S0-AVW-466

Construction Databook

Construction Databook

Sidney M. Levy

McGraw-Hill
New York San Francisco Washington, D.C. Auckland Bogotá
Caracas Lisbon London Madrid Mexico City Milan
Montreal New Delhi San Juan Singapore
Sydney Tokyo Toronto

Library of Congress Cataloging-in-Publication Data

Levy, Sidney M.
Construction databook / Sidney M. Levy.
p. cm.
ISBN 0-07-038365-0
1. Building—Handbooks, manuals, etc. I. Title.
TH151 .L48 1998
690—dc21 98-40478
CIP

McGraw-Hill

A Division of The ***McGraw-Hill*** *Companies*

4 5 6 7 8 9 0 DOC/DOC 0 3 2 1 0

ISBN 0-07-038365-0

The sponsoring editor for this book was Larry Hager, the editing supervisor was Andrew Yoder, and the production supervisor was Pamela Pelton. It was set in ITC Century Light by Jana Fisher through the services of Barry E. Brown (Broker—Editing, Design and Production).

Printed and bound by R. R. Donnelley & Sons Company

This book is printed on recycled, acid-free paper containing a minimum of 50% recycled, de-inked material.

Contents

Introduction *vii*

Section 1: Sitework *1*
Section 2: Site Utilities *33*
Section 3: Concrete *71*
Section 4: Masonry *97*
Section 5: Structural Steel, Joists, and Metal Decks *137*
Section 6: Wood and Lumber Products *181*
Section 7: Plywood, Composite Wood Products, High-Pressure Laminates *223*
Section 8: Roofing *249*
Section 9: Fireproofing *287*
Section 10: Sealants *301*
Section 11: Acoustics/Sound Control *319*
Section 12: Doors and Windows *337*
Section 13: Finish Hardware *387*
Section 14: Drywall, Metal Framing, and Plaster *419*
Section 15: Flooring *457*
Section 16: Painting *483*
Section 17: Elevators—Dumbwaiters *507*
Section 18: Plumbing *523*
Section 19: Fire Protection *553*
Section 20: Heating, Ventilating, and Air Conditioning *573*
Section 21: Electrical *615*
Section 22: Metrification *643*
Section 23: Useful Tables, Charts, and Formulas *661*

Index *679*

Introduction

The *Construction Databook* provides the builder, project manager, construction superintendent, and design consultants and facility owners with a one-source reference guide to the most commonly encountered construction materials and components. Valuable information ranging from acoustics to mechanical systems, with scores of other topics in between, is included in this one-easy-to-access single volume.

Product specifications, installation, and maintenance data on hundreds of building products is included in this book along with other sections containing useful tables and formulas. An explanation of the on-going metrification movement includes charts for both "soft" and "hard" conversions.

The *Construction Databook* contains diagrams and schematics of numerous plumbing and HVAC installations to acquaint the reader with the most frequently installed systems and the equipment required to operate them.

Selected sections of the National Electric Code (NEC) are included in the *Databook* to supplement the reader's knowledge of acceptable electrical installation techniques. Much of the material in this volume has been gleaned from manufacturer's sources and trade association supplied information; some of the manufacturer supplied data may be proprietary in nature but generally is similar to products made by other vendors. This one-source *Construction Databook* should prove invaluable for office and field based construction and design professionals, since it contains, in one volume, answers to so many of the design and product application questions that arise during a construction project.

The *Databook* will provide the architect, builder, engineer, and owner with a handy, easy-to-access source of building material specifications and weights per foot, formulas, conversion factors, diagrammatic data and installation instructions.

How many times during project meetings, field visits, or in conversions with specialty contractors is it convenient to have a concise source of product data and specifications readily at hand? The *Databook* was written with those needs in mind.

I selected the construction components and specifications that, in my forty year experience in the construction industry, appear to be those for which reference material is so often required and, of course, always needed "yesterday".

And many manufacturers and related trade organizations are often eager to furnish additional information or more specific information if requested.

I hope you find the *Construction Databook* a worthwhile addition to your construction library.

Sidney M. Levy

Section

1

Sitework

Contents

1.1.1 Site investigations
1.1.2 Glossary of terms
1.1.3 Soil classification systems
1.1.3.1 Definition of soil by grain size
1.1.4 U.S./Metric sieve sizes
1.1.5 Interpreting soil-test boring logs
1.1.5.1 Classification terminology used in conjunction with test borings
1.1.5.2 OSHA soil classifications
1.1.6 OSHA simple slope, single, and multiple bench diagrams
1.1.6.1 OSHA simple slope and vertical- sided trench-excavation diagrams
1.1.7 Caissons
1.1.8 Piles (types)
1.1.8.1 Basic parts of a typical pile-driving rig
1.1.8.2 Pile-driving rig with fixed-lead system
1.1.8.3 Pile-driving rig with swing-lead system
1.1.8.4 Lead types
1.1.8.5 Pile-driving rig in batter configurations
1.1.8.6 Pile-driving hammer types
1.1.8.7 Sheet piles
1.1.8.8 Typical sheet pile sections
1.1.8.9 Steel H Piles (typical sections and specifications)
1.1.8.10 Typical soldier pile detail (rock bearing)
1.1.8.11 Typical braced soldier pile with deadman
1.1.8.12 Typical soldier pile drilled into rock (concrete filled)
1.1.8.13 Typical soldier pile with wood lagging, steel walers, and bracing
1.1.9 Dewatering
1.1.9.1 Diagrammatical display of water flow from wellpoints
1.1.9.2 Drawdown (diagramatical)
1.2.0 Protecting foundations from ground water
1.3.0 Hand signals for boom-equipment operators
1.4.0 Avoiding bituminous paving pitfalls
1.4.1 Calculating the amount of bitumastic paving per 100 linear feet and per mile
1.4.1.1 Linear feet of paving covered by one ton of material
1.4.1.2 Linear meters of asphalt covered by one ton of material
1.4.2 Cubic meters of asphalt for various widths/depths of paving (metric)
1.4.3 Tons of asphalt required per mile for various widths/pounds per yard
1.4.4 Asphalt paving block specifications

1.1.1 Investigation

Site work involves working with various types of soils and dealing with the unexpected—even in the presence of extensive soil test borings.

Even prior to commencing construction, a thorough investigation of the site, both visually and after a review of available geo-technical reports, the contractor will be more prepared for what lies ahead.

1. Does a visual inspection of the site reveal any clues to the composition and consistency of the soil?
2. Are there rock outcroppings? If so, what is the nature of the rock?
3. Is there any indication of the presence of ground water close to the surface of the site?
4. Do any remains of abandoned subsurface structures appear in areas where excavation will be required?
5. Do any structures require demolition in areas where new structures are to be built or where underground utilities are to be installed?
6. Are any utilities absent that might be required during construction (i.e., water, electric power, telephone lines, sanitary and storm sewers, or gas mains)? Are any of these utilities in areas where new construction will be required and are to be relocated?
7. What do the soil test borings reveal?

Analyzing a typical soil test boring should start with a look at the consistency of the soil, as reported on the report, the presence or absence of rock or any other underground obstructions, the level at which water was observed, and the blow count (an indication of soil-bearing capacity). The blows per foot also reveal the plasticity of the soil.

1.1.2 Glossary of Terms

Aeolian deposits Wind-deposited materials such as sand dunes or other silty-type materials.

Alluvium Material deposited by streams that might no longer exist, or form existing floodplains.

Aquifer A geological formation that provides water in sufficient quantities to create a spring or well.

Bank-run gravel Often called *run-of-bank gravel*. Gravel as it is excavated from a bank in its natural state.

Base course A layer of material selected to provide a subgrade for some load-bearing structure (such as paving) or to provide for drainage under a structure above.

Binder A material that will pass a No. 40 U.S. standard sieve.

Boulder A rock fragment with a diameter larger than 12 inches (304.8 mm).

Cemented soil Soil in which particles are held together by a chemical agent, such as calcium carbonate.

Clay A mineral soil consisting of particles less than 0.002 mm in equivalent diameter; a soil textural class, or a fine-grained soil with more than 50 percent passing through a No. 200 sieve that has a high plasticity index in relation to its liquid limit.

Cobble A rock fragment, generally oblong or rounded, with an average dimension ranging from 3 inches (75 mm) to 12 inches (305 mm).

Cohesion Shear resistance of soil at zero normal stress.

Cohesionless soil A soil when air dried in an unconfined space, has little cohesion when submerged.

Cohesive soil A soil when in an unconfined state, has considerable strength when air dried and submerged.

Compaction A process to decrease voids between soil particles when subjected to the forces applied by special equipment.

Density The mass of solid particles in a sample of soil or rock.

Dry soil Soil that does not exhibit visible signs of moisture content.

Fines Clay-sized particles (less than 0.002 mm).

Fissured soil Soil material that has a tendency to break along definite planes of fracture with little resistance.

Glacial till Unstratified glacial materials deposited by the movement of ice and consisting of sand, clay, gravel, and boulders in any proportion.

Granular soil Gravel, sand, or silt with little or no clay content. It has no cohesive strength, cannot be molded with moist, and crumbles easily when dry.

Gravel Round or semi-round particles of rock that will pass through a 3-inch (76.2 mm) sieve and be retained by a No. 4 U.S. standard sieve (approximately ¼ inch, (6.35 mm). It is also defined as an aggregate, consisting of particles that range in size from ¼ inch (6.35 mm) to 3 inches (76.2 mm).

Hardpan Soil that has become rock-like because of the accumulation of cementing minerals, such as calcium carbonate, in the soil.

Layered system Two or more distinctly different soil or rock types arranged in layers.

Loess A uniform aeolian deposit of silty material having an open structure and relatively high cohesion because of the cementation of clay or marl.

Marl Calcareous clay that contains 35 to 65 percent calcium carbonate.

Optimum moisture content Water content at which a soil can be compacted to a maximum-unit dry-unit weight.

Organic clay/soil/silt Clay/soil/silt with high organic content.

Perched water table A water table of generally limited area that appears above the normal free-water elevation.

Plastic A property of soil that allows the soil to be deformed or molded without cracking or causing an appreciable volume change.

Plastic limit Water content at which a soil will just begin to crumble when rolled into a cylinder approximately ⅛ inch (3.17 mm) in diameter.

Relative compaction The dry unit weight of soil, compared to the maximum unit weight obtained in a laboratory compaction test, expressed as a ratio.

Specific gravity The ratio of the weight in air of a given volume of solids at a stated temperature to the weight in air of an equal volume of distilled water at the stated temperature.

1.1.3 Soil Classification Systems

Soils can be classified in several different methods and categories. The Tyler System uses opening per lineal inch of wire screen to determine particle size. For example, according to this system, a No. 20 mesh, has 20 openings per lineal inch of screen, which equates to a sieve size of 0.0328 inches (0.833 mm).

The Unified Soil Classification System, the most widely used classification system, uses letters to designate soil types within three major groups: coarse-grained, fine-grained, and highly organic soils.

- *Coarse-grained soil* Includes gravel, sands, and mixtures of the two. The letter *G* denotes gravel and the letter *S* denotes sand. In mixtures, the first letter indicates the primary constituent, e.g., GS. Both gravel and sand are further divided into four groups:
  - *Well graded* Designated by the letter *W*.
  - *Poorly graded* Designated by the letter *P*.
  - *Dirty with plastic fines* Designated *P*
  - *Dirty with nonplastic silty fines* If it will pass through a No. 200 sieve, it is designated by the letter *M*.

The coefficient of uniformity (Cu) is computed from data taken from a grain size distribution curve.

- *Fine-grained soils* These soils are further divided into inorganic silts (M), inorganic clays (c), and organic silts or clays (O). Each group is further divided into soils having liquid limits lower than 50 percent (L) and those with liquid limits higher than 50 percent (H). For example, an inorganic silt with liquid limit lower than 50 percent would be designated *ML*.
- *Highly Organic soils* This group is identified by the letters *Pt*, for peat, which is characteristic of materials in this grouping.

1.1.3.1 Definition of Soil By Grain Size

Sieve size	Corresponding soil classification
12" (304.8 mm) or more	Boulders
3" (76.2 mm) to 12" (304.8 mm)	Cobbles
¾" (19.05 mm) to 3" (76.2 mm)	Coarse gravel
No. 4 to ¾" (19.05 mm)	Fine gravel
No. 4 to No. 10	Coarse sand
No. 10 to No. 40	Medium sand
No. 40 to No. 200	Fine sand
Passing through No. 200	Silt and clay fines

1.1.4 U.S.A./Metric Sieve Sizes

This chart shows the various sieve-size openings and their metric conversions.

U.S.A. Sieve Series and Equivalents—A.S.T.M. E-11-87

Sieve Designation		Sieve Opening		Nominal Wire Diameter	
Standard (a)	Alternative	mm	in (approx. equivnts.)	mm	in (approx. equivnts.)
125 mm	5"	125	5.00"	8.00	.3150"
106 mm	4.24"	106	4.24"	6.40	.2520"
100 mm	4"(b)	100	4.00"	6.30	.2480"
90 mm	3.5"	90	3.50"	6.08	.2394"
75 mm	3"	75	3.00"	5.80	.2283"
63 mm	2.5"	63	2.50"	5.50	.2165"
53 mm	2.12"	53	2.12"	5.15	.2028"
50 mm	2"(b)	50	2.00"	5.05	.1988"
45 mm	1.75"	45	1.75"	4.85	.1909"
37.5 mm	1.5"	37.5	1.50"	4.59	.1807"
31.5 mm	1.25"	31.5	1.25"	4.23	.1665"
26.5 mm	1.06"	26.5	1.06"	3.90	.1535"
25.0 mm	1"(b)	25.0	1.00"	3.80	.1496"
22.4 mm	7/8"	22.4	0.875"	3.50	.1378"
19.0 mm	3/4"	19.0	0.750"	3.30	.1299"
16.0 mm	5/8"	16.0	0.625"	3.00	.1181"
13.2 mm	.530"	13.2	0.530"	2.75	.1083"
12.5 mm	1/2"(b)	12.5	0.500"	2.67	.1051"
11.2 mm	7/16"	11.2	0.438"	2.45	.0965"
9.5 mm	3/8"	9.5	0.375"	2.27	.0894"
8.0 mm	5/16"	8.0	0.312"	2.07	.0815"
6.7 mm	.265"	6.7	0.265"	1.87	.0736"
6.3 mm	1/4"(b)	6.3	0.250"	1.82	.0717"
5.6 mm	No. 3-1/2(c)	5.6	0.223"	1.68	.0661"
4.75 mm	No. 4	4.75	0.187"	1.54	.0606"
4.00 mm	No. 5	4.00	0.157"	1.37	.0539"
3.35 mm	No. 6	3.35	0.132"	1.23	.0484"
2.80 mm	No. 7	2.80	0.11"	1.10	.0430"
2.36 mm	No. 8	2.36	0.0937"	1.00	.0394"
2.00 mm	No. 10	2.00	0.0787"	.900	.0345"
1.70 mm	No. 12	1.70	0.0661"	.810	.0319"
1.40 mm	No. 14	1.40	0.0555"	.725	.0285"
1.18 mm	No. 16	1.18	0.0469"	.650	.0256"
1.00 mm	No. 18	1.00	0.0394"	.580	.0228"
850 µm	No. 20	0.850	0.0331"	.510	.0201"
710 µm	No. 25	0.710	0.0278"	.450	.0177"
660 µm	No. 30	0.600	0.0234"	.390	.0154"
500 µm	No. 35	0.500	0.0197"	.340	.0134"
425 µm	No. 40	0.425	0.0165"	.290	.0114"
355 µm	No. 45	0.355	0.0139"	.247	.0097"
300 µm	No. 50	0.300	0.0117"	.215	.0085"
250 µm	No. 60	0.250	0.0098"	.180	.0071"
212 µm	No. 70	0.212	0.0083"	.152	.0060"
180 µm	No. 80	0.180	0.0070"	.131	.0052"
150 µm	No. 100	0.150	0.0059"	.110	.0043"
125 µm	No. 120	0.125	0.0049"	.091	.0036"
106 µm	No. 140	0.106	0.0041"	.076	.0030"
90 µm	No. 170	0.090	0.0035"	.064	.0025"
75 µm	No. 200	0.075	0.0029"	.053	.0021"
63 µm	No. 230	0.063	0.0025"	.044	.0017"
53 µm	No. 270	0.053	0.0021"	.037	.0015"
45 µm	No. 325	0.045	0.0017"	.030	.0012"
38 µm	No. 400	0.038	0.0015"	.025	.0010"
32 µm	No. 450		0.00126"	.0011	
25 µm	No. 500		0.00098"	.001	
20 µm	No. 635		0.00079"	.0008	

(a) These standard designations correspond to the values for test sieve apertures recommended by the International Standards Organization Geneva, Switzerland.
(b) These sieves are not in the fourth root of 2 Series, but they have been included because they are in common usage.
(c) These numbers (3-1/2 to 400) are the approximate number of openings per linear inch but it is preferred that the sieve be identified by the standard designation in millimeters or microns (1000 microns = 1 mm.)

1.1.5 Interpreting Soil-Test Boring Logs

Engineers and Scientists
SML Geotechnical

Stratford Court
Uncasville,CT

Boring No. 1
Page 1 of 1
File No. 50512
Chkd. By: JMB

Boring Co. ABC Drilling
Foreman Tom Jones
Rep.
Date Start 1/8/91 End 1/8/91
Location NE section of proposed building
GS.Elev. 143 ± Datum NGVD

	Casing	Sampler
Type		Split Spoon
I.D./O.D.		1 3/8"/2"
Hammer Wt.		140 lbs.
Hammer Fall		30 in.
Other	2¼" HSA	AX

Groundwater Readings

Date	Time	Depth	Casing	Stab. Time
1/8/91	1420	*	out	none

DEPTH	CASNG BLOWS	No.	Pen./Rec.	Depth (Ft.)	Blows/6"	Field Testing (ppm)	SAMPLE DESCRIPTION & CLASSIFICATION	Stratum Description	REMKS	Equipment Installed
		S-1	24/18	0-2.0	1-3-5-8		Top 6": Loose, brown TOPSOIL. Next 6": Loose, brown, fine SAND, little Silt. Bottom 6": Medium dense, brown, fine to medium SAND, little fine Gravel, trace Silt.	TOPSOIL; 0.5' LOOSE SILTY SAND; 1.0' MEDIUM DENSE SAND		- - - - - -
5		S-2	18/12	5.0-6.5	16-28-70		Very dense, grey and brown, fine to coarse SAND, some fine to coarse Gravel, trace Silt.	5.0' VERY DENSE SAND		
10		S-3	5/5	10.0-10.4	100/5"		Very dense, grey, weathered ROCK.	10.0' VERY DENSE WEATHERED ROCK		
15		C-1	60/34	13.5-18.5	7 min/ft.		GNEISS (RQD=0)	13.5' BEDROCK	1	
					8					
					8					
					7					
					6					
20								18.5' E.O.B.		
25										

REMARKS
1. Auger refusal at 13.5'.
* Water used for coring altered the water level.
Note: Ground surface elevation is interpolated from topographic plan. Borings were located in the field by taping from existing site features.

Stratification lines represent approximate boundaries between soil types, transitions may be gradual. Water level readings have been made at times and under conditions stated. Fluctuations of groundwater may occur due to factors other than those present at the time measurements were made.

Boring No. 1

Typical auger boring, spoon sampling report.

The blow count reveals that it took:

1 blow to drive a 140-pound hammer six inches.
3 blows were required to drive a 140-pound hammer 12 inches.
5 blows were required to drive a 140-pound hammer 18 inches.
8 blows were required to drive a 140-pound hammer to a depth of 24 inches.
16 blows were required to drive the hammer to a depth of five feet.
100 blows were required to drive the hammer to 10 feet five inches.

As far as water level is concerned, the use of water during the coring operation did not allow the Geotech to ascertain ground water levels with certainty. When nonwater coring operations are used, ground water levels are so indicated.

1.1.5.1 Classification Terminology Used in Conjunction with Test Borings

COMPONENT GRADATION TERMS

MATERIAL	FRACTION	SIEVE SIZE
GRAVEL	COARSE	3/4" TO 3"
	FINE	NO. 4 TO 3/4"
SAND	COARSE	N0. 10 TO NO. 4
	MEDIUM	NO. 40 TO NO. 10
	FINE	NO. 200 TO NO. 40
FINES		PASSING NO. 200

FINES FRACTION

PLASTICITY	PI	NAME	SMALLEST THREAD DIA ROLLED
NON-PLASTIC	0	SILT	NONE
SLIGHT	1-5	Clayey SILT	1/4"
LOW	5-10	SILT & CLAY	1/8"
MEDIUM	10-20	CLAY & SILT	1/16"
HIGH	20-40	Silty CLAY	1/32"
VERY HIGH	>40	CLAY	1/64"

RELATIVE DENSITY OR CONSISTENCY TERMS

NON-PLASTIC SOILS		PLASTIC SOILS	
BLOWS/FT	DENSITY	BLOWS/FT	PLASTIC SOILS
0-4	V. LOOSE	<2	V. SOFT
4-10	LOOSE	2-4	SOFT
10-30	M. DENSE	4-8	M. STIFF
30-50	DENSE	8-15	STIFF
>50	V. DENSE	15-30	V. STIFF
		>30	HARD

PROPORTIONAL TERMS

PROPORTIONAL TERM	PERCENT BY WEIGHT
AND	35-50
SOME	20-35
LITTLE	10-20
TRACE	1-10

BEDROCK WEATHERING CLASSIFICATION

GRADE	SYMBOL	DIAGNOSTIC FEATURES
Fresh	F	No visible signs of decomposition or discoloration. Rings under hammer impact.
Slightly Weathered	WS	Slight discoloration inwards from open fractures, otherwise similar to F.
Moderately Weathered	WM	Discoloration throughout. Weaker minerals such as feldspar decomposed. Strength somewhat less than fresh rock but cores cannot be broken by hand or scraped by knife. Texture preserved.
Highly Weathered	WH	Most minerals somewhat decomposed. Specimens can be broken by hand with effort or shaved with knife. Core stones present in rock mass. Texture becoming indistinct but fabric preserved.
Completely Weathered	WC	Minerals decomposed to soil but fabric and structure preserved (Saprolite). Specimens easily crumbled or penetrated.
Residual Soil	RS	Advanced state of decomposition resulting in plastic soils. Rock fabric and structure completely destroyed. Large volume change.

1.1.5.2 OSHA Soil Classification

OSHA uses a soil-classification system as a means of categorizing soil and rock deposits in a hierarchy of stable rock, Type A soil, Type B soil, and Type C soil, in decreasing order of stability. Maximum allowable slopes are set forth, according to the soil or rock type.

Soil or rock type	Maximum allowable slope for excavation less than 20 feet
Stable rock	Vertical (90 degrees)
Type A soil	3/4:1 (53 degrees)
Type B soil	1:1 (45 degrees)
Type C soil	1½:1 (34 degrees)

A short-term maximum allowable slope of 1½ H:1V (63°) is allowed in excavations in Type A soil that are 12 ft (3.67 m) or less in depth. Short-term maximum allowable slopes for excavations greater than 12 ft (3.67 m) in depth shall be ¾ H:1V (53°)

Note: Consult OSHA for definition of *short-term.*

Type A: A cohesive soil with an unconfined compressive strength of 1.5 tons per square foot (144 kPa) or greater. Cohesive soils can be categorized as silty, clay, sandy clay, clay loam, and cemented soils. No soil is classified Type A if:

1. The soil is fissured.
2. The soil is subject to vibration from heavy traffic, or pile driving.
3. The soil has previously been disturbed.
4. The soil is part of a sloped, layered system, where the layers dip into the excavation on a slope of 4 horizontal to 1 vertical.
5. The material is subject to other factors that tend to make it less stable.

Type B: A cohesive soil with an unconfined compressive strength of greater than 0.5 tons per square foot (48 kPa), but not less than 1.5 tons per square foot (144 kPa). This classification applies to cohesionless soils, including angular gravel (similar to crushed rock), silt, silt loam, sandy loam, and in some cases, silty clay loam and sandy clay loam. This classification also applies to previously disturbed soils, except those that would be classified as Type C or soil that meets the unconfined compressive strength or cementation requirements for Type A, but is fissured or subject to vibration, dry rock that is not stable, or material that is part of a sloped, layered system, where the layers dip into the excavation on a slope less steep that 4 horizontal to 1 vertical, but only if the material would otherwise be classified Type B.

Type C: A cohesive soil with an unconfined compressive strength of 0.5 tons per square foot (48 kPa) or less, generally consisting of granular soils (including gravel, sand and loamy sand, submerged soil, soil form which water freely seeps, submerged rock that is not stable, or material in a sloped, layered system, where the layers dip into the excavation on a slope of 4 horizontal to 1 vertical (or steeper).

OSHA, in 1926.652 Appendix B, lists standards, interpretations, and illustrations of simple, single, multiple benches, and the use of trench support and shield systems for 20-foot (maximum) excavation depths. OSHA pages 186.8 and 186.9 of Appendix B contain diagrams that depict benched excavations for various types of excavations. For a complete explanation of excavations and trench-protection requirements, refer to the entire text of OSHA CFR 1926.652 in Appendix B.

1.1.6 OSHA Simple Slope, Single, and Multiple Bench Diagrams

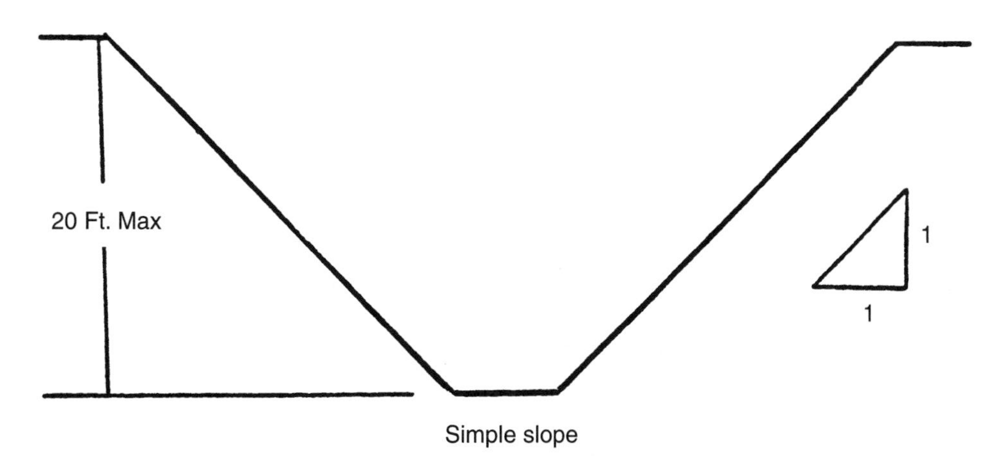

Simple slope

2. All benched excavations 20 feet or less in depth shall have a maximum allowable slope of 1:1 and maximum bench dimensions as follows:

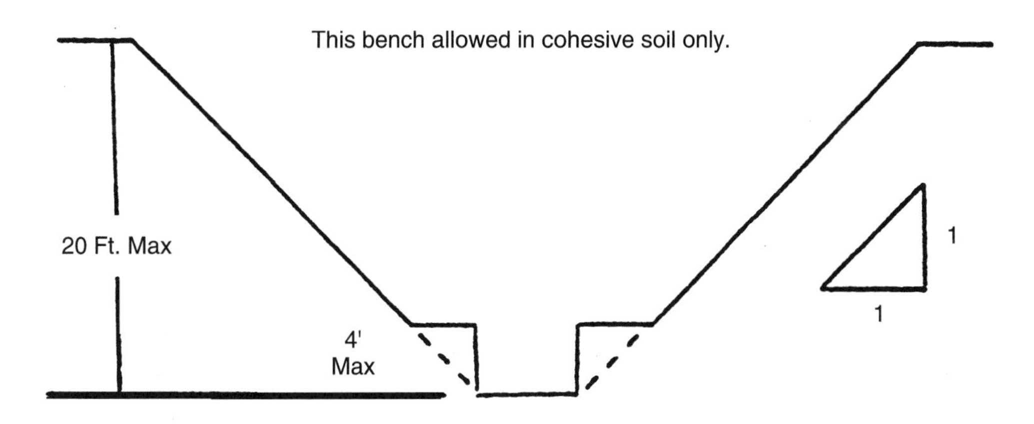

Single Bench

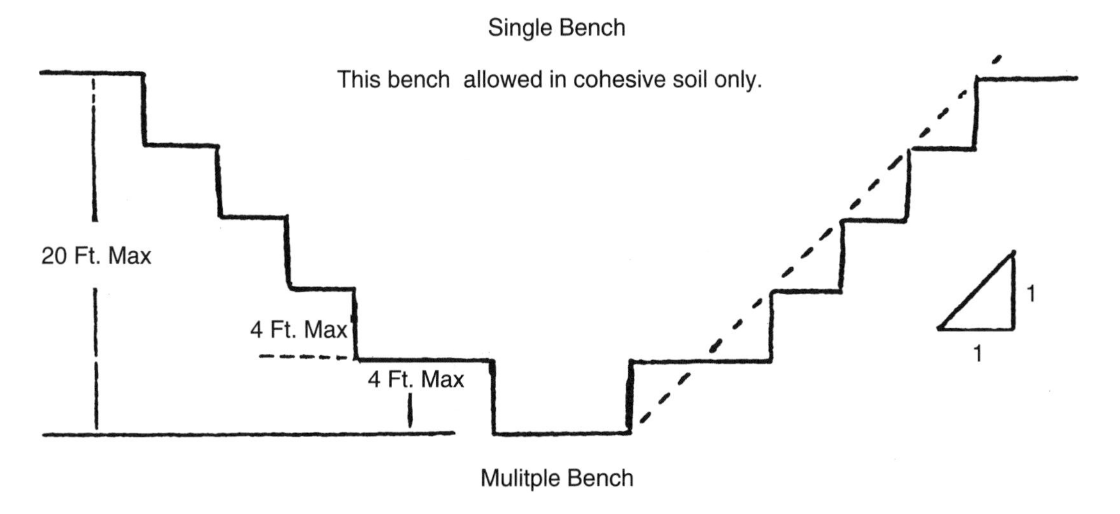

Mulitple Bench

3. All excavations 20 feet or less in depth which have vertically sided lower portions shall be shielded or supported to a height at least 18 inches above the top of the vertical slide. All such excavations shall have a maximum allowable slope of 1:1.

1.1.6.1 OSHA Simple Slope and Vertical-Sided Trench-Excavation Diagrams

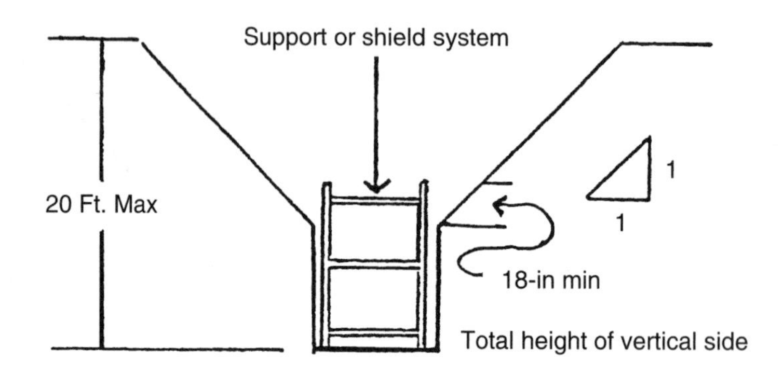

Vertically Sided Lower Portion

4. All other sloped excavations shall be in accordance with the other options permitted in § 1926.652(b).

B-1.3 Excavations Made in Type C-Soil

1. All simple slope excavations 20 feet or less in depth shall have a maximum allowable slope of 1½:1.

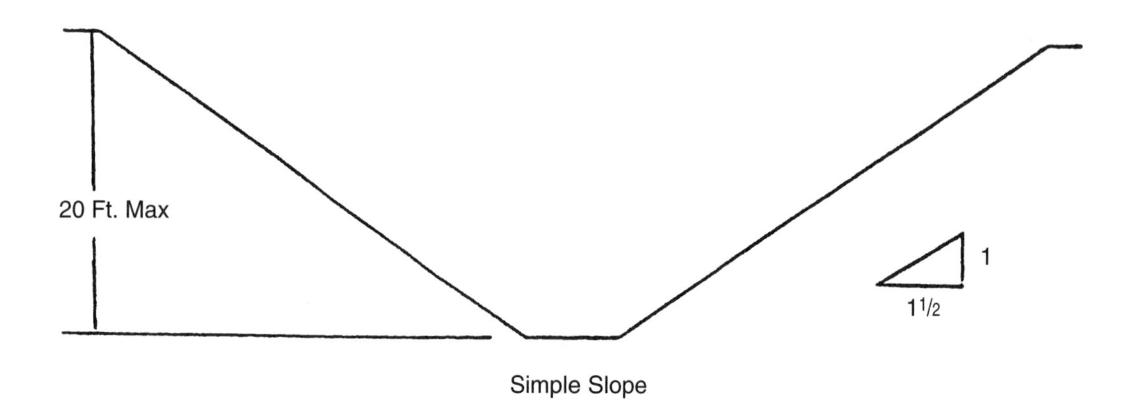

Simple Slope

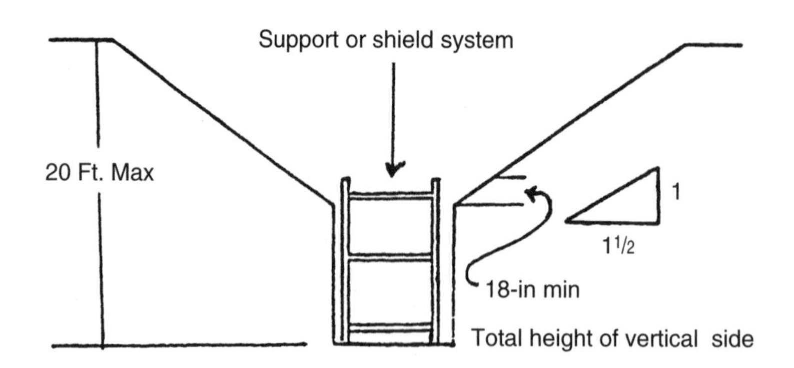

Vertical Sided Lower Portion

3. All other sloped excavations shall be in accordance with the other options permitted in § 1926.652(b).

B-1.4 Excavations Made in Layered Soils

1. All excavations 20 feet or less in depth made in layered soils shall have a maximum allowable slope for each layer as set forth below.

1.1.7 Caissons

Caissons are typically drilled or augered holes, into which a metal casing is installed; caissons are often referred to as *drilled* or *bored* piers. Typically, caissons fall into three groups: straight shaft, belled, or rock socketed. Installing caissons involves excavation, generally by a boring machine, to the depth required to meet either end bearing or friction, lowering a metal casing into the excavated hole, and filling the casing with reinforced concrete. When a belled caisson is constructed, a special belling tool is inserted in the bottom of the excavated shaft to create an enlarged base. In the case of a rock-socketed caisson, the boring machine will have a cutting shoe on the end to bore into the rock-bearing surface. A rock socket is then drilled into the rock to accept an H pile before filling the caisson with concrete.

The slurry method of installing caissons, utilizing bentonite, is also used when proper soil and ground-water conditions prevail.

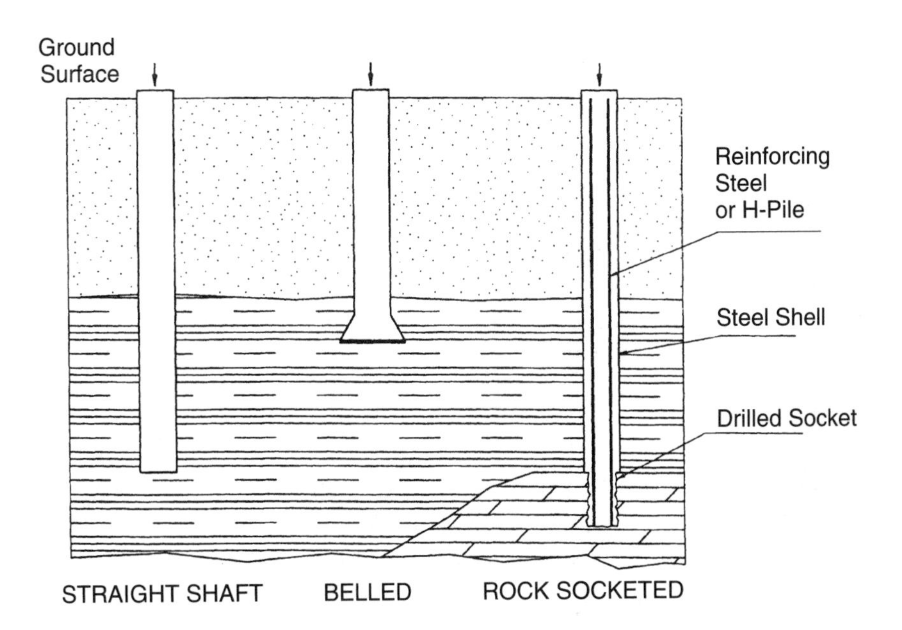

1.1.8 Piles (Types)

Pile-driving rigs are used to install various types of piles, as well as sheet piling. The power source used to operate the pile-driving hammer can be by steam boiler, a compressed air source, or by an electrohydraulic system. All pile-driving rigs share similar components, some of which are more complex than others. Figure 1.8.1 reveals a typical pile-driving rig with all parts identified.

Pile types

- *Timber* Made of whole trees and driven with the small end down. The wooden piles are treated with an American Wood Preservers Association (AWPA) approved preservative to help resist decay and attack by insects.
- Steel Either H piles confirming to American Institute of Steel Construction (AISC) specifications for HP shapes, pipe piles that meet ASTM A252, or pile shells of relatively light-gauge corrugated steel driven with an internal steel mandrel. These types of piles are considered non-load bearing and serve generally as a form to protect the concrete as it cures.

- *Concrete piles* Precast concrete piles and cast-in-place concrete piles that utilize driven corrugated shells as their forming material.
- *Soldier piles* HP piles driven in soil, to bedrock, or to concrete-filled shafts, depending upon the depth of toe-in required soldier piles are used as support for wood lagging in lieu of sheet piling.

1.1.8.1 The Basic Parts of a Typical Pile-Driving Rig

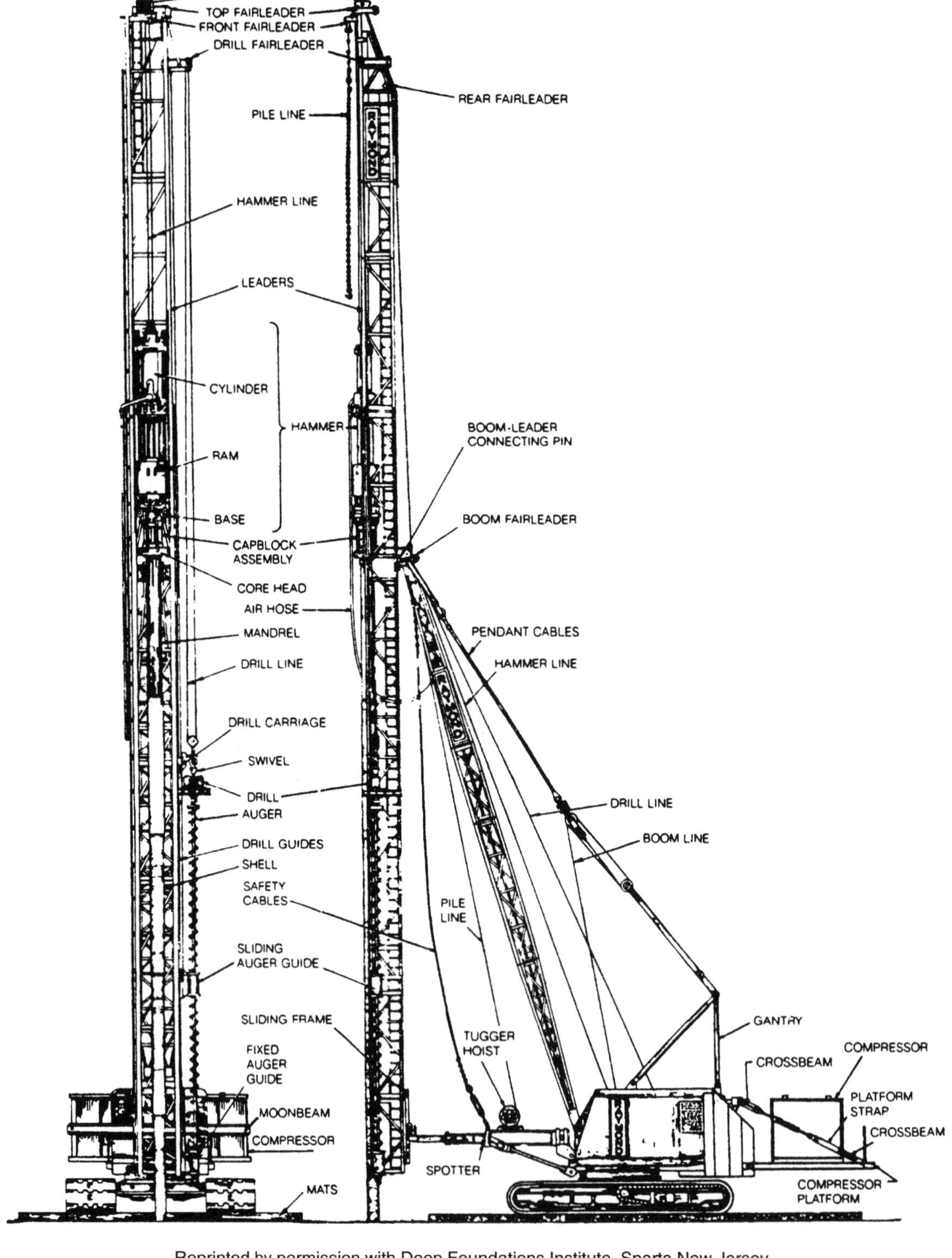

Reprinted by permission with Deep Foundations Institute, Sparta New Jersey

1.1.8.2 Pile-Driving Rig with Fixed-Lead System

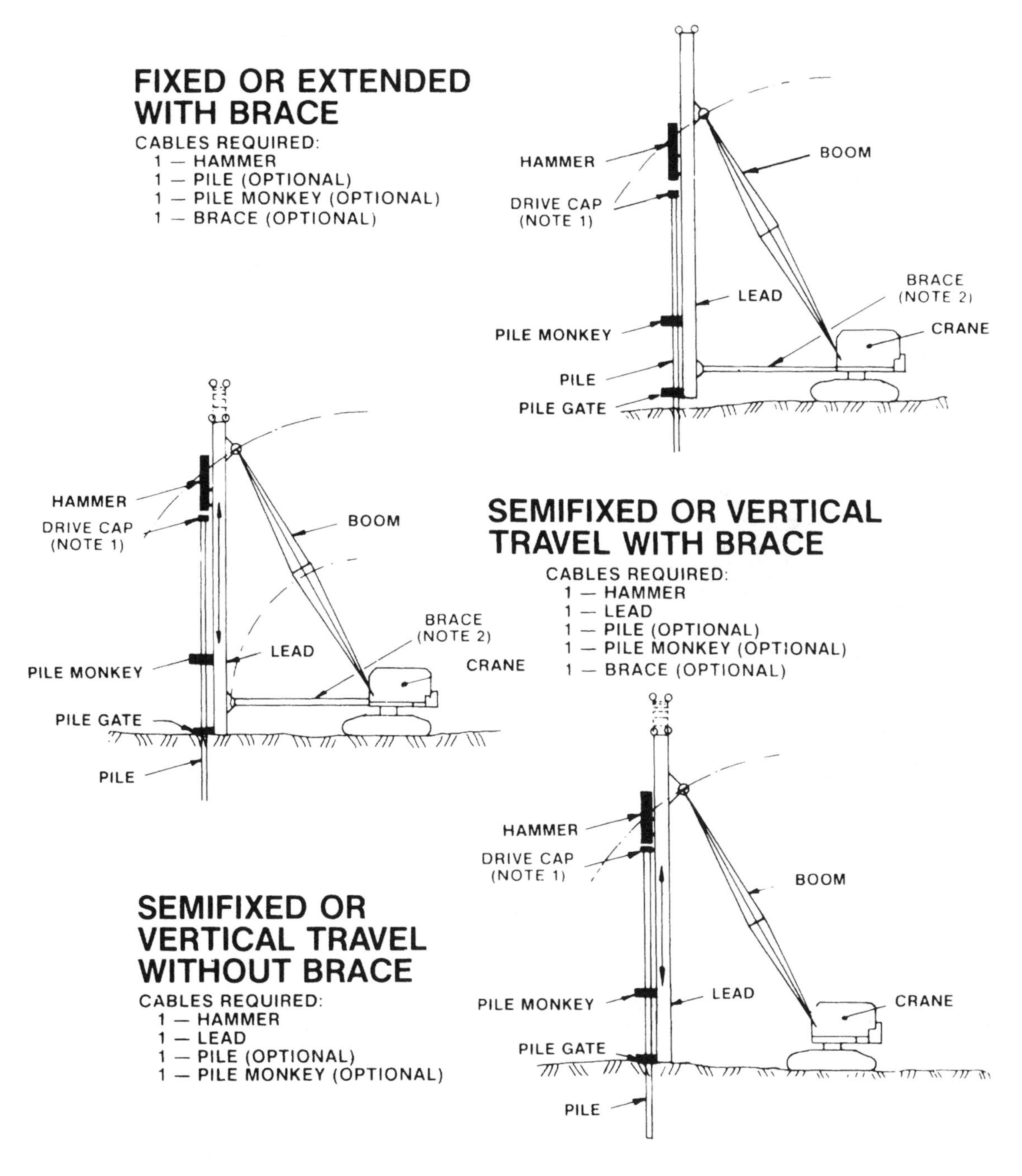

NOTE 1. Also called anvil block, bonnet, cap, driving head, follow cap, helmet, hood, rider cap.

NOTE 2. Also called A-frame, apron, bottom brace, bottom strut, kicker, parallelogram, platform, spider, spotter, spreader, spreader bars.

1.1.8.3 Pile-Driving Rig with Swing-Lead System

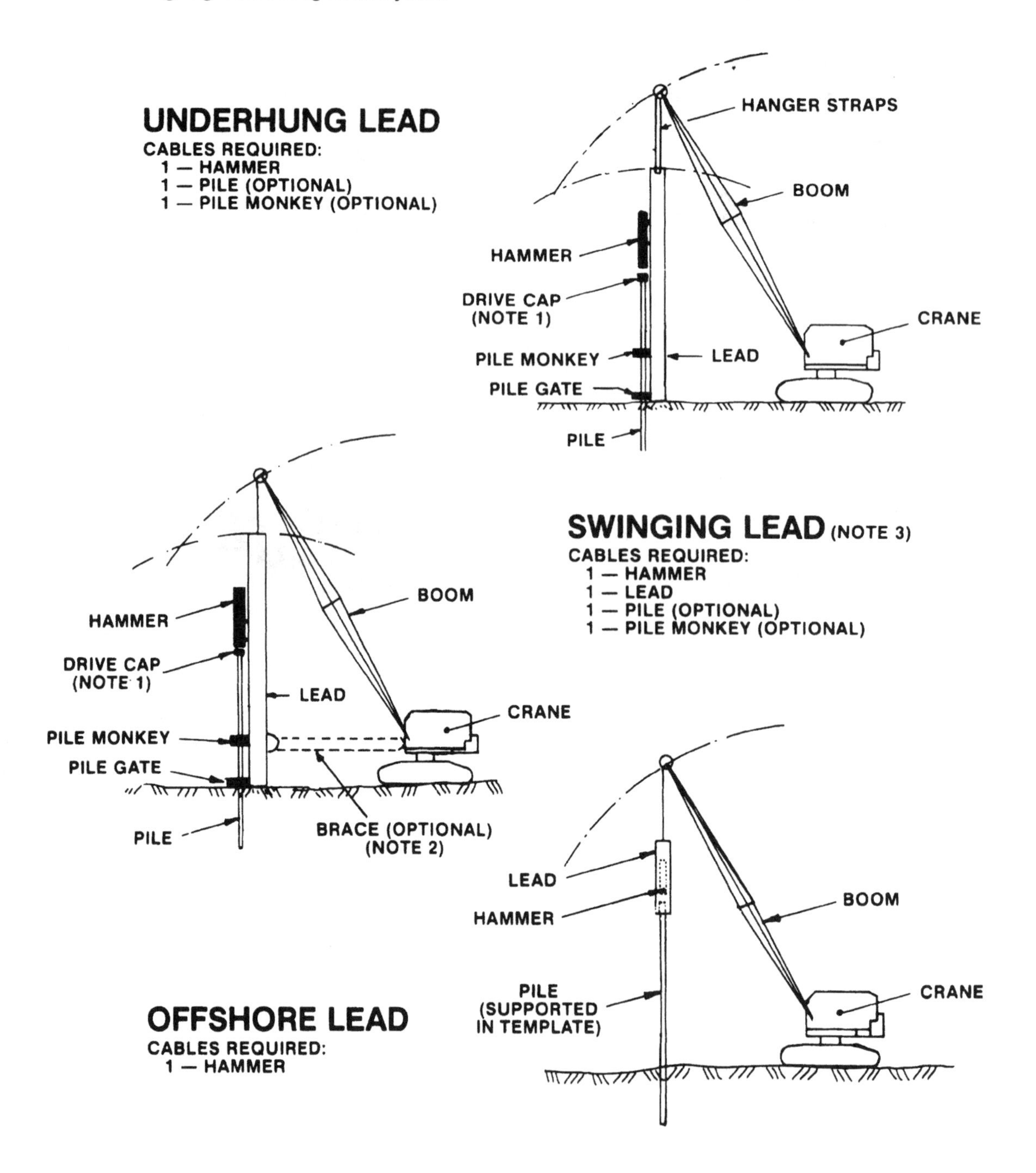

NOTE 1. Also called anvil block, bonnet, cap, driving head, follow cap, helmet, hood, rider cap.

NOTE 2. Also called A-frame, apron, bottom brace, bottom strut, kicker, parallelogram, platform, spider, spotter, spreader, spreader bars.

NOTE 3. Also called hanging lead.

1.1.8.4 Lead Types

H-BEAM

Also Called Spud Lead. Wide Flange. Monkey Stick. European Lead

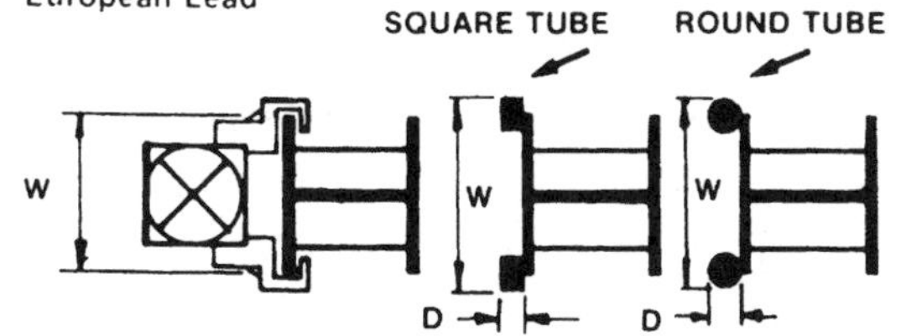

GUIDE PROFILES & APPROX. WEIGHTS:
10" W — 50 lb./ft.
12" W — 60-80 lb./ft.
14" W — 80-120 lb./ft.
21" W — 145 lb./ft.
3" D x 15¾" W — 80 lb./ft.
2¾" Dia. x 15¾" W — 80-130 lb./ft.

PIPE LEAD

Also Called Monkey Stick, Spud Lead, European Lead or Pogo Stick

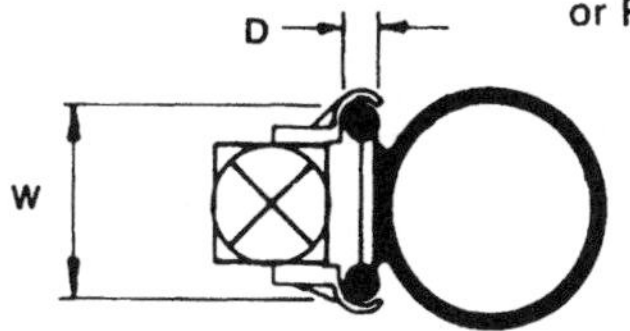

GUIDE PROFILES & APPROX. WEIGHTS:
2" Dia. x 11" W — 30-60 lb./ft.
2¾" Dia. x 15¾" W — 70-125 lb./ft.

BOX LEAD

Also Called "U" Lead. Steam Lead

WITH OR WITHOUT PLATFORM

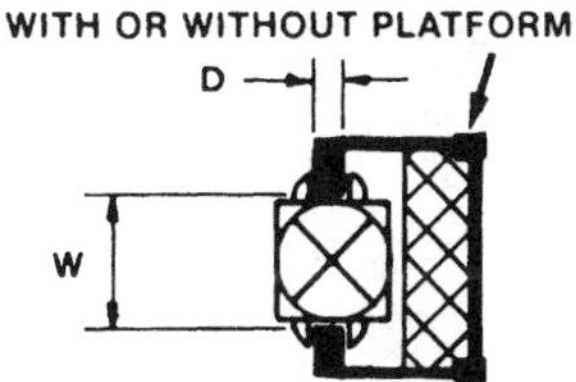

GUIDE PROFILES & APPROX. WEIGHTS:
6", 7" or 8" D x 20" W — 65-90 lb./ft.
8" or 9" D x 26" W — 80-160 lb./ft.
8", 9" or 10" D x 32" W — 130-210 lb./ft.
10" or 11" D x 37" W — 200-280 lb./ft.
11" D x 56" W — 210-300 lb./ft.

OFFSHORE LEAD

Also Called Chuck Lead, Can, Rope Suspended Lead or Pogo Stick

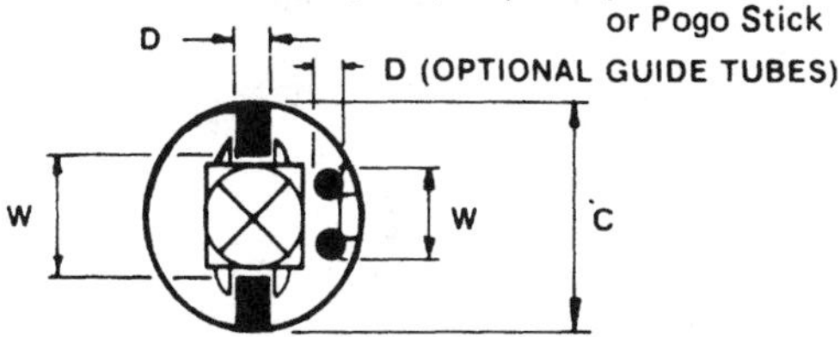

GUIDE PROFILES & APPROX. WEIGHTS:
Diesel Hammers —
6" D x 20" W x 24" C — 4,000 lb.
8" D x 26" W x 36" C — 6,500 lb.
10" D x 32" W x 48" C — 10,000 lb.
2" Dia. x 11" W x 24" C — 4,000 lb.
2¾" Dia. x 15¾" W x 36" C — 6,500 lb.
Air/Steam Hammers —
10" D x 54" W x 48" C — 16,850 lb.
14" D x 80" W x 72" C — 33,700 lb.
19" D x 88" W x 72" C — 68,000 lb.
22" D x 144" W x 120" C — 137,000 lb.
Weights can vary due to hammer size and thickness of drive plate.

TRUSS LEAD

Also Called Monkey Stick. Spud Lead. European Lead

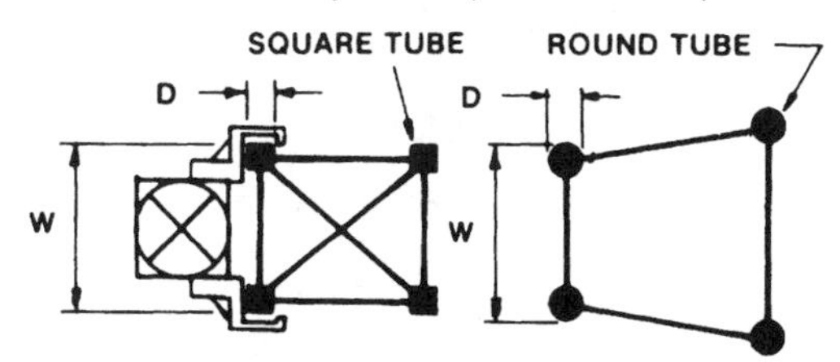

GUIDE PROFILES & APPROX. WEIGHTS:
3" D x 15¾" W — 60-100 lb./ft.
3" D x 21½" W — 60 lb./ft.
3" D x 28½" W — 100 lb./ft.
5" D x 28½" W — 150-190 lb./ft.
2" Dia. x 11" W — 30 lb./ft.
2¾" Dia. x 15¾" W — 70-125 lb./ft.

TRIANGULAR LEAD

Also Called Monkey Stick. Spud Lead. European Lead

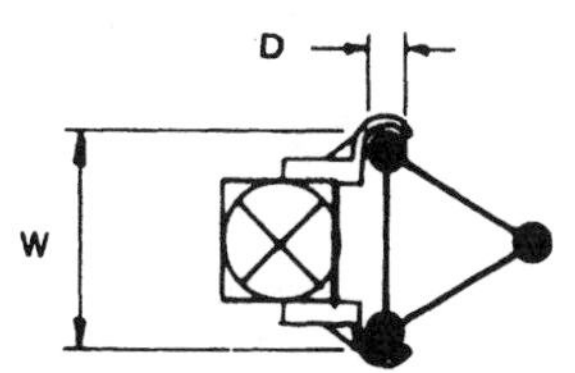

GUIDE PROFILES & APPROX. WEIGHTS:
2" Dia. x 11" W — 20 lb./ft.
2¾" Dia. x 15¾" W — 40 lb./ft.

Reprinted by permission with Deep Foundations Institute, Sparta New Jersey

1.1.8.5 Pile-Driving Rig in Batter Configurations

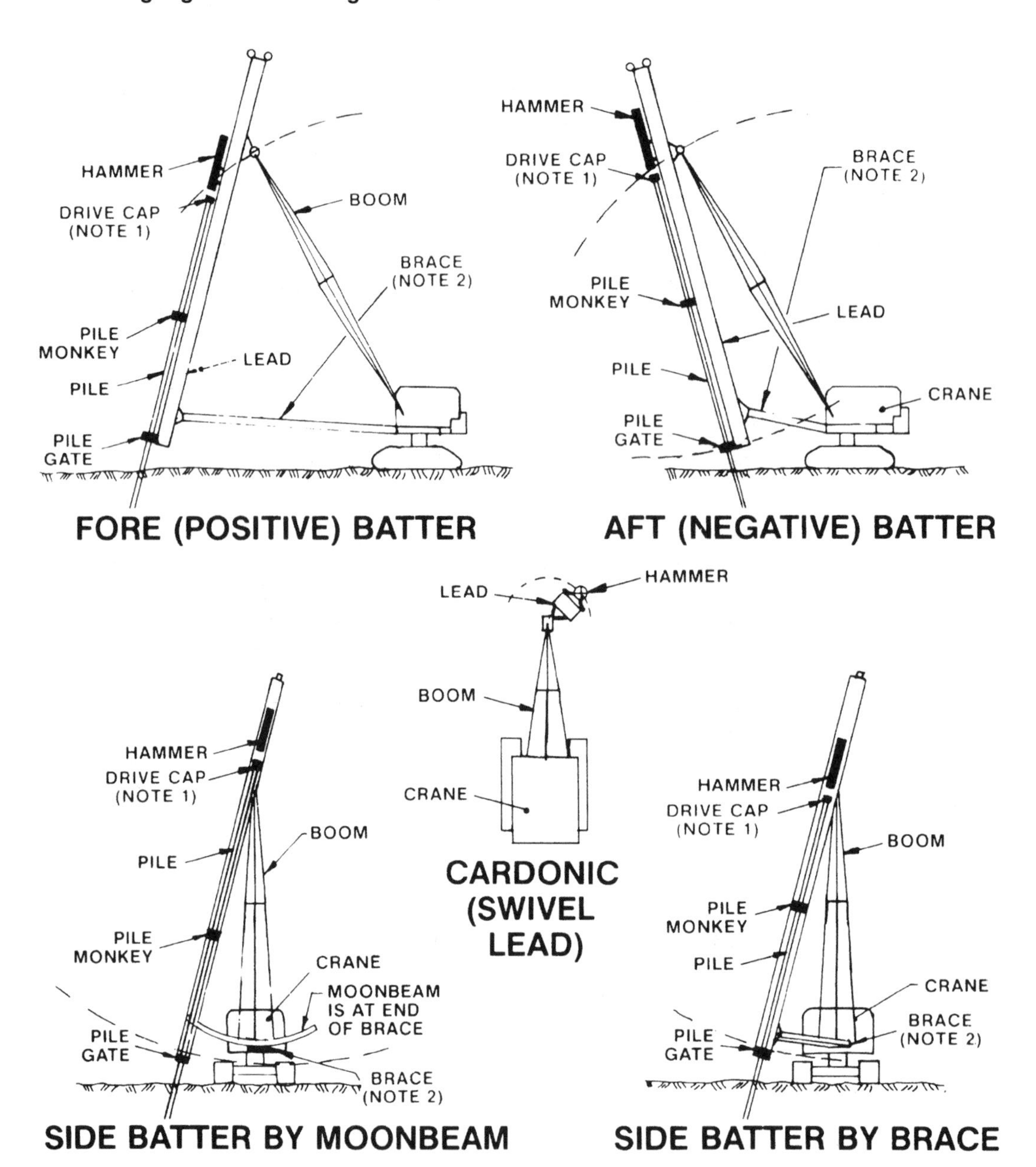

NOTE 1. Also called anvil block, bonnet, cap, driving head, follow cap, helmet, hood, rider cap.

NOTE 2. Also called A-frame, apron, bottom brace, bottom strut, kicker, parallelogram, platform, spider, spotter, spreader, spreader bars.

Reprinted by permission with Deep Foundations Institute, Sparta New Jersey

1.1.8.6 Pile-Driving Hammer Types

1. *The drop hammer* Rarely used, except for installing compacted-concrete piles.
2. *Single-acting hammers* Powered by steam or air pressure, which is used to raise the hammer ram for each down stroke. Gravity and the weight of the hammer deliver the kinetic energy required to drive the pile.
3. *Double-acting hammers* Generally powered by compressed air or hydraulics, which provides the power to raise the hammer ram and accelerate its fall.

4. *Vibratory hammers* Paired, oscillating rotating weights connected to the pile delivers anywhere from 0 to 2000 vibrations per minute at low frequency or from 0 to 8000 vibrations per minute for hihigh-frequency hammers to drive the pile to design depth. This type hammer is effective only in granular or cohesiveless soils.

1.1.8.7 Sheet Piles

When deep excavations are required in the near vicinity of existing structures, steel sheet piling or soldier beams with wood lagging are generally used to create a stable excavate. Where soil conditions permit, the sheet piling or soldier beams are driven to sufficient depth to allow for ample toe-in to support the sheeting or soldier beams once excavation reaches the required depth. If insufficient depth of soil is available, often in the case of soldier beams, they will be drilled into rock and supported with bracing.

1.1.8.8 Typical Sheet Pile Sections

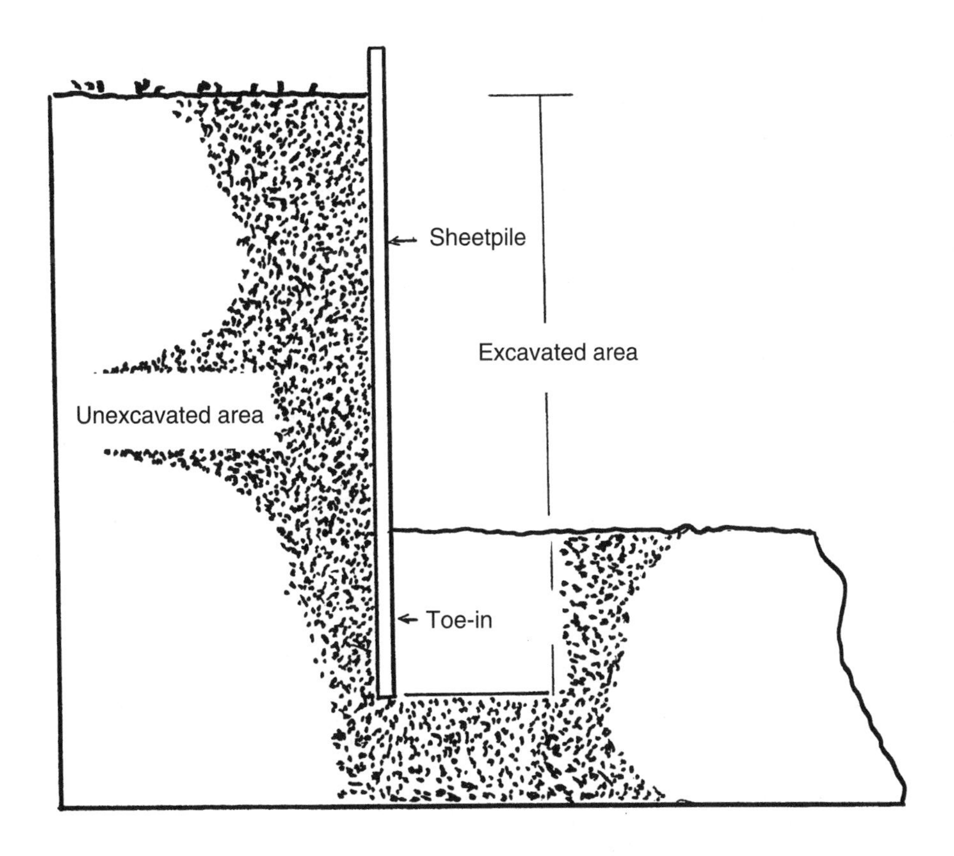

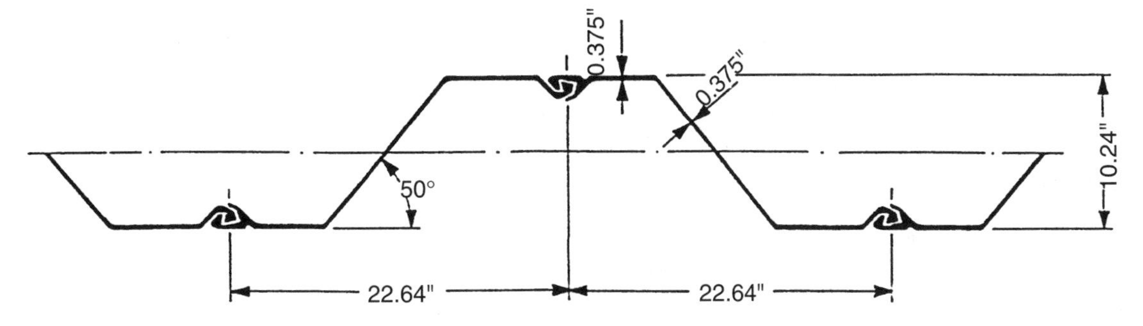

1.1.8.9 Steel H-Piles (Typical Sections and Specifications)

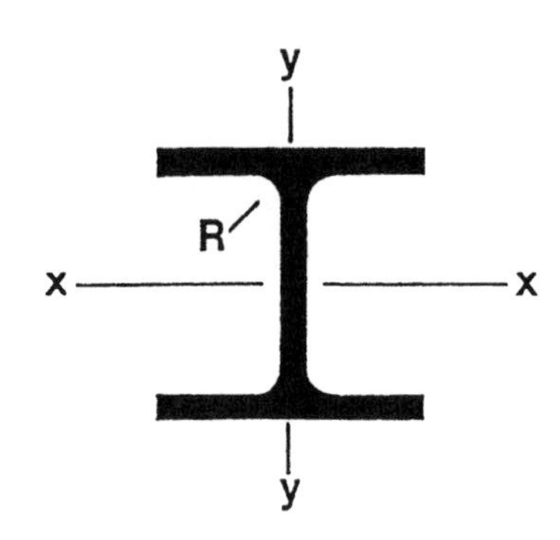

Properties for Design:

Designation and Nominal Size	Weight Per Foot	Area	Depth	Flange Width	Flange Thickness	Web Thickness	Fillet Rad. R	Surface Area	Axis x-x I	Axis x-x S	Axis x-x r	Axis y-y I	Axis y-y S	Axis y-y r
inch.	lb.	$inch^2$	inch	inch	inch	inch	inch	$ft.^2/ft.$	$inch^4$	$inch^3$	inch	$inch^4$	$inch^3$	inch
HP14 14 x 14½	117	34.4	14.21	14.885	.805	.805	.60	1024	1220	172	5.96	443	59.5	3.59
	102	30.0	14.01	14.785	.705	.705	.60	1017	1050	150	5.92	380	51.4	3.56
	89	26.1	13.83	14.695	.615	.615	.60	1010	904	131	5.88	326	44.3	3.53
	73	21.4	13.61	14.585	.505	.505	.60	1002	729	107	5.84	261	35.8	3.49
HP12 12 x 12	84	24.6	12.28	12.295	.685	.685	.60	860	650	106.0	5.14	213	34.6	2.94
	74	21.8	12.13	12.215	.610	.605	.60	850	569	93.8	5.11	186	30.4	2.92
	63	18.4	11.94	12.125	.515	.515	.60	844	472	79.1	5.06	153	25.3	2.88
	53	15.5	11.78	12.045	.435	.435	.60	838	393	66.8	5.03	127	21.1	2.86
HP10 10 x 10	57	16.8	9.99	10.225	.565	.565	.50	707	294	58.8	4.18	101	19.7	2.45
	42	12.4	9.70	10.075	.420	.415	.50	696	210	43.4	4.13	71.7	14.2	2.41
HP8 8 x 8	36	10.6	8.02	8.155	.445	.445	.40	565	119	29.8	3.36	40.3	9.88	1.95

H-Pile Specifications:

ASTM Grades	Yld. Point	Ten. Str.	Advantages
A-36	36,000	70,000	Basic Specification.
A-572 GR50	50,000	65,000	Higher yield.
A-690	50,000	70,000	2-3 times corrosion resistance in splash zone.

1.8.10 Typical Soldier Pile Detail (Rock Bearing)

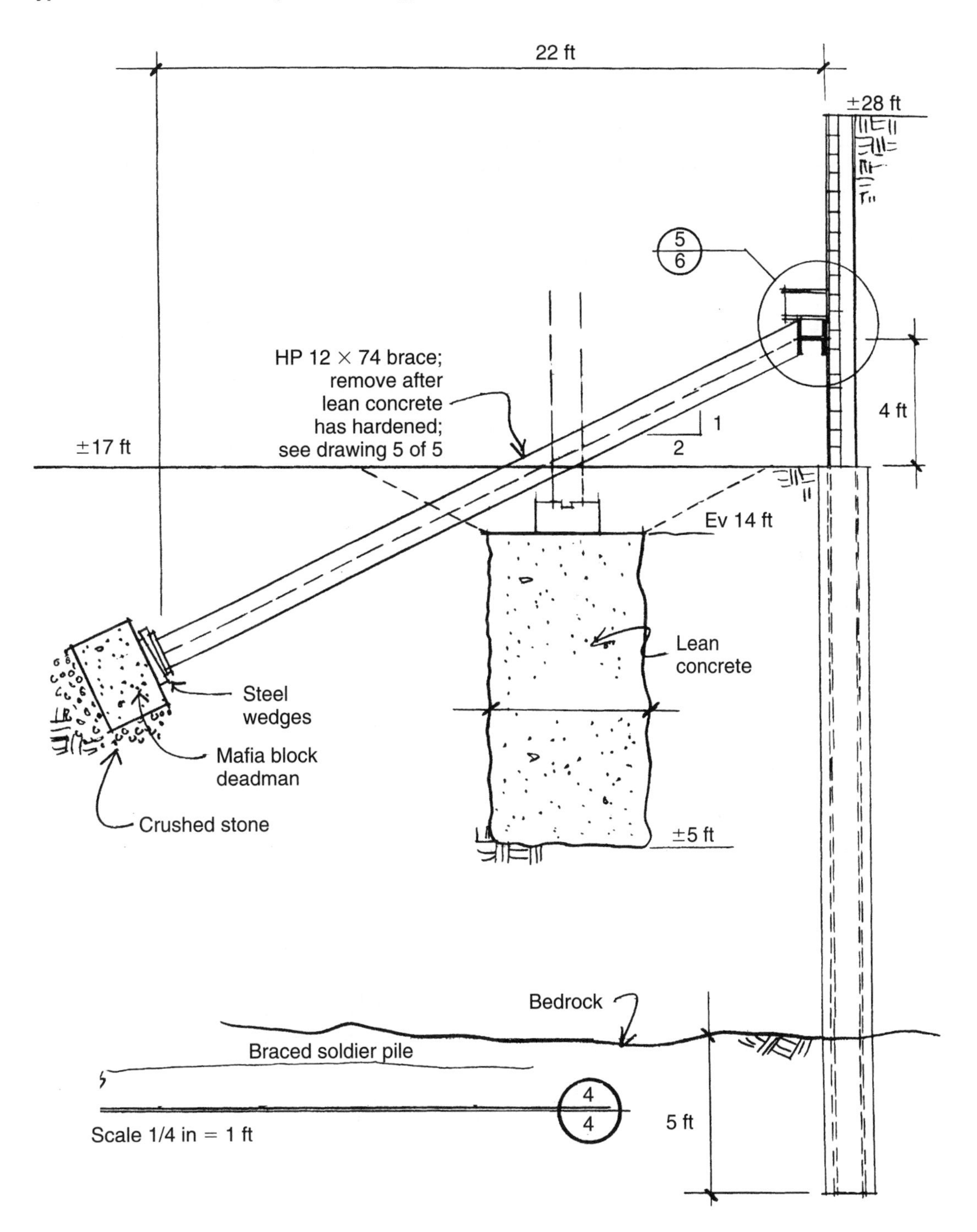

1.8.11 Typical Braced Soldier Pile with Deadman

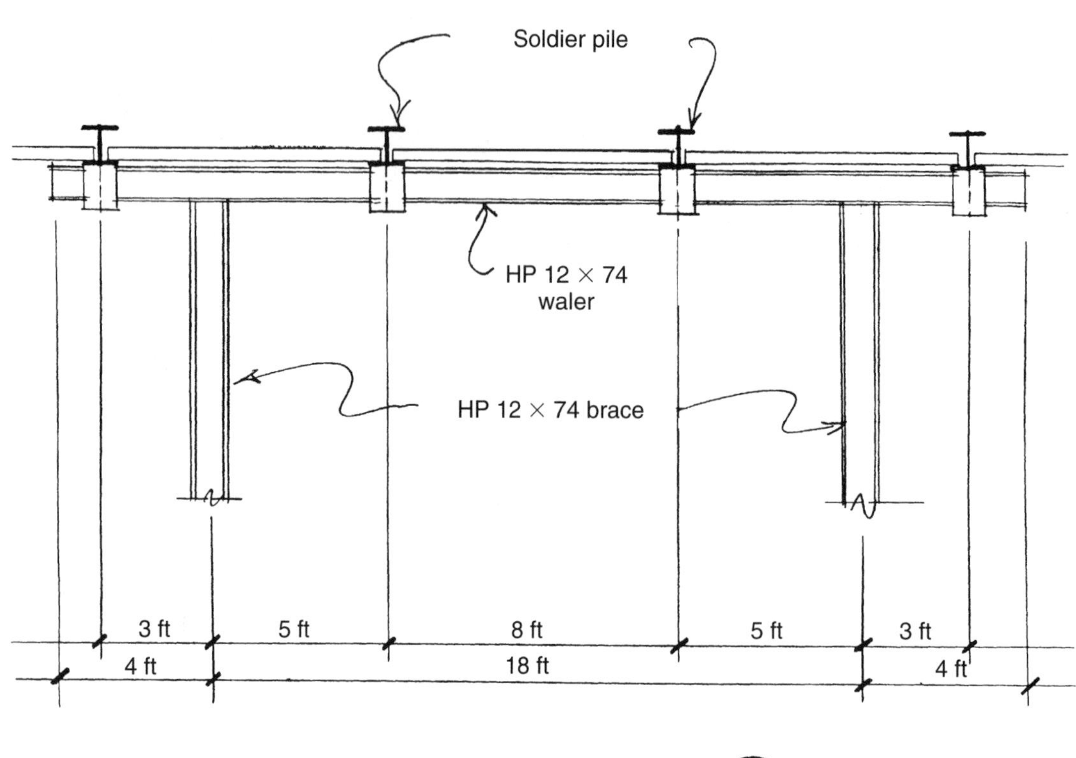

Bracing plan

Scale 1/4 = 1 ft

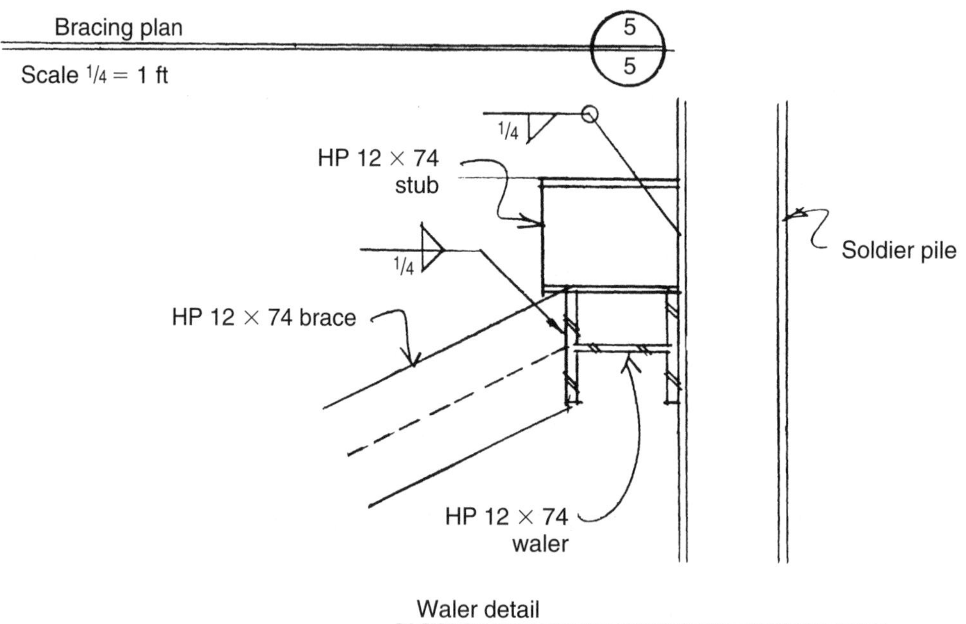

Waler detail

3/4 in = 1 ft

1.1.8.12 Typical Soldier Pile Drilled into Rock (Concrete Filled)

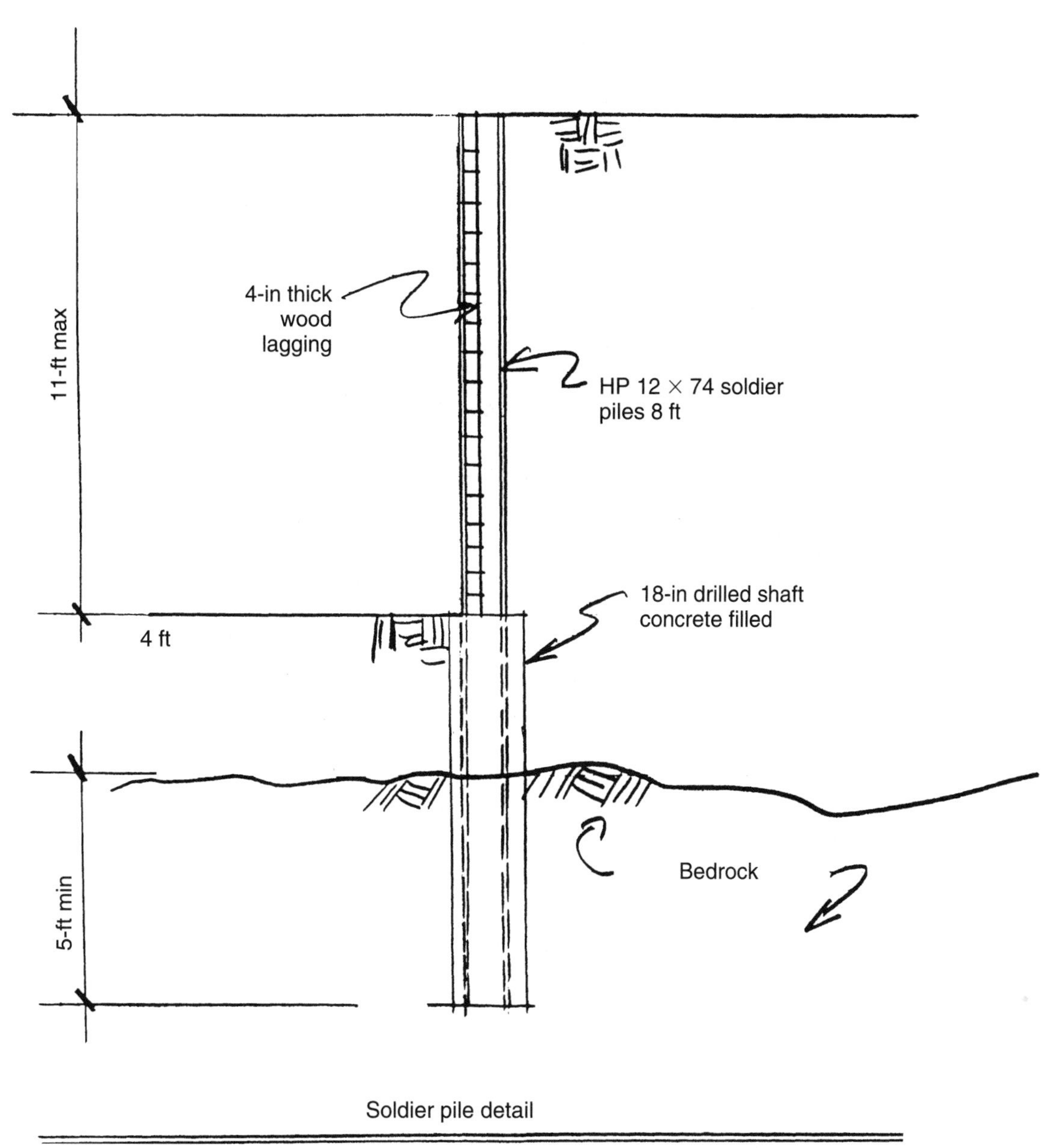

Soldier pile detail

Scale: 1/4 in = 1 ft

1.1.8.13 Typical Soldier Pile with Wood Lagging, Steel Walers, and Bracing

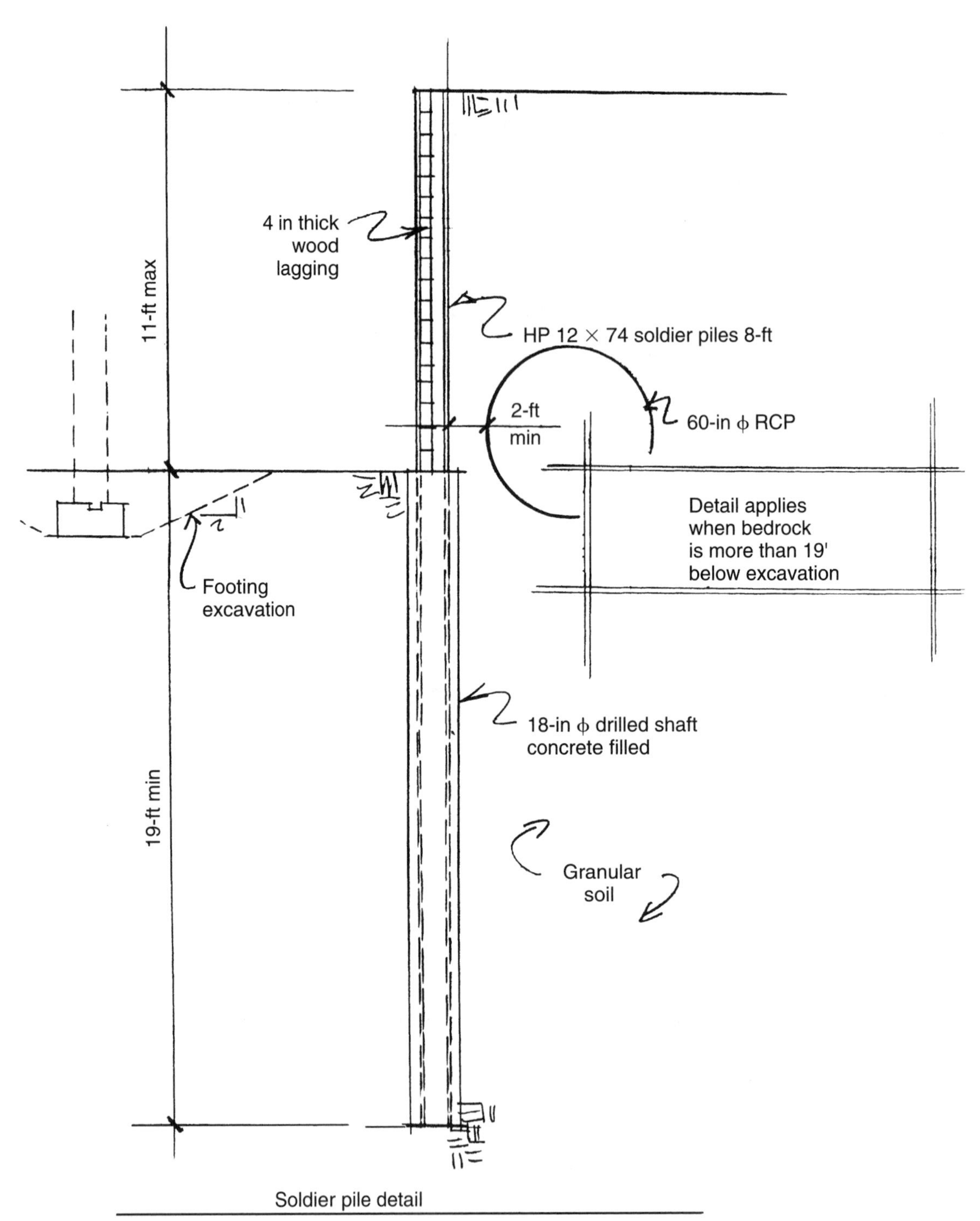

Soldier pile detail

Scale: 1/4 in = 1 ft

1.1.9 Dewatering

The lateral movement of water through soils is through the most pervious (porous) layer, but vertical movement is dictated by the least pervious or permeable soil layer. Dewatering can be as simple as excavating a pit, filling it partially with crushed stone, and inserting a pump. Or it can be as complex as creating a series of wells connected to a header pipe, which, in turn, is connected to a turbine pump operating under net suction conditions to draw the water to the surface.

These systems of creating well points are all based upon the need to draw down the water level by creating a radial flow from the aquifer, which will generally produce a funnel effect within the water table. This can be accomplished via single or multiple well points, but the drawdown procedure is basically the same; the utilization of a casing that contains a screened or slotted opening to which either a positive displacement or vacuum pump is attached to bring the ground water to the surface, taking into account loss of head through the screened/slotted head plus the depth of the casing and pump.

1.1.9.1 Diagrammatical Display of Water Flow from Wellpoints

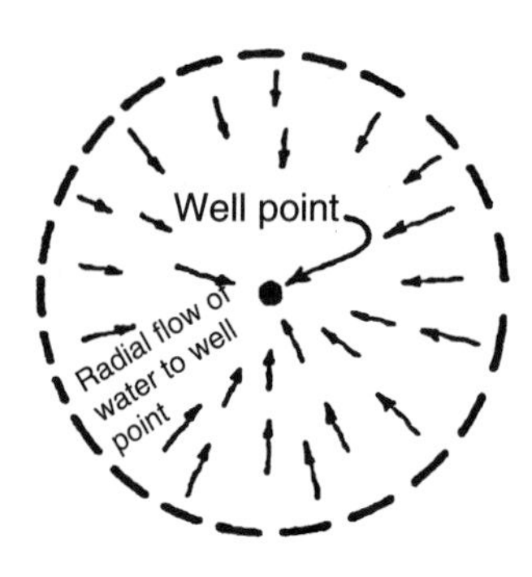

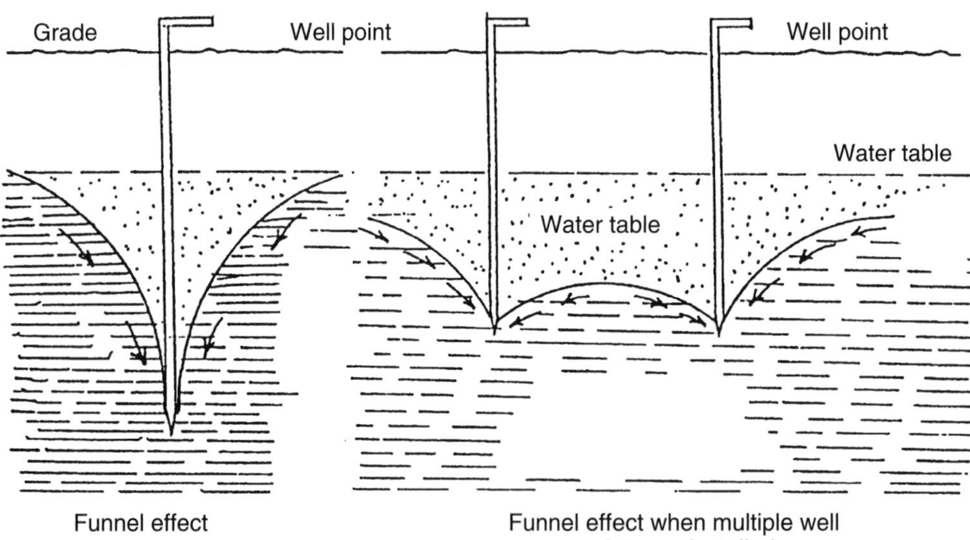

1.1.9.2 Drawdown (Diagrammatical)

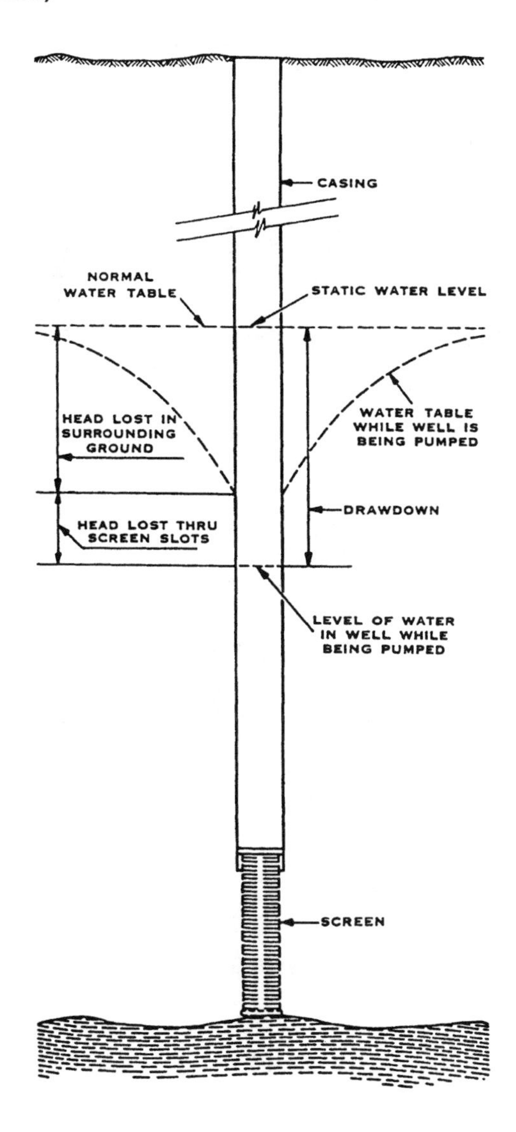

1.2.0 Protecting Foundations From Ground Water

Where foundations are being constructed and ground water is encountered it is standard procedure to protect the walls either fully or partially below grade, if that space is to be occupied. A prefabricated drainage system incorporating bitumastic coating on the exterior of the wall, a cupsated plastic drainage mat installed vertically and protected by a geotextile, will allow water to drop to a perforated footing drain to divert water away from the wall.

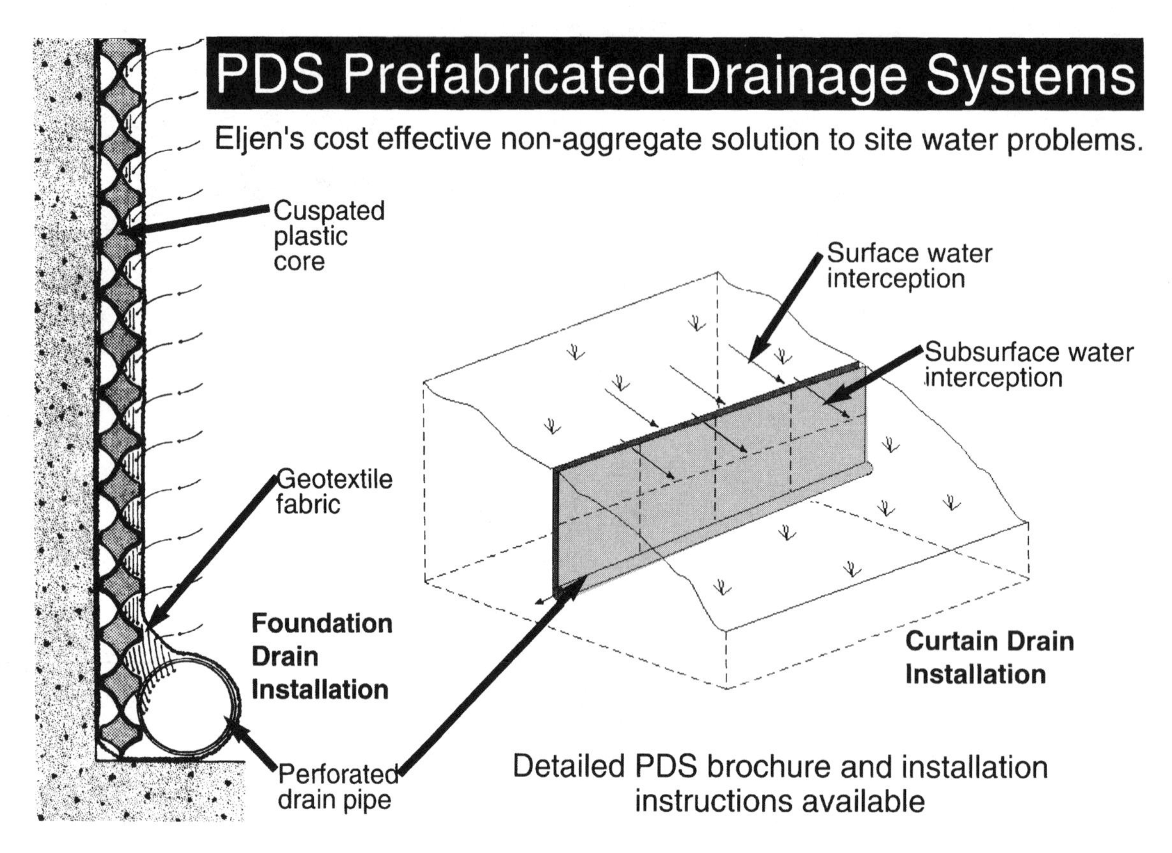

By permission of Eljen Corporation, Storrs, Connecticut

1.3.0 Hand Signals for Boom-Equipment Operators

1.4.0 Avoiding Bituminous Paving Pitfalls

POSSIBLE CAUSES OF DEFICIENCIES IN PLANT-MIX PAVEMENTS

Aggregates Too Wet	Inadequate Bunker Separation	Aggregate Feed Gates Not Properly Set	Over-rated Dryer Capacity	Dryer set too Steep	Improper Dryer Operation	Temp. Indicator Out of Adjustment	Aggregate Temperatures Too High	Worn Out Screens	Faulty Screen Operation	Bin Overflows Not Functioning	Leaky Bins	Segregation of Aggregates in Bins	Carryover in Bins Due to Overloading Screens	Aggregate Scales Out of Adjustment	Improper Weighing	Feed of Mineral Filler Not Uniform	Insufficient Aggregates in Hot Bins	Improper Weighing Sequence	Insufficient Asphalt	Too Much Asphalt	Faulty Distribution of Asphalt to Aggregates	Asphalt Scales Out of Adjustment	Asphalt Meter Out of Adjustment	Undersize or Oversize Batch	Mixing time not proper	Improperly Set or Worn Paddles	Faulty Dump Gate	Asphalt and Aggregate Feed Not Synchronized	Occasional Dust Shakedown in Bins	Irregular Plant Operation	Faulty Sampling	Types of Deficiencies That May Be Encountered in Producing Hot Plant-Mix Paving Mixtures.
		A												B	B				A	A	A	B	C	B	B	B		C			A	Asphalt Content Does Not Check Job Mix Formula
	A	A						B	B	B	B	A	A	B	B	B	A							B		B	B	C	B		A	Aggregate Gradation Does not Check Job Mix Formula
	A	A							B	B	B	A	A	B	B	B	A							B	B			C	B		A	Excessive Fines in Mix
A			A	A	A	A	A																							A		Uniform Temperatures Difficult to Maintain
											B			B	B									B								Truck Weights Do Not Check Batch Weights
														B	B					A	A	B	C	B		B		C				Free Asphalt on Mix in Truck
																		B									B					Free Dust on Mix in Truck
A			A	A	A	A													A		A	B	C	B	B	B		C		A		Large Aggregates Uncoated
									B	B	A	A	A	B	B	B	A	B			A	B	C		B	B	B	C	B	A		Mixture in Truck Not Uniform
																		B			A			B	B					A		Mixture in Truck Fat on One Side
					A															A	A	B	C	B				C		A		Mixture Flattens in Truck
		A			A	A	A																							A		Mixture Burned
A			A	A	A	A			B				A						A			B	C	B				C		A		Mixture Too Brown or Gray
														B	B	B	A			A	A	B	C	B				C		A		Mixture Too Fat
					A	A	A																							A		Mixture Smokes in Truck
A			A	A	A	A																								A		Mixture Steams in Truck
					A	A	A												A										A	A		Mixture Appears Dull in Truck

A—Applies to Batch and Drum Mix Plants. B—Applies to Batch Plants only. C—Applies to Drum Mix Plants only.

POSSIBLE CAUSES OF IMPERFECTIONS IN FINISHED PAVEMENTS

Insufficient or Non-Uniform Tack Coat	Improperly Cured Prime or Tack Coat	Mixture Too Coarse	Excess Fines in Mixture	Insufficient Asphalt	Excess Asphalt	Improperly Proportioned Mixture	Unsatisfactory Batches in Load	Excess Moisture in Mixture	Mixture Too Hot or Burned	Mixture Too Cold	Poor Spreader Operation	Spreader in Poor Condition	Inadequate Rolling	Over-Rolling	Rolling Mixture When Too Hot	Rolling Mixture When Too Cold	Roller Standing on Hot Pavement	Overweight Rollers	Too High Amplitude — Vibratory Roller	Improper Frequency — Vibratory Roller	Vibration or When Reversing or Stopped	Overrolling — Vibratory Roller	Excessive Moisture in Subsoil	Excessive Prime Coat or Tack Coat	Excessive Hand Raking	Labor Careless or Unskilled	Excessive Segregation in Laying	Operating Finishing Machine Too Fast	*Types of Pavement Imperfections That May Be Encountered In Laying Plant Mix Paving Mixtures.*
					X	X	X																	X					Bleeding
				X				X	X																				Brown, Dead Appearance
					X	X	X																	X			X		Rich or Fat Spots
		X	X			X	X			X	X	X	X	X	X	X									X	X	X	X	Poor Surface Texture
X	X	X				X	X			X	X	X	X		X	X	X	X	X	X	X	X			X	X	X	X	Rough Uneven Surface
		X		X		X	X			X	X	X	X			X									X	X	X		Honeycomb or Raveling
		X								X	X	X	X		X	X									X	X	X		Uneven Joints
			X		X	X				X			X		X	X	X	X			X	X				X			Roller Marks
X	X		X		X	X	X	X			X	X			X			X				X			X				Pushing or Waves
			X	X		X								X	X			X	X			X	X						Cracking (Many Fine Cracks)
														X				X				X	X						Cracking (Large Long Cracks)
		X				X				X	X	X		X	X			X	X		X	X							Rocks Broken by Roller
		X		X		X			X	X	X	X															X	X	Tearing of Surface During Laying
X	X		X		X	X		X		X			X	X		X		X	X			X	X	X					Surface Slipping on Base

By permission of the Asphalt Institute, Lexington, Kentucky

1.4.1 Calculating the Amount of Bitumastic Paving Per 100 Linear Feet and Per Mile

Width, Feet	Depth—Inches											
	1	2	3	4	5	6	7	8	9	10	11	12
1	0.31	0.62	0.93	1.23	1.54	1.85	2.16	2.47	2.78	3.09	3.40	3.70
2	0.62	1.23	1.85	2.47	3.09	3.70	4.32	4.94	5.56	6.17	6.79	7.41
3	0.93	1.85	2.78	3.70	4.63	5.56	6.48	7.41	8.33	9.26	10.20	11.10
4	1.23	2.47	3.70	4.94	6.17	7.41	8.64	9.88	11.10	12.30	13.60	14.80
5	1.54	3.09	4.63	6.17	7.72	9.26	10.80	12.30	13.90	15.40	17.00	18.50
6	1.85	3.70	5.56	7.41	9.26	11.10	13.00	14.80	16.70	18.50	20.40	22.20
7	2.16	4.32	6.48	8.64	10.80	13.00	15.10	17.30	19.40	21.60	23.80	25.90
8	2.47	4.94	7.41	9.88	12.30	14.80	17.30	19.80	22.20	24.70	27.20	29.60
9	2.78	5.56	8.33	11.10	13.90	16.70	19.40	22.20	25.00	27.80	30.60	33.30
10	3.09	6.17	9.26	12.30	15.40	18.50	21.60	24.70	27.80	30.90	34.00	37.00
20	6.17	12.30	18.50	24.70	30.90	37.00	43.20	49.40	55.60	61.70	67.90	74.10
30	9.26	18.50	27.80	37.00	46.30	55.60	64.80	74.10	83.30	92.60	102.00	111.00
40	12.30	24.70	37.00	49.40	61.70	74.10	86.40	98.80	111.00	123.00	136.00	148.00
50	15.40	30.90	46.30	61.70	77.20	92.60	108.00	123.00	139.00	154.00	170.00	185.00
60	18.50	37.00	55.60	74.10	92.60	111.00	130.00	148.00	167.00	185.00	204.00	222.00
70	21.60	43.20	64.80	86.40	108.00	130.00	151.00	173.00	194.00	216.00	238.00	259.00
80	24.70	49.40	74.10	98.80	123.00	148.00	173.00	198.00	222.00	247.00	272.00	296.00
90	27.80	55.60	83.30	111.00	139.00	167.00	194.00	222.00	250.00	278.00	306.00	333.00
100	30.90	61.70	92.60	123.00	154.00	185.00	216.00	247.00	278.00	309.00	340.00	370.00

Width, Feet	1	2	3	4	5	6	7	8	9	10	11	12
1	16.30	32.60	48.90	65.20	81.50	97.50	114.00	130.00	147.00	163.00	179.00	196.00
2	32.60	65.20	97.80	130.00	163.00	196.00	228.00	261.00	293.00	326.00	359.00	391.00
3	48.90	97.80	147.00	196.00	244.00	293.00	342.00	391.00	440.00	489.00	538.00	587.00
4	65.20	130.00	196.00	261.00	326.00	391.00	456.00	521.00	587.00	652.00	717.00	782.00
5	81.50	163.00	244.00	326.00	407.00	489.00	570.00	652.00	733.00	815.00	896.00	978.00
6	97.80	196.00	293.00	391.00	489.00	587.00	684.00	782.00	880.00	978.00	1,076.00	1,173.00
7	114.00	228.00	342.00	456.00	570.00	684.00	799.00	913.00	1,027.00	1,141.00	1,255.00	1,369.00
8	130.00	261.00	391.00	521.00	652.00	782.00	913.00	1,043.00	1,173.00	1,304.00	1,434.00	1,564.00
9	147.00	293.00	440.00	587.00	733.00	880.00	1,027.00	1,173.00	1,320.00	1,467.00	1,613.00	1,760.00
10	163.00	326.00	489.00	652.00	815.00	978.00	1,141.00	1,304.00	1,467.00	1,630.00	1,793.00	1,956.00
20	326.00	652.00	978.00	1,304.00	1,630.00	1,956.00	2,281.00	2,607.00	2,933.00	3,259.00	3,585.00	3,911.00
30	489.00	978.00	1,467.00	1,956.00	2,440.00	2,933.00	3,422.00	3,911.00	4,440.00	4,889.00	5,378.00	5,867.00
40	652.00	1,304.00	1,956.00	2,607.00	3,259.00	3,911.00	4,563.00	5,215.00	5,867.00	6,519.00	7,170.00	7,822.00
50	815.00	1,630.00	2,444.00	3,259.00	4,074.00	4,889.00	5,704.00	6,519.00	7,333.00	8,148.00	8,963.00	9,778.00
60	978.00	1,956.00	2,933.00	3,911.00	4,889.00	5,867.00	6,844.00	7,822.00	8.800.00	9,778.00	10,756.00	11,733.00
70	1,141.00	2,281.00	3,422.00	4,563.00	5,704.00	6,844.00	7,985.00	9,126.00	10,267.00	11,407.00	12,548.00	13,689.00
80	1,304.00	2,607.00	3,911.00	5,215.00	6,519.00	7,822.00	9,126.00	10,430.00	11,733.00	13,037.00	14,341.00	15,644.00
90	1,467.00	2,933.00	4,400.00	5,867.00	7,333.00	8,800.00	10.267.00	11,733.00	13,200.00	14,667.00	16,133.00	17,600.00
100	1,630.00	3,259.00	4,889.00	6,519.00	8,148.00	9,778.00	11,407.00	13,037.00	14,667.00	16,296.00	17,926.00	19,556.00

NOTE: Formulas used for calculations: Per 100 Lin Ft: $q=\left(\frac{D}{36}\right)\left(\frac{W}{3}\right)\left(\frac{100}{3}\right)=0.3086\ DW$

Per Mile: $q=\left(\frac{D}{36}\right)\left(\frac{W}{3}\right)\left(\frac{5{,}280}{3}\right)=16.2963\ DW$

where: q = Quantity of material, cubic yards
D = Depth, inches
W = Width, feet
L = Length

1.4.1.1 Linear Feet of Paving Covered by One Ton of Material

	Width—Metres											
kg/m2	2	2.5	3	3.5	4	4.5	5	5.5	6	6.5	7	7.5
5	100.0	80.0	66.7	57.1	50.0	44.4	40.0	36.4	33.3	30.8	28.6	26.7
10	50.0	40.0	33.3	28.6	25.0	22.2	20.0	18.2	16.7	15.4	14.3	13.3
15	33.3	26.7	22.2	19.0	16.7	14.8	13.3	12.1	11.1	10.3	9.5	8.9
20	25.0	20.0	16.7	14.3	12.5	11.1	10.0	9.1	8.3	7.7	7.1	6.7
25	20.0	16.0	13.3	11.4	10.0	8.9	8.0	7.3	6.7	6.2	5.7	5.3
30	16.7	13.3	11.1	9.5	8.3	7.4	6.7	6.1	5.6	5.1	4.8	4.4
35	14.3	11.4	9.5	8.2	7.1	6.3	5.7	5.2	4.8	4.4	4.1	3.8
40	12.5	10.0	8.3	7.1	6.3	5.6	5.0	4.5	4.2	3.8	3.6	3.3
45	11.1	8.9	7.4	6.3	5.6	4.9	4.4	4.0	3.7	3.4	3.2	3.0
50	10.0	8.0	6.7	5.7	5.0	4.4	4.0	3.6	3.3	3.1	2.9	2.7
60	8.3	6.7	5.6	4.8	4.2	3.7	3.3	3.0	2.8	2.6	2.4	2.2
70	7.1	5.7	4.8	4.1	3.6	3.2	2.9	2.6	2.4	2.2	2.0	1.9
80	6.3	5.0	4.2	3.6	3.1	2.8	2.5	2.3	2.1	1.9	1.8	1.7
90	5.6	4.4	3.7	3.2	2.8	2.5	2.2	2.0	1.9	1.7	1.6	1.5
100	5.0	4.0	3.3	2.9	2.5	2.2	2.0	1.8	1.7	1.5	1.4	1.3
200	2.5	2.0	1.7	1.4	1.3	1.1	1.0	0.9	0.8	0.8	0.7	0.7
300	1.7	1.3	1.1	1.0	0.8	0.7	0.7	0.6	0.6	0.5	0.5	0.4
400	1.3	1.0	0.8	0.7	0.6	0.6	0.5	0.5	0.4	0.4	0.4	0.3
500	1.0	0.8	0.7	0.6	0.5	0.4	0.4	0.4	0.3	0.3	0.3	0.3
600	0.8	0.7	0.6	0.5	0.4	0.4	0.3	0.3	0.3	0.3	0.2	0.2

NOTE: Formula used for calculation: $L = \frac{1000}{RW}$

where: L = Linear metres covered by one megagram of material
R = Rate of spread, kg/m^2
W = Width of spread, metres

By permission of the Asphalt Institute, Lexington, Kentucky

1.4.1.2 Linear Meters of Asphalt Covered by One Ton of Material

	Width—Feet									
lb/yd2	8	9	10	11	12	13	14	15	16	17
20	112.5	100.0	90.0	81.8	75.0	69.2	64.3	60.0	56.3	52.9
25	90.0	80.0	72.0	65.5	60.0	55.4	51.4	48.0	45.0	42.4
30	75.0	66.7	60.0	54.5	50.0	46.2	42.9	40.0	37.5	35.3
35	64.3	57.1	51.4	46.8	42.9	39.6	36.7	34.3	32.1	30.3
40	56.3	50.0	45.0	40.9	37.5	34.6	32.1	30.0	28.1	26.5
45	50.0	44.4	40.0	36.4	33.3	30.8	28.6	26.7	25.0	23.5
50	45.0	40.0	36.0	32.7	30.0	27.7	25.7	24.0	22.5	21.2
60	37.5	33.3	30.0	27.3	25.0	23.1	21.4	20.0	18.8	17.6
70	32.1	28.6	25.7	23.4	21.4	19.8	18.4	17.1	16.1	15.1
80	28.1	25.0	22.5	20.5	18.8	17.3	16.1	15.0	14.1	13.2
90	25.0	22.2	20.0	18.2	16.7	15.4	14.3	13.3	12.5	11.8
100	22.5	20.0	18.0	16.4	15.0	13.8	12.9	12.0	11.3	10.6
150	15.0	13.3	12.0	10.9	10.0	9.2	8.6	8.0	7.5	7.1
200	11.3	10.0	9.0	8.2	7.5	6.9	6.4	6.0	5.6	5.3
250	9.0	8.0	7.2	6.5	6.0	5.5	5.1	4.8	4.5	4.2
300	7.5	6.7	6.0	5.5	5.0	4.6	4.3	4.0	3.8	3.5
400	5.6	5.0	4.5	4.1	3.8	3.5	3.2	3.0	2.8	2.6
500	4.5	4.0	3.6	3.3	3.0	2.8	2.6	2.4	2.2	2.1
600	3.7	3.3	3.0	2.7	2.5	2.3	2.1	2.0	1.9	1.8
700	3.2	2.9	2.6	2.3	2.1	2.0	2.0	1.7	1.6	1.5
800	2.8	2.5	2.3	2.0	1.9	1.7	1.6	1.5	1.4	1.3
900	2.5	2.2	2.0	1.8	1.7	1.5	1.4	1.3	1.2	1.2
1,000	2.3	2.0	1.8	1.6	1.5	1.4	1.3	1.2	1.1	1.1
1,100	2.0	1.8	1.6	1.5	1.4	1.3	1.2	1.1	1.0	1.0

NOTE: Formula used for calculation: $L = \frac{2{,}000(9)}{RW} = \frac{18{,}000}{RW}$

where: L = Linear feet covered by one ton of material
R = Rate of spread, lb/ft^2
W = Width of spread, feet

By permission of the Asphalt Institute, Lexington, Kentucky

1.4.2 Cubic Meters of Asphalt for Various Widths/Depths of Paving (Metric)

	Width, Metres	Depth—Millimetres											
		20	40	60	80	100	120	140	160	180	200	220	240
PER 50 LINEAR METRES	1	1.00	2.00	3.00	4.00	5.00	6.00	7.00	8.00	9.00	10.00	11.00	12.00
	2	2.00	4.00	6.00	8.00	10.00	12.00	14.00	16.00	18.00	20.00	22.00	24.00
	3	3.00	6.00	9.00	12.00	15.00	18.00	21.00	24.00	27.00	30.00	33.00	36.00
	4	4.00	8.00	12.00	16.00	20.00	24.00	28.00	32.00	36.00	40.00	44.00	48.00
	5	5.00	10.00	15.00	20.00	25.00	30.00	35.00	40.00	45.00	50.00	55.00	60.00
	6	6.00	12.00	18.00	24.00	30.00	36.00	42.00	48.00	54.00	60.00	66.00	72.00
	7	7.00	14.00	21.00	28.00	35.00	42.00	49.00	56.00	63.00	70.00	77.00	84.00
	8	8.00	16.00	24.00	32.00	40.00	48.00	56.00	64.00	72.00	80.00	88.00	96.00
	9	9.00	18.00	27.00	36.00	45.00	54.00	63.00	72.00	81.00	90.00	99.00	108.00
	10	10.00	20.00	30.00	40.00	50.00	60.00	70.00	80.00	90.00	100.00	110.00	120.00
	15	15.00	30.00	45.00	60.00	75.00	90.00	105.00	120.00	135.00	150.00	165.00	180.00
	20	20.00	40.00	60.00	80.00	100.00	120.00	140.00	160.00	180.00	200.00	220.00	240.00
	25	25.00	50.00	75.00	100.00	125.00	150.00	175.00	200.00	225.00	250.00	275.00	300.00
	30	30.00	60.00	90.00	120.00	150.00	180.00	210.00	240.00	270.00	300.00	330.00	360.00

By permission of the Asphalt Institute, Lexington, Kentucky

1.4.3 Tons of Asphalt Required Per Mile for Various Widths/Pounds Per Yard

	Width—Feet														
lb/yd²	1	2	3	4	5	6	7	8	9	10	20	30	40	50	60
10	2.9	5.9	8.8	11.7	14.7	17.6	20.5	23.5	26.4	29.3	58.7	88.0	117.3	146.7	176.0
20	5.9	11.7	17.6	23.5	29.3	35.2	41.1	46.9	52.8	58.7	117.3	176.0	234.7	293.3	352.0
30	8.8	17.6	26.4	35.2	44.0	52.8	61.6	70.4	79.2	88.0	176.0	264.0	352.0	440.0	527.9
40	11.7	23.5	35.2	46.9	58.7	70.4	82.1	93.9	105.6	117.3	234.7	352.0	469.3	586.7	704.0
50	14.7	29.3	44.0	58.7	73.3	88.0	102.7	117.3	132.0	146.7	293.3	440.0	586.7	733.3	880.0
60	17.6	35.2	52.8	70.4	88.0	105.6	123.2	140.8	158.4	176.0	352.0	528.0	704.0	880.0	1,056.0
70	20.5	41.1	61.6	82.1	102.7	123.2	143.7	164.3	184.8	205.3	410.7	616.0	821.3	1,026.7	1,232.0
80	23.5	46.9	70.4	93.9	117.3	140.8	164.3	187.7	211.2	234.7	469.3	704.0	938.7	1,173.3	1,408.0
90	26.4	52.8	79.2	105.6	132.0	158.4	184.8	211.2	237.6	264.0	528.0	792.0	1,056.0	1,320.0	1,584.0
100	29.3	58.7	88.0	117.3	146.7	176.0	205.3	234.7	264.0	293.3	586.7	880.0	1,173.3	1,466.7	1,760.0
200	58.7	117.3	176.0	234.7	293.3	352.0	410.7	469.3	528.0	586.7	1,173.3	1,760.0	2,346.7	2,933.3	3,520.0
300	88.0	176.0	264.0	352.0	440.0	528.0	616.0	704.0	792.0	880.0	1,760.0	2,640.0	3,520.0	4,400.0	5,280.0
400	117.3	234.7	352.0	469.3	586.7	704.0	821.3	938.7	1,056.0	1,173.3	2,346.7	3,520.0	4,693.3	5,866.7	7,040.0
500	146.7	293.3	440.0	586.7	733.3	880.0	1,026.7	1,173.3	1,320.0	1,466.7	2,933.3	4,400.0	5,866.7	7,333.3	8,800.0
600	176.0	352.0	528.0	704.0	880.0	1,056.0	1,232.0	1,408.0	1,584.0	1,760.0	3,520.0	5,280.0	7,040.0	8,800.0	10,560.0
700	205.3	410.7	616.0	821.3	1,026.7	1,232.0	1,437.3	1,642.7	1,848.0	2,053.3	4,106.7	6,160.0	8,213.3	10,266.7	12,320.0
800	234.7	469.3	704.0	938.7	1,173.3	1,408.0	1,642.7	1,877.3	2,112.0	2,346.7	4,693.3	7,040.0	9,386.7	11,733.3	14,080.0
900	264.0	528.0	792.0	1,056.0	1,320.0	1,584.0	1,848.0	2,112.0	2,376.0	2,640.0	5,280.0	7,920.0	10,560.0	13,200.0	15,840.0
1,000	293.3	586.7	880.0	1,173.3	1,466.7	1,760.0	2,053.3	2,346.7	2,640.0	2,933.3	5,866.7	8,800.0	11,733.3	14,666.7	17,600.0

NOTE: Formula used for calculation: $w = \left(\frac{W}{3}\right)\left(\frac{5280}{3}\right)\left(\frac{R}{2000}\right) = 0.2933\ RW$

where: w = Weight of material, tons per mile
R = Rate of application, lb/yd²
W = Width of application, feet

By permission of the Asphalt Institute, Lexington, Kentucky

1.4.4 Asphalt Paving Block Specifications

Typical Applications	Thickness of Unit Recommended
Industrial Floors	1½″, 2″ or 2½″
Warehouse, Baggage and Express Room Floors	1½″ or 2″
Traffic Aisles and Loading Platforms	1½″ or 2″
Piers and Docks	1½″ or 2″
Roof Decks—Parking or Storage	1½″
Roof Decks and Balconies—Recreational	1¼″ or 1½″
Airport, Hangars, Runways, Aprons	1½″, 2″ or 2½″
Ramps and Bridge Approaches	2½″ or 3″
Streets, Roads, Bridges, Viaducts	2½″ or 3″
Waterproofing Protection Courses	1¼″
Estate, Residential and Institutional Driveways	2″ {Hexagonal or Rectangular
Walks, Courts, Plazas and Terraces	2″ {Hexagonal or Rectangular

To convert in. to mm multiply by 25.4.

Size	Pounds per Block	Pounds per sq. ft.	Pounds per sq. yd.	Net Tons per Thousand Blocks	Number of Blocks		Per M Blocks	
					per sq. ft.	per sq. yd.	sq. ft.	sq. yd.
5″×12″×1¼″	6.6	15.6	140	3.30	2.4	21.6	423	47
5″×12″×1½″	7.9	18.7	168	3.95	2.4	21.6	423	47
5″×12″×2″	10.6	25.1	226	5.30	2.4	21.6	423	47
5″×12″×2½″	13.1	31.0	279	6.55	2.4	21.6	423	47
5″×12″×3″	16.0	37.8	340	8.00	2.4	21.6	423	47
6″×12″×1¼″	7.9	15.6	140	3.95	2.0	18.0	505	56
6″×12″×1½″	9.5	18.8	169	4.75	2.0	18.0	505	56
6″×12″×2″	12.67	25.1	226	6.335	2.0	18.0	505	56
6″×12″×2½″	15.8	31.3	282	7.90	2.0	18.0	505	56
6″×12″×3″	19.0	37.6	338	9.50	2.0	18.0	505	56
8″ Hex* 1¼″	5.97	15.5	140	2.985	2.6	23.4	385	43
8″ Hex* 1½″	7.12	18.5	167	3.56	2.6	23.4	385	43
8″ Hex* 2″	9.53	24.8	223	4.765	2.6	23.4	385	43
8″×8″×1¼″	7.04	15.8	142	3.52	2.24	20.2	446	50
8″×8″×1½″	8.4	18.8	169	4.20	2.24	20.2	446	50
8″×8″×2″	11.2	25.1	226	5.60	2.24	20.2	446	50

*Hexagonal blocks
For metric conversion factors refer to Chapter III.

By permission of the Asphalt Institute, Lexington, Kentucky

Section

2

Site Utilities

Contents

2.1.0 Installation of piping (general)
2.1.1 Bedding materials and terminology
2.1.1.1 Pipe-zone bedding materials
2.1.2 Setting up batter boards
2.1.3 Trench preparation for concrete pipe
2.1.4 Trench requirements for rigid and thermoplastic pipe
2.1.4.1 Trench width consideration
2.1.5 Assembling a trench shield
2.1.5.1 Determining trench shield size
2.1.6 Reinforced concrete pipe (RCP) specification ASTMC76 T&G joints
2.1.6.1 Reinforced concrete pipe (RCP) specifications ASTMC14, ASTMC76 (bell and spigot)
2.1.6.2 Installation procedures for bell and spigot concrete pipe
2.1.7 Trench recommendations for cast-iron soil pipe
2.1.7.1 Depth of cover for cast-iron soil pipe/rated working pressures
2.1.7.2 Equivalent sizes for cast-iron soil pipe
2.1.7.3 Ring test crushing loads on cast-iron soil pipe
2.1.7.4 Slopes required for self-cleaning cast-iron pipe
2.1.7.5 Typical pipe-joining methods for cast-iron soil pipe
2.1.8 Ductile iron-pipe dimensions and weight for push/on mechanical joints
2.1.8.1 Ductile iron-pipe specification and assembly tips
2.1.8.2 Ductile iron specifications (3" to 36" pipe)
2.1.9 Schedule 40/80 PVC pipe specifications
2.1.9.1 Deflection of thermoplastic pipe
2.1.9.2 Expansion and contraction of PVC pipe
2.2.0 Corrugated steel-pipe specifications (12" to 144" diameter)
2.2.1 Corrugated steel-pipe specifications (arch height of cover limits, 3" × 1")
2.2.1.1 Corrugated steel-pipe specifications (arch height of cover limits, 5" × 1")
2.2.1.2 Corrugated aluminum/galvanized steel-pipe specifications
2.2.2 Expansion characteristics of various metal/thermoplastic pipes
2.3.0 Testing of underground pipe installations
2.3.1 Diagram of infiltration tests
2.3.2 Diagram of exfiltration tests
2.4.0 Storm and sanitary manhole schematics with sections
2.4.1 Storm sewer manhole components
2.5.0 Casting for sanitary and storm manholes

2.1.0 Installation of Piping (General)

A great deal of construction activity involves the installation or rerouting of underground utilities (sanitary and storm sewers, domestic water lines and fire mains, electrical and telecommunications services, and natural gas lines). The nature and variety of these installations vary substantially from site to site, but the basic materials of construction generally do not.

Underground site utility work consists of the installation of conduits of various sizes and materials of construction to carry these utilities; the basic piping materials are either reinforced concrete pipe, thermoplastics, cast iron or ductile iron, lightweight aluminum or steel and corrugated metal pipe.

Installation of these types of pipes have several things in common:

1. *Excavation and pipe laying* Depending upon the type of soil and the width and depth of excavation, either "open cut" will be used or a trench cut (utilizing sheet piling) or a trench box (to avoid collapse of the walls of the excavate).
2. *Bedding material* Depending upon the type of pipe being installed and the nature of the subsoil, off-site bedding materials might be required, not only to place under the pipe, but for initial backfill
3. Compaction of the soil above the pipe will also depend on the depth of the excavate, the soil conditions, and the percentage compaction required.

2.1.1 Bedding and Backfill Materials For Site Utility Work and the Pipe Zone

To discuss the backfill procedures for underground pipes, it is necessary to understand pipe zone terminology.

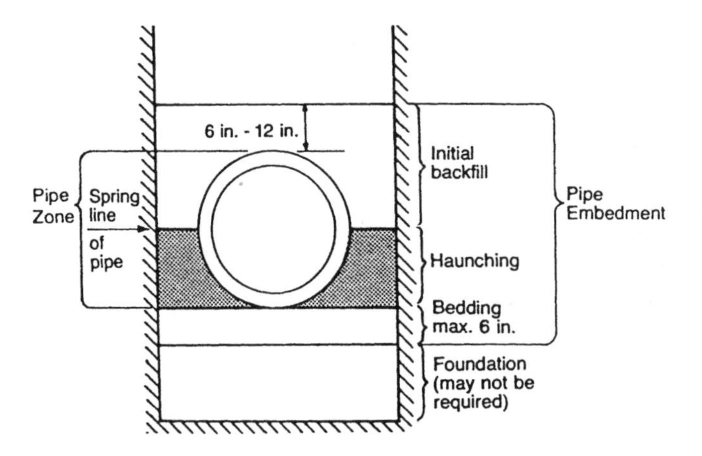

- *Foundation* Might not be required if the trench bottom is stable and will support a rigid pipe without causing deviation in grade or such flexing of the pipe that will create flexural failure.
- *Bedding* This material is required to bring the trench bottom up to grade and to provide uniform longitudinal support. Sand is often used for this purpose.
- *Haunching* The material used in this zone will supply structural support for the pipe and prevent it from deflecting (if it is a flexible pipe) or having joint misalignment when further backfilling and compaction above occurs.
- *Initial backfill* Material placed 6 to 12 inches above the spring line will only provide additional side support, most of the support coming from compaction of the soil in the haunching area.

2.1.1.1 Pipe-Zone Bedding Materials

- *Class I* Angular stone, graded from ¼" (6.4mm) to ½" (12.7 mm), including crushed stone, crushed shells, and cinders.

- *Class II* Coarse sand with a maximum particle size of 1½" (38.1 mm), including various graded sands and gravel containing small percentages of fines. Soil type GW, SP, SM, and C * (See the unified soil classification listing in Chapter 1).
- *Class III* Fine sand and clayey gravel, including fine sand, sand-clay mixtures, and gravel-clay mixes. Soil types GM, GC, SM, and SC are included in this class.
- *Class IV* Silt, silty clays (including inorganic clays), and silts of medium to high plasticity and liquid limits. Soil types MH, ML, CH, and CL are included in this class.
- *Class V* Soils not recommended for bedding, haunching, or initial backfill consisting of organic silts, organic clays and peat, and other highly organic materials.

Common sense, experience, and OSHA regulations will dictate the precautions required during site utilities excavation. OSHA Handbook *Title 29 of the Code of Federal Regulations* (29 CFR Part 1926) is to be referred to for detailed regulations regarding excavation and trenching operations. OSHA *Construction Industry Digest* (OSHA 2202) is a pocket-sized digest of basic applicable standards, including excavation and trenching. This handy booklet can be obtained by calling the local U.S. Department of Labor office.

2.1.2 Setting Up Batter Boards and Determining Trench Widths and Depths

The figure shows how to set up a batter boards so as to establish a center line of underground pipe.

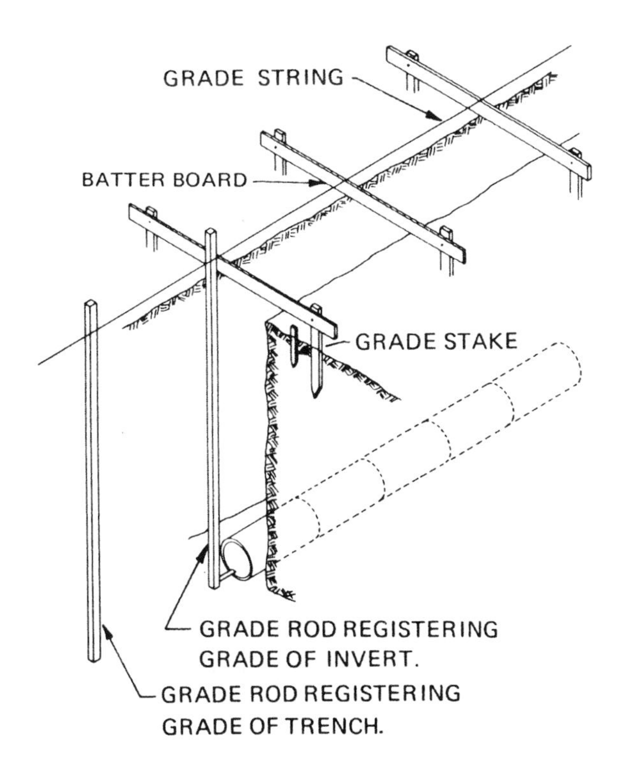

2.1.3 Trench Preparation for Concrete Pipe

Trench width required to install concrete pipe of varying pipe diameters (metric conversion).

Pipe, diameter, in (mm)	Trench width, ft	Pipe diameter, in (in)	Trench width, ft
4	1.6	60	8.5
6	1.8	66	9.2
8	2.0	72	10.0
10	2.3	78	10.7
12	2.5	84	11.4
15	3.0	90	12.1
18	3.4	96	12.9
21	3.8	102	13.6
24	4.1	108	14.3
27	4.5	114	14.9
33	5.2	120	15.6
36	5.6	126	16.4
42	6.3	132	17.1
48	7.0	138	17.8
54	7.8	144	18.5
100	0.47	150	25.0
150	0.54	165	28.0
200	0.60	180	30.0
250	0.68	195	32.0
300	0.80	210	34.0
375	0.91	225	36.0
450	1.02	240	39.0
525	1.10	255	41.0
600	1.20	270	43.0
675	13.0	285	45.0
825	16.0	300	48.0
900	17.0	315	50.0
1050	19.0	330	52.0
1200	21.0	345	54.0
1350	23.0	360	56.0

NOTE: Trench widths based on 1.25 Bc + 1 ft where Bc is the outside diameter of the pipe in inches, and + 300 where Bc is the outside diameter of the pipe in millimeters.

2.1.4 Trench Requirements for Rigid and Thermoplastic Pipe

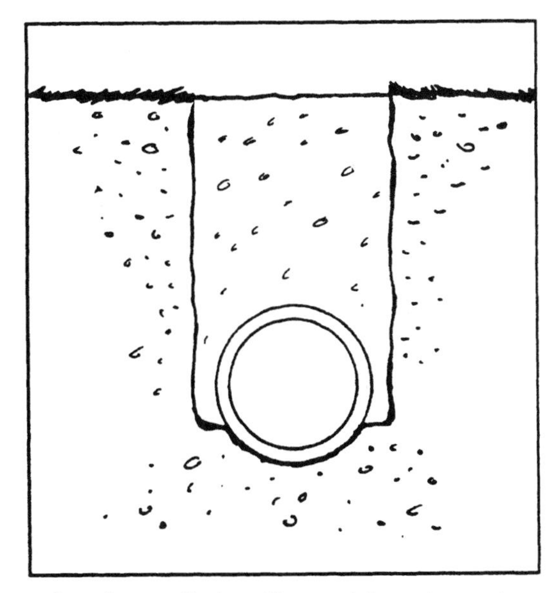

Cast-iron soil pipe No special requirements for trench width needed.

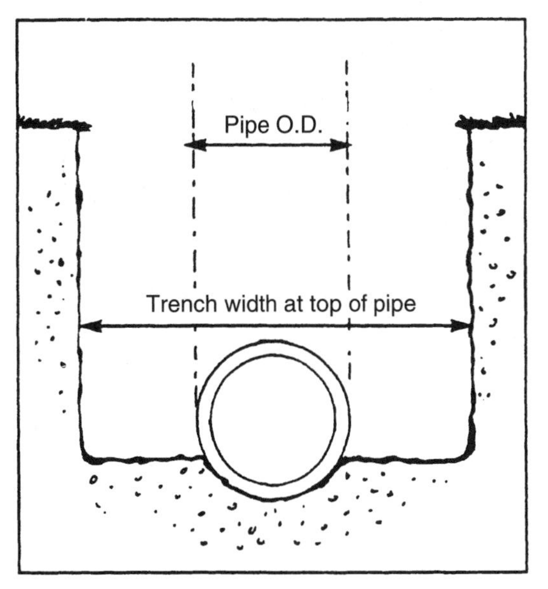

Thermoplastic pipe Special requirements, trench width must be 1.25 × O.D. of pipe plus 12 inches.

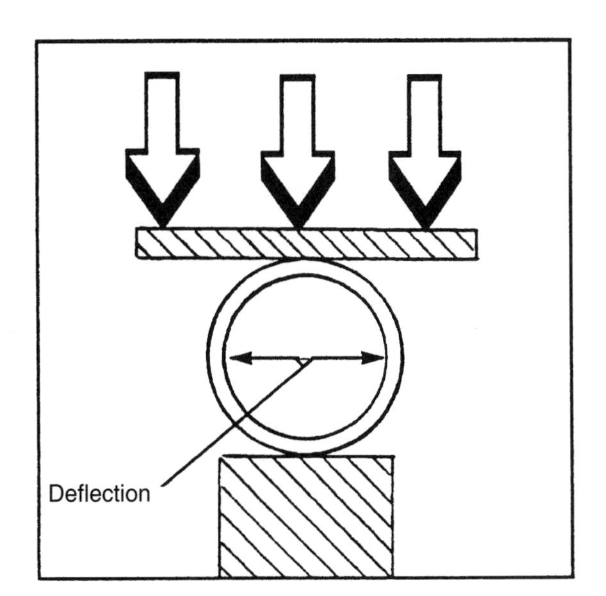

Cast-iron soil pipe No special bedding required unless installations are in rock.

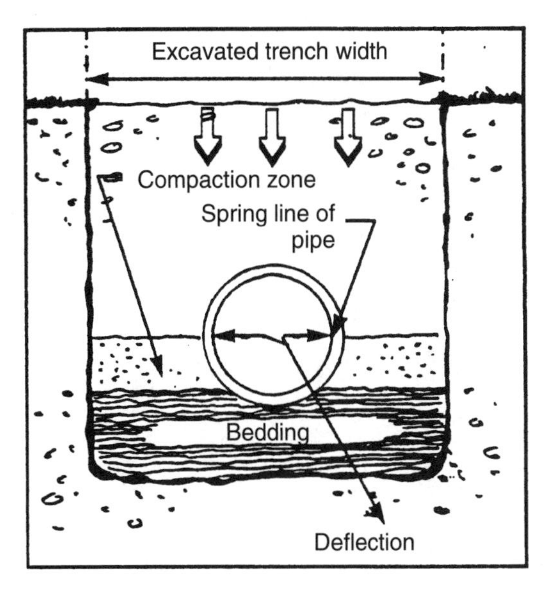

Thermoplastic pipe Special bedding requirements per ASTM D 2321-89.

*Reprinted by permission of the Cast Iron Soil Pipe Institute

2.1.4.1 Trench Width Considerations

- *Cast-iron soil* This is a rigid pipe that does not depend on sidefill stiffness; therefore, the trench can be as narrow as an installer requires in order to join the pipe sections together and complete the joint connections.
- *Thermoplastic pipe* This flexible pipe requires sidewall stiffness in the trench to limit deflections. ASTM D2321-89 recommends that the trench width be as wide as the outside diameter of the pipe being installed plus 16 inches (400 mm). An alternative to this formula is to multiply the outside diameter of the pipe by 1.25 and add 12 inches (300 mm). For example, a 6-inch (150 mm) pipe has an outside diameter of 6.625 inches (165.6 mm) and would require a 20-inch (500 mm) wide trench. The added width of the trench for "flexible" pipe is to allow for compaction equipment to operate in the "compaction zone" on each side of the flexible pipe to create sidewall stiffness.

2.1.5 Assembling a Trench Shield

Assembly

Lay side panel flat on ground with collar sockets up . . .

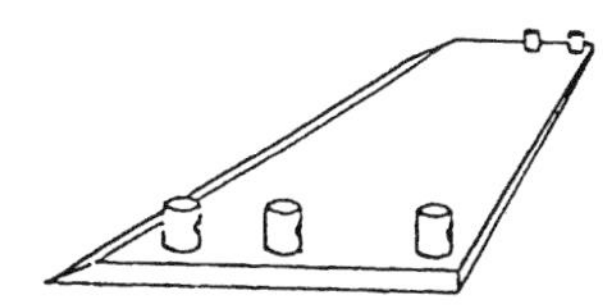

Place spreader pipe and/or plate onto collars or into brackets and pin in place. Secure pins with keepers. A minimum of 2 spreader units are required at each end of trench shield.

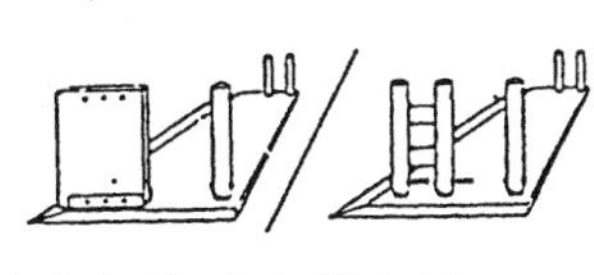

Lower second sidewall onto spreaders and pin.

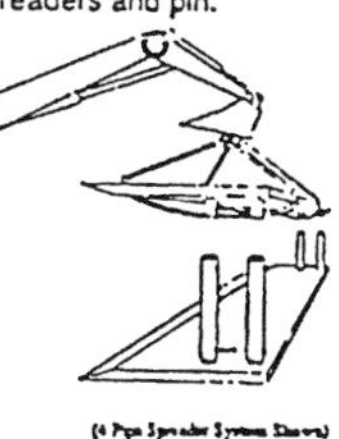

Stand trench shield in upright position and prepare for installation.

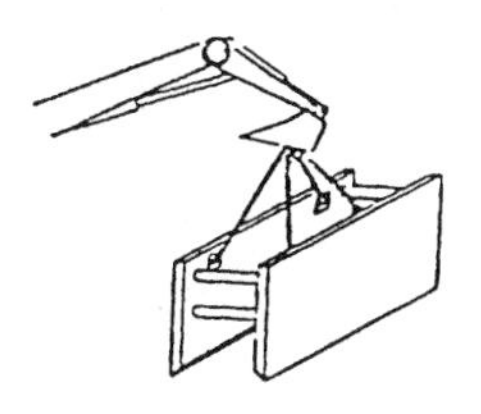

Using a trench shield in stable soil

Excavate to grade just slightly wider than the trench shield. Dig walls vertical to a minimum of 18" below the top of the shield. Slope soil above shield according to OSHA regulations. Install shield in trench.

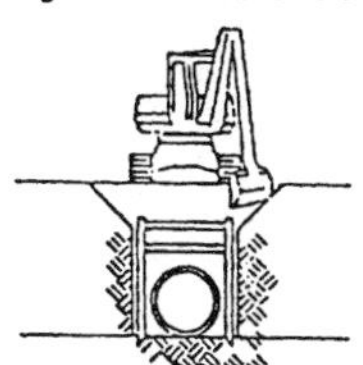

Excavate in front of the trench shield.

Pull shield forward by front top spreader pipe or with pulling eyes. (Pulling eyes should be used with spreaders wider than 72" or when soil pressure is severe enough to cause spreader to deflect).

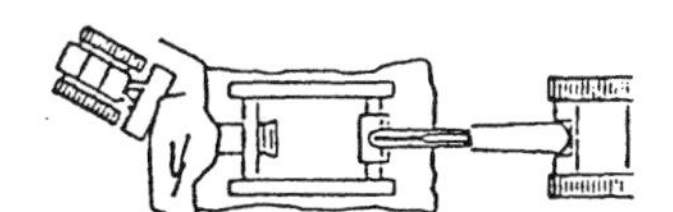

Using a shield in unstable soil

Excavate until soil begins to crumble beyond desired trench width. Place shield on line of excavation.

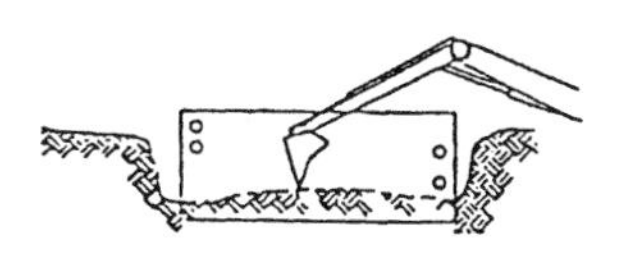

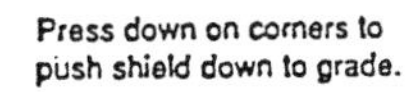

Press down on corners to push shield down to grade.

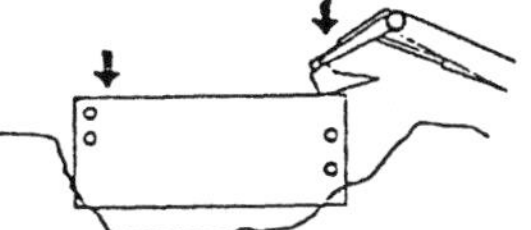

Pull shield forward and up on appropriate angle.

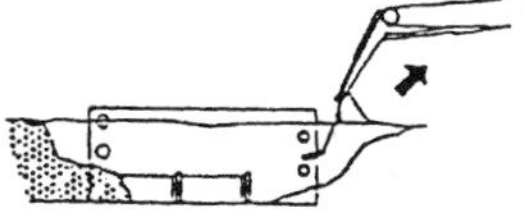

Excavate soil within the shield and repeat previous process.

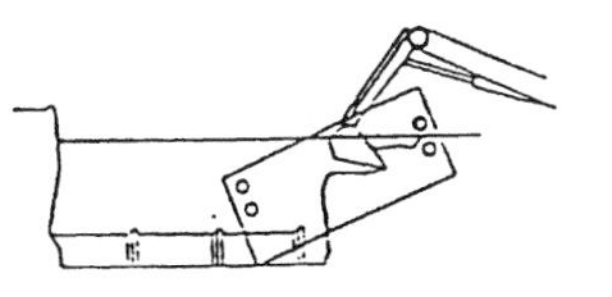

Using shields for patchwork, repairs, or tie-ins

- Center shield over work area.
- Lay soil at ends back according to OSHA regulations or use manufacturer's designed end plates to protect from cave-ins.

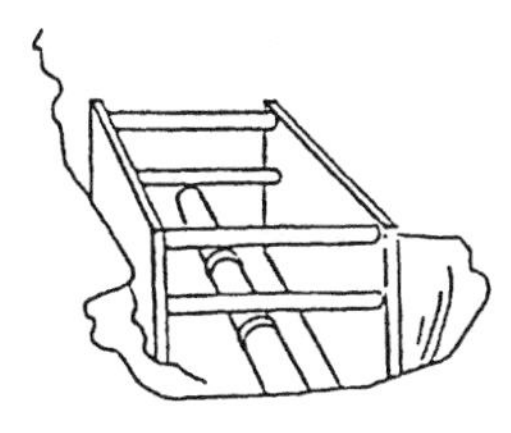

Manhole box with corner end plates

Corner end plates help prevent loose material from running into the end of the shield. Soil at ends should be sloped according to OSHA regulations.

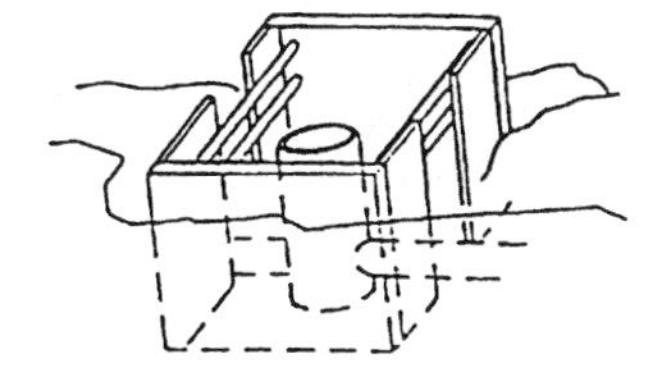

Using 4-sided shields

When using shields as protection during manhole assembly work, insure that proper end panels are used, or lay soil at the ends back according to OSHA regulations.

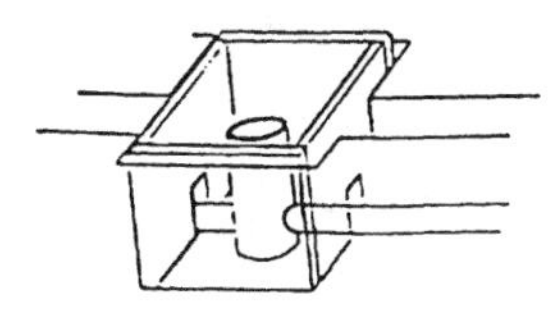

By permission Efficiency Productions, Inc., Lansing, Michigan

2.1.5.1 Determining Trench Shield Size

If the company does not own a trench box, but plans to rent one, certain data, shown below, must be given to the rental company to ensure that the proper size box is ordered to fit the job at hand.

To size a trench shield

Depth of cut ________

Soil Conditions*

Type A (25#) ________
Type B (45#) ________
Type C (60#) ________
Hydrostatic ________

Outside pipe diameter ________

(Shield 12 in wider than pipe OD)

Pipe length ________

(Shield 2 to 4 ft longer)

Bucket width ________

(Inside shield: 12 in less than shield)

(Outside shield: 4 in more than shield)

Machine lift capacity ________

(1.5 times shield weight at 20-ft radius at grade)

* Soil conditions refer to OSHA classifications. (See Sec. 1.1.5.2 for a full explanation of Type A, B, and C soils.)

By permission Efficiency Production, Lansing, Michigan

2.1.6 Reinforced Concrete Pipe (RCP) Specification ASTMC76 T&G Joints

ASTM C 76

Reinforced Concrete Culvert, Storm Drain and Sewer Pipe, Tongue and Groove Joints

	WALL A		WALL B		WALL C	
Internal Diameter, inches	Minimum Wall Thickness, inches	Approximate Weight, pounds per foot	Minimum Wall Thickness, inches	Approximate Weight, pounds per foot	Minimum Wall Thickness, inches	Approximate Weight, pounds per foot
12	1¾	79	2	93	–	–
15	1⅞	103	2¼	127	–	–
18	2	131	2½	168	–	–
21	2¼	171	2¾	214	–	–
24	2½	217	3	264	3¾	366
27	2⅝	255	3¼	322	4	420
30	2¾	295	3½	384	4¼	476
33	2⅞	336	3¾	451	4½	552
36	3	383	4	524	4¾	654
42	3½	520	4½	686	5¼	811
48	4	683	5	867	5¾	1011
54	4½	864	5½	1068	6¼	1208
60	5	1064	6	1295	6¾	1473
66	5½	1287	6½	1542	7¼	1735
72	6	1532	7	1811	7¾	2015
78	6½	1797	7½	2100	8¼	2410
84	7	2085	8	2409	8¾	2660
90	7½	2395	8½	2740	9¼	3020
96	8	2710	9	3090	9¾	3355
102	8½	3078	9½	3480	10¼	3760
108	9	3446	10	3865	10¾	4160

Large Sizes of Pipe Tongue and Groove Joint

Internal Diameter Inches	Internal Diameter Feet	Wall Thickness Inches	Approximate Weight, pounds per foot
114	9½	9½	3840
120	10	10	4263
126	10½	10½	4690
132	11	11	5148
138	11½	11½	5627
144	12	12	6126
150	12½	12½	6647
156	13	13	7190
162	13½	13½	7754
168	14	14	8339
174	14½	14½	8945
180	15	15	9572

These tables are based on concrete weighing 150 pounds per cubic foot and will vary with heavier or lighter weight concrete.

2.1.6.1 Reinforced Concrete Pipe Specifications ASTMC14, ASTMC76 (Bell and Spigot)

ASTM C 14—Nonreinforced Sewer and Culvert Pipe, Bell and Spigot Joint.						
	CLASS 1		CLASS 2		CLASS 3	
Internal Diameter, inches	Minimum Wall Thickness, inches	Approx. Weight, pounds per foot	Minimum Wall Thickness, inches	Approx. Weight, pounds per foot	Minimum Wall Thickness, inches	Approx. Weight, pounds per foot
4	5/8	9.5	3/4	13	7/8	15
6	5/8	17	3/4	20	1	24
8	3/4	27	7/8	31	1 1/8	36
10	7/8	37	1	42	1 1/4	50
12	1	50	1 3/8	68	1 3/4	90
15	1 1/4	80	1 5/8	100	1 7/8	120
18	1 1/2	110	2	160	2 1/4	170
21	1 3/4	160	2 1/4	210	2 3/4	260
24	2 1/8	200	3	320	3 3/8	350
27	3 1/4	390	3 3/4	450	3 3/4	450
30	3 1/2	450	4 1/4	540	4 1/4	540
33	3 3/4	520	4 1/2	620	4 1/2	620
36	4	580	4 3/4	700	4 3/4	700

ASTM C 76—Reinforced Concrete Culvert, Storm Drain and Sewer Pipe, Bell and Spigot Joint.				
	WALL A		WALL B	
Internal Diameter, inches	Minimum Wall Thickness inches	Approximate Weight, pounds per foot	Minimum Wall Thickness, inches	Approximate Weight, pounds per foot
12	1 3/4	90	2	110
15	1 7/8	120	2 1/4	150
18	2	160	2 1/2	200
21	2 1/4	210	2 3/4	260
24	2 1/2	270	3	330
27	2 5/8	310	3 1/4	390
30	2 3/4	360	3 1/2	450

2.1.6.2 Installation Procedures for Bell and Spigot Concrete Pipe

Clean Bell

Step 1

Carefully clean all dirt and foreign substances from the jointing surface of the bell or groove end of pipe.

Improperly prepared bell jointing surface may prevent homing of the pipe.

Clean Spigot

Step 2

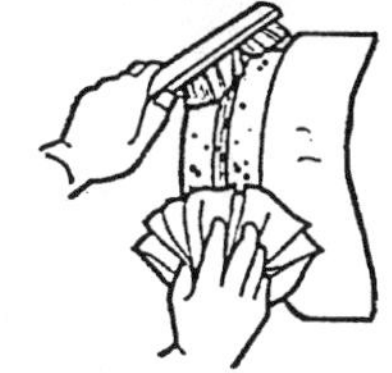

Carefully clean spigot or tongue end of pipe, including the gasket recess.

Improperly prepared spigot and gasket recess may prevent gasket from sealing correctly.

Lubricate Bell

Step 3

Lubricate bell jointing suface liberally. Use a brush, cloth, sponge or gloves to cover entire inside surface. Only approved lubricant should be used.

A bell not lubricated or improperly lubricated may cause gasket to roll and possibly damage the bell.

Lubricate Spigot

Step 4

Lubricate the spigot or tongue end of pipe, especially the gasket recess.

Gasket may twist out of recess if lubricant in recess is lacking or insufficient.

Lubricate Gasket

Step 5

Lubricate the gasket thoroughly before it is placed on the spigot or tongue.

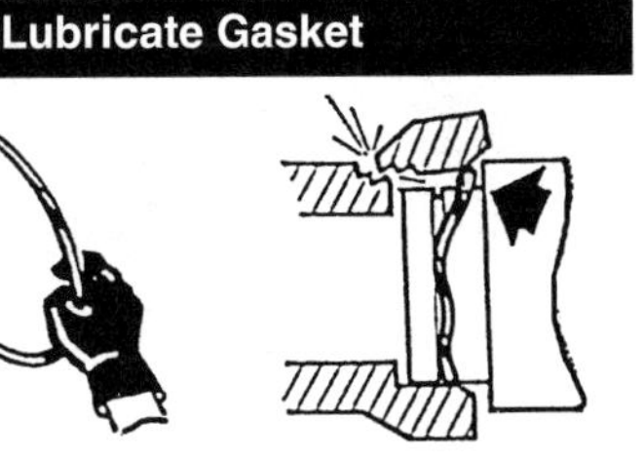

Excessive force will be required to push the pipe to the home position if gasket is not well lubricated.

Install Gasket

Step 6

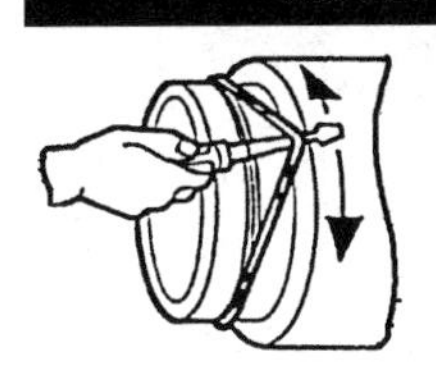

Fit the gasket carefully. Equalize the rubber gasket stretch by running a smooth, round object, inserted between gasket and spigot, around the entire circumference several times.

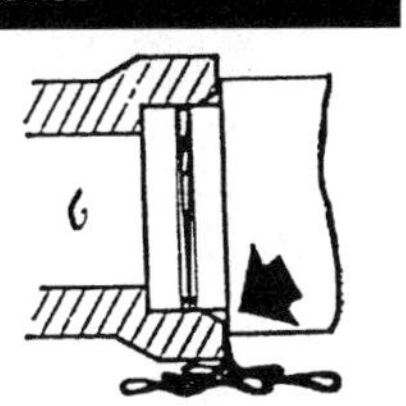

Unequal stretch could cause bunching of gasket and may cause leaks in the joint or crack the bell.

When jointing small diameter pipe, a chain or cable is wrapped around the barrel of the pipe a few feet behind the tongue or spigot and fastened with a grab hook or other suitable connecting device. A lever assembly is anchored to the installed pipe, several sections back from the last installed section, and connected by means of a chain or cable to the grab hook on the pipe to be installed. By pulling the lever back, the tongue or spigot of the pipe being jointed is pulled into the bell or groove of the last installed pipe section. To maintain close control over the alignment of the pipe, a laying sling can be used to lift the pipe section slightly off the bedding foundation.

Large diameter pipe can be jointed by placing a *dead man blocking* inside the installed pipe, several sections back from the last installed section, which is connected by means of a chain or cable to a *strong back* placed across the end of the pipe section being installed. The pipe is pulled home by lever action similar to the external assembly.

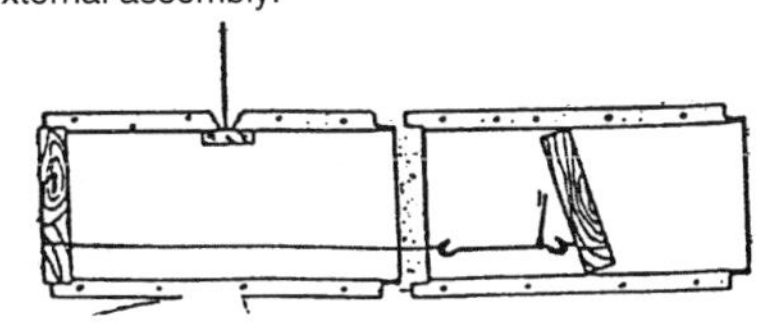

By permission American Concrete Pipe Assoc., Irving, Tx.

2.1.7 Trenching Recommendations for Cast-Iron Soil Pipe

PIPE SIZE		1 1/2"			2"			3"			4"			5"		
Depth of Cover	Trench Width	12"	18"	24"	12"	18"	24"	12"	18"	24"	12"	18"	24"	18"	24"	36"
	EL	57	57	57	75	75	75	111	111	111	144	144	144	177	177	177
2.0′	TL	77	77	77	192	192	192	365	365	365	614	614	614	864	864	864
	L	134	134	134	267	267	267	476	476	476	758	758	758	1041	1041	1041
	EL	72	72	72	95	95	95	140	140	140	184	184	184	226	226	226
2.5′	TL	44	44	44	141	141	141	211	211	211	387	387	387	563	563	563
	L	116	116	116	236	236	236	351	351	351	571	571	571	789	789	789
	EL	86	86	86	115	115	115	169	169	169	223	223	223	250	275	275
3.0′	TL	35	35	35	106	106	106	141	141	141	282	282	282	422	422	422
	L	121	121	121	221	221	221	310	310	310	505	505	505	672	697	697
	EL	101	101	101	134	134	134	199	199	199	262	262	262	276	324	324
3.5′	TL	26	26	26	67	67	67	96	96	96	192	192	192	288	288	288
	L	127	127	127	201	201	201	295	295	295	454	454	454	564	612	612
	EL	116	116	116	154	154	154	228	228	228	298	301	301	373	373	373
4.0′	TL	21	21	21	48	48	48	80	80	80	144	144	144	232	232	232
	L	137	137	137	202	202	202	308	308	308	442	445	445	605	605	605
	EL	131	131	131	173	173	173	258	258	258	318	341	341	422	422	422
4.5′	TL	18	18	18	40	40	40	72	72	72	120	120	120	196	196	196
	L	149	149	149	213	213	213	330	330	330	438	461	461	618	618	618
	EL	145	145	145	193	193	193	287	287	287	336	380	380	471	471	471
5.0′	TL	16	16	16	32	32	32	64	64	64	96	96	96	160	160	160
	L	161	161	161	225	225	225	351	351	351	432	476	476	631	631	631
	EL	160	160	160	213	213	213	317	317	317	351	419	419	520	520	520
5.5′	TL	14	14	14	30	30	30	60	60	60	88	88	88	140	140	140
	L	174	174	174	243	243	243	377	377	377	439	507	507	660	660	660
	EL	175	175	175	232	232	232	346	346	346	365	458	458	569	569	569
6.0′	TL	13	13	13	29	29	29	56	56	56	80	80	80	120	120	120
	L	188	188	188	261	261	261	402	402	402	445	539	539	689	689	689
	EL	189	189	189	252	252	252	375	375	375	376	497	497	618	618	618
6.5′	TL	12	12	12	28	28	28	54	54	54	76	76	76	112	112	112
	L	201	201	201	280	280	280	429	429	429	452	573	573	730	730	730
	EL	204	204	204	271	271	271	387	405	405	387	537	537	667	667	667
7.0′	TL	11	11	11	27	27	27	52	52	52	72	72	72	104	104	104
	L	215	215	215	298	298	298	439	457	457	459	609	609	771	771	771
	EL	219	219	219	291	291	291	396	396	396	396	576	576	716	716	716
7.5′	TL	10	10	10	26	26	26	50	50	50	68	68	68	96	96	96
	L	229	229	229	317	317	317	446	446	446	464	644	644	812	812	812
	EL	234	234	234	311	311	311	404	464	464	404	615	615	765	765	765
8.0′	TL	10	10	10	26	26	26	48	48	48	64	64	64	88	88	88
	L	244	244	244	337	337	337	452	512	512	468	679	679	853	853	853

Note: All O.D.'s are based on service weight nominal O.D.'s (ASTM A-74).

By permission Cast Iron Soil Pipe Institute

2.1.7 Trenching Recommendations for Cast-Iron Soil Pipe (Continued)

PIPE SIZE Depth of Cover		4"	6"	8"	10"	12"
3.0′	EL	392	465	538	611	685
	TL	282	563	774	986	1232
	L	674	1028	1312	1597	1917
3.5′	EL	439	523	607	692	777
	TL	192	384	576	736	896
	L	631	907	1183	1428	1673
4.0′	EL	482	576	672	768	865
	TL	144	320	480	624	752
	L	626	896	1152	1392	1617
4.5′	EL	521	626	732	839	947
	TL	120	272	417	536	648
	L	641	898	1148	1375	1595
5.0′	EL	556	671	788	906	1025
	TL	96	240	352	448	544
	L	652	911	1140	1354	1569
5.5′	EL	589	713	840	969	1099
	TL	88	192	312	392	488
	L	677	905	1152	1361	1587
6.0′	EL	618	752	889	1028	1168
	TL	80	160	272	336	432
	L	698	912	1161	1364	1600
6.5′	EL	645	788	934	1083	1234
	TL	76	148	248	308	396
	L	721	936	1182	1391	1630
7.0′	EL	670	821	976	1135	1296
	TL	72	136	224	280	360
	L	742	957	1200	1415	1656

EL = Earth Load In Pounds
TL = Truck Load In Pounds
L = Total Load In Pounds

By permission Cast Iron Soil Pipe Institute

2.1.7.1 Depth of Cover for Cast-Iron Soil Pipe/Rated Working Pressures

Size in.	Pressure class,[1] psi	Nominal thickness, in.	Laying Conditions				
			Type 1 trench[2]	Type 2 trench	Type 3 trench	Type 4 trench	Type 5 trench
			Max. depth of cover,[ft]				
3	350	0.25	78	88	99	100§	100[4]
4	350	0.25	53	61	69	85	100[4]
6	350	0.25	26	31	37	47	65
8	350	0.25	16[5]	20	25	34	50
10	350	0.26	11	15	19	28	45
12	350	0.28	10[-5]	15	19	28	44
14	250	0.28		11[4]	15	23	36
	300	0.30		13	17	26	42
	350	0.31		14	19	27	44
16	250	0.30		11**	15	24	34
	300	0.32		13	17	26	39
	350	0.34		15	20	28	44
18	250	0.31		10[4]	14	22	31
	300	0.34		13	17	26	36
	350	0.36		15	19	28	41
20	250	0.33		10	14	22	30
	300	0.36		13	17	26	35
	350	0.38		15	19	28	38
24	200	0.33		8[4]	12	17	25
	250	0.37		11	15	20	29
	300	0.40		13	17	24	32
	350	0.43		15	19	28	37
30	150	0.34		—	9	14	22
	200	0.38		8[4]	12	16	24
	250	0.42		11	15	19	27
	300	0.45		12	16	21	29
	350	0.49		15	19	25	33
36	150	0.38		—	9	14	21
	200	0.42		8[4]	12	15	23
	250	0.47		10	14	18	25
	300	0.51		12	16	20	28
	350	0.56		15	19	24	32

1. Ductile-iron pipe is adequate for the rated working pressure indicated for each nominal size plus a surge allowance of 100 psi. Calculations are based on a 2.0 safety factor times the sum of working pressure and 100 psi surge allowance. (See ANSI/AWWA C150/A21.50 for design formula.) Ductile-iron pipe for working pressures higher than 350 psi is available.

2. For pipe 14 in. and larger, consideration should be given to the use of laying conditions other than Type 1.

3. An allowance for a single H-20 truck with 1.5 impact factor is included for all depths of cover.

4. Calculated maximum depth of cover exceeds 100 ft.

5. Minimum allowable depth of cover exceeds 100 ft.

Reprinted by permission from Atlantic State Cast Iron Pipe Company, Phillipsburg, New Jersey

2.1.7.2 Cast Iron Soil Pipe Equivalents

	1½	2	3	4	5	6	8	10	12	15
1½	1	1.8	4	7.1	10.8	15.7	28.	44.4	63.4	100
2		1	2.3	4	6.1	8.8	15.8	25	35.6	56.3
3			1	1.8	2.7	3.9	7	11.1	15.8	25
4				1	1.5	2.2	3.9	6.3	8.9	14.1
5					1	1.4	2.6	4.1	5.8	9.2
6						1	1.8	2.8	4.	6.4
8							1	1.6	2.3	3.6
10								1	1.4	2.3
12									1	1.6
15										1

EXAMPLE: A 4" cast iron soil pipe is equivalent to how many 2" cast iron soil pipe? In the vertical column under 4", and opposite 2", read the equivalent which is 4: This means that four 2" cast iron soil pipe are the equivalent of one 4" cast iron soil pipe in inside cross-sectional area.

By permission Cast Iron Soil Pipe Institute

2.1.7.3 Ring Test Crushing Loads on Cast-Iron Soil Pipe

NO-HUB				SERVICE WEIGHT				EXTRA HEAVY			
Pipe Size In.	Nominal O.D. (D_o)	Nominal Thickness (1)	Ring Crushing Load* (w)	Pipe Size In.	Nominal O.D. (D_o)	Nominal Thickness (1)	Ring Crushing Load* (w)	Pipe Size In.	Nominal O.D. (D_o)	Nominal Thickness (1)	Ring Crushing Load* (w)
1½	1.90	.16	8328	—	—	—	—	—	—	—	—
2	2.35	.16	6617	2	2.30	.17	7680	2	2.38	.19	9331
3	3.35	.16	4542	3	3.30	.17	5226	3	3.50	.25	10885
4	4.38	.19	4877	4	4.30	.18	4451	4	4.50	.25	8324
5	5.30	.19	3999	5	5.30	.18	3582	5	5.50	.25	6739
6	6.30	.19	3344	6	6.30	.18	2997	6	6.50	.25	5660
8	8.38	.23	3674	8	8.38	.23	3674	8	8.62	.31	6546
10	10.56	.28	4317	10	10.50	.28	4342	10	10.75	.37	7465
				12	12.50	.28	3632	12	12.75	.37	6259
				15	15.88	.36	4727	15	15.88	.44	7097

*Pounds per linear foot

By permission Cast Iron Soil Pipe Institute

2.1.7.4 Slopes Required for Self-Cleaning Cast-Iron Pipe

Slopes required to obtain self-cleaning velocities of 2.0 and 2.5 ft./sec. (based on Mannings Formula with N = 0.012)

Pipe Size (In.)	Velocity (Ft./Sec.)	¼ FULL		½ FULL		¾ FULL		FULL	
		Slope (Ft./Ft.)	Flow (Gal./Min.)	Slope (Ft./Ft.)	Flow (Gal./Min.)	Slope (Ft./Ft.)	Flow (Gal./Min.)	Slope (Ft./Ft.)	Flow (Gal./Min.)
2.0	2.0	0.0313	4.67	0.0186	9.34	0.0148	14.09	0.0186	18.76
	2.5	0.0489	5.84	0.0291	11.67	0.0231	17.62	0.0291	23.45
3.0	2.0	0.0178	10.77	0.0107	21.46	0.0085	32.23	0.0107	42.91
	2.5	0.0278	13.47	0.0167	26.82	0.0133	40.29	0.0167	53.64
4.0	2.0	0.0122	19.03	0.0073	38.06	0.0058	57.01	0.0073	76.04
	2.5	0.0191	23.79	0.0114	47.58	0.0091	71.26	0.0114	95.05
5.0	2.0	0.0090	29.89	0.0054	59.79	0.0043	89.59	0.0054	119.49
	2.5	0.0141	37.37	0.0085	74.74	0.0067	11.99	0.0085	149.36
6.0	2.0	0.0071	43.18	0.0042	86.36	0.0034	129.54	0.0042	172.72
	2.5	0.0111	53.98	0.0066	107.95	0.0053	161.93	0.0066	214.90
8.0	2.0	0.0048	77.20	0.0029	154.32	0.0023	231.52	0.0029	308.64
	2.5	0.0075	96.50	0.0045	192.90	0.0036	289.40	0.0045	385.79
10.0	2.0	0.0036	120.92	0.0021	241.85	0.0017	362.77	0.0021	483.69
	2.5	0.0056	151.15	0.0033	302.31	0.0026	453.46	0.0033	604.61
12.0	2.0	0.0028	174.52	0.0017	349.03	0.0013	523.55	0.0017	698.07
	2.5	0.0044	218.15	0.0026	436.29	0.0021	654.44	0.0026	872.58
15.0	2.0	0.0021	275.42	0.0012	550.84	0.0010	826.26	0.0012	1101.68
	2.5	0.0032	344.28	0.0019	688.55	0.0015	1032.83	0.0019	1377.10

By permission Cast Iron Soil Pipe Institute

2.1.7.5 Typical Pipe-Joining Methods for Cast-Iron Pipe

Note lead and oakum will be found on older piping installations only.

(a) Typical hubless coupling

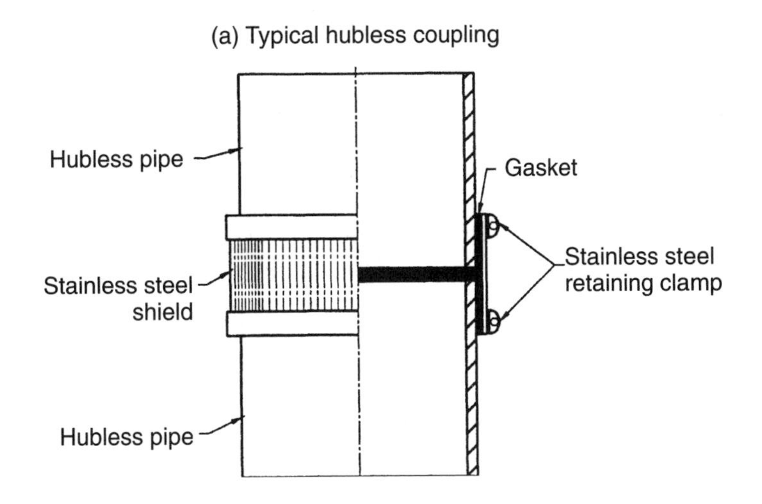

(b) Compression joint

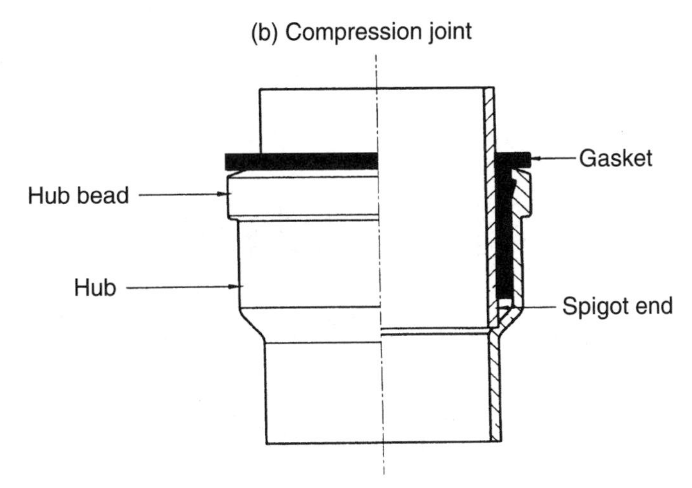

(c) Lead and oakum joint

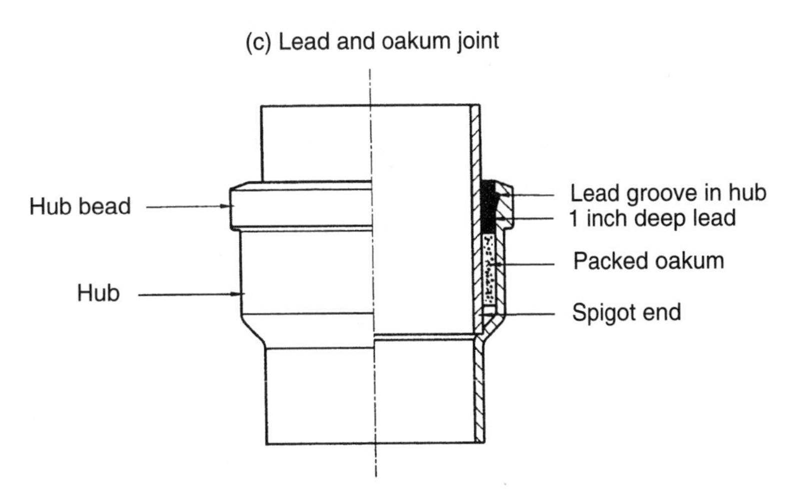

By permission Cast Iron Soil Pipe Institute

2.1.8 Ductile Iron Pipe Dimensions/Weights for Push on Mechanical Joint Pipe

Standards Applicable to Ductile Iron Pipe and Fittings

Thickness Design of Ductile Iron Pipe	ANSI/AWWA C150/A21,50
Ductile Iron Pipe for Water and Other Liquids	ANSI/AWWA C151/A21, 51 FEDERAL WWP421D, Grade C
Ductile Iron Pipe for Gravity Flow Service	ANSI/ASTM A746
Ductile Iron Fittings for Water and Other Liquids 30 in through 36 in	ANSI/AWWA C110/A21.10
Ductile Iron Compact Fittings 3 in through 24 in	ANSI/AWWA C153/A21.53
Flanged Fittings	ANSI/AWWA C110/A21.10 ANSI B16-1
Ductile Iron Pipe with Threaded Flanges	ANSI/AWWA C115/21.15
Castings and Linings:	
Asphaltic	ANSI/AWWA C151/A21.51 ANSI/AWWA C110/A21.10 ANSI/AWWA C153/A21.53
Cement Lining	ANSI/AWWA C104/A21.4
Various Epoxy Linings and Casings	MANUFACTURER'S STANDARD
Exterior Polyethylene Encasement	ANSI/AWWA C105/A21.5
Joints—Pipe and Fittings	
Push-on and Mechanical Rubber-Gasket joints	ANSI/AWWA C111/A21.11 FEDERAL WWP421D
Flanged	ANSI/AWWA C115/A21.15 ANSI B16.1
Grooved and Shouldered	ANSI/AWWA C606
Pipe Threads	ANSI B2.1
Installation	ANSI/AWWA C600

Mechanical Joint Pipe, Maximum Allowable Joint Deflection

Size of Pipe, in	Y, max. Joint Deflection	X, Deflection Inches per 18-ft. Length	Approx. Radius, Feet of Curve, Produced by Succession of Joints 18 ft. Length
3	8° 18'	35*	140*
4	8° 18'	35*	140*
6	7° 7'	30*	160*
48	5° 21'	20	195
10	5° 21'	20	195
12	5° 21'	20	195
14	3° 35'	13.5	285
16	3° 35'	13.5	285
18	3° 0'	11	340
20	3° 0'	11	340
24	2° 23'	9	450

* 20-ft length.

Laying Conditions

Type 1†*	Type 2	Type 3	Type 4	Type 5
Flat-bottom trench.† Loose backfill.	Flat-bottom trench.† Backfill lightly consolidated to centerline of pipe.	Pipe bedded in 4-in. minimum loose soil.‡ Backfill lightly consolidated to top of pipe.	Pipe bedded in sand, gravel or crushed stone to depth of ⅛ pipe diameter, 4-in. minimum. Backfill compacted to top of pipe. (Approximately 80% Standard Proctor, AASHTO§ T-99).	Pipe bedded in compacted granular material to centerline of pipe. Compacted granular or select** material to top of pipe. (Approximately 90% Standard Proctor, AASHTO§-T-99)

Notes: Consideration of the pipe-zone embedment conditions included in this figure may be influenced by factors other than pipe strength. For additional information on pipe bedding and backfill, see ANSI/AWWA C600.

* For nominal pipe sizes 14 in and larger, consideration should be given to the use of laying conditions other than Type 1.

† Flat bottom is defined as undisturbed earth.

‡ Loose soil or select material is defined as native soil excavated from the trench, free of rocks, foreign materials, and frozen earth.

§ American Association of State Highway and Transportation Officials, 444 N. Capitol St. N.W., Suite 225, Washington, DC 20001.

2.1.8 Ductile Iron Pipe Dimensions/Weights for Push on Mechanical Joint Pipe (Continued)

Push-on Joint Pipe,
Maximum Deflection Full Length Pipe

Size of pipe, in	Max. joint deflection, degrees	Deflection, in		Approximate radius feet of curve, produced by succession of joints, 3 in. same as 4 in	
		18 ft length	20 ft length	18 ft length	20 ft length
3	5		21		230
4	5		21		230
6	5	19	21	206	230
8	5	19		206	
10	5	19		206	
12	5	19		206	
14	5	19		206	
16	5	19		206	
18	5	19		206	
20	5	19		206	
24	5	19		206	
30	5	19		206	
36	5	19		206	

*20-ft length.

2.1.8 Ductile Iron Pipe Dimensions/Weights for Push on Mechanical Joint Pipe (Continued)

Dimensions and Weights for Special Classes of Push-on Joint and Mechanical Joint Ductile Iron Pipe

Pipe manufactured in accordance with ANSI'AWWA C151/A21.51–91 under method of design outlined in ANSI/AWWA C150/A21.30

					Push-on joint			Mechanical joint		
Size in	Thickness class	Thickness, in	OD,* in	Weight of barrel, lb/ft	Weight of bell, lb	Weight per length,† lb	Ave. Weight per foot‡ lb	Weight of bell, lb	Weight per length,† lb	Ave. weight per foot‡/lb
3	51	0.25	3.96	8.9	9	185	9.4	11	190	9.4
3	52	0.28	3.96	9.9	9	205	10.4	11	210	10.4
3	53	0.31	3.96	10.9	9	225	11.4	11	230	11.2
3	54	0.34	3.96	11.8	9	245	12.2	11	245	12.2
3	55	0.37	3.96	12.8	9	265	13.2	11	265	13.2
3	56	0.40	3.96	13.7	9	265	14.2	11	285	14.2
4	51	0.26	4.80	11.3	11	235	11.8	16	240	12.1
4	52	0.29	4.80	12.6	11	265	13.2	16	270	13.4
4	53	0.32	4.80	13.8	11	285	14.4	16	290	14.6
4	54	0.35	4.80	15.0	11	310	15.6	16	315	15.8
4	55	0.38	4.80	16.1	11	335	16.6	16	340	16.9
4	56	0.41	4.80	17.3	11	355	17.8	16	360	18.1
6	50	0.25	6.90	16.0	15	305	16.8	18	305	17.0
6	51	0.28	6.90	17.8	15	335	18.6	18	340	18.8
6	52	0.31	6.90	19.6	15	370	20.4	18	370	20.6
6	53	0.34	6.90	21.4	15	400	22.2	18	405	22.4
6	54	0.37	6.90	23.2	15	435	24.0	18	435	24.2
6	55	0.40	6.90	25.0	15	465	25.8	18	470	26.0
6	56	0.43	6.90	26.7	15	495	27.5	18	500	27.7
8	50	0.27	9.05	22.8	22	430	24.0	24	435	24.1
8	51	0.30	9.05	25.2	22	475	26.4	24	480	26.5
8	52	0.33	9.05	27.7	22	530	28.9	24	525	29.0
8	53	0.36	9.05	30.1	22	565	31.3	24	565	31.4
8	54	0.39	9.05	32.5	22	605	33.7	24	610	33.8
8	55	0.42	9.05	34.8	22	650	36.0	24	650	36.1
8	56	0.45	9.05	37.2	22	690	38.4	24	695	38.5
10	50	0.29	11.10	30.1	29	570	31.7	31	575	31.8
10	51	0.32	11.10	33.2	29	625	34.8	31	630	34.9
10	52	0.35	11.10	36.2	29	680	37.8	31	685	37.9
10	53	0.38	11.10	39.2	29	735	40.8	31	735	40.9
10	54	0.41	11.10	42.1	29	785	43.7	31	790	43.9
10	55	0.44	11.10	45.1	29	840	46.7	31	845	46.8
10	56	0.47	11.10	48.0	29	895	49.6	31	895	49.7
12	50	0.31	13.20	38.4	35	725	40.3	37	730	40.5
12	51	0.34	13.20	42.0	35	790	43.9	37	795	44.1
12	52	0.37	13.20	45.6	35	835	47.5	37	860	47.7
12	53	0.40	13.20	49.2	35	920	51.1	37	925	51.3
12	54	0.43	13.20	52.8	35	985	54.7	37	985	54.9
12	55	0.46	13.20	56.3	35	1050	58.2	37	1050	58.4
12	56	0.49	13.20	59.9	35	1115	61.8	37	1115	62.0
14	50	0.33	15.30	47.5	60	915	50.8	61	915	50.9
14	51	0.36	15.30	51.7	60	990	55.0	61	990	55.1
14	52	0.39	15.30	55.9	60	1065	59.2	61	1065	59.3
14	53	0.42	15.30	60.1	60	1140	63.4	61	1145	63.5
14	54	0.45	15.30	64.2	60	1215	67.5	61	1215	67.5
14	55	0.48	15.30	68.4	60	1290	71.7	61	1290	71.8
14	56	0.51	15.30	72.5	60	1365	75.8	61	1365	75.9
16	50	0.34	17.40	55.8	68	1070	59.6	74	1080	59.9
16	51	0.37	17.40	60.6	68	1160	61.4	74	1165	64.7
16	52	0.40	17.40	65.4	68	1245	69.2	74	1250	69.5
16	53	0.43	17.40	70.1	68	1330	71.9	74	1335	74.2
16	54	0.46	17.40	74.9	68	1415	78.7	74	1420	79.0
16	55	0.49	17.40	79.7	68	1505	83.5	74	1510	83.8
16	56	0.52	17.40	84.4	68	1585	88.2	74	1595	88.5

2.1.8 Ductile Iron Pipe Dimensions/Weights for Push on Mechanical Joint Pipe (Continued)

18	50	0.35	19.50	64.4	78	1235	68.7	85	1245	69.1
18	51	0.35	19.50	69.3	78	1335	74.1	85	1340	74.5
18	52	0.41	19.50	75.2	78	1430	79.5	85	1440	79.9
18	53	0.44	19.50	80.6	78	1530	84.9	85	1535	85.3
18	54	0.47	19.50	86.0	78	1625	90.3	85	1635	90.7
18	55	0.50	19.50	91.3	78	1720	95.6	85	1730	96.0
18	56	0.53	19.50	96.7	78	1820	101.0	85	1825	101.4
20	50	0.36	21.60	73.5	87	1410	78.3	98	1420	78.9
20	51	0.39	21.60	79.5	87	1520	84.3	98	1530	84.9
20	52	0.42	21.60	85.5	87	1625	90.3	98	1635	90.9
20	53	0.45	21.60	91.5	87	1735	96.3	98	1745	96.9
20	54	0.48	21.60	97.5	87	1840	102.3	98	1855	102.9
20	55	0.51	21.60	103.4	87	1950	108.2	98	1960	108.8
20	56	0.54	21.60	109.3	87	2053	114.1	98	2065	114.7
24	50	0.38	25.80	92.9	103	1775	98.7	123	1795	99.7
24	51	0.41	25.80	100.1	103	1905	105.9	123	1925	106.9
24	52	0.44	25.80	107.3	103	2035	113.1	123	2055	114.1
24	53	0.47	25.80	114.4	103	2165	120.2	123	2180	121.2
24	54	0.50	25.80	121.6	103	2295	127.4	123	2310	128.4
24	55	0.53	25.80	128.8	103	2425	134.6	123	2440	135.6
24	56	0.56	25.80	135.9	103	2550	141.7	123	2570	142.7
30	50	0.39	32.00	118.5	170	2305	127.9		...	
30	51	0.43	32.00	130.5	170	2520	139.9		...	
30	52	0.47	32.00	142.5	170	2735	151.9		...	
30	53	0.51	32.00	154.4	170	2950	163.8		...	
30	54	0.55	32.00	166.3	170	3165	175.7		...	
30	55	0.59	32.00	178.2	170	3180	187.6		...	
30	56	0.63	32.00	190.0	170	3590	199.4		...	
36	50	0.43	38.30	156.5	239	3055	169.8		...	
36	51	0.48	38.30	174.5	239	3380	187.8		...	
36	52	0.53	38.30	192.4	239	3700	205.7		...	
36	53	0.58	38.30	210.3	239	4025	223.6		...	
36	54	0.63	38.30	228.1	239	4345	241.4		...	
36	55	0.68	38.30	245.9	239	4665	259.2		...	
36	56	0.73	38.30	263.7	239	4985	277.0		...	

*Tolerances of OD of spigot end: 3–12 in, ±0.6 in, 1–24 in., +0.05 in, –0.08 in, 30–36 in, †0.08 in, –0.06 in.

†Including bell; calculated weight of pipe rounded off to nearest 5 lb

‡Including bell; average weight per foot, based on calculated weight of pipe before rounding.

3–4-in nominal 20-ft laying length; 6-in nominal 18 or 20 ft; 8–36-in nominal 18-ft laying length.

2.1.8.1 Ductile Iron-Pipe Specifications and Assembly Tops

Push-on Restrained Joint Pipe

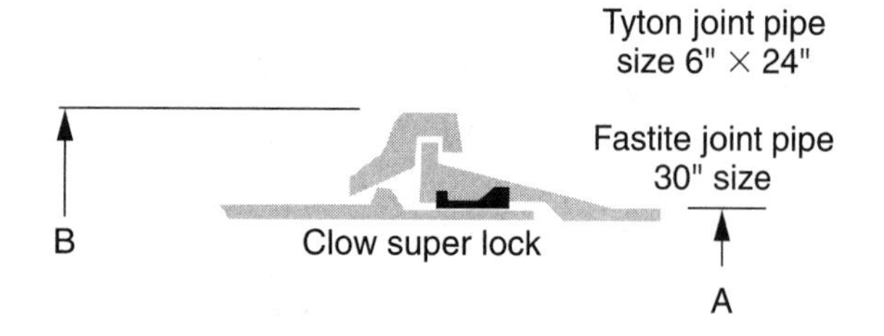

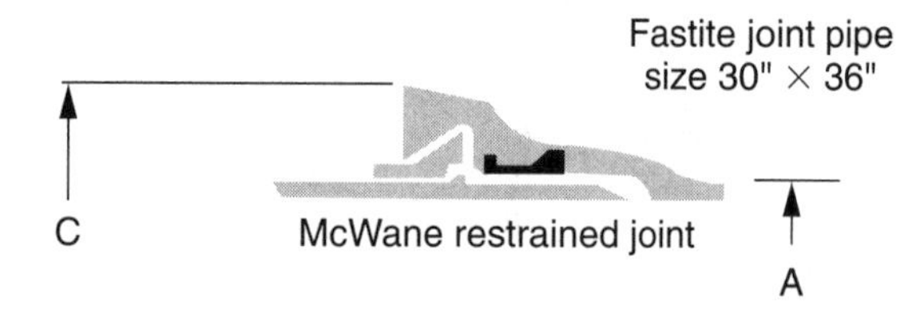

Nominal pipe size, in	Pressure rating,* psi	Joint deflection Degrees	Joint deflection Inches in 18 ft	A, pipe OD, in	B, retainer OD, in	C, bell OD, in
6	350	4	15	6.90	11.75	
8	350	4	15	9.05	14.38	
10	350	4	15	11.10	16.75	
12	350	4	15	13.20	19.13	
14	350	3	11	15.30	21.75	
16	350	3	11	17.40	24.00	
18	350	3	11	19.50	26.38	
20	350	3	11	21.60	28.63	
24	350	3	11	25.80	33.75	
30	250	3	7	32.00	40.13	38.75
36	250	3	7	38.30		45.63

*In the 14-in and larger sizes pressure rating limited to the rating of the pipe barrel thickness selected.

All BND Socket Joint Pipe

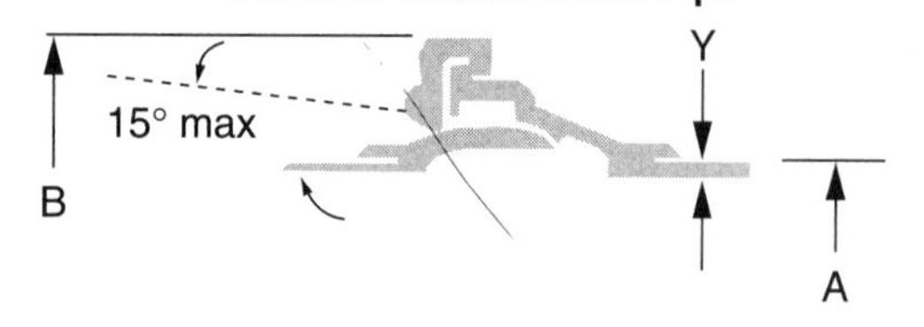

ASSEMBLY OF FIELD CUT TYPE

When pipe are cut in the field, the cut end may be readily conditioned so that it can be used to make up the next joint. The outside of the cut end should be beveled about $\frac{1}{4}$-in at an angle of about 30° (Figure 1). This can be quite easily done, with a coarse file or a portable grinder. The operation removes any sharp, rough edges which otherwise might injure the gasket.

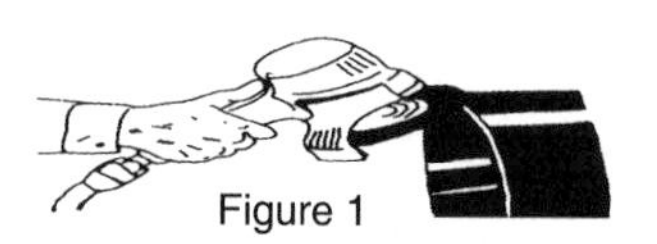
Figure 1

When ductile iron pipe 14 in and larger is to be cut in the field, the material should be ordered as "*gauged full length.*" Pipe that is gauged full "length" is specially marked to avoid confusion. The ANSI/AWWA standard for ductile iron pipe requires factory gauging of the spigot end. Accordingly, pipe selected for field cutting should also be field-gauged in the location of the cut and found to be within the tolerances shown in Table 1. In the field a mechanical joint gland can be used a gauging device.

Table 1. Suitable Pipe Diameters, for Field Cuts and Restrained Joint Field Fabrication

Nominal pipe size, in	Min. pipe diameter, in	Max. pipe diameter, in	Min. pipe circumference, in	Max. pipe circumference, in
3	3.90	4.02	$12\frac{1}{4}$	$12\frac{5}{8}$
4	4.74	4.86	$14\frac{29}{32}$	$15\frac{9}{32}$
6	6.84	6.96	$21\frac{1}{2}$	$21\frac{7}{8}$
8	8.99	9.11	$28\frac{1}{4}$	$28\frac{5}{8}$
10	11.04	11.16	$34\frac{11}{16}$	$35\frac{1}{16}$
12	13.14	13.26	$41\frac{9}{32}$	$41\frac{21}{32}$
14	15.22	15.35	$47\frac{13}{16}$	$48\frac{7}{32}$
16	17.32	17.45	$54\frac{13}{32}$	$54\frac{15}{16}$
18	19.42	19.55	61	$61\frac{13}{32}$
20	21.52	21.65	$67\frac{19}{32}$	68
24	25.72	25.85	$80\frac{13}{16}$	$81\frac{7}{32}$
30	31.94	32.08	$100\frac{11}{32}$	$100\frac{25}{32}$
36	38.24	38.38	$120\frac{1}{8}$	$120\frac{9}{16}$

Table based on ANSI/AWWA C151/A21.51 guidelines for push-on joints.

THE BACKHOE METHOD OF ASSEMBLY

A backhoe may be used to assemble pipe of intermediate and larger sizes. The plain end of the pipe should be carefully guided by hand into the bell of the previously assembled pipe. The bucket of the backhoe may then be used to push the pipe until fully seated. A timber header should be used between the pipe and backhoe bucket to avoid damage to the pipe.

2.1.8.1 Ductile Iron-Pipe Specifications and Assembly Tops (Continued)

Size, in	Thickness: Class (A21.51)	Thickness: in	A, pipe OD, in	B, retainer OD, in	Full-length weight,* lb: As shipped	Under water: Full of air	Under water: Full of water	Safe end pull (lb)
6	55	0.40	6.90	13⅞	545	240	465	50,000
8	55	0.42	9.05	16⅞	770	240	655	70,000
10	55	0.44	11.10	19⅛	1005	200	860	95,000
12	55	0.46	13.20	22	1270	155	1080	120,000
14	56	0.51	15.30	24½	1655	160	1410	145,000
16	56	0.52	17.40	27	1990	45	1685	165,000
18	56	0.53	19.50	30	2375	–70	2015	195,000
	58+	0.59			2560	110	2170	
20	56	0.54	21.60	32¾	2810	–200	2375	210,000
	59+	0.63			3110	100	2635	
24	56	0.56	25.80	38¼	3700	–620	3100	260,000
	62+	0.74			4415	95	3715	
30	58	0.71	32.00	46¼	5855	–900	4920	330,000
	61+	0.83			6435	–180	5360	
36	57	0.78	38.30	54½	8145	–1300	6880	400,000
	59+	0.88			8725	–725	7330	

*Weights are for 18 ft 0 in laying lengths. Nominal full lengths vary by size.

Pipe, bell, ball, and retainer are ductile iron.

Dimensions and weights subject to manufacturing tolerances.

6–24-in pressure rating: 350 psi.

30–36-in pressure rating: 250 psi.

+Thickness required to overcome buoyancy.

2.1.8.2 Ductile Iron-Pipe Specifications for 3" to 36" Pipe

Nominal Thickness for Standard Pressure Classes of Ductile-Iron Pipe

Size, in	Outside diameter, in	Pressure class* 150	200	250	300	350
		Nominal thickness, in				
3	3.96	—	—	—	—	0.25†
4	4.80	—	—	—	—	0.25†
6	6.90	—	—	—	—	0.25†
8	9.05	—	—	—	—	0.25†
10	11.10	—	—	—	—	0.26
12	13.20	—	—	—	—	0.28
14	15.30	—	—	0.28	0.30	0.31
16	17.40	—	—	0.30	0.32	0.34
18	19.50	—	—	0.31	0.34	0.36
20	21.60	—	—	0.33	0.36	0.38
24	25.80	—	0.33	0.37	0.40	0.43
30	32.00	0.34	0.38	0.42	0.45	0.49
36	38.30	0.38	0.42	0.47	0.51	0.56

*Pressure classes are defined as the rated water pressure of the pipe in psi. The thicknesses shown are adequate for the rated water working pressure plus a surge allowance of 100 psi. Calculations are based on a minimum yield strength of 42,000 and a 2.0 safety factor times the sum of the working pressure and 100 psi surge allowance.

†Calculated thicknesses for these sizes and pressure ratings are less than those shown above. Presently these are the lowest nominal thicknesses available in these sizes.

Note: Per ANSI/AWWA C150/A21.50 the thicknesses above include the 0.06" service allowance and the casting tolerance listed below by size ranges:

Size, in	Casting tolerances, in
3–8	–0.05
10–12	–0.06
14–36	–0.07

Standard Dimensions and Weights of 3 in through 36 in Push-on-Joint Ductile Iron Pipe

Size, in	Pressure class	Thickness, in	Outside diameter,* in	18-ft laying length: Weight per length,† lb	Ave. weight lb/ft
3§	350	0.25	3.96	185	9.2
4§	350	0.25	4.80	225	11.3
6§	350	0.25	6.90	300	16.6
8	350	0.25	9.05	395	22.0
10	350	0.26	11.10	510	28.4
12	350	0.28	13.20	655	36.4
14	250	0.28	15.30	770	42.9
	300	0.30	15.30	825	45.8
	350	0.31	15.30	850	47.2
16	250	0.30	17.40	940	52.3
	300	0.32	17.40	1000	55.5
	350	0.34	17.40	1060	58.8
18	250	0.31	19.50	1090	60.5
	300	0.34	19.50	1185	65.9
	350	0.36	19.50	1250	65.9
20	250	0.33	21.60	1290	71.6
	300	0.36	21.60	1395	77.6
	350	0.38	21.60	1470	81.6
24	200	0.33	25.80	1550	86.1
	250	0.37	25.80	1725	95.8
	300	0.40	25.80	1855	103.0
	350	0.43	25.80	1985	110.2
30	150	0.34	32.00	2000	111.2
	200	0.38	32.00	2220	123.2
	250	0.42	32.00	2435	135.2
	300	0.45	32.00	2595	144.2
	350	0.49	32.00	2810	156.1
36	150	0.38	38.30	2675	148.7
	200	0.42	38.30	2935	163.1
	250	0.47	38.30	3260	181.1
	300	0.51	38.30	3520	195.5
	350	0.56	38.30	3840	213.4

*Tolerance of OD of spigot end: 3–12 in, ±0.06 in, 14–24 in, +0.05-in, –0.08 in, 30–36 in, +0.08 in, –0.06 in.

†Including bell; calculated weight of pipe rounded off to nearest 5 lbs

‡Including bell; average weight, per foot, based on calculated weight of pipe before rounding.

§Available in 20-ft lengths.

2.1.8.2 Ductile Iron-Pipe Specifications for 3" to 36" Pipe

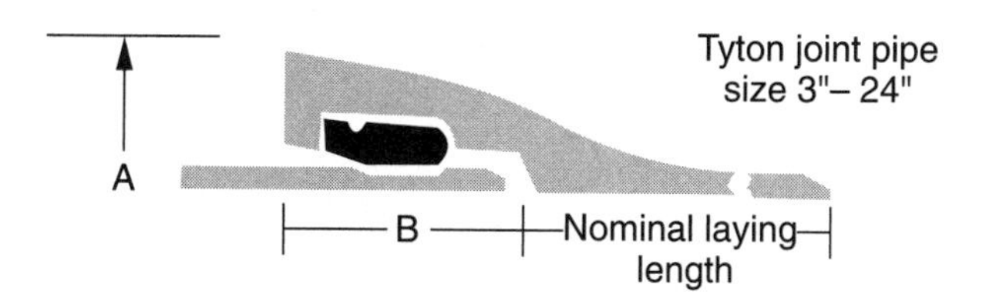

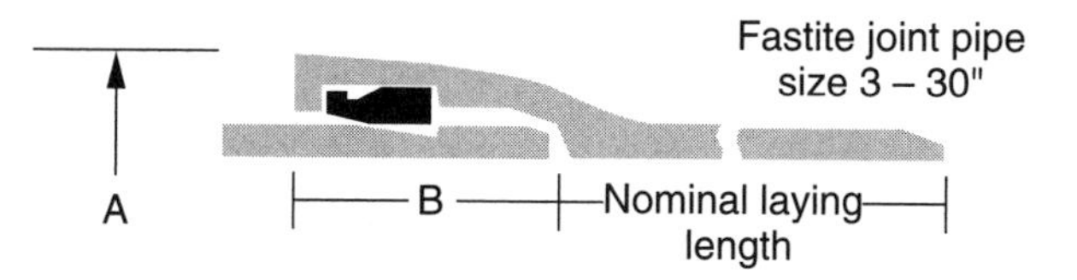

Standardized Mechanical Joint Pipe
Joint dimension and weight

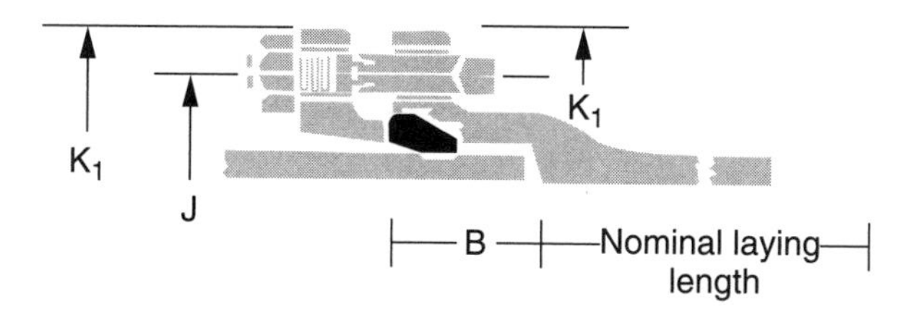

Size, in	Pipe thickness,* in		Outside diameter, in	Dimensions, in	
	From	To		A	B
3	0.25	0.40	3.96	5.80	3.00
4	0.25	0.41	4.80	6.86	3.15
6	0.25	0.43	6.90	8.75	3.38
8	0.25	0.45	9.05	11.05	3.69
10	0.26	0.47	11.10	13.15	3.75
12	0.28	0.49	13.20	15.30	3.75
14	0.28	0.51	15.30	17.85	5.00
16	0.30	0.52	17.40	20.00	5.00
18	0.31	0.53	19.50	22.10	5.00
20	0.33	0.54	21.60	24.25	5.00
24	0.33	0.56	25.80	28.50	5.00
30	0.34	0.63	32.00	34.95	6.50
36	0.38	0.73	38.30	41.37	6.50

*3–4-in nominal 20-ft laying length; 6-in nominal 18 or 20 ft; 8–36-in nominal 18-ft laying length.

Dimensions subject to manufacturing tolerances.

Size inches	Pipe thickness inches		Outside diameter inches	Dimension in inches				Bolts			Bell weight pounds	Gland bolts gasket weight pounds
	From	To		B	J	K*	K*	No	Size inches	Length inches		
3	0.25	0.40	3.96	2.50	6.19	7.62	7.69	4	⅝	3	11	7
4	0.26	0.41	4.80	2.50	7.50	9.06	9.12	4	¾	3½	16	10
6	0.25	0.43	6.90	2.50	9.50	11.06	11.12	6	¾	3½	18	16
8	0.27	0.45	9.05	2.50	11.75	13.31	13.37	6	¾	4	24	25
10	0.29	0.47	11.10	2.50	14.00	15.62	15.62	8	¾	4	31	30
12	0.31	0.49	13.20	2.50	16.25	17.88	17.88	8	¾	4	37	40
14	0.33	0.51	15.30	3.50	18.75	20.25	20.25	10	¾	4½	61	45
16	0.34	0.52	17.40	3.50	21.00	22.50	22.50	12	¾	4½	74	55
18	0.35	0.53	19.50	3.50	23.25	24.75	24.75	12	¾	4½	85	65
20	0.36	0.54	21.60	3.50	25.50	27.00	27.00	14	¾	4½	98	85
24	0.38	0.56	25.80	3.50	30.00	31.50	31.50	16	¾	5	123	105

*3"–4" nominal 20' laying length. –6" nominal 18' or 20'
8"–24" nominal 18' laying length.
Dimensions subject to manufacturing tolerances.

Reprinted by permission from Atlantic State Cast Iron Pipe Company, Phillipsburg, New Jersey

2.1.9 Schedule 40/80 PVC Pipe Specifications

Schedule 40/80 pipe size: inside/outside dimensions, weight per foot for UL-rated PVC pipe.

J-M SCH. 40 CONDUIT
U.L. Listed

RIGID NON-METALIC CONDUIT FOR USE IN BOTH ABOVE GROUND AND UNDERGROUND INSTALLATIONS

Schedule 40 Conduit — Rated for 90°C Conductors

Size	Part Number	Avg. OD	Nom. ID	Min. Wall	Approx. Wt. 100/Ft	Ft. Per Bundle	Feet Per Lift	Price/ 100 Ft
1/2	40050	.840	.622	.109	18	100	6000	18.01
3/4	40075	1.050	.824	.113	24	100	4400	24.45
1	40100	1.315	1.049	.133	33	100	3600	35.32
1¼	40125	1.660	1.380	.140	45	50	3300	47.83
1½	40150	1.900	1.610	.145	56	50	2250	57.69
2	40200	2.375	2.067	.154	74	50	1400	76.35
2½	40250	2.875	2.469	.203	126	10	900	122.25
3	40300	3.500	3.068	.216	163	10	880	158.90
3½	40350	4.000	3.548	.226	197	10	630	190.17
4	40400	4.500	4.026	.237	234	10	480	224.77
5	40500	5.563	5.047	.258	319	10	230	319.35
6	40600	6.625	6.065	.280	411	10	220	410.11

Schedule 40 is furnished in standard 10' lengths with one bell end.
20 ft. lengths are available upon request.

J-M SCH. 80 CONDUIT
U.L. Listed

RIGID NON-METALIC CONDUIT FOR USE IN BOTH ABOVE GROUND AND UNDERGROUND INSTALLATIONS

Schedule 80 Conduit — Rated for 90°C Conductors

Size	Part Number	Avg. OD	Nom. ID	Min. Wall	Approx. Wt. 100/Ft	Ft. Per Bundle	Feet Per Lift	Price/ 100 Ft
1/2	80050	.840	.546	.147	22	100	6000	23.35
3/4	80075	1.050	.742	.154	30	100	4400	31.35
1	80100	1.315	.957	.179	42	100	3600	44.93
1¼	80125	1.660	1.278	.191	60	50	3300	62.72
1½	80150	1.900	1.500	.200	72	50	2250	74.57
2	80200	2.375	1.939	.218	98	10	1400	102.85
2½	80250	2.875	2.323	.276	151	10	900	157.30
3	80300	3.500	2.900	.300	213	10	880	209.90
4	80400	4.500	3.826	.337	310	10	480	305.50
5	80500	5.563	4.813	.375	430	10	230	440.42
6	80600	6.625	5.761	.432	590	10	220	583.00

Schedule 80 is furnished in standard 10' lengths with one bell end.
20 ft. lengths are available upon request.

By permission J-M Manufacturing Company, Inc., Livingston, N.J.

2.1.9.1 Deflection in Thermoplastic Pipe

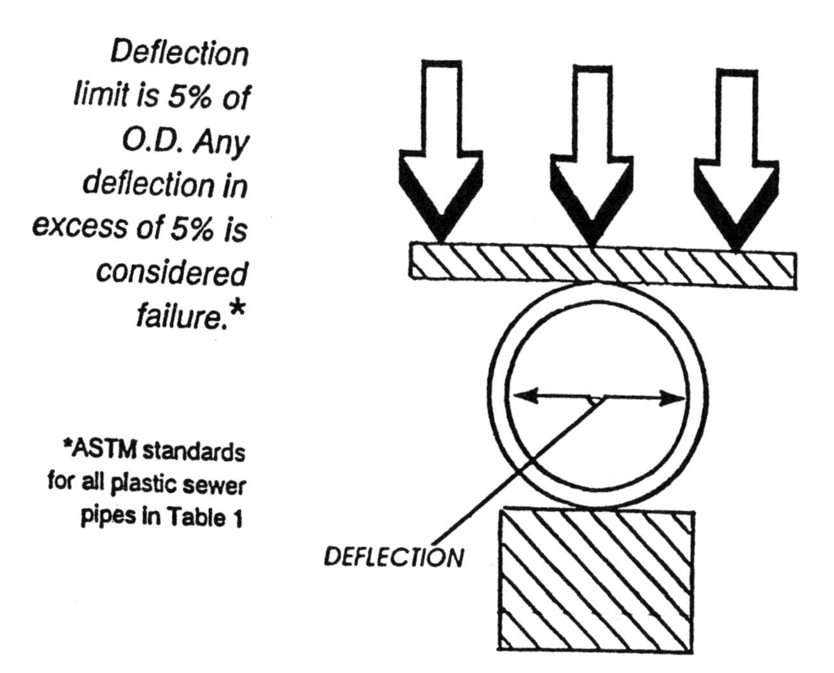

Reprinted by permission of the Cast Iron Soil Pipe Institute

2.1.9.2 Expansion and Contraction of PVC Pipe

PVC non-metallic conduit will expand and contract with temperature variations. When it is necessary to allow for movement of PVC conduit because of temperature changes, the amount of movement can be determined from the chart below. The coefficient of thermal expansion of J-M PVC conduit is 3.0×10^{-5} in/in°F. If major temperature variations are expected the use of expansion joints should be considered and should be installed, in accordance with the engineer's design.

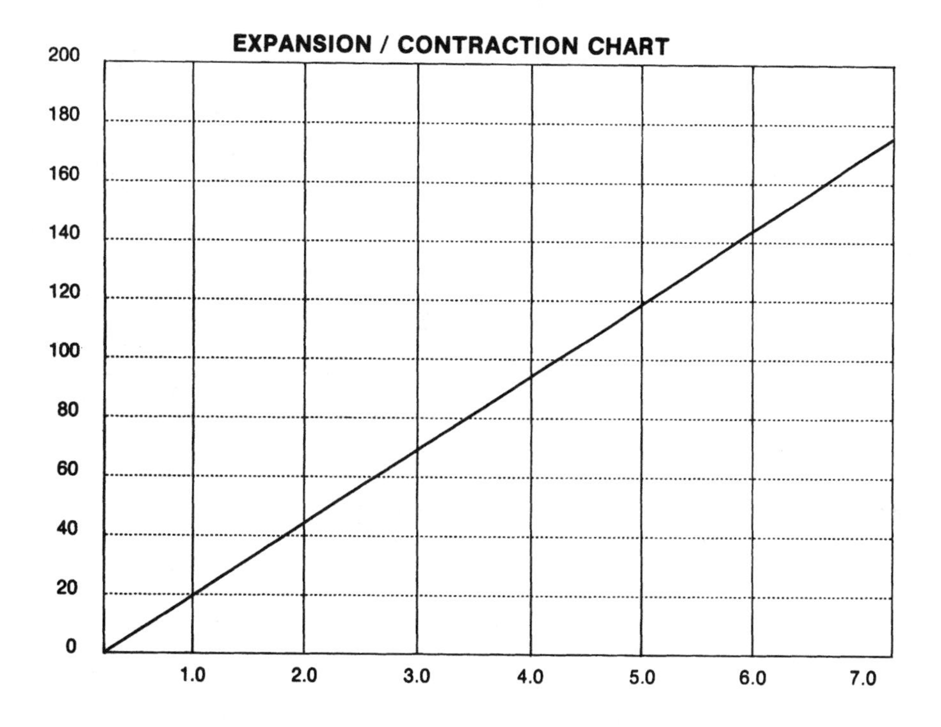

CHANGE IN LENGTH (INCHES) PER 100-FOOT LENGTH OF DUCT

By permission J-M Manufacturing Company, Inc., Livingston, N.J.

2.2.0 Corrugated Steel Pipe Specifications (12" to 144" Diameter)

Approximate Weight/Foot CONTECH Corrugated Steel Pipe

(Estimated Average Weights—Not for Specification Use)

2⅔" x ½" Corrugation

Inside Diameter in.	Specified Thickness in.	Galvanized & ALUMINIZED	Full Coated	Coated & PAVED-INVERT	SMOOTH-FLO	HEL-COR CL
12	0.052	8	10	13		
	0.064	10	12	15		
	0.079	12	14	17		
15	0.052	10	13	16	26	
	0.064	12	15	18	28	
	0.079	15	18	21	31	
18	0.052	12	16	19	31	
	0.064	15	19	22	34	
	0.079	18	22	25	37	
21	0.052	14	18	23	36	
	0.064	17	21	26	39	
	0.079	21	25	30	43	
24	0.052	15	20	26	41	
	0.064	19	24	30	45	65
	0.079	24	29	35	50	69
	0.109	33	38	44	59	77
30	0.052	20	26	32	51	
	0.064	24	30	36	55	82
	0.079	30	36	42	60	87
	0.109	41	47	53	72	96
36	0.052	24	31	39	50	
	0.064	29	36	44	65	98
	0.079	36	43	51	75	104
	0.109	49	56	64	90	116
	0.138	62	69	77	100	127
42	0.052	28	36	45	71	
	0.064	34	42	51	77	114
	0.079	42	50	59	85	121
	0.109	57	65	74	100	135
	0.138	72	80	89	115	149
48	0.064	38	48	57	85	128
	0.079	48	58	67	95	138
	0.109	65	75	84	112	154
	0.138	82	92	101	129	170
	0.168	100	110	119	147	186
54	0.079	54	65	76	105	156
	0.109	73	84	95	124	173
	0.138	92	103	114	143	191
	0.168	112	123	134	163	209
60	0.109	81	92	106	140	192
	0.138	103	114	128	162	212
	0.168	124	135	149	183	232
66	0.109	89	101	117	160	211
	0.138	113	125	141	180	233
	0.168	137	149	165	210	255
72	0.138	123	137	154	210	254
	0.168	149	163	180	236	278
78	0.168	161	177	194	260	302
84	0.168	173	190	208	270	325
90	0.168	186	204	224	289	348
96	0.168	198	217	239	309	371

3" x 1" or 5" x 1" Corrugation

Inside Diameter in.	Specified Thickness in.	Galvanized & ALUMINIZED	Full Coated	Coated & PAVED-INVERT	SMOOTH-FLO	HEL-COR CL
54	0.064	50	66	84	138	197
	0.079	61	77	95	149	207
	0.109	83	100	118	171	226
	0.138	106	123	140	194	245
	0.168	129	146	163	217	264
60	0.064	55	73	93	153	218
	0.079	67	86	105	165	229
	0.109	92	110	130	190	251
	0.138	118	136	156	216	272
	0.168	143	161	181	241	293
66	0.064	60	80	102	168	240
	0.079	74	94	116	181	252
	0.109	101	121	143	208	276
	0.138	129	149	171	236	299
	0.168	157	177	199	264	322
72	0.064	66	88	111	183	262
	0.079	81	102	126	197	275
	0.109	110	132	156	227	301
	0.138	140	162	186	257	326
	0.168	171	193	217	288	351
78	0.064	71	95	121	198	
	0.079	87	111	137	214	298
	0.109	119	143	169	246	326
	0.138	152	176	202	279	353
	0.168	185	209	235	312	380
84	0.064	77	102	130	213	
	0.079	94	119	147	230	321
	0.109	128	154	182	264	351
	0.138	164	189	217	300	379
	0.168	199	224	253	335	409
90	0.064	82	109	140	228	
	0.079	100	127	158	246	
	0.109	137	164	195	283	376
	0.138	175	202	233	321	406
	0.168	213	240	271	359	438
96	0.064	87	116	149	242	
	0.079	107	136	169	262	
	0.109	147	176	209	302	401
	0.138	188	217	250	343	433
	0.168	228	257	290	383	467
102	0.064	93	124	158	258	
	0.079	114	145	179	279	
	0.109	155	186	220	320	426
	0.138	198	229	263	363	460
	0.168	241	272	306	406	496
108	0.079	120	153	188	295	
	0.109	165	198	233	340	
	0.138	211	244	279	386	487
	0.168	256	289	324	431	525
114	0.079	127	162	199	312	
	0.109	174	209	246	359	
	0.138	222	257	294	407	514
	0.168	271	306	343	456	554
120	0.109	183	220	259	378	
	0.138	234	271	310	429	541
	0.168	284	321	360	479	583
126	0.138	247	285	326	452	
132	0.138	259	299	342	474	
	0.168	314	354	397	529	
138	0.138	270	312	357	495	
	0.168	328	370	415	553	
144	0.168	344	388	435	579	

Note: Smooth-Flo is fully lined with asphalt to provide added hydraulic efficiency Hel-Cor has helical corrugations. Inside Dimensions on various types of corrugated metal drainage pipes.

By permission CONTECH Construction Products Inc. Middletown, Ohio

2.2.1 Corrugated Steel-Pipe Specifications (Arch Height Cover Limits, 3" × 1")

H 20 and H 25 Live Load

Size		Minimum Specified Thickness, Inches*	Minimum Cover, Inches	Maximum Cover, Feet
Equivalent Pipe Diameter	Span x Rise, Inches			2 Tons/Ft.² Corner Bearing Pressure
48	53 x 41	0.079	12	25
54	60 x 46	0.079	15	25
60	66 x 51	0.079	15	25
66	73 x 55	0.079	18	24
72	81 x 59	0.079	18	21
78	87 x 63	0.079	18	20
84	95 x 67	0.079	18	20
90	103 x 71	0.079	18	20
96	112 x 75	0.079	21	20
102	117 x 79	0.109	21	19
108	128 x 83	0.109	24	19
114	137 x 87	0.109	24	19
120	142 x 91	0.138	24	19

E 80 Live Load

Size		Minimum Specified Thickness, Inches*	Minimum Cover, Inches	Maximum Cover, Feet
Equivalent Pipe Diameter	Span x Rise, Inches			2 Tons/Ft.² Corner Bearing Pressure
48	53 x 41	0.079	24	25
54	60 x 46	0.079	24	25
60	66 x 51	0.079	24	25
66	73 x 55	0.079	30	24
72	81 x 59	0.079	30	21
78	87 x 63	0.079	30	18
84	95 x 67	0.079	30	18
90	103 x 71	0.079	36	18
96	112 x 75	0.079	36	18
102	117 x 79	0.109	36	17
108	128 x 83	0.109	42	17
114	137 x 87	0.109	42	17
120	142 x 91	0.138	42	17

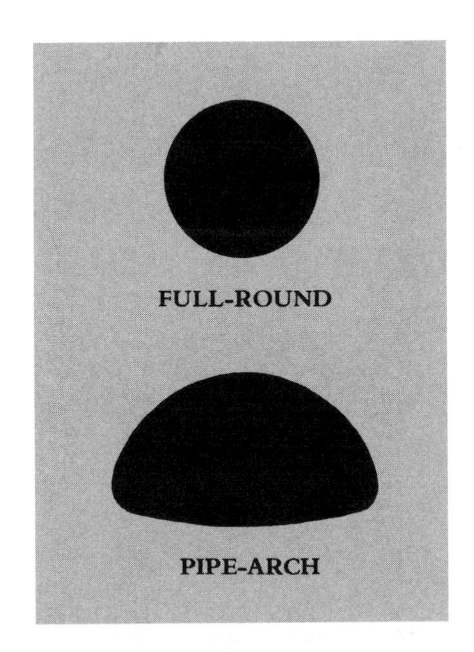

Reference Specifications

Material	Galvanized Steel	ASTM A 929 and AASHTO M218
	ALUMINIZED STEEL Type 2	ASTM A 929 and AASHTO M274
	FIBER-BONDED Steel	ASTM A 885
	Polymer-Coated Steel	ASTM A 742 and AASHTO M246
Pipe	Steel (Galvanized and ALUMINIZED STEEL Type 2)	ASTM A 760 and AASHTO M36
	Steel (Polymeric)	ASTM A 762 and AASHTO M245
Coating/Lining	Asphalt and Concrete	ASTM A 849 and AASHTO M190
Design	Steel	ASTM A 796 and AASHTO Standard Specification for Highway Bridges, Section 12
Installation	Steel	ASTM A 798 and AASHTO Standard Specification for Highway Bridges, Section 26

Reprinted by permission of CONTECH Construction Products, Inc., Middletown, Ohio

2.2.1.1 Corrugated Steel Pipe Specifications (Arch Height Cover Limits, 5" × 1")

H 20 and H 25 Live Load

Size: Equivalent Pipe Diameter	Size: Span x Rise, Inches	Minimum Specified Thickness, Inches*	Minimum Cover, Inches	Maximum Cover, Feet: 2 Tons/Ft.² Corner Bearing Pressure
72	81 x 59	0.109	18	21
78	87 x 63	0.109	18	20
84	95 x 67	0.109	18	20
90	103 x 71	0.109	18	20
96	112 x 75	0.109	21	20
102	117 x 79	0.109	21	19
108	128 x 83	0.109	24	19
114	137 x 87	0.109	24	19
120	142 x 91	0.138	24	19

E 80 Live Load

Size: Equivalent Pipe Diameter	Size: Span x Rise, Inches	Minimum Specified Thickness, Inches*	Minimum Cover, Inches	Maximum Cover, Feet: 2 Tons/Ft.² Corner Bearing Pressure
72	81 x 59	0.109	30	21
78	87 x 63	0.109	30	18
84	95 x 67	0.109	30	18
90	103 x 71	0.109	36	18
96	112 x 75	0.109	36	18
102	117 x 79	0.109	36	17
108	128 x 83	0.109	42	17
114	137 x 87	0.109	42	17
120	142 x 91	0.138	42	17

*Some 3" x 1" and 5" x 1" minimum gages shown for pipe-arch are due to manufacturing limitations.

Construction loads

For temporary construction vehicle loads, an extra amount of **compacted cover** may be required over the top of the pipe. The height-of-cover shall meet the minimum requirements shown in the table below. The use of heavy construction equipment necessitates greater protection for the pipe than finished grade cover minimums for normal highway traffic.

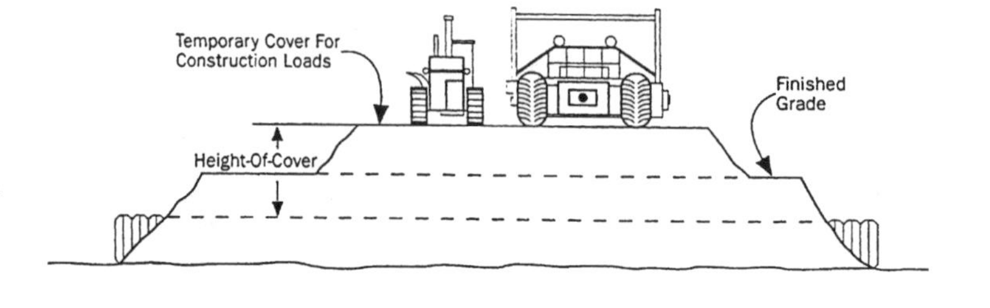

General Guidelines for Minimum Cover Required for Heavy Off-Road Construction Equipment

Pipe Span, Inches	Minimum Cover (feet) for Indicated Axle Loads (kips): 18-50	50-75	75-110	110-150
12-42	2.0	2.5	3.0	3.0
48-72	3.0	3.0	3.5	4.0
78-120	3.0	3.5	4.0	4.0
126-144	3.5	4.0	4.5	4.5

** Minimum cover may vary, depending on local conditions. The contractor must provide the additional cover required to avoid damage to the pipe. Minimum cover is measured from the top of the pipe to the top of the **maintained** construction roadway surface.*

Corrugated steel pipe is used extensively to rehabilitate failing reinforced concrete pipe.

2.2.1.2 Corrugated Aluminum/Galvanized Steel Pipe Specifications

Aluminized Steel or Galvanized Steel Corrugated Metal Pipe

Diameter (Inches)	Weight (Pounds/Lineal Foot)		
	Specified Thickness and Gage		
	(.064″) 16	(.079″) 14	(.109″) 12
18	14.9	18.3	
21	17.4	21.4	29.1
24	19.9	24.4	35.9
30	24.9	30.5	41.5
36	29.8	36.7	49.9
42	34.8	42.8	58.3
48	39.8	48.9	66.6
54	44.8	55.0	74.9
60	49.8	61.2	83.3
66		67.2	91.5
72		73.2	99.6
78		79.3	107.9
84			116.1
90			124.4
96			132.6
102			139.8

Aluminum Corrugated Pipe

Diameter (Inches)	Weight (Pounds/Lineal Foot)			
	Specified Thickness and Gage			
	(.060″) 16	(.075″) 14	(.105″) 12	(.135″) 10
18	5.2	6.4		
21	6.0	7.5	10.5	
24	6.9	8.6	12.0	
30	8.6	10.7	15.0	
36	10.3	12.8	18.0	22.4
42	12.1	15.0	21.0	26.2
48		17.1	24.0	29.9
54		19.3	27.0	33.7
60			30.0	37.4
66			33.0	41.1
72			36.0	44.8
78				48.5
84				52.2

By permission CONTECH Construction Products Inc., Middletown, Ohio

2.2.2 Expansion Characteristics of Various Metal/Thermoplastic Pipe

Expansion: Allowances for expansion and contraction of building materials are important design considerations. Material selection can create or prevent problems. Cast iron is in tune with building reactions to temperature. Its expansion is so close to that of steel and masonry that there is no need for costly expansion joints and special offsets. That is not always the case with other DWV materials.

Thermal expansion of various materials.

Material	Inches per inch 10^{-6} X per °F	Inches per 100′ of pipe per 100°F.	Ratio-assuming cast iron equals 1.00
Cast iron	6.2	0.745	1.00
Concrete	5.5	0.66	.89
Steel (mild)	6.5	0.780	1.05
Steel (stainless)	7.8	0.940	1.26
Copper	9.2	1.11	1.49
PVC (high impact)	55.6	6.68	8.95
ABS (type 1A)	56.2	6.75	9.05
Polyethylene (type 1)	94.5	11.4	15.30
Polyethylene (type 2)	83.3	10.0	13.40

Here is the *actual* increase in length for 50 feet of pipe and 70° temperature rise.

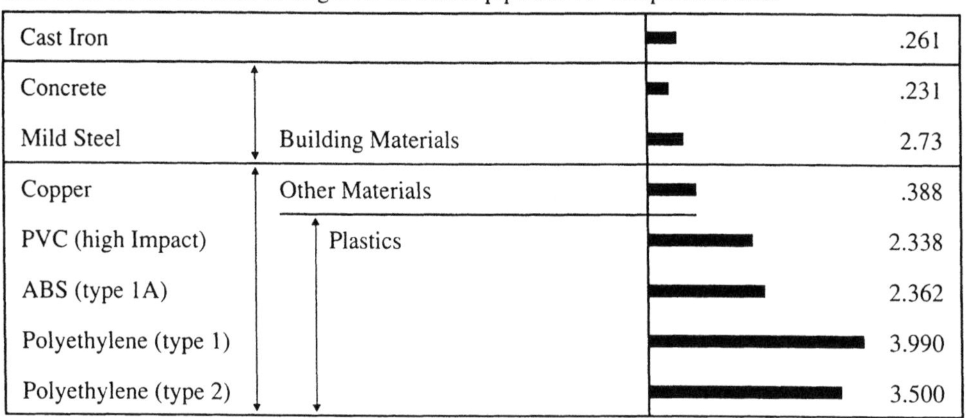

By permission Cast Iron Soil Pipe Institute

2.3.0 Testing of Underground Pipe Installations

- *Infiltration test* This test measures the integrity of the pipe to withstand infiltration from ground water. It is conducted in accordance with ASTM C 969M (C969). This test is only applicable if the water table is at least 2 feet (600 mm) above the crown of the pipe.

  The infiltration test is usually performed on storm and sanitary lines, and conducted between two adjacent manholes. All service laterals, stubs, and fittings are plugged or capped at the connection to the test pipe section. A V-notch weir or other suitable measuring device is installed in the inlet pipe to the downstream manhole; infiltrating water is then allowed to build up and level off behind the weir until steady, uniform flow is obtained. When this action occurs over the weir, leakage is determined by direct reading the graduations on the weir or converting the flow quantity to gallons per minute per unit length of pipe per unit of time.

- *Exfiltration test* This test results in subjecting the entire system to a pressure test and is generally used on small-diameter sanitary lines and follows ASTM C969M (C969) procedures. Tests are performed between manholes and the test section is filled with water through the upstream manhole. Once the test section is filled with water and allowed to stand for an adequate period of time, the water level in the upstream manhole is brought up to the proper test level. After a set period of time, the water elevation should be measured from the same reference point and the loss of water during the test period calculated. Or the water should be restored to its original level and the amount of water used to do so accounted for to determine the leakage.

Illustrations of infiltration and exfiltration tests follow and include a chart converting feet head to pounds per square inch (the metric equivalent is kilopascals).

2.3.1 Diagram of Infiltration Tests

The following test shows plugged underground lines.

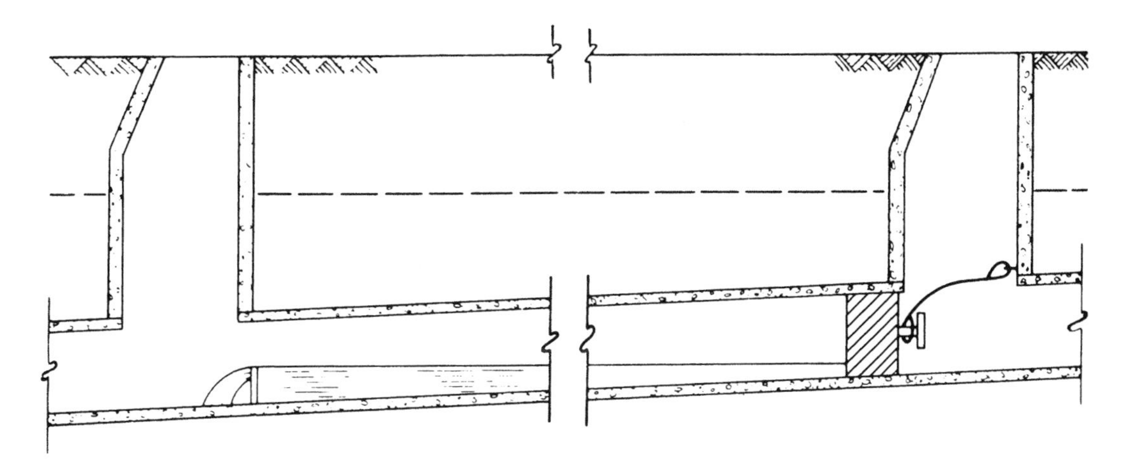

FEET HEAD OF WATER INTO PRESSURE, POUNDS PER SQUARE INCH							
Feet Heed	Lbs. per Square inch	Feet Head	Lbs. per Square inch	Feet Head	Lbs. per Square inch	Feet Head	Lbs. per Square Inch
1	.43	20	8.66	75	32.48	160	69.29
2	.87	25	10.83	80	34.65	170	73.63
3	1.30	30	12.99	85	36.81	180	77.96
4	1.73	35	15.16	90	38.98	190	82.29
5	2.17	40	17.32	95	41.14	200	86.62
6	2.60	45	19.40	100	43.31	225	97.45
7	3.03	50	21.65	110	47.64	250	108.27
8	3.40	55	23.82	120	51.97	275	119.10
9	3.90	60	25.99	130	56.30	300	129.93
10	4.33	65	28.15	140	60.63	325	140.75
15	6.50	70	30.32	150	64.96	350	151.58

Chart of conversion of feet head of water to pounds per square inch

FEET HEAD OF WATER INTO KILOPASCALS (kN/m²)							
Feet Head	kPa	Feet Head	kPa	Feet Head	kPa	Feet Head	kPa
1	2.99	20	59.77	75	224.14	160	478.16
2	5.98	25	74.71	80	239.08	170	508.05
3	8.96	30	89.66	85	254.02	180	537.93
4	11.95	35	104.60	90	268.97	190	567.81
5	14.94	40	119.54	95	283.91	200	597.70
6	17.93	45	134.48	100	298.85	225	672.41
7	20.92	50	149.43	110	328.73	250	747.13
8	23.91	55	164.37	120	358.62	275	821.84
9	26.90	60	179.31	130	388.51	300	896.55
10	29.89	65	194.25	140	418.39	325	971.26
15	44.83	70	209.20	150	448.28	350	1046.00

Metric conversion of feet head chart

2.3.2 Diagram of Exfiltration Tests

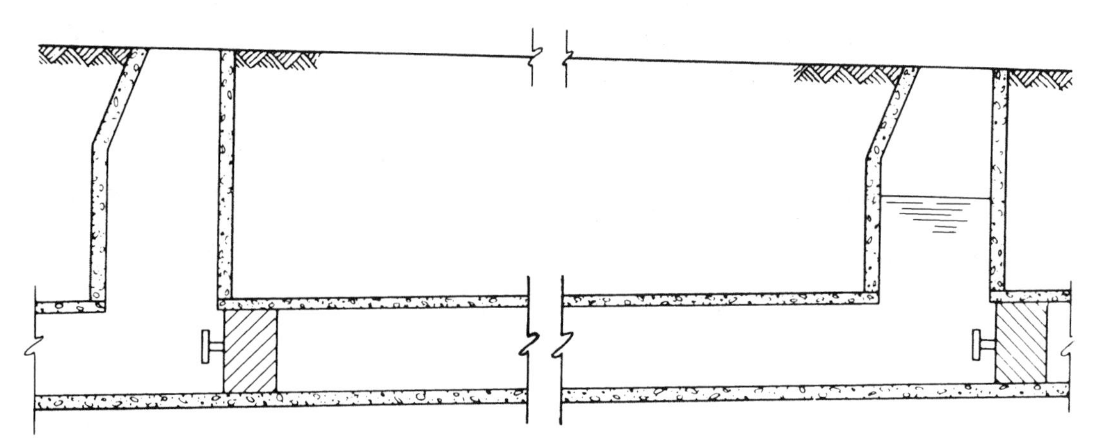

Per permission American Concrete Pipe Association, Irving, Tx.

2.4.0 Storm and Sanitary Manhole Schematics with Sections

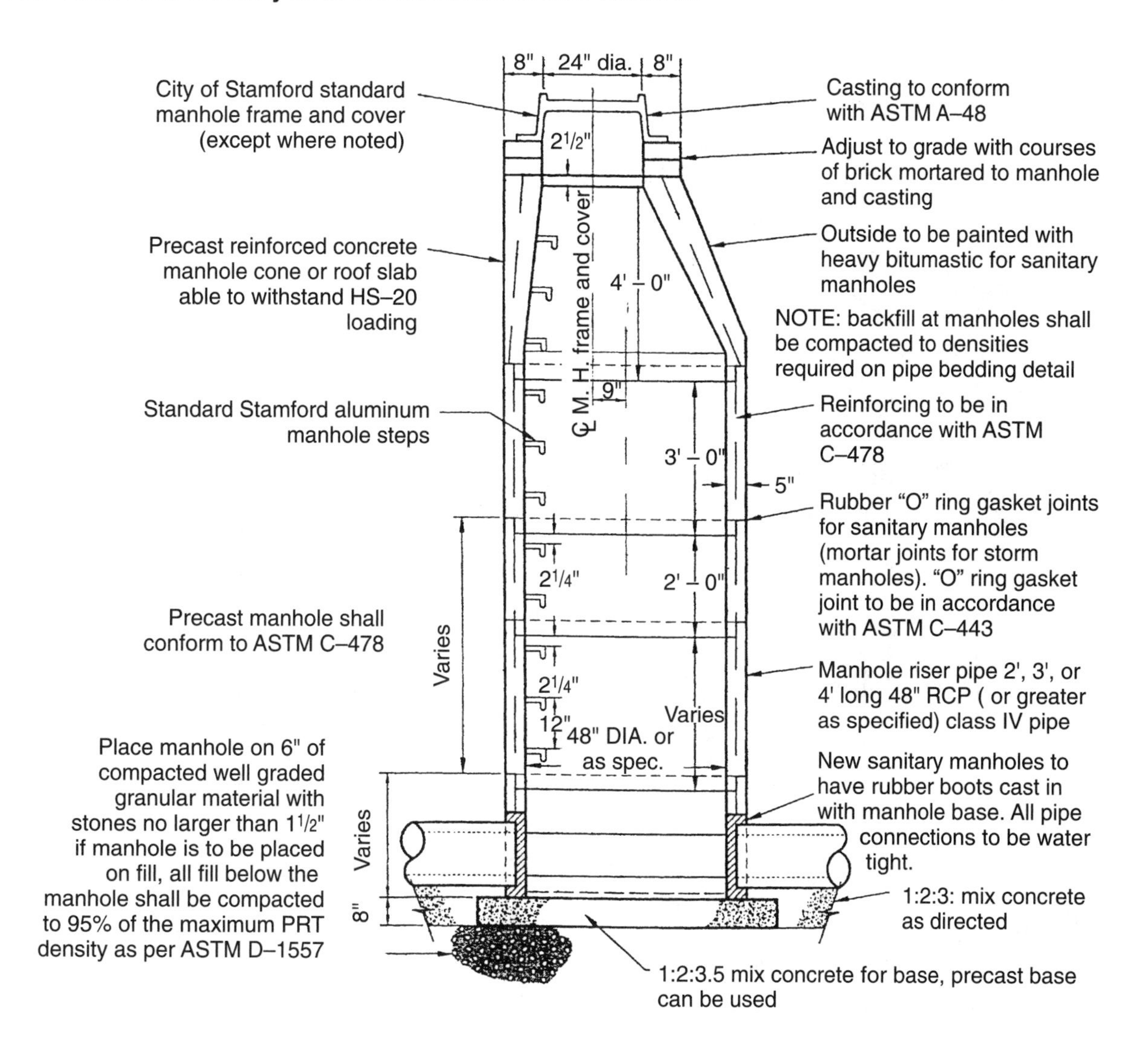

2.4.1 Storm Sewer Manhole Components

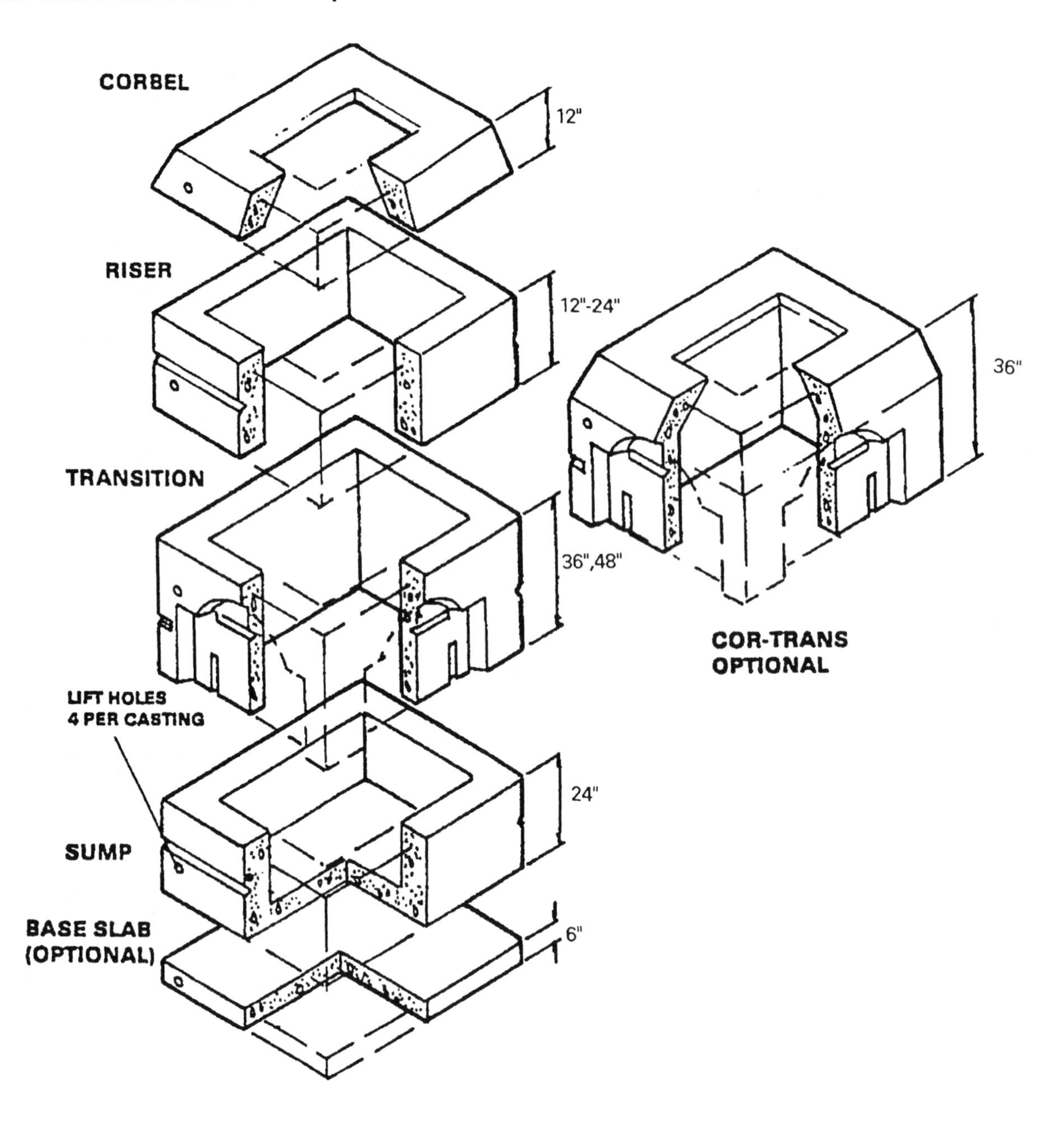

2.5.0 Castings for Sanitary and Storm Manholes

Gray iron exhibits excellent corrosion resistance, as well as excellent compressive strength.

Ductile iron is the material of choice when gray iron castings do not have enough load-bearing capacities or where impact resistance is a factor. Ductile iron castings are used where loads greater than H20 are required, such as locations subject to fork truck traffic, airports, and container ports.

Mating ductile iron lids with Class-35B gray iron frames is often a cost-effective method of providing covers in most situations other than H20.

CASTINGS FOR DIFFERING TRAFFIC CONDITIONS

FOUR STRENGTH CATEGORIES

Extra Heavy Duty: For airport and concentrated loads. These applications require special castings to accommodate uniformly distributed loads of 100 to over 225 psi. In selecting castings for heavy concentrated loads, advise loading conditions: total wheel load, tire or wheel contact area, and tire pressure. We will confirm your choice as being suitable for the purpose intended.

Heavy duty: Castings in this category are generally suitable for highway traffic or H20 wheel loads of 16,000 lb. Many of the castings are suitable for much heavier loading.

Medium duty: Use these castings for driveway, parks, ramps, and similar installations where wheel loads will not exceed 2000 lb.

Light Duty: These castings are recommended for sidewalks, terraces, and very light traffic.

Typical Specifications and Mechanical Properties of Gray Iron and Ductile Iron

Gray Iron

Class no.	Tensile strength, psi	Specifications
30	30,000	ASTM A48-83
35	35,000	ASTM A48-83 AASHTO M105-82
40	40,000	ASTM A48-83
45	45,000	ASTM A48-83

Ductile Iron

Grade	Tensile strength, psi	Yield strength, psi	Elongation %	Specifications
60-40-18	60,000	40,000	18 min.	ASTM A536-80 SAE J434C
65-45-12	65,000	45,000	12 to 20	ASTM A536-80 SAE J434C
80-55-06	80,000	55,000	6 to 12	ASTM A536-80 SAE J434C
100-70-03	100,000	70,000	3 to 10	ASTM A536-80 SAE J434C
CLASS A	60,000	45,000	15 min.	MIL-1-24137 CLASS A

Specifications and Mechanical Properties for Nonferrous Metals

Aluminum

ASTM no.	Alloy no.	Physical properties: Tensile, psi	Yield, psi	Elongation, %
B26	713.0	32,000	22,000	3
B26	319.0	23,000	13,000	1.5

Bronze

ASTM no.	Alloy no.	Physical properties: Tensile, psi	Yield, psi	Elongation, %
B584	C87200-12A	45,000	18,000	20
B584	C86300	110,000	60,000	12

Section

3

Concrete

Contents

3.1.0 History
3.1.1 General properties
3.2.0 Portland cement as a major component
3.2.1 High early cement
3.2.1.2 How cement content affects shrinkage
3.2.1.3 Effect of cement/water content on shrinkage
3.3.0 Control joints
3.3.1 Maximum spacing of control units
3.3.1.1 Dowel spacing
3.4.0 Admixtures
3.5.0 Chloride content in the mixing water
3.6.0 Guidelines for mixing small batches of concrete
3.7.0 Recommended slumps
3.8.0 Forms or cast-in-place concrete
3.8.1 Maximum allowable tolerances for form work
3.8.2 Release agents for forms
3.8.3 Principal types of commercially available form ties
3.9.0 Curing of concrete
3.9.1 Curing procedures
3.9.2 Curing times at 50 degrees
3.10.0 Concrete-reinforcing bar-size/weight chart
3.10.1 Material specifications for reinforcing bars
3.10.2 Mechanical/chemical requirements for reinforcing bars
3.10.3 Welded-wire fabric (WWF)
3.10.3.1 Common styles of welded-wire fabric
3.10.3.2 Welded-wire fabric (WWF)
3.10.4 Recommended industry practices for rebar fabrication
3.10.4.1 Recommended industry practices for rebar fabrication (continued)
3.10.4.2 Recommended industry practices for refar fabrication (continued)
3.10.4.3 Recommended industry practices for rebar fabrication (continued)
3.11.0 Reinforcng bar supports (typical types)
3.11.1 Typical wire size and geometry of bar supports
3.11.2 Typical types of bar supports for precast concrete
3.11.3 All-plastic bar supports
3.11.4 Sequence of placing bar supports (two-way flat plate slab)
3.11.4.1 Sequence of placing bar supports (two-way flat slab)
3.11.5 Bar supports on corrugated steel forms
3.12.0 Notes on the metrication of reinforcing steel

3.1.0 History

Concrete is an ancient materal of construction, first used during the Roman Empire, which extended from about 20 BC to 200 AD. The word concrete is derived from the Roman *concretus*, meaning to grow together. Although this early mixture was made with lime, cement, and a volcanic ash material called *pozzolana*, concrete today is a sophisticated material in which exotic constituents can be added and, with computer-controlled batching, can produce a product capable of achieving 50,000 psi compressive strength.

The factors contributing to a successful batch of concrete are:

- Precise measurement of water content.
- Type, size, and amount of cement and aggregate.
- Type, size, and location of reinforcement within the concrete pour to compensate for the lack of tensile strength basic in concrete.
- Proper curing procedures during normal, hot or cold weather conditions.

3.1.1 General Properties

With some exceptions, the two most widely used concrete mixtures are:

- Normal-weight (stone) concrete with a dry weight of 145 psf (6.93kPa).
- Lightweight concrete (LWC) with a weight of approximately 120 psf (5.74 KPa). Extra light concrete, with weights as low as 80 psf (3.82 kPa), can be achieved with the use of special aggregates.

Other Types of Concrete

- *Lightweight Insulating* Containing perlite, vermiculite, and expanded polystyrene, which is used as fill over metal roof decks, in partitions, and in panel walls.
- *Cellular* Contains air or gas bubbles suspended in mortar and either no coarse aggregates or very limited quantities are included in the mixture. Used where high insulating properties are required.
- *Shot-crete or Gunite* The method of placement characterizes this type of concrete, which is applied via pneumatic equipment. Typical uses are swimming pools, shells, or domes, where formwork would be complicated because of the shape of the structure.
- *Ferrocement* Basically a mortar mixture with large amounts of light-gauge wire reinforcing. Typical uses include bins, boat hulls, and other thin, complex shapes.

3.2.0 Portland Cement as a Major Component

Different types of portland cement are manufactured to meet specific purposes and job conditions

- Type I is a general-purpose cement used in pavements, slabs, and miscellaneous concrete pads and structures.
- Type IA is used for normal concrete, to which an air-entraining admixture is added.
- Type II creates a moderate sulfur-resistant product that is used where concrete might be exposed to groundwater that contains sulfates.
- Type IIA is the same as Type II, but is suited for an air-entrainment admixture.
- Type III is known as *high early strength* and generates high strength in a week or less.
- Type IIIA is high early, to which is added an air-entrainment admixture.
- Type IV cement produces low heat of hydration and is often used in mass pours, such as dam construction or thick mat slabs.
- Type V is a high sulfate-resistant cement that finds application in concrete structures exposed to high sulfate-containing soils or groundwater.

- White Portland cement is generally available in Type I or Type III only and gains its white color from the selection of raw materials containing negligible amounts of iron and magnesium oxide. White cement is mainly used as a constituent in architectural concrete.

3.2.1 High Early Cement

High early cement does exactly what its name implies: it provides higher compressive strength at an earlier age. Although Type III or Type IIIA cement can produce high early strength, there are other ways to achieve the same end result:

- Add more cement to the mixture (600 lbs (272 Kg) to 1000 lbs (454 Kg).
- Lower the water content (0.2 to 0.45) by weight.
- Raise the curing temperature after consultation with the design engineer.
- Introduce an admixture into the design mix.
- Introduce microsilica, also known as *silica fume* to the design mix.
- Cure the cast-in-place concrete by autoclaving (steam curing).
- Provide insulation around the formed, cast-in-place concrete to retain heat of hydration.

3.2.1.2 How Cement Contents Affects Shrinkage

When low slumps, created in conjunction with minimum water requirements, are used with correct placement procedures, the shrinkage of concrete will be held to a minimum. Conversely, high water content and high slumps will increase shrinkage. A study at the Massachusetts Institute of Technology, as reported by the Portland Cement Association, indicated that for every 1% increase in mixing water, shrinkage of concrete increased by 2%. This study produced the following chart, showing the correlation of water and cement content to shrinkage.

3.2.1.3 Effect of Cement/Water Content on Shrinkage

Cement Content	Concrete Composition					Water cement	Slump	Shrinkage
Bags/cubic yard	Cement	Water	Air	Aggregate	Water + air	Ratio by Weight	(inches)	(av. 3 × 3 × 10" prism)
4.99	0.089	0.202	0.017	0.692	0.219	0.72	3.3	0.0330
5.99	0.107	0.207	0.016	0.670	0.223	0.62	3.6	0.330
6.98	0.124	0.210	0.014	0.652	0.224	0.54	3.8	0.0289
8.02	0.143	0.207	0.015	0.635	0.223	0.46	3.8	0.0300

3.3.0 Control Joints

Thermal shrinkage will occur and the object of control joints, sometimes referred to as construction joints, is to avoid the *random cracking* that often comes about when a concrete slab dries and produces excess tensile stress. Control joint spacing depends upon the slab thickness, aggregate size, and water content, as reported by the Portland Cement Association in their article "Concrete Floors on Concrete," second edition, 1983.

3.3.1 Maximum Spacing of Control Joints

Slab Thickness	Slump of 4–6 inches (101.6 mm–152.4 mm)		
	Max. size aggregate less than ¾ inches (19.05 mm)	Max. size aggregate larger than ¾ inches	Slump less than 4 inches (101.6 mm)
4" (101.6 mm)	8' (2.4 m)	10' (3.05 m)	12' (3.66 m)
5" (126.9 mm)	10' (3.05 m)	13' (3.96 m)	15' (4.57 m)
6" (152.4 mm)	12' (3.66 m)	15' (4.57 m)	18' (5.49 m)
7" (177.8 mm)	14' (4.27 m)	18' (5.49 m)	21' (6.4 m)
8" (203.1 mm)	16' (4.88 m)	20' (6.1 m)	24' (7.32 m)
9" (228.6 mm)	18' (5.49 m)	23' (7.01 m)	27' (8.23 m)
10" (253.9 mm)	20' (6.1 mm)	25' (7.62 m)	30' (9.14 m)

The term *control joint* is often used as being synonymous with *construction joint*, however, there is a difference between the two. A *control joint* is created to provide for movement in the slab and induce cracking at that point, whereas a *construction joint* is a bulkhead that ends that day's slab pour. When control joints are created by bulkheading off a slab pour, rather than saw-cutting after the slab has been poured, steel dowels are often inserted in the bulkhead to increase load transfer at this joint.

3.3.1.1 Dowel Spacing

Slab Depth in. (mm)	Diameter (bar number)	Total length in. (mm)	Spacing in. (mm) center to center
5" (126.9 mm)	#5	12 in. (304.8 mm)	12 in. (304.8 mm)
6" (152.4 mm)	#6	14 in. (355.6 mm)	12 in. (304.8 mm)
7" (177.8 mm)	#7	14 in. (355.6 mm)	12 in. (304.8 mm)
8" (203.1 mm)	#8	14 in. (355.6 mm)	12 in. (304.8 mm)
9" (228.6 mm)	#9	16 in. (406.4 mm)	12 in. (304.8 mm)
10" (253.9 mm)	#10	16 in. (406.4 mm)	12 in. (304.8 mm)

3.4.0 Admixtures

Although concrete is an extremely durable product, it faces deterioration from various sources: chemical attack, permeation by water and/or gases from external sources, cracking because of chemical reaction (known as *heat of hydration*), corrosion of steel reinforcement, freeze/thaw cycles, and abrasion. Much of the deterioration caused by these internal and exterior factors can be drastically delayed by the addition of a chemical admixture to the ready-mix concrete.

Admixtures are chemicals developed to make it easier for a contractor to produce a high-quality concrete product. Some admixtures retard curing, some accelerate it; some create millions of microscopic bubbles in the mixture; others allow a substantial reduction in water content, but still permit the concrete to flow like thick pea soup.

- *Water-reducing admixtures* Improve strength, durability, workability of concrete. Available in normal range and high range.
- *High-range water-reducing admixture* Also known as superplasticizer, it allows up to 30% reduction in water content with no loss of ultimate strength, but it creates increased flowability. It is often required where reinforcing steel is placed very close together in intricate forms.
- *Accelerating admixtures* They accelerate the set time of concrete, thereby reducing the protection time in cold weather, allowing for earlier stripping of forms. Accelerating admixtures are available in both chloride- and nonchloride-containing forms. Nonchloride is required if concrete is to be in contact with metal and corrosion is to be avoided.
- *Retarder admixtures* Retards the setting time, a desirable quality during very hot weather.
- *Air-entraining admixtures* Creates millions of microscopic bubbles in the cured concrete, allowing for expansion of permeated water, which freezes and is allowed to expand into these tiny bubbles, thereby resisting hydraulic pressures caused by the formation of ice.
- *Fly ash* When added to the concrete mixture, it creates a more dense end product, making the concrete extremely impermeable to water, which affords more protection to steel reinforcement contained in the pour. The addition of fly ash can increase ultimate strength to as much as 6500 psi (44.8 MPa), in the process, making the concrete more resistant to abrasion.
- *Silica fume* Also known as microsilica, it consists of 90 to 97% silicon dioxide, containing various amounts of carbon that are spherical in size and average about 0.15 microns in size. These extremely fine particles disperse into the spaces around the cement grains and create a uniform, dense microstructure that produces concrete with ultra-high compressive strengths, in the nature of 12,000 (82.73 MPa) to 17,000 psi (117.20 MPa).
- *Multifilament or fibrillated fibers* This material is not a chemical admixture per se, but several manufacturers of concrete chemical additives also sell containers of finely chopped synthetic fibers, generally polypropylene, which, when added to the ready-mix concrete, serves as secondary reinforcement and prevent cracks.

3.5.0 Chloride Content in the Mixing Water

Excessive chloride ions in mixing water can contribute to accelerated reinforcing-steel corrosion and should be a concern when evaluating a mix design. Maximum water-soluble chloride ions, in various forms of concrete (as a percentage), should not exceed the following:

- Prestressed concrete 0.06%
- Reinforced concrete exposed to chloride in service (e.g., garbage slab) 0.15%
- Reinforced concrete that will be dry and/or protected from moisture infiltration 1.00%
- Other reinforced concrete 0.30%

3.6.0 Guidelines For Mixing Small Batches of Concrete (By Weight)

Max. size aggregate	Cement (lbs/Kg)	Wet-fine aggregate (lbs/Kg)	Wet coarse aggregate (lbs/Kg)	Water (lbs/Kg)
⅜" (9.52 mm)	29 lbs (13.15 Kg)	59 lbs (26.76 Kg)	46 lbs (20.87 Kg)	11 lbs (4.99 Kg)
½" (12.6 mm)	27 lbs (12.25 Kg)	53 lbs (24.04 Kg)	55 lbs (24.95 Kg)	11 lbs (4.99 Kg)
¾" (19.05 mm)	25 lbs (11.34 Kg)	47 lbs (21.32 Kg)	65 lbs (29.66 Kg)	10 lbs (4.54 Kg)
1" (25.39 mm)	24 lbs (10.89 Kg)	45 lbs (20.41 Kg)	70 lbs (31.75 Kg)	10 lbs (4.54 Kg)
1½" (37.99 mm)	23 lbs (10.43 Kg)	43 lbs (19.50 Kg)	75 lbs (34.02 Kg)	9 lbs (4.08 Kg)

Guidelines For Mixing Small Batches Of Concrete (By Volume)

Max size aggregate	Cement	Wet-fine aggregate	Wet-coarse aggregate	Water
⅜" (9.52 mm)	1	2½	1½	½
½" (12.6 mm)	1	2½	2	½
¾" (19.05 mm)	1	2½	2½	½
1" (25.29 mm)	1	2½	2¾	½
1½" (37.99 mm)	1	2½	3	½

3.7.0 Recommended Slumps

The Portland Cement Association recommends the following slumps:

Component	Max. slump (inches)	Min. slump (inches)
Footings (reinforced or not)	3	1
Foundation walls	3	1
Substructure walls	3	1
Caissons	3	1
Beams and reinforced walls	4	1
Building columns	4	1
Pavements and slabs	3	1
Mass concrete	2	1

3.8.0 Forms For Cast-In-Place Concrete

Many different types of forms are on the market: wood, steel, aluminum, and fiberglass. Each has its advantage and disadvantage; however, some items (form ties and form-release materials) are common to all forms. Also, numerous types and configurations of form liners are available, primarily for architectural concrete use.

3.8.1 Maximum Allowable Tolerances for Form Work

The American Concrete Institute (ACI), in their ACI 347 Manual, include recommended maximum allowable tolerances for various types of cast-in-place and precast concrete, for example:

- *Maximum variations from plumb* In column and wall surfaces in any 10 feet (3.05 m) of length ¼ inch (6.35 mm)
- *Maximum for entire length*
  ½ inch (12.7 mm)
- *Maximum variations form established position in plan shown on drawings—walls*
  ¾ inch (19.05 mm)
- *Variations in cross-sectional dimensions of beams/slab-wall thickness*
  Minus: ⅛ inch (3.175 mm)
  Plus: ¼ inch (6.35 mm)

3.8.2 Release Agents for Forms

A number of commercially available form release agents are on the market and some contractors use their own formula, but precautions are necessary, in some instances, to protect the form material:

Form face material Release agent comments and precautions.

Wood forms Oils penetrate wood and extend its life.

Unsealed plywood Apply a liberal amount of release agent several days before using, then wipe off, so only a thin layer remains prior to placing concrete.

Sealed/overlaid plywood Do not use diesel oil or motor oil on HDO/MDO plywood. Products containing castor oil can discolor concrete.

Steel Use a product with a rust inhibitor.

Aluminum Avoid products that contain wax or paraffin.

Glass-fiber reinforced Follow the form manufacturer's recommendation to avoid damage to forms.

Rigid plastic forms Follow the form manufacturer's recommendations to avoid damage to forms.

Elastomeric liners These often do not require release agents, but using the proper agent can prolong life. When deep textures are required, release agents should be used. Follow the manufacturer's recommendations to avoid damage to forms.

Foam expanded plastic liners Petroleum-based agents can dissolve thew foam. These liners are generally "one-time" use only.

Rubber liners/molds Do not use petroleum, mineral oil, or solvent-based form oils to avoid damage to liner.

Concrete molds Avoid chemically active release agents and avoid match-cast or slab-on-slab work when the casting surface used as the form is only a few days old.

Controlled-permeability forms No release agent required.

Plaster waste molds Pretreat the mold with shellac or some other type of waterproof coating. Yellow cup grease (thinned) is an effective release agent.

3.8.3 Principal Types of Commercially Available Form Ties

	Type of Tie	Typical Working Loads In Tension* (lb.)	Notes/Comments
One-Piece Ties	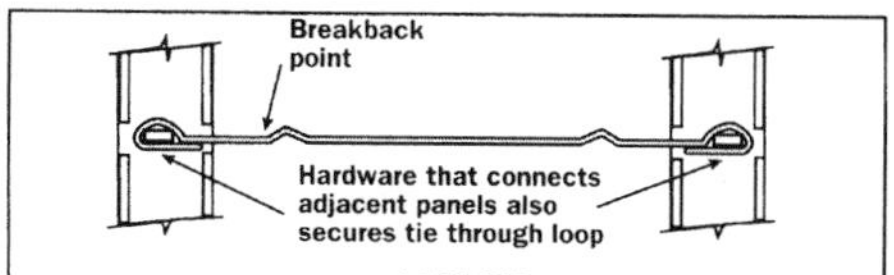 LOOP TIE (Breakback point; Hardware that connects adjacent panels also secures tie through loop)	Standard: 2,250 Heavy: 3,000	Shown with manufactured panel; also used with combination lock and bearing-plate hardware in job-built forms.
One-Piece Ties	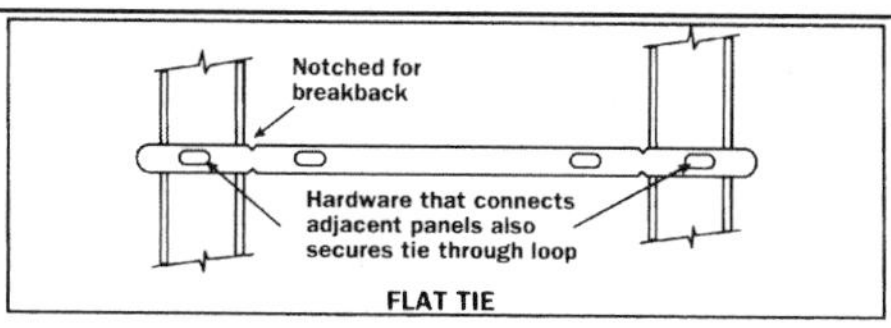 FLAT TIE (Notched for breakback; Hardware that connects adjacent panels also secures tie through loop)	Standard: 2,250 Heavy: 3,000	Also available for 1,500-pound loads.
One-Piece Ties	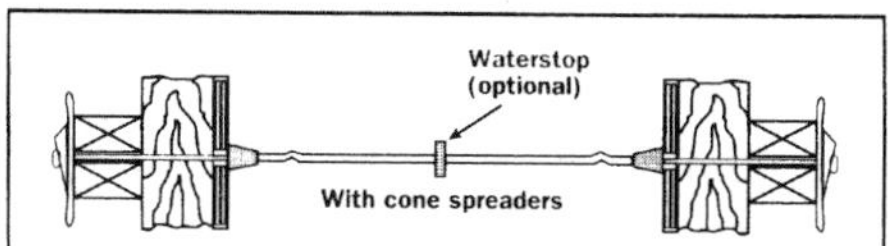 SNAP TIE (Waterstop (optional); With cone spreaders)	Standard: 2,250 Heavy: 3,000-3,200	Shown with cone spreaders; also available with washer spreaders.
One-Piece Ties	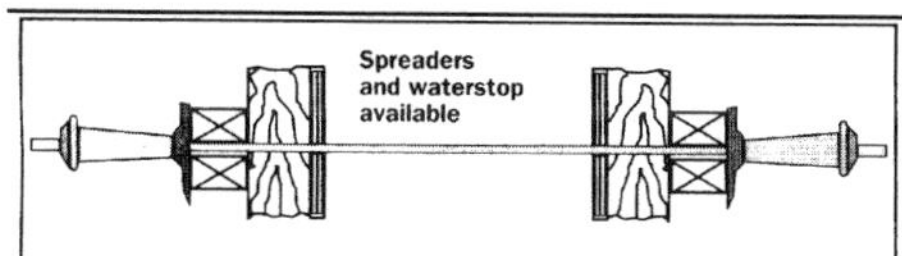 FIBERGLASS TIE (Spreaders and waterstop available)	3,000; 7,500; and 25,000, with diameters of 0.3, 0.5, and 1 inch, respectively	Available in 10- and 12-foot pieces for cutting to any desired length. Spreaders available.
One-Piece Ties	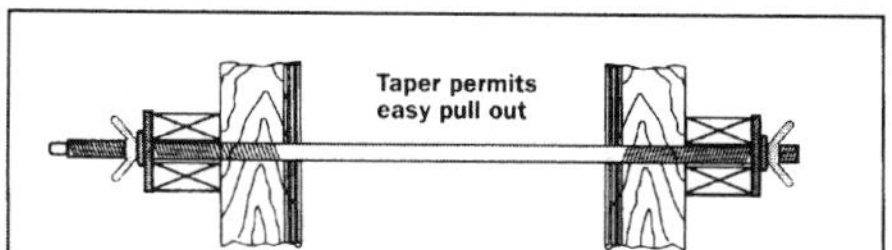 TAPER TIE (Taper permits easy pull out)	7,500-64,000, depending on diameter and grade of steel	Completely reusable; grease before installation to facilitate removal. No spreaders included.
	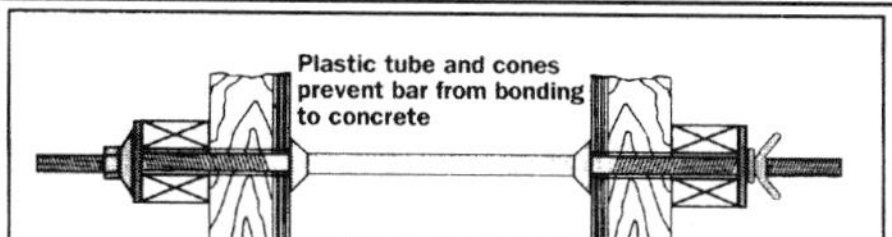 THREADED BAR TIE (Plastic tube and cones prevent bar from bonding to concrete)	7,000-69,000, with diameters from ½ inch to 1½ inches	Stock up to 50 feet long can be cut to required length. Plastic sleeve makes it removable.
Internally Disconnecting Ties	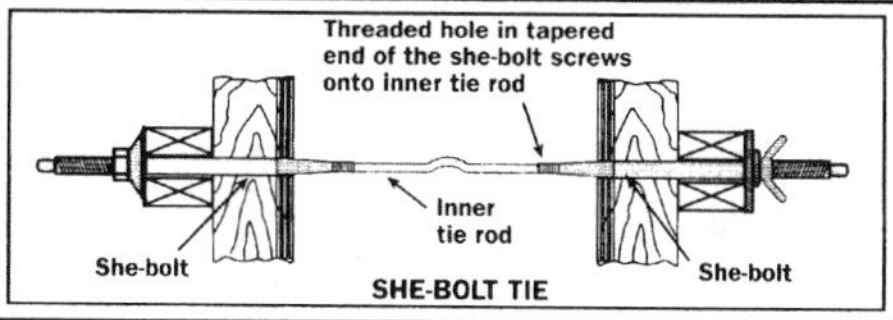 SHE-BOLT TIE (Threaded hole in tapered end of the she-bolt screws onto inner tie rod; Inner tie rod; She-bolt; She-bolt)	5,000-64,000	No internal spreader. External spreader bracket available.
Internally Disconnecting Ties	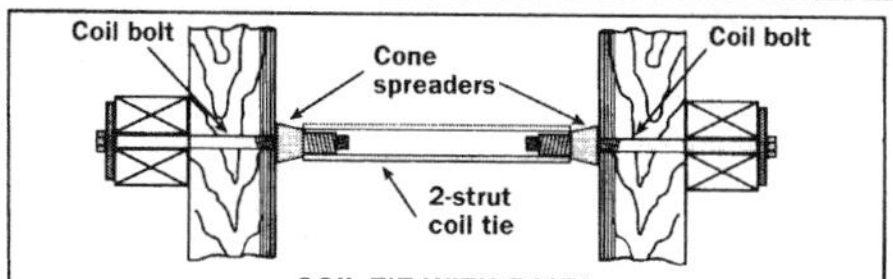 COIL TIE WITH BOLTS (Coil bolt; Coil bolt; Cone spreaders; 2-strut coil tie)	Two-strut: 4,500-64,000 Four-strut: 18,000-27,000	Shown with cone spreader, but can be used as combination tie/spreader where it is not necessary to keep the tie ends at the back of the wall face.

* Based on manufacturers' data, using a 2-1 factor of safety. Wide working-load ranges indicate a range of form-tie diameters and grades of steel.

By permission Aberdeen's Concrete Construction

3.9.0 Curing of Concrete

To attain design strength, curing is a crucial part of the cast-in-place concrete process in order that the proper amount of moisture content and ambient temperature is maintained immediately following the placement of the concrete. The optimum curing cycle will take into account the prevention or replenishment of moisture content from the concrete and the maintenance of a favorable temperature for a specific period of time. During winter months, temporary protection and heat is required in conjunction with the curing process and during summer months; moisture replenishment becomes an integral part of the curing process.

3.9.1 Curing Procedures

1. Apply a membrane-curing compound—either by spraying or rolling on the surface immediately after the troweling process on slabs has ceased, or on walls, columns, beams, after the forms have been removed.
2. Curing by water in other than cold-weather conditions is acceptable, as long as it is continuous.
3. Waterproof paper, applied directly over the concrete surface after it has received a spray of water, is often effective.
4. Damp burlap, free of foreign substances that could leach out and stain the concrete, is also a proven curing procedure, as long as the burlap is kept moist.
5. Polyethylene sheets can be used as a blanket in much the same manner as waterproof paper, as long as its edges are lapped and sealed properly.
6. Damp sand or straw is also used on occasion, when nothing else is available. These materials must also be sprayed from time to time to maintain the moisture content.

The length of curing depends upon a number of factors, including the type of cement used and ambient temperatures. The following can be used as a guideline to determine the length of curing time.

3.9.2 Curing at 50 Degrees F – Air Entrained Concrete

Percentage design strength required	Type cement used in mix		
	I	II	III
50%	6	9	3
65%	11	14	5
85%	21	28	16
95%	29	35	26

Curing at 70 Degrees F (21 Degrees C.) Days – Air Entrained Concrete

Percentage design strength required	Type cement used in mix		
	I	II	III
50%	4	6	3
65%	8	10	4
85%	16	18	12
95%	23	24	20

3.10.0 Concrete-Reinforcing Bar-Size/Weight Chart

Because of concrete's low resistance to shear and tensile strength, the type configuration and placement of reinforcement is crucial to achieve the project's design criteria. The most common form of concrete reinforcement is the deformed reinforcing bar and welded wire fabric. The most commonly used reinforcing bars are set forth in the following chart.

BAR SIZE DESIGNATION	WEIGHT	NOMINAL DIMENSIONS–ROUND SECTIONS		
	POUNDS PER FOOT	DIAMETER INCHES	CROSS-SECTIONAL AREA-SQ INCHES	PERIMETER INCHES
#3	.376	.375	.11	1.178
#4	.668	.500	.20	1.571
#5	1.043	.625	.31	1.963
#6	1.502	.750	.44	2.356
#7	2.044	.875	.60	2.749
#8	2.670	1.000	.79	3.142
#9	3.400	1.128	1.00	3.544
#10	4.303	1.270	1.27	3.990
#11	5.313	1.410	1.56	4.430
#14	7.650	1.693	2.25	5.320
#18	13.600	2.257	4.00	7.090

3.10.1 Material Specifications for Reinforcing Bars

Identification Marks*—ASTM Standard Rebars

The ASTM specifications for billet-steel, rail-steel, axle-steel and low-alloy reinforcing bars (A615, A616, A617 and A706, respectively) require identification marks to be rolled into the surface of one side of the bar to denote the Producer's mill designation, bar size, type of steel, and minimum yield designation. Grade 60 bars show these marks in the following order.

1st—Producing Mill (usually a letter)
2nd—Bar Size Number (#3 through #11, #14, #18)
3rd—Type of Steel:

S for Billet (A615)
W for Low-Alloy (A706)

I for Rail (A616)

I R for Rail meeting Supplementary Requirements S1 (A616)

A for Axle (A617)

4th—Minimum Yield Designation

Minimum yield designation is used for Grade 60 and Grade 75 bars only. Grade 60 bars can either have one single longitudinal line (grade line) or the number 60 (grade mark). Grade 75 bars can either have two grade lines or the grade mark 75.

A grade line is smaller and is located between the two main ribs which are on opposite sides of all bars made in the United States. A grade line must be continued through at least 5 deformation spaces, and it may be placed on the same side of the bar as the other markings or on the opposite side. A grade mark is the 4th mark on the bar.

Grade 40 and 50 bars are required to have only the first three identification marks (no minimum yield designation).

VARIATIONS: Bar identification marks may also be oriented to read horizontally (at 90° to those illustrated). Grade mark numbers may be placed within separate consecutive deformation spaces to read vertically or horizontally.

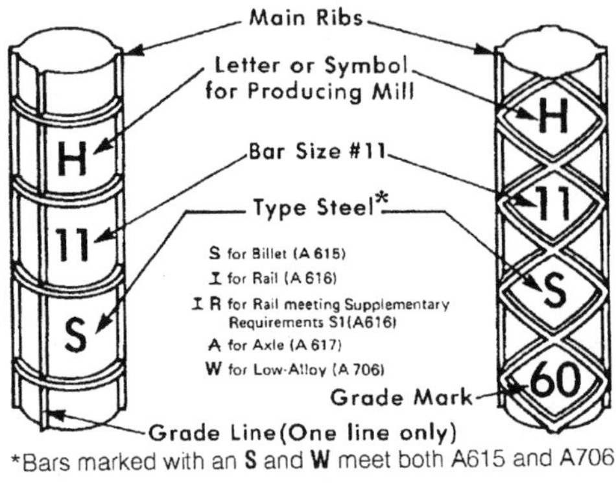

*Bars marked with an **S** and **W** meet both A615 and A706

GRADE 60

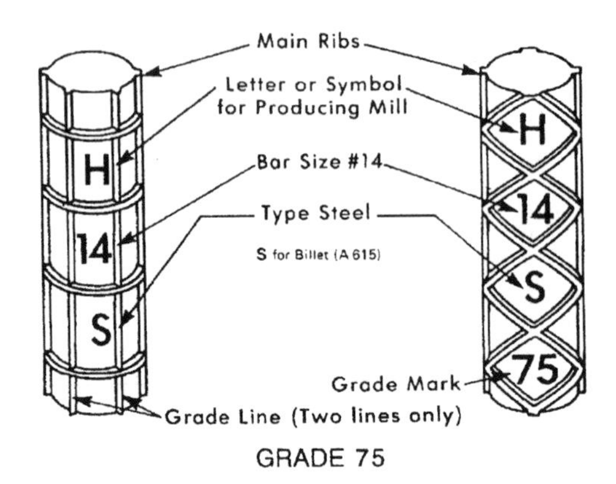

GRADE 75

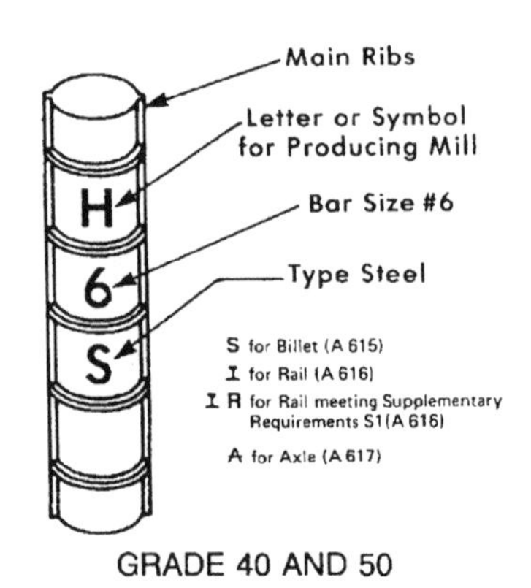

GRADE 40 AND 50

By permission Concrete Reinforcing Steel Institute, Schramsburg, Illinois

3.10.2 Mechanical/Chemical Requirements for Reinforcing Bars

Type of Steel and ASTM Designation	Bar Nos. Range	Grade[1]	Minimum[2] Yield Strength, psi	Minimum Tensile Strength, psi	Minimum Percentage Elongation in 8 in.	Cold Bend Test[3] Pin Diameter (d=nominal diameter of specimen)
Billet-Steel A615	3-6	40	40,000	70,000	#3 . . . . 11 #4, #5, #6 12	#3, #4, #5 . . . . 3½d #6 . . . . 5d
	3-11, 14, 18	60	60,000	90,000	#3, #4, #5, #6 . . . . 9 #7, #8 . . . . 8 #9, #10, #11, #14, #18. . . . 7	#3, #4, #5 . . . . 3½d #6, #7, #8 . . . . 5d #9, #10, #11 . . . . 7d #14, #18 (90°). . . . 9d
	6-11, 14, 18	75	75,000	100,000	#6, #7, #8 . . . . 7 #9, #10, #11, #14, #18. . . . 6	#6, #7, #8 . . . . 5d #9, #10, #11 . . . . 7d #14, #18 (90°). . . . 9d
Low-Alloy Steel A706	3-11, 14, 18	60	60,000[4]	80,000[5]	#3, #4, #5, #6 . . . . 14 #7, #8, #9, #10, #11 . . . . 12 #14, #18 10	#3, #4, #5 . . . . 3d #6, #7, #8 . . . . 4d #9, #10, #11 . . . . 6d #14, #18 . . . . 8d

[1] Minimum yield designation.
[2] Yield point or yield strength. See ASTM specifications.
[3] Test bends 180° unless noted otherwise.
[4] Maximum yield strength 78,000 psi (ASTM A706 only).
[5] Tensile strength shall not be less than 1.25 times the actual yield strength (ASTM A706 only).
* For the mechanical requirements of rail-steel and axle-steel bars, see ASTM specifications A616 and A617, respectively.

DEFORMATION REQUIREMENTS FOR STANDARD ASTM DEFORMED REINFORCING BARS

Size No.	Maximum Average Spacing, in.	Minimum Average Height, in.	Maximum[1] Gap, in.
3	0.262	0.015	0.143
4	0.350	0.020	0.191
5	0.437	0.028	0.239
6	0.525	0.038	0.286
7	0.612	0.044	0.334
8	0.700	0.050	0.383
9	0.790	0.056	0.431
10	0.889	0.064	0.487
11	0.987	0.071	0.540
14	1.185	0.085	0.648
18	1.58	0.102	0.864

[1] Chord of 12.5% of nominal perimeter

CHEMICAL COMPOSITION REQUIREMENTS FOR STANDARD ASTM DEFORMED REINFORCING BARS

Type of Steel and ASTM Designation	Condition*	Element									
		Carbon (C)	Manganese (Mn)	Phosphorus (P)	Sulfur (S)	Silicon (Si)	Copper (Cu)	Nickel (Ni)	Chromium (Cr)	Molybdenum (Mo)	Vanadium (V)
Billet-Steel A615	1	X	X	X	X						
	2			0.06%							
	3			0.075%							
Low-Alloy Steel A706	1	X	X	X	X	X	X	X	X	X	X
	2	0.30%	1.50%	0.035%	0.045%	0.50%					
	3	0.33%	1.56%	0.043%	0.053%	0.55%					

*CONDITION DEFINITIONS:
1 Analysis required of these elements for each heat.
2 Maximum allowable chemical content for each heat.
3 Maximum allowable chemical content for finished bar.

By permission Concrete Reinforcing Steel Institute, Schramsburg, Illinois

3.10.3 Welded Wire Fabric (WWF)

Cross-sectional area and weight of welded wire fabric

Wire Size Number		Nominal Diameter, in.	Nominal Weight, lbs/ft	Area Per Width (in.²/ft) for Various Spacings (in.)						
Plain	Deformed			2	3	4	6	8	12	16
W45	D45	0.757	1.53	2.70	1.80	1.35	0.90	0.68	0.45	0.34
W31	D31	0.628	1.05	1.86	1.24	0.93	0.62	0.47	0.31	0.23
W20	D20	0.505	0.680	1.2	0.80	0.60	0.40	0.30	0.20	0.15
W18	D18	0.479	0.612	1.1	0.72	0.54	0.36	0.27	0.18	0.14
W16	D16	0.451	0.544	0.96	0.64	0.48	0.32	0.24	0.16	0.12
W14	D14	0.422	0.476	0.84	0.56	0.42	0.28	0.21	0.14	0.11
W12	D12	0.391	0.408	0.72	0.48	0.36	0.24	0.18	0.12	0.09
W11	D11	0.374	0.374	0.66	0.44	0.33	0.22	0.17	0.11	0.08
W10.5		0.366	0.357	0.63	0.42	0.32	0.21	0.16	0.11	0.08
W10	D10	0.357	0.340	0.60	0.40	0.30	0.20	0.15	0.10	0.08
W9.5		0.348	0.323	0.57	0.38	0.29	0.19	0.14	0.095	0.07
W9	D9	0.338	0.306	0.54	0.36	0.27	0.18	0.14	0.090	0.07
W8.5		0.329	0.289	0.51	0.34	0.26	0.17	0.13	0.085	0.06
W8	D8	0.319	0.272	0.48	0.32	0.24	0.16	0.12	0.080	0.06
W7.5		0.309	0.255	0.45	0.30	0.23	0.15	0.11	0.075	0.06
W7	D7	0.299	0.238	0.42	0.28	0.21	0.14	0.11	0.070	0.05
W6.5		0.288	0.221	0.39	0.26	0.20	0.13	0.097	0.065	0.05
W6	D6	0.276	0.204	0.36	0.24	0.18	0.12	0.090	0.060	0.05
W5.5		0.265	0.187	0.33	0.22	0.17	0.11	0.082	0.055	0.04
W5	D5	0.252	0.170	0.30	0.20	0.15	0.10	0.075	0.050	0.04
W4.5		0.239	0.153	0.27	0.18	0.14	0.090	0.067	0.045	0.03
W4	D4	0.226	0.136	0.24	0.16	0.12	0.080	0.060	0.040	0.03
W3.5		0.211	0.119	0.21	0.14	0.11	0.070	0.052	0.035	0.03
W3		0.195	0.102	0.18	0.12	0.090	0.060	0.045	0.030	0.02
W2.9		0.192	0.099	0.17	0.12	0.087	0.058	0.043	0.029	0.02
W2.5		0.178	0.085	0.15	0.10	0.075	0.050	0.037	0.025	0.02
W2.1		0.162	0.070	0.13	0.84	0.063	0.042	0.031	0.021	0.02
W2		0.160	0.068	0.12	0.080	0.060	0.040	0.030	0.020	0.02
W1.5		0.138	0.051	0.090	0.060	0.045	0.030	0.022	0.015	0.01
W1.4		0.134	0.048	0.084	0.056	0.042	0.028	0.021	0.014	0.01

Notes:

1. The above listing of plain and deformed wire sizes represents wires normally selected to manufacture welded wire fabric to specific areas of reinforcement. Wire sizes other than those listed above, including larger sizes, may be available if the quantity required is sufficient to justify manufacture.
2. The nominal diameter of a deformed wire is equivalent to the diameter of a plain wire having the same weight per foot as the deformed wire.
3. The ACI Building Code requirements for tension development lengths and tension lap splice lengths of welded wire fabric are not included herein. These design requirements are covered in *Reinforcement: Anchorages, Lap Splices and Connections* available from CRSI. For additional information, see *Manual of Standard Practice—Structural Welded Wire Fabric* and *Structural Detailing Manual,* both published by the Wire Reinforcement Institute.

By permission Concrete Reinforcing Steel Institute, Schramsburg, Illinois

3.10.3.1 Common Styles of Welded-Wire Fabric

This specification includes requirements for the epoxy-coating material; surface preparation of the steel prior to application of the coating; the method of application of the coating; limits on coating thickness; and acceptance tests to ensure that the coating was properly applied. Small spots of coating damage might occur during handling and processing of the coated wire or WWF. All damaged areas of coating must be repaired (touched-up) with patching material.

Certain styles of welded wire fabric as shown in Table 1 have been recommended by the Wire Reinforcement Institute as common styles. WWF manufacturers can meet specific steel area requirements when ordered for designated projects, or, in some localities, can be available from inventory.

Style Designation (W = Plain, D = Deformed)	Steel Area (in.²/ft)		Approximate Weight (lbs per 100 sq ft)
	Longitudinal	Transverse	
4 x 4-W1.4 x W1.4	0.042	0.042	31
4 x 4-W2.0 x W2.0	0.060	0.060	43
4 x 4-W2.9 x W2.9	0.087	0.087	62
4 x 4-W/D4 x W/D4	0.120	0.120	86
6 x 6-W1.4 x W1.4	0.028	0.028	21
6 x 6-W2.0 x W2.0	0.040	0.040	29
6 x 6-W2.9 x W2.9	0.058	0.058	42
6 x 6-W/D4 x W/D4	0.080	0.080	58
6 x 6-W/D4.7 x W/D4.7	0.094	0.094	68
6 x 6-W/D7.4 x W/D7.4	0.148	0.148	107
6 x 6-W/D7.5 x W/D7.5	0.150	0.150	109
6 x 6-W/D7.8 x W/D7.8	0.156	0.156	113
6 x 6-W/D8 x W/D8	0.160	0.160	116
6 x 6-W/D8.1 x W/D8.1	0.162	0.162	118
6 x 6-W/D8.3 x W/D8.3	0.166	0.166	120
12 x 12-W/D8.3 x W/D8.3	0.083	0.083	63
12 x 12-W/D8.8 x W/D8.8	0.088	0.088	67
12 x 12-W/D9.1 x W/D9.1	0.091	0.091	69
12 x 12-W/D9.4 x W/D9.4	0.094	0.094	71
12 x 12-W/D16 x W/D16	0.160	0.160	121
12 x 12-W/D16.6 x W/D16.6	0.166	0.166	126

*Many styles may be obtained in rolls.

By permission Concrete Reinforcing Steel Institute, Schramsburg, Illinois

3.10.3.2 Welded-Wire Fabric Identification

Welded-wire fabric rolls can be manufactured in any lengths, up to the maximum weight per roll that is convenient for handling. The lengths of rolls vary with the individual manufacturing practices of different producers. Typical lengths are 100, 150, and 200 ft. *Sheet* or *roll length* is defined as the length, tip to tip, of longitudinal wires. This length should be a whole multiple of the transverse wire spacing.

The sum of the two end overhangs on either sheets or rolls should be equal to one transverse wire spacing. Unless otherwise specified, each end overhang equals one half of a transverse spacing.

Epoxy-coated wire and welded-wire fabric are being used in reinforced-concrete construction as a corrosion-protection system. Coated fabric is also used in reinforced earth construction, such as mechanically stabilized embankments.

The ASTM A884 specification covers the epoxy coating of plain and deformed steel wire, and plain and deformed steel welded-wire fabric. The specification defines two classes of coating thickness: Class A is intended for use as reinforcement in concrete and Class B as reinforcement in earth. Class A coating thickness after curing should be 7 to 17 mils, inclusive, and Class B has a minimum thickness of 18 mils.

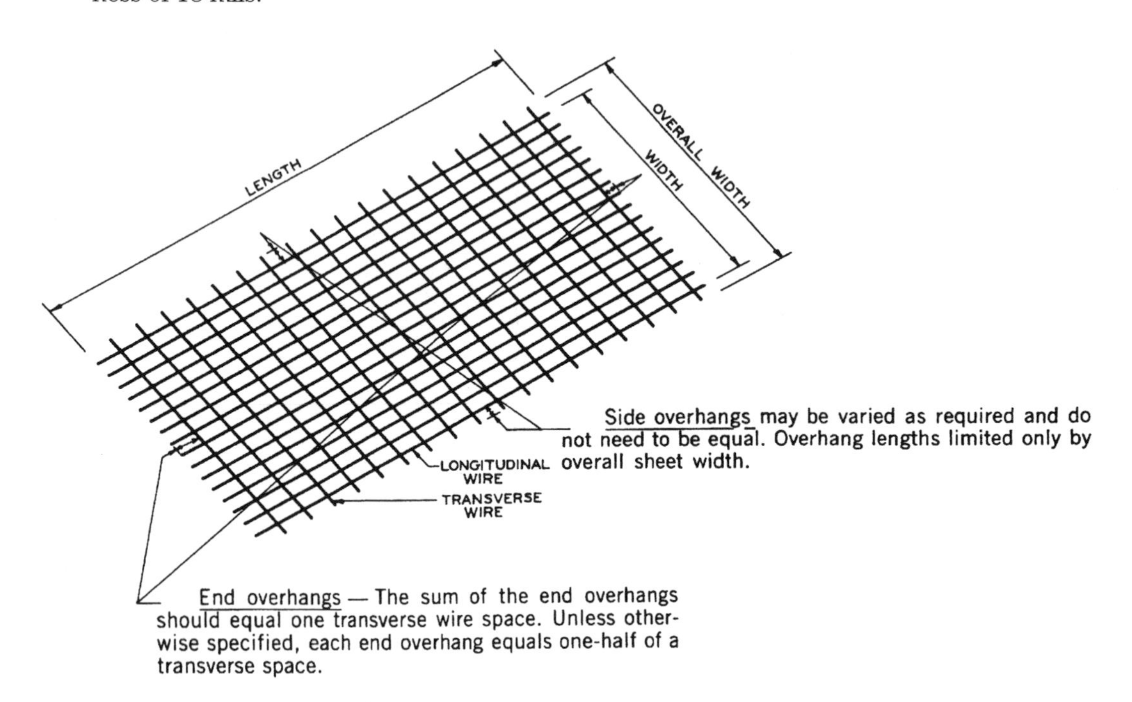

Industry Method of Designating Style:
Example — WWF 6x12—W16xW8

Longitudinal wire spacing . . . . 6"	Longitudinal wire size W16
Transverse wire spacing . . . . 12"	Transverse wire size W8

By permission Concrete Reinforcing Steel Institute, Schramsburg, Illinois

3.10.4 Recommended Industry Practices for Rebar Fabrication

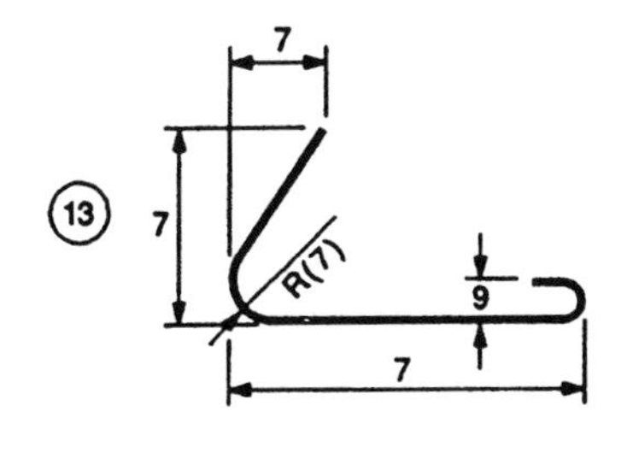

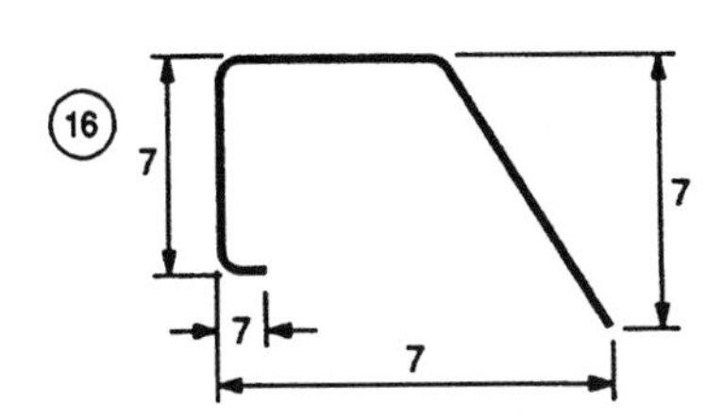

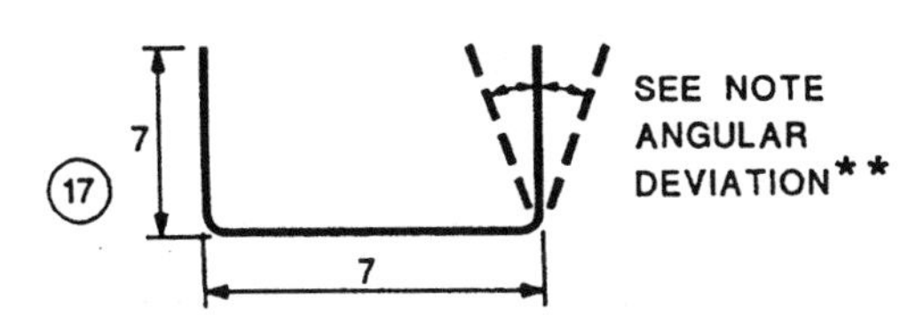

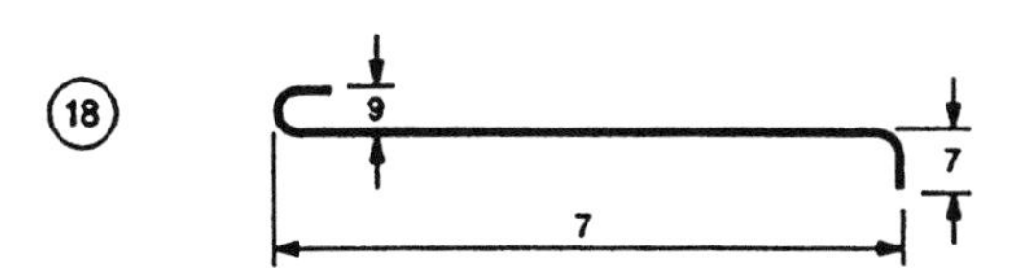

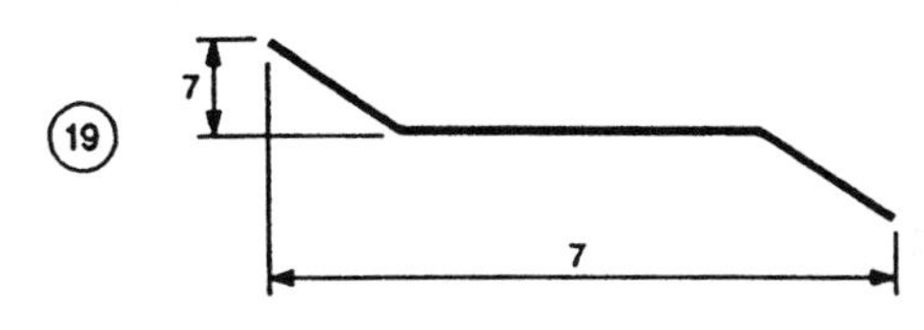

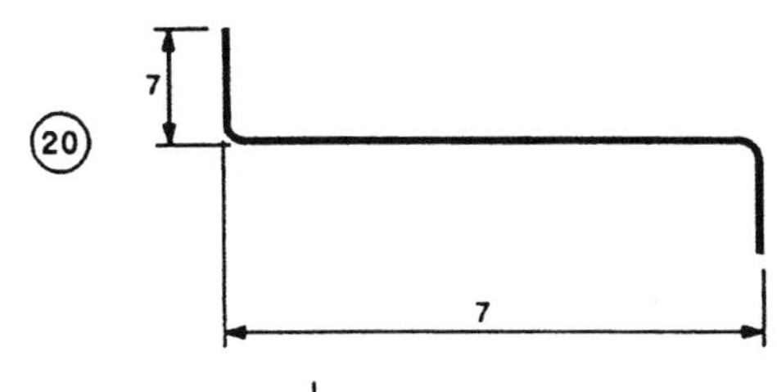

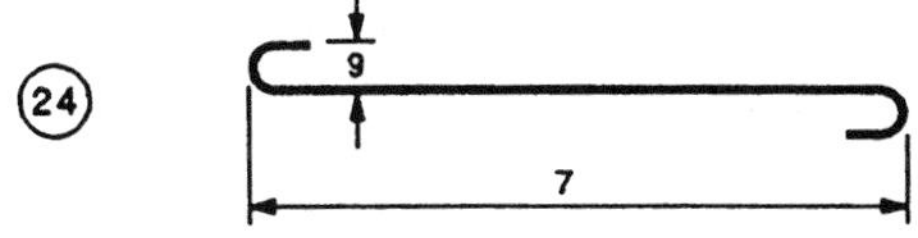

TOLERANCE SYMBOLS

Symbol	#14	#18
7	±2½ in.	±3½ in.
8	±2 in.	±2 in.
9	±1½ in.	±2 in.
10 = 2% x "O" dimension, ≥	± 2½ in. min.	±3½ in. min.

All tolerances single plane and as shown. Also, note end cutting deviations on page 7–2.

*Saw-cut both ends—overall length ±½".

**Angular Deviation—maximum ±2½" or ±½"/ft on all 90° hooks and bends.

***If application of positive tolerance to Type 9 results in a chord length ≥ the arc or bar length, the bar may be shipped straight.

By permission Concrete Reinforcing Steel Institute, Schramsburg, Illinois

3.10.4.1 Recommended Industry Practices for Rebar Fabrication (Continued)

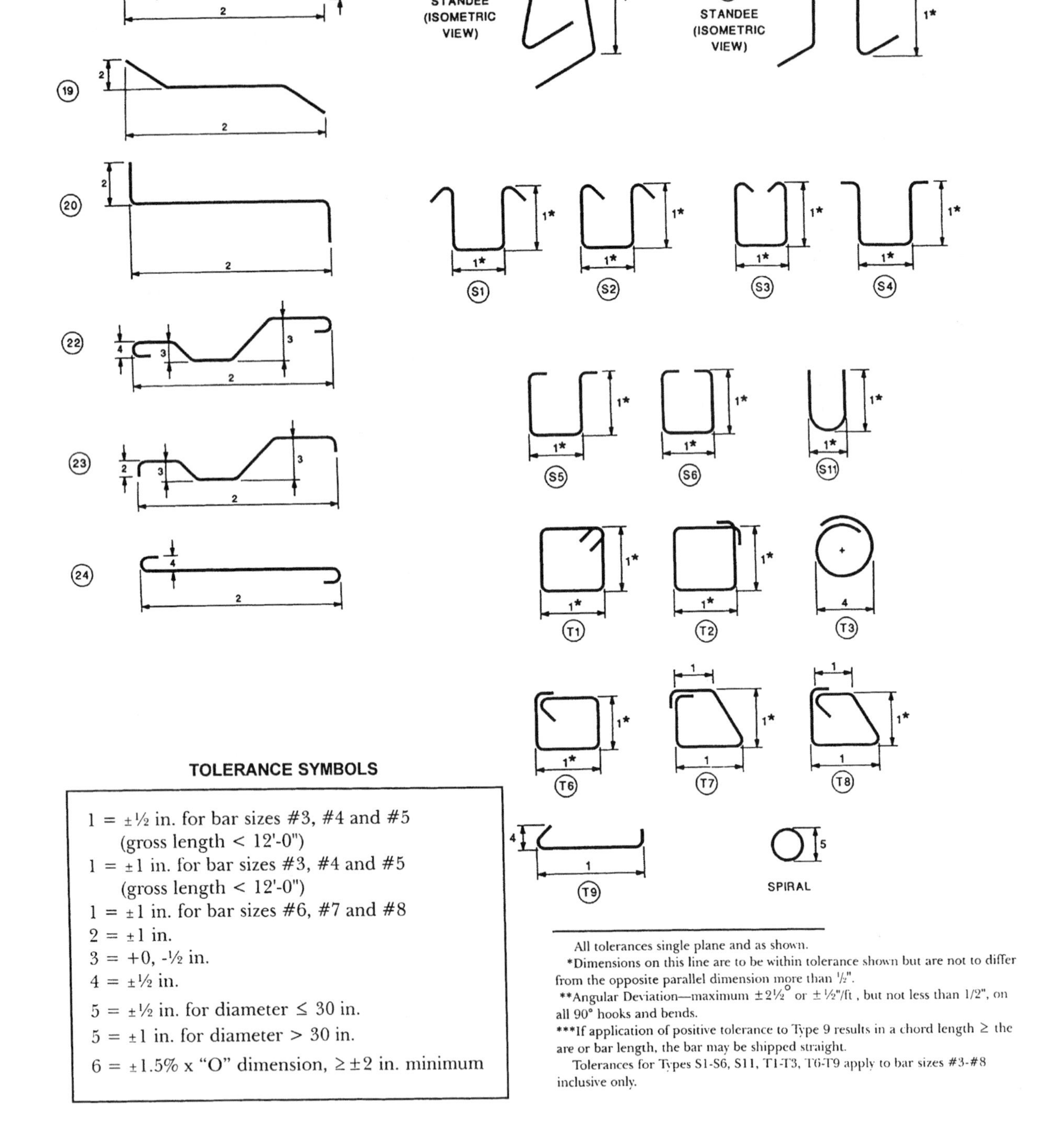

TOLERANCE SYMBOLS

1 = ±½ in. for bar sizes #3, #4 and #5 (gross length < 12'-0")
1 = ±1 in. for bar sizes #3, #4 and #5 (gross length < 12'-0")
1 = ±1 in. for bar sizes #6, #7 and #8
2 = ±1 in.
3 = +0, -½ in.
4 = ±½ in.
5 = ±½ in. for diameter ≤ 30 in.
5 = ±1 in. for diameter > 30 in.
6 = ±1.5% x "O" dimension, ≥ ±2 in. minimum

All tolerances single plane and as shown.

*Dimensions on this line are to be within tolerance shown but are not to differ from the opposite parallel dimension more than ½".

**Angular Deviation—maximum ±2½° or ±½"/ft, but not less than 1/2", on all 90° hooks and bends.

***If application of positive tolerance to Type 9 results in a chord length ≥ the are or bar length, the bar may be shipped straight.

Tolerances for Types S1-S6, S11, T1-T3, T6-T9 apply to bar sizes #3-#8 inclusive only.

By permission Concrete Reinforcing Steel Institute, Schramsburg, Illinois

3.10.4.2 Recommended Industry Practices for Rebar Fabrication (Continued)

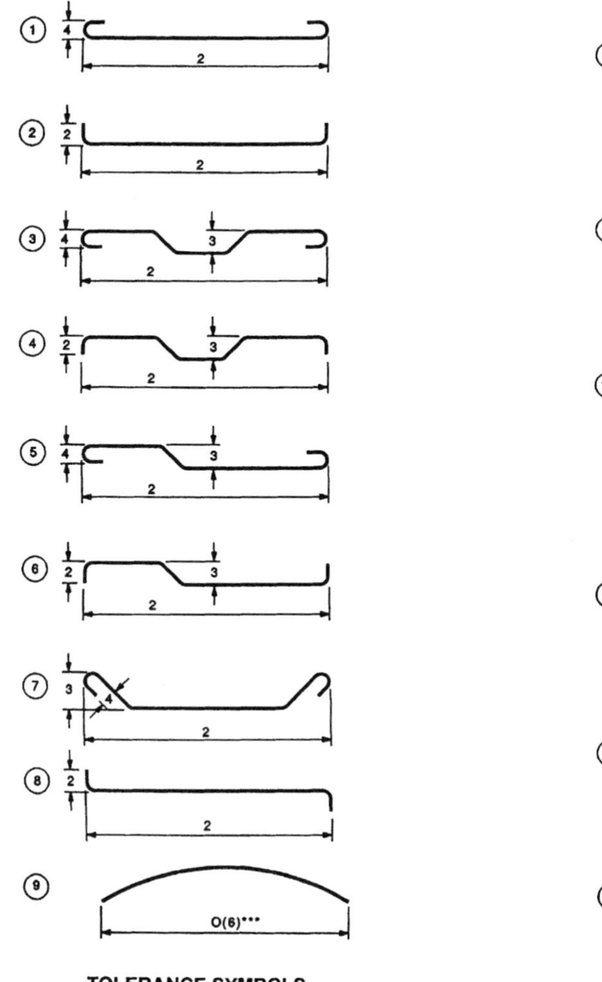

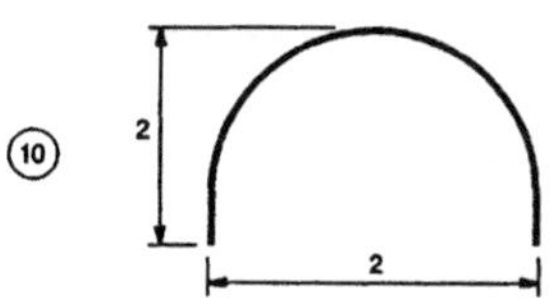

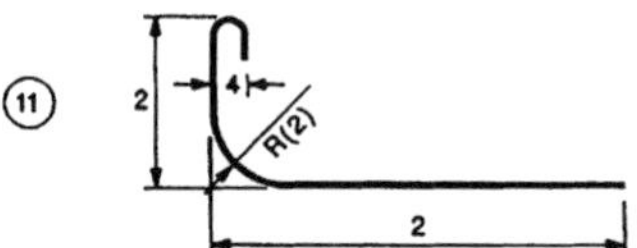

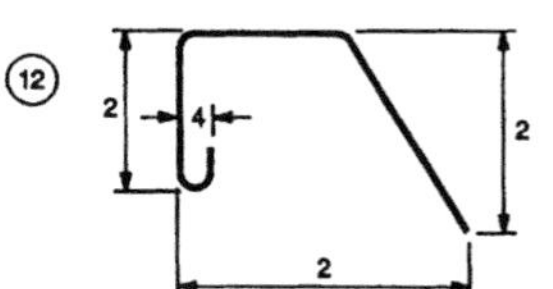

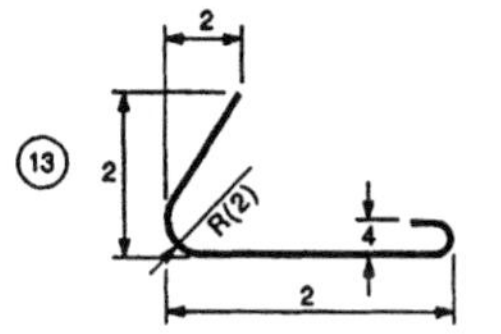

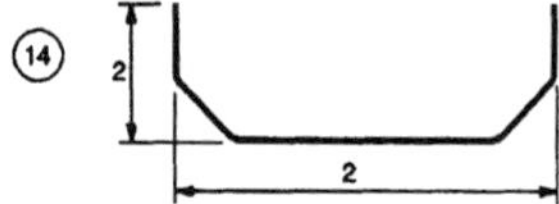

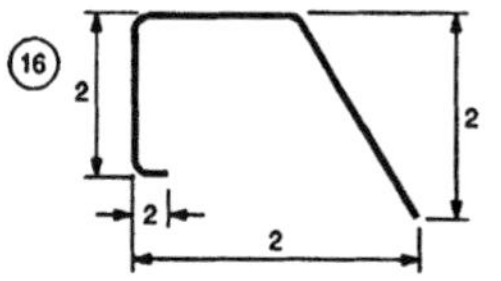

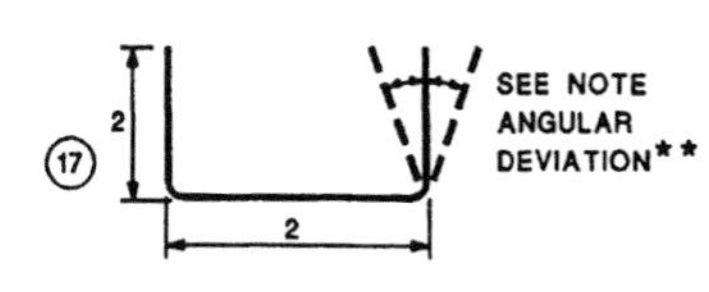

TOLERANCE SYMBOLS

1 = ±½ in. for bar sizes #3, #4 and #5 (gross length < 12'-0")
1 = ±1 in. for bar sizes #3, #4 and #5 (gross length < 12'-0")
1 = ±1 in. for bar sizes #6, #7 and #8
2 = ±1 in.
3 = +0, -½ in.
4 = ±½ in.
5 = ±½ in. for diameter ≤ 30 in.
5 = ±1 in. for diameter > 30 in.
6 = ±1.5% x "O" dimension, ≥ ±2 in. minimum

All tolerances single plane and as shown.

*Dimensions on this line are to be within tolerance shown but are not to differ from the opposite parallel dimension more than ½".

**Angular Deviation—maximum ±2½° or ±½"/ft, but not less than 1/2", on all 90° hooks and bends.

***If application of positive tolerance to Type 9 results in a chord length ≥ the arc or bar length, the bar may be shipped straight.

Tolerances for Types S1-S6, S11, T1-T3, T6-T9 apply to bar sizes #3-#8 inclusive only.

3.10.4.3 Recommended Industry Practices for Rebar Fabrication (Continued)

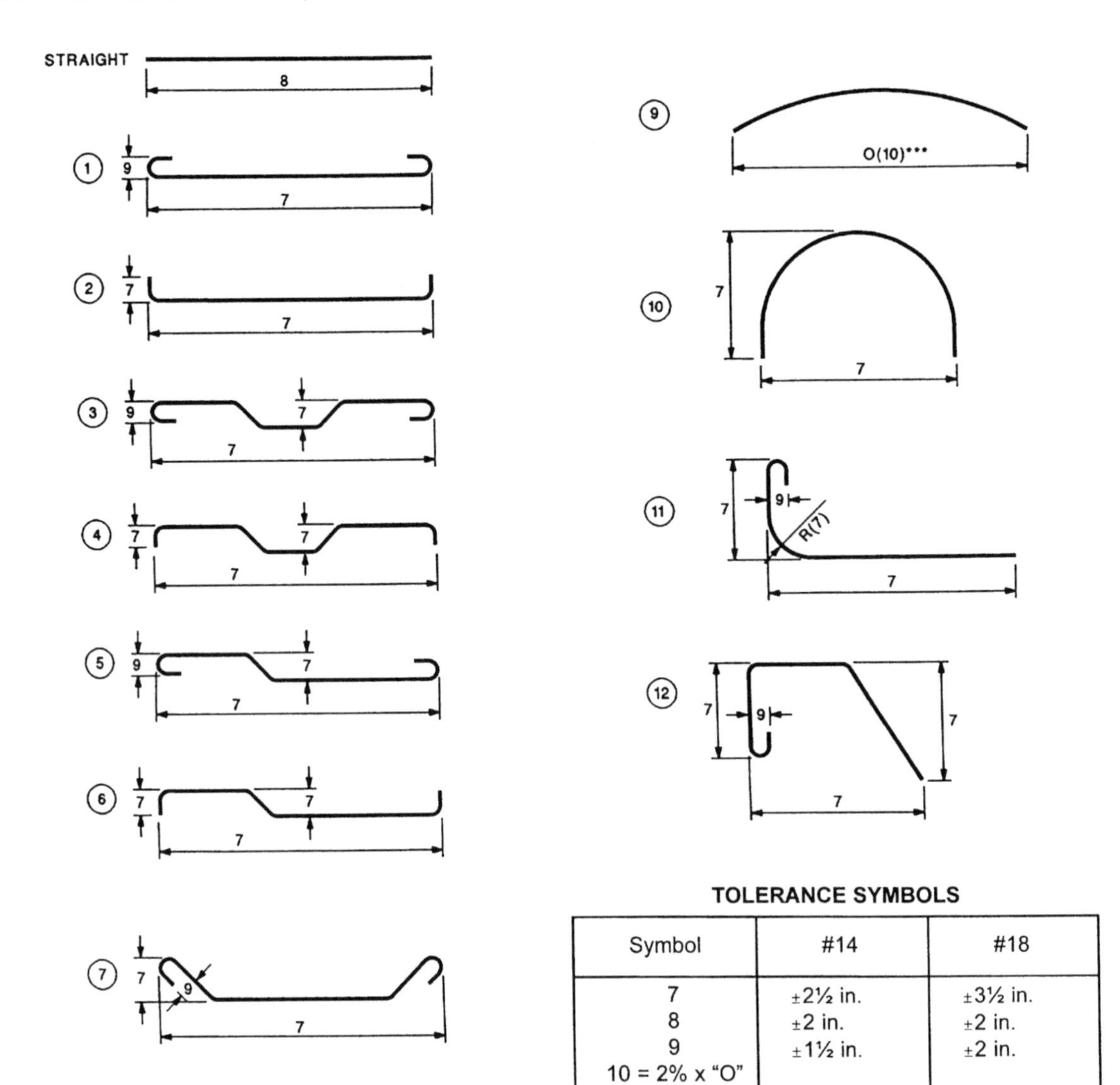

TOLERANCE SYMBOLS

Symbol	#14	#18
7	±2½ in.	±3½ in.
8	±2 in.	±2 in.
9	±1½ in.	±2 in.
10 = 2% x "O" dimension, ≥	± 2½ in. min.	±3½ in. min.

All tolerances single plane and as shown. Also, note end cutting deviations in Section 7.

*Saw-cut both ends—overall length ±½".

**Angular Deviation—maximum ±21/2" or ±1/2"/ft on all 90° hooks and bends.

***If application of positive tolerance to Type 9 results in a chord length ≥ the arc or bar lngth, the bar may be shipped straight.

By permission Concrete Reinforcing Steel Institute, Schramsburg, Illinois

3.11.0 Reinforcing Bar Supports (Typical Types)

SYMBOL	BAR SUPPORT ILLUSTRATION	BAR SUPPORT ILLUSTRATION PLASTIC CAPPED OR DIPPED	TYPE OF SUPPORT	TYPICAL SIZES
SB	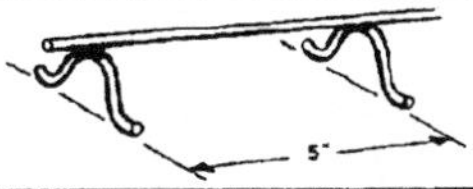 5"	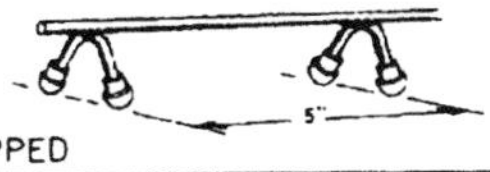 CAPPED 5"	Slab Bolster	¾, 1, 1½, and 2 in. heights in 5 ft and 10 ft lengths
SBU*	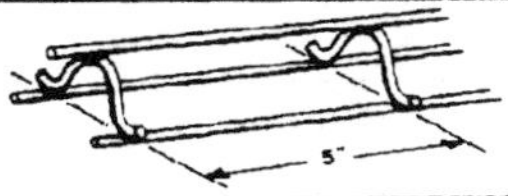 5"		Slab Bolster Upper	Same as SB
BB	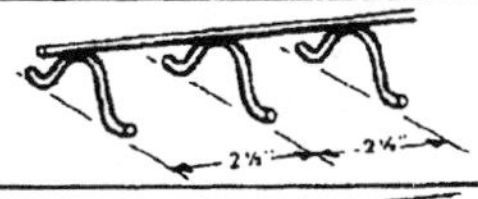 2½" 2½"	 CAPPED 2½" 2½"	Beam Bolster	1, 1½, 2 to 5 in. heights in increments of ¼ in. in lengths of 5 ft
BBU*	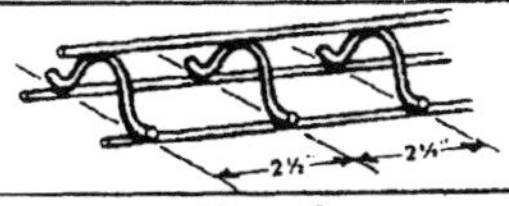 2½" 2½"		Beam Bolster Upper	Same as BB
BC	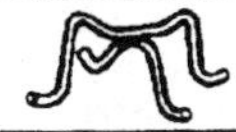	DIPPED 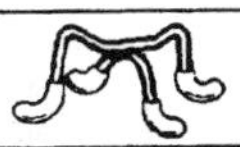	Individual Bar Chair	¾, 1, 1½, and 1¾ in. heights
JC	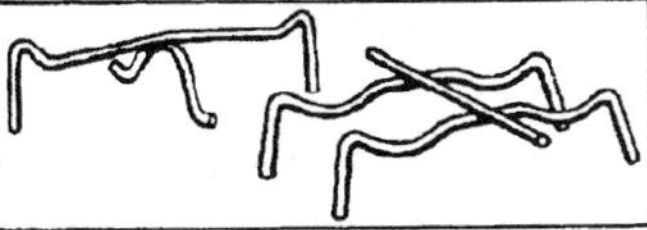	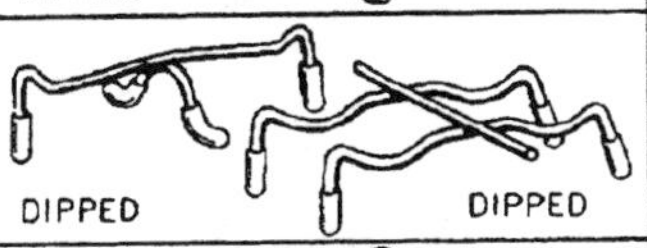 DIPPED DIPPED	Joist Chair	4, 5, and 6 in. widths and ¾, 1 and 1½ in. heights
HC	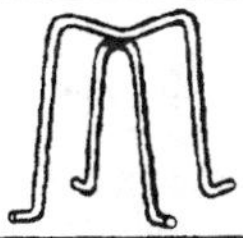	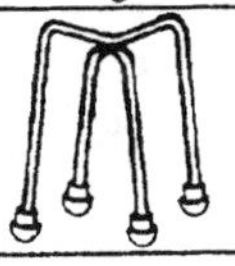 CAPPED	Individual High Chair	2 to 15 in. heights in increments of ¼ in.
HCM*			High Chair for Metal Deck	2 to 15 in. heights in increments of ¼ in.
CHC	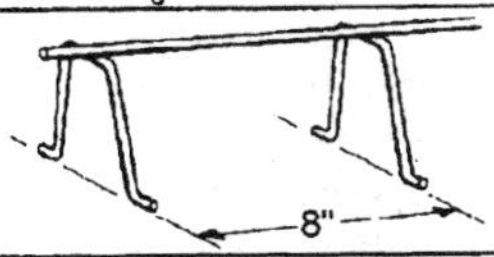 8"	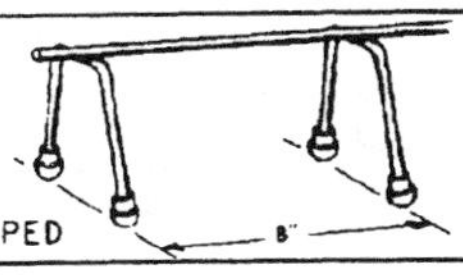 CAPPED 8"	Continuous High Chair	Same as HC in 5 ft and 10 ft lengths
CHCU*	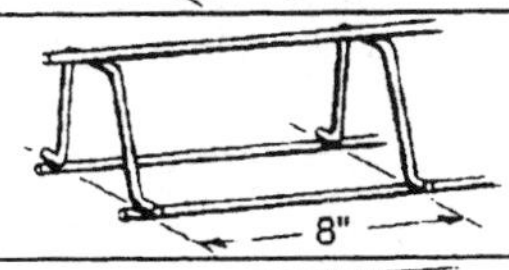 8"		Continuous High Chair Upper	Same as CHC
CHCM*	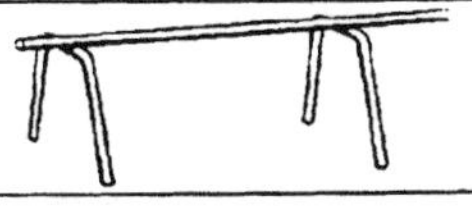		Continuous High Chair for Metal Deck	Up to 5 in. heights in increments of ¼ in.
JCU**	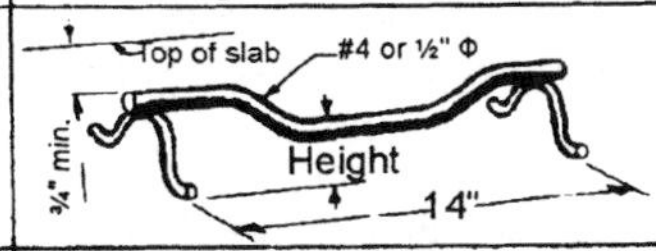 Top of slab #4 or ½" Φ ¾" min. Height 14"	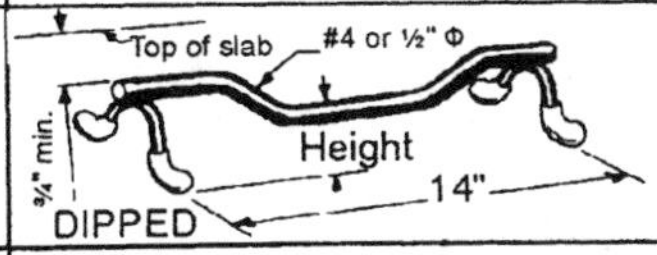 Top of slab #4 or ½" Φ ¾" min. Height 14" DIPPED	Joist Chair Upper	14 in. span heights —1 thru +3½ in. vary in ¼ in. increments
CS	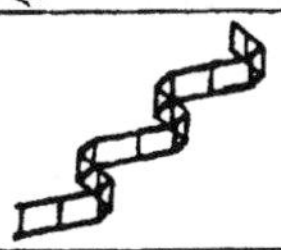		Continuous Support	1½ to 12 in. in increments of ¼ in. in lengths of 6'-8"

*Usually available in Class 3 only, except on special order.
**Usually available in Class 3 only, with upturned or end bearing legs.

By permission Concrete Reinforcing Steel Institute, Schramsburg, Illinois

3.11.1 Typical Wire Size and Geometry of Bar Supports

SYMBOL	NOMINAL HEIGHTS[2]	TYPICAL WIRE SIZES CARBON STEEL			USUAL GEOMETRY
		TOP[3]	LEGS	RUNNER	
SB	All	4 ga.	6 ga.	N/A	Legs spaced 5 in. on center.
SBU	All	4 ga.	6 ga.	7 ga.	Same as SB
BB	Up to 1½ in. incl Over 1½ in. to 2 in. incl Over 2 in. to 3½ in. incl Over 3½ in.	7 ga. 7 ga. 4 ga. 4 ga.	7 ga. 7 ga. 4 ga. 4 ga.	N/A N/A N/A N/A	Legs spaced 2½ in. on center.
BBU	Up to 2 in. incl Over 2 in.	7 ga. 4 ga.	7 ga. 4 ga.	7 ga. 4 ga.	Same as BB.
BC	All	N/A	7 ga.	N/A	—
JC	All	N/A	6 ga.	N/A	—
HC	2 in. to 3½ in. incl Over 3½ in. to 5 in. incl Over 5 in. to 9 in. incl Over 9 in. to 15 in. incl	N/A N/A N/A N/A	4 ga. 4 ga. 2 ga. 0 ga.	N/A N/A N/A N/A	Legs at 20 deg or less with vertical. When height exceeds 12 in., legs are reinforced with welded cross wires or encircling wires.
HCM	2 in. to 5 in. incl Over 5 in. to 9 in. incl Over 9 in. to 15 in. incl	N/A N/A N/A	4 ga. 2 ga. 0 ga.	N/A N/A N/A	Same as HC. The longest leg will govern the size of wire to be used.
CHC	2 in. to 3½ in. incl Over 3½ in. to 5 in. incl Over 5 in. to 9 in. incl Over 9 in. to 15 in. incl	2 ga. 2 ga. 2 ga. 2 ga.	4 ga. 4 ga. 2 ga. 0 ga.	N/A N/A N/A N/A	Legs at 20 deg or less with vertical. All legs 8¼ in. on center maximum, with leg within 4 in. of end of chair, and spread between legs not less than 50% of nominal height.
CHCU	2 in. to 5 in. incl Over 5 in. to 9 in. incl Over 9 in. to 15 in. incl	2 ga. 2 ga. 2 ga.	4 ga. 2 ga. 0 ga.	4 ga. 4 ga. 4 ga.	Same as CHC
CHCM	Up to 2 in. incl Up to 2 in. incl Over 2 in. to 5 in. incl	4 ga. 2 ga. 2 ga.	6 ga. 4 ga. 4 ga.	N/A N/A N/A	With 4 ga. top wire, maximum leg spacing is 5 in. on center. With 2 ga. top wire, maximum spacing is 10 in. on center.
JCU	−1 in. to +3½ in. incl (Measured from form to top of middle portion of saddle bar) in ¼ in. increments.	#4 bar or ½ in. dia	2 ga.	N/A	Legs spaced 14 in. on center. Maximum height of JCU at support legs should be slab thickness minus ¾ in.
CS	1½ in. to 7 in. incl 5 in. to 12 in. incl 7½ in. to 12 in. incl	8 ga. 6 ga. 4 ga.	8 ga. 6 ga. 4 ga.	8 ga. 6 ga. 4 ga.	Legs spaced 6 in. on center, 4 in. on center at bend point. Middle runner used for heights over 7 in.

[1]Wire sizes are American Steel & Wire gauges.

[2]The nominal height of the bar support is taken as the distance from the bottom of the leg, sandplate or runner wire to the bottom of the reinforcement. Variations of ± ⅛ in. from the stated nominal height are generally permitted by usual construction specifications for tolerances.

[3]Top wire on continuous supports may be straight or corrugated, at the option of the Manufacturer.

By permission Concrete Reinforcing Steel Institute, Schramsburg, Illinois

3.11.2 Typical Types of Bar Supports for Precast Concrete

SYMBOL	BAR SUPPORT ILLUSTRATION	TYPE OF SUPPORT	TYPICAL SIZES	DESCRIPTION
PB		Plain Block	A—3/4" to 6" B—2" to 6" C—2" to 48"	Used when placing rebar off grade and formwork. When "C" dimension exceeds 16" a piece of rebar should be cast inside block.
WB		Wired Block	A—3/4" to 4" B—2" to 3" C—2" to 3"	Generally 16 ga. tie wire is cast in block, commonly used against vertical forms or in positions necessary to secure the block by tying to the rebar.
TWB		Tapered Wired Block	A—3/4" to 3" B—3/4" to 2½" C—1¼" to 3"	Generally 16 ga. tie wire is cast in block, commonly used where minimal form contact is desired.
CB		Combination Block	A—2" to 4" B—2" to 4" C—2" to 4" D—fits #3 to #5 bar	Commonly used on horizontal work.
DB		Dowel Block	A—3" B—3" to 5" C—3" to 5" D—hole to accommodate a #4 bar	Used to support top mat from dowel placed in hole. Block can also be used to support bottom mat.
DSSS		Side Spacer - Wired	Concrete cover, 2" to 6"	Used to align the rebar cage in a drilled shaft.* Commonly 16 ga. tie wires are cast in spacer. Items for 5" to 6" cover have 9 ga. tie wires at top and bottom of spacer.
DSBB		Bottom Bolster - Wired	Concrete cover, 3" to 6"	Used to keep the rebar cage off of the floor of the drilled shaft.* Item for 6" cover is actually 8" in height with a 2" shaft cast in the top of the bolster to hold the vertical bar.
DSWS		Side spacer for drilled shaft applications	Concrete cover, 3" to 6"	Generally used to align rebar in a drilled shaft. Commonly manufactured with two sets of 12 ga. annealed wires, assuring proper clearance from the shaft wall surface.

*Also known as a pier, caisson or cast-in-drilled hole.

By permission Concrete Reinforcing Steel Institute, Schramsburg, Illinois

3.11.3 All-Plastic Bar Supports

Epoxy-coated reinforcing bars have become a widely used corrosion-protection system for reinforced concrete structures. Compatible types of bar supports should be used to support epoxy-coated reinforcing bars. The purpose of the compatible types of bar supports is to minimize damage to the coating on the bars during field placing of the coated bars, and not to introduce a potential source of corrosion at, and in close proximity to the point of contact of the bar supports with the coated bars. CRSI recommends:

1. Wire bar supports should be coated entirely with dielectric material, such as epoxy or plastic, compatible with concrete, for a distance of at least 2 in. from the point of contact with the epoxy-coated reinforcing bars, or;
2. Bar supports should be made of dielectric material. If precast concrete blocks with embedded tie wires or precast concrete doweled blocks are used, the wires or dowels should be epoxy coated or plastic coated; or;
3. Reinforcing bars that are used as support bars should be epoxy coated. In walls reinforced with epoxy-coated bars, spreader bars, where specified by the architect/engineer, should be epoxy-coated. Proprietary combination bar clips and spreaders that are used in walls with epoxy-coated reinforcing bars should be made of corrosion-resistance material or coated with dielectric material.

SYMBOL	BAR SUPPORT ILLUSTRATION	TYPE OF SUPPORT	TYPICAL SIZES	DESCRIPTION
BS		Bottom Spacer	Heights, ¾" to 6"	Generally for horizontal work. Not recommended for ground or exposed aggregrate finish.
BS-CL		Bottom Spacer	Heights, ¾" to 2"	Generally for horizontal work, provides bar clamping action. Not recommended for ground or exposed aggregate finish.
HC		High Chair	Heights, ¾" to 5"	For use on slabs or panels.
HC-V		High Chair, Variable	Heights 2½" to 6¼"	For horizonatal and vertical work. Provides for different heights.
WS		Wheel Spacer	Concrete Cover ⅜" to 3"	Generally for vertical work. Bar clamping action and minimum contact with forms. Applicable for column reinforcing steel.
DSWS		Side Spacer for drilled shaft applications	Concrete Cover 2½" to 6"	Generally used to align rebar in a drilled shaft.* Two piece wheel that closes and locks on to the stirrup or spiral assuring proper clearance from the shaft wall surface.
VLWS		Locking Wheel Spacer for all vertical applications	Concrete Cover ¾" to 6"	Generally used in both drilled shaft and vertical applications where excessive loading occurs. Surface spines provide minimal contact while maintaining required tolerance.

*Also known as a pier, caisson or cast-in-drilled hole.

By permission Concrete Reinforcing Steel Institute, Schramsburg, Illinois

3.11.4 Sequence of Placing Bar Supports (Two-Way Flat Plate Slab)

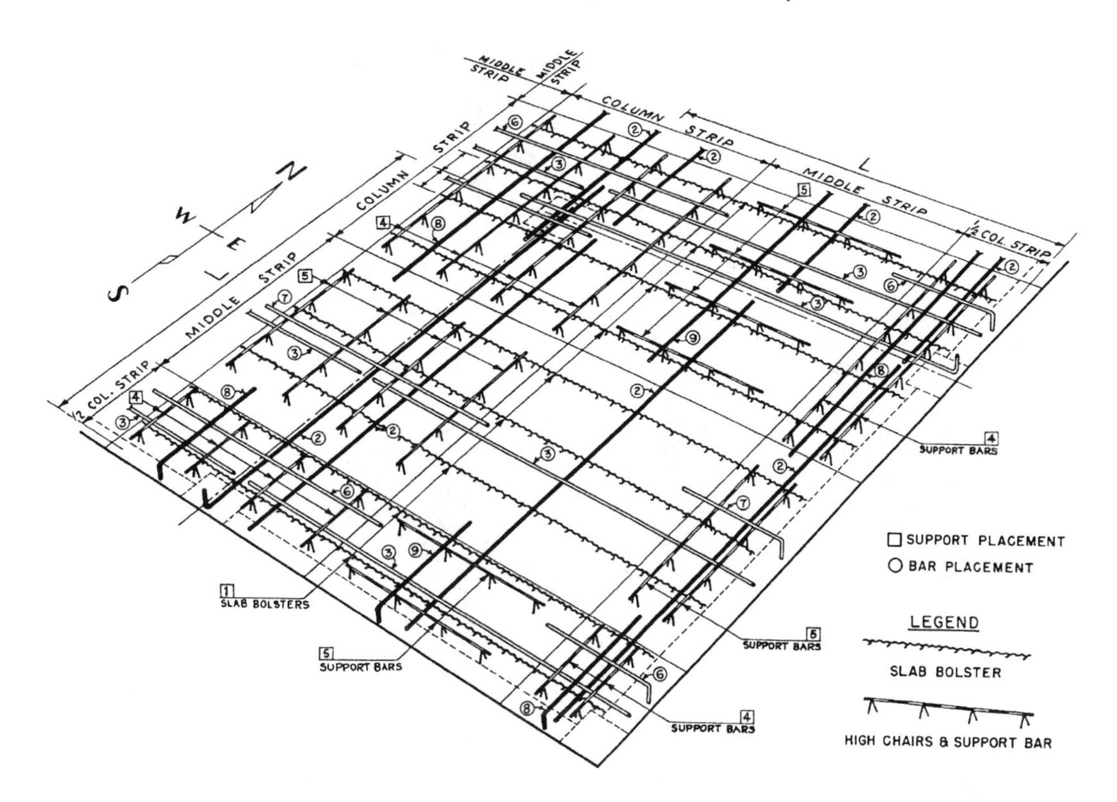

☐ 1. Place continuous lines of slab bolsters in E-W direction at 4'-0" maximum o.c. between columns.
○ 2. Set N-S bottom bars in column and middle strips.*
○ 3. Set E-W bottom bars in column and middle strips.*
☐ 4. Place 3 or more rows of #4 support bars (length 0.5L) at 4'-0" maximum o.c. on high chairs at 3'-0" maximum o.c. in N-S direction at each column head.
☐ 5. Place 3 or more rows of #4 support bars (length approx. 0.4L) at 4'-0" maximum o.c. on high chairs at 3'-0" maximum o.c. between columns lengthwise in N-S and E-W column strips.
○ 6. Set E-W top column strip bars at column heads.
○ 7. Set E-W top middle strip bars.
○ 8. Set N-S top column strip bars at column heads.
○ 9 Set N-S top middle strip bars.

Note 1: This sequence is used when the Architect/Engineer specifies the outmost layer direction. In this case the N-S bars are closest to the bottom and top of slab.

Note 2: Placing practices in certain regions may prefer to substitute individual bar supports in lieu of slab bolsters.

Note 3: Refer to Section 6.2 for use of various types and materials of bar supports.

*For structural integrity; the ACI 318 Building Code requires that all common strip bottom bars must be made continuous with adjacent spans. If bars must be spliced, use a Class A tension splice located at the support. Two of these rebars must pass through the column core and be placed within the column reinforcement. Note that, in the illustration above, these bars have been hooked at the exterior support and that the slab bolsters were extended.

By permission Concrete Reinforcing Steel Institute, Schramsburg, Illinois

3.11.4.1 Sequence of Placing Bar Supports (Two-Way Flat Slab)

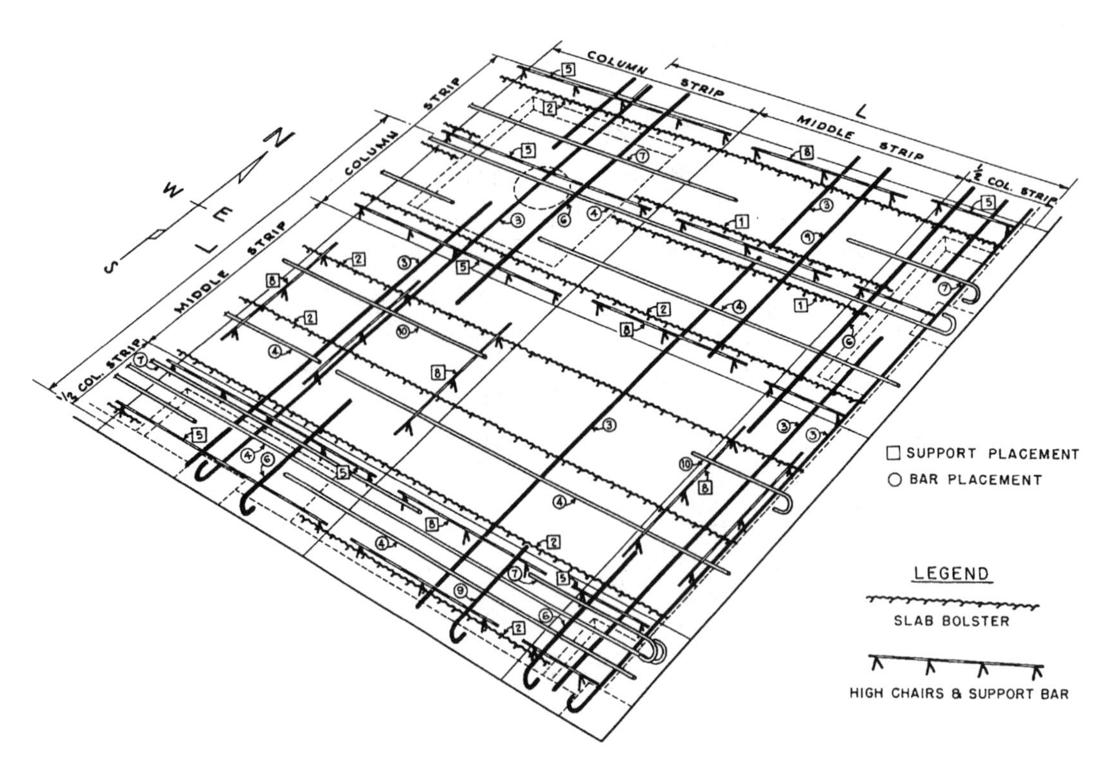

□ 1. Place a single line of slab bolsters in E-W direction on each side adjacent to column centerline between drop panels.

□ 2. Place continuous lines of slab bolsters in E-W direction at 4'-0" maximum o.c. between drop panels. Begin spacing 3" outside drop panels. Add one E-W slab bolster at slab edges between drop panels.

○ 3. Set N-S bottom bars, column and middle strips.*

○ 4. Set E-W bottom bars, column and middle strips.*

□ 5. Place 3 rows of #4 support bars (length 0.5L) on high chairs at 3'-0" maximum o.c. in E-W direction at each column head. Tie middle support bar to column verticals.

○ 6. Set N-S column strip top bars.

○ 7. Set E-W column strip top bars.

□ 8. Place 3 or more rows of #4 support bars (length 0.32L) at 4'-0" maximum o.c. in N-S and E-W column strips, parallel to the strips. Place 2 rows at all slab edges.

○ 9. Set N-S top bars in middle strips.

○ 10. Set E-W top bars in middle strip.

By permission Concrete Reinforcing Steel Institute, Schramsburg, Illinois

3.11.5 Bar Supports on Corrugated Steel Forms

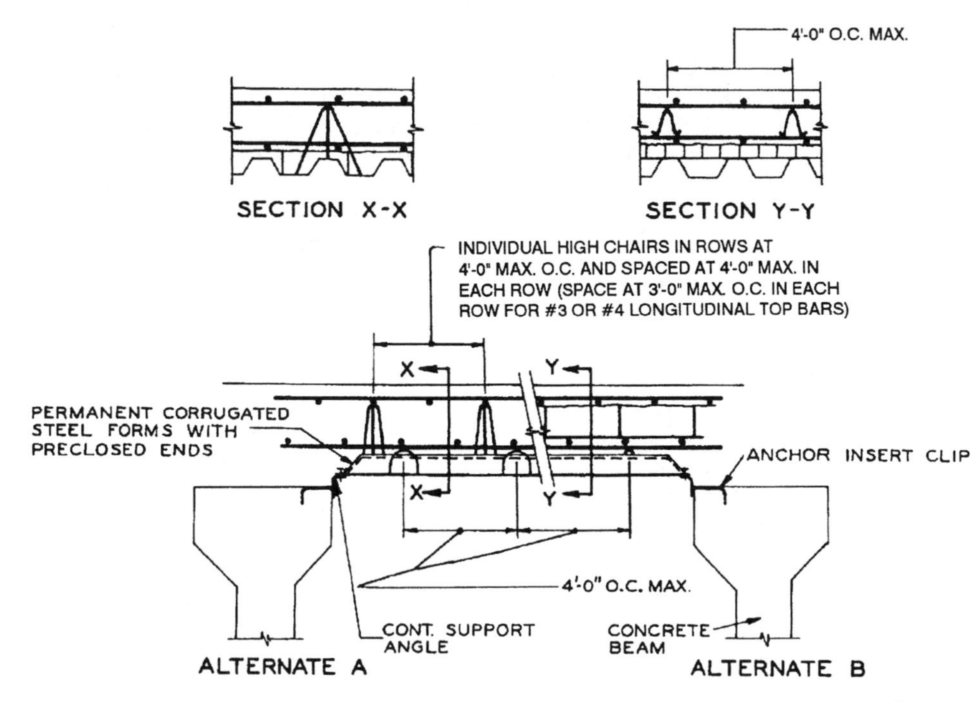

WIRE BAR SUPPORTS ON PERMANENT STEEL FORMS

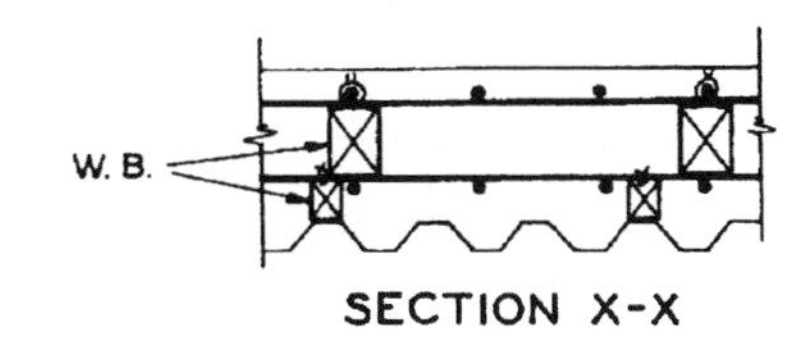

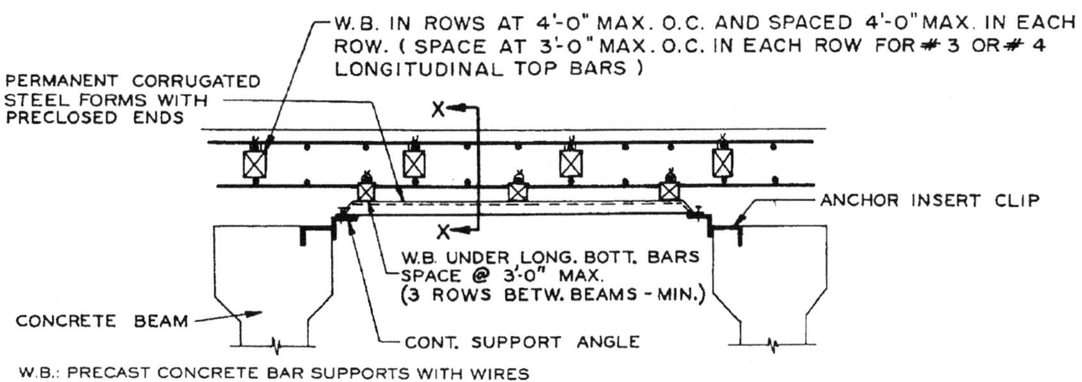

PRECAST CONCRETE BAR SUPPORTS ON PERMANENT STEEL FORMS

By permission Concrete Reinforcing Steel Institute, Schramsburg, Illinois

3.12.0 Notes on the Metrication of Reinforcing Steel

Drawing Scales

Metric drawing scales are expressed in nondimensional ratios. Nine scales are preferred (1:1, 1:5, 1:10, 1:20, 1:50, 1:100, 1:200, 1:500 and 1:1000). Three others have limited usage (1:2, 1:25 and 1:250). A comparison between inch-foot and metric scales follows:

DRAWING SCALES

INCH-FOOT SCALE	RATIO	METRIC SCALE		REMARKS
		PRE-FERRED	OTHER	
FULL SIZE	1:1	1:1		No change
HALF SIZE	1:2		1:2	No change
4" = 1'-0" 3" = 1'-0"	1:3 1:4	1:5		Close to 3" scale
2" = 1'-0" 1-1/2" = 1-0" 1"= 1'-0"	1:6 1:8 1:12	1:10		Between 1" and 1½" scale
3/4" = 1'-0" 1/2" = 1'-0"	1:16 1:24	1:20	1:25	Between 1/2" and 3/4" scales Very close to 1/2" scale
3/8" = 1'-0" 1/4" = 1'-0" 1" = 5'-0" 3/16" = 1'-0"	1:32 1:48 1:60 1:64	1:50		Close to 1/4" scale
1/8" = 1'-0" 1" = 10'-0" 3/32" = 1'-0"	1:96 1:120 1:128	1:100		Very close to 1/8" scale
1/16" = 1'-0"	1:196	1:200		Very close to 1/16" scale
1" = 20'-0"	1:240		1:250	Very close to 1"=20'-0" scale
1" = 30'-0" 1/32" = 1'-0" 1" = 40'-0"	1:360 1:384 1:480	1:500		Very close to 1"=40'-0" scale
1" = 50'-0" 1" = 60'-0" 1" = 1 chain 1" = 80'-0"	1:600 1:720 1:792 1:960	1:1000		Very close to 1"=80'-0" scale

Metric Units Used On Drawings

- Use only one unit of measure on a drawing. Except for large scale site drawings, the unit should be the millimeter (mm).
- Delete unit symbols but provide an explanatory note ("All dimensions are shown in millimeters" or "All dimensions are shown in meters").
- Whole numbers should indicate millimeters; decimal numbers taken to three places should indicate meters.
- Where modules are used, the recommended basic module is 100 mm, which is similar to the 4-inch module in building construction (4 inches = 101.6 mm).

Drawing Sizes

The ISO "A" series drawing sizes are preferred metric sizes for design drawings. There are five "A" series sizes:

A0 = 1189 x 841 mm (46.8 x 33.1 in.)

A1 = 841 x 594 mm (33.1 x 23.4 in.)

A2 = 594 x 420 mm (23.4 x 16.5 in.)

A3 = 420 x 297 mm (16.5 x 11.7 in.)

A4 = 297 x 210 mm (11.7 x 8.3 in.)

A0 is the basic drawing size with an area of one square meter. Smaller sizes are obtained by halving the long dimension of the previous size. All "A" series sizes have a height to width ratio of one to the square root of 2.

Of course, metric drawings may be made on any size paper.

Rounding and Conversion

- When converting numbers from inch-pounds to metric, round the metric value to the same number of digits as there were in the inch-pound number (11 miles equals 17.699 km, which rounds to 18 km).
- Convert mixed inch-pound units (feet and inches, pounds and ounces) to the smaller inch-pound unit before converting to metric and rounding.
- "Rounding down" from multiples of 4 inches to multiples of 100 mm makes dimensions exactly 1.6 percent smaller and areas about 3.2 percent smaller. About 3/16 inch is lost in every linear foot.
- In a "soft" conversion, an inch-pound measurement is mathematically converted to its exact (or nearly exact) metric equivalent. With "hard" conversion, a new rounded, rationalized metric number is created that is convenient to work with and remember (1 inch = 25.4 mm (soft) = 25 mm (hard)).

By permission Concrete Reinforcing Steel Institute, Schramsburg, Illinois

Section

4

Masonry

Contents

4.0.0 History of masonry
4.1.0 Mortar
4.1.1 Mortar types
4.1.2 Mortar additives
4.1.3 Mortar testing
4.1.4 Compressive strengths of masonry, based upon type of mortar
4.1.5 Compressive strength of mortar made with various types of cement
4.1.6 Allowable compressive stresses for masonry
4.1.7 Foundation wall construction (depth of unbalanced back fill)
4.2.0 Brick sizes (nomenclature)
4.2.1 Other brick sizes
4.2.2 Modular/nonmodular brick sizes (illustrated)
4.2.3 Brick positions in a wall
4.2.4 Traditional bond patterns (illustrated)
4.2.5 Traditional bond patterns explained
4.2.6 Brick arches (illustrated)
4.3.0 Estimating concrete masonry
4.3.1 Horizontal brick coursing
4.3.2 Nominal height of brick and block walls by coursing
4.4.0 Typical Atlas brick construction
4.4.1 Brick orientation (illustrated)
4.4.2 Corner, beam, and jamb details
4.4.3 Pilaster/parapet wall details
4.4.4 Flashing details
4.4.5 Flashing and caulking details at brick-relieving angles
4.4.6 Miscellanous flashing details
4.4.7 Pilaster details
4.4.8 Corbeling limitations
4.4.9 Wall elevation sections
4.4.10 Bearing areas, running bond at intersections
4.5.0 Grout strengths/proportions by weight and volume
4.6.0 Tile-wall systems
4.6.1 Standard tile-cladding shapes
4.7.0 Glass block (typical sill details)
4.7.1 Glass block (typical head and jamb details)
4.7.2 Glass block (typical panel anchor details)
4.8.0 Masonry reinforcement (types of ties)
4.8.1 Masonry reinforcement (materials and physical properties of bars/wire)
4.8.2 Wall anchorage details
4.8.3 Truss and Ladur reinforcement
4.8.4 Masonry wall ties
4.8.5 Masonry veneer anchors
4.8.6 Seismic masonry veneer anchors
4.8.7 Seismic masonry ladur and comb reinforcement
4.9.0 Investigating unstable masonry conditions to prevent failures
4.9.1 Fire-resistance ratings of various concrete masonry units and assemblies

4.0.0 History of Masonry

The first recorded brick masonry units were made by the Egyptians in 10,000 BC and the Romans used brick in many of their structures 2000 years ago. At the Great Pyramid of Giza in Egypt is the first recorded use of mortar. Brick manufacture and use occurred in the mid 1600s and was patterned on English methods and practices. It was not until 1930, however, that cavity wall construction (as we know it today) was introduced into the United States from Europe as a means of controlling moisture. This method provides a physical separation between the inner and outer wythes to serve as a drainage cavity for water, which would be expelled through weep holes in the outer wythe.

Masonry today is primarily devoted to the construction of brick, block, structural clay products, and natural and cast stone. Walls can be basically categorized as load bearing or nonload bearing walls, cavity walls, veneer walls, and solid walls. No matter the type of material used or the method by which the masonry wall is constructed, two components remain crucial: mortar and wall reinforcement.

4.1.0 Mortar

Mortar is the bonding agent that holds all of the masonry units together. Bond strength is the crucial element that differs from its close relative, concrete, where compressive strength is the most important physical property.

Mortar serves four functions:

1. It bonds the masonry units together and seals the space between them.
2. It allows for dimensional variations in the masonry units while still maintaining a high degree of levelness.
3. It bonds to the reinforcing steel in the wall.
4. It provides an added decorative effect to the wall inasmuch as various colors or tooled joints can be introduced.

4.1.1 Mortar Types

- *Type M* High compressive strength (2500 psi average), containing greater durability than other types. Therefore, it is generally recommended for unreinforced masonry walls below grade.
- *Type S* Reasonable high compressive strength (1800 psi average) and having great tensile bond strength. It is usually recommended for reinforced masonry walls, where maximum flexural strength is required.
- *Type N* Mid-range compressive strength (750 psi average) and suitable for general above-grade masonry construction for parapets and chimneys.
- *Type O* Low compressive strength (350 psi average) and suitable for interior nonload-bearing masonry walls.
- *Type K* Very low compressive strength (75 psi average) and occasionally used for interior nonload-bearing walls, where permitted by local building codes.

Workability or plasticity of the mortar is an essential characteristic of proper mortar mixes. The mortar must have both cohesive and adhesive qualities when it makes contact with the masonry units. Hardness or high strength is not necessarily a measure of durability. Mortar that is stronger than the masonry units to which it is applied might not "give," thereby causing stress to be relieved by the masonry units. This could result in these units cracking or spalling.

4.1.2 Mortar Additives

Like concrete, mortar admixtures can be added for many reasons:

- *Accelerators* To speed up the setting time by 30 to 40% and increase the 24-hour strength. Some accelerators contain calcium chloride and are not acceptable to the architect/engineer.

- *Retarders* Extends the board life of the mortar by as much as 4 to 5 hours. It slows down the set time of mortar when temperatures exceed 70 degrees F.
- *Integral water repellents* It reduces water absorption and is useful when a single wythe wall will be exposed to the elements.
- *Bond modifiers* Improves adhesion to block. It is particularly useful when glass block walls are being built.
- *Corrosion inhibitors* Used in marine environments where salt air could penetrate the mortar and begin to corrode any wall reinforcement.

4.1.3 Mortar Testing

Mortar testing is performed by the "prism" test method, in accordance with ASTM E 447, Method B. The compressive strength is the average strength of three prisms.

4.1.4 Compressive Strength of Masonry, Based Upon Type of Mortar

Net area compressive strength of concrete masonry units, psi (MPa)		Net area compressive strength of masonry, psi[1] (MPa)
Type M or S mortar	Type N mortar	
1250 (8.6)	1300 (9.0)	1000 (6.9)
1900 (13.1)	2150 (14.8)	1500 (10.3)
2800 (19.3)	3050 (21.0)	2000 (13.8)
3750 (25.8)	4050 (27.9)	2500 (17.2)
4800 (33.1)	5250 (36.2)	3000 (20.1)

[1] For units of less than 4 in. (102 mm) height, 85 percent of the values listed.

4.1.5 Compressive Strength of Mortars Made with Various Types of Cement

Type of cement	Minimum compressive strength, psi				ASTM designation
	1 day	3 days	7 days	28 days	
Portland cements					C 150-85
I	—	1800	2800	4000*	
IA	—	1450	2250	3200*	
II	—	1500	2500	4000*	
	—	1000†	1700†	3200*†	
IIA	—	1200	2000	3200*	
	—	800†	1350†	2560*†	
III	1800	3500	—	—	
IIIA	1450	2800	—	—	
IV	—	—	1000	2500	
V	—	1200	2200	3000	
Blended cements					C 595-85
I(SM), IS, I(PM), IP	—	1800	2800	3500	
I(SM)-A, IS-A, I(PM)-A, IP-A	—	1450	2250	2800	
IS(MS), IP(MS)	—	1500	2500	3500	
IS-A(MS), IP-A(MS)	—	1200	2000	2800	
S	—	—	600	1500	
SA	—	—	500	1250	
P	—	—	1500	3000	
PA	—	—	1250	2500	
Expansive cement					C 845-80
E-1	—	—	2100	3500	
Masonry cements					C 91-83a
N	—	—	500	900	
S	—	—	1300	2100	
M	—	—	1800	2900	

*Optional requirement.
†Applicable when the optional heat of hydration or chemical limit on the sum of C_3S and C_3A is specified.

Note: When low or moderate heat of hydration is specified for blended cements (ASTM C 595), the strength requirement is 80% of the value shown.

By permission from the Masonry Society, ACI, ASCE from their manual *Building Code Requirements for Masonry Structures*

4.1.6 Allowable Compressive Stresses for Masonry

Construction; compressive strength of unit, gross area, psi (MPa)	Allowable compressive stresses[1] gross cross-sectional area, psi (MPa)	
	Type M or S mortar	Type N mortar
Solid masonry of brick and other solid units of clay or shale; sand-lime or concrete brick:		
8000 (55.1) or greater	350 (2.4)	300 (2.1)
4500 (31.0)	225 (1.6)	200 (1.4)
2500 (17.2)	160 (1.1)	140 (0.97)
1500 (10.3)	115 (0.79)	100 (0.69)
Grouted masonry, of clay or shale; sand-lime or concrete:		
4500 (31.0) or greater	225 (1.6)	200 (1.4)
2500 (17.2)	160 (1.1)	140 (0.97)
1500 (8.3)	115 (0.79)	100 (0.69)
Solid masonry of solid concrete masonry units:		
3000 (20.7) or greater	225 (1.6)	200 (1.4)
2000 (13.8)	160 (1.1)	140 (0.97)
1200 (8.3)	115 (0.79)	100 (0.69)
Masonry of hollow load bearing units:		
2000 (13.8) or greater	140 (0.97)	120 (0.83)
1500 (10.3)	115 (0.79)	100 (0.69)
1000 (6.9)	75 (0.52)	70 (0.48)
700 (4.8)	60 (0.41)	55 (0.38)
Hollow walls (noncomposite masonry bonded[2])		
Solid units:		
2500 (17.2) or greater	160 (1.1)	140 (0.97)
1500 (10.3)	115 (0.79)	100 (0.69)
Hollow units	75 (0.52)	70 (0.48)
Stone ashlar masonry:		
Granite	720 (5.0)	640 (4.4)
Limestone or marble	450 (3.1)	400 (2.8)
Sandstone or cast stone	360 (2.5)	320 (2.2)
Rubble stone masonry		
Coursed, rough, or random	120 (0.83)	100 (0.69)

Net area compressive strength of units, psi (MPa)	Moduli of elasticity[1] E_{mv} psi x 10^6 (MPa x 10^3)	
	Type N mortar	Type M or S mortar
6000 (41.3) and greater	—	3.5 (24)
5000 (34.5)	2.8 (19)	3.2 (22)
4000 (27.6)	2.6 (18)	2.9 (20)
3000 (20.7)	2.3 (16)	2.5 (17)
2500 (17.2)	2.2 (16)	2.4 (17)
2000 (13.8)	1.8 (12)	2.2 (15)
1500 (10.3)	1.5 (10)	1.6 (11)

[1] Linear interpolation permitted.

4.1.7 Foundation Wall Construction (Depth of Unbalanced Back Fill)

Wall construction	Nominal wall thickness, in. (mm)	Maximum depth of unbalanced backfill, ft (m)
Hollow unit masonry	8 (203) 10 (254) 12 (305)	5 (1.53) 6 (1.83) 7 (2.14)
Solid unit masonry	8 (203) 10 (254) 12 (305)	5 (1.53) 7 (2.14) 7 (2.14)
Fully grouted masonry	8 (203) 10 (254) 12 (305)	7 (2.14) 8 (2.44) 8 (2.44)

By permission from the Masonry Society, ACI, ASCE from their manual *Building Code Requirements for Masonry Structures*

4.2.0 Brick Sizes (Nomenclature)

MODULAR BRICK SIZES								
Unit Designation	Nominal Dimensions, in.			Joint Thickness[2], in.	Specified Dimensions[3], in.			Vertical Coursing
	w	h	l		w	h	l	
Modular	4	2⅔	8	⅜ ½	3⅝ 3½	2¼ 2¼	7⅝ 7½	3C = 8 in.
Engineer Modular	4	3⅕	8	⅜ ½	3⅝ 3½	2¾ 2 13/16	7⅝ 7½	5C = 16 in.
Closure Modular	4	4	8	⅜ ½	3⅝ 3½	3⅝ 3½	7⅝ 7½	1C = 4 in.
Roman	4	2	12	⅜ ½	3⅝ 3½	1⅝ 1½	11⅝ 11½	2C = 4 in.
Norman	4	2⅔	12	⅜ ½	3⅝ 3½	2¼ 2¼	11⅝ 11½	3C = 8 in.
Engineer Norman	4	3⅕	12	⅜ ½	3⅝ 3½	2¾ 2 13/16	11⅝ 11½	5C = 16 in.
Utility	4	4	12	⅜ ½	3⅝ 3½	3⅝ 3½	11⅝ 11½	1C = 4 in.
NON-MODULAR BRICK SIZES								
Standard				⅜ ½	3⅝ 3½	2¼ 2¼	8 8	3C = 8 in.
Engineer Standard				⅜ ½	3⅝ 3½	2¾ 2 13/16	8 8	5C = 16 in.
Closure Standard				⅜ ½	3⅝ 3½	3⅝ 3½	8 8	1C = 4 in.
King				⅜	3 3	2¾ 2⅝	9⅝ 9⅝	5C = 16 in.
Queen				⅜	3	2¾	8	5C = 16 in.

[1] 1 in. = 25.4 mm; 1 ft = 0.3 m
[2] Common joint sizes used with length and width dimensions. Joint thicknesses of bed joints vary based on vertical coursing and specified unit height.
[3] Specified dimensions may vary within this range from manufacturer to manufacturer.

Reprinted by permission: Brick Institute of America, Reston, Virginia

4.2.1 Other Brick Sizes

MODULAR BRICK SIZES							
Nominal Dimensions, in.			Joint Thickness[2], in.	Specified Dimensions[3], in.			Vertical Coursing
w	h	l		w	h	l	
4	6	8	⅜ ½	3⅝ 3½	5⅝ 5½	7⅝ 7½	2C = 12 in.
4	8	8	⅜ ½	3⅝ 3½	7⅝ 7½	7⅝ 7½	1C = 8 in.
6	3⅕	12	⅜ ½	5⅝ 5½	2¾ 2¹³⁄₁₆	11⅝ 11½	5C = 16 in.
6	4	12	⅜ ½	5⅝ 5½	3⅝ 3½	11⅝ 11½	1C = 4 in.
8	4	12	⅜ ½	7⅝ 7½	3⅝ 3½	11⅝ 11½	1C = 4 in.
8	4	16	⅜ ½	7⅝ 7½	3⅝ 3½	15⅝ 15½	1C = 4 in.
NON-MODULAR BRICK SIZES							
			⅜	3 3	2¾ 2⅝	8⅝ 8⅝	5C = 16 in.

[1] 1 in. = 25.4 mm; 1 ft = 0.3 m
[2] Common joint sizes used with length and width dimensions. Joint thicknesses of bed joints vary based on vertical coursing and specified unit height.
[3] Specified dimensions may vary within this range from manufacturer to manufacturer.

Reprinted by permission: Brick Institute of America, Reston, Virginia

4.2.2 Modular and Nonmodular Brick Sizes

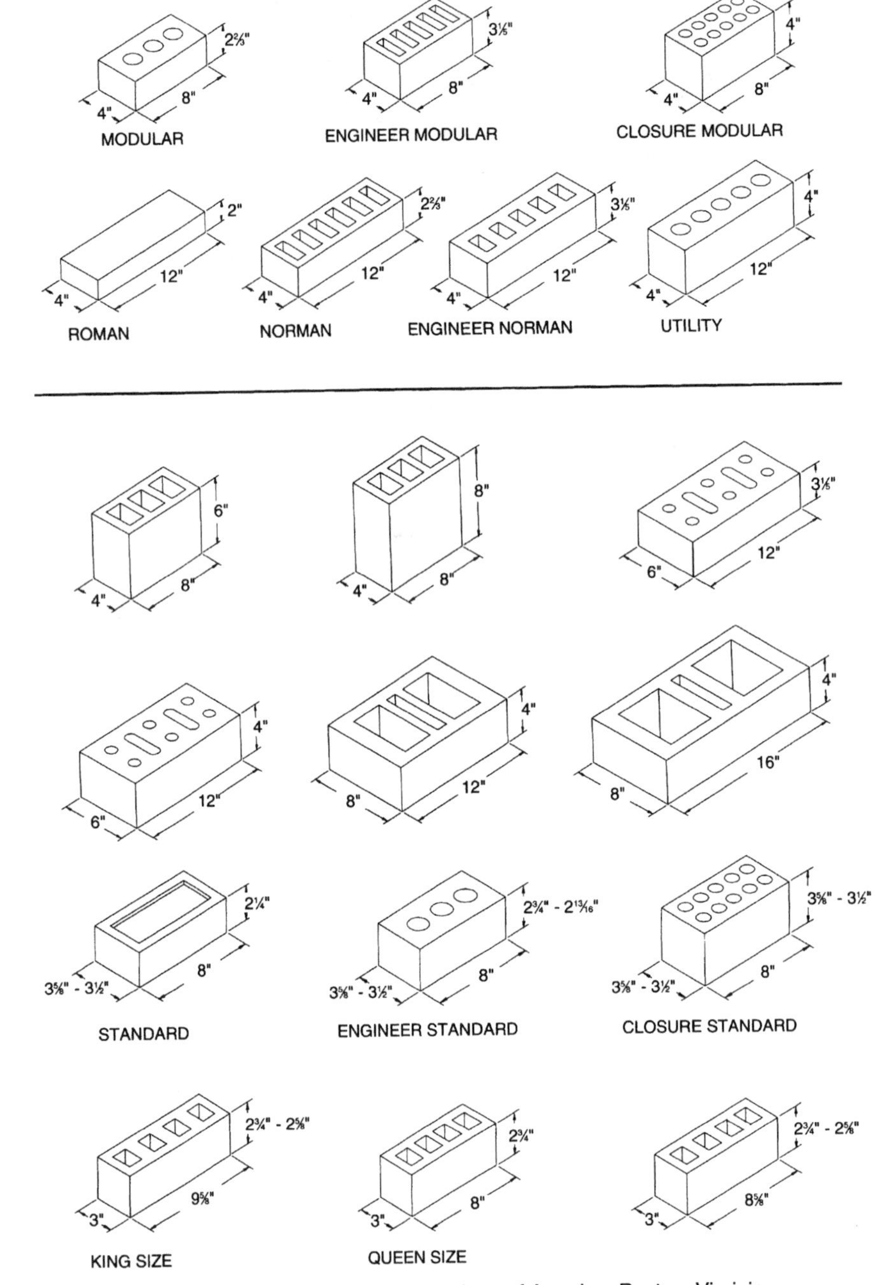

Reprinted by permission: Brick Institute of America, Reston, Virginia

4.2.3 Brick Positions in a Wall

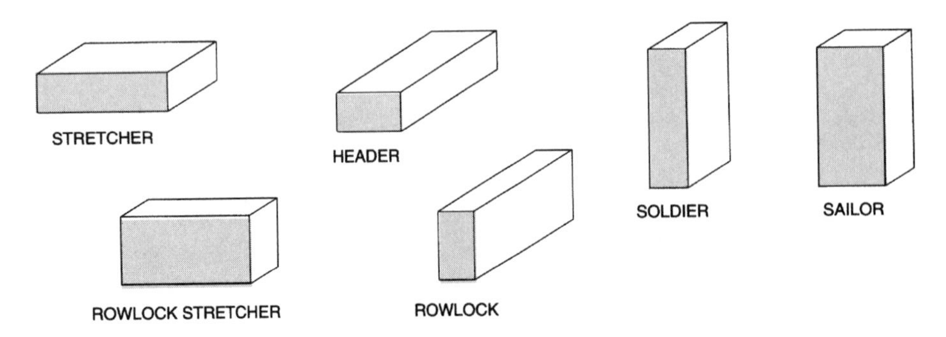

4.2.4 Traditional Bond Patterns (Illustrated)

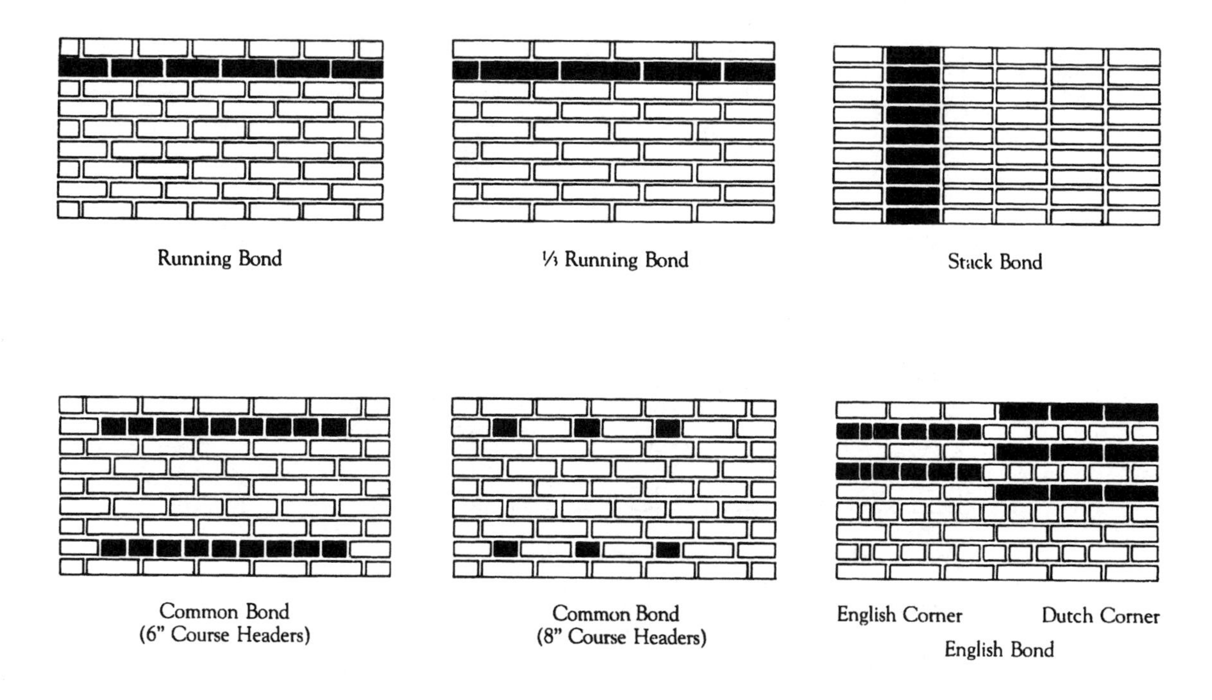

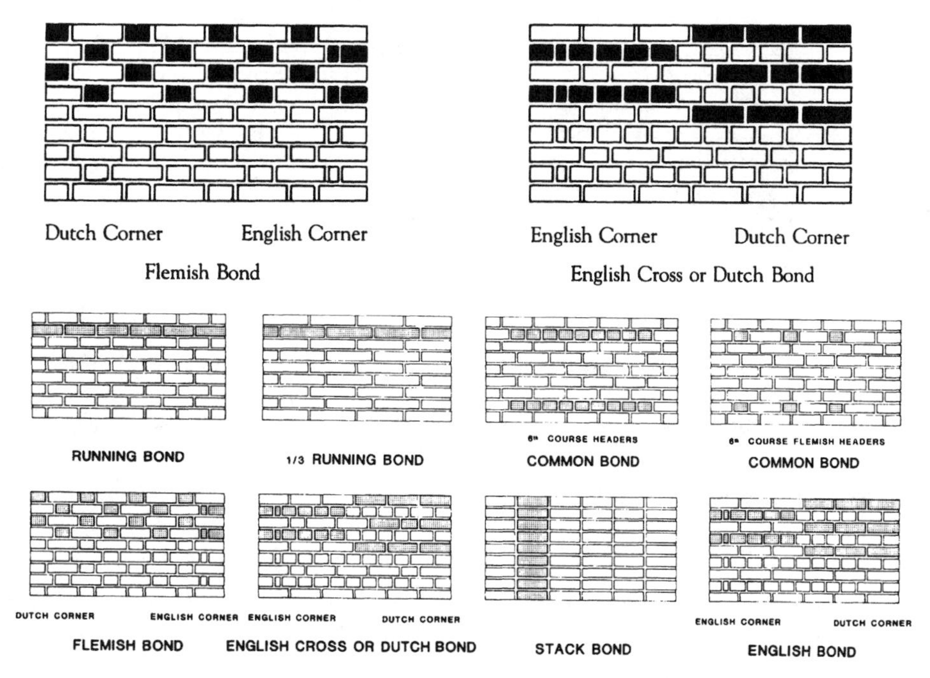

Reprinted by permission: Brick Institute of America, Reston, Virginia

4.2.4 Traditional Bond Patterns (Illustrated) (Continued)

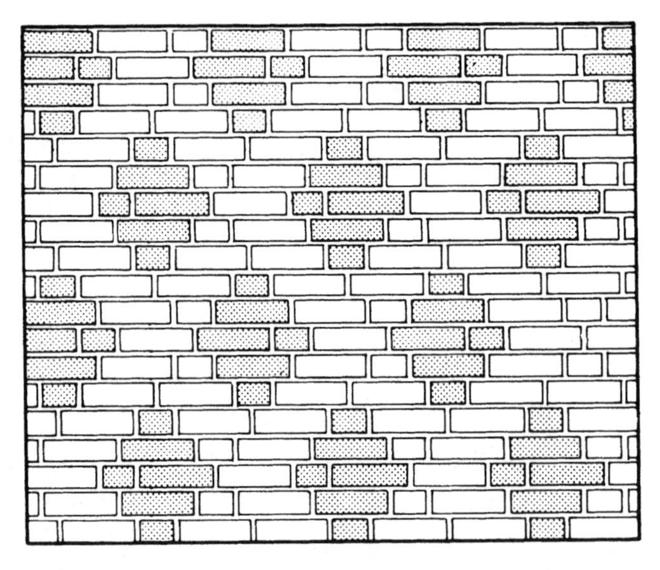

Double Stretcher Garden Wall Bond with Units in Diagonal Lines

Garden Wall Bond with Units in Dovetail Fashion

Reprinted by permission: Brick Institute of America, Reston, Virginia

4.2.5 Traditional Bond Patterns Explained

Standard patterns for brick walls are:

- *Running bond* The simplest of all brick structures, this pattern consists of all stretchers. Metal ties are used when this type of wall is used in cavity-wall or veneer-wall construction.
- *Common or American bond* A variation of the running bond, this pattern introduces a course of full-length headers at regular intervals, generally every fifth, sixth, or seventh course.
- *English bond* This pattern consists of alternate courses of headers and stretchers. The headers are centered on the stretchers and joints between stretchers in all courses are aligned vertically.
- *English cross or Dutch bond* This is a variation on the English Bond, but it differs in that vertical joints between the stretchers in alternate courses do not align vertically.

- *Flemish bond* Each course of brick consists of alternate stretchers and headers. Headers in alternate courses are centered over the stretchers in the intervening courses. Half brick or "snapped" headers can be used where structural bonding between two whythes is not required.
- *Block or stacked bond* There is no overlapping of units because all vertical joints are aligned. Generally, this patterned wall is bonded to the backing with rigid steel ties and reinforcement in the horizontal mortar joints.

4.2.6 Brick Arches (Illustrated)

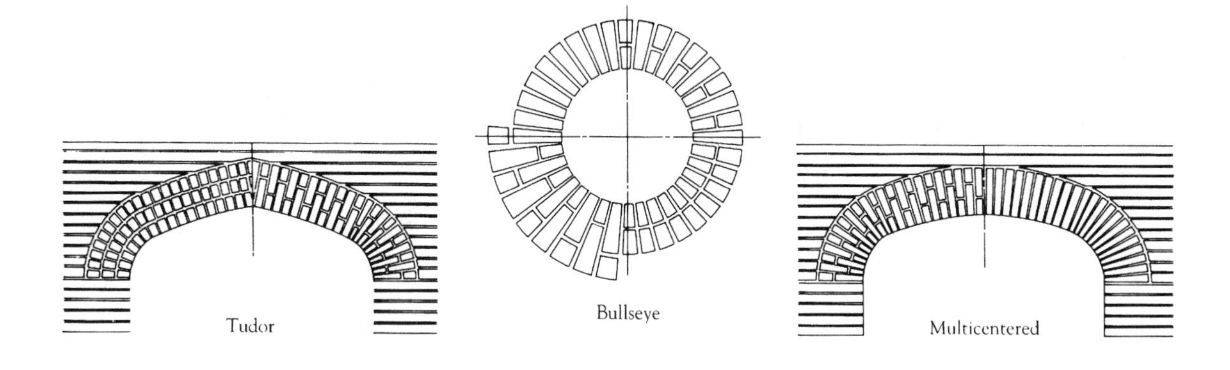

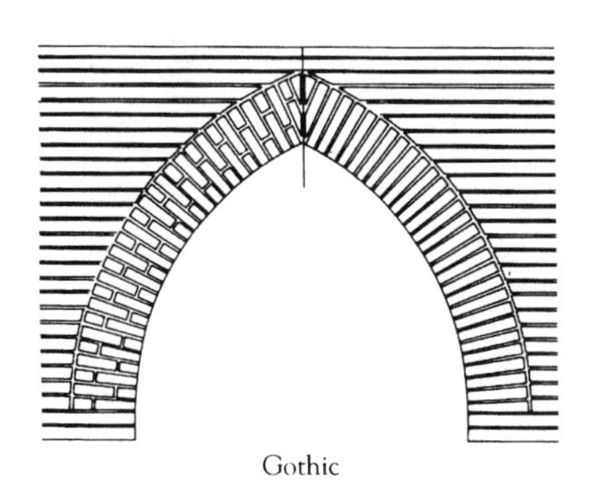

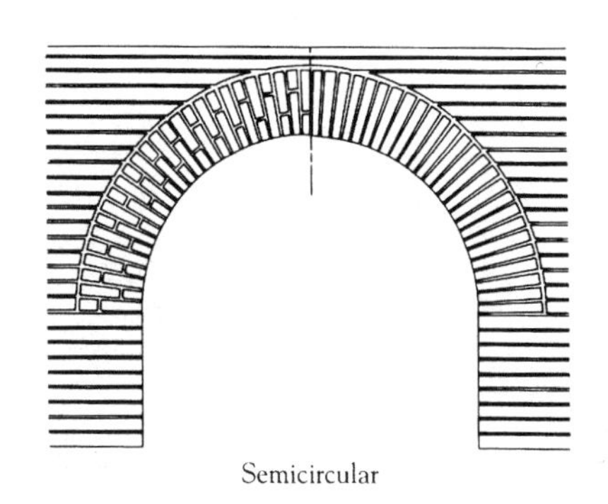

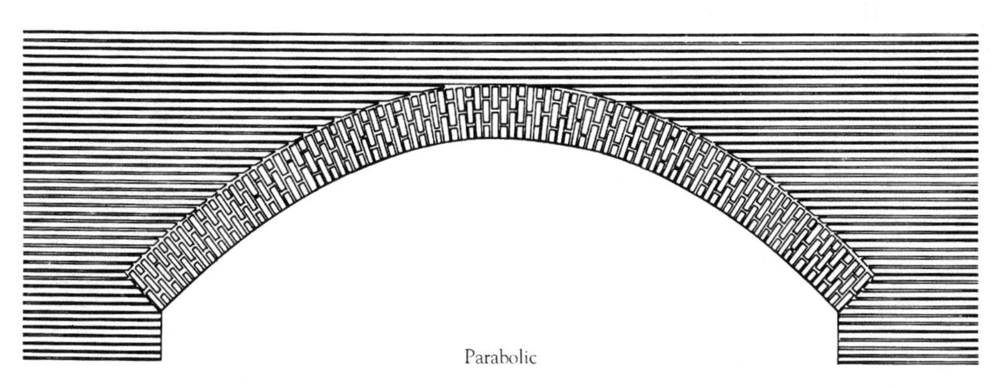

Reprinted by permission: Brick Institute of America, Reston, Virginia

4.3.0 Estimating Concrete Masonry

Estimating Concrete Masonry

NOMINAL LENGTH OF CONCRETE MASONRY WALLS BY STRETCHERS

(Based on units 15⅝" long and half units 7⅝" long with ⅜" thick head joints)

LENGTH OF WALL	NO. OF UNITS	LENGTH OF WALL	NO. OF UNITS	LENGTH OF WALL	NO. OF UNITS	LENGTH OF WALL	NO. OF UNITS	LENGTH OF WALL	NO. OF UNITS	LENGTH OF WALL	NO. OF UNITS
0'-8"	½	20'-8"	15½	40'-8"	30½	60'-8"	45½	80'-8"	60½	100'-8"	75½
1'-4"	1	21'-4"	16	41'-4"	31	61'-4"	46	81'-4"	61	101'-4"	76
2'-0"	1½	22'-0"	16½	42'-0"	31½	62'-0"	46½	82'-0"	61½	102'-0"	76½
2'-8"	2	22'-8"	17	42'-8"	32	62'-8"	47	82'-8"	62	102'-8"	77
3'-4"	2½	23'-4"	17½	43'-4"	32½	63'-4"	47½	83'-4"	62½	103'-4"	77½
4'-0"	3	24'-0"	18	44'-0"	33	64'-0"	48	84'-0"	63	104'-0"	78
4'-8"	3½	24'-8"	18½	44'-8"	33½	64'-8"	48½	84'-8"	63½	104'-8"	78½
5'-4"	4	25'-4"	19	45'-4"	34	65'-4"	49	85'-4"	64	105'-4"	79
6'-0"	4½	26'-0"	19½	46'-0"	34½	66'-0"	49½	86'-0"	64½	106'-0"	79½
6'-8"	5	26'-8"	20	46'-8"	35	66'-8"	50	86'-8"	65	106'-8"	80
7'-4"	5½	27'-4"	20½	47'-4"	35½	67'-4"	50½	87'-4"	65½	107'-4"	80½
8'-0"	6	28'-0"	21	48'-0"	36	68'-0"	51	88'-0"	66	108'-0"	81
8'-8"	6½	28'-8"	21½	48'-8"	36½	68'-8"	51½	88'-8"	66½	108'-8"	81½
9'-4"	7	29'-4"	22	49'-4"	37	69'-4"	52	89'-4"	67	109'-4"	82
10'-0"	7½	30'-0"	22½	50'-0"	37½	70'-0"	52½	90'-0"	67½	110'-0"	82½
10'-8"	8	30'-8"	23	50'-8"	38	70'-8"	53	90'-8"	68	110'-8"	83
11'-4"	8½	31'-4"	23½	51'-4"	38½	71'-4"	53½	91'-4"	68½	111'-4"	83½
12'-0"	9	32'-0"	24	52'-0"	39	72'-0"	54	92'-0"	69	112'-0"	84
12'-8"	9½	32'-8"	24½	52'-8"	39½	72'-8"	54½	92'-8"	69½	112'-8"	84½
13'-4"	10	33'-4"	25	53'-4"	40	73'-4"	55	93'-4"	70	113'-4"	85
14'-0"	10½	34'-0"	25½	54'-0"	40½	74'-0"	55½	94'-0"	70½	114'-0"	85½
14'-8"	11	34'-8"	26	54'-8"	41	74'-8"	56	94'-8"	71	114'-8"	86
15'-4"	11½	35'-4"	26½	55'-4"	41½	75'-4"	56½	95'-4"	71½	115'-4"	86½
16'-0"	12	36'-0"	27	56'-0"	42	76'-0"	57	96'-0"	72	116'-0"	87
16'-8"	12½	36'-8"	27½	56'-8"	42½	76'-8"	57½	96'-8"	72½	116'-8"	87½
17'-4"	13	37'-4"	28	57'-4"	43	77'-4"	58	97'-4"	73	117'-4"	88
18'-0"	13½	38'-0"	28½	58'-0"	43½	78'-0"	58½	98'-0"	73½	118'-0"	88½
18'-8"	14	38'-8"	29	58'-8"	44	78'-8"	59	98'-8"	74	118'-8"	89
19'-4"	14½	39'-4"	29½	59'-4"	44½	79'-4"	59½	99'-4"	74½	119'-4"	89½
20'-0"	15	40'-0"	30	60'-0"	45	80'-0"	60	100'-0"	75	120'-0"	90

NOMINAL HEIGHT OF CONCRETE MASONRY WALLS BY COURSES

(Based on units 7⅝" high and ⅜" thick mortar joints)

HEIGHT OF WALL	NO. OF UNITS	HEIGHT OF WALL	NO. OF UNITS	HEIGHT OF WALL	NO. OF UNITS	HEIGHT OF WALL	NO. OF UNITS
0'-8"	1	8'-8"	13	16'-8"	25	24'-8"	37
1'-4"	2	9'-4"	14	17'-4"	26	25'-4"	38
2'-0"	3	10'-0"	15	18'-0"	27	26'-0"	39
2'-8"	4	10'-8"	16	18'-8"	28	26'-8"	40
3'-4"	5	11'-4"	17	19'-4"	29	27'-4"	41
4'-0"	6	12'-0"	18	20'-0"	30	28'-0"	42
4'-8"	7	12'-8"	19	20'-8"	31	28'-8"	43
5'-4"	8	13'-4"	20	21'-4"	32	29'-4"	44
6'-0"	9	14'-0"	21	22'-0"	33	30'-0"	45
6'-8"	10	14'-8"	22	22'-8"	34	30'-8"	46
7'-4"	11	15'-4"	23	23'-4"	35	31'-4"	47
8'-0"	12	16'-0"	24	24'-0"	36	32'-0"	48

HOW TO USE THESE TABLES

The tables on this page are an aid to estimating and designing with standard concrete masonry units. The following are examples of how they can be used to advantage.

Example:

Estimate the number of units required for a wall 76' long and 12' high.

From table: 76' = 57 units
12' = 18 courses

57 x 18 = 1026 = No. masonry units required.

Example:

Estimate the number of units required for a foundation 24' x 30' = 11 courses high.

2 (24 + 30) = 108' = distance for a foundation

From table: 108' = 81 units

81 x 11 = 891 = No. masonry units required.

This table can also be useful in the layout of a building on a modular basis to eliminate cutting of units. Example: If design calls for a wall 41' long it can be found from the table that making this wall 41'-4", will eliminate cutting units and consequent waste. Example: If the distance between two openings has been tentatively established at 2'-9", consulting the table will show that 2'-8" dimension would eliminate cutting of units.

4.3.1 Horizontal Brick Coursing

Number of Units	Unit Length					
	Nominal Dimensions, in.		Specified Dimensions, in.			
	8	12	8		8⅝	9⅝
			½ in. jt.	⅜ in. jt.	⅜ in. jt.	⅜ in. jt.
1	0' - 8"	1' - 0"	0' - 8½"	0' - 8⅜"	0' - 9"	0' - 10"
2	1' - 4"	2' - 0"	1' - 5"	1' - 4¾"	1' - 6"	1' - 8"
3	2' - 0"	3' - 0"	2' - 1½"	2' - 1⅛"	2' - 3"	2' - 6"
4	2' - 8"	4' - 0"	2' - 10"	2' - 9½"	3' - 0"	3' - 4"
5	3' - 4"	5' - 0"	3' - 6½"	3' - 5⅞"	3' - 9"	4' - 2"
6	4' - 0"	6' - 0"	4' - 3"	4' - 2¼"	4' - 6"	5' - 0"
7	4' - 8"	7' - 0"	4' - 11½"	4' - 10⅝"	5' - 3"	5' - 10"
8	5' - 4"	8' - 0"	5' - 8"	5' - 7"	6' - 0"	6' - 8"
9	6' - 0"	9' - 0"	6' - 4½"	6' - 3⅜"	6' - 9"	7' - 6"
10	6' - 8"	10' - 0"	7' - 1"	6' - 11¾"	7' - 6"	8' - 4"
11	7' - 4"	11' - 0"	7' - 9½"	7' - 8⅛"	8' - 3"	9' - 2"
12	8' - 0"	12' - 0"	8' - 6"	8' -4½"	9' - 0"	10' - 0"
13	8' - 8"	13' - 0"	9' - 2½"	9' - 0⅞"	9' - 9"	10' - 10"
14	9' - 4"	14' - 0"	9' - 11"	9' - 9¼"	10' - 6"	11' - 8"
15	10' - 0"	15' - 0"	10' - 7½"	10' - 5⅝"	11' - 3"	12' - 6"
16	10' - 8"	16' - 0"	11' - 4"	11' - 2"	12' - 0"	13' - 4"
17	11' - 4"	17' - 0"	12' - 0½"	11' - 10⅜"	12' - 9"	14' - 2"
18	12' - 0"	18' - 0"	12' - 9"	12' - 6¾"	13' - 6"	15' - 0"
19	12' - 8"	19' - 0"	13' - 5½"	13' - 3⅛"	14' - 3"	15' - 10"
20	13' - 4"	20' - 0"	14' - 2"	13' - 11½"	15' - 0"	16' - 8"
21	14' - 0"	21' - 0"	14' - 10½"	14' - 7⅞"	15' - 9"	17' - 6"
22	14' - 8"	22' - 0"	15' - 7"	15' - 4¼"	16' - 6"	18' - 4"
23	15' - 4"	23' - 0"	16' - 3½"	16' - 0⅝"	17' - 3"	19' - 2"
24	16' - 0"	24' - 0"	17' - 0"	16' - 9"	18' - 0"	20' - 0"
25	16' - 8"	25' - 0"	17' - 8½"	17' - 5⅜"	18' - 9"	20' - 10"
26	17' - 4"	26' - 0"	18' - 5"	18' - 1¾"	19' - 6"	21' - 8"
27	18' - 0"	27' - 0"	19' - 1½"	18' - 10⅛"	20' - 3"	22' - 6"
28	18' - 8"	28' - 0"	19' - 10"	19' - 6½"	21' - 0"	23' - 4"
29	19' - 4"	29' - 0"	20' - 6½"	20' - 2⅞"	21' - 9"	24' - 2"
30	20' - 0"	30' - 0"	21' - 3"	20' - 11¼"	22' - 6"	25' - 0"
31	20' - 8"	31' - 0"	21' - 11½"	21' - 7⅝"	23' - 3"	25' - 10"
32	21' - 4"	32' - 0"	22' - 8"	22' - 4"	24' - 0"	26' - 8"
33	22' - 0"	33' - 0"	23' - 4½"	23' - 0⅜"	24' - 9"	27' - 6"
34	22' - 8"	34' - 0"	24' - 1"	23' - 8¾"	25' - 6"	28' - 4"
35	23' - 4"	35' - 0"	24' - 9½"	24' - 5⅛"	26' - 3"	29' - 2"
36	24' - 0"	36' - 0"	25' - 6"	25' - 1½"	27' - 0"	30' - 0"
37	24' - 8"	37' - 0"	26' - 2½"	25' - 9⅞"	27' - 9"	30' - 10"
38	25' - 4"	38' - 0"	26' - 11"	26' - 6¼"	28' - 6"	31' - 8"
39	26' - 0"	39' - 0"	27' - 7½"	27' - 2⅝"	29' - 3"	32' - 6"
40	26' - 8"	40' - 0"	28' - 4"	27' - 11"	30' - 0"	33' - 4"
41	27' - 4"	41' - 0"	29' - 0½"	28' - 7⅜"	30' - 9"	34' - 2"
42	28' - 0"	42' - 0"	29' - 9"	29' - 3¾"	31' - 6"	35' - 0"
43	28' - 8"	43' - 0"	30' - 5½"	30' - 0⅛"	32' - 3"	35' - 10"
44	29' - 4"	44' - 0"	31' - 2"	30' - 8½"	33' - 0"	36' - 8"
45	30' - 0"	45' - 0"	31' - 10½"	31' - 4⅞"	33' - 9"	37' - 6"
46	30' - 8"	46' - 0"	32' - 7"	32' - 1¼"	34' - 6"	38' - 4"
47	31' - 4"	47' - 0"	33' - 3½"	32' - 9⅝"	35' - 3"	39' - 2"
48	32' - 0"	48' - 0"	34' - 0"	33' - 6"	36' - 0"	40' - 0"
49	32' - 8"	49' - 0"	34' - 8½"	34' - 2⅜"	36' - 9"	40' - 10"
50	33' - 4"	50' - 0"	35' - 5"	34' - 10¾"	37' - 6"	41' - 8"
100	66' - 8"	100' - 0"	70' - 10"	69' - 9½"	75' - 0"	83' - 4"

[1] 1 in. = 25.4 mm; 1 ft = 0.3 m

Reprinted by permission: Brick Institute of America, Reston, Virginia

4.3.2 Nominal Height of Brick and Block Walls by Coursing

COURSES	REGULAR 4 2¼" bricks + 4 equal joints =					MODULAR 3 bricks + 3 joints =	CONCRETE BLOCKS	
	10" ¼" joints	10½" ⅜" joints	11" ½" joints	11½" ⅝" joints	12" ¾" joints	8"	3⅝" blocks ⅜" joints	7⅝" blocks ⅜" joints
1	2½"	2⅝"	2¾"	2⅞"	3"	2 11/16"	4"	8"
2	5"	5¼"	5½"	5¾"	6"	5 5/16"	8"	1'4"
3	7½"	7⅞"	8¼"	8⅝"	9"	8"	1'0"	2'0"
4	10"	10½"	11"	11½"	1'0"	10 11/16"	1'4"	2'8"
5	1'0½"	1'1⅛"	1'1¾"	1'2⅜"	1'3"	1'1 5/16"	1'8"	3'4"
6	1'3"	1'3¾"	1'4½"	1'5¼"	1'6"	1'4"	2'0"	4'0"
7	1'5½"	1'6⅜"	1'7¼"	1'8⅛"	1'9"	1'6 11/16"	2'4"	4'8"
8	1'8"	1'9"	1'10"	1'11"	2'0"	1'9 5/16"	2'8"	5'4"
9	1'10½"	1'11⅝"	2'0¾"	2'1⅞"	2'3"	2'0"	3'0"	6'0"
10	2'1"	2'2¼"	2'3½"	2'4¾"	2'6"	2'2 11/16"	3'4"	6'8"
11	2'3½"	2'4⅞"	2'6¼"	2'7⅝"	2'9"	2'5 5/16"	3'8"	7'4"
12	2'6"	2'7½"	2'9"	2'10½"	3'0"	2'8"	4'0"	8'0"
13	2'8½"	2'10⅛"	2'11¾"	3'1⅜"	3'3"	2'10 11/16"	4'4"	8'8"
14	2'11"	3'0¾"	3'2½"	3'4¼"	3'6"	3'1 5/16"	4'8"	9'4"
15	3'1½"	3'3⅜"	3'5¼"	3'7⅛"	3'9"	3'4"	5'0"	10'0"
16	3'4"	3'6"	3'8"	3'10"	4'0"	3'6 11/16"	5'4"	10'8"
17	3'6½"	3'8⅝"	3'10¾"	4'0⅞"	4'3"	3'9 5/16"	5'8"	11'4"
18	3'9"	3'11¼"	4'1½"	4'3¾"	4'6"	4'0"	6'0"	12'0"
19	3'11½"	4'1⅞"	4'4¼"	4'6⅝"	4'9"	4'2 11/16"	6'4"	12'8"
20	4'2"	4'4½"	4'7"	4'9½"	5'0"	4'5 5/16"	6'8"	13'4"
21	4'4½"	4'7⅛"	4'9¾"	5'0⅜"	5'3"	4'8"	7'0"	14'0"
22	4'7"	4'9¾"	5'0½"	5'3¼"	5'6"	4'10 11/16"	7'4"	14'8"
23	4'9½"	5'0⅜"	5'3¼"	5'6⅛"	5'9"	5'1 5/16"	7'8"	15'4"
24	5'0"	5'3"	5'6"	5'9"	6'0"	5'4"	8'0"	16'0"
25	5'2½"	5'5⅝"	5'8¾"	5'11⅞"	6'3"	5'6 11/16"	8'4"	16'8"
26	5'5"	5'8¼"	5'11½"	6'2¾"	6'6"	5'9 5/16"	8'8"	17'4"
27	5'7½"	5'10⅞"	6'2¼"	6'5⅝"	6'9"	6'0"	9'0"	18'0"
28	5'10"	6'1½"	6'5"	6'8½"	7'0"	6'2 11/16"	9'4"	18'8"
29	6'0½"	6'4⅛"	6'7¾"	6'11⅜"	7'3"	6'5 5/16"	9'8"	19'4"
30	6'3"	6'6¾"	6'10½"	7'2¼"	7'6"	6'8"	10'0"	20'0"
31	6'5½"	6'9⅜"	7'1¼"	7'5⅛"	7'9"	6'10 11/16"	10'4"	20'8"
32	6'8"	7'0"	7'4"	7'8"	8'0"	7'1 5/16"	10'8"	21'4"
33	6'10½"	7'2⅝"	7'6¾"	7'10⅞"	8'3"	7'4"	11'0"	22'0"
34	7'1"	7'5¼"	7'9½"	8'1¾"	8'6"	7'6 11/16"	11'4"	22'8"
35	7'3½"	7'7⅞"	8'0¼"	8'4⅝"	8'9"	7'9 5/16"	11'8"	23'4"
36	7'6"	7'10½"	8'3"	8'7½"	9'0"	8'0"	12'0"	24'0"
37	7'8½"	8'1⅛"	8'5¾"	8'10⅜"	9'3"	8'2 11/16"	12'4"	24'8"
38	7'11"	8'3¾"	8'8½"	9'1¼"	9'6"	8'5 5/16"	12'8"	25'4"
39	8'1½"	8'6⅜"	8'11¼"	9'4⅛"	9'9"	8'8"	13'0"	26'0"
40	8'4"	8'9"	9'2"	9'7"	10'0"	8'10 11/16"	13'4"	26'8"
41	8'6½"	8'11⅝"	9'4¾"	9'9⅞"	10'3"	9'1 5/16"	13'8"	27'4"
42	8'9"	9'2¼"	9'7½"	10'0¾"	10'6"	9'4"	14'0"	28'0"
43	8'11½"	9'4⅞"	9'10¼"	10'3⅝"	10'9"	9'6 11/16"	14'4"	28'8"
44	9'2"	9'7½"	10'1"	10'6½"	11'0"	9'9 5/16"	14'8"	29'4"
45	9'4½"	9'10⅛"	10'3¾"	10'9⅜"	11'3"	10'0"	15'0"	30'0"
46	9'7"	10'0¾"	10'6½"	11'0¼"	11'6"	10'2 11/16"	15'4"	30'8"
47	9'9½"	10'3⅜"	10'9¼"	11'3⅛"	11'9"	10'5 5/16"	15'8"	31'4"
48	10'0"	10'6"	11'0"	11'6"	12'0"	10'8"	16'0"	32'0"
49	10'2½"	10'8⅝"	11'2¾"	11'8⅞"	12'3"	10'10 11/16"	16'4"	32'8"
50	10'5"	10'11¼"	11'5½"	11'11¾"	12'6"	11'1 5/16"	16'8"	33'4"

4.4.0 Typical Atlas Brick Construction

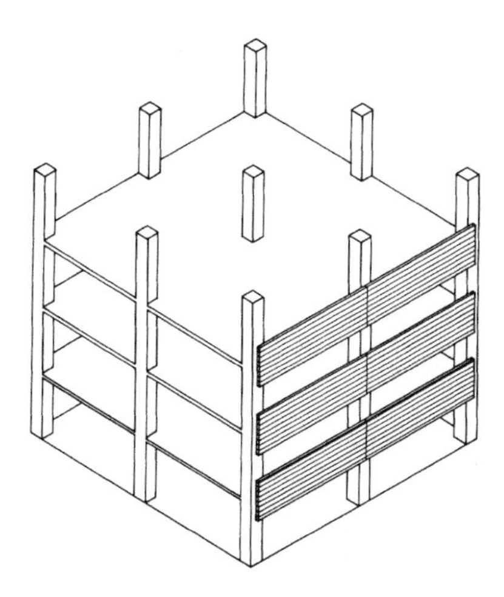

1. PREFABRICATED* PANEL CURTAIN WALL SYSTEM

Description: Panels are "hung" from the structural frame to provide the curtain wall. All loads are transferred to the frame or load bearing system.

*The panels may be prefabricated, or laid-in-place.

Advantages: 1) Essentially a veneer system, without expensive back-up or exposed supporting steel angles required.

2) Allows frame structure and curtain wall fabrication to proceed independently.

3) Prefabrication allows off-site masonry construction for "tight" jobsites.

Applications: Most economical where there is a significant amount of repetitive design elements (i.e. spandrels, soffits, lintels, or column cover elements). Brick panels can be the entire exterior cladding, or be used in conjunction with other systems where convenient (load bearing, structural skin, pre-cast concrete systems, etc.). Panels are adaptable to any construction form.

Prefabricated panels also allow a high degree of aesthetic flexibility.

2. STRUCTURAL "SKIN" (CURTAIN WALL)

Description: The building structure is a load bearing moment-resting space frame. Reinforced, grouted Atlas Brick is supported at the foundation, and tied laterally to the building frame.

Advantages: 1) Allows independent construction of the load bearing, moment frame and brick skin, requiring less trade coordination.

2) Eliminates traditional veneer support angles and back-up wall systems.

3) Provides a more structurally stable cladding system than traditional unreinforced masonry (particularly in earthquake areas).

Applications: Universally applied on single- or multi-story buildings, wherever a frame structure is used and the economic and aesthetic demands of exposed face brick is desired.

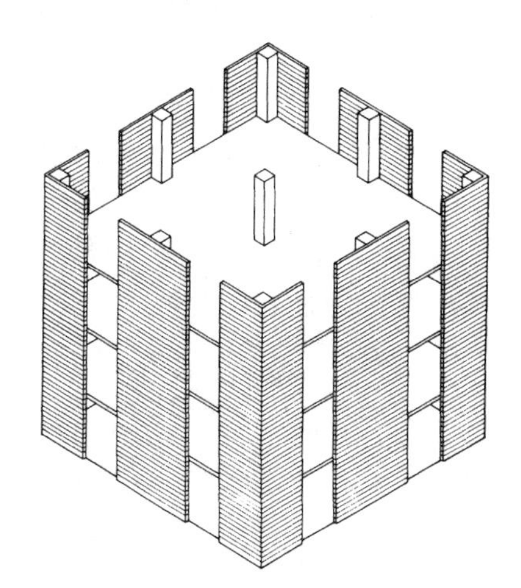

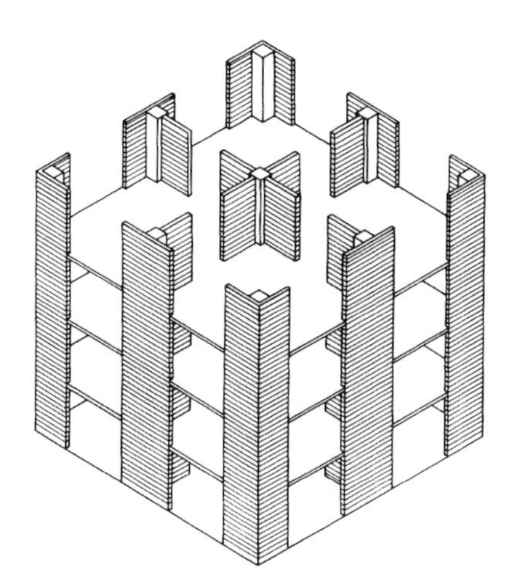

3. DUAL FRAMING SYSTEM

Description: This system uses a load bearing space frame that is designed to carry the gravity loads as well as 25% of the shear load.

Reinforced, grouted Atlas Brick walls serve as the shear resisting elements, and are designed to carry the full lateral load.

Advantages: 1) Allows independent construction of frame and shear wall systems. Amount of trade coordination is decreased.

2) The complexity of the frame construction is decreased since only 25% of the shear load is transferred through the frame connections.

Applications: Used on any structure where there is frame and shear wall construction acting together to resist design loads.

4. LOAD BEARING SHEAR WALL SYSTEM

Description: All gravity dead loads, live loads, and lateral loadings due to earthquake or wind are resisted by the reinforced grouted Atlas Brick walls, in conjunction with the structural floor diaphragm.

Advantages: Economy results from multiple use of structural elements. The brick walls serve as:

1) structure
2) space partitions (finished walls)
3) fire separations
4) sound partitions
5) exterior finish

Applications: Used on single & multi-story structures where there are a number of walls that can carry the vertical and horizontal loads, especially apartment buildings, hotels, single story structures like warehouses, shopping centers, etc.

4.4.1 Brick Orientation (Illustrated)

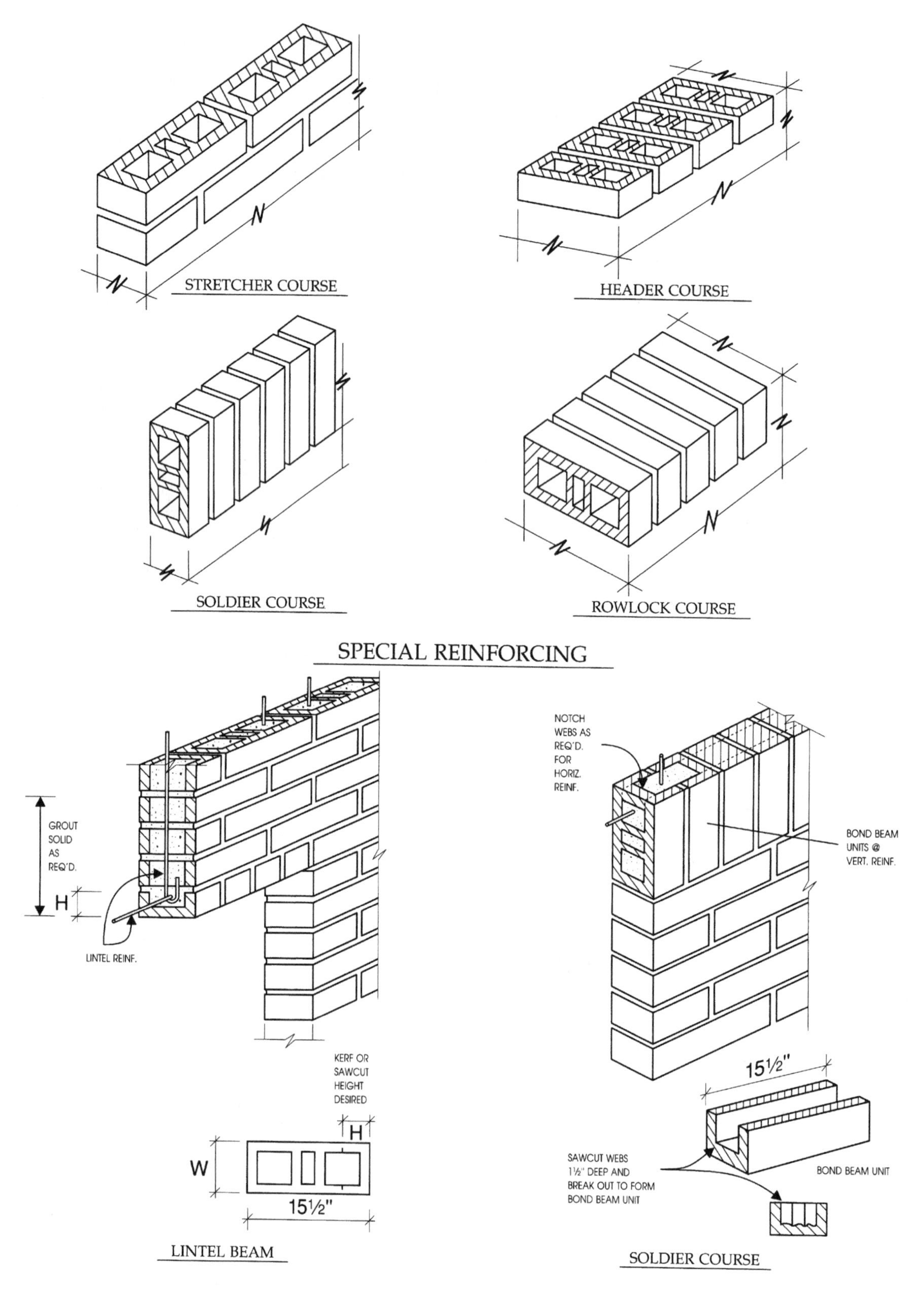

Reprinted with permission from Interstate Brick, West Jordan, Utah

4.4.2 Corner, Beam, and Jamb Details

CORNER DETAILS

W (Wall Width)	L (Corner Unit Length)
3½"	11½"
5½"	13½"
7½"	15½" (Reg. Stretcher Unit, No Knifed Corner Req'd.)

W (Wall Width)	A	B	C
3½"	1¼"	9¼"	7½"
5½"	2 1/16"	10 1/16"	7½"
7½"	2⅞"	10⅞"	7½"

1/2" MORTAR JOINT TYPICAL
STANDARD STRETCHER UNIT
KNIFED CORNER UNIT

90° CORNER

TYPICAL STRETCHER UNIT
SPECIAL 45° CORNER UNIT

45° CORNER

BEAM AND JAMB OPTIONS

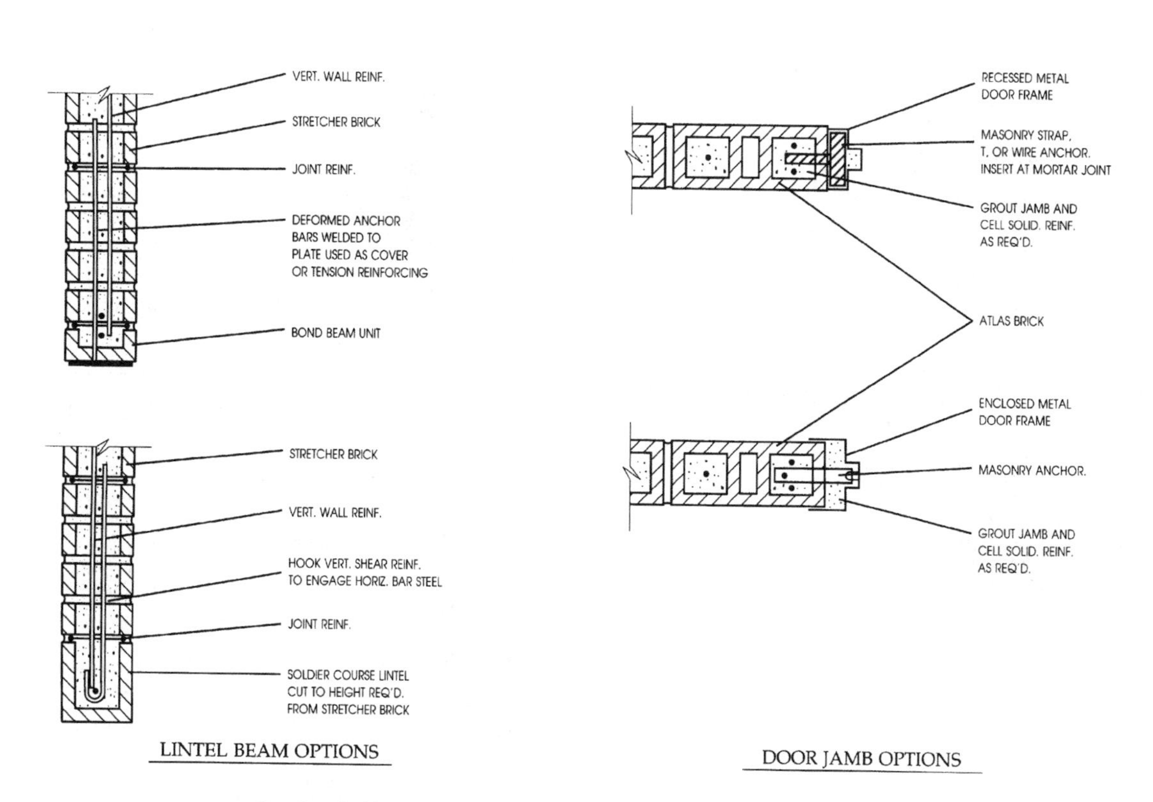

LINTEL BEAM OPTIONS

DOOR JAMB OPTIONS

Reprinted with permission from Interstate Brick, West Jordan, Utah

4.4.3 Pilaster and Parapet Wall Details

"INTEGRAL PILASTER" CONCEPT

RIGID INSULATION

CAVITY FOR INSULATION
NOTE:
IF RIGID INSULATION IS NOT USED, GROUT DAMS (NOT SHOWN) ARE NECESSARY TO FORM PILASTERS.

INTEGRAL GROUT PILASTER

IN SEISMIC ZONES 2, 3 AND 4 WHEN PILASTERS ARE SPACED MORE THAN 48" APART, INTERMEDIATE "MINIMUM" STEEL IS REQUIRED.

PILASTER SPACING (AS REQ'D BY DESIGN)

VERT. REINF. STEEL

FIELD-CUT "NOTCH" IN ATLAS BRICK FOR GROUT (4" ATLAS SHOWN)

HORIZ. JOINT REINF. (AS REQ'D BY DESIGN)

INTEGRAL PILASTER WALL

Description: a "Hybrid' system using a double wythe, insulated cavity wall and grouted pilasters or piers forming the structural tie between the brick wythes.

Advantages: Combines the appearance of two-faced brick construction while allowing a cavity of rigid board insulation to fulfill specific high insulation requirements. The interior and exterior brick mass also aids in the thermal performance of this wall system.

Applications: This wall system can be used in load bearing applications (gymnasium, auditoriums, etc.) or anywhere a highly insulated (high R value) double face, structural brick "sandwich" wall is required.

Note: This Atlas system requires a complete rational analysis to determine particular application acceptability. Local codes may or may not provide for this structural system. A complete detail of this wall is described in the paper, "High, Thin Brick Walls That Can Beat The Energy Crunch," Donald A. Wakefield, P.E., 5th International Brick Masonry Conference, Washington, D.C., October, 1979.

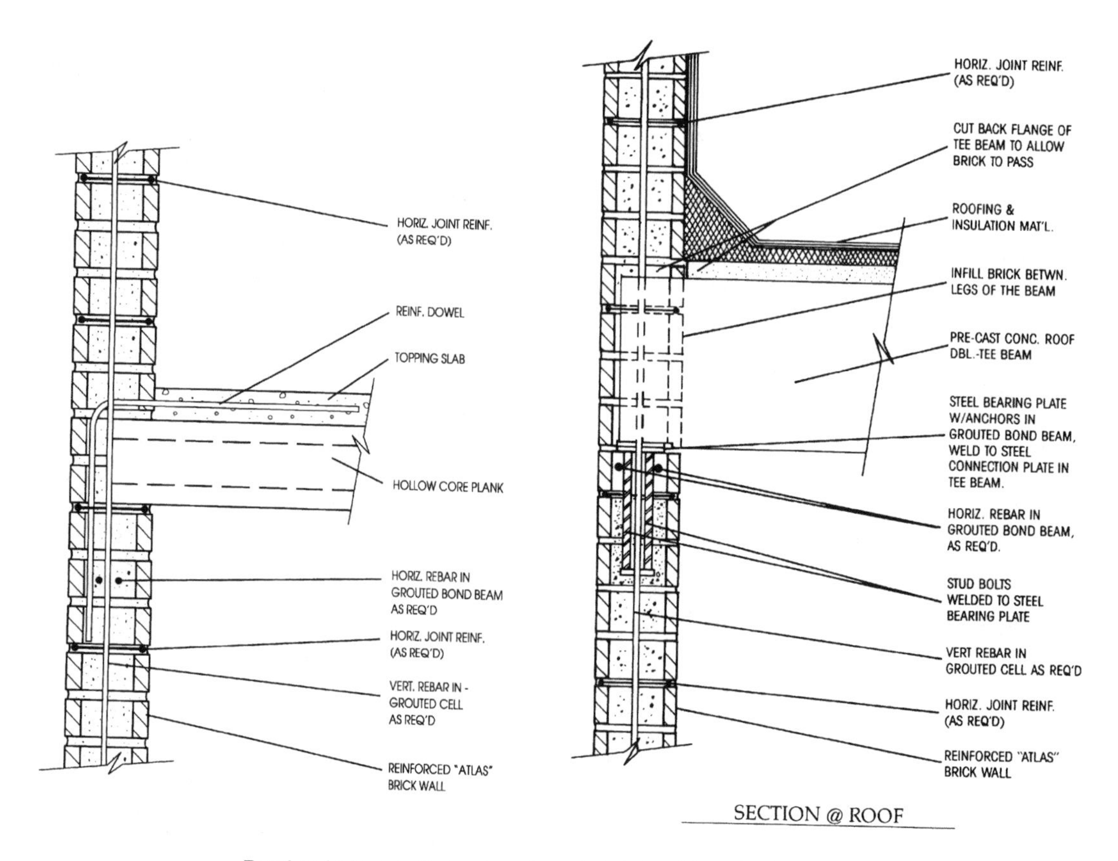

Reprinted with permission from Interstate Brick, West Jordan, Utah

4.4.4 Flashing Details

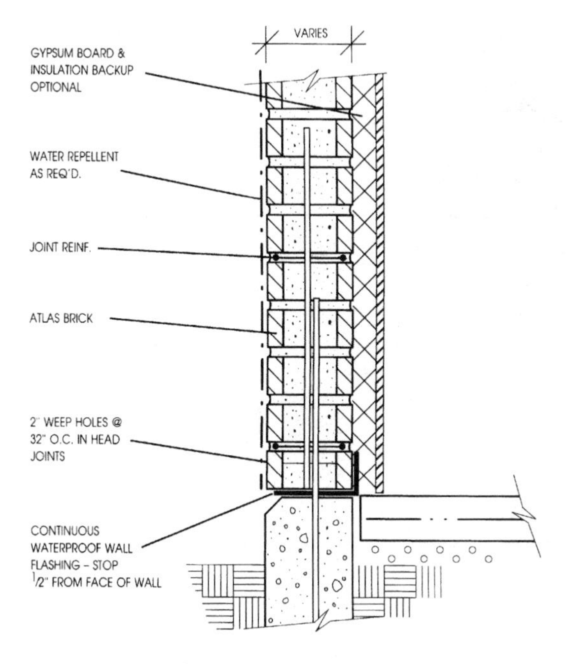

SINGLE UNIT BASE WITH FLASHING

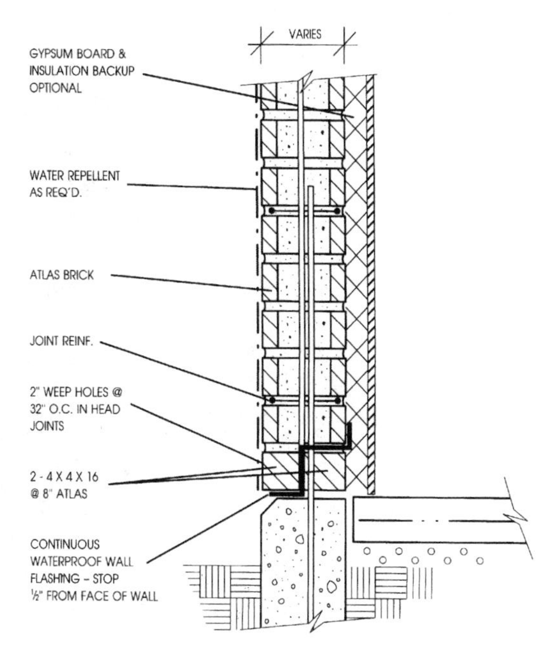

DOUBLE UNIT BASE WITH FLASHING

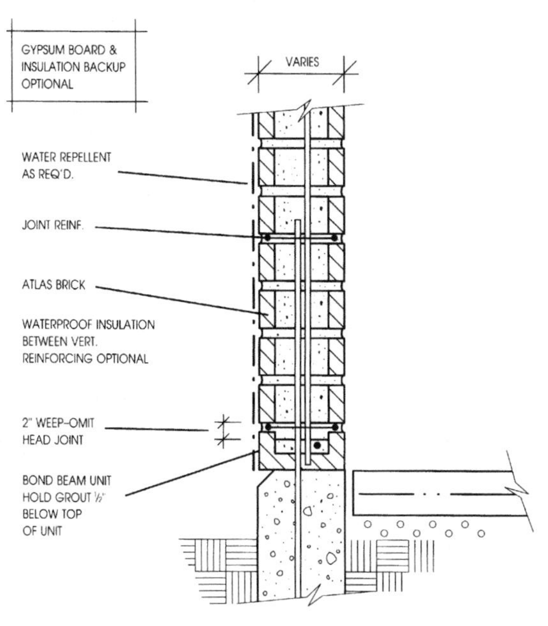

SINGLE UNIT BASE W/O FLASHING

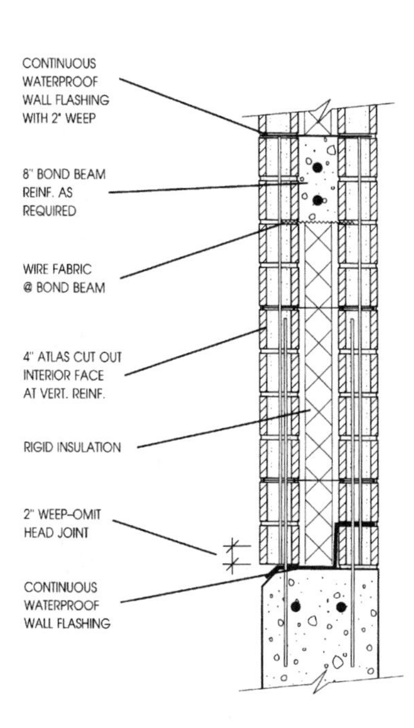

INTEGRAL PILASTER WITH FLASHING

Reprinted with permission from Interstate Brick, West Jordan, Utah

4.4.5 Flashing and Caulking Details at Brick-Relieving Angles

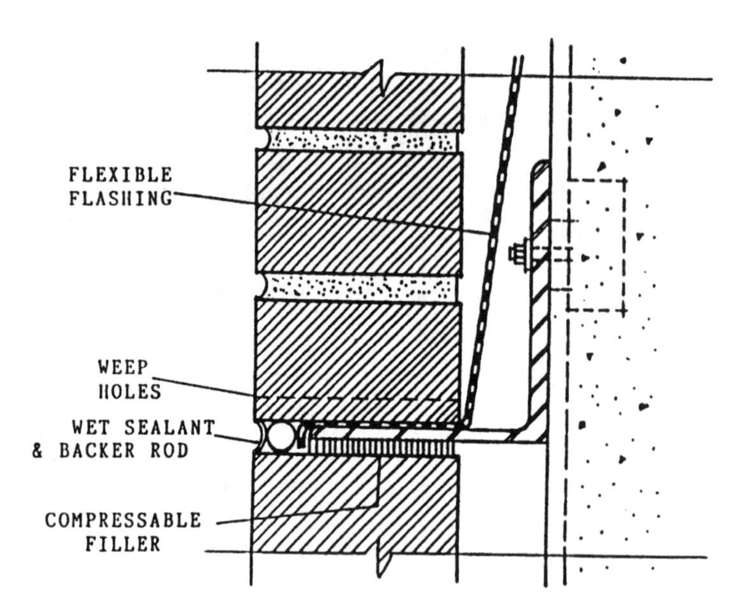

Flexible flashing terminated behind wet sealant & backer rod.

4.4.6 Miscellaneous Flashing Details

Metal flashing details for cavity and block back-up brick wall.

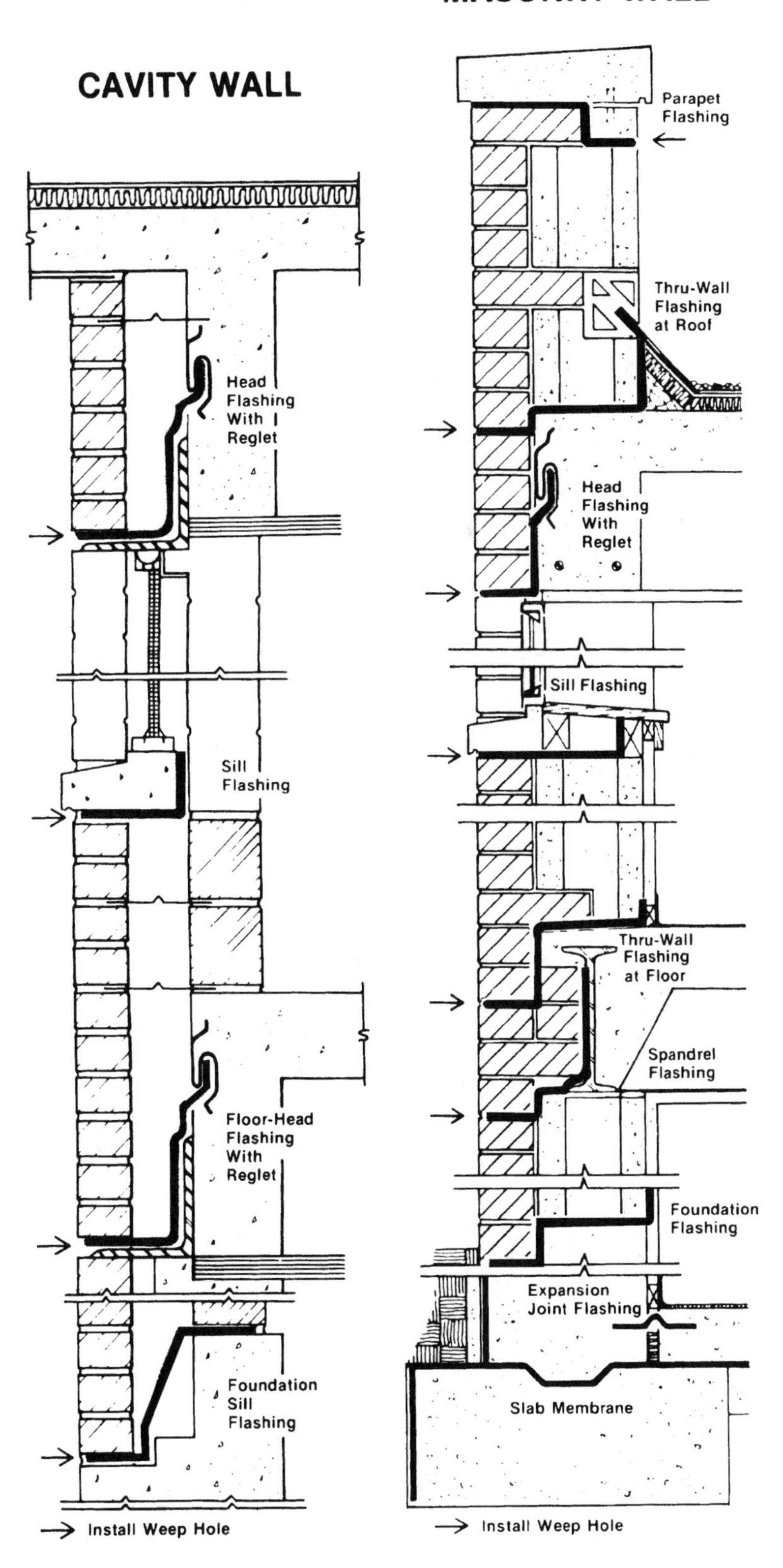

By permission AFCO Products, Inc., Somerville, MA

4.4.7 Pilaster Details

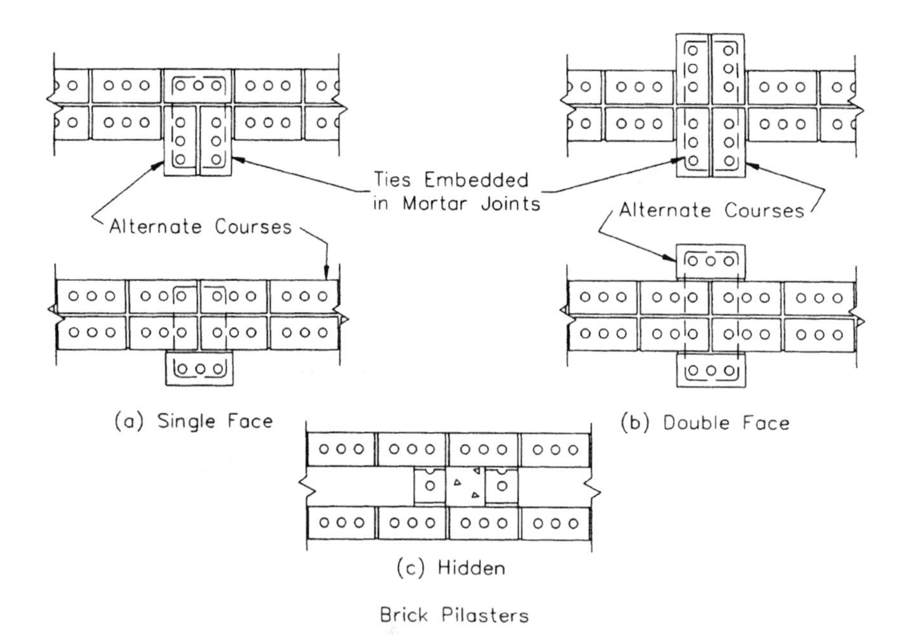

Brick Pilasters

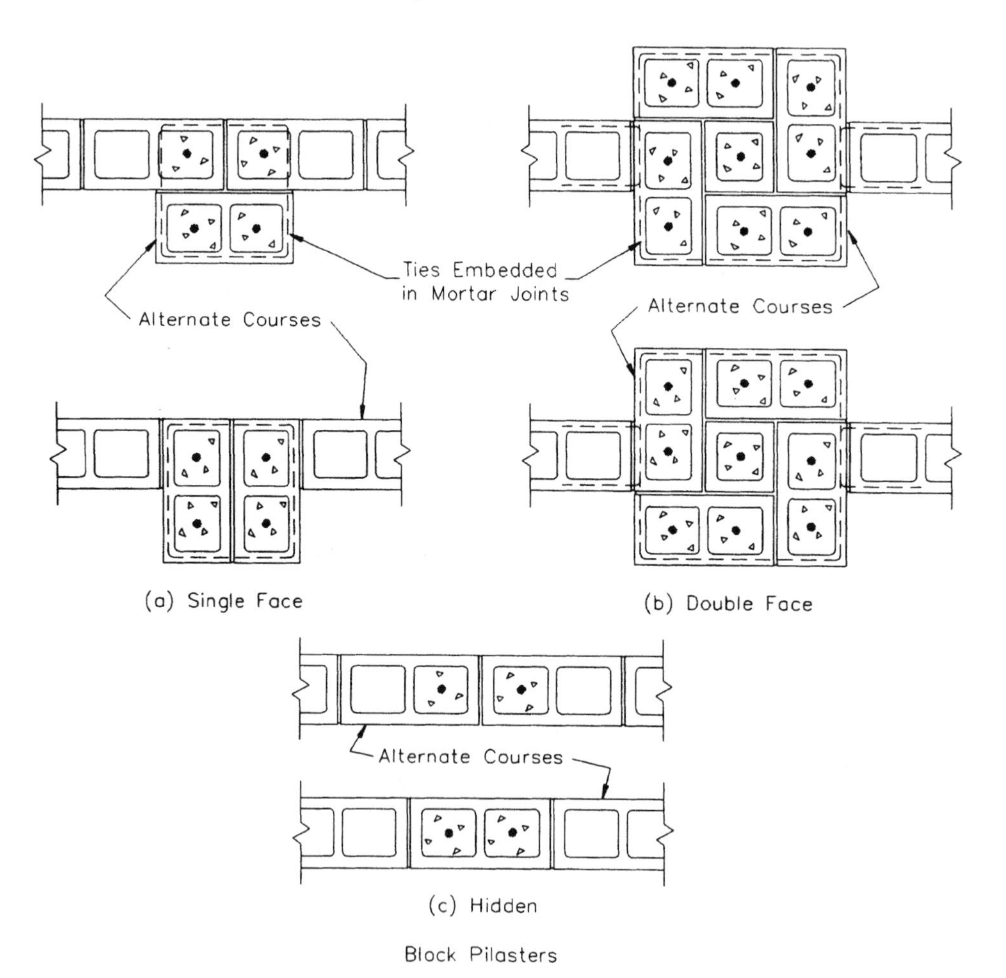

Block Pilasters

By permission from the Masonry Society, ACI, ASCE from their manual *Building Code Requirements for Masonry Structures*

4.4.8 Corbeling Limitations

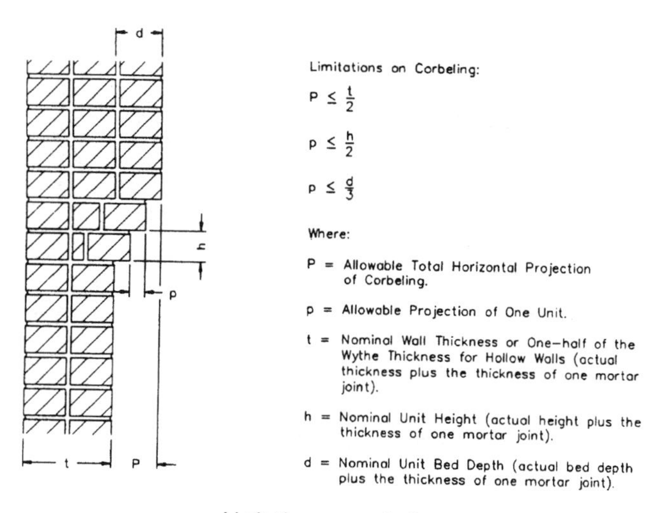

Limitations on corbeling

By permission from the Masonry Society, ACI, ASCE from their manual *Building Code Requirements for Masonry Structures*

4.4.9 Wall-Elevation Sections

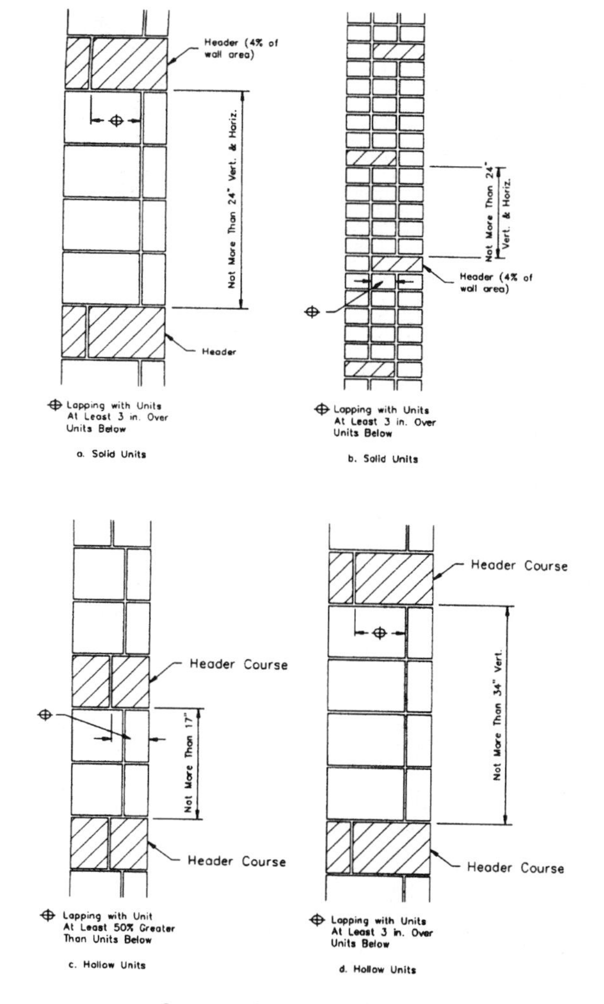

Cross section of wall elevations

By permission from the Masonry Society, ACI, ASCE from their manual *Building Code Requirements for Masonry Structures*

4.4.10 Bearing Areas, Running Bond at Intersections

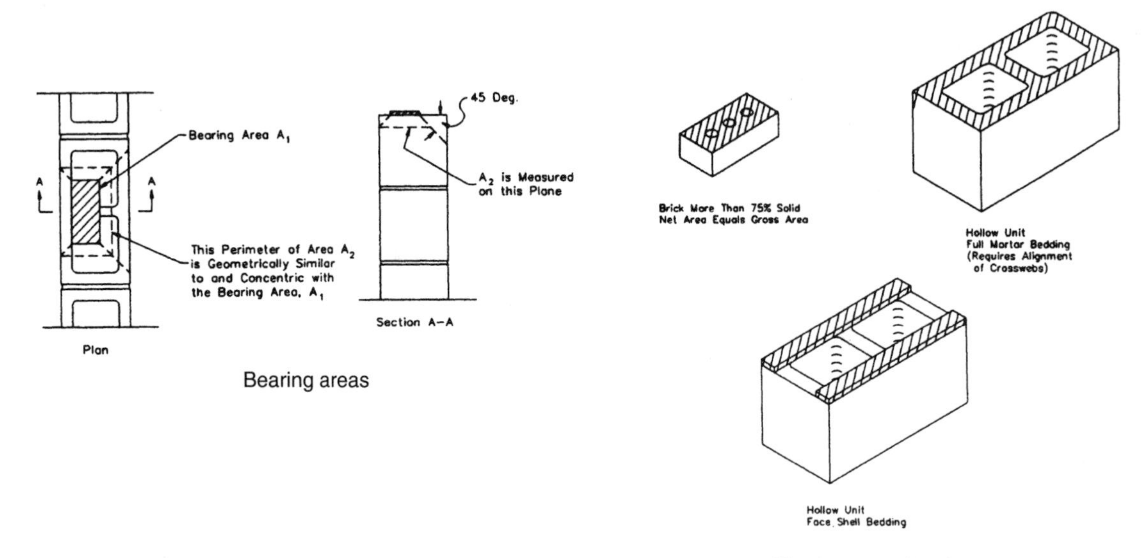

Bearing areas

Net cross-sectional areas

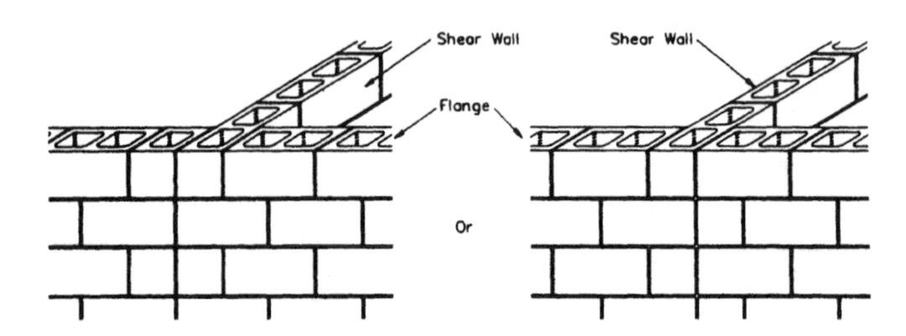

Running bond lap at intersection

By permission from the Masonry Society, ACI, ASCE from their manual *Building Code Requirements for Masonry Structures*

4.5.0 Grout Strengths/Proportions by Weight and Volume

Grout strengths

Grout type	Location	Compressive strength, psi (MPa)			Reference
		Low	Mean	High	
Coarse	Lab	1965(13.5)	3106(21.4)	4000(27.6)	2.11
Coarse	Lab	3611(24.9)	4145(28.6)	4510(31.1)	2.12
Coarse	Field	5060(34.9)	5455(37.6)	5940(41.0)	2.13

Grout proportions by volume

Grout type	Cement	Lime	Aggregate, damp, loose[1]	
			Fine	Coarse
Fine	1	$0\text{-}^1/_{10}$	$2^1/_4$ to 3	
Coarse	1	$0\text{-}^1/_{10}$	$2^1/_4$ to 3	1 to 2

[1]Times the sum of the volumes of the cementitious materials.

Reprinted by permission: Brick Institute of America, Reston, Virginia

4.6.0 Tile-Wall Systems

Innovative wall systems, utilizing thin tile as wall coverings, provide exciting design opportunities in today's competitive building market. Various concepts (see schematics), either prefabricated as panels in the factory or set-in-place on site, offer numerous wall-system options. Design assistance and cost analysis are available through local tile contractors or panel fabricators.

Tile Cladding Benefits:

- Design freedom
- Lightweight construction
- Quick installation
- Economical in-place cost
- Durability and fire resistance
- Increased insulation value
- All-weather construction

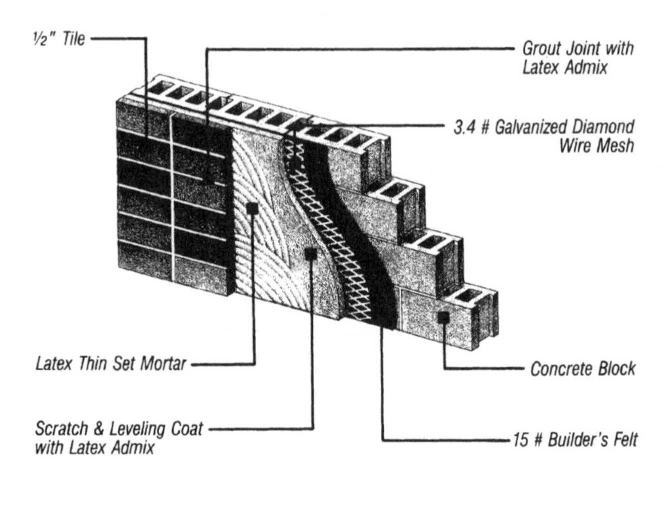

MORTAR BED SYSTEM

Pre-Fabricated or On-Site Panelization

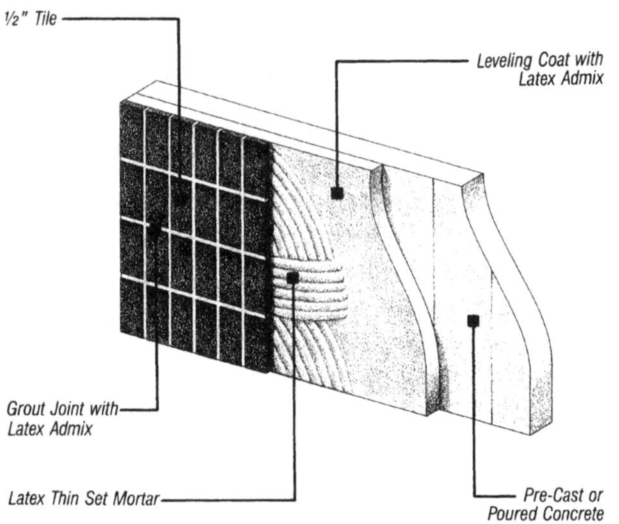

PANEL SYSTEM

Pre-Fabricated or On-Site Panelization

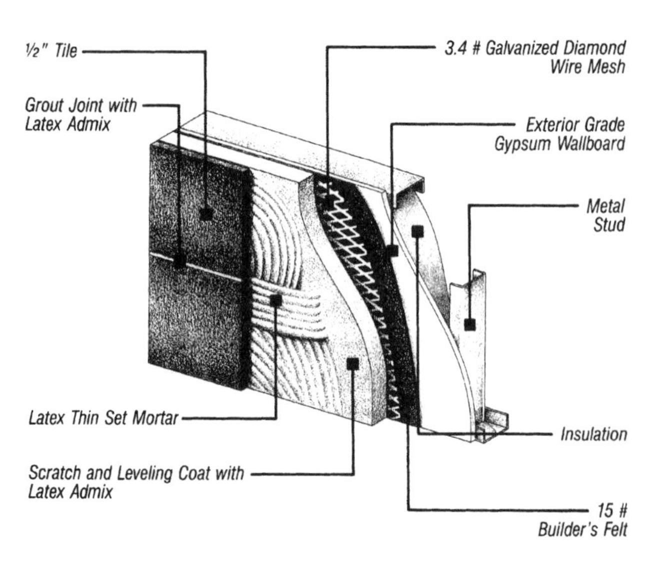

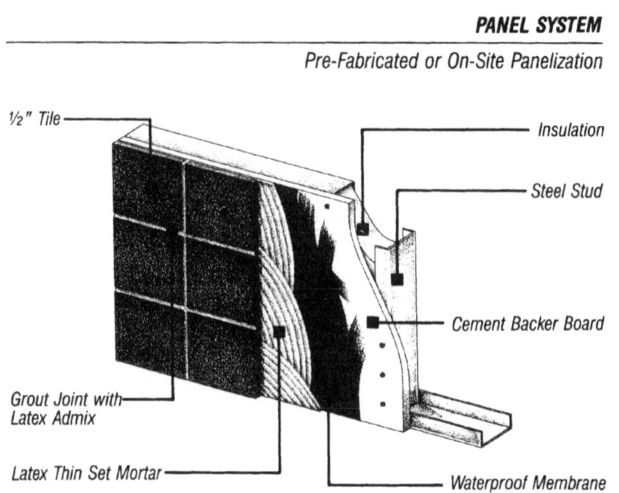

By permission Endicott Clay Products Co., Fairbury, Nebraska

4.6.1 Standard Tile-Cladding Shapes

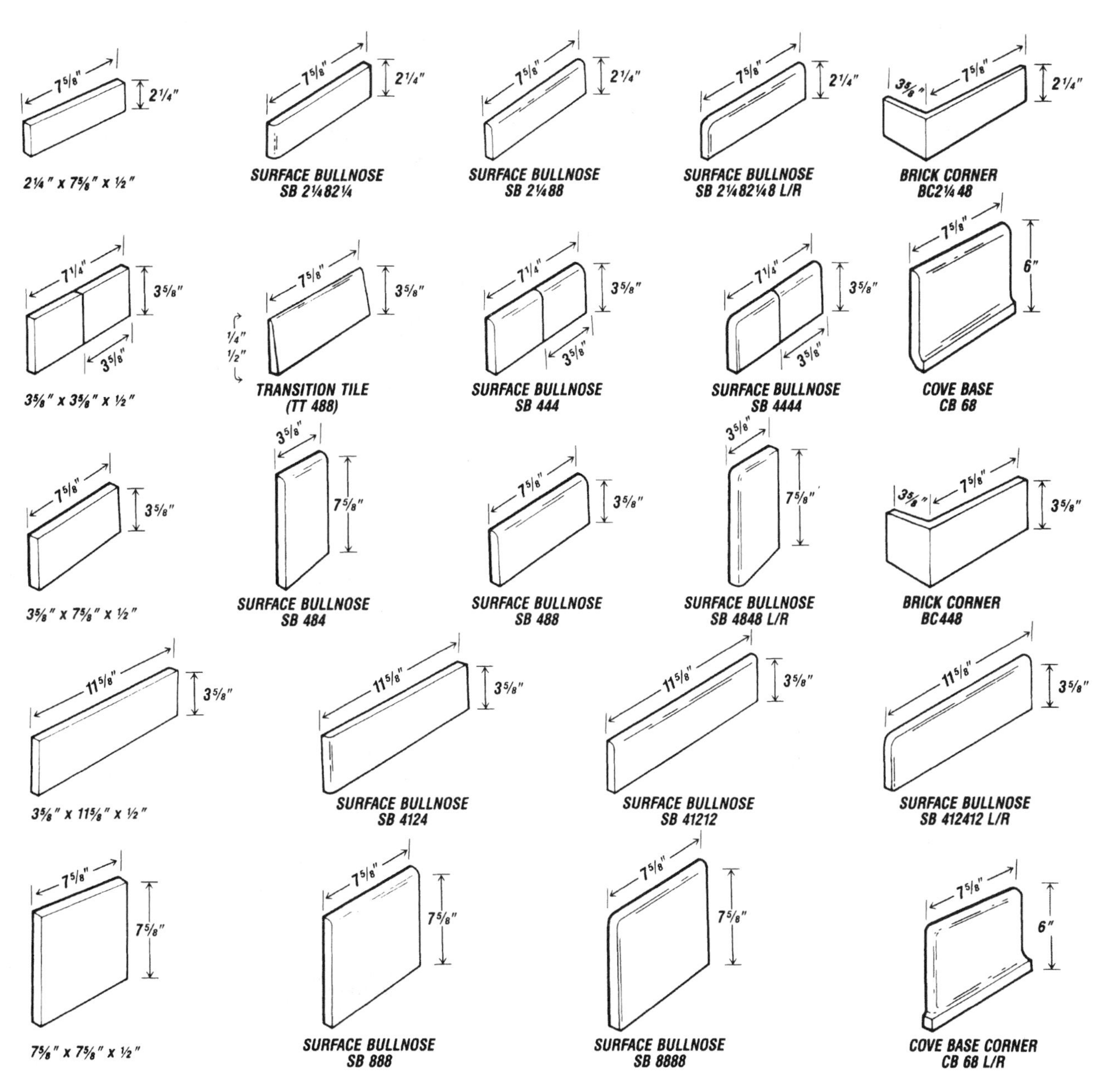

By permission Endicott Clay Products Co., Fairbury, Nebraska

4.7.0 Glass Block (Typical Sill Details)

Glass block is often used in Building Construction, however, installation details vary considerably from brick- or block-wall construction.

Typical Sill Details

Exterior Openings

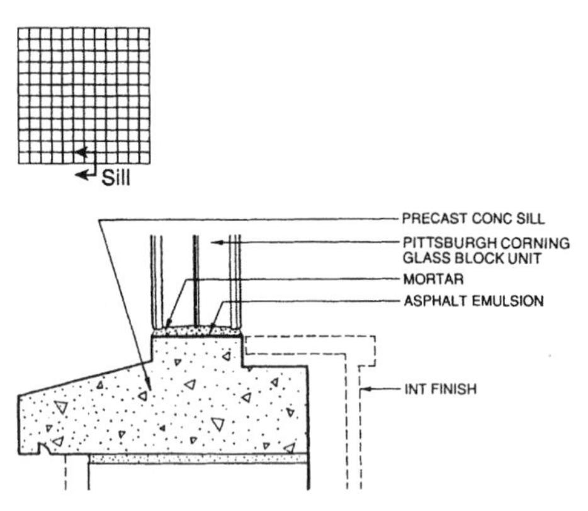

Sill - Glass Block in CMU Wall (PCD 006) Fire Rated

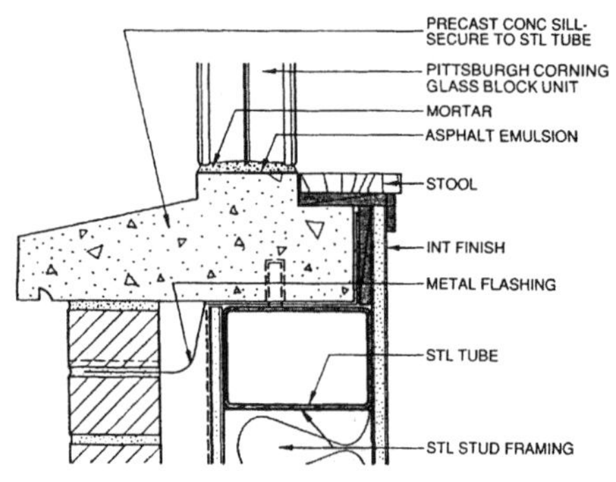

Sill - Glass Block in Steel Stud Wall With Brick Veneer (PCD 063)

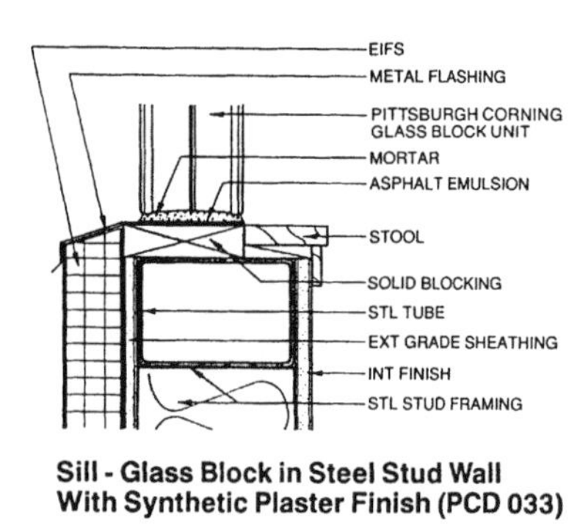

Sill - Glass Block in Steel Stud Wall With Synthetic Plaster Finish (PCD 033)

Typical Mortared Stiffener Details

250 Sq. Ft. Panels

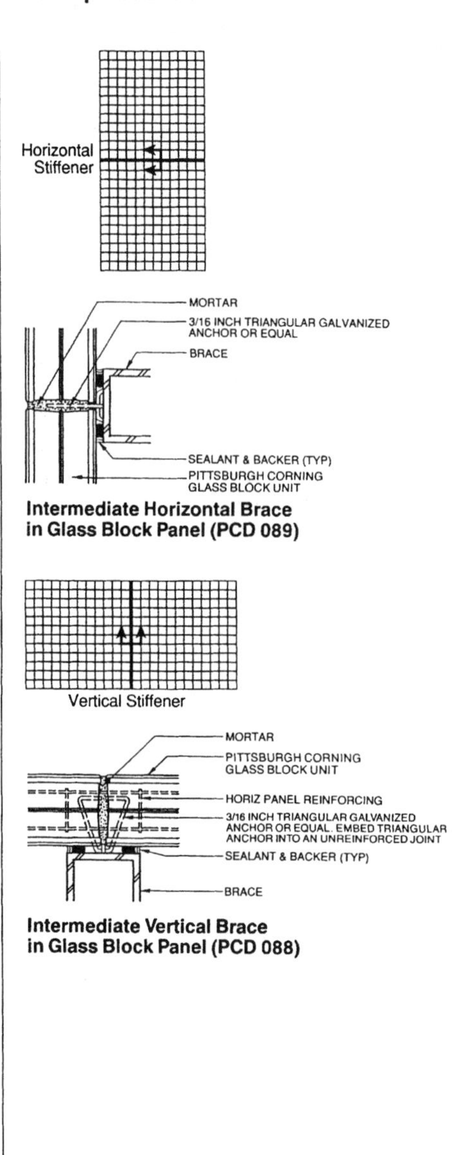

Intermediate Horizontal Brace in Glass Block Panel (PCD 089)

Intermediate Vertical Brace in Glass Block Panel (PCD 088)

By permission Pittsburgh Corning Glass Block, Pittsburgh, Pennsylvania

4.7.1 Glass Block (Typical Head and Jamb Details)

Typical Head Details
Exterior Openings

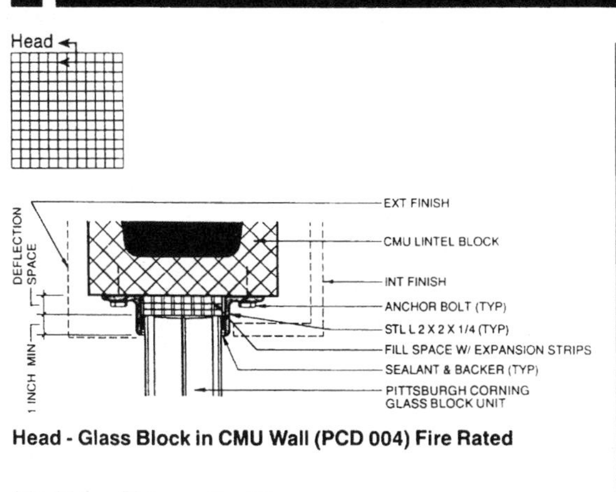

Head - Glass Block in CMU Wall (PCD 004) Fire Rated

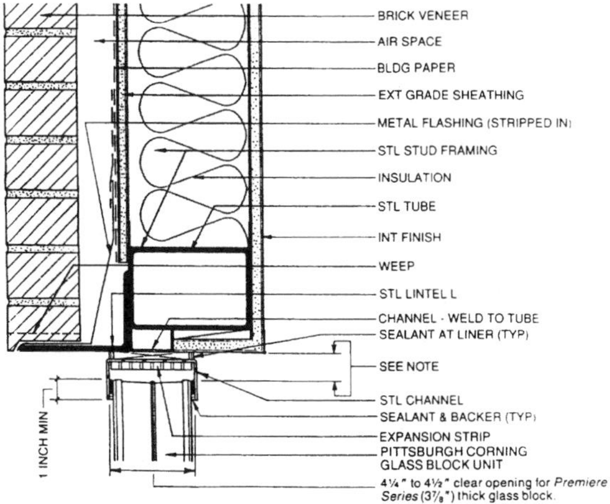

NOTE: This dimension is determined by the anticipated deflection of the structural member above the glass block.

Head - Glass Block in Steel Stud Wall With Brick Veneer (PCD 061)

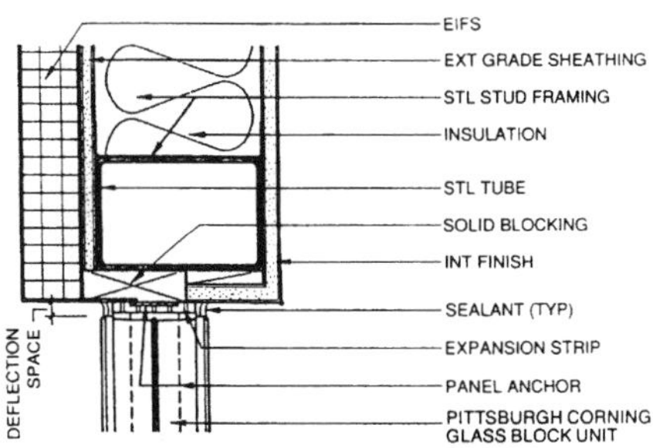

Head - Glass Block in Steel Stud Wall With Synthetic Plaster Finish (PCD 031)

Typical Jamb Details
Exterior Openings

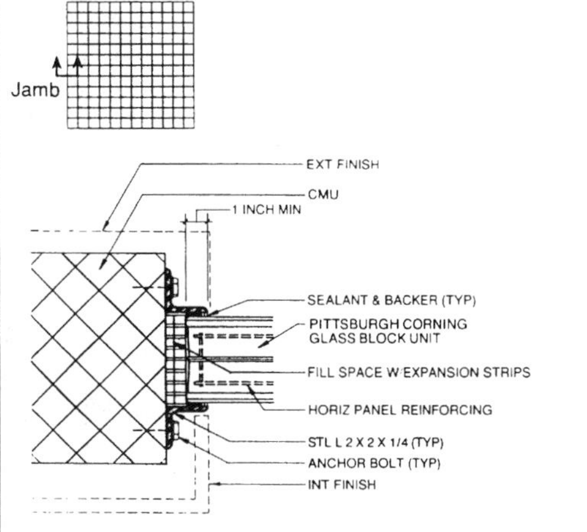

Jamb - Glass Block in CMU Wall (PCD 005) Fire Rated

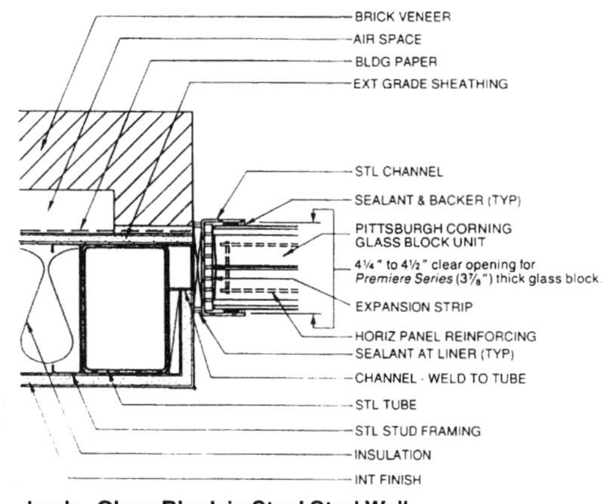

Jamb - Glass Block in Steel Stud Wall With Brick Veneer (PCD 062)

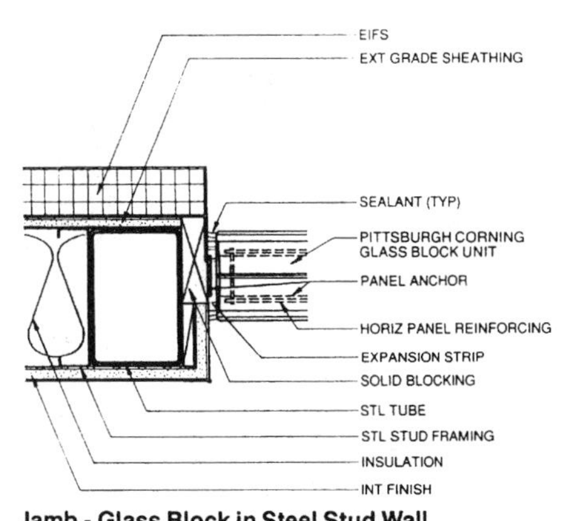

Jamb - Glass Block in Steel Stud Wall With Synthetic Plaster Finish (PCD 032)

By permission Pittsburgh Corning Glass Block, Pittsburgh, Pennsylvania

4.7.2 Glass Block (Typical Panel Anchor Details)

Panel Anchor Construction

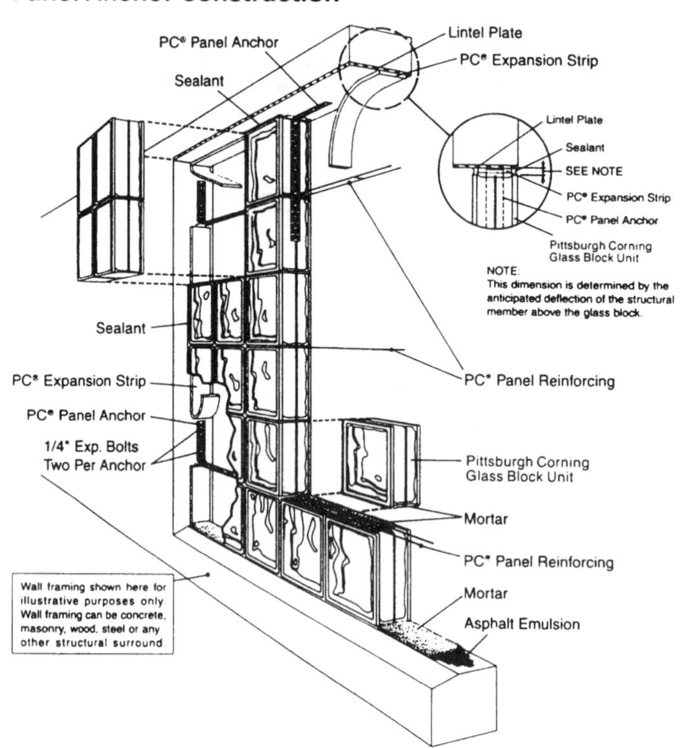

Channel-Type Restraint Construction

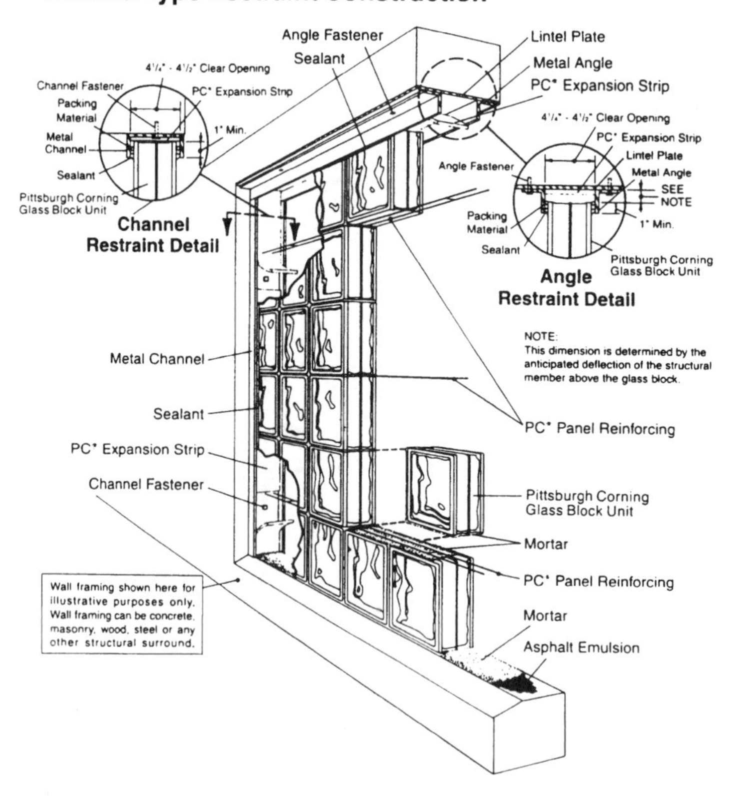

By permission Pittsburgh Corning Glass Block, Pittsburgh, Pennsylvania

4.8.0 Masonry Reinforcement (Types of Ties)

Whenever a double wythe wall is constructed or a cavity wall containing a masonry veneer is built, anchors, ties, or reinforcement is required to stabilize the two components. Seismic requirements add other components to the conventional masonry wall reinforcement to stabilize the structure in case of a seismic event.

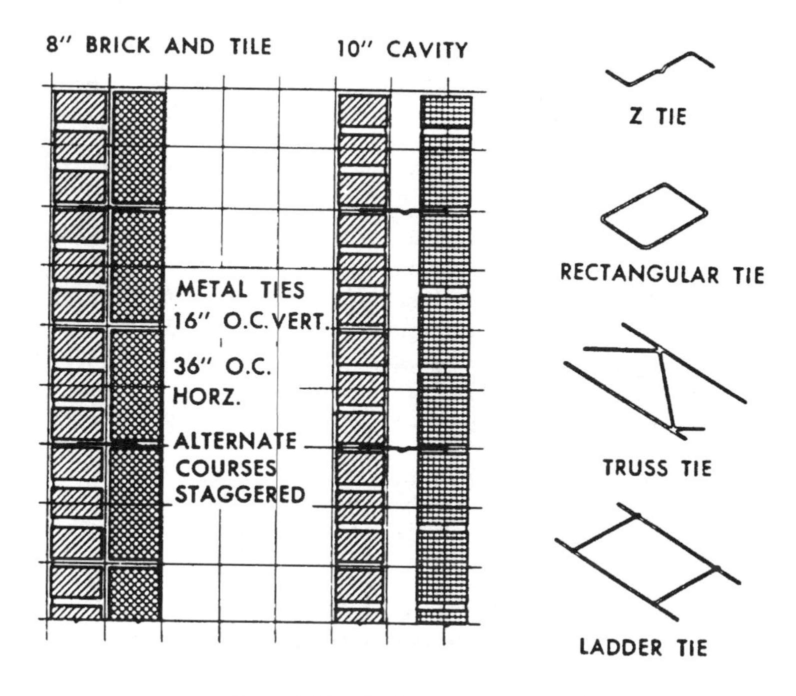

Metal-Tied Masonry Walls

Reprinted by permission: Brick Institute of America, Reston, Virginia

4.8.1 Masonry Reinforcement (Materials and Physical Properties of Bars/Wire)

Reinforcement and metal accessories

ASTM specification	Material	Use	Yield strength, ksi (MPa)	ASTM yield stress, MPa
A 36	Structural steel	Connectors	36 (248)	250
A 82	Steel wire	Joint reinforcement, ties	70 (483)	485
A 167	Stainless steel	Bolts, reinforcement, ties	30 (207)	205
A 185	Steel wire	Wire fabric, ties	75 (517)	485
A 307	Carbon steel	Connectors	60 (414)	
A 366	Carbon steel	Connectors	—	
A 496	Steel wire	Reinforcement	75 (517)	485
A 497	Steel wire fabric	Reinforcement, wire fabric	70 (483)	485
A 615	Billet steel	Reinforcement	40,60 (276, 414)	300,400
A 616	Rail steel	Reinforcement	50,60 (345, 414)	350,400
A 617	Axle steel	Reinforcement	40,60 (276, 414)	300,400
A 706	Low alloy steel	Reinforcement	60 (414)	

Physical properties of steel reinforcing wire and bars

Designation		Diameter, in. (mm)	Area, in.² (mm²)	Perimeter, in. (mm)
Wire				
W1.1 (11 gage)		0.121 (3.07)	0.011 (7.10)	0.380 (9.65)
W1.7 (9 gage)		0.148 (3.76)	0.017 (11.0)	0.465 (11.8)
W2.1 (8 gage)		0.162 (4.12)	0.020 (12.9)	0.509 (12.9)
W2.8 (3/16 wire)		0.187 (4.75)	0.027 (17.4)	0.587 (14.9)
W4.9 (1/4 wre)		0.250 (6.35)	0.049 (31.6)	0.785 (19.9)
Bars	Metric			
#3		0.375 (9.53)	0.11 (71.0)	1.178 (29.92)
	10	0.445 (11.3)	0.16 (100)	1.398 (35.5)
#4		0.500 (12.7)	0.20 (129)	1.571 (39.90)
#5	15	0.625 (15.9)	0.31 (200)	1.963 (49.86)
#6		0.750 (19.1)	0.44 (284)	2.456 (62.38)
	20	0.768 (19.5)	0.47 (300)	2.413 (61.3)
#7		0.875 (22.2)	0.60 (387)	2.749 (69.83)
	25	0.992 (25.2)	0.76 (500)	3.118 (79.2)
#8		1.000 (25.4)	0.79 (510)	3.142 (79.81)
#9		1.128 (28.7)	1.00 (645)	3.544 (90.02)
	30	1.177 (29.9)	1.09 (700)	3.697 (93.9)
#10		1.270 (32.2)	1.27 (819)	3.990 (101.3)
	35	1.406 (35.7)	1.55 (1000)	4.417 (112.2)
#11		1.410 (35.8)	1.56 (1006)	4.430 (112.5)

Wire size	*Minimum number of ties required*
W1.7	one wall tie per $2^2/_3$ ft² (0.25 m²) of wall
W2.8	one wall tie per $4^1/_2$ ft² (0.42 m²) of wall

4.8.2 Wall Anchorage Details

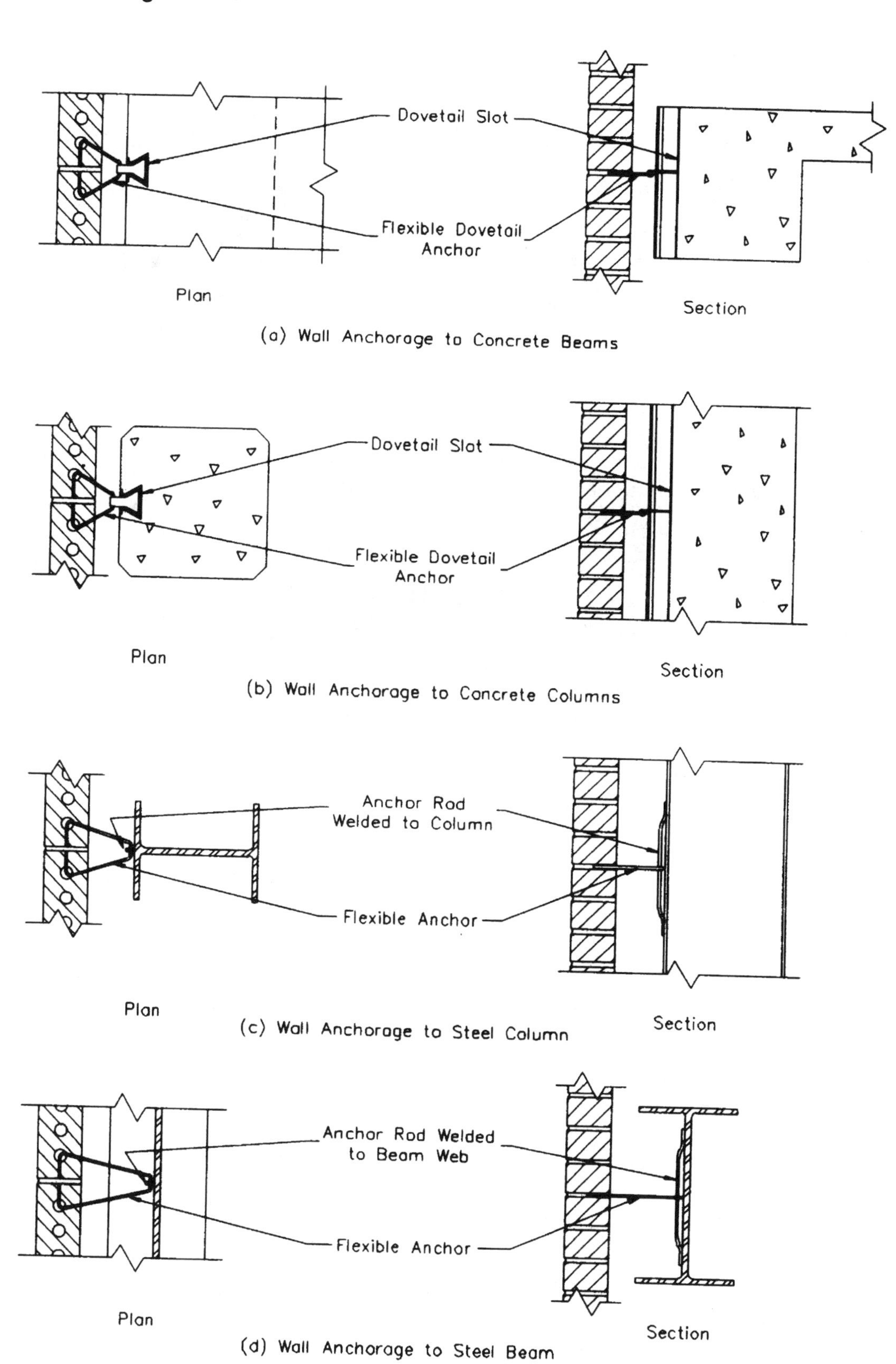

By permission from the Masonry Society, ACI, ASCE from their manual *Building Code Requirements for Masonry Structures*

4.8.3 Truss and Ladur Reinforcement

DUR-O-WAR® TRUSS

D/A 310 TRUSS

D/A 310 TR TRI-ROD

D/A 310 DSR DOUBLE SIDE ROD

LADUR TYPE®

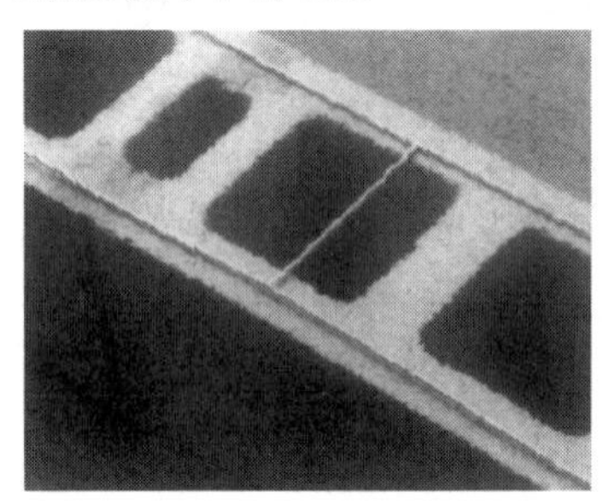

D/A 320 LADUR

D/A 320 TR TRI-ROD

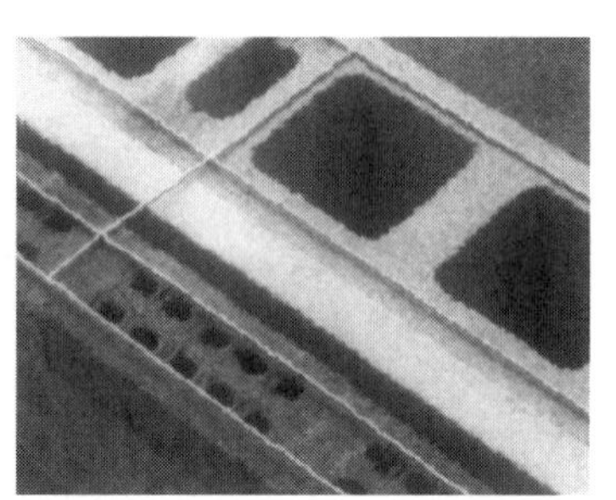

D/A 320 DSR DOUBLE SIDE ROD

INSTALLATION - TRUSS AND LADUR

Use at least one longitudinal side rod for each bed joint. Out-to-out spacing of the side rods is approximately 2" (50mm) less than the nominal thickness of the wall or wythe in which the reinforcement is placed.

Splices

Side rods should be lapped 6" (150mm) at splices in order to provide adequate continuity of the reinforcement when subjected to normal shrinkage stresses.

Centering and Placement

Place joint reinforcement directly on masonry and place mortar over wire to form bed joint. This applies to both truss type (shown) and ladur type.

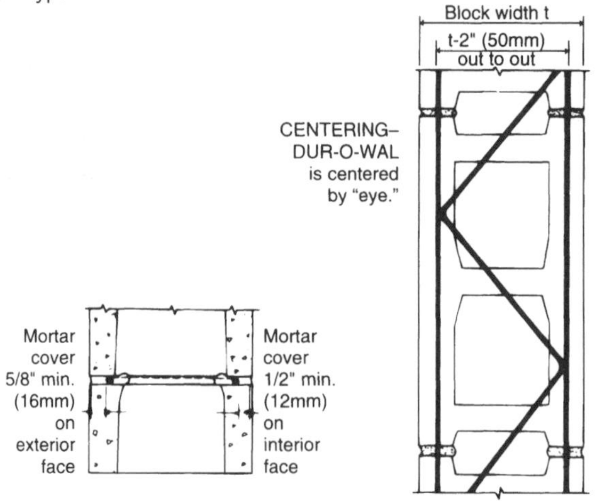

By permission from Dur-O-Wall, Inc., Arlington Heights, Illinois

4.8.4 Masonry Wall Ties

D/A 5801

Recommended for non-insulated cavity/walls. The channel base plate is secured to the back-up and has a 1-1/4" (30mm) slot for coursing adjustability. The 3/16" (5mm) triangular wire tie is mortared in the veneer. Hot dipped galvanized and stainless steel finishes are available.

D/A 5431

Recommended for reconstructing brick wythes of composite walls. The 14 gauge (1.9 mm) corrugated strap has a 1-1/4" (30mm) of adjustability. The tie is mortared in place with the new brick wythe. Shear lugs accommodate seismic ladur or pencil rod. Hot dipped galvanized and stainless steel finishes are available.

D/A 5213S with Seismic Ladur

By permission from Dur-O-Wall, Inc., Arlington Heights, Illinois

4.8.5 Masonry Veneer Anchors

D/A 213

D/A 207 WITH D/A 701

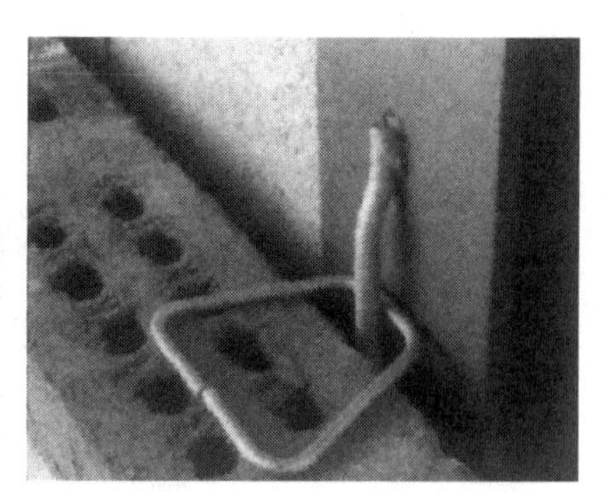

D/A 709 WITH D/A 701

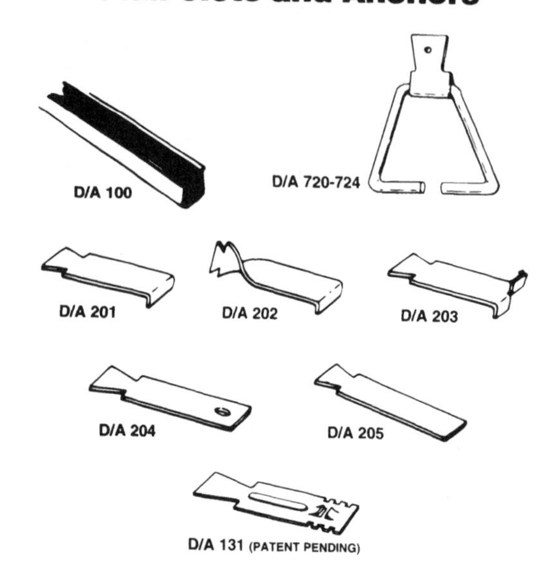

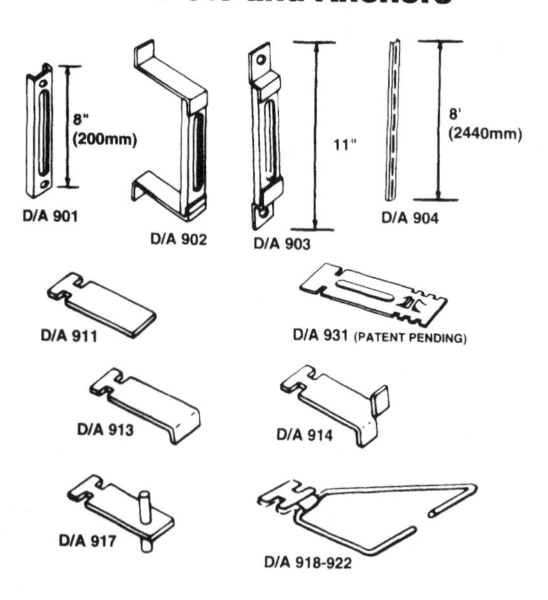

By permission from Dur-O-Wall, Inc., Arlington Heights, Illinois

4.8.6 Seismic Masonry Veneer Anchors

Seismic Veneer Anchoring Application

DUR-O-WAL's seismic veneer anchors are designed to meet performance criteria as defined by building codes. These anchors can be used for tieing brick veneers to wood stud, steel studs, steel framing, masonry, brick and concrete, They are fabricated with shear lugs that accommodate 9 gauge veneer reinforcement. The connectors are individually mounted and are easily installed.

Seismic Veneer Anchors (patented)

This anchor has the same plate and pintle design as Seismic Dur-O-Eye. The plate is engineered to be attached to the face of a CMU or concrete (D/A 5213) steel stud, wood stud or steel frame (D/A 213S) rather than embedded in mortar. The pintle shear lugs hold pencil rod or Seismic Ladur in place for greater pull out stress resistance and ductility. Adjusts 1-1/4" (30mm) up or down to allow for different course heights and allows at least 1/2" (13mm) horizontal in-plane movement to accommodate expansion and contraction. A hot dipped galvanized finish (1.5 oz. zinc per sq. ft.) (458g/m^2) is standard, and 304 stainless steel is available. DUR-O-WAL recommends the use of two screws for stud applications, either the D/A 807 for steel, D/A 808 for wood, or D/A 995, or a special 1/4" (6mm) expansion bolt for concrete or masonry retrofit applications (D/A 5213).

D/A 931 Seismic Channel Slot Anchor Assembly (patent pending)

Engineered for use with standard channel slots. Pencil rod or Seismic Ladur fits inside shear lug for positive placement without the need for special clips.

D/A 431 Seismic Strap Anchor (patent pending)

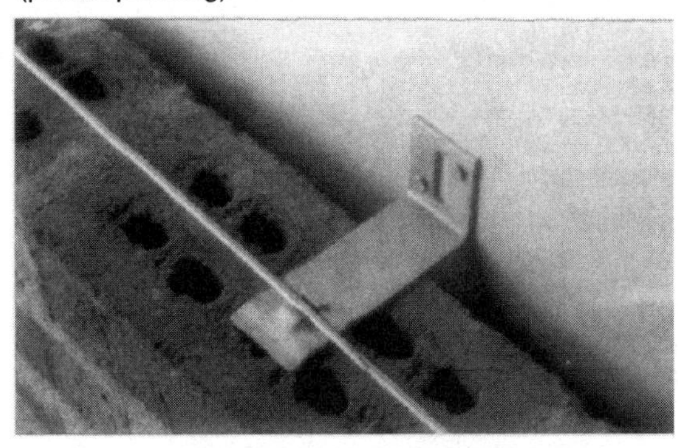

A special 14 ga. (1.9mm) adjustable seismic corrugated veneer anchor with two shear lugs, which is engineered for use with pencil rod or Seismic Ladur to resist out of plane movement and afford greater ductility in seismic zones 3 and 4 or Seismic Performance Categories D and E can be nailed or screwed to wood stud backup (D/A 808).

D/A 131 Seismic Dovetail Anchor Assembly (patent pending)

Specially designed tie with shear lug locks for pencil rod or Seismic Ladur to assure positive positioning and reinforcement without the need for special clips. Engineered to fit standard dovetail slots with 5/8" (16mm) throat opening.

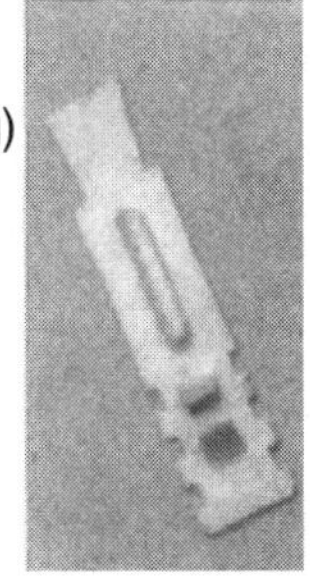

By permission from Dur-O-Wall, Inc., Arlington Heights, Illinois

4.8.7 Seismic Masonry Ladur and Comb Reinforcement

D/A 360 S SEISMIC LADUR-EYE

D/A 370 S SEISMIC DUR -O-EYE

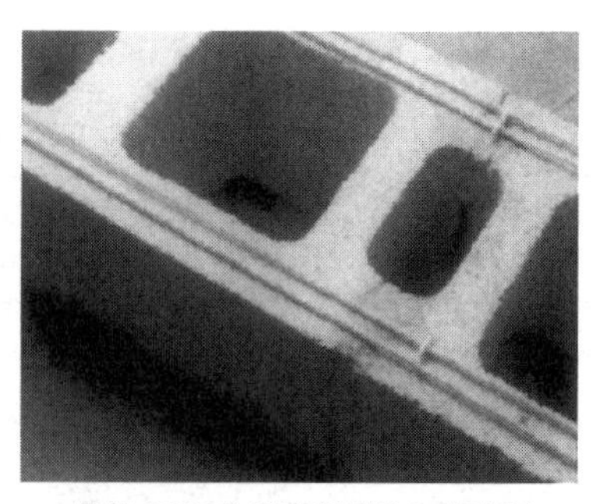
D/A 320 S SEISMIC LADUR

D/A 5213/Seismic 5213S

Recommended for brick cavity walls with or without insulation. Dual leg 3/16" (5mm) pintle adjusts vertically 1-1/4" (30mm), up or down. The plate projects off the back-up wall to accommodate insulation, or bridge cavities. Hot dip galvanized, and stainless steel finishes available.

Seismic Comb (patent pending)

Masonry confinement reinforcement located in horizontal mortar joint to improve seismic performance of shear walls. Provides the Vertical Rebar confinement requirements in Section 2108.2.5.6 (1994). Made with 3/16" diameter wire conforming to ASTM A82. A hot dipped galvanized finish (1.5 oz., 458 g/m^2, zinc per square foot), per ASTM A153, is standard. Available for 6" (150mm), 8" (200mm), 10" (250mm) and 12" (300mm) hollow masonry units.

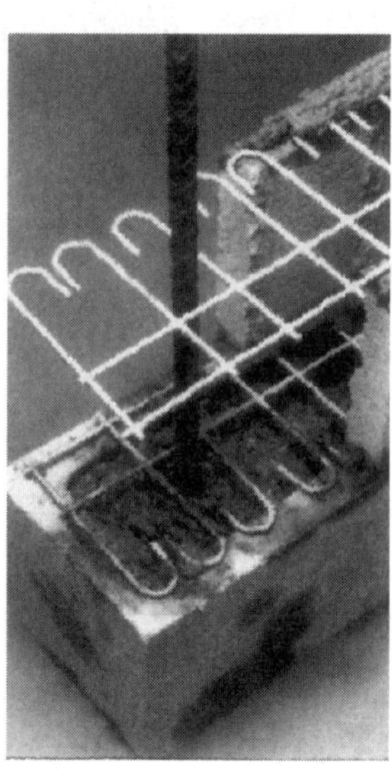

By permission from Dur-O-Wall, Inc., Arlington Heights, Illinois

4.9.0 Investigating Unstable Masonry Conditions to Prevent Failures

Although masonry walls are extremely durable, "old age" and neglect can take its toll on even the most durable structure. When inspecting a masonry facade for potential problems and restoration, a number of contributing factors must be considered. Often, it is necessary to cut out a small section of wall in the area/areas where failures are suspected.

The following checklist will aid in this investigation:

1. When initially built, were all ties and anchors installed as required?
2. Were the ties properly installed (e.g., embedded adequately in the bed joint and connected to the backup correctly)
3. Does there appear to be excessive differential wall movement caused by thermal movement, settlement, or freeze/thaw conditions?
4. Were the proper size and type of ties/anchors used to avoid stresses that exceed the facade materials' capacity?
5. Were the proper type of expansion and control joints installed at the proper distances?
6. Have the ties, anchors, fasteners, relieving angles, and lintels corroded because of moisture being trapped? Is there accelerated corrosion from chlorides or has galvanic action taken place because of a combination of carbon steel anchors in contact with dissimilar materials?
7. Has excessive water penetrated the wall system from any poorly maintained parapet flashings or roof-coping flashings?

8. Have the caulk joints been allowed to deteriorate?
9. Have the weep holes been caulked when maintenance caulking was performed and have the lintels been caulked at the point where brick is bearing on them?
10. Have the mortar joints deteriorated and not been tuckpointed during routine maintenance inspections?

4.9.1 Fire Resistance Ratings of Various Concrete Masonry Units and Assemblies

Listed is the minimum required equivalent thickness of concrete masonry assembly (inches and centimeters, metric in parenthesis)

Aggregate type in the CMU	4 hour	3 hour	2 hour	1.5 hours	1 hour	0.75 hours	0.5 hours
Calcareous or siliceous gravel	6.2 (15.75)	5.3 (13.46)	4.2 (10.67)	3.6 (9.14)	2.8 (7.11)	2.4 (6.09)	2.0 (5.08)
Limestone, cinders, slag	5.9 (14.99)	5.0 (12.7)	4.0 (10.16)	3.4 (8.73)	2.7 (6.86)	2.3 (5.84)	1.9 (4.82)
Expanded clay, shale or slate	5.1 (12.95)	4.4 (11.17)	3.6 (9.14)	3.3 (8.38)	2.6 (6.6)	2.2 (5.59)	1.8 (4.57)
Expanded slag pumice	4.7 (11.94)	4.0 (10.16)	3.2 (8.13)	2.7 (6.86)	2.1 (5.33)	1.9 (4.82)	1.5 (3.81)

Reinforced Concrete Masonry Columns

Minimum column dimensions inches/centimeters and fire-resistance rating

1 hour	2 hours	3 hours	4 hours
(8 inches) (20.32)	10 inches (25.4)	12 inches (30.48)	14 inches (35.56)

Reinforced Concrete Masonry Lintels

Minimum longitudinal reinforcing cover (inches/centimeters)

Nominal lintel	Fire-resistance rating			
Width (inches and centimeters)	1 hour	2 hurs	3 hours	4 hours
6 inches (15.24)	1½	2	—	—
8 inches (20.32)	1½	1½	1¾	3
10 inches or more (25.4 cm or more)	1½	1½	1½	1¾

Equivalent Thickness of Concrete Masonry Units

Nominal width	Based on typical hollow units		Based on percent solid 75%	100%
4 (10.16)	2.68 (6.8)	[73.8]	2.72 (6.91)	3.62 (9.19)
6 (15.24)	3.09 (7.85)	[55.0]	4.22 (10.72)	5.62 (14.27)
8 (20.32)	4.04 (10.26)	[53.0]	5.72 (14.53)	7.62 (19.35)
10 (25.4)	4.98 (12.65)	[51.7]	7.22 (18.34)	9.62 (24.43)
12 (30.48)	5.66 (14.38)	[48.7]	8.72 (22.15)	11.62 (29.51)

(*Note*: Values in brackets are percent solid values based on typical two-core concrete masonry units. Numbers in parenthesis are metric equivalents, in centimeters, to inch dimensions.

Section

5

Structural Steel, Joists, and Metal Decks

Contents

5.0.0 History of steel and grades of structural steel
5.1.0 Surface areas/box areas of "W" shapes (W4 to W12)
5.1.1 Surface areas/box areas of "W" shapes (W12 to W18)
5.1.2 Surface areas/box areas of "W" shapes (W18 to W36)
5.2.0 Standard mill practices (camber)
5.2.1 Standard mill practices ("W" shape tolerances)
5.3.0 Suggested beam-framing details
5.3.1 Suggested column base plate details
5.3.2 Suggested structural steel erection details (miscellaneous)
5.4.0 Welded joints (standard symbols)
5.5.0 Threaded fasteners (bolt head shapes)
5.5.1 Threaded fasteners (weight of bolts)
5.5.2 Threaded fasteners (weight of ASTM A325 /A490 bolts)
5.5.3 Properties of heavy hex nuts and indentifying marks
5.5.4 Bolt diameters and standard hole dimensions
5.5.5 Capscrews/bolts/heavy hex nut markings
5.5.6 Dimensions of finished hex nuts
5.5.7 Dimensions of finished hex bolts
5.5.8 Tension control (TC) bolt-installation procedures
5.5.9 Tru-Tension (TC) bolt-assembly specifications
5.6.0 Major characteristics of joist series
5.6.1 General information on K series joists
5.6.2 Standard specifications for open web joists (K series)
5.6.3 K series joists (top chord extensions and extended end)
5.6.4 General information (LH- and DLW-series joists)
5.6.5 LH and DLW series longspan details
5.7.0 Joist girders (what are they?)
5.7.1 Joist girder notes and connection details
5.7.2 Joist girder moment connection details
5.7.3 Specifying joist girders
5.8.0 Recommended maximum spans for steel decking
5.8.1 Methods of lapping steel deck
5.8.2 Noncomposite and composite deck details
5.8.3 Pour-stop selection table
5.8.4 Cellular floor-deck and form-deck profiles
5.8.5 Composite floor-deck and roof-deck profiles
5.8.6 Reinforcing openings in steel decks
5.8.7 Example of 6" penetration in steel deck
5.8.8 Maximum spans for roof deck
5.9.0 Fire-resistance ratings for roof decks
5.9.1 Floor-ceiling fire-resistance ratings with steel joist

5.0.0 History of Steel and Grades of Structural Steel

Iron was produced by primitive man by placing iron ore and charcoal in a clay pot and building a fire in the pot, using a crude bellows to provide the forced draft that deposited iron at the bottom. It was not until the mid-1800s that Henry Bessemer, an English metalurgist, developed a process whereby forced air was introduced into the iron-refining procedure raising the temperature of the crucible so that impurities in the molten pig iron were burned away. In the process, a more malleable metal, steel, was created.

Various minerals and metals are added to molten steel nowadays to enhance certain characteristics:

- *Nickel* Improves the hardenability of steel and increases impact strength at low temperatures.
- *Sulfur* Increases machinability.
- *Manganese* Increases strength and hardness.
- *Carbon* The principal hardening agent in steel.
- *Molydenum* Prevents brittleness.
- *Vanadium* Gives steel a fine grain structure and improves the fatigue values.
- *Silicon* Improves strength. It is a deoxidizer.
- *Phosphorous* Improves the machinability of high-sulfur steels and imparts some resistance to corrosion.

ASTM Structural Steel Specifications

ASTM designation	Steel type
A36	Carbon
A529	Carbon
A441	High strength (low alloy)
A572 grade (includes 42, 50, 60, 65)	High strength (low alloy)
A242	Corrosion resistant, high strength Low alloy
A588	Corrosion resistant, high strength Low alloy
A852	Quenched and tempered (low alloy) (Plates only)
A514	Quenched and tempered alloy (Plates only)

5.1.0 Surface Area/Box Areas of "W" Shapes (W4 to W12)

Designation	Case A	Case B	Case C	Case D
W 12x 58	4.39	5.22	2.87	3.70
x 53	4.37	5.20	2.84	3.68
W 12x 50	3.90	4.58	2.71	3.38
x 45	3.88	4.55	2.68	3.35
x 40	3.86	4.52	2.66	3.32
W 12x 35	3.63	4.18	2.63	3.18
x 30	3.60	4.14	2.60	3.14
x 26	3.58	4.12	2.58	3.12
W 12x 22	2.97	3.31	2.39	2.72
x 19	2.95	3.28	2.36	2.69
x 16	2.92	3.25	2.33	2.66
x 14	2.90	3.23	2.32	2.65
W 10x112	4.30	5.17	2.76	3.63
x100	4.25	5.11	2.71	3.57
x 88	4.20	5.06	2.66	3.52
x 77	4.15	5.00	2.62	3.47
x 68	4.12	4.96	2.58	3.42
x 60	4.08	4.92	2.54	3.38
x 54	4.06	4.89	2.52	3.35
x 49	4.04	4.87	2.50	3.33
W 10x 45	3.56	4.23	2.35	3.02
x 39	3.53	4.19	2.32	2.98
x 33	3.49	4.16	2.29	2.95
W 10x 30	3.10	3.59	2.23	2.71
x 26	3.08	3.56	2.20	2.68
x 22	3.05	3.53	2.17	2.65
W 10x 19	2.63	2.96	2.04	2.38
x 17	2.60	2.94	2.02	2.35
x 15	2.58	2.92	2.00	2.33
x 12	2.56	2.89	1.98	2.31

Designation	Case A	Case B	Case C	Case D
W 8x67	3.42	4.11	2.19	2.88
x58	3.37	4.06	2.14	2.83
x48	3.32	4.00	2.09	2.77
x40	3.28	3.95	2.05	2.72
x35	3.25	3.92	2.02	2.69
x31	3.23	3.89	2.00	2.67
W 8x28	2.87	3.42	1.89	2.43
x24	2.85	3.39	1.86	2.40
W 8x21	2.61	3.05	1.82	2.26
x18	2.59	3.03	1.79	2.23
W 8x15	2.27	2.61	1.69	2.02
x13	2.25	2.58	1.67	2.00
x10	2.23	2.56	1.64	1.97
W 6x25	2.49	3.00	1.57	2.08
x20	2.46	2.96	1.54	2.04
x15	2.42	2.92	1.50	2.00
W 6x16	1.98	2.31	1.38	1.72
x12	1.93	2.26	1.34	1.67
x 9	1.90	2.23	1.31	1.64
W 5x19	2.04	2.45	1.28	1.70
x16	2.01	2.43	1.25	1.67
W 4x13	1.63	1.96	1.03	1.37

Case A: Shape perimeter, minus one flange surface.
Case B: Shape perimeter.
Case C: Box perimeter, equal to one flange surface plus twice the depth.
Case D: Box perimeter, equal to two flange surfaces plus twice the depth.

By permission of the American Institute of Steel Construction, Chicago, Ill.

5.1.1 Surface Area/Box Areas of "W" Shapes (W12 to W18)

Designation	Case A	Case B	Case C	Case D
W 18x 46	4.41	4.91	3.52	4.02
x 40	4.38	4.88	3.48	3.99
x 35	4.34	4.84	3.45	3.95
W 16x100	5.28	6.15	3.70	4.57
x 89	5.24	6.10	3.66	4.52
x 77	5.19	6.05	3.61	4.47
x 67	5.16	6.01	3.57	4.43
W 16x 57	4.39	4.98	3.33	3.93
x 50	4.36	4.95	3.30	3.89
x 45	4.33	4.92	3.27	3.86
x 40	4.31	4.89	3.25	3.83
x 36	4.28	4.87	3.23	3.81
W 16x 31	3.92	4.39	3.11	3.57
x 26	3.89	4.35	3.07	3.53
W 14x730	7.61	9.10	5.23	6.72
x665	7.46	8.93	5.08	6.55
x605	7.32	8.77	4.94	6.39
x550	7.19	8.62	4.81	6.24
x500	7.07	8.49	4.68	6.10
x455	6.96	8.36	4.57	5.98
W 14x426	6.89	8.28	4.50	5.89
x398	6.81	8.20	4.43	5.81
x370	6.74	8.12	4.36	5.73
x342	6.67	8.03	4.29	5.65
x311	6.59	7.94	4.21	5.56
x283	6.52	7.86	4.13	5.48
x257	6.45	7.78	4.06	5.40
x233	6.38	7.71	4.00	5.32
x211	6.32	7.64	3.94	5.25
x193	6.27	7.58	3.89	5.20
x176	6.22	7.53	3.84	5.15
x159	6.18	7.47	3.79	5.09
x145	6.14	7.43	3.76	5.05

Designation	Case A	Case B	Case C	Case D
W 14x132	5.93	7.16	3.67	4.90
x120	5.90	7.12	3.64	4.86
x109	5.86	7.08	3.60	4.82
x 99	5.83	7.05	3.57	4.79
x 90	5.81	7.02	3.55	4.76
W 14x 82	4.75	5.59	3.23	4.07
x 74	4.72	5.56	3.20	4.04
x 68	4.69	5.53	3.18	4.01
x 61	4.67	5.50	3.15	3.98
W 14x 53	4.19	4.86	2.99	3.66
x 48	4.16	4.83	2.97	3.64
x 43	4.14	4.80	2.94	3.61
W 14x 38	3.93	4.50	2.91	3.48
x 34	3.91	4.47	2.89	3.45
x 30	3.89	4.45	2.87	3.43
W 14x 26	3.47	3.89	2.74	3.16
x 22	3.44	3.86	2.71	3.12
W 12x336	5.77	6.88	3.92	5.03
x305	5.67	6.77	3.82	4.93
x279	5.59	6.68	3.74	4.83
x252	5.50	6.58	3.65	4.74
x230	5.43	6.51	3.58	4.66
x210	5.37	6.43	3.52	4.58
W 12x190	5.30	6.36	3.45	4.51
x170	5.23	6.28	3.39	4.43
x152	5.17	6.21	3.33	4.37
x136	5.12	6.15	3.27	4.30
x120	5.06	6.09	3.21	4.24
x106	5.02	6.03	3.17	4.19
x 96	4.98	5.99	3.13	4.15
x 87	4.95	5.96	3.10	4.11
x 79	4.92	5.93	3.07	4.08
x 72	4.89	5.90	3.05	4.05
x 65	4.87	5.87	3.02	4.02

Case A: Shape perimeter, minus one flange surface.
Case B: Shape perimeter.
Case C: Box perimeter, equal to one flange surface plus twice the depth.
Case D: Box perimeter, equal to two flange surfaces plus twice the depth.

By permission of the American Institute of Steel Construction, Chicago, Ill.

5.1.2 Surface Area/Box Areas of "W" Shapes (W18 to W36)

Designation	Case A	Case B	Case C	Case D
W 36x300	9.99	11.40	7.51	8.90
x280	9.95	11.30	7.47	8.85
x260	9.90	11.30	7.42	8.80
x245	9.87	11.20	7.39	8.77
x230	9.84	11.20	7.36	8.73
W 36x210	8.91	9.93	7.13	8.15
x194	8.88	9.89	7.09	8.10
x182	8.85	9.85	7.06	8.07
x170	8.82	9.82	7.03	8.03
x160	8.79	9.79	7.00	8.00
x150	8.76	9.76	6.97	7.97
x135	8.71	9.70	6.92	7.92
W 33x241	9.42	10.70	7.02	8.34
x221	9.38	10.70	6.97	8.29
x201	9.33	10.60	6.93	8.24
W 33x152	8.27	9.23	6.55	7.51
x141	8.23	9.19	6.51	7.47
x130	8.20	9.15	6.47	7.43
x118	8.15	9.11	6.43	7.39
W 30x211	8.71	9.97	6.42	7.67
x191	8.66	9.92	6.37	7.62
x173	8.62	9.87	6.32	7.57
W 30x132	7.49	8.37	5.93	6.81
x124	7.47	8.34	5.90	6.78
x116	7.44	8.31	5.88	6.75
x108	7.41	8.28	5.84	6.72
x 99	7.37	8.25	5.81	6.68
W 27x178	7.95	9.12	5.81	6.98
x161	7.91	9.08	5.77	6.94
x146	7.87	9.03	5.73	6.89
W 27x114	6.88	7.72	5.39	6.23
x102	6.85	7.68	5.35	6.18
x 94	6.82	7.65	5.32	6.15
x 84	6.78	7.61	5.28	6.11

Designation	Case A	Case B	Case C	Case D
W 24x162	7.22	8.30	5.25	6.33
x146	7.17	8.24	5.20	6.27
x131	7.12	8.19	5.15	6.22
x117	7.08	8.15	5.11	6.18
x104	7.04	8.11	5.07	6.14
W 24x 94	6.16	6.92	4.81	5.56
x 84	6.12	6.87	4.77	5.52
x 76	6.09	6.84	4.74	5.49
x 68	6.06	6.80	4.70	5.45
W 24x 62	5.57	6.16	4.54	5.13
x 55	5.54	6.13	4.51	5.10
W 21x147	6.61	7.66	4.72	5.76
x132	6.57	7.61	4.68	5.71
x122	6.54	7.57	4.65	5.68
x111	6.51	7.54	4.61	5.64
x101	6.48	7.50	4.58	5.61
W 21x 93	5.54	6.24	4.31	5.01
x 83	5.50	6.20	4.27	4.96
x 73	5.47	6.16	4.23	4.92
x 68	5.45	6.14	4.21	4.90
x 62	5.42	6.11	4.19	4.87
W 21x 57	5.01	5.56	4.06	4.60
x 50	4.97	5.51	4.02	4.56
x 44	4.94	5.48	3.99	4.53
W 18x119	5.81	6.75	4.10	5.04
x106	5.77	6.70	4.06	4.99
x 97	5.74	6.67	4.03	4.96
x 86	5.70	6.62	3.99	4.91
x 76	5.67	6.59	3.95	4.87
W 18x 71	4.85	5.48	3.71	4.35
x 65	4.82	5.46	3.69	4.32
x 60	4.80	5.43	3.67	4.30
x 55	4.78	5.41	3.65	4.27
x 50	4.76	5.38	3.62	4.25

Case A: Shape perimeter, minus one flange surface.
Case B: Shape perimeter.
Case C: Box perimeter, equal to one flange surface plus twice the depth.
Case D: Box perimeter, equal to two flange surfaces plus twice the depth.

5.2.0 Standard Mill Practices (Camber)

All beams are straightened after rolling to meet sweep and camber tolerances listed hereinafter for W shapes and S shapes. The following data refers to the subsequent cold cambering of beams to produce a predetermined dimension.

The maximum lengths that can be cambered depend on the length to which a given section can be rolled, with a maximum of 100 feet. The following table outlines the maximum and minimum induced camber of W shapes and S shapes.

MAXIMUM AND MINIMUM INDUCED CAMBER

Sections Nominal Depth in.	Specified Length of Beam, ft.				
	Over 30 to 42, incl.	Over 42 to 52, incl.	Over 52 to 65, incl.	Over 65 to 85, incl.	Over 85 to 100, incl.
	Max. and Min. Camber Acceptable, in.				
W shapes, 24 and over	1 to 2, incl.	1 to 3, incl.	2 to 4, incl.	3 to 5, incl.	3 to 6, incl.
W shapes, 14 to 21, incl. and S shapes, 12 in. and over	$^3/_4$ to $2^1/_2$, incl.	1 to 3, incl.	2 to 4, incl.	$2^1/_2$ to 5, incl.	Inquire

Consult the producer for specific camber and/or lengths outside the above listed available lengths and sections.

Mill camber in beams of less depth than tabulated should not be specified.

A single minimum value for camber, within the ranges shown above for the length ordered, should be specified.

Camber is measured at the mill and will not necessarily be present in the same amount in the section of beam as received due to release of stress induced during the cambering operation. In general, 75% of the specified camber is likely to remain.

Camber will approximate a simple regular curve nearly the full length of the beam, or between any two points specified.

Camber is ordinarily specified by the ordinate at the mid-length of the portion of the beam to be curved. Ordinates at other points should not be specified.

Although mill cambering to achieve reverse or other compound curves is not considered practical, fabricating shop facilities for cambering by heat can accomplish such results as well as form regular curves in excess of the limits tabulated above. Refer to Effect of Heat on Steel, Part 6 of this Manual, for further information.

CAMBER ORDINATE TOLERANCES

Lengths	Plus Tolerance	Minus Tolerance
50 ft. and Less	$^1/_2$ inch	0
Over 50 ft.	$^1/_2$ inch plus $^1/_8$ inch for each 10 ft. or fraction thereof in excess of 50 ft.	0

5.2.1 Standard Mill Practices ("W" Shape Tolerances)

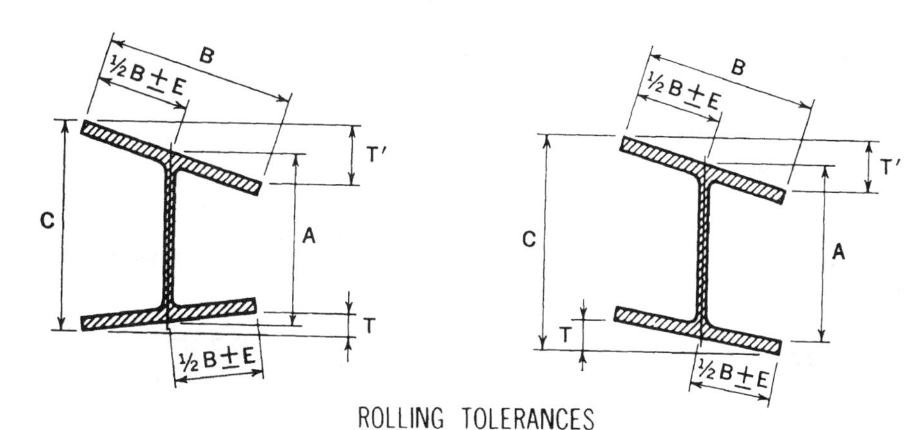

ROLLING TOLERANCES

Section Nominal Size, in.	A, Depth, in. Over Theoretical	A, Depth, in. Under Theoretical	B, Flg. Width, in. Over Theoretical	B, Flg. Width, in. Under Theoretical	T + T′, Flanges, Out of Square, max, in.	[a]E, Web off Center, max, in.	C, Max. Depth at any Cross-Section over Theoretical Depth, in.
To 12, incl.	1/8	1/8	1/4	3/16	1/4	3/16	1/4
Over 12	1/8	1/8	1/4	3/16	5/16	3/16	1/4

[a] Variation of 5/16-in. max. for sections over 426 lb./ft.

CUTTING TOLERANCES

W Shapes	Variations from Specified Length for Lengths Given, in. 30 ft. and Under: Over	30 ft. and Under: Under	Over 30 ft.: Over	Over 30 ft.: Under
Beams 24 in. and under in nominal depth	3/8	3/8	3/8 plus 1/16 for each additional 5 ft. or fraction thereof	3/8
Beams over 24 in. nom. depth; all columns	1/2	1/2	1/2 plus 1/16 for each additional 5 ft. or fraction thereof	1/2

OTHER TOLERANCES

Area and Weight Variation: ±2.5% theoretical or specified amount.

Ends Out-of-Square: 1/64 in. per in. of depth, or of flange width if it is greater than the depth.

Camber and Sweep:

Sizes	Length	Permissible Variation, in. Camber	Permissible Variation, in. Sweep
Sizes with flange width equal to or greater than 6 in.	All	$\frac{1}{8}$ in. × $\frac{\text{(total length, ft.)}}{10}$	
Sizes with flange width less than 6 in.	All	$\frac{1}{8}$ in. × $\frac{\text{(total length, ft.)}}{10}$	$\frac{1}{8}$ in. × $\frac{\text{(total length, ft.)}}{5}$
[b] Certain sections with a flange width approx. equal to depth & specified on order as columns	45 ft. and under	$\frac{1}{8}$ in. × $\frac{\text{(total length, ft.)}}{10}$ with 3/8 in. max.	
	Over 45 ft.	$\frac{3}{8}$ in. + $\left[\frac{1}{8}\text{ in.} \times \frac{\text{(total length, ft.} - 45)}{10}\right]$	

[b] Applies only to: W 8 x 31 and heavier, W 12 x 65 and heavier, W 10 x 49 and heavier, W 14 x 90 and heavier. If other sections are specified on the order as columns, the tolerance will be subject to negotiation with the manufacturer.

By permission of the American Institute of Steel Construction, Chicago, Ill.

5.3.0 Suggested Details Beam-Framing

WELDED MOMENT SPLICES

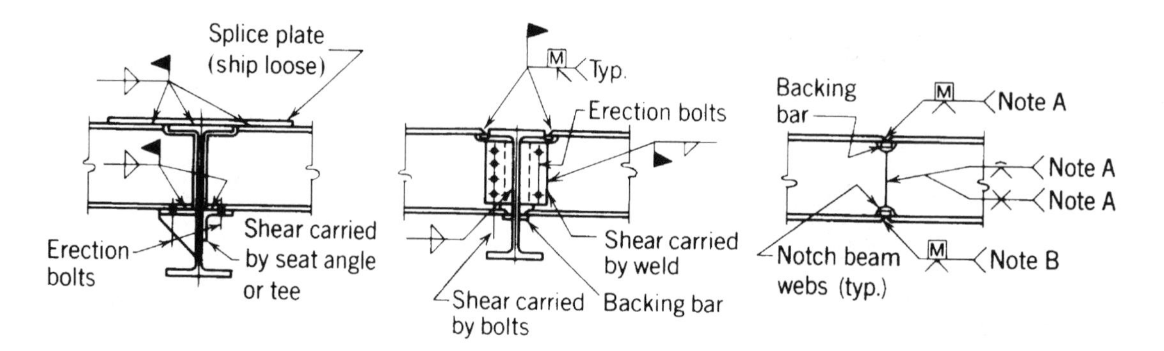

Note A: Joint preparation depends on thickness of material and welding process.
Note B: Invert this joint preparation if beam cannot be turned over.

MOMENT SPLICE AT RIDGE (FIELD BOLTED)

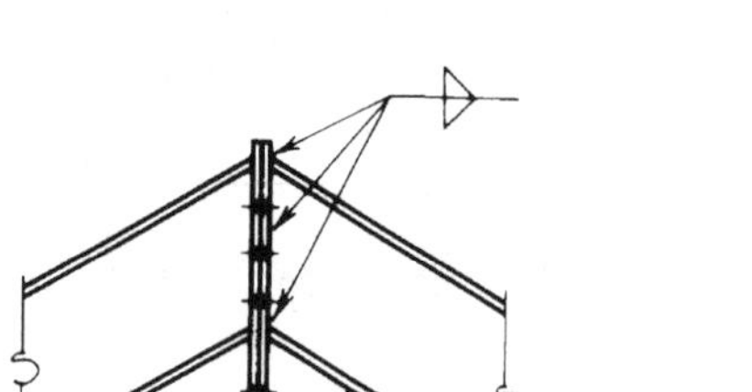

*BEAM OVER COLUMN (WITH CONTINUITY)

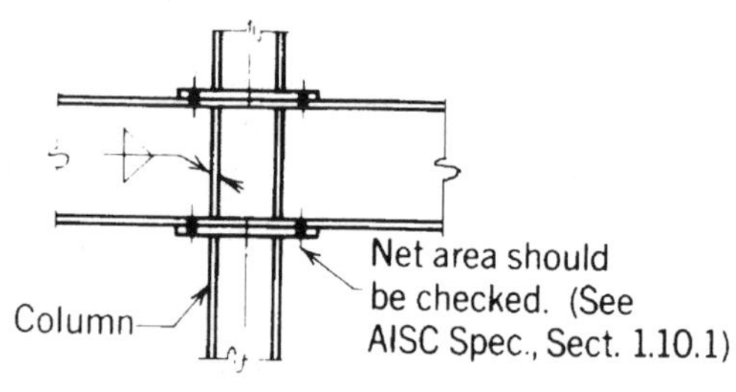

*For Plastic Design see Spec. Sect. 2.6.

5.3.1 Suggested Column Base Plate Details

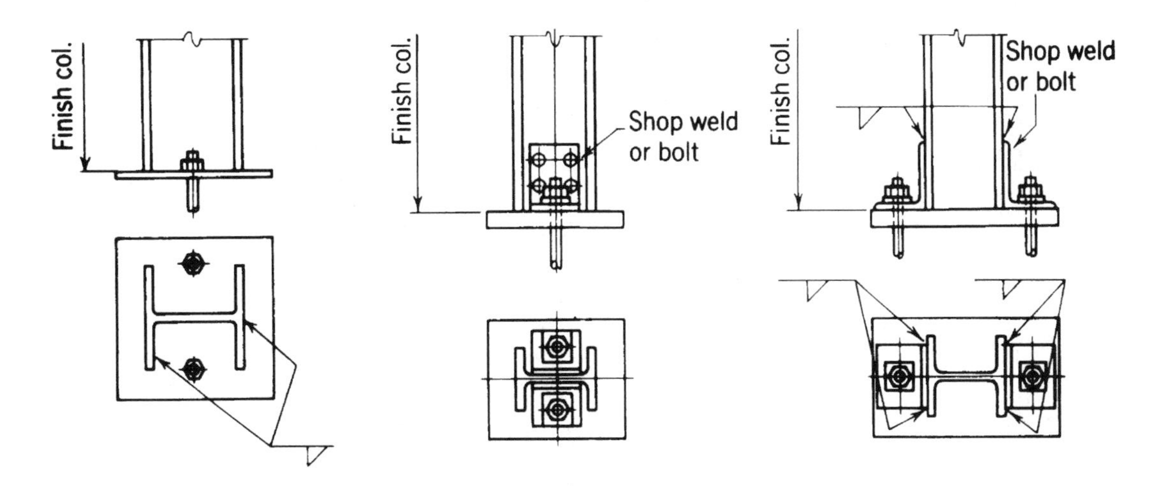

Base plate detailed and shipped loose when required.

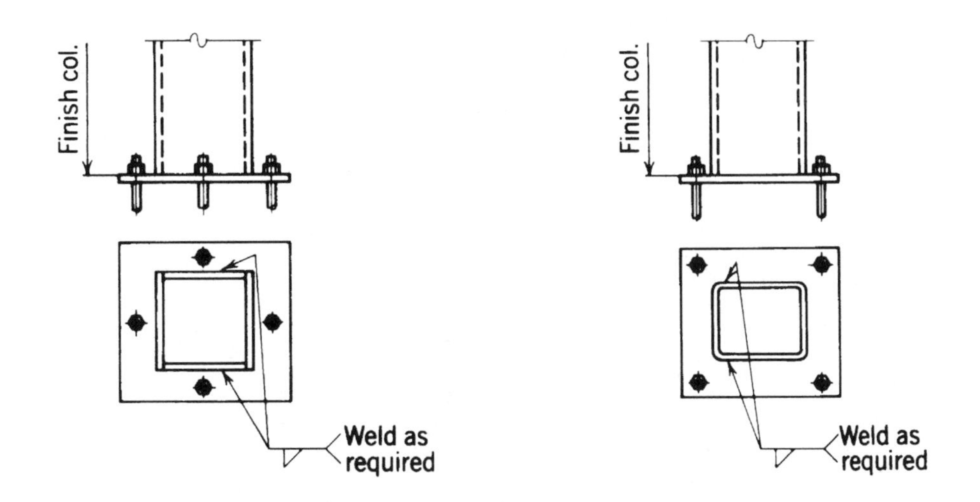

Notes: 1. Hole sizes for anchor bolts are normally made oversize to facilitate erection as follows:
Bolts ¾ to 1″○—5/16″ oversize
Bolts 1 to 2″○—½″ oversize
Bolts over 2″○—1″ oversize

2. The stability of a column with its loading should be considered at all stages of erection and its base designed accordingly for anchors and base plate.

By permission of the American Institute of Steel Construction, Chicago, Ill.

5.3.2 Suggested Structural Steel Erection Details (Miscellaneous)

SHELF ANGLES WITH ADJUSTMENT

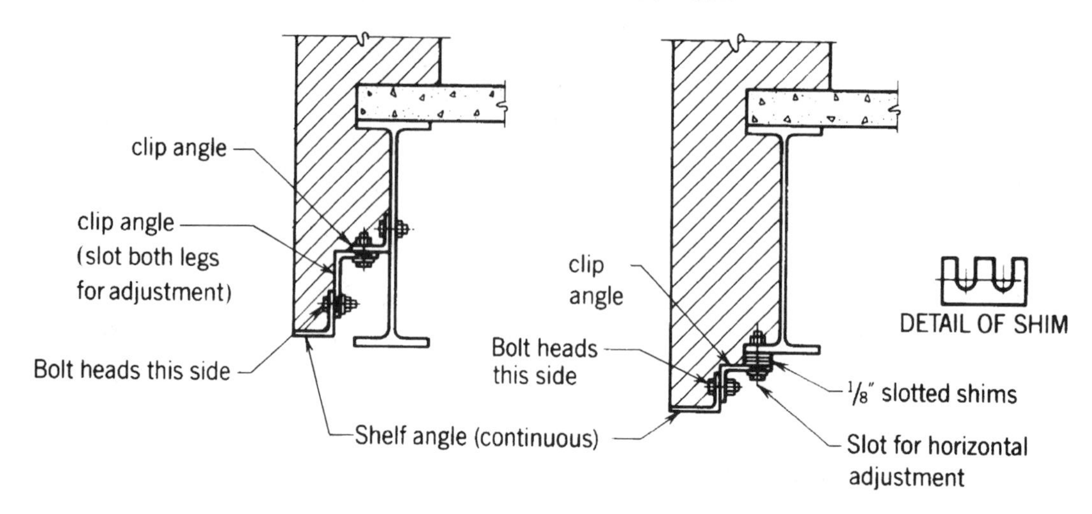

Notes: Horizontal adjustment is made by slotted holes; vertical adjustment may be made by slotted holes or by shims.
For tolerance allowance in alignment, see AISC Code of Standard Practice.

TIE RODS AND ANCHORS

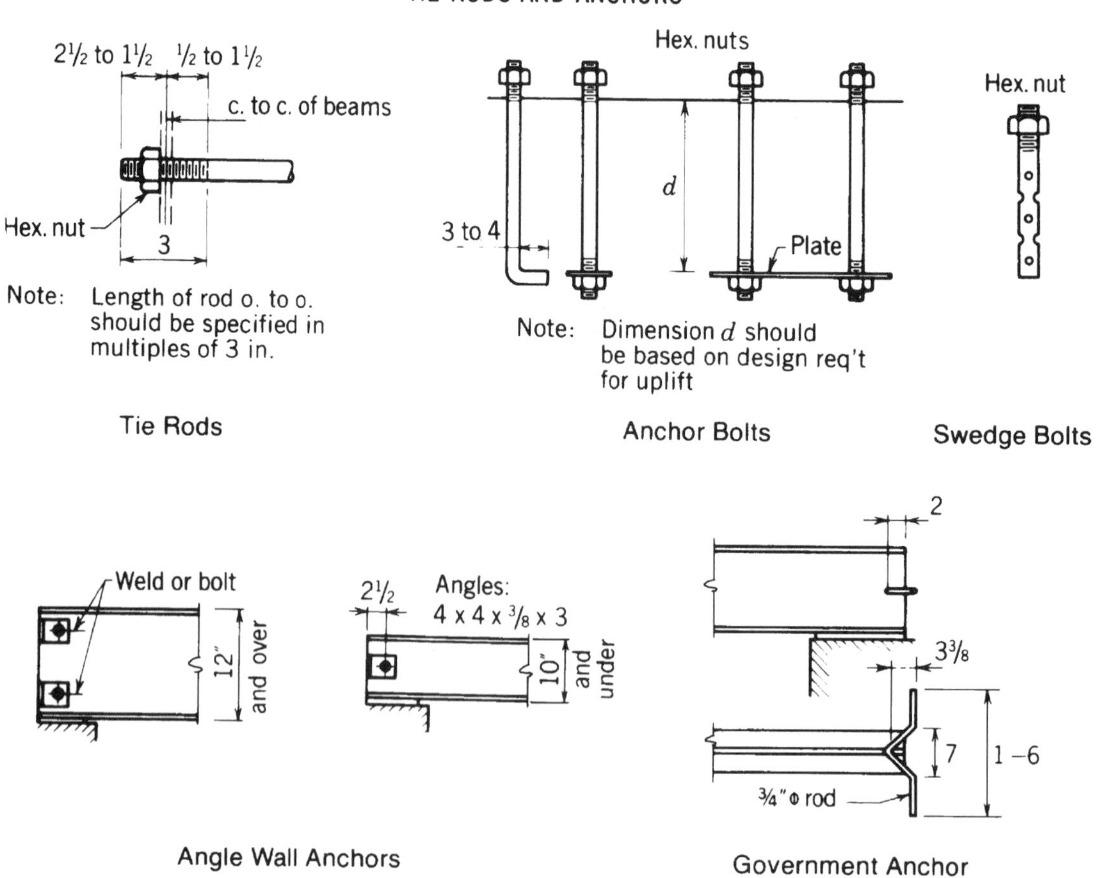

By permission of the American Institute of Steel Construction, Chicago, Ill.

5.4.0 Welded Joints (Standard Symbols)

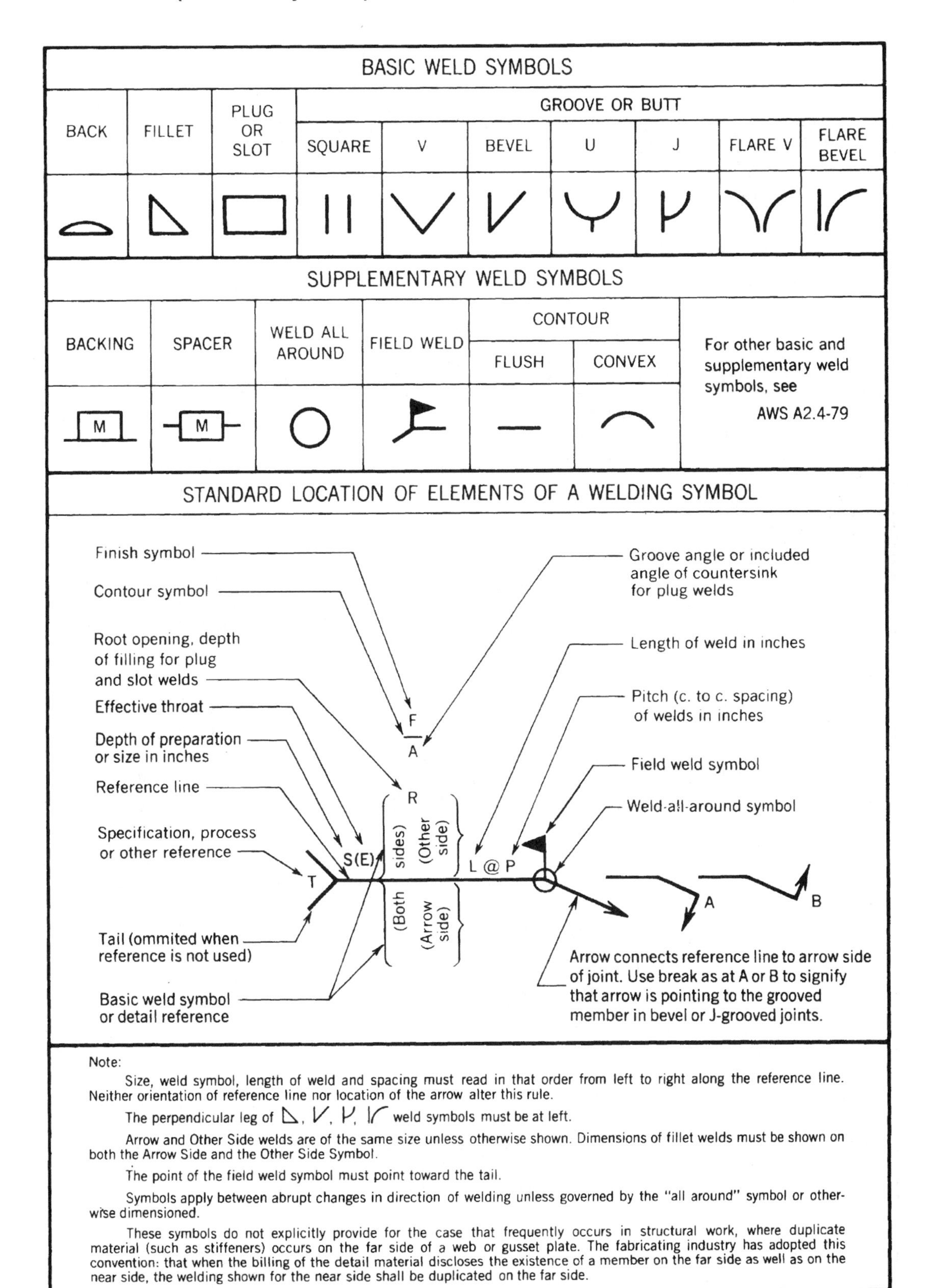

By permission of the American Institute of Steel Construction, Chicago, Ill.

5.5.0 Threaded Fasteners (Bolt Head Shapes)

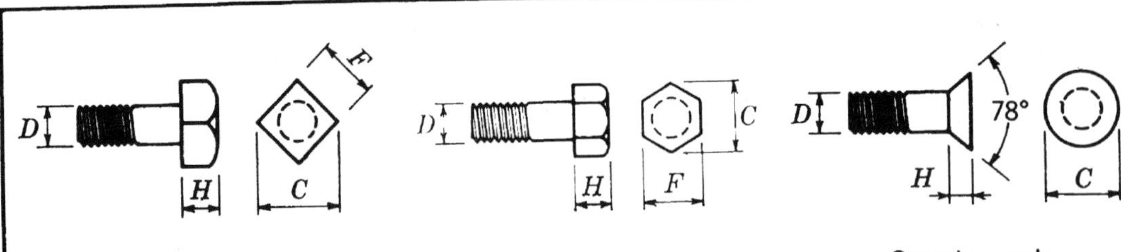

Square Hex Countersunk

Bolt head dimensions, rounded to nearest 1/16 inch, are in accordance with ANSI B18.2.1—1972 (Square and Hex) and ANSI 18.5—1971 (Countersunk)

Standard Dimensions for Bolt Heads

Diam. of Bolt D	Square			Hex			Heavy Hex			Countersunk	
	Width F	Width C	Height H	Width F	Width C	Height H	Width F	Width C	Height H	Diam. C	Height H
In.	In.	In.	In.	In.	In.	In.	In.	In.	In.	In.	In.
1/4	3/8	1/2	3/16	7/16	1/2	3/16	...	...	...	1/2	1/8
3/8	9/16	13/16	1/4	9/16	5/8	1/4	...	...	...	11/16	3/16
1/2	3/4	1 1/16	5/16	3/4	7/8	3/8	7/8	1	3/8	7/8	1/4
5/8	15/16	1 5/16	7/16	15/16	1 1/16	7/16	1 1/16	1 1/4	7/16	1 1/8	5/16
3/4	1 1/8	1 9/16	1/2	1 1/8	1 5/16	1/2	1 1/4	1 7/16	1/2	1 3/8	3/8
7/8	1 5/16	1 7/8	5/8	1 5/16	1 1/2	9/16	1 7/16	1 11/16	9/16	1 9/16	7/16
1	1 1/2	2 1/8	11/16	1 1/2	1 3/4	11/16	1 5/8	1 7/8	11/16	1 13/16	1/2
1 1/8	1 11/16	2 3/8	3/4	1 11/16	1 15/16	3/4	1 13/16	2 1/16	3/4	2 1/16	9/16
1 1/4	1 7/8	2 5/8	7/8	1 7/8	2 3/16	7/8	2	2 5/16	7/8	2 1/4	5/8
1 3/8	2 1/16	2 15/16	15/16	2 1/16	2 3/8	15/16	2 3/16	2 1/2	15/16	2 1/2	11/16
1 1/2	2 1/4	3 3/16	1	2 1/4	2 5/8	1	2 3/8	2 3/4	1	2 11/16	3/4
1 3/4	...	...	...	2 5/8	3	1 3/16	2 3/4	3 3/16	1 3/16	...	...
2	...	...	...	3	3 7/16	1 3/8	3 1/8	3 5/8	1 3/8	...	...
2 1/4	...	...	...	3 3/8	3 7/8	1 1/2	3 1/2	4 1/16	1 1/2	...	...
2 1/2	...	...	...	3 3/4	4 5/16	1 11/16	3 7/8	4 1/2	1 11/16	...	...
2 3/4	...	...	...	4 1/8	4 3/4	1 13/16	4 1/4	4 15/16	1 13/16	...	...
3	...	...	...	4 1/2	5 3/16	2	4 5/8	5 5/16	2	...	...
3 1/4	...	...	...	4 7/8	5 5/8	2 3/16	...	...	...	...	...
3 1/2	...	...	...	5 1/4	6 1/16	2 5/16	...	...	...	...	...
3 3/4	...	...	...	5 5/8	6 1/2	2 1/2	...	...	...	...	...
4	...	...	...	6	6 15/16	2 11/16	...	...	...	...	...

For dimensions for high strength bolts, refer to "Specifications for Structural Joints Using ASTM A325 or A490 Bolts" in Part 5 of this manual.
Countersunk head bolts may be ordered with slotted or socket head.

5.5.1 Threaded Fasteners (Weight of Bolts)

Length Under Head Inches	Diameter of Bolts in Inches								
	¼	⅜	½	⅝	¾	⅞	1	1⅛	1¼
1	2.38	6.11	13.0	24.1	38.9	...	...	...	...
1¼	2.71	6.71	14.0	25.8	41.5	...	...	...	...
1½	3.05	7.47	15.1	27.6	44.0	67.3	95.1	...	...
1¾	3.39	8.23	16.5	29.3	46.5	70.8	99.7	...	...
2	3.73	8.99	17.8	31.4	49.1	74.4	104	143	...
2¼	4.06	9.75	19.1	33.5	52.1	77.9	109	149	...
2½	4.40	10.5	20.5	35.6	55.1	82.0	114	155	206
2¾	4.74	11.3	21.8	37.7	58.2	86.1	119	161	213
3	5.07	12.0	23.2	39.8	61.2	90.2	124	168	221
3¼	5.41	12.8	24.5	41.9	64.2	94.4	129	174	229
3½	5.75	13.5	25.9	44.0	67.2	98.5	135	181	237
3¾	6.09	14.3	27.2	46.1	70.2	103	140	188	246
4	6.42	15.1	28.6	48.2	73.3	107	145	195	254
4¼	6.76	15.8	29.9	50.3	76.3	111	151	202	262
4½	7.10	16.6	31.3	52.3	79.3	115	156	208	271
4¾	7.43	17.3	32.6	54.4	82.3	119	162	215	279
5	7.77	18.1	33.9	56.5	85.3	123	167	222	288
5¼	8.11	18.9	35.3	58.6	88.4	127	172	229	296
5½	8.44	19.6	36.6	60.7	91.4	131	178	236	304
5¾	8.78	20.4	38.0	62.8	94.4	136	183	242	313
6	9.12	21.1	39.3	64.9	97.4	140	188	249	321
6¼	9.37	21.7	40.4	66.7	100	143	193	255	329
6½	9.71	22.5	41.8	68.7	103	147	198	262	337
6¾	10.1	23.3	43.1	70.8	106	151	204	269	345
7	10.4	24.0	44.4	72.9	109	156	209	275	354
7¼	10.7	24.8	45.8	75.0	112	160	214	282	362
7½	11.0	25.5	47.1	77.1	115	164	220	289	371
7¾	11.4	26.3	48.5	79.2	118	168	225	296	379
8	11.7	27.0	49.8	81.3	121	172	231	303	387
8½	...	28.6	52.5	85.5	127	180	241	316	404
9	...	30.1	55.2	89.7	133	189	252	330	421
9½	...	31.6	57.9	93.9	139	197	263	343	438
10	...	33.1	60.6	98.1	145	205	274	357	454
10½	...	34.6	63.3	102	151	213	284	371	471
11	...	36.2	66.0	106	157	221	295	384	488
11½	...	37.7	68.7	110	163	230	306	398	505
12	...	39.2	71.3	115	170	238	316	411	522
12½	...	...	74.0	119	176	246	327	425	538
13	...	...	76.7	123	182	254	338	439	556
13½	...	...	79.4	127	188	263	349	452	572
14	...	...	82.1	131	194	271	359	466	589
14½	...	...	84.8	135	200	279	370	479	605
15	...	...	87.5	140	206	287	381	493	622
15½	...	...	90.2	144	212	296	392	507	639
16	...	...	92.9	148	218	304	402	520	656
Per Inch Additional	1.3	3.0	5.4	8.4	12.1	16.5	21.4	27.2	33.6

Bolt is Square Bolt, ANSI B18.2.1—72 and nut is Hex Nut, ANSI B18.2.2—72. This table conforms to weight standards adopted by the Industrial Fasteners Institute.

5.5.2 Threaded Fasteners (Weight of ASTM A325 or A490 Bolts)

Heavy hex structural bolts with heavy hex nuts in pounds per 100

Length Under Head Inches	Diameter of Bolt in Inches								
	1/2	5/8	3/4	7/8	1	1 1/8	1 1/4	1 3/8	1 1/2
1	16.5	29.4	47.0	. . .	. . .	. . .	. . .	. . .	. . .
1 1/4	17.8	31.1	49.6	74.4	104	. . .	. . .	. . .	. . .
1 1/2	19.2	33.1	52.2	78.0	109	148	197	. . .	. . .
1 3/4	20.5	35.3	55.3	81.9	114	154	205	261	333
2	21.9	37.4	58.4	86.1	119	160	212	270	344
2 1/4	23.3	39.8	61.6	90.3	124	167	220	279	355
2 1/2	24.7	41.7	64.7	94.6	130	174	229	290	366
2 3/4	26.1	43.9	67.8	98.8	135	181	237	300	379
3	27.4	46.1	70.9	103	141	188	246	310	391
3 1/4	28.8	48.2	74.0	107	146	195	255	321	403
3 1/2	30.2	50.4	77.1	111	151	202	263	332	416
3 3/4	31.6	52.5	80.2	116	157	209	272	342	428
4	33.0	54.7	83.3	120	162	216	280	353	441
4 1/4	34.3	56.9	86.4	124	168	223	289	363	453
4 1/2	35.7	59.0	89.5	128	173	230	298	374	465
4 3/4	37.1	61.2	92.7	133	179	237	306	384	478
5	38.5	63.3	95.8	137	184	244	315	395	490
5 1/4	39.9	65.5	98.9	141	190	251	324	405	503
5 1/2	41.2	67.7	102	146	196	258	332	416	515
5 3/4	42.6	69.8	105	150	201	265	341	426	527
6	44.0	71.9	108	154	207	272	349	437	540
6 1/4	. . .	74.1	111	158	212	279	358	447	552
6 1/2	. . .	76.3	114	163	218	286	367	458	565
6 3/4	. . .	78.5	118	167	223	293	375	468	577
7	. . .	80.6	121	171	229	300	384	479	589
7 1/4	. . .	82.8	124	175	234	307	392	489	602
7 1/2	. . .	84.9	127	179	240	314	401	500	614
7 3/4	. . .	87.1	130	183	246	321	410	510	626
8	. . .	89.2	133	187	251	328	418	521	639
8 1/4	. . .	. . .	. . .	192	257	335	427	531	651
8 1/2	. . .	. . .	. . .	196	262	342	435	542	664
8 3/4	. . .	. . .	. . .	. . .	. . .	. . .	444	552	676
9	. . .	. . .	. . .	. . .	. . .	. . .	453	563	689
Per inch additional add	5.5	8.6	12.4	16.9	22.1	28.0	34.4	42.5	49.7
For each 100 plain round washers add	2.1	3.6	4.8	7.0	9.4	11.3	13.8	16.8	20.0
For each 100 beveled square washers add	23.1	22.4	21.0	20.2	19.2	34.0	31.6	. . .	. . .
This table conforms to weight standards adopted by the Industrial Fasteners Institute, 1965, updated for washer weights.									

By permission of the American Institute of Steel Construction, Chicago, Ill.

5.5.3 Properties of Heavy Hex Nuts and Identifying Marks

Grade	Proof Load Stress	Rockwell Hardness
A563 Grade C & C3	144,000 PSI	B78-C38
A563 Grade DH & DH3	175,000 PSI	C24-C38
A19 Grade 2H	175,000 PSI	C24-C38

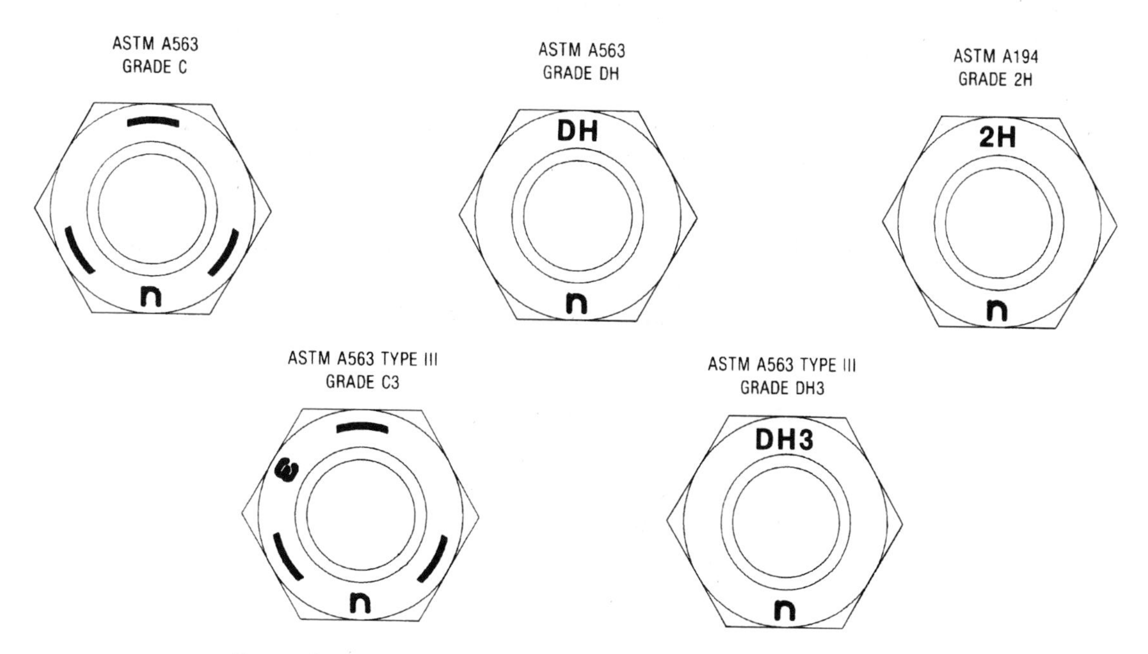

By permission of Nucor Fastener division of Nucor Corp., St. Joe, Indiana

5.5.4 Bolt Diameters and Standard Hole Dimensions

AISC/LRFD (ASTM A325-A490)				ISO/TC 167 (ASTM A325M - A490M)	
Bolt Diameter		Hole		Bolt Diameter	Hole
in	mm	in	mm	mm	mm
1/2	12.7	9/16	14.3	-	-
5/8	15.9	11/16	17.5	M16	18
3/4	19.0	13/16	20.6	-	-
-	-	-	-	M20	22
7/8	22.2	15/16	23.8	M22	24
-	-	-	-	M24	26
1	25.4	1 1/16	27.0	-	-
1 1/8	28.6	1 3/16	30.2	M27	30
1 1/4	31.8	1 5/16	33.3	M30	33
1 3/8	34.9	1 7/16	36.5	-	-
-	-	-	-	M36	39
1 1/2	38.1	1 9/16	39.7	-	-

Standard Hole Diameters

Metric Bolt mm	U.S. Substitution inch
M16	5/8
M22	7/8
M27	1 1/8
M30	1 1/4

Suggested Permissible Bolt Substitutions

By permission of Nucor Fastener division of Nucor Corp., St. Joe, Indiana

5.5.5 Capscrews/Bolts/Heavy Hex Nut Identifying Marks

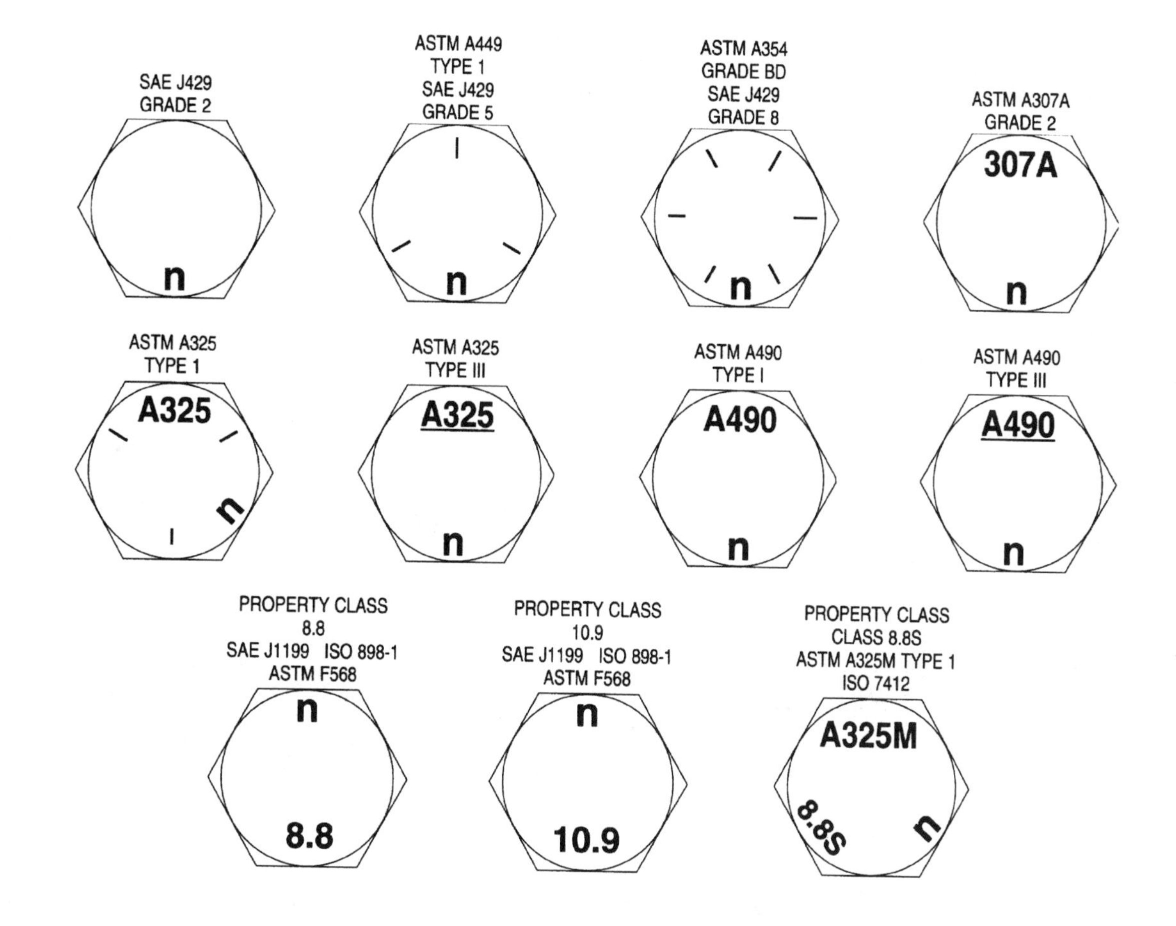

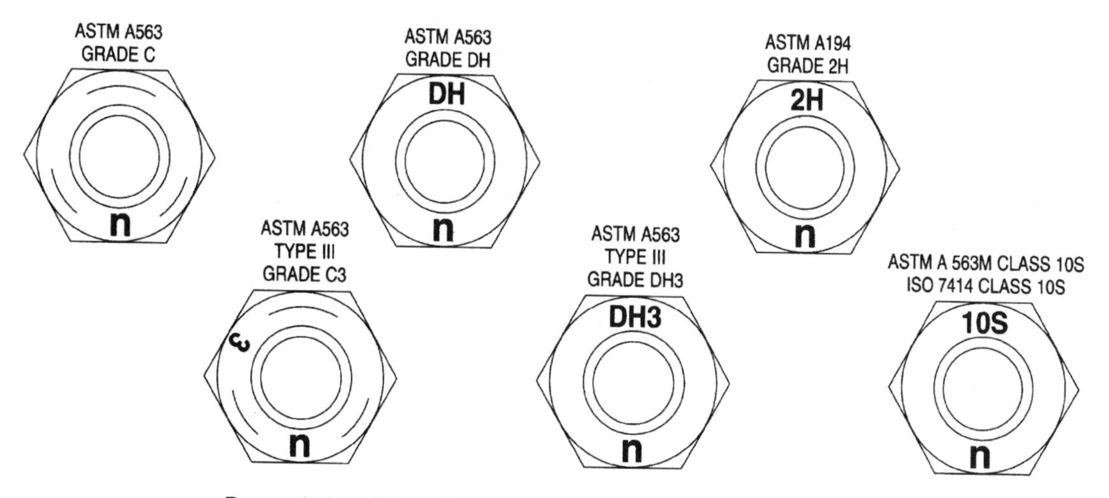

By permission of Nucor Fastener division of Nucor Corp., St. Joe, Indiana

5.5.6 Dimensions of Finished Hex Nuts

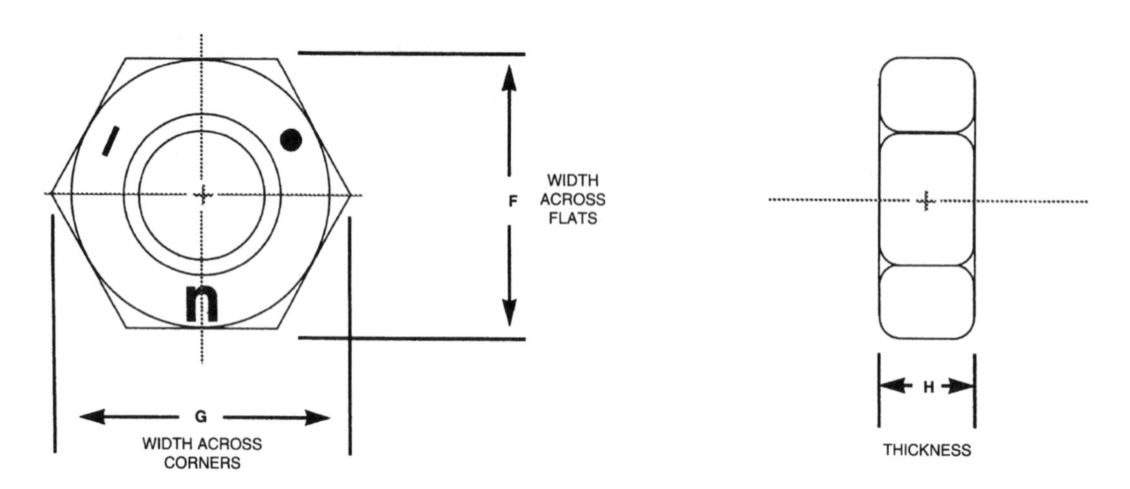

Table 2
DIMENSIONS OF FINISHED HEX NUTS

Nominal Size or Basic Major Diam. of Thread		Width Across Flats			Width Across Corners		Thickness Hex Nuts		
		Basic	Max.	Min.	Max.	Min.	Basic	Max.	Min.
1/4	0.2500	7/16	0.438	0.428	0.505	0.488	7/32	0.226	0.212
5/16	0.3125	1/2	0.500	0.489	0.577	0.557	17/64	0.273	0.258
3/8	0.3750	9/16	0.562	0.551	0.650	0.628	21/64	0.337	0.320
7/16	0.4375	11/16	0.688	0.675	0.794	0.768	3/8	0.385	0.365
1/2	0.5000	3/4	0.750	0.736	0.866	0.840	7/16	0.448	0.427
9/16	0.5625	7/8	0.875	0.861	1.010	0.982	31/64	0.496	0.427
5/8	0.6250	15/16	0.938	0.922	1.083	1.051	35/64	0.559	0.535
3/4	0.7500	1-1/8	1.125	1.088	1.299	1.240	41/64	0.665	0.617
7/8	0.8750	1-5/16	1.312	1.269	1.516	1.447	3/4	0.776	0.724
1	1.0000	1-1/2	1.500	1.450	1.732	1.653	55/64	0.887	0.831
1-1/8	1.1250	1-11/16	1.688	1.631	1.949	1.859	31/32	0.999	0.939
1-1/4	1.2500	1-7/8	1.875	1.812	2.165	2.066	1-1/16	1.094	1.030
1-3/8	1.3750	2-1/16	2.062	1.994	2.382	2.273	1-11/64	1.206	1.136
1-1/2	1.5000	2-1/4	2.250	2.175	2.598	2.480	1-9/32	1.317	1.245

(ANSI B18.2.2-1987)

By permission of Nucor Fastener division of Nucor Corp., St. Joe, Indiana

5.5.7 Dimensions of Finished Hex Bolts

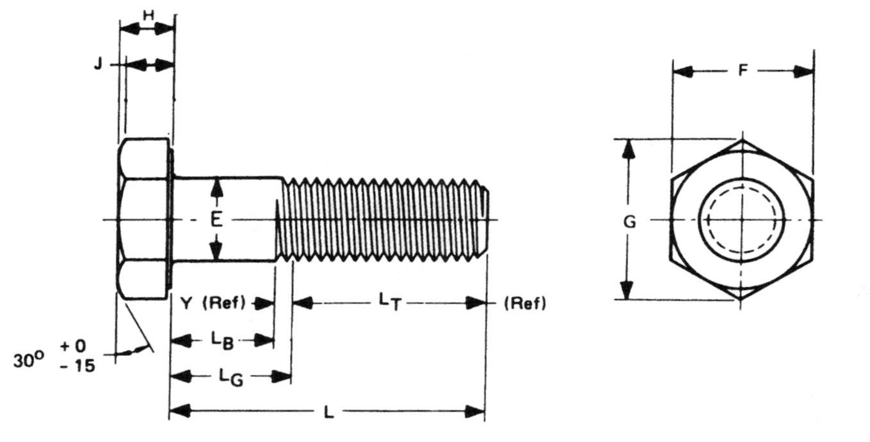

Nominal Size Or Basic Product Diameter	E Body Diameter		F Width Across Flats			G Width Across Corners		H Height			J Wrenching Height	L_T Thread Length For Screw Lengths 6 in. and Shorter	L_T Thread Length For Screw Lengths Over 6 in.	Y Transition Thread Length	Runout of Bearing Surface FIM
	Max	Min	Basic	Max	Min	Max	Min	Basic	Max	Min	Min	Basic	Basic	Max	Max
1/4 0.2500	0.2500	0.2450	7/16	0.438	0.428	0.505	0.488	5/32	0.163	0.150	0.106	0.750	1.000	0.250	0.010
5/16 0.3125	0.3125	0.3065	1/2	0.500	0.489	0.577	0.557	13/64	0.211	0.195	0.140	0.875	1.125	0.278	0.011
3/8 0.3750	0.3750	0.3690	9/16	0.562	0.551	0.650	0.628	15/64	0.243	0.226	0.160	1.000	1.250	0.312	0.012
7/16 0.4375	0.4375	0.4305	5/8	0.625	0.612	0.722	0.698	9/32	0.291	0.272	0.195	1.125	1.375	0.357	0.013
1/2 0.5000	0.5000	0.4930	3/4	0.750	0.736	0.866	0.840	5/16	0.323	0.302	0.215	1.250	1.500	0.385	0.014
9/16 0.5625	0.5625	0.5545	13/16	0.812	0.798	0.938	0.910	23/64	0.371	0.348	0.250	1.375	1.625	0.417	0.015
5/8 0.6250	0.6250	0.6170	15/16	0.938	0.922	1.083	1.051	25/64	0.403	0.378	0.269	1.500	1.750	0.455	0.017
3/4 0.7500	0.7500	0.7410	1 1/8	1.125	1.100	1.299	1.254	15/32	0.483	0.455	0.324	1.750	2.000	0.500	0.020
7/8 0.8750	0.8750	0.8660	1 5/16	1.312	1.285	1.516	1.465	35/64	0.563	0.531	0.378	2.000	2.250	0.556	0.023
1 1.0000	1.0000	0.9900	1 1/2	1.500	1.469	1.732	1.675	39/64	0.627	0.591	0.416	2.250	2.500	0.625	0.026
1 1/8 1.1250	1.1250	1.1140	1 11/16	1.688	1.631	1.949	1.859	11/16	0.718	0.658	0.461	2.500	2.750	0.714	0.029
1 1/4 1.2500	1.2500	1.2390	1 7/8	1.875	1.812	2.165	2.066	25/32	0.813	0.749	0.530	2.750	3.000	0.714	0.033
1 3/8 1.3750	1.3750	1.3630	2 1/16	2.062	1.994	2.382	2.273	27/32	0.878	0.810	0.569	3.000	3.250	0.833	0.036
1 1/2 1.5000	1.5000	1.4880	2 1/4	2.230	2.175	2.598	2.480	1 5/16	0.974	0.902	0.640	3.250	3.500	0.833	0.039

By permission of Nucor Fastener division of Nucor Corp., St. Joe, Indiana

5.5.8 Tension Cntrol (TC) Bolt Installation Procedures

Tru-Tension™ Fasteners are designed to be installed with various types of lightweight portable electric wrenches specifically intended for use with this style of structural fastener. They can be utilized for any applications where A325 and A490 bolts are specified. The installation tool has an inner socket, which engages the spline tip of the bolt spline, and when the tension is sufficient in the fastener, the spline tip simply twists off, leaving the tightened bolt correctly installed in the connection.

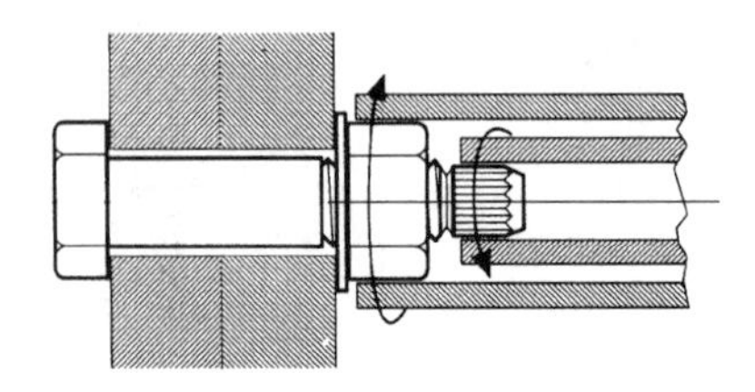

INSTALLATION PROCEDURES

1

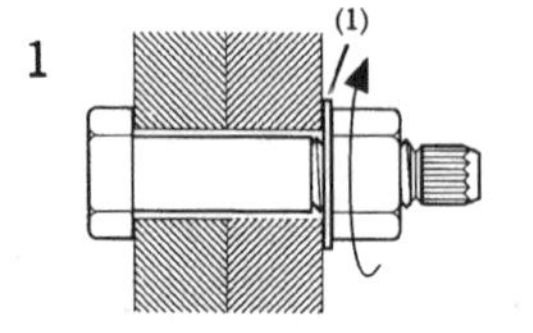

Place the bolt into the connection with the washer (1) under the nut. Finger tighten the nut.

2

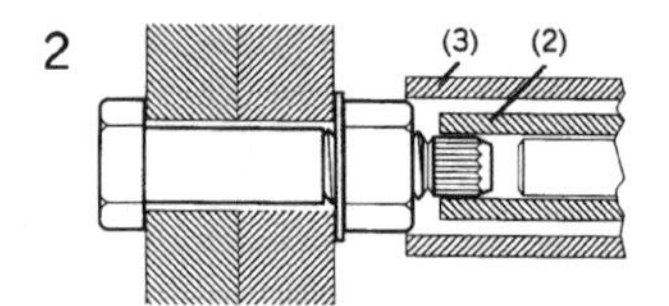

Fit inner socket (2) over the grooved spline and push the wrench slightly then engage the outer socket (3) over the nut.

3

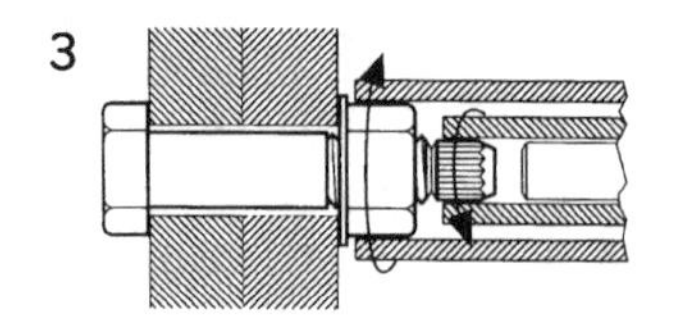

Start the wrench. The outer socket rotates the nut relative to the bolt during tightening, and the bolt will be tightened until the required bolt tension is reached. At this point the splined tip shears off.

4

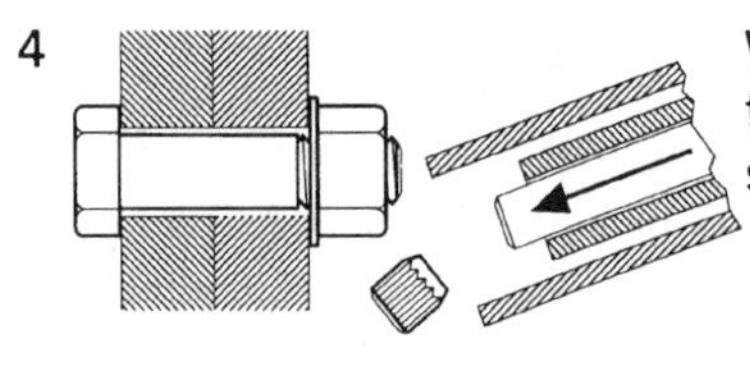

When the installation is complete remove the socket from the nut and depress the ejection lever to discharge the sheared spline from the inner socket of the wrench.

Note: Particularly when installing multiple rows of bolts or where uneven steel contact is encountered, the fasteners should be preloaded to snug tight conditions prior to final tightening. This method will prevent interactions between bolts as additional bolts are tightened. As always, fasteners should be tightened in sequence from the most rigid section out. As with all high-strength structural fasteners, Tru-Tension™ fasteners should be stored in their sealed metal kegs until ready for use. Opened cans should be stored indoors protected from the elements to prevent environmental contamination (rain, dirt, etc.).

By permission of Nucor Fastener division of Nucor Corp., St. Joe, Indiana

5.5.9 Tru-Tension (TC) Bolt-Assembly Specifications

DETERMINATION OF TRU-TENSION™ LENGTH

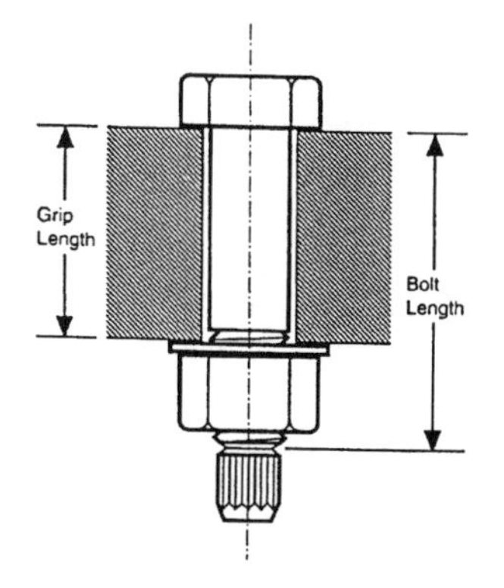

Bolt Size In.	To Determine Required Bolt Length Add to Grip, In.
5/8	7/8
3/4	1
7/8	1 1/8
1	1 1/4
1 1/8	1 1/2

STRUCTURAL FASTENER TENSION

Fastener Test Tension Required for Slip-critical Connections and Connections Subject to Direct Tension

Nominal Bolt Size, Inches	Minimum Tension[2] in 1000's of Pounds (kips)	
	A325 Bolts	A490 Bolts
5/8	20.0	25.2
3/4	29.4	36.8
7/8	41.0	51.5
1	53.6	67.2
1 1/8	58.8	84.0

[2]Equal to 70 percent of specified minimum tensile strengths of bolts (as specified in ASTM Specifications for test of full size A325 and A490 bolts with UNC threads loaded in axial tension) rounded to nearest 100 lbs. (Includes 5% per AISC spec.)

TRU-TENSION™ ASSEMBLY WEIGHTS

A325 and A490 ASSEMBLIES

(Assembly: Bolt 1, Nut 1, and Washer 1)

Nominal bolt size	5/8″			3/4″			7/8″			1″		
Length (Inches)	Net Weight Per 100 Pieces (lb.)	Container Quantity (pcs.)	Net Container Weight (lb.)	Net Weight Per 100 Pieces (lb.)	Container Quantity (pcs.)	Net Container Weight (lb.)	Net Weight Per 100 Pieces (lb.)	Container Quantity (pcs.)	Net Container Weight (lb.)	Net Weight Per 100 Pieces (lb.)	Container Quantity (pcs.)	Net Container Weight (lb.)
1½	39.3	500	197	61.7	320	197						
1¾	41.1	450	185	64.8	300	194	93.8	210	197			
2	43.2	420	182	67.9	280	190	98.0	200	196	133.0	140	186
2¼	45.3	400	182	71.1	270	192	102.3	190	194	138.6	140	194
2½	47.4	380	180	74.2	250	186	106.6	180	192	166.4	110	183
2¾	49.5	360	178	77.3	250	193	110.8	180	199	149.7	130	195
3	51.6	320	165	80.5	240	193	115.1	170	196	155.3	120	186
3¼	53.6	300	161	83.6	230	192	119.3	160	203	160.9	120	193
3½	55.7	300	167	86.7	220	191	123.6	150	185	166.4	110	183
3¾	57.8	290	168	89.9	200	180	127.9	140	179	172.0	110	189
4	59.9	280	168	93.0	190	177	132.1	140	185	177.6	100	178
4¼				96.1	180	173	136.3	130	177	183.2	100	183
4½	64.1	270	173	99.3	180	179	140.7	120	169	188.7	100	189
4¾				102.4	170	174	144.9	120	174	194.3	90	175
5	68.3	250	171	105.5	160	169	149.2	110	165	199.9	90	180
5¼				108.7	140	153	153.4	110	169	205.4	90	185
5½				111.8	130	145	157.7	100	158	211.0	80	169
5¾				114.9	130	150	162.0	100	162	216.6	80	173
6				118.1	120	142	166.2	90	150	222.1	80	178

By permission of Nucor Fastener division of Nucor Corp., St. Joe, Indiana

5.6.0 Major Characteristics of Joist Series**

MAJOR CHARACTERISTICS OF JOIST SERIES **

K Series
Min. Fy=50000 psi
Depths 8" thru 30"
Spans to 60'-0

CS Series
Min. Fy=50000 psi
Depths 10" thru 30"
Spans 20'-0 thru 60'-0

LH Series
Min. Fy=50000 psi
Depths 18" thru 48"
Spans to 96'-0

DLH Series
Min. Fy=50000 psi
Depths 52" thru 72"
Spans to 144'-0

SLH Series
Min. Fy=50000 psi
Depths from 80"
Spans - Contact Vulcraft

JOIST GIRDER Series
Min. Fy=50000 psi
Depths as required
Spans as required

** Some design and/or delivery requirements may dictate yield strength other than that shown below.

By permission of Nucor Research and Development, Norfolk, Nebraska

5.6.1 General Information on K Series Joists

- Economical
- High strength
- *Design* Vulcraft K Series open web steel joists are designed in accordance with specifications of the Steel Joist Institute.
- SJI spans to 60' 0"
- *Paint* Vulcraft joists receive a shop-coat of rust-inhibitive primer, whose performance characteristics conform to those of the Steel Joist Institute specifications 3.3

Standing Beam Bridging

The bridging table was developed to support the top chords against lateral movement during the construction period. It is then intended that the floor or roof deck will laterally support the top chords under a full loading condition by meeting the provisions of Section 5.8 of the specifications. Most standing-seam roof systems will not adequately brace the top chords laterally with the number of rows as required by the bridging table. We, therefore, recommend that when standing-seam roof systems are specified, the specifying engineer employ a note to have the joist manufacturer to check the system and to provide bridging as required to adequately brace the top cords against lateral movement under a full-loading condition.

Uplift Bridging

Where uplift forces caused by wind are a design requirement, these forces must be indicated on the structural drawings in terms of net uplift in pounds per square foot or pounds per lineal foot. When these loads are specified, they must be considered in the design of joists and bridging. A single line of bottom chord bridging must be provided near the first bottom cord panel points whenever uplift from wind load is a design consideration.

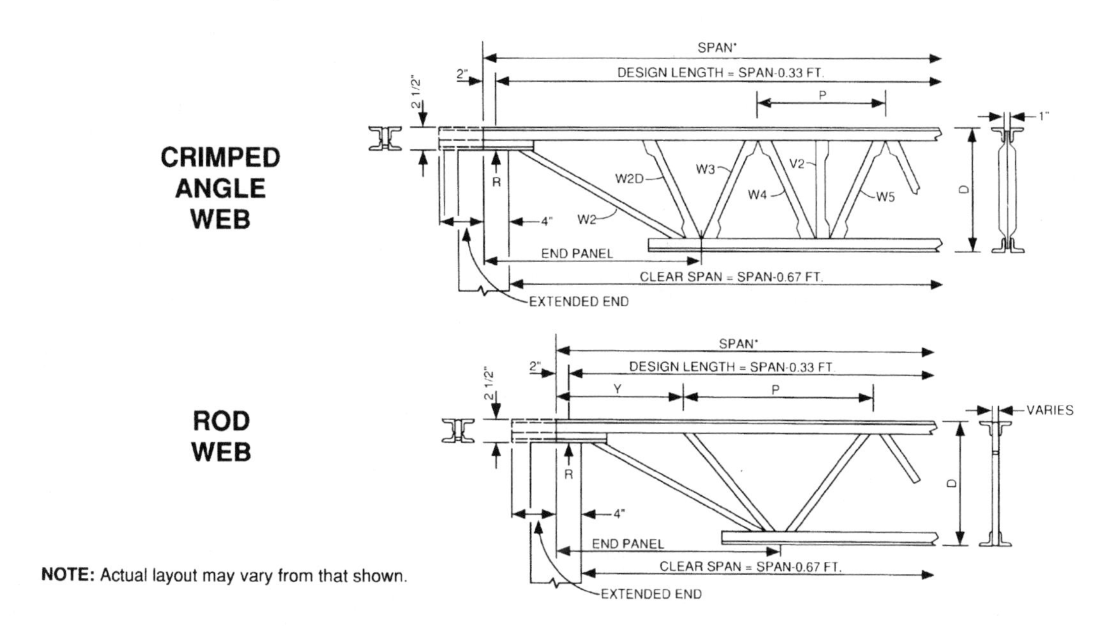

NOTE: Actual layout may vary from that shown.

Number of Rows of Bridging***
Distances are Span Lengths

Section Number	1 Row	2 Rows	3 Rows	4 Rows**	5 Rows***
#1	Up thru 16'	Over 16' thru 24'	Over 24' thru 28'		
#2	Up thru 17'	Over 17' thru 25'	Over 25' thru 32'		
#3	Up thru 18'	Over 18' thru 28'	Over 28' thru 38'	Over 38' thru 40'	
#4	Up thru 19'	Over 19' thru28'	Over 28' thru 38'	Over 38' thru 48'	
#5	Up thru 19'	Over 19' thru 29'	Over 29' thru 39'	Over 39' thru 50'	Over 50' thru 52'
#6	Up thru 19'	Over 19' thru 29'	Over 29' thru 39'	Over 39' thru 51'	Over 51' thru 56'
#7	Up thru 20'	Over 20' thru 33'	Over 33' thru 45'	Over 45' thru 58'	Over 58' thru 60'
#8	Up thru 20'	Over 20' thru 33'	Over 33' thru 45'	Over 45 thru 58'	Over 58' thru 60'
#9	Up thru 20'	Over 20' thru 33'	Over 33' thru 46'	Over 46' thru 59'	Over 59' thru 60'
#10	Up thru 20	Over 20' thru 37'	Over 37' thru 51'	Over 51' thru 60'	
#11	Up thru 20'	Over 20' thru 38'	Over 38' thru 53'	Over 53' thru 60'	
#12	Up thru 20'	Over 20' thru 39'	Over 39' thru 53'	Over 53' thru 60'	

*Last digit(s) of joist designation shown in Load Tables.
**Where 4 or 5 rows of bridging are required, a row nearest the midspan of the joist shall be diagonal bridging with bolted connections at chords and intersection.
***See Section 5.11 of the specifications.

Sizes of Horizontal Bridging

Size	Maximum Joist Spacing
L 1 x 1 x 7/64	5'-0"
L 1 1/4 x 1 1/4 x 7/64	6'-3"
L 1 1/2 x 1 1/2 x 7/64	7'-6"
L 1 3/4 x 1 3/4 x 1/8	8'-9"
L 2 x 2 x 1/8	10'-0"

By permission of Nucor Research and Development, Norfolk, Nebraska

5.6.2 Standard Specifications for Open Web Joists (K Series)

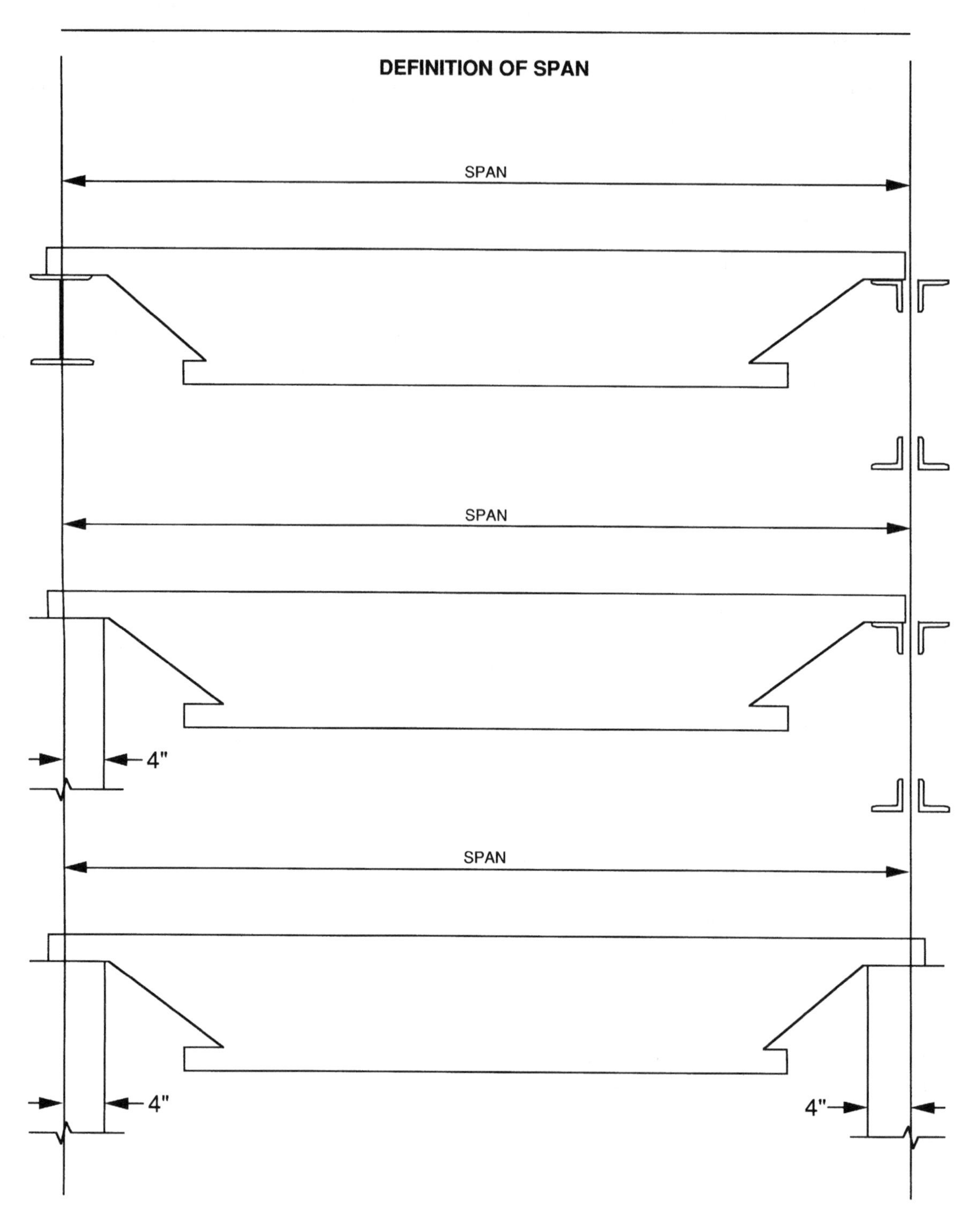

[DESIGN LENGTH = SPAN – 0.33 FT.]

By permission of the Steel Joist Institute, Myrtle Beach, South Carolina

5.6.3 K Series Open Web Steel Joists

Top Chord Extensions and Extended Ends

Joist extensions are commonly furnished to support a variety of overhang conditions. The two types are pictured. The first is the top chord extension or "S" type, which has only the top chord angles extended. The second is the extended end or "R" type in which the standard 2½" end-bearing depth is maintained over the entire length of the extension. The "S" type extension is so designated because of its simple nature whereas the "R" type involves reinforcing the top chord angles. The specifying authority should be aware that an "S" type is more economical and should be specified whenever possible.

The following load tables for K-series top chord extensions and extended ends have been developed as an aid to the specifying authority. The black number in the tables is the maximum allowable uniform load in pounds per linear foot. The blue number is the uniform load, which will produce an approximate deflection of $L_1/240$, where L_1 is the length of the extension. The load tables are applicable for uniform loads only. If there are concentrated loads/and or non-uniform loads, a loading diagram must be provided by the specifying authority on the contract drawings. In cases where it is not possible to meet specific job requirements with a 2½" deep "R" type extension (refer to "S" and "I" values in the Extended End Load Table), the depth of the extension must be increased to provide greater load-carrying capacity. If the loading diagram for any condition is not shown, the joist manufacturer will design the extension to support the uniform load indicated in the K-Series Joist Load Table for the span of the joist.

When top chord extensions or extended ends are specified, the allowable deflection and the bracing requirements must be considered by the specifying authority.

Note that an "R" type extension must be specified when building details dictate a 2½" depth at the end of the extension. In the absence of specific instructions, the joist manufacturer could provide either type.

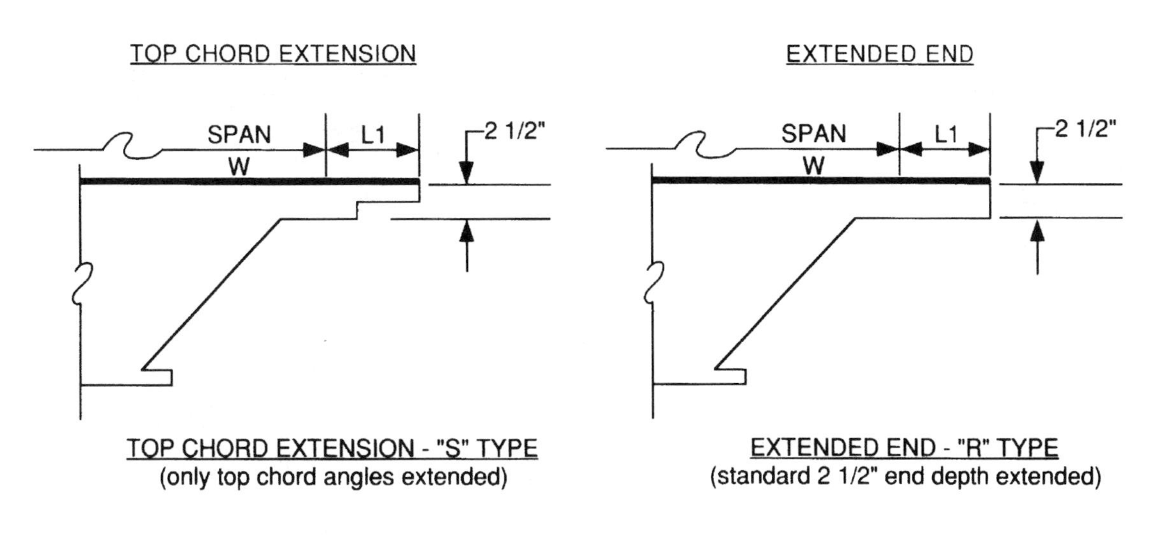

TOP CHORD EXTENSION - "S" TYPE
(only top chord angles extended)

EXTENDED END - "R" TYPE
(standard 2 1/2" end depth extended)

W = Uniform Load
LI = Length of Extension
SPAN = See Page 20 for Definition of SPAN

By permission of the Steel Joist Institute, Myrtle Beach, South Carolina

5.6.4 General Information (LH and DLH Series Joists)

- High strength
- Economical
- *Design* Vulcraft LH and DLH series long-span steel joists are designed in accordance with the specifications of the Steel Joist Institute.
- Roof spans to 144'
- Floor spans to 120'
- *Paint* Vulcraft joists receive a shop-coat of rust inhibitive primer whose performance characteristics conform to those of the Steel Joist Institute specification 102.4.

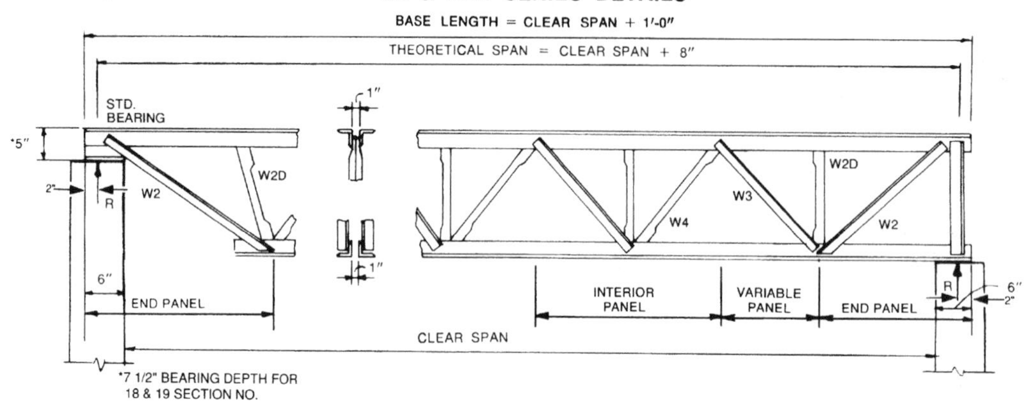

LH & DLH TABLE
MINIMUM BEARING LENGTHS

Joist Type	On Masonry	On Concrete	On Steel
LH 02 thru 17 DLH 10 thru 19	6"*	6"*	4"
MINIMUM BEARING PLATE WIDTHS			
LH 02 thru LH 12 DLH 10 thru DLH 12	9"*	9"*	
LH 13 thru LH 17 DLH 13 thru DLH 19	12"*	12"*	

*See Sect. 104.4 on page 43.

BRIDGING SPACING

Section No.*	Min. Bolt Diameter**	Maximum Spacing of Lines of Bridging
LH 02 to 09, incl.	3/8''	11'-0"
DLH 10	3/8''	14'-0"
LH 10 to 14, incl.	3/8''	16'-0"
DLH 11 to 14, incl.	3/8''	16'-0"
LH 15 to 17, incl.	1/2''	21'-0"
DLH 15 to 17, incl.	1/2''	21'-0"
DLH 18 to 19, incl.	5/8''	26'-0"

*Last two digits of joint designation shown in load table.
**Size required due to requirements as indicated for bolted x bridging connections in Section 104.5(e). Minimum A307 Bolt required for connection.

JOIST SPACING FOR BRIDGING ANGLE SIZE

DIAGONAL BRIDGING CHART
Bridging Angle Size

DEPTH	L1x1x7/64	L1¼x1¼x7/64	L1½x1½x7/64	L1¾x1¾x1/8	L2x2x1/8
18	6'- 5"	8'- 2"	9'-10"	11'- 6"	
20	6'- 5"	8'- 1"	9'-10"	11'- 6"	
24	6'- 4"	8'- 1"	9'- 9"	11'- 5"	
28	6'- 2"	8'- 0"	9'- 8"	11'- 5"	
32	6'- 1"	7'-10"	9'- 7"	11'- 4"	13'- 0"
36		7'- 9"	9'- 6"	11'- 3"	12'-11"
40		7'- 7"	9'- 5"	11'- 2"	12'-10"
44		7'- 5"	9'- 3"	11'- 0"	12'- 9"
48		7'- 3"	9'- 1"	10'-11"	12'- 8"
52			9'- 0"	10'- 9"	12'- 7"
56			8'-10"	10'- 8"	12'- 5"
60			8'- 7"	10'- 6"	12'- 4"
64			8'- 5"	10'- 4"	12'- 2"
68			8'- 2"	10'- 2"	12'- 0"
72			8'- 0"	10'- 0"	11'-10"

†HORIZONTAL BRIDGING CHART
Bridging Angle Size

DEPTH	L1x1x7/64	L1¼x1¼x7/64	L1½x1½x7/64	L1¾x1¾x1/8	L2x2x1/8
ALL DEPTHS	5'- 0"	6'- 3"	7'- 6"	8'- 9"	10'- 0"

† See specification section 104.5 for the proper use of horizontal bridging.

NOTES:
1. Special designed LH and DLH can be supplied in longer lengths. See SLH Series Page 47.
2. Additional bridging may be required when joists support standing seam roof decks. The specifying engineer should require that the joist manufacturer check the system and provide bridging as required to adequately brace the joists against lateral movement. For bridging requirements due to uplift pressures refer to sect. 104.12.

By permission of Nucor Research and Development, Norfolk, Nebraska

5.6.5 LH and DLH Series Longspan Steel Joists

Standard Types

Longspan steel joists can be furnished with either underslung or square ends, with parallel chords, or with single- or double-pitched top chords to provide sufficient slope for roof drainage.

The Longspan joist designation is determined by its nominal depth at the center of the span, except for offset double-pitched joists, where the depth should be given at the ridge. A part of the designation should be either the section number or the total design load over the design live load (TL/LL given in plf). All pitched joists will be cambered in addition to the pitch.

Nonstandard Types

The following joists can also be suppled by Vulcraft, however, the district sales office or manufacturing facility nearest you should be contacted for any limitations in depth or length that they might have.

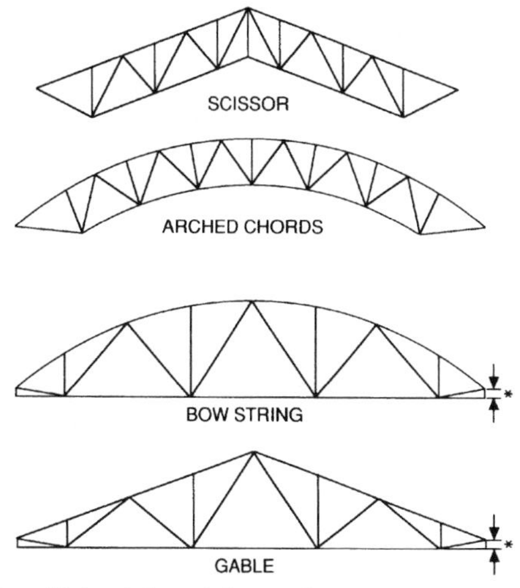

*Contact Vulcraft for minimum depth at ends.

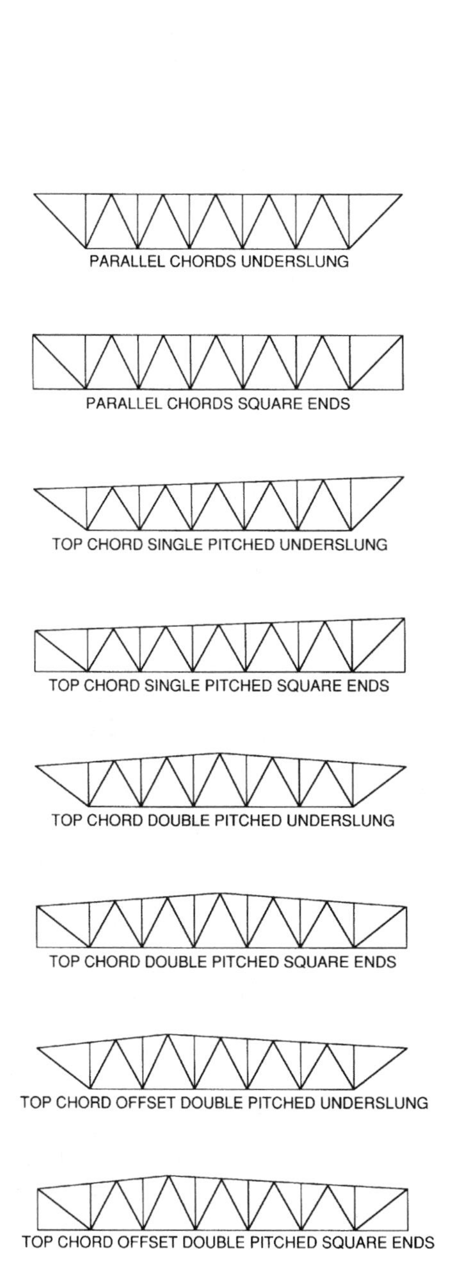

CAMBER FOR STANDARD TYPES

LH & DLH series joists shall have camber in accordance with the following table:**

Top Chord Length	Approx. Camber
20'	1/4"
30'	3/8"
40'	5/8"
50'	1"
60'	1 1/2"
70'	2"
80'	2 3/4"
90'	3 1/2"
100'	4 1/4"
110'	5"
120'	6"
130'	7"
140'	8"
144'	8 1/2"

** NOTE: If full camber is not desired near walls or other structural members please note on the structural drawings.

By permission of Nucor Research and Development, Norfolk, Nebraska

5.6.5 LH and DLH Series Longspan Steel Joists (Continued)

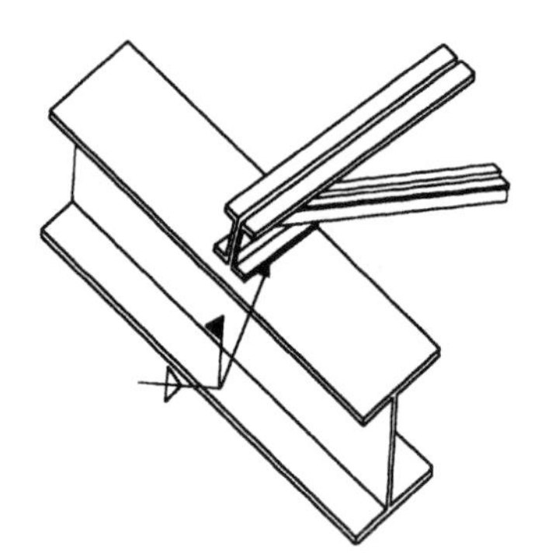

ANCHORAGE TO STEEL
SEE SJI SPECIFICATION
104.4 (b) AND 104.7 (b).

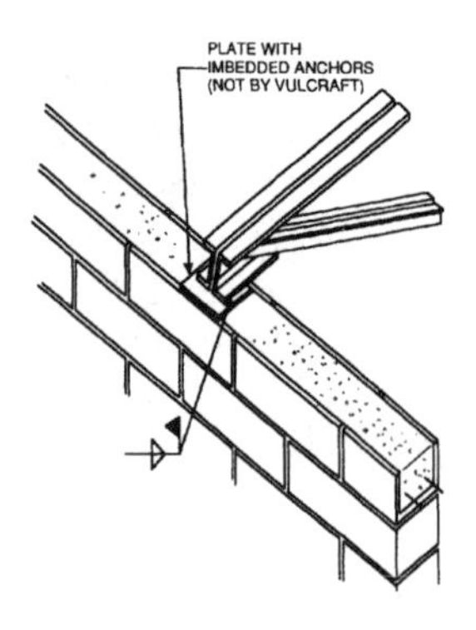

ANCHORAGE TO MASONRY
SEE SJI SPECIFICATION
104.4 (a) AND 104.7 (a).

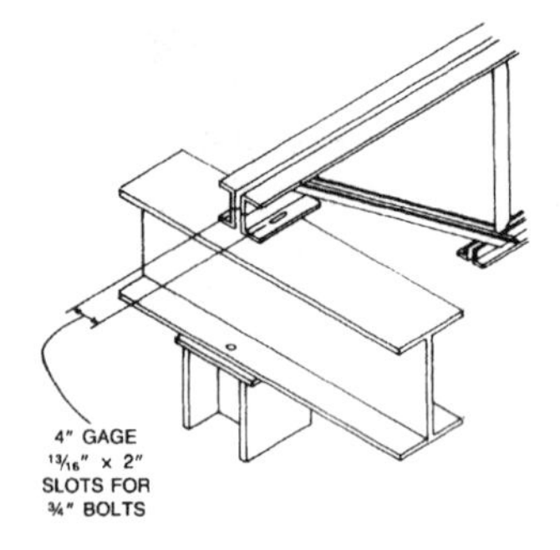

BOLTED CONNECTION (c)
Typically required at columns

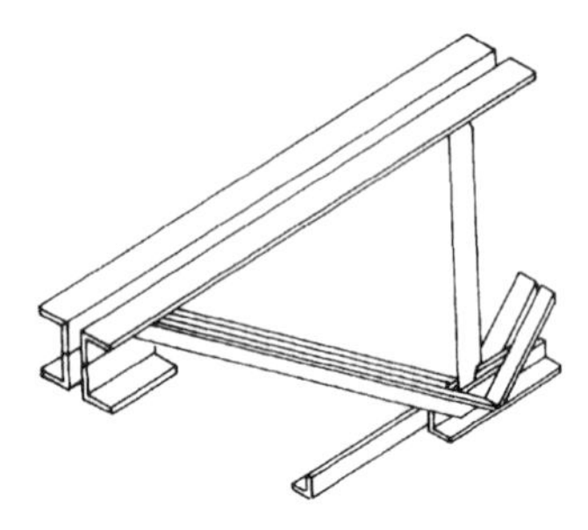

CEILING EXTENSION

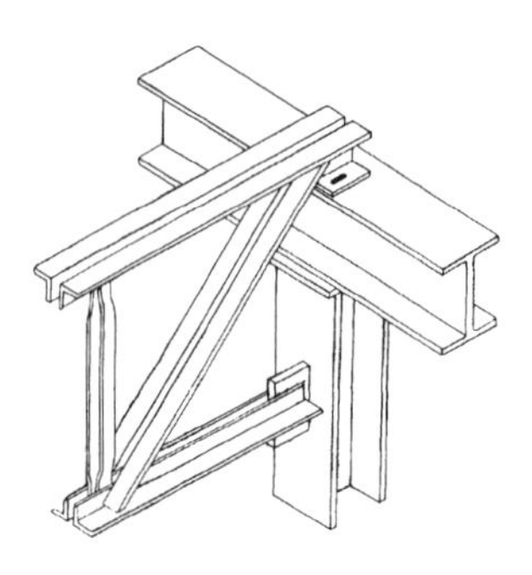

BOTTOM CHORD STRUT

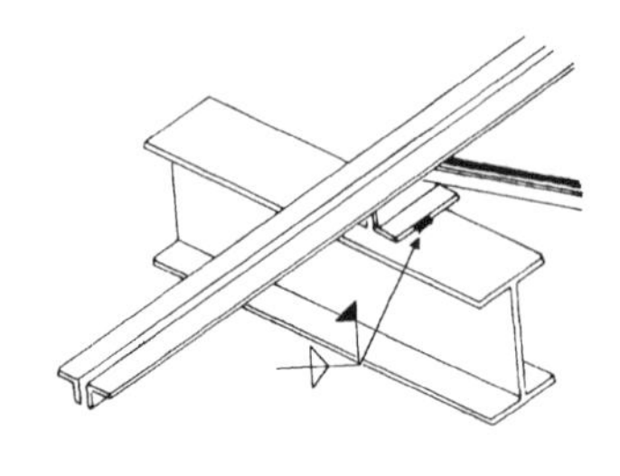

TOP CHORD EXTENSION (a)

(a) Extended top chords or full depth cantilever ends require the special attention of the specifying engineer.

The magnitude and location of the design loads to be supported, the deflection requirements, and the proper bracing shall be clearly indicated on the structural drawings.

(b) See SJI Specification - Section 105 for Handling and Erection of LH and DLH Joists.

(c) The Occupational Safety and Health Administration Standards (OSHA), Paragraph 1910.12 refers to Paragraph 1518.751 of "Construction Standards" which states:

"In steel framing, where bar joists are utilized, and columns are not framed in at least two directions with structural steel members, a bar joist shall be field-bolted at columns to provice lateral stability during construction."

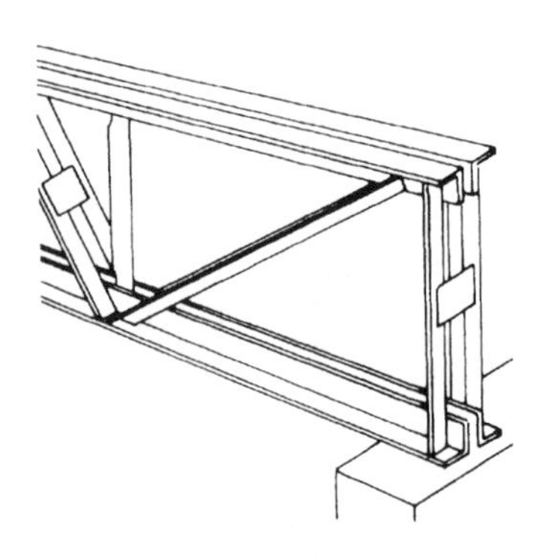

SQUARE END
See SJI Specification 104.5 (f)

By permission of Nucor Research and Development, Norfolk, Nebraska

5.7.0 Joist Girders (What Are They?)

Joist girders are primarily framing members. The design is simple span supporting equally spaced concentrated loads from open-web steel joists. These concentrated loads are considered to act at the panel points of the joist girder. Joist girders are designed to allow for the efficient use of steel in longer spans for primary framing members.

The following weight tables list joist girders from 20" to 96" deep and spans up to 100 feet. (For depths and lengths not listed, contact Vulcraft.) The depth designation is determined by the nominal depth at the center of the span, except for offset double-pitched girders, where the depth is determined at the ridge.

The standard configuration of a joist girder is a parallel chord with underslung ends and bottom chord extensions. (Joist girders can be furnished in other configurations.) The standard depth of bearing for joist girders is 6* inches at the end of the bearing seat.

The standard method of connecting girders to columns is two ¾" diameter A325 bolts. A loose connection of the lower chord to the column or other support is required during erection in order to stabilize the lower chord laterally and to help brace the joist girder against overturning. Caution: If a rigid connection of the bottom chord is to be made to column or other support, it is to be made only after the application of the dead loads. The joist girder is then no longer simply supported and the system must be investigated for continuous frame action by the specifying engineer.

Joist girders along the perimeter, with joists coming in from one side only, and those with unbalanced loads must be designed so that the reactions pass through the center of the joist girder.

The weight tables list the approximate weight per linear foot for a joist girder supporting the panel point loads given by the specifying engineer. Note: The weight of the joist girder must be included in the panel point load.

For calculating the approximate deflection or checking ponding the following formula can be used in determining the approximate moment of inertia of the joist girder.

$$I_{JG} = 0.027\,NPLd$$

Where N = number of joist spaces, P = panel point load in kips, L = joist girder length in feet, and d = effective depth of the joist girder in inches. Contact Vulcraft if a more exact joist girder moment of inertia must be known.

*Increase seat depth to 7½ inches if weight of joist girder appears to the right of the stepped blue lines in the weight tables.

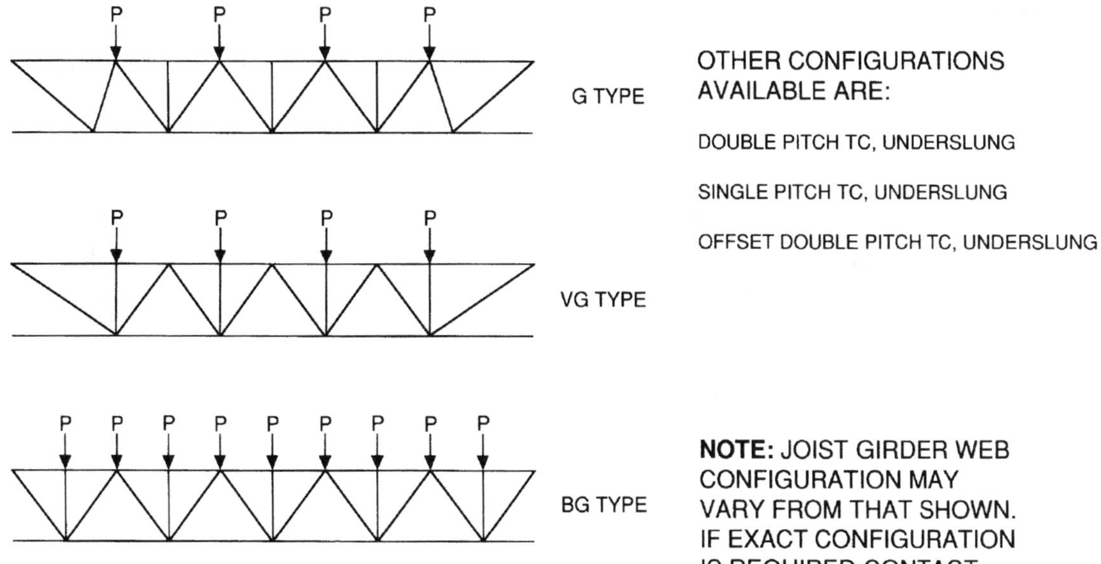

By permission of Nucor Research and Development, Norfolk, Nebraska

5.7.1 Joist Girder Notes and Connection Details

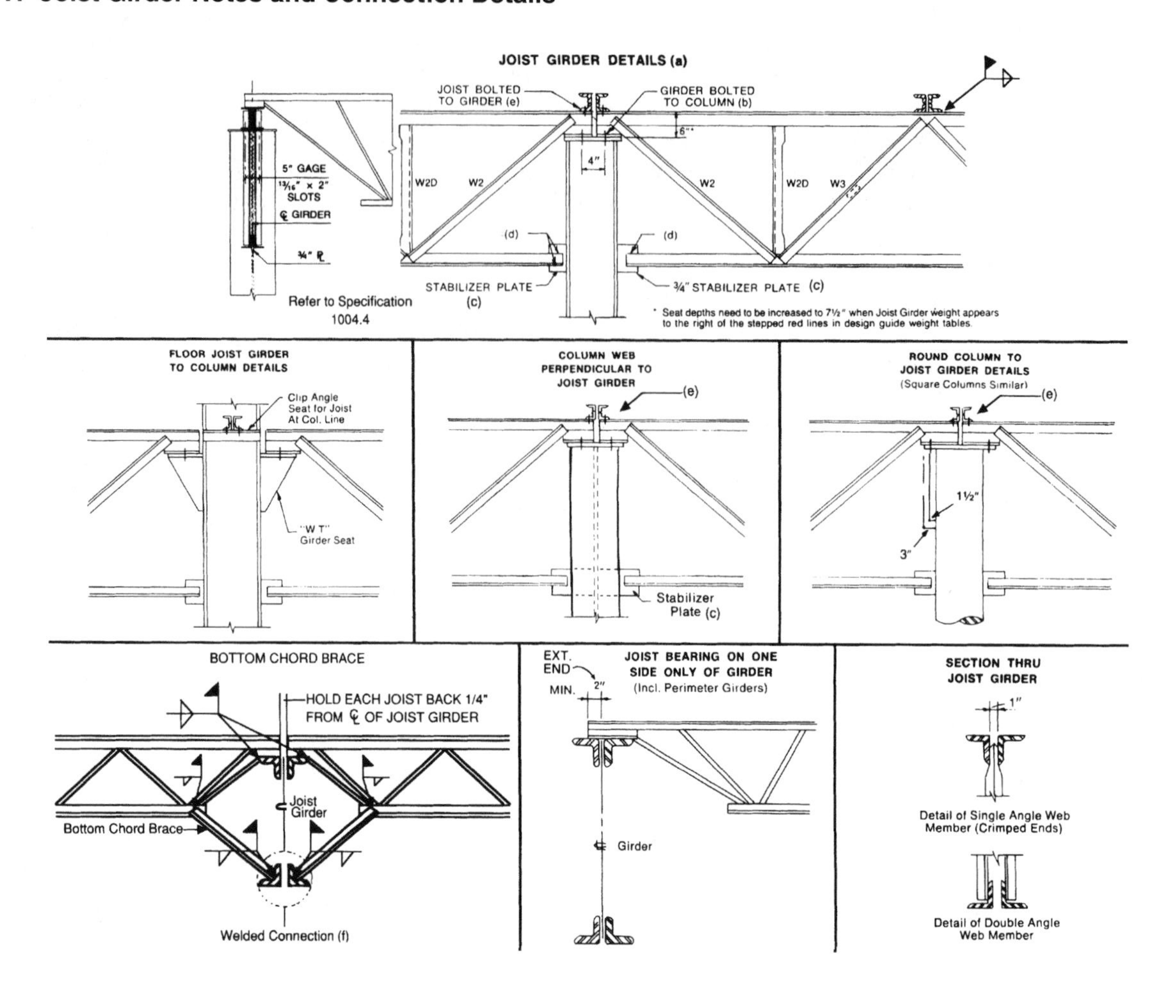

(a) All Joist Girder dimensions shown are subject to change when required by the physical size of large Joist Girders. If changes are necessary Vulcraft will so note on the placing plans.

(b) The standard connection for Joist Girders to columns is $^{13}/_{16}$ inch slots for ¾ inch bolts in girder bearings. The girder erection bolts are by others. If the specifying engineer wishes to use the Joist Girder bearing to transmit horizontal loads, he should specify the required amount of weld to connect the Joist Girder seat to the column. For additional information see the section of this catalog "JOIST GIRDERS IN MOMENT RESISTIVE FRAMES."

(c) Stabilizer plates between bottom chord angles stabilize the bottom chord laterally and brace the Joist Girder against overturning during erection. (Refer to 1004.4)

(d) Joist Girder bottom chord struts do not require welding to the stabilizer plate unless required by design to transmit horizontal forces. When welding is required, the amount of weld should be specified by the specifying engineer. UNLESS OTHERWISE SPECIFIED, BOTTOM CHORD STRUTS SHOULD NOT BE WELDED.

(e) Joists are connected to the girder by welding except that the joists at (or nearest) the column shall be bolted (O.S.H.A. Sec. 1910.12 Construction Standards Sec 1518.751).

(f) The l/ry of the bottom chord of the Joist Girder cannot exceed 240. For STANDARD Joist Girders, the specifying engineer can use the "Joist Girder Bottom Chord Brace Chart" in conjunction with the "Design Guide Weight Table/Joist Girders, G Series" to select the correct number of bottom chord braces. Joist Girders which must resist uplift, end moments, or axial bottom chord forces may require additional braces.

By permission of Nucor Research and Development, Norfolk, Nebraska

5.7.2 Joist Girder Moment Connection Details

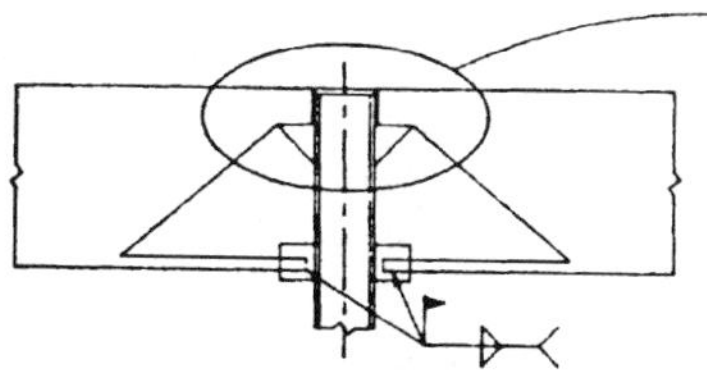

Presented below are five suggested details for a moment resistive connection involving roof Joist Girders. Similar details could also be utilized for longspan or even shortspan joists with end moments. In all cases, the bottom chord is to be connected to the column with a vertical stabilizer plate which is to be sized to carry the required load and obtain required weld (use 6 × 6 × ¾ plate minimum for Joist Girders).

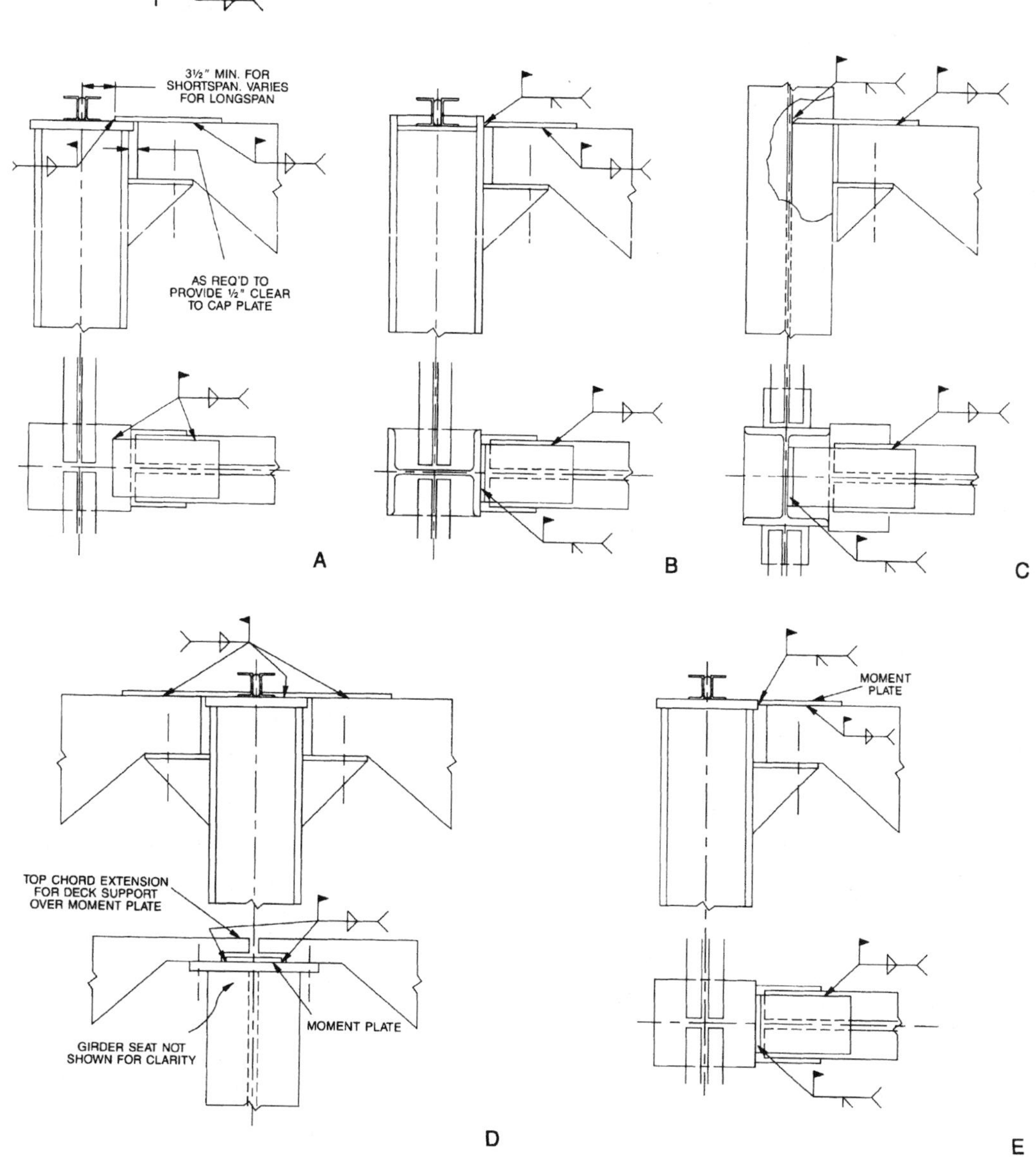

NOTES:
(1) Connections type B & C would also be recommended for floor girder details.
(2) Where a backer bar is required for groove welds, additional clearance must be provided when determining girder hold back dimension.
(3) Similar details would apply at other types of columns.
(4) Additional stiffener plates as required not shown for clarity.
(5) In all details, moment plate design and material is not by Vulcraft.

By permission of Nucor Research and Development, Norfolk, Nebraska

5.7.3 Specifying Joist Girders

For a given joist girder span, the designer first determines the number of joist spaces. Then the panel point loads are calculated and depth is selected. The following tables gives the Joist Girder weight per linear foot for various depths and loads.

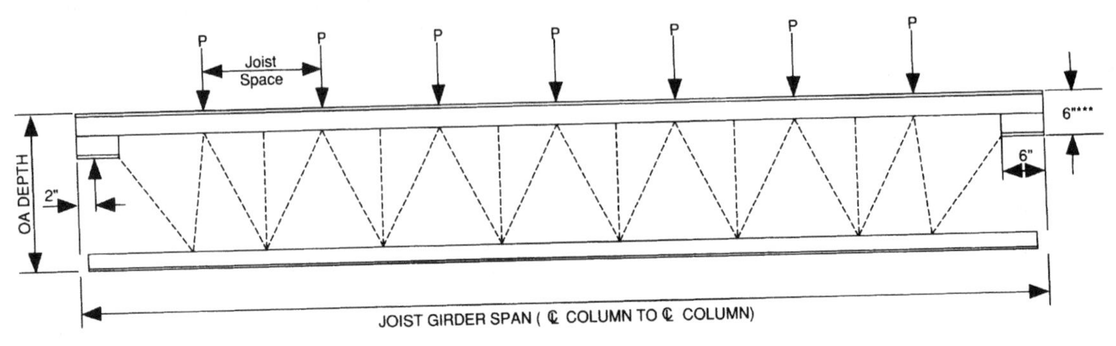

Example: Given : 50'-0 x 40'-0 bay   Joists spaced on 6'-3 centers
Live Load = 20 psf
Dead Load = 15 psf *
Total Load = 35 psf

Note: Web configuration may vary from that shown. Contact Vulcraft if exact lay-out must be known.

* Includes the approximate Joist Girder weight in panel point loads.
**See page 59 for other Girder Types.
*** Increase to 7 1/2" if weight of Joist Girder is to right of stepped lines in the weight tables.

1. Determine number of actual joist spaces (N). In this example, N = 8
2. Joist Selection
   a) Span = 40'-0
   b) T.L. = 6.25 x 35 = 219 plf
   c) from K-Series load tables select a 22K7 (T.L. = 231 > 219; L.L. = 185 > 125) 123 x 1.5 = 185 (l/240 limit applies since ceiling is not plastered)
3. Joist Girder Selection
   a) compute the concentrated load at top chord panel points P = 219 x 40 = 8,760 lbs. = 8.8 kips (use 9K for depth selection) Live load deflection rarely governs in Joist Girder selection because of their depth.
   b) Select girder depth
   The 50'-0 span 8 panel Joist Girder table on page 72 indicates that the rule of about one inch of depth for each foot of span is a good compromise of limited depth and economy. Therefore select depth of 48 inches.
   c) the Joist Girder will then be designated 48G8N8.8K
   d) the Joist Girder table shows the weight for a 48G8N9K is 43 pounds per lineal foot
   e) total weight of this Joist Girder system per square foot is:
   Joists 9.7 plf/6.25 = 1.55
   Girder 43 plf/40 = 1.07
   2.62 psf
4. For rectangular bays check economy with joists and girders spanning the opposite way
   Joists (26K10) 13.8 plf/6.67 = 2.07
   Girder (40G6N12K) 41 plf/50 = .82
   2.89 psf

NOTES:
1. When it is required to have joists bear only at vertical web members to gain space for duct work, the Joist Girder should be labeled as a "VG" in lieu of a "G".
2. The following tables serve as a design guide only. Odd size joist girder lengths, depths, kip loadings, and panel lengths are available.
3. Based on tests by Underwriters Laboratories Inc., Vulcraft Joist Girders have been approved for use in designs P231, G256, G514, N732, N754 and N736 as primary framing members. For additional fire resistance information, see FIRE RATING SECTION on page 83 and the Underwriters Laboratories Fire Resistance Directory.

By permission of Nucor Research and Development, Norfolk, Nebraska

5.8.0 Recommended Maximum Spans for Steel Decking

Recommended Maximum Spans for Construction and Maintenance Loads Standard 1½-Inch and 3-Inch Roof Deck

	Type	Span Condition	Span Ft.-In.	Maximum Recommended Spans Roof Deck Cantilever
Narrow Rib Deck	NR22	1	3'-10"	1'-0"
	NR22	2 or more	4'-9"	
	NR20	1	4'-10"	1'-2"
	NR20	2 or more	5'-11"	
	NR18	1	5'-11"	1'-7"
	NR18	2 or more	6'-11"	
Intermediate Rib Deck	IR22	1	4'-6"	1'-2"
	IR22	2 or more	5'-6"	
	IR20	1	5'-3"	1'-5"
	IR20	2 or more	6'-3"	
	IR18	1	6'-2"	1'-10"
	IR18	2 or more	7'-4"	
Wide Rib Deck	WR22	1	5'-6"	1'-11"
	WR22	2 or more	6'-6"	
	WR20	1	6'-3"	2'-4"
	WR20	2 or more	7'-5"	
	WR18	1	7'-6"	2'-10"
	WR18	2 or more	8'-10"	
Deep Rib Deck	3DR22	1	11'-0"	3'-6"
	3DR22	2 or more	13'-0"	
	3DR20	1	12'-6"	4'-0"
	3DR20	2 or more	14'-8"	
	3DR18	1	15'-0"	4'-10"
	3DR18	2 or more	17'-8"	

Type (gage)	Design Thickness		Minimum Thickness	
	In.	mm	In.	mm
28	0.0149	0.38	0.014	0.35
26	0.0179	0.45	0.017	0.43
24	0.0238	0.60	0.023	0.57
22	0.0295	0.75	0.028	0.71
20	0.0358	0.91	0.034	0.86
18	0.0474	1.20	0.045	1.14
16	0.0598	1.52	0.057	1.44

Finishes available are:

1. Galvanized (Conforming to ASTM A924-94 and or ASTM A653-94);
2. Uncoated (Black);
3. Painted with a shop coat of primer paint (one or both sides).

The uncoated finish is, by custom, referred to as "black" by some users and manufacturers; the use of the word "black" does not refer to paint color on the product.

5.8.1 Methods of Lapping Steel Deck

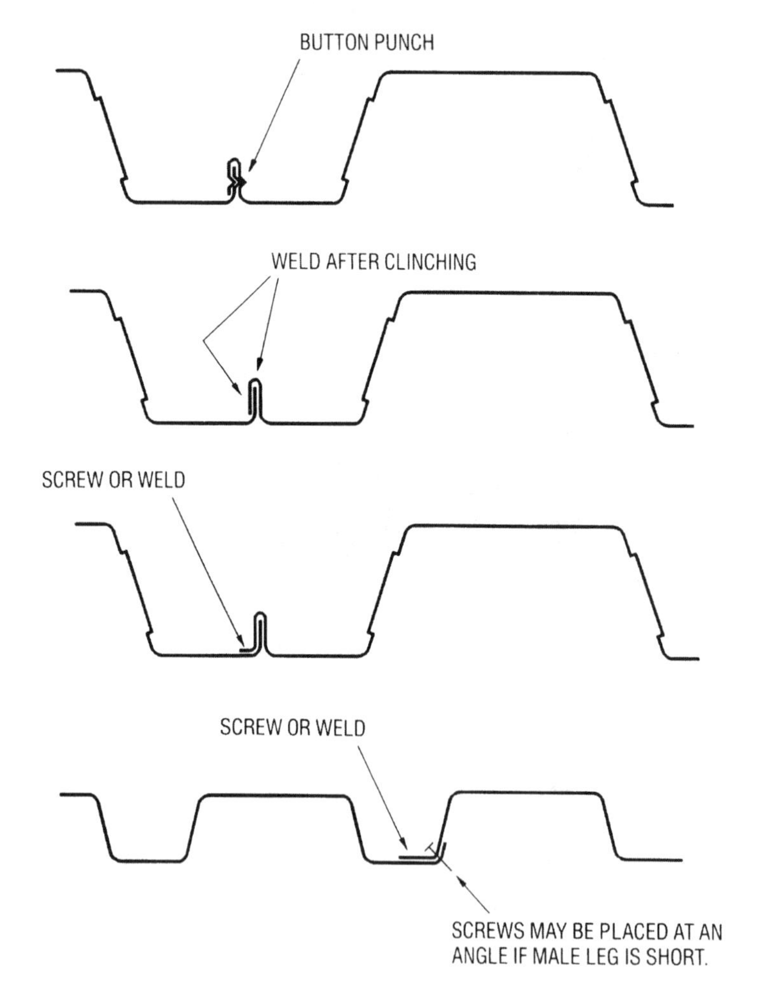

5.8.2 Noncomposite and Composite Deck Details

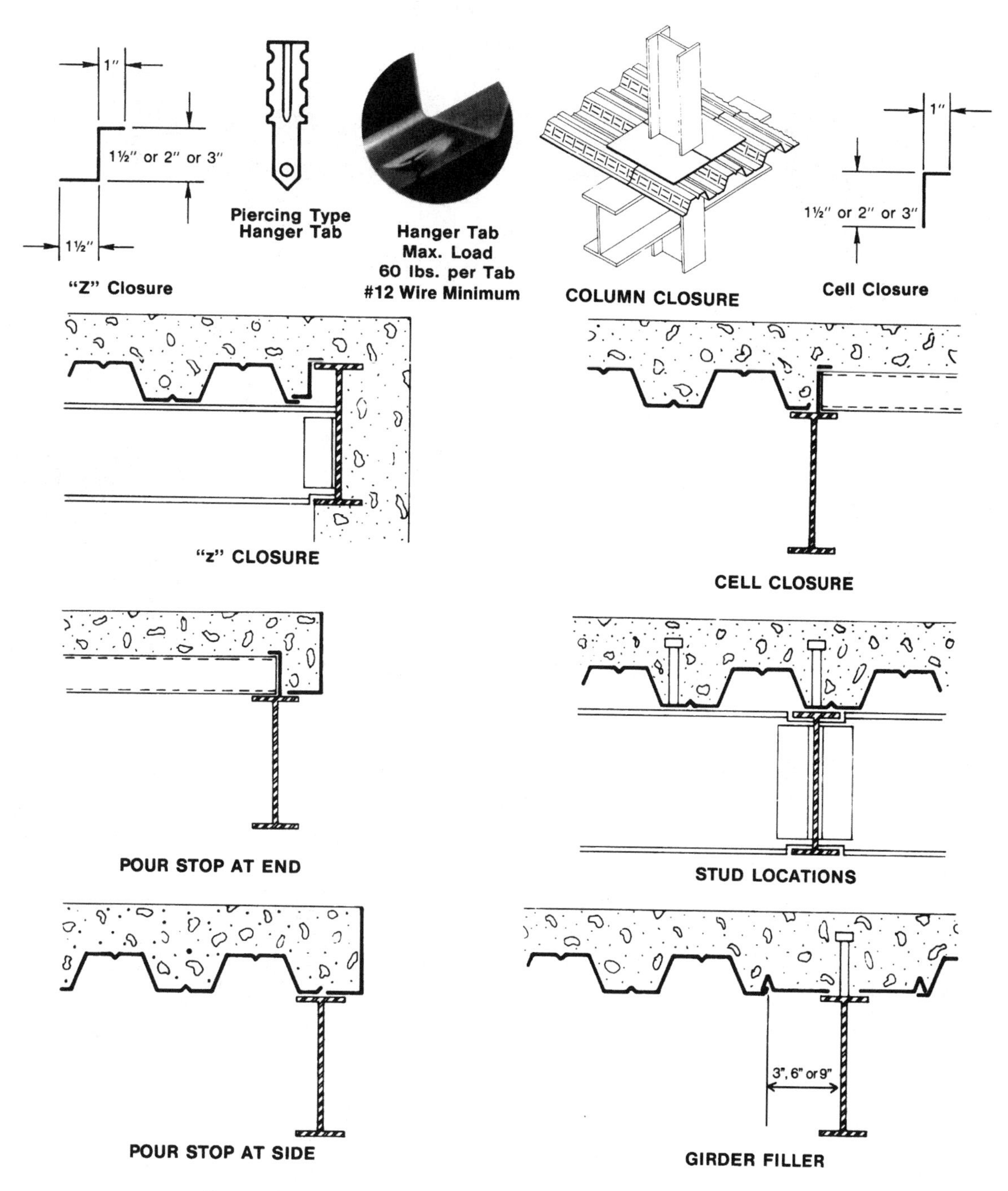

By permission of Nucor Research and Development, Norfolk, Nebraska

5.8.3 Pour-Stop Selection Table

Allowable cantilever of metal deck where pour stops are required.

SELECTION TABLE

SLAB DEPTH (Inches)	OVERHANG (INCHES)												
	0	1	2	3	4	5	6	7	8	9	10	11	12
						POUR STOP TYPES							
4.00	20	20	20	20	18	18	16	14	12	12	12	10	10
4.25	20	20	20	18	18	16	16	14	12	12	12	10	10
4.50	20	20	20	18	18	16	16	14	12	12	12	10	10
4.75	20	20	18	18	16	16	14	14	12	12	10	10	10
5.00	20	20	18	18	16	16	14	14	12	12	10	10	
5.25	20	18	18	16	16	14	14	12	12	12	10	10	
5.50	20	18	18	16	16	14	14	12	12	12	10	10	
5.75	20	18	16	16	14	14	12	12	12	12	10	10	
6.00	18	18	16	16	14	14	12	12	12	10	10	10	
6.25	18	18	16	14	14	12	12	12	12	10	10		
6.50	18	16	16	14	14	12	12	12	12	10	10		
6.75	18	16	14	14	14	12	12	12	10	10	10		
7.00	16	16	14	14	12	12	12	12	10	10	10		
7.25	16	16	14	14	12	12	12	10	10	10			
7.50	16	14	14	12	12	12	12	10	10	10			
7.75	16	14	14	12	12	12	10	10	10	10			
8.00	14	14	12	12	12	12	10	10	10				
8.25	14	14	12	12	12	10	10	10	10				
8.50	14	12	12	12	12	10	10	10					
8.75	14	12	12	12	12	10	10	10					
9.00	14	12	12	12	10	10	10						
9.25	12	12	12	12	10	10	10						
9.50	12	12	12	10	10	10							
9.75	12	12	12	10	10	10							
10.00	12	12	10	10	10	10							
10.25	12	12	10	10	10								
10.50	12	12	10	10	10								
10.75	12	10	10	10									
11.00	12	10	10	10									
11.25	12	10	10										
11.50	10	10	10										
11.75	10	10											
12.00	10	10											

TYPES	DESIGN THICKNESS
20	0.0358
18	0.0474
16	0.0598
14	0.0747
12	0.1046
10	0.1345

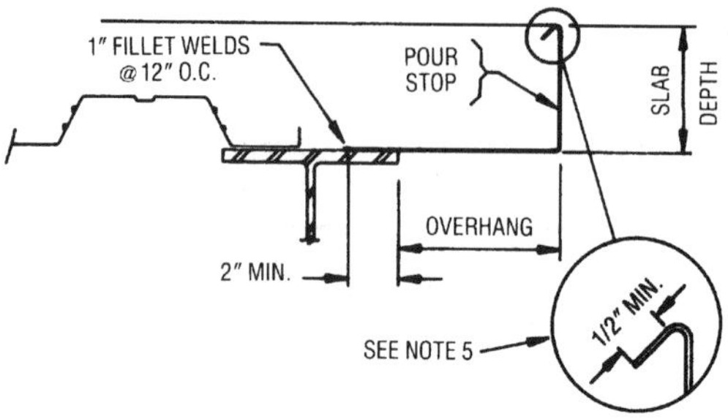

1" FILLET WELDS @ 12" O.C.
POUR STOP
SLAB DEPTH
OVERHANG
2" MIN.
SEE NOTE 5
1/2" MIN.

NOTES: The above Selection Table is based on following criteria:

1. Normal weight concrete (150PCF).
2. Horizontal and vertical deflection is limited to 1/4" maximum for concrete dead load.
3. Design stress is limited to 20 KSI for concrete dead load temporarily increased by one-third for the construction live load of 20 PSF.
4. Pour Stop Selection Table does not consider the effect of the performance, deflection, or rotation of the pour stop support which may include both the supporting composite deck and/or the frame.
5. Vertical leg return lip is recommended for type 16 and lighter.
6. This selection is not meant to replace the judgement of experienced Structural Engineers and shall be considered as a reference only.

SDI reserves the right to change any information in this selection without notice.

5.8.4 Cellular Floor-Deck and Form-Deck Profiles

Cellular Floor Deck Profiles	Name	Nominal Thickness Range	Weight Range	Comments
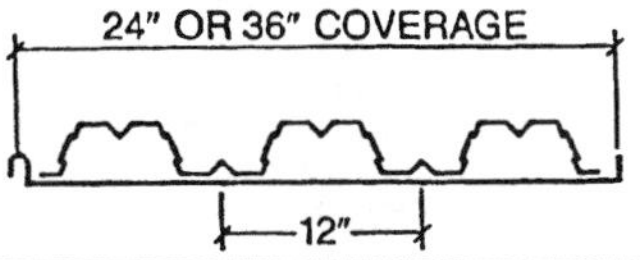 24" OR 36" COVERAGE; 12"	3" x 12" Composite Cellular	.03" to .06"	4 psf to 7 psf	Bottom plate may be perforated for acoustical.
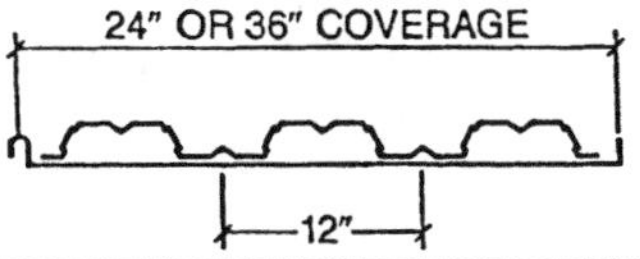 24" OR 36" COVERAGE; 12"	2" x 12" Composite Cellular	.03" to .06"	4 psf to 7 psf	Bottom plate may be perforated for acoustical.
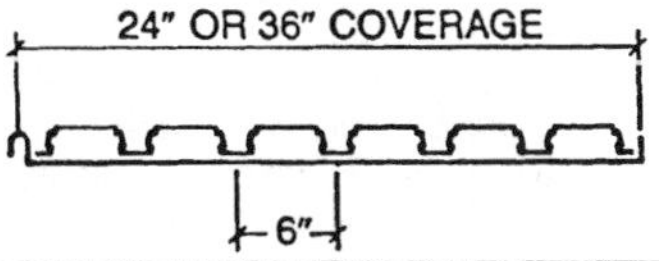 24" OR 36" COVERAGE; 6"	1½" x 6" Composite Cellular	.03" to .06"	4 psf to 7 psf	May also be used as roof deck. Bottom plate may be perforated for acoustical.
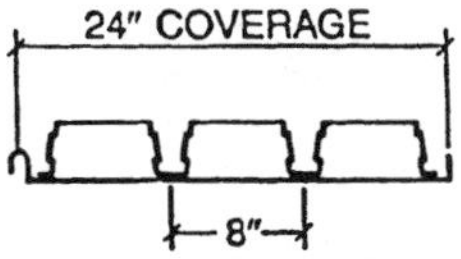 24" COVERAGE; 8"	3" x 8" Composite Cellular	.03" to .06"	4 psf to 7 psf	May also be used as roof deck. Bottom plate may be perforated for acoustical.

Form Deck Profiles	Name	Nominal Thickness Range	Weight Range	Comments
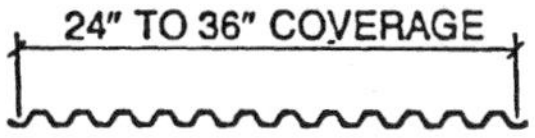 24" TO 36" COVERAGE	9/16" x Varies Form Deck	.014" to .030"	0.8 psf to 1.5 psf	Standard form deck. Used as centering.
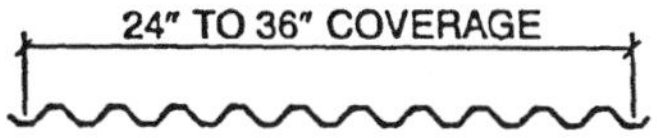 24" TO 36" COVERAGE	15/16" x Varies Form Deck	.017" to .040"	1.0 psf to 2.0 psf	Heavy duty form deck. Used as centering.
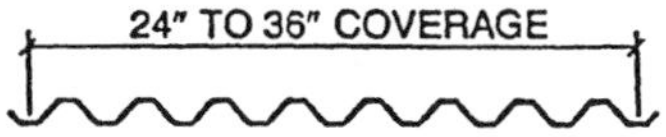 24" TO 36" COVERAGE	1 5/16" x Varies Form Deck	.017" to .047"	1.0 psf to 2.8 psf	Extra heavy duty form deck. Used as centering.
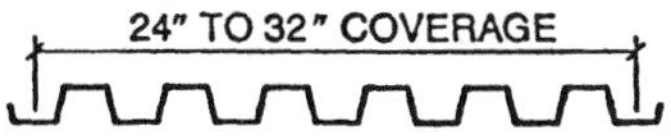 24" TO 32" COVERAGE	1½" or 2" x Varies Form Deck	.023" to .047"	1.4 psf to 2.8 psf	Super duty form deck. Used as centering.

Note: All profiles may be used as roof deck (for a patented assembly)

5.8.5 Composite Floor-Deck and Roof-Deck Profiles

Composite Floor Deck Profiles	Name	Nominal Thickness Range	Weight Range	Comments
36″ OR 24″ COVERAGE; 12″	1½″ x 12″ 2 x 12″ 3″ x 12″ Composite	.03″ to .06″	2 psf to 4 psf	Embossment patterns will vary from manufacturer to manufacturer. Side laps are flat adjustable or button punchable.
24½″ COVERAGE; 6⅛″	2″ x 12″ Composite	.03″ to .06″	2 psf to 4 psf	
36″ OR 30″ COVERAGE; 6″	1½″ x 6″ Composite	.03″ to .06″	2 psf to 4 psf	Embossment patterns will vary from manufacturer to manufacturer. Side laps are flat adjustable or button punchable.
24″ COVERAGE; 8″	3″ x 8″ Composite	.03″ to .06″	2 psf to 4 psf	Embossment patterns will vary from manufacturer to manufacturer. Side laps are flat adjustable or button punchable. This profile is not generally suitable for use with shear studs.

Roof Deck Profiles	Name	Nominal Thickness Range	Weight Range	Comments
36″ OR 30″ COVERAGE; 2½″ NOM.; 6″; 1¾″ MIN.	1½″ x 6″ Wide Rib (WR)	.03″ to .06″	2 psf to 4 psf	May be referred to as "B" deck. Sidelaps may be flat adjustable or button punchable. Acoustical deck will have perforated webs.
36″ OR 30″ COVERAGE; 1¾″ NOM.; 6″; ½″ MIN.	1½″ x 6″ Intermediate Rib (IR)	.03″ to .06″	2 psf to 4 psf	May be referred to as "F" deck.
36″ OR 30″ COVERAGE; 1″ NOM.; 6″; ⅜″ MIN.	1½″ x 6″ Narrow Rib (NR)	.03″ to .06″	2 psf to 4 psf	May be referred to as "A" deck.
24″ COVERAGE; 2¾″ NOM.; 8″; 1½″ MIN.	3″ x 8″ Deep Rib (DR)	.03″ to .06″	2 psf to 4 psf	May be referred to as "N" deck. Sidelaps may be flat adjustable or button punchable. Acoustical deck will have perforated webs.

5.8.6 Reinforcing Openings in Steel Decks

Methods of cutting and reinforcing penetrations through decking.

SUMP REINFORCING AT END OF DECK

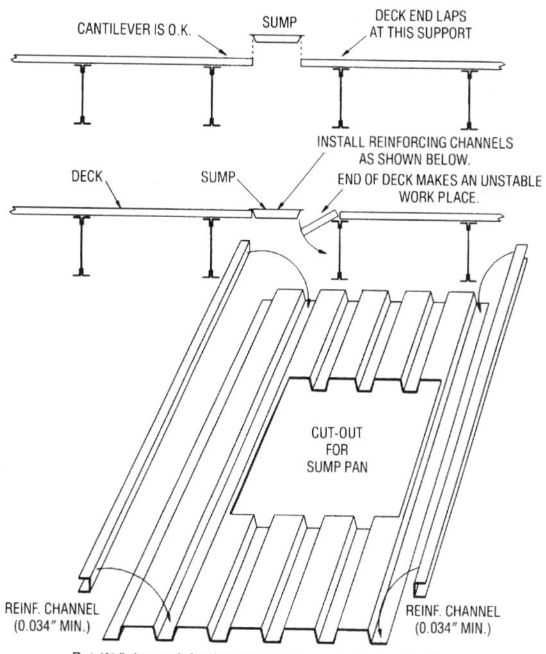

Put 1½" deep reinforcing channels in each rib at each side of opening (flush with top of deck). Channels span between joists. Attach flanges of sump pan to channels.

Burn holes in deck side laps, caused by welded side lap attachments, are spaced far enough apart not to cause problems. Burn holes near intermediate supports are unlikely to cause much loss of strength unless a total area greater than a 6" diameter hole is removed. These burn holes are usually caused by the welder searching for the unseen structural member; therefore, the use of chalk lines is recommended.

Distributed small dents, such as those caused by foot traffic, will not cause a structural problem; but if the denting covers a large percentage of the job, the insulation board will be better attached with mechanical fasteners rather than by adhesives. The designer must approve any change in fastening.

Vigilance should be maintained to detect and correct any "soft" spots in roofs that could cause insulation boards to crack under foot loading.

EXAMPLES OF DETAILS FOR OPENINGS

DETAILS FOR OPENINGS TO 2′-0″ PERPENDICULAR TO DECK

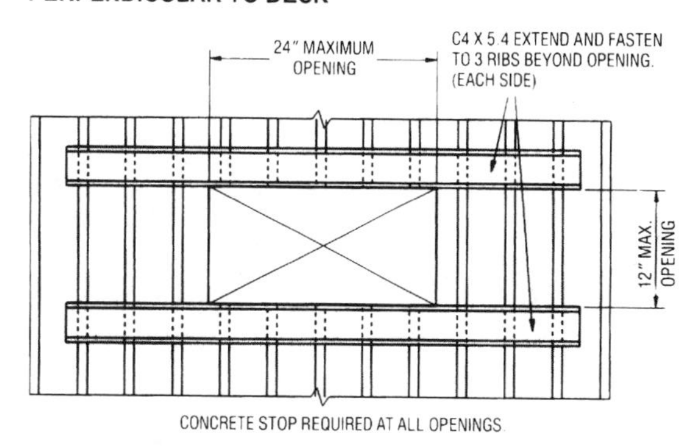

DETAILS FOR OPENINGS TO 12″ PERPENDICULAR TO DECK

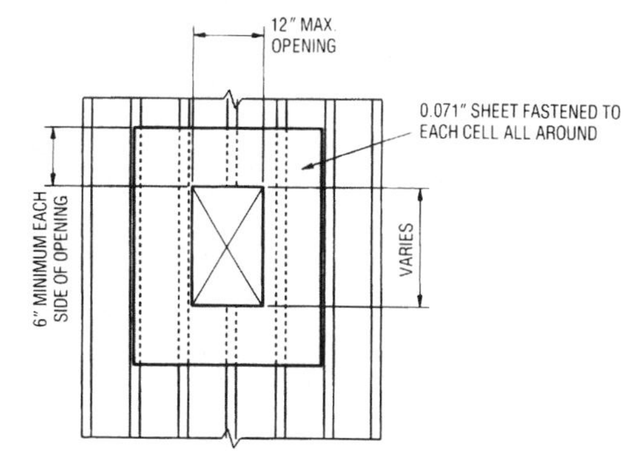

NOTE:
For holes 6″ ∅ or less no reinforcing or minimum 0.045″ plate required, depending on location.

5.8.7 Example of 6" Penetration in Steel Deck

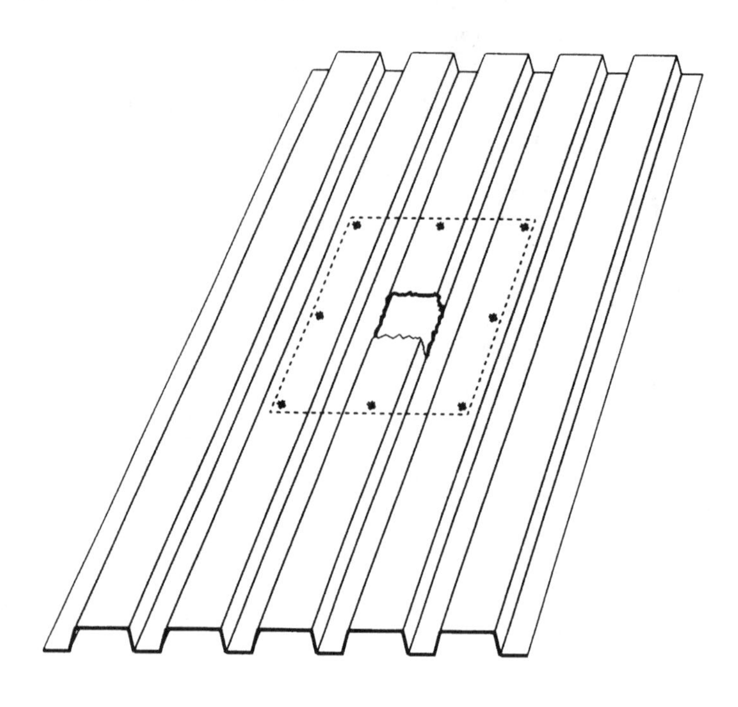

SUGGESTED SCHEDULE:

One Rib Removed (6″ Diameter)	No Reinforcing Or 0.045″ Plate (Min.)
8″ Diameter	0.045″ Plate (Min.)
8″ to 13″ Diameter	0.057″ Plate (Min.)
Over 13″	Frame Opening* (Design By Project Engineer)

*Check cantilever ability of deck

5.8.8 Maximum Spans for Roof Deck

Recommended Maximum Spans for Construction and Maintenance Loads Standard for 1½ Inch and 3 Inch Roof Deck

	Type	Span Condition	Span Ft.-In.	Span Meters	Max. Recommended Spans Roof Deck Cantilever Ft.-In.	Max. Recommended Spans Roof Deck Cantilever Meters
Narrow Rib Deck	NR22	1	3′-10″	1.15 m	1′-0″	.30 m
	NR22	2 or more	4′-9″	1.45 m		
	NR20	1	4′-10″	1.45 m	1′-2″	.35 m
	NR20	2 or more	5′-11″	1.80 m		
	NR18	1	5′-11″	1.80 m	1′-7″	.45 m
	NR18	2 or more	6′-11″	2.10 m		
Intermediate Rib Deck	IR22	1	4′-6″	1.35 m	1′-2″	.35 m
	IR22	2 or more	5′-6″	1.65 m		
	IR20	1	5′-3″	1.60 m	1′-5″	.40 m
	IR20	2 or more	6′-3″	1.90 m		
Wide Rib Deck	WR22	1	5′-6″	1.65 m	1′-11″	.55 m
	WR22	2 or more	6′-6″	1.75 m		
	WR20	1	6′-3″	1.90 m	2′-4″	.70 m
	WR20	2 or more	7′-5″	2.25 m		
	WR18	1	7′-6″	2.30 m	2′-10″	.85 m
	WR18	2 or more	8′-10″	2.70 m		
Deep Rib Deck	3DR22	1	11′-0″	3.35 m	3′-5″	1.05 m
	3DR22	2 or more	13′-0″	3.95 m		
	3DR20	1	12′-6″	3.80 m	3′-11″	1.20 m
	3DR20	2 or more	14′-8″	4.45 m		
	3DR18	1	15′-0″	4.55 m	4′-9	1.45 m
	3DR18	2 or more	17′-8″	5.40 m		

Construction and maintenance loads:
SPANS are governed by a maximum stress of 26 ksi (180 MPa) and a maximum deflection of L/240 with a 200 pound (0.89 kN) concentrated load at midspan on a 1′-0″ (300 mm) wide section of deck. If the designer contemplates loads of greater magnitude, spans shall be decreased or the thickness of the steel deck increased as required.

All loads shall be distributed by appropriate means to prevent damage to the completed assembly during construction.

STEEL DECK CANTILEVER EXAMPLE: WR22

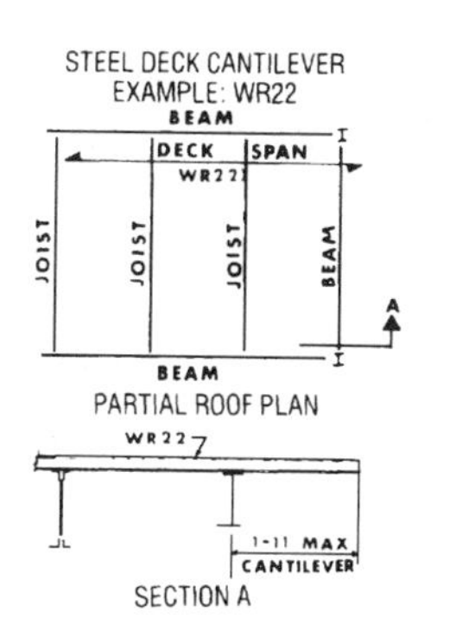

Cantilever loads:
Construction phase load of 10 psf (0.48 kPa) on adjacent span and cantilever, plus 200 pound load (0.89 kN) at end of cantilever with a stress limit of 26 ksi (180 MPa).

Service load of 45 psf (2.15 kPa) on adjacent span and cantilever, plus 100 pound load (0.44 kN) at end of cantilever with a stress limit of 20 ksi (140 MPa).

Deflection limited to L/240 of adjacent span for interior span and deflection at end of cantilever to L/240 of overhang.

Notes:

1. Adjacent span: Limited to those spans shown in Section 3.4 of Roof Deck Specifications. In those instances where the adjacent span is less than 3 times the cantilever span, the individual manufacturer should be consulted for the appropriate cantilever span.
2. Sidelaps must be attached at end of cantilever and at a maximum of 12 inches (300 mm) on center from end.
3. No permanent suspended loads are to be supported by the steel deck.
4. The deck must be completely attached to the supports and at the sidelaps before any load is applied to the cantilever.

4. Installation & Site Storage

4.1 Site Storage: Steel deck shall be stored off the ground with one end elevated to provide drainage, and shall be protected from the elements with a waterproof covering, ventilated to avoid condensation.

5.9.0 Fire Resistance Ratings for Roof Decks

FIRE RESISTANCE RATINGS

2-Hour Rating with Directly-Applied Protection

Illustration refers to UL Design P801 using a sprayed mineral fiber insulation. See also UL Designs P701, 711, and P805

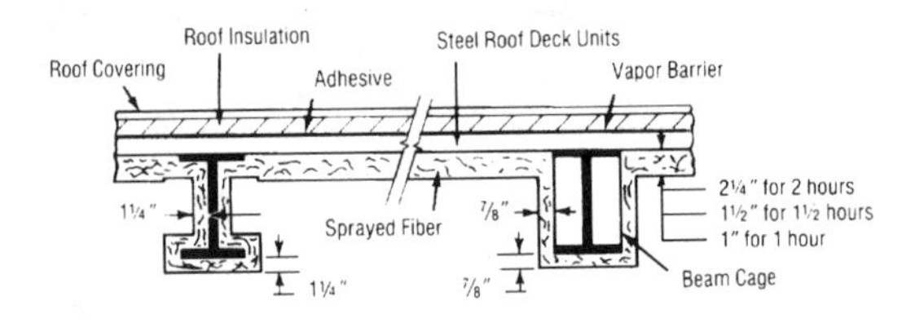

2-Hour Rating with Metal Lath and Plaster Ceiling

Illustration refers to UL Design P404. See also UL Design P409.

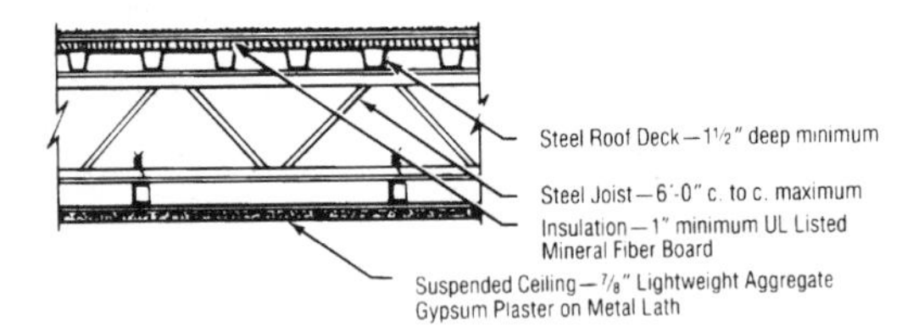

Other 2-Hour Ratings

Although standard roof deck sections were not used for the following tests, it is the opinion of persons knowledgeable in fire test procedures that galvanized steel roof deck with a minimum depth of 1½ inches and a 0.0295-inch design thickness can be used without decreasing the fire resistance of the assembly. In each case, the assembly was tested using either a steel form unit with a minimum depth of 9⁄16 inch or a steel floor deck essentially identical to products marketed as roof deck. The authorities having jurisdiction should be consulted before substituting steel roof deck in the following assemblies:

UL Designs P215 and P219: accoustical ceiling systems. 2 inches vermiculite concrete on special roof topping mixture on steel deck.

UL Design P902: no ceiling required. 2¾ inches cellular concrete on steel deck.

1-Hour Ratings with Suspended Acoustical Ceiling

Illustration refers to UL Design P201. See also UL Designs P204, P210, P211, P224, P232, P235, P238, and P243, and Factory Mutual Roof-Ceiling Construction 3-1 hour.

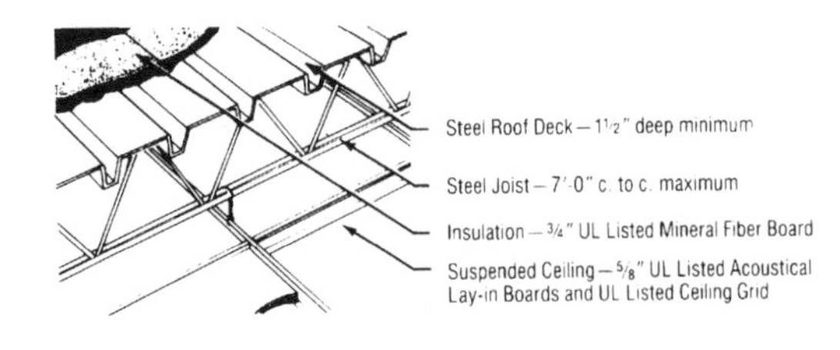

5.9.1 Floor Ceiling Fire-Resistance Ratings with Steel Joist

FLOOR-CEILING ASSEMBLIES WITH MEMBRANE PROTECTION

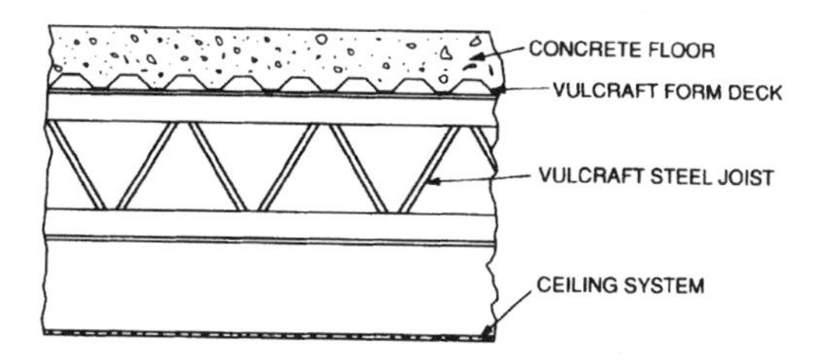

RESTRAINED ASSEMBLY RATING	TYPE OF PROTECTION SYSTEM	CONCRETE TYPE & THICKNESS ABOVE DECK		MINIMUM JOIST SIZE	MAX. JOIST SPACING SEE NOTE 2	U.L. DESIGN NUMBER	UNRESTRAINED BEAM RATING
1 HR	EXPOSED GRID	2 1/2"	NW	8K1	4'-0	G253	1 HR
		2 1/2"	NW	8K1	6'-0	G256	1, 2, 3 HR
1 1/2 HRS	EXPOSED GRID	3"	NW	10K1	4'-0	G203	1 1/2, 2 HR
		2 1/2"	NW	10K1	4'-0	G228	1 1/2, 2 HR
		2"	NW	10K1	4'-0	G229	1 1/2, 2, 3 HR
		2 1/2"	NW	10K1	4'-0	G243	1 1/2, 2 HR
2 HRS	CONCEALED GRID	2 1/2"	NW	10K1	4'-0	G008	2 HR
		2 1/2"	NW	10K1	4'-0	G018	---
		2 1/4"	NW	8K1	4'-0	G023	2 HR
		2 1/2"	NW	10K1	4'-0	G028	---
		2 1/2"	NW	8K1	4'-0	G031	3 HR
		2 1/2"	NW	10K1	4'-0	G036	3 HR
		2 1/4"	NW	10K1	4'-0	G037	2 HR
	EXPOSED GRID	3"	NW	8K1	4'-0	G203	1 1/2, 2 HR
		2 1/2"	NW	10K1	4'-0	G204	2 HR
		2 1/2"	NW	10K1	4'-0	G208	2 HR
		3"	NW	10K1	4'-0	G209	---
		2 1/2"	NW	10K1	4'-0	G211	---
		3"	NW	8K1	4'-0	G212	2 HR
		2 1/2"	NW	10K1	4'-0	G213	2, 3 HR
		2 1/2"	NW	10K1	4'-0	G227	3 HR
		2 1/2"	NW	10K1	4'-0	G228	1 1/2, 2 HR
		2"	NW	10K1	4'-0	G229	1 1/2, 2, 3 HR
		2 1/2"	NW	10K1	4'-0	G243	1 1/2, 2 HR
		3"	NW	8K1	4'-0	G244	2 HR
		2 1/2"	NW	10K1	4'-0	G250	2 HR
		2 1/2"	NW	10K1	4'-0	G255	1, 2, 3 HR
		2 1/2"	NW	8K1	6'-0	G256	1, 2, 3 HR
		2 1/2"	NW	8K1	4'-0	G258	2, 3 HR
	GYPSUM BOARD	2 1/2"	NW	8K1	4'-0	G523	2, 3 HR
		2 1/2"	NW, LW	10K1	4'-0	G529	3 HR
3 HRS	CONCEALED GRID	3 1/2"	NW	8K1	4'-0	G033	3 HR
		3 1/4"	NW	10K1	4'-0	G036	3 HR
	EXPOSED GRID	3"	NW	10K1	4'-0	G213	2, 3 HR
		3 1/4"	NW	10K1	4'-0	G229	1 1/2, 2, 3 HR
		3 1/2"	NW	8K1	4'-0	G256	1, 2, 3 HR
	GYPSUM BOARD	2 3/4"	NW	10K1	4'-0	G529	3 HR
4 HRS	METAL LATH	2 1/2"	NW	12K5	4'-0	G401	---

By permission of Nucor Research and Development, Norfolk, Nebraska

Section

6

Wood and Lumber Products

Contents

6.0.0 Introduction to softwoods, hardwoods, and lumber terminology
6.1.0 Introduction to Western Wood Products Association (WWPA) and Southern Pine Inspection Bureau (SPIB)
6.2.0 American Lumber Standards Committee (ALSC) and wood preservatives
6.2.1 ALSC pressure-treated wood-stamp markings
6.2.2 ALSC registered trademarks
6.3.0 Moisture content in lumber
6.4.0 WWPA guide to understanding grade stamps
6.4.1 Species of wood included in WWPA jurisdiction
6.4.2 Species identification and facsimile grade stamps
6.4.3 Design values for various species of Western wood products
6.4.4 Adjustment factors for base values of Western Wood products
6.4.5 Additional adjustment factors for Western wood dimension lumber
6.5.0 Standard sizes for Western wood finish and selects (dry lumber)
6.5.1 Standard sizes for Western wood common boards, studs, and patterns
6.5.2 Western wood scaffolding sizes and design values
6.6.0 American Softwood Standards for boards and timbers
6.6.1 American Softwood Standards for shiplap and centermatch lumber
6.6.2 American Softwood standards for worked lumber
6.6.3 American Softwood Standards for siding (19% moisture content)
6.6.4 American Softwood Standards for finish, floor, and ceiling partition lumber
6.7.0 Specifying Southern pine lumber (grade stamp markings)
6.7.1 Southern pine span tables for joists
6.7.2 Southern pine span tables for wet-service joists and rafters
6.7.3 Spans for various Southern pine species
6.7.4 Extent of notching of structural pine framing members
6.7.5 Southern pine rafter spans and birdsmouth data
6.7.6 Conversion diagram for Southern pine rafters
6.8.0 Properties of sections of Southern pine framing members
6.8.1 Standard sizes of Southern pine dimension lumber, boards, and decking
6.9.0 Southern pine header load tables and connection details
6.9.1 Southern pine rafter framing details
6.9.2 Southern pine floor joist framing details
6.9.3 Additional Southern pine joist framing details
6.10.0 Southern Pine Inspection Bureau grading rules for decking
6.10.1 Southern Pine Inspection Bureau grading rules for finish and boards
6.10.2 Southern Pine Inspection Bureau grading rules for 2" dimensions
6.11.0 Southern pine wood-preservative retention standards
6.12.0 Knots and how to measure them
6.13.0 Commercial names of principal softwood species
6.14.0 Lumber industry abbreviations

The numerous species of wood can be divided into two basic classifications: softwood and hardwood. These classifications do not necessarily refer to the hardness or softness of the species, but rather by the type of tree from which the wood is taken.

6.0.0 Introduction to Softwoods, Hardwoods, and Lumber Terminology

Hardwood comes from trees that shed their leaves at the end of a growing season (such as oak, hickory, chestnut, elm, maple, and birch). Softwoods, on the other hand, are trees, such as evergreens, that do not shed their leaves (cedar, pine, hemlock, larch, and spruce, for example). Hardwoods are generally used for flooring, furniture, cabinetry, and millwork. Softwoods find wide application as framing members, although some species of pine are used as shelving or are incorporated into various types of millwork.

The characteristics of wood vary from tree to tree as well as from section to section within a tree. Therefore, some method is required to select and grade pieces of lumber cut from a tree to form some degree of uniformity. Then organizations were established to set the standards for various grades of lumber. They have the authority to inspect member mills to ensure that the buyer receives the quality they bargain for.

6.1.0 Introduction to Western Wood Products Association (WWPA) and Southern Pine Inspection Bureau (SPIB)

The Western Wood Products Association (WWPA) was formed around 1900. By 1924, various other grading associations in the United States developed product standards with the assistance of the U.S. Department of Commerce. The WWPA, headquartered in Portland, Oregon establishes standards of size and levels of quality for a variety of western softwoods. Its inspectors regularly visit member mills to ensure that the quality and production of these mills meet pre-established standards. Only then is the mill allowed to stamp their product with the approved WWPA certification. Softwood lumber is further classified according to extent of manufacture:

- *Rough lumber* Lumber that has not been dressed, but only sawn edged and trimmed to the extent of showing saw marks on all four sides.
- *Dressed or surfaced lumber* Lumber that has been run through a surfacing machine to achieve a smooth and uniform surface on one side (S1S), two sides (S2S), one edge (S1E), two edges (S2E), all four sides (S4S), or any combination thereof.
- *Worked lumber* Lumber that, in addition to being dressed or surfaced, has been matched, shiplapped or tongue and grooved.
- *Resawn lumber* Lumber that is dressed before resawing and not afterward. Uniformity of thickness does not characterize resawn lumber.

The Southern Pine Inspection Bureau (SPIB) in Pensacola, Florida establishes the grading rules for four principle species of Southern pine: longleaf (*pinus palustris*), slash (*pinus elliottii*), shortleaf (*pinus echinata*), and loblolly (*pinus eaeda*). A few other species of negligible or less importance to the construction industry are also included.

6.2.0 American Lumber Standards Committee (ALSC) and Wood Preservatives

The American Lumber Standard Committee (ALSC) also stamps lumber and is administered by the U.S. Department of Commerce. The ALSC provides supervisory inspections for pressure-treated wood products and has established a series of abbreviations for the various types of wood preservatives in use today.

CCA	chromated copper arsenate
ACA	ammoniacal copper arsenate
ACZA	ammoniacal copper zine arsenate
ACC	acid copper chromate
ACQ	ammoniacal copper quat. type-B
COPPER NAP	copper naphthenate
PENTA	pentachlorophenol
CREOSOTE	creosote and/or solutions
BORATE	borates

6.2.1 ALSC Pressure-Treated Wood-Stamp Markings

ACCREDITED AGENCIES FOR SUPERVISORY AND LOT INSPECTION OF PRESSURE TREATED WOOD PRODUCTS March 1996

Agencies accredited by the Board of Review of the American Lumber Standard Committee, Incorporated and typical quality marks.

Interpreting a Quality Mark

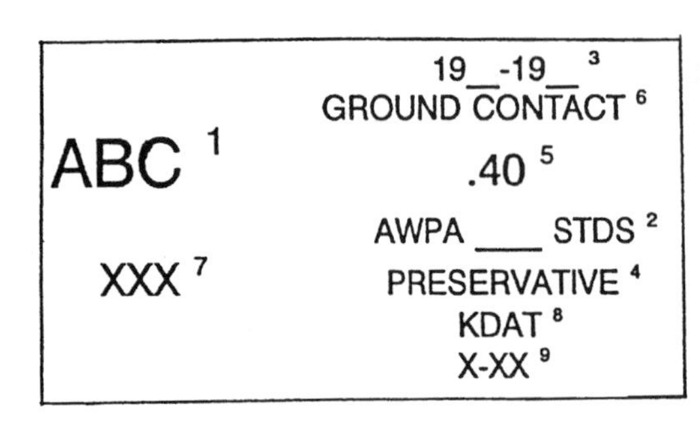

1 - The identifying symbol, logo or name of the accredited agency.
2 - The applicable American Wood Preservers' Association (AWPA) commodity standard.
3 - The year of treatment if required by AWPA standard.
4 - The preservative used, which may be abbreviated.
5 - The preservative retention.
6 - The exposure category (e.g. Above Ground, Ground Contact, etc.).
7 - The company name and location of home office; or company name and number; or company number.
8 - If applicable, moisture content after treatment.
9 - If applicable, length, and/or class.

As specified below for particular agencies, some or all of the following American Wood Preservers' Association commodity standards are used by American Lumber Standard Committee, Incorporated accredited agencies which supervise facilities which pressure treat wood products:

C1 All Timber Products--Preservative Treatment by Pressure Processes
C2 Lumber, Timbers, Bridge Ties and Mine Ties--Preservative Treatment by Pressure Processes
C3 Piles--Preservative Treatment by Pressure Processes
C4 Poles--Preservative Treatment by Pressure Processes
C5 Fence Posts--Preservative Treatment by Pressure Processes
C6 Crossties and Switch Ties-Preservative Treatment by Pressure Process
C9 Plywood--Preservative Treatment by Pressure Processes
C15 Wood for Commercial-Residential Construction--Preservative Treatment by Pressure Processes
C17 Playground Equipment Treated with Inorganic Preservatives--Preservative Treatment by Pressure Processes
C18 Standard for Pressure Treated Material in Marine Construction
C22 Lumber and Plywood for Permanent Wood Foundations--Preservative Treatment by Pressure Processes
C23 Round Poles and Posts used in Building Construction--Preservative Treatment by Pressure Processes
C24 Sawn Timber Piles Used for Residential and Commercial Building
C25 Sawn Crossarms-Preservative Treatment by Pressure Process
C28 Standard for Preservative Treatment of Structural Glued Laminated Members and Laminations Before Gluing of Southern Pine, Pacific Coast Douglas Fir, Hemfir and Western Hemlock by Pressure Processes
C31 Lumber Used Out of Contact With the Ground and Continuously Protected from Liquid Water--Treatment by Pressure Processes
C33 Standard for Preservative Treatment of Structural Composite Lumber by Pressure Processes
C34 Shakes and Shingles-Preservative Treatment by Pressure Processes

6.2.2 ALSC Registered Trademarks

There are twenty-five agencies certified by the American Lumber Standard Committee (ALSC). The ALSC program is based on Voluntary Product Standard PS 20-94 and is administered by the Department of Commerce. Each agency has a registered trade-mark which is an integral part of the grade-mark applied to lumber graded under each agency's supervision. Copies of a brochure printed by the ALSC entitled "ALSC Certified Agencies and Typical Grade-Marks" can be obtained at no charge through the ALSC, P. O. Box 210, Germantown, MD 20874. Your personnel should be familiar with the species of lumber used and the agencies providing service for that species. A copy of the ALSC brochure should be available to your personnel at all times.

An example of an ALSC certified agency grade-mark and the information that a certified grade-mark must contain:

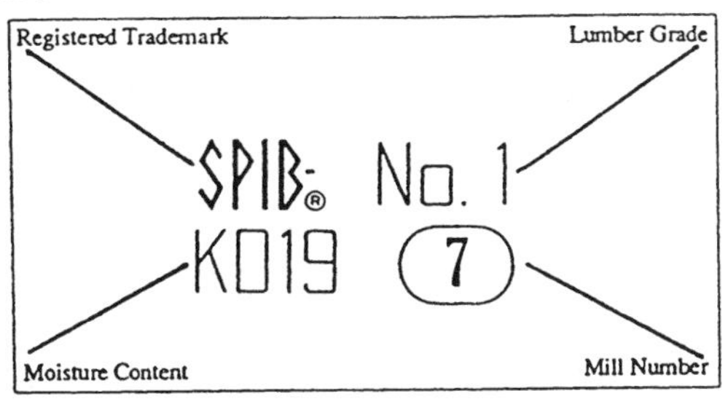

1. Agency logo or species of lumber bearing the stamp. In the case of the Southern Pine Inspection Bureau, the agency logo identifies the species Southern Pine.

2. Grade of Lumber.

3. Moisture content of lumber at the time of dressing if dressed lumber is involved. The moisture content designation is required on lumber in thickness less than 5 inches.

   KD-15 -- Kiln dried to 15% max. moisture content
   KD-19 -- Kiln dried to 19% max. moisture content
   S-DRY -- Kiln or Air dried to 19% max. moisture content
   S-GRN -- Indicates moisture content in excess of 19% and should be applied to all green lumber from 2-1/2" to 4-1/2" nominal thickness

4. Mill Identification Number.

SPIB will provide accurate information concerning ALSC approved grade-marks upon request.

SPIB: Jim Loy, (904) 434-2611

ALSC: John McDaniel, (301) 972-1700

Updated: 9/95

6.3.0 Moisture Content in Lumber

Both the WWPA and the SPIB have similar standards to designate moisture content in the lumber bearing their grading stamps. The moisture content of lumber is the weight of water contained in the lumber, expressed as a percentage of weight of the wood from which some water has been removed. Dry lumber is defined as having a moisture content of 19% or less; lumber with a moisture content in excess of 19% is classified as unseasoned lumber.

When standard-sized dry lumber is grade-stamped, the grade stamp will indicate the condition of "seasoning" as either MC15, KD15, S-DRY, or KD.

- *MC-15* Lumber surfaced with a moisture content of 15% or less.
- *KD-15* Kiln-dried lumber, surfaced, with a moisture content of 15% or less (kiln-dried lumber is lumber that has been heat-seasoned in a chamber to produce a predetermined moisture content).
- *S-DRY* Lumber surfaced with a moisture content of 19% or less.
- *KD* Kiln-dried lumber with a moisture content of 19% or less.
- *S-GRN* Unseasoned lumber with a moisture content in excess of 19%.

It is important to note that restrictions on moisture content apply at the time of shipment, as well as the time when it was surfaced. When lumber is shipped on open conveyances where it is susceptible to picking up moisture, the seller is relieved of any moisture content restrictions as long as the buyer is notified of the method of shipment (e.g., open-to-the-weather trucks, rail cars, or even ships) and agrees to this method of shipment.

6.4.0 WWPA Guide to Understanding Grade Stamps

Integrity of Grade Stamp

Western Wood Products Association is the largest association of lumber manufacturers in the United States. WWPA members and grading service subscribers are located in the 12 western states: Arizona, California, Colorado, Idaho, Montana, Nevada, New Mexico, Oregon, South Dakota, Utah, Washington and Wyoming. The Association's Quality Standards Department supervises lumber grading by maintaining a highly competent staff of lumber inspectors who regularly check the quality of mill production, including visual grade requirements of glued products and machine stress-rated lumber.

The Association's ***Grading Rules for Western Lumber*** establishes standards of size and levels of quality in conformance with the American Softwood Lumber Standard PS 20-94. The Association is certified as a rules writing and inspection agency by the Board of Review, American Lumber Standard Committee. The Association is approved to provide mill supervisory services under its rules and the rules of the West Coast Lumber Inspection Bureau, the Redwood Inspection Service, the National Lumber Grades Authority for Canadian Lumber and the NGR portion of the Southern Pine Inspection Bureau Rules. In addition, WWPA is approved to supervise finger-jointed and machine stress-rated lumber.

Interpreting Grade Marks

Most grade stamps, except those for rough lumber or heavy timbers, contain 5 basic elements:

a. **WWPA certification mark.** Certifies Association quality supervision. ® is a registered trademark.

b. **Mill identification.** Firm name, brand or assigned mill number. WWPA can be contacted to identify an individual mill whenever necessary.

c. **Grade designation.** Grade name, number or abbreviation.

d. **Species identification.** Indicates species by individual species or species combination. Species identification marks for groups to which design values are assigned are:

SPF^S WEST WOODS WEST CDR

e. **Condition of seasoning.** Indicates condition of seasoning at time of surfacing:

MC-15 — KD-15	15% maximum moisture content
S-DRY — KD	19% maximum moisture content
S-GRN —	over 19% moisture content (unseasoned)

Inspection Certificate

When an inspection certificate issued by the Western Wood Products Association is required on a shipment of lumber and specific grade marks are not used, the stock is identified by an imprint of the Association mark and the number of the shipping mill or inspector.

Grade Stamp Facsimiles

WWPA uses a set of marks similar to the randomly selected examples shown on the reverse side, to identify lumber graded under its supervision.

Species Combinations

The species groupings for dimension lumber products are shown left and explained in the second box on the reverse side. When alternative species combinations, as shown in the third box on the reverse side, are used for structural applications, design values are controlled by the species with the lowest strength value within the combination.

By permission of Western Wood Products Association

6.4.1 Species of Wood Included in WWPA Jurisdiction

Species or Species Combination	Mark
Douglas Fir and Larch Douglas Fir Western Larch	DOUG. FIR-L
Douglas Fir-South Lumber manufactured from Douglas Fir grown in Arizona, Colorado, Nevada, New Mexico and Utah.	
Hem-Fir California Red Fir, Grand Fir, Noble Fir, Pacific Silver Fir, White Fir and Western Hemlock	
Spruce-Pine-Fir (South) Engelmann Spruce, Sitka Spruce, Lodgepole Pine, Balsam Fir, Jack Pine, Red Pine, and Eastern Spruces The SPF[s] grouping is used by all U.S. rule writing agencies that write grading rules for certain Spruces, Pines and Firs. In the United States the SPF[s] mark can be used on any one of these species or combinations thereof.	SPF[S]
Western Cedars Incense, Western Red, Alaska and Port Orford Cedar	
Western Woods Any combination of western softwood species except Redwood and Western Cedars.	

Assigned design values for the following species combination are the same as those shown for Western Woods.

White Woods Any true firs, spruces, hemlocks or pines.	

By permission of Western Wood Products Association

6.4.2 Species Identification and Facsimile Grade Stamps

Species Identification

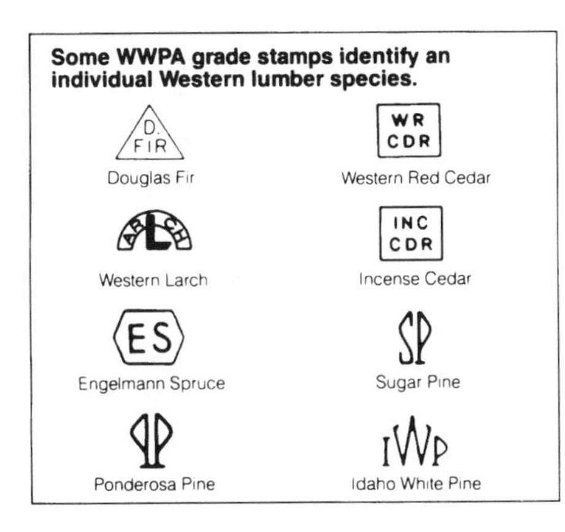

A number of Western lumber species have similar performance properties and are marketed with a common species designation. These species groupings are used for lumber to which design values are assigned.

Stamp	Species	Stamp	Species
DOUG. FIR-L	Douglas Fir and Larch	D FIR S	Douglas Fir South*
HEM FIR	California Red Fir, Grand Fir, Noble Fir, Pacific Silver Fir, White Fir and Western Hemlock	SPF S	Engelmann and Sitka Spruce, Lodgepole Pine (and Eastern firs and spruces)
WEST WOODS	Alpine Fir, Ponderosa Pine, Sugar Pine, Idaho White Pine and Mountain Hemlock, plus any of the species in the other groupings except Western Cedars	WEST CDR	Incense, Western Red, Port Orford and Alaska Cedar

*Lumber manufactured from Douglas Fir grown in Arizona, Colorado, Nevada, New Mexico and Utah.

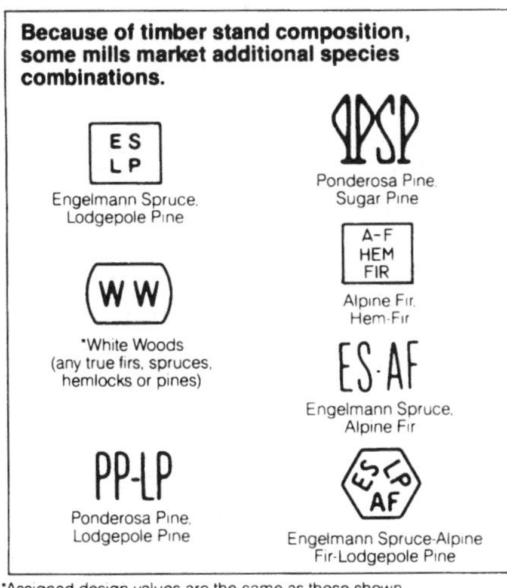

*Assigned design values are the same as those shown for Western Woods.

Facsimiles of Typical Grade Stamps

Dimension Grades

12 WWP® 2 S-DRY D. FIR

12 CONST WWP® S-DRY SPF S

12 STAND & BTR WWP® S-DRY D FIR S

Glued Products

12 STUD WWP® S-DRY VERTICAL USE ONLY WEST WOODS CERT GLUED JNTS

12 WWP® 1 S-DRY CERT EXT JNTS HEM FIR

Finish Grade — Graded Under WCLIB Rules

12 C & BTR WWP® VG S-DRY WCLB RULES D. FIR

Cedar Grades

12 CLEAR MC15 WWP® VG HEART WR CDR

12 WWP® A MC15 WEST CDR

Commons

12 2&BTR COM WWP® S-DRY PP

12 4 COM WWP® S-DRY ES

12 STERLING WWP® S-DRY IWP

Machine Stress-Rated Products

MACHINE RATED WWP® 12 S-DRY D FIR 1650 Fb 1.5E

MACHINE RATED WWP® 12 S-DRY D FIR 1650Fb 1020Ft 1.5E

Finish & Select Grades

12 C & BTR SEL WWP® MC 15 PP

12 PRIME WWP® MC 15 D FIR

12 D SEL WWP® MC 15 SP

Decking

12 SEL DECK WWP® MC 15 INC CDR

12 PATIO 1 WWP® S-DRY PP

6.4.3 Design Values for Various Species of Western Wood Products

Species or Group	Grade	Extreme Fiber Stress in Bending "Fb" Single	Tension Parallel to Grain "Ft"	Horizontal Shear "Fv"	Compression Perpendicular "Fc⊥"	Compression Parallel to Grain "Fc//"	Modulus of Elasticity "E"
Douglas Fir-Larch	Select Structural	1450	1000	95	625	1700	1,900,000
Douglas Fir	No. 1 & Btr.	1150	775	95	625	1500	1,800,000
Western Larch	No. 1	1000	675	95	625	1450	1,700,000
	No. 2	875	575	95	625	1300	1,600,000
	No. 3	500	325	95	625	750	1,400,000
	Construction	1000	650	95	625	1600	1,500,000
	Standard	550	375	95	625	1350	1,400,000
	Utility	275	175	95	625	875	1,300,000
	Stud	675	450	95	625	825	1,400,000
Douglas Fir-South	Select Structural	1300	875	90	520	1550	1,400,000
Douglas Fir South	No. 1	900	600	90	520	1400	1,300,000
	No. 2	825	525	90	520	1300	1,200,000
	No. 3	475	300	90	520	750	1,100,000
	Construction	925	600	90	520	1550	1,200,000
	Standard	525	350	90	520	1300	1,100,000
	Utility	250	150	90	520	875	1,000,000
	Stud	650	425	90	520	825	1,100,000
Hem-Fir	Select Structural	1400	900	75	405	1500	1,600,000
Western Hemlock	No. 1 & Btr.	1050	700	75	405	1350	1,500,000
Noble Fir	No. 1	950	600	75	405	1300	1,500,000
California Red Fir	No. 2	850	500	75	405	1250	1,300,000
Grand Fir	No. 3	500	300	75	405	725	1,200,000
Pacific Silver Fir	Construction	975	575	75	405	1500	1,300,000
White Fir	Standard	550	325	75	405	1300	1,200,000
	Utility	250	150	75	405	850	1,100,000
	Stud	675	400	75	405	800	1,200,000
Spruce-Pine-Fir (South)	Select Structural	1300	575	70	335	1200	1,300,000
Western Species:	No. 1	850	400	70	335	1050	1,200,000
Engelmann Spruce	No. 2	750	325	70	335	975	1,100,000
Sitka Spruce	No. 3	425	200	70	335	550	1,000,000
Lodgepole Pine	Construction	850	375	70	335	1200	1,000,000
	Standard	475	225	70	335	1000	900,000
	Utility	225	100	70	335	650	900,000
	Stud	575	250	70	335	600	1,000,000
Western Cedars	Select Structural	1000	600	75	425	1000	1,100,000
Western Red Cedar	No. 1	725	425	75	425	825	1,000,000
Incense Cedar	No. 2	700	425	75	425	650	1,000,000
Port Orford Cedar	No. 3	400	250	75	425	375	900,000
Alaska Cedar	Construction	800	475	75	425	850	900,000
	Standard	450	275	75	425	650	800,000
	Utility	225	125	75	425	425	800,000
	Stud	550	325	75	425	400	900,000
Western Woods	Select Structural	875	400	70	335	1050	1,200,000
Any of the species in the first four species groups above plus any or all of the following:	No. 1	650	300	70	335	925	1,100,000
Idaho White Pine	No. 2	650	275	70	335	875	1,000,000
Ponderosa Pine	No. 3	375	175	70	335	500	900,000
Sugar Pine	Construction	725	325	70	335	1050	1,000,000
Alpine Fir	Standard	400	175	70	335	900	900,000
Mountain Hemlock	Utility	200	75	70	335	600	800,000
	Stud	500	225	70	335	550	900,000

*Design values in pounds per square inch.

By permission of Western Wood Products Association

6.4.4 Adjustment Factors for Base Values of Western Wood Products

SIZE FACTORS (C_F) — Table A

Apply to Dimension Lumber Base Values

Grades	Nominal Width (depth)	F_b 2" & 3" thick nominal	F_b 4" thick nominal	F_t	$F_{c//}$	Other Properties
Select Structural, No. 1 & Btr., No. 1, No. 2 & No. 3	2", 3" & 4"	1.5	1.5	1.5	1.15	1.0
	5"	1.4	1.4	1.4	1.1	1.0
	6"	1.3	1.3	1.3	1.1	1.0
	8"	1.2	1.3	1.2	1.05	1.0
	10"	1.1	1.2	1.1	1.0	1.0
	12"	1.0	1.1	1.0	1.0	1.0
	14" & wider	0.9	1.0	0.9	0.9	1.0
Construction & Standard	2", 3" & 4"	1.0	1.0	1.0	1.0	1.0
Utility	2" & 3"	0.4	—	0.4	0.6	1.0
	4"	1.0	1.0	1.0	1.0	1.0
Stud	2", 3" & 4"	1.1	1.1	1.1	1.05	1.0
	6" & wider	1.0	1.0	1.0	1.0	1.0

REPETITIVE MEMBER FACTOR (C_r) — Table B

Apply to Size-adjusted F_b

Where 2" to 4" thick lumber is used repetitively, such as for joists, studs, rafters and decking, the pieces side by side share the load and the strength of the entire assembly is enhanced. Therefore, where three or more members are adjacent or are not more than 24" on center and are joined by floor, roof or other load distributing elements, the F_b value can be increased 1.15 for repetitive member use.

REPETITIVE MEMBER USE
$F_b \times 1.15$

DURATION OF LOAD ADJUSTMENT (C_D) — Table C

Apply to Size-adjusted Values

Wood has the property of carrying substantially greater maximum loads for short durations than for long durations of loading. Tabulated design values apply to normal load duration. (Factors do not apply to MOE or $F_{c\perp}$).

LOAD DURATION	FACTOR
Permanent	0.9
Ten Years (Normal Load)	1.0
Two Months (Snow Load)	1.15
Seven Day	1.25
One Day	1.33
Ten Minutes (Wind and Earthquake Loads)	1.6
Impact	2.0

Confirm load requirements with local codes. Refer to Model Building Codes or the National Design Specification for high-temperature or fire-retardant treated adjustment factors.

HORIZONTAL SHEAR ADJUSTMENT (C_H) — Table D

Apply to F_v Values

Horizontal shear values published in Table 1 are based upon the maximum degree of shake, check or split that might develop in a piece. When the actual size of these characteristics is known, the following adjustments may be taken.

2" THICK LUMBER

For convenience, the table below may be used to determine horizontal shear values for any grade of 2" thick lumber in any species when the length of split or check is known and any increase in them is not anticipated.

When length of split on wide face is:	Multiply Tabulated Fv value by:
No split	2.00
1/2 of wide face	1.67
3/4 of wide face	1.50
1 of wide face	1.33
1½ of wide face or more	1.00

3" and THICKER LUMBER

Horizontal shear values for 3" and thicker lumber also are established as if a piece were split full length. When specific lengths of splits are known and any increase in them is not anticipated, the following adjustments may be applied.

When length of split on wide face is:	Multiply Tabulated Fv value by:
No split	2.00
1/2 of narrow face	1.67
1 of narrow face	1.33
1½ of narrow or more	1.00

BASE VALUE EQUATIONS

The basic difference between using BASE VALUES and the design values that were published for dimension lumber prior to the results of the In-Grade Testing Program, is that BASE VALUES must be adjusted for SIZE before conditions of use. The table below shows how the adjustments are applied to BASE VALUES.

BASE VALUE EQUATIONS

Apply to Dimension Lumber Values in Table 1

Base Value	x	Size Adjustment Factor	x	Routine Adjustment Factors	x	Special Use Factors	=	Design Value
F_b	x	C_F	x	$C_D \times C_r$	x	$C_M \times C_R \times C_t \times C_{fu}$	=	F'_b
F_t	x	C_F	x	C_D	x	$C_M \times C_R \times C_t$	=	F'_t
F_v			x	$C_D \times C_H$	x	$C_M \times C_R \times C_t$	=	F'_v
$F_{c\perp}$*					x	$C_M \times C_R \times C_t$	=	$F'_{c\perp}$
$F_{c//}$	x	C_F	x	C_D	x	$C_M \times C_R \times C_t$	=	$F'_{c//}$
E					x	$C_M \times C_R \times C_t$	=	E'

* For $F_{c\perp}$ value of 0.02" deformation basis, see Table F.

Note:
- C_F = Size Factor
- C_r = Repetitive Member Factor
- C_H = Horizontal Shear
- C_D = Duration of Load
- C_{fu} = Flat Use Factor
- C_M = Wet Use Factor
- C_R = Fire Retardant Factor, refer to the National Design Specification
- C_t = Temperature Factor, refer to the National Design Specification

The following adjustment factors are shown in the WWPA Product Use Manual:

Flat Use Factors (C_{fu})	(Table E)
Adjustments for Compression Perpendicular to Grain ($C_{c\perp}$)	(Table F)
Wet Use Factors (C_M)	(Table G)

By permission of Western Wood Products Association

6.4.5 Additional Adjustment Factors for Western Wood Dimension Lumber

FLAT USE FACTORS (C_{fu}) — Table E

Apply to Size-adjusted F_b

NOMINAL WIDTH	NOMINAL THICKNESS	
	2" & 3"	4"
2" & 3"	1.00	—
4"	1.10	1.00
5"	1.10	1.05
6"	1.15	1.05
8"	1.15	1.05
10" & wider	1.20	1.10

ADJUSTMENTS FOR COMPRESSION PERPENDICULAR-TO-GRAIN ($C_{c\perp}$) — Table F

For Deformation Basis of 0.02"
Apply to $F_{c\perp}$ Values

Design values for compression perpendicular-to-grain ($F_{c\perp}$) are established in accordance with the procedures set forth in ASTM Standards D 2555 and D 245. ASTM procedures consider deformation under bearing loads as a serviceability limit state comparable to bending deflection because bearing loads rarely cause structural failures. Therefore, ASTM procedures for determining compression perpendicular-to-grain values are based on a deformation of 0.04" and are considered adequate for most classes of structures. Where more stringent measures need to be taken in design, the following formula permits the designer to adjust design values to a more conservative deformation basis of 0.02":

$$Y_{02} = 0.73\, Y_{04} + 5.60$$

EXAMPLE: Douglas Fir-Larch: $Y_{04} = 625$ psi
$Y_{02} = 0.73\,(625) + 5.60 = 462$ psi

WET USE FACTORS (C_M) — Table G

Apply to Size-adjusted Values

The design values shown in the accompanying tables are for routine construction applications where the moisture content of the wood does not exceed 19%. When use conditions are such that the moisture content of dimension lumber will exceed 19%, the Wet Use Adjustment Factors below are recommended:

PROPERTY		ADJUSTMENT FACTOR
F_b	Extreme Fiber Stress in Bending	0.85*
F_t	Tension Parallel-to-Grain	1.0
F_c	Compression Parallel-to-Grain	0.8**
F_v	Horizontal Shear	0.97
$F_{c\perp}$	Compression Perpendicular-to-Grain	0.67
E	Modulus of Elasticity	0.9

*Wet Use Factor 1.0 for size-adjusted F_b not exceeding 1150 psi.
**Wet Use Factor 1.0 for size-adjusted F_c not exceeding 750 psi.

SPECIAL DIMENSION LUMBER

Grades/End Uses - There are two categories of Special Dimension Lumber grades. Design values are shown in Tables 2 and 3.

a. Structural Decking - 2x4 through 4x12
b. Machine Stress-Rated Lumber (MSR) - nominal 2" and less in thickness, 2" and wider

STRUCTURAL DECKING

Grades/End Uses - Standard decking patterns, in nominal 2" single T&G and 3" and 4" double T&G, are available in vee or eased joints to meet most architectural design requirements. For diagrams of available patterns and sizes, order WWPA's *Standard Patterns* (G-16).

While known and used as "roof decking," the load-bearing capacities of structural decking also make it useful as floor decking and solid sidewall construction. Published design values need to be adjusted for depth effect. Refer to Tables 2 and H below.

Decking spans are provided in Table 10, page 15.

STRUCTURAL DECKING DESIGN VALUES* — Table 2

2" to 4" thick, 4" to 12" wide
USE WITH ADJUSTMENTS, TABLES C, G, H
For Flatwise Use Only

		DRY or MC 15			
		Extreme Fiber Stress in Bending "F_b"		Compression Perpendicular "$F_{c\perp}$"	Modulus of Elasticity "E"
Species	Grade	Single Member	Repetitive Member		
Douglas Fir-Larch	Sel.	1750	2000	625	1,800,000
	Com.	1450	1650	625	1,700,000
Douglas Fir-South	Sel.	1750	1900	520	1,400,000
	Com.	1400	1600	520	1,300,000
Hem-Fir	Sel.	1400	1600	405	1,500,000
	Com.	1150	1350	405	1,400,000
SPFS	Sel.	1150	1350	335	1,400,000
	Com.	950	1100	335	1,200,000
Western Cedars	Sel.	1250	1450	425	1,100,000
	Com.	1050	1200	425	1,000,000
Western Woods	Sel.	1150	1300	335	1,200,000
	Com.	950	1100	335	1,100,000

*Design values in pounds per square inch.
See Table 1 (p. 6) for compression perpendicular-to-grain ($F_{c\perp}$) values.

ADJUSTMENT FACTORS FOR DEPTH EFFECT — Table H

For all widths of Structural Decking
Apply to Dimension Lumber Base Values

Decking bending design values may be adjusted for thickness as shown below because the bending values shown in Table 2 are based on a 4" thick member loaded flatwise.

NOMINAL THICKNESS		
2"	3"	4"
1.10	1.04	1.00

ADJUSTMENTS FOR STRUCTURAL DECKING — Checklist 2

- ☐ Duration of Load (C_D) — Table C, page 7
- ☐ Wet Use Factor (C_M) (only when appropriate) — Table G, page 9
- ☐ Depth Effect — Table H, page 9

6.5.0 Standard Sizes for Western Wood Finish and Selects (Dry Lumber)

The metric dimensions listed in these rules are calculated at 25.4 millimeters (mm) times the actual dimension in inches, rounded to the nearest millimeter. In case of a dispute on size measurements, the conventional (inch) method of measurement shall take precedence.

STANDARD SIZES for FINISH DRY LUMBER

Thicknesses			Widths		
Nominal	Surfaced		Nominal	Surfaced	
	Inch	mm (1)		Inch	mm (1)
3/8"	5/16	8	2"	1 1/2	38
1/2"	7/16	11	3"	2 1/2	64
5/8"	9/16	14	4"	3 1/2	89
3/4"	5/8	16	5"	4 1/2	114
1"	3/4	19	6"	5 1/2	140
1 1/4"	1	25	7"	6 1/2	165
1 1/2"	1 1/4	32	8" and wider	3/4 off nominal	19 off nominal
1 3/4"	1 3/8	35			
2"	1 1/2	38			
2 1/2"	2	51			
3"	2 1/2	64			
3 1/2"	3	76			
4"	3 1/2	89			

(1) See Section 723.00.

STANDARD SIZES for SELECTS DRY LUMBER

Thicknesses			Widths		
Nominal	Surfaced		Nominal	Surfaced	
	Inch	mm (1)		Inch	mm (1)
4/4	3/4	19	2"	1 1/2	38
5/4	1 5/32	29	3"	2 1/2	64
6/4	1 13/32	36	4"	3 1/2	89
7/4	1 19/32	40	5"	4 1/2	114
8/4	1 13/16	46	6"	5 1/2	140
9/4	2 3/32	53	7"	6 1/2	165
10/4	2 3/8	60	8" and wider	3/4 off nominal	19 off nominal
11/4	2 9/16	65			
12/4	2 3/4	70			
16/4	3 3/4	95			

(1) See Section 723.00.

By permission of Western Wood Products Association

6.5.1 Standard Sizes for Western Wood Common Boards, Studs, and Battens

STANDARD SIZES for COMMON BOARDS (Including Thick Lumber Shipped Under Board Rules) DRY LUMBER

Thicknesses			Widths		
Nominal	Surfaced		Nominal	Surfaced	
	Inch	mm (1)		Inch	mm (1)
3/4	5/8	16	2"	1½	38
4/4	¾	19	3"	2½	64
5/4	1 5/32	29	4"	3½	89
6/4	1 13/32	36	5"	4½	114
7/4	1 19/32	40	6"	5½	140
8/4	1 13/16	46	7"	6½	165
9/4	2 3/32	53	8" and wider	¾ off nominal	19 off nominal
10/4	2 3/8	60			
11/4	2 9/16	65			
12/4	2¾	70			
16/4	3¾	95			

(1) See Section 723.00.

Surfaced square size shall be governed by thickness.
At manufacturer's option, dry 4/4 may be 25/32"
Standard lengths are 6' and longer in multiples of 1.

STANDARD SIZES for STUDS

Nominal	Surfaced Dry		Surfaced Unseasoned	
	Inch	mm (1)	Inch	mm (1)
Thicknesses				
2"	1½	38	1 9/16	40
3"	2½	64	2 9/16	65
4"	3½	89	3 9/16	90
Widths				
2"	1½	38	1 9/16	40
3"	2½	64	2 9/16	65
4"	3½	89	3 9/16	90
5"	4½	114	4 5/8	117
6"	5½	140	5 5/8	143
8" and wider	¾ off nominal	19 off nominal	½ off nominal	13 off nominal

(1) See Section 723.00.

BATTENS

All Species

Standard widths are:

Nominal	Net	
	Inch	mm (1)
Flat Battens—3"	¼ x 2½	6 x 64
O.G. Battens—2"	¾ x 1¾	19 x 44
O.G. Battens—2½"	¾ x 2¼	19 x 57
O.G. Battens—3"	¾ x 2½	19 x 64

(1) See Section 723.00.

By permission of Western Wood Products Association

6.5.2 Western Wood Scaffolding Sizes and Design Values

SCAFFOLD PLANK
Douglas Fir and Larch
1¼" and Thicker
8" and Wider

There are two grades of Scaffold Plank: SCAFFOLD NO. 1 and SCAFFOLD NO. 2. Design Values for Douglas Fir and Larch are as follows:

Design Values—For Flatwise Use*

Thickness	Grade	Extreme Fiber Stress in Bending (Fb) in psi	Modulus of Elasticity (E) in psi
2" & less	No. 1	2350	1,800,000
	No. 2	2200	1,800,000
These values apply to dry use conditions. For wet use conditions, these values shall be multiplied by 0.86 for Fb and 0.97 for E.			
3"	No. 1	1800	1,600,000
	No. 2	1650	1,600,000
These values apply to both dry and wet use conditions.			

*See Sections 100.000 through 170.00 for information about these values.

Other species may be graded under these rules and design values for them may be obtained from the Association. All pieces are FOHC and the face showing the more serious characteristics are used to determine the grade. Knot size is determined by the average diameter of the largest knot showing on either wide face. Knots showing on narrow faces are permitted if they displace no more of the cross section than knots on wide faces, except spike knots across the full width are not permitted.

Scaffold plank is usually ordered unseasoned and grades are based on rough lumber. Scaffold plank is full sawn, except an occasional piece may be ⅛" scant in thickness or ¼" scant in width.

By permission of Western Wood Products Association

6.6.0 American Softwood Standards for Boards and Timbers

The thicknesses apply to all widths and all widths apply to all thicknesses. Sizes are given in inches and millimeters. Metric units are based on actual size. See B2, Appendix B for rounding rule for metric units.

Item	Thicknesses					Face Widths				
	Nominal Inch	Minimum Dressed				Nominal Inch	Minimum Dressed			
		Dry[a]		Green[a]			Dry[a]		Green[a]	
		Inch	mm	Inch	mm		Inch	mm	Inch	mm
						2	1-1/2	38	1-9/16	40
						3	2-1/2	64	2-9/16	65
						4	3-1/2	89	3-9/16	90
						5	4-1/2	114	4-5/8	117
	1	3/4	19	25/32	20	6	5-1/2	140	5-5/8	143
						7	6-1/2	165	6-5/8	168
Boards[b]	1-1/4	1	25	1-1/32	26	8	7-1/4	184	7-1/2	190
						9	8-1/4	210	8-1/2	216
	1-1/2	1-1/4	32	1-9/32	33	10	9-1/4	235	9-1/2	241
						11	10-1/4	260	10-1/2	267
						12	11-1/4	286	11-1/2	292
						14	13-1/4	337	13-1/2	343
						16	15-1/4	387	15-1/2	394
						2	1-1/2	38	1-9/16	40
						3	2-1/2	64	2-9/16	65
	2	1-1/2	38	1-9/16	40	4	3-1/2	89	3-9/16	90
	2-1/2	2	51	2-1/16	52	5	4-1/2	114	4-5/8	117
	3	2-1/2	64	2-9/16	65	6	5-1/2	140	5-5/8	143
Dimension	3-1/2	3	76	3-1/16	78	8	7-1/4	184	7-1/2	190
	4	3-1/2	89	3-9/16	90	10	9-1/4	235	9-1/2	241
	4-1/2	4	102	4-1/16	103	12	11-1/4	286	11-1/2	292
						14	13-1/4	337	13-1/2	343
						16	15-1/4	387	15-1/2	394
Timbers	5 & thicker	1/2 off	13 off	1/2 off	13 off	5 & wider	1/2 off	13 off	1/2 off	13 off

[a] See 2.7 and 2.11 for the definitions of dry and green lumber.

[b] Boards less than the minimum thickness for nominal 1-inch but 5/8 inch (16 mm) or greater thickness dry (11/16 inch (17 mm) green) shall be regarded as ALS lumber, but such boards shall be marked to show the size and condition of seasoning at the time of dressing. They shall also be distinguished from nominal 1-inch boards on invoices and certificates.

6.6.1 American Softwood Standards for Shiplap and Centermatch Lumber

The thicknesses apply to all widths and all widths apply to all thicknesses. Sizes are given in inches and millimeters. Metric units are based on actual size. See B2, Appendix B for rounding rule for metric units.

Item	Thicknesses					Face Widths				
	Nominal Inch	Minimum Dressed				Nominal Inch	Minimum Dressed			
		Dry[a]		Green[a]			Dry[a]		Green[a]	
		Inch	mm	Inch	mm		Inch	mm	Inch	mm
Shiplap, 3/8-inch (10 mm) lap	1	3/4	19	25/32	20	4	3-1/8	79	3-3/16	81
						6	5-1/8	130	5-1/4	133
						8	6-7/8	175	7-1/8	181
						10	8-7/8	225	9-1/8	232
						12	10-7/8	276	11-1/8	283
						14	12-7/8	327	13-1/8	333
						16	14-7/8	378	15-1/8	384
Shiplap, 1/2 inch (13 mm) lap	1	3/4	19	25/32	20	4	3	76	3-1/16	78
						6	5	127	5-1/8	130
						8	6-3/4	171	7	178
						10	8-3/4	222	9	229
						12	10-3/4	273	11	279
						14	12-3/4	324	13	330
						16	14-3/4	375	15	381
Centermatch, 1/4 inch (6 mm) tongue	1	3/4	19	25/32	20	4	3-1/8	79	3-3/16	81
	1-1/4	1	25	1-1/32	26	5	4-1/8	105	4-1/4	108
	1-1/2	1-1/4	32	1-9/32	33	6	5-1/8	130	5-1/4	133
						8	6-7/8	175	7-1/8	181
						10	8-7/8	225	9-1/8	232
						12	10-7/8	276	11-1/8	283
2 inch (51 mm) D & M, 3/8 inch (10 mm) tongue	2	1-1/2	38	1-9/16	40	4	3	76	3-1/16	78
						6	5	127	5-1/8	130
						8	6-3/4	171	7	178
						10	8-3/4	222	9	229
						12	10-3/4	273	11	279
2 inch (51 mm) Shiplap, 1/2 inch (13 mm) lap	2	1-1/2	38	1-9/16	40	4	3	76	3-1/16	78
						6	5	127	5-1/8	130
						8	6-3/4	171	7	178
						10	8-3/4	222	9	229
						12	10-3/4	273	11	279

National Institute of Standards and Technology

6.6.2 American Softwood Standards for Worked Lumber

The thicknesses apply to all widths and all widths apply to all thicknesses. Sizes are given in inches and millimeters. Metric units are based on actual size. See B2, Appendix B for rounding rule for metric units.

Thicknesses[a]					Face Widths				
Nominal Inch	Minimum Dressed				Nominal Inch	Minimum Dressed			
	Dry		Green			Dry		Green	
	Inch	mm	Inch	mm		Inch	mm	Inch	mm
Tongue and Grooved									
2-1/2	2	51	2-1/16	52	4	3	76	3-1/16	78
3	2-1/2	64	2-9/16	65	6	5	127	5-1/8	130
3-1/2	3	76	3-1/16	78	8	6-3/4	171	7	178
4	3-1/2	89	3-9/16	90	10	8-3/4	222	9	229
4-1/2	4	102	4-1/16	103	12	10-3/4	273	11	279
Shiplap									
2-1/2	2	51	2-1/16	52	4	3	76	3-1/16	78
3	2-1/2	64	2-9/16	65	6	5	127	5-1/8	130
3-1/2	3	76	3-1/16	78	8	6-3/4	171	7	178
4	3-1/2	89	3-9/16	90	10	8-3/4	222	9	229
4-1/2	4	102	4-1/16	103	12	10-3/4	273	11	279
Grooved-for-Splines									
2-1/2	2	51	2-1/16	52	4	3-1/2	89	3-9/16	90
3	2-1/2	64	2-9/16	65	6	5-1/2	140	5-5/8	143
3-1/2	3	76	3-1/16	78	8	7-1/4	184	7-1/2	190
4	3-1/2	89	3-9/16	90	10	9-1/4	235	9-1/2	241
4-1/2	4	102	4-1/16	103	12	11-1/4	286	11-1/2	292

In worked lumber of nominal 2-inch and greater thickness, the tongue shall be 3/8 inch (10 mm) wide in tongued-and-grooved lumber and the lap shall be 1/2 inch (13 mm) wide in shiplapped lumber, with the over-all widths 3/8 inch (10 mm) and 1/2 inch (13 mm) wider, respectively, than the face widths shown in the above table. Double tongued-and-grooved decking shall be manufactured with a 3/8 inch (10 mm) or 5/16 inch (8 mm) wide tongue.

[a] See Table 3 for information on nominal 2-inch dimension.

National Institute of Standards and Technology

6.6.3 American Softwood Standards for Siding (19% Moisture Content)

The thicknesses apply to all widths and all widths apply to all thicknesses. Sizes are given in inches and millimeters. Metric units are based on actual size. See B2, Appendix B for rounding rule.

Item	Thicknesses			Face Widths		
	Nominal[a] Inch	Minimum Dressed		Nominal Inch	Minimum Dressed	
		Inch	mm		Inch	mm
Bevel Siding	1/2 9/16 5/8 3/4 1	7/16 butt, 3/16 tip 15/32 butt, 3/16 tip 9/16 butt, 3/16 tip 11/16 butt, 3/16 tip 3/4 butt, 3/16 tip	11 butt, 5 tip 12 butt, 5 tip 14 butt, 5 tip 17 butt, 5 tip 19 butt, 5 tip	4 5 6 8 10 12	3-1/2 4-1/2 5-1/2 7-1/4 9-1/4 11-1/4	89 114 140 184 235 286
Bungalow Siding	3/4	11/16 butt, 3/16 tip	17 butt, 5 tip	8 10 12	7-1/4 9-1/4 11-1/4	184 235 286
Rustic and Drop Siding (shiplapped, 3/8 inch (10 mm) lap)	5/8 1	9/16 23/32	14 18	4 5 6	3 4 5	76 102 127
Rustic and Drop Siding (shiplapped, 1/2 inch (13 mm) lap)	5/8 1	9/16 23/32	14 18	4 5 6 8 10 12	2-7/8 3-7/8 4-7/8 6-5/8 8-5/8 10-5/8	73 98 124 168 219 270
Rustic and Drop Siding (dressed and matched)	5/8 1	9/16 23/32	14 18	4 5 6 8 10	3-1/8 4-1/8 5-1/8 6-7/8 8-7/8	79 105 130 175 225

[a] For lumber of less than nominal 1-inch thickness, the board measure count is based on the nominal surface dimensions (width by length). Otherwise, the nominal inch units of designated thicknesses and widths in this table are the same as the board measure or count sizes. Lumber shall be measured by board or cubic measure.

National Institute of Standards and Technology

6.6.4 American Softwood Standards for Finish, Floor, and Ceiling Partition Lumber

The thicknesses apply to all widths and all widths apply to all thicknesses except as modified. Sizes are given in inches and millimeters. Metric units are based on actual size. See B2, Appendix B for rounding rule for metric units.

Item	Thicknesses			Face Widths		
	Nominal[a] Inch	Minimum Dressed		Nominal Inch	Minimum Dressed	
		Inch	mm		Inch	mm
Finish	3/8	5/16	8	2	1-1/2	38
	1/2	7/16	11	3	2-1/2	64
	5/8	9/16	14	4	3-1/2	89
	3/4	5/8	16	5	4-1/2	114
	1	3/4	19	6	5-1/2	140
	1-1/4	1	25	7	6-1/2	165
	1-1/2	1-1/4	32	8	7-1/4	184
	1-3/4	1-3/8	35	9	8-1/4	210
	2	1-1/2	38	10	9-1/4	235
	2-1/2	2	51	11	10-1/4	260
	3	2-1/2	64	12	11-1/4	286
	3-1/2	3	76	14	13-1/4	337
	4	3-1/2	89	16	15-1/4	387
Flooring[b]	3/8	5/16	8	2	1-1/8	29
	1/2	7/16	11	3	2-1/8	54
	5/8	9/16	14	4	3-1/8	79
	1	3/4	19	5	4-1/8	105
	1-1/4	1	25	6	5-1/8	130
	1-1/2	1-1/4	32			
Ceiling[b]	3/8	5/16	8	3	2-1/8	54
	1/2	7/16	11	4	3-1/8	79
	5/8	9/16	14	5	4-1/8	105
	3/4	11/16	17	6	5-1/8	130
Partition[b]	1	23/32	18	3	2-1/8	54
				4	3-1/8	79
				5	4-1/8	105
				6	5-1/8	130
Stepping[b]	1	3/4	19	8	7-1/4	184
	1-1/4	1	25	10	9-1/4	235
	1-1/2	1-1/4	32	12	11-1/4	286
	2	1-1/2	38			

[a] For lumber of less than nominal 1-inch thickness, the board measure count is based on the nominal surface dimensions (width by length). Otherwise, the nominal inch units of designated thicknesses and widths in this table are the same as the board measure or count sizes. Lumber shall be measured by board or cubic measure.

[b] In tongued-and-grooved flooring and in tongued-and-grooved and shiplapped ceiling of 5/16 inch (8 mm), 7/16 inch (11 mm), and 9/16 inch (14 mm) dressed thicknesses, the tongue or lap shall be 3/16 inch (5 mm) wide, with the over-all widths 3/16 inch (5 mm) wider than the face widths shown in the above table. In all other worked lumber of dressed thicknesses of 5/8 inch (16 mm) to 1-1/4 inches (32 mm), the tongue shall be 1/4 inch (6 mm) wide or wider in tongued-and-grooved lumber, and the lap shall be 3/8 inch (10 mm) wide or wider in shiplapped lumber, and the over-all widths shall be not less than the dressed face widths shown in the above table plus the width of the tongue or lap.

National Institute of Standards and Technology

6.7.0 Specifying Southern Pine Lumber (Grade Stamp Markings)

The Southern Pine Inspection Bureau is the rules writing agency for the Southern Pine Industry. For your grade-marked Southern Pine orders specify the SPIB logo for quality.

Typical facsimiles of the approved SPIB registered grade-marks are displayed below:

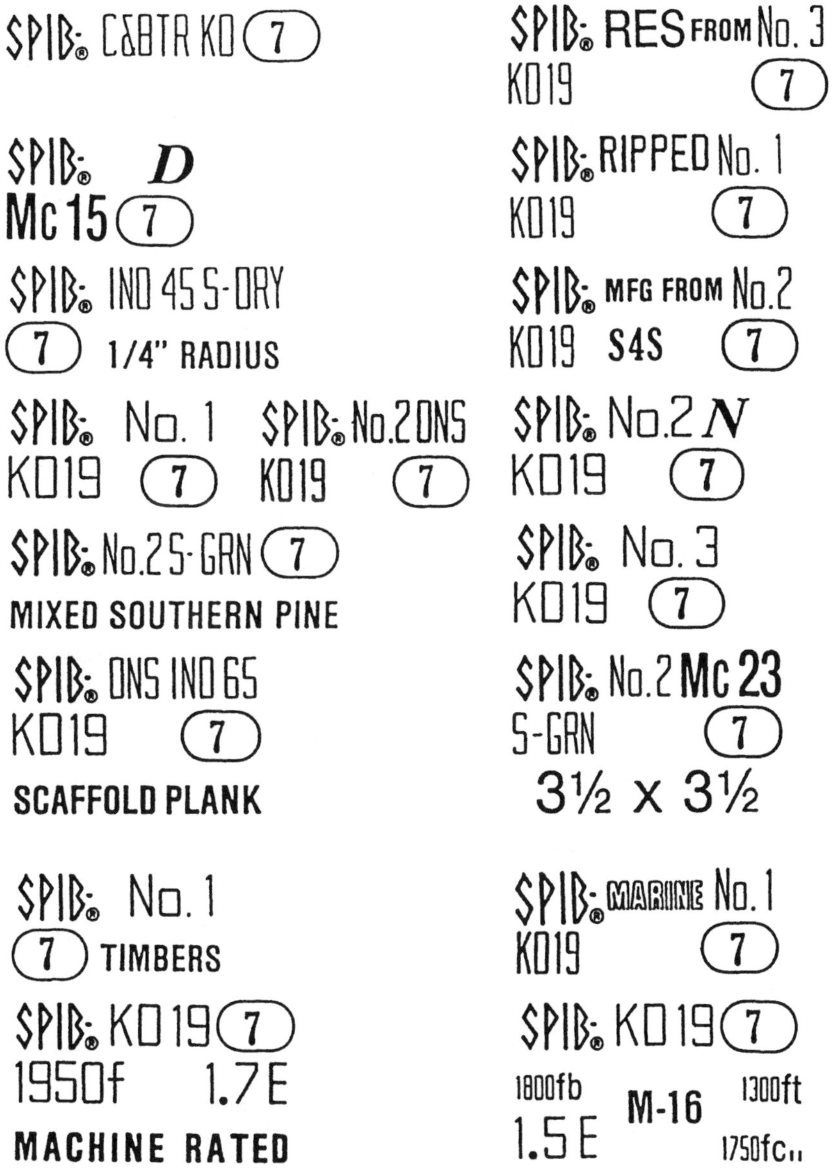

SPIB® C&BTR KD (7)

SPIB® D
Mc 15 (7)

SPIB® IND 45 S-DRY
(7) 1/4" RADIUS

SPIB® No. 1
KD19 (7)

SPIB® No.2 DNS
KD19 (7)

SPIB® No.2 S-GRN (7)
MIXED SOUTHERN PINE

SPIB® DNS IND 65
KD19 (7)
SCAFFOLD PLANK

SPIB® No. 1
(7) TIMBERS

SPIB® KD 19 (7)
1950f 1.7E
MACHINE RATED

SPIB® RES FROM No. 3
KD19 (7)

SPIB® RIPPED No. 1
KD19 (7)

SPIB® MFG FROM No.2
KD19 S4S (7)

SPIB® No.2 N
KD19 (7)

SPIB® No. 3
KD19 (7)

SPIB® No.2 Mc 23
S-GRN (7)
3½ x 3½

SPIB® MARINE No. 1
KD19 (7)

SPIB® KD 19 (7)
1800fb M-16 1300ft
1.5 E 1750fc⊥

*Timbers 5" x 5" and larger are not required to be dry unless specified.

Before specifying, consult current editions of the SPIB Standard Grading Rules and/or SPIB Special Product Rules. Please feel free to contact the SPIB office for further information concerning your specifications. Our telephone number is: (904) 434-2611.

By permission of Southern Pipe Inspection Bureau

6.7.1 Southern Pine Span Tables for Joists

SOUTHERN PINE SPAN TABLES

Maximum spans given in feet and inches
Inside to inside of bearings

Tables 5 thru 11 are abbreviated span tables for the most commonly available grades of Southern Pine lumber. For other grades, loading conditions and spacings, refer to *Maximum Spans for Southern Pine Joists and Rafters* published by the Southern Pine Council.

These spans are based on *1993 AF&PA Span Tables for Joists and Rafters*, and *1994 SPIB Standard Grading Rules for Southern Pine Lumber.* Except for Table 8, they are intended for use in covered structures or where the moisture content in use does not exceed 19 percent for an extended period of time.

Floor Joists

Design Criteria: Deflection – limited to span in inches divided by 360 (live load only).
Strength – based on 30, 40, or 50 pounds per square foot (psf) live load, plus 10 psf dead load.

		Size (inches) and Spacing (inches on center)											
		2 x 6			2 x 8			2 x 10			2 x 12		
Grade	Live Load	12″oc	16″oc	24″oc	12″oc	16″oc	24″oc	12″oc	16″oc	24″oc	12″oc	16″oc	24″oc
No. 1	30 psf	12 – 0	10 – 11	9 – 7	15 – 10	14 – 5	12 – 7	20 – 3	18 – 5	16 – 1	24 – 8	22 – 5	19 – 6
	40 psf	10 – 11	9 – 11	8 – 8	14 – 5	13 – 1	11 – 5	18 – 5	16 – 9	14 – 7	22 – 5	20 – 4	17 – 5
	50 psf	10 – 2	9 – 3	8 – 1	13 – 5	12 – 2	10 – 8	17 – 1	15 – 6	13 – 4	20 – 9	18 – 10	15 – 11
No. 2	30 psf	11 – 10	10 – 9	9 – 4	15 – 7	14 – 2	12 – 4	19 – 10	18 – 0	14 – 8	24 – 2	21 – 1	17 – 2
	40 psf	10 – 9	9 – 9	8 – 6	14 – 2	12 – 10	11 – 0	18 – 0	16 – 1	13 – 2	21 – 9	18 – 10	15 – 4
	50 psf	9 – 11	9 – 1	7 – 9	13 – 1	11 – 11	10 – 0	16 – 9	14 – 8	12 – 0	19 – 10	17 – 2	14 – 0
No. 3	30 psf	10 – 5	9 – 1	7 – 5	13 – 3	11 – 6	9 – 5	15 – 8	13 – 7	11 – 1	18 – 8	16 – 2	13 – 2
	40 psf	9 – 4	8 – 1	6 – 7	11 – 11	10 – 3	8 – 5	14 – 0	12 – 2	9 – 11	16 – 8	14 – 5	11 – 10
	50 psf	8 – 6	7 – 5	6 – 0	10 – 10	9 – 5	7 – 8	12 – 10	11 – 1	9 – 1	15 – 3	13 – 2	10 – 9

Ceiling Joists – Drywall Ceiling

Design Criteria: Deflection – limited to span in inches divided by 240 (live load only).
Strength – based on 10 or 20 pounds per square foot (psf) live load, plus 5 or 10 psf dead load.

		Size (inches) and Spacing (inches on center)											
		2 x 4			2 x 6			2 x 8			2 x 10		
Grade	Live Load	12″oc	16″oc	24″oc	12″oc	16″oc	24″oc	12″oc	16″oc	24″oc	12″oc	16″oc	24″oc
No. 1	10 psf	12 – 8	11 – 6	10 – 0	19 – 11	18 – 1	15 – 9	26 – 0	23 – 10	20 – 10	26 – 0*	26 – 0*	26 – 0*
	20 psf	10 – 0	9 – 1	8 – 0	15 – 9	14 – 4	12 – 6	20 – 10	18 – 11	15 – 11	26 – 0*	23 – 2	18 – 11
No. 2	10 psf	12 – 5	11 – 3	9 – 10	19 – 6	17 – 8	15 – 6	25 – 8	23 – 4	20 – 1	26 – 0*	26 – 0*	24 – 0
	20 psf	9 – 10	8 – 11	7 – 8	15 – 6	13 – 6	11 – 0	20 – 1	17 – 5	14 – 2	24 – 0	20 – 9	17 – 0
No. 3	10 psf	11 – 7	10 – 0	8 – 2	17 – 1	14 – 9	12 – 1	21 – 8	18 – 9	15 – 4	25 – 7	22 – 2	18 – 1
	20 psf	8 – 2	7 – 1	5 – 9	12 – 1	10 – 5	8 – 6	15 – 4	13 – 3	10 – 10	18 – 1	15 – 8	12 – 10

Floor Joists – Heavy Live Loads

Design Criteria: Deflection – limited to span in inches divided by 360 (live load only).
Strength – based on 75, 100, 125 or 150 pounds per square foot (psf) live load, plus 10 psf dead load.

		Size (inches) and Spacing (inches on center)											
		2 x 6			2 x 8			2 x 10			2 x 12		
Grade	Live Load	12″oc	16″oc	24″oc	12″oc	16″oc	24″oc	12″oc	16″oc	24″oc	12″oc	16″oc	24″oc
No. 1	75 psf	8 – 10	8 – 1	7 – 1	11 – 8	10 – 8	9 – 3	14 – 11	13 – 7	11 – 3	18 – 2	16 – 4	13 – 4
	100 psf	8 – 1	7 – 4	6 – 5	10 – 8	9 – 8	8 – 4	13 – 7	12 – 1	9 – 10	16 – 6	14 – 5	11 – 9
	125 psf	7 – 6	6 – 10	5 – 11	9 – 10	9 – 0	7 – 6	12 – 7	10 – 11	8 – 11	15 – 0	13 – 0	10 – 7
	150 psf	7 – 1	6 – 5	5 – 6	9 – 3	8 – 5	6 – 11	11 – 7	10 – 0	8 – 2	13 – 9	11 – 11	9 – 9
No. 2	75 psf	8 – 8	7 – 11	6 – 6	11 – 6	10 – 4	8 – 5	14 – 3	12 – 4	10 – 1	16 – 8	14 – 5	11 – 9
	100 psf	7 – 11	7 – 0	5 – 9	10 – 5	9 – 1	7 – 5	12 – 6	10 – 10	8 – 10	14 – 8	12 – 8	10 – 4
	125 psf	7 – 4	6 – 4	5 – 2	9 – 6	8 – 2	6 – 8	11 – 4	9 – 9	8 – 0	13 – 3	11 – 5	9 – 4
	150 psf	6 – 9	5 – 10	4 – 9	8 – 8	7 – 6	6 – 2	10 – 5	9 – 0	7 – 4	12 – 2	10 – 6	8 – 7
No. 3	75 psf	7 – 2	6 – 2	5 – 1	9 – 1	7 – 11	6 – 5	10 – 9	9 – 4	7 – 7	12 – 10	11 – 1	9 – 1
	100 psf	6 – 4	5 – 5	4 – 5	8 – 0	6 – 11	5 – 8	9 – 5	8 – 2	6 – 8	11 – 3	9 – 9	7 – 11
	125 psf	5 – 8	4 – 11	4 – 0	7 – 3	6 – 3	5 – 1	8 – 6	7 – 5	6 – 0	10 – 2	8 – 10	7 – 2
	150 psf	5 – 3	4 – 6	3 – 8	6 – 8	5 – 9	4 – 8	7 – 10	6 – 9	5 – 7	9 – 4	8 – 1	6 – 7

* The listed maximum span has been limited to 26′ - 0″ based on material availability. Check sources of supply for lumber longer than 20 .

By permission of Southern Pine Council

6.7.2 Southern Pine Span Tables for Wet-Service Joist and Rafters

Wet-Service Floor Joists

Design Criteria: Deflection–limited to span in inches divided by 360 (live load only).
Strength–based on 40 or 60 pounds per square foot (psf) live load, plus 10 psf dead load.

		Size (inches) and Spacing (inches on center)											
		2 x 6			2 x 8			2 x 10			2 x 12		
Grade	Live Load	12"oc	16"oc	24"oc	12"oc	16"oc	24"oc	12"oc	16"oc	24"oc	12"oc	16"oc	24"oc
No. 1	40 psf	10–7	9–7	8–5	13–11	12–8	11–1	17–9	16–2	13–5	21–7	19–8	16–1
	60 psf	9–3	8–5	7–4	12–2	11–1	9–7	15–6	13–11	11–4	18–10	16–7	13–7
No. 2	40 psf	10–4	9–5	7–10	13–8	12–5	10–1	17–5	15–10	13–2	21–2	18–10	15–4
	60 psf	9–1	8–1	6–8	11–11	10–6	8–7	15–2	13–7	11–1	18–4	15–11	13–0
No. 3	40 psf	9–4	8–1	6–7	11–11	10–3	8–5	14–0	12–2	9–11	16–8	14–5	11–10
	60 psf	7–11	6–10	5–7	10–0	8–8	7–1	11–10	10–3	8–5	14–1	12–3	10–0

Rafters – Drywall or No Finished Ceiling – Construction Load (C_D = 1.25)[1]

Design Criteria: Deflection–limited to span in inches divided by 240 or 180 (live load only).
Strength–based on 20 pounds per square foot (psf) live load, plus 10 psf dead load.

		Size (inches) and Spacing (inches on center)											
		2 x 6			2 x 8			2 x 10			2 x 12		
Grade	Deflection	12"oc	16"oc	24"oc	12"oc	16"oc	24"oc	12"oc	16"oc	24"oc	12"oc	16"oc	24"oc
No. 1	240	15–9	14–4	12–6	20–10	18–11	16–6	26–0*	24–1	21–1	26–0*	26–0*	25–2
	180	17–4	15–9	13–9	22–11	20–10	17–9	26–0*	25–10	21–1	26–0*	26–0*	25–2
No. 2	240	15–6	14–1	12–3	20–5	18–6	15–10	26–0*	23–2	18–11	26–0*	26–0*	22–2
	180	17–0	15–1	12–4	22–5	19–5	15–10	26–0*	23–2	18–11	26–0*	26–0*	22–2
No. 3	240	13–6	11–8	9–6	17–2	14–10	12–2	20–3	17–7	14–4	24–1	20–10	17–0
	180	13–6	11–8	9–6	17–2	14–10	12–2	20–3	17–7	14–4	24–1	20–10	17–0

Rafters – Drywall Ceiling – Snow Load (C_D = 1.15)[1]

Design Criteria: Deflection–limited to span in inches divided by 240 (live load only).
Strength–based on 30 or 40 pounds per square foot (psf) live load, plus 10 psf dead load.

		Size (inches) and Spacing (inches on center)											
		2 x 6			2 x 8			2 x 10			2 x 12		
Grade	Live Load	12"oc	16"oc	24"oc	12"oc	16"oc	24"oc	12"oc	16"oc	24"oc	12"oc	16"oc	24"oc
No. 1	30 psf	13–9	12–6	10–11	18–2	16–6	14–5	23–2	21–1	17–6	26–0*	25–7	20–10
	40 psf	12–6	11–5	9–11	16–6	15–0	13–1	21–1	19–2	15–8	25–7	22–10	18–8
No. 2	30 psf	13–6	12–3	10–2	17–10	16–2	13–2	22–3	19–3	15–9	26–0*	22–7	18–5
	40 psf	12–3	11–2	9–1	16–2	14–5	11–10	19–11	17–3	14–1	23–4	20–2	16–6
No. 3	30 psf	11–2	9–8	7–11	14–3	12–4	10–1	16–10	14–7	11–11	20–0	17–4	14–2
	40 psf	10–0	8–8	7–1	12–9	11–0	9–0	15–1	13–0	10–8	17–11	15–6	12–8

Rafters – No Finished Ceiling – Snow Load (C_D = 1.15)[1]

Design Criteria: Deflection–limited to span in inches divided by 180 (live load only).
Strength–based on 30 or 40 pounds per square foot (psf) live load, plus 10 psf dead load.

		Size (inches) and Spacing (inches on center)											
		2 x 4			2 x 6			2 x 8			2 x 10		
Grade	Live Load	12"oc	16"oc	24"oc	12"oc	16"oc	24"oc	12"oc	16"oc	24"oc	12"oc	16"oc	24"oc
No. 1	30 psf	9–8	8–9	7–8	15–2	13–9	11–9	20–0	18–0	14–9	24–9	21–5	17–6
	40 psf	8–9	8–0	7–0	13–9	12–6	10–6	18–2	16–2	13–2	22–2	19–2	15–8
No. 2	30 psf	9–6	8–7	7–1	14–5	12–6	10–2	18–8	16–2	13–2	22–3	19–3	15–9
	40 psf	8–7	7–9	6–4	12–11	11–2	9–1	16–8	14–5	11–10	19–11	17–3	14–1
No. 3	30 psf	7–7	6–7	5–4	11–2	9–8	7–11	14–3	12–4	10–1	16–10	14–7	11–11
	40 psf	6–9	5–10	4–9	10–0	8–8	7–1	12–9	11–0	9–0	15–1	13–0	10–8

*The listed maximum span has been limited to 26'-0" based on material avaliability. Check sources of supply for lumber longer than 20'.

(1) C_D = duration of load factor. See page 12 for additional information on adjustment factors.

By permission of Southern Pine Council

6.7.3 Spans for Various Southern Pine Species

Species and Grade	40 psf live load, 10 psf dead load, ℓ/360				30 psf live load, 10 psf dead load, ℓ/360			
	2x10		2x12		2x10		2x12	
	16" o.c.	24" o.c.	16" o.c.	24" o.c.	16" o.c.	24" o.c.	16" o.c.	24" o.c.
SP No. 1	16'-9"	14'-7"	20'-4"	17'-5"	18'-5"	16'-1"	22'-5"	19'-6"
DFL No. 1	16'-5"	13'-5"	19'-1"	15'-7"	18'-5"	15'-0"	21'-4"	17'-5"
SP No. 2	16'-1"	13'-2"	18'-10"	15'-4"	18'-0"	14'-8"	21'-1"	17'-2"
HF No. 1	16'-0"	13'-1"	18'-7"	15'-2"	17'-8"	14'-8"	20'-10"	17'-0"
SPF Nos. 1 & 2	15'-4"	12'-7"	17'-10"	14'-7"	17'-2"	14'-0"	19'-11"	16'-3"
DFL No. 2	15'-4"	12'-7"	17'-10"	14'-7"	17'-2"	14'-0"	19'-11"	16'-3"
HF No. 2	15'-2"	12'-5"	17'-7"	14'-5"	16'-10"	13'-10"	19'-8"	16'-1"
SP No. 3	12'-2"	9'-11"	14'-5"	11'-10"	13'-7"	11'-1"	16'-2"	13'-2"
DFL No. 3	11'-8"	9'-6"	13'-6"	11'-0"	13'-0"	10'-8"	15'-1"	12'-4"
HF No. 3	11'-8"	9'-6"	13'-6"	11'-0"	13'-0"	10'-8"	15'-1"	12'-4"
SPF No. 3	11'-8"	9'-6"	13'-6"	11'-0"	13'-0"	10'-8"	15'-1"	12'-4"

Note: These spans were calculated using published design values and are for comparison purposes only. They include the repetitive member factor, C_r=1.15, but do not include composite action of adhesive and sheathing. Spans may be slightly different than other published spans due to rounding. SP=Southern Pine, DFL=Douglas Fir–Larch, HF=Hem–Fir, SPF=Spruce–Pine–Fir.

6.7.4 Extent of Notching of Structural Pine Framing Members

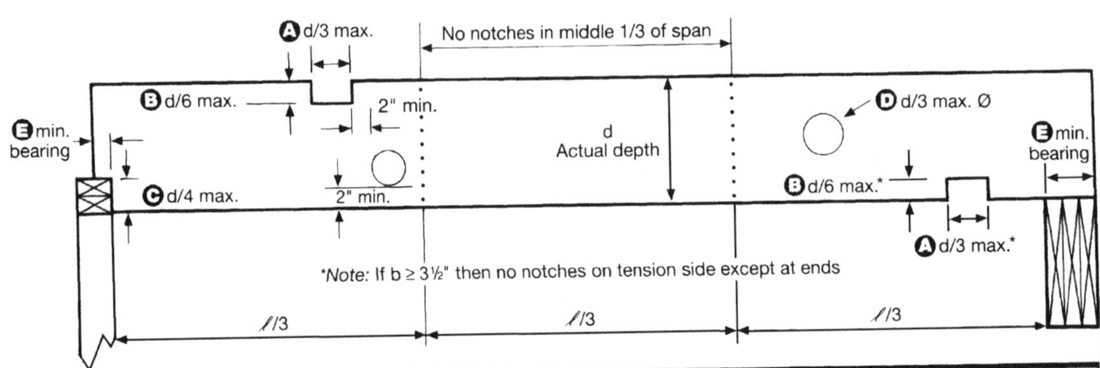

Joist Size	A Maximum Notch Length	B Maximum Notch Depth	C Maximum End Notch Depth	D Maximum Hole Diameter	E Minimum Bearing Length[6]	
2 x 6	1-13/16"	7/8"	1-3/8"	1-13/16"	1-1/2"	3"
2 x 8	2-3/8"	1-3/16"	1-13/16"	2-3/8"	1-1/2"	3"
2 x 10	3-1/16"	1-1/2"	2-5/16"	3-1/16"	1-1/2"	3"
2 x 12	3-3/4"	1-7/8"	2-13/16"	3-3/4"	1-1/2"	3"

6 Minimum bearing: 1-1/2" on wood or steel; 3" bearing on masonry.

By permission of Southern Pine Council

6.7.5 Southern Pine Rafter Spans and Birdsmouth Data

Maximum Span Comparisons for Rafters

Southern Pine also demonstrates its strength and performance leadership for rafters. For more detailed rafter span information, see *Southern Pine Maximum Spans for Joists and Rafters.*

Species and Grade	30 psf live, 15 psf dead, ℓ/180, C_D=1.15, 6 on 12 slope						20 psf live, 10 psf dead, ℓ/240, C_D=1.25, 3 on 12 slope					
	2x6		2x8		2x10		2x6		2x8		2x10	
	16" o.c.	24" o.c.	16" o.c.	24" o.c.	16" o.c.	24" o.c.	16" o.c.	24" o.c.	16" o.c.	24" o.c.	16" o.c.	24" o.c.
SP No. 1	13'-6"	11'-1"	17'-0"	13'-11"	20'-3"	16'-6"	14'-4"	12'-6"	18'-11"	16'-6"	24'-1"	21'-1"
DFL No. 1	12'-0"	9'-10"	15'-3"	12'-5"	18'-7"	15'-2"	14'-4"	12'-6"	18'-11"	15'-10"	23'-9"	19'-5"
SP No. 2	11'-9"	9'-7"	15'-3"	12'-5"	18'-2"	14'-10"	14'-1"	12'-3"	18'-6"	15'-10"	23'-2"	18'-11"
HF No. 1	11'-9"	9'-7"	14'-10"	12'-1"	18'-1"	14'-9"	13'-9"	12'-0"	18'-1"	15'-6"	23'-1"	18'-11"
DFL No. 2	11'-3"	9'-2"	14'-3"	11'-8"	17'-5"	14'-3"	14'-1"	11'-9"	18'-2"	14'-10"	22'-3"	18'-2"
SPF Nos.1&2	11'-3"	9'-2"	14'-3"	11'-8"	17'-5"	14'-3"	13'-5"	11'-9"	17'-9"	14'-10"	22'-3"	18'-2"
HF No. 2	11'-1"	9'-1"	14'-0"	11'-6"	17'-2"	14'-0"	13'-1"	11'-5"	17'-3"	14'-8"	21'-11"	17'-11"
SP No. 3	9'-1"	7'-5"	11'-7"	9'-6"	13'-9"	11'-3"	11'-8"	9'-6"	14'-10"	12'-2"	17'-7"	14'-4"
DFL No. 3	8'-6"	6'-11"	10'-9"	8'-10"	13'-2"	10'-9"	10'-10"	8'-10"	13'-9"	11'-3"	16'-9"	13'-8"
HF No. 3	8'-6"	6'-11"	10'-9"	8'-10"	13'-2"	10'-9"	10'-10"	8'-10"	13'-9"	11'-3"	16'-9"	13'-8"
SPF No. 3	8'-6"	6'-11"	10'-9"	8'-10"	13'-2"	10'-9"	10'-10"	8'-10"	13'-9"	11'-3"	16'-9"	13'-8"

Note: These spans were calculated using published design values and are for comparison purposes only. They include the repetitive member factor, C_r=1.15, but do not include composite action of adhesive and sheathing. Spans may be slightly different than other published spans due to rounding. SP=Southern Pine, DFL=Douglas Fir–Larch, HF=Hem–Fir, SPF=Spruce–Pine–Fir. C_D = load duration factor.

Cutting a Rafter Birdsmouth

A common roof framing technique is to use a rafter birdsmouth cut for the connection of the rafter to the top plate of the exterior wall. The following steps, tables and figures detail the birdsmouth cut.

Instructional Steps for Cutting a Rafter Birdsmouth

(see diagrams to right)

1. Determine the rafter length.
   Ex: Run = 20', slope = 4.
   Rafter length = 21'-1" using Table 2.
2. Measure Θ *(from Table 1)* at top edge of rafter.
3. Draw the building line.
4. Draw 2/3 width line from top edge of rafter.
5. Use square to draw seat cut line from bottom edge of rafter to intersect building line.

Note: The birdsmouth notch should be limited to 1/3 the rafter width to maintain 2/3 of the rafter section.

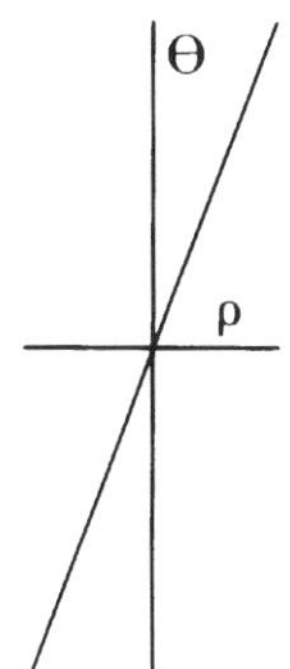

Table 1

Slope	3	4	5	6	7	8	9	10	11	12
θ°	14	18	23	27	30	34	37	40	43	45
ρ°	76	72	67	63	60	56	53	50	47	45

Layout lines for a common rafter

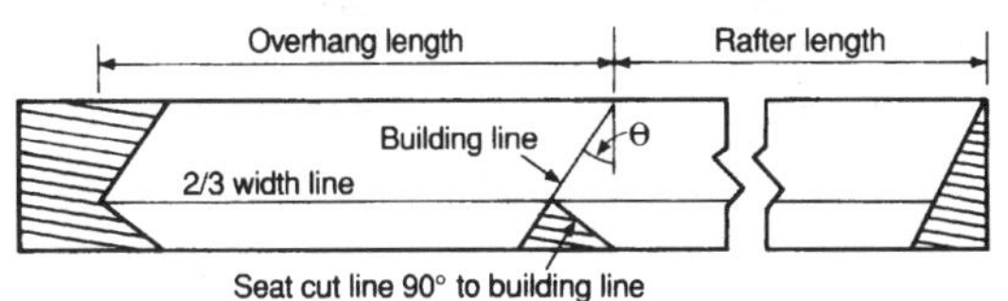

Table 2

Description	Rafter length = Slope factor times run									
Slope	3	4	5	6	7	8	9	10	11	12
Slope factor	1.031	1.054	1.083	1.118	1.158	1.202	1.250	1.302	1.357	1.414

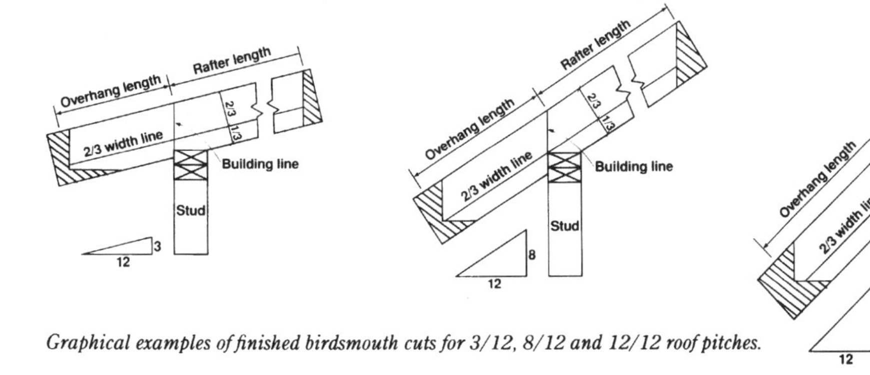

Graphical examples of finished birdsmouth cuts for 3/12, 8/12 and 12/12 roof pitches.

By permission of Southern Pine Council

6.7.6 Conversion Diagram for Southern Pine Rafters

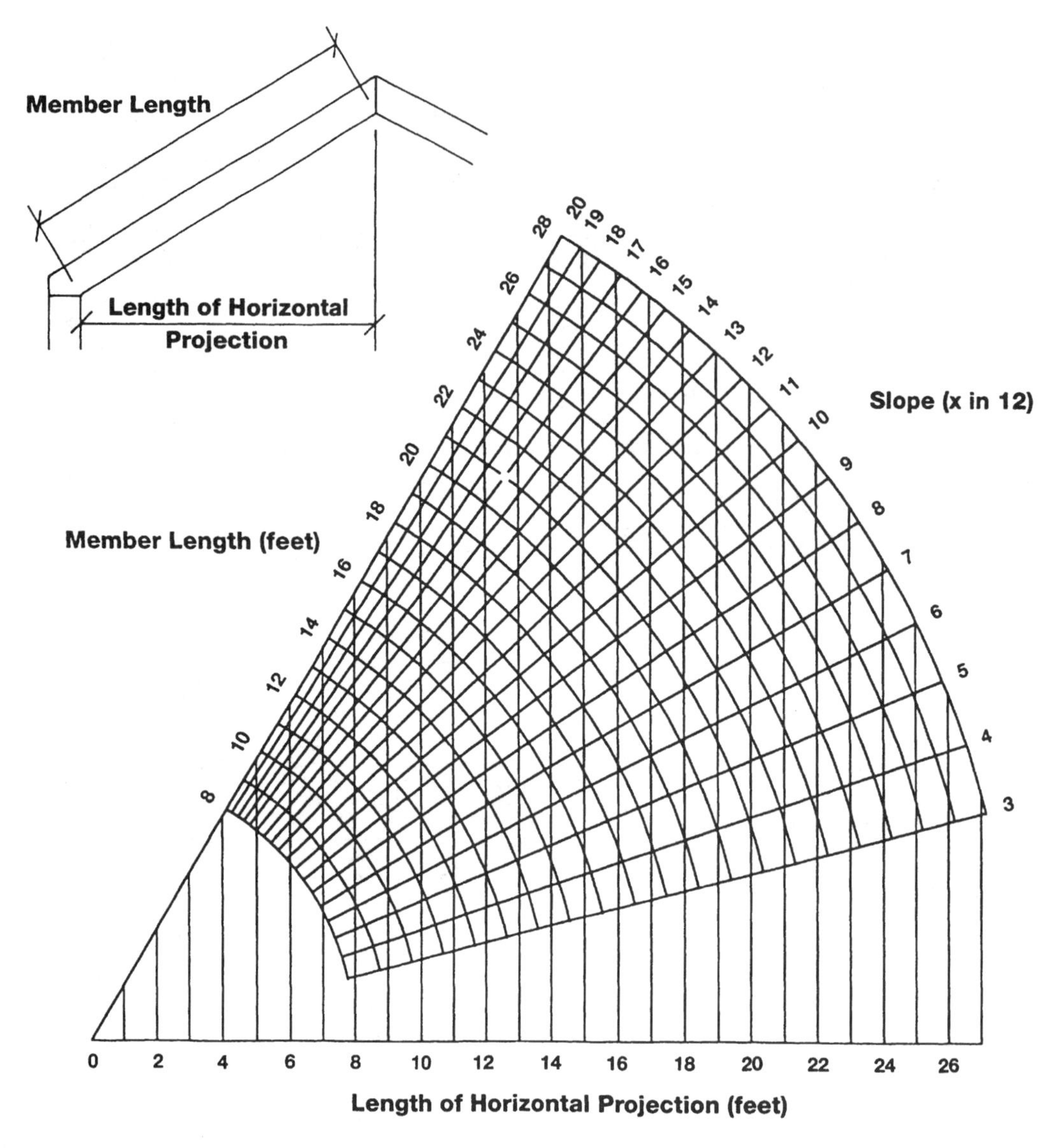

To use the diagram, select the known horizontal distance and follow the vertical line to its intersection with the radial line of the specified slope. Then proceed along the arc to read the sloping distance. In some cases it may be desirable to interpolate between the one-foot separations. The diagram also may be used to find the horizontal distance corresponding to a given sloping distance, or to find the slope when the horizontal and sloping distances are known.

Example: With a roof slope of 8 in 12, and a horizontal distance of 20 feet, the sloping distance may be read as 24 feet.

By permission of Southern Pine Council

6.8.0 Properties of Sections of Southern Pine Framing Members

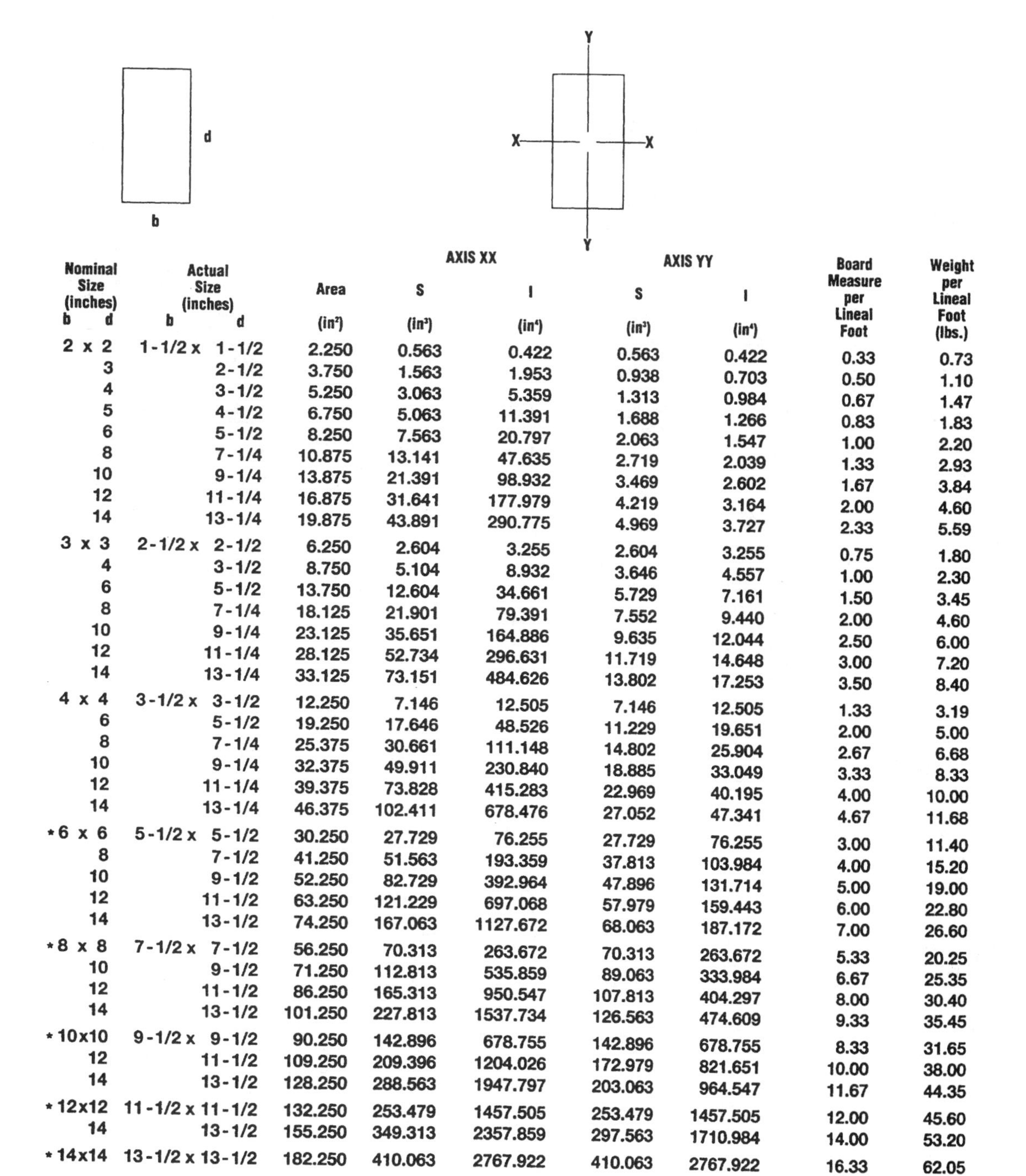

Nominal Size (inches) b d	Actual Size (inches) b d	Area (in²)	AXIS XX S (in³)	AXIS XX I (in⁴)	AXIS YY S (in³)	AXIS YY I (in⁴)	Board Measure per Lineal Foot	Weight per Lineal Foot (lbs.)
2 x 2	1-1/2 x 1-1/2	2.250	0.563	0.422	0.563	0.422	0.33	0.73
3	2-1/2	3.750	1.563	1.953	0.938	0.703	0.50	1.10
4	3-1/2	5.250	3.063	5.359	1.313	0.984	0.67	1.47
5	4-1/2	6.750	5.063	11.391	1.688	1.266	0.83	1.83
6	5-1/2	8.250	7.563	20.797	2.063	1.547	1.00	2.20
8	7-1/4	10.875	13.141	47.635	2.719	2.039	1.33	2.93
10	9-1/4	13.875	21.391	98.932	3.469	2.602	1.67	3.84
12	11-1/4	16.875	31.641	177.979	4.219	3.164	2.00	4.60
14	13-1/4	19.875	43.891	290.775	4.969	3.727	2.33	5.59
3 x 3	2-1/2 x 2-1/2	6.250	2.604	3.255	2.604	3.255	0.75	1.80
4	3-1/2	8.750	5.104	8.932	3.646	4.557	1.00	2.30
6	5-1/2	13.750	12.604	34.661	5.729	7.161	1.50	3.45
8	7-1/4	18.125	21.901	79.391	7.552	9.440	2.00	4.60
10	9-1/4	23.125	35.651	164.886	9.635	12.044	2.50	6.00
12	11-1/4	28.125	52.734	296.631	11.719	14.648	3.00	7.20
14	13-1/4	33.125	73.151	484.626	13.802	17.253	3.50	8.40
4 x 4	3-1/2 x 3-1/2	12.250	7.146	12.505	7.146	12.505	1.33	3.19
6	5-1/2	19.250	17.646	48.526	11.229	19.651	2.00	5.00
8	7-1/4	25.375	30.661	111.148	14.802	25.904	2.67	6.68
10	9-1/4	32.375	49.911	230.840	18.885	33.049	3.33	8.33
12	11-1/4	39.375	73.828	415.283	22.969	40.195	4.00	10.00
14	13-1/4	46.375	102.411	678.476	27.052	47.341	4.67	11.68
*6 x 6	5-1/2 x 5-1/2	30.250	27.729	76.255	27.729	76.255	3.00	11.40
8	7-1/2	41.250	51.563	193.359	37.813	103.984	4.00	15.20
10	9-1/2	52.250	82.729	392.964	47.896	131.714	5.00	19.00
12	11-1/2	63.250	121.229	697.068	57.979	159.443	6.00	22.80
14	13-1/2	74.250	167.063	1127.672	68.063	187.172	7.00	26.60
*8 x 8	7-1/2 x 7-1/2	56.250	70.313	263.672	70.313	263.672	5.33	20.25
10	9-1/2	71.250	112.813	535.859	89.063	333.984	6.67	25.35
12	11-1/2	86.250	165.313	950.547	107.813	404.297	8.00	30.40
14	13-1/2	101.250	227.813	1537.734	126.563	474.609	9.33	35.45
*10x10	9-1/2 x 9-1/2	90.250	142.896	678.755	142.896	678.755	8.33	31.65
12	11-1/2	109.250	209.396	1204.026	172.979	821.651	10.00	38.00
14	13-1/2	128.250	288.563	1947.797	203.063	964.547	11.67	44.35
*12x12	11-1/2 x 11-1/2	132.250	253.479	1457.505	253.479	1457.505	12.00	45.60
14	13-1/2	155.250	349.313	2357.859	297.563	1710.984	14.00	53.20
*14x14	13-1/2 x 13-1/2	182.250	410.063	2767.922	410.063	2767.922	16.33	62.05

**Note:* Properties are based on minimum dressed green size which is 1/2 inch off nominal in both b and d dimensions.

By permission of Southern Pine Council

6.8.1 Standard Sizes of Southern Pine Dimension Lumber, Boards, and Decking

Based on 1994 SPIB Grading Rules

	Thickness (inches)			Width (inches)		
	Nominal	Dressed Dry	Dressed Green	Nominal	Dressed Dry	Dressed Green
Dimension Lumber and Timbers, dressed[1]	2	1-1/2		2	1-1/2	
	2-1/2	2	2-1/16	3	2-1/2	2-9/16
	3	2-1/2	2-9/16	4	3-1/2	3-9/16
	3-1/2	3	3-1/16	5	4-1/2	4-5/8
	4	3-1/2	3-9/16	6	5-1/2	5-5/8
				8	7-1/4	7-1/2
				10	9-1/4	9-1/2
				12	11-1/4	11-1/2
				14	13-1/4	13-1/2
				16	15-1/4	15-1/2
				18	17-1/4	17-1/2
				20	19-1/4	19-1/2
	Timbers 5″ & thicker	1/2″ off nominal	1/2″ off nominal	5″ & wider	1/2″ off nominal	1/2″ off nominal
	Nominal	Dressed		Nominal	Dressed	
Boards, S4S	1	3/4[2]		2	1-1/2	
	1-1/4	1		3	2-1/2	
	1-1/2	1-1/4		4	3-1/2	
				5	4-1/2	
				6	5-1/2	
				7	6-1/2	
				8	7-1/4	
				9	8-1/4	
				10	9-1/4	
				11	10-1/4	
				12	11-1/4	
				over 12	3/4″ off nominal	
	Nominal	Dressed		Nominal	Dressed	
Finish, dry	3/8	5/16		2	1-1/2	
	1/2	7/16		3	2-1/2	
	5/8	9/16		4	3-1/2	
	3/4	5/8		5	4-1/2	
	1	3/4		6	5-1/2	
	1-1/4	1		7	6-1/2	
	1-1/2	1-1/4		8	7-1/4	
	1-3/4	1-3/8		9	8-1/4	
	2	1-1/2		10	9-1/4	
	2-1/2	2		11	10-1/4	
	3	2-1/2		12	11-1/4	
	3-1/2	3		14	13-1/4	
	4	3-1/2		16	15-1/4	
	Nominal	Dressed		Nominal	Dressed	
Radius Edge Decking	1-1/4	1		4	3-1/2	
				5	4-1/2	
				6	5-1/2	

(1) Dimension Lumber 2″ thick and less than 14″ wide is required to be dry with a moisture content of 19% or less. Heavy Dimension Lumber (2 x 14 and wider, 2-1/2″ thick by all widths, and 3 x 3 and larger) and Timbers are not required to be dry unless specified. Thicknesses apply to their corresponding widths as squares and wider, except a thickness to 1-9/16″ applies to nominal 2″ in widths of 14″ and wider if dressed green. (In 2″ Dimension, widths over 12″ are not customary stock sizes, so 2 x 14 and wider sizes are usually produced only on special order.)

(2) Boards less than the minimum dressed thickness for 1″ nominal but which are 5/8″ or greater thickness dry may be regarded as American Standard Lumber, but such Boards shall be marked to show the size and condition of seasoning at the time of dressing. They shall also be distinguished from 1″ Boards on invoices and certificates.

By permission of Southern Pine Council

6.9.0 Southern Pine Header Load Tables and Connection Details

Maximum Load Comparisons for Headers *(plf)*

Total load / live load

Clear Opening	Size	SP No. 1	DFL No. 1	SP No. 2	HF No. 1	DFL No. 2	SPF Nos. 1&2	HF No. 2
18' (two-car garage)	2-2x10	107/ 85	90 / 85	85 / 80	86 / 75	79	79 / 70	77 / 65
	2-2x12	153 / 152	122	118	116	106	107	104
9' (single-car garage)	2-2x10	440	375	356	354	329	326	317
	2-2x12	560*	502	487	467*	439	436*	423*
6' (window opening)	2-2x10	740	782	740	617*	733	576	617*
	2-2x12	974*	1029*	974*	812*	973*	757*	812*

Note: This table is for comparison purposes only. Values shown are the maximum uniformly distributed loads in pounds per lineal foot (plf) that can be applied to the header in addition to its own weight. When different, total load deflection limit = ℓ/240 (left) and live load deflection limit = ℓ/360 (right); otherwise these values are the same. The load duration factor, C_D = 1.00. SP = Southern Pine, DFL = Douglas Fir-Larch, HF = Hem-Fir, SPF = Spruce-Pine-Fir.

*Requires two trimmers (3" bearing); all others require one trimmer (1.5" bearing).

Header Connection Details

The key to header performance is the manner in which they are connected. The graphical examples below provide guidance on the types of connections that can be used in the field.

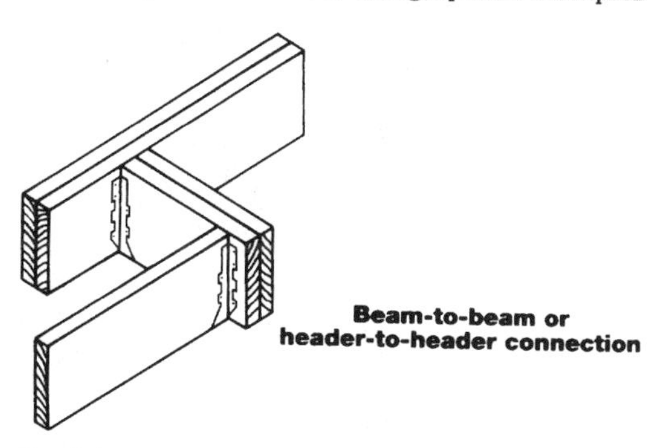

Note: Follow code or connector manufacturer requirements for nailing schedules and allowable loads for headers and connections.

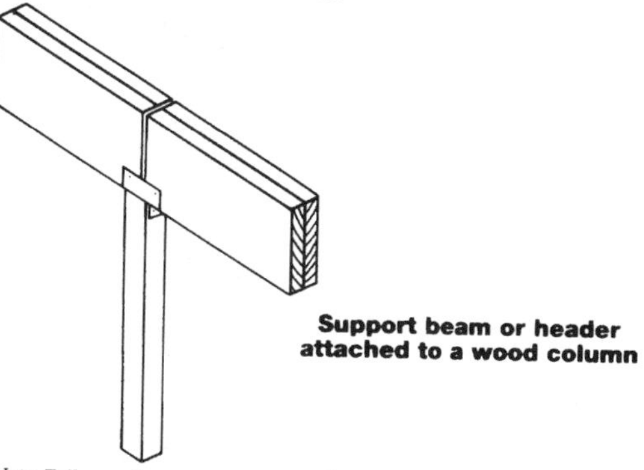

Note: Follow code or connector manufacturer requirements for nailing schedules and allowable loads for headers and connections.

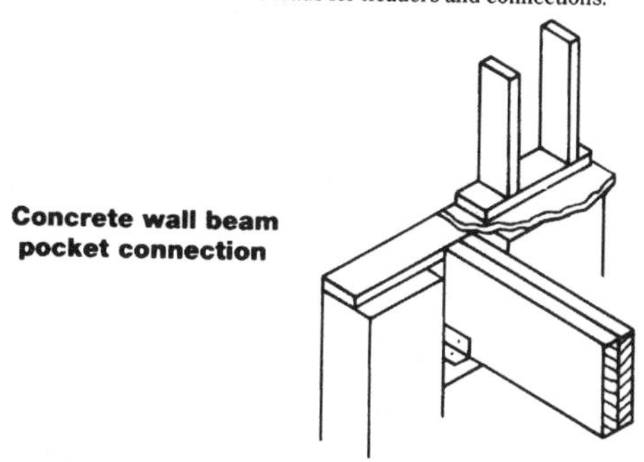

Note: Follow code requirements for wood in contact with concrete and bearing support connections.

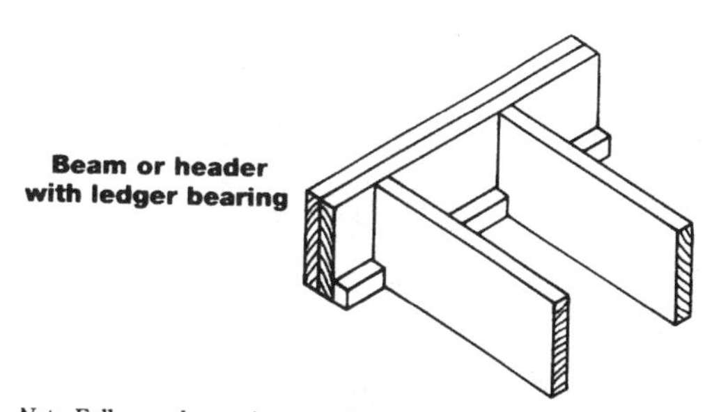

Note: Follow code requirements for nailing schedules for joist-to-header and ledger-to-header connections.

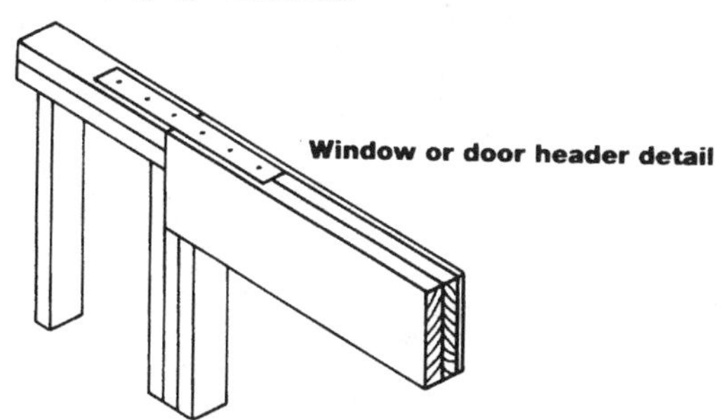

Note: Follow code requirements for nailing schedules, allowable loads, proper straps and proper bearing connections.

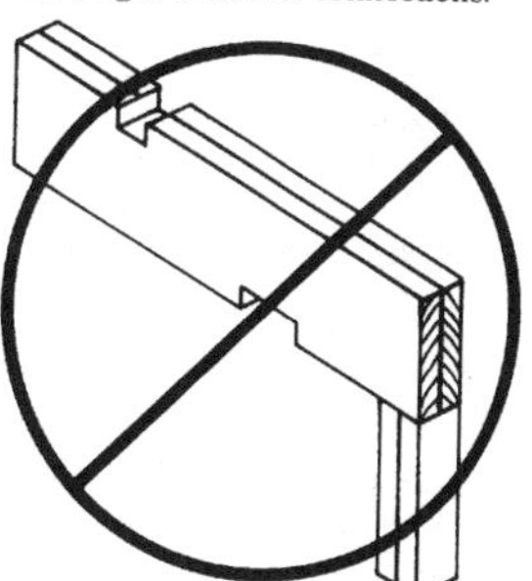

Caution: *Do not cut, drill or notch beams or headers.*

6.9.1 Southern Pine Rafter Framing Details

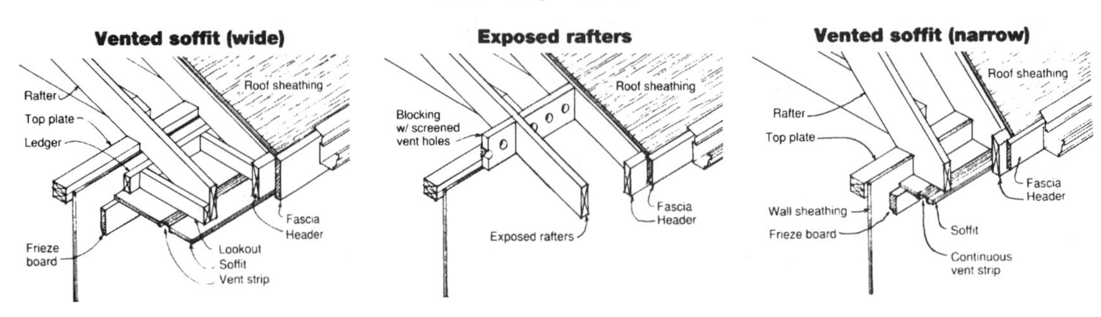

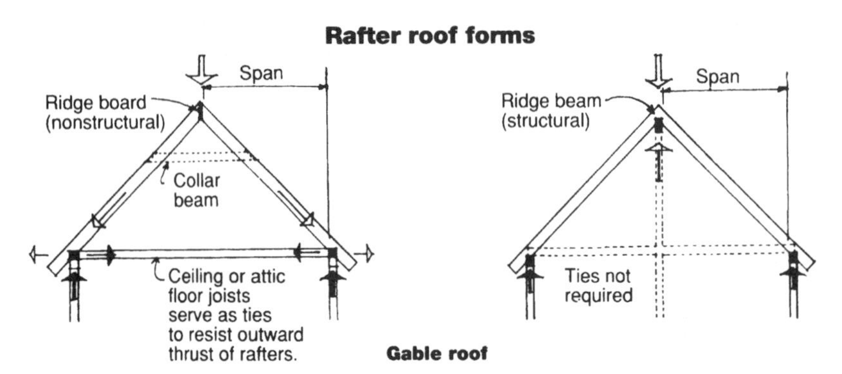

Typical roof framing types are shown in the figures above. The arrows show the flow of force on the roof framing members.

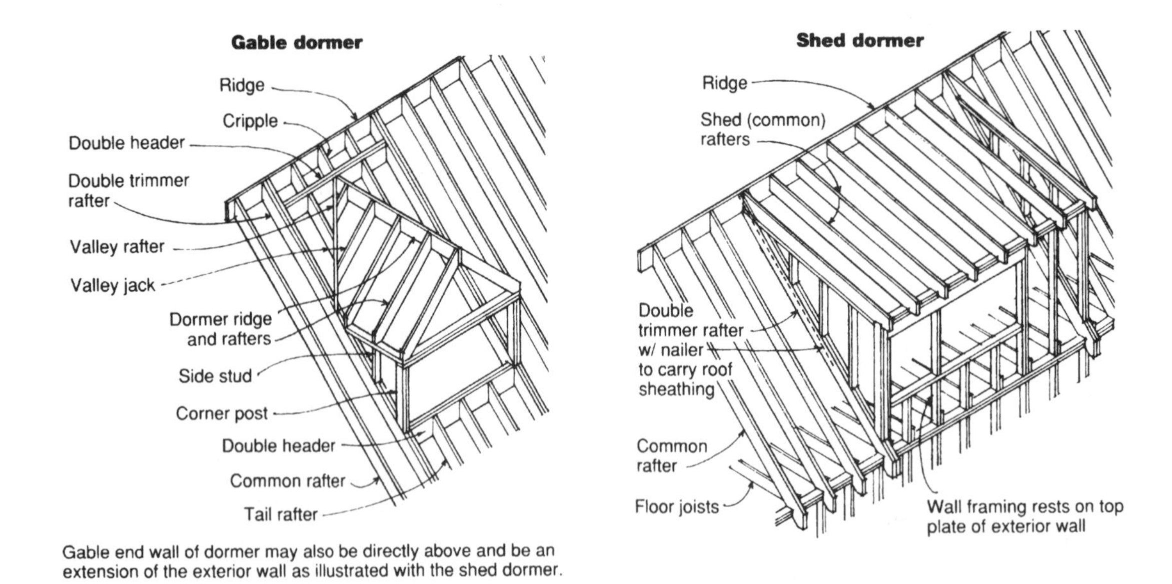

Dormers are framed into the roof system to add style to the roof and provide light for the attic space or upper floor living area.

By permission of Southern Pine Council

6.9.2 Southern Pine Floor Joist Framing Details

Lateral Support

Typically, joists are laterally supported by:
1) a rim joist applied to both ends of the joist to provide stability and to prevent rotation, and
2) sheathing attached to the top of the joist to provide compression edge support. No additional lateral support is required for most common joist applications. There are some conditions, however, where additional lateral support provided by blocking, bridging or cross bracing will be required. For example, the need for lateral support becomes greater as the depth to breadth (d/b) ratio of a joist increases.

The model building codes and the *National Design Specification (NDS)® for Wood Construction* provide additional guidance on lateral support requirements. The local building code, however, will determine the lateral support required for a particular building. Examples of blocking and cross bracing are shown to the right.

Solid blocking

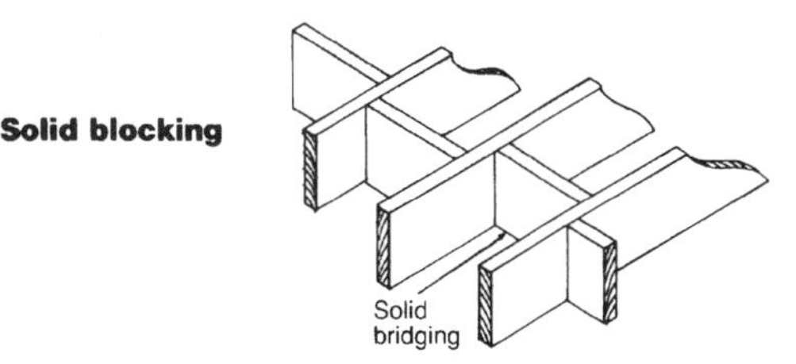

Wood or steel cross bracing

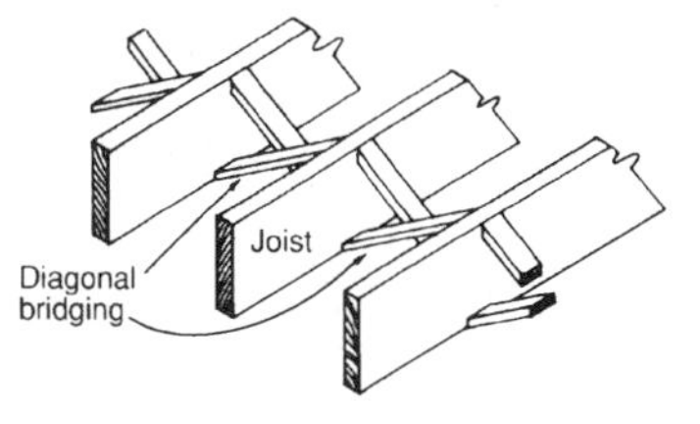

Floor Joist Framing Details

Beam support conditions

Top of joist flush with beam*

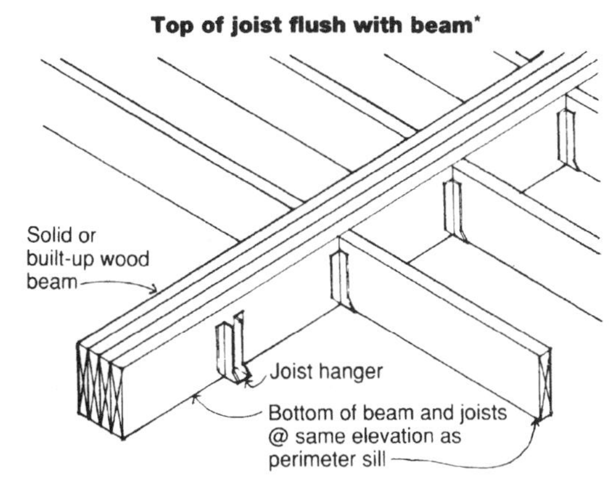

*Use only with well-seasoned lumber

Ledger bearing on steel beam

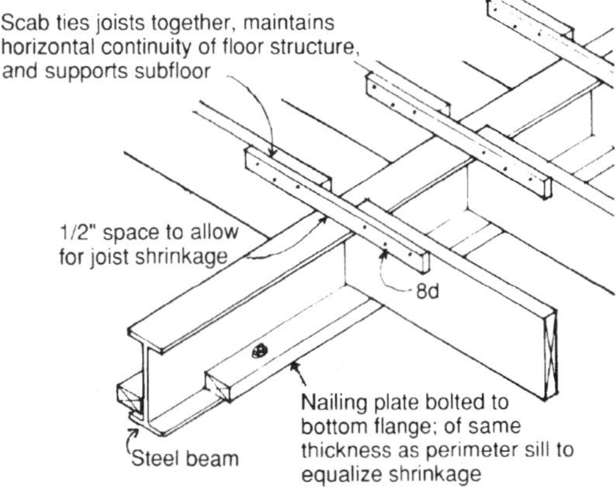

Ledger bearing

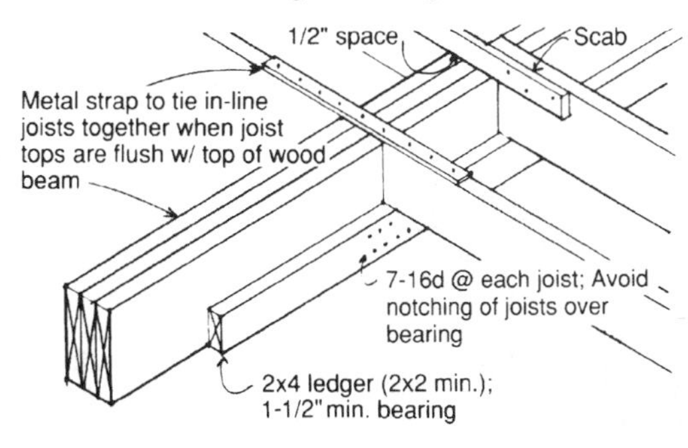

Glued-laminated beam bearing

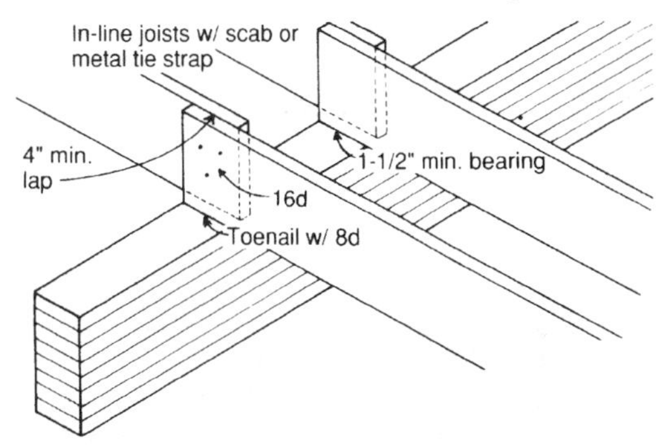

By permission of Southern Pine Council

6.9.3 Additional Floor Joist Framing Details

Additional Floor Joist Framing Details

Floor projections

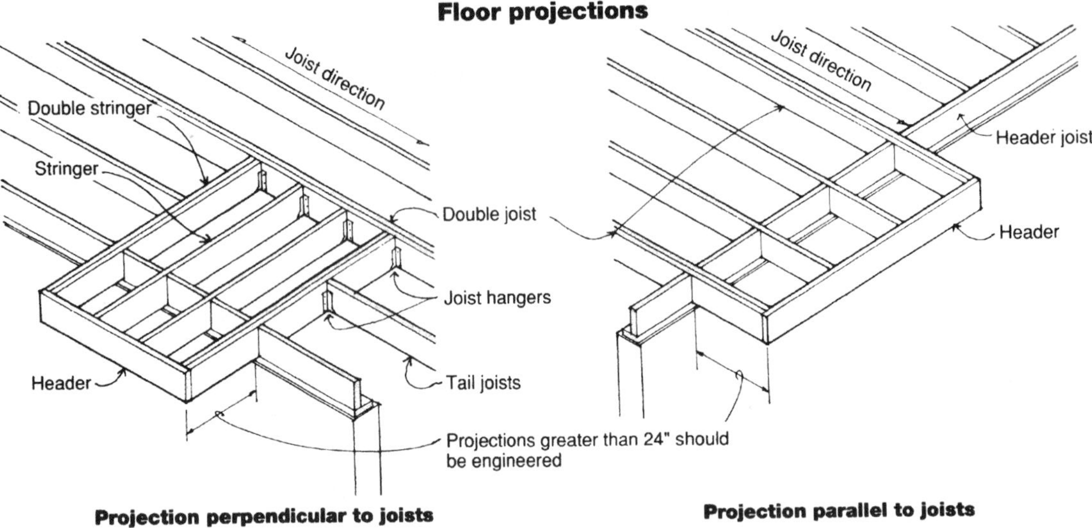

Balconies, room extensions, and bay windows can easily be added to a floor plan.

Floor openings

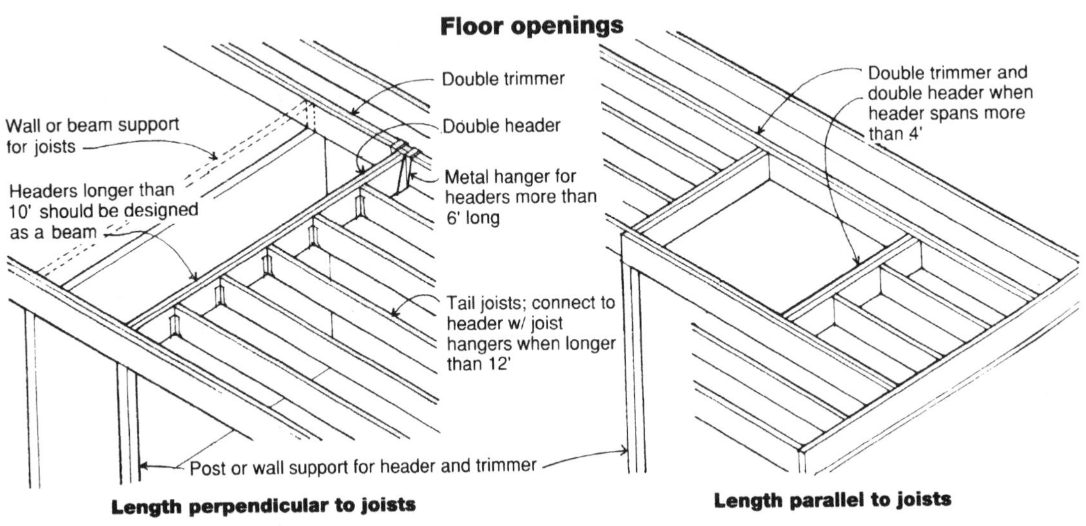

Stairwell and chimney openings are also easily framed. The versatility of joist construction even allows some of these changes to be made when construction is in progress for maximum design flexibility.

Floor Performance

Spans given in the previous joist span comparison table meet all model code requirements. However, meeting the minimum code requirements may not always be sufficient to satisfy the customer. A stricter deflection criteria, such as $l/480$, can be used to provide a more solid-feeling floor system. In addition to this, floor performance may be enhanced by: glue-nailing the floor sheathing to the joists; using thicker sheathing material (e.g., 3/4" versus 5/8" plywood); and/or using 12" o.c. spacing versus 16" o.c. Most important is the proper installation of the joists—making sure walls and girders are level and nailing the sheathing to joists accurately.

Floor vibrations can also occur in some floors. Continuous solid blocking or cross bracing can improve the floor's vibration performance. Vibrations can also be minimized by attaching a gypsum ceiling directly to the bottom of the joists where no ceiling previously existed.

6.10.0 Southern Pine Inspection Bureau Grading Rules for Decking

SPIB GRADING RULES FOR RADIUS EDGE DECKING			
DEFECTS	PREMIUM	STANDARD	REVERSE SIDE STANDARD
Checks	1/32" wide x 10" long. If through - 1/32" wide by width of piece	Surface - not limited Through - 2 times width	No limit
Compression Wood	None if readily identifiable and damaging form		
Decay	None	None	1/3 width by 1/3 thickness
Firm Red Heart	25% of face	No limit	No limit
Holes	One 1/4" hole every 4'	Well scattered - 1/4"	No limit
Knots	Sound-Firm-Encased-Pith, if tight and well spaced: 4" width - 1" 6" width - 1-1/2" Decayed knots with serious pits and cavities - 1/2 the sound knot size. All knots in any 4' may not be over twice the maximum size knot	Sound-Firm-Encased-Pith, if tight and well spaced: 4" width - 1-1/2" 6" width - 2-1/2" Decayed knots with serious pits and cavities: 4" width - 1" 6" width - 1-1/2"	No limit except to 55% of the cross section as noted directly below
	Premium: Knot average on front and back cannot be larger than 45% of the cross section using the displacement method (between lines parallel to the edges)		
	Well-Spaced - The sum of all knots in any 6" of length may not be larger than twice the size of maximum size knot.		
Manufacture	Same as No.2 Boards STD E	Same as No.2 Boards STD E	No limit
Pitch	Medium	Medium	Heavy
Pitch Pockets	Medium	Medium	No limit
Pitch Streaks	1/6 width by 1/3 length	1/6 width by 1/3 length	No limit
Pith	No limit if sound	No limit if sound	No limit
Shakes	Surface - 1/32" wide Through at end - Width of piece	Surface - 1/8" wide Through at end - Two times width, but not over 1/6 length	No limit
Splits	Width of piece	Two times width	
Skips	1/64" for 6" or equivalent Same on edge	1/32" on 10% face 1/16" on edge (Hit or Miss)	1/32" full length
Slope of Grain	1" in 8"	1" in 8"	1" in 8"
Stain	Medium	Medium	Heavy
Wane	1/8" deep by 1/2" wide Reverse side - 1/2" deep by 1" wide	1/8" deep by 1/2" wide	1/2" deep by 1" wide
Crook For 6" width	10 12 14 16 18 5/8 7/8 1-1/8 1-3/8 1-1/2	8 10 12 14 16 18 5/8 1 1-1/4 1-1/2 1-3/4 2	
Loosened Grain	1/8" separation	1/8" separation	No limit
Offset	1/16" deep	1/16" deep	1/16" deep
Bites	1/16" deep	1/16" deep	1/16" deep

By permission of Southern Pine Council

6.10.1 Southern Pine Inspection Bureau Grading Rules for Finish and Boards

SPIB GRADING RULES FOR FINISH AND BOARDS				
DEFECTS	C&BTR	D	NO. 2	NO. 3
Compression Wood	None	Not Limited	Not Limited	Not Limited
Firm Red Heart	25% Face Not Limited if Otherwise B&B	Not Limited	Not Limited	Not Limited
Decay	None	None	Heart Center Only 1/2" Wide by 1/4 Length	Allowed if Suitable for Nailing Throughout
Holes	1/16" Limited to 6 per/ft	1/16" Limited to 12 per/ft	1/4" Not Limited One 1" Per Piece	1-1/2" in 1x4 and 1x6 1/4 W in 1x8 1x10 1x12
Knots	Sound or Firm and Tight 3/4"-All Widths 1 - 1-1/2" in 6" Width 2 - 1-1/2" in 8, 10 & 12" All Knots in Any 4 Ft Must Not Exceed Twice Diameter of Maximum Knot Allowed	Decayed 3/4"- All Widths if Smooth and Even with surface Sound, Firm, Encased & Pith 3/4" - 1x4 1-1/2" - 1x6 2" - 1x8, 1x10 & 1x12 All Knots Must Be Tight-All Knots in Any 4 Ft Must Not Exceed Twice Diameter of Maximum Knot Allowed	No.1 Knot Size* 1x4 - 2-1/2" 1-1/2" 1x6 - 3" 2" 1x8 - 3-1/2" 2" 1x10 - 4" 2-1/2" 1x12 - 4-1/2" 2-1/2" * Decayed and Hollow Knots Limited to No. 1 Sizes - Knot Holes & Loose Knots Same as Holes	Not Limited In-Size Except Must be Able to Handle Without Breaking Loose Knots and Through Openings in Hollow Knots Limited Same as Holes
Pith	3/4 of Square Inch	1/6 Length	Not Limited	Not Limited
Stain	15% Face - Medium	25% Face - Medium	Not Limited if Medium	Not Limited if Medium
Pitch	Light - Medium if B&B	Medium, if C - Heavy (1/4W x 1/2L)	Not Limited	Not Limited
Pitch Streak	1/6 Width x 1/3 Length	1/6 Width x 1/3 Length If C-1/4 Width x 1/2 Length	Not Limited Worm-Eaten Area=Knot Size	Not Limited
Pitch Pocket	1/4" x 2" All Widths (Small) 1-3/8" x 4" in 1x6 (Medium) 2-3/8" x 4" in 1x8, 1x10 and 1x12	All Widths 3/8" x 4" (Med) 1x6 - 1 Large (4 sq") 1x8, 10, 12 - 2 Large (4 sq") If C: Through Pocket 3/8x4"	Not Limited	Not Limited
Shakes	1/32" Wide by Width of Piece - None Through	1/32" Wide - None Through	1/4 Length if Close Fitting	1/2 Length if Close Fitting
Skips	Face and Edge 1/64" for 6" or Equivalent	1/32" on 10% Face 1/16" on Edge (Full Length) If C - 1/32" Scant in Width for Each Inch of Width	1/32" on 25% Face 1/16" on Edge (Full Length) 10% of Pieces up to 1/8" Scant in Width	1/16" Full Length 1/8" on 10% of Pieces 1/4" in Width if Not Over 2 Ft Long
Split	Width of Piece	Twice Width But Cannot Exceed 1/6 Length	1/4 Length if Close Fitting	1/2 Length if Close Fitting
Checks	Surface - 1/32" x 10" Through - 1/32" x Width of Piece	Surface - 1/16" x 20" Through - 1/32" x 10" If C - Through 1/16" x 20"	Through - 1/4 Length if Close Fitting	If Through 1/2 Length
Wane	Face: 1/8" Deep x 1/2" Wide Reverse Side 1/4" Deep x 1/8 Width x 1/3 Length	Face: 1/4" Deep x 1/8 Width x 1/3 Length - Reverse: 1/2" Deep x 1/4 Width - Not Exceed 2" Wide	1/4 Width or 2" Wide - 1/8" of Wood on Edge - Sharp Edge for 8" on Occasional Piece	Face: 1/3 Width Reverse 3/4 Width Sharp Edge

6.10.2 Southern Pine Inspection Bureau Grading Rules for 2" Dimensions

CHARACTERISTICS		SEL STR	NO.1	NO.2	NO.3
COMPRESSION WOOD		← NOT ALLOWED IN DAMAGING FORM FOR THE GRADE CONSIDERED →			
SLOPE OF GRAIN		1" in 12"	1" in 10"	1" in 8"	1" in 4"
DECAY		Not Permitted	Not Permitted	Heart center, 1/3 thickness X 1/3 width	Heart Center, 1/3 cross section. Must not destroy nailing edge. See para. 710(e)
HOLES		Same as unsound knots	Same as unsound knots	See chart below	See chart below
KNOTS		Edge / Centerline / Unsound Knots	Edge / Centerline / Unsound Knots	Edge / Centerline / Holes	Edge / Centerline / Holes
	2x4	3/4" / 7/8" / 3/4"	1" / 1-1/2" / 1"	1-1/4" / 2" / 1-1/4"	1-3/4" / 2-1/2" / 1-3/4"
	2x5	1" / 1-1/2" / 7/8"	1-1/4" / 1-7/8" / 1-1/8"	1-5/8" / 2-3/8" / 1-3/8"	2-1/4" / 3" / 1-7/8"
	2x6	1-1/8" / 1-7/8" / 1"	1-1/2" / 2-1/4" / 1-1/4"	1-7/8" / 2-7/8" / 1-1/2"	2-3/4" / 3-3/4" / 2"
	2x8	1-1/2" / 2-1/4" / 1-1/4"	2" / 2-3/4" / 1-1/2"	2-1/2" / 3-1/2" / 2"	3-1/2" / 4-1/2" / 2-1/2"
	2x10	1-7/8" / 2-5/8" / 1-1/4"	2-1/2" / 3-1/4" / 1-1/2"	3-1/4" / 4-1/4 / 2-1/2"	4-1/2" / 5-1/2" / 3"
	2x12	2-1/4" / 3" / 1-1/4"	3" / 3-3/4" / 1-1/2"	3-3/4" / 4-3/4" / 3"	5-1/2" / 6-1/2" / 3-1/2"
		Sound, firm, encased, pith, tight & well spaced. One hole or equivalent smaller holes per 4 lin. ft.	Sound, firm, encased, pith, tight & well spaced. One hole or equivalent smaller holes per 3 lin. ft.	Well spaced knots of any quality. One hole or equivalent smaller holes per 2 lin. ft.	Well spaced knots of any quality. One hole or equivalent smaller holes per 1 lin. ft.
SHAKES		← Ends: same as splits → Elsewhere: 2' surface; none through		Ends: same as splits Elsewhere: surface 3' or 1/4 length; 2' through	1/6 length if through at edges or ends, elsewhere through shakes 1/3 length
CHECKS		← Surface seasoning checks not limited Through checks at ends limited as splits →			
SKIPS		← Hit and miss in 10% of the pieces. See para. 720(f) →		Hit and miss. 5% of the pieces may be hit or miss or heavy skip for 2'. See para. 720(e, f, and g)	Hit or miss. 10% of the pieces may have heavy skip. See para. 720(e and g)
SPLITS		← Equal to the width →		Equal to 1-1/2 times the width	Equal to 1/6 length
WANE		← 1/4 thickness x 1/4 width x full length or equivalent; must not exceed 1/2 thickness x 1/3 width for up to 1/4 length. Also see para. 750. →		1/3 thickness x 1/3 width x full length or equivalent; must not exceed 2/3 thickness x 1/2 width for up to 1/4 length. Also see para. 750.	1/2 thickness x 1/2 width x full length or equivalent; must not exceed 7/8 thickness or 3/4 width for up to 1/4 length. Also see para. 750.
BOW		← 10' /1-3/8"; 12' /1-1/2"; 14' /2"; 16' /2-1/2" →		10' /1-1/2"; 12' /2"; 14' /2-1/2"; 16'/ 3-1/4"	10' /2-3/4"; 12' /3"; 14' /4"; 16' /5"
CROOK	Size	← 10' / 12' / 14' / 16' →		10' / 12' / 14' / 16'	10' / 12' / 14' / 16'
	2x4	3/8" / 1/2" / 5/8" / 3/4"		1/2" / 11/16" / 7/8" / 1"	3/4" / 1" / 1-1/4" / 1-1/2"
	2x6	5/16" / 7/16" / 9/16" / 11/16"		7/16" / 5/8" / 3/4" / 7/8"	5/8" / 7/8" / 1-1/8" / 1-3/8"
	2x8	1/4" / 13/32" / 1/2" / 9/16"		3/8" / 1/2" / 5/8" / 3/4"	1/2" / 13/16" / 1" / 1-1/8"
	2x10	7/32" / 3/8" / 7/16" / 1/2"		1/4" / 7/16" / 1/2" / 5/8"	7/16" / 3/4" / 7/8" / 1"
	2x12	3/16" / 9/32" / 3/8" / 7/16"		3/16" / 3/8" / 3/8" / 1/2"	3/8" / 9/16" / 3/4" / 7/8"

DENSE GRAIN: Requires 6 rings/inch & 1/3 summerwood or 4 rings/inch & 1/2 summerwood.
EXCEPTIONALLY LIGHT WEIGHT PIECES: Should not be placed in No.2N and higher grades (Exceptionally light weight pieces have less than 15% summerwood).

By permission of Southern Pine Inspection Bureau

6.11.0 Southern Pine Wood-Preservative Retention Standards

	Waterborne Preservatives[2]				Creosote and Oilborne Preservatives[3]			
	Ammoniacal Copper Arsenate (ACA)	Ammoniacal Copper Zinc Arsenate (ACZA)	Chromated Copper Arsenate (CCA)	AWPA Standard(s)	Creosote	Creosote-Petroleum	Creosote Solutions	Pentachlorophenol (Penta)
	Retention Assay of Treated Wood – lbs./cu.ft.							
Lumber, Timbers & Plywood								
Above Ground	0.25	0.25	0.25	C2/C9	8[5]	8[5]	8[5]	0.40
Soil & Freshwater use	0.40	0.40	0.40	C2/C9	10[5]	10[5]	10[5]	0.50
Permanent Wood Foundation (PWF)	0.60	0.60	0.60	C22	**NR**	**NR**	**NR**	**NR**
Saltwater use	2.5	2.5	2.5	C2/C9	25	**NR**	25	**NR**
Piles								
Land or freshwater use & foundations	0.80	0.80	0.80	C3	12	12	12	0.60
Marine								
Prevalent Marine Organism								
Teredo only	2.5[4] and 1.5	2.5[4] and 1.5	2.5[4] and 1.5	C18	20	**NR**	20	**NR**
Pholads only	**NR**	**NR**	**NR**	C18	20	**NR**	20	**NR**
Limnoria tripunctata only	2.5[4] and 1.5	2.5[4] and 1.5	2.5[4] and 1.5	C18	**NR**	**NR**	**NR**	**NR**
Sphaeroma terebrans or for both pholads and limnoria tripunctata use a dual treatment								
First treatment	1.0	1.0	1.0	C18	–	–	–	–
Second treatment	–	–	–	C18	20	–	20	–
Poles								
Utility								
Normal	0.60	0.60	0.60	C4	7.5	7.5	7.5	0.38
Severe service conditions (high incidence of decay and termite attack)	0.60	0.60	0.60	C4	9.0	9.0	9.0	0.45
Building Construction – Round	0.60	0.60	0.60	C23	9.0[5]	**NR**	**NR**	0.45
Posts								
Commercial-Residential								
Fence								
Round, half-round, and quarter-round	0.40	0.40	0.40	C5	8[5]	8[5]	8[5]	0.40
Sawn four sides	0.40	0.40	0.40	C2	10[5]	10[5]	10[5]	0.50
Highway Construction								
Fence, Guide, Sign, and Sight Posts								
Round, half-round, and quarter-round	0.40	0.40	0.40	C14	8	8	8	0.40
Sawn four sides	0.40	0.40	0.40	C14	10	10	10	0.50
Guardrail and Spacer Blocks								
Round	0.50	0.50	0.50	C14	10	10	10	0.50
Sawn four sides	0.60	0.60	0.60	C14	12	12	12	0.60

NR – Not Recommended

(1) AWPA Standards detail plant operating procedures for pressure treatment of wood. These Standards include minimum vacuum, pressure, penetration requirements, and maximum steaming parameters. AWPA also details minimum retention requirements, sampling zones for assay and maximum redrying temperature allowance for each preservative, commodity, and wood species. For a copy of the AWPA Standards booklet, please write to the American Wood Preservers' Association, P.O. Box 286, Woodstock, Maryland 21163-0286. **(2)** ACA, ACZA and CCA are the most commonly available waterborne preservatives. Ammoniacal Copper Quat (ACQ) – Type B, Copper Citrate and CDDC are also approved by AWPA as waterborne preservatives for Southern Pine as lumber, timbers, and ties. **(3)** Copper Naphthenate is also approved by AWPA as an oilborne preservative for specific wood species and applications excluding saltwater use. **(4)** The assay retentions are based on two assay zones – 0 to 0.5 inch, and 0.5 to 2.0 inches. **(5)** Not recommended where cleanliness and freedom from odor are necessary.

By permission of Southern Pine Council

6.12.0 Knots and How to Measure Them

Measuring of knots for Southern Pine lumber.

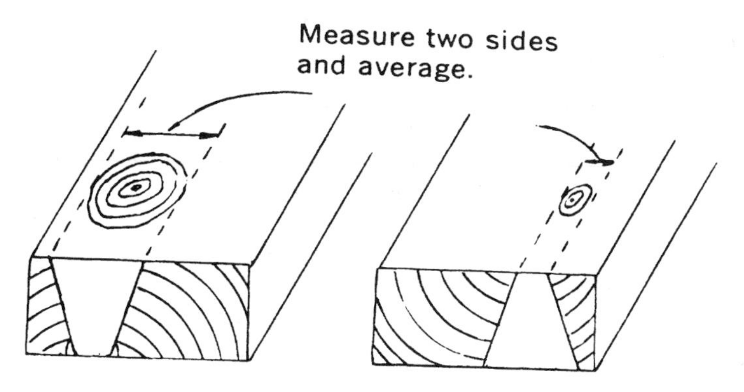

GRADE	KNOT SIZE ON NARROW FACE	KNOT MAY EXTEND INTO PIECE
No. 1	1-1/2"	1/2 of width (Fig. 1)
	1"	3/4 of width (Fig. 2)
No. 2	1-1/2"	Slightly less than 3/4 of width.
No. 3	1-1/2"	Slightly less than width

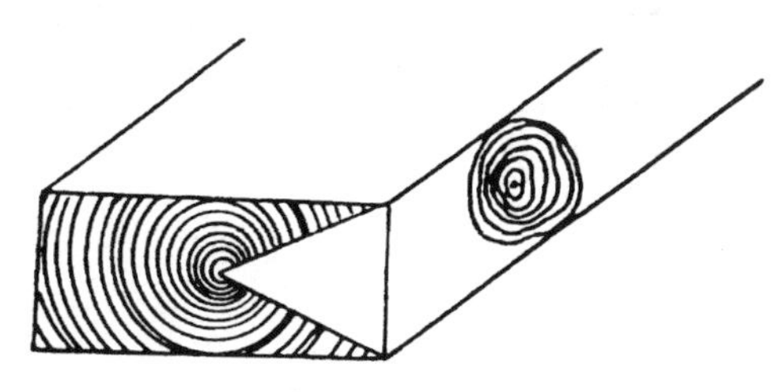

Figure 1

1-1/2" Narrow face knot extending 1/2 of width.

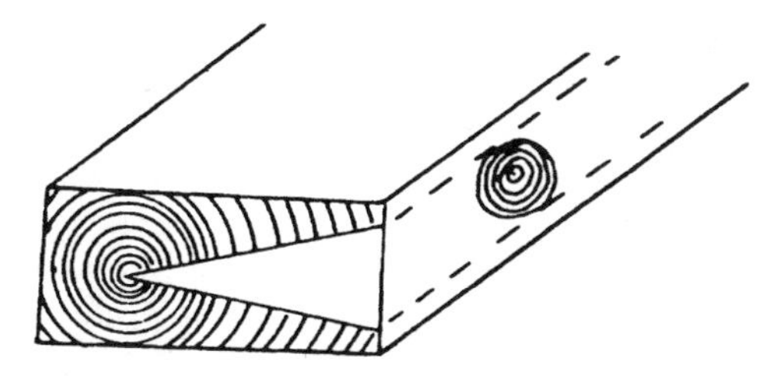

Figure 2

1" Narrow face knot extending 3/4 of width.

By permission of Southern Pine Council

6.13.0 Commercial Names of the Principal Softwood Species

Commercial Species or Species Group Names[8]	*Official Common Tree Names*[9]	*Botanical Names*
CEDAR:		
Alaska Cedar	Alaska-cedar	Chamaecyparis nootkatensis
Incense Cedar	incense-cedar	Libocedrus decurrens
Port Orford Cedar	Port-Orford-cedar	Chamaecyparis lawsoniana
Eastern Red Cedar	eastern redcedar	Juniperus virginiana
	southern redcedar	J. silicicola
Western Red Cedar	western redcedar	Thuja plicata
Northern White Cedar	northern white-cedar	T. occidentalis
Southern White Cedar	Atlantic white-cedar	Chamaecyparis thyoides
CYPRESS:[10]		
Baldcypress	baldcypress	Taxodium distichum
Pond cypress	pondcypress	T. distichum var. nutans
FIR:		
Balsam Fir[11]	balsam fir	Abies balsamea
	Fraser fir	A. fraseri
Douglas Fir[12]	Douglas-fir	Pseudotsuga menziesii
Noble Fir	noble fir	Abies procera
White Fir	subalpine fir	A. lasiocarpa
	California red fir	A. magnifica
	grand fir	A. grandis
	noble fir	A. procera
	Pacific silver fir	A. amabilis
	white fir	A. concolor

6.13.0 Commercial Names of the Principal Softwood Species (Continued)

Commercial Species or Species Group Names[8]	*Official Common Tree Names*[9]	*Botanical Names*
HEMLOCK:		
Eastern Hemlock	Carolina hemlock	Tsuga caroliniana
	eastern hemlock	T. canadensis
Mountain Hemlock	mountain hemlock	T. mertensiana
West Coast Hemlock	western hemlock	T. heterophylla
JUNIPER:		
Western Juniper	alligator juniper	Juniperus deppeana
	Rocky Mountain juniper	J. scopulorum
	Utah juniper	J. osteosperma
	western juniper	J. occidentalis
LARCH:		
Western Larch	western larch	Larix occidentalis
PINE:		
Jack Pine	jack pine	Pinus banksiana
Limber Pine	limber pine	P. flexilis
Lodgepole Pine	lodgepole pine	P. contorta
Norway Pine	red pine	P. resinosa
Pitch Pine	pitch pine	P. rigida
Ponderosa Pine	ponderosa pine	P. ponderosa
Radiata/Monterey Pine	Monterey pine	P. radiata
Sugar Pine	sugar pine	P. lambertiana
Whitebark Pine	whitebark pine	P. albicaulis
Idaho White Pine	western white pine	P. monticola
Northern White Pine	eastern white pine	P. strobus
Longleaf Pine[13]	longleaf pine	P. palustris
	slash pine	P. elliottii
Southern Pine (Major)	loblolly pine	P. taeda
	longleaf pine	P. palustris
	shortleaf pine	P. echinata
	slash pine	P. elliottii
Southern Pine (Minor)	pond pine	P. serotina
	Virginia pine	P. virginiana
	sand pine	P. clausa
	spruce pine	P. glabra
REDWOOD:		
Redwood	redwood	Sequoia sempervirens
SPRUCE:		
Eastern Spruce	black spruce	Picea mariana
	red spruce	P. rubens
	white spruce	P. glauca
Engelmann Spruce	blue spruce	P. pungens
	Engelmann spruce	P. engelmannii
Sitka Spruce	Sitka spruce	P. sitchensis
TAMARACK:		
Tamarack	tamarack	Larix laricina
YEW:		
Pacific Yew	Pacific yew	Taxus brevifolia

6.14 Lumber Industry Abbreviations

These abbreviations are commonly used for softwood lumber, although all of them are not necessarily applicable to all species. Additional abbreviations which are applicable to a particular region or species shall not be used unless included in certified grading rules.

Abbreviations are commonly used in the forms indicated, but variations such as the use of upper- and lower-case type, and the use or omission of periods and other forms of punctuation are not required.

AD	Air-dried
ADF	After deducting freight sides
ALS	American Softwood Lumber Standard
AV or AVG	Average
Bd	Board
Bd ft	Board foot or feet
Bdl	Bundle
Bev	Beveled
B/L	Bill of lading
BM	Board Measure
Btr	Better
B&B or B&Btr	B and better
B&S	Beams and stringers
CB1S	Center bead one side
CB2S	Center bead two sides
CF	Cost and freight
CG2E	Center groove two edges
CIF	Cost, insurance, and freight
CIFE	Cost, insurance, freight, and exchange
Clg	Ceiling
Clr	Clear
CM	Center matched
Com	Common
CS	Caulking seam
Csg	Casing
Cu Ft	Cubic foot or feet
CV1S	Center Vee one side
CV2S	Center Vee two sides
D&H	Dressed and headed
D&M	Dressed and matched
DB Clg	Double-beaded ceiling (E&CB1S)
DB Part	Double-beaded partition (E&CB2S)
DET	Double end trimmed
Dim	Dimension
Dkg	Decking
D/S or D/Sdg	Drop siding
EB1S	Edge bead one side
EB2S	Edge bead two sides
E&CB1S	Edge and center bead one side
E&CB2S	Edge and center bead two sides
E&CV1S	Edge and center Vee one side
E&CV2S	Edge and center Vee two sides
EE	Eased edges
EG	Edge (vertical) grain
EM	End matched

6.14 Lumber Industry Abbreviations (Continued)

EV1S	Edge Vee one side
EV2S	Edge Vee two sides
Fac	Factory
FAS	Free alongside (named vessel)
FBM	Foot or board measure
FG	Flat (slash) grain
Flg	Flooring
FOB	Free on board (named point)
FOHC	Free of heart center or centers
FOK	Free of knots
Frt	Freight
Ft	Foot or feet
GM	Grade marked
G/R or G/Rfg	Grooved roofing
HB	Hollow back
H&M	hit-and-miss
H or M	hit-or-miss
Hrt	Heart
Hrt CC	Heart cubical content
Hrt FA	Heart facial area
Hrt G	Heart girth
IN	Inch or inches
J&P	Joists and planks
KD	Kiln-dried
Lbr	Lumber
LCL	Less than carload
LFT or Lin Ft	Linear foot or feet
Lgr	Longer
Lgth	Length
Lin	Linear
Lng	Lining
M	Thousand
MBM	Thousand (feet) board measure
MC	Moisture content
Merch	Merchantable
Mldg	Moulding
mm	Millimeter
No	Number
N1E	Nosed one edge
N2E	Nosed two edges
Og	Ogee
Ord	Order
Par	Paragraph
Part	Partition
Pat	Pattern
Pc	Piece
Pcs	Pieces
PE	Plain end
PO	Purchase order
P&T	Post and timbers
Reg	Regular
Res	Resawed or resawn
Rfg	Roofing
Rgh	Rough

6.14 Lumber Industry Abbreviations (Continued)

R/L	Random lengths
R/W	Random widths
R/W&L	Random widths and lengths
Sdg	Siding
Sel	Select
S&E	Side and Edge (surfaced on)
SE Sdg	Square edge siding
SE & S	Square edge and sound
S/L or S/LAP	Shiplap
SL&C	Shipper's load and count
SM or Std M	Standard matched
Specs	Specifications
Std	Standard
Stpg	Stepping
Str or Struc	Structural
S1E	Surfaced one edge
S1S	Surfaced one side
S1S1E	Surfaced one side and one edge
S1S2E	Surfaced one side and two edges
S2E	Surfaced two edges
S2S	Surfaced two sides
S2S1E	Surfaced two sides and one edge
S2S&CM	Surfaced two sides and center matched
S2S&SM	Surfaced two sides and standard matched
S4S	Surfaced four sides
S4S&CS	Surfaced four sides and caulking seam
T&G	Tongued and grooved
VG	Vertical grain
Wdr	Wider
Wt	Weight

Section

7

Plywood, Composite Wood Products, High-Pressure Laminates

7.0.0 American Plywood Association (APA) grading guidelines
7.1.0 Plywood types and typical applications
7.1.1 APA-registered trademarks explained
7.1.2 Plywood veneer grades
7.2.0 Exposure ratings (exposure 1 and 2)
7.3.0 Plywood species group numbers
7.3.1 Chart of classification of species
7.4.0 Variety of surface textures available for APA-rating siding
7.5.0 Plywood panel dimensions (U.S. customary and metric)
7.6.0 Span tables for plywood sheathing and subfloors
7.7.0 Recommended spans for roof sheathing and fastening schedules
7.8.0 Typical plywood sheathing construction
7.8.1 Composite wood products
7.8.1.1 Medium-Density Fiberboard (MDF)
7.8.1.2 Hardboard (compressed fiberboard)
7.8.1.3 Cellulosic fiberboard (softboard)
7.8.1.4 Oriented Strand Board (OSB)
7.8.1.5 Waferboard
7.8.1.6 Laminated Veneer Lumber (LVL)
7.8.1.7 Parallel-Strand Lumber (PSL)
7.8.1.8 Oriented Strand Lumber (OSL)
7.8.1.9 Com-Ply
7.9.0 Moisture content of particleboard and the impact on warpage
7.9.1 Plywood underlayment span tables and glue/nailed fastening recommendations
7.9.2 Ideal fabrication conditions chart
7.9.3 Moisture content zones in the U.S.
7.9.4 Dimensional changes in Medium-Density Fiberboard (MDF) and Industrial-Grade Particle Board (PBI)
7.10.0 APA-rated sturdi-floor subfloor and floor framing for hardwood floors
7.11.0 High-Pressure Laminate (HPL) Q&A
7.11.1 HPL tips for avoiding panel warpage
7.11.2 HPL stress crack avoidance
7.11.3 HPL post-forming counter tops
7.11.4 HPL post-forming counter tops (manual techniques)
7.12.0 Common post-forming problems
7.13.0 Low-Pressure Laminates (LPL)
7.14.0 APA specifications for roof sheathing

Used as sheathing, flooring, in the production of cabinetry and millwork, plywood and composite wood products play a key role in the construction industry.

7.0.0 American Plywood Association (APA) Grading Guidelines

The American Plywood Association, headquartered in Tacoma, Washington, establishes grades and specifications for plywood products. The National Particleboard Association, located in Gaithersburg, Maryland, is the authority on composite wood products.

Plywood

Similar to the grading agencies for Western wood products and Southern pine lumber, the American Plywood Association (APA) provides the industry with specification guidelines and grade stamps by which to identify these grades. The term *grade* can apply to the type of veneer being used or the use for which the panel is best suited.

7.1.0 Plywood Types and Typical Applications

Where interior usage for cabinetry, shelving, built-ins, and so forth is required, APA-Sanded and Touch-Sanded designations apply:

- *A-A* For use where appearance on both sides is important.
- *A-B* For use where appearance on only one side is important, but where two solid sides are required.
- A-C For use where appearance on one side is important in exterior applications, such as soffits, truck lining, and structural uses.
- *A-D* For use where appearance on one side is important in interior applications, such as paneling and partitions.
- *B-B* Utility panel with two sides. Interior use-primarily: limited exterior use.
- *B-C* Utility panel for farm-service work, box cars, and truck linings for exterior use.
- *B-D* Utility panel for backing, sides of built-ins, separator boards, and slip sheets for interior and exterior use.
- *C-C plugged* For use as an underlayment over structural subfloor, pallet fruit bins, and for use in areas to be covered by carpet.
- *C-D plugged* For open soffits, cable reels, walkways, interior, or protected applications. Not to be used as underlayment.
- *Underlayment* For application over structural subfloor, it provides a smooth surface for carpet and, touch sanded, for resilient floors.

Specialty Panels

- *APA high-density overlay (HDO)* Manufactured with a semi-opaque resin-fiber overlay on both sides. It is used for concrete forms, industrial bins, and exhaust ducts.
- *APA marine* Plywood made only with Douglas fir or Western larch have highly restrictive limitations on core gaps and face repairs. As the name implies, it is ideal for boat hulls and other marine uses.
- *APA B-B plyform Class 1* Used for concrete formwork and designed for multi-use applications.
- *APA medium-density overlay MDO* Made with a smooth, opaque, resin-treated fiber overlay, producing an ideal base for finish painting, signs, and shelving.
- *APA decorative* Plywood with a rough-sawn, brushed, and grooved surface for interior accent walls, paneling, exhibit displays, etc.
- *APA plyron* Plywood with a hardboard face adhered on both sides, for countertops, cabinet doors, and shelving.

Plyform Exterior-grade plywood used for concrete forms

B-B plyform It has a smooth, solid surface. It can be re-used many times.

B-C EXT Sanded panel used where only one smooth side is needed.

HDO plyform High-density overlay with hard, semi-opaque resin-fiber finish. Resists abrasion up to 200 re-uses. Requires a release agent.

Structural 1 plyform Stronger and stiffer than B-B and HDO. Recommended for high-pressure applications.

7.1.1 APA-Registered Trademarks Explained

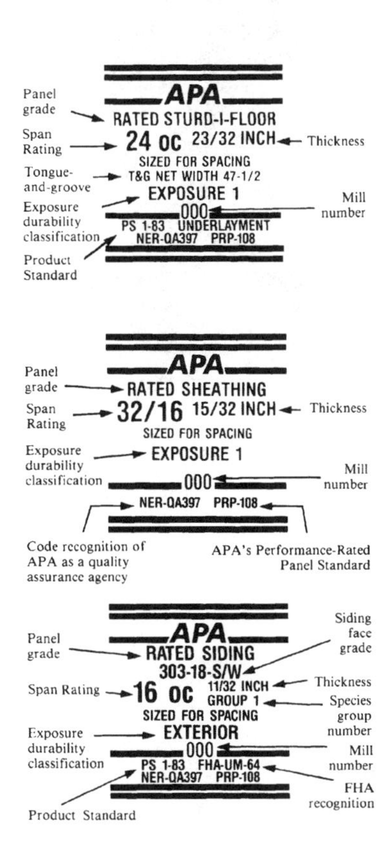

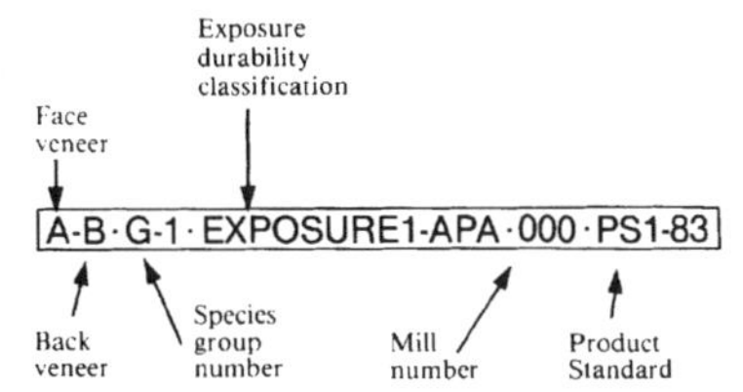

APA PERFORMANCE STANDARDS

APA performance standards are the result of new manufacturing technology that makes possible the manufacture of structural panel products from wood by-products and species not provided for in *U.S. Product Standard PS 1-83*. APA performance standards deal exclusively with how a product must perform in a designated application rather than from what or how the product must be manufactured.

Panels produced under APA performance standards — called APA Performance Rated Panels — must meet several performance baseline requirements according to the panel's designated end use. These performance requirements include uniform and concentrated static and impact load capacity, fastener-holding ability, racking resistance, dimensional stability, and bond durability.

In addition to conventional veneer plywood, APA performance standards encompass such other panel products as composites, waferboard and oriented strand board. (See APA Performance Rated Panels," page 8.)

For complete performance testing and qualification information, write APA for ***PRP-108, Performance Standards and Policies for Structural-Use Panels***, Form E445.

GRADE

The term "grade" may refer to *panel* grade or to *veneer* grade. Panel grades are generally identified in terms of the veneer grade used on the face and back of the panel (e.g., A-B, B-C, etc.), or by a name suggesting the panel's intended end use (e.g., APA Rated Sheathing, Underlayment, etc.).

Veneer grades define veneer appearance in terms of natural unrepaired growth characteristics and allowable number and size of repairs that may be made during manufacture. The highest quality veneer is "A,"(1) the lowest "D." The minimum grade of veneer permitted in Exterior plywood is "C." "D" veneer is used only in panels intended for interior use or for applications protected from permanent exposure to the weather.

EXPOSURE DURABILITY

APA trademarked panels may be produced in four exposure durability classifications — Exterior, Exposure 1, Exposure 2, and Interior.

Exterior panels have a fully waterproof bond and are designed for applications subject to permanent exposure to the weather or to moisture.

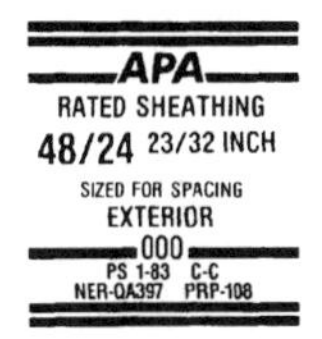

(1) Some manufacturers also produce a premium "N" grade (natural finish) veneer, available only on special order.

By permission of APA, The Engineered Wood Association, Tacoma, Washington

7.1.2 Plywood Veneer Grades

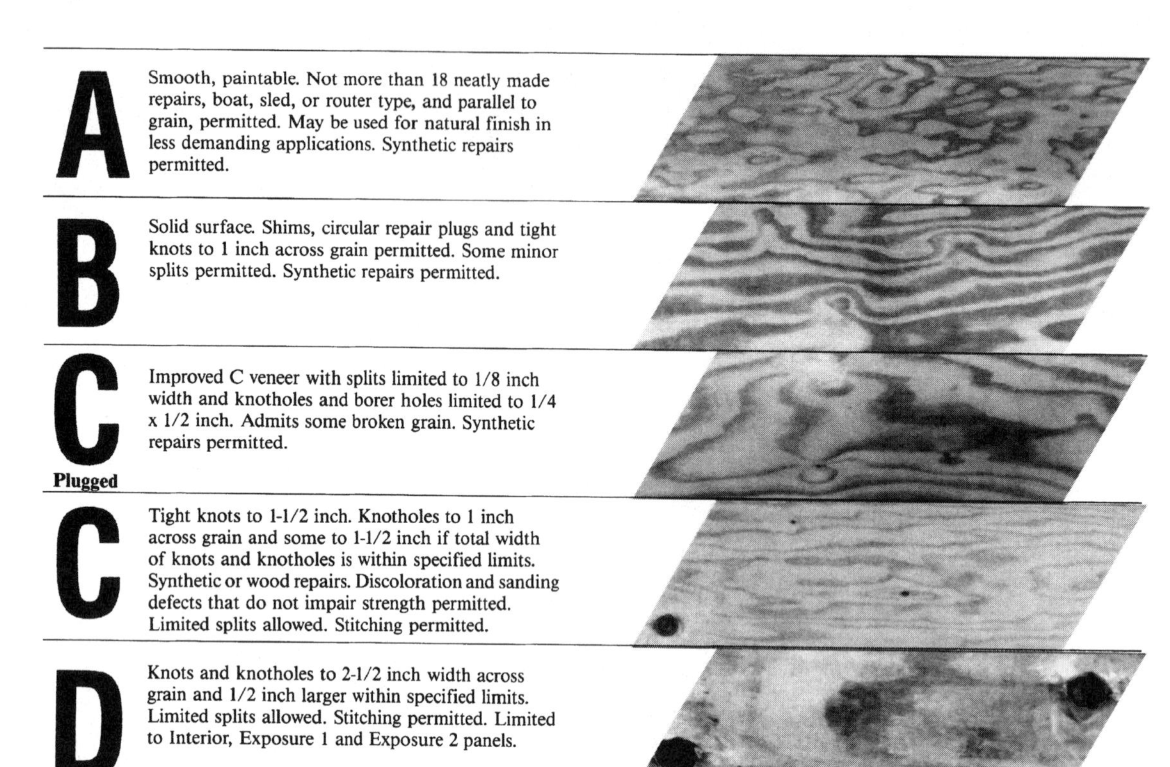

By permission of APA, The Engineered Wood Association, Tacoma, Washington

7.2.0 Exposure Ratings (Exposure 1 and 2)

Exposure 1 is for exterior use and has a fully waterproof bond designed for applications where the plywood will be permanently exposed to the weather or to moisture. Plywood so designated is stamped Exposure 1. Exposure 2 is for protected construction applications and is constructed with intermediate glue. This product is identified as Exposure 2 on the ADA grade stamp.

7.3.0 Plywood Species Group Numbers

Plywood manufactured in accordance with U.S. Product Standard (PS) 183 can be made of more than 70 species of wood and these species are divided into 5 groups. Group 1 is the strongest and stiffest and Group 5 the least strong and least stiff.

7.3.1 Chart of Classification of Species

Group 1	Group 2	Group 3	Group 4	Group 5
Apitong	Cedar, Port Orford	Alder, Red	Aspen	Basswood
Beech, American	Cypress	Birch, Paper	Bigtooth	Poplar, Balsam
Birch	Douglas-Fir 2[a]	Cedar, Alaska	Quaking	
Sweet	Fir	Fir, Subalpine	Cativo	
Yellow	Balsam	Hemlock, Eastern	Cedar	
Douglas-Fir 1[a]	California Red	Maple	Incense	
Kapur	Grand	Bigleaf	Western Red	
Keruing	Noble	Pine	Cottonwood	
Larch, Western	Pacific Silver	Jack	Eastern	
Maple, Sugar	White	Lodgepole	Black (Western Poplar)	
Pine	Hemlock, Western	Ponderosa	Pine	
Caribbean	Lauan	Spruce	Eastern White	
Ocote	Almon	Redwood	Sugar	
Pine, South.	Bagtikan	Spruce		
Loblolly	Mayapis	Engelmann		
Longleaf	Red	White		
Shortleaf	Tangile			
Slash	White			
Tanoak	Maple, Black			
	Mengkulang			
	Meranti, Red[b]			
	Mersawa			
	Pine			
	Pond			
	Red			
	Virginia			
	Western White			
	Spruce			
	Black			
	Red			
	Sitka			
	Sweetgum			
	Tamarack			
	Yellow-Poplar			

(a) Douglas-Fir from trees grown in the states of Washington, Oregon, California, Idaho, Montana, Wyoming, and the Canadian Provinces of Alberta and British Columbia shall be classed as Douglas-Fir No. 1. Douglas-Fir from trees grown in the states of Nevada, Utah, Colorado, Arizona and New Mexico shall be classed as Douglas-Fir No. 2.

(b) Red Meranti shall be limited to species having a specific gravity of 0.41 or more based on green volume and oven dry weight.

7.4.0 Variety of Surface Textures Available on APA-Rated Siding

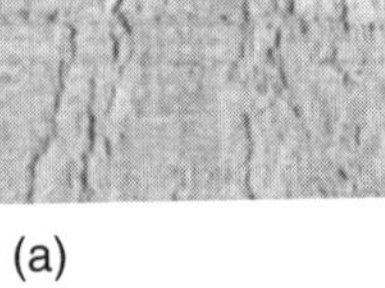

(a)

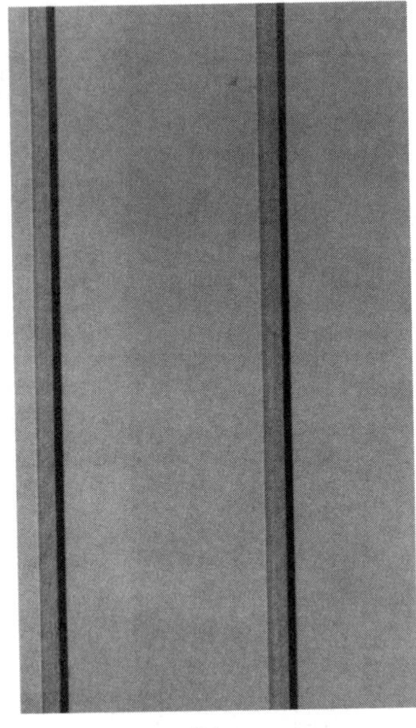

(b)

(c)

COM-PLY®

APA Rated Siding composite panel with rough-sawn veneer faces bonded to solid, reconstituted structural wood core. Available with grooves typically 4″ or 8″ oc, similar to Texture 1-11; or 1-1/2″-wide grooves spaced 12″ oc, similar to reverse board-and-batten pattern. Available in 19/32″, and 5/8″ thicknesses. Long edges shiplapped for continuous pattern. Available with Douglas-fir or cedar veneer faces.

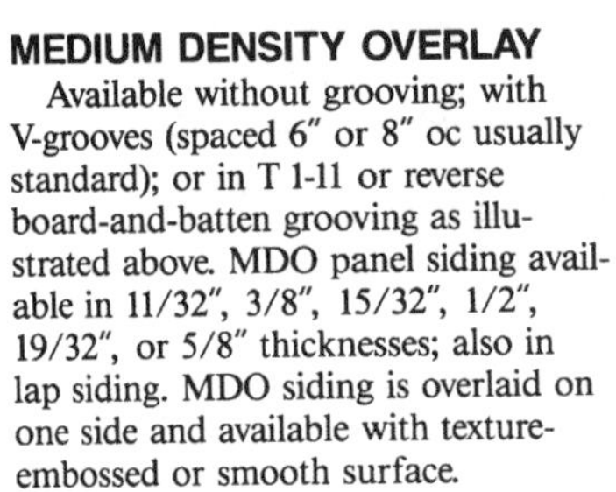

MEDIUM DENSITY OVERLAY

Available without grooving; with V-grooves (spaced 6″ or 8″ oc usually standard); or in T 1-11 or reverse board-and-batten grooving as illustrated above. MDO panel siding available in 11/32″, 3/8″, 15/32″, 1/2″, 19/32″, or 5/8″ thicknesses; also in lap siding. MDO siding is overlaid on one side and available with texture-embossed or smooth surface.

BRUSHED

Brushed or relief-grain textures accent the natural grain pattern to create striking surfaces. Generally available in 11/32″, 3/8″, 15/32″, 1/2″, 19/32″, and 5/8″ thicknesses. Available in Douglas-fir, cedar and other species.

7.4.0 Variety of Surface Textures Available on APA-Rated Siding (Continued)

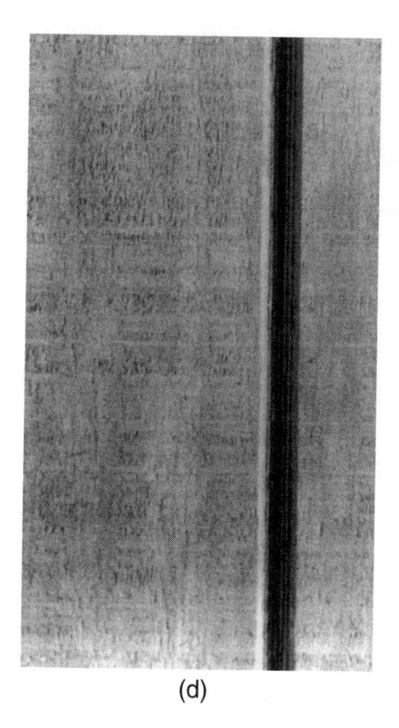

(d)

APA TEXTURE 1-11

Special Rated Siding 303 panel with shiplapped edges and parallel grooves 1/4″ deep, 3/8″ wide; grooves 4″ or 8″ oc are standard. Other spacings may be available on special order. T 1-11 is available only in 19/32″ and 5/8″ thicknesses. Rough-sanded panel shown above. Also available with scratch-sanded, overlaid, brushed and other surfaces. Available in Douglas-fir, cedar, redwood, southern pine and other species.

(e)

ROUGH SAWN

Manufactured with a slight, rough-sawn texture running across panel. Available without grooves, or with grooves of various styles; in lap sidings, as well as in panel form. Generally available in 11/32″, 3/8″, 15/32″, 1/2″, 19/32″ and 5/8″ thicknesses. Rough sawn also available in kerfed (shown) with grooves typically 4″ oc in multiples of 2″, Texture 1-11, reverse board-and-batten, channel groove and V-groove (15/32″, 1/2″, 19/32″, or 5/8″ thick). Available in Douglas-fir, redwood, cedar, southern pine and other species.

By permission of APA, The Engineered Wood Association, Tacoma, Washington

7.5.0 Plywood Panel Dimensions (U.S. Customary and Metric)

Metric Conversions

Metric equivalents of nominal thicknesses and common sizes of APA Rated Siding products are tabulated below. (1 inch = 25.4 millimeters):

APA RATED SIDING NOMINAL THICKNESS

in.	mm
11/32	8.7
3/8	9.5
7/16	11.1
15/32	11.9
1/2	12.7
19/32	15.1
5/8	15.9

PANEL SIDING NOMINAL DIMENSIONS (Width x Length)

ft.	mm	m (approx.)
4 x 8	1219 x 2438	1.22 x 2.44
4 x 9	1219 x 2743	1.22 x 2.74
4 x 10	1219 x 3048	1.22 x 3.05

LAP SIDING NOMINAL DIMENSIONS (Width x Length)

in. x ft.	mm	m (approx.)
6 x 16	152.4 x 4877	0.15 x 4.88
8 x 16	203.2 x 4877	0.20 x 4.88
12 x 16	304.8 x 4877	0.30 x 4.88

7.6.0 Span Tables fo Plywood Sheathing and Subfloors

Wood Structural Panel Sheathing[a][c] – Panel Continuous Over 2 or More Spans

PANEL SPAN RATING	MAXIMUM STUD SPACING (inches)	MAXIMUM FASTENER SPACING (inches)[b]	
		PANEL EDGES (when over framing)	INTERMEDIATE (each stud)
12/0, 16/0, 20/0 or Wall-16 oc	16	6	12
24/0, 24/16, 32/16 or Wall-24 oc	24	6	12

(a) When wood structural panel is used, building paper and diagonal wall bracing are not required.
(b) Use fastener recommended by metal-framing manufacturer.
(c) See requirements for nailable panel sheathing when exterior covering is to be nailed to sheathing.

Recommended Uniform Floor Live Loads for APA RATED STURD-I-FLOOR and APA RATED SHEATHING with Long Dimension Perpendicular to Supports.

STURD-I-FLOOR SPAN RATING	SHEATHING SPAN RATING	MAXIMUM SPAN (in.)	ALLOWABLE LIVE LOADS (psf)[a]						
			JOIST SPACING (in.)						
			12	16	20	24	32	40	48
16 oc	24/16, 32/16	16	185	100					
20 oc	40/20	20	270	150	100				
24 oc	48/24	24	430	240	160	100			
32 oc	60/32	32		430	295	185	100		
48 oc		48			460	290	160	100	55

(a) 10 psf dead load assumed. Live load deflection limit is l/360.
Note: Shaded joist spacing meet Code Plus recommendations.

APA Panel Subflooring (APA RATED SHEATHING)[a]

PANEL SPAN RATING	MINIMUM PANEL THICKNESS (in.)	MAXIMUM SPAN (in.)	MAXIMUM FASTENER SPACING (in.)[c][e]	
			SUPPORTED PANEL EDGES	INTERMEDIATE SUPPORTS
24/16	7/16	16	6	12
32/16	15/32	16[b]	6	12
40/20	19/32	20[b][d]	6	12
48/24	23/32	24	6	12
60/32	7/8	32	6	12

(a) For subfloor recommendations under ceramic tile, refer to APA Design/Contruction Guide: Residential and Commercial. For subfloor recommendations under gypsum concrete, contact manufacturer of floor topping.
(b) Span may be 24 inches if 3/4-inch wood strip flooring is installed at right angles to joists.
(c) Use fastener recommended by metal-framing manufacturer.
(d) Span may be 24 inches if a minimum 1-1/2 inches of lightweight concrete is applied over panels.
(e) Other code-approved fasteners may be used.

7.7.0 Recommended Spans for Roof Sheathing and Fastening Schedules

Recommended Uniform Roof Live Loads for APA RATED SHEATHING[c] and APA RATED STURD-I-FLOOR with Long Dimension Perpendicular to Supports[e]

PANEL SPAN RATING	MINIMUM PANEL THICKNESS (in.)	MAXIMUM SPAN(in.)		ALLOWABLE LIVE LOADS (psf)[d]							
				SPACING OF SUPPORTS CENTER-TO-CENTER (in.)							
		WITH EDGE SUPPORT [a]	WITHOUT EDGE SUPPORT	12	16	20	24	32	40	48	60
APA RATED SHEATHING[c]											
12/0	5/16	12	12	30							
16/0	5/16	16	16	70	30						
20/0	5/16	20	20	120	50	30					
24/0	3/8	24	20[b]	190	100	60	30				
24/16	7/16	24	24	190	100	65	40				
32/16	15/32	32	28	325	180	120	70	30			
40/20	19/32	40	32	—	305	205	130	60	30		
48/24	23/32	48	36	—	—	280	175	95	45	35	
60/32	7/8	60	48	—	—	—	305	165	100	70	35
APA RATED SHEATHING[f]											
16 oc	19/32	24	24	185	100	65	40				
20 oc	19/32	32	32	270	150	100	60	30			
24 oc	23/32	40	36	—	240	160	100	50	30	25	
32 oc	7/8	48	48	—	—	295	185	100	60	40	
48 oc	1-3/32	60	48	—	—	—	290	160	100	65	40

(a) Tongue-and-groove edges, panel edge clips (one midway between each support, except two equally spaced between supports 48 inches on center), lumber blocking, or other. For low slope roofs, see Table 5.
(b) 24 inches for 15/32-inch and 1/2-inch panels.
(c) Includes apa Rated Sheathing/ceiling deck.
(d) 10 psf dead load assumed.
(e) Applies to panels 24 inches or wider.
(f) Also applies to C-C Plugged grade plywood.
Note: Shaded support spacing meet Code Plus recommendations.

Recommended Maximum Spans for APA Panel Roof Decks for Low Slope Roofs[a]
(Long panel dimension perpendicular to supports and continuous over two or more spans.)

Grade	Minimum Nominal Panel Thickness (in.)	Minimum Span Rating	Maximum Span (in.)	Panel Clips Per Span[b] (number)
apa rated sheathing	15/32	32/16	24	1
	19/32	40/20	32	1
	23/32	48/24	48	2
	7/8	60/32	60	2

(a) Low slope roofs are applicable to built-up, single-ply and modified bitumen roofing systems. For guaranteed or warranted roofs contact membrane manufacturer for acceptable deck.
(b) Edge support may also be provided by tongue-and-groove edges or solid blocking.

FASTENER SCHEDULES

When attaching wood structural panels to metal decking, the main purpose of the fasteners is to keep the panels flat. The fastener schedule should be at least the same as if the panel was applied to framing that is spaced in accordance with the panel's Span Rating. For example, a 32/16 span rated sheathing panel should have fasteners spaced at 6 inches on center along the 4-foot ends, and at no more than 32 inches on center by 12 inches on center across the width of the panel (28 fasteners per panel). If wind uplift is a consideration, additional fasteners may be required.

Recommended Minimum Fastening Schedule for APA Panel Roof Sheathing (Increased fastener schedules may be required in high wind or seismic zones.)

Panel Thickness[b] (in.)	Fasteners[c]	
	Maximum Spacing (in.)	
	Panel Edges	Intermediate
5/16-1	6	12[a]
1-1/8	6	12[a]

(a) For spans 48 inches or greater, space fasteners 6 inches at all supports.
(b) For stapling asphalt shingles to 5/16-inch and thicker panels, use staples with a 15/16-inch minimum crown width and a 1-inch leg length. Space according to shingle manufacturer's recommendations.
(c) Use fastener recommended by metal-framing manufacturer.

7.8.0 Typical Plywood Sheathing Construction

APA Rated Sheathing easily meets building code requirements for bending and racking strength without diagonal straps. Building paper is not required over wall sheathing, except under stucco, and under brick veneer where required by local building code. Rated Sheathing provides an excellent nail base for exterior siding. For information on installing exterior panel siding over nailable sheathing, refer to the *APA Design/Construction Guide: Residential & Commercial*, Form E30.

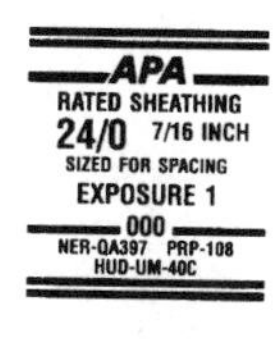

APA
RATED SHEATHING
32/16 15/32 INCH
SIZED FOR SPACING
EXPOSURE 1
000
PS 1-83 C-D
NER-QA397 PRP-108

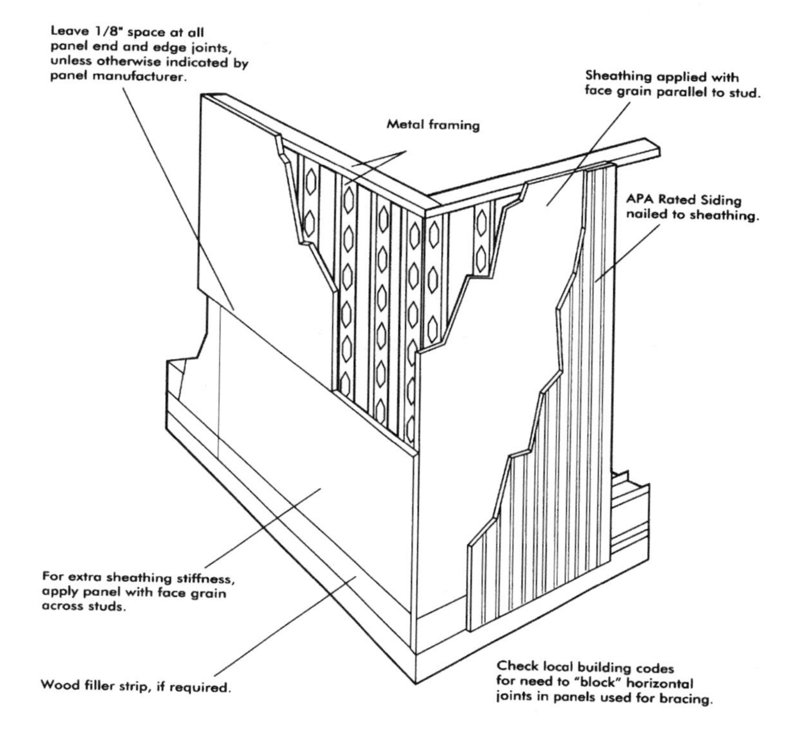

Wood Structural Panel Sheathing(a)(c) – Panel Continuous Over 2 or More Spans

PANEL SPAN RATING	MAXIMUM STUD SPACING (inches)	MAXIMUM FASTENER SPACING (inches)(b)	
		PANEL EDGES (when over framing)	INTERMEDIATE (each stud)
12/0, 16/0, 20/0 or Wall-16 oc	16	6	12
24/0, 24/16, 32/16 or Wall-24 oc	24	6	12

(a) When wood structural panel is used, building paper and diagonal wall bracing are not required.
(b) Use fastener recommended by metal-framing manufacturer.
(c) See requirements for nailable panel sheathing when exterior covering is to be nailed to sheathing.

By permission of APA, The Engineered Wood Association, Tacoma, Washington

7.8.1 Composite Wood Products

Along with lumber and plywood, within the past 40 years, a new wood product has gained wide acceptance in the industry, composite wood products. These products are panels and laminated materials made up of small pieces of wood glued together, oftentimes with plastic fillers. These products are frequently referred to as *engineered wood products*.

7.8.1.1 Medium-Density Fiberboard (MDF)

Dry-formed panels manufactured from lignocellulosis fibers, combined with a synthetic resin or other suitable binder.

- *Available thicknesses:* 3/16" (4.74 mm) to 1½" (38.1 mm) (3", 76.2 mm, is available on special order).
- *Density:* 40 to 50 pounds/cubic foot (641 to 801 kg/cubic meter).
- *Uses:* Moldings or millwork where it replaces solid wood.

7.8.1.2 Hardboard (Compressed Fiberboard)

A board manufactured from interfelted lignocellulosis fibers, consolidated under heat and pressure to form a dense material.

- *Available thicknesses* Typically ⅛" (12.7 mm) to ½" (38.1 mm).
- *Density* 45 to 70 pounds/cubic foot (705 to 112 kg/cubic meter).
- *Uses* Exterior siding, peg board, decorative wall paneling, underlayment, drawer bottoms, furniture backs, and simulated wood shingles and shakes.

7.8.1.3 Cellulosic Fiberboard (Softboard)

Made from wood fibers, recycled paper, bagasse (a plant residue, such as from sugar cane), and other agricultural by products.

- *Available thicknesses* Typically ½" (12.7 mm) to 2" (50.8 mm).
- *Density* Typically 10 to 25 pounds/cubic foot (160 to 400 kg/cubic meter).
- *Uses* Wall sheathing, roof insulation, and sound insulation.

7.8.1.4 Oriented Strand Board (OSB)

This material evolved from waferboard and is constructed of strands of softwood or hardwood ½" (12.7 mm) wide by 3" (76.2 mm) to 4'6" (1.37 m) in length.

- *Available thicknesses* Typically ¼" (6.4 mm) to 1⅛" (28.6 mm).
- *Density* 36 to 44 pounds/cubic foot (577 to 705 kg/cubic meter).
- *Uses* Interchangeably used in structural applications in the same way as plywood. Phenolic paper overlaid OSB is used for siding.

7.8.1.5 Waferboard

Similar to OSB, except that it is composed of large flakes of wood bonded together and generally made from low-density hardwoods, such as aspen. Once used a great deal as sheathing, it has largely been replaced by OSB.

7.8.1.6 Laminated Veneer Lumber (LVL)

Primarily a structural member made of veneer laid up in one grain direction and made in billets 27" (68.6 cm) to 50" (127 cm) wide and 1½" (38.1 mm) or 1¾" (44.5 mm) thick. Produced under pressure to cure the adhesives, mostly phenolic glues. This material is nondestructively tested to ensure consistent strength. TrusJoist MacMillan uses this material as flanges in their I-joists.

7.8.1.7 Parallel-Strand Lumber (PSL)

These products are made of oriented strands of waste softwood veneer. The ½" (12.7 mm) wide by 37" (94 cm) long strands are oriented and laid up into a mat, which is processed through a microwave-heating system into billets of 11" (279 mm) × 18" (457 mm) or 11" (279 mm) × 14" (355 mm). These billets are sawn into lengths and thicknesses, as required. PSL members are used where high-strength lumber or timber materials are required. TrusJoist MacMillan's Parallam is a PSL product.

7.8.1.8 Oriented Strand Lumber (OSL)

OSL is made with nominal 12" (300 mm) long strands and pressed in a steam-injection press machine to produce uniform density throughout. This material, developed by McMillan Bloedel, Ltd., is also used in joist construction.

7.8.1.9 Com-Ply

Com-Ply is a material developed by the USDA Forest Service in the 1970s and composed of random or oriented wood flakes or particles sandwiched between two layers of veneer. One or more layers of veneer are also placed on the faces or edges of the lumber. This material is not widely used today.

7.9.0 Moisture Content of Particleboard and the Impact on Warpage

When used as a substrate for plastic laminate facings, these particleboard and fiberboard panels are subject to warpage if not stored properly. Warpage can also occur when an unbalanced laminated panel is produced—one with a face sheet of high-pressure laminate, but no backer sheet. Moisture content building up in the unfaced panel causes stresses to accumulate. When these stresses become excessive and are no longer equally balanced, cracks can occur in the laminate. This unbalance can occur because of a number of factors:

- Selection of laminate other than HPL, such as a wood veneer.
- The environment in which laminating is to occur.
- Conditioning (or lack thereof) of each component of the assembly.
- Product design problems.
- Installation procedures.

Unusually moist or dry conditions should be avoided in both the storage of the substrate and the laminating environment.

7.9.1 Plywood Underlayment Span Tables and Glue/Nailed Fastening Recommendations

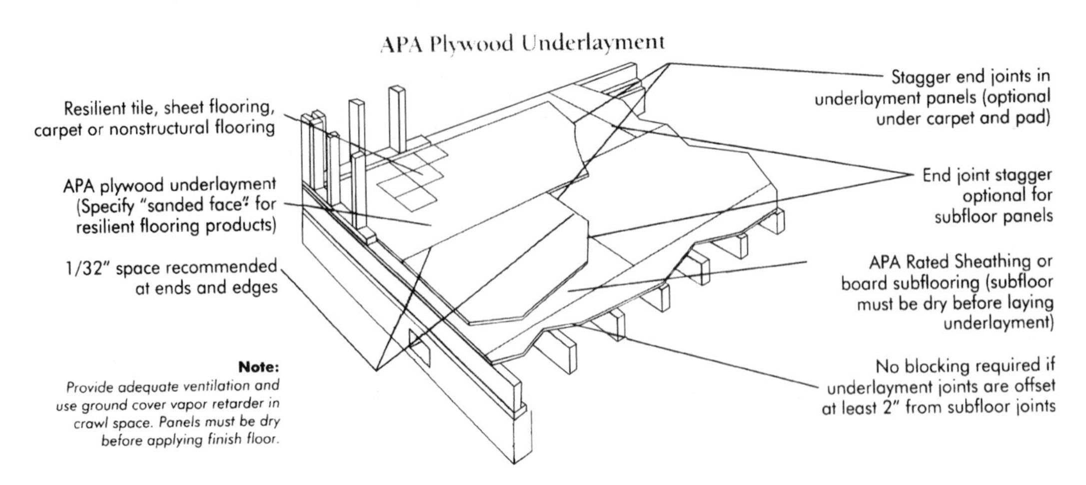

APA Plywood Underlayment[a]

Plywood Grades[b]	Application	Minimum Plywood Thickness (in.)	Fastener Size and Type	Maximum Fastener Spacing (in.)[e]	
				Panel Edges[d]	Intermediate
APA UNDERLAYMENT APA C-C Plugged EXT APA RATED STURD-I-FLOOR (19/32 in. or thicker)	Over smooth subfloor	1/4	3d (1-1/4 in.) ring- or screw-shank nails min. 12-1/2 gage (0.099 in.) shank dia.[c]	3	6 each way
	Over lumber subfloor or uneven surfaces	11/32		6	8 each way

(a) For underlayment recommendations under ceramic tile, refer to Table 7.
(b) In areas to be finished with resilient floor coverings such as tile or sheet vinyl, or with fully-adhered carpet, specify Underlayment, C-C Plugged or veneer-faced STURD-I-FLOOR with "sanded face." Underlayment A-C, Underlayment B-C, Marine EXT or sanded plywood grades marked "Plugged Crossbands Under Face," "Plugged Crossbands (or Core)," "Plugged Inner Plies" or "Meets Underlayment Requirements" may also be used under resilient floor coverings.
(c) Use 4d (1½ in.) ring- or screw-shank nails, minimum 12½ gauge (0.099 in.) shank diameter, for underlayment panels 19/32 in. to ¾ in. thick.
(d) Fasten panels ⅜ in. from panel edges.
(e) Fasteners for 5-ply plywood underlayment panels and for panels greater than ½ in. thick may be spaced 6 in. on center at edges and 12 in. each way intermediate.

APA Rated Sturd-I-Floor[a]

Span Rating (Maximum Joist Spacing) (in.)[g]	Panel Thickness[b] (in.)	Fastening: Glue-Nailed[c]			Fastening: Nailed-Only		
		Nail Size and Type	Spacing (in.) Supported Panel Edges	Spacing (in.) Intermediate Supports	Nail Size and Type	Spacing (in.) Supported Panel Edges	Spacing (in.) Intermediate Supports
16	19/32, 5/8	6d ring- or screw-shank[d]	12	12	6d ring- or screw-shank	6	12
20	19/32, 5/8	6d ring- or screw shank[d]	12	12	6d ring- or screw-shank	6	12
24	23/32, 3/4	6d ring- or screw-shank[d]	12	12	6d ring- or screw-shank	6	12
	7/8	8d ring- or screw-shank[d]	12	12	8d ring- or screw-shank	6	12
32	7/8	8d ring- or screw-shank[d]	6	12	8d ring- or screw-shank	6	12
48	1-3/32, 1-1/8	8d ring- or screw-shank[e]	6	(f)	8d ring- or screw-shank	6	(f)

APA Panel Subflooring (APA Rated Sheathing)[a][f]

Panel Span Rating (f)	Minimum Panel Thickness (in.)	Maximum Span (in.)	Nail Size & Type[e]	Nail Spacing (in.) Supported Panel Edges	Nail Spacing (in.) Intermediate Supports
24/16	7/16	16	6d common	6	12
32/16	15/32, 1/2	16[b]	8d common[c]	6	12
40/20	19/32, 5/8	20[b][d]	8d common	6	12
48/24	23/32, 3/4	24	8d common	6	12
60/32[g]	7/8	32	8d common	6	12

(a) For subfloor recommendations under ceramic tile, refer to Table 7. For subfloor recommendations under gypsum concrete, contact manufacturer of floor topping.
(b) Span may be 24. in. if ¾-in. wood strip flooring is installed at right angles to joists.
(c) 6d common nail permitted if panel is ½ in. or thinner.
(d) Span may be 24 in. if a minimum 1½ in. of lightweight concrete is applied over panels.
(e) Other code-approved fasteners may be used.
(f) For Code Plus Floors, see pages 29–30 for requirements.
(g) Check with supplier for availability.

By permission of APA, The Engineered Wood Association, Tacoma, Washington

7.9.2 Ideal Fabrication Conditions Chart

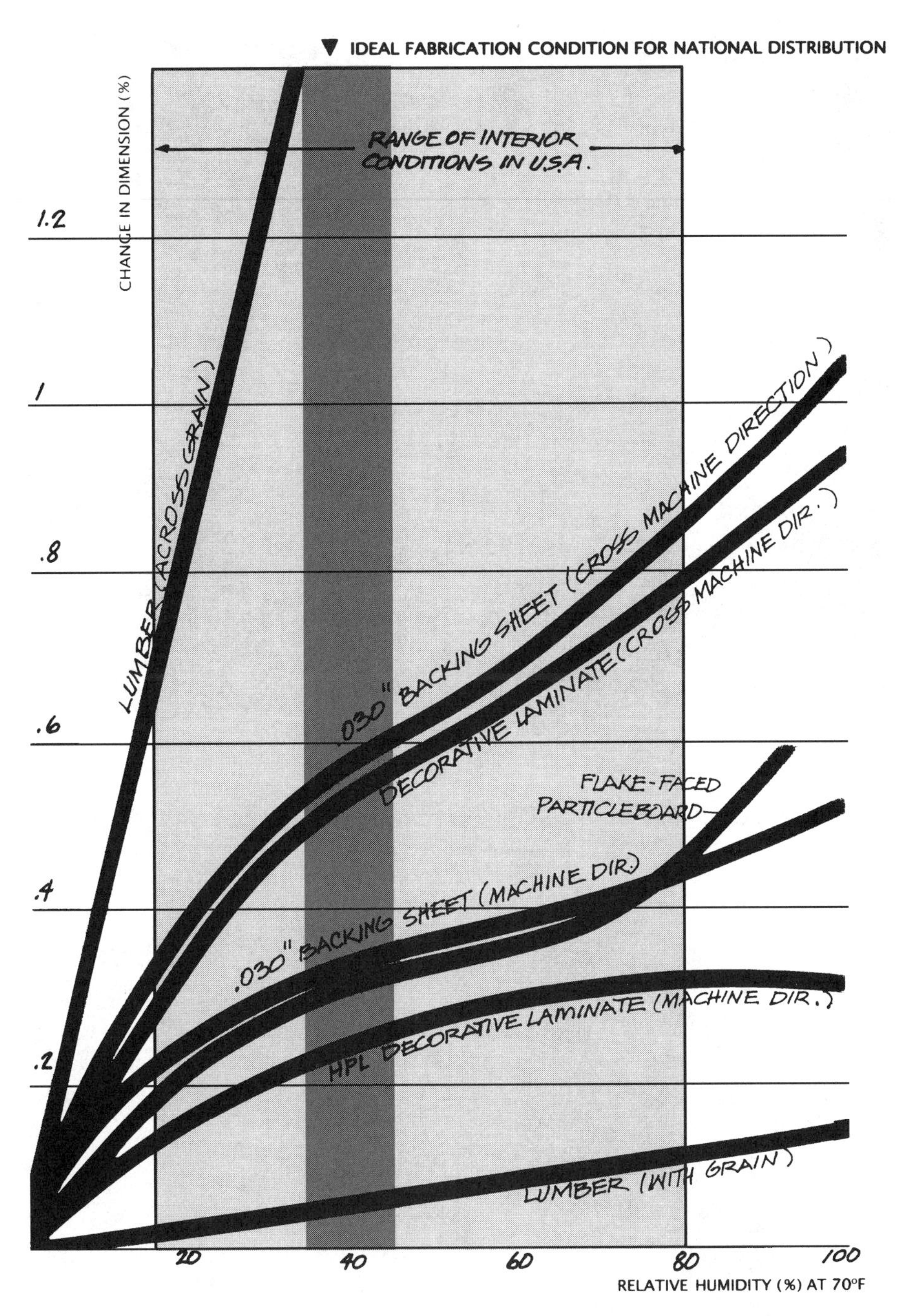

Reprinted by permission of National Particleboard Association, Gaithersburg, Maryland

7.9.3 Moisture Content Zones in the U.S.

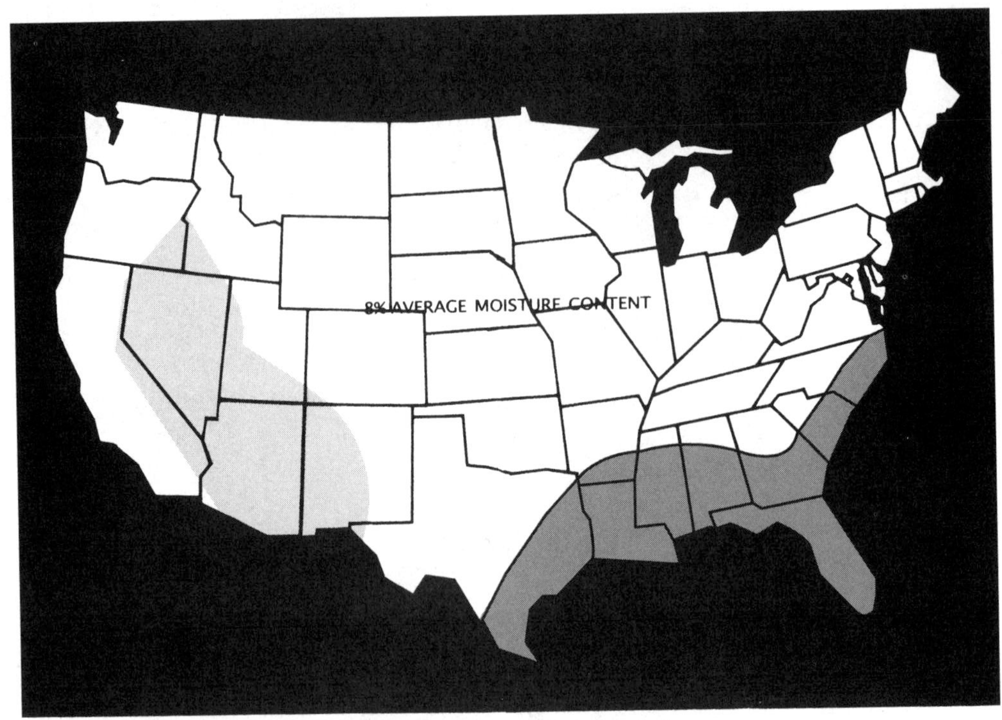

Approximate equilibrium moisture content zones for wood-based products. Values may vary with local and seasonal conditions.

7.9.4 Dimensional Changes in Medium-Density Fiberboard (MDF) and Industrial-Grade Particle Board (PBI)

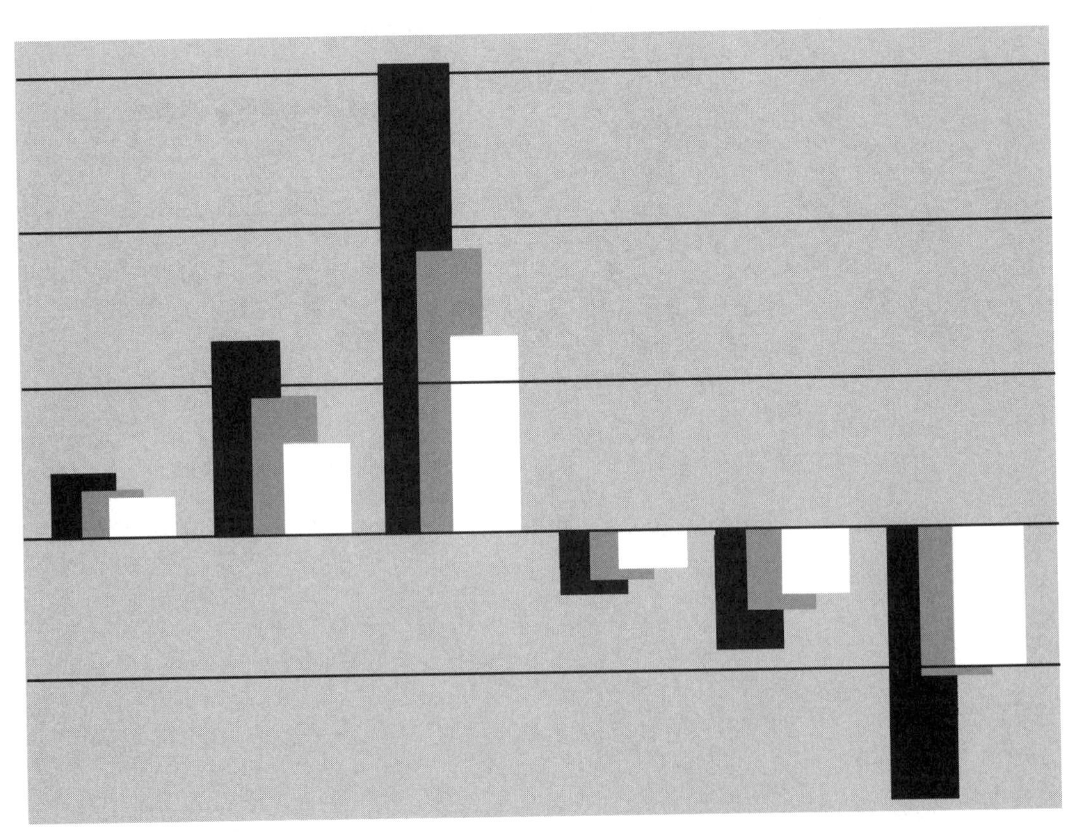

Reprinted by permission of National Particleboard Association, Gaithersburg, Maryland

7.10.0 APA-Rated Sturdi-Floor Subfloor and Floor Framing for Hardwood Floors

APA Rated Sturd-I-Floor 16, 20, 24, 32 and 48 oc

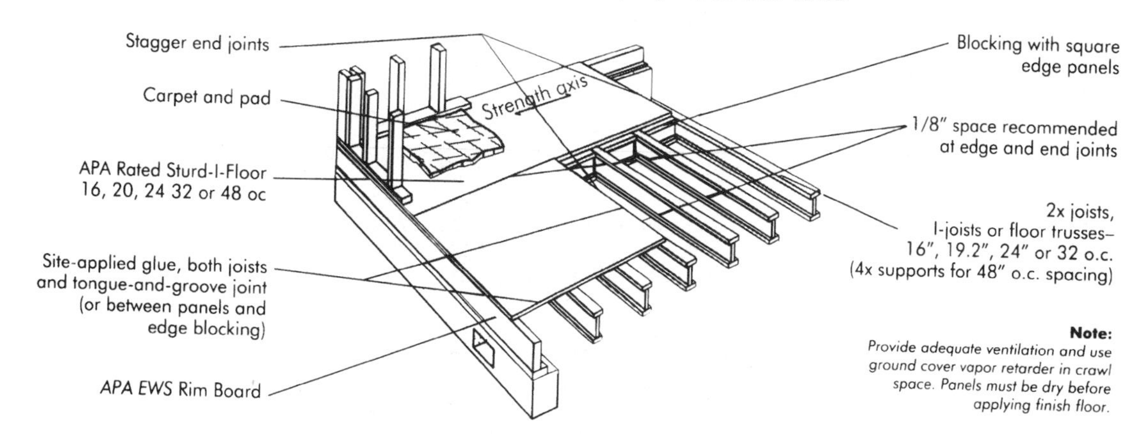

Subflooring and Spacing of Floor Framing for Hardwood Flooring

APA Rated Sheathing or Sturd-I-Floor Span Rating	Spacing (in.) of Floor Framing	
	Maximum Spacing	Code Plus Spacing
40/20, 20 oc*	19.2	12
48/24, 24 oc	24	19.2
32 oc	32	24
48 oc	48	32

* The National Oak Flooring Manufacturers Association (NOFMA) and the National Wood Floor Association (NWFA) both recommend the use of 23/32" minimum thickness OSB or plywood as a subfloor material.

By permission of APA, The Engineered Wood Association, Tacoma, Washington

7.11.0 High-Pressure Laminate (HPL) Q&A

Q. What is HPL?

A. High pressure laminate is a thermoset paper/plastic composite, where decorative papers impregnated with melamine are consolidated over phenolic impregnated kraft papers at high temperature and pressure to form a homogeneous laminate.

Q. What is the difference between horizontal, vertical, and postforming grades of HPL?

A. Horizontal grade 10/HGS is thicker, .050," and not intended to be post-formed to a tight radius. Horizontal surfaces include countertops, vanity tops, store fixtures, window sills, desks, table tops, convector covers, furniture and casework.
Vertical grade 55/VGS is thinner, .030," and does not have the impact resistance of a horizontal grade. Vertical surfaces include wall panels, elevator cabs, toilet compartments, etc.
Postforming is available in both horizontal grade 12/HGP, .042," and vertical grade 20/VGP, .030." Postforming is designed for tight inside and outside bends.

Q. What do the letters following the grade number mean?

A. Formica Corporation being a worldwide manufacturer utilizes the International Organization for Standardization (ISO) nomenclature. Examples are:

Grade 10/HGS ....Horizontal, General Purpose, Standard
Grade 12/HGP ....Horizontal, General Purpose, Postformable
Grade 20/VGP ....Vertical, General Purpose, Postformable
Grade 55/VGS ....Vertical, General Purpose, Standard
Grade 50/HGF ....Horizontal, General Purpose, Fire Rated
Grade 32/VGF ....Vertical, General Purpose, Fire Rated
Grade 72/CLS ....Cabinet Liner, Standard

Q. What causes expansion and contraction of laminates after fabrication? How can this be prevented?

A. High pressure laminate is a wood, paper product and like all wood products moves with changes in humidity. Laminates expand in high humidity and contract in low humidity. Laminate and core should be conditioned at 45% to 50% R.H. at least 48 hours prior to laminating. Pick a substrate that moves at the same dimensional change rate as HPL such as medium density fiberboard (MDF) or 45# industrial grade particleboard.

Q. What causes stress cracking? How can it be eliminated?

A. Excessive diminensional movement of the laminate can cause stress, especially on inside corners, which is relieved by the cracking. To eliminate cracking: acclimate the laminate and core, minimize cross directional dimensions, use the thickest laminate possible for the application, use the strongest adhesive possible for the application, and rout inside corners (⅛" minimum).

By permission of Formica Corporation, Cincinnati, Ohio

7.11.0 High-Pessure Laminate (HPL) Q&A (Continued)

Q. What causes laminated panels to warp?

A. Panel warpage is caused by a difference of movement between the laminate and the substrate. To minimize warpage, acclimate the laminate and core prior to bonding. Panels that require flatness should be balanced by bonding the same grade of laminate to both sides.

Q. Can HPL be used for exterior applications?

A. No.

Q. Can surface scratches be repaired?

A. No. Melamine is one of the hardest plastics known, but it can be scratched. Like glass, melamine scratches white, thus scratches are more apparent on dark solid colors. Because the finish is pressed into the laminate, it is impossible to repair. Superficial scratches can be hidden with the use of furniture polish.

Q. Do laminates fade?

A. Laminates will fade if exposed to direct sunlight. Bright chromatics fade easier than earthtones. All FORMICA® brand laminate colors surpass industry fade requirements.

Q. Can you resurface laminate over laminate?

A. Yes, self edge or flat surfaces can be resurfaced. Follow the recommended procedures in our Fabrication Data Sheet titled "Resurfacing Laminated Assemblies With FORMICA® brand products."

Q. Can laminates be painted?

A. Yes. However, the laminate surface has trace amounts of release agent which prevents paint adhesion. Lightly sanding the surface removes this agent and provides tooth for the paint. Epoxy paint adheres the best. Painted surfaces do not have the durability of laminate.

Q. What adhesives should be used to bond laminates?

A. FORMICA® brand contact adhesives are available in brush, spray, flammable, and non-flammable formulations. Resorcinols, ureas, and PVAc (white glue) type adhesives yield stronger bonds.

Q. How should laminate be cleaned?

A. There is a thin layer of melamine resin on the surface of HPL, which is very hard and stain resistant, but there are many modern household reagents that will attack it. Do not use acids, alkalies, bleaches, or abrasive cleansers on laminate. Surfaces should be cleaned with a clean, soft cotton cloth and mild detergent such as Pine-Sol®.

FORMICA is a registered trademark of Formica Corporation.
PINE-SOL is a registered trademark of American Cyanamid Company.

By permission of Formica Corporation, Cincinnati, Ohio

7.11.1 HPL Tips for Avoiding Panel Warpage

Causes of Panel Warpage

Laminate clad panels are susceptible to warpage if they are not physically restrained or balanced. Balanced panel construction equalizes the forces acting on both sides of the core material. If for any reason, these forces become unbalanced, warpage can result.

Warpage of wood product panel assemblies (e.g. laminate clad particleboard or MDF) is attributed to the differences in dimensional movement between the face and back laminates and the core or substrate material. This movement and its subsequent stresses are caused by the expansion or contraction of paper fibers in the laminate skins and wood fibers in wood composite cores as they respond to relative humidity changes. The stress and dimensional movement, generated within a laminate skin, is transmitted to the core through its glue line. The forces involved are tremendous and, if they are not properly considered in the panel design, warpage can result.

The use of laminates and substrates that have different strengths and/or dimensional movement potentials is not the only cause of warpage. Exposing one side of a panel assembly to different humidity conditions than the other side can also cause warpage. For example, a "balanced" panel will warp if one side is exposed to air conditioning and the other is against a damp, below grade wall (e.g. basement wall without a proper moisture barrier).

Tips for Avoiding Panel Warpage

1) All panel components should be acclimated to the same environment prior to assembly. This will ensure that one component will not be contracting while the other is expanding due to subsequent relative humidity changes. In addition, under extreme conditions, materials that have not been properly acclimated to the same condition prior to fabrication, can buckle or delaminate as well as warp. Proper preconditioning of materials can also help to minimize shrink-back or laminate growth problems on machined edges.

2) For critical applications requiring a well balanced assembly (doors, etc.), the same laminate or skin should be applied on both sides. Less critical applications may only require a cabinet liner or phenolic backer. Small components and mechanically restrained panels (countertops, etc.), on the other hand, may not need balancing sheets.

3) Thick panels warp less than thin panels due to increased rigidity and the geometry of the forces involved. For critical applications the thickest core material permissible should be selected to help minimize warpage.

4) Laminates expand and contract twice as much in their cross-grain direction as they do in their grain (parallel with the sanding lines) direction. Always align the sanding lines of the front and back laminates in the same direction and, wherever possible, align the grain direction of the laminate with the longest panel dimension. It is also advisable to align the grain and cross-grain directions of the laminates with that of the substrate.

Note: When multiple panels are viewed together, keep all laminate components aligned in the same direction to minimize visual changes in color or gloss due to the directionality of the underlying surface paper and laminate finish.

7.11.2 HPL Stress Crack Avoidance

Causes of stress cracking

Stress cracking of high pressure laminate is caused by the concentration or build-up of stresses in a particular area of a laminated assembly. When this stress becomes greater than that which the laminate can withstand, a stress crack will occur. If such stresses are allowed to concentrate around a cut-out or other such fabrication detail, one or more cracks can characteristically radiate from the sharper corners of the cut-out, where, for mechanical reasons, the laminate is weakest.

These stresses can be caused by external mechanical forces but are generally caused by the normal dimensional movements of the laminated assembly as it reacts to the surrounding environment. As with all wood based products, high pressure laminates and their substrates react to humidity changes. Under moist conditions laminated assemblies gain moisture and expand dimensionally. When this same assembly is subjected to dry conditions, however, this moisture is lost and shrinkage results. If the laminate shrinks more than the substrate, stress cracking of the laminate surface can occur in certain areas.

Techniques for controlling stress cracking

The occurrence of stress cracking can be greatly minimized by using fabrication techniques and practices which recognize and moderate the dimensional movement and associated stresses that can develop within a laminated assembly. These techniques and practices consist of: preconditioning, proper substrate selection, obtaining a good bond, proper inside corner fabrication, proper seam placement and good installation practices.

Preconditioning

Prior to the fabrication, allow the laminate and substrate to acclimate for at least 48 hours to the same ambient conditions. Optimum conditions are approximately 75°F and a relative humidity of 45 to 55%. Provision should be made for the circulation of air around the components.

Substrate selection

FORMICA® brand laminate and COLORCORE® brand surfacing material should be bonded to either a MDF (Medium Density Fiberboard) or a 45 lb. density industrial grade particleboard (CS 236-66: Type 1, Grade B, Class 2). The dimensional change properties of these substrates, being similar to that of high pressure laminate, greatly reduces the potential for stress cracking when the assembly is subjected to low humidity conditions.

Plywood substrates should be avoided, whenever possible, for use with FORMICA® brand laminate and should never be used as a substrate for COLORCORE® brand surfacing materials. Because of its cross ply construction, plywood expands and shrinks less than either of these laminate grades. This results in greater stress built up within the laminate and thereby increases the chance of stress cracking.

Adhesive bond

The quality and nature of the bond between the laminate and the substrate is also an important factor to consider when trying to minimize stress cracking. Basically, the stronger and more rigid the bond, the less are the chances for stress cracking.

Contact adhesives, by their nature, are elastomeric and therefore transfer less of the stress to the substrate. Assemblies made with contact adhesives, therefore, are less crack resistant than those fabricated with rigid or semi-rigid adhesives. If contact adhesives are used they should be properly applied and fused to obtain the strongest possible bond.

Rigid and semi-rigid adhesives such as resorcinal, ureas and PVAc (white glues) transfer stresses directly to the substrate. Assemblies fabricated with these adhesives are more crack resistant.

By permission of Formica Corporation, Cincinnati, Ohio

7.11.2 HPL Stress Crack Avoidance (Continued)

The stress crack performance of assemblies using contact adhesive can be greatly improved if a PVAc (white glue) is used at all inside corners as illustrated below. Note: If the assembly is to be water resistant, a catalyzed PVAc glue should be used.

A. The cutout area of the laminate and substrate assembly is masked prior to applying the contact adhesive.
B. Once the contact adhesive has been applied and dried, the masking is removed and a PVAc glue is applied.
C. The laminate and substrate are then joined and nip rolled together to fuse the contact adhesive. The masked off area is then clamped until the adhesive sets. This usually takes about one hour.

(See attached figures)

Figure A

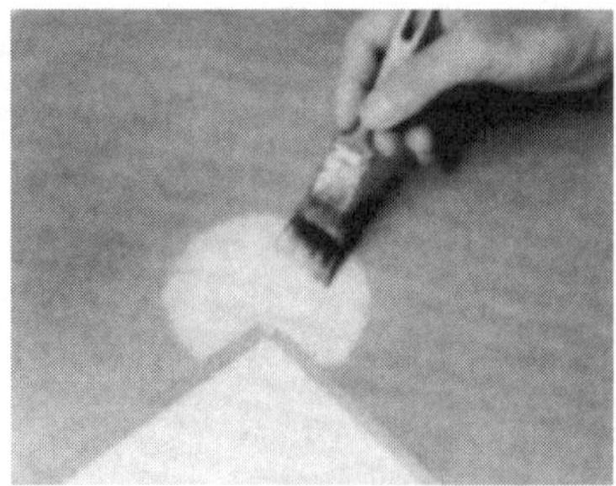

Figure B

Figure C

Inside corner fabrication

The inside corners of all cutouts must be radiused as large as possible (1/8" minimum) to minimize stress cracking. A radiused corner created by a 1/4" diameter router bit is normally used. All edges and inside corners should be filed smooth and free of any chips or nicks.

Seam placement

Another effective means of minimizing the chances of stress cracking is to plan the placement of seams to reduce the number of inside corners. Examples of proper seam positions are shown in the following illustrations.

(See attached illustration)

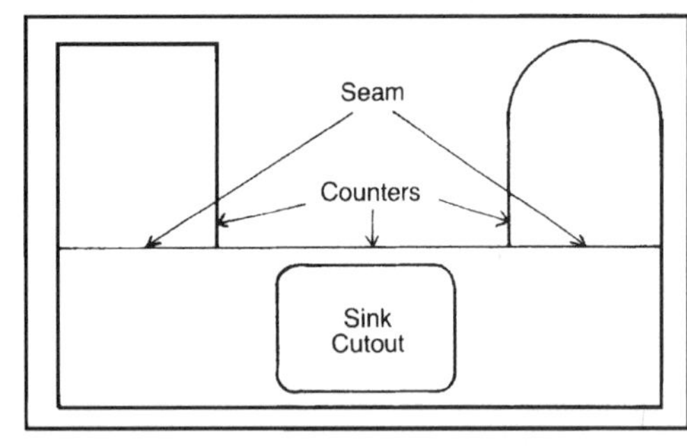

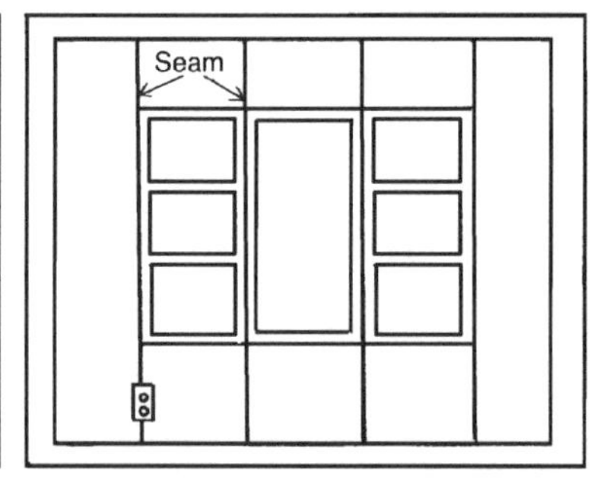

Installation

Install the laminated assembly with sufficient clearance at pipes, electrical boxes, panel edges, etc. to allow for normal dimensional movement. Sinks, louvers, drop in ranges, etc. should fit easily into openings without binding. Do not install a panel or laminated assembly by force fitting. Panels should be installed in a flat plane by shimming, as necessary, to avoid mechanical stresses caused by bending or twisting.

Summary

1. Precondition laminate and substrate for a minimum of 48 hours prior to fabrication. Optimum conditions are approximately 75°F and 45 to 55% relative humidity.
2. Select the proper substrate...MDF or 45lb. density particleboard. Plywood should not be used with COLOR-CORE® brand surfacing material.
3. Obtain a good bond. Assemblies bonded with rigid or semi-rigid adhesives are more crack resistant than those assembled with contact adhesives.
4. Radius inside corners as large as possible, 1/8" minimum.
5. Plan the placement of seams to minimize inside corners.
6. Provide sufficient clearance at sinks, electrical boxes, range cutouts, etc. to allow for dimensional movement. Do not force fit. Do not induce mechanical stresses.

By permission of Formica Corporation, Cincinnati, Ohio

7.11.3 HPL Post-Forming CounterTops

CONDITIONS AFFECTING POSTFORMING

Successful postforming is easily accomplished by using various techniques which recognize and moderate the common variables associated with postforming. These techniques incorporate: preconditioning, temperature control, elimination of drafts and proper equipment adjustment and maintenance.

PRECONDITIONING

Postforming grade laminate is slightly hygroscopic; that is, it is capable of losing or absorbing moisture from the atmosphere. Therefore, if it is exposed to dry air conditions, a loss of moisture can result that adversely affects its postforming properties. To assure proper postforming performance, FORMICA® brand postforming grade laminate should be preconditioned prior to use for at least 48 hours at 70°F and 50% relative humidity. Small shop areas can be economically humidified with portable humidifier units. Larger areas may require specific recommendations from a HVAC equipment supplier.

Remember, when seasonal changes approach, preconditioning practices should be observed to maintain consistent postforming conditions inside the shop, regardless of the atmospheric conditions outside. This is especially important during the winter months when dry air conditions often exist.

TEMPERATURE CONTROL

The optimum postforming temperature for FORMICA® brand laminate is at or near 325°F. Lower temperatures may cause cracking while higher temperatures may cause gloss changes, blistering and/or cracking. If either occurs, alter the surface temperature accordingly. On most equipment this can be accomplished by adjusting the power input to the heater, the heater height or the line speed.

To determine the surface temperature of laminated plastic there are two primary techniques which can facilitate equipment set-up.

One relatively simple technique involves the use of temperature indicators such as TEMPILAQ® Temperature Indicating Liquid or TEMPILSTIK® Temperature Indicating Crayons to facilitate equipment set-up. These are available from the Tempil Division of Big Three Industries, Inc., 2901 Hamilton Blvd., South Plainfield, NJ 07080 (phone: 201-757-8300).

Another effective method of monitoring and measuring the laminate surface temperature is to use a non-contact infrared thermometer. One unit that we have found to be particularly useful is a Model D500-RS remote sensor Microscanner from Exergen Corp., 1 Bridge Street, Newton, MA 02158 (phone: 800-422-3006). A unit of this type is recommended for larger shops.

ELIMINATE DRAFTS

Avoid open windows or doors near the postforming operation. Sudden drafts over the heated laminate surface can drop its temperature below optimum conditions and cause cracking or crazing. This is especially important during cold weather when cold blasts from open doors, etc. can happen unexpectedly. The use of temporary or permanent partitions to eliminate drafts is often required.

EQUIPMENT INSPECTION

Commercial or custom built postforming equipment will perform efficiently and properly only if it is in good working condition. All equipment should, therefore, be inspected periodically. Automatic timers may malfunction. Heating elements may develop hot spots or fail to heat up. Guides or stops may loosen. Rollers may become misaligned or worn. Planned periodic inspection of all critical components will help avoid costly material damage and loss of valuable production time.

By permission of Formica Corporation, Cincinnati, Ohio

7.11.4 HPL Post-Forming Counter Tops (Manual Techniques)

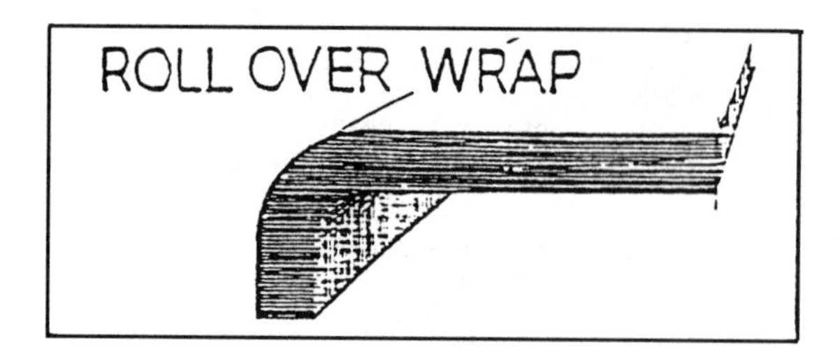

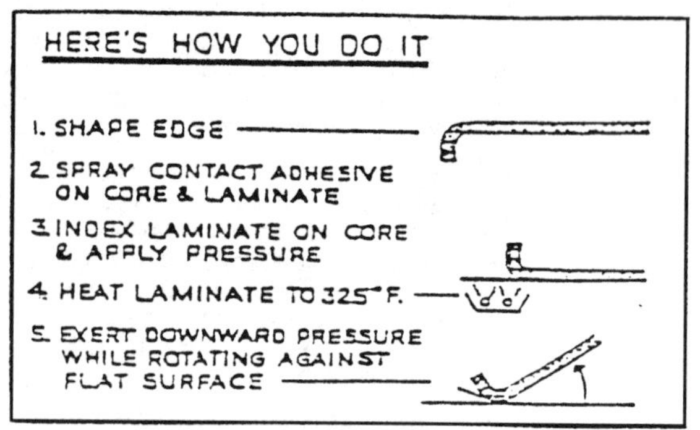

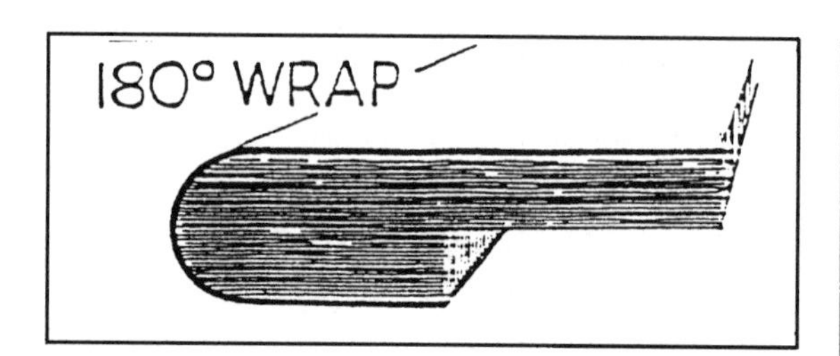

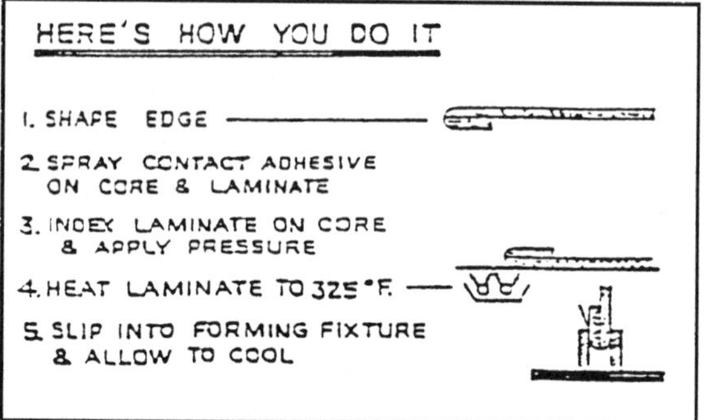

By permission of Formica Corporation, Cincinnati, Ohio

7.12.0 Common Post-Forming Problems

SYMPTOM	PROBLEM	CAUSE	CORRECTION
Cracking, crazing	Heat source	Insufficient heat	Increase heat or rate of heat-up
		Improper heater position	Adjust heater to focus on bend area
	Cores	Irregular radius	Sand core
		Poor machining	Check cutter alignment
		Cold cores	Store at 65°F +
		Contaminated or dusty cores	Clean prior to forming
		Radius too tight	Increase radius
	Equipment	Poor alignment	Align equipment
		Dirty equipment	Clean equipment
	Laminate	Wrong grade	Use proper grade
		Dry conditions	Humidify storage area
Blisters	Heat source	Too much heat	Reduce heat
Glueline delamination	Heat source	Insufficient heat to soften laminate	Increase heat
		Too much heat	Reduce heat
	Core	Radius too tight	Increase radius
	Equipment	Poor alignment	Align equipment
	Adhesive	Insufficient adhesive	Increase spread rate
		Improper adhesive	Consult manufacturer
	Drying oven	Insufficient dry time	Increase drying time or oven temperature
Gloss change	Heat source	Too much heat	Reduce heat

IMPORTANT NOTICE

The information and statements herein are believed to be reliable but are not to be construed as a warranty or representation for which Formica Corporation assumes legal responsibility. Users should undertake sufficient verification and testing to determine the suitability for their own particular purposes of any information or products referred to herein. NO WARRANTY OF FITNESS FOR A PARTICULAR PURPOSE IS MADE.

By permission of Formica Corporation, Cincinnati, Ohio

7.13.0 Low-Pressure Laminates (LPL)

Sometimes referred to as *saturated paper laminates*, these LPLs can take the form of solid-color decorative papers that have been saturated with either a melamine, a phenolic or a polyester resin. These low-pressure laminates are wood based and will shrink and expand in the presence of moisture or the lack of moisture. Although the contractor will generally purchase these kinds of panel materials from a manufacturer, it is helpful to have knowledge of the factors involving successful assembly of these products.

- During assembly, the press should be loaded and closed as quickly as possible.
- Hot boards should be stacked flat and well supported while cooling. Rapid cooling is to be avoided.
- The volatile material in the papers should be retained and not dried out.
- The press platen temperature and conditions for the proper curing of both sides must be set when using different papers.
- When using phenolic papers with elevated temperatures and extended press times, degradation of the substrate must be avoided. Proper cooling of these panels is essential.

7.14.0 APA Specifications for Roof Sheathing

The Code Plus Roof

1 Apply APA Rated Sheathing with a minimum Span Rating of 32/16 to roof framing spaced 24 inches o.c., or;

2 Apply APA Rated Sheathing with a minimum Span Rating of 24/16 to roof framing spaced 16 inches o.c.

3 Provide adequate ventilation to meet or exceed local code requirements.

4 For panels up to 1 inch thick use 8d nails. Use 8d deformed shank or 10d common nails for panels 1-3/32 inch or thicker. Space nails 6 inches o.c. at supported edges and 12 inches o.c. at intermediate supports. (See Tables 16 and 17 for details.)

5 Space panels 1/8-inch at all ends and edges.

Recommended Maximum Spans for Code Plus Roofs
(APA Rated Sheathing with Long Dimension Perpendicular to Supports)[a]

Panel Span Rating	Minimum Panel Thickness (in.)	Maximum Span (in.)	Allowable Live Loads (psf) Spacing of Supports Center-to-Center (in.)			
			12	16	24	32
24/16	7/16	16	190	100		
32/16	15/32, 1/2	24		180	70	
40/20	19/32, 5/8	24			130	
48/24	23/32, 3/4	32			175	95

(a) Applies to panels 24 in. or wider. For narrower panels, additional edge support is recommended. (See *APA Technical Note R275* for details.)

Shaded areas meet Code Plus requirements.

Recommended Uniform Roof Live Loads for APA Rated Sheathing[c] and APA Rated Sturd-I-Floor with Long Dimension Perpendicular to Supports[e]

Panel Span Rating	Minimum Panel Thickness (in.)	Maximum Span (in.)		Allowable Live Loads (psf)[d] Spacing of Supports Center-to-Center (in.)			
		With Edge Support[a]	Without Edge Support	12	16	24	48
APA RATED SHEATHING[c]							
12/0	5/16	12	12	30			
16/0	5/16	16	16	70	30		
20/0	5/16	20	20	120	50		
24/0	3/8	24	20[b]	190	100	30	
24/16	7/16	24	24	190	100	40	
32/16	15/32, 1/2	32	28	325	180	70	
40/20	19/32, 5/8	40	32	—	305	130	
48/24	23/32, 3/4	48	36	—	—	175	35
60/32[g]	7/8	60	48	—	—	305	70

By permission of APA, The Engineered Wood Association, Tacoma, Washington

Section

8

Roofing

Contents

8.0.0 Most frequently used types of roofing
8.0.1 Built-up membrane roofing
8.0.2 Fluid-applied membrane roofs
8.0.3 Single-ply membrane roofs
8.0.4 Metal sheet and metal panel roofs
8.0.5 Shingles, shakes, and tile roofs
8.1.0 Roof flashings
8.1.1 Flashing types and locations
8.2.0 3-ply built-up roof on approved insulation
8.2.1 3-ply built-up roof on nailable deck
8.2.2 3-ply built-up roof on lightweight fill insulated deck
8.3.0 4-ply gravel surface built-up roof over insulation, inclines to 3" per foot
8.3.1 4-ply smooth surface built-up roof over insulation, inclines to 3" per foot
8.4.0 3- and 4-ply hot-mopped modified bitumen roofs
8.5.0 Built-up roof-flashing details
8.5.1 Built-up roof-flashing details (continued)
8.6.0 Single-ply membrane securement data
8.6.1 Single-ply membrane securement data (continued)
8.6.2 Basic wind-speed map
8.6.3 Single-ply membrane splicing cement guide
8.6.4 Single-ply membrane ballasted roof stone specifications
8.7.0 Single-ply membrane curb flashing details
8.7.1 Single-ply membrane reglet and cap flashing details.
8.7.2 Single-ply membrane curb and vertical pipe flashing details
8.7.3 Single-ply membrane counterflashing /vertical termination flashing details
8.7.4 Single-ply membrane expansion-joint details
8.7.5 Single-ply membrane box gutter/roof drain flashing details
8.7.6 Single-ply membrane acceptable roof deck chart
8.8.0 Single-ply membrane Underwriters Laboratories specifications
8.9.0 Single-ply membrane Roofing Preventative Maintenance Guidelines
8.9.1 Investigation of leaks on a ballasted single-ply membrane roof
8.10.0 A typical fire vent for BUR and SPM roofs
8.10.1 A typical roof hatch where a ladder is used for access
8.10.2 Typical roof hatch where a ships ladder is used for access
8.10.3 Typical roof hatch installation where stairs are used for access
8.11.0 Copper and lead-coated copper roofing material sizes and weights
8.12.0 Standard sizes and exposure to weather for slate roof tiles
8.12.1 Slate roof installation procedures
8.13.0 Cedar shingle/shake installation diagrams
8.13.1 Cedar shingle-grade label facsimiles
8.13.2 Cedar shingle and shake installation and maintenance tips
8.14.0 A checklist to deflect or avoid roof leaks

8.0.0 Most Frequently Used Types of Roofing

8.0.1 Built-Up Membrane Roofing

All BURs share three basic components: felts, bitumens, and protective caps. The felts, asphalt-impregnated, fiberglass-reinforced membrane sheets are designed to act in concert with the bitumens (a semi-solid asphalt or coal tar pitch material) to create a moisture-resistant surface. The cap, weathering-grade asphalt embedded with mineral granules or gravel to protect the built-up roof from the elements is the third element in this assembly.

Built-up roofs can be subdivided into three categories:

1. *Smooth surface* BUR without any gravel topping. These roofs are lightweight, easy to inspect, and, if leaks occur, make it simple to determine the source of the leak.
2. *Gravel surface* BUR with a stone-aggregate spread over its entire surface after a flood coat of bitumen as been applied to protect the membrane from the elements. Gravel-surfaced BURs are limited to those roofs with slopes of 3 inches or less.
3. *Mineral surface* BUR with a top sheet of weathering-grade asphalt embedded with mineral granules to protect the surface from the elements.

8.0.2 Fluid-Applied Membrane Roofs

Fluid-applied roofs can be installed with either hot or cold materials. This type of roof installation requires a stable substrate, such as a cast-in-place concrete deck. When applied over concrete, which must meet certain moisture content standards, a prime coat is first sprayed or rolled on. This is generally followed by the installation of a nylon or fiberglass mat mopped directly onto the primed concrete surface after which a top coat is applied by roller or spray. The fluid applied membrane makes it easy to spot leaks, which might occur if cracks appear in the substrate and the nylon/fiberglass mat cannot bridge the gap. The liquid-applied roof is often used where free-form roofs are constructed.

8.0.3 Single-Ply Membrane Roofs

The advent of man-made elastomeric materials, such polyvinyl chloride (PVC) and ethylene propylene diene monomer (EPDM), ushered in the era of single-ply membrane roofs. Elastic, flexible, easy to install, ozone and ultraviolet-ray resistant, these wide-width sheets (some as wide as 40 feet) provide a roof membrane with significantly fewer seams that is very cost effective, long-lived and relatively easy to repair, if damaged.

A variation on the single-ply membrane roof is the IRMA roof (Inverted Roof Membrane Assembly), where the single-ply membrane is placed directly on the roof deck and rigid insulation, protection board, and aggregate ballast is placed on top. The membrane nestles protected from the elements and from roof traffic that could damage the membrane.

8.0.4 Metal Sheet and Metal Panel Roofs

Metals of various alloys (such as lead, tern, zinc, and copper) have been used for hundreds of years and are still popular today, primarily for aesthetic reasons or when historic restorations are being undertaken. Formed metal roofing should not be installed on sloped roofs with a pitch less than 1½ inches in one foot.

8.0.5 Shingles, Shakes, and Tile Roofs

These materials are actually watershedding materials, rather than waterproofing materials, and rely upon roof pitch to rapidly drain the water from the surface on the roof. Slopes of 3 to 4 inches per foot are recommended before selecting any of these materials. Wood shingles and wood shakes require installation where air can circulate behind them so that they can dry out after becoming wet. Slate shingles are expensive to purchase and install, but are extremely long lasting. This material is generally specified when restoration work is being undertaken. Porcelain enamel tiles or clay tiles are frequently used in certain parts of the country where mission or Spanish-style roofs are popular, such as the Southwest.

8.1.0 Roof Flashings

- *Gravel stops* Gravel stops are metal flashing attached to the edge of the roof to protect and secure the edge of the roof membrane. When gravel is placed on the roof, the profile of the gravel stop is such that it prevents the gravel from rolling or washing over the edge of the roof.
- *Copings* Similar in nature to gravel stops, except that they are placed on top of perimeter parapet walls to secure the roof's base flashing.
- *Base flashings* Generally flexible materials that provide watertight integrity between the horizontal roof membrane and some vertical surface. Base flashing can also be made of metal and require either a reglet or counterflashing on the vertical surface to ensure watertight conditions.
- *Counter flashings* Flashings that act as a shield to cover the seamed base flashing below. They are generally constructed of aluminum, copper, lead, or stainless steel.
- *Pipe and conduit flashings* Whenever a mechanical or electrical pipe or conduit penetrates the roof surface, some form of flashing must be installed to seal off this penetration. Factory-supplied "boots" or shop-fabricated "pitch pockets" are used to seal off these roof surfaces.
- *Roof drain flashings* When installed in a roof, generally at a low point in the roof surface where water tends to accumulate, special care is required where these flashings are installed. Usually installed by the plumbing contractor, roof drains can be purchased with flashings specially designed for that purpose.
- *Roof vent flashings* Roof vents installed through the roof surface require "boots" that can be purchased or fabricated for that purpose.
- *Pitch pockets* The "pocket" is usually formed of aluminum or copper and is fastened to the roof deck, which encloses a pipe or series of pipes that penetrate the roof surface. This pocket or dam is then filled with pitch, a black viscous tar that "cold" flows to seal the spaces around the penetrations. Pitch pockets require periodic inspections to ensure that the pitch levels are maintained.
- *Expansion joint covers* When a large expanse of roof is constructed, allowance must be made for expansion and subsequent contraction. Various types of bellow or slip-joint expansion joints can be installed, and (depending on the configuration) might require additional flashing to make them watertight.
- *Ridge flashings* Where the valley and eaves are created in a roof, flashings must be installed. Generally, this occurs when shingled roofs are installed, whether wood, tile, or slate.

8.1.1 Flashing Types and Locations

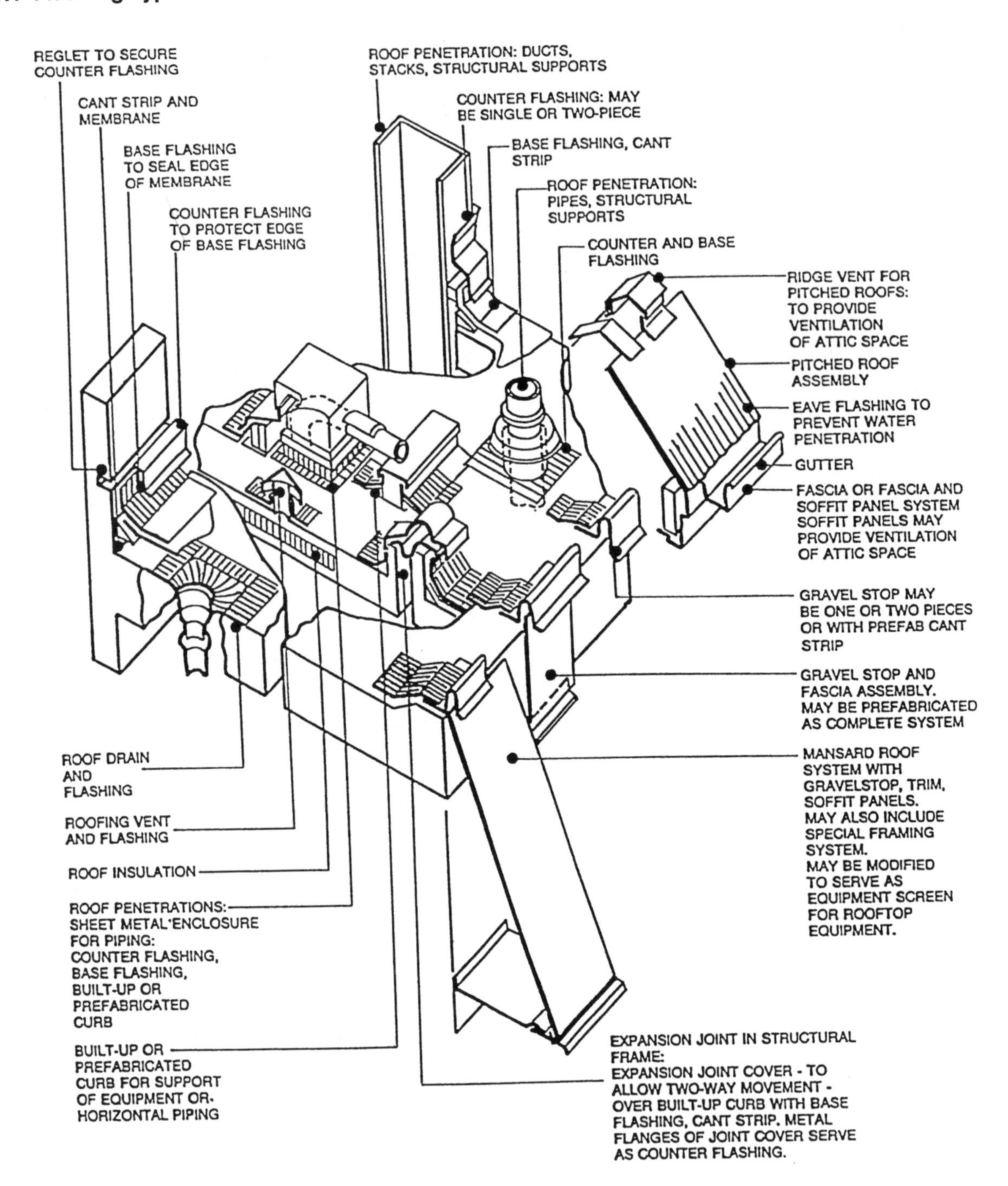

8.2.0 3-Ply Built-Up Roof on Approved Insulation

Specification 3GIG (Alternate)

Three Ply
Gravel Surfaced
Fiber Glass Built-Up Roof

For use over Schuller insulation, approved decks or other approved insulations, on inclines of up to 3" per foot (250 mm/m)

For Regions 1, 2 and 3
(Not acceptable in all locations, consult Schuller Technical Service Specialist)

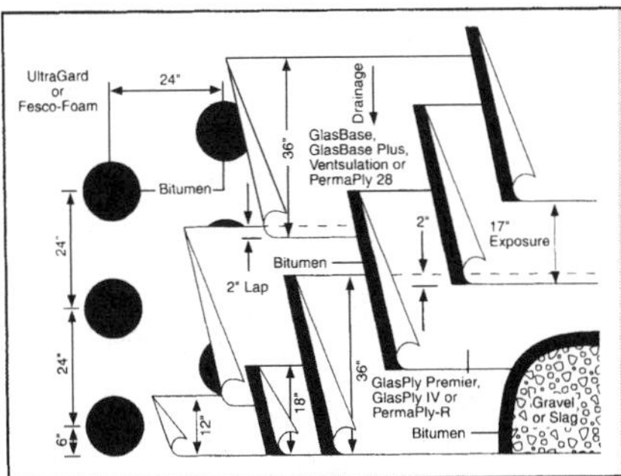

Materials per 100 sq. ft. of Roof Area

Felts:
Ventsulation, GlasBase, PermaPly 28 or GlasBase Plus .... 1 ply
GlasPly Premier, PermaPly-R or GlasPly IV .... 2 plies

Bitumen (Interply):

Incline per foot	Bitumen	Nominal Weight
Up to ½"	170°F, Type II, Flat	53 lbs.
½" to 3"	190°F, Type III, Steep or 220°F, Type IV, Special Steep	53 lbs.
0 to 3"	PermaMop	53 lbs.

Surfacing:
Flood coat of bitumen .... 60 lbs.
Gravel .... 400 lbs.
or Slag .... 300 lbs.

Aggregate density, size and coverage will determine actual weight.
Approximate installed weight: 434 - 622 lbs.

General

This specification is for use over any type of approved structural deck which is not nailable and which offers a suitable surface to receive the roof. Poured and pre-cast concrete decks require priming with Schuller Concrete Primer prior to application of hot bitumen.

This specification is also for use over Schuller roof insulations or other approved rigid roof insulations, which are not nailable and which offer a suitable surface to install the roof. Specific written approval is required for any roof insulation not manufactured or supplied by Schuller. Insulation should be installed in accordance with the appropriate Schuller Insulation Specification detailed in the current Schuller Commercial/Industrial Roofing Systems Manual. This specification can also be used in certain reroofing situations. Refer to the "Reroofing" section of the Schuller Commercial/Industrial Roofing Systems Manual. This specification is not to be used directly over poured or pre-cast gypsum or lightweight, insulating concrete fills.

Design and installation of the deck and/or substrate must result in the roof draining freely and to outlets numerous enough and so located as to remove water promptly and completely. Areas where water ponds for more than 24 hours are unacceptable and are not eligible to receive a Schuller Roofing Systems Guarantee.

Note: All general instructions contained in the current Schuller Commercial/Industrial Roofing Systems Manual should be considered part of this specification.

Flashings

Flashing details can be found in the "Bituminous Flashings" section of the Schuller Commercial/Industrial Roofing Systems Manual.

Application

Note: On roof decks with slopes up to 1" per foot (83.3 mm/m), the roofing felts may be installed either perpendicular or parallel to the roof incline. On slopes over 1" per foot (83.3 mm/m), refer to **Paragraph 6.11** of this section for special requirements.

Using Ventsulation, GlasBase, GlasBase Plus or PermaPly 28, start with a 12" (305 mm) width (the use of a specific base sheet may be a condition of Guarantee). The following base sheet courses should be applied full width, lapping the side laps 2" (51 mm) and the end laps 4" (102 mm) over the preceding felts. Set each felt firmly into spot moppings of hot bitumen (within ±25°F [±14°C] of the EVT). The spot moppings should be applied by machine at the rate of approximately 7 lbs. per square (0.3 kg/m²). The spots should be approximately 12" (305 mm) in diameter and 24" (610 mm) o.c. Each row should be staggered from the previous one.

Using GlasPly Premier, PermaPly-R, or GlasPly IV, apply a piece 18" (457 mm) wide, then over that, a full width piece. The following felts are to be applied full width overlapping the preceding felts by 19" (483 mm) so that at least 2 plies of felt cover the base felt/substrate at all locations. Install each felt so that it is firmly and uniformly set, without voids, into the hot bitumen (within ±25°F [±14°C] of the EVT) applied just before the felt at a nominal rate of 23 lbs. per square (1.1 kg.m²) over the entire surface. Installation over porous substrates such as roof insulation may require up to 33 lbs. per square (1.6 kg/m²) of hot bitumen.

Surfacing

Flood the surface with the appropriate bitumen at an approximate rate of 60 lbs. per square (2.9 kg/m²). Into the hot bitumen, embed an acceptable gravel at a rate of 400 lbs. per square (19.5 kg/m²) or an acceptable slag at a rate of 300 lbs. per square (14.6 kg/m²). Aggregate must be installed so that there is complete coverage across the entire surface and at least 50% of the aggregate is solidly adhered in the hot bitumen. Aggregate should meet the requirements of ASTM D 1863.

Asphalt should meet the requirements of ASTM D 312. The contractor must provide a Schuller confirmation number for asphalt on jobs which require a Guarantee. Check with a Schuller Technical Service Specialist for special requirements in hot climates.

By permission of Schuller Roofing Systems, Denver, Colorado

8.2.1 3-Ply Built-Up Roof on Nailable Deck

Specification 3GNG

Three Ply
Gravel Surfaced
Fiber Glass Built-Up Roof

For use over wood or other nailable decks on inclines of up to 3" per foot (250 mm/m)

For Regions 2 and 3

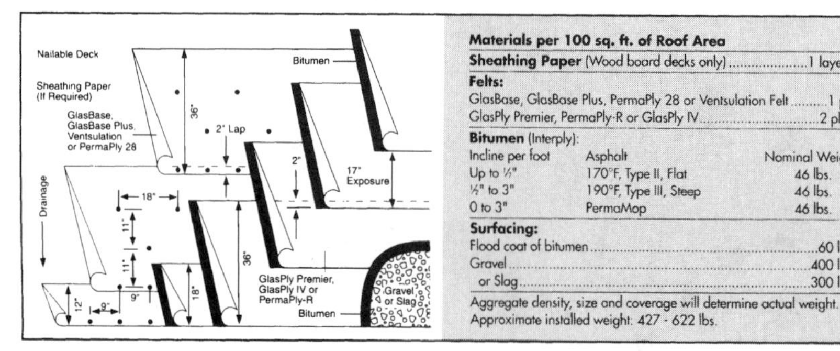

Materials per 100 sq. ft. of Roof Area

Sheathing Paper (Wood board decks only) .... 1 layer

Felts:
GlasBase, GlasBase Plus, PermaPly 28 or Ventsulation Felt .... 1 ply
GlasPly Premier, PermaPly-R or GlasPly IV .... 2 plies

Bitumen (Interply):

Incline per foot	Asphalt	Nominal Weight
Up to ½"	170°F, Type II, Flat	46 lbs.
½" to 3"	190°F, Type III, Steep	46 lbs.
0 to 3"	PermaMop	46 lbs.

Surfacing:
Flood coat of bitumen .... 60 lbs.
Gravel .... 400 lbs.
or Slag .... 300 lbs.

Aggregate density, size and coverage will determine actual weight.
Approximate installed weight: 427 - 622 lbs.

General

This specification is for use over any type of approved structural deck (without insulation) which can receive and adequately retain nails or other types of mechanical fasteners that may be recommended by the deck manufacturer. Examples of such decks are wood and plywood. This specification is not for use directly over lightweight, insulating concrete decks.

Design and installation of the deck and/or substrate must result in the roof draining freely and to outlets numerous enough and so located as to remove water promptly and completely. Areas where water ponds for more than 24 hours are unacceptable and are not eligible to receive a Schuller Roofing Systems Guarantee.

Note: All general instructions contained in the current Schuller Commercial/Industrial Roofing Systems Manual should be considered part of this specification.

Flashings

Flashing details can be found in the "Bituminous Flashings" section of the Schuller Commercial/Industrial Roofing Systems Manual.

Application

Over wood board decks, one ply of sheathing paper must be used under the base felt and on top of the wood board deck.

Note: On roof decks with slopes up to 1" per foot (83.3 mm/m), the roofing felts may be installed either perpendicular or parallel to the roof incline. On slopes over 1" per foot (83.3 mm/m), refer to **Paragraph 6.11** of this section for special requirements.

Using GlasBase, GlasBase Plus, Ventsulation, or PermaPly 28, start with a 12" (305 mm) width (a specific base sheet may be a condition of Guarantee). The following base sheet courses are to be applied full width, lapping the preceding felt 2" (51 mm) on the side laps and 4" (102 mm) on the end laps. Nail the side laps 9" (229 mm) o.c. Down the longitudinal center of each felt, place two rows of nails spaced approximately 11" (279 mm) apart, with the nails staggered on approximately 18" (457 mm) centers. Use nails or fasteners appropriate to the type of deck with 1" (25 mm) minimum diameter caps. For additional fastener information, refer to the Fastener Data in the "Roof Deck" section of the current Schuller Commercial/Industrial Roofing Systems Manual.

Using GlasPly Premier, PermaPly-R, or GlasPly IV, apply a piece 18" (457 mm) wide, then over that, a full width piece. The following felts are to be applied full width overlapping the preceding felts by 19" (483 mm) so that at least 2 plies of felt cover the base felt/substrate at all locations. Install each felt so that it is firmly and uniformly set, without voids, into the hot bitumen (within ±25°F (±14°C) of the EVT) applied just before the felt at a nominal rate of 23 lbs. per square (1.1 kg.m²) over the entire surface. Installation over porous substrates such as roof insulation may require up to 33 lbs. per square (1.6 kg/m²) of hot bitumen.

Surfacing

Flood the surface with the appropriate bitumen at an approximate rate of 60 lbs. per square (2.9 kg/m²). Into the hot bitumen, embed an acceptable gravel at a rate of 400 lbs. per square (19.5 kg/m²) or an acceptable slag at a rate of 300 lbs. per square (14.6 kg/m²). Aggregate must be installed so that there is complete coverage across the entire surface and at least 50% of the aggregate is solidly adhered in the hot bitumen. Aggregate should meet the requirements of ASTM D 1863.

Asphalt should meet the requirements of ASTM D 312. The contractor must provide a Schuller confirmation number for asphalt on jobs which require a Guarantee. Check with a Schuller Technical Service Specialist for special requirements in hot climates.

8.2.2 3-Ply Built-Up Roof on Lightweight Fill Insulated Deck

Specification 3GLG-CT

Three Ply
Gravel Surfaced
Fiber Glass Built-Up Roof

For use over approved, lightweight, insulating fill decks on inclines of up to ¼" per foot (20.8 mm/m)

For Regions 2 and 3

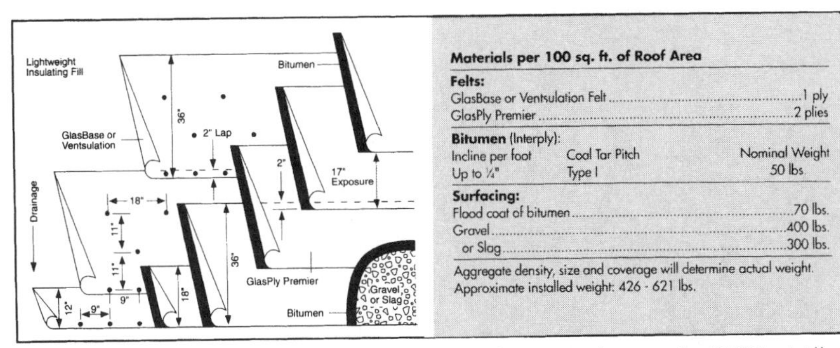

Materials per 100 sq. ft. of Roof Area

Felts:
GlasBase or Ventsulation Felt .... 1 ply
GlasPly Premier .... 2 plies

Bitumen (Interply):

Incline per foot	Coal Tar Pitch	Nominal Weight
Up to ¼"	Type I	50 lbs.

Surfacing:
Flood coat of bitumen .... 70 lbs.
Gravel .... 400 lbs.
or Slag .... 300 lbs.

Aggregate density, size and coverage will determine actual weight.
Approximate installed weight: 426 - 621 lbs.

General
This specification is for use over any type of approved, lightweight, insulating concrete fill deck (without insulation) which can receive and adequately retain nails or other types of mechanical fasteners as may be recommended by the deck manufacturer. Examples of such decks are Zonolite, Celcore and Elastizell. Schuller Ventsulation Felt is recommended over any wet fill deck and may be required as a condition of guarantee.

Design and installation of the deck and/or substrate must result in the roof draining freely and to outlets numerous enough and so located as to remove water promptly and completely. Areas where water ponds for more than 24 hours are unacceptable and are not eligible to receive a Schuller Roofing Systems Guarantee.

Note: All general instructions contained in the current Schuller Commercial/Industrial Roofing Systems Manual should be considered part of this specification.

Flashings
Flashing details can be found in the "Bituminous Flashings" section of the Schuller Commercial/Industrial Roofing Systems Manual.

Application
Over wood board decks, one ply of sheathing paper must be used under the base felt and on top of the wood board deck.

Note: On roof decks with slopes up to ¼" per foot (20.8 mm/m), the roofing felts may be installed either perpendicular or parallel to the roof incline.

DANGER: Coal tar is considered a hazard by inhalation, ingestion and skin contact. The International Agency for Research on Cancer (IARC) has classified coal tar as an agent which is carcinogenic to humans (Group 1). Schuller does not make or sell a coal tar pitch waterproofing agent, and does not recommend its use. Alternative materials, such as asphalt should be utilized.

Using GlasBase or Ventsulation, start with a 12" (305 mm) width (a specific base sheet may be a condition of Guarantee). The following base sheet courses are to be applied full width, lapping the preceding felt 2" (51 mm) on the side laps and 4" (102 mm) on the end laps. Nail the side laps 9" (229 mm) o.c. Down the longitudinal center of each felt, place two rows of nails spaced approximately 11" (279 mm) apart, with the nails staggered on approximately 18" (457 mm) centers. Use nails or fasteners appropriate to the type of deck with 1" (25 mm) minimum diameter caps. For additional fastener information, refer to the Fastener Data in the "Roof Deck" section of the current Schuller Commercial/Industrial Roofing Systems Manual.

Using GlasPly Premier, apply a piece 18" (457 mm) wide, then over that, a full width piece. The following felts are to be applied full width overlapping the preceding felts by 19" (483 mm) so that at least 2 plies of felt cover the base felt/substrate at all locations. Install each felt so that it is firmly and uniformly set, without voids, into the hot bitumen (within ±25°F [±14°C] of the EVT) applied just before the felt at a nominal rate of 25 lbs. per square (1.2 kg.m²) over the entire surface. Installation over porous substrates such as roof insulation may require up to 33 lbs. per square (1.6 kg/m²) of hot bitumen.

Surfacing
Flood the surface with the appropriate bitumen at an approximate rate of 70 lbs. per square (3.4 kg/m²). Into the hot bitumen, embed an acceptable gravel at a rate of 400 lbs. per square (19.5 kg/m²) or an acceptable slag at a rate of 300 lbs. per square (14.6 kg/m²). Aggregate must be installed so that there is complete coverage across the entire surface and at least 50% of the aggregate is solidly adhered in the hot bitumen. Aggregate should meet the requirements of ASTM D 1863.

Coal Tar Pitch must meet the requirements of ASTM D 450, Type I and be certified as such by the manufacturer, in writing.

By permission of Schuller Roofing Systems, Denver, Colorado

8.3.0 4-Ply Gravel Surface Built-Up Roof Over Insulation, Inclines to 3" Per Foot

Specification 4GNG-CT

Four Ply
Gravel Surfaced
Fiber Glass Built-Up Roof

For use over wood or other nailable decks on inclines of up to ¼" per foot (20.8 mm/m)

For Regions 1, 2 and 3

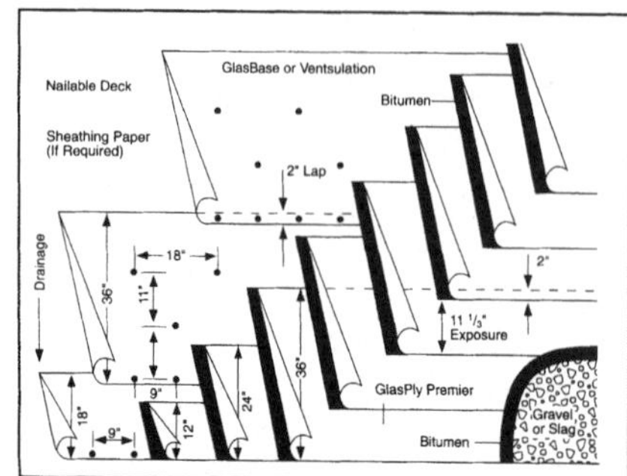

Materials per 100 sq. ft. of Roof Area

Sheathing Paper:
Any wood decks .... 1 layer

Felts:
GlasBase or Ventsulation Felt .... 1 ply
GlasPly Premier .... 3 plies

Bitumen (Interply):

Incline per foot	Coal Tar Pitch	Nominal Weight
Up to ¼"	Type I	75 lbs.

Surfacing:
Flood coat of bitumen .... 70 lbs.
Gravel .... 400 lbs.
or Slag .... 300 lbs.

Aggregate density, size and coverage will determine actual weight.
Approximate installed weight: 446 - 652 lbs.

General

This specification is for use over any type of approved structural deck (without insulation) which can receive and adequately retain nails or other types of mechanical fasteners that may be recommended by the deck manufacturer. Examples of such decks are wood and plywood. This specification is not for use directly over lightweight, insulating concrete decks.

Design and installation of the deck and/or substrate must result in the roof draining freely and to outlets numerous enough and so located as to remove water promptly and completely. Areas where water ponds for more than 24 hours are unacceptable and are not eligible to receive a Schuller Roofing Systems Guarantee.

Note: All general instructions contained in the current Schuller Commercial/Industrial Roofing Systems Manual should be considered part of this specification.

Flashings

Flashing details can be found in the "Bituminous Flashings" section of the Schuller Commercial/Industrial Roofing Systems Manual.

Application

Over wood board decks, one ply of sheathing paper must be used under the base felt and on top of the wood board deck.

Note: On roof decks with slopes up to ¼" per foot (20.8 mm/m), the roofing felts may be installed either perpendicular or parallel to the roof incline.

DANGER: Coal tar is considered a hazard by inhalation, ingestion and skin contact. The International Agency for Research on Cancer (IARC) has classified coal tar as an agent which is carcinogenic to humans (Group 1). Schuller does not make or sell a coal tar pitch waterproofing agent, and does not recommend its use. Alternative materials, such as asphalt should be utilized.

Using Ventsulation or GlasBase, start with an 18" (457 mm) width (the use of a specific base sheet may be a condition of Guarantee). The following base sheet courses are to be applied full width, lapping the preceding felt 2" (51 mm) on the side laps and 4" (102 mm) on the end laps. Nail the side laps 9" (229 mm) o.c. Down the longitudinal center of each felt, place two rows of nails spaced approximately 11" (279 mm) apart, with the nails staggered on approximately 18" (457 mm) centers. Use nails or fasteners appropriate to the type of deck with 1" (25 mm) minimum diameter caps. For additional fastener information, refer to the Fastener Data in the "Roof Deck" section of the current Schuller Commercial/Industrial Roofing Systems Manual.

Using GlasPly Premier, PermaPly-R, or GlasPly IV, apply a piece 12" (305 mm) wide, then over that, one 24" (610 mm) wide, then over both, a full width piece. The following felts are to be applied full width, overlapping the preceding felts by 24⅔" (627 mm) so that at least 3 plies of felt cover the base felt/substrate at all locations. Install each felt so that it is firmly and uniformly set, without voids, into the hot bitumen (within ±25°F [±14°C] of the EVT) applied just before the felt at a nominal rate of 25 lbs. per square (1.2 kg/m²) over the entire surface. Installation over porous substrates such as roof insulation may require up to 33 lbs. per square (1.6 kg/m²) of hot bitumen.

Surfacing

Flood the surface with the appropriate bitumen at an approximate rate of 70 lbs. per square (3.4 kg/m²). Into the hot bitumen, embed an acceptable gravel at a rate of 400 lbs. per square (19.5 kg/m²) or an acceptable slag at a rate of 300 lbs. per square (14.6 kg/m²). Aggregate must be installed so that there is complete coverage across the entire surface and at least 50% of the aggregate is solidly adhered in the hot bitumen. Aggregate should meet the requirements of ASTM D 1863.

Coal Tar Pitch must meet the requirements of ASTM D 450, Type I and be certified as such by the manufacturer, in writing.

By permission of Schuller Roofing Systems, Denver, Colorado

8.3.1 4-Ply Smooth Surface Built-Up Roof Over Insulation, Inclines to 3" Per Foot

Specification 4GIG (Alternate)

Four Ply
Gravel Surfaced
Fiber Glass Built-Up Roof

For use over Schuller insulation, approved decks or other approved insulations, on inclines of up to 3" per foot (250 mm/m)

For Regions 1, 2 and 3
(Not acceptable in all locations, consult Schuller Technical Service Specialist)

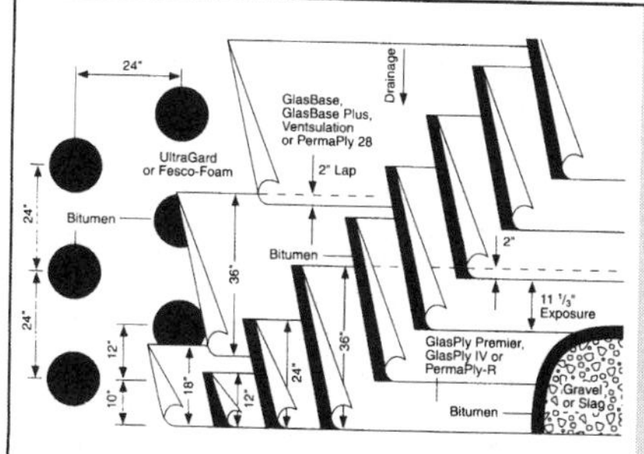

Materials per 100 sq. ft. of Roof Area

Felts:		
GlasBase, GlasBase Plus, PermaPly 28 or Ventsulation Felt		1 ply
GlasPly Premier, PermaPly-R or GlasPly IV		3 plies
Bitumen (Interply):		
Incline per foot	Bitumen	Nominal Weight
Up to ½"	170°F, Type II, Flat	75 lbs.
½" to 3"	190°F, Type III, Steep or 220°F, Type IV, Special Steep	75 lbs.
0 to 3"	PermaMop	75 lbs.
Surfacing:		
Flood coat of bitumen		60 lbs.
Gravel		400 lbs.
or Slag		300 lbs.

Aggregate density, size and coverage will determine actual weight.
Approximate installed weight: 460 - 665 lbs.

General
This specification is for use over any type of approved structural deck which is not nailable and which offers a suitable surface to receive the roof. Poured and pre-cast concrete decks require priming with Schuller Concrete Primer prior to application of hot bitumen.

This specification is also for use over Schuller roof insulations or other approved rigid roof insulations, which are not nailable and which offer a suitable surface to install the roof. Specific written approval is required for any roof insulation not manufactured or supplied by Schuller. Insulation should be installed in accordance with the appropriate Schuller Insulation Specification detailed in the current Schuller Commercial/Industrial Roofing Systems Manual. This specification can also be used in certain reroofing situations. Refer to the "Reroofing" section of the Schuller Commercial/Industrial Roofing Systems Manual. This specification is not to be used directly over poured or pre-cast gypsum or lightweight, insulating concrete fills.

Design and installation of the deck and/or substrate must result in the roof draining freely and to outlets numerous enough and so located as to remove water promptly and completely. Areas where water ponds for more than 24 hours are unacceptable and are not eligible to receive a Schuller Roofing Systems Guarantee.

Note: All general instructions contained in the current Schuller Commercial/Industrial Roofing Systems Manual should be considered part of this specification.

Flashings
Flashing details can be found in the "Bituminous Flashings" section of the Schuller Commercial/Industrial Roofing Systems Manual.

Application
Note: On roof decks with slopes up to 1" per foot (83.3 mm/m), the roofing felts may be installed either perpendicular or parallel to the roof incline. On slopes over 1" per foot (83.3 mm/m), refer to **Paragraph 6.11** of this section for special requirements.

Using Ventsulation, GlasBase, GlasBase Plus or PermaPly 28, start with an 18" (457 mm) width (the use of a specific base sheet may be a condition of Guarantee). The following base sheet courses should be applied full width, lapping the side laps 2" (51 mm) and the end laps 4" (102 mm) over the preceding felts. Set each felt firmly into spot moppings of hot bitumen (within ±25°F [±14°C] of the EVT). The spot moppings should be applied by machine at the rate of approximately 7 lbs. per square (0.3 kg/m²). The spots should be approximately 12" (305 mm) in diameter and 24" (610 mm) o.c. Each row should be staggered from the previous one.

Using GlasPly Premier, PermaPly-R, or GlasPly IV, apply a piece 12" (305 mm) wide, then over that, one 24" (610 mm) wide, then over both, a full width piece. The following felts are to be applied full width, overlapping the preceding felts by 24 2/3" (627 mm) so that at least 3 plies of felt cover the base felt/substrate at all locations. Install each felt so that it is firmly and uniformly set, without voids, into the hot bitumen (within ±25°F [±14°C] of the EVT) applied just before the felt at a nominal rate of 23 lbs. per square (1.1 kg/m²) over the entire surface. Installation over porous substrates such as roof insulation may require up to 33 lbs. per square (1.6 kg/m²) of hot bitumen.

Surfacing
Flood the surface with the appropriate bitumen at an approximate rate of 60 lbs. per square (2.9 kg/m²). Into the hot bitumen, embed an acceptable gravel at a rate of 400 lbs. per square (19.5 kg/m²) or an acceptable slag at a rate of 300 lbs. per square (14.6 kg/m²). Aggregate must be installed so that there is complete coverage across the entire surface and at least 50% of the aggregate is solidly adhered in the hot bitumen. Aggregate should meet the requirements of ASTM D 1863.

Asphalt should meet the requirements of ASTM D 312. The contractor must provide a Schuller confirmation number for asphalt on jobs which require a Guarantee. Check with a Schuller Technical Service Specialist for special requirements in hot climates.

By permission of Schuller Roofing Systems, Denver, Colorado

8.4.0 3- and 4-Ply Hot-Mopped Modified Bitumen Roofs

Specification 3CIG/3FIG/3PIG

Three Ply Hot Mopped Modified Bitumen Gravel Surfaced Roofing System

For use over Schuller insulation, approved decks, or other approved insulations on inclines up to 3" per foot (250 mm/m)

For Regions 1, 2 and 3

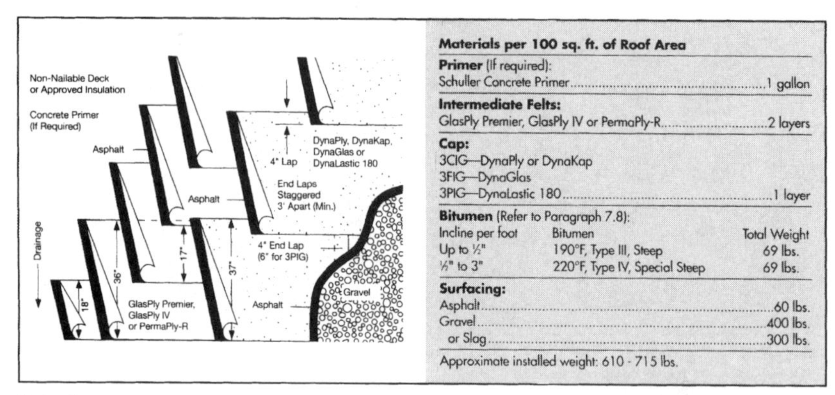

Materials per 100 sq. ft. of Roof Area

Primer (If required):
Schuller Concrete Primer .... 1 gallon

Intermediate Felts:
GlasPly Premier, GlasPly IV or PermaPly-R .... 2 layers

Cap:
3CIG—DynaPly or DynaKap
3FIG—DynaGlas
3PIG—DynaLastic 180 .... 1 layer

Bitumen (Refer to Paragraph 7.8):

Incline per foot	Bitumen	Total Weight
Up to ½"	190°F, Type III, Steep	69 lbs.
½" to 3"	220°F, Type IV, Special Steep	69 lbs.

Surfacing:
Asphalt .... 60 lbs.
Gravel .... 400 lbs.
or Slag .... 300 lbs.

Approximate installed weight: 610 - 715 lbs.

General

This specification is for use over any type of approved structural deck which is not nailable and which provides a suitable surface to receive the roof. Poured and pre-cast concrete decks require priming with Schuller Concrete Primer prior to application of hot bitumen.

This specification is also for use over Schuller roof insulations, or other approved roof insulations which are not nailable and which provide a suitable surface to receive the roof. Specific written approval is required for any roof insulation that is not supplied by Schuller. Insulation should be installed in accordance with the appropriate Schuller Insulation Specification detailed in the Schuller Commercial/Industrial Roofing Systems Manual. This specification can also be used in certain reroofing situations. Refer to the "Reroofing" section of the Schuller Commercial/Industrial Roofing Systems Manual. This specification is not to be used directly over gypsum, either poured or pre-cast, or lightweight, insulating concrete decks or fills.

Design and installation of the deck and/or roof substrate must result in the roof draining freely, to outlets numerous enough and so located as to remove water promptly and completely. Areas where water ponds for more than 24 hours are unacceptable and will not be eligible for a Schuller Roofing System Guarantee.

Note: All general instructions contained in the current Schuller Commercial/Industrial Roofing Systems Manual shall be considered part of this specification.

Flashings

Flashing details can be found in the "Bituminous Flashings" section of the Schuller Commercial/Industrial Roofing Systems Manual.

Application

On roof decks with slopes up to ½" per foot (41.6 mm/m), the roofing felts and modified bitumen sheets may be installed either perpendicular or parallel to the roof incline.

Roll an 18" (457 mm) wide piece of one of the intermediate felts listed into a full mopping of bitumen. Over that, apply a full width piece. The remaining felts are to be applied full width, overlapping the preceding felts by 19" (483 mm), so that at least 2 plies of felt cover the substrate at all locations. Apply a full width piece of one of the cap sheets listed into a full mopping of bitumen. Subsequent sheets are to be applied in the same manner, with 4" (102 mm) side and end laps over the preceding sheets (6" [152 mm] end laps for DynaLastic products).

Apply all felts so that they are firmly and uniformly set, without voids, into the hot bitumen. Bitumen temperature should be at the Equiviscous Temperature (EVT), ±25°F (±14°C), at the point of application. All felt edges shall be well sealed. The bitumen shall be applied just before the felt, at a nominal rate of 23 lbs. per square (1.1 kg/m²). When applying over insulations, more than 23 lbs. per square (1.1 kg/m²) of bitumen may be needed due to the absorbency of the insulation. For modified bitumen sheets, the bitumen temperature shall be at a minimum of 400°F (204°C), or at the EVT, whichever is higher, when the sheet is set into it. This higher temperature maximizes the bonding of the modified bitumen sheet.

For cold weather application techniques, refer to **Paragraph 7.31.**

Steep Slope Requirements

Special procedures are required on inclines over ½" per foot (41.6 mm/m). Refer to **Paragraph 7.29.**

Surfacing

Flood the surface with the appropriate bitumen, depending on the roof slope, at an approximate rate of 60 lbs. per square (2.9 kg/m²). Embed an acceptable gravel at the rate of approximately 400 lbs. per square (19.5 kg/m²), or an acceptable slag at a rate of approximately 300 lbs. per square (14.6 kg/m²), into the hot bitumen. Aggregate must be installed so that there is complete coverage across the entire surface and at least 50% of the aggregate is solidly adhered in the hot bitumen.

8.4.0 3- and 4-Ply Hot-Mopped Modified Bitumen Roofs (Continued)

Specification 4CID/4FID/4PID

**Four Ply Hot Mopped
Modified Bitumen
Mineral Surfaced Roofing System**

For use over Schuller insulation, approved decks, or other approved insulations on inclines up to 3" per foot (250 mm/m)

For Regions 1, 2 and 3

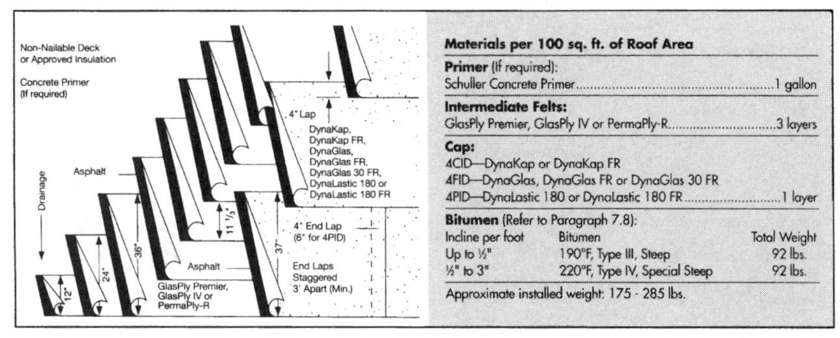

Materials per 100 sq. ft. of Roof Area

Primer (If required):
Schuller Concrete Primer........1 gallon

Intermediate Felts:
GlasPly Premier, GlasPly IV or PermaPly-R........3 layers

Cap:
4CID—DynaKap or DynaKap FR
4FID—DynaGlas, DynaGlas FR or DynaGlas 30 FR
4PID—DynaLastic 180 or DynaLastic 180 FR........1 layer

Bitumen (Refer to Paragraph 7.8):

Incline per foot	Bitumen	Total Weight
Up to ½"	190°F, Type III, Steep	92 lbs.
½" to 3"	220°F, Type IV, Special Steep	92 lbs.

Approximate installed weight: 175 - 285 lbs.

General
This specification is for use over any type of approved structural deck which is not nailable and which provides a suitable surface to receive the roof. Poured and pre-cast concrete decks require priming with Schuller Concrete Primer prior to application of hot bitumen.

This specification is also for use over Schuller roof insulations, or other approved roof insulations which are not nailable and which provide a suitable surface to receive the roof. Specific written approval is required for any roof insulation that is not supplied by Schuller. Insulation should be installed in accordance with the appropriate Schuller Insulation Specification detailed in the Schuller Commercial/Industrial Roofing Systems Manual. This specification can also be used in certain reroofing situations. Refer to the "Reroofing" section of the Schuller Commercial/Industrial Roofing Systems Manual. This specification is not to be used directly over gypsum, either poured or pre-cast, or lightweight, insulating concrete decks or fills.

Design and installation of the deck and/or roof substrate must result in the roof draining freely, to outlets numerous enough and so located as to remove water promptly and completely. Areas where water ponds for more than 24 hours are unacceptable and will not be eligible for a Schuller Roofing System Guarantee.

Note: All general instructions contained in the current Schuller Commercial/Industrial Roofing Systems Manual shall be considered part of this specification.

Flashings
Flashing details can be found in the "Bituminous Flashings" section of the Schuller Commercial/Industrial Roofing Systems Manual.

Application
On roof decks with slopes up to ½" per foot (41.6 mm/m), the roofing felts and modified bitumen sheets may be installed either perpendicular or parallel to the roof incline.

Roll a 12" (305 mm) wide piece of one of the intermediate felts listed into a full mopping of bitumen. Over that, apply one 24" (610 mm) wide. Over both, apply a full width piece. The remaining felts are to be applied full width, overlapping the preceding felts by 24⅔" (627 mm), so that at least 3 plies of felt cover the substrate at all locations.

Apply a full width piece of one of the cap sheets listed into a full mopping of bitumen. Subsequent sheets are to be applied in the same manner, with 4" (102 mm) side and end laps over the preceding sheets (6" [152 mm] end laps for DynaLastic products).

Apply all felts so that they are firmly and uniformly set, without voids, into the hot bitumen. Bitumen temperature should be at the Equiviscous Temperature (EVT), ±25°F (±14°C), at the point of application. All felt edges shall be well sealed. The bitumen shall be applied just before the felt, at a nominal rate of 23 lbs. per square (1.1 kg/m²). When applying over insulations, more than 23 lbs. per square (1.1 kg/m²) of bitumen may be needed due to the absorbency of the insulation. For modified bitumen sheets, the bitumen temperature shall be at a minimum of 400°F (204°C), or at the EVT, whichever is higher, when the sheet is set into it. This higher temperature maximizes the bonding of the modified bitumen sheet.

For cold weather application techniques, refer to **Paragraph 7.31.**

Steep Slope Requirements
Special procedures are required on inclines over ½" per foot (41.6 mm/m). Refer to **Paragraph 7.29.**

Surfacing
No additional surfacing is required.

By permission of Schuller Roofing Systems, Denver, Colorado

8.5.0 Built-Up Roof-Flashing Details

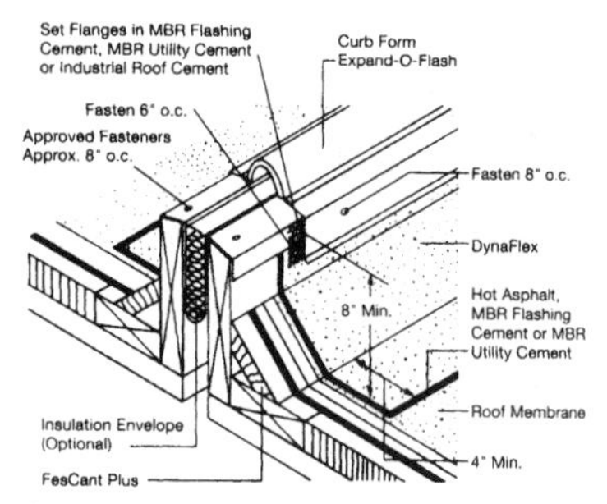

Specification DFE-7

Expansion Joint Cover: Application of the base flashing is outlined in Specification DFE-1 (NLB). Install and splice Expand-O-Flash in accordance with the installation instructions provided with the product.

Prefabricated intersections, as well as horizontal-to-vertical transitions, are available to complete the Expand-O-Flash installation. Refer to Section 12 on "Roofing Accessories" in the current Schuller Commercial/Industrial Roofing Systems Manual.

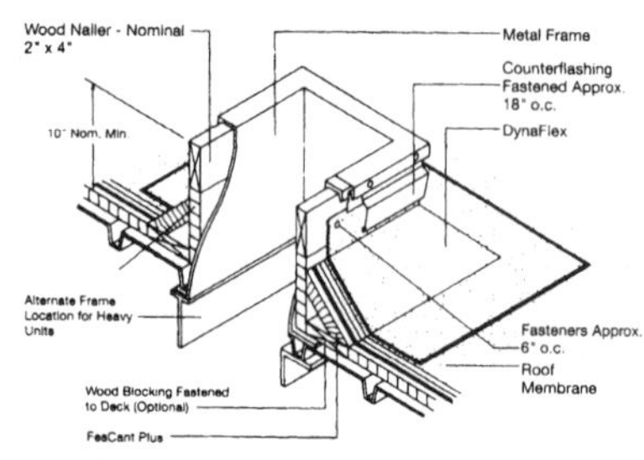

Specification DFE-8

Prefabricated Curb: Refer to Flashing Specification DFE-1 (NLB) for detailed instructions on application of the base flashing. Base flashing felts should extend as far up the prefabricated curb as practicable, but not less than 8" (203 mm). Install the flashing receiver and metal counterflashing in accordance with the prefabricated curb manufacturer's specifications and details, or in accordance with the DFE-8 detail.

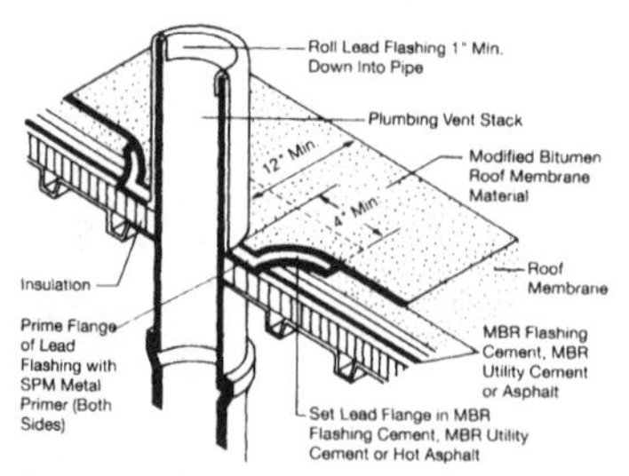

Specification DFE-9

Plumbing Vent Flashing: Prime both sides of the flange of the lead boot with SPM Metal Primer. Set the flange into a bed of MBR Flashing Cement, MBR Utility Cement, or a mopping of hot bitumen. Cover the flange with a layer of modified bitumen membrane sheet, set in MBR Flashing Cement, MBR Utility Cement, or hot bitumen. Roll the top edge of the lead boot down into the pipe a minimum of 1" (25 mm). Minimum weight of lead sheet: 2½ lbs. per square foot (12.2 kg/m²).

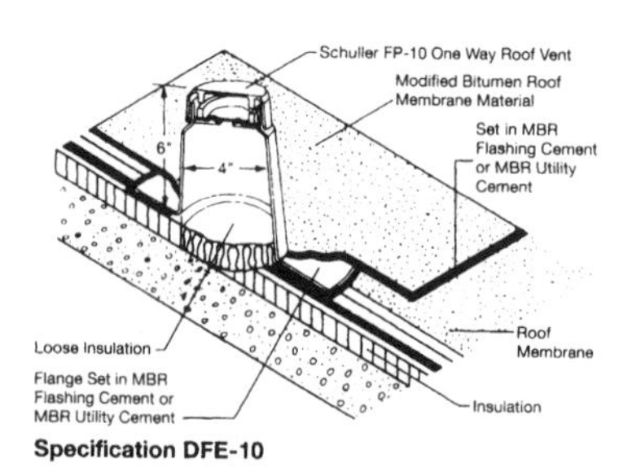

Specification DFE-10

FP-10 One Way Roof Vent: Cut a 5" (127 mm) diameter hole in membrane. Remove all or part of the insulation, as necessary to facilitate venting; replace with loose insulation to prevent possible condensation. Apply a layer of MBR Flashing Cement or MBR Utility Cement around the 5" (127 mm) hole and press the vent flange into place. Flash in the vent with a layer of modified bitumen membrane sheet, set in MBR Flashing Cement or MBR Utility Cement.

Note: Hot asphalt may be used in lieu of the MBR Flashing or Utility Cements to set and flash in the vent, however, do not mix the two methods of application.

For load-bearing masonry parapet construction with nailing facilities

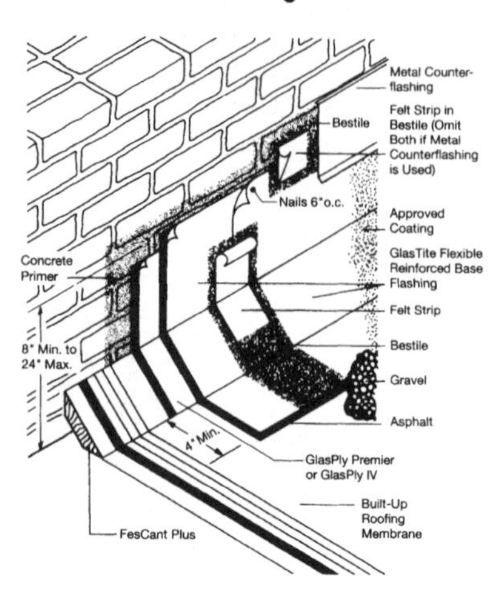

For non load-bearing construction, using roof-to-wall expansion joint cover

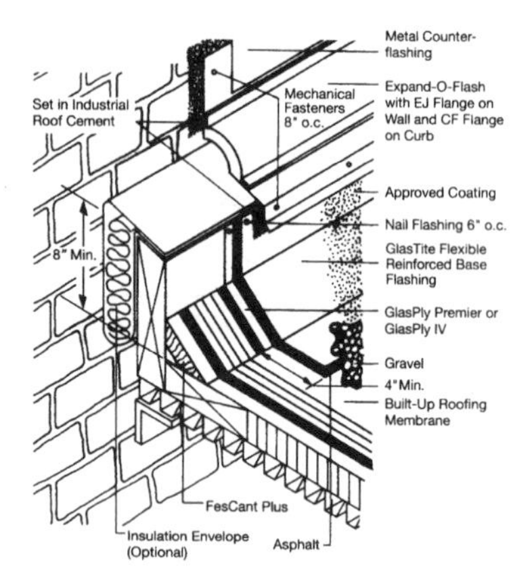

For load-bearing masonry parapet construction with no nailing facilities

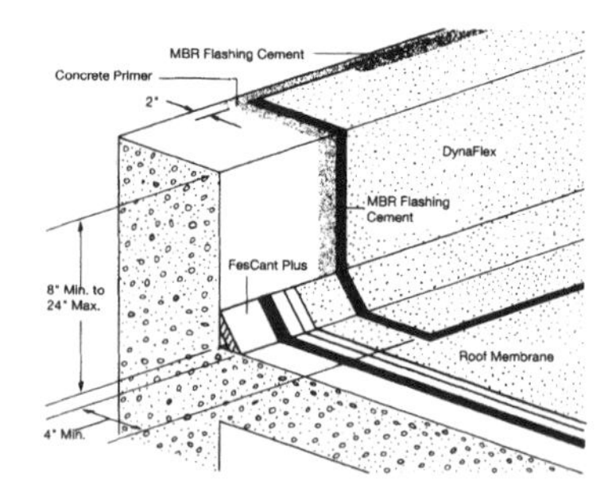

8.5.1 Built-Up Roof-Flashing Details (Continued)

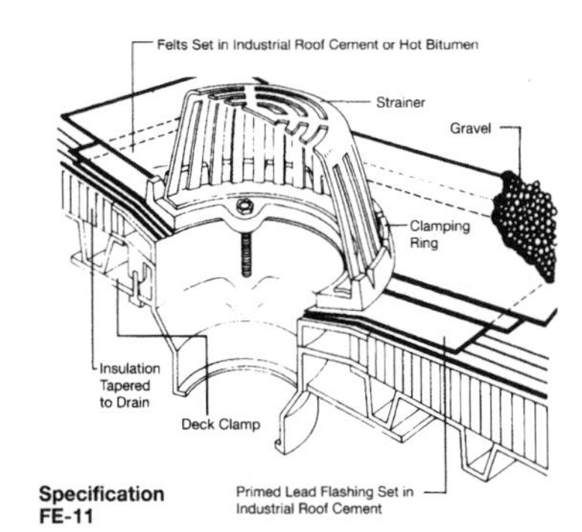

Specification FE-11

Flashing to Metal Drain: Run membrane plies to edge of drain opening. Prime both sides of a 30" (762 mm) square (minimum) piece of lead flashing (minimum 2½ lb./sq. ft. [12.2 kg/m²]) with Schuller Concrete Primer and apply to the roof surface in Industrial Roof Cement. Cover the lead flashing with 2 plies of GlasPly Premier, GlasPly IV, or PermaPly-R, set in Industrial Roof Cement or hot bitumen. Flashing felts should extend 4" and 6" (102 mm and 152 mm) beyond the edge of the lead flashing, in all directions. The membrane plies, lead flashing, and flashing felts should all extend under the clamping ring. Attach the clamping ring and tighten uniformly.

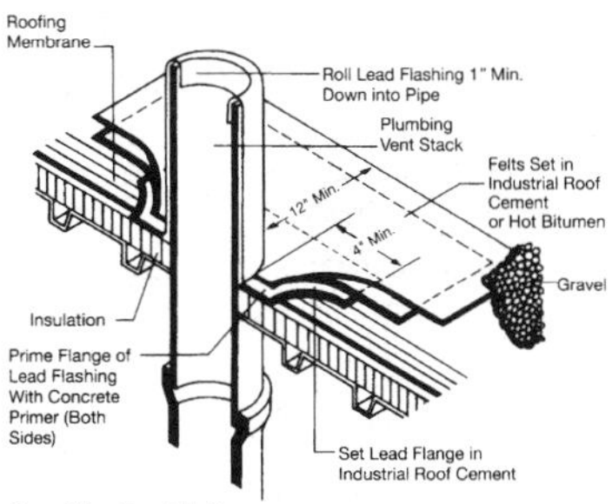

Specification FE-9

Plumbing Vent Flashing: Prime both sides of the flange of the lead boot with Schuller Concrete Primer and set into a bed of Industrial Roof Cement. Cover with 2 layers of GlasPly Premier, GlasPly IV, or PermaPly-R, set in Industrial Roof Cement or hot bitumen. Roll top edge of lead boot down into pipe. Minimum weight of lead sheet: 2½ lbs. per square foot (12.2 kg/m²).

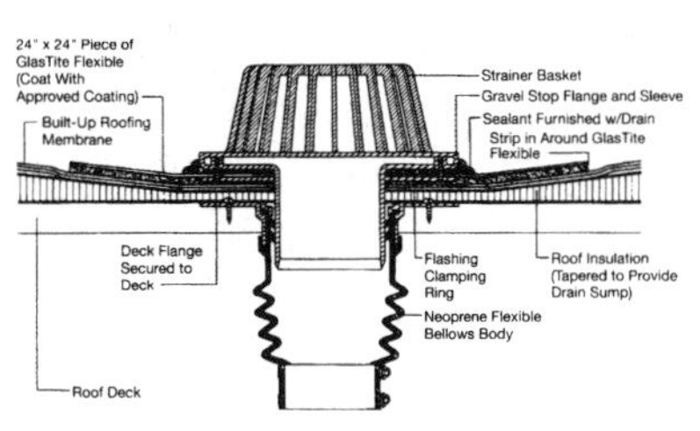

Specification FE-12

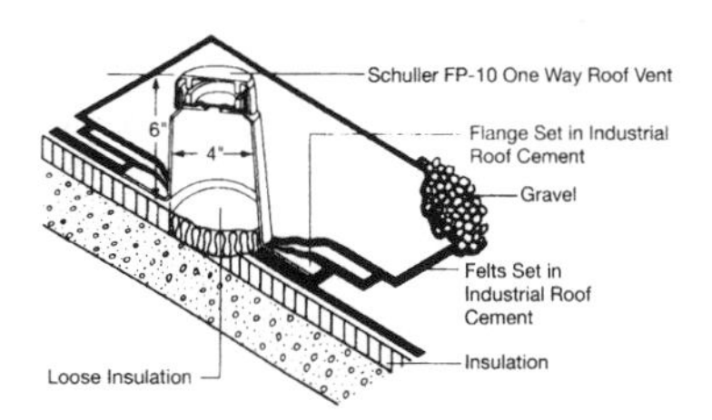

Specification FE-10

FP-10 One-Way Roof Vent: Cut a 5" (127 mm) diameter hole in the membrane. Remove all or part of the insulation, as necessary to facilitate venting; replace with loose insulation to prevent possible condensation. Apply a layer of Industrial Roof Cement around the 5" (127 mm) hole and press the vent flange into place. Flash in the vent with 2 plies of GlasPly Premier, GlasPly IV, or PermaPly-R, set in Industrial Roof Cement. One FP-10 Vent should be used per 10 squares of roof area.

Note: Hot asphalt may be used in lieu of Industrial Roof Cement to set and strip in the vent, however, do not mix the two methods of application.

Roof Edge Details

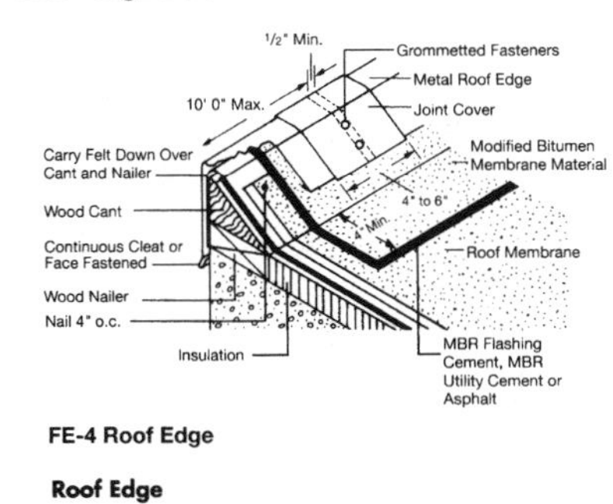

FE-4 Roof Edge

Roof Edge

Roof edges and gravel stops

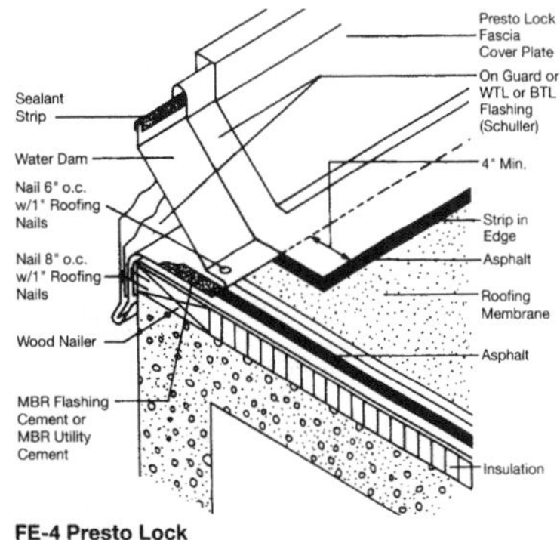

FE-4 Presto Lock

Presto Lock Fascia and Flashing System

By permission of Schuller Roofing Systems, Denver, Colorado

8.6.0 Single-Ply Membrane Securement Data

A. The following charts indicate the required number of perimeter membrane sheets, width of field membrane sheets and required fastening density for Carlisle's Sure-Seal/Brite-Ply Mechanically-Fastened Roofing System. The chart is categorized by deck type and includes four different wind zones which are identified on the "Basic Wind Speed Map" at the end of this section.

B. To determine appropriate securement requirements, identify project wind zone from the map and select the chart based on project deck type. The building height is then used to determine membrane securement requirements for the project.

Wind Zone	Deck Type (1)	Building Height	# of Perimeter Sheets	Field Membrane Width	Fastening Density (Field & Perimeter Sheets)
Zone 1 79 MPH or Less (126 km/h)	Steel and Lightweight Concrete over Steel	0' - 75' (23 m)	1	10' (3 m)	12" (31 cm) O.C.
		76' - 150' (23.2 - 46 m)	1	10'	6" (15.5 cm) O.C.
			1	7' (2.1 m)	12" O.C.
	Structural Concrete	0' - 75' (23 m)	1	10'	12" O.C.
		76' - 150' (23.2 - 46 m)	2	10'	12" O.C.
	Plywood, Wood Planks (2) or Oriented Strand Board	0' - 50' (15.2 m)	1	10'	12" O.C.
		51' - 150' (15.5 - 46 m)	2	10'	6" O.C.
	Gypsum and Fibrous Cement	0' - 75' (23 m)	2	10'	6" O.C.
			2	7'	12" O.C.
Zone 2 80 - 89 MPH (128-142 km/h)	Steel and Lightweight Concrete over Steel	0' - 75' (23 m)	1	10'	6" O.C.
			1	7'	12" O.C.
		76' - 100" (23.2 - 30.5 m)	2	10'	6" O.C.
			2	7"	12" O.C.
		101' - 150' (30.8 - 46 m)	2	10'	6" O.C.
	Structural Concrete	0' - 75' (23 m)	2	10'	12" O.C.
		76' - 150' (23.2 - 46 m)	2	10'	6" O.C.
			2	7'	12" O.C.
	Plywood, Wood Planks (2) or Oriented Strand Board	0' - 50' (15.2 cm)	2	10'	6" O.C.
			2	7'	12" O.C.
		51' - 150' (15.5 - 46 m)	4	10'	6" O.C.
	Gypsum and Fibrous Cement	0' - 50' (15.2 m)	2	10'	6" O.C.
			2	7'	12" O.C.

By permission of The Carlisle Corporation, Carlisle, Pennsylvania

8.6.1 Single-Ply Membrane Securement Data (Continued)

Wind Zone	Deck Type (1)	Building Height	# of Perimeter Sheets	Field Membrane Width	Fastening Density (Field & Perimeter Sheets)
Zone 3 90-99 MPH (3) (144-158 km/h)	Steel and Lightweight Concrete over Steel	0' - 40' (12.2 m)	2	10' (3 m)	6" (15.5 cm) O.C.
			2	7' (2.1 m)	12" (31 cm) O.C.
		41' - 75' (12.5 - 30.5 m)	2	10'	6" (15.5 cm) O.C.
		76' - 100' (23.2 - 30.5 m)	2	7' (2.1 m)	6" O.C.
	Structural Concrete	0' - 75' (23 m)	2	10'	6" O.C.
			2	7'	12" O.C.
		76' - 150' (23.2 - 46 m)	3	7'	12" O.C.
			2	10'	6" O.C.
	Plywood, Wood Planks (2) or Oriented Strand Board	0' - 100' (30.5 m)	2	7'	6" O.C.
	Gypsum and Fibrous Cement	0' - 75' (23 m)	2	7'	6" O.C.
Zone 4 100 MPH (160 km/h) or Greater	Steel and Lightweight Concrete over Steel	0' - 100' (30.5 m)	2	7'	6" O.C.
	Structural Concrete	0' - 150' (46 m)	1	7'	6" O.C.
	Plywood, Wood Planks (2) or Oriented Strand Board	NOT ACCEPTABLE			
	Gypsum and Fibrous Cement	NOT ACCEPTABLE			

Notes:

(1) Refer to "Attachment I", Pullout Values/Withdrawal Resistance Criteria, for roof deck/pullout requirements and the required Carlisle Fastener.

(2) On plywood or wood plank decks, if pullout tests exceed 425 pounds (192 kg) per fastener, the membrane securement requirements for steel decks may be followed providing the pullout tests are submitted to Carlisle for approval.

(3) Those areas located between wind zone contours of 90-100 MPH (144 - 160 km/h) that are within 20 miles (32 km) of the coastline shall be considered as a Zone 4 Wind Zone.

C. The fastening criteria shown above does not necessarily reflect Factory Mutual approvals. For specific requirements when a Factory Mutual rating is required, refer to the Carlisle *Code Approval Guide* which is published separately

By permission of The Carlisle Corporation, Carlisle, Pennsylvania

8.6.2 Basic Wind-Speed Map

This map is based on ASCE 7-88, formerly ANSI A 58.1–1982.

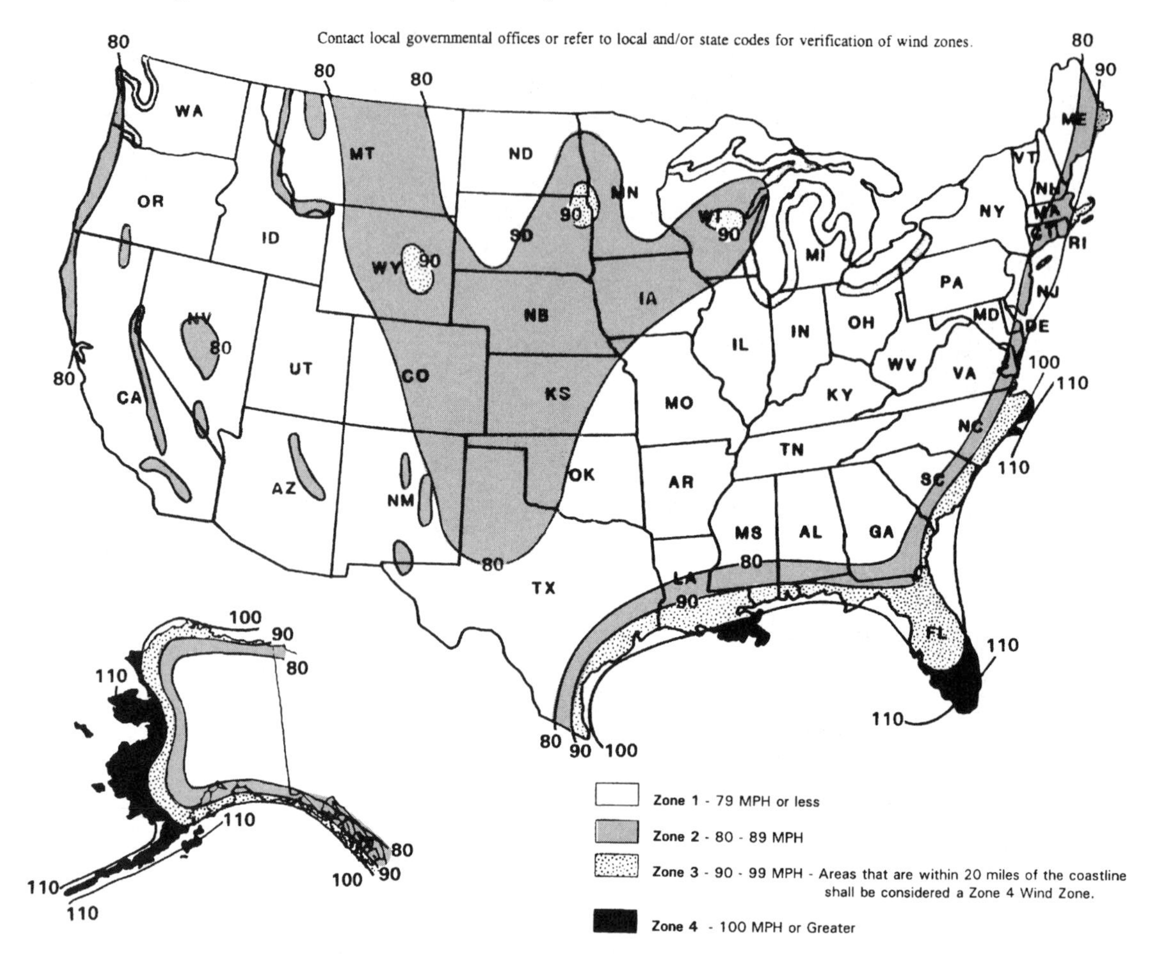

By permission of The Carlisle Corporation, Carlisle, Pennsylvania

8.6.3 Single-Ply Membrane Splicing Cement Guide

One gallon of Splicing Cement, applied in a medium, relatively even coat, will achieve the approximate coverage rates as listed:

Linear Feet	Splice Width
150 feet	3 inches
120 feet	4 inches
100 feet	5 inches
85 feet	6 inches
75 feet	7 inches

Note: The above coverage rates have been calculated to include the application of Splicing Cement 1 inch beyond the splice width on both mating surfaces of the membrane.

FOR CURED-TO-CURED MEMBRANE SPLICES ONLY:

a. While the Splicing Cement is drying, apply a bead of In-Seam Sealant™ no less than 1/8 inch and no more than 1/4 inch wide within 1/2 inch of the inside edge of the bottom membrane sheet.

Note: When minimum 6 inch wide membrane splices incorporate Sure-Seal HP Purlin Fasteners and HP Locking Seam Plates, the In-Seam Sealant shall be applied along the center line used to locate fastening plates (approximately 3 inches from the edge of the membrane sheet). At the Fastening Plates, apply the In-Seam Sealant around the edge of the plate which is nearest the outside edge of the top membrane sheet. Refer to Detail MR-2-B.

Approximately 75 linear feet of coverage per tube can be achieved when a 5/32 inch diameter bead of In-Seam Sealant is applied.

b. Maintain a continuous bead of In-Seam Sealant on all membrane splices.

c. During splice cleaning procedures, Sure-Seal HP Splice Wipes contaminated with In-Seam Sealant cannot be reused for the application of Splice Cleaner.

Allow the cement to dry until it is tacky but will not string or stick to a dry finger touch and will not move when pushed with a dry finger.

Roll the top membrane sheet onto the mating surface. Take care not to stretch or wrinkle the membrane sheet to avoid a fishmouth in the field splice.

Assemble the seam with hand pressure by wiping toward the splice edge.

Immediately roll the splice with a 2 inch wide steel roller, using positive pressure, toward the outer edge of the splice. DO NOT ROLL PARALLEL TO THE SPLICE EDGE. On a completed splice, the In-Seam Sealant must remain evident and must be sensitive to the touch.

8.6.4 Single-Ply Membrane Ballasted Roof Stone Specifications

Rounded Water-Worn Gravel must be applied over the EPDM membrane at the minimum rate of **1000 pounds (488 kg/10 m²)** per square and must be evenly distributed to maintain an average of 10 pounds per square foot (approximately 48.8 kg/m²).

ASTM D 448 SIZE NUMBER	MINIMUM COVERAGE RATE (pounds per square) (kg/10 m²)	AVERAGE COVERAGE RATE (pounds per square foot continuously distributed)	Average kg/m² (continuously distributed)
4 (1-1/2 inch (3.8 cm) nominal diameter)	1000 (488)	10	48.8
3 (2 inch (5 cm) nominal diameter)	1000 (488)	10	48.8
24 (2-1/2 inch (6.4 cm) nominal diameter)	1000 (488)	10	48.8
2 (2-1/2 inch (6.4 cm) nominal diameter)	1300 (634)	13	63.4
1 (3-1/2 inch (8.9 cm) nominal diameter)	1300 (634)	13	63.4

Standard sizes of coarse aggregate - Based on ASTM D448

Size Number	1	2	24	3	4
Nominal Size Square Openings	3-1/2" (8.9 cm) to 1-1/2" (3.8 cm)	2-1/2" (6.4 cm) to 1-1/2" (3.8 cm)	2-1/2" (6.4 cm) to 3/4" (1.9 cm)	2" (5 cm) to 1" (2.5 cm)	1-1/2" (3.8 cm) to 3/4" (1.9 cm)
Amounts Passing Each Lab Sieve (Square Opening), Percent (%)					
4" (10 cm)	100				
3-1/2" (8.9 cm)	90 to 100				
3" (8 cm)		100	100		
2-1/2" (6.4 cm)	25 to 60	90 to 100	90 to 100	100	
2" (5 cm)		35 to 70		90 to 100	100
1-1/2" (3.8 cm)	0 to 15	0 to 15	25 to 60	35 to 70	90 to 100
1" (2.5 cm)				0 to 15	20 to 55
3/4" (1.9 cm)	0 to 5	0 to 5	0 to 10		0 to 15
1/2" (1.3 cm)			0 to 5	0 to 5	
3/8" (1 cm)					0 to 5

By permission of The Carlisle Corporation, Carlisle, Pennsylvania

8.7.0 Single-Ply Membrane Curb Flashing Details

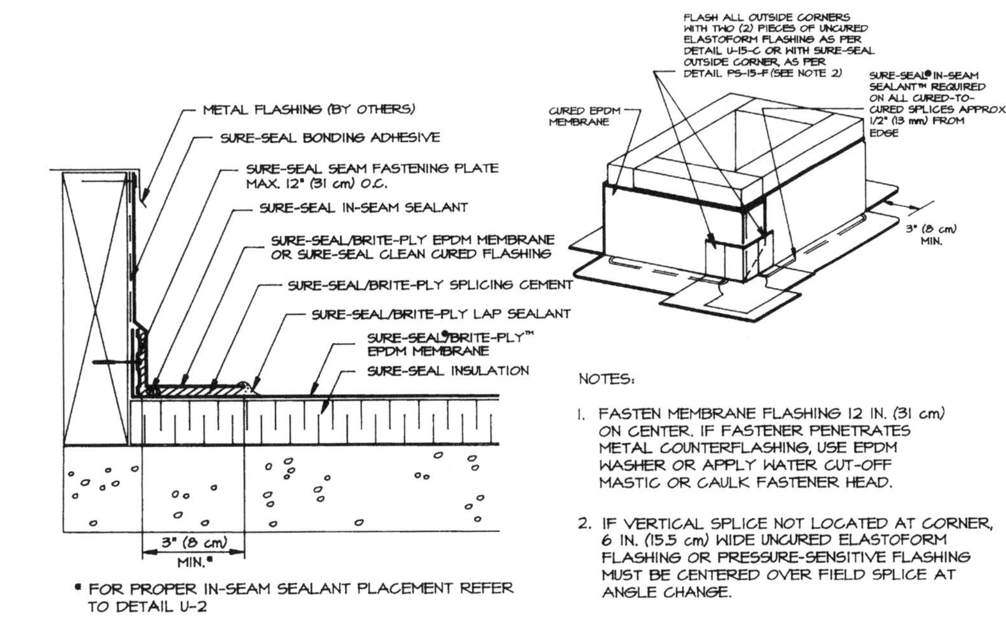

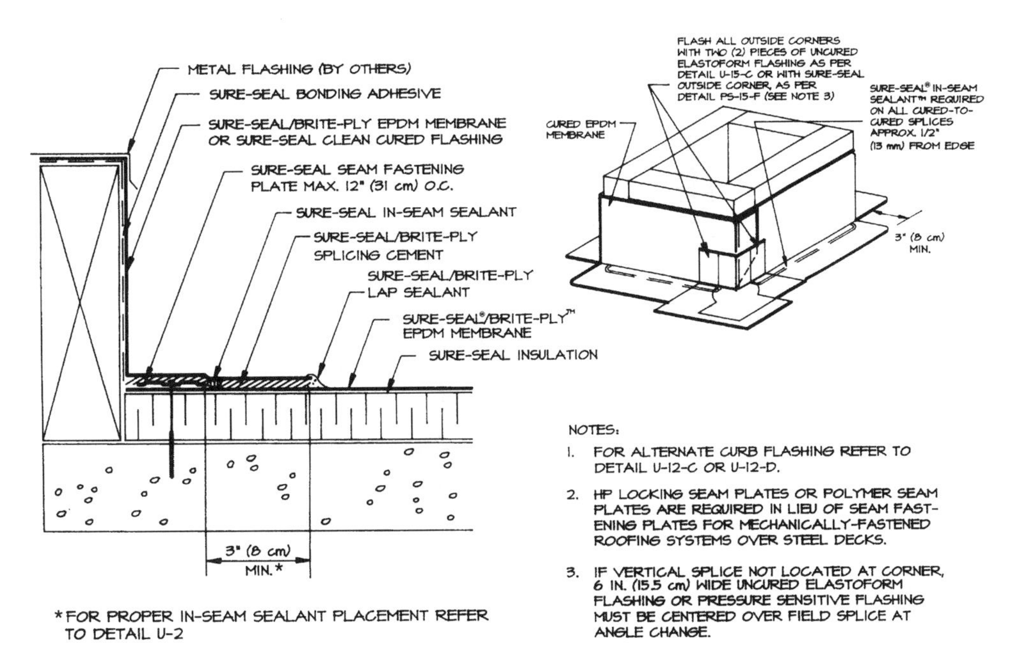

By permission of The Carlisle Corporation, Carlisle, Pennsylvania

8.7.1 Single-Ply Membrane Reglet and Cap Flashing Details

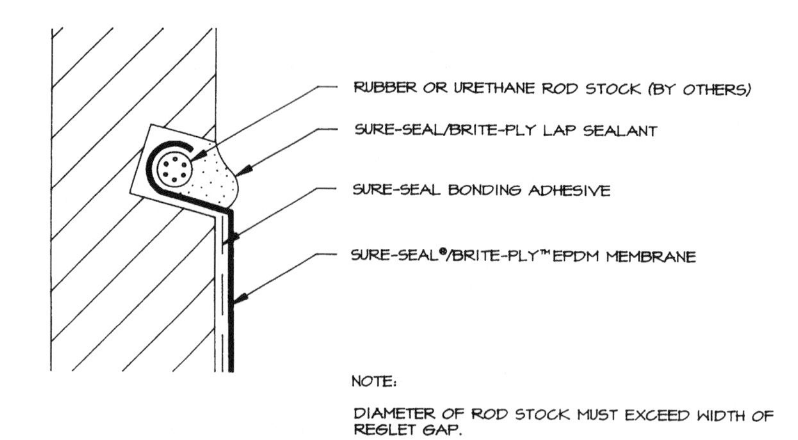

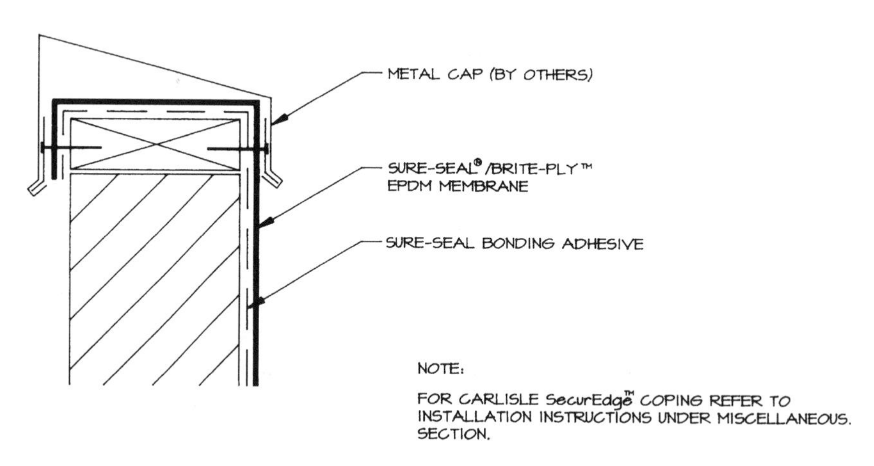

By permission of The Carlisle Corporation, Carlisle, Pennsylvania

8.7.2 Single-Ply Membrane Curb and Vertical Pipe Flashing Details

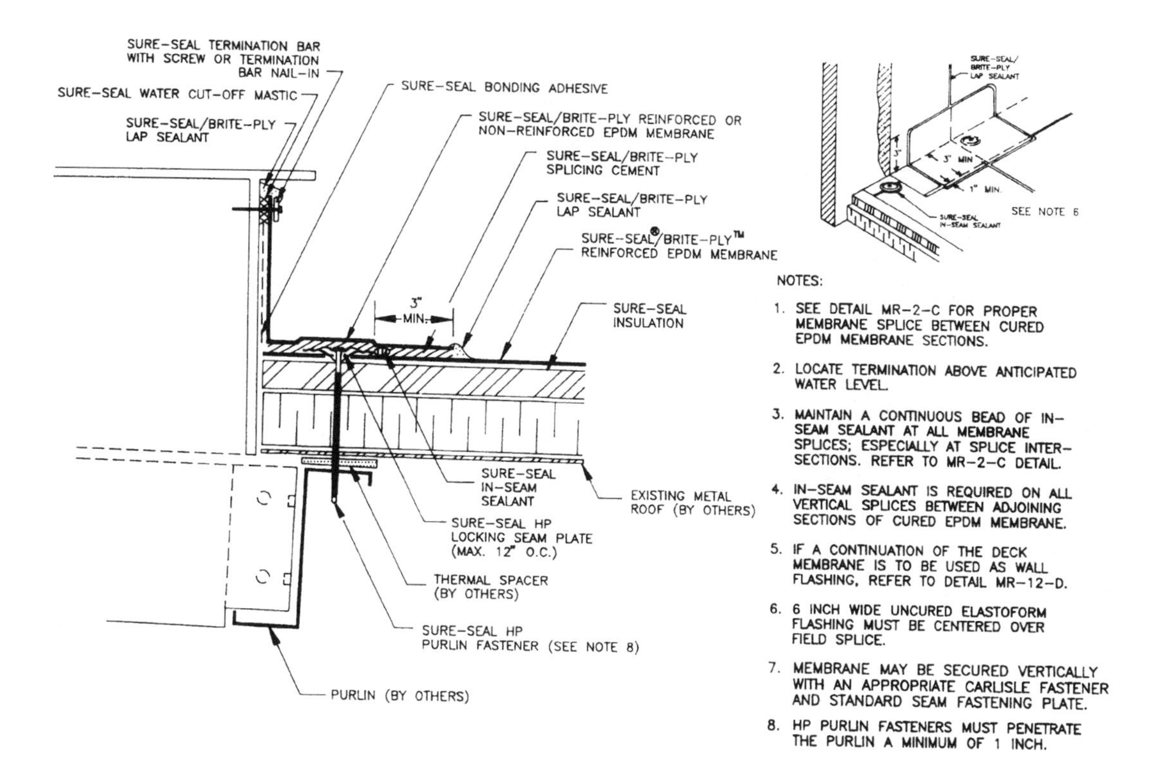

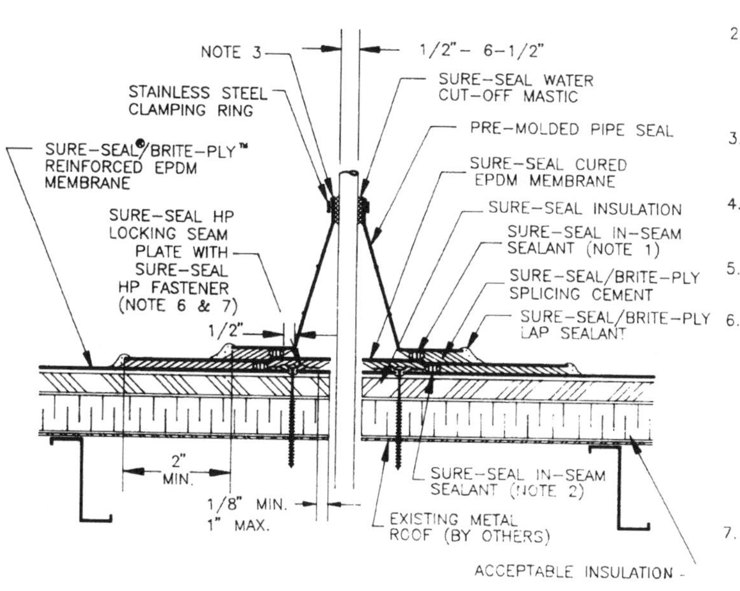

NOTES:

1. APPLY IN-SEAM SEALANT 1/2 INCH FROM THE INSIDE EDGE OF THE PIPE SEAL FLANGE.
2. IN-SEAM SEALANT MUST BE LOCATED A MAXIMUM OF 1/2 INCH AWAY FROM THE HP LOCKING SEAM PLATE AND SHALL BE CONTINUOUS AROUND THE PIPE.
3. PRE-MOLDED PIPE SEAL MUST HAVE INTACT RIB AT THE TOP EDGE REGARDLESS OF PIPE DIAMETER.
4. HP LOCKING SEAM PLATES ARE ALSO REQUIRED AROUND FIELD FABRICATED PIPE SEALS (U-14 DETAILS).
5. SEE SPECIFICATION FOR OTHER DETAILS REQUIRING HP LOCKING SEAM PLATES.
6. SPACING SHALL BE A MAXIMUM OF 6 INCHES ON CENTER INTO 26 AND 28 GAUGE METAL ROOFING. SPACING CAN BE A MAXIMUM OF 12 INCHES ON CENTER INTO 24 GAUGE METAL ROOFING PROVIDING THE PULLOUT VALUES IDENTIFIED IN THIS SPECIFICATION ARE ACHIEVED.
   A MINIMUM OF 3 HP FASTENERS AND SURE-SEAL HP LOCKING SEAM PLATES ARE REQUIRED.
7. HP FASTENER MUST PENETRATE THE EXISTING ROOF A MINIMUM OF 3/4 INCH.

By permission of The Carlisle Corporation, Carlisle, Pennsylvania

8.7.3 Single-Ply Membrane Counterflashing/Vertical Termination Flashing Details

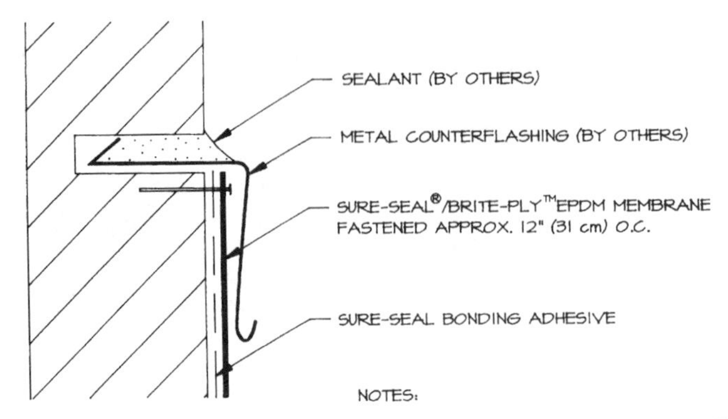

NOTES:

1. IF FASTENER PENETRATES METAL COUNTERFLASHING, USE EPDM WASHER OR APPLY WATER CUT-OFF MASTIC OR CAULK FASTENER HEAD.
2. FOR 15 YEAR WARRANTY, A CARLISLE TERMINATION BAR (SEE DETAIL U-9-H) MUST BE INSTALLED BEHIND THE COUNTERFLASHING.

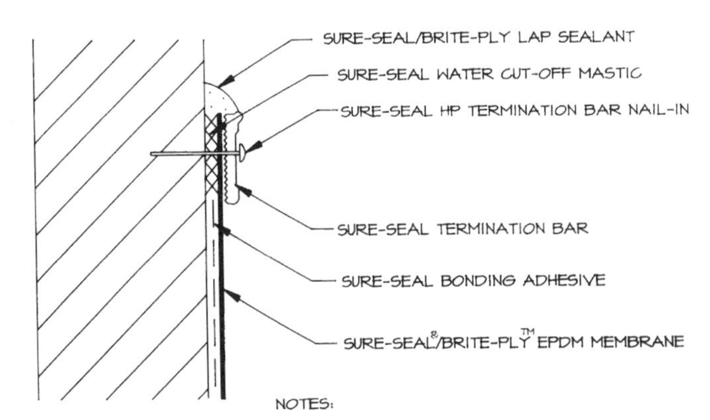

NOTES:

1. APPLY ON HARD SMOOTH SURFACE ONLY; NOT FOR USE ON WOOD.
2. WATER CUT-OFF MASTIC MUST BE HELD UNDER CONSTANT COMPRESSION.
3. DO NOT WRAP COMPRESSION TERMINATION AROUND CORNERS.
4. ALLOW 1/4 INCH (6 mm) MIN. TO 1/2 INCH (13 mm) MAX. SPACING BETWEEN CONSECUTIVE LENGTHS OF TERMINATION BAR.
5. TERMINATION BAR BY OTHERS MUST BE 1/8" X 1" (.3 X 2.5 cm) MINIMUM.

By permission of The Carlisle Corporation, Carlisle, Pennsylvania

8.7.4 Single-Ply Membrane Expansion-Joint Details

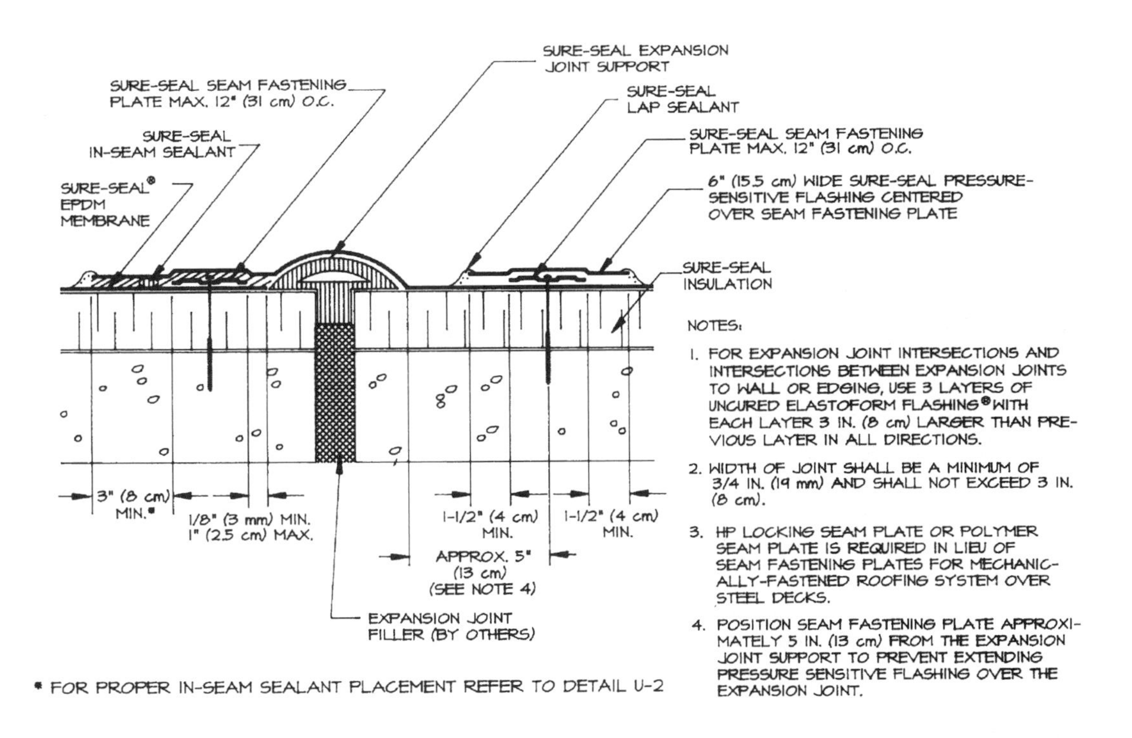

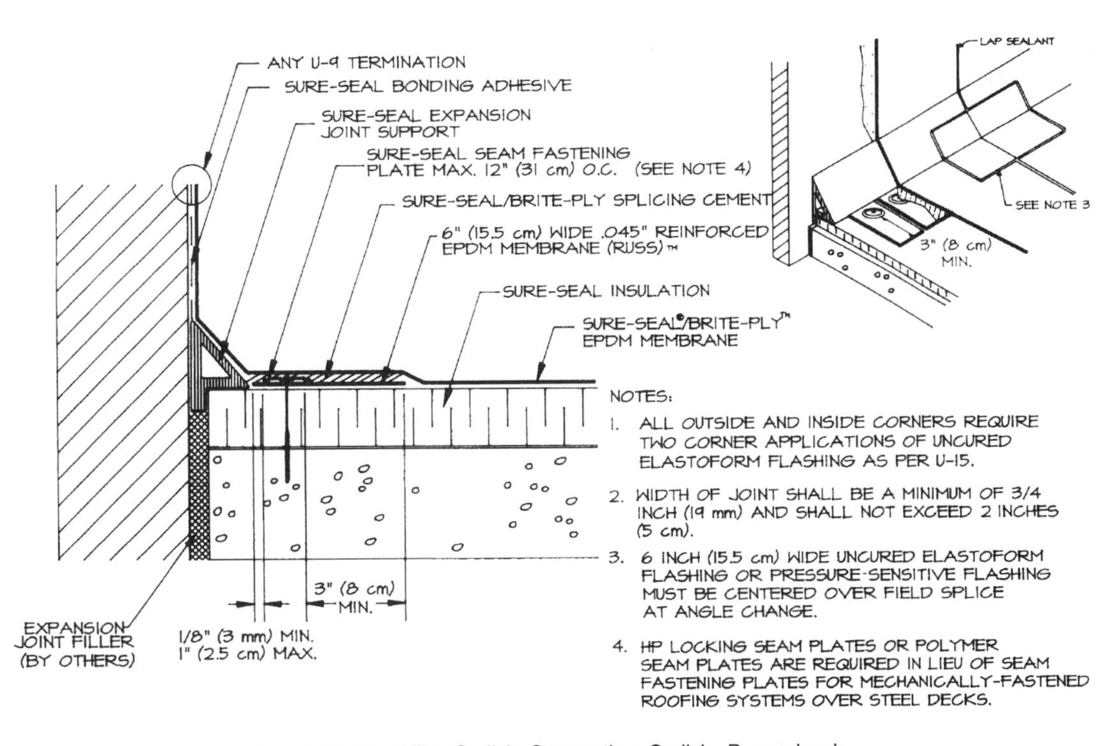

By permission of The Carlisle Corporation, Carlisle, Pennsylvania

8.7.5 Single-Ply Membrane Box Gutter/Roof Drain Flashing Details

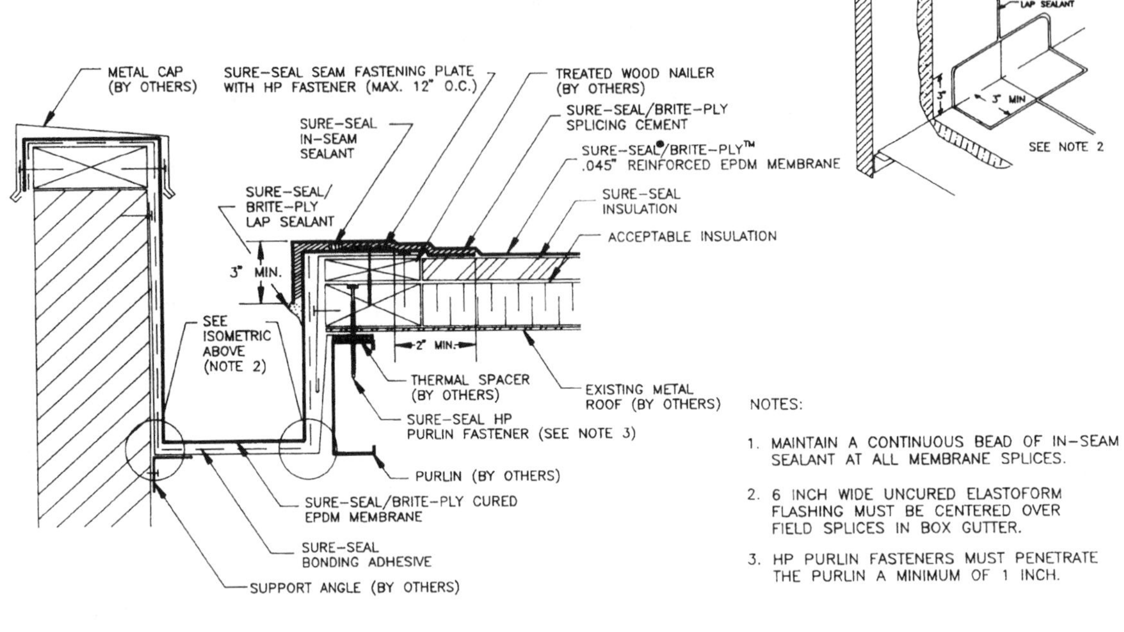

FOR DRAINS WITH TAPERED INSULATION AT DRAIN SUMP GREATER THAN 3 INCHES (8 cm) TO ONE (31 cm) HORIZONTAL FOOT

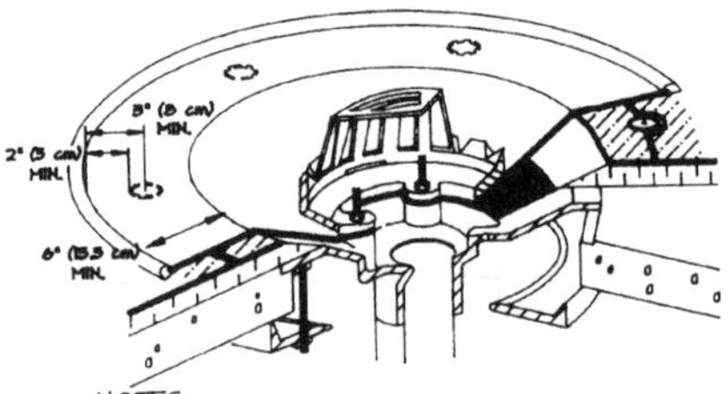

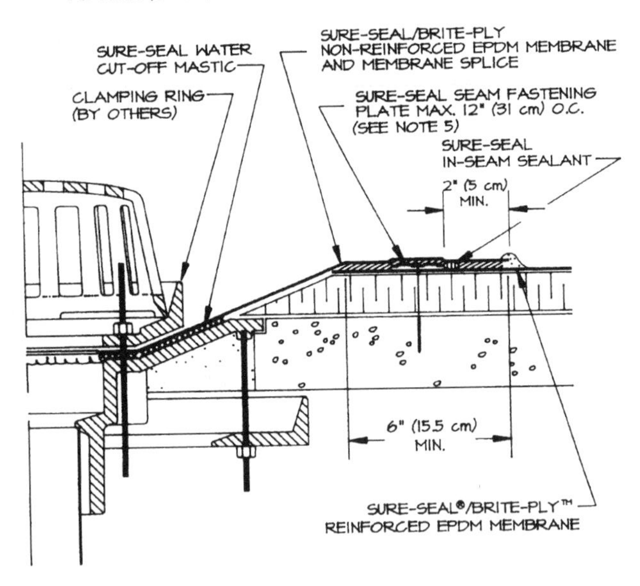

NOTES:

1. REINFORCED EPDM MEMBRANE MUST BE FASTENED WITH SEAM FASTENING PLATES NO MORE THAN 12 INCHES (31 cm) ON CENTER WHEN THE TAPERED INSULATION AT THE DRAIN SUMP IS GREATER THAN 3 INCHES (8 cm) TO THE HORIZONTAL FOOT. CUT REINFORCED MEMBRANE EVEN WITH TOP EDGE OF THE DRAIN SUMP.

   WHEN TAPERED INSULATION AT THE DRAIN SUMP IS LESS THAN 3 INCHES (8 cm) TO THE HORIZONTAL FOOT, REFER TO U-6-B DETAIL.
2. USE NON-REINFORCED EPDM MEMBRANE AS A SURFACE SPLICE AND EXTEND INTO DRAIN CLAMPING RING.
3. LOCATE EDGE OF THE SURFACE SPLICE OUT OF THE DRAIN SUMP AT LEAST 6 INCHES (15.5 cm) IN ALL DIRECTIONS ONTO THE HORIZONTAL MEMBRANE.
4. INSULATION TAPER SHALL NOT BE STEEPER THAN 6 INCHES (15.5 cm) (VERTICAL) IN 12 INCHES (31 cm) (HORIZONTAL).
5. HP LOCKING SEAM PLATES OR POLYMER SEAM PLATES ARE REQUIRED IN LIEU OF SEAM FASTENING PLATES OVER STEEL DECKS. REFER TO DETAIL MFS-2-B.

WHEN A SQUARE OR RECTANGULAR SECTION OF NON-REINFORCED EPDM MEMBRANE IS USED AS A SURFACE SPLICE, ROUND THE CORNERS OF THE NON-REINFORCED MEMBRANE FOR PROPER SPLICE.

By permission of The Carlisle Corporation, Carlisle, Pennsylvania

8.7.6 Single-Ply Membrane Acceptable Roof Deck Chart

1. Proper decking shall be provided by the building owner. The building owner or its designated representative must have a registered engineer investigate the building structure to ensure its ability to withstand the total weight of this roofing system, as well as construction loads and live loads, in accordance with all applicable codes. The specifier must also designate the maximum allowable weight and location for material loading and storage on the roof.
2. Acceptable decks, minimum pullouts, and approved Carlisle Fasteners:

Deck Type	Minimum Pullout	Approved Fastener	Minimum Penetration
Steel, 22 gauge or heavier	425 pounds	HP Fastener	3/4 Inch
Lightweight Insulating Concrete over Steel	360 pounds	HP Fastener	3/4 Inch
Structural Concrete, rated 3,000 psi or greater	800 pounds	HP Concrete Spike or HP Fastener [1]	1-1/4 Inches
Wood Planks and Plywood (minimum 15/32 inch thick APA Grade CDX)	360 pounds	HP Fastener [2]	1 Inch (Maximum 1-1/2 inches on wood planks)
Oriented Strand Board (OSB) (minimum 7/16 inch thick APA Rated non-veneer)	360 pounds	HP Woodie Fastener [3]	1-1/2 Inches
Cementitious Wood Fiber and Gypsum	300 pounds	HP Lightweight Deck Fastener	1-1/2 Inches

Notes:
1. HP Fasteners over 6 inches in length are not recommended for use on concrete decks.
2. If the minimum pullout into plywood decks cannot be achieved, a trial test should be conducted with the HP Woodie Fastener to determine acceptability (refer to Note 3 below).
3. A maximum of 1-1/2 inch thick insulation can be specified in conjunction with HP Woodie Fasteners.

If toggle bolts are specified for membrane securement, contact Carlisle for requirements.

3. Withdrawal resistance tests are strongly suggested to determine the suitability of a roof deck. Cementitious wood fiber, gypsum, lightweight insulating concrete over steel and oriented strand board (regardless of thickness), or plywood (less than 5/8 inch in thickness) must be tested. If the minimum pullout requirements cannot be achieved, Carlisle may be contacted for options concerning an appropriate roofing system.

8.8.0 Single-Ply Membrane Underwriters Laboratories Specifications

The following information highlights the Underwriters Laboratories (UL) and Factory Mutual (FM) code ratings achieved with Carlisle's Sure-Weld Mechanically Fastened Roofing System:

Underwriters Laboratories

UL Class "A"

Deck Type	Insulation	Thickness	Maximum Slope
Non-Combustible and Combustible (For combustible decks, gypsum wallboard must be installed beneath the insulations listed) (1) (2)	Carlisle HP Recovery Board	1/2"-3"	1"
	Carlisle HP Recovery Board/Polyisocyanurate	1/2" Min./Any	1"
	Carlisle HP Recovery Board/Polystyrene	1/2" Min./Any	1"
	Carlisle Polyisocyanurate HP, HP-N or HP-W	Any	1/2"
Combustible	Gypsum Board Gypsum Board/Polyisocyanurate Gypsum Board/Polystyrene	1/2" 1/2"/Any 1/2"/Any	2" 2" 2"

UL Class "B"

Deck Type	Insulation (3)	Thickness	Maximum Slope
Combustible	Carlisle Polyisocyanurate HP, HP-N, HP-W	2" Min.	1/2"
	Carlisle Polyisocyanurate/G2 Base Sheet (4)	1-1/2" Min./ G2 Base	1/2"
	HP Recovery Board Board/Polyisocyanurate	1/2" Min./ 1-1/2" Min.	1"
	HP Recovery Board/Polyisocyanurate/G2 Base Sheet (4)	1/2" Min./1" Min./G2 Base	1"
	HP Recovery Board/G2 Base Sheet (4)	1" Min./G2 Base	1"

Notes:

(1) Minimum 1/2 inch thick gypsum wallboard can be a classified or unclassified material with a minimum weight of 1.84 pounds per square foot. 1/4 inch thick Georgia Pacific Dens-Deck or Sound Deadening Board with a minimum weight of 1.09 pounds per square foot may be substituted for 1/2 inch thick gypsum wallboard.

(2) On Retrofit/No Tearoff projects, where the existing roof is Class A rated, the gypsum board can be eliminated. Existing roofs which are Class B or C rated will require the use of gypsum board to achieve a Class A rating, otherwise, the new roofing system will retain the existing UL rating.

(3) Insulation joints (bottom layer) are to be staggered a minimum of 6 inches from joints in wood deck.

(4) Acceptable G2 base sheets can be one of the following; Celotex Type G2 Vaporbar GB, GAF Gafglas No. 75 Base Sheet, Manville Glasbase, Owens Corning Perma Ply No. 28 or Tamko Glass Base.

By permission of The Carlisle Corporation, Carlisle, Pennsylvania

8.9.0 Single-Ply Membrane Roofing Preventative Maintenance Guidelines

Periodic maintenance to the roofing system will help to address those locations where moisture could infiltrate and cause damage. It is imperative that the building owner recognizes the importance of preventative maintenance in an effort to increase the life expectancy of the roofing system beyond the warranty period.

Preventative Maintenance

The following is a list of general care and maintenance requirements for Carlisle Roofing Systems. These maintenance items will help attain maximum performance from the roofing system.

- *Provide proper drainage* Keep the roof surface clean of leaves, twigs, paper or accumulated dirt at drain areas to avoid clogged drains. Excessive ponding of water on the surface of the membrane will increase the probability of moisture entering the structure in the event of a puncture or cut in the membrane.
- *Avoid degrading the membrane.*

Do not expose the membrane to the following materials because of possible degradation of the membrane:

- Liquids that contain petroleum products
- Solvents
- Grease used for lubricating roof top units
- Oils (new or old) used for air conditioning or compressor units
- Kitchen wastes or other animal fats
- Chemicals

Catch pans and proper drainage of these pans or other means of containment can be used for membrane protection. Prolonged exposure to these materials will cause swelling and possible degradation of the membrane if the spills are not removed.

8.9.1 Investigation of Leaks on a Ballasted Single-Ply Membrane Roof

DO NOT USE RAKES OR SHOVELS FOR BALLAST REMOVAL AS DAMAGE TO THE EPDM MAY RESULT. A GRAVEL PUSHER OR PUSH BROOM MUST BE USED.

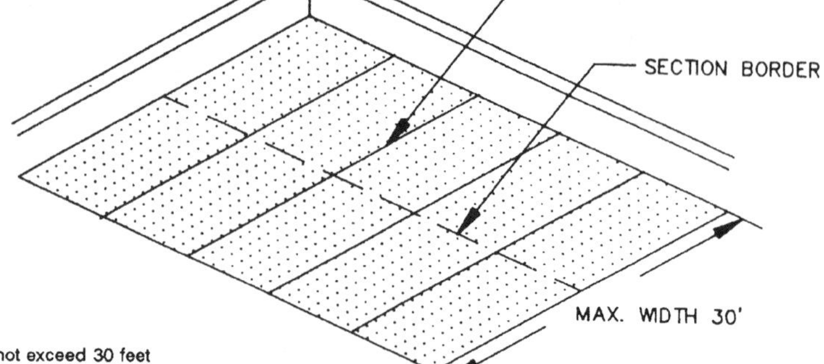

Step #1

It is recommended that the area to be investigated not exceed 30 feet in width for more practical removal of the ballast. If the area to be investigated is larger than 30 feet in width, divide the area into equal sections with each section not wider than 30 feet. Each area to be investigated should be laid out with the factory splices of the membrane crossing the narrow direction (or the width) of the area to be investigated.

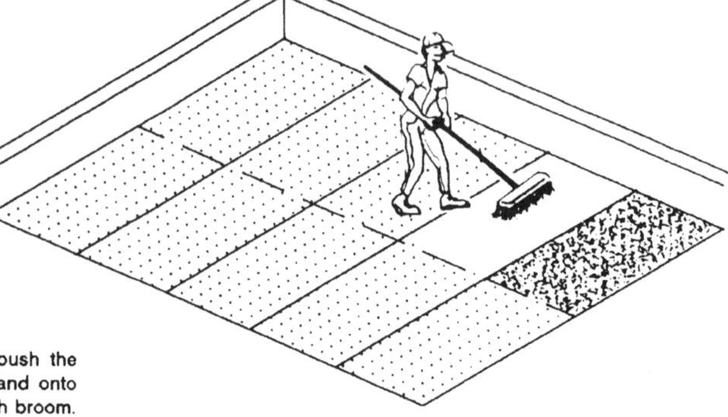

Step #2

Beginning at one end of the section being investigated, push the ballast out of one section between a set of factory splices and onto the adjoining section, as shown, with a gravel pusher or push broom. The area of the ballast to be removed should be approximately half the width of the section between the factory splices. Investigate the exposed EPDM membrane for signs of leaks as outlined in Paragraph C, Leak Investigation.

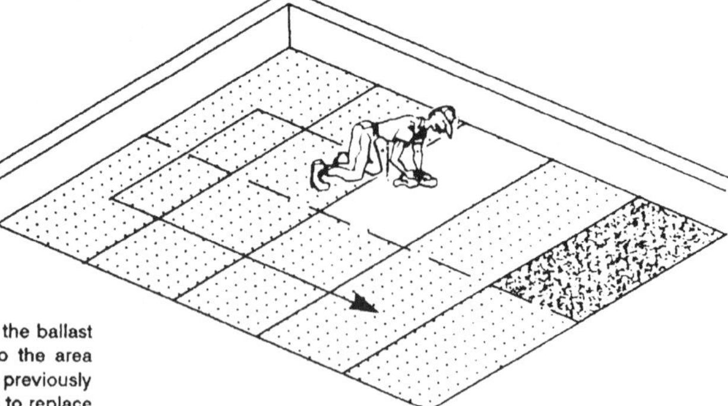

Step #3

To continue ballast removal and leak investigation, remove the ballast from the second factory splice section by pushing it onto the area previously investigated. When applying the ballast onto the previously investigated area, completely cover the exposed membrane to replace the ballast which was removed for the previous investigation.

Continue ballast removal and leak investigation in consecutive sections by removing the ballast from the next area and place it onto the area already investigated.

8.9.1 Investigation of Leaks on a Ballasted Single-Ply Membrane Roof (Continued)

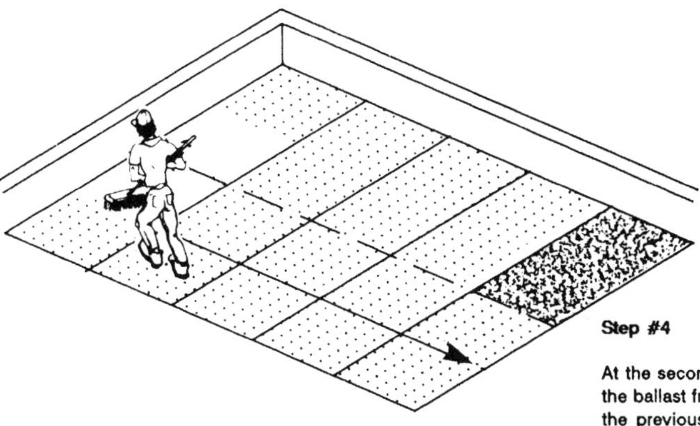

Step #4

At the second half of the area to be investigated, continue to remove the ballast from the section between factory splices by pushing it onto the previously exposed membrane at the end of the first half of the area being investigated.

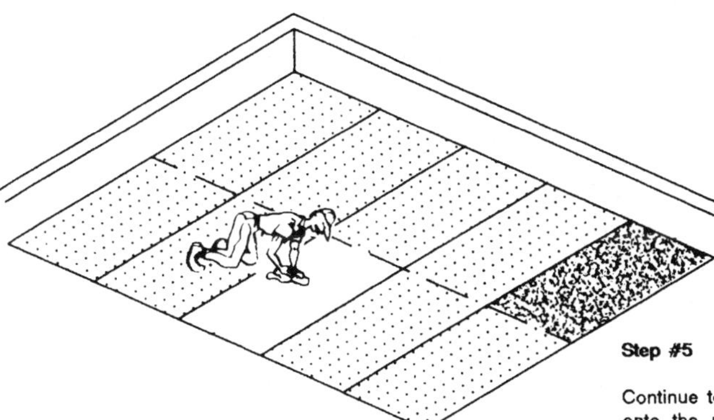

Step #5

Continue to remove ballast from consecutive sections by pushing it onto the previously exposed membrane at the adjoining section already investigated. Continue removing and replacing ballast along the second half of the area being investigated until the last section is exposed.

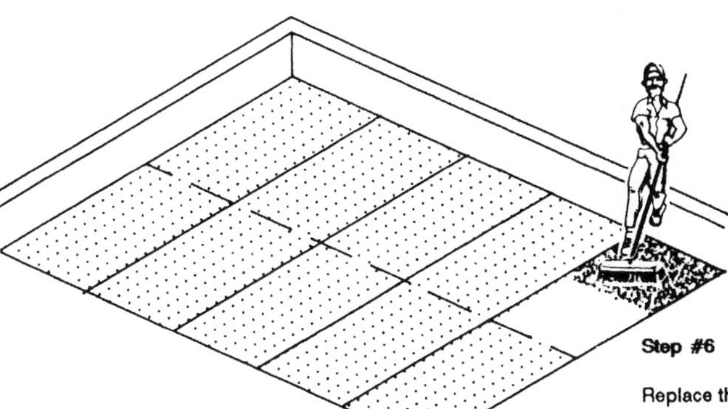

Step #6

Replace the ballast at the last section with half the ballast from the first section where ballast removal initially began to expose the final factory splice section. After investigating the final section, replace the ballast.

Continue the procedures across all sections of the roof (maximum 30 feet wide) until the leak has been found.

Use of the ballast removal steps, outlined above, avoids the double movement of ballast except at the first section.

By permission of The Carlisle Corporation, Carlisle, Pennsylvania

8.10.0 A Typical Fire Vent with Inside Pull Release Cable and Fusible Link

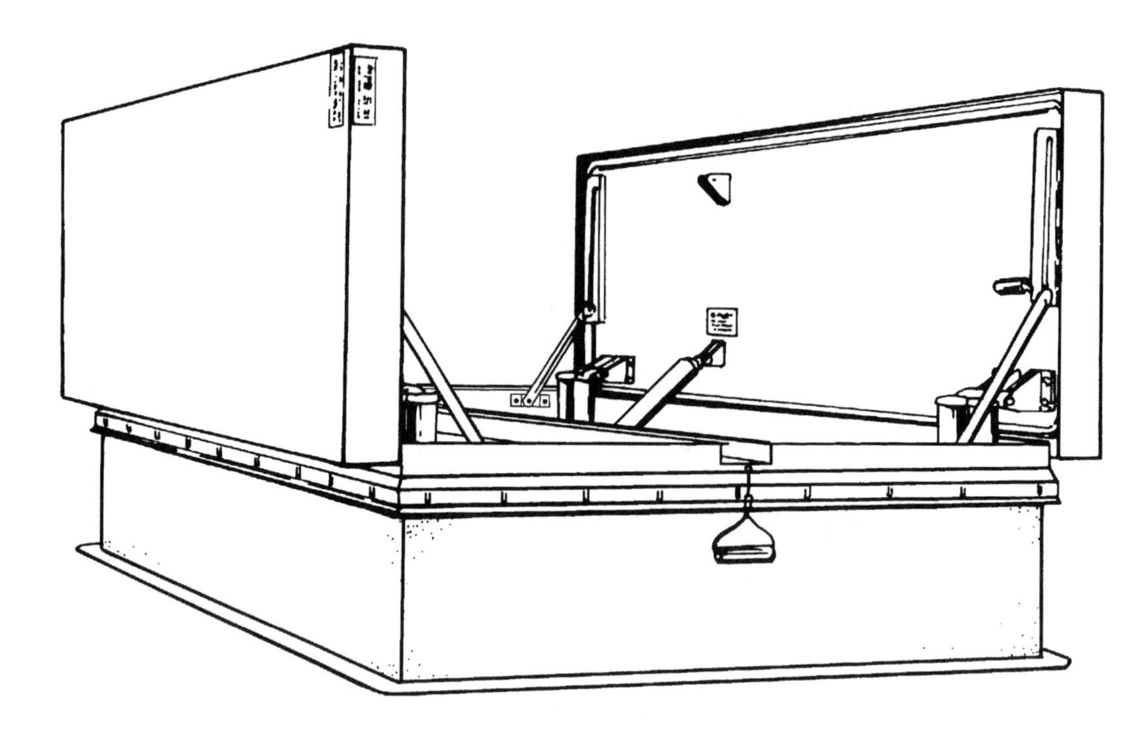

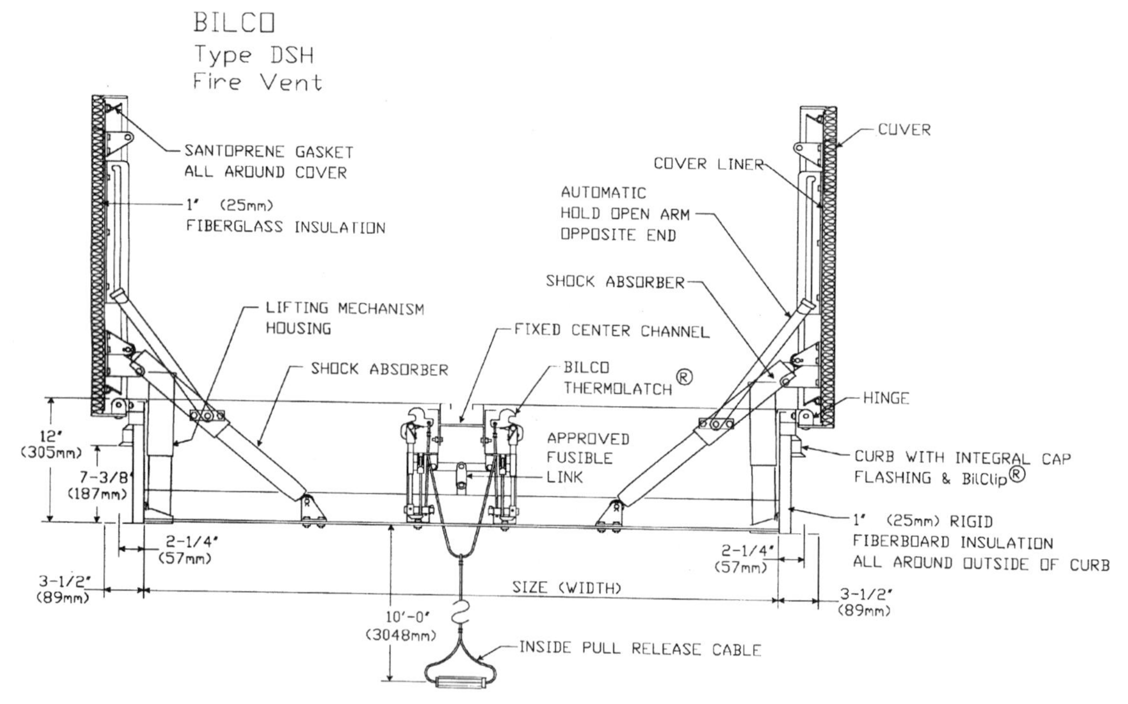

By permission of the Bilco Company, New Haven, CT

8.10.1 A Typical Roof Hatch Where a Ladder is Used For Access

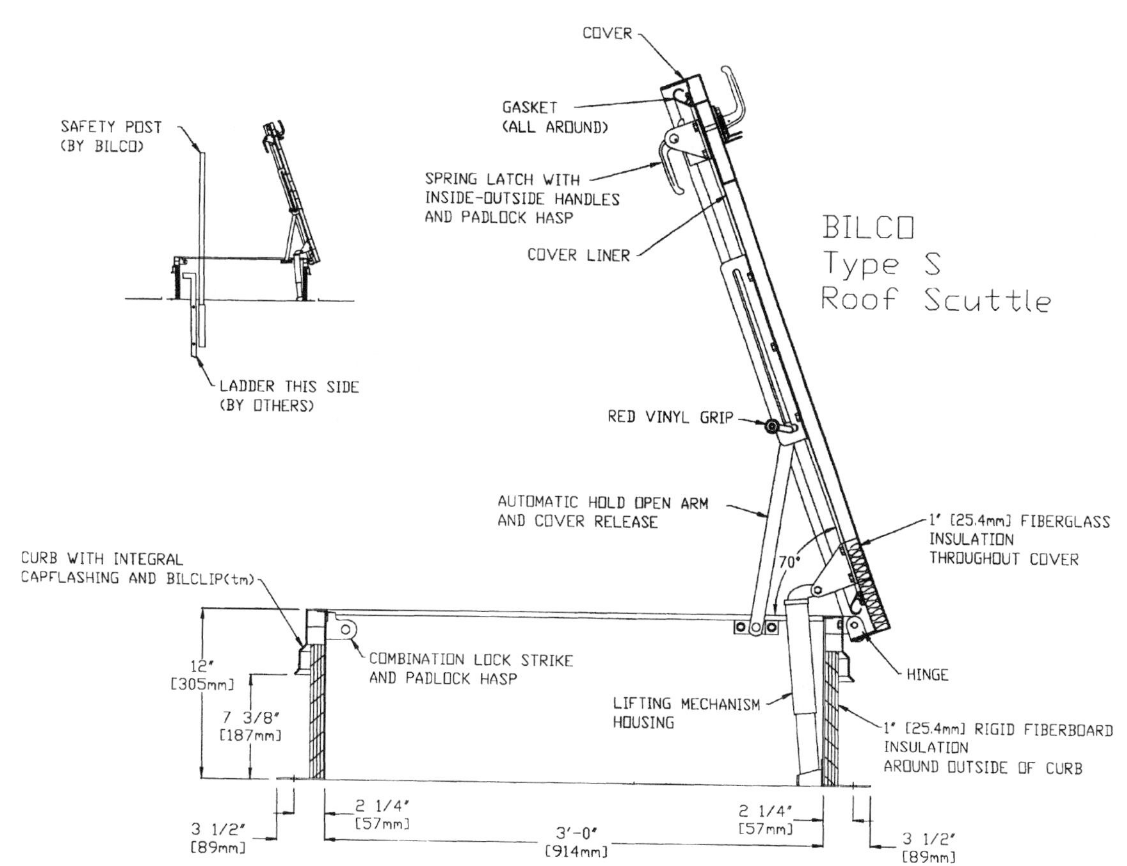

By permission of The Bilco Company, New Haven, CT

8.10.2 Typical Roof Hatch Where a Ships Ladder is Used for Access

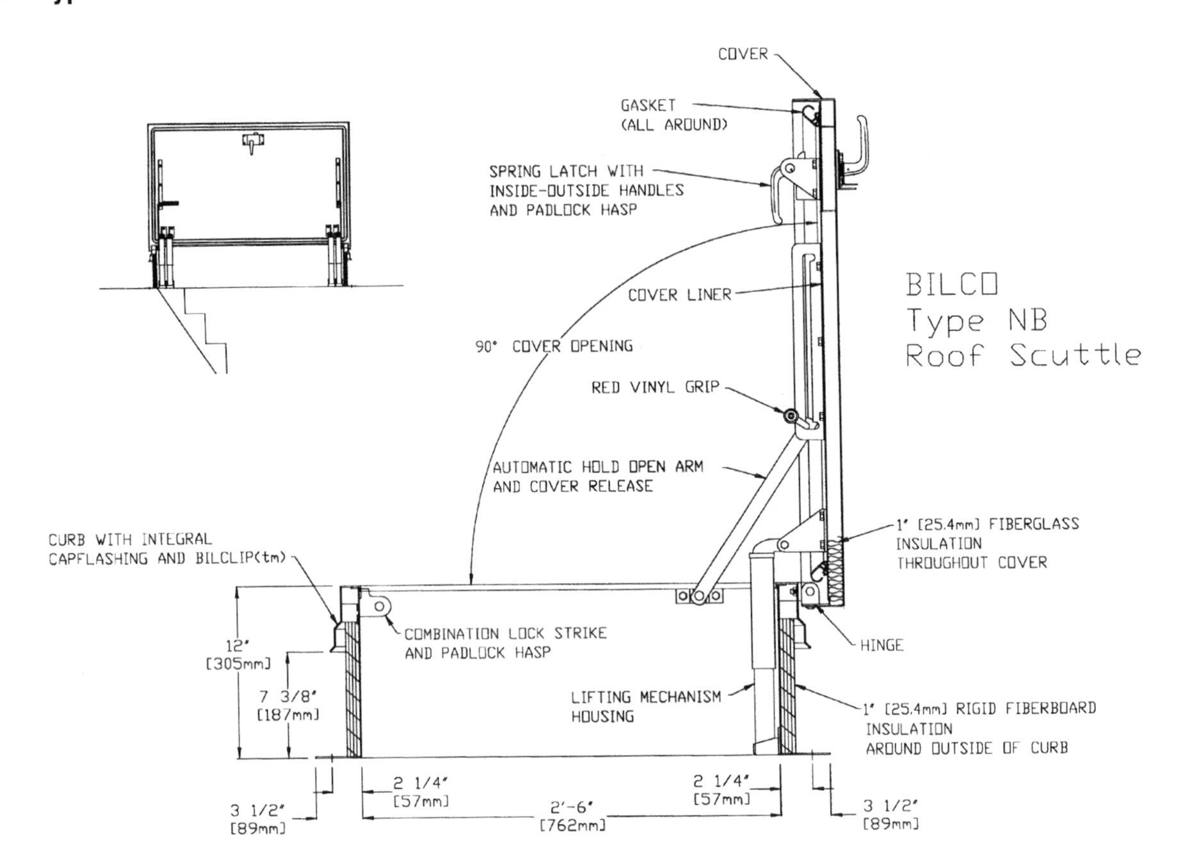

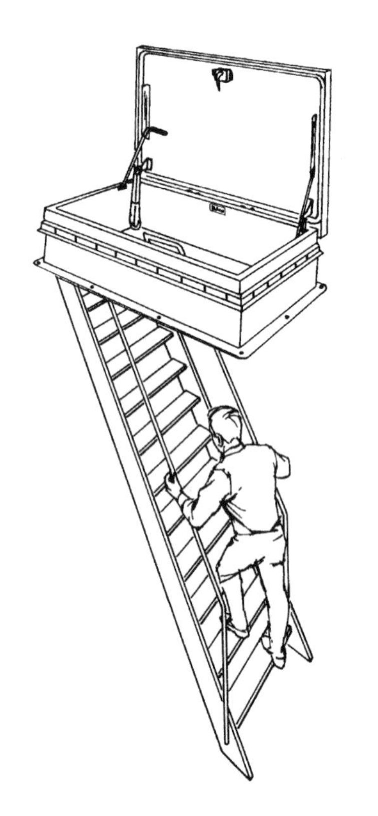

By permission of The Bilco Company, New Haven, CT

8.10.3 Typical Roof Hatch Installation Where Stairs Are Used for Access

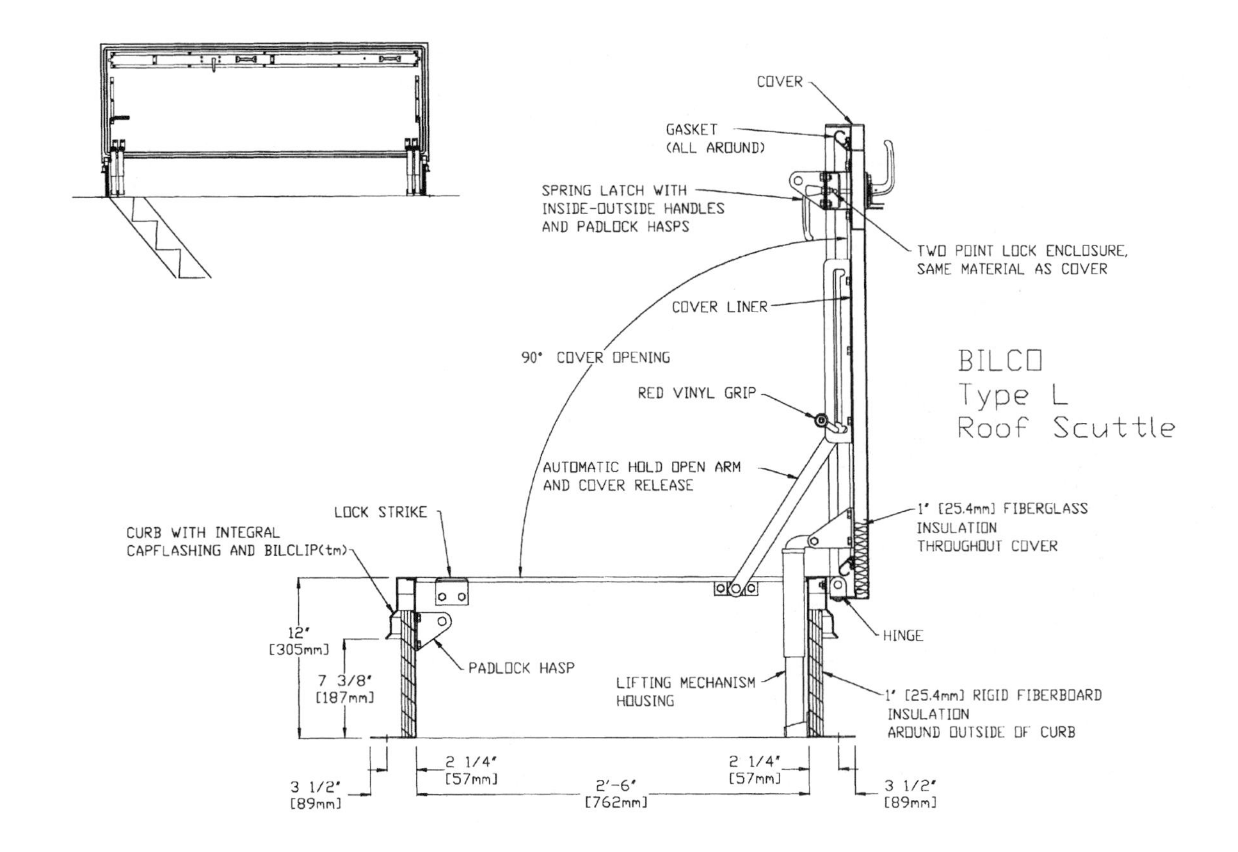

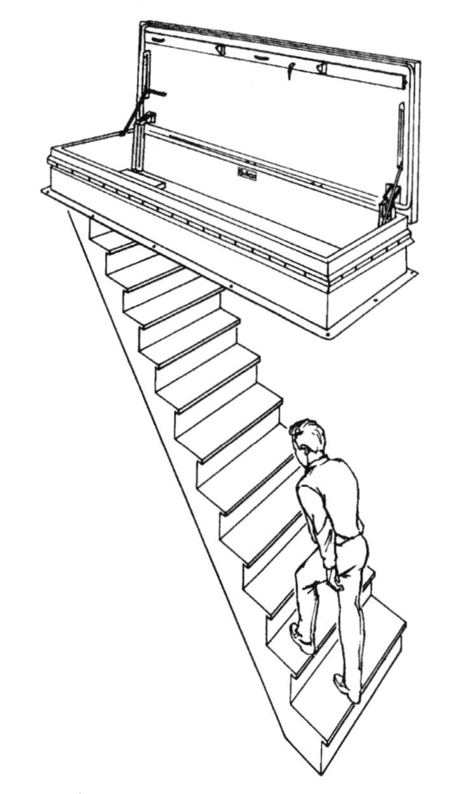

By permission of The Bilco Company, New Haven, CT

8.11.0 Copper and Lead-Coated Copper Roofing Material Sizes and Weights

Manufactured in accordance with ASTM B 370

COLD ROLLED COPPER SHEET

SIZES			COLD ROLLED			LEAD COATED	
			POUNDS PER SHEET	POUNDS PER CASE	SHEETS PER CASE	POUNDS PER CASE	SHEETS PER CASE
* 12 oz.	(.0162)	36 x 96	17.8	1068	60		
12 oz.	(.0162)	36 x 120	22.3	1070	48		
16 oz.	(.0216)	24 x 96	15.8	1027	65	1027	65
16 oz.	(.0216)	24 x 120	19.8	990	50	990	50
16 oz.	(.0216)	30 x 96	19.8	990	50	990	50
16 oz.	(.0216)	30 x 120	24.7	988	40		
16 oz.	(.0216)	36 x 96	23.7	1042	44	1042	44
16 oz.	(.0216)	36 x 120	29.7	1069	36	1069	36
20 oz.	(.0270)	24 x 96	19.9	1054	53		
20 oz.	(.0270)	24 x 120	24.9	1070	43		
20 oz.	(.0270)	30 x 96	24.9	1070	43		
20 oz.	(.0270)	30 x 120	31.1	1088	35		
20 oz.	(.0270)	36 x 96	29.9	1046	35	1046	35
20 oz.	(.0270)	36 x 120	37.3	1044	28	1044	28
24 oz.	(.0323)	36 x 96	35.6	1068	30		
24 oz.	(.0323)	36 x 120	44.5	1112	25		
32 oz.	(.0431)	36 x 96	47.3	1040	22		
32 oz.	(.0431)	36 x 120	59.1	1063	18		
48 oz.	(.0646)	36 x 96	70.9	1063	15		
48 oz.	(.0646)	36 x 120	88.7	1064	12		

COLD ROLLED COPPER COILS

GAUGE	WIDTH	COIL ID	COIL WT.
16 oz. (.0216)	9 15/16"	16"	1500/2000#
16 oz. (.0216)	9 7/8"	16"	1500/2000#
16 oz. (.0216)	10 1/2"	16"	1500/2000#
16 oz. (.0216)	11 5/8"	16"	1500/2000#
16 oz. (.0216)	11 3/4"	16"	1500/2000#
16 oz. (.0216)	11 7/8"	16"	1500/2000#
16 oz. (.0216)	13 1/8"	16"	1500/2000#
16 oz. (.0216)	13 3/4"	16"	1500/2000#
16 oz. (.0216)	15"	16"	1500/2000#
16 oz. (.0216)	18"	20"	1000/1200#
16 oz. (.0216)	20"	20"	1000/1200#
16 oz. (.0216)	24"	20"	1000/1200#

SOFT COPPER ROLLS

SIZE AND GAUGE	NO. OF ROLLS PER BOX	NET WEIGHT PER BOX
6" x 16 oz. (.0216)	5	500
7" x 16 oz. (.0216)	5	500
8" x 16 oz. (.0216)	5	500
10" x 16 oz. (.0216)	5	500
12" x 16 oz. (.0216)	5	500
14" x 16 oz. (.0216)	5	500
16" x 16 oz. (.0216)	5	500
18" x 16 oz. (.0216)	5	500
20" x 16 oz. (.0216)	5	500
24" x 16 oz. (.0216)	5	500

* Weight of a square foot of material is equal to the above identified ounces.

8.12.0 Standard Sizes and Exposure to Weather for Slate Roof Tiles

SCHEDULE OF STANDARD ROOFING SIZES

Standard thickness 3/16 inch; Other thicknesses available.

Sizes of slate, in.	No. in each sq.	Exposed when laid 3 in. lap	Approximate nails needed per square	
			LBS.	OZS.
24x16	86	10-1/2 in.	1	0
24x14	98	10-1/2 in.	1	2
24x13	106	10-1/2 in.	1	3
24x11	125	10-1/2 in.	1	7
24x12	114	10-1/2 in.	1	5
22x14	108	9-1/2 in.	1	4
22x13	117	9-1/2 in.	1	5
22x12	126	9-1/2 in.	1	7
22x11	138	9-1/2 in.	1	9
22x10	152	9-1/2 in.	1	12
20x14	121	8-1/2 in.	1	6
20x13	132	8-1/2 in.	1	8
20x12	141	8-1/2 in.	1	10
20x11	154	8-1/2 in.	1	12
20x10	170	8-1/2 in.	1	15
20x 9	189	8-1/2 in.	2	3
18x14	137	7-1/2 in.	1	9
18x13	148	7-1/2 in.	1	11
18x12	160	7-1/2 in.	1	13
18x11	175	7-1/2 in.	2	0
18x10	192	7-1/2 in.	2	3
18x 9	213	7-1/2 in.	2	7
16x14	160	6-1/2 in.	1	13
16x12	184	6-1/2 in.	2	2
16x11	201	6-1/2 in.	2	5
16x10	222	6-1/2 in.	2	8
16x 9	246	6-1/2 in.	2	13
16x 8	277	6-1/2 in.	3	2
14x12	218	5-1/2 in.	2	8
14x11	238	5-1/2 in.	2	11
14x10	261	5-1/2 in.	3	3
14x 9	291	5-1/2 in.	3	5
14x 8	327	5-1/2 in.	3	12
14x 7	374	5-1/2 in.	4	4
12x10	320	4-1/2 in.	3	10
12x 9	355	4-1/2 in.	4	1
12x 8	400	4-1/2 in.	4	9
12x 7	457	4-1/2 in.	5	3
12x 6	533	4-1/2 in.	6	1
10x 8	515	3-1/2 in.	5	14
10x 7	588	3-1/2 in.	7	4
10x 6	686	3-1/2 in.	7	13

By permission: Buckingham Slate, Arvonia, Virginia

8.12.1 Slate Roof Installation Procedures

SLATE

(A) Slate shall be Genuine Unfading BUCKINGHAM-VIRGINIA SLATE as furnished by the Buckingham-Virginia Slate Corporation, 1 Main Street, P.O. Box 8, Arvonia, Virginia 23004-008, of the following sizes and thicknesses:
(B) All slate shall be hard, dense, sound rock, punched for two nails each. No cracked slate shall be used. All exposed corners shall be practically full. No broken corners on covered ends which sacrifice nailing strength or the laying of a water-tight roof will be allowed.

SLATING

(A) The entire surface of all roofs, unless otherwise specified, and all other surfaces so indicated on the drawings, shall be covered with slate as herein specified, in a proper and watertight manner.
(B) The slate shall project 2" at the eaves and from 1/2" to 1" as directed at all gable ends, and shall be laid in horizontal courses with 3" headlap, and each course shall break joints with the preceding one by at least 3". Slates at the eaves or cornice line shall be doubled using same thickness slate for under-eaves at first exposed course. Under eave slate to be approximately 3" longer than exposure of first course.
(C) Wood can't strip at eaves to be furnished by others.
(D) Slates overlapping sheet metal work shall have the nails so placed as to avoid puncturing the sheet metal. Exposed nails shall be permissible only in top courses where unavoidable.
(E) Neatly fit slate around aH pipes, ventilators, and other vertical surfaces.
(F) Nails shall not be driven so far as to produce a strain on the slate.
(G) Cover all exposed nail heads with elastic cement. Hip slates and ridge slates shall be laid in elastic cement spread thickly over unexposed surface of under courses of slate, nailed securely in place and carefully pointed with elastic cement.
(H) Build in and place all flashing pieces, snow-guards, etc., furnished by the sheet metal contractor and cooperate with him in doing the work of flashing. (If roofing contractor has the flashing and sheet metal work under his contract, change this paragraph to suit.)
(I) Upon completion, all slate must be sound, whole, clean, and the roof shall be left watertight and neat in every respect, and subject to the architect's approval.

ROOFING FELT

(A) On all surfaces to be covered with slate, furnish and lay genuine asphalt saturated rag felt of an approved equal, not less in weight than that commercially known as "30 pound" felt or equal.
(B) Felt shall be laid in horizontal layers with joints lapped towards the eaves at least 2", and well secured along laps and at ends as necessary to properly hold the felt in place and protect the structure until covered with the slate. All felt shall be preserved unbroken, tight, and whole.
(C) Felt shall lap all hips and ridges at least 12" to form double thickness and shall be lapped 2" over the metal of any valleys or built-in gutters.

NAILS

(A) All slate shall be fastened with two large head slaters' hard copper wire nails, cut copper, cut brass or cut yellow metal slating nails to be inserted as desired of sufficient length to adequately penetrate the roof boarding. (Gauge or weight of nails should be inserted.)
(B) (In event the nailing base is other than wood, change the above paragraph to suit material used.)

HIPS

(A) All hips shall be laid to form "Fantail", "Saddle", "Mitred", "Boston", (to be inserted as desired.)

RIDGES

(A) All ridges to be laid to form "Comb", "Saddle", "Strip Saddle", (to be inserted as desired.) The nails of the combing slate shall pass through the joints of the slate below.
(B) The combing slate shall be laid with the same exposure as the next course down. (If desired, the combing slate sloping away from the direction of prevailing storms may project 1" above the combing slate on opposite side of ridge.)

VALLEYS

(A) All valleys shall be laid to form "Closed", "Open", "Round", (to be inserted as desired.)

FLASH & SHEET METAL WORK

(Specifications for flashing and sheet metal work to be inserted her if included under this specification.

By permission: Buckingham Slate, Arvonia, Virginia

8.13.0 Cedar Shingle/Shake Installation Diagrams

Shingle Exposure

PITCH	Maximum exposure recommended for roofs								
	Length								
	No. 1 Blue Label			No. 2 Red Label			No. 3 Black Label		
	16"	18"	24"	16"	18"	24"	16"	18"	24"
3/12 to 4/12	3¾"	4¼"	5¾"	3½"	4"	5½"	3"	3½"	5"
4/12 and steeper	5"	5½"	7½"	4"	4½"	6½"	3½"	4"	5½"

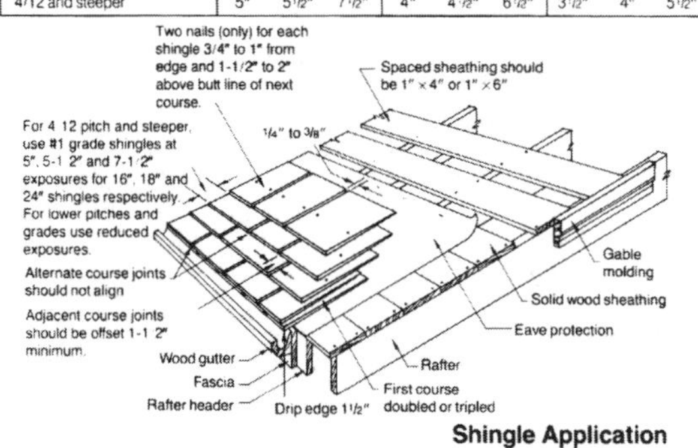

Shingle Application

Shake Exposure

PITCH	Maximum exposure recommended for roofs	
	Length	
	18"	24"
4/12 and steeper	7½"	10"(a)
(a) 24" x 3/8" handsplit shakes limited to 7-1/2" maximum weather exposure (5" per UBC)		

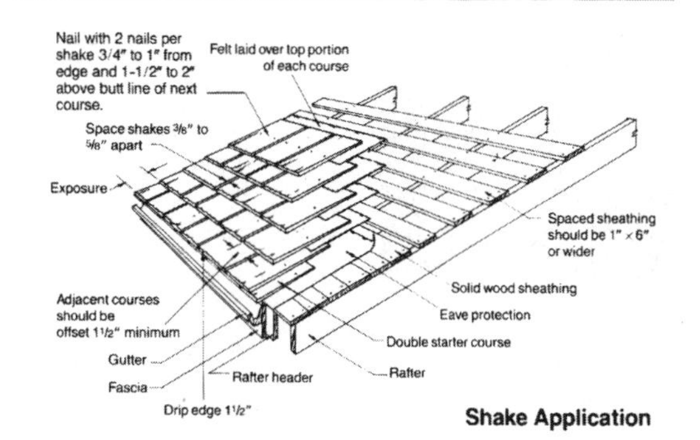

Shake Application

8.13.1 Cedar Shingle-Grade Label Facsimiles

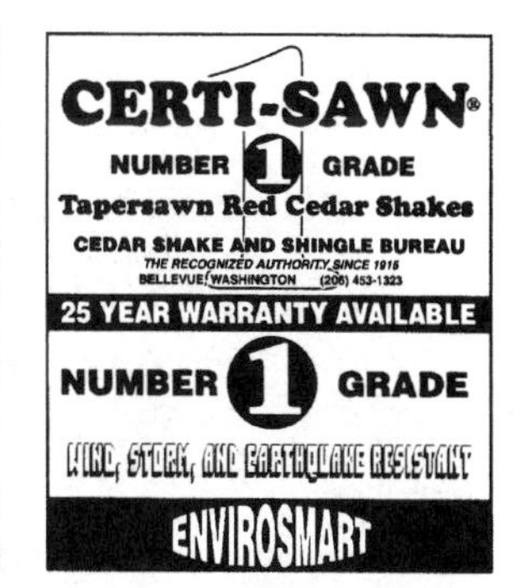

CEDAR SHAKE & SHINGLE BUREAU

By permission of Cedar Shake and Shingle Bureau, Bellevue, WA.

8.13.2 Cedar Shingle and Shake Installation and Maintenance Tips

SOME BASIC MAINTENANCE TIPS TO FOLLOW

Using a CEDAR SHAKE AND SHINGLE BUREAU **"APPROVED MAINTENANCE TECHNICIAN"** will help to ensure your safety and professional workmanship.

If you're doing it yourself BE CAREFUL! Use extreme caution on the roof, on slippery surfaces, around power lines, ladders and equipment.

SHOES

Wear suitable footwear. Tennis shoes will provide traction and will minimize damage to the shakes or shingles.

TRIM OVERHANGING BRANCHES

This will prevent debris and moss from clogging the valleys and gutters and from keeping the roof wet or damp. It will also eliminate roof damage in wind storms.

RUN LEADERS TO THE GROUND

Run downspouts (leaders) to the ground onto splashblocks slanting away from the foundation, or directly to another gutter below, never onto a lower roof surface.

CLEAN GUTTERS AND DOWNSPOUTS

Spring and Fall

NEVER BLOCK OFF ROOF VENTILATION

Such as louvers, ridge vents or soffit vents, even in winter. One of the most critical factors in roof durabililty is proper ventilation. Without it, heat and moisture build-up in the attic area and can cause rafters and sheathing to rot, roofing to buckle, and insulation to lose its effectiveness. Also, ice dams frequently occur when attics are not properly ventilated.

STEP RIGHT

Avoid walking on a cedar roof that is hot from the sun.

Never walk or stand on the lower end (butt end) of the shake or shingle to avoid cracking or weakening. Shingles and tapersawn shakes lay flat and therefore will not crack as easily. Carefully place your feet directly against the butt end of the row above.

SWEEP CLEAN

Remove debris (branches, pine needles, leaves, etc.). Leaving them on the roof retains moisture and encourages decay. This accumulation could also impede the run-off rain water which then could result in leaks. Being careful not to damage the shakes or shingles, clean them by using a stiff broom or brush. Remove foreign matter from the spaces (keyways) between the individual shake or shingle.

MOLD AND MILDEW

Mold and mildew can be killed and cleaned temporarily from wood roofs with the following solution:

MOLD AND MILDEW CONTROL FORMULA

3 ounces trisodium phosphate (TSP)
1 ounce detergent (e.g. Tide)
1 quart 5% sodium hypochlorite (Clorox)
3 quarts of warm water

This solution should be applied undiluted, and the surface scrubbed with a soft brush. When the surface is clean, it should be rinsed thoroughly with fresh water.

Care should be taken not to spray vegetation. If it does happen, rinse the plants thoroughly with fresh water.

MOSS CONTROL

In dry weather, control of moss can be accomplished be spraying or brushing the roof with a 10% solution of zinc sulfate. The moss absorbs the zinc oxide and eventually can be swept off the roof.

A solution of household bleach (sodium hypochlorite) mixed in a ratio of one part bleach to four parts water should prove to be equally effective.

Caution should be exercised in the use of all chemicals because of their high toxicity. Generally there is no hazard to plants provided that the chemicals do not contact the surrounding soil. If this should happen, either by direct contact or by the chemicals running through a septic tank and into the soil, no vegetation may grow for some time since the soil may be sterilized.

ZINC OR COPPER STRIPS FOR MOSS CONTROL

The use of these strips nailed at the ridge cap can be effective for moss control. These strips should run the full length of the roof and have a portion exposed to the weather. The reaction between the rain water and the zinc or copper forms a mild chemical solution that is carried down over the roof and retards formation of moss, fungus and mildew.

REPLACE AND BLEND

Replace all broken or missing shakes, shingles and ridge capping. New replacements can be made to blend in with the rest of the roof by dipping or spraying them with a 50% solution of baking soda and water.

(1 lb. of baking soda dissolved in 1/2 gallon of water).

The shakes and shingles will turn a weathered gray color after 4 or 5 hours in sunlight. This is a chemical reaction and is permanent.

REPLACING A MISSING OR DAMAGED SHAKE OR SHINGLE

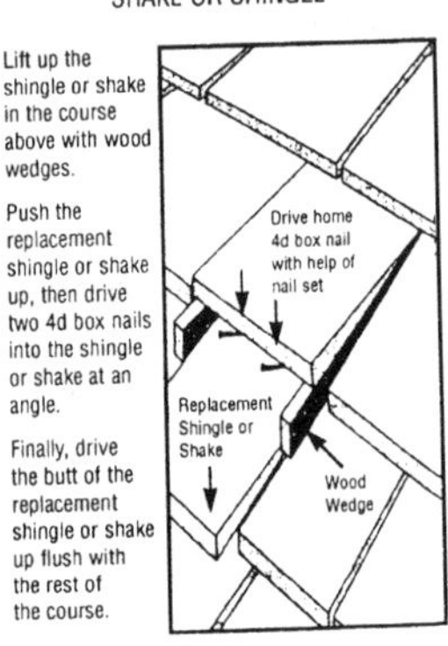

By permission of Cedar and Shake Bureau, Bellevue, Washington

8.14.0 A Checklist to Detect or Avoid Roof Leaks

The source of a leak is not necessarily directly above the appearance of water penetration on the inside of a building. Water has a tendency to travel by the forces of gravity or to be forced into a certain path by high winds. Careful inspection of the roof and all flashings is sometimes necessary to detect a leak; planned inspections by the owner might uncover a potential problem so that repairs can be affected.

1. Most leaks occur at the perimeter of the building because this is where more movement occurs, except at structural expansion joints. This area requires frequent inspection or "first look" if a leak has been reported.
2. Roof penetrations, those at roof drains or roof curbs or around roof accessories or pipe/conduit flashings, would be the next best place to inspect.
3. Parapet walls, exposed in two sides, might experience greater temperature variations and subsequent expansion and contraction activity, giving rise to tears in the flashing and leaks.
4. Equipment supports are frequently sources of roof leaks. Roof insulation attached to the outside surface of structural steel supports could act as a thermal bridge and increase the potential for condensation build up.
5. Tears or splits in the membrane itself, caused by workers working on the roof and abusing the surface, is another area of investigation. Servicing of roof-top equipment where oils and lubricants are used can also result in leaks because of the oils or lubricants being carelessly spilled on the roof membrane and dissolving a portion of the membrane.

Section

9

Fireproofing

Contents

9.0.0 Fireproofing or fire resisting?
9.1.0 Four accepted methods to fireproof steel
9.1.1 Spray- or trowel-on "dry" or "wet" systems
9.2.0 Fireproofing terminology
9.3.0 Typical spray fireproofing specifications
9.4.0 Spray fireproofing guide for dry mix applications
9.5.0 UL/ULC fire-resistance ratings chart (dry and wet mixes)
9.6.0 Standard physical-performance properties for spray-applied materials
9.7.0 Column fireproofing utilizing gypsum drywall (two and three-hour ratings)
9.8.0 Two-hour fire-rated drywall column enclosure (UL Design X518)
9.9.0 Two-three-hour drywall column enclosure (UL Design X518, X515)
9.10.0 Three-hour drywall column enclosure at precast concrete panel (UL Design U904)
9.11.0 Three-hour drywall column enclosure at 12" block wall corner (UL Design X-515)
9.12.0 Three-hour concrete column enclosure (traffic area)
9.13.0 Three-hour masonry column enclosure

9.0.0 Fireproofing or Fire Resisting

Fireproofing, in many cases, might better be referred to as fire resistance because the materials applied, mainly to structural steel systems, are meant to protect these systems from collapsing when exposed to the presence of fire for specific periods of time (one, two, three, or four hours). In other cases, the term *fire retardant* is more applicable, particularly when applied to flammable or combustible materials, like wood. In this case, fire retardancy provides a limit to the flame spread, fuel contribution, and smoke development that would have occurred if the combustible surface had not been treated with a fire-resistive coating.

9.1.0 Four Accepted Methods to Fireproof Steel

1. Spray or trowel on materials of a cementitious or mineral fiber nature.
2. Concrete encase structural steel columns or beams, or increase the thickness of concrete-suspended slabs on metal deck.
3. Apply specific numbers of layers of gypsum drywall onto the structural steel members.
4. Spray, brush, or roll on a water- or solvent-based intumescent material or mastic.

9.1.1 Spray- or Trowel-On "Dry" or "Wet" Systems

Spray- or trowel-on "dry" or "wet" cementitious or mineral fiber materials are the most prevalent forms of structural-steel fireproofing and are generally divided into two groupings (Type I and Type II).

- *Type I* A factory-mixed cementitious materials with a minimum density of 15/14 pounds per cubic foot (240 kg/cubic meter).
- *Type II* A factory, mixed, asbestos-free, mineral fiber material with inorganic binders, having a minimum applied dry density of 15 pounds per cubic foot (240 kg/cubic meter). If this system is used, it is generally followed by a water overspray to press any loose fibers and allow the binders to migrate and produce a firm surface.

9.2.0 Fireproofing Terminology

- *Air erosion* Resistance of spray fireproofing to dusting, flaking, sifting, and delamination because of air movement across its surface. ASTME-859-82/GSA sets the performance quality for air erosion; it is to be 0.025 gm/ft^2 maximum.
- *Bond strength* The ability of the spray fireproofing to resist pulling away from the steel substrate. The higher the bond strength, the lower the chance for cohesive or adhesive failure. ASMTE E-736-80 refers to bond strength and sets 200 lbs/ft^2 as the minimum bond strength.
- *Damageability* The resistance to physical abuse from abrasion, impact penetration, and compression. Two tests conducted by the City of San Francisco developed two standards and one test by ASTM provides the third:
  - ~ *Impact penetration* Six cubic centimeters maximum (City of San Francisco).
  - ~ *Abrasion resistance* 22 cubic centimeters maximum (City of San Francisco).
  - ~ *Compression* 500 pounds per square foot minimum (ASTM E-761-80).
- *Dry mix* It has no requirement to pre-mix with water or other additives. It can be applied in its original state by air under pressure. Water is introduced at the spray nozzle prior to application. The dry mix is quick and easy to apply.
- *Wet mix* The product is mixed with water to provide a slurry that is applied under high pressure through a nozzle. Although it is frequently referred to as *cementitious*, many manufacturer's products do not contain cement. This type of application provides cost-effective, fire-resistant performance per unit thickness.

9.3.0 Typical Spray Fireproofing Specifications

Physical Performance Characteristics: Fireproofing material shall meet the following physical performance standards:

1. Dry Density: The field density shall be measured in accordance with ASTM Standard E 605. Minimum average density shall be that listed in the UL Fire Resistance Directory *for each rating indicated,* ICBO Evaluation Report as required by the authority having jurisdiction, or minimum average 240 kg/cubic meter (15 pcf), whichever is greater.
2. Deflection: Material shall not crack or delaminate from the surface to which it is applied when tested in accordance with ASTM E 759.
3. Bond Impact: Material subject to impact tests in accordance with ASTM E 760 shall not crack or delaminate from the surface to which it is applied.
4. Bond Strength: Fireproofing, when tested in accordance with ASTM E 736, shall have a minimum average bond strength of 9.6 KPa (200 psf) and a minimum individual bond strength of 7.2 KPa (150 psf).
5. Air Erosion: Maximum allowable total weight loss of the fireproofing material shall be .05 gms/square meter (.005 grams/ft^2) when tested in accordance with ASTM E 859. Sample surface shall be "as applied" (not pre-purged) and the total reported weight loss shall be the total weight loss over a 24 hour period from the beginning of the test.
6. High Speed Air Erosion: Materials to be used in plenums or ducts shall exhibit no continued erosion after 4 hours at an air speed of 12.7 m/s (47 km/h) [2500 ft./min. (29 mph)] when tested in accordance with UMC Standard 6-1 and ASTM E 859.
7. Compressive Strength: The fireproofing shall not deform more than 10% when subjected to compressive forces of 57 KPa (1200 psf) when tested in accordance with ASTM E 761.
8. Corrosion Resistance: Fireproofing applied to steel shall be tested in accordance with ASTM E 937 and shall not promote corrosion of steel.
9. Abrasion Resistance: No more than 15 cm^3 shall be abraded or removed from the fireproofing substrate when tested in accordance with the test methods developed by the City of San Francisco, Bureau of Building Inspection.
10. Impact Penetration: The fireproofing material shall not show a loss of more than 6 cm^3 when subjected to impact penetration tests in accordance with the test methods developed by the City of San Francisco, Bureau of Building Inspection.
11. Surface Burning Characteristics: Material shall exhibit the following surface burning characteristics when tested in accordance with ASTM E 84:
    Flame Spread 0
    Smoke Development 0
12. Resistance to Mold: The fireproofing material shall be formulated at the time of manufacturing with a mold inhibitor. Fireproofing material shall be tested in accordance with ASTM G 21 and shall show resistance to mold growth for a period of 21 days for general use and 60 days for materials to be installed in plenums.
13. Combustibility: Material shall have a maximum total heat release of 20 MJ/m^2 and a maximum 125 kw/m^2 peak rate of heat release 600 seconds after insertion when tested in accordance with ASTM E 1354 at a radiant heat flux of 75kw/m^2 with the use of electric spark ignition. The sample shall be tested in the horizontal orientations.

Primed/Painted Substrates and Metal Decking.

Cross Reference Sec. 05100 Structural Steel and Section 05300 Metal Decking.

Primed/Painted Substrates: Fireproofing obtains its maximum bond when applied to unprimed/unpainted structural steel. Priming of interior structural steel is generally unnecessary and is not recommended by the steel industry. Primers add to the cost of the structure and may adversely affect the fire-resistance rating and the bond of the fireproofing to the substrate. Grace recommends that the structural steel specification include the following: "Interior structural steel to receive application of spray-applied fireproofing shall be free of primer and paint."

Currently, no primer/paint is specifically listed by Underwriters Laboratories Inc. for use with interior structural steel. According to the UL's Fire Resistance Directory, primer/paint removal, bond strength tests, mechanical attachment, bonding agents, or combination thereof may be required to maintain a fire resistive rating. Contact your Grace Representative for more information. Please note that there are limited UL approvals for primed/painted metal decks and joist element.

Metal Decking: Rolling compounds or lubricants are commonly used in the manufacture of steel decking. These compounds may impair proper adhesion of fireproofing to the substrate. Lubricants are available which, when used in appropriate quantities, will not adversely affect the bond of fireproofing to steel deck surfaces. Grace recommends that Section 05300 Metal Decking states: "Steel Deck manufacturer shall supply decking free of amounts of lubricants or oils which would impair the adhesion of spray-applied fireproofing."

9.4.0 Spray Fireproofng Guide for Dry Mix Applications

This is an abbreviated guide and is not intended as a substitute for the CAFCO® Application and Installation Manual. All applicators should thoroughly review the Application and Installation Manual prior to applying this product.

PREFERRED NOZZLE: 2-1/2" (65 mm) I.D. High output Air/Water nozzle, made by Hydra-Cone. The use of an expander sleeve is recommended to provide an even spray pattern. A 10 to 20 cfm (280 to 570 liters/min) AIR COMPRESSOR providing 60 psi (4.1 kg/cm²) air pressure at the nozzle is required.

ACCEPTABLE NOZZLES: 2-1/2" (65 mm) I.D. RA-9 Airless or 2" (50 mm) I.D. RA-6 Airless nozzles, made by Hydra-Cone. The use of an expander sleeve is recommended to provide an even spray pattern.
2-1/2" (65 mm) I.D. Boss 8 and 6 jet Airless nozzles, made by Contractors Consulting Service.

UNACCEPTABLE NOZZLE: 2" or 2-1/2" (50 or 65 mm) I.D. Hydra-Cone (Center Stem Jet), made by Hydra-Cone.

RECOMMENDED EQUIPMENT: Unisul - All Pneumatic Fireproofing Machines
Contractors Consulting Service - All BOSS Machines

MACHINE SETTINGS: Unisul - Carding boxes or slide gates should be set at 6 to 8.
BOSS -discs should be set at position 8. When feeding material, empty only one bag of material into machine hopper at a time. When the hopper is 1/4 full, empty next bag into the hopper.

WATER RATIO: 1.2 to 1 water to material ratio, by weight. Water pressure should be a minimum of 60 psi (4.1 kg/cm²) as measured at the nozzle. Refer to the CAFCO Application and Installation Manual for methods to determine water flow rate and material feed rate.

WATER BOOSTER PUMP: IT IS MANDATORY THAT A WATER BOOSTER PUMP WITH A 55 GAL.(200 LITER) MINIMUM RESERVOIR TANK BE USED TO INSURE PROPER WATER PRESSURE AND VOLUME.

HOSE SET-UP: ***TRANSFER HOSE*** must be smooth interior, rubber or plastic with a 2-1/2" (65 mm) or 3" (75 mm) Inside Diameter (I.D.). It must be reinforced to resist kinking or cracking and must resist static build up. The maximum transfer hose length, not including standpipe, is 250 ft. (75 m).

LIGHTWEIGHT FLEX HOSE (WHIP HOSE) must be rubber or plastic with a 2" (50 mm) or 2-1/2" (65 mm) Inside Diameter. It must be lightweight and flexible to allow mobility at the nozzle and must resist static build up. The maximum whip hose length is 25 ft. (8 m).

By permission of Isolatex International, Stanhope, New Jersey

9.4.0 Spray Fireproofing Guide for Dry Mix Applications (Continued)

NOZZLE DISTANCE:	18" to 24" (450 to 600 mm) from the substrate.
SURFACE PREPARATION:	Ensure surfaces are clean and free of dirt, oil , grease, loose mill scale, paints/primers (other than those approved) and any other materials that may impair adhesion. For applications to primed steel, contact the Isolatek International Technical Department. Note: See CAFCO Application and Installation Manual for use of CAFCO BOND-SEAL on various substrates.
APPLICATION TEMPERATURE:	Maintain a minimum substrate and ambient temperature of 40°F (4°C) prior to, during and a minimum of 24 hours after application.
VENTILATION:	Provide a minimum of 3 complete air exchanges per hour until the material is dry.
WATER OVERSPRAY:	IT IS MANDATORY THAT THE **BLAZE-SHIELD** II BE OVERSPRAYED WITH WATER BEFORE THE END OF THE WORK DAY.

NOTE: Only the listed equipment, nozzles and procedures are approved for applying CAFCO BLAZE-SHIELD II Deviations from any of these recommendations will result in product not meeting claims as published in Isolatek's literature. For complete details, refer to the CAFCO Application and Installation Manual. **This guide is not a substitute for the CAFCO Application and Installation Manual.**

By permission of Isolatex International, Stanhope, New Jersey

9.5.0 UL/ULC Fire-Resistance Ratings Chart (Dry and Wet Mixes)

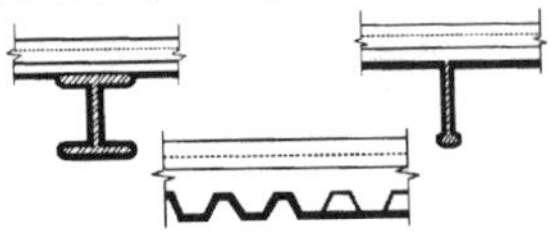

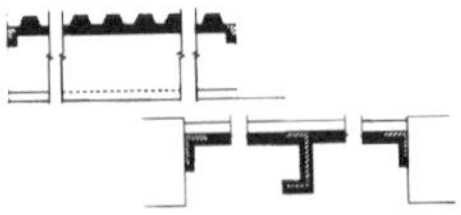

Dry Mix Fire Protection

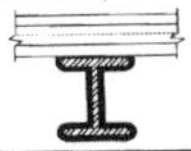

Assembly	BLAZE-SHIELD® DC/F, II Rating (Hr.) 1	1½	2	3	4	UL, ULC Design	BLAZE-SHIELD® HP Rating (Hr.) 1	1½	2	3	4	UL, ULC Design
Floor Assemblies (Protected)	•	•	•	•		D832	•	•	•	•		D832
	•	•	•	•	•	D858	•	•	•	•		G801
	•	•	•	•		D859	•	•	•	•		F816 ▲
	•	•	•	•	•	D860						
	•	•	•	•		G801						
					•	F801 ▲						
	•	•	•	•		F816 ▲						
	•	•	•	•		F819 ▲						
Floor Assemblies (Unprotected)	•	•	•	•		D902	•	•	•	•		D902
	•	•	•			F904 ▲	•	•	•			F904 ▲
Beam Only Floors	•	•	•	•		N815	•	•	•	•	•	N816
	•	•	•	•	•	N816	•	•	•	•		N826
	•	•	•	•	•	N823	•	•	•	•	•	O804 ▲
	•	•	•	•		N826						
	•	•	•			N830						
	•		•			N802 ▲						
	•	•	•	•	•	O804 ▲						
Roof Assemblies Protected (with board insulation)	•	•	•			P801	•	•	•			P801
	•	•	•			P814	•	•	•			P814
	•	•	•			P819	•	•	•			P819
	•	•	•			P825	•					R805 ▲
	•					R805 ▲						
Roof Assemblies Unprotected (with insulating concrete)	•	•	•			P908	•	•	•			P907
	•	•	•			P922	•	•	•			P922
Beam Only Roofs	•	•	•	•		S801	•	•	•	•		S801
	•	•	•			S802 *	•	•	•			S802 *
	•	•	•	•	•	S805 *	•	•	•	•	•	S805 *
	•	•	•			S806	•	•	•			S806
Columns: Wide Flange, Pipe and Tube	•	•	•	•	•	X829	•	•	•	•	•	X829
	•	•	•	•	•	X827	•	•	•	•	•	X827
	•	•	•	•		Z808 ▲						
	•	•	•			Z802 ▲						
	•	•	•	•		Z803 ▲						
Nonbearing Wall	•	•	•	•	•	U804	•	•	•	•		U804
	•	•	•	•	•	W801 ▲	•	•	•	•	•	W801 ▲

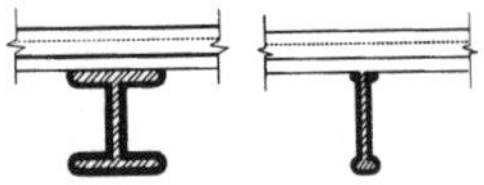

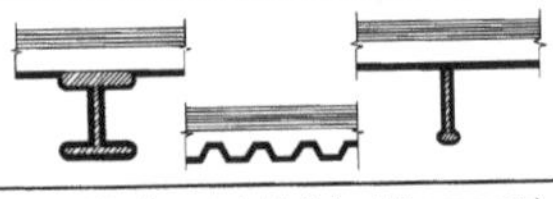

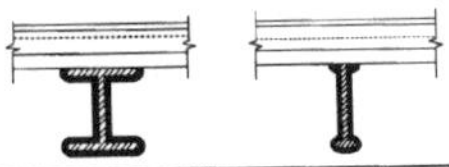

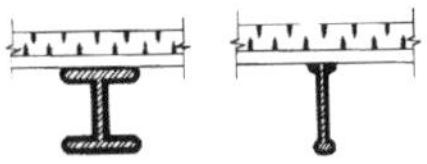

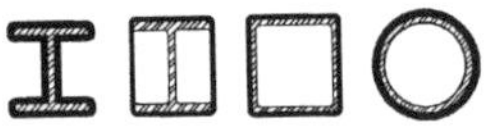

Wet Mix Fire Protection

Assembly	Cafco® 300, 400 Rating (Hr.) 1	1½	2	3	4	UL, ULC (1) Design	Cafco® 800 Rating (Hr.) 1	1½	2	3	4	UL, ULC Design
Floor Assemblies (Protected)	•	•	•	•		D759	•	•	•	•	•	D744
	•	•	•	•		G705						
	•	•	•	•		D860						
Floor Assemblies (Unprotected)	•	•	•	•		D902	•	•	•	•		D902
Beam Only Floors	•	•	•	•	•	N759	•	•	•	•	•	N742
	•	•	•	•		N761	•	•	•	•		N760
	•	•	•	•	•	O708 ▲	•	•	•	•	•	O707 ▲
Roof Assemblies Protected (with board insulation)	•	•	•	•		P719						
	•	•	•			P723						
Roof Assemblies Unprotected (with insulating concrete)	•	•	•			P908						
	•	•	•			P922						
Beam Only Roofs	•	•	•	•	•	S721	•	•	•	•		S720
	•	•	•	•	•	S729						
Columns: Wide Flange, Pipe and Tube	•	•	•	•	•	X790	•	•	•	•	•	X764
	•	•	•	•	•	Z715 ▲	•	•	•	•		X767
					•	Z716 ▲	•	•	•	•	•	X768
							•	•	•	•	•	XR703
Nonbearing Wall												

(1) ULC designs listed are for Cafco 300 only.

★ Requires material on underside of deck.

▲ ULC design

By permission of Isolatex International, Stanhope, New Jersey

9.6.0 Standard Physical-Performance Properties for Spray-Applied Materials

CHARACTERISTIC	ASTM STANDARD	LOW DENSITY	MEDIUM DENSITY	HIGH DENSITY
Surface Burning Characteristics	E84	Flame.......0 Smoke......0	Flame.......0 Smoke......0	Flame.......0 Smoke......0
Density	E605	15 lb/ft^3 (240 kg/m^3)	22 lb/ft^3 (352 kg/m^3)	40 lb/ft^3 (640 kg/m^3)
Cohesion / Adhesion (Bond Strength)	E736	150 lb/ft^2 (7.2 kPa)	434 lb/ft^2 (20.8 kPa)	1,000 lb/ft^2 (48.1 kPa)
Deflection	E759	No cracks or delaminations	No cracks or delaminations	No cracks or delaminations
Bond Impact	E760	No cracks or delaminations	No cracks or delaminations	No cracks or delaminations
Compressive Strength	E761	750 lb/ft^2 (35.9 kPa)	7340 lb/ft^2 (351 kPa)	43,200 lb/ft^2 (2068 kPa)
Air Erosion Resistance	E859	< 0.025 g/ft^2	< 0.025 g/ft^2	< 0.025 g/ft^2
Corrosion Resistance	E937, Mil Std 810	Does not promote corrosion of steel	Does not promote corrosion of steel	Does not promote corrosion of steel
Combustibility	E136, E1354	Noncombustible	Noncombustible	Noncombustible

By permission of Isolatex International, Stanhope, New Jersey

9.7.0 Column Fireproofing Utilizing Gypsum Drywall (Two- and Three-Hour Ratings)

Column Fireproofing

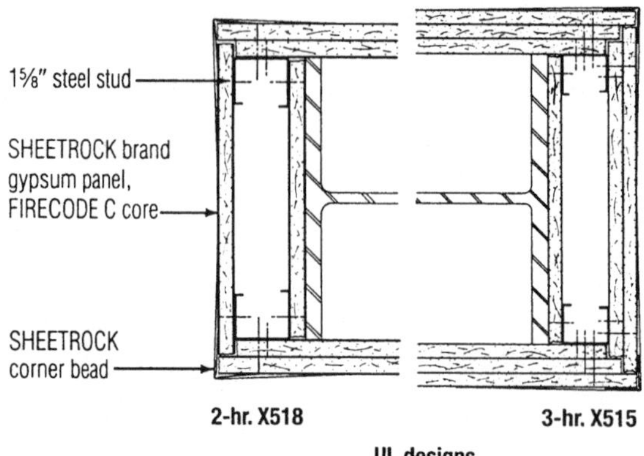

UL designs

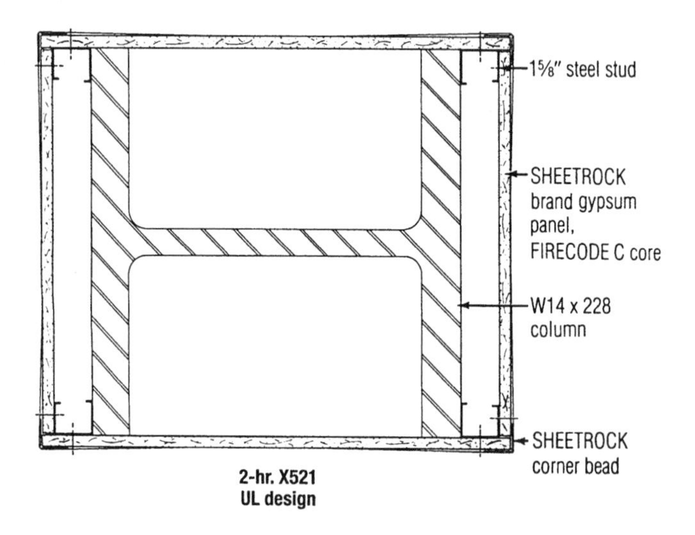

2-hr. X521
UL design

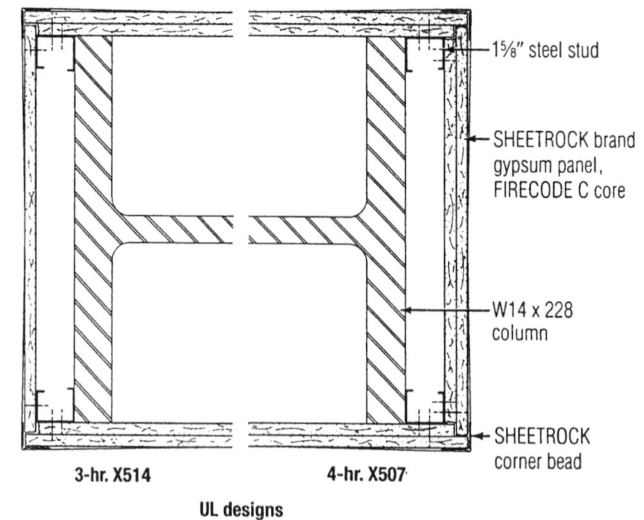

UL designs

By permission United States Gypsum Corp., Chicago, Illinois

9.8.0 Two-Hour Fire-Rated Drywall Column Enclosure (UL Design X518)

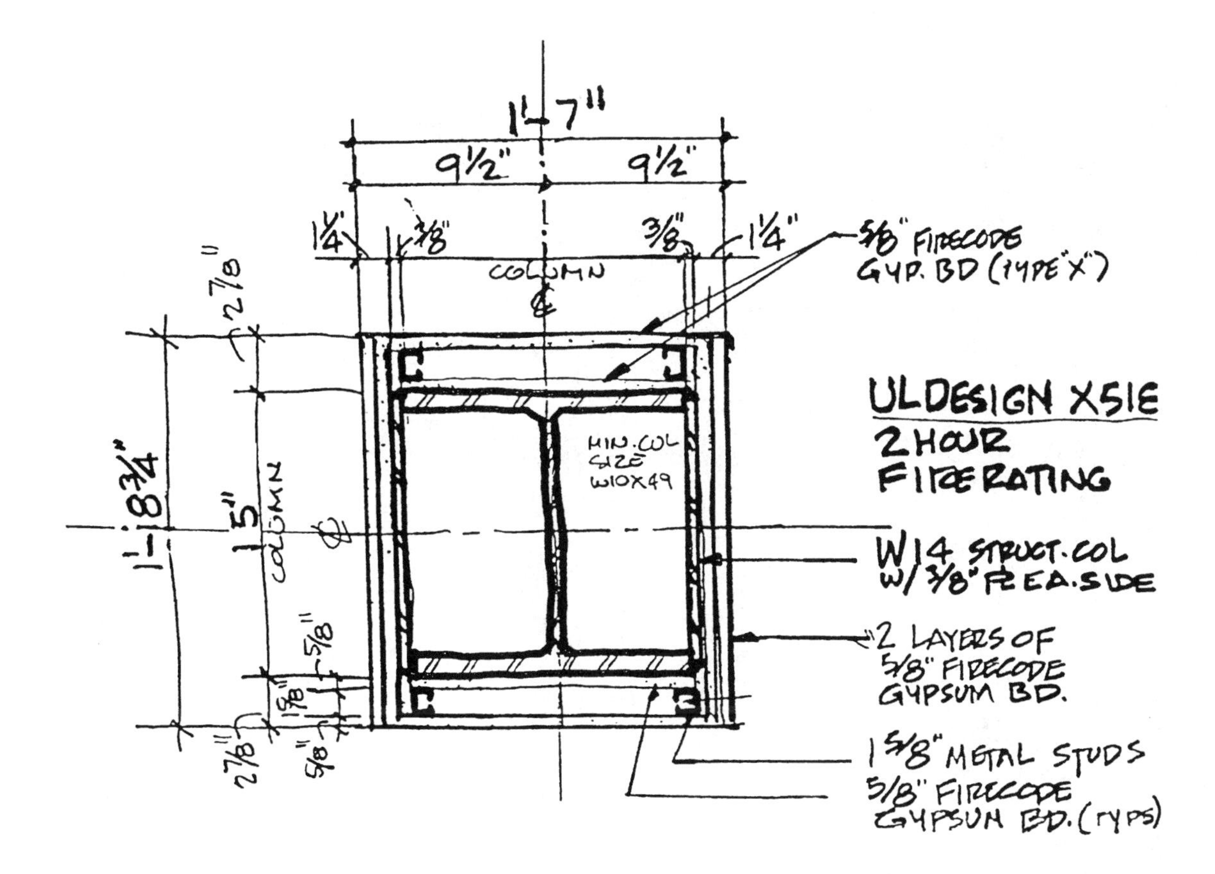

9.9.0 Two-Three-Hour Drywall Column Enclosure (UL Design X518, X515)

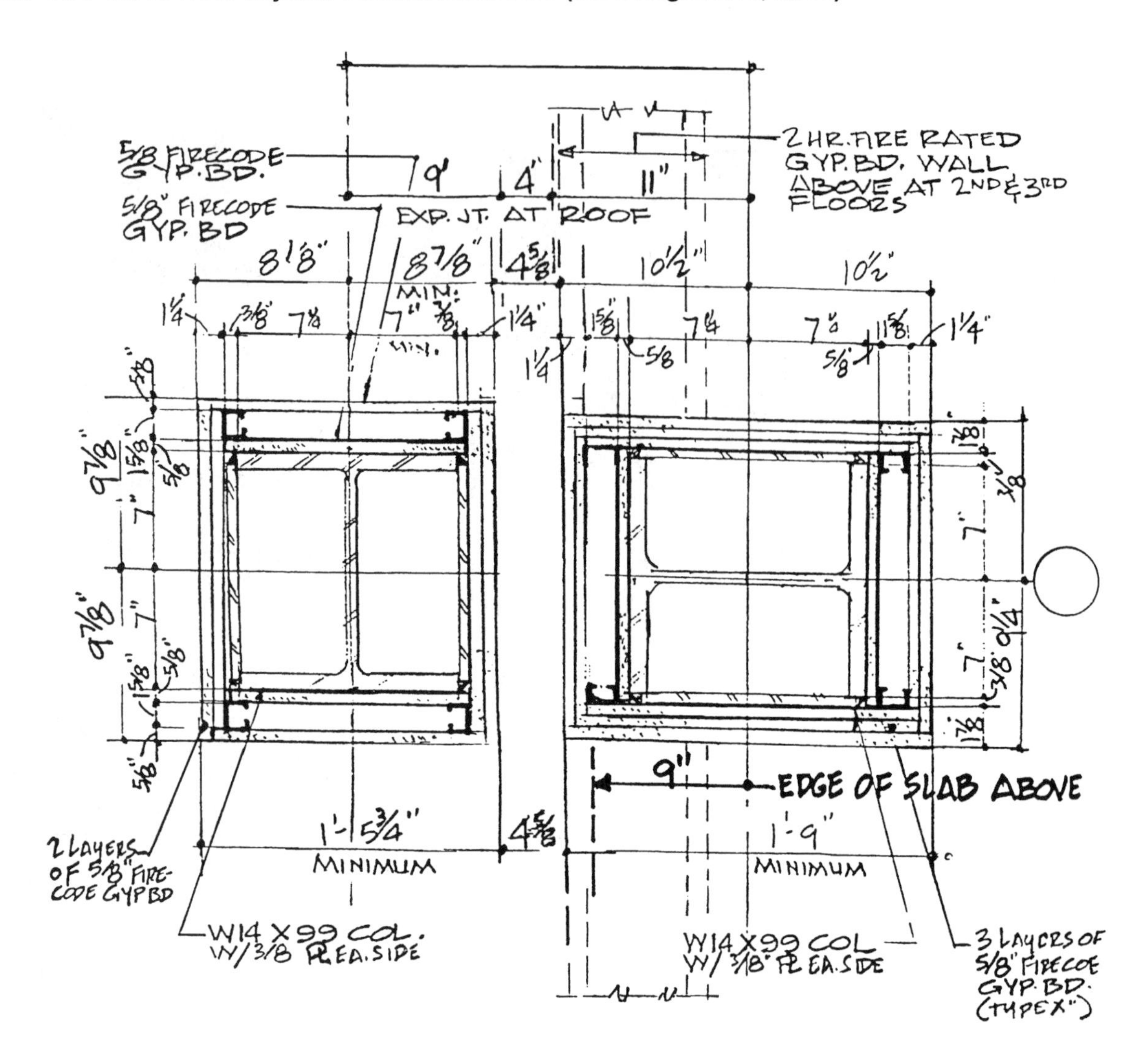

9.10.0 Three-Hour Drywall Column Enclosure at Precast Concrete Panel (UL Design U904)

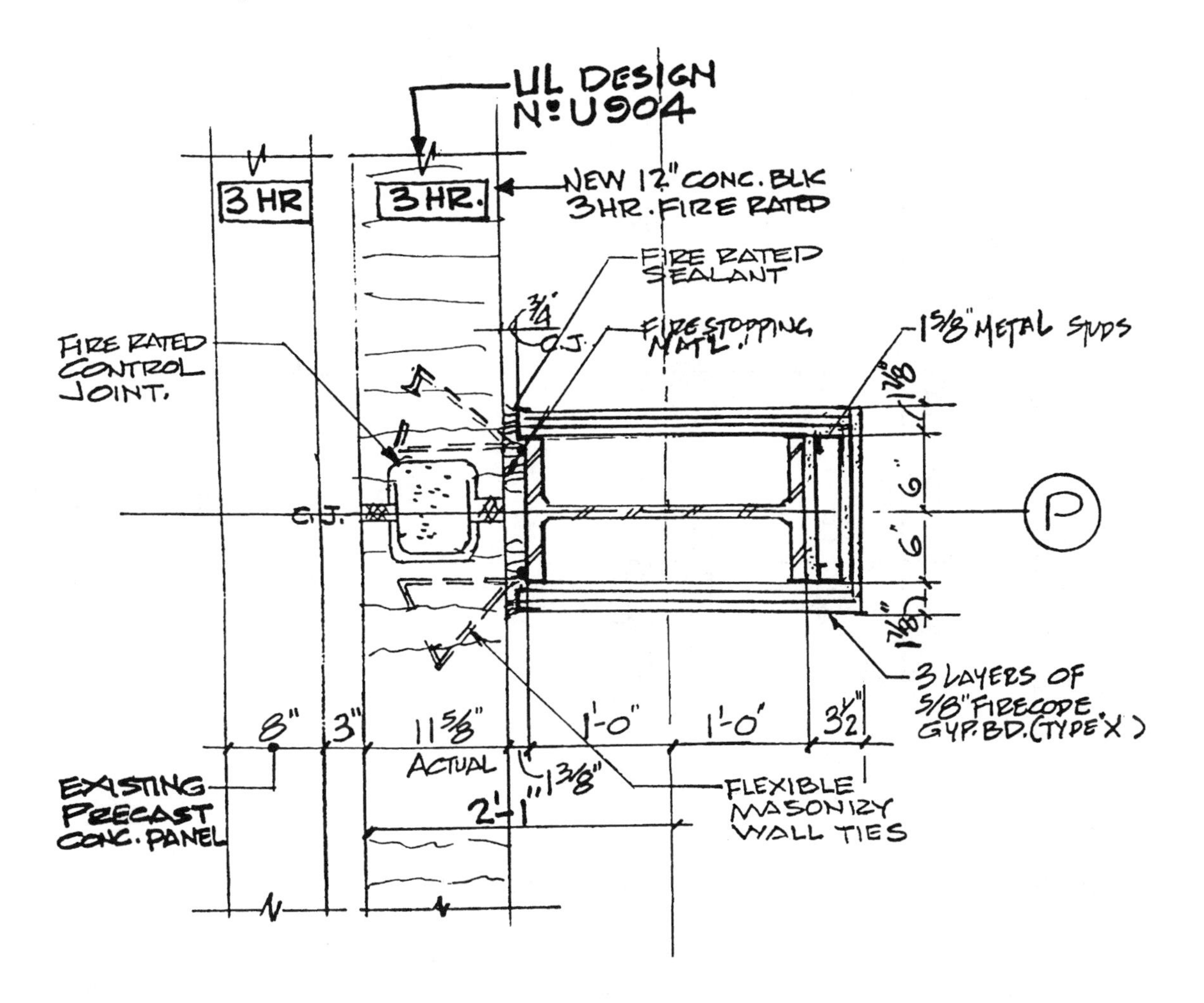

9.11.0 Three-Hour Drywall Column Enclosure at 12" Block Wall Corner (UL Design X-515)

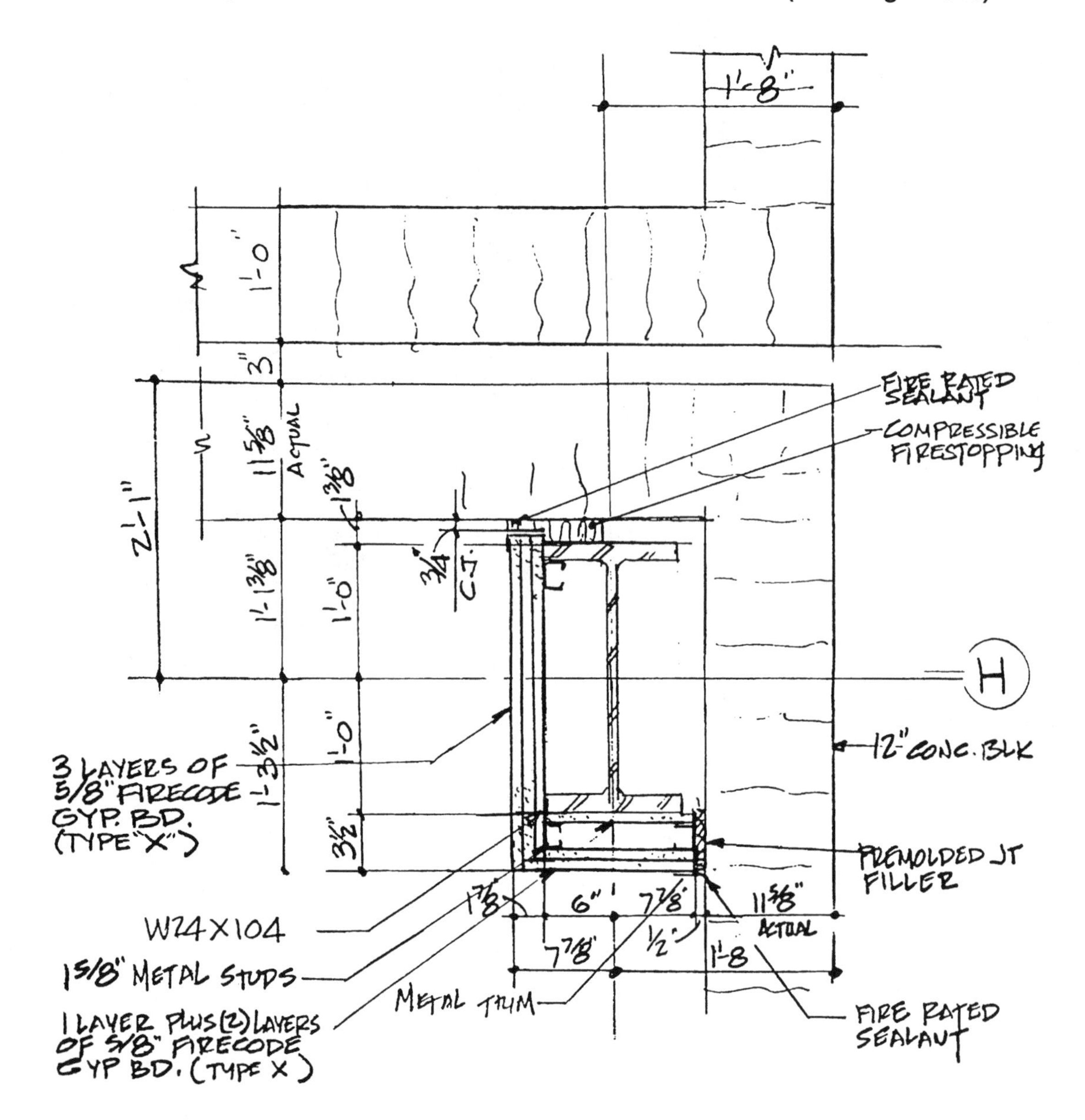

9.12.0 Three-Hour Concrete Column Enclosure (Traffic Area)

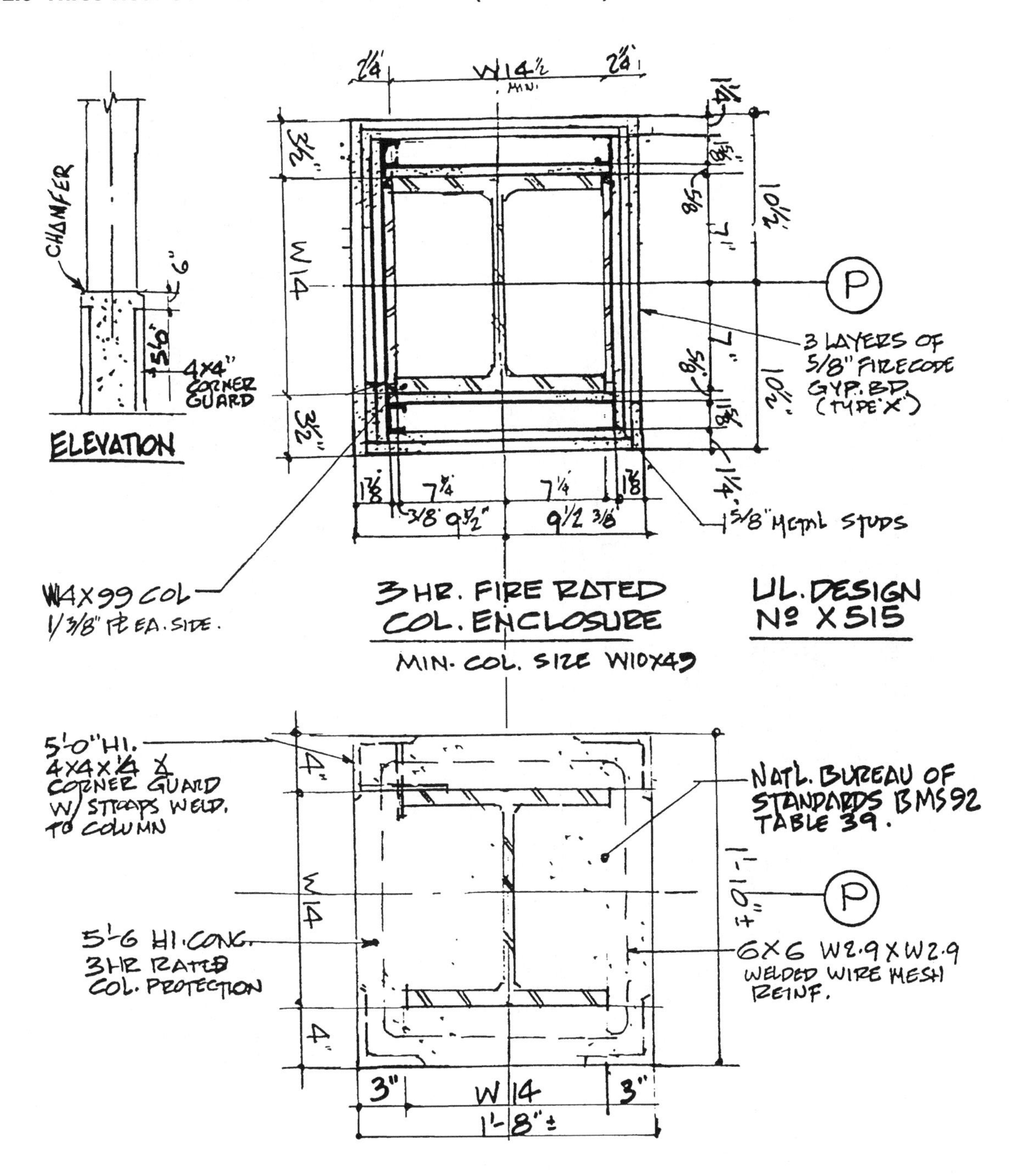

9.13.0 Three-Hour Masonry Column Enclosure

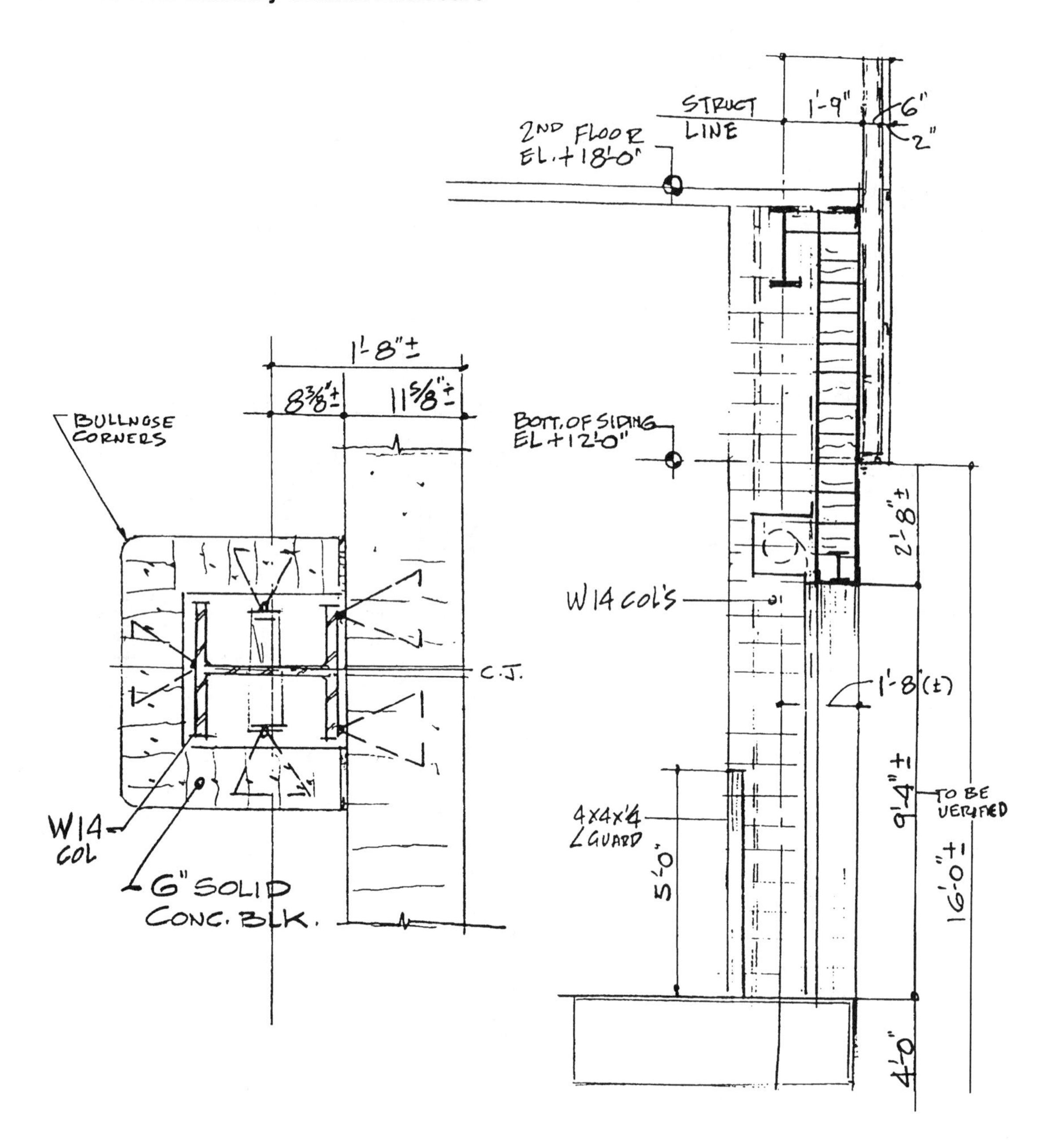

Section

10

Sealants

Contents

10.0.0 Sealants as joint-filling compounds
10.1.0 Proper application of sealants
10.2.0 Typical properties of non-cementitious vs. cementitious repair materials
10.3.0 Advantages/disadvantages of various sealants
10.4.0 Properties of various sealant materials
10.5.0 Temperature vs. sealant performance
10.6.0 Dow Corning silicone-sealant designs /UL ratings/estimating requirements
10.7.0 Typical butt joints and other joint details
10.8.0 Typical exterior-wall air-seal applications
10.9.0 Acceptable/unacceptable air-seal applications
10.10.0 Adhesion test procedures
10.11.0 Proper parapet wall-sealants diagrams
10.12.0 When is it time to repoint? Mortar joint details
10.13.0 Inspection of mortar joints to determine water-resistant integrity
10.14.0 Steps taken to repoint properly
10.15.0 Guidelines for waterproof back-up of wythes in masonry cavity wall
10.16.0 Diagram of a typical composite waterproofing system
10.17.0 Parking garage inspection checklists

The category "sealants" spans a wide range of construction activities and applications—from preventing water and moisture from infiltrating into below-grade structures to maintaining the watertight integrity of the entire superstructure.

This section deals primarily with caulking and sealant compounds: selection and application, and, secondarily, curtain wall and masonry sealants.

10.0.0 Sealants as Joint-Filling Compounds

These materials generally fall into one of three categories:

- *Dynamic joints* Joints that exhibit changes due to movement from expansion, contraction, isolation and loadings.
- *Static joints* Joints that exhibit little or no movement, such as masonry mortar joints. However, no joint in a building is truly static because all materials exhibit some movement from temperature changes and load factors.
- *Butt joints* Joints that have opposing faces that contract and expand and place a sealant in compression, tension, and can also exhibit shear from extreme loading forces or seismic events.

10.1.0 Proper Application of Sealants

The key to proper application of any sealant begins with proper surface preparation, which can vary considerably from one material to another. Most manufacturers go to great lengths to provide detailed surface preparation and application procedures, which are often ignored by the applicator, resulting in either poor performance or outright failure.

The following general guidelines are to be augmented by the manufacturer's instructions for the sealant and surface selected:

- *Concrete and masonry* Concrete can have the most variable surface conditions of any product because of variations in curing conditions, moisture content, finishing techniques, additives, hardeners, curing compounds, and form-release agents. Concrete and masonry surfaces can exhibit weak surface layers because of laitance present in concrete and the potential for spalling in masonry structures. Surfaces contaminated by laitance, hardeners, curing compounds and form-release materials can be sandblasted or wire brushed to remove these contaminants.

  Newly placed concrete or masonry must be allowed to cure before applying sealants. If these surfaces, once cured, become wet from rain, they should be allowed to dry at least 24 hours in good drying weather before sealant or primer application. Because most sealant manufacturers do not recommend applying their products in temperatures below 40 degrees F, frost is a problem. Under these conditions, an application of isopropyl alcohol or methyl ethyl ketone will cause surface moisture to evaporate and a sealant can be quickly applied before frost forms again.
- *Stone* These surfaces generally provide good sealant adhesion. However, some material (such as granite, limestone, and marble) should be primed before a sealant is applied. If the surface area of the stone appears to be flaking or dust, it must be cleaned by either water blasting, sandblasting, or wire brushing before priming and sealant application.
- *Glass and porcelain surfaces* These surfaces are excellent substrates for sealants once their surfaces are cleaned of contaminants and oils. Methyl ethyl ketone or alcohol is an ideal cleaner.
- *Painted and laquered surfaces* Depending on where these surfaces are located and their exposure to the weather, sealants should not be applied to flaking painted or laquered surfaces. Sound painted/laquered surfaces should first be cleaned by wiping with a solvent to remove oil and dust. It is preferable to do a test section to ensure that the solvent does not "lift" the painted surface.
- *Rigid plastic materials* Solvents will clean these surfaces adequately. However, the manufacturer of the fiberglass, acrylic, or other plastic compound should be consulted to determine which solvents will not permanently damage the plastic surface.
- *Flexible plastics and elastomers* These materials are difficult for sealants to adhere to. Test applications of a solvent, such as VM&P naptha, should be applied to determine if it is harmful to the plastic or elastomer.

- *Aluminum with a mill finish* A good degreasing solvent, such as trichloroethane or xylene, will clean these surfaces properly. A rub down with fine steel wool or fine emery cloth might permit better adhesion.
- *Aluminum with an anodized finish* This surface generally provides an excellent surface for sealant application. However, it should be wiped down with methyl ethyl ketone or xylene to remove any surface contaminants.
- *Copper* Copper can oxidize and this patina must be removed by either sanding or rubbing with steel wool. Copper is not compatible with many sealants; the sealant manufacturer or distributor should be contacted for the proper selection.
- *Lead* Though not used extensively as a new material, lead is often encountered in restoration work. It is difficult to obtain adhesion to a lead surface—even after cleaning with xylene or methyl ethyl ketone. Seek the manufacturer's recommendation.
- *Steel* Most steel surfaces to be caulked will have been painted, and procedures for any painted surface will apply. For unpainted steel surfaces, the steel must be free from rust, oil, and other surface contaminants. Abrade the surface by sandblasting or wire brushing down to a sound surface, clean with a solvent, and then apply the caulking.
  - *Stainless steel* This is another difficult surface for adhesion purposes. Primers are often recommended along with solvent cleaning of the surface.
  - *Galvanized steel* New galvanized surfaces present more difficult surfaces for adhesion than weathered galvanized surfaces. Once again, consultation with the sealant manufacturer is recommended.

10.2.0 Typical Properties of Non-Cementitious vs. Cementitious Repair Materials

Property	Epoxy	Polyester	MMA	Cement	Latex-Cement
Compressive strength	High	High	Moderate	Moderate	Moderate
Adhesion:					
Dry surfaces	Excellent	Variable	Very good	Fair-good	Good-VG
Wet surfaces	Excellent *(some)*	Poor	Poor	Good	Very good
Shrinkage	Minimal (<1%)	High (8%)	Moderate	Moderate	Low-Moderate
Thermal coefficient of expansion	High (14×10^{-6}/°F)	----------	Very high (40×10^{-6}/°F)	Moderate (8×10^{-6}/°F)	Moderate (8×10^{-6}/°F)
Modulus of elasticity	Variable (low mod used for masonry)	Low to medium (variable)	Medium	Medium	Low-medium
Permeability	*Permeability controlled by proper aggregate:binder ratios*			Good	Good
Appearance: Color	Yellows in sun	Yellows in sun	Non-yellowing	May fade	Resists fading
wet/dry	*Do not develop same wet/dry appearances as natural stone*			Good w/d	Good w/d
Common uses	Welding cracks Consolidation Rebonding Terra cotta repair	Marble analogs Consolidation Concrete repair	Impregnation Consolidation Civil engineering	Patching Grouting Coating	Patching Coating Rebonding
Safety/handling	Sensitizer Corrosive hardeners	Irritating odor Mod. toxicity	Irritating odor Flammable	Dust (silica) Alkaline (cement)	Dust (silica) Alkaline

Reprinted by permission from the Sealant, Waterproofing, and Restoration Institute, Kansas City, Missouri

10.3.0 Advantages/Disadvantages of Various Sealants

Sealant Type *(Typical Cost)**	Key Attributes	Disadvantages
ORGANICS Butyl, Acrylic, & Solvent Acrylic [6-8 ¢/ft]	paintability	very low movement high shrinkage poor weatherability
POLYSULFIDEs [10-12 ¢/ft]	chemical resistance abrasion resistance paintability below grade applications	modulus changes with temperature compression set potential new formulations unproven old formulations contained PCBs
POLYURETHANES [10-15 ¢/ft]	color flexibility limited life "self cleaning" paintability	reversion with heat, humidity, UV modulus changes with temperatures poor application in cold temperatures
SILICONES Acetoxy (vinegar smell) [9-12¢/ft]	optically clear available field proven history long shelf-life Antifungal formulations	incompatible with reflective glass, concrete, some metals adhesion to fluoropolymer paint abrasion resistance
NEUTRAL [11-20¢/ft]	20+ year lifetime largest range of modules the only sealant for structural adhesionmodulus stability at various temperatures field proven history	abrasion resistance overplastisized formulations and stain adjacent surfaces

**Average Pacific Northwest contractor cost per foot based on a 1/4" x 1/4" joint*

10.4.0 Properties of Various Sealant Materials

Properties of Interest	General Purpose Epoxy	Novolac Epoxy	Polymer Alloys	Polyester	Vinyl Ester	Acrylic	Poly-urethane	Water-based Urethane
Alkali Resistance	Excellent	Excellent	Excellent	Poor to Fair	Good to Very Good	Very Good	Very Good	Very Good
Acid Resistance	Good	Excellent 98% H_2SO_4	Excellent 98% H_2SO_4	Good	Very Good to Excellent	Good	Good	Good to Very Good
Solvent Resistance	Fair to Good	Good to Very Good	Excellent	Fair to Good	Excellent	Poor to Good	Fair to Good	Excellent 200
Physical Properties	Hard, Tough & Rigid	Hard, Tough & Rigid	Hard, Tough & Rigid	Hard, Tough & Rigid	Hard, Tough & Rigid	Hard Scratch-Resistant, Tough	Durable Scratch-Resistant, Tough	Hard, Tough & Rigid
Flexibility	Good	Good	Good	Good	Good	Very Good to Excellent	Excellent	Good
Impact Resistance	Good	Good	Good	Good	Good	Very Good to Excellent	Excellent	Good
Abrasion Resistance	Good	Good	Good	Good	Good	Very Good	Excellent	Good
UV Resistance	Fair	Fair	Fair	Good	Fair	Excellent	Fair to Excellent	Very Good
Preferred Application Temperatures	40°-110°F	50°-110°F	50°-110°F	50°-110°F	50°-110°F	-20°-90°F	40°-110°F	50°-110°F
Moisture Tolerance *(During Application)*	Very Good	Very Good	Very Good	Poor	Poor	Poor	Poor	Good to Very Good
V.O.C's *(Volatile Organic Compounds)*	Very Low to None	None	None	—	—	High (MMA)	Very Low To None	Very Low

Reprinted by permission from the Sealant, Waterproofing, and Restoration Institute, Kansas City, Missouri

10.5.0 Temperature Vs. Sealant Performance

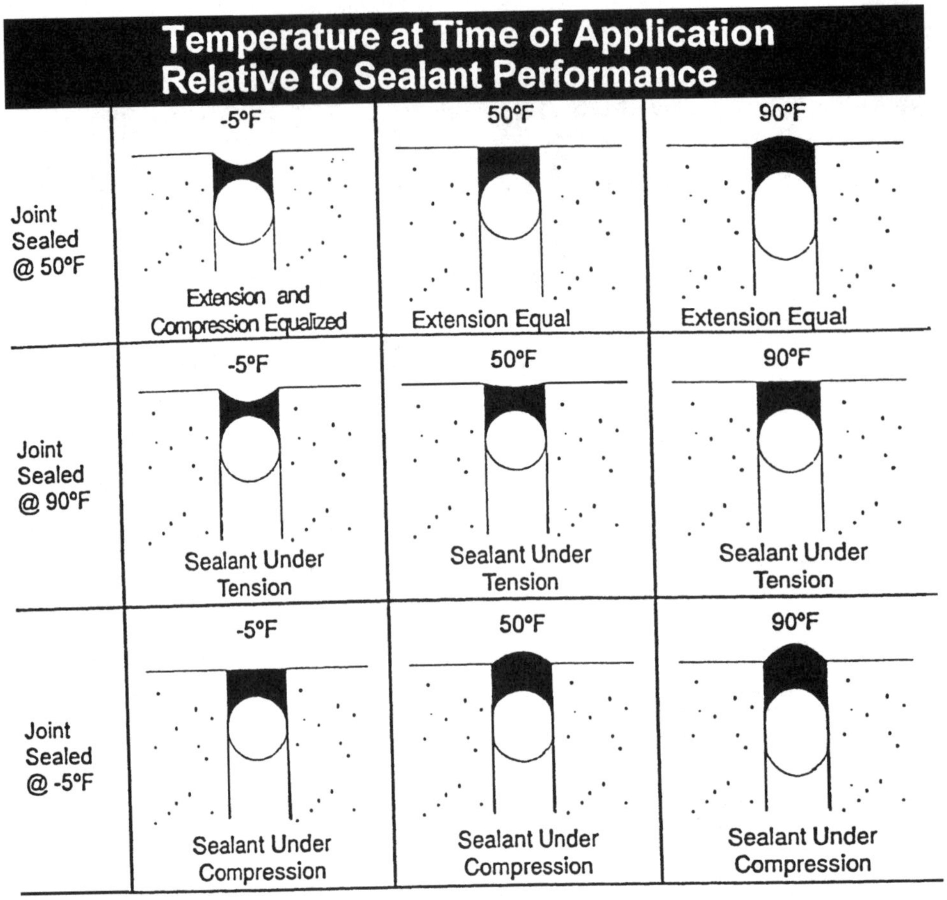

Temperature at Time of Application Relative to Sealant Performance

	-5°F	50°F	90°F
Joint Sealed @ 50°F	Extension and Compression Equalized	Extension Equal	Extension Equal
Joint Sealed @ 90°F	Sealant Under Tension	Sealant Under Tension	Sealant Under Tension
Joint Sealed @ -5°F	Sealant Under Compression	Sealant Under Compression	Sealant Under Compression

	Canada and Northern USA		Southern USA	
	°F	°C	°F	°C
A. Estimated highest building surface temperature	155	68	180	82
B. Estimated lowest building surface temperature	-45	-43	-20	-29
Maximum temperature differential controlling joint movement (A-B)	200	111	200	111

10.6.0 Dow Corning Silicone-Sealant Designs/UL Ratings/Estimating Requirements

FIGURE 1: RECOMMENDED JOINT DESIGN

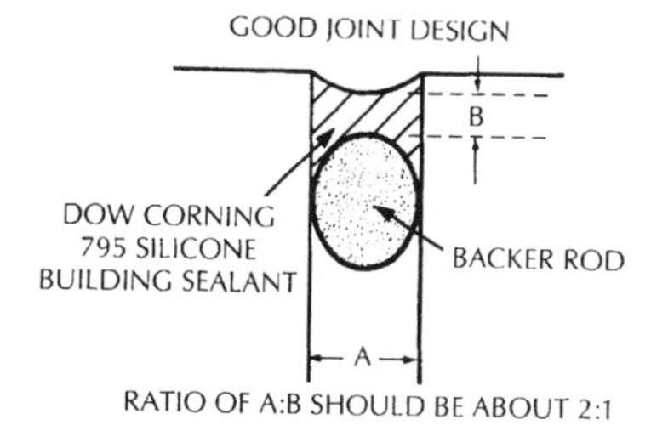

TABLE I: UL FIRE RESISTANCE RATING FOR JOINTS USING DOW CORNING 795 SILICONE BUILDING SEALANT

Maximum Joint Width, inches	Exterior Joint Sealant Thickness, inches	Forming Material	Forming Material Thickness (Item 2), inches	Rating, hours
1	1/2	Mineral Wool	3	2
1	1 1/2[1]	Backer Rod	–	2

[1]This is not a typical joint design. Cure time for such a design will be considerably lengthened, but the sealant will still perform. This is not a recommended design for a joint requiring ±50 percent movement.

FIGURE 2: BOND BREAKER TAPE

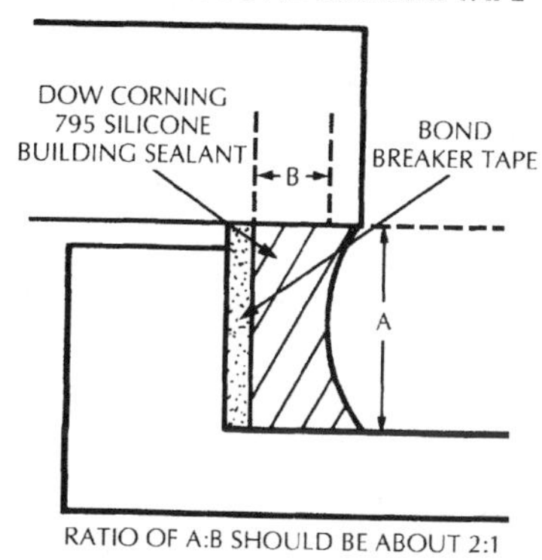

TABLE II: ESTIMATING REQUIREMENTS

Linear Feet per Gallon of DOW CORNING 795 Silicone Building Sealant for Various Joint Sizes

DEPTH, Inches	WIDTH, Inches: 1/4	3/8	1/2	5/8	3/4	1	2	3
1/8	616	411	307	—	—	—	—	—
3/16	411	275	205	164	—	—	—	—
1/4	307	205	154	123	103	—	—	—
3/8	—	137	103	82	68	51	25	17
1/2	—	—	77	62	51	39	19	12

FIGURE 3: EXTERIOR JOINT SEALING CONFIGURATIONS AND FIRE RATINGS

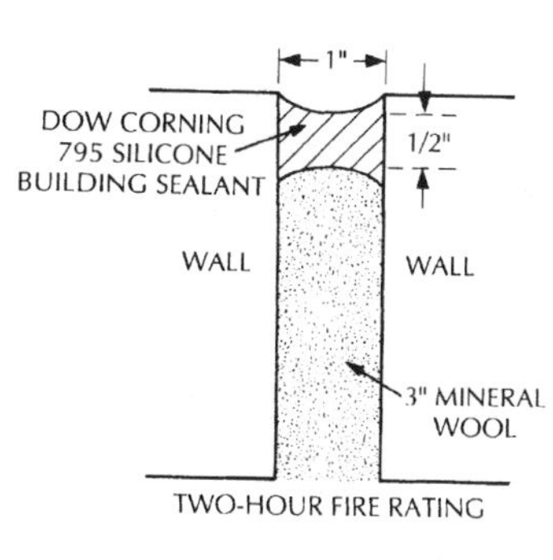

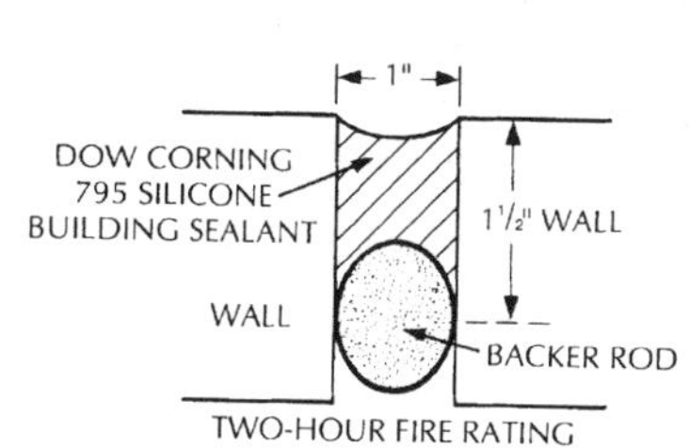

FIGURE 4: EXAMPLES OF TYPICAL GLAZING DETAILS

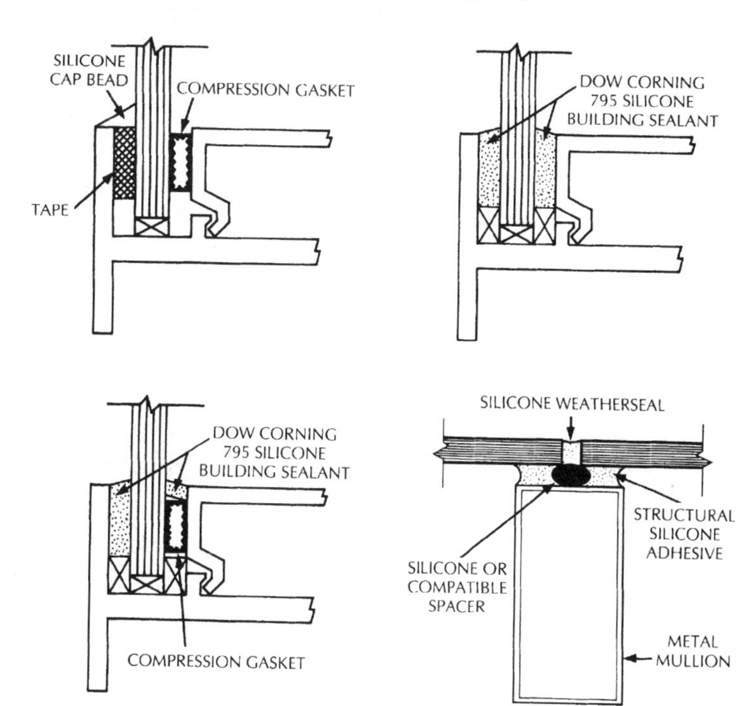

Reprinted by permission from the Sealant, Waterproofing, and Restoration Institute, Kansas City, Missouri

10.7.0 Typical Butt Joints and Other Joint Details

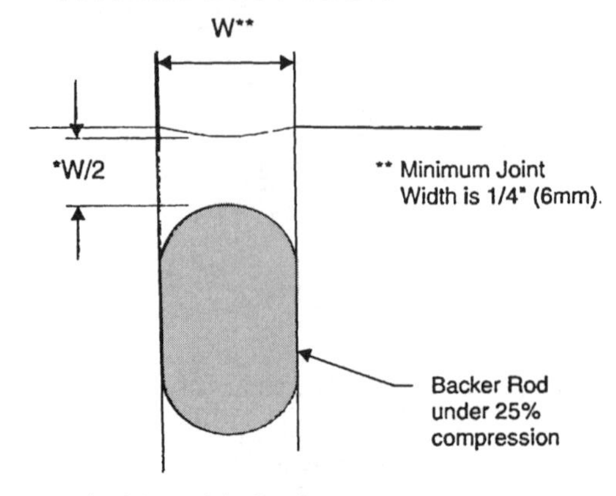

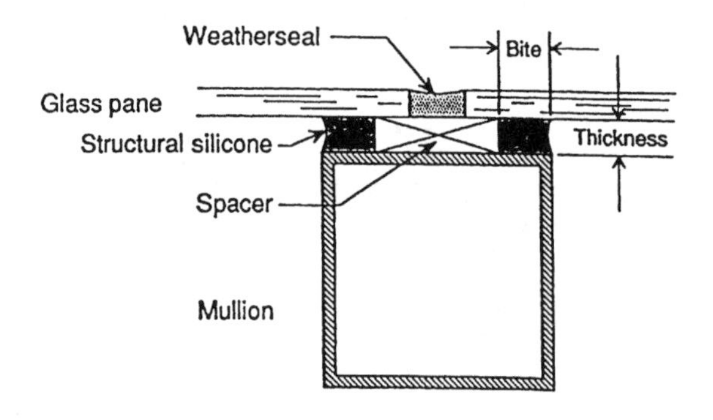

Structural glazing joints sealed with silicone sealant.

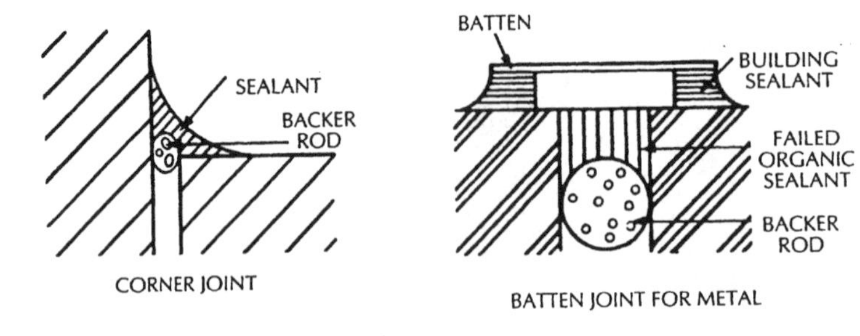

10.8.0 Typical Exterior-Wall Air-Seal Applications

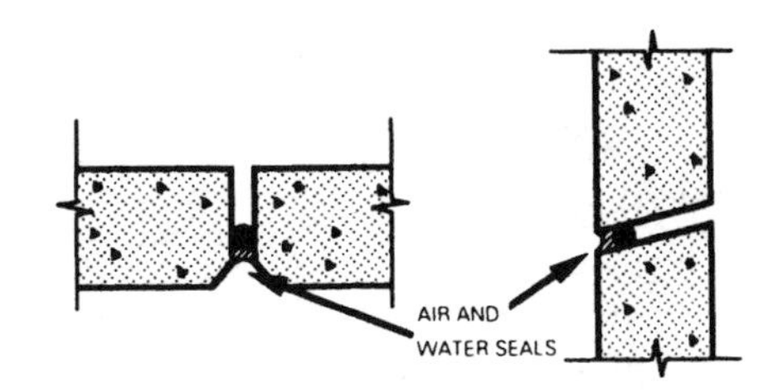

PRECAST WALL PANEL WITH ONE-STAGE JOINTS

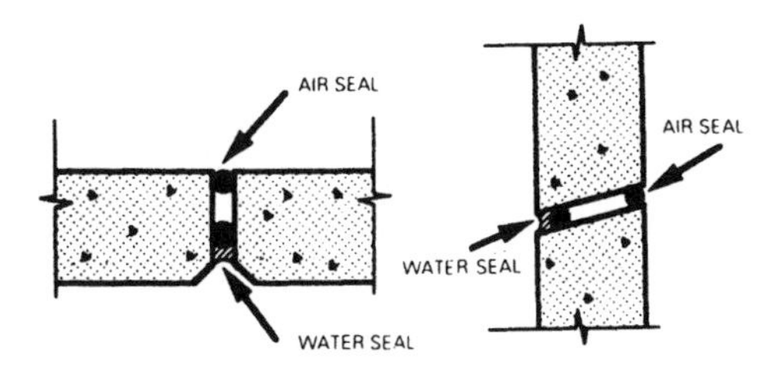

PRECAST WALL PANEL WITH TWO-STAGE JOINTS

AIR SEAL
20 SLOPE
AIR SEAL
VENTED
RAIN SEAL
EXTERIOR
INTERIOR
VENTED
RAIN SEAL

TWO-STAGE PRESSURE-EQUALIZED JOINTS

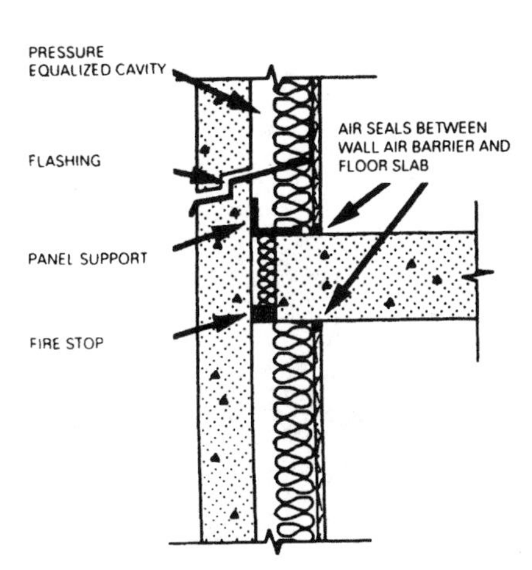

PRECAST CONCRETE PRESSURE-EQUALIZED RAIN SCREEN

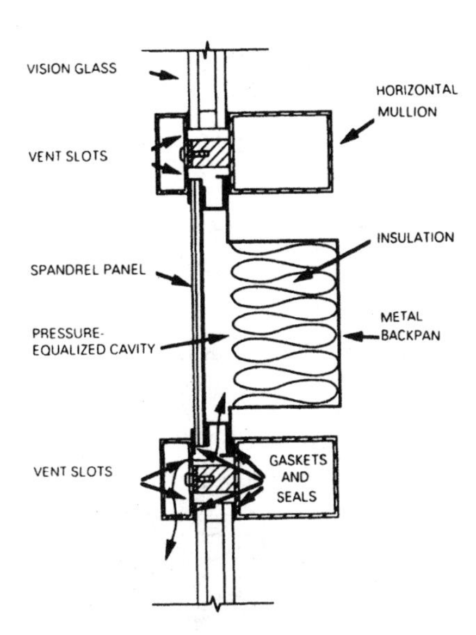

PRESSURE-EQUALIZED CURTAIN WALL MULLIONS

10.9.0 Acceptable/Unacceptable Air-Seal Applications

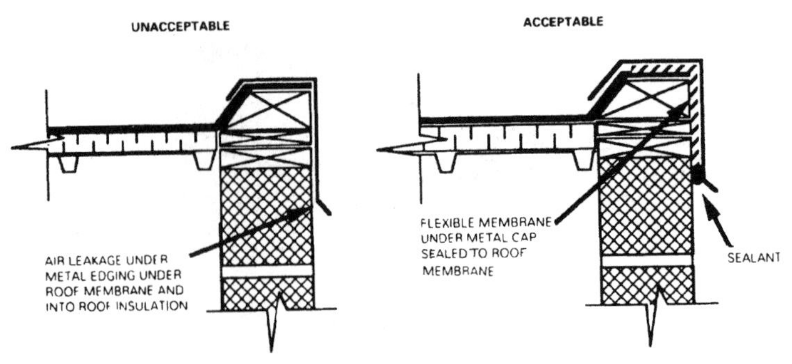

NOTE WALL INSULATION AIR BARRIER AND VAPOR RETARDER NOT SHOWN

AIR LEAKAGE AT ROOF EDGE

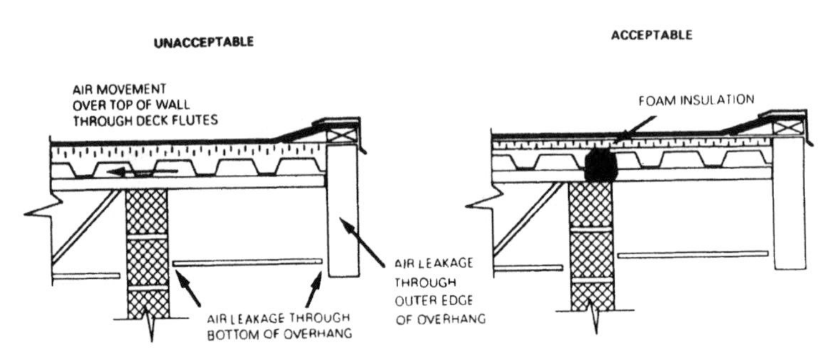

NOTE WALL INSULATION AIR BARRIER AND VAPOR RETARDER NOT SHOWN

AIR LEAKAGE AT ROOF OVERHANG

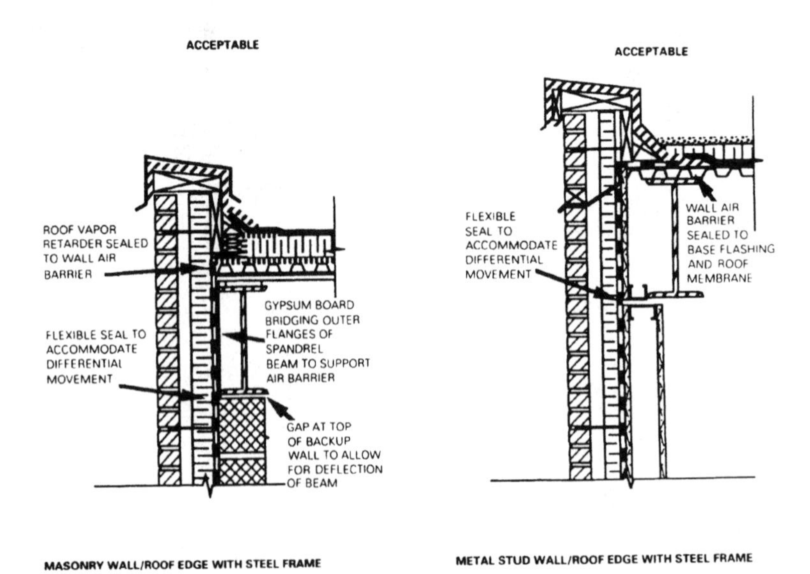

MASONRY WALL/ROOF EDGE WITH STEEL FRAME

METAL STUD WALL/ROOF EDGE WITH STEEL FRAME

Reprinted by permission from the Sealant, Waterproofing, and Restoration Institute, Kansas City, Missouri

10.10.0 Adhesion Test Procedures

Recently, Bill Walter needed a field test for adhesion. He had a substrate with limited surface integrity and wanted to know if his surface preparation would be adequate.

My answer was that he should prepare the surface on a test piece, then follow the procedure outlined below for either low or high modulus sealants (depending on his choice).

Bill thought the test was handy and of probable value to many applicators and contractors, so I am passing it on through the APPLICATOR.

As a check for adhesion, a hand pull test may be run on the job site after the sealant is fully cured. (Usually within 14 to 21 days).

The hand pull test procedure is as follows:

1. Make a knife cut horizontally from one side of the joint to the other.
2. Make two vertical cuts approximately 2 inches long, at the sides of the joint, meeting the horizontal cut at the top of the 2 inch cuts.
3. Place a 1 inch mark on the sealant tab as shown in the picture below.

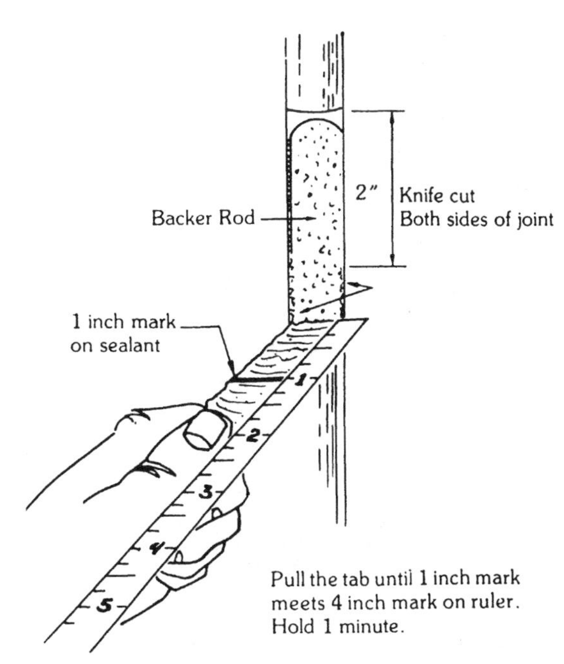

4. Grasp the 2 inch piece of sealant firmly between the finger just above the 1 inch mark and pull at a 90° angle. Hold a ruler along side the extending sealant.
5. If the 1 inch mark on the sealant can be pulled 3 inches to the 4 inch mark on the ruler (300% elogation) and held with no failure of sealant (the sealant is not pulling away from the walls of the joint), the sealant will perform in 50% joint expansion.
6. Sealant may be replaced in test area easily, merely by applying more sealant in the same manner it was originally installed (assuming good adhesion was obtained). Care should be taken to assure that the new sealant is in contact with the original and that the original sealant surfaces are clean, so that good bond between the new and old sealant will be obtained.

NOTE: Adhesion may be adversely affected by:

1. Moisture in or on the substrate during sealant application and cure.
2. Contaminated or weak surfaces.
3. Poor application technique.

NOTE: If the test is done on a flat surface, a test piece like that below is recommended.

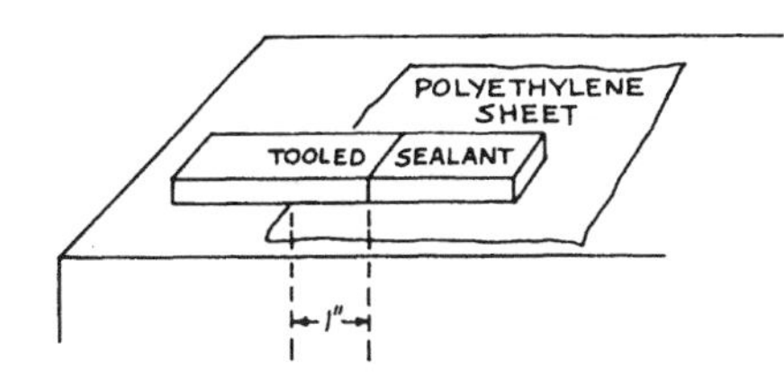

¼ inch deep, 1 inch wide, approximately 4 inches long
Pull at 90° holding at 1 inch mark.
No under cutting is needed since she sealants generally do not adhere well to polyethylene.
After cure, proceed starting at s
No under cutting is needed since the sealants generally do not adhere well to polyethylene.
After cure, proceed starting at step #3 above.

NOTE: It is often desirable to submerge the test piece in water for one day or seven days and repeat the test starting at step #4. Whether one day or seven days is chosen depends on the climate or environment where the sealant is expected to be used.

As a check for adhesion, a hand pull test may be run on the job site after the sealant is fully cured (usually within fourteen to 21 days).

The hand pull test procedure is as follows;

1. Make a knife cut horizontally from one side of the joint to the other.
2. Make two vertical cuts approximately 2 inches long, at the sides of the joint, meeting the horizontal cut at the top of the 2 inch cuts.
3. Grasp the 2 inch piece of sealant firmly between the fingers and pull down at a 90° angle or more, and try to pull the uncut sealant out of the joint.
4. If adhesion is acceptable, the sealant should tear cohesively in itself before releasing adhesively from the substrate.
5. Sealant may be replaced in test area easily merely by applying more sealant in the same manner it was originally installed (assuming good adhesion was obtained). Care should be taken to assure that the new sealant is in contact with both surfaces.

Reprinted by permission from the Sealant, Waterproofing, and Restoration Institute, Kansas City, Missouri

10.11.0 Proper Parapet Wall-Sealants Diagrams

The Best Moisture Escape Routes

1. Ventilate the cavity for walls to breathe.
2. Install weep holes and/or clean existing weep holes that might have become clogged.
3. Correct improperly installed flashing and/or install additional flashing at problems areas.

The Best Barriers to Water Entry

1. Create water infiltration barriers, such as cap flushing.
2. Install adequate expansion and control joints to accommodate expansion due to thermal movement, moisture absorption and freeze-thaw cycles.
3. Replace spalled brick.
4. Repoint deteriorating joints.

A word of caution: When replacing glazed brick, do not use corner brick in any location other than corners. With its two glazed sides, corner brick will fail to provide a proper bond on one side.

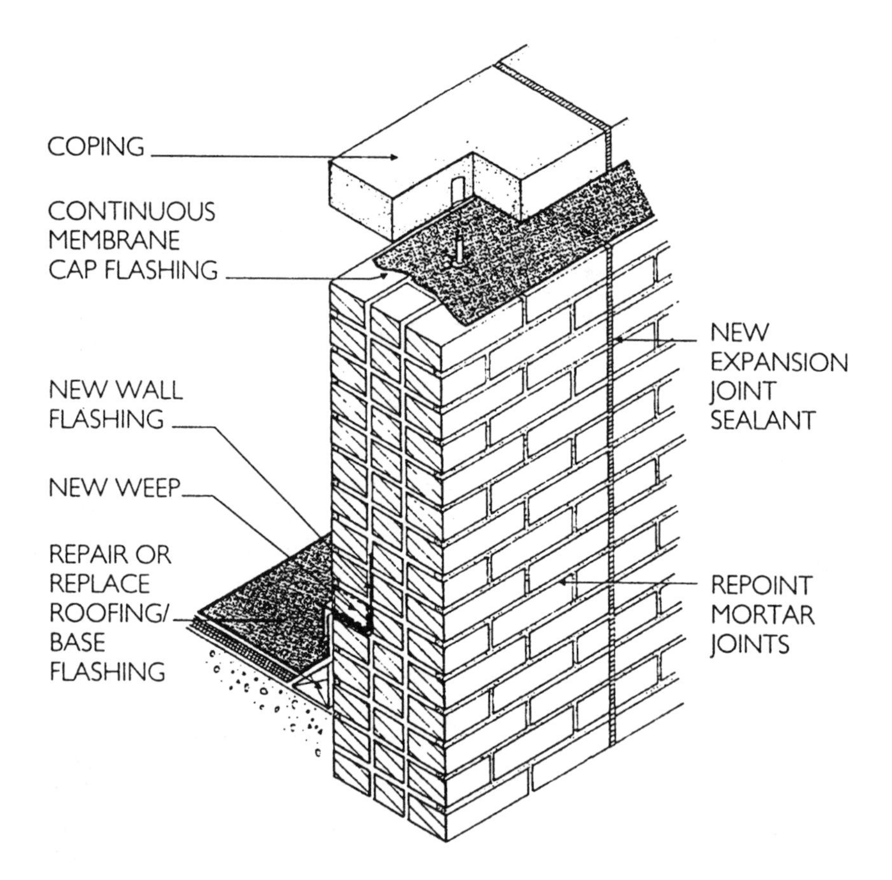

10.12.0 When Is It Tme to Repoint? Mortar Joint Details

You know it's time to repoint when:

- Mortar has eroded to expose the brick behind the glazed face.
- Mortar has crumbled from the joint.
- Hairline cracks have appeared in the mortar.
- The bond between the mortar and the glazed brick is broken.

Strategies for maintaining mortar joints include the following:

1. Remove the old mortar by cutting out to a depth of at least ⅝"; remove more if necessary to eliminate unsound mortar.
2. Clean joints of old mortar, dust, and dirt prior to repointing.
3. Avoiding damaging brick edges when removing old mortar.
4. Use a mix ratio of 1 part Portland cement: 1 to 1½ parts hydrated lime: 6 parts sand for a flexible, but durable mortar.
5. Day and evening temperatures should be above 40° F during repointing; the area of work should be protected from the weather when not being worked on.
6. All excess mortar, smears, and droppings should be cleaned up before the mortar sets.
7. Joint configuration must be designed so that the mortar meets the top edge of the glaze and the joint easily sheds water.

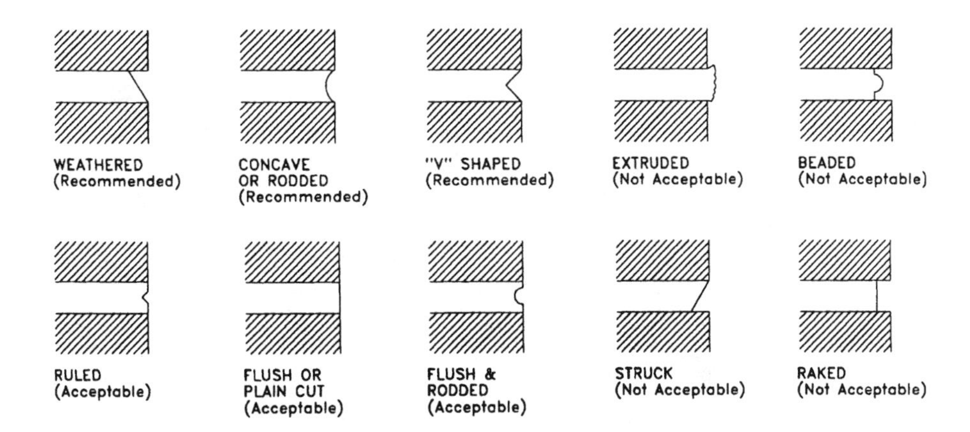

10.13.0 Inspection of Mortar Joints to Determine Water-Resistant Integrity

- Has the mortar eroded to the point where a large portion of the underside of the brick above and below is visible? If so, it is time to repoint.
- Has the mortar begun to crumble from the joint? If so, it is time to repoint.
- Have hairline cracks formed in the mortar? If so, it is time to repoint.
- Is the bond between the mortar and brick broken? If so, it is time to repoint.

10.1.4.0 Steps Taken to Repoint Properly

1. Cut out old mortar to a depth of at least ⅝ inch. Remove more if a sound surface has not been found at that depth.
2. Avoid damaging the edges of the bricks while cleaning out the old mortar joint.
3. Clean out dust and dirt from the old joint.

4. Mix up a batch of mortar with the following proportions:
   1 part Portland cement
   1 to 1½ parts hydrated lime
   6 parts sand
5. Repointing should not take place when both day and night temperatures are below 40 degrees F.
6. Clean off excess mortar, drips, etc. before the mortar sets up.
7. The proper selection mortar-joint configuration will help to prevent a recurrence of premature failure.

Recommended and not recommended joint profiles follow.

10.15.0 Guidelines For Waterproof Back-Up of Wythes in Masonry Cavity Walls

* Don't neglect the need for properly installed flashing and adequately spaced weep holes. Waterproofing on the outer surface of a backup wythe is not intended to work by itself, but must work integrally with other details to prevent water leakage.

* The surface on which the waterproofing is to be applied should be clean and smooth, with all mortar projections cut flush.

* Most waterproofing must be applied within a specific temperature range. For example, some coatings should not be applied below 50 degrees F or above 95 degrees F. Always check manufacturers' recommendations.

* Use adequate safety protection, as needed: a face shield or protective goggles, an approved respirator, gloves, etc. Check product labels for safety information, proper application technique, and cautionary advice.

* Always request and read the manufacturer-supplied Material Safety Data Sheet (MSDS) for all products. It also is a good idea make sure that products comply with all Volatile Organic Compound (VOC) regulations.

* Dispose of all empty containers in accordance with federal, state, and local regulations.

* When spray-applying a waterproof coating, be aware that high winds may make it difficult to get a consistent application and may even blow the spray to neighboring areas where it can damage exposed surfaces and foilage.

* Protect vegetation and painted areas from spills or overspray.

* For sealers or coatings that must be mixed with water, use only clean water free of any contaminants; make sure the mixing container also is clean.

10.16.0 Diagram of a Typical Composite Waterproofing System

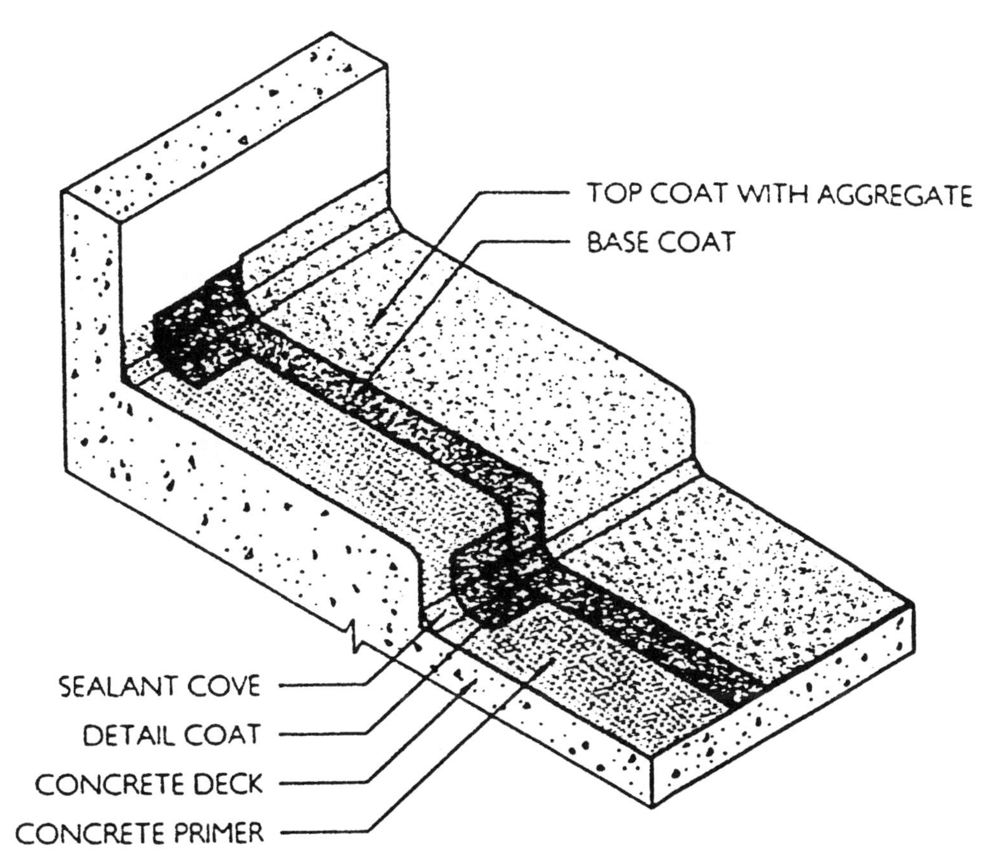

Reprinted by permission of Sealant Waterproofing & Restoration Institute, Kansas City, Mo.

10.17.0 Parking Garage Inspection Checklists

Inspected by		Date
Name of Structure		
Address		
Owner		
Construction Type		
Age of Structure		
Approximate Square Footage		
Number of Levels		Vehicle Capacity
Overhead Clearance		
Usage (Light, Moderate, Heavy)		
Previous Repairs	Type	Location

Instructions

This checklist is designed for use in quick, walk-through surveys of existing parking structures. It is not intended for thorough, in-depth investigations.

Each level of the parking garage should be surveyed separately, with observations for each level recorded on a separate copy of the checklist.

Conditions to be Checked

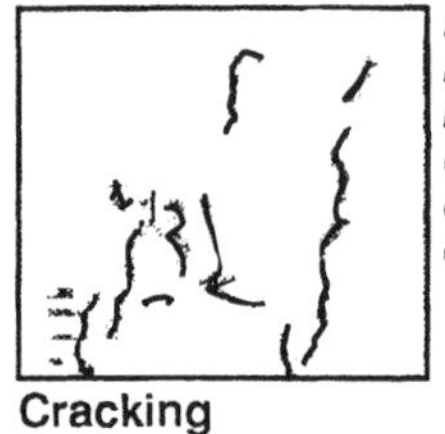

Fissures resulting in moisture and chloride entry into concrete.

Cracking

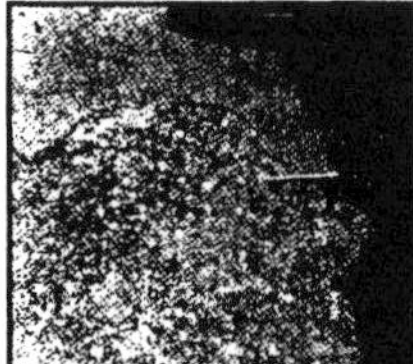

Loss of surface of concrete caused by freeze-thaw cycle and inadequate air entrainment.

Scaling

Potholes resulting from corrosion induced stress.

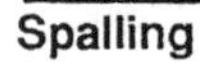

Spalling

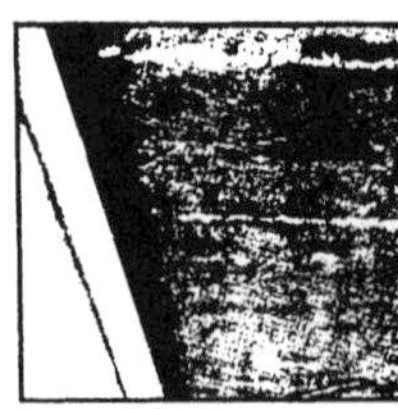

Water migration through concrete slab leading to corrosion of reinforcing steel and spalling of concrete.

Leaching

Tell-tale signs including ponding, staining and damage to floor below.

Leaking

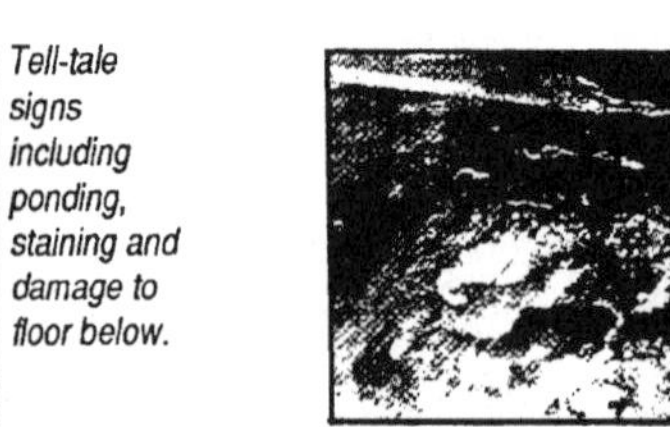

Condition caused by concrete deterioration resulting in corrosion of reinforcing steel.

Exposed Reinforcing Steel

10.17.0 Parking Garage Inspection Checklists (Continued)

Level

	Problems			Locations	Comments
	major	minor	none		
I. Concrete Slab					
A. Floor (Top of Slab)					
Concrete					
Cracking					
Scaling					
Spalling/Delamination					
Potholes					
Leaching					
Water Stains					
Unevenness of deck					
Structural/Reinforcing Steel					
Exposed Rebars					
Corrosion					
Slab Protection					
Membrane					
Sealer					
B. Ceiling (Underside of Slab)					
Concrete					
Cracking					
Scaling					
Spalling/Delamination					
Leaching					
Water Stains					
Structural/Reinforcing Steel					
Exposed Rebars					
Corrosion					
II. Expansion Joints/Control Joints					
A. Freeze/Thaw Damage					
B. Damage from Traffic or Snow Plows					
C. Joint Failure					
D. Bearing Pads					
III. Drainage					
A. Floor Drains					
B. Ponding					
IV. Beams and Girders					
A. Concrete					
Cracking					
Horizontal					
Vertical					
Diagonal					
Scaling					
Spalling/Delamination					
Leaching					
Water Stains					

10.16.0 Parking Garage Inspection Checklists (Continued)

	Problems major	minor	none	Locations	Comments
B. Structural/Reinforcing Steel					
Exposed Rebars					
Corrosion					
V. Support Columns					
A. Concrete					
Cracking					
Scaling					
Spalling/Delamination					
Leaching					
Water Stains					
B. Structural/Reinforcing Steel					
Exposed Rebars					
Corrosion					
C. Out-of-Plumb					
VI. Walls					
A. Concrete					
Cracking					
Horizontal					
Vertical					
Diagonal					
Scaling					
Spalling/Delamination					
Leaching					
Water Stains					
Sealants					
B. Structural/Reinforcing Steel					
Exposed Rebars					
Corrosion					
C. Out-of-Plumb					
VII. Spandrels and Guard Rails					
A. Concrete					
Cracking					
Scaling					
Spalling/Delamination					
Leaching					
Water Stains					
Sealants					
B. Structural/Reinforcing Steel					
Exposed Rebars					
Corrosion					
C. Out-of-Plumb					

Section

11

Acoustics/Sound Control

Contents

11.0.0 What is sound?
11.1.0 Sound and the office environment
11.2.0 Sound rating systems
11.2.1 STC ratings
11.2.2 Common STC ratings
11.2.3 Decibel levels of common noises
11.3.0 Sound control (general factors that affect acoustical control)
11.4.0 Dos and don'ts for drywall partitions
11.5.0 Typical STC ratings for various types of concrete and masonry walls/floors
11.5.1 Dos and Don'ts (illustrated)
11.6.0 Estimated wood floor sound performance
11.7.0 The challenge of tv/stereo
11.8.0 Controlling octave band transmission with sound-attenuation blankets
11.9.0 STC ratings for various partition types
11.10.0 Suggested STC ratings and construction
11.11.0 STC ratings or 2" to 6" concrete slabs and various STC-rated ceiling assemblies
11.12.0 The effect of acoustical doors on STC ratings
11.13.0 Noise-muffling qualities of various types of plumbing pipe materials
11.14.0 Plumbing installation acoustical considerations
11.15.0 Duct systems and acoustical considerations
11.16.0 Composite wall/electrical box installations
11.17.0 Electrical transformers and increased decibel levels

Acoustics is the science of sound and vibration. The control of sound and vibration transmission within a structure involves architectural design and structural, mechanical, and electrical engineering considerations. The end result of a building where acoustical and vibration control is taken into account during design and where these considerations are carried out by the contractor results in the creation of an environment in which people can live and work more comfortably and productively.

11.0.0 What Is Sound?

Sound is a vibration that occurs at various frequencies in an elastic medium. It is generated at a source and it travels through either a gaseous, liquid, or solid environment. Sound-pressure levels are represented in *decibels*—a ratio of intensity of sound, as measured to an intensity equivalent to the threshold of hearing. Changes in decibel levels do not follow arithmetic progressions (e.g., a change in 10-db pressure will result in the perception of hearing sound twice as loud). However, a change of 3 db, up or down, will be barely perceptible. Resistance to sound transmission varies with different frequencies. The span of human hearing ranges from 15 Hertz (Hz) to 20,000 Hz. Sound transmission coefficient factors (STC) are tested at frequencies in the 125- to 4000-Hz range.

11.1.0 Sound and the Office Environment

The American Society of Interior Designers (ASID) hired the Yankelovich Partners in 1996 to determine if noise-level reduction was of major concern to office workers. Seventy percent of the respondents indicated that their productivity would increase if they worked in a less-noisy environment. Changes in the work place have resulted in a noisier office environment today, brought about by:

- Higher work-station densities.
- Increased use of speaker phones.
- Increased use of video conferencing and the resultant higher levels of noise concentrated in a central area.
- Team conferencing and more frequent crosstalk occurring in an open office environment among divider panels not suited to absorb noise effectively.
- The proliferation of computer screens throughout the workplace and the tendency to increase screen size, thereby creates even larger hard-surface areas.

11.2.0 Sound Rating Systems

Various rating systems have been devised to qualify acoustical design. Although many such systems exist, five basic systems are most often encountered by the contractor:

- *STC (Sound Transmission Coefficient)* It evaluates the effectiveness of construction components in isolation speech sound sources.
- *MTC (Music/Mechanical Transmission Class)* It is used to measure low-frequency sound. The higher the number, the better the acoustic quality of the wall between the source and adjacent areas.
- *dBa (decibel level)* The loudness level that is most often used to weigh human response to sound.
- *RC* It evaluates the constant background noise in a space from a source, such as an air-handling unit.
- IIC (Impact Insulation Class) Impact sound transmission is produced when a structural element is set into vibration by direct impact (for example, when someone walks on a concrete floor above an occupied area). The higher the IIC, the better the impact noise control.

Other acoustical terms are also important:

- *Frequency band* A division of audible sound relating to convenient sections or octaves.

- *Noise-reduction coefficient* An arithmetic average, to the nearest 0.05, of four sound-absorption coefficients. The ratio of the sound-absorbing relationship of a material at four specific frequencies, compared to the effectiveness of a perfectly sound-absorbing material at the same frequency.

11.2.1 STC Ratings

It is important to remember that STC ratings apply only to those sounds that have the same frequency spectrum or sound profile as those produced by the human voice. One way to remember this is to think of STC as "speech transmission class." STC ratings are applicable when audible sound remains within the range of 125 Hz; machinery, HVAC equipment, and high-fidelity recordings occupy the frequency from 20 Hz to 20,000 Hz and must be dealt within a different manner than STC ratings. The higher the STC, the greater the sound barrier required.

11.2.2 Common STC Ratings

- *STC-25* Normal speech can be heard clearly through a barrier.
- *STC 30* Loud speech can be heard and clearly understood. However, normal speech can be heard, but not easily understood.
- *STC 35* Loud speech can be heard, but is difficult to understand.
- *STC 42* Loud speech can be heard, but only faintly.
- *STC 45* Normal speech cannot be heard
- *STC 46 to 50* Loud speech cannot be heard; other loud sounds can barely be heard.

Sound from the source drops off over the distance traveled to reach a partition. As sound travels through a room, sound levels are affected by the surfaces that the sound contacts. Some common acoustic coefficients are (with 1.0 being the highest, absorbing more sound):

Acoustic tile	0.8
Audience of people	0.8
Carpet and pad	0.6
Cloth upholstered seats	0.6
Fabric	0.3
Glass	0.09
Gypsum drywall	0.05
Concrete	0.02
Tile	0.01

11.2.3 Decibel Levels of Common Noises

Rustling of leaves	10 dB
Empty room	20 dB
Inside bedroom, quiet conversation	30 dB
Private office	40 dB
General office area	50 dB
Face-to-face conversation	60 dB
Bathroom/television	70 dB
Inside speeding automobile	80 dB
Hi-fi stereo	90 dB
Noisy party/symphony orchestra	100 dB
Elevated train	120 dB
Jet aircraft	140 dB

11.3.0 Sound Control (General Factors That Affect Acoustical Control)

Sound is divided into two basic types, according to origin: airborne (conversation, music, and street noise) and structure borne (footsteps on a hard surface, telephone ringing, and vibration from machinery rigidly attached to the structure).

The following methods, used individually, or in conjunction with each other, are used to control both airborne and structure-borne sound.

- *Mass* Thicker floor slabs and/or demising partitions, and inertia pads used in conjunction with the vibration isolation of mechanical equipment.
- *Decoupling* Vibration isolators for mechanical equipment, resilient channels attached to either wood or metal studs, or separated rows of studs, foam-backed carpeting, or resilient flooring.
- *Absorption* Using such materials as sound-soak panels, fiberglass batts, or sound-attenuation blankets.
- *Sealants* Use of flexible acoustical sealant to close off open areas, where ducts, electrical and mechanical conduits, and wiring devices have penetrated floors, ceilings, and partitions.

11.4.0 Dos and Don'ts For Drywall Partitions

United States Gypsum Company, in various articles in their *Form & Function* magazine, set forth the following helpful hints:

- *Perimeter seals* Don't use standard weather caulking, which has a tendency to harden and lose the resiliency required for proper sealing. Don't use drywall tape and joint compound that could crack as various building structural components deflect under load. Don't place caulking under the runner track, but place it to fill the perimeter gap between the gypsum board faces and the surrounding floor, wall, and ceiling elements. This is accomplished by placing a heavy bead of caulking adjacent to the runner prior to installing the gypsum board.
- *Penetrations* Do offset electrical/telecommunication penetrations through a demising wall by at least one stud cavity. Do seal the back and sides of any such outlet boxes with acoustical sealant. Apply this acoustical sealant around all ductwork penetrating demising walls
- *Metal-resilient components* Resilient channel installed where screws are of sufficient length to penetrate the resilient channel, but not penetrate the surface beyond will decouple and isolate the wall or ceiling components. Don't use screws any longer than those recommended by the manufacturer of the resilient channel. Do allow the channel to float upon installation and maintain a minimum ¼-inch clearance between it and the adjacent assembly.

11.5.0 Typical STC Ratings For Various Types of Concrete and Masonry Walls/Floors

Concrete Masonry Units, Brick, and Concrete Walls	
4-inch (51 mm) CMU, brick, or concrete wall	37–42
6-inch (76 mm) CMU, brick, or concrete wall	42–46
8-inch (102 cm) CMU, brick, or concrete wall	47–51
12-inch (153 mm) CMU, brick, or concrete wall	52–56

Concrete floors	
4-inch (51mm slabs)	41
6-inch (76 mm) slabs	46
8-inch (102 mm) slabs	51

If a resilient suspended ceiling is attached to the underside of a concrete slab, the STC rating will increase by approximately 12. If sleepers are attached to the upper surface of a concrete slab, the STC rating will improve (approximately) by 7.

11.5.1 Do's and Don'ts (Illustrated)

The following dos and don'ts are illustrative of several methods to prevent the transmission of sound from one partitioned area to the next.

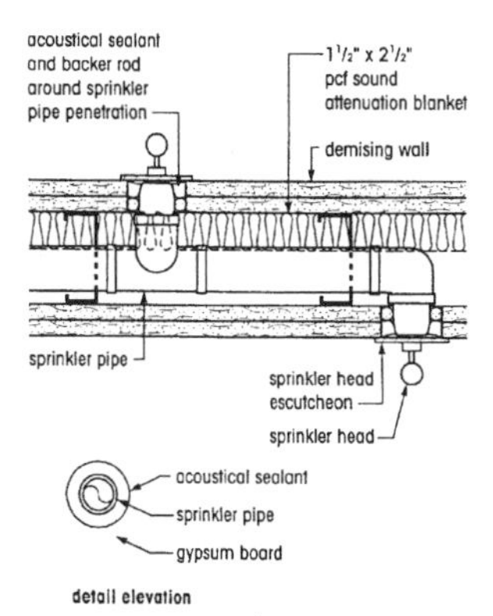

Sprinklers also need acoustical treatment to prevent sound leaks. Offset sprinkler heads on opposite sides of dimising walls by at least one stud cavity.

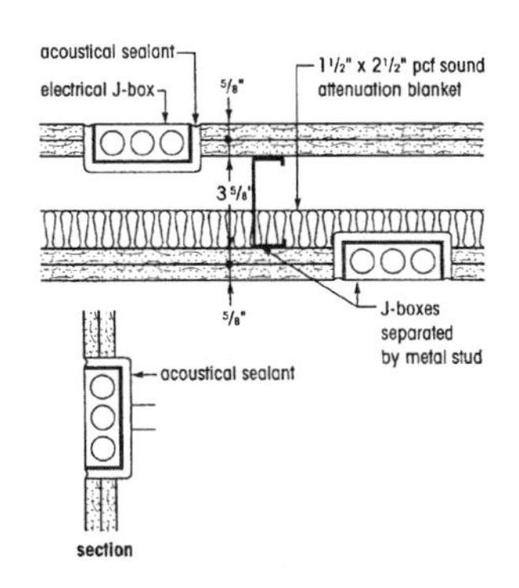

Penetrations for electrical boxes must be acoustically sealed to prevent sound transmission through the wall.

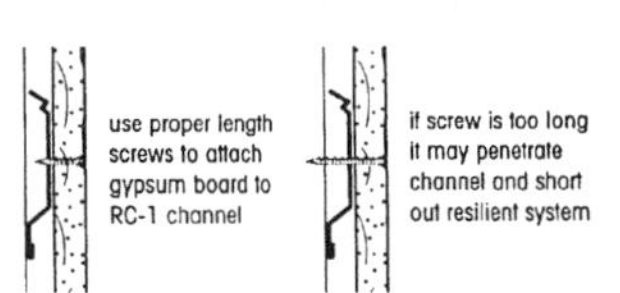

The use of the wrong length of screw can "short out" acoustical separation provided by RC-1 Resilient Channels.

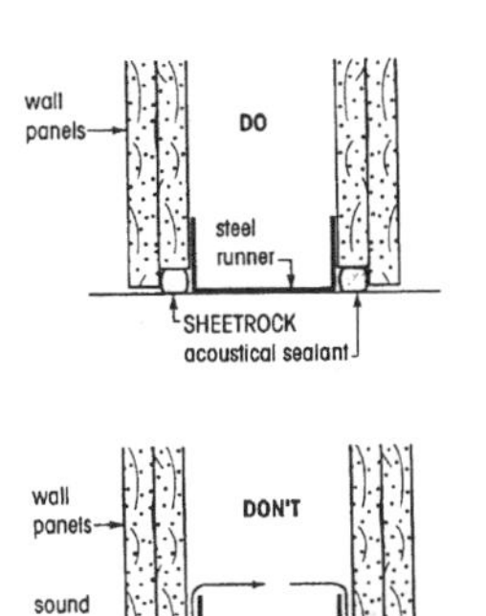

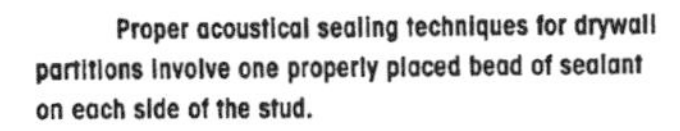

Proper acoustical sealing techniques for drywall partitions involve one properly placed bead of sealant on each side of the stud.

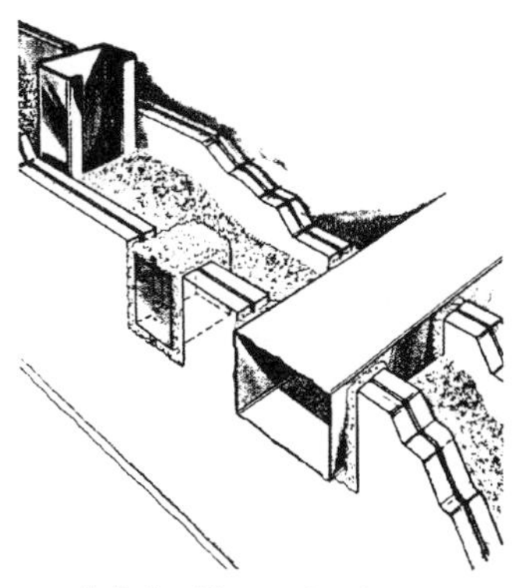

Application of SHEETROCK Acoustical Sealant around ducts effectively seals the wall to reduce sound transmission.

11.6.0 Estimated Wood Floor Sound Performance

Sound transmission and impact-insulation characteristics of a wood-floor assembly can be calculated by adding various components to the basic floor assembly. For example, to the basic wood-floor assembly with an STC frequency of 36, add resilient channel (STC 10) plus 1/2" sound-deadening board (STC 1) for a total assembly rating of STC 47.

Description	STC frequency IIC	Low frequency
Basic wood floor (wood joist, ¾" decking, ⅝" gypsum board attached directly to ceiling	36	33
Add cushioned vinyl/linoleum	0	2
Add non-cushioned vinyl/linoleum	0	0
Add ½" parquet flooring	0	1
Add ¾" Gypcrete	7–8	1
Add 1½" lightweight concrete	7–8	1
Add ½" sound-deadening board	1	5

Description	STC frequency IIC	Low frequency
Add R-19 batt insulation	2	0
Add R-11 batt insulation	1	0
Add 3" mineral wood insulation	1	0
Add resilient channel	10	8
Add resilient channel with insulation	13	15
Add an extra layer of ⅝" gypsum board	0–2	2–4
Carpet and padding	0	20–25

Source: Southern Pine Council, Kenner, Louisiana

11.7.0 The Challenge of TV/Stereo

Equipment Frequency Spectrums

The sound spectrums produced by five types of sound equipment that can be used in hotel guest rooms are compared in the graphs in Fig. 1. Music is the Source, and it is reproduced at 75 dBA. Fig. 1A shows the sound-pressure level in the octave centered at 250 Hz (middle "C" is 256 Hz). Fig. 1b shows the level in the 125-Hz octave and Fig. 1c, the 63-Hz octave. The top source, a typical hotel portable mono (monophonic, monaural) TV, is used as the basic reference source because the industry has so much experience with the success or failure of their isolation systems with this equipment.

It can be seen in Fig. 1a that all the equipment easily reproduces the energy in the 250-Hz octave band. The differences begin in the 125-Hz octave (Fig. 1b). A top of the line, 1988 27-in. portable stereo TV performs about the same as a standard portable mono TV in the 125-Hz octave. The console TV and full-range sound system (bass controls set on flat) are 4 or 5 dB louder in this frequency range. A full-range system with controls set to boost bass will be at least 10 dB louder than the portable mono set.

The most significant differences in performance occur in the 63-Hz octave band. The sound produced in the 63-Hz octave band by a typical portable mono TV generally is insignificant. The portable stereo TV is 10 dB louder and the full-range system (bass boost) can easily be 35 dB louder than the mono portable! The amount of sound isolation required at 125 Hz and lower increases as the equipment capabilities to accurately reproduce the recorded music is improved. High-quality stereo equipment, including the portable stereo TV, also produce significantly more sound energy in the 2000-Hz octave band. This fidelity improvement could cause some speech-intrusion problems where they might not have previously existed because the portable mono TV produces little sound at 2000 Hz and above.

a. 75 dBA Music in 250 Hz Octave Band.

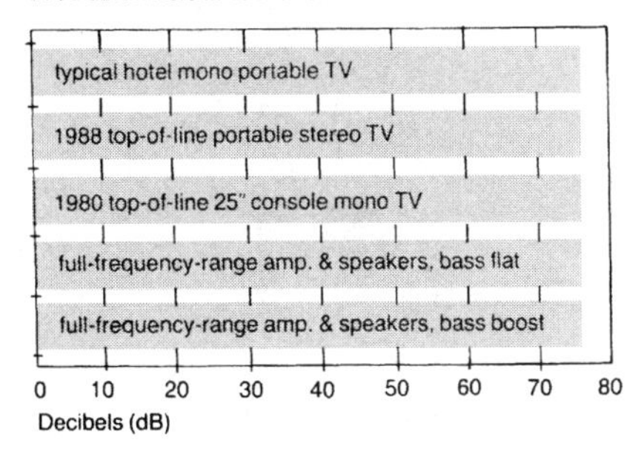

b. 75 dBA Music in 125 Hz Octave Band.

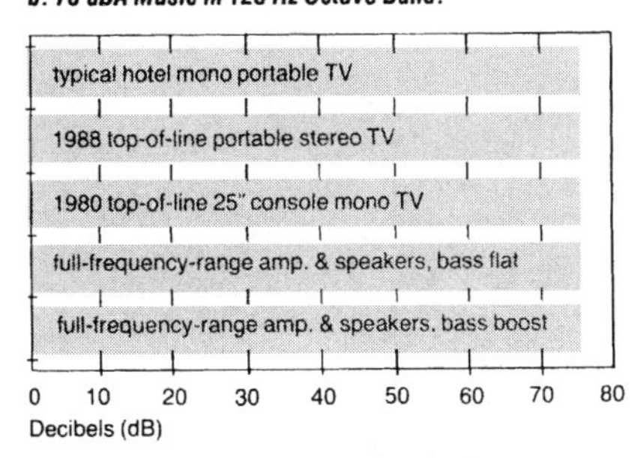

c. 75 dBA Music in 63 Hz Octave Band.

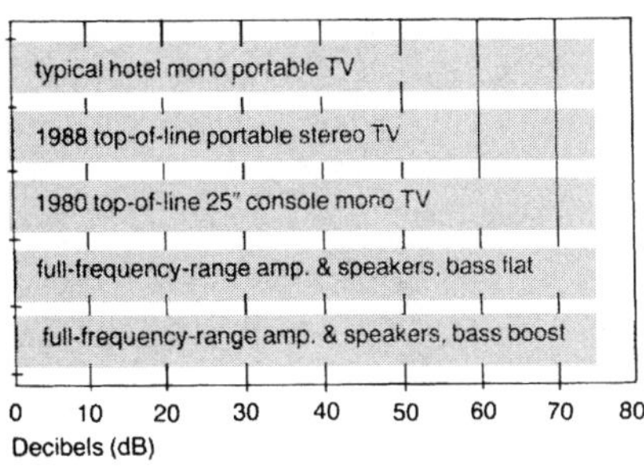

Soft background music	60 dBA
Normal speech effort (3 ft. from talker)	62 dBA
Loud speech	77 dBA
Fairly loud TV (typical playback level)	75 dBA
Minimum for serious listening to orchestra music, below minimum for rock listeners	80 dBA
Loud orchestra music, moderately loud rock music	90 dBA
Extremely loud orchestra music; loud rock music	95 dBA
Controlled hard rock concert (not unusual in teenager's bedroom)	100 dBA
Uncontrolled hard rock concert	115 dBA

Reprinted by permission of *Form & Function* Magazine, published by USG Corporation

RC (Room Criteria) curves

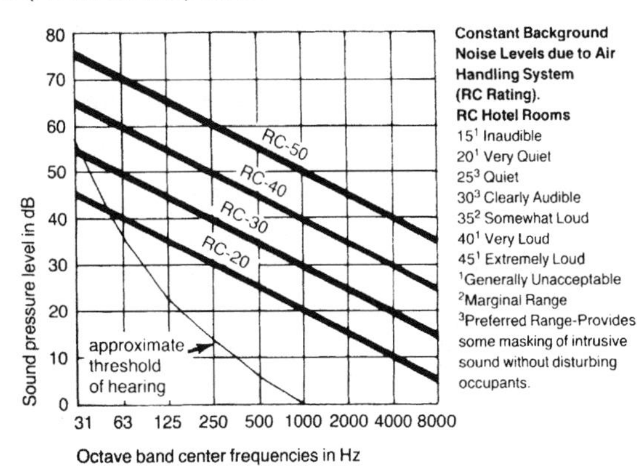

11.7.0 The Challenge of TV/Stereo (Continued)

Conclusions

The quality of TV sound has improved significantly during the last few years with the playback equipment, rather than the broadcast or recorded signal, the factor usually limiting the frequency range reproduced. The newer portable stereo TVs extend the frequency range about an octave lower and an octave higher than the typical portable mono TV of the past. The frequency range of stereo TV (broadcast or VCR), albums, cassette tapes, and CDs are similar when played back through a high wattage, full-frequency-range stereo audio system. There might be issues of the quality of sound, but the quantity can be very similar.

It should be expected that stereo TVs will require partition systems with MTC ratings of 4 to 5 points higher than the partition systems used with the older mono systems to achieve about the same degree of acoustical privacy. The table shows that reasonable results can be achieved with STC-50/MTC-45 isolation with the portable mono TV. An STC-54/MTC-50 is required for similar privacy from a stereo TV. Special, high-performance designs are needed when full-frequency-range systems are installed in luxury hotels.

Sound Isolating Partition	Laboratory Sound Rating (assumed as field achieved)		Typical Mono Hotel TV or Radio: 75 dBA Loudness*		Portable High-Quality Stereo TV: 75 dBA Loudness*		Full Range Stereo with Bass Control Flat: 85 dBA Loudness*	
	STC	MTC	Speech	Music	Speech	Music	Speech	Music
A 2-1/2-in. steel studs, single-layer 5/8" SHEETROCK FIRECODE "C" Gypsum Panels, 1-1/2" THERMAFIBER Sound Attenuation Fire Blankets (SAFB) in cavity	44	40	2	2	1	1	0	0
B Same as "A" but with double-layer of panels on one side	51	45	3	3	3–	2	1	0
C Same as "A" but with double-layer of panels on both sides	54	50	4	4	4	3	2+	2–
D 3-5/8-in. steel studs, single 5/8" layer of panels, 3" SAFB in cavity	48	44	3–	2	2	2	1–	0
E Same as "D" but with double-layer of panels on one side	53	51	3+	3	3	3–	2+	2–
F Same as "D" but with double-layer of panels on both sides	57	54	4+	4	4+	4	3	2
G 3-1/2-in, 20-ga. steel studs, RC-1 Resilient Channels, single and double-layer 1/2-in. SHEETROCK FIRECODE "C" Gypsum Panels, 3" SAFB in cavity	55	49	4	4	4	3	3–	2–
H Same as "G" but with double-layer of panels on both sides	60	54	5	5	5	4	3+	2+
I Same as "G" but with double and triple layers of panels	61	56	5	5	5	5	4	3
J USG Double Wall System, 3" SAFB in cavity	60	57	5	5	5	5	4	3

A B C D E F G H I J

***Key to Rating System**

0 Intrusive music or speech significantly above background noise
1 Over 50% sentence intelligibility, music clearly audible
2 About 50% sentence intelligibility, music audible
3 About 0 to 10% sentence intelligibility, music barely audible
4 Speech or music sound may be just perceptible with careful listening
5 Speech and music generally inaudible with careful listening

United States Gypsum Company
101 South Wacker Drive
Chicago, Illinois 60606-4385
A Subsidiary of USG Corporation

Reprinted by permission of *Form & Function* magazine, published by USG Corporation

11.8.0 Controlling Octave Band Transmission with Sound-Attenuation Blankets

Partition design (insulation density only variable in each test)	Octave band center frequencies in Hz* 125	250	500	1000	2000	4000	STC Improvement
1. SAFB - 5-in. GF - 6-in. GB - 5/8-in.	dB 1.0	dB 1.8	dB 2.9	dB 1.0	dB 6.2	dB 4.6	4
2. SAFB - 3-in. GF - 3-1/2-in. GB - 5/8-in.	−0.6	1.3	1.3	0.4	3.8	2.3	2
3. SAFB - 3-in. GF - 3-1/2-in. GB - 5/8-in.	−0.6	0.4	1.8	0.6	3.3	2.9	2
4. SAFB - 3-in. GF - 3-in. GB - 1/2-in.	0.5	2.6	1.0	1.3	3.1	2.6	2
5. SAFB - 3-in. GF - 3-1/2-in. GB - 5/8-in.	0.1	0.2	1.2	0	1.6	2.6	2
6. SAFB - 1-1/2-in. GF - 1-1/2-in. GB - 1/2-in.	−0.5	2.0	1.8	2.4	3.0	2.0	0
7. SAFB - 3-in. GF - 3-1/2-in. GB - 5/8-in.	2.5	2.7	2.4	4.4	5.2	3.0	3

*Octave band data is derived from 1/3-octave band data reported to nearest decibel. Conversion from 1/3-octaves to octaves is rounded to nearest 0.1 decibel.
Test Reference Numbers: 1. RAL-TL84-139/TL83-230 2. RAL-TL84-147/TL84-144 3. RAL-TL84-148/TL84-145
4. USG 71508/71405 5. USG 830507/830509 6. USG 71413/71404 7. USG 830436/830501

Reprinted by permission of *Form & Function* magazine, published by USG Corporation

11.9.0 STC Ratings for Various Partition Types

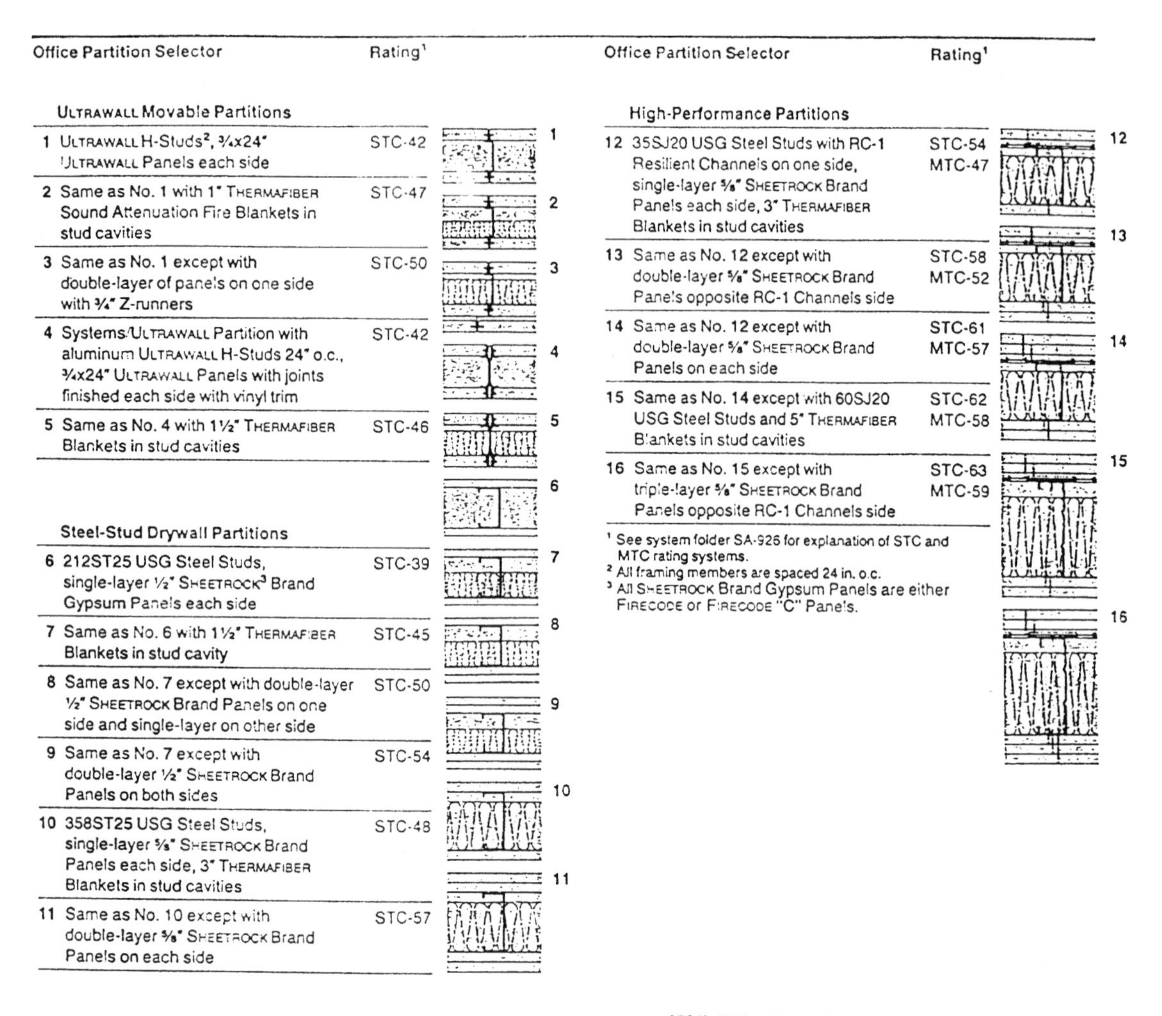

Office Partition Selector	Rating[1]
ULTRAWALL Movable Partitions	
1 ULTRAWALL H-Studs[2], 3/4x24" ULTRAWALL Panels each side	STC-42
2 Same as No. 1 with 1" THERMAFIBER Sound Attenuation Fire Blankets in stud cavities	STC-47
3 Same as No. 1 except with double-layer of panels on one side with 3/4" Z-runners	STC-50
4 Systems/ULTRAWALL Partition with aluminum ULTRAWALL H-Studs 24" o.c., 3/4x24" ULTRAWALL Panels with joints finished each side with vinyl trim	STC-42
5 Same as No. 4 with 1 1/2" THERMAFIBER Blankets in stud cavities	STC-46
Steel-Stud Drywall Partitions	
6 212ST25 USG Steel Studs, single-layer 1/2" SHEETROCK[3] Brand Gypsum Panels each side	STC-39
7 Same as No. 6 with 1 1/2" THERMAFIBER Blankets in stud cavity	STC-45
8 Same as No. 7 except with double-layer 1/2" SHEETROCK Brand Panels on one side and single-layer on other side	STC-50
9 Same as No. 7 except with double-layer 1/2" SHEETROCK Brand Panels on both sides	STC-54
10 358ST25 USG Steel Studs, single-layer 5/8" SHEETROCK Brand Panels each side, 3" THERMAFIBER Blankets in stud cavities	STC-48
11 Same as No. 10 except with double-layer 5/8" SHEETROCK Brand Panels on each side	STC-57

Office Partition Selector	Rating[1]
High-Performance Partitions	
12 35SJ20 USG Steel Studs with RC-1 Resilient Channels on one side, single-layer 5/8" SHEETROCK Brand Panels each side, 3" THERMAFIBER Blankets in stud cavities	STC-54 MTC-47
13 Same as No. 12 except with double-layer 5/8" SHEETROCK Brand Panels opposite RC-1 Channels side	STC-58 MTC-52
14 Same as No. 12 except with double-layer 5/8" SHEETROCK Brand Panels on each side	STC-61 MTC-57
15 Same as No. 14 except with 60SJ20 USG Steel Studs and 5" THERMAFIBER Blankets in stud cavities	STC-62 MTC-58
16 Same as No. 15 except with triple-layer 5/8" SHEETROCK Brand Panels opposite RC-1 Channels side	STC-63 MTC-59

[1] See system folder SA-926 for explanation of STC and MTC rating systems.
[2] All framing members are spaced 24 in. o.c.
[3] All SHEETROCK Brand Gypsum Panels are either FIRECODE or FIRECODE "C" Panels.

STC ratings for walls containing cracks or small openings.

Opening or Crack Size (Area in 100 ft.² Wall)	STC of Wall with No Openings								
	20	25	30	35	40	45	50	55	60
	STC of Wall Containing Cracks or Openings								
144.0 in.²	17	19	20	20	20	20	20	20	20
72.0 in.²	18	21	22	23	23	23	23	23	23
36.0 in.²	19	23	25	26	26	26	26	26	26
18.0 in.²	20	24	27	29	29	29	29	29	29
9.0 in.²	20	25	28	30	31	32	32	32	32
4.5 in.²	20	25	29	32	34	35	35	35	35
2.25 in.²	20	25	29	33	36	37	38	38	38
1.0 in.²	20	25	30	34	38	40	41	41	41
0.5 in.²	20	25	30	35	39	42	44	44	44
0.25 in.²	20	25	30	35	39	43	46	47	47
0.125 in.²	20	25	30	35	40	44	47	49	50
0.063 in.²	20	25	30	35	40	45	48	51	53

STC limitations imposed on composite constructions by various duct arrangements.

Description of Supply or Return System Serving Adjacent Spaces	Approx. Max. Rating
Supply air via common unlined branch duct	STC-30
Supply air via separate unlined branch duct connected to common unlined main duct	STC-35
Return air through ceiling to common plenum	STC-40
Supply air via common duct with 1-in. thick acoustical lining, min. 10 ft. and two elbows between room outlets	STC-45
Return air through ceiling to common plenum utilizing 3-ft. section of duct with 1-in. acoustical lining and one lined elbow, open ends of duct boots min. 6 ft. apart.	STC-50

Reprinted by permission of *Form & Function* magazine, published by USG Corporation

11.10.0 Suggested STC Ratings and Construction

Suggested minimum STC ratings for various types of composite office construction.

Space Relationship	Background Noise Level RC-30	RC-35
Executive office to executive office	STC-50	STC-45
Executive office to private office	STC-50	STC-45
Executive office to secretary	STC-45	STC-40
Conference room to private office	STC-45	STC-40
Private office to private office	STC-40	STC-35
Private office to secretary	STC-35	STC-30

STC ratings of partitions with door or window occupying 21% of composite area.

STC of wall only \ STC of door or window only	14	16	18	20	22	24	26	28	30	32	34	36	38	40	42	44
36	21	23	24	26	28	30	31	33	34	35	36	36	36	36	37	37
38	21	23	25	26	28	30	32	33	35	36	37	38	38	38	38	39
40	21	23	25	27	28	30	32	34	35	37	38	39	39	40	40	41
42	21	23	25	27	29	30	32	34	36	37	39	40	41	41	42	42
44	21	23	25	27	29	30	32	34	36	38	39	41	42	43	43	44
46	21	23	25	27	29	31	33	34	36	39	40	41	43	44	45	45
48	21	23	25	27	29	31	33	35	36	38	40	42	43	45	46	47
50	21	23	25	27	29	31	33	35	37	38	40	42	44	45	47	48
52	21	23	25	27	29	31	33	35	37	39	40	42	44	46	47	49
54	21	23	25	27	29	31	33	35	37	39	41	43	44	46	48	49
56	21	23	25	27	29	31	33	35	37	39	41	43	45	46	48	50

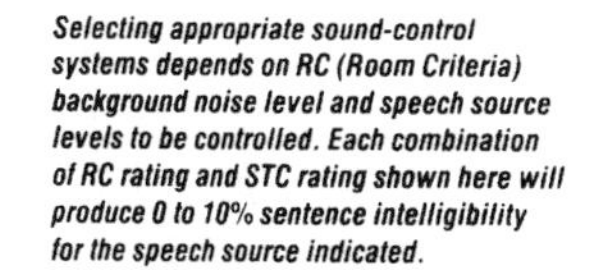

Selecting appropriate sound-control systems depends on RC (Room Criteria) background noise level and speech source levels to be controlled. Each combination of RC rating and STC rating shown here will produce 0 to 10% sentence intelligibility for the speech source indicated.

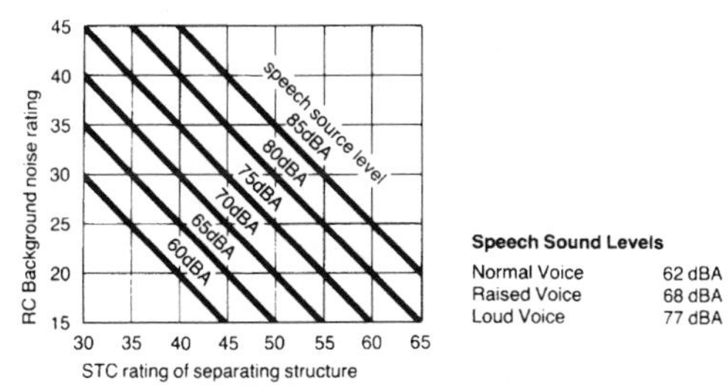

Room Criteria background noise-level curves in 10 dB increments.

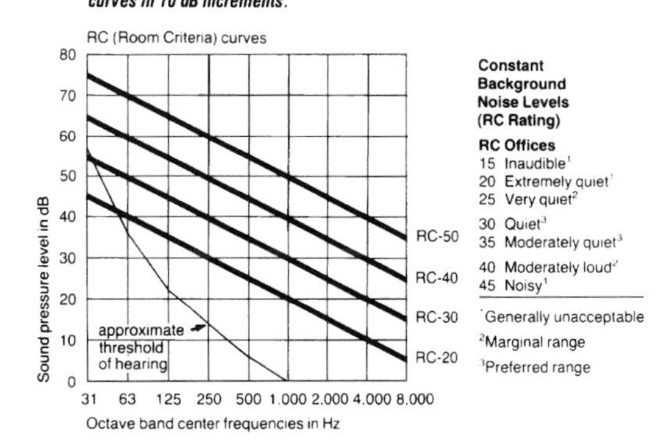

Constant Background Noise Levels (RC Rating)

RC Offices
15 Inaudible[1]
20 Extremely quiet[1]
25 Very quiet[2]
30 Quiet[3]
35 Moderately quiet[3]
40 Moderately loud[2]
45 Noisy[1]

[1]Generally unacceptable
[2]Marginal range
[3]Preferred range

Reprinted by permission of *Form & Function* magazine, published by USG Corporation

11.11.0 Ratings of 2" to 6" Concrete Slabs and Various STC-Rated Ceiling Assemblies

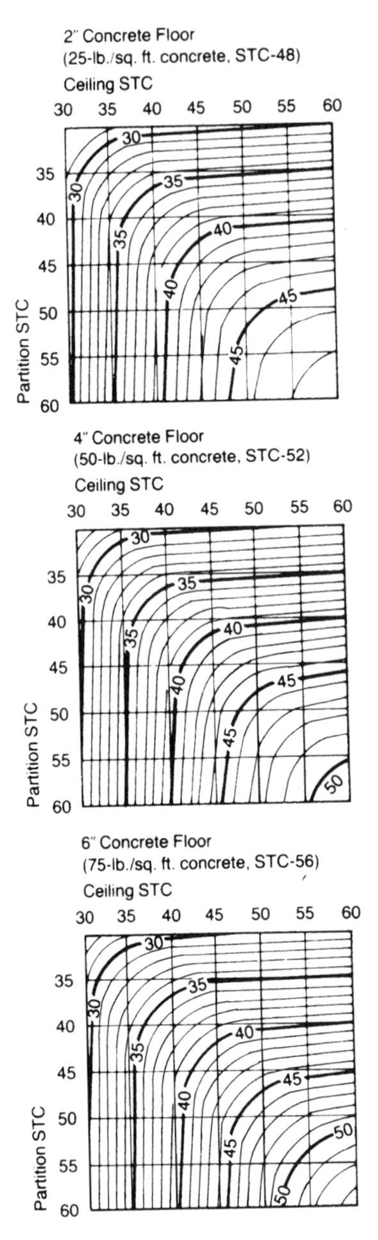

Composite office construction STC graphs for wall/ceiling systems compared to concrete floor systems of 2 in., 4 in. and 6 in. Note that wall, floor and ceiling areas are considered approximately equal for these graphs and that flanking sound paths are not considered.

Reprinted by permission of *Form & Function* magazine, published by USG Corporation

11.12.0 The Effect of Acoustical Doors on STC Ratings

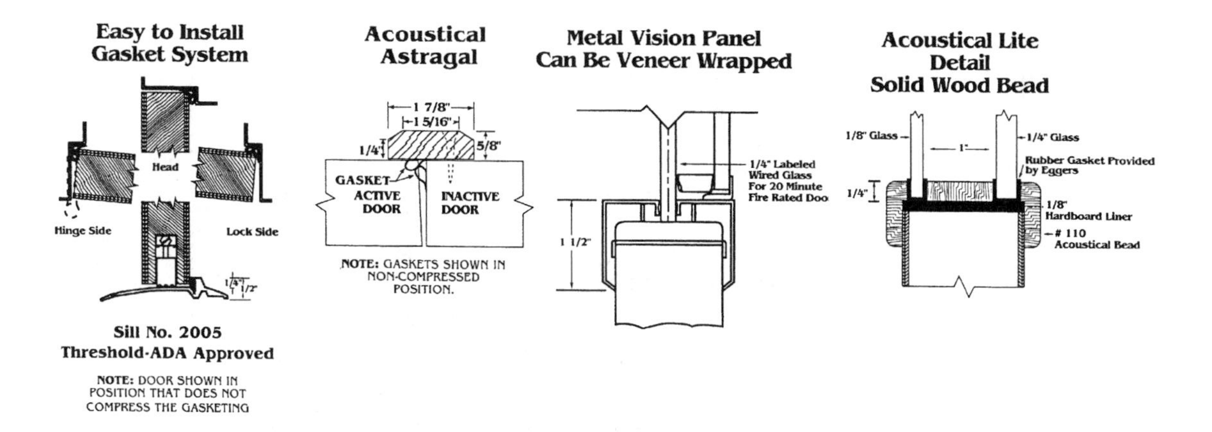

ACOUSTICAL DOORS (TECHNICAL INFORMATION)

STANDARDS (meet or exceed)	AWI Section 1300, NWWDA I.S. 1-A.
TEST METHODS	ASTM E90-90.
SIZES	
MAXIMUM OVERALL	4' 0" x 10' 0". 20 minute limited to 4' 0" x 8' 0" singles; 8' 0" x 8' 0" pairs. 45 minute limited to 4' 0" x 8' 0" singles.
THICKNESS	1 3/4", 2 1/4" & 1 3/8".
FACES	All availble species, sketch face, plastic, medium density overlay.
EDGESTRIP	
TOP & BOTTOM RAILS	STC 28, 31, 33, 36–1 1/8" min. option hardwood glued to core. STC 37 through 53–5" glued to core.
VERTICAL STILES	STC 31 and 36– stave 5/8", particle 1 3/8" matching or compatible to face veneer. Glued to core. STC 37 through 53–1 9/16" glued to core. STC 35 through 53–veneer edged with matching veneer to face veneer.
FLOOR SEAL	All ratings have a center-mounted drop seal.
GASKETING	All gasketing is supplied with door and can be installed in standard stopped, hollow metal frame.
LITES	All lites subject to the following for warranty: no less than 5" between adjacent cutouts such as hardware, lites, etc. Total area not to exceed 40% of door area or 50% of door height. Except STC 39 full lite.
FINISHING	Gardall II, polyurethane, primed, painted, sealed, as specified.
PREMACHINING	Prefitting, mortised for appropriate hardware.
APPLIED MOULDING	Allowable on one or both faces. 3/4" high max. by 3" wide max.
MATCHING TRANSOM–PAIRS	Virtually unlimited in standard veneers.
WARRANTY	
INTERIOR	Life of orginal installation.
EXTERIOR	Not recommended– A special STC 31 through 51 door for residences around airports is available with a five-year warranty. Certain geographical locations subject to special installation requirements.

Reprinted by permission of Eggers Industries, Twin Rivers, WI

11.13.0 The Noise-Muffling Qualities of Various Types of Plumbing Risers

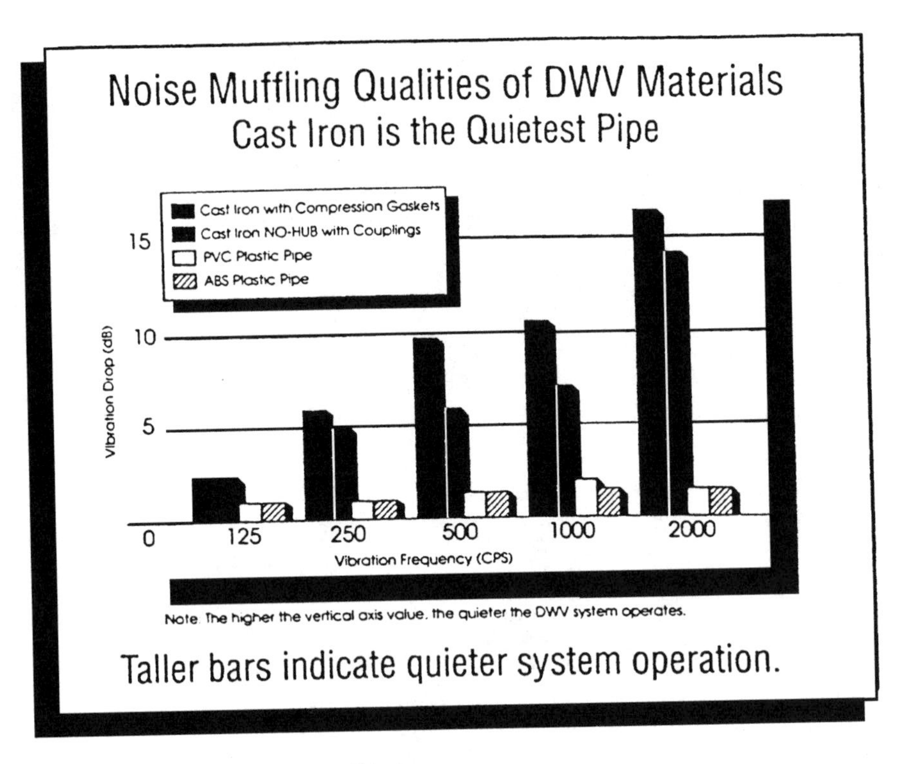

Note: DWV = Drainage, Waste, and Vents

By permission of Cast Iron Soil Pipe Institute

11.14.0 Plumbing Installation Acoustical Considerations

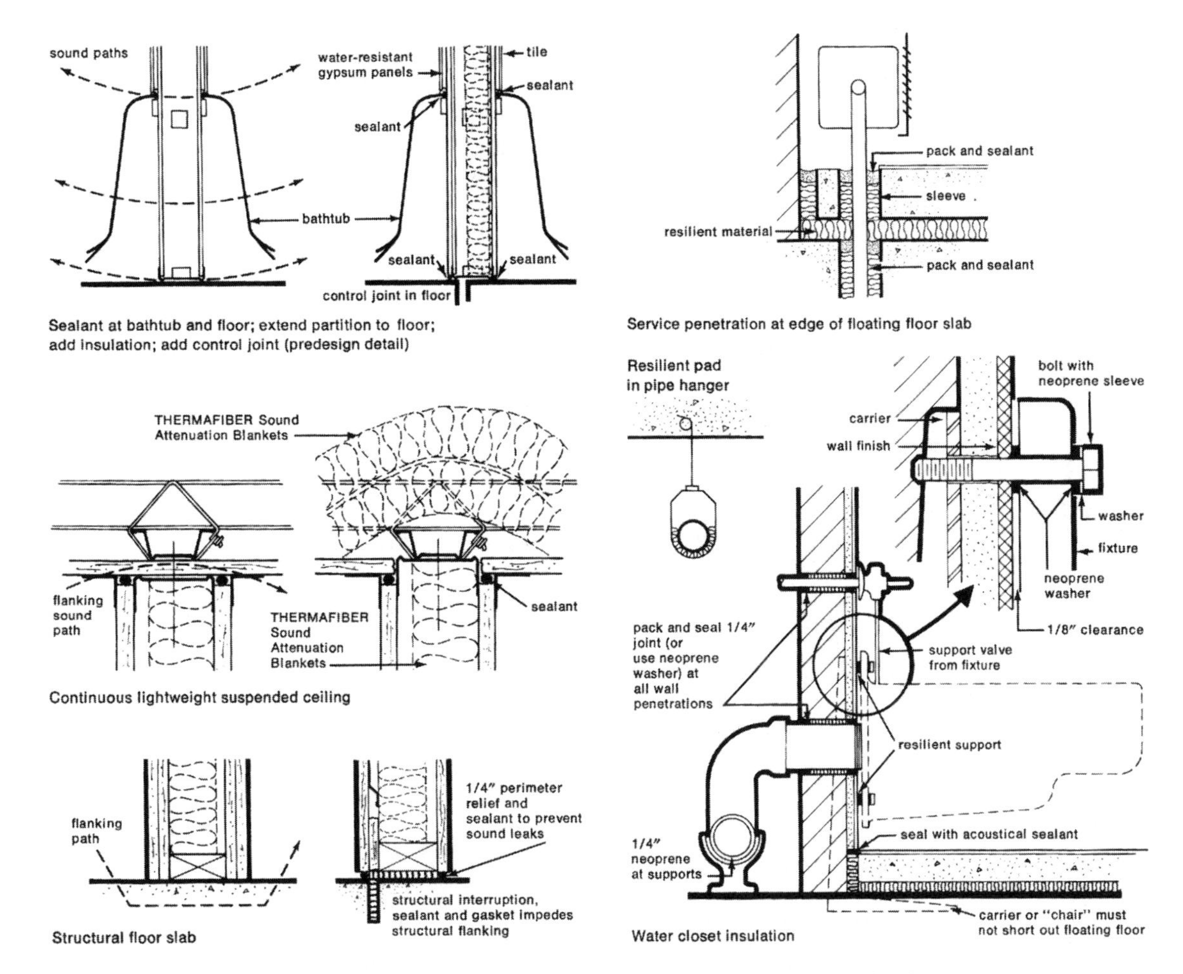

Reprinted by permission of *Form & Function* magazine, published by USG Corporation

11.15.0 Duct Systems and Acoustical Considerations

Duct systems in both commercial and residential buildings can be constructed of metal or fiberglass, lined or wrapped with insulating materials. Not only is noise generated by the actual flow of air through the duct system, but noise is generated or can be controlled by the type of material from which the ductwork is constructed.

Description	Octave Band Frequency (Hz)					
	125	250	500	1000	2000	4000
Bare sheet metal*	0.1	0.1	0.1	0.1	0.1	0.1
Wrapped sheet metal*	0.2	0.2	0.2	0.2	0.2	0.2
Lined sheet metal* (one inch thick)	0.3	0.7	1.9	5.3	4.8	2.3
Fiberglass duct (one inch thick)	0.4	1.4	3.3	3.9	5.0	3.7

**1978 ASHRAE Transactions*, Vol. 84, Part 1, p. 122

11.16.0 Composite Wall/Electrical Box Installations

STC limitations imposed on composite constructions by flanking walls and window.

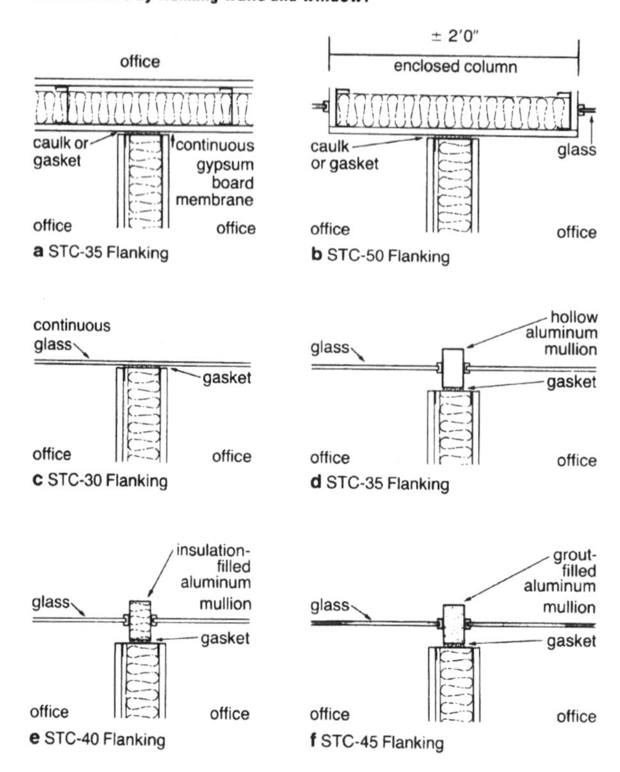

Acoustical details for installing electrical boxes in sound-rated walls. Note that an acoustician usually must develop specific job details for walls rated over STC-60.

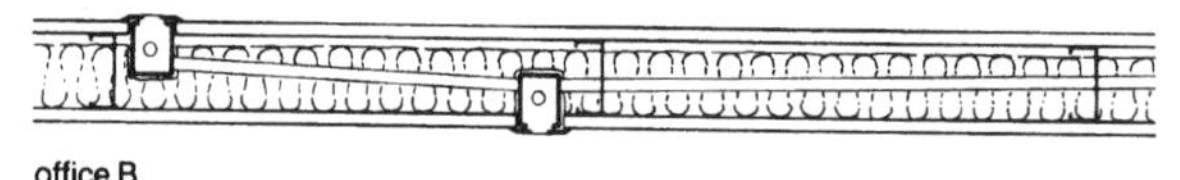

a Caulked boxes back to back or side to side in same stud cavity. Arrangement should not be used with walls rated above STC-40.

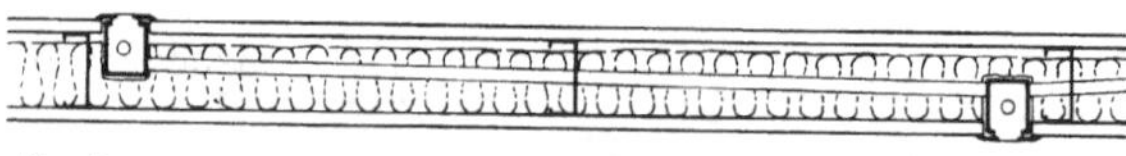

b Caulked boxes side to side in separate stud cavities, min. 36 in. apart. Arrangement suitable for walls rated up to STC-50.

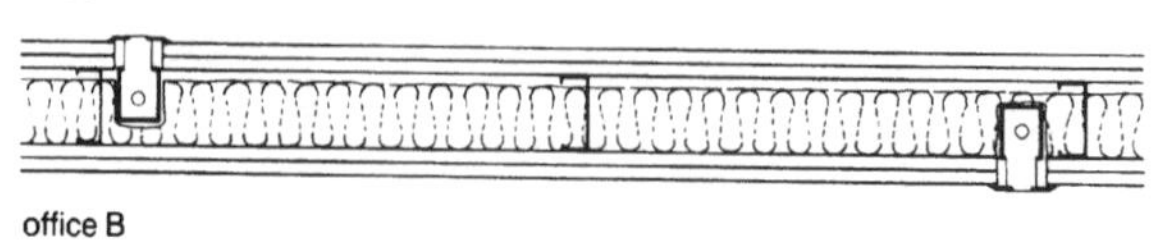

c Caulked boxes, min. 36 in. apart, conduit from overhead or beneath floor. Arrangement suitable for walls rated up to STC-60.

11.17.0 Electrical Transformers and Increased Decibel (dBA) Levels

When locating office space adjacent to electrical equipment rooms or electrical closets where sizable electrical transformers are installed, precautions should be taken in wall construction to avoid or lessen the transmission of excessive decibel levels to these areas.

Listed are the transformer ratings and their corresponding decibel sound output.

Transformer rating	Decibel sound output
9	40
15	42
30	42
45	42
75	45
112–½	45
150	45
225	49
300	49
500	53

Section

12

Doors and Windows

Contents

12.0.0 Hollow metal doors and frames
12.0.1 Classifications of hollow metal doors
12.0.2 Standard opening sizes for hollow metal doors
12.0.3 Hardware locations and reinforcing required for hollow metal doors and frames
12.0.4 Metal thickness of hollow metal doors
12.1.0 Dealing with hollow metal door installation problems
12.1.1 Frame loose in drywall partition
12.1.2 Frame loose in drywall partition (another condition)
12.1.3 Improper door/frame clearances
12.1.4 Door binding and sagging (hinge problems)
12.1.5 Springing a twisted door
12.1.6 Springing a twisted door (another method)
12.1.7 Reswagging hinges
12.1.8 Hinge binding against rabbet
12.1.9 Thermal bow in a hollow metal door
12.2.0 UL label off a fire-rated door?
12.2.1 UL label off a fire-rated frame?
12.3.0 Hollow metal door paint problems
12.4.0 Wood veneer doors, stave lumber core (specifications/grades)
12.4.1 Wood veneer doors, particleboard core (specifications/grades)
12.4.2 Wood veneer doors, mineral core (specifications/grades)
12.5.0 Appearance of standard wood veneer cuts
12.5.1 Matching of individual veneer skins
12.6.0 Laminate-faced particleboard core doors (specifications/grades)
12.6.1 Laminate-faced mineral core doors (specifications/grades)
12.7.0 Wood door construction details
12.8.0 Fire-rated wood door construction
12.8.1 Fire-rated, sound retardant, lead-lined, and electrostatic shield doors
12.9.0 Data required to order pre-machined wood doors
12.9.1 Hardware and special reinforcing requirements
12.9.2 Wood door glazing and louver options
12.10.0 Installation of exterior wood swinging doors
12.11.0 Warp tolerance and telegraphing tolerances for wood doors
12.12.0 How to Store, Handle, Finish, Install, and Maintain Wood Doors
12.13.0 Aluminum door types/sectional dimensions
12.13.1 Aluminum revolving doors
12.14.0 Windows (Aluminum, Wood, Steel, and Plastic)
12.15.0 Window performance grades and ANSI and NWWDA standards for wood windows
12.16.0 Effect of glazing selections on heat gain
12.17.0 NWWDA air-infiltration standards
12.18.0 Steps required to order wood-clad windows

12.19.0 Low-E glazing (illustration)
12.20.0 Thermal movement and frame deflection
12.21.0 Defining fixed and hinged portions of french door assemblies
12.22.0 Aluminum window wall (stick-built construction)
12.23.0 Aluminum window wall (shear block fabrication)
12.24.0 Aluminum window wall (screwspline fabrication)
12.25.0 Sloped glazing and skylight configurations

Numerous configurations of doors and and windows are in use in residential, commercial, and industrial construction today.

Sliding, revolving, folding, and vertical rise doors are specified in some projects, but it is the rare construction project that does not include swinging doors—either wood or metal or laminate clad. This section emphasizes these later three types.

The materials of construction for windows include: wood, steel and aluminum, vinyl, fiberglass, and combinations of these materials. However, the availability of different window configurations allow for a multitude of fenestration configurations: single and double hung, fixed lights, casements, sliders, awnings, and pivots to meet specific architectural designs. This section is devoted to general window design and materials of construction.

12.0.0 Metal Doors and Frames

Commonly referred to as hollow metal, these doors and frames are available in many standard sizes and configurations and any number of custom design variations. The design and classification standards are established by the Steel Door Institute (SDI) for grades, sizes, metal gauges, and hardware locations.

12.0.1 Classifications of Hollow Metal Doors

Grade I–Standard-duty 1⅜" and 1¾" (Level C)
- Model 1 . . . . .Full Flush Design
- Model 2 . . . . .Seamless Design

Grade II–Heavy-duty 1¾" (Level B)
- Model 1 . . . . .Full Flush Design
- Model 2 . . . . .Seamless Design

Grade III–Extra Heavy-duty 1¾" (Level A)
- Models 1 & 1A . . . . .Full Flush Design
- Models 2 & 2A . . . . .Seamless Design
- Model 3 . . . . .Stile and Rail - Flush panel

By permission of Steel Door Institute (SDI), Cleveland, Ohio

12.0.2 Standard Opening Sizes For Hollow Metal Doors

STANDARD OPENING SIZE

Opening Widths	Opening Heights 1 3/4 " Doors							1 3/8 " Doors	
2'0"	6'8"	7'0"	7'2"	7'10"	8'0"	8'10"	10'0"	6'8"	7'0"
2'4"	6'8"	7'0"	7'2"	7'10"	8'0"	8'10"	10'0"	6'8"	7'0"
2'6"	6'8"	7'0"	7'2"	7'10"	8'0"	8'10"	10'0"	6'8"	7'0"
2'8"	6'8"	7'0"	7'2"	7'10"	8'0"	8'10"	10'0"	6'8"	7'0"
2'10"	6'8"	7'0"	7'2"	7'10"	8'0"	8'10"	10'0"	6'8"	7'0"
3'0"	6'8"	7'0"	7'2"	7'10"	8'0"	8'10"	10'0"	6'8"	7'0"
3'4"	6'8"	7'0"	7'2"	7'10"	8'0"	8'10"	10'0"		
3'6"	6'8"	7'0"	7'2"	7'10"	8'0"	8'10"	10'0"		
3'8"	6'8"	7'0"	7'2"	7'10"	8'0"	8'10"	10'0"		
3'10"	6'8"	7'0"	7'2"	7'10"	8'0"	8'10"	10'0"		
4'0"	6'8"	7'0"	7'2"	7'10"	8'0"	8'10"	10'0"		

Doors

Nominal Design Clearances

The nominal clearance between the door and frame head and jambs shall be ⅛" in the case of both single swing and pairs of doors.

The nominal clearance between the meeting edges of pairs of doors can range from ⅛" to ¼" (⅛" for fire rated doors).

The nominal clearance at the bottom shall be ¾".

The nominal clearance between the face of the door and door stop shall be 1/16".

All clearances are subject to a tolerance of + or –1/32".

Construction Features-Full Flush and Seamless

Door Faces

Full Flush Faces

Form each door face from a single sheet of steel of a thickness as defined by Table II. There shall be no visible seams on the surface of the faces.

By permission of Steel Door Institute (SDI), Cleveland, Ohio

12.0.3 Hardware Locations and Reinforcing Required for Hollow Metal Doors and Frames

HARDWARE REINFORCING GAGES

HARDWARE TYPES [4]		MINIMUM GAGE		MINIMUM THICKNESS [1]	
		DOOR	FRAME	DOOR	FRAME
HINGES	1 3/8" DOORS	12 [2]	12 [2]	.093	.093
	1 3/4" DOORS	10 [2,3]	10 [2]	.123	.123
MORTISE LOCKS & DEADBOLTS		14 [2]	14 [2]	.067	.067
BORED OR CYLINDRICAL LOCKS		14 [2]	14 [2]	.067	.067
FLUSH BOLTS		14	14	.067	.067
SURFACE BOLTS		14	14	.067	.067
SURFACE APPLIED CLOSERS		14	14	.067	.067
HOLD OPEN ARMS		14	14	.067	.067
PULL PLATES & BARS		16	— —	.053	— —
SURFACE PANIC DEVICES		14	14	.067	.067
FLOOR CHECKING HINGES		7	7	.167	.167
PIVOT HINGES		7	7	.167	.167
KICK & PUSH PLATES		REINFORCING IS NOT REQUIRED			

(1) THE MINIMUM STEEL THICKNESS FOR EACH SPECIFIC GAGE ARE DERIVED FROM PUBLISHED FIGURES OF UNDERWRITERS LABORATORIES.(TABLE III).

(2) A THINNER GAGE OF STEEL MAY BE EMPLOYED AS LONG AS THE TAPPED HOLES, USED FOR MOUNTING THE HARDWARE, ARE EXTRUDED TO PRODUCE AN EQUIVALENT NUMBER OF THREADS THAT WOULD BE PROVIDED USING THE GAGE OF MATERIAL INDICATED.

(3) IF THE REINFORCING IS ANGULAR OR CHANNEL SHAPED, 12 GAGE IS PERMISSIBLE

(4) WHEN REINFORCEMENT IS OMITTED AND THROUGH-BOLTING IS REQUIRED, THE USE OF SPACERS OR SEX-BOLTS SHALL BE PART OF THE SPECIFICATION.

HARDWARE LOCATIONS

Locks, Latches, Roller Latches, and Double Handle Sets		40 5/16" to Centerline of Lock Strike from Bottom of Frame. (Refer to Appendix "C" for Additional Information)
Rim and Mortise Panic Devices		
Cylindrical and Mortise Deadlocks		48 " to Centerline of Strike from Bottom of Frame
Push Plates		Centerline of 45" from the Bottom of Frame
Pull Plates		Centerline of Grip @ 42" from the Bottom of Frame
Combination Push Bar		Centerline of 42" from Bottom of Frame
Hospital Arm Pull		Centerline of Lower Base is 45" from Bottom of Frame with Grip Open at Bottom
Hinges	Top	Up to 11 3/4" from Rabbet Section of Frame to Centerline of Hinge
	Bottom	Up to 13" from Bottom of Frame to Centerline of Hinge
	Intermediate	Equally Spaced Between Top and Bottom Hinges

By permission of Steel Door Institute (SDI), Cleveland, Ohio

12.0.4 Metal Thickness of Hollow Metal Doors

TABLE II
METAL THICKNESS/DOORS

GRADE	MODEL	FULL FLUSH OR SEAMLESS		STILES AND RAILS	
		MSG NO.*	MINIMUM THICKNESS	MSG NO.*	MINIMUM THICKNESS
I	1	20	0.032		
	2	20	0.032		
II	1	18	0.042		
	2	18	0.042		
III	1	16	0.053		
	1A	14	0.067		
	2	16	0.053		
	2A	14	0.067		
	3	18	0.042	16	0.053

**Nominal inch equivalent is based on Manufacturers Standard Gage and is Subject to Normal tolerances*

Gage vs Minimum Metal Thickness. The minimum steel thicknesses for each specific gage are derived from the published figures of Underwriters Laboratories Inc. Those limits are shown in Table III.

Table III

MSG* NO.	UNCOATED	HOT DIPPED			ELECTROLYTIC		
		A40	A60	G60	"A"	"B"	"C"
12	.093	.093	.093	.093	.093	.093	.093
14	.067	.067	.067	.067	.067	.067	.067
16	.053	.053	.053	.053	.053	.053	.053
18	.042	.042	.042	.042	.042	.042	.042
20	.032	.032	.032	.032	.032	.032	.032

**Nominal inch equivalent is based on Manufacturers Standard Gage and is Subject to Normal tolerances*

By permission of Steel Door Institute (SDI), Cleveland, Ohio

12.1.0 Dealing With Hollow Metal Door Installation Problems

Whether the hollow metal frames are "set up and welded" or "knocked down" (KD), if they are not properly stored and installed in metal-framed drywall partitions or masonry openings, problems will arise, if not during actual construction, certainly during the post-construction period. Although the contractor might be diligent in supervising and inspecting the installation of hollow metal doors and frames, by their own forces or by a subcontractor, improper storage or less-than-adequate installation procedures can result in problems that require corrective action. Many of these problem installations can be corrected without total removal of either the door or frame.

12.1.1 Frame Loose in Drywall Partition

FRAME LOOSE ON DRYWALL

Frame manufacturers closely control the dimensions which their frames are manufactured to. Since automated equipment is used, these dimensions are easily repeated from piece to piece. The majority of cases will reveal that the overall wall thickness has not been properly maintained. Wall thickness conditions can easily vary from undersize to oversize. The thickness should be checked if possible, to verify the wall's compliance with the Job Specification.

Frames installed in drywall walls can use two different anchoring methods as follows

* WELDED OR SNAP-IN STEEL OR WOOD STUD ANCHORS

Some frames use welded or snapped in steel or wood stud anchors. These frames are installed prior to the drywall material being attached to the studs. In this case, the drywall can either be "butted-up" against the return of the frame or be "tucked in" behind the return of the frame. Only in the installation where the drywall is "tucked in" behind the return can there be a condition where the frame is loose on the drywall. Refer to Figure 1 and Figure 2. This gap could be uniform along the entire length (height) of the jamb or could be only in certain areas. Since the frame cannot be removed, the only available options are to caulk the gap or cover it with trim.

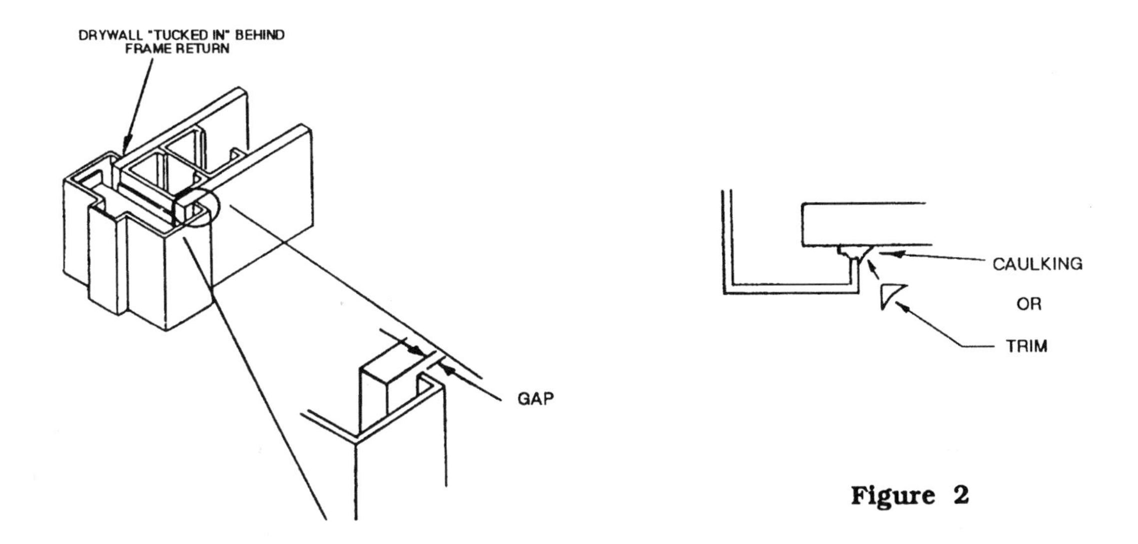

Figure 1

Figure 2

By permission of Steel Door Institute (SDI), Cleveland, Ohio

12.1.2 Frame Loose in Drywall Partition (Another Condition)

This condition should be reported to the appropriate jobsite personnel. The condition can be corrected by putting a bearing plate on each side of the corner and compressing the internal steel studs with a clamp, refer to Figure 8. However, the responsibility for correcting this condition belongs to the sub-contractor responsible for the actual wall construction.

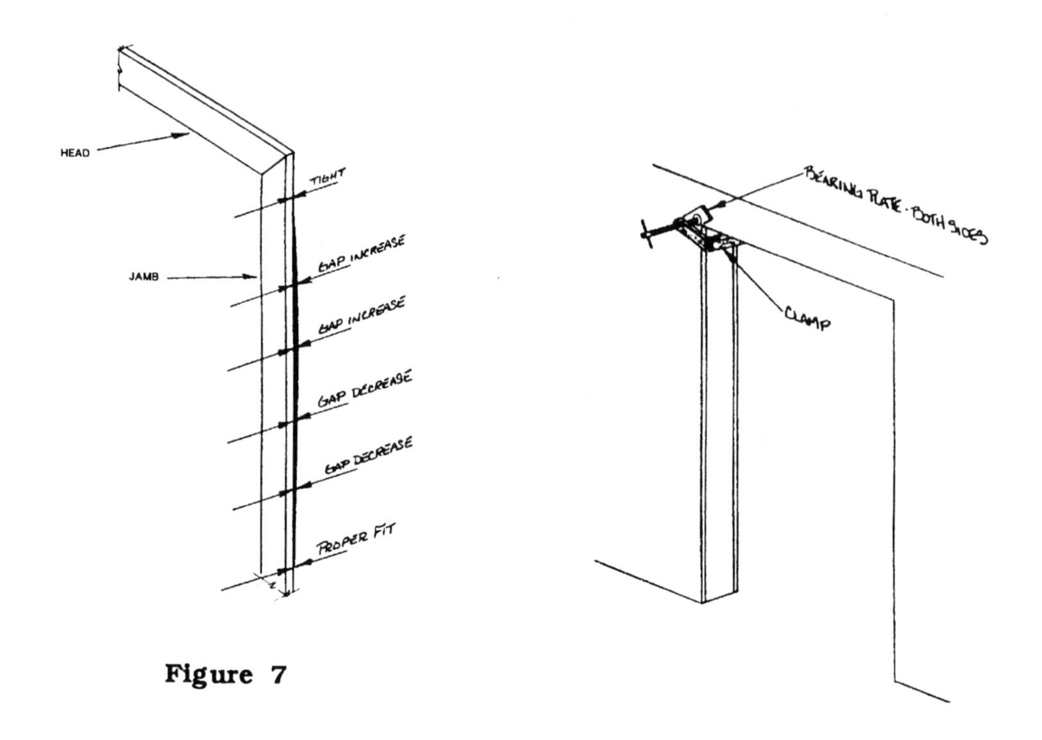

Figure 7

Figure 8

C) The third condition is different from the first two which talk about the "fit" of the frame over the wall thickness. The third condition is that of compression anchors which have not been tightened. The drywall frame would then be loose across the width of the opening and move from side to side against the rough opening.

The frame should be plumbed, square and secure in the opening by properly adjusting the compression anchors following the manufacturers instructions.

By permission of Steel Door Institute (SDI), Cleveland, Ohio

12.1.3 Improper Door/Frame Clearances

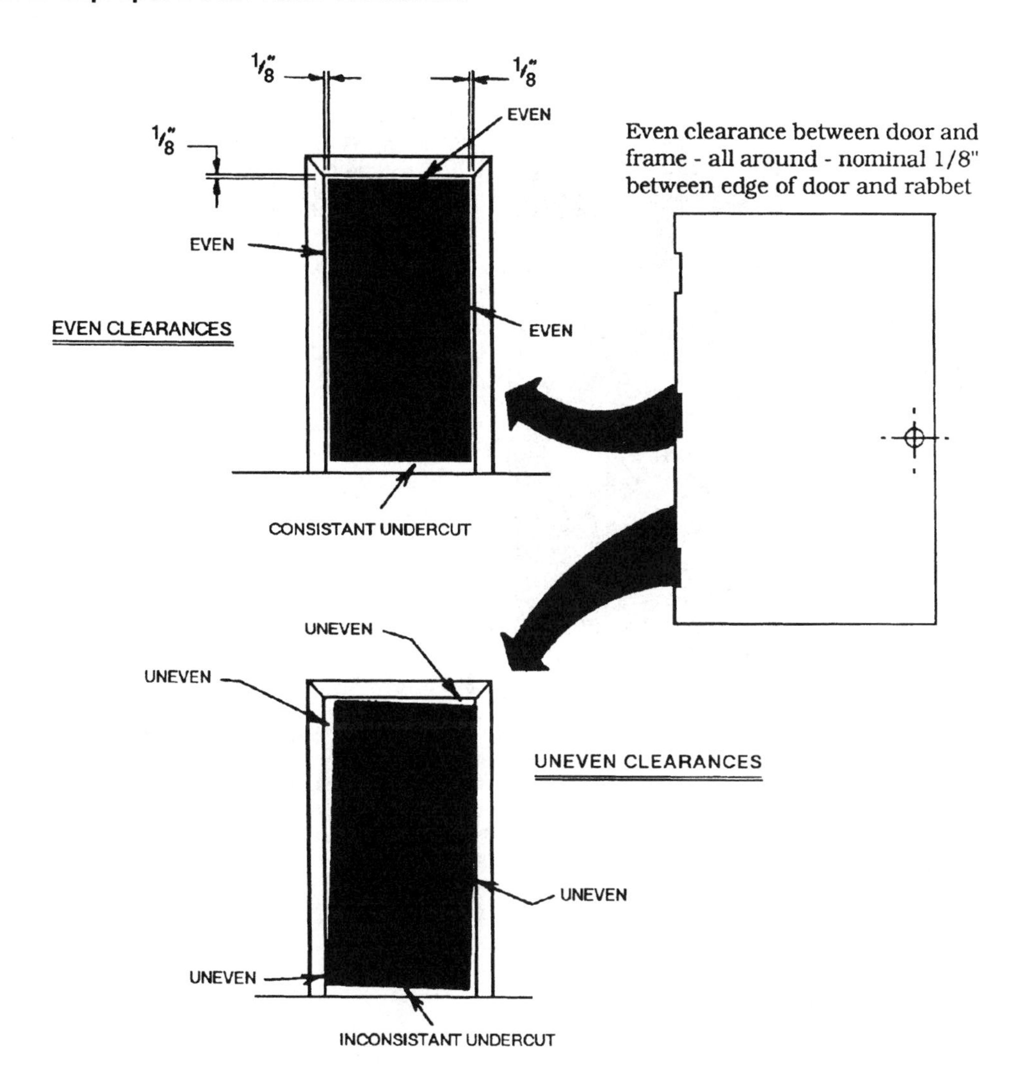

A door and frame are both the same geometric figure, that is, a rectangle. One rectangle, the door, must open and close within another rectangle, the frame. To do so, the clearance between the two must be properly maintained. All standard steel door and frame manufacturers closely hold tolerances which result in a nominal clearance between the door and frame of 1/8 inch. If this clearance is not maintained, an interference will develop and/or hardware misalignment may occur.

From this, it can be seen that proper installation is extremely important in establishing clearances and will prevent a multitude of potential problems from developing.

The Steel Door Institute has many publications which were developed to establish industry standards and assist in specifying as well as installing standard steel doors and frames. One publication, SDI-105 will be of assistance regarding the erection and installation of standard steel frames.

By permission of Steel Door Institute (SDI), Cleveland, Ohio

12.1.4 Door Binding and Sagging (Hinge Problems)

Is Door Sagging ?

If sag cannot be corrected and door and frames returned to plumb relationship, repositioning or shimming the strike may relieve this condition. Additionally filing the strike will compensate for minor mis-alignment (refer to section "Lock fits too tight in strike")

Is your door binding ?

Frames which are out of plumb are frequently the cause of faulty operation of locksets and binding of bolts in the strike. Check carefully.

Are hinges loose ?

If hinge screw will not remain tight, the screw can be held in place by the use of a "locktite" type product which prevents the screw from loosening. Additionally, "Nylok" type fasteners can be used to replace the normal machine screws.

Are hinges worn ?

If excessive wear has occured on hinge knuckles, door will not be held tightly. Replace hinges.

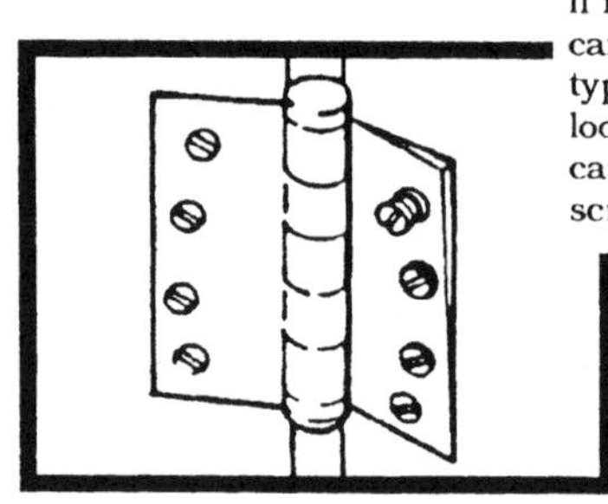

Are Hinges Properly Swaged?

The hinge manufacturers specifications should be checked to determine what the proper hinge swage should be.

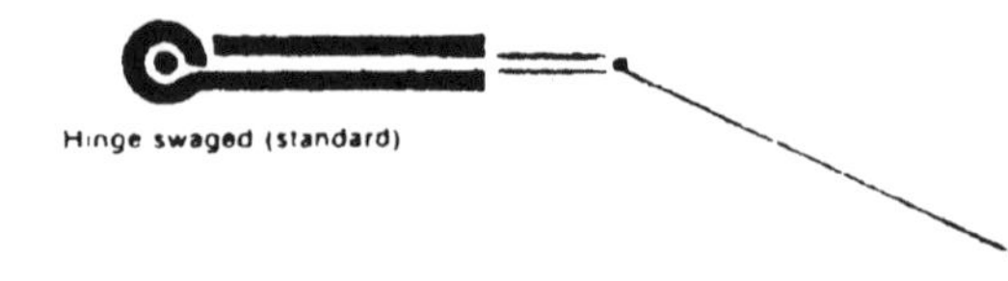

Swaging is a slight offset of the hinge leaf at the barrel which permits the leaves to come together.

Standard swaging of standard weight and heavy weight full mortise hinges when closed to the parallel position provides a 1/16" clearance between leaves.

By permission of Steel Door Institute (SDI), Cleveland, Ohio

12.1.5 Springing a Twisted Door

TWISTED DOOR

It is possible, in some cases, to "spring" the door back to (or much closer to) its ideal position of being parallel with the imaginary plane across the faces of the frame. This can usually be done with the door remaining in the frame. A piece of wood blocking must be placed between the door and frame and pressure is then applied at the twisted area to "spring" the door. However, caution should be exercised on drywall installations since the frame could possibly work loose from the wall, particularly on slip on drywall type frames.

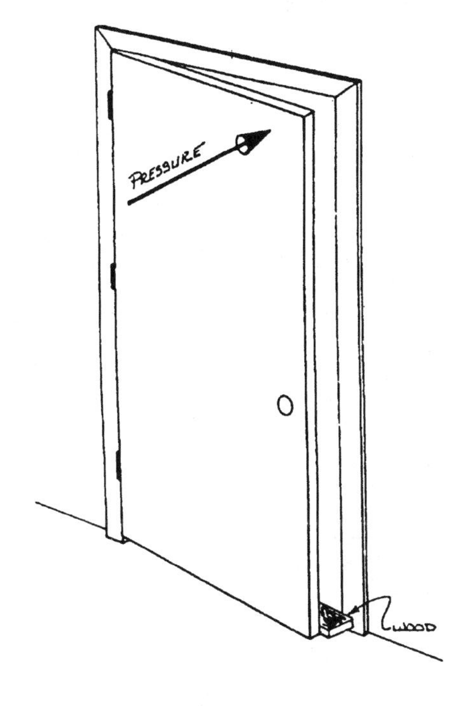

Twisted door, with top lock area of door "breaking—through" the imaginary plane. Place wood block on floor, between door and frame as shown. Apply pressure to top lock area as shown to "spring" door back into position. Remove wood block, close door and check condition. Repeat if necessary.

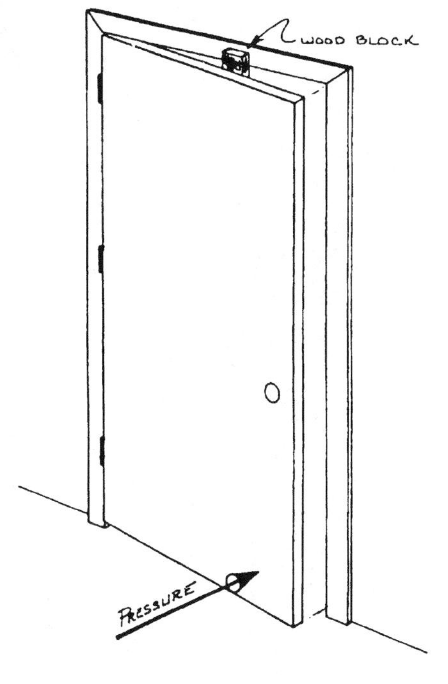

Twisted door, with bottom lock area of door "breaking-through" the imaginary plane. Place wood block between frame head and door as shown. Apply pressure to the bottom lock area as shown to "spring" door back into position. Remove wood block, close door and check condition. Repeat if necessary.

By permission of Steel Door Institute (SDI), Cleveland, Ohio

12.1.6 Springing a Twisted Door (Another Method)

TWISTED DOOR

An alternate method can also be used which will allow the door to remain in the opening. This might be appropriate in drywall installations as previously mentioned. Although the example shown reflects the top half of the door, this method could be used on the bottom half of the door as well.

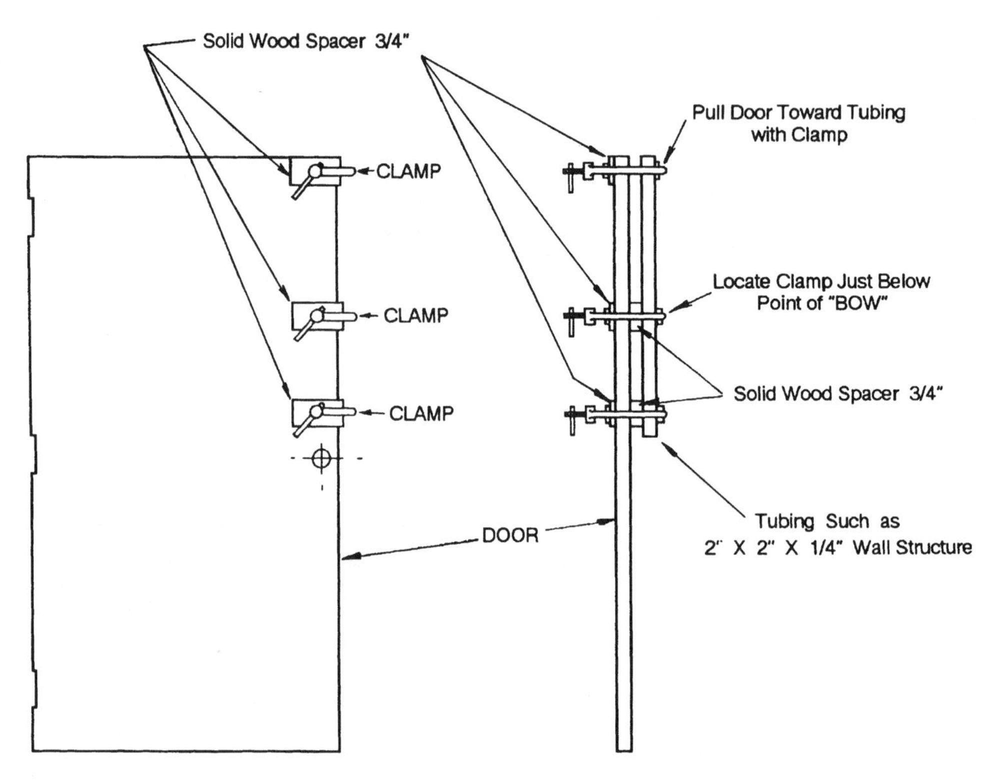

By permission of Steel Door Institute (SDI), Cleveland, Ohio

12.1.7 Reswagging Hinges

The following example shows how a hinge leaf can be reswaged to correct minor improper door / frame clearances. This particular method allows the reswaging to be accomplished while the door remains in the opening and the hinge leaves remain on the door and frame. The example shows a top hinge reswaged to correct a sag type condition. However, any of the hinges can be reswaged in this manner to compensate for conditions opposite to that of a sag condition.

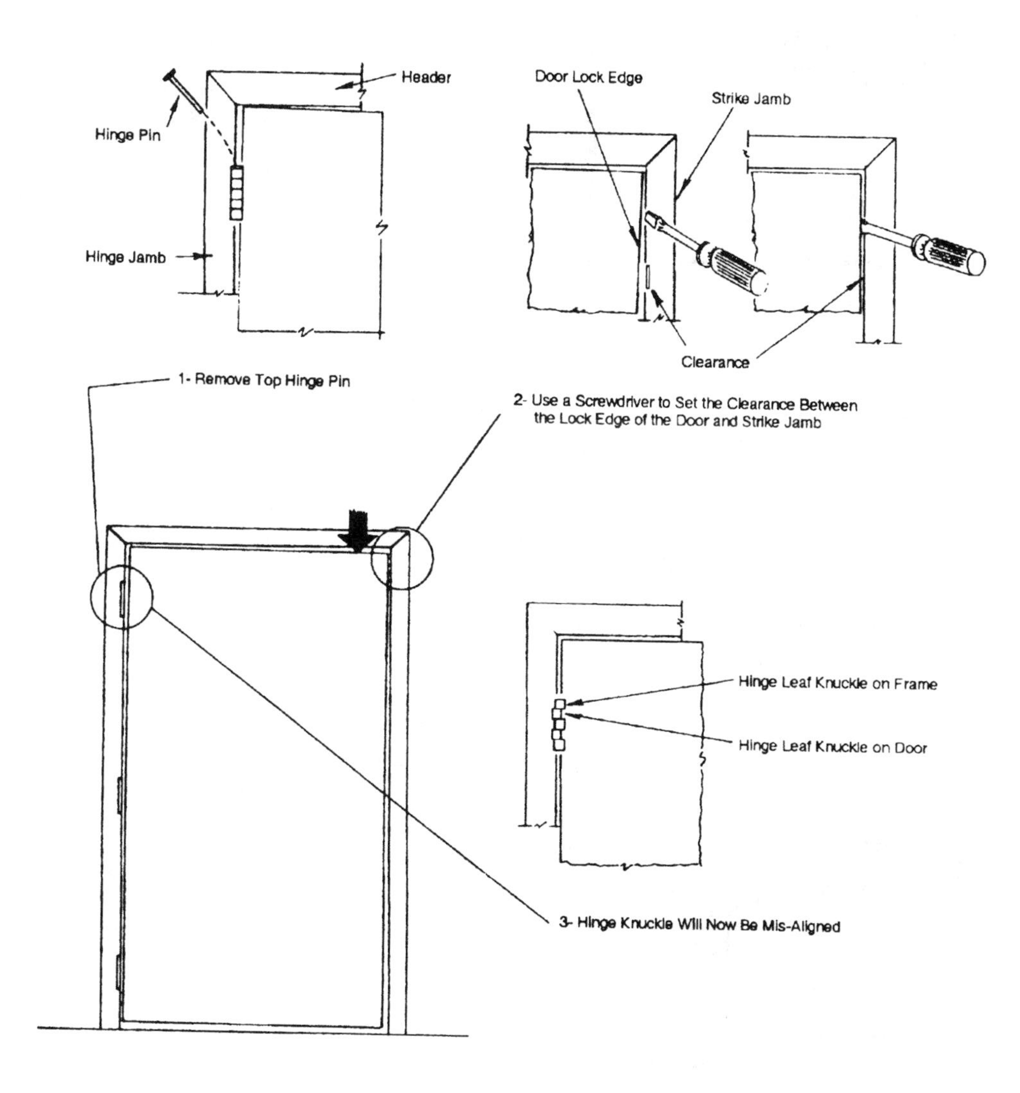

By permission of Steel Door Institute (SDI), Cleveland, Ohio

12.1.8 Hinge Binding Against Rabbet

Normally, hinge bind is found between the door and rabbet. There are several ways of shimming which will move the door in different directions. The following guidelines should be used in shim applications.

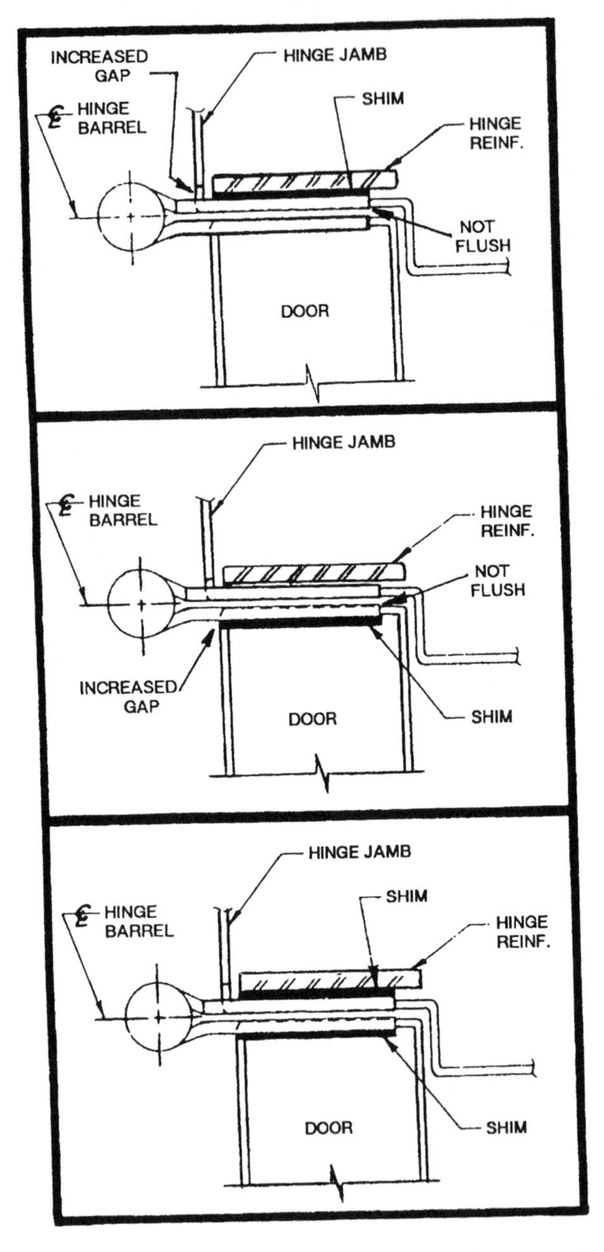

1. A shim can be placed between the frame hinge reinforcement and the hinge leaf. This will move the door towards the strike jamb. However, the hinge notch face gap will be increased and the hinge leaf surface will not be flush with the rabbet surface.

2. A shim can be placed between the door and the hinge leaf. This will also move the door towards the strike jamb. However, an increased gap will be created by the shim and the hinge leaf surface will not be flush with the backset surface on the door.

3. To minimize the gaps and the hinge leaf surfaces not being flush in #1 and #2 above, two shims can be used. These two shims would be half the thickness as those used in #1 or #2. This would minimize gaps and out of flush conditions.

By permission of Steel Door Institute (SDI), Cleveland, Ohio

12.1.9 Thermal Bow in a Hollow Metal Door

The entire door/frame surface should then be lightly sanded and "feathered" into any heavily sanded areas. The entire surface should then be re-prime painted.

* For products which are finish painted, the affected areas should be adequately sanded. If necessary the area should be sanded to bare metal. The entire re maining finish painted area should then be lightly sanded and "feathered" into any heavily sanded areas. If bare metal is showing these areas should be re-prime painted and lightly sanded to "feather" into the lightly sanded finish painted areas. The product should then be re-finish painted.

**In all cases, when the door is being prepared for top, finish coat painting the surface should be cleaned. Use the same solvent that will be used to thin topcoat paint and thoroughly clean all surfaces to be painted.

THERMAL BOW

Installers should be aware of a condition known as Thermal Bow. Thermal Bow is a temporary condition which may occur in metal doors due to the inside-outside temperature differential. This is more common when the direct rays of the sun are on a door surface. This condition is temporary, and to a great extent depends on the door color, door construction, length of exposure, temperature, etc. This condition can often be alleviated by painting the exposed surface a light color. In some cases of extreme cold, this condition may occur in reverse.

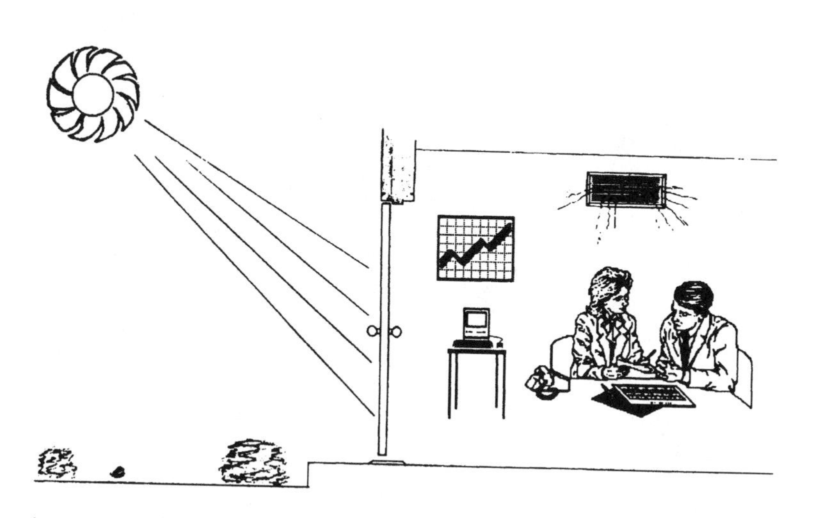

By permission of Steel Door Institute (SDI), Cleveland, Ohio

12.2.0 UL Label Off Fire-Rated Door?

Fire-rated doors are an important element of compliance with building codes and fire-protection standards. Consequently, proper control of the labels that are attached to the doors is top priority for the manufacturer, code official, and labeling agency. The manufacturer must account for every label used and the label can only be applied at the manufacturer's facility or at an authorized distributor of the manufacturer. These are the only places at which a label can be affixed to product. Once the product is in the field, whether it's installed or not, even the manufacturer is not allowed to attach labels unless a representative of the labeling agency has inspected the product for compliance with the manufacturer's procedures. As you can see, not just anyone can attach labels to doors in the field and not just anyone can be in possession of fire-rating labels. Only authorized individuals can be in possession of fire-rating labels. Only authorized individuals can be in possession of and attach labels to fire rated products in the field. Anything other than this is illegal!

All labels on fire-rated doors are located in the same place. Be sure that you are looking for the label in the right location. The label will be located on the hinge edge of the door between the top and middle hinge. If the label is not present, you should contact the distributor who provided the door. They, in turn, will initiate the appropriate action to correct the missing-label problem.

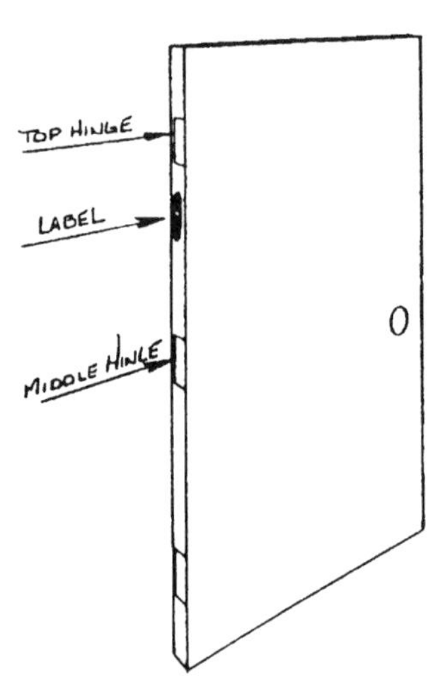

By permission of Steel Door Institute (SDI), Cleveland, Ohio

12.2.1 UL Label Off Fire-Rated Frame?

Like doors, fire-rated frames are an important element of compliance with building codes and fire-protection standards. Consequently, proper control of the labels that are attached to the frame is top priority for the manufacturer, code official, and labeling agency. The manufacturer must account for every label used and the label can only be applied at the manufacturer's facility or at an authorized distributor of the manufacturer. These are the only places that a label can be affixed to the product by the manufacturer. Once the product is in the field, whether it is installed or not, even the manufacturer is not allowed to attach labels unless a representative of the labeling agency has inspected the product for compliance with the manufacturer's procedures. As you can see, not just anyone can attach labels to frames in the field and not just anyone can be in possession of fire-rating labels. Only authorized individuals can be in possession of and attach labels to fire-rated products in the field. Anything other than this is illegal!

All labels on fire-rated frames are located in the same place. However, it should be noted that some frames have an embossed label, rather than the surface-attached label. The embossed label is actually "stamped" into the frame rabbet. Whether the label is surface-attached or embossed, it is located in the same place, on the hinge jamb between the top and middle hinge. If the label or embossment is not present, you should contact the distributor who provided the frame. They, in turn, will initiate the appropriate action to correct the missing-label problem.

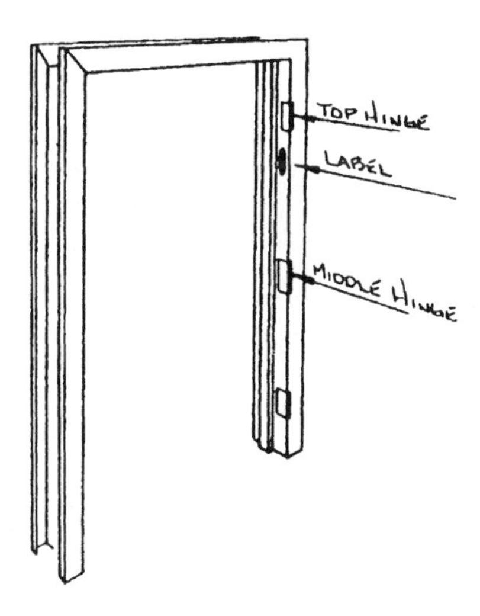

By permission of Steel Door Institute (SDI), Cleveland, Ohio

12.3.0 Hollow Metal Door Paint Problems

Paint Peeling to Bare Metal

Two conditions must be considered when evaluating paint peeling to bare metal.

Primer Paint Only

If the product is only primer painted, then poor adhesion between the primer and bare metal has occurred. This can usually be attributed to inadequate surface preparation before priming. The bare metal must be adequately prepared to ensure good primer paint adhesion.

The door should be completely sanded, washed with an appropriate solvent and re-primed. The sanding and washing operations should provide an adequate surface to ensure good primer adhesion.

Primer Paint and Top (Finish) Coat

The failure could be caused by either poor surface preparation before priming or the use of a non-compatible finish paint, which has reacted with the primer and lifted all paint to bare metal. In either case, the corrective measures would be the same. The door should be completely sanded and washed with an appropriate solvent. The door should then be reprimed. Lightly sand the primer coat, wipe, and finish paint with a compatible top coat.

In all cases, when the door is being prepared for top, finish-coat painting, the surface should be cleaned. Use the same solvent that will be used to thin top-coat paint and thoroughly clean all surfaces to be painted.

Paint in Tapped Holes

Both hollow metal doors and frames have various holes that are drilled and tapped. These holes are in various components, such as reinforcements. All of the components are brought together as an assembly prior to the painting operation.

There are a variety of painting methods which manufacturers can use. Some of these methods could result in a paint build up in the tapped holes of the reinforcements. The build up could, occasionally, make installation of the screw difficult. The build up should be removed to make screw installation easier and assure that the screws are properly sealed.

The best method of cleaning the tapped holes is to use an actual thread tap which matches the screw thread. It will easily cut through and clean the paint build up by simply running the tap in and out of the hole. If the build up is not as great and extra screws are available (or can be obtained) the screw can be run in and out of the hole to clean minor build up prior to final screw installation.

Water Stain Damage

Water stain damage is a direct result of improper storage of prime-painted products. If the product is still in prime-painted products. If the product is still in primer (no finish coat has been applied), the condition is easily detectable:

- Initially, the water stain appears as a discoloration or variance in sheen or gloss in the primer. A specific area or areas can be distinctly noticed, which look and possibly feel different from the rest of the product.
- If the water stain has existed for a considerable length of time and was caused by enough water, rust will start to appear through the discolored areas.

If the product has had a finish coat of paint applied, water stain damage can cause failure of the finish coat as well.

- This condition can be detected by finish-paint failure randomly on the door, as well as the appearance of uniform rust development in those areas. In some cases, the finish paint will show good adhesion in those areas, but will also show a uniform layer of rust developing through the finish paint.

These conditions can be attributed to improperly stored prime products that were exposed to water.

- For products that are prime only, the affected areas should be adequately sanded. If necessary, the area should be sanded to bare metal.

By permission of Steel Door Institute (SDI), Cleveland, Ohio

12.4.0 Wood Veneer Doors, Stave Lumber Core (Specifications and Grades)

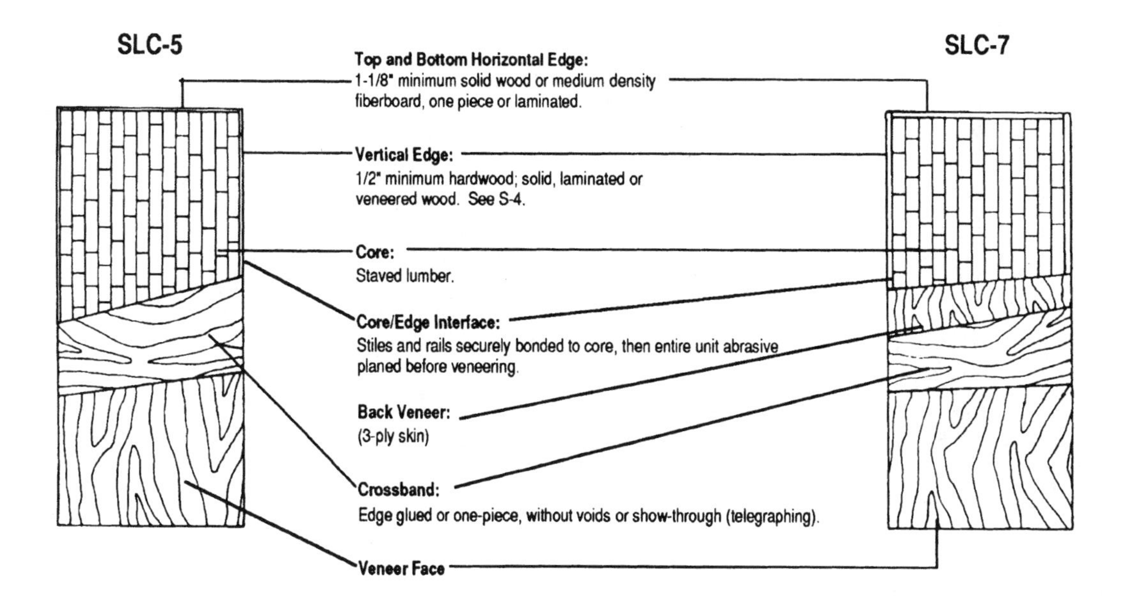

GRADES:	PREMIUM	CUSTOM	ECONOMY
Veneer face:	Minimum 1/50". A Grade. Edge glued joints.	Minimum 1/50". A Grade. Edge glued joints.	Mill option. B Grade. Mill option.
Veneer match:	Book, slip or random. Center, balanced or running.[1] Pair and set match. Door and transom match.	Book, slip or random. Running. Pair and set selected for similar color and grain. Transom selected for similar color only.	No match. No match. No match. No match.
Vertical edges:	Same species visible surface. Sanded ease. No visible joints allowed.	Compatible species visible surface. Sanded ease. Visible joints allowed on hinge edge.[3]	Mill option. Mill option.
Lights[2], louvers[4] and moulding:	Same species lumber, or veneered, or metal vision frames.	Compatible species lumber or metal vision frames.	Mill option.
Transoms:	Bottom horizontal edge runs full width. Matching species lumber or veneered.	Bottom horizontal edge runs full width. Compatible species lumber or veneered.	No match.

[1]Veneer match to be selected by architect.
[2]Maximum 1,296 sq. in. for 20-minute doors.
[3]Visible joints allowed on both edges for opaque finish.
[4]Louvers not allowed in 20-minute doors.

NOTE: 9 ply door constructions are available and may be specified when evaluated or approved by the design professional.

NOTE: Due to scarcity of Birch lumber, Birch faced doors may use compatible species edges, lights, and moulding in premium grade.

12.4.1 Wood Veneer Doors, Particleboard Core (Specifications and Grades)

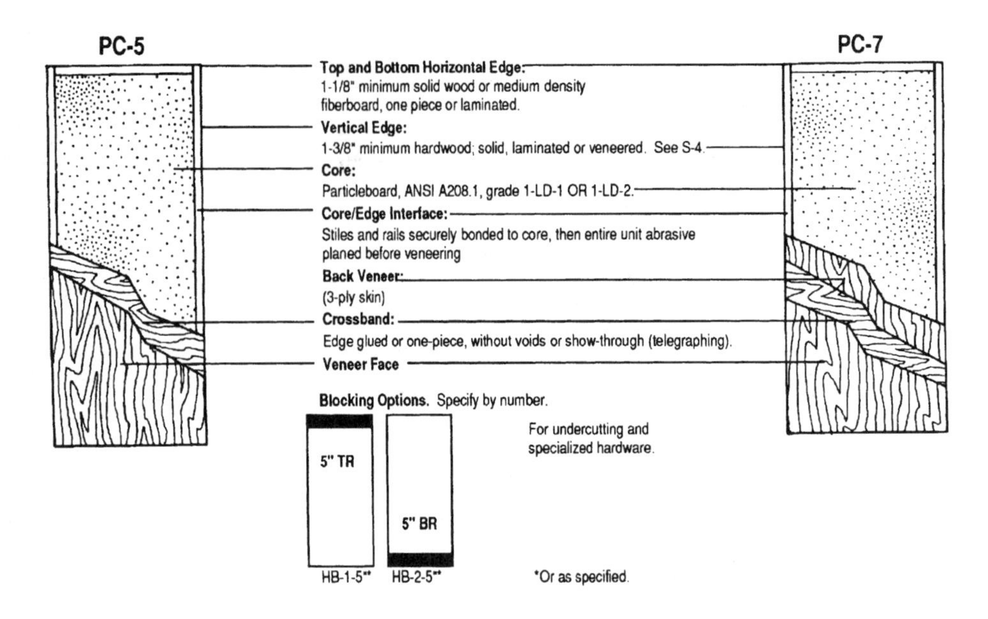

GRADES:	PREMIUM	CUSTOM	ECONOMY
Veneer face:	Minimum 1/50". A Grade. Edge glued joints.	Minimum 1/50". A Grade. Edge glued joints.	Mill option. B Grade. Mill option.
Veneer match:	Book, slip or random. Center, balanced or running.[1] Pair and set match. Door and transom match.	Book, slip or random. Running. Pair and set selected for similar color and grain. Transom selected for similar color only.	No match. No match. No match. No match.
Vertical edges:	Same species visible surface. Sanded ease. No visible joints allowed.	Compatible species visible surface. Sanded ease. Visible joints allowed on hinge edge.[3]	Mill option. Mill option. Mill option.
Lights,[2] louvers[4] and moulding:	Same species lumber, or veneered, or metal vision frames.	Compatible species lumber or metal vision frames.	Mill option.
Transoms:	Bottom horizontal edge runs full width. Matching species lumber or veneered.	Bottom horizontal edge runs full width. Compatible species lumber or veneered.	No match.

[1]Veneer match to be selected by architect.
[2]Maximum 1,296 sq. in. for 20-minute doors.
[3]Visible joints allowed on both edges for opaque finish.
[4]Louvers not allowed in 20-minute doors.

NOTE: 9 ply door constructions are available and may be specified when evaluated or approved by the design professional

NOTE: Due to scarcity of Birch lumber, Birch faced doors may use compatible species edges, lights, and moulding in premium grade.

By permission of National Wood Window and Door Association, Des Plaines, Illinois

12.4.2 Wood Veneer Doors - Mineral Core - Specifications and Grades

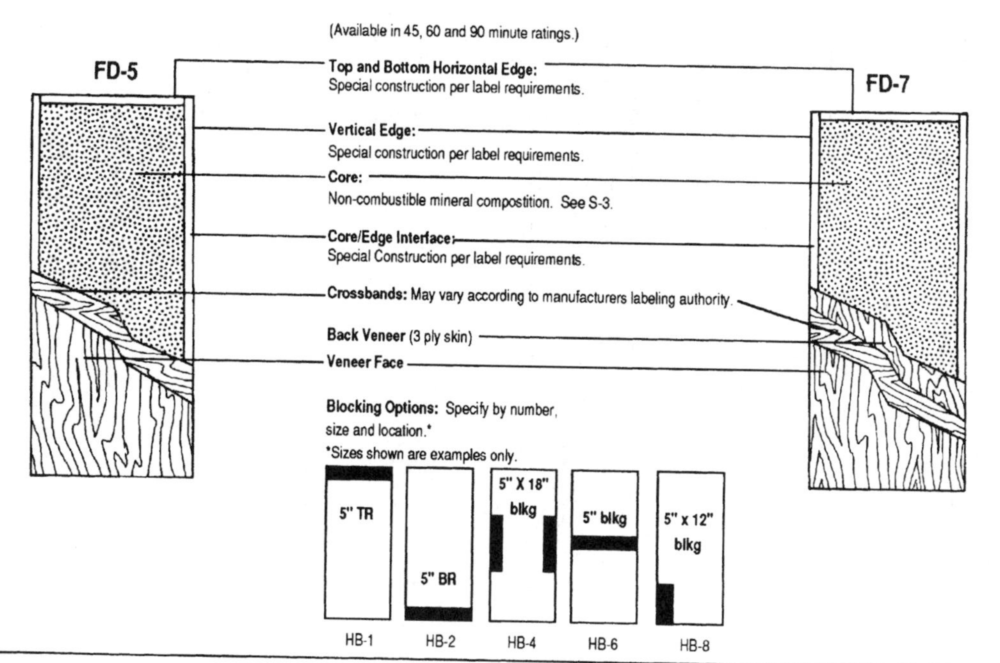

GRADES:	PREMIUM	CUSTOM	ECONOMY
Veneer face:	Minimum 1/50". A Grade. Edge glued joints.	Minimum 1/50". A Grade. Edge glued joints.	Mill option. B Grade. Mill option.
Veneer match:	Book, slip or random. Center, balanced or running.[1] Pair and set match. Door and transom match.	Book, slip or random. Running. Pair and set selected for similar color and grain. Transom selected for similar color only.	No match. No match. No match. No match.
Vertical edges:	Same species visible surface. Sanded ease. No visible joints allowed.	Compatible species visible surface. Sanded ease. Visible joints allowed on hinge edge.[3]	Mill option. Mill option.
Lights,[2] louvers,[4] and moulding:	Special construction per label requirements.	Special construction per label requirements.	Mill option.
Transoms:	Bottom horizontal edge runs full width. Special construction per label requirements.	Bottom horizontal edge runs full width. Special construction per label requirements.	No match.

[1]Veneer match to be selected by architect.
[2]Maximum 100 sq. in. for 60- and 90-minute rated doors.
Maximum 1,296 sq. in. for 45-minute rated doors.
[3]Visible joints allowed on both edges for opaque finish.
[4]Fusible link louvers are allowed in 45-, 60- and 90-minute doors.

NOTE: 9 Ply door constructions are available and may be specified when evaluated or approved by the design professional.

NOTE: Due to scarcity of Birch lumber, Birch faced doors may use compatible species edges, lights, and molding in premium grade.

By permission of National Wood Window and Door Association, Des Plaines, Illinois

12.5.0 Appearance of Standard Wood Veneer Cuts

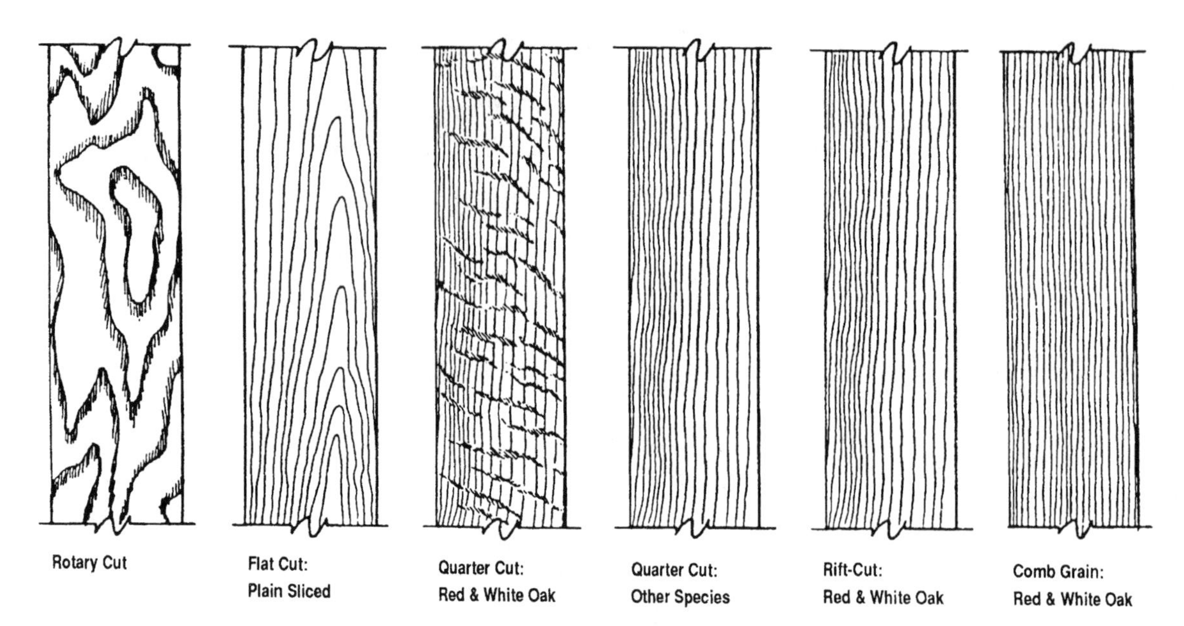

Veneer Cuts

The way in which a log is cut, in relation to the annual growth rings, determines the appearance of veneer. The beauty of veneer is in the natural variations of texture, grain, figure, color, and the way it is assembled on a door face.

Faces will have the natural variations in grain inherent in the species and cut. Natural variations of veneer grain and pattern will vary from these illustrations.

Rotary

This cut follows the log's annual growth rings, providing a general bold random appearance.

Flat Cut (Plain Sliced)

Slicing is done parallel to a line through the center of the log. Cathedral and straight grained patterns result. The individual pieces of veneer are kept in the order they are sliced, permitting a natural grain progression when assembled as veneer faces.

Quarter Cut

A series of stripes is produced. These stripes vary in width from species to species. Flake is a characteristic of this cut in red and white oak.

Rift-Cut (only in Red & White Oak)

The cut slices slightly across the medullary rays, accentuating the vertical grain and minimizing the "flake." Rift grain is restricted to red and white oak.

Comb Grain (only in Red & White Oak)

Limited availability. This is a rift-cut veneer distinguished by the tightness and straightness of the grain along the entire length of the veneer. Slight angle in the grain is allowed. Comb grain is restricted to red and white oak. See section G-11 for maximum grain slope. There are occasional cross bars and flake is minimal.

12.5.1 Matching of Individual Veneer Skins

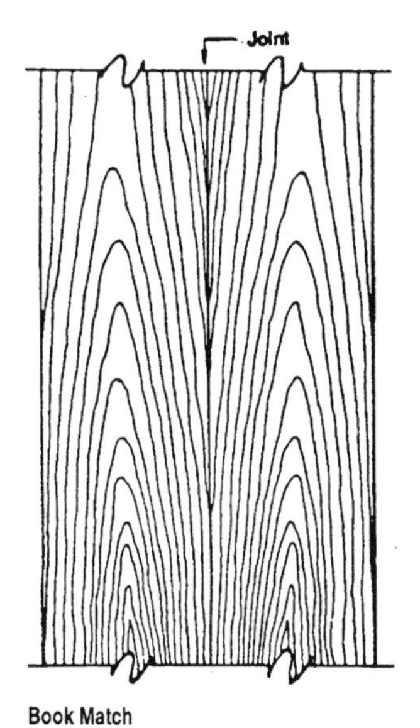

Book Match

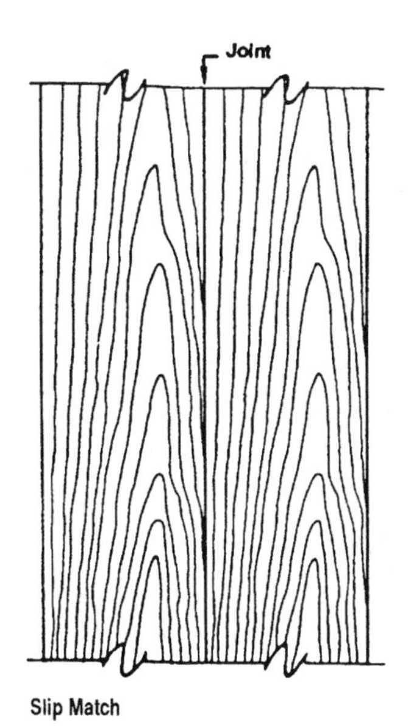

Slip Match

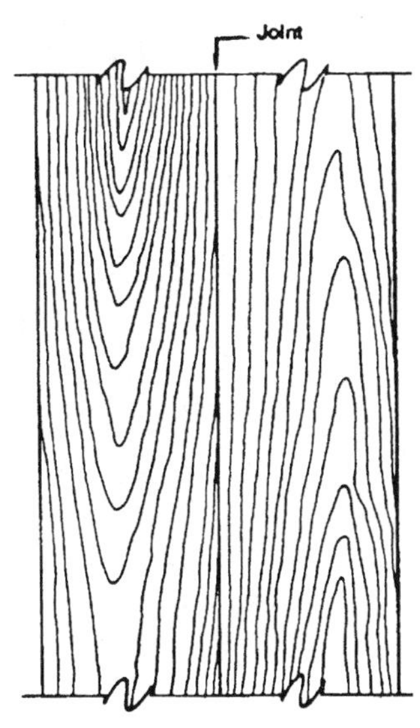

Random Match

Book Match

The most commonly used match in the industry. Every other piece of veneer is turned over so adjacent pieces are opened like two adjacent pages in a book. The veneer joints match and create a mirrored image pattern at the joint line, yielding a maximum continuity of grain. Book matching is used with rotary, plain sliced, quarter, rift cut or comb grain veneers.

Barber Pole Effect in Book Match

Because the "tight" and "loose" faces alternate in adjacent pieces of veneer, they may accept stain differently, and this may yield a noticeable color variation called barber poling. See slip match for further information on color variation. Barber pole can be minimized through proper sanding and finishing techniques.

Slip Match

Adjoining pieces of veneer are placed in sequence without turning over every other piece. The grain figure repeats, but joints won't show mirrored effect. Slip matching is often used in quarter cut, rift-cut and comb grain veneers to eliminate the barber pole effect. However, it may cause a sloping appearance of the veneer, especially in larger veneers.

Pleasing Match

A face containing components which provides a pleasing overall appearance. The grain of the various components need not be matched at the joints. Sharp color contrasts at the joints of the components are not permitted.

Random Match

A random selection of individual pieces of veneer from one or more logs. Produces a "board-like" appearance. It is most commonly used in opaque finish grades.

Note to Specifiers:
The matching of veneers at a joint line must be specified.

By permission of National Wood Window and Door Association, Des Plaines, Illinois

12.6.0 Laminate-Faced Particleboard Core Doors (Specifications and Grades)

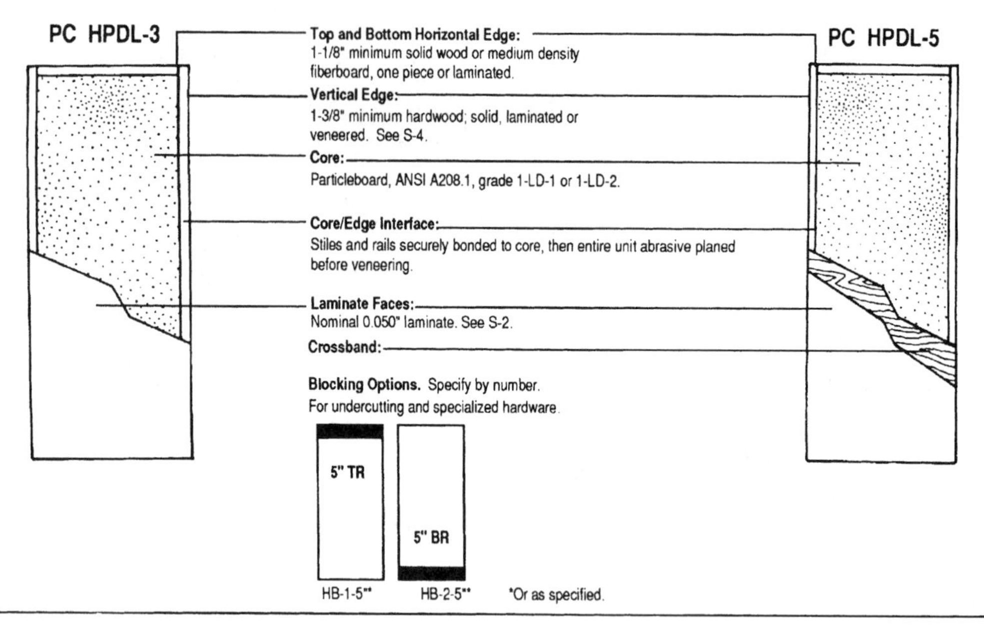

GRADES:	PREMIUM	CUSTOM	ECONOMY
Laminate faces:	Nominal 0.050" high pressure laminate.[1] See S-2.	Nominal 0.050" high pressure laminate.[1] See S-2.	Nominal 0.050" high pressure laminate.[1] See S-2.
Vertical edges for woodgrain patterns:	Matching 0.050" laminate, or lumber or veneer for transparent finish.[2] No visible joints allowed.	Matching 0.050" laminate, or lumber or veneer for transparent finish.[2] Joints allowed on hinge edge.	Mill option.
Vertical edges for solid colors:[4]	Matching 0.050" laminate or close grain hardwood for paint finishes.[2] No visible joints allowed.[6]	Matching 0.050" laminate or close close grain hardwood for paint finishes.[2] No visible joints allowed.[6]	Mill option.
Lights,[3] louver,[5] and moulding for woodgrain patterns:	Compatible species, lumber or veneer with transparent finish,[2] or primed metal vision frame or louver.	Compatible species lumber or veneer for finishing or primed metal vision frame or louver.	Mill option.
Lights,[3] louvers,[5] and moulding for solid colors:[4]	Close grain hardwood for paint finish or primed vision frame or louver	Close grain hardwood for paint finish, or primed vision frame or louver.	Mill option.
Transom-bottom horizontal edges for woodgrain patterns:	Matching 0.050" laminate, or designated species, lumber or veneer for transparent finish.[2] No visible joints allowed.	Matching 0.050" laminate or compatible species lumber or veneer for finishing. No visible joints allowed.	Mill option.
Transom-bottom horizontal edges for solid colors:[4]	Matching 0.050" laminate, or close grain hardwood for paint finish.	Matching 0.050" laminate, or close grain hardwood for paint finish.	Mill option.

[1]Pair matching not available.
[2]Species and stain for entire wood trim package to be selected by architect.
[3]Maximum 1,296 sq. in. for 20-minute rated doors.
[4]Includes other non-wood patterns.
[5]Louvers not allowed in 20-minute doors.
[6]Visible joints allowed on both edges if for opaque finish.

By permission of National Wood Window and Door Association, Des Plaines, Illinois

12.6.1 Laminate-Faced Mineral Core Doors (Specifications/Grades)

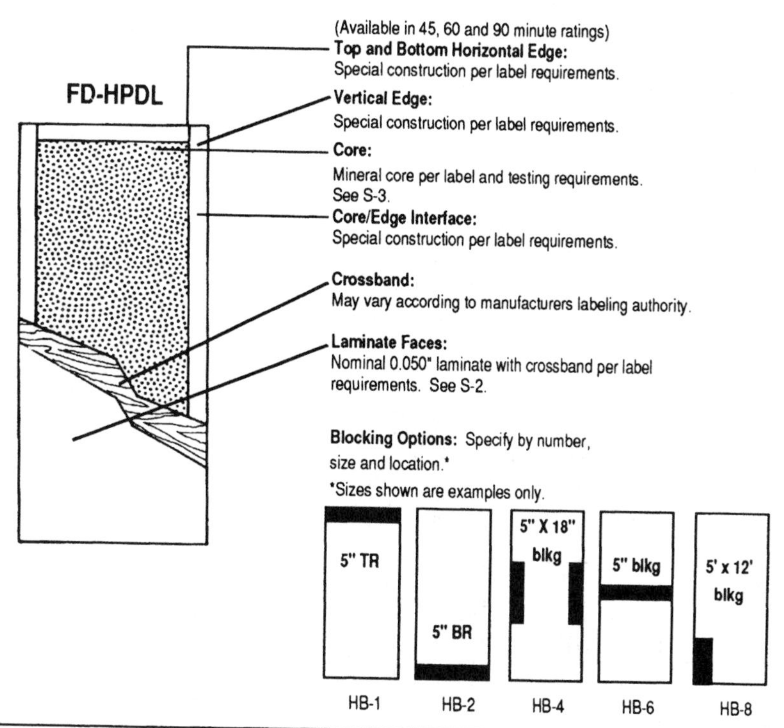

GRADES:	PREMIUM	CUSTOM	ECONOMY
Laminate faces:	Nominal 0.050" high pressure laminate[1] with crossband per label requirements.	Nominal 0.050" high pressure laminate[1] with crossband per label requirements.	Nominal 0.050" high pressure laminate[1] with crossband per label requirements.
Vertical edges for woodgrain patterns:	Matching 0.050" laminate, or lumber or veneer for transparent finish.[2] No visible joints allowed.	Matching 0.050" laminate or lumber or veneer for transparent finish.[2] Joints allowed on hinge edge.	Mill option.
Vertical edges for solid colors:[4]	Matching 0.050" laminate; or close grain hardwood or special construction for paint finishes.[2] No visible joints allowed.[5]	Matching 0.050" laminate or close grain hardwood or special construction for paint finishes.[2] No visible joints allowed.[5]	Mill option.
Lights,[3] louvers, and moulding for woodgrain patterns:	Special construction per label requirements.	Special construction per label requirements.	Mill option.
Lights,[3] louvers, and mouldings for solid colors:[4]	Special construction per label requirements.	Special construction per label requirements.	Mill option.
Transom-bottom horizontal edges for woodgrain patterns:	Special construction per label requirements.	Special construction per label requirements.	Mill option.
Transom-bottom horizontal edges for solid colors:[4]	Special construction per label requirements.	Special construction per label requirements.	Mill option.

[1]Pair matching not available.
[2]Species and stain for entire wood trim package to be selected by architect.
[3]Maximum 100 sq. in. for 60- and 90-minute rated doors. Maximum 1,296 sq. in. for 45-minute rated doors.
[4]Includes other non-wood patterns.
[5]Visible joints allowed on both edges if for opaque finish.

By permission of National Wood Window and Door Association, Des Plaines, Illinois

12.7.0 Wood Door Construction Details

General Moulding Requirements

- Species shall match or be compatible with face veneer or laminate.
- Specify transparent or opaque finish.
- Moulding shall be free of open defects, shake, splits, or doze.
- Moulding must be smooth and free of visible knife, saw, or sanding marks.
- Specify following options.

Meeting Edge Options — Specify by number

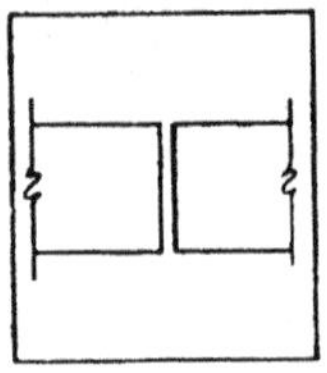

Option E1: Meeting Edge (No Bevel)

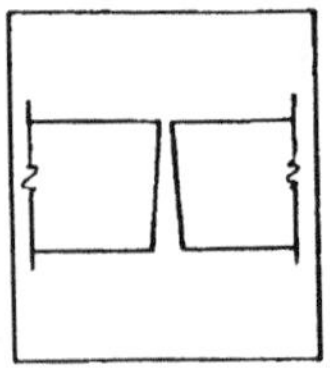

Options E2: Meeting Edge (Bevel)

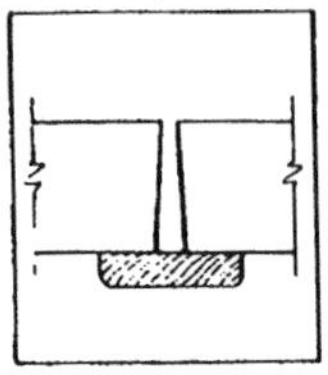

Option E3: Flat Astragal

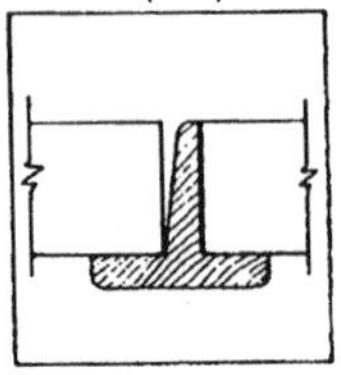

Option E4: Tee Astragal

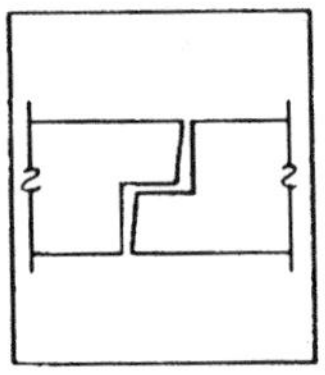

Option E5: Rabbeted

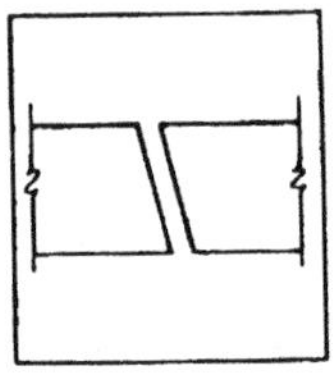

Option E6: Parallel Bevel (Double Egress)

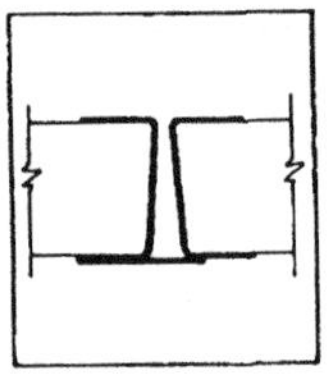

Option E7: Metal Edge Guard and Astragal

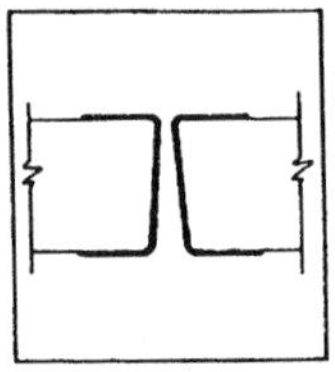

Option E8: Metal Edge Guard

Dutch Door Options — Specify by number

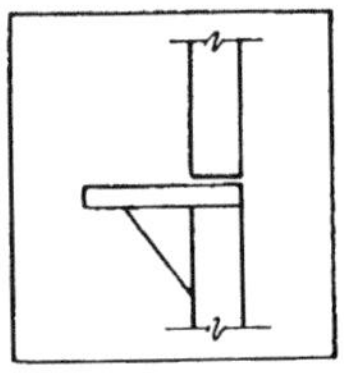

Option D1: One side shelf

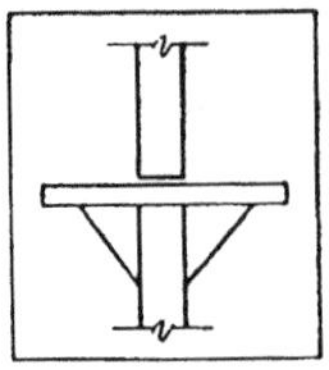

Option D2: Two side Shelf

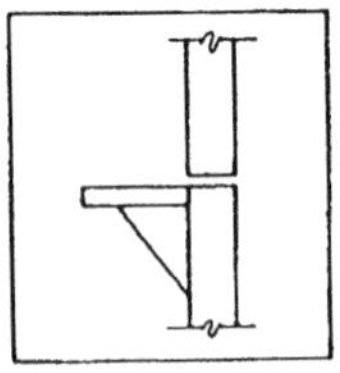

Option D3: 20-Minute shelf

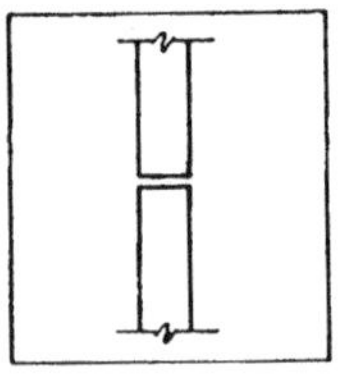

Option D4: No shelf

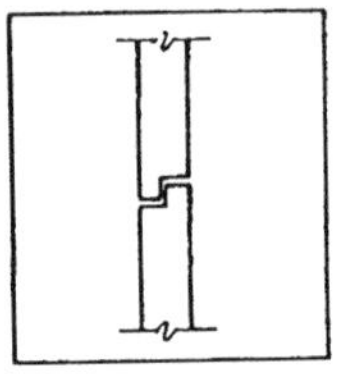

Option D5: Rabbeted meeting rails.

Transom Meeting Edge Options

Specify by number

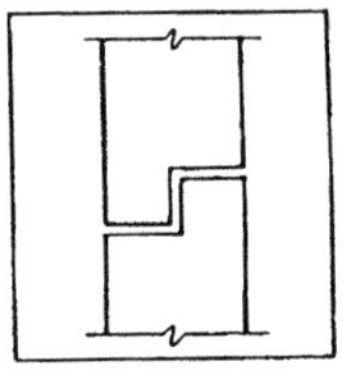

Option T1: Rabbeted

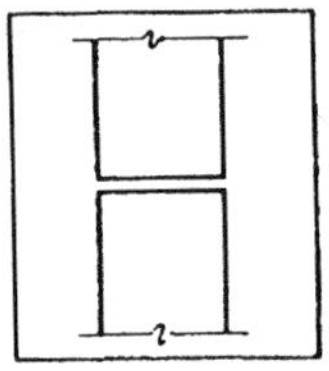

Option T2: Non-rabbeted

Note to Specifiers:
Options E1, E2, E5, E6, E7, E8 available for fire doors per individual manufacturer's approval. Some may require fire-retardant treated edges.

By permission of National Wood Window and Door Association, Des Plaines, Illinois

12.8.0 Fire-Rated Wood Door Construction

Fire-rating construction for wood doors with large lites.

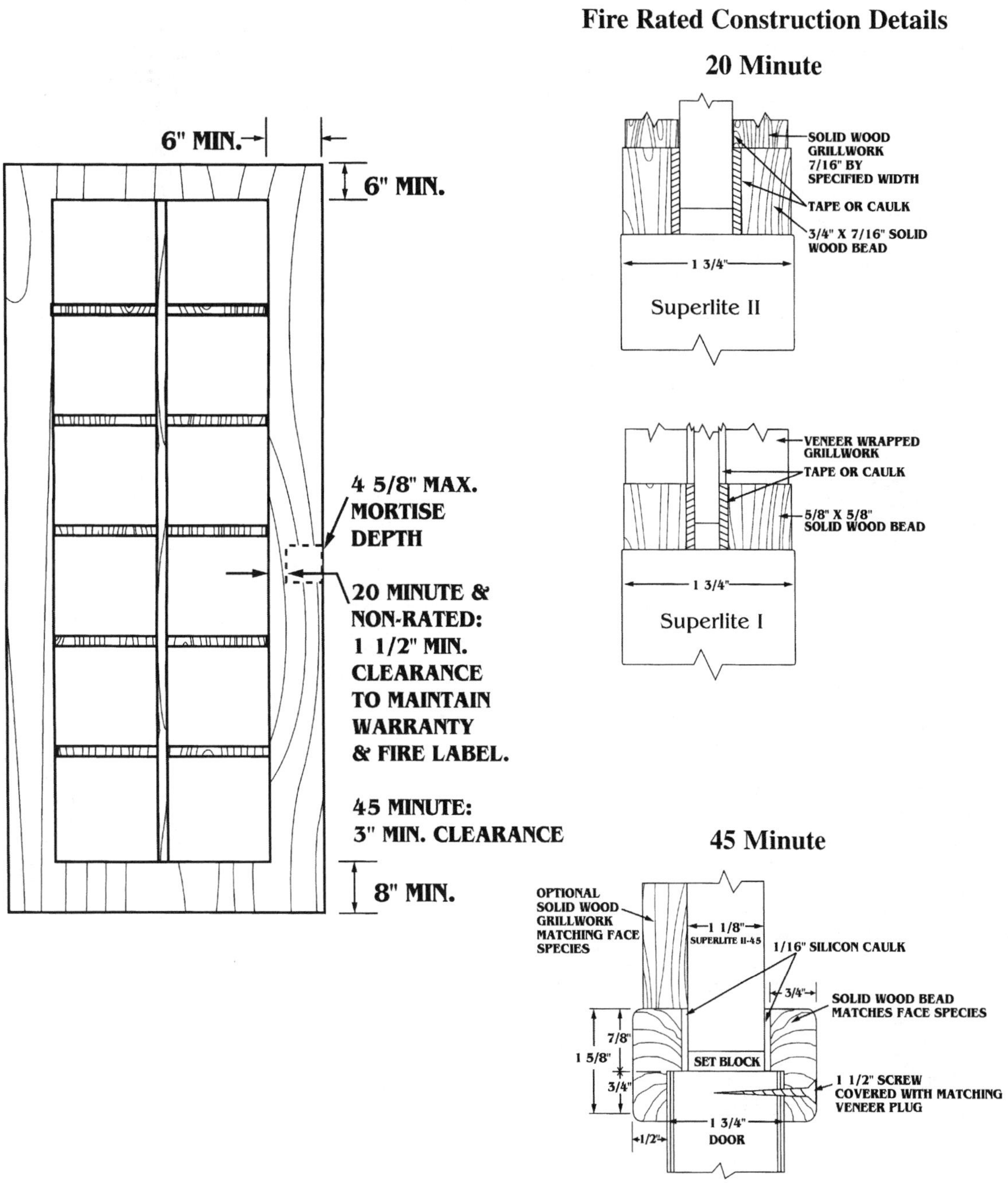

By permission of Eggers Industries, Twin Rivers, WI

12.8.1 Fire-Rated, Sound-Retardant, Lead Lined, Electrostatic Shield Doors

Fire Door Ratings and Openings Classifications

The Model Codes have established a fire door rating and operating classification system for use in protecting door openings in fire resistive rated wall constructions. The Fire Door Ratings table describes these doors. The Fire Door classifications table provides the relationship of the fire resistive ratings of doors and the use and rating of the wall in which the door opening is installed.

All fire doors must meet the requirements of ASTM E-152 and bear certifying labels of an independent testing agency approved by the building official.

Installation is required to be in accordance with the National Fire Protection Association's Publications NFPA 80, "Fire Doors and Windows," and NFPA 101, "Life Safety Code."

Labeling and Listing

The Model Codes require fire doors to be labeled. Essentially, a label indicates the rating and use of a door. It is a permanent identifying mark attached to the door by the manufacturer. A testing organization provides random unannounced inspection of the production of the fire door. The manufacturer, by labeling the door, indicates compliance with the standard fire test for fire doors and NFPA 80. In addition to the door, the door frame and hardware are required to be labeled for use with a specific fire door. All fire doors must be self-closing and self-latching.

Fire Door Ratings

LABEL	RATING	DESCRIPTION	WALL RATING
20-minute	1/3 hr.	For smoke and draft control between rooms or office corridors.	1 hr.
45-minute	3/4 hr.	In corridor and room partitions.	1 hr.
60-minute	1 hr.	In one-hour enclosures in vertical exitways.	1 hr.
90-minute	1-1/2 hr.	In two-hour enclosures in vertical exitways.	2 hr.

Special Function Doors

Sound Retardant (Acoustical):

Sound Transmission Class (STC) ratings are prescribed in ASTM Standard E-90. Door thickness may exceed 1-3/4". 1-3/4" doors with gasketing can provide varying STC ratings. These doors generally have cores with a damping compound which prevents the faces from vibrating in unison. Consult manufacturer for special stop, gasketing and automatic bottom seal requirements. Contact NWWDA for 1989 Acoustical Test conducted by Warnock Hersey, document #495-0015.

> ***Note to Specifiers:***
> *Specify the Sound Transmission Class (STC) required.*

Lead Lined (X-ray):

These doors are manufactured with a continuous lead sheet from edge to edge in the center of the door or between the crossbanding and the core.

> ***Note to Specifiers:***
> *Specify the thickness of the lead which determines the shielding rating.*

Bullet Resistant:

These doors are manufactured with special materials which resist penetration by shots of various calibers. Resistance may be rated as resistant to medium power, high-power or super-power small arms and high-power rifles.

Electrostatic Shield:

These doors are manufactured with wire mesh either in the center of the core or between the crossbanding and the core. The mesh is grounded with electrical leads through the hinges to the frame.

> ***Note to Specifiers:***
> *Specify the number and location of electrical leads.*

12.9.0 Data Required to Order Pre-Machined Wood Doors

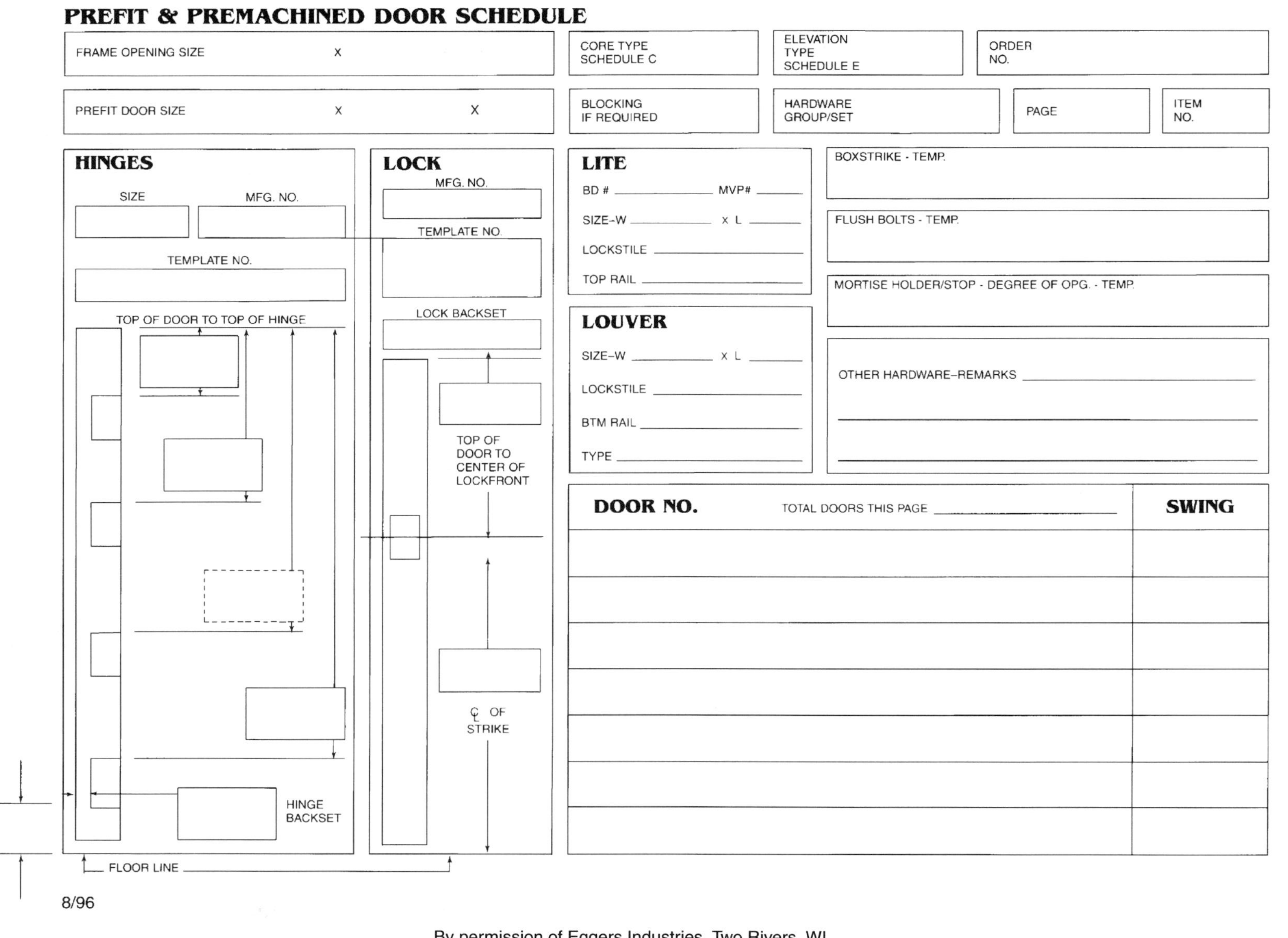
PREFIT & PREMACHINED DOOR SCHEDULE

FRAME OPENING SIZE X
CORE TYPE SCHEDULE C
ELEVATION TYPE SCHEDULE E
ORDER NO.

PREFIT DOOR SIZE X X
BLOCKING IF REQUIRED
HARDWARE GROUP/SET
PAGE
ITEM NO.

HINGES
SIZE
MFG. NO.
TEMPLATE NO.
TOP OF DOOR TO TOP OF HINGE
HINGE BACKSET
FLOOR LINE

LOCK
MFG. NO.
TEMPLATE NO.
LOCK BACKSET
TOP OF DOOR TO CENTER OF LOCKFRONT
℄ OF STRIKE

LITE
BD # ______ MVP# ______
SIZE–W ______ X L ______
LOCKSTILE ______
TOP RAIL ______

LOUVER
SIZE–W ______ X L ______
LOCKSTILE ______
BTM RAIL ______
TYPE ______

BOXSTRIKE - TEMP.
FLUSH BOLTS - TEMP.
MORTISE HOLDER/STOP - DEGREE OF OPG. - TEMP.
OTHER HARDWARE–REMARKS ______

DOOR NO.
TOTAL DOORS THIS PAGE ______
SWING

8/96

By permission of Eggers Industries, Two Rivers, WI

12.9.1 Hardware and Special Reinforcing Requirements

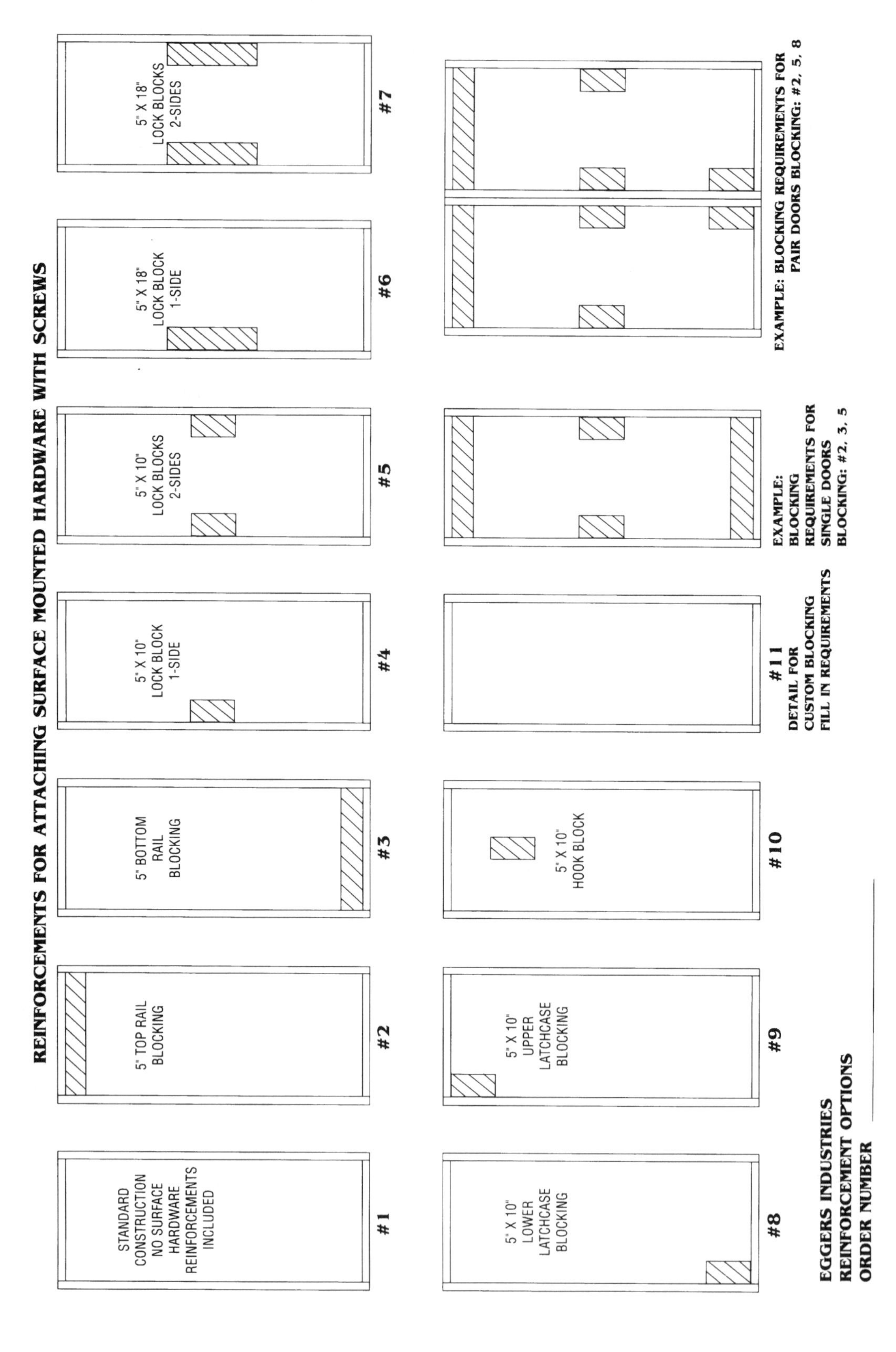

By permission of Eggers Industries, Twin Rivers, WI

12.9.2 Wood Door Glazing and Louver Options

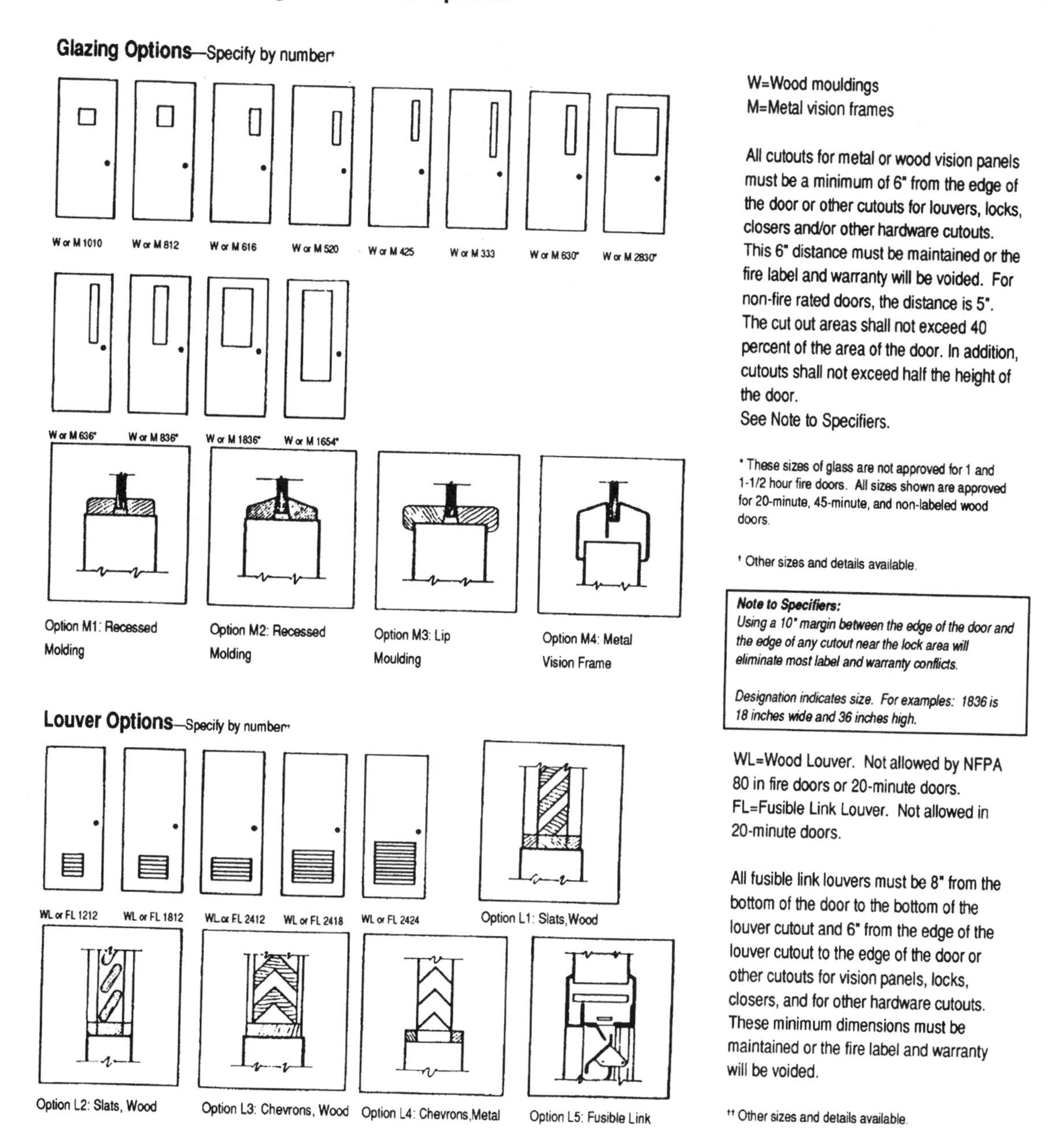

By permission of National Wood Window and Door Association, Des Plaines, Illinois

12.10.0 Installation of Exterior Wood Swinging Doors

- Measure the rough opening for size, out of plumb, and out of square.
- Check the existing sub-sill and ensure that it is level.
- Review the manufacturer's installation tolerances and instructions for proper dimensions.
- In the absence of any manufacturer's information, the rough opening should be no more than 1½ inches wider and no more than 1 inch higher than the outside dimensions of the door frame jamb.
- The rough opening should be no more than ⅛ inch out of plumb over the height of the opening.
- The sub-sill should be capable of being leveled to within 1⁄16 inch over the width of the opening, but not sloped to the interior of the structure.

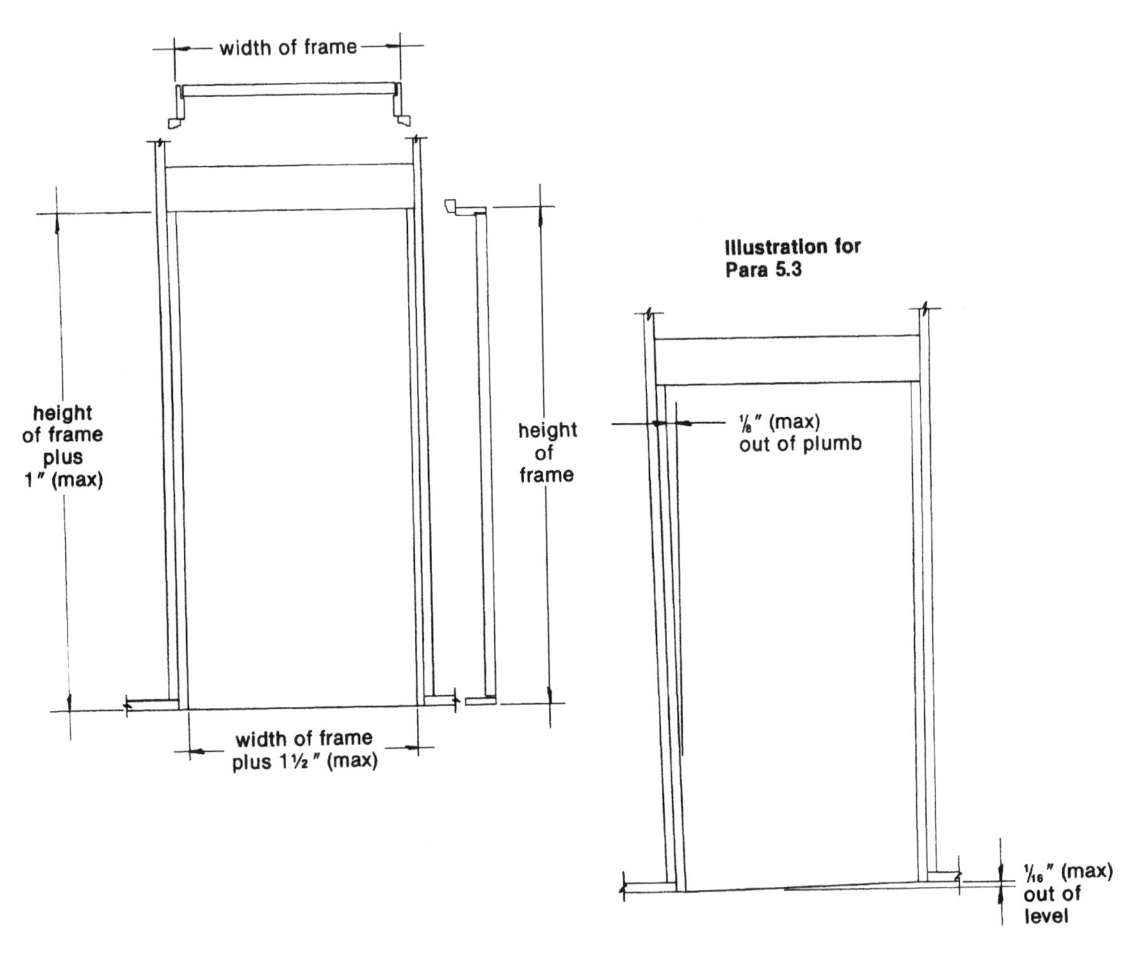

By permission of National Wood Window and Door Association, Des Plaines, Illinois

12.11.0 Warp Tolerance and Telegraphing Tolerances for Wood Doors

Warp

Warp is any distortion in the door itself, and it does not refer to the door in relation to the frame or the jamb in which it is hung. Warp is measured by placing a straight edge or a taut string on the concave face and determining the maximum distance from the straight edge or string to the door face. The accompanying table and drawing illustrate the standard and test.

Door Thickness	Door Size	Warp a defect when maximum deviation exceeds
1-3/8" [35 mm]	3'-0" x 7'-0" or smaller [900 x 2100 mm]	1/4" [6 mm]
1-3/4" [44 mm] or thicker	3'-6" x 7'-0" or smaller [900 x 2100 mm]	1/4" [6 mm]
1-3/4" [44 mm] or thicker	Larger than 3'-6" x 7'-0" [900 x 2100 mm]	1/4" [6 mm] in any 3'-6" x 7'-0" section [900 x 2100 mm]
NOTE: 1-3/8" doors are not recommended for sizes in excess of 3'-0" x 7'-0"		

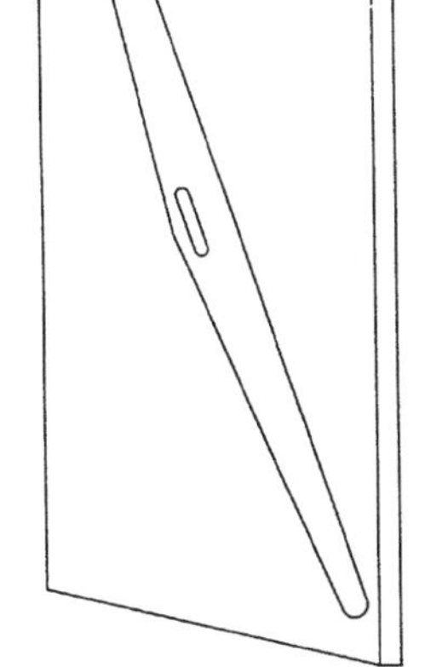

Illustration of Warp Test

Show-through or Telegraphing

Telegraphing is any distortion in the face veneer of a door caused by variations in thickness between the core materials and/or the vertical or horizontal edge bands. In any grade, variation from a true plane in excess of 0.010" in any three inch span is considered a defect. The accompanying drawing illustrates the typical condition. (The selection of high gloss finishes should be avoided because they tend to accentuate natural variations in material and construction.)

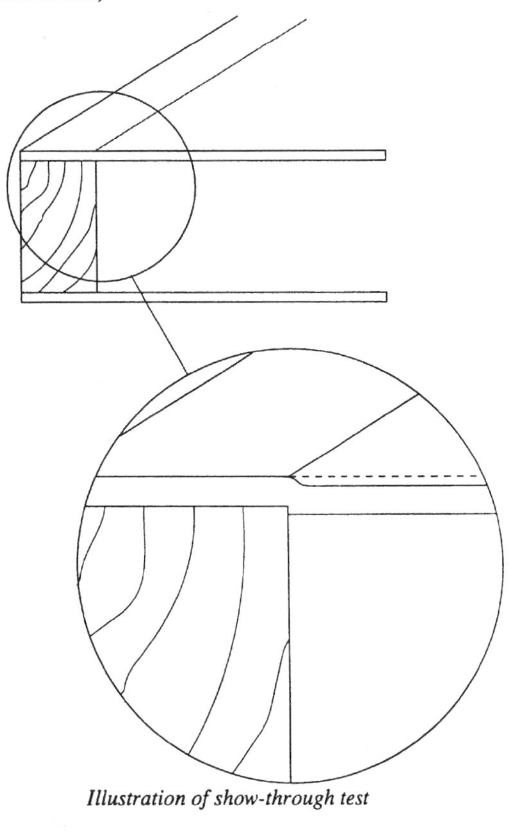

Illustration of show-through test

By permission of National Wood Window and Door Association, Des Plaines, Illinois

12.12.0 How to Store, Handle, Finish, Install, and Maintain Wood Doors

Installation

1. The utility or structural strength of the doors must not be impaired when fitting to the opening, in applying hardware, in preparing for lights, louvers, plant-ons, or other detailing.
2. Use two hinges for solid-core doors up to 60 inches in height, three hinges up to 90 inches in height, and an additional hinge for every additional 30 inches of height or portion thereof. Interior hollow-core doors weighing less than 50 pounds and not over 7'6" in height can be hung on two hinges. Use heavy weight hinges on doors over 175 lbs. Pivot hardware can be used in lieu of hinges. Consult the hinge or pivot hardware manufacturer with regard to weight and size of hinges or pivots required.
3. Clearances between top and hinge door edges and door frame should be a minimum of ⅛" (3.2 mm). For a single door latch edge, the clearance should be ⅛" (3.2 mm). For a pair of doors, the meeting edge clearance should be 1/16" (1.6 mm) per leaf. The bottom edge should be ¾" (19 mm)

maximum from the top of a non-combustible floor and ⅜" (10 mm) maximum from the top of a non-combustible sill.

4. All hardware locations, preparations, and methods of attachment must be appropriate for the specific door construction. Templates for specific hardware preparation are available from hardware manufacturers or their distributors.
5. When light or louver cutouts are made for exterior doors, they must be protected to prevent water from entering the door core.
6. Pilot holes must be drilled for all screws that act as hardware attachments. Threaded to the head screws are preferable for fastening hardware to non-rated doors and are required on fire-rated doors.
7. In fitting for height, do not trim the top or bottom edge by more than ¾ inches unless accommodated by additional blocking. Trimming of fire-rated doors must be in accordance wih NFPA 80.
8. Doors and door frames should be installed plumb, square, and level.

Cleaning and Touchup

1. Inspect all wood doors prior to hanging them on the job. Repair noticeable marks or defects that might have occurred from improper storage and handling.
2. Field repairs and touchups are the responsibility of the installing contractor upon completion of initial installation. Field touchups shall include the filling of exposed nail or screw holes, re-finishing raw surfaces resulting from job fitting, repairing job-inflicted scratches and mars, and final cleaning of finished surfaces.
3. When cleaning door surfaces, use a non-abrasive commercial cleaner designed for cleaning wood door or paneling surfaces that do not leave a film residue that would build-up or affect the surface gloss of the door finish.

Adjustment and Maintenance

1. Ensure that all doors swing freely and do not bind in their frame. Adjust the finish hardware for proper alignment, smooth operation and proper latching without unnecessary force or excessive clearance.
2. Review with the owner/owner's representative how to periodically inspect all doors for wear, damage, and natural deterioration.
3. Review with the owner/owner's representative how to periodically inspect and adjust all hardware to ensure that it continues to function as it was originally intended.
4. Finishes on exterior doors could deteriorate because of exposure to the environment. To protect the door, it is recommended that the condition of the exterior finish be inspected at least once a year and re-finished as needed.

Storage and handling

1. Store doors flat on a level surface in a dry, well-ventilated building. Doors should not come in contact with water. Doors should be kept at least 3½" off the floor and should have protective coverings under the bottom door and over the top. Covering should protect doors from dirt, water, and abuse, but allow for air circulation under and around the stack.
2. Avoid exposure of interior doors to direct sunlight. Certain species (e.g., cherry, mahogany, walnut, and teak) in an unfinished state are more susceptible to discoloration if exposed to sunlight or some forms of artificial light. To protect doors from light damage after delivery, opaque wrapping of individual doors could be specified.
3. Do not subject interior doors to extremes of heat and/or humidity. Do not allow doors to come in contact with water. Prolonged exposure could cause damage. Buildings where humidity and temperature are controlled provide the best storage facilities (recommended conditions 25%–55% RH and 50° F to 90° F).

4. Do not install doors in buildings that have wet plaster or cement unless they have been properly finished. Do not store doors in buildings with excessive moisture content. HVAC systems should be operating and balanced.
5. Doors should always be handled with clean hands or while wearing clean gloves.
6. Doors should be lifted and carried when being moved, not dragged across one another.

Finishing

1. Wood is hygroscopic and dimensionally influenced by changes in moisture content caused by changes within its surrounding environment. To ensure uniform moisture exposure and dimensional control, all surfaces must be finished equally.
2. Doors should not be considered ready for finishing when initially received. Before finishing, remove all handling marks, raised grain, scuffs, burnishes, and other undesirable blemishes by block sanding all surfaces in a horizontal position with a 120- 150- or 180-grit sandpaper. Solid-core flush doors, because of their weight, naturally compress the face veneer grain while in the stack. Therefore, sanding of the overall surface will be required to open the veneer grain to receive a field applied finish evenly. To avoid cross-grain scratches, sand with the grain.
3. Certain species of wood, particularly oak, might contain extractives that react unfavorably with foreign materials in the finishing system. Eliminate the use of steel wool on bare wood, rusty containers or other contaminants in the finishing system.
4. A thinned coat of sanding sealer can be applied prior to staining to promote a uniform finish and avoid sharp contrasts in color or a blotchy appearance. Door manufacturers are not responsible for the final appearance of field-finished doors. It is expected that the painting contractor will make adjustments, as needed, to achieve desired results.
5. All exposed wood surfaces must be sealed, including top and bottom rails. Cutouts for hardware in exterior doors must be sealed prior to installation of hardware and exposure to weather.
6. Dark-colored finishes should be avoided on all surfaces if the door is exposed to direct sunlight, in order to reduce the chance of warping or veneer checking.
7. Water-based coatings on unfinished wood could cause veneer splits, highlight joints, and raise wood grain. If used on exterior doors, the coating should be an exterior-grade product. When installed in exterior applications, doors must be properly sealed and adequately protected from the elements. Please follow the finish manufacturer's recommendations regarding the correct application and use of these products.
8. Be sure that the door surface being finished is satisfactory in both smoothness and color after each coat. Allow adequate drying time between coats. Desired results are best achieved by following the finish manufacturer's recommendations. Do not finish doors until a sample of the finish has been approved.
9. Certain wood fire doors have fire-retardant salts impregnated into various wood components that make the components more hygroscopic than normal wood. When exposed to high-moisture conditions, these salts will concentrate on exposed surfaces and interfere with the finish. Before finishing the treated wood, reduce the moisture content below 11% and remove the salt crystals with a damp cloth followed by drying and light sanding. For further information on fire doors, see the NWWDA publication regarding *Installing, Handling & Finishing Fire Doors.*

12.13.0 Aluminum Door Types/Sectional Dimensions

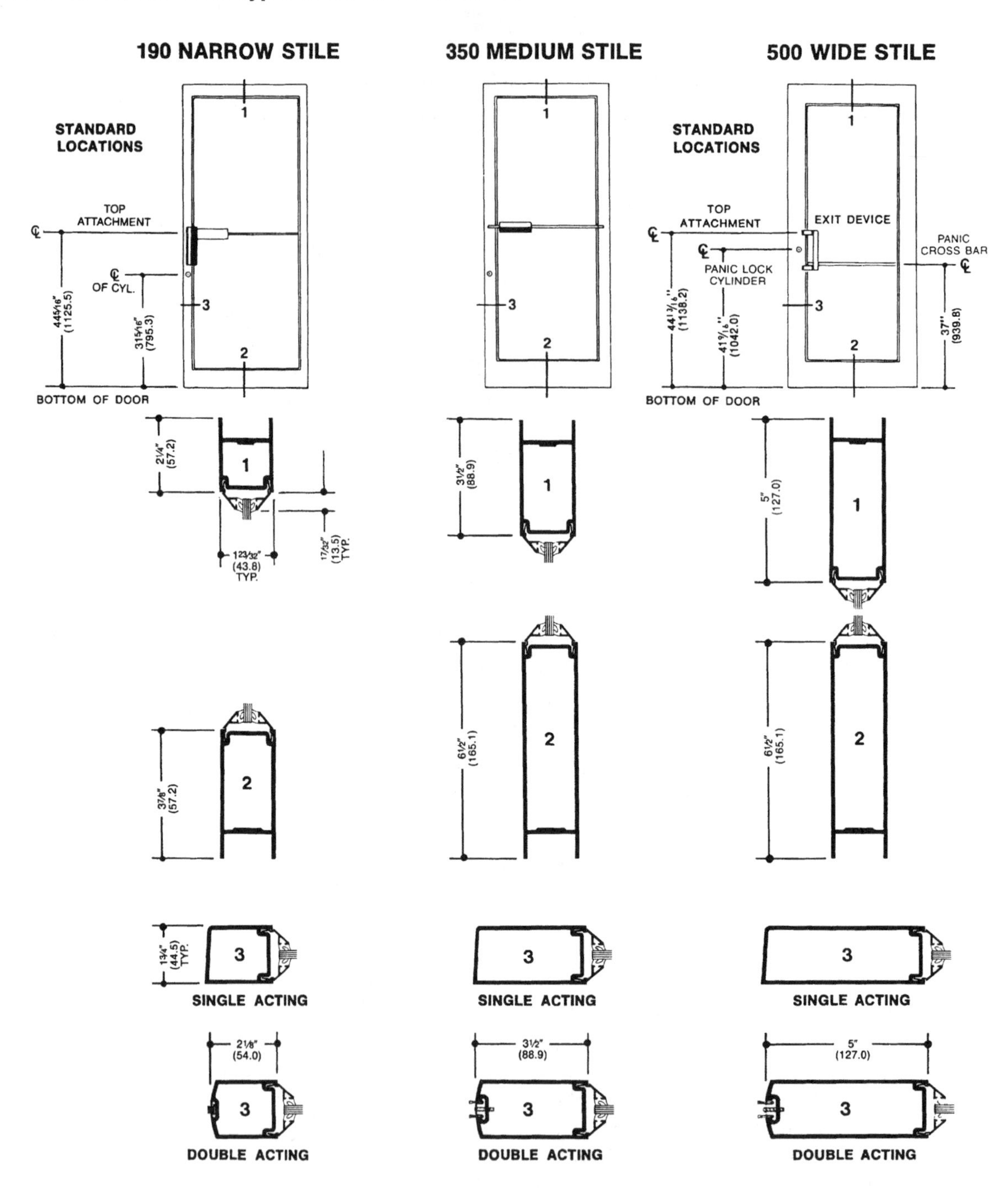

By permission of Kawneer Corporation, Norcross, Georgia

12.13.1 Aluminum Revolving Doors

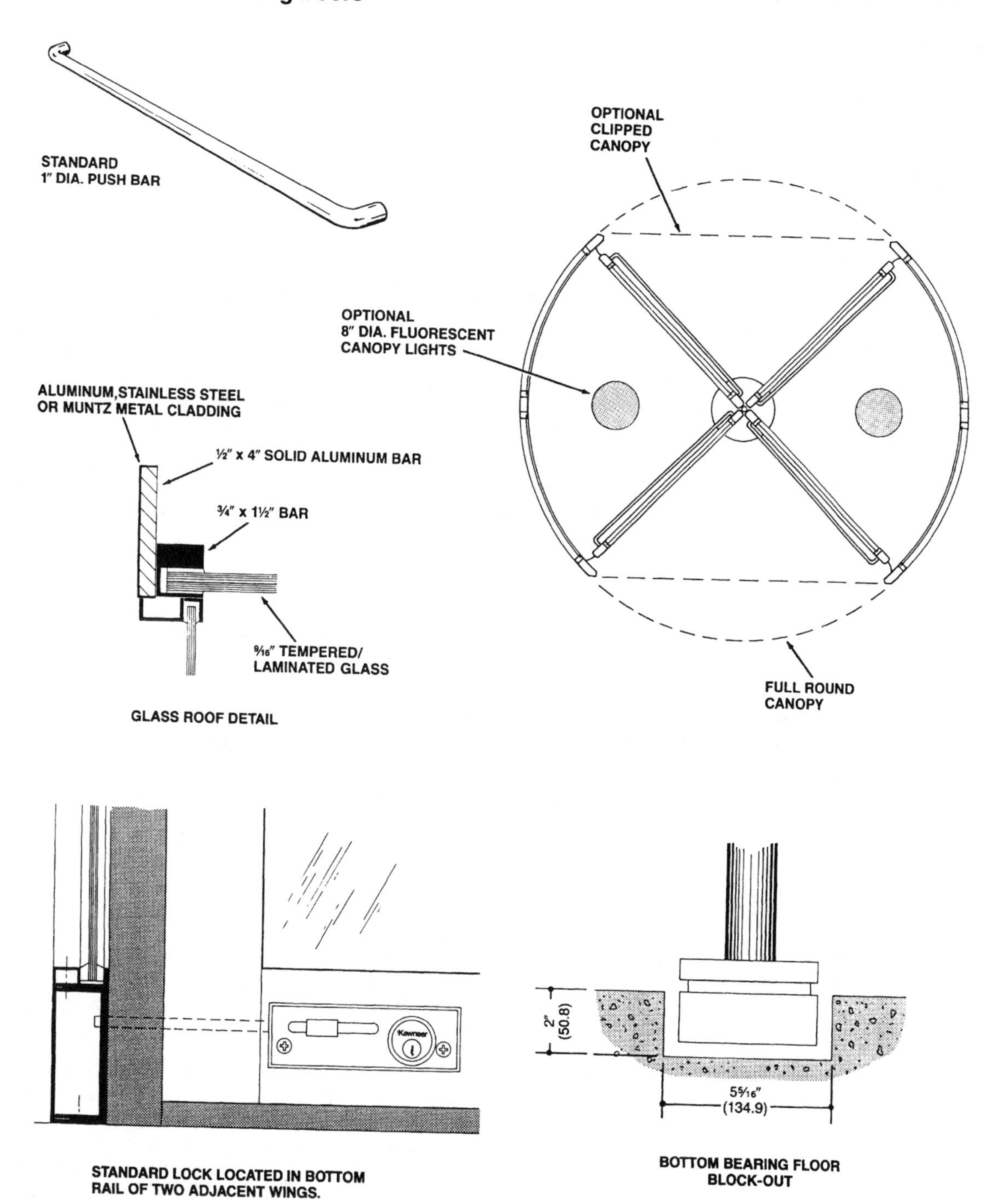

By permission of Kawneer Corporation, Norcross, Georgia

12.14.0 Windows (Aluminum, Wood, Steel, and Plastic

Aluminum Windows

According to ANSI/AAMA-101, aluminum used in the manufacture of windows must meet the following specifications:

- *Yield strength* 16,000 psi (110.24 MPa)
- *Tensile strength* 22,000 psi (151.6 MPa)
- *Coefficient of thermal expansion* 13 × 10 to the –6-inch.inch/(2.45 cm) degree Fahrenheit (to convert F to C, subtract 32 and divide by 1.8)

Aluminum windows are susceptible to corrosion if their painted or anodized surfaces are exposed to the environment. Unless airborne contaminants are removed periodically by washing, they will attract and hold moisture. In combination with pollutants, over time, the exposed painted or anodized metal surface will be attached.

Aluminum is an excellent heat and cold transmitter. Without a thermal break in the window frame, it will always present a cold interior surface during winter months. Aluminum window components tend to expand and contract rapidly in response to temperature changes, causing stresses on improperly installed glazing. If these stresses become excessive, cracks will develop in the glazed section However, aluminum windows are very cost effective; are manufactured in a wide range of sizes, configurations, and colors; and are generally maintenance-free, compared to wood windows.

Steel Windows

These windows are usually constructed of hot-rolled, #12 steel and are classified by the minimum combined weight of the outside frame and vent member.

- *Residential grade* Minimum 2.0 pounds (0.9 kilogram) with maximum 1 inch (2.54 cm) from front to back. The maximum dimension is 6½ feet (1.98 meters) and the maximum spacing of mullions is 3½ feet (1.07 meters).
- *Standard grade* Minimum 3.0 pounds per lineal foot (1.36 kilograms per 30.48 cm) with a maximum of 1¼ inches (3.17 cm) front to back, ¾ inch (1.9 cm) vertical muntin required in projected vents over 4½ feet (1.37 meters) wide. The maximum glazed area is 60 square feet (5.58 square meters) and a maximum dimension is 10 feet (3.05 meters). For combined units, a maximum mullion spacing of 6½ feet (1.98 meters) is permitted.
- *Heavy intermediate grade* Minimum of 3.5 pounds per lineal foot (1.58 kilograms per 30.48 cm) with a maximum of 1⁵⁄₁₆ inches (3.33 cm) from front to back, ¾ inches (1.90 cm) vertical muntin in projected vents over 5 feet (1.52 meters). The maximum glazed area is 84 squre feet (7.8 square meters). For combined units, a maximum spacing of mullions is 6½ feet (1.98 meters).
- *Heavy custom grade* Minimum 4.2 pounds per lineal foot (1.91 kilograms per 30.48 cm) with a maximum of 1½ inches (3.8 cm) from front to back of the ventilator and the supporting frame.

Steel windows exhibit great strength, allowing for large glazed areas. Thermal expansion is minimal, but thermal breaks in the frames are required to prevent the transmission of heat and cold from exterior to interior areas. These windows require periodic maintenance to ensure the integrity of their protective coatings to prevent rusting of their components.

Plastic/Vinyl Windows

Vinyl windows are manufactured to ASTM D4216 specifications that require the minimum properties of the polyvinychloride (PVC) to have an impact resistance of 0.65 four pounds per inch (0.045 kilograms per square centimeter) of notch, a tensile strength of 5000 psi (34.5 Mpa), a modulus of elasticity in tension of 0.29×10^6, deflection temperature under load at 140 degrees F (77 degrees C) and a coefficient of expansion of less than 2.2 × 10 to the minus 5th inch (2.54 cm)/inch (2.54 cm)/degree Fahrenheit (to convert F to C, subtract 32 and divide by 1.8).

Vinyl windows can be manufactured in many textures and colors, including wood-finish look-alikes. Although stabilizers are added to the vinyl compound, some dark colors have been known to fade or distort when exposed to strong sunlight for extended periods of time. Vinyl windows are difficult to refinish if damaged or if the color fades. Vinyl windows exhibit excellent thermal properties, do not expand or contract to any noticeable degree when subject to heat or cold and are relatively maintenance free and cost effective.

Wood Windows

Wood windows offer beauty and warmth, as well as exhibiting excellent thermal qualities. Protection from the elements and condensation requires that both interior and exterior surfaces are either painted or otherwise sealed to prevent wood rot. Several manufacturers offer aluminum or vinyl cladding to minimize exterior maintenance.

12.15.0 Window Performance Grades and ANSI and NWWDA Standards for Wood Windows

Grades of Performance

	Pass	Grade 20	Grade 40	Grade 60
Preliminary (Design) Load: (Minimum test pressure sustained without damage, psf)	13.3	13.3	26.6	40
Operating Force (Pounds of force)		25	30	35
Air Infiltration: (Maximum infiltration at test pressure)	0.34	0.34	0.25	0.10
Water Penetration: (Minimum test pressure sustained without leakage, psf)	2.86	2.86	4.43	6.24

Grades of Performance*
(Metric Units)

	Grade 20	Grade 40	Grade 60
Preliminary Load: (Minimum test pressure sustained without damage, Pa)	638	1277	1920
Operating Force (Newtons)	111	133	156
Air Infiltration: (Maximum air infiltration in cfm at 1.56 psf test pressure)	5.26×10–4	3.81×10–4	1.55×10–4
Water Penetration: (Minimum test pressure sustained without leakage, Pa)	137	215	300
Structural Performance: (Minimum test pressure sustained without damage, Pa)	960	1920	2880

*The loads and levels prescribed in this table are actual quantities to be applied or measured during testing and do not include consideration of safety factors.

(Reprinted from NWWDA I.S. 2–87)

American National Standards Institute (ANSI)
National Wood Window and Door Association (NWWDA)
Standards for Wood Window Units I.S. 2–87

	Air Infiltation ASTM E-283	Water Infiltration ASTM E-547	Physical Load ASTM E-330
Grade 20- Suitable for residential construction	At an air pressure of 1.56 PSF (25 MPH), not more than .34 cubic feet per minute (CFM) per lineal foot of sash crack perimeter	No water shall pass beyond interior of unit in a 15 minute test, 5 gals. per hour per sq. ft. (equals 8" of rain per hour), under air pressure of 2.86 PSF (34 MPH)	Positive–20 PSF (89) MPH is applied to the exterior of the window, held for 10 seconds and released. Negative – Same as above as applied to the interior of the window and released. No glass breakage, no hardware damage nor deformation shall result in malfunction. Residual deflection to any member shall not exceed .4% of its span.
Grade 40- Suitable for light commercial construction	Same as above, not more than .25 cubic feet per minute	Same as above under air pressure of 4.43 PSF (42 MPH)	Same as above with positive and negative testing done under 40 PSF (126.5 MPH)
Grade 60- Suitable for heavy commercial construction	Same as above, not more than .10 cubic feet per minute	Same as above under air pressure of 6.24 PSF (50 MPH)	Same as above with positive and negative testing done under 60 PSF (154.9 MPH)

By permission of Marvin Windows and Doors, Warroad, Minnesota

12.16.0 Effect of Glazing Selections on Heat Gain

Heat Gain and Performance Data

Heat Gain Data

In areas of the U.S. where cooling is the major energy cost, glazing may be the most important factor in energy-saving. That's because cooling costs are based almost solely on heat gains transmitted through the glass. The accompanying table is used to show maximum heat gain by type of glass.

Clear	Heat Gain†	Tinted Grey/Bronze	Heat Gain†	Medium Performance Reflective	Heat Gain†
Single-pane 3/32" or 1/8"	214	Single-pane grey 3/16" (for comparison only)	165	Single-pane bronze (for comparison only)	106
Single-pane 3/16" (for comparison only)	208	Single pane bronze 3/16" (for comparison only)	168		
Double-pane (for comparison only)	186				
Double-pane high-performance insulating	113	Double-pane high-1 performance sun insulating			

12.17.0 NWWDA Air-Infiltration Standards

Operating force refers to maximum amount of force, expressed in pounds, required to open and close a window unit.

Air Infiltration • Testing

NWWDA I.S. 2.93- DP Ratings	DP15	DP20	DP25	DP30	DP35	DP40
Design pressure (psf)	15	20	25	30	35	40
Structural test pressure (psf)	22.5	30	37.5	45	52.5	60
Water infiltration (psf)	2.86	3.00	3.75	4.50	5.25	6.00
Air infiltration @ 1.57 psf (cfm/ft^2)	.37	.37	.25	.25	.25	.15
Operating force (lb)	25	25	30	30	30	35

NWWDA I.S.3 (old I.S. 2-87)	Grade 20	Grade 40	Grade 60
Design pressure (psf)	13.3	26.6	40.0
Structural test pressure (psf)	20	40	60
Water infiltration (psf)	2.86	4.43	6.24
Air infiltration @ 1.57 psf (cfm/ft^2)*	.34	.25	.10
Operating force (lb)	25	30	35

*Note: I.S. 2-87 - air infiltration @ 1.57 psf (cfm/lin. ft. of crack)

Note: Windows had been previously rated by the structural test pressure attained (e.g., Grade 20, 40, and 60); however, units today are rated by using design pressure (DP) ratings.

By permission of Anderson Corporation, Bayport, Minnesota

12.18.0 Steps Required To Order Wood/Clad Windows

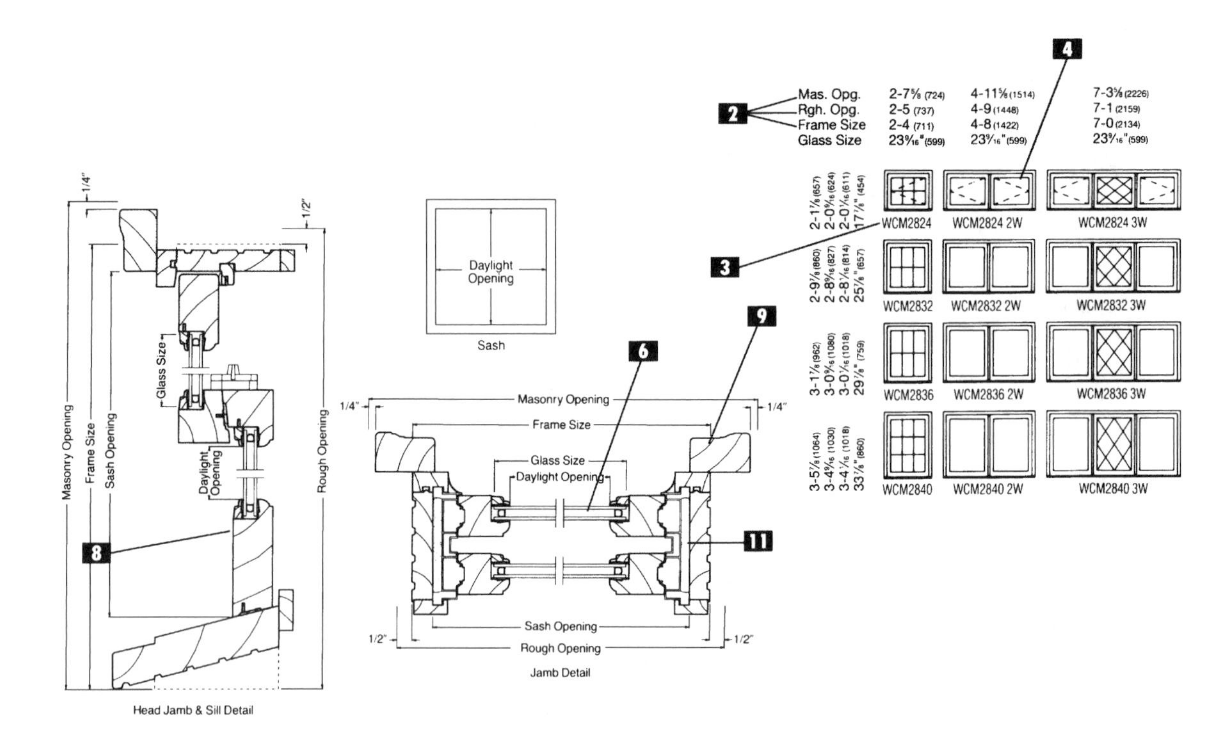

Items to consider when placing an order for windows

1. Select style and material (wood, wood/clad, etc.).
2. Determine product size by using the rough opening, masonry opening, and frame size.
3. Identify manufacturer's unit number.
4. Specify operation.
5. Specify screens, if required.
6. Specify any glazing options.
7. Specify interior wood finish (bare or factory primed).
8. Specify exterior wood finish (bare, factory primed, and clad).
9. Specify color of hardware options, any drips, metal accessory items.
10. Specify type of exterior casing.
11. Specify jamb width.
12. Select any additional options.

By permission of Marvin Windows and Doors, Warroad, Minnesota

12.19.0 Low-E Glazing (Illustration)

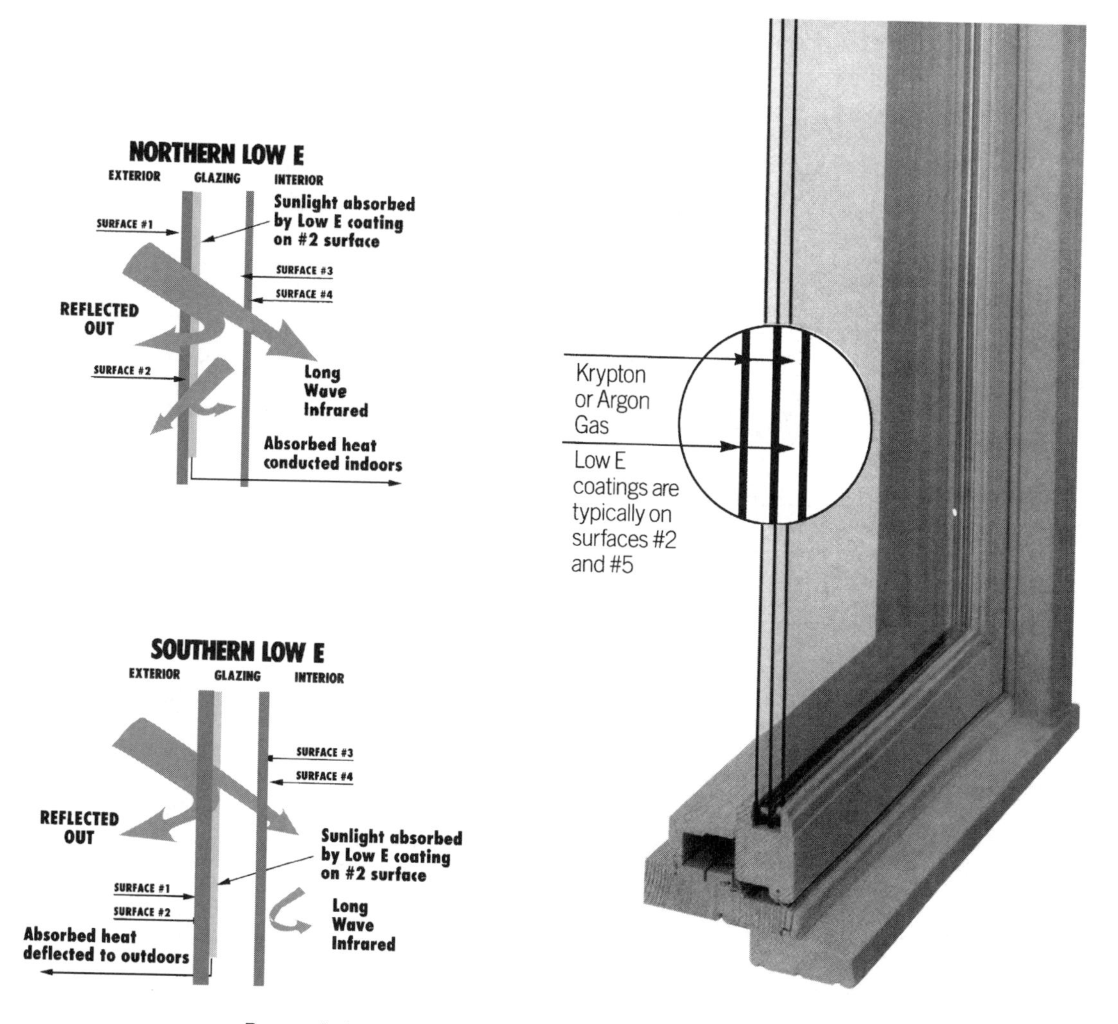

By permission of Marvin Windows and Doors, Warroad, Minnesota

12.20.0 Thermal Movement and Frame Deflection

THERMAL MOVEMENT, in frame and glass

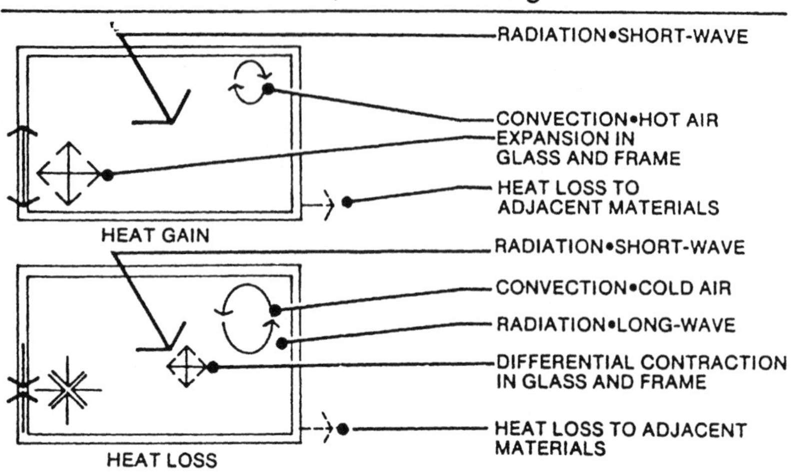

DEFLECTION, vertical framing members

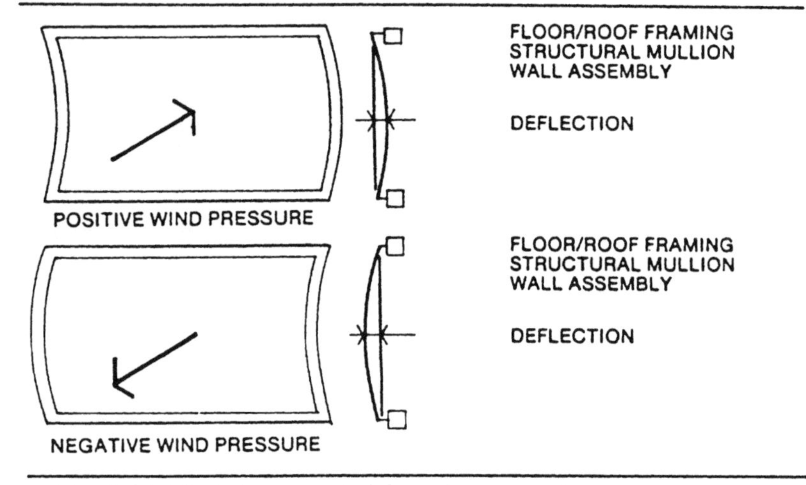

DEFLECTION, horizontal framing members

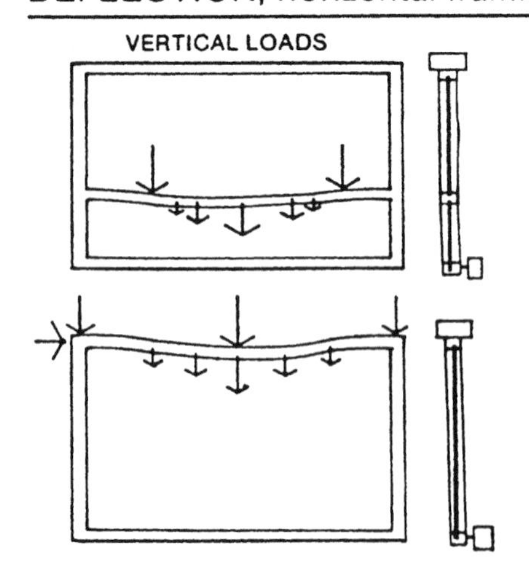

DEFLECTION IN LINTEL
DEFLECTION IN MULLION
EXCESSIVE DEFLECTION MAY IMPOSE STRESS ON GLASS

TWIST IN MULLION DUE TO VERTICAL OR LATERAL LOADS•MAY RESULT IN LATERAL PRESSURE ON GLASS

DEFLECTION IN MULLION DUE TO WEIGHT OF GLASS MAY IMPOSE STRESSES ON GLASS IN LOWER FRAME

12.21.0 Defining Fixed and Hinged Portions of French Door Assemblies

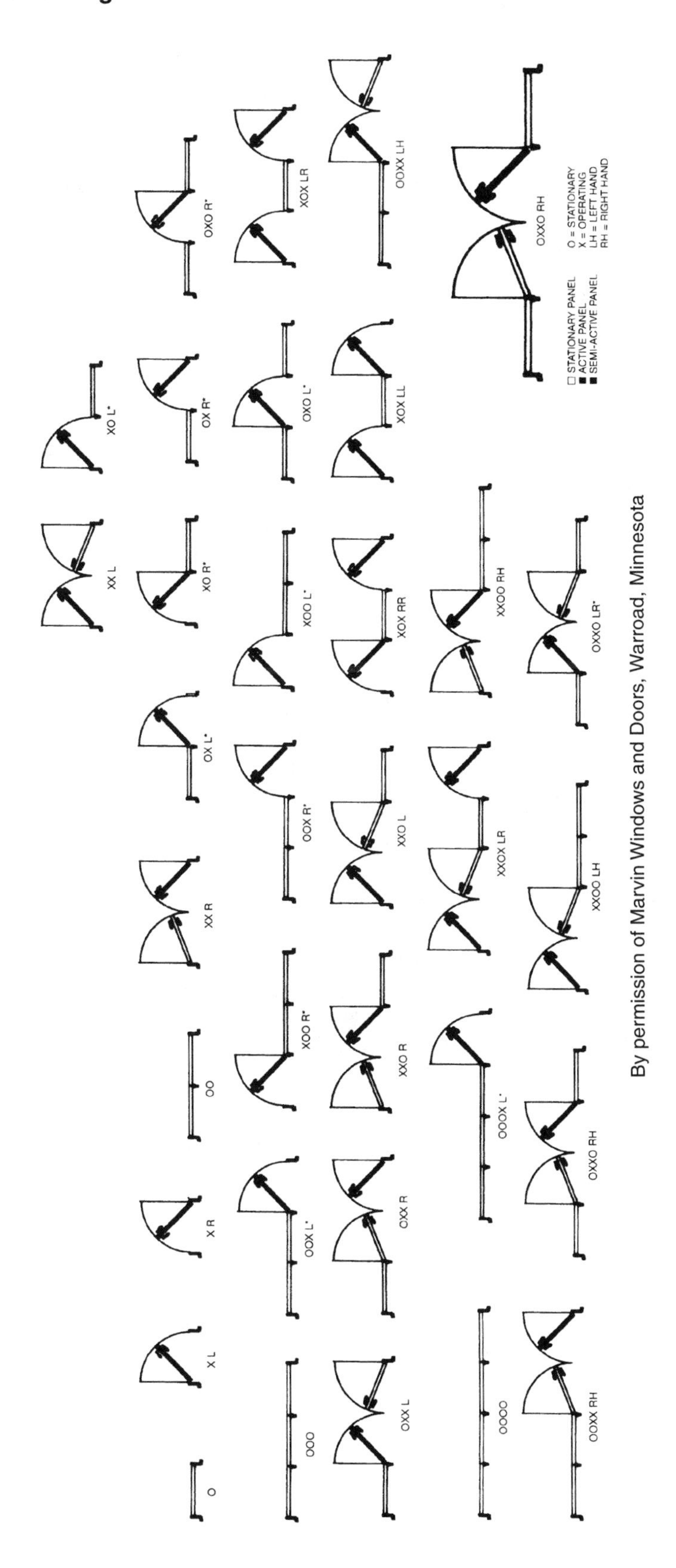

By permission of Marvin Windows and Doors, Warroad, Minnesota

12.22.0 Aluminum Window Wall (Stick-Built Construction)

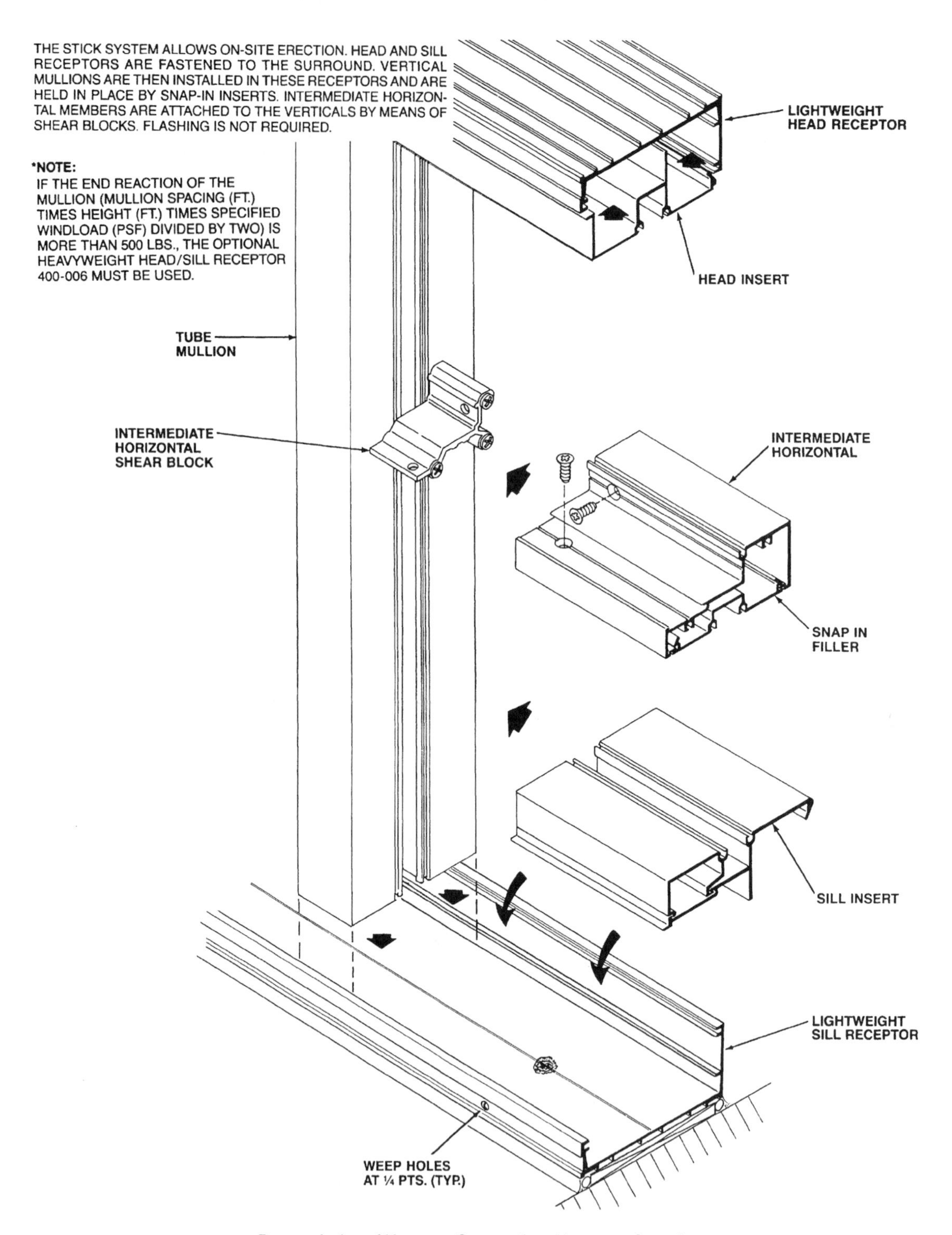

By permission of Kawneer Corporation, Norcross, Georgia

12.23.0 Aluminum Window Wall (Shear Block Fabrication)

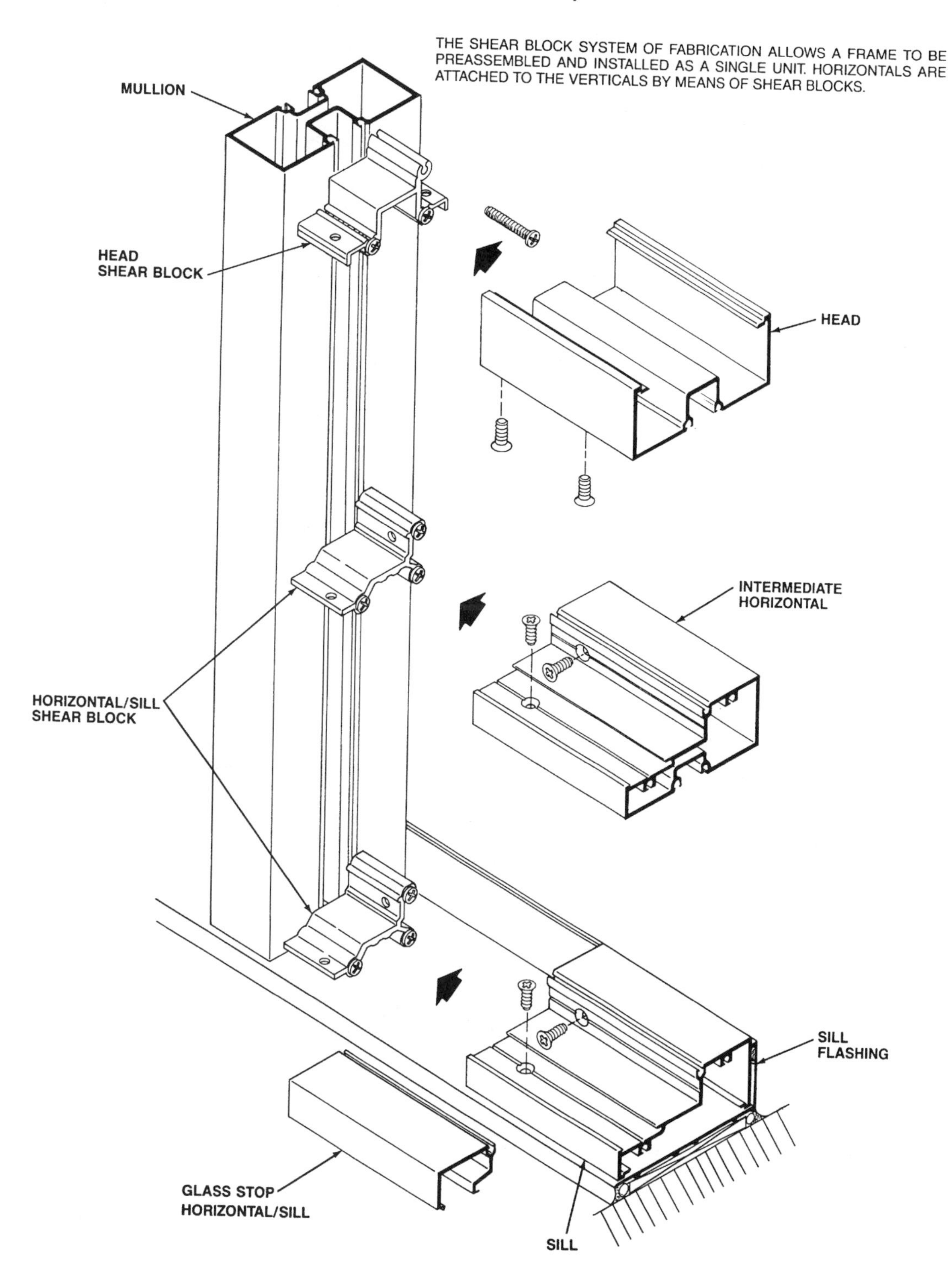

By permission of Kawneer Corporation, Norcross, Georgia

12.24.0 Aluminum Window Wall (Screwspline Fabrication)

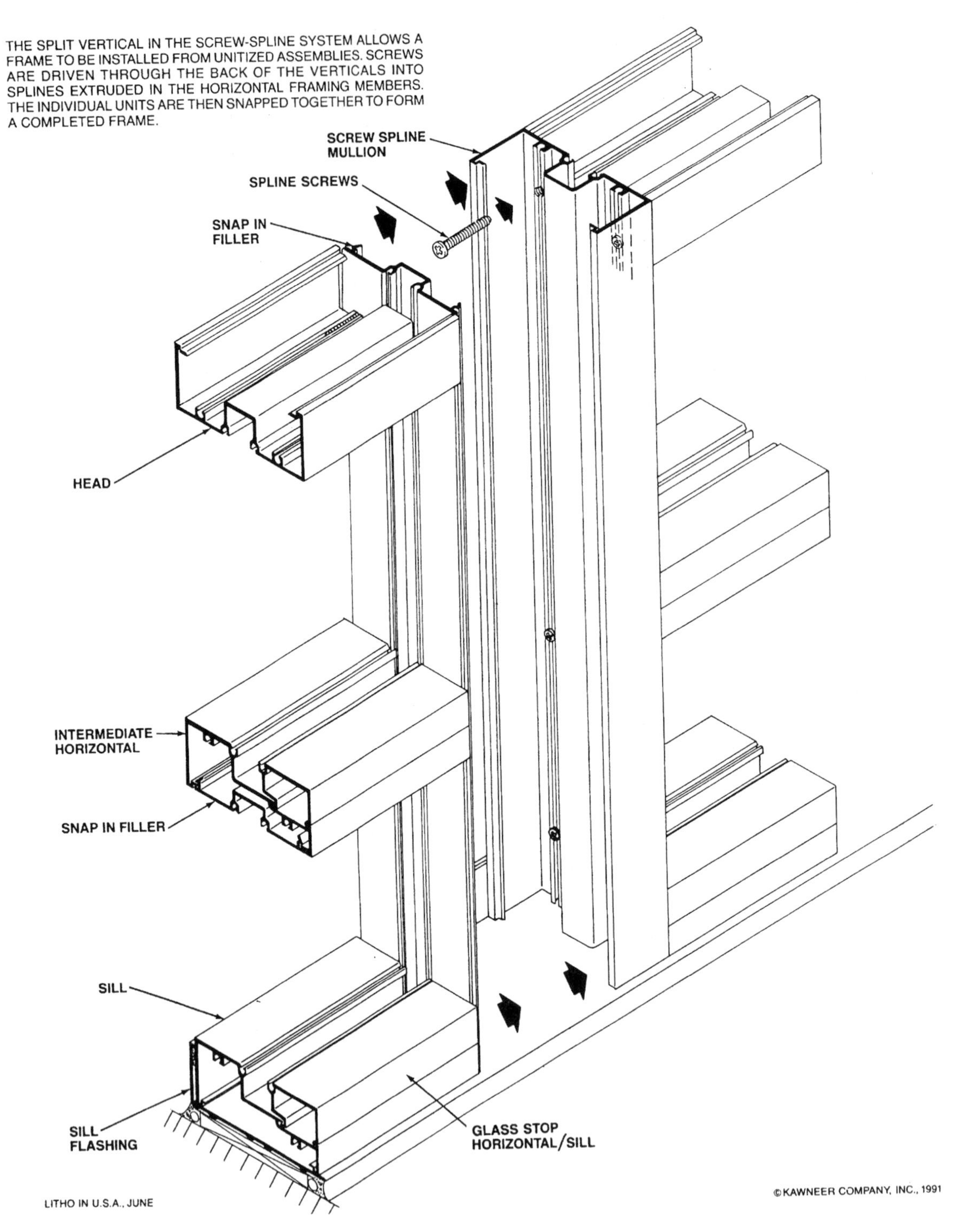

By permission of Kawneer Corporation, Norcross, Georgia

12.25.0 Sloped Glazing and Skylight Configurations

1400 S.S. is a flexible and economical slope/skylight system suited for short span, low rise applications. The following general notes and illustrations are useful in determining the proper application of 1400 S.S. If the structural requirements of a particular project exceed the design limits outlined here, consider 1600 S.G.

1. It is imperative to consult local, national, and state codes in regard to safety glazing and combined load requirements. Acrylic and polycarbonate infills should not be used in 1400 S.S. sections.
2. 1400 S.S. is intended for application to Kawneer Tri-fab 450, Tri-fab 451, Isoglaze 450 and Nucore framing systems. The height of vertical storefront materials beneath 1400 S.S. should not exceed 10′-0″ with 4′-0″ modules at a 30 PSF windload. For applications which exceed these limits consult your Kawneer Representative.
3. 1400 S.S. aluminum sections are intended for straight sloping surfaces only. Curving or bending of sections is not recommended.
4. The structural properties of 1400 S.S. should not be supplemented by the addition of structural steel or by applying it to a steel grid. For increased structural capability, see 1600 S.G.
5. 1400 S.S. accommodates slope angles from a minimum of 20° to a maximum of 45°. (Slope angles of less than 20°, regardless of the system selected, make proper drainage difficult, contribute to pooling and staining at purlin pressure plates and may lead to other maintenance problems.) For angles greater then 45° see 1600 S.G.
6. Structural silicone glazed (S.S.G.) purlins are subject to unique structural considerations when compared with conventional purlins; therefore, it is important to consult the loading charts for S.S.G. purlins (see page F2-11).
7. Maintain a minimum DLO of 6″ between hip/valley rafters and adjacent rafter at head/eave intersection.
8. Intermediate rafters are required at intersection of hip rafter and ridge.

1400 S.S. DESIGN CONFIGURATIONS

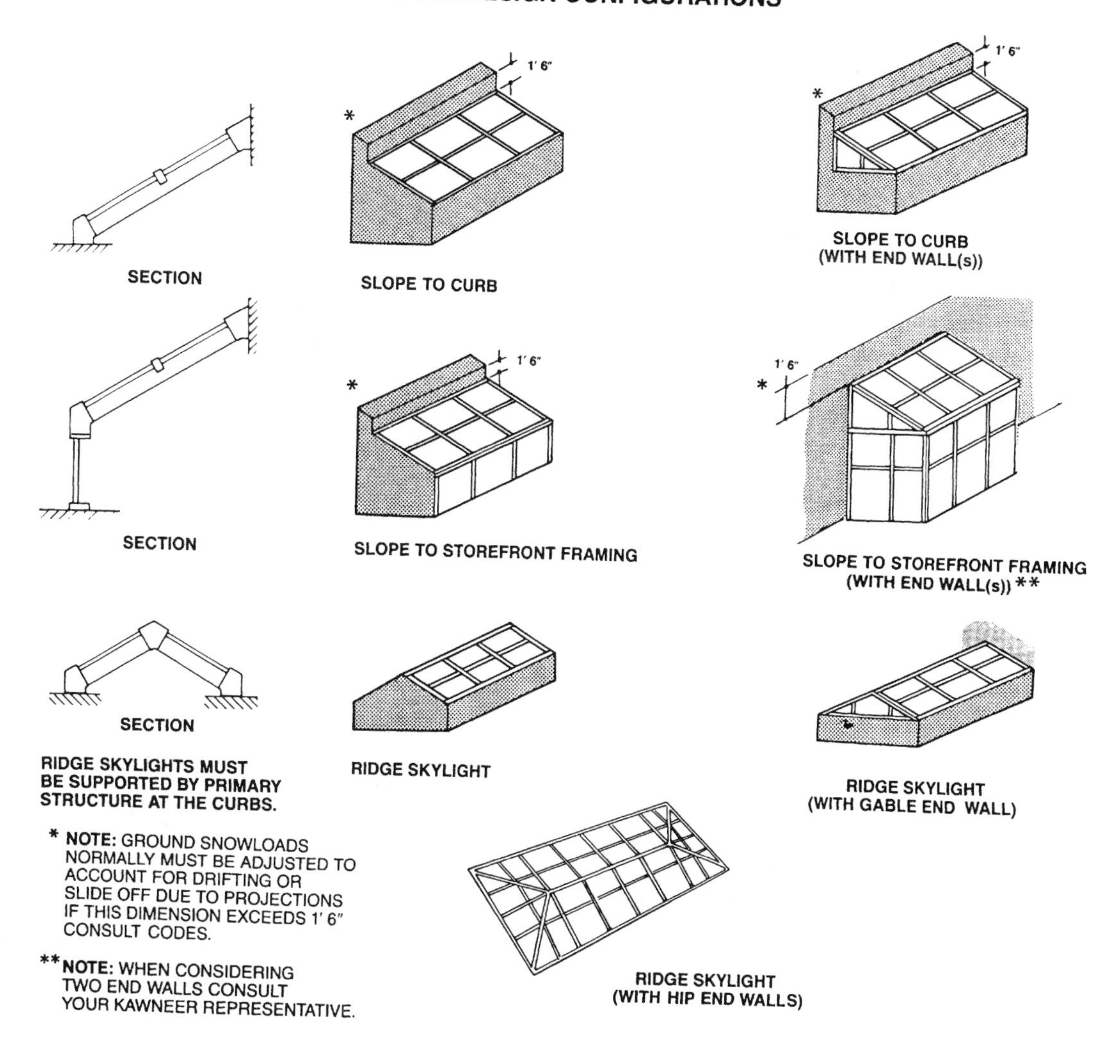

By permission of Kawneer Corporation, Norcross, Georgia

Section

13

Finish Hardware

Contents

13.0.0 Introduction to contents
13.1.0 Door hinges (types and illustrations)
13.2.0 Lockset and latchset configurations and functions
13.3.0 Heavy-duty mortise cases, hubs, and spring cartridges
13.4.0 Strikes (illustrated)
13.5.0 Door knob designs
13.6.0 Lever handle designs
13.6.1 Lever handle designs (forged and wrought)
13.7.0 Turn levers
13.8.0 Mortise cylinders
13.8.1 Rosette and blocking rings for cylinders
13.8.2 Miscellaneous cams for mortise cylinders
13.9.0 Deadbolts, spindles, security fasteners, and guard bolts
13.10.0 Construction key systems
13.10.1 Removal core cylinders
13.11.0 Panic devices (concealed/surface-applied vertical rod devices)
13.11.1 Panic devices (mortise lock devices)
13.11.2 Panic devices (rim devices; conventional and enclosed push-bar type)
13.11.3 Panic devices (rim devices, other types of pushes)
13.11.4 Panic devices (outside trim)
13.12.0 Standard keying terms, codes, and designations
13.13.0 Finish symbols and descriptions of these finishes
13.14.0 Recommended number of hinges and frequency of operations
13.15.0 ASTM specifications applicable to finish hardware requirements

13.0.0 Introduction to Contents

Finish hardware selections and specifications span a wide range of functions, materials of construction and decorative requirements. The information contained in this section touches on hardware mainstays: locksets, latchsets with trim and cylinders, hinges (butts), panic devices, and informative specification tables. Although much of this information was furnished by two manufacturers, it remains very much generic in nature.

13.1.0 Door Hinges (Types and Illustrations)

The butts are available in a wide range of metals

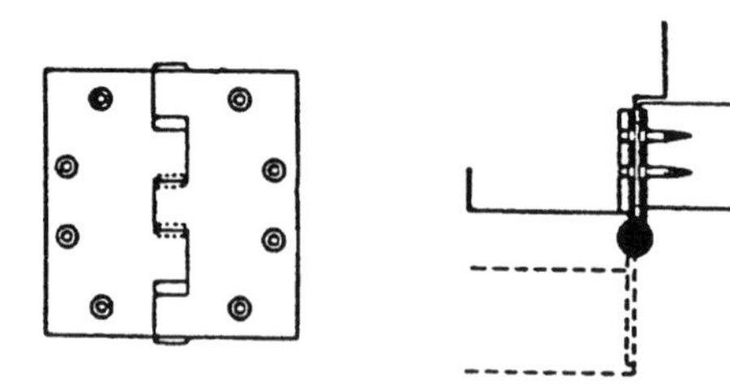

These butts have two equal square-edged leaves; one is mortised into the door and the other into the frame. It is available in standard, heavy, or extra heavy weight.

Half Surface

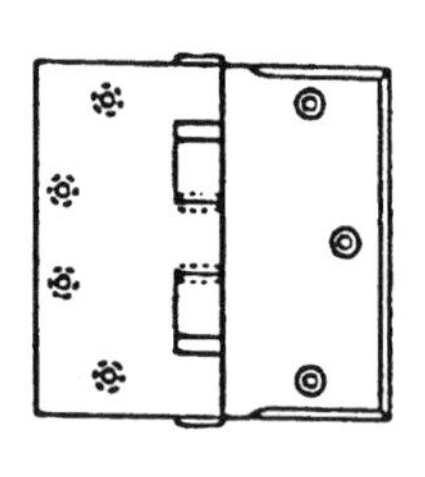

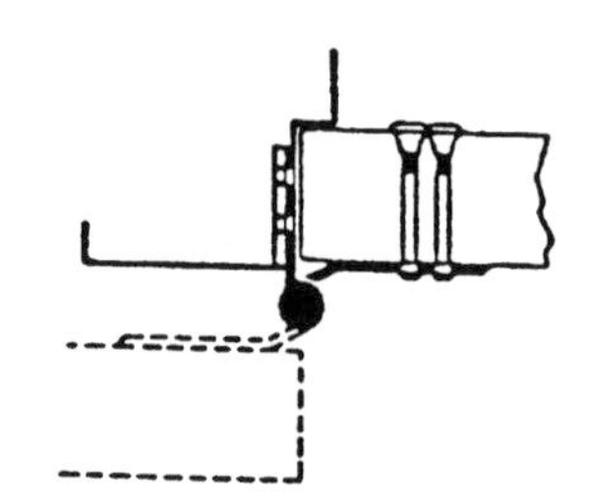

These butts have two equal leaves; one is square-edged and the other is bevel-edged; the square edge is mortised into the frame, the bevel edge is surface mounted on the door. It is available in standard and heavy weight.

Half Mortise

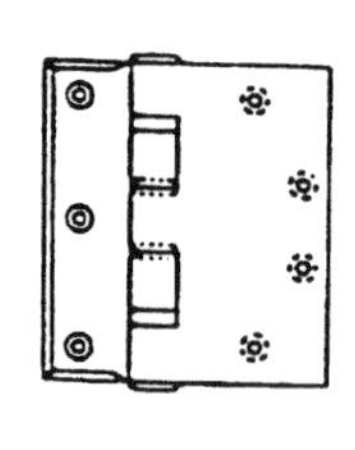

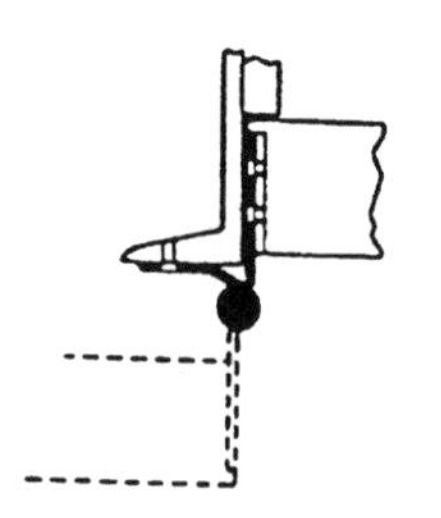

These butts have two equal leaves; one is square edged and the other is bevel edged; the square edge is mortised into the door edge and the bevel edge is mounted on the frame. It is available in standard and heavy weight.

Full Surface

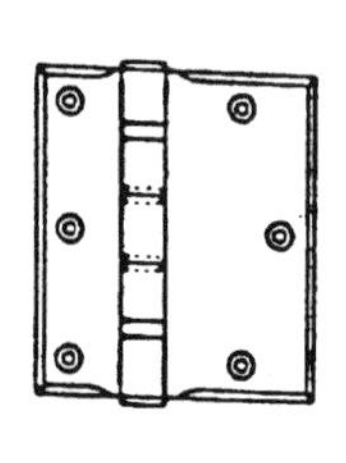

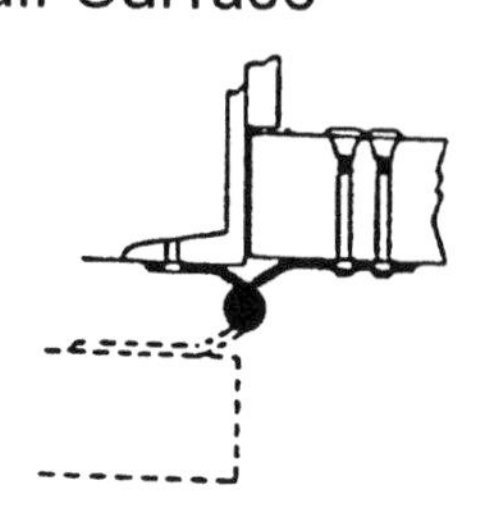

Two bevel-edged leave butts are of unequal size; one is mounted on the frame, the other on the door.

13.1.0 Door Hinges (Types and Illustrations) (Continued)

All of the above butts are general available in sizes referring to their height: 41⁄2" (11.43 cm), 5" (12.7 cm), and 6" (15.24 cm).

Special Butts

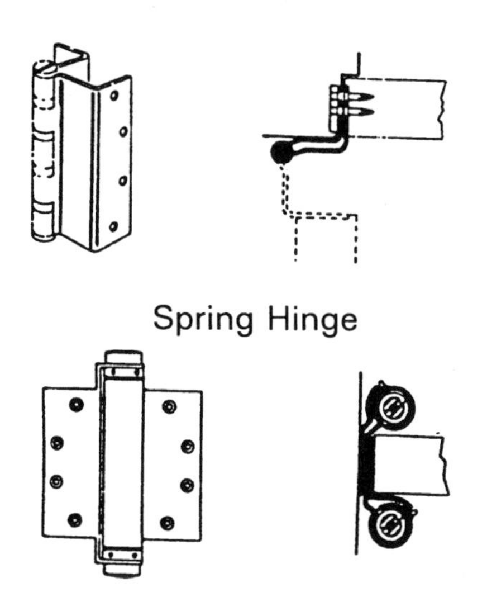

Spring Hinge

Swing clear/full mortise are also available in half-surface, half-mortise, and full-surface configurations. These types of butts provide an unobstructed clear frame opening when door is in the 90-degree open position. It is available in either a single- or double-acting configuration, usually mortised into the door and frame, providing closing action without a separate closer.

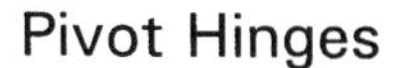

Pivot Hinges

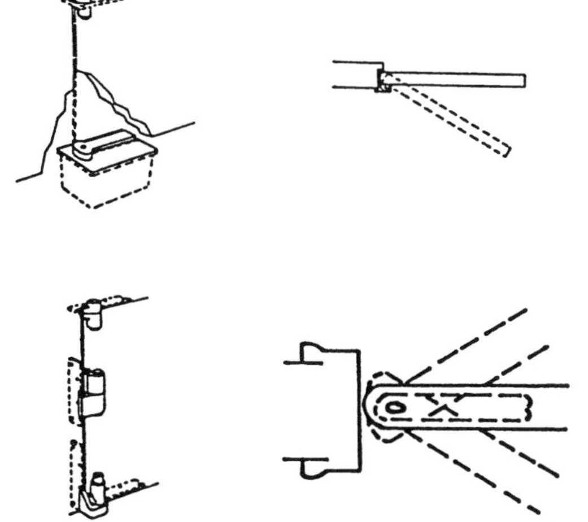

Offset pivot hinges are mortised into the top and bottom edges of the door and into the frame jamb at the top and bottom. These hinges can also be mortised into the floor and the top of the frame. Center pivot hinges are attached to the top and bottom edges of the door and either into the top and bottom of the frame or into the floor and the top of the frame. Fully mortised into the edge of the door and frame, the hinge portion is not visible when the door is closed, except when the Paumelle or Olive Knuckle hinge is used, the olive-shaped portion is visible as an architectural feature.

Invisible Hinges

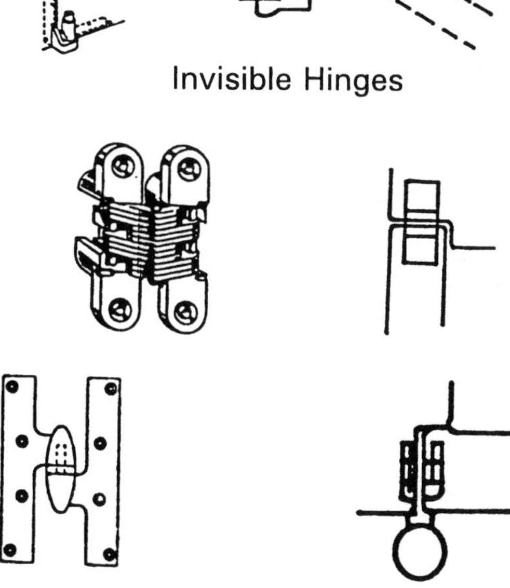

13.2.0 Lockset and Latchset Configurations and Functions

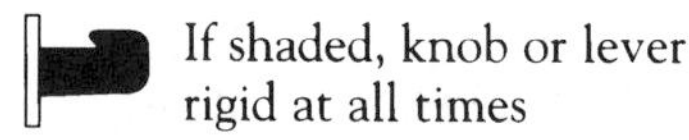

16 PUBLIC ENTRANCE

- Guardbolt deadlocks latch bolt
- Latchbolt retracted by either lever unless outside lever is locked by key inside
- Key outside retracts latch bolt when outside lever is locked

ANSI	8200 Levers	7800 Knobs
F09	8216	7816

17 UTILITY

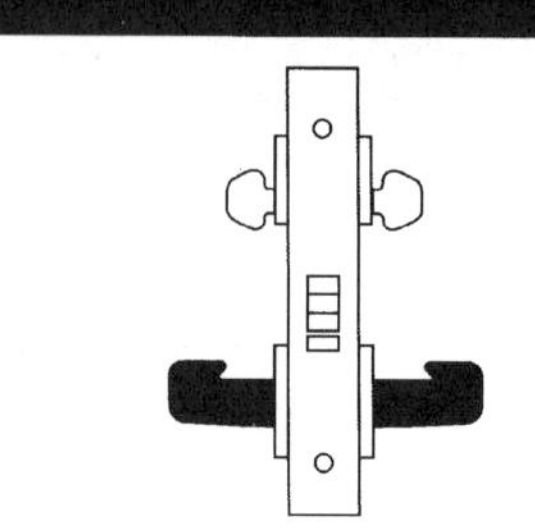

- Guardbolt deadlocks latchbolt
- Both levers rigid at all times
- Latchbolt retracted by key either side

ANSI	8200 Levers	7800 Knobs
*	8217	7817

* ANSI does not list this function

20 DEADLOCK

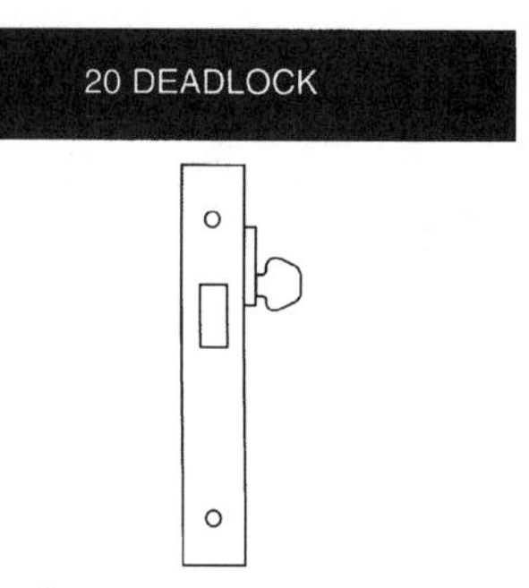

- Deadbolt operated from outside by key
- No inside operation

ANSI	8200	7800
F18	8220	N/A

21 DEADLOCK

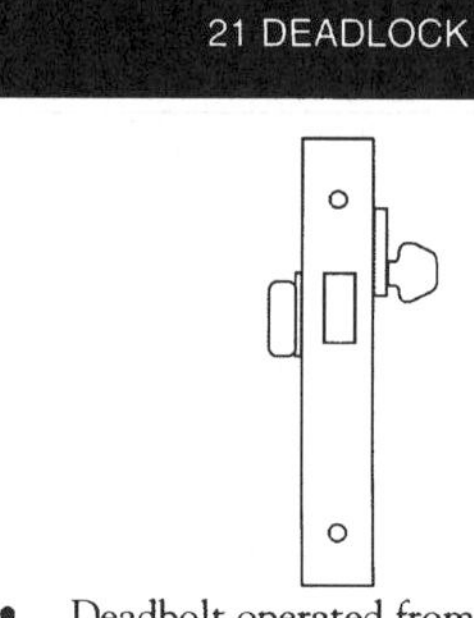

- Deadbolt operated from outside by key
- Deadbolt operated from inside by turn lever

ANSI	8200	7800
F17	8221	N/A

22 DEADLOCK

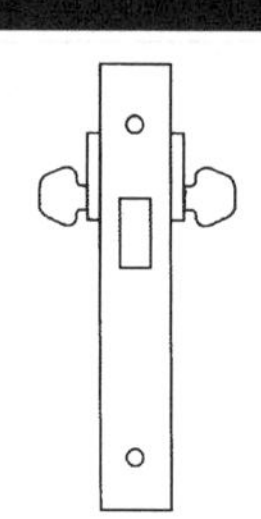

- Deadbolt operated from either side by key

ANSI	8200	7800
F16	8222	N/A

23 CLASSROOM DEADLOCK

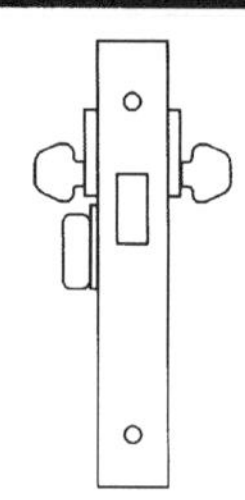

- Deadbolt operated from either side by key
- Deadbolt retracted by turn lever inside, but turn lever will not project it

ANSI	8200	7800
*	8223	N/A

* ANSI does not list this function

By permission: Sargent Manufacturing Company, New Haven, Connecticut

13.2.0 Lockset and Latchset Configurations and Functions (Continued)

If shaded, knob or lever rigid at all times

24 AND 25 DORMITORY AND STOREROOM

- Latchbolt retracted by either lever
- Deadbolt operated from outside by key, inside by turn lever
- When deadbolt is projected, outside lever is automatically locked
- 25 function only – Operating inside lever automatically retracts latchbolt and deadbolt unlocks outside lever

ANSI	8200 Levers	7800 Knobs
F21	8224	7824
F13	8225	7825

26 STOREROOM LOCK

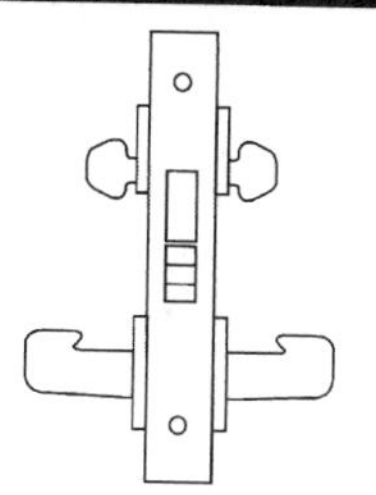

- Latchbolt retracted by either lever
- Deadbolt operated from either side by key
- Inside lever does not retract deadbolt

ANSI	8200 Levers	7800 Knobs
F14	8226	7826

31 UTILITY

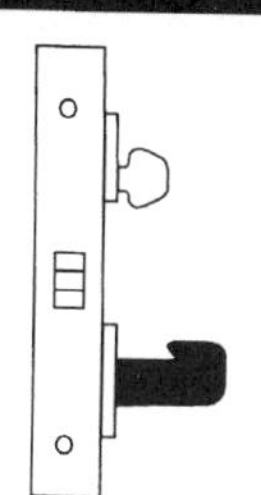

- Guardbolt deadlocks latchbolt
- Latchbolt retracted by key outside
- Outside lever rigid at all times
- No inside cylinder or trim

ANSI	8200 Levers	7800 Knobs
*	8231	7831

* ANSI does not list this function

35 STOREROOM

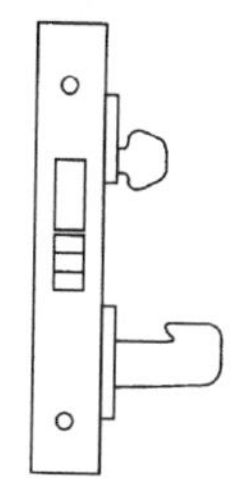

- Deadbolt operated from outside by key
- Latchbolt retracted by outside lever except when deadbolt is projected, outside lever is automatically locked
- No inside cylinder or trim

ANSI	8200 Levers	7800 Knobs
*	8235	7835

* ANSI does not list this function

36 CLOSET

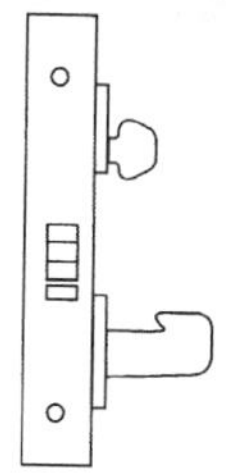

- Guardbolt deadlocks latch bolt
- Latchbolt retracted by outside lever
- Key outside lock or unlocks outside lever
- No inside cylinder or trim

ANSI	8200 Levers	7800 Knobs
*	8236	7836

* ANSI does not list this function

37 CLASSROOM LOCK

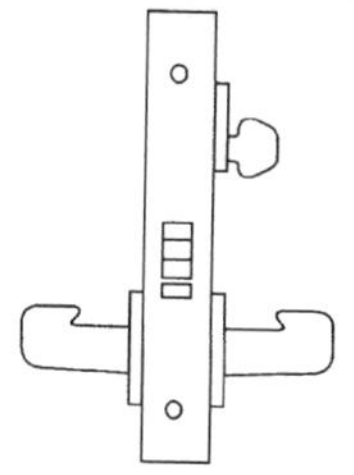

- Guardbolt deadlocks latchbolt
- Latchbolt retracted by either lever unless outside lever is locked by key
- Key outside locks or unlocks outside lever
- Inside lever always operative
- Latchbolt can be retracted by key in locked position

ANSI	8200 Levers	7800 Knobs
F05	8237	7837

By permission: Sargent division ASSA ABLOY, New Haven, Connecticut

13.2.0 Lockset and Latchset Configurations and Functions (Continued)

If shaded, knob or lever rigid at all times

03 CLASSROOM DEADLOCK

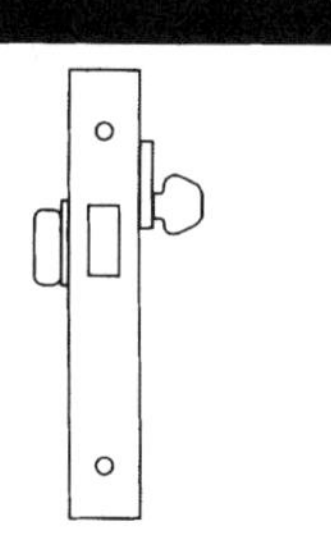

- Deadbolt operated from outside by key
- Turn lever inside retracts deadbolt only, but will not project it

ANSI	8200 Levers	7800 Knobs
*	8203	N/A

* ANSI does not list this function

04 STOREROOM OR SERVICE

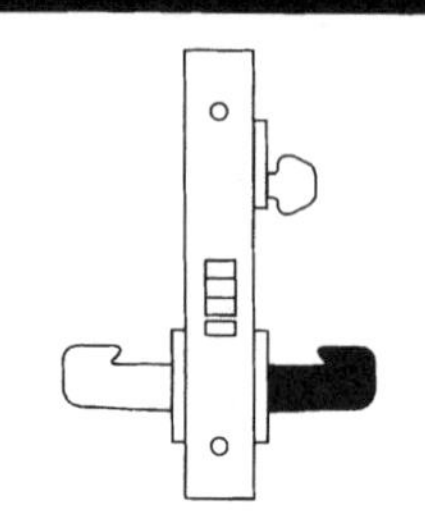

- Guardbolt deadlocks latchbolt
- Latchbolt retracted by lever inside or key outside
- Outside lever rigid at all times

ANSI	8200 Levers	7800 Knobs
F07	8204	7804

05 OFFICE

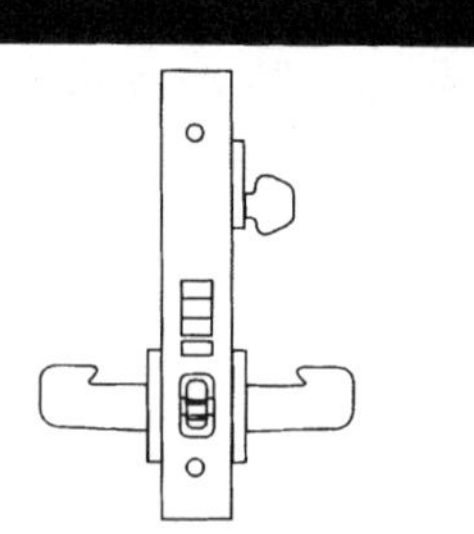

- Guardbolt deadlocks latchbolt
- Latchbolt retracted by either lever unless outside lever is locked by toggle in lock front
- Key outside retracts latch bolt when outside lever is locked

ANSI	8200 Levers	7800 Knobs
F04	8205	7805

06 STOREROOM OR SERVICE

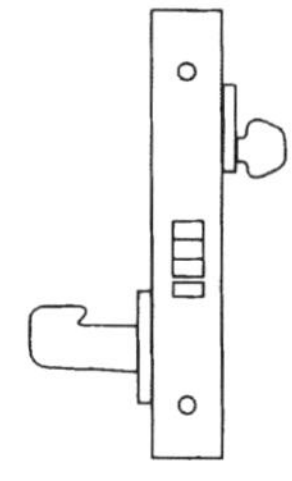

- Guardbolt deadlocks latchbolt
- Latchbolt retracted by lever inside or key outside
- Outside cylinder only

ANSI	8200 Levers	7800 Knobs
*	8206	7806

* ANSI does not list this function

13 EXIT LATCH

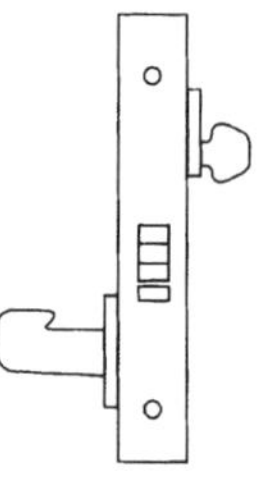

- Guardbolt deadlocks latchbolt
- Latchbolt retracted by lever inside
- No outside trim

ANSI	8200 Levers	7800 Knobs
*	8213	7813

* ANSI does not list this function

15 PASSAGE

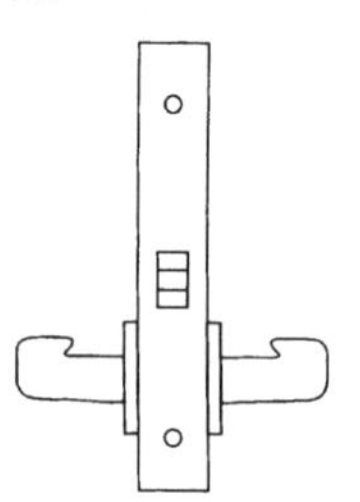

- Latchbolt retracted by either lever

ANSI	8200 Levers	7800 Knobs
F01	8215	7815

By permission: Sargent division ASSA ABLOY, New Haven, Connecticut

13.3.0 Heavy-Duty Mortise Cases, Hubs, and Spring Cartridges

HEAVY DUTY CASE, CAP & INSIDE FRONT

High impact strength is achieved through increased thickness of the case and cap to .109 and the inside front to .125

SIMPLE REVERSIBILITY

Instructions on each lockbody facilitates quick field reversibility for rehanding, using a standard screwdriver. Rehanding is done without opening the lockbody.

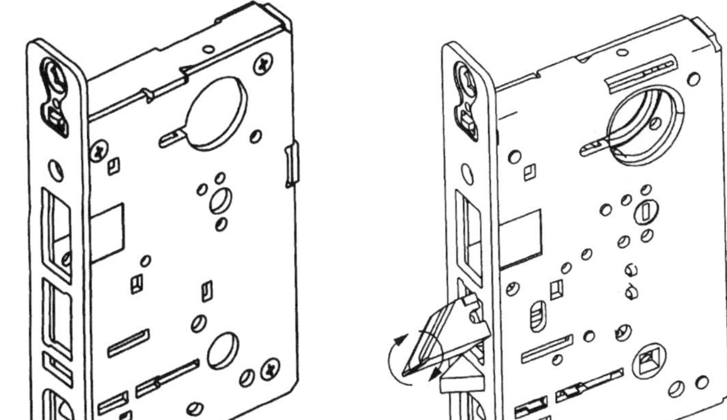

STAINLESS STEEL LATCHBOLT

Stainless steel ¾" one-piece anti-friction, reversible latchbolt.

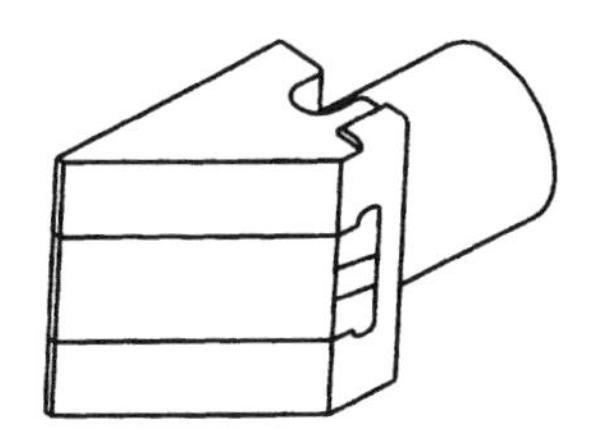

HEAVY DUTY HUBS & SPRING CARTRIDGE

Heavy duty hubs and spring cartridge provide superior strength and cycle life of the lock. Stainless steel hubs are available for institutional requirements.

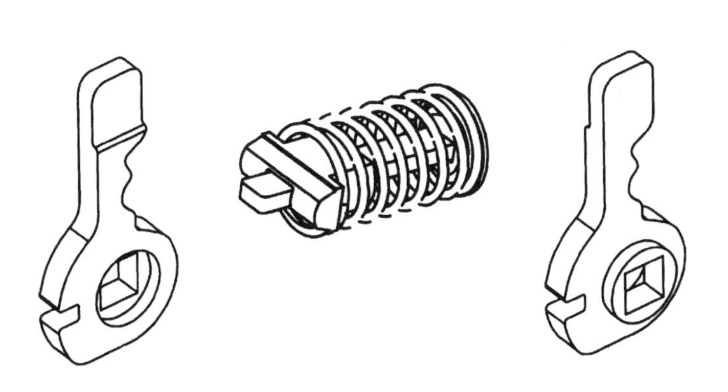

UNIVERSAL STRIKE

Universal, non-handed, curved-lip strike to simplify ordering and installation. Wrought box strike furnished standard.

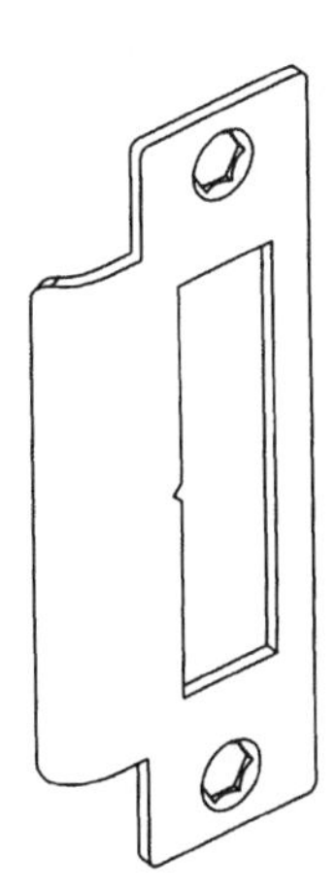

By permission: Sargent Manufacturing Company, New Haven, Connecticut

13.4.0 Strikes (Illustrated)

All sets are packed standard with a universal non-handed curved lip ANSI 4⅞" strikes. See chart below for part number and Lip Lengths. Standard is 1¼" lip length, Part Number 82-0110.

STRIKES — 82-0108

Part No.	Lip Length
82-0109	1⅛"
82-0110	1¼"
82-0111	1⅜"
82-0112	1⅝"
82-0113	1⅞"
82-0114	2⅛"
82-0115	2⅜"
82-0116	2⅝"
82-0117	2⅞"

To order strikes separately, give strike part number and finish. Strikes ordered separately are furnished with wood screws and without strike boxes.

STRIKES — 877 OBS

Door Thickness	Part No. RHRB	Part No. LHRB
1¾"	82-0332	82-0333
2"	82-0334	82-0335
2¼"	82-0336	82-0337
2½"	82-0338	82-0339
2¾"	82-0340	82-0341
3"	82-0342	82-0343

82-0108 SERIES STRIKES

4⅞" 123.8mm

3/32" 2.4mm THICK

1¼" 31.8mm

78-0034 DEADBOLT STRIKE

4⅞" 123.8mm

3/32" 2.4mm THICK

1¼" 31.8mm

877 OPEN BACK STRIKE

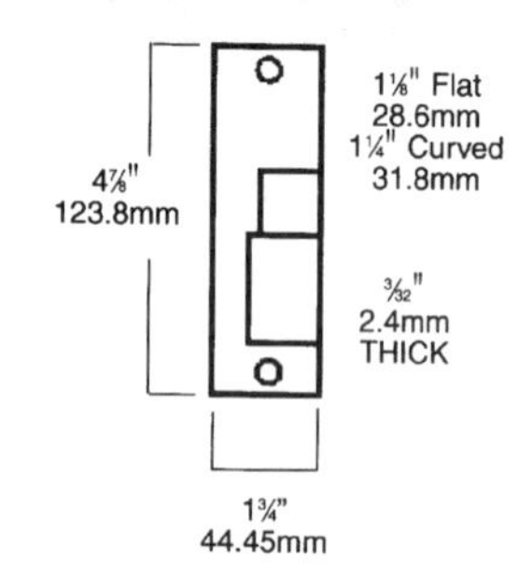

82-0229 FLAT LIP STRIKE

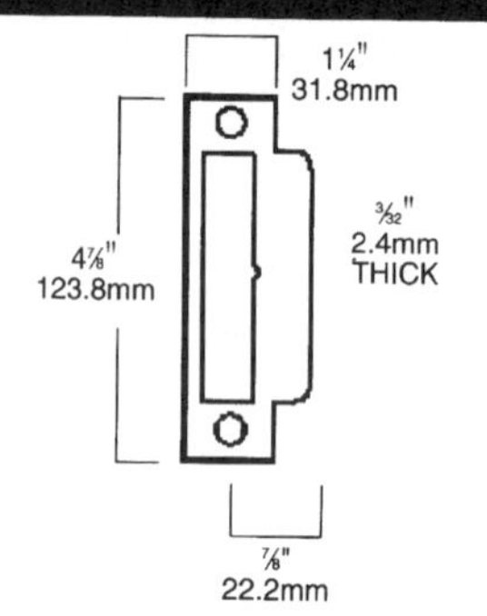

For use on pairs of doors

HOTEL LOCK INDICATORS (PREFIX 50-)

Available in 7700 and 8100 Series only. Wrought brass, bronze. For 50 function only. To order add 50-prefix to lockset designation. (Example: 50-7850 or 50-8250) Not available for narrow front locksets (1-prefix)

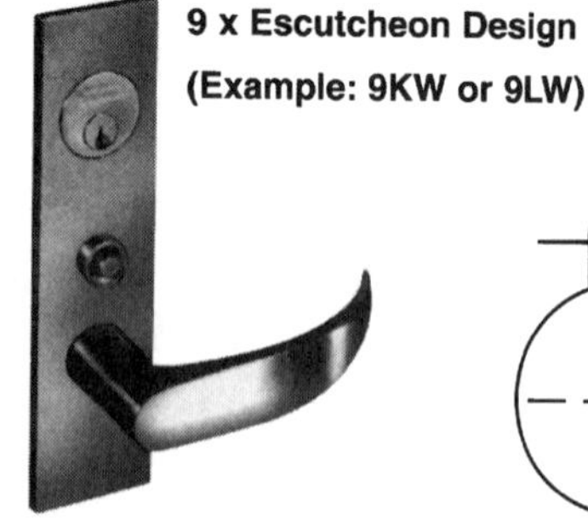

9 x Escutcheon Design

(Example: 9KW or 9LW)

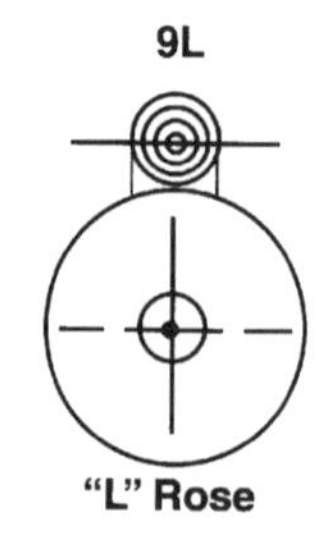

"L" Rose

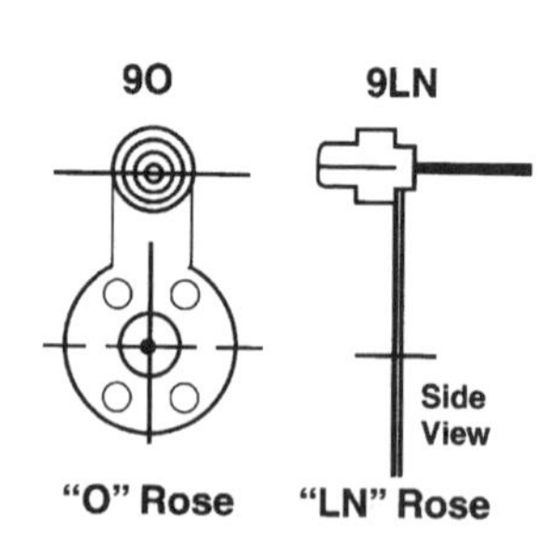

"O" Rose "LN" Rose

WROUGHT STRIKE BOX

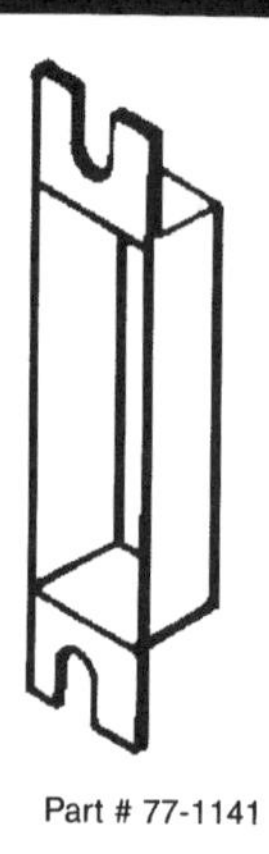

Part # 77-1141

By permission: Sargent Manufacturing Company, New Haven, Connecticut

13.5.0 Door Knob Designs

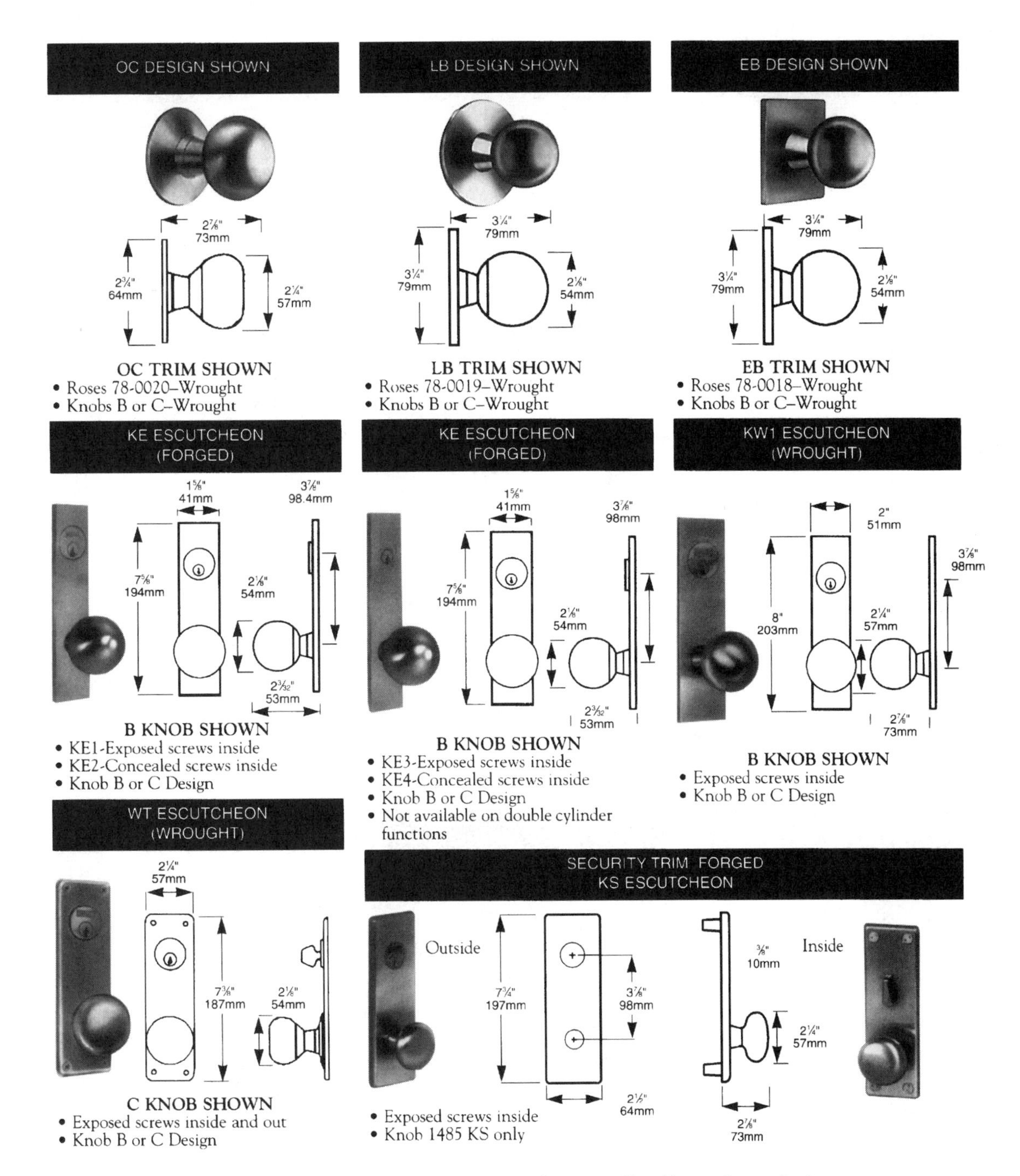

By permission: Sargent Manufacturing Company, New Haven, Connecticut

13.6.0 Lever Handle Designs

Any lever design can be used with any rose design.

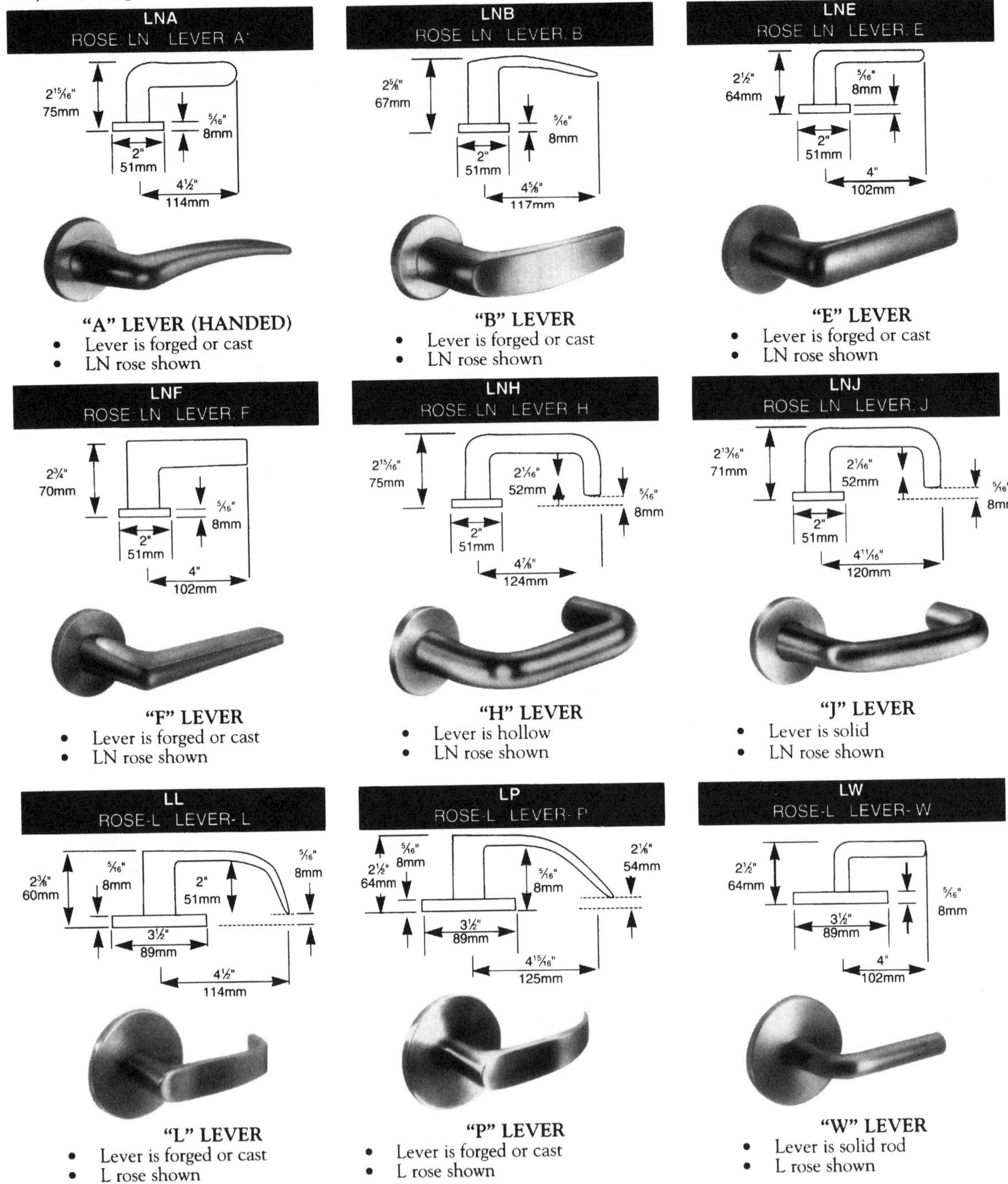

By permission: Sargent Manufacturing Company, New Haven, Connecticut

13.6.1 Lever Handle Designs (Forged and Wrought)

Any lever design can be used with any escutcheon design. LS Security Trim available with "L" lever only.

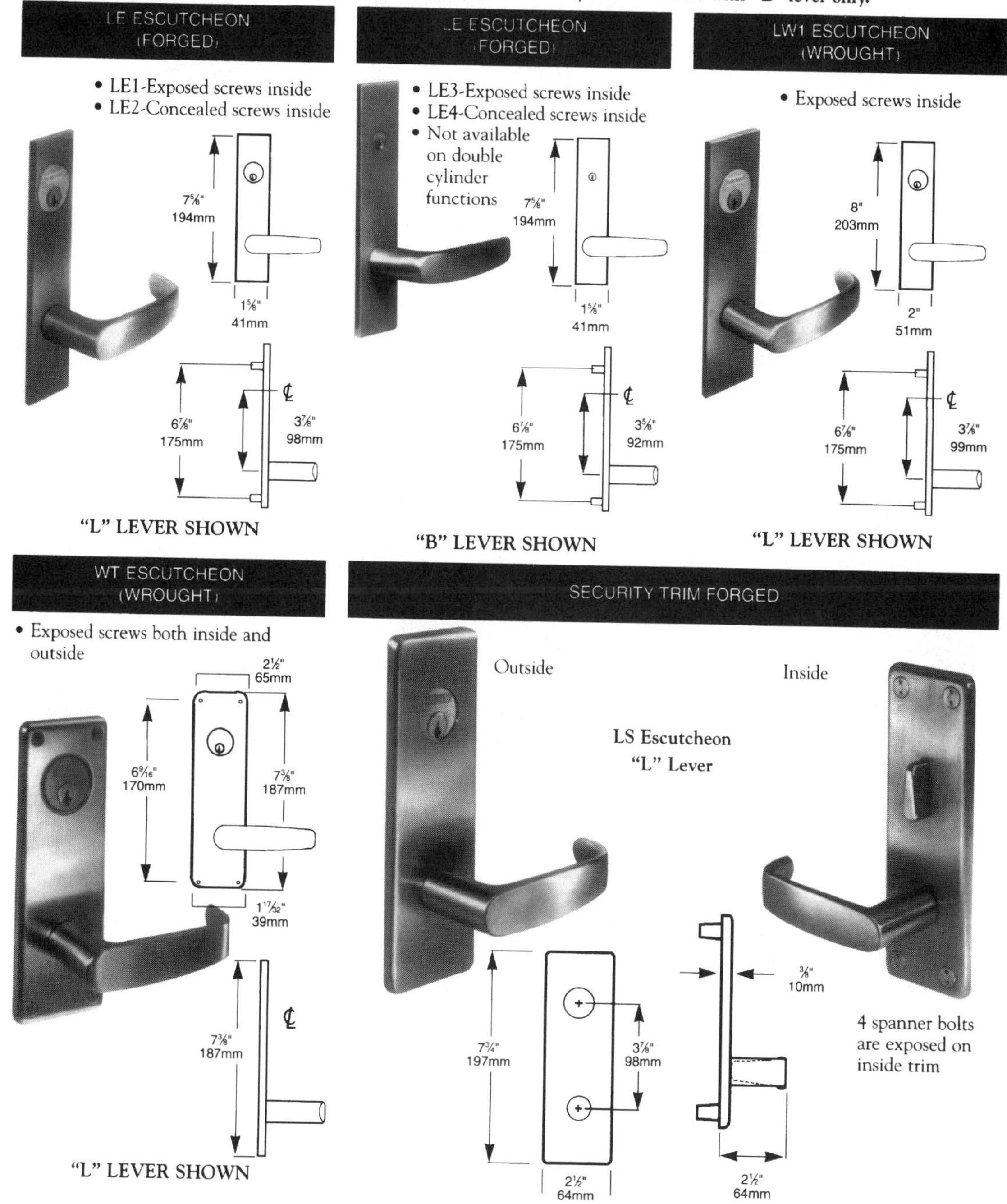

By permission: Sargent Manufacturing Company, New Haven, Connecticut

13.7.0 Turn Levers

130 W

Brass, bronze or stainless steel. Wrought plate, wrought lever. Furnished with flat spindle. Finishes: 3, 4, 9, 10, 10B, 10BL, 20D, 26, 26D, 32, 32D. Used with 7800 locksets.

130 KB

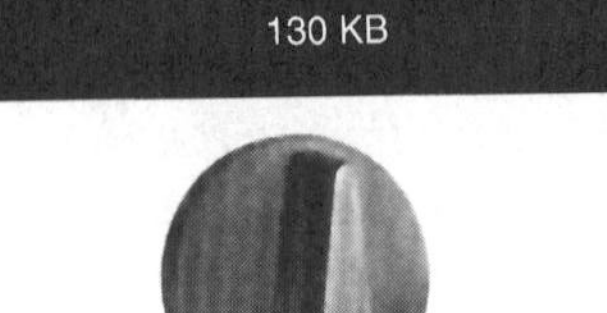

Forged brass, bronze or stainless steel. Furnished with flat spindle. Finishes: 3, 4, 9, 10, 10B, 10BL, 20D, 26, 26D, 32, 32D. Used with 8200 rose trim.

130 KA

Forged brass, bronze or stainless steel. Furnished with flat spindle. Finishes: 3, 4, 9, 10, 10B, 10BL, 20D, 26, 26D, 32, 32D. Used with 78-0018E rose.

184 W

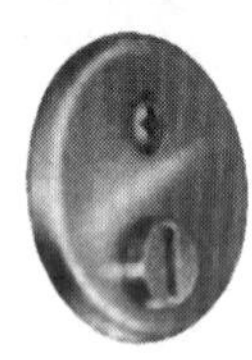

Brass, bronze or stainless steel plate. Brass or stainless steel button. Furnished with flat spindle. For 65 function with sectional trim using 130W Turn Lever. Finishes: 3, 4, 9, 10, 10B, 10BL, 20D, 26, 26D, 32, 32D.

184 KB

Brass, bronze or stainless steel. Furnished with flat spindle. For 65 function using 130 KB Turn Lever. Furnished with flat spindle. Finishes: 3, 4, 9, 10, 10B, 10BL, 20D, 26D, 32, 32D.

184 KA

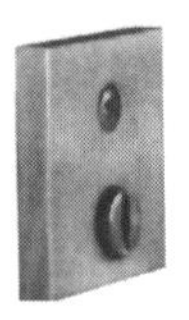

Brass, bronze or stainless steel. Furnished with flat spindle. For 65 function using 130 KA Turn Lever. Finishes: 3, 4, 9, 10, 10B, 10BL, 20D, 26D, 32, 32D.

RABBETED DOOR KIT NO. 677

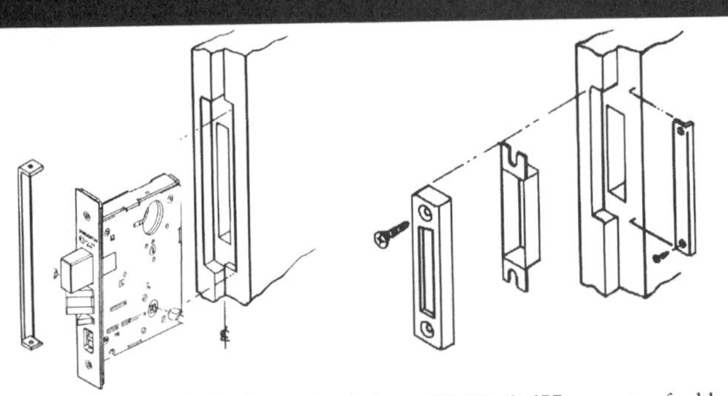

Kit adapts 7800 and 8200 locks for rabbeted doors. Kit No.2-677 consists of rabbeted strip and strike. Kit furnished standard for 1¼" doors. Specify when using 2¼" thick doors. Finishes: 3, 4, 9, 10, 10B, 10BL, 20D, 26, 26D.

OUTSIDE EMERGENCY RELEASE KEY

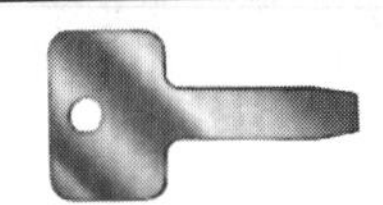

No. 14-0057. Carbon steel.

126 T-TURN

(USED IN LIEU OF KNOB FOR 7800 SERIES LOCKSETS)

Brass or bronze. Finishes: 3, 4, 9, 10, 10B, 10BL, 20D, 26, 26D. Spindle engages hub on one side of lockset only.

By permission: Sargent Manufacturing Company, New Haven, Connecticut

13.8.0 Mortise Cylinders

40 SERIES TYPE CYLINDER

Cylinder: Brass

Cap: Brass, Bronze and Stainless Steel.

Finishes: 3, 4, 9, 10, 10B, 10BL, 20D, 26, 26D, 32, 32D.

Furnished standard with No. 97 Compression Ring. See Function table on page 6 for cam required.

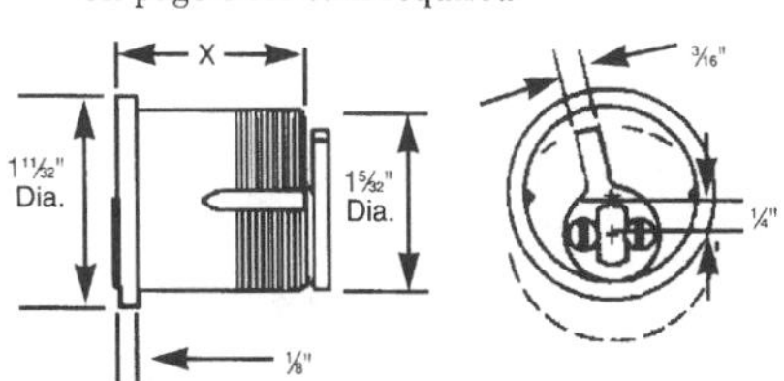

Length under cylinder head										
No.	41	42	43	44	46	48	50	52	54	56
Length Including cam (dim.x)	1 1/8"	1 1/4"	1 3/8"	1 1/2"	1 3/4"	2"	2 1/4"	2 1/2"	2 3/4"	3"

HOTEL TYPE MORTISE CYLINDER (PREFIX 50-) 50-40 SERIES

Cylinder: Brass

Cap: Brass, Bronze and Stainless Steel.

Finishes: 4, 15

For use only with Sargent Escutcheon Trim (KE, LE) See Function table for cam required.

Length under cylinder head				
No.	50-41	50-42	50-43	50-44
Length including cam (dim.x)	1 1/8"	1 1/4"	1 3/8"	1 1/2"

MORTISE CYLINDERS, EXPOSED BARREL ONLY 78-40 SERIES

Cylinder: Brass

Cap: Brass, Bronze and Stainless Steel.

Finishes: 4, 15

For use only with Sargent Escutcheon Trim (KE, LE) See Function table for cam required.

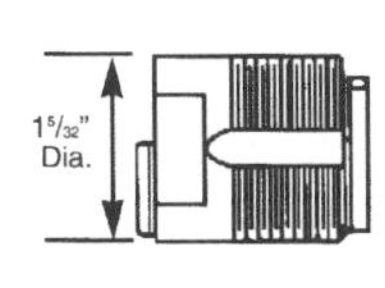

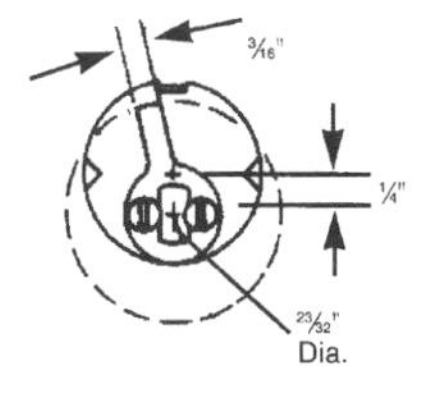

Door Thickness	Cylinder
1 3/4"	6 pin; single cylinder function only
2"	6 or 7 pin; single cylinder functions only
2 1/4"	6 or 7 pin; all lock functions

By permission: Sargent Manufacturing Company, New Haven, Connecticut

13.8.1 Rosette and Blocking Rings for Cylinders

NO. 1KB ROSETTE

Cast Brass, Bronze and Stainless Steel

Finishes: 3, 4, 10, 10B, 10BL, 20D, 26, 26D, 32, 32D

1 1/2" diameter, 5/16" projection

No. 1 KB-2 7/16" projection

No. 1 KB-3 9/16" projection

NO. 97 ROSETTE

Brass, Bronze and Stainless Steel
Finishes: 3, 4, 10, 10B, 10BL, 20D, 26, 26D, 32, 32D

1 11/16" diameter

9/32" projection

NO. 90 BLOCKING RING

Brass, Bronze and Stainless Steel

Finishes: 3, 4, 10, 10B, 10BL, 20D, 26, 26D, 32, 32D

1 3/8" diameter; 1/16", 1/8", 3/16", 1/4", 3/8" projections.
Specify projection required when ordering.

HOW TO FIND YOUR ROSETTE AND BLOCKING RING REQUIREMENTS

Rosettes and ring requirements are coded by letters A through L in the table to the right. Each letter represents a rosette or ring as listed below. As an example, a 41 cylinder for a 1 3/8" door using KE escutcheon trim for a single cylinder function would require (A) a No. 97 Rosette; and (H) a No. 90 1/8" Blocking Ring.

A. No. 97 Rosette (includes spring)
B. No. 97-02 Spring
C. No rosette or ring required
D. 1KB Rosette (includes spring)
E. 1KB-2 Rosette (includes spring)
F. 1KB-3 Rosette (includes spring)
G. No. 90 1/16" Blocking Ring
H. No. 90 1/8" Blocking Ring
J. No. 90 3/16" Blocking Ring
K. No. 90 1/4" Blocking Ring
L. No. 90 3/8" Blocking Ring

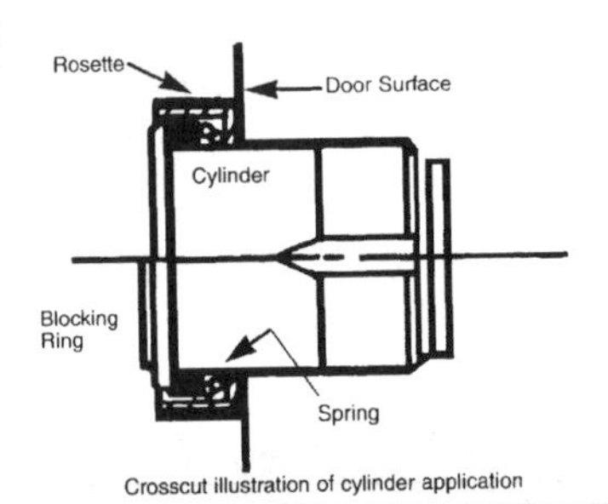

Crosscut illustration of cylinder application

ESCUTCHEON TRIM

Trim	KE, KW,			LE, LW		
Function	Single Cylinder			Double Cylinder		
Door	1⅜	1¾	2¼	1⅜	1¾	2¼
41 Cylinder 5 or 6 pin	B H	C	– J	B H H	B	–
42 Cylinder 5, 6 or 7 pin	B K	C	C K	B J K	B	C
43 Cylinder 5, 6, or 7 pin	B L	B J	C	B J L	B L	B H

SECTIONAL TRIM

Trim	KS, LS					
Function	Single Cylinder			Double Cylinder		
Door	1	1¾	2¼	1⅜	1¾	2¼
41 Cylinder 5 or 6 pin	–	A	–	–	A H	–
42 Cylinder 5, 6 or 7 pin	–	A	A	–	A K	A
43 Cylinder 5, 6, or 7 pin	– K	A	A L	– H	A L	A J

Trim	KS, LS					
Function	Single Cylinder			Double Cylinder		
Door	1	1¾	2¼	1⅜	1¾	2¼
41 Cylinder 5 or 6 pin	E H	D	–	F H J	E H	–
42 Cylinder 5, 6 or 7 pin	F K	D G	D	F J K	F K	D
43 Cylinder 5, 6, or 7 pin	F L	E J	D	F J L	F H	E H

By permission: Sargent Manufacturing Company, New Haven, Connecticut

13.8.2 Miscellaneous Cams for Mortise Cylinders

MORTISE ADJUSTABLE FRONT CYLINDER

Cylinder: Brass
Cap: Brass, Bronze and Stainless Steel.
Finishes: 3, 4, 9, 10, 10B, 10BL, 20D, 26, 26D, 32, 32D.
No cylinder ring or rosette is required. Spring action front adjust 1/16" to assure proper mounting of cylinder. Available with three different cap heights to accommodate various door and trim thicknesses. Can be used with most trim except escutcheon designs.

			Cylinder No.			
No.	Cap Size	Dim A	35-41	35-42	35-43	35-44
Length	5	5/16"	31/32"	1 3/32"	1 7/32"	1 11/32"
Including	7	7/16"	31/32"	1 3/32"	1 7/32"	
cam (dim.x)	9	9/16"	23/32"	27/32"	31/32"	1 3/32"

Also available for use with hotel function locks, for ordering cylinders only. Example: 35-50-7-2-US26D.

35-40 SERIES CYLINDER-CAP HEIGHT REQUIREMENTS

When ordering cylinders, use the following suffixes as required:

Suffix 2	Hotel Cam
Suffix 3	Short Cam
Suffix 5	5/16" Cap Height
Suffix 7	7/16" Cap Height
Suffix 9	9/16" Cap Height

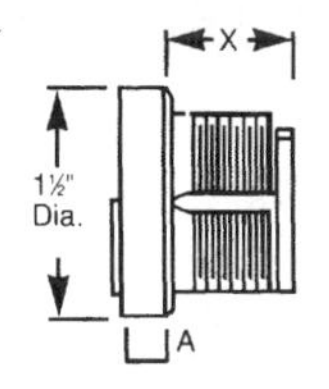

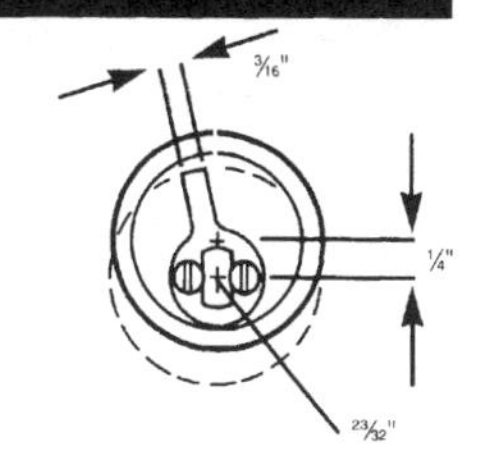

Trim Function	All Designs — Sectional Trim					
	Single Cyl			Double Cyl		
	1⅜"	1¾"	2¼"	1⅜"	1¾"	2¼"
35-41 Cylinder 6 PIN	7/16"	5/16"	—	9/16"	7/16"	—
35-42 Cylinder 6 or 7 PIN	—	7/16"	5/16"	—	9/16"	5/16"
35-43 Cylinder 6 or 7 PIN	—	—	5/16"	—	—	7/16"

CAMS AND APPLICATIONS

Application	All Functions Except 92 and 16 Functions inside Cylinder and 50 Function Hotel Cylinder	92 and 16 Functions inside and all Functions with 88 Prefix
Standard Cams	13-0660	13-0661
Construction Cylinder Cam required for 41 size only	13-0662	13-0663
Slotted Cam required for a 6 pin key on a 5 pin cylinder or 7 pin key on a 6 pin cylinder	13-0664	13-0665

APPLICATIONS

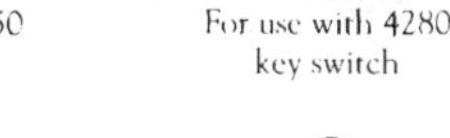

For use with 50 function

13-2045 Hotel Cylinder Cam

For use with 4280 key switch

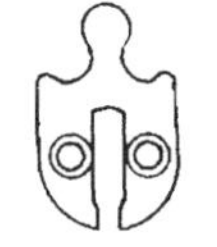

13-0921 Standard Cloverleaf Cam

MISCELLANEOUS CAMS

Manufacturer and Lock Number	Desc.	Cam No.	Cam with Cylinder
Adams Rite 1850	Offset Cam	13-0512	47-41-101
Adams Rite 4050	Offset Cam	13-0513	47-142-101
Adams Rite 4150	Offset Cam	13-0513	47-142-101
Adams Rite 4250	Offset Cam	13-0513	47-142-101
Adams Rite 4350	Offset Cam	13-0513	47-142-101
Adams Rite 4070	Offset Cam	13-0513	47-142-101
Schlage "L" Latch	Cam	13-0938	–

13-0512

13-0513

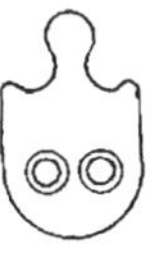

13-0938

By permission: Sargent Manufacturing Company, New Haven, Connecticut

13.9.0 Deadbolts, Spindles, Security Fasteners, and Guard Bolts

STAINLESS STEEL DEADBOLT

Stainless steel deadbolt with hardened steel rollers has a 1" throw.

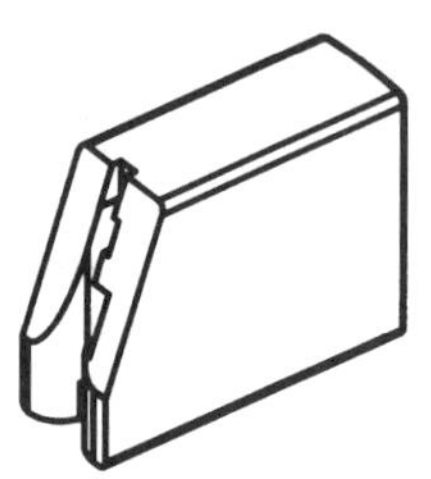

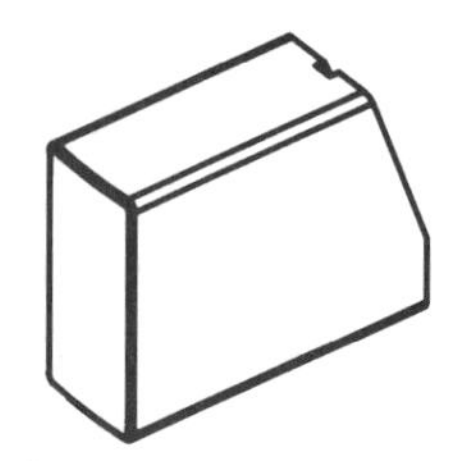

SPINDLES

Durable spindle design provides security and integrity of the lockbody by shearing off under extreme loads while preventing any damage to the lockbody. Inside and outside spindles operate independently.

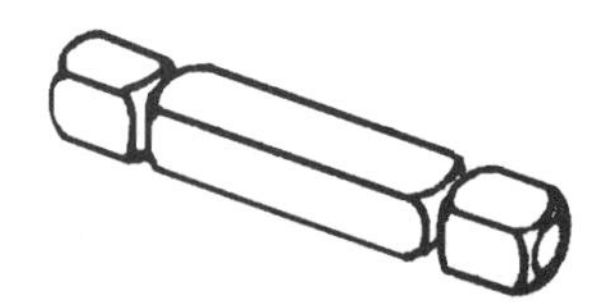

SECURITY HARDWARE

Six-lobe security screws (prefix 36-) or spanner head (prefix 37-) are available.

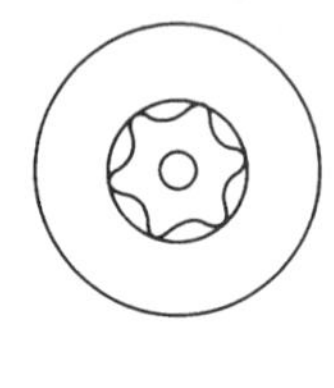

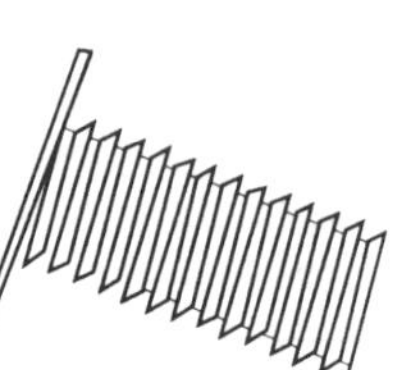

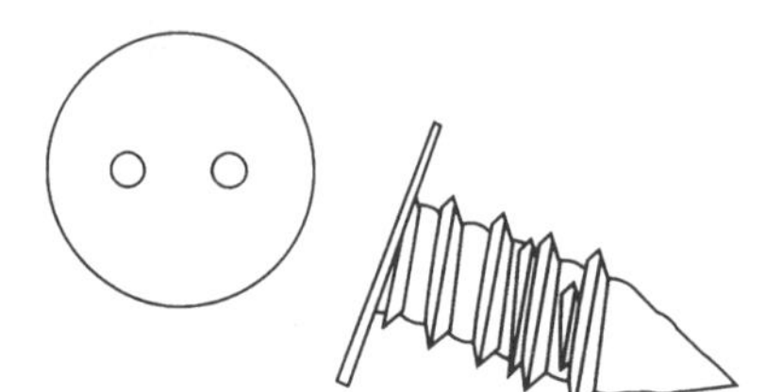

STAINLESS STEEL GUARD BOLT

Stainless steel, non-handed guardbolt.

THRU-BOLTED TRIM

Mortise lock trim is thru-bolted to ensure proper alignment and security. Greater torque resistance.

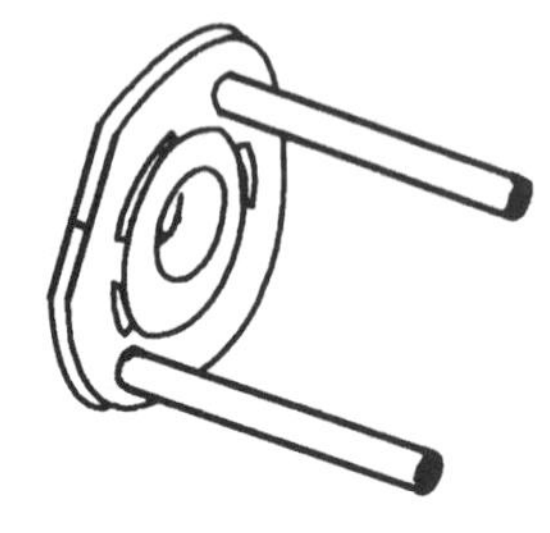

TRIM CONVERTIBILITY

Outside trim levers or knobs can be easily disassembled by unscrewing the retaining nut to separate the rose/escutcheon from the lever or knob.

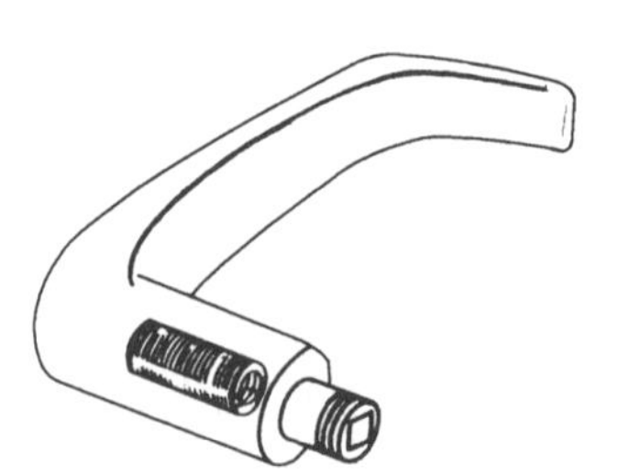

By permission: Sargent Manufacturing Company, New Haven, Connecticut

13.10.0 Construction Key Systems

CONSTRUCTION KEY SYSTEMS (PREFIX-21)

The Sargent construction keying system protects the building owner by providing temporary masterkeying during the construction period. Regular day and masterkeys are retained by the distributor and cannot be duplicated or obtained by unauthorized personnel during construction. Temporary keys become inoperative when the regular keys are turned over to the building owner.

Orders for this system must show individual item numbers for each lock, and where room or opening numbers are known, they also must appear with each lockset.

To order prefix 21, see Sargent cylinder catalog for more details.

MORTISE CYLINDER TURN LEVER 124 SERIES

Cylinder: Brass
Turn Lever: Brass, Bronze, Aluminum
Cap: Brass, Bronze and Stainless Steel
Finishes: 3, 4, 9, 10, 10B, 10BL, 20D, 26, 26D, 32, 32D
Furnished standard with No. 97 Compression Ring

Length under cylinder head					
No.	124-41	124-42	124-43	124-44	124-46
Length Including Cam (dim. x)	1⅛"	1¼"	1⅜"	1½"	1¾"

Cam No.	Description
124-1	Sizes 41 through 46 x cam for standard cylinder
124-3	Sizes 41 through 46 x cam for 16 and 37
124-8	Sizes 41 through 46 x cam for inside cylinder or 7892
124-101	Sizes 41 through 46 x cam for Adams Rite 1850 Lock

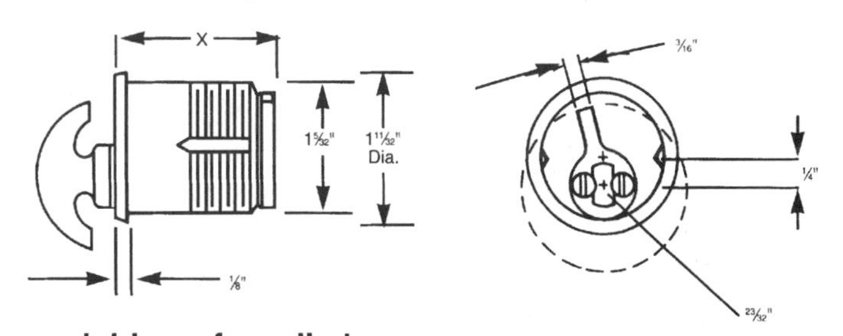

Pins and drivers for cylinder

	ORDERING NO.	PIN NO.	LENGTH
Bottom Plus	13-0064	1	.170
	13-0065	2	.190
	13-0066	3	.210
	13-0067	4	.230
	13-0068	5	.250
	13-0069	6	.270
	13-0070	7	.290
	13-0071	8	.310
	13-0072	9	.330
	13-0073	10	.350
Master Plus/Drivers	13-0051	2	.040
	13-0052	3	.060
	13-0053	4	.080
	13-0054	5	.100
	13-0055	6	.120
	13-0056	7	.140
	13-0057	8	.160
	13-0058	9	.180
	13-0058	10	.200

By permission: Sargent Manufacturing Company, New Haven, Connecticut

13.10.1 Removable Core Cylinders

REMOVABLE CORE CYLINDERS

Sargent removable core offers security and convenience by making keying changes a simple matter. Rekeying and transferring keying to another door is facilitated because it is no longer necessary to disassemble the lock. A special control key releases the locking cam of the cylinder core and allows immediate removal of the core. Virtually unlimited key changes are possible, and removable core cylinders can be master keyed or grand master keyed. Removable core is available across the Sargent line of padlocks, deadlocks, bored locks and exit devices.

CONSTRUCTION REMOVABLE CORE KEYED CYLINDERS (PREFIX 64-)

The Sargent removable construction core system protects the security of an owner's masterkey system during the period of construction. It is used throughout the construction period in lieu of the permanent masterkeyed cores. This prevents the keys to the permanent system from becoming available to unauthorized persons. Upon completion of the building, the temporary construction cores are removed and replaced with the permanent removable cores which are inoperative by the construction keys. During the construction period, locks can be furnished with returnable cylinders, or plastic disposable cores. Temporary cylinders (64 prefix) are installed only in doors which must be locked during construction. The disposable plastic core (prefix 60) is recommended for all non-essential locking doors of the construction period.

It will be the distributor's responsibility to:

- Deliver all permanent cores to the job site
- Remove all the temporary cores and install permanent cores
- Inspect each lockset to ensure satisfactory operation of permanent cores
- Deliver to building owner all day, master and control keys for the permanent system
- Return all temporary cores to New Haven on a return goods authorization (RGA)

REMOVABLE CORES ONLY SERIES 6300

For all locksets 6-pin only
Finishes: 4, 15
When ordering, give all pertinent keying information

DISPOSABLE PLASTIC CORE

May be used for those doors that do not require locking during the construction period. (prefix 60-)
These cores are ordered with 60-7805-OB-26D

OLD STYLE REMOVABLE CORE

Available for existing systems only
Permanent Removable Cores (Prefix 51-)
If ordering for existing construction key system, give all pertinent keying information.

Old Style Construction Removable Core (Prefix 52-)
A separate order for permanent cores, with all necessary keying information and item numbers, for identical purposes, should accompany the lockset order. The permanent cores will be shipped directly to the distributor and not to the job site.

Mortise Type Cylinder Series 6340 & 50-6343 (for Hotels)
Brass
Cap: Brass, Bronze and Stainless Steel
Finishes: 3, 4, 9, 10, 10B, 10B, 20D, 26, 26D, 32, 32D.
Furnished standard with No. 97 compression ring.
Cam is permanently staked to body and cannot be changed in field.
50-6343 is available in C series keyways only for use in hotel function locks.

Removable Core Mortise Cylinders Series 040 and 1400
140-6 pin system Cylinder: Brass

Length under cylinder head				
No.	041	042	043	044
Length Including Cam	10⅛"	1¼"	1⅜"	1½"

REMOVABLE CORE CAMS

13-0542 Standard Sargent Cam

13-0806 92 and 16 Standard Sargent Cam

13-0832 Adams Rite 1850 4710

13-0833 Adams Rite 4050's

13-0922 4280 Switch Lock

13-0928 Schlage "L" Latch

Straight X1-3000

By permission: Sargent Manufacturing Company, New Haven, Connecticut

13.11.0 Panic Devices (Concealed/Surface-Applied Vertical Rod Devices)

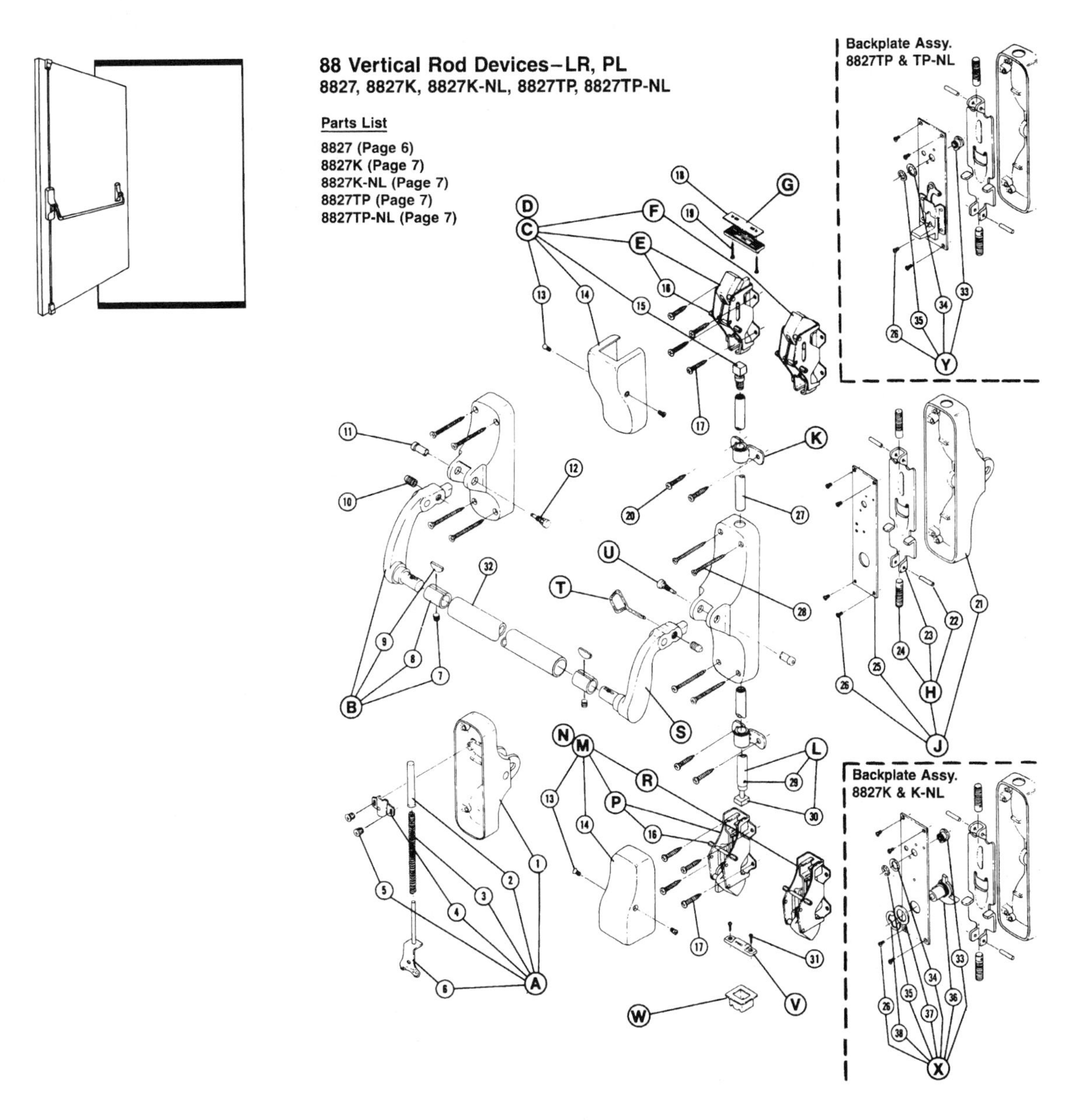

By permission: Von Duprin Exit Devices - Ingersoll-Rand, Inc., Indianapolis, Indiana

13.11.0 Panic Devices (Concealed/Surface-Applied Vertical Rod Devices) (Continued)

Parts List–8827 Vertical Rod Device–LR, PL

Reference Illustration Page 5

Item	Qty.	Part No.	Description	
A	1	101713	End Case Assy.–Brz.	X
A	1	101714	End Case Assy.–Alum.	X
A	1	101715	End Case Assy.–St. Stl	X
B	1	100729	Lever Arm Assy.–LH–Brz.	X
B	1	100746	Lever Arm Assy.–LH–Alum.	X
B	1	101487	Lever Arm Assy.–LH–St. Stl.	X
C	1	101754	Top Latch Case Assy.–LR–Brz.	X
C	1	101755	Top Latch Case Assy.–LR–Alum.	X
C	1	101756	Top Latch Case Assy.–LR–St. Stl.	X
D	1	101760	Top Latch Case Assy.–PL–Brz.	X
D	1	101761	Top Latch Case Assy.–PL–Alum.	X
D	1	101762	Top Latch Case Assy.–PL–St. Stl.	X
E	1	101750	Top Latch Bolt & Chassis Assy.–LR	
F	1	101752	Top Latch Bolt & Chassis Assy.–PL	
G	1	030298	299 Strike Assy.–Top	X
H	1	102701	Rod Control Assy.	
J	1	101707	Center Case Assy.–Brz.	X
J	1	101708	Center Case Assy.–Alum.	X
J	1	101709	Center Case Assy.–St. Stl.	X
K	2	101648	Rod Guide Assy.–Brz.	X
K	2	101774	Rod Guide Assy.–Alum.	X
K	2	101775	Rod Guide Assy.–St. Stl.	X
L	1	101777	Bottom Vertical Rod Assy.–Brz.	X
L	1	103234	Bottom Vertical Rod Assy.–St. Stl.	X
M	1	101757	Bottom Latch Case Assy.–LR–Brz.	X
M	1	101758	Bottom Latch Case Assy.–LR–Alum.	X
M	1	101759	Bottom Latch Case Assy.–LR–St. Stl.	X
N	1	101763	Bottom Latch Case Assy.–PL–Brz.	X
N	1	101764	Bottom Latch Case Assy.–PL–Alum.	X
N	1	101765	Bottom Latch Case Assy.–PL–St. Stl.	X
P	1	101751	Bottom Latch Bolt & Chassis Assy.–LR	
R	1	101753	Bottom Latch Bolt & Chassis Assy.–PL	
S	1	100744	Lever Arm Assy.–RH–Brz.	X
S	1	100747	Lever Arm Assy.–RH–Alum.	X
S	1	101486	Lever Arm Assy.–RH–St. Stl.	X
T	1	103868	Wedge Tite Package	
U	1	103865	Lever Arm Axle Package	
V	1	030602	248L4 Strike Assy.	
W	1	030659	304L Strike Assy.	
1	1	960517	End Case–Brz.	X
1	1	960518	End Case–Alum.	X
1	1	960519	End Case–St. Stl.	X
2	1	951081	Spring Tube	
3	1	953630	Spring	
4	1	951910	Spring Tube Bracket	
5	2	963094	#8-32 × 5/16″ PPHMS–Thd. Form.	
6	1	107682	Spring Stop Sub-Assy.	

Item	Qty.	Part No.	Description	
7	2	963851	Wedge Tite Adaptor Screw (5/16″–18 × 5/8″ Set Scr.)	
8	2	965676	Tube Attaching Ring	
9	2	956520	Attaching Ring Wedge	
10	1	968485	Dog Screw	
11*	2	969573	88 Axle–Female	
12*	2	969572	88 Axle–Male	
13	4	956010	#8-32 × 1/4″ FPHMS–Brz.	X
13	4	956011	#8-32 × 1/4″ FPHMS–St. Stl.	X
14	2	960548	Top & Bottom Latch Case Cover–Brz.	X
14	2	960585	Top & Bottom Latch Case Cover–Alum.	X
14	2	960586	Top & Bottom Latch Case Cover–St. Stl.	X
15	1	960577	Top Rod Connector–LR	
15	1	967341	Top Rod Connector–PL	
16	4	960652	Special Chassis Nut	
17	8	965288	#10-12 × 10-24 × 1″ PBHCS	
18	1	945521	Adjusting Shim	
19	2	965289	#10-12 × 10-24 × 1 1/4″ OPHCS	X
20***	4	965286	#10-12 × 10-24 × 1″ PTHCS–Brz.	X
20***	4	965287	#10-12 × 10-24 × 1″ PTHCS–St. Stl.	X
21	1	960421	Center Case–Brz.	X
21	1	960422	Center Case–Alum.	X
21	1	960423	Center Case–St. Stl.	X
22	2	963193	1/8″ × 11/16″ Lg. Spirol Pin	
23	1	961229	Rod Control	
24	2	961283	Rod Adaptor	
25	1	952530	Back Plate	
26	4	963096	#8-32 × 3/8″ Lg. FPHMS–Thd. Form.	
27	1	961628	Top Vertical Rod–Brz.	X
27	1	960581	Top Vertical Rod–St. Stl.	X
28**	8	965291	#10-12 × 10-24 × 2″ OPHCS–Brz.	X
28**	8	965292	#10-12 × 10-24 × 2″ OPHCS–St. Stl.	X
29	1	963585	1/8″ × 1/2″ Roll Pin	
30	1	960578	Bottom Rod Connector	
31	2	963008	#10-24 × 1/2″ Lg. OPHMS	
32	1	060275	Cross Bar Tube–Std.–27 1/2″ Lg.	X
32	1	061275	Cross Bar Tube–Knurled–27 1/2″ Lg.	X
32	1	060295	Cross Bar Tube–Std.–29 1/2″ Lg.	X
32	1	061295	Cross Bar Tube–Knurled–29 1/2″ Lg.	X
32	1	060360	Cross Bar Tube–Std.–36″ Lg.	X
32	1	061360	Cross Bar Tube–Knurled–36″ Lg.	X
32	1	060420	Cross Bar Tube–Std.–42″ Lg.	X
32	1	061420	Cross Bar Tube–Knurled–42″ Lg.	X
32	1	060500	Cross Bar Tube–Std.–(Longer than 42″)	X
32	1	061500	Cross Bar Tube–Knurled–(Longer than 42″)	X
*	1	103865	Lever Arm Axle Package	
**	1	900500	Mtg. Screw Package	X
***	1	900534	Mtg. Screw Package	X

X designates items that are finished.

Note: For ordering parts provide the part number, description, total quantity and finish required.

By permission: Von Duprin Exit Devices - Ingersoll-Rand, Inc., Indianapolis, Indiana

13.11.1 Panic Devices (Mortise Lock Devices)

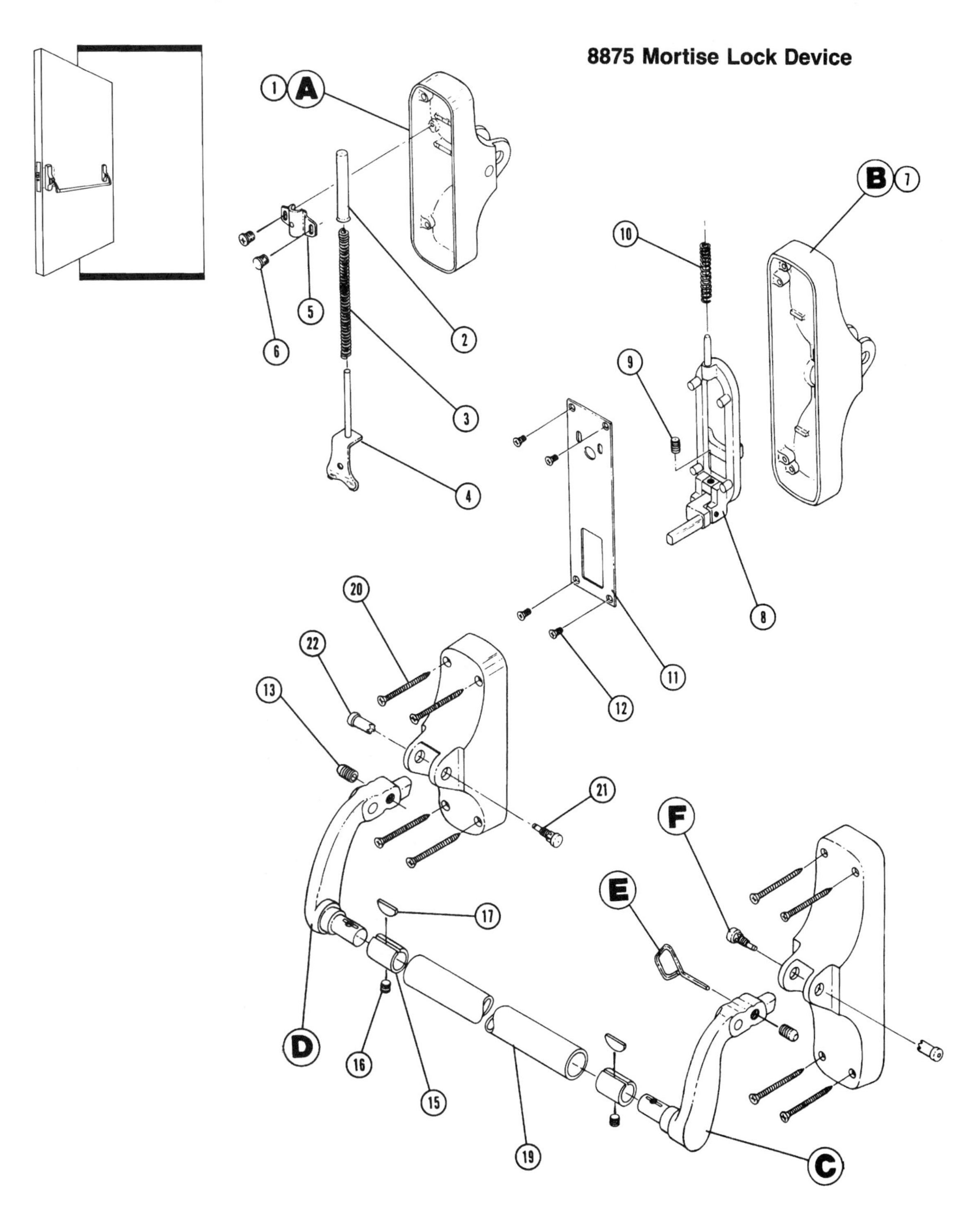

By permission: Von Duprin Exit Devices - Ingersoll-Rand, Inc., Indianapolis, Indiana

13.11.1 Panic Devices (Mortise Lock Devices) (Continued)

Parts List–8875-F Mortise Lock Device

Item	Qty.	Part No.	Description	
A	1	101716	End Case Assy.–Brz.	X
A	1	101717	End Case Assy.–Alum.	X
A	1	101718	End Case Assy.–St. Stl.	X
B	1	101695	Center Case Assy.–RH–Brz.	X
B	1	101696	Center Case Assy.–RH–Alum.	X
B	1	101697	Center Case Assy.–RH–St. Stl.	X
B	1	101698	Center Case Assy.–LH–Brz.	X
B	1	101699	Center Case Assy.–LH–Alum.	X
B	1	101700	Center Case Assy.–LH–St. Stl.	X
C	1	109867	Lever Arm Assy.–RH–Brz.	X
C	1	109868	Lever Arm Assy.–RH–Alum.	X
C	1	109869	Lever Arm Assy.–RH–St. Stl.	X
D	1	109858	Lever Arm Assy.–LH–Brz.	X
D	1	109859	Lever Arm Assy.–LH–Alum.	X
D	1	109860	Lever Arm Assy.–LH–St. Stl.	X
F	1	103865	Lever Arm Axle Package	
1	1	960912	End Case–Brz.	X
1	1	960913	End Case–Alum.	X
1	1	960914	End Case–St. Stl.	X
2	1	951081	Spring Tube	
3	1	953630	Spring	
4	1	107682	Spring Stop Guide Assy.	
5	1	951910	Spring Tube Bracket	
6	2	963094	#8-32 × 5/16″ PPHMS–Thd. Form.	
7	1	960915	Center Case–Brz.	X
7	1	960916	Center Case–Alum.	X
7	1	960917	Center Case–St. Stl.	X
8	1	104367	Latch Control Finger Assy.–RH	
8	1	100693	Latch Control Finger Assy.–LH	
9	1	951780	Flat Point Set Screw	
10	1	958643	Spring	
11	1	960756	Back Plate	
12	4	963096	#8-32 × 3/8″ PFHMS–Self Tap	
15	2	965676	Tube Attaching Ring	
16	2	963851	Wedge Tite Adaptor Screw	
17	2	956520	Attaching Ring Wedge	
19	1	060275	Cross Bar Tube–Std.–27½″ Lg.	
19	1	061275	Cross Bar Tube–Knurled–Std.–27½″ Lg.	
19	1	060295	Cross Bar Tube–Std.–29½″ Lg.	
19	1	061295	Cross Bar Tube–Knurled–Std.–29½″ Lg.	
19	1	060360	Cross Bar Tube–Std.	
19	1	061360	Cross Bar Tube–Knurled–Std.–36″ Lg.	
19	1	060420	Cross Bar Tube–Std.–42″ Lg.	
19	1	061420	Cross Bar Tube–Knurled–Std.–42″ Lg.	
19	1	060500	Cross Bar Tube–Custom–(Lger. than 42″)	
19	1	061500	Cross Bar Tube–Custom–Knurled–(Lger. than 42″)	
20	8	965291	#10-12 × 10-24 × 2″ OPHCS–Brz.	
20	8	965292	#10-12×10-24×2″ OPHCS–St. Stl.–Alum.	
21**	2	969572	88 Axle–Male	
22**	2	969573	88 Axle–Female	
*	1	900500	Mounting Screw Package	
**	1	103865	Lever Arm Axle Package	

X designates items that are finished.

Note: For ordering parts provide the part number, description, total quantity and finish required.

By permission: Von Duprin Exit Devices - Ingersoll-Rand, Inc., Indianapolis, Indiana

13.11.2 Panic Devices (Rim Devices Conventional and Enclosed Push-Bar Type)

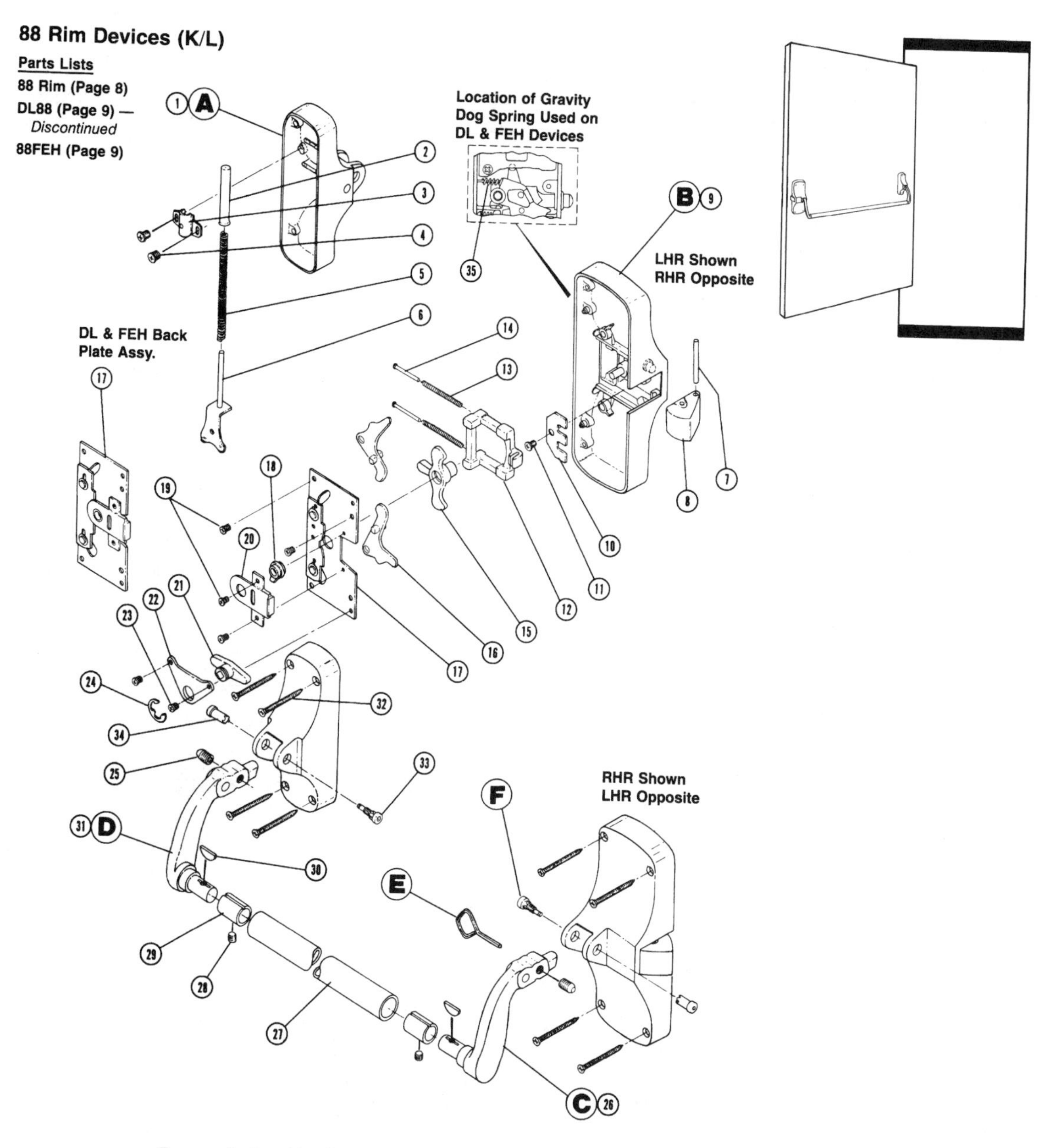

By permission: Von Duprin Exit Devices - Ingersoll-Rand, Inc., Indianapolis, Indiana

13.11.2 Panic Devices (Rim Devices Conventional and Enclosed Push-Bar Type) (Continued)

Parts List–88 Rim Device–K/L

Reference Illustration–Page 7

Item	Qty.	Part No.	Description	
A	1	101713	End Case Assy.–Brz.	X
A	1	101714	End Case Assy.–Alum.	X
A	1	101715	End Case Assy.–St. Stl.	X
B	1	104154	Center Case Assy.–RH–Brz.	X
B	1	104155	Center Case Assy.–RH–Alum.	X
B	1	104156	Center Case Assy.–RH–St. Stl.	X
C	1	100744	Lever Arm Assy.–RH	X
C	1	100747	Lever Arm Assy.–RH	X
C	1	101486	Lever Arm Assy.–RH	X
D	1	100729	Lever Arm Assy.–LH	X
D	1	100746	Lever Arm Assy.–LH	X
D	1	101487	Lever Arm Assy.–LH	X
E	1	103868	Wedge Tite Key	
F	1	103865	Lever Arm Axle Pack	
1	1	960517	End Case–Brz.	X
1	1	960518	End Case–Alum.	X
1	1	960519	End Case–St. Stl.	X
2	1	951081	Spring Tube	
3	1	951910	Spring Tube Bracket	
4	2	963094	#8-32 × 5/16" PPHMS–Thd. Form.	
5	1	953630	Spring	
6	1	107682	Spring Stop Sub-Assy.	
7	1	945501	Latch Bolt Axle	
8	1	100751	Latch Bolt Sub-Assy.	X
9	1	101653	Center Case Sub-Assy.–Brz.–LH & RH	X
9	1	101654	Center Cash Sub-Assy.–Alum.–LH & RH	X
9	1	101655	Center Case Sub-Assy.–St. Stl.–LH & RH	X
10	1	960675	Latch Retainer Plate	
11	1	963098	#10-24 × 5/16" Lg. PPHMS–Thd. Form.	
12	2	952280	Latch Tail	
13	2	966762	Latch Tail Spring	
14	2	963099	1/8" × 1 1/4" Lg. Rd. Hd. Rivet	
15	1	951960	Master Cam	
16	1	965420	Knob Cam Lift	
17	1	104150	Back Plate Sub-Assy.	
18	1	965506	K Cylinder Cam	

Item	Qty.	Part No.	Description	
19	5	963094	#8-32 × 5/16" Lg. PPHMS–Thd. Form.	
20	1	965503	Auxiliary Back Plate	X
21	1	951950	Knob Cam	
22	1	965418	Knob Cam Bracket	
23	1	963712	#8-32 × 3/8" Lg. PPHMS–Thd. Form.	
24	1	963103	Snap Ring–Truarc #5133-62	
25	2	968485	Dog Screw	
26	1	101444	Lever Arm & Adaptor Assy.–RH–Brz.	X
26	1	101446	Lever Arm & Adaptor Assy.–RH–Alum.	X
26	1	101479	Lever Arm & Adaptor Assy.–RH–St. Stl.	X
27	1	060275	Cross Bar Tube–Std.–27 1/2" Lg.	
27	1	061275	Cross Bar Tube–Knurled–27 1/2" Lg.	
27	1	060295	Cross Bar Tube–Std.–29 1/2" Lg.	
27	1	061295	Cross Bar Tube–Knurled–29 1/2" Lg.	
27	1	060360	Cross Bar Tube–Std.–36" Lg.	
27	1	061360	Cross Bar Tube–Knurled–36" Lg.	
27	1	060420	Cross Bar Tube–Std.–42" Lg.	
27	1	061420	Cross Bar Tube–Knurled–42" Lg.	
27	1	060500	Cross Bar Tube–Custom– (Longer than 42")	
27	1	061500	Cross Bar Tube–Custom–Knurled– (Longer than 42")	
28	2	963851	Wedge Tite Adapt. Screw (5/16"-18 × 5/8" Set Scr.)	
29	2	965676	Tube Attaching Ring	
30	2	956520	Attaching Ring Wedge	
31	1	101445	Lever Arm & Adaptor Assy.–LH–Brz.	X
31	1	101447	Lever Arm & Adaptor Assy.–LH–Alum.	X
31	1	101480	Lever Arm & Adaptor Assy.–LH–St. Stl.	X
32*	8	965291	#10-12 × 10-24 × 2 OPHCS–Brz.	
32*	8	955292	#10-12 × 10-24 × 2 OPHCS–St. Stl.	
33**	2	969572	88 Axle–Male	
34**	2	969573	88 Axle–Female	
*	1	900500	Mounting Screw Package	X
**	1	103865	Lever Arm Axle Package	

X designates items that are finished.

Note: For ordering parts provide the part number, description, total quantity and finish required.

By permission: Von Duprin Exit Devices - Ingersoll-Rand, Inc., Indianapolis, Indiana

13.11.3 Panic Devices (Rim Devices, Other Types of Pushes)

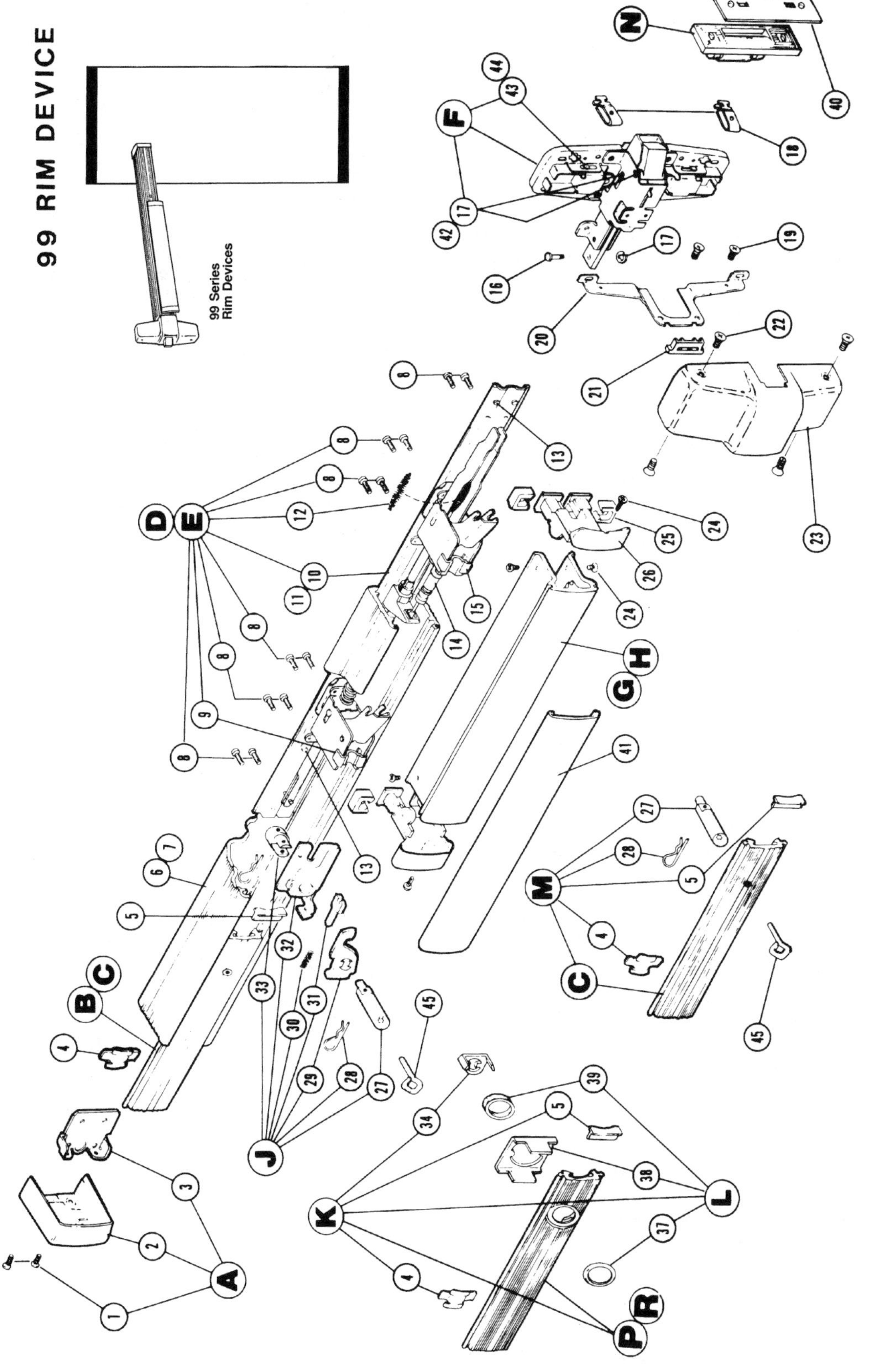

By permission: Von Duprin Exit Devices - Ingersoll-Rand, Inc., Indianapolis, Indiana

13.11.3 Panic Devices (Rim Devices, Other Types of Pushes) (Continued)

Parts List–99 Rim Device

Item	Qty.	Part No.	Description	
A	1	110283	Mechanism End Cap	X
B	1	110671	Std. Cover Plate Assy. 2′6″-3′0″ Door	X
C	1	110672	Std. Cover Plate Assy. 3′1″-4′0″ Door	X
D	1	110760	Base Plate Assy. 2′6″-3′0″ Door	
E	1	110761	Base Plate Assy. 3′1″-4′0″ Door	
F	1	110831	Center Case Assy. (Less Cover)	
G	1	108823	Push Bar Sub-Assy. 2′6″-3′0″ Door	
H	1	108824	Push Bar Sub-Assy. 3′1″-4′0″ Door	
J	1	107970	Dogging Sub-Assy.	
K	1	050115	CD Conversion Kit	
L	1	107813	Cylinder Hardware Assy.	
M	1	050114	Std. Conversion Kit	
N	1	030298	299 Strike	
P	1	110673	CD Cover Plate Assy. 2′6″-3′0″ Door	X
R	1	110674	CD Cover Plate Assy. 3′1″-4′0″ Door	X
1	2	963911	#8-32 × ½″ OPHMS Thd. Cut	
2	1	970266	Mechanism Case End Cap	X
3	1	970151	Mechanism Case Mounting Bracket	
4	1	967093	Anti-Rattle Spring	
5	1	966268	Cover Bearing Insert	
6	1	968144	Mechanism Case 2′6″-3′0″ Door	X
7	1	968146	Mechanism Case 3′1″-4′0″ Door	X
8	12	963094	#8-32 × 5⁄16″ PPHMS	
9	1	970079	Bearing Strip	
10	1	970033	Base Plate–2′6″-3′0″ Door	
11	1	970034	Base Plate–3′1″-4′0″ Door	
12	1	968555	Latch Return Spring	
13	4	970183	Rubber Bumper	
14	1	969613	Shock Absorber	
15	1	969612	Holder	
16	1	969520	Control Link Pin	
17	3	964066	Retaining Ring (Truarc #T5304-18)	
18	2	968101	Cover Retaining Clip	

Parts List–99 Rim Device

Item	Qty.	Part No.	Description	
19	2	964166	#12-24 × ½″ PPHMS-T/Cut Type T	
20	1	969841	Bracket	
21	1	969400	Cover Bearing Insert	
22	4	964041	#8-18 × ½″ FPHSMS-Type A	X
23	1	969398	99 Rim Center Case Cover (Zinc)	X
23	1	970082	99 Rim Center Case Cover (Brs.)	X
24	6	964041	#8-18 × ½″ FPHMS (Undercut AB)	
25	4	968496	Push Bar Guide	
26	2	968650	Push Bar End Cap	X
27	1	968112	Dogging Shaft	
28	2	963909	#22 Hitch Pin	
29	1	969941	Dogging Hook	
30	1	966384	Dogging Spring	
31	1	968115	Dogging Spring Guide	
32	1	968117	Dogging Housing	
33	1	968114	Dogging Axle	
34	1	968116	CD Actuator Arm	
37	1	961267	Cylinder Collar	X
38	1	967032	Cylinder Locating Washer	
39	1	959010	Cylinder Lock Nut	
40	1	945521	Adjusting Shim	
41	1	067016	Std. Push Bar Trim–2′6″-3′0″ Door	X
41	1	067023	Std. Push Bar Trim–3′1″-4′0″ Door	X
41	1	067116	Knurled Push Bar Trim–2′6″-3′0″ Door	X
41	1	067123	Knurled Push Bar Trim–3′1″-4′0″ Door	X
41	1	067216	Embossed Push Bar Trim–2′6″-3′0″ Door	X
41	1	067223	Embossed Push Bar Trim–3′1″-4′0″ Door	X
42	2	969467	Latch Bolt Pin	
43	2	967448	Latch Link Pin	
44	2	964085	Retaining Ring (Truarc T5304-15)	
45	1	959066	Special Hex Key	
	1	900561	Device Mounting Screw Package	
	1	900263	299 Strike Mounting Screw Package	X

X Designates items that are finished.

By permission: Von Duprin Exit Devices - Ingersoll-Rand, Inc., Indianapolis, Indiana

13.11.4 Panic Devices (Outside Trim)

Parts List – 991K-R&V Trim

Item#	Qty	Part#	Description	Mat'l	Fin. (X)
		047011	991K-R&V Trim	Brs	X
		047013	991K-R&V Trim (Knurled Knob)	Brs	X
A	1	110537	Orbit Knob Assy	Brs	X
A	1	110538	Orbit Knob Assy (Knurled)	Brs	X
B	1	110539	991K-R&V Trim Plate Assy	Brs	X
1	4	969546	Hex Stud	Stl	
2	2	969551	Pinion Gear	Stl	
3	1	969544	Straight Key	Stl	
4	1	963767	Retaining Ring (Truarc #5100-43)	Stl	
5	1	969537	Slider Rack	Stl	
6	2	969012	Spring	Stl	
7	1	969535	Slider	Stl	
8	1	963925	#10 Internal Tooth Lock Washer	Stl	
9	1	964086	#10-24 x 3/8" Button Hd. Cap Screw	Stl	
10	1	969545	Rim & Vertical Finger	Stl	
11	1	963638	1/4-20 x 3/4" Soc. Set Screw	Stl	
12	1	968201	Cylinder Retaining Cup	Stl	
		900837	Mounting Screw Package – 1¾ DR	Stl	
		900838	Mounting Screw Package – 2¼ DR	Stl	

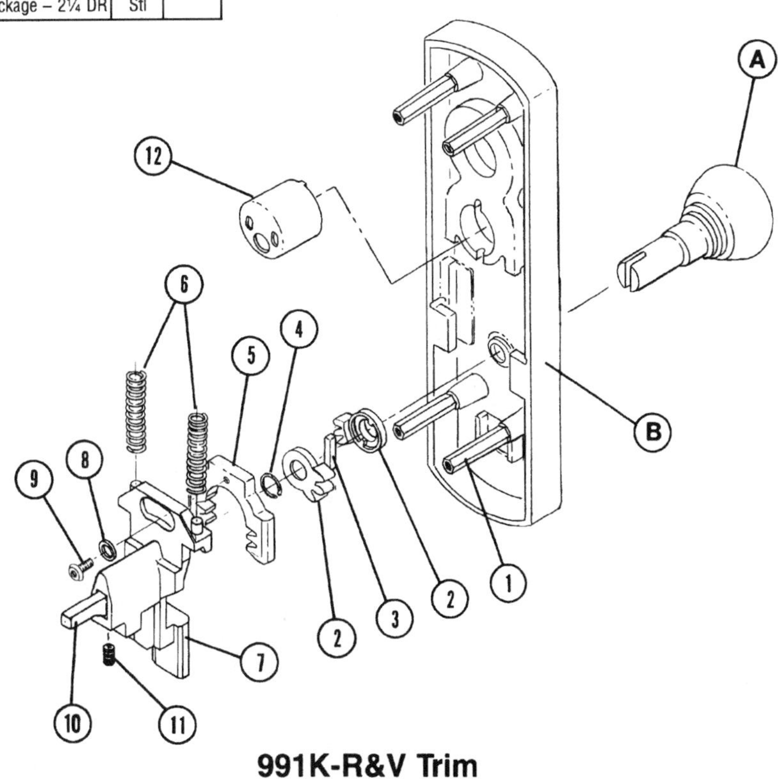

991K-R&V Trim

By permission: Von Duprin Exit Devices - Ingersoll-Rand, Inc., Indianapolis, Indiana

13.11.4 Panic Devices (Outside Trim) (Continued)

Parts List – 992L-BE-M Lever Trim

Item#	Qty	Part#	Description	Mat'l	Fin. (X)
A	1	110792	#01 Lever Assy	Brs	X
A	1	111895	#01 Lever Assy	Brz	X
A	1	111667	#02 Lever Assy	Brs	X
A	1	111668	#02 Lever Assy	Brz	X
A	1	110791	#03 Lever Assy	Brs	X
A	1	111643	#03 Lever Assy	Brz	X
A	1	110793	#06 Lever Assy	Brs	X
A	1	111644	#06 Lever Assy	Brz	X
A	1	111665	#07 Lever Assy	Brs	X
A	1	111666	#07 Lever Assy	Brz	X
A	1	111681	#12 Lever Assy – RHR	Brs	X
A	1	111683	#12 Lever Assy – LHR	Brs	X
A	1	111682	#12 Lever Assy – RHR	Brz	X
A	1	111684	#12 Lever Assy – LHR	Brz	X
A	1	111663	#17 Lever Assy	Brs	X
A	1	111664	#17 Lever Assy	Brz	X
B	1	110798	Trim Plate Sub-Assy	Brs	X
C	1	110767	Slider Assy	Stl	
1	4	969546	Hex Stud	Stl	
2	1	969558	Shear Pin	Stl	
3	1	969548	Turnpiece Cam	Stl	
4	1	964097	Retaining Ring – Truarc #5101-43	Stl	
5	2	969504	Lift Spring	S.S.	
6	2	964086	#10-24 x ⅜" Cap Screw	Stl	
7	2	963925	#10 Int. Tooth Lock Washer	Stl	
8	1	970120	Adjustment Shim	S.S.	
9	1	969536	Slider Yoke	Stl	
10	1	969535	Slider	Stl	
	1	900837	Mounting Screw Package – 1¾ DR	Stl	
	1	900838	Mounting Screw Package – 2¼ DR	Stl	

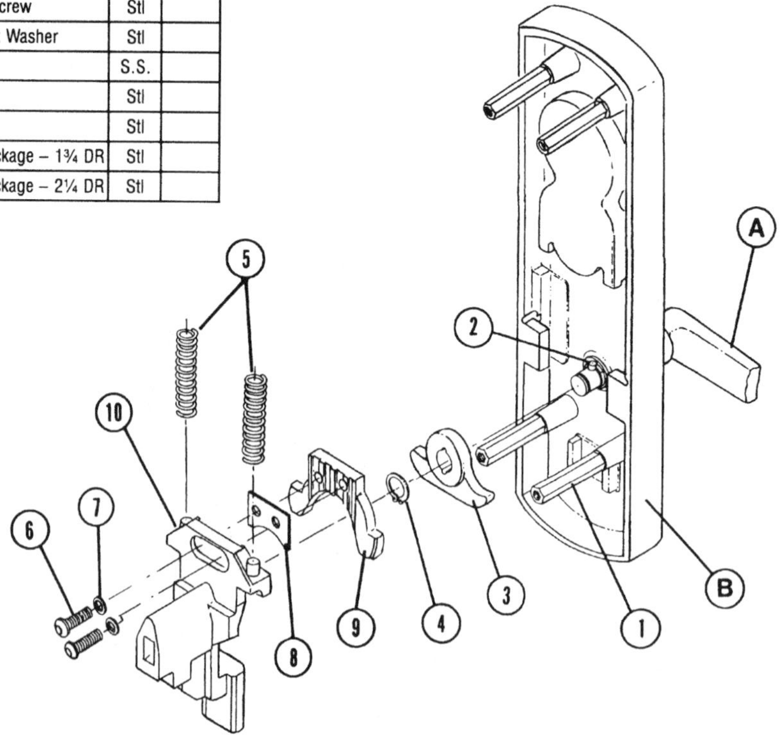

992L-BE-M Lever Trim

By permission: Von Duprin Exit Devices - Ingersoll-Rand, Inc., Indianapolis, Indiana

13.12.0 Standard Keying Terms, Codes, and Designations

Standard keying terms

Keys and Terms	Abbreviation	Definition
Change key	—	Individual lock key.
Keyed Differently	KD	Each lock is set to a different key combination.
Keyed alike	KA	Two or more locks set to the same key combination, KA2, KA3 KA4 etc.
Masterkey	MKD	Operates any given quantity of cylinders with different key changes.
Grand Masterkey	GMKD	Operates all individual locks already operated by two or more masterkeys.
Great grand Masterkey	GGMKD	Operates all locks under the various masterkeys and grand masterkeys already established.
Emergency key	EMKD	Operates hotel locks having shut out feature which blocks entry by all other keys.
Construction key	CK	Operates all cylinders designated for a temporary period during construction.
Control key	—	Key to remove active core of cylinders.
Keyway	—	Broaching in cylinder barrel.
Key section	—	Sidewarding in cylinder match broaching in barrel.
"To operate"	—	**Identifying a key** or keys to operate other cylinders having different key change (Note: Never use phrase "to pass" or "subject to").
"To be operated by"	—	**Identifying a cylinder** to be operated by one or more individual keys other hand its own key (Note: Never use phrase "to pass" or "subject to").
Lock out key	—	Permits hotel management to lock door against use of all other keys except emergency key.
Single keyed	SKDI	Cylinders operated by their change key only (in a master, grand or great grand masterkey system).

Standard keying code

1. Use two letters for both masterkeyed and grand masterkeyed systems.
   a. Masterkey systems have change key numbers prefixed. Example: 1AA, 2AA
   b. Grand masterkey systems have change key numbers suffixed. Example: AA1, AA2
2. Each change combination has a different number affixed to letter symbol. Every keyed different cylinder must be listed with a different number. Cylinders in keyed alike groups will have the same number affixed.
3. Letter symbol only (A) indicates to be operated by grand masterkey only, no change key (Single letter would not be used at all in simple masterkey system).
4. Two letter symbol only (AA) indicates to be operated by: AA masterkey and A grand masterkey only, no change key.
5. Symbol A1, A2; these are changes under the "A" grand masterkey only.
6. Symbol GGM1, GGM2; these are changes under great grand masterkey only.
7. Symbol 1AA, 2AA, etc. used in great grand masterkey system. The change numbers are prefixed on all locks operated by masterkeys under great grand masterkey only—no grand masterkey operates these locks.
8. Symbol SKD1, SKD2, etc.—single keyed—used for locks in a master, grand master or great grand masterkey system. These locks operated by their change keys only (not masterkeyed, grand masterkeyed, etc.).

Examples

SIMPLE MASTERKEY SYSTEM

Masterkey AA
Change key 1AA
2AA
3AA

GRAND MASTERKEY SYSTEM

Grand Masterkey A

MasterkeyAA	MasterkeyAB	MasterkeyAC
Change key AA1	AB1	AC1
AA2	AB2	AC2
AA3	AB3	AC3

13.13.0 Finish Symbols and Descriptions of These Finishes

(McKinney, BHMA, U.S. Government Codes.)

McKINNEY	DESCRIPTION	BASE BHMA	U.S. FINISH	MATERIAL
P	Primed for Painting	600	USP	Steel
2C	Zinc Plated, Commercial	602	US2C	Steel
2G	Zinc Plated, Government Specification	603	US2G	Steel
3	Bright Brass	605 632	US3	Brass Steel
4	Dull Brass	606 633	US4	Brass Steel
5	Dull Brass, Oxidized	609 638	US5	Brass Steel
7	Brass, Nickel Oxidized, Bright Relieved	610 636	US7	Brass Steel
9	Bright Bronze	611 637	US9	Bronze Steel
10	Dull Bronze	612 638	US10	Bronze Steel
10B	Antique Bronze, Oiled	613 640	US10B	Bronze Steel
11	Dull Bronze, Oxidized	616 643	US11	Bronze Steel
14	Bright Nickel Plated	618 645	US14	Brass, Bronze Steel
15	Dull Nickel Plated	619 646	US15	Brass, Bronze Steel
15A	Nickel Oxidized, Relieved	620 647	US15A	Brass, Bronze Steel
17A	Half Polished Iron, Smooth	621 648	US17A	Brass, Bronze Steel
20	Statuary Bronze, Light	623 649	US20	Bronze Steel
20A	Statuary Bronze, Dark	624 650	US20A	Bronze Steel
26	Bright Chromium	625 651	US26	Brass, Bronze Steel
26D	Dull Chromium	626 652	US26D	Brass, Bronze Steel
32	Polished Stainless Steel	629	US32	Stainless Steel Series 300
32D	Dull Stainless Steel	630	US32D	Stainless Steel Series 300
AP	Aluminum Powder Coat	—	—	Steel
BZ	Zinc Plated — Buffed Bright	—	—	Steel
DZ	Zinc Plated — Dull	—	—	Steel
D2	Co-Laq, Medium	—	—	Steel
D3	Co-Laq, Dark	—	—	Steel
D4	Co-Laq, Black	—	—	Steel
PG	Powdered Gold	—	—	Steel
PW	Powdered White	—	—	Steel
PB	Powdered Beige	—	—	Steel
PN	Powdered Neutral	—	—	Steel
3C	US3, with Clear Powder Coat	—	—	Steel
4C	US4, with Clear Powder Coat	—	—	Steel

Finishes on McKinney hinges comply with U.S. standards. Where a special finish or a matched finish is required, a sample should be submitted. McKinney rust resisting finish is specified by prefixing S to catalog number.

13.4.0 Recommended Number of Hinges and Frequency of Operations

RECOMMENDED NUMBER OF HINGES PER DOOR, EITHER WOOD OR METAL

Door Height, In. (mm)	Number of Hinges per Door
Up to 60 (1524)	2
60 to 90 (1524 to 2286)	3
90 to 120 (2286 to 3048)	4

RECOMMENDED SIZE OF HINGES PER DOOR, EITHER WOOD OR METAL

Door		Hinge	
Thickness In. (mm)	Width In. (mm)	Height In. (mm)	Gauge
1 3/8 (35)	up to 36 (914)	3 1/2 (89)	.119
1 3/8 (35)	over 36 (914)	4 (102)	.129
1 3/4 (44)	up to 36 (914)	4 1/2 (114)	*.134
1 3/4 (44)	over 36 - 48 (914 - 1219)	5 (127)	*.134
1 3/4 (44)	over 48 (1219)	6 (152)	*.160
2-2 1/2 (51-64)	up to 42 (1067)	5 (127) HW	.190
2-2 1/2 (51-64)	over 42 (1067)	6 (152) HW	.190

*Heavy hinges should be used on all extra heavy doors or those exposed to high frequency use! Five knuckle heavy weight hinges are four bearing. The following gauges of metal apply:
Heavy weight 4-1/2" (114) high = .180" gauge
Heavy weight 5" (127) high = .190" gauge
Heavy weight 6" (152) high = .190" gauge
Note: Five knuckle 8" (203) high hinges have six bearings.

EXPECTED FREQUENCY OF DOOR OPERATION

One Cycle = one complete opening and closing.

Installation Type	Expected Frequency Daily	Expected Frequency Yearly	
Commercial			
Commercial store entrance	5,000	1,500,000	High
Office building entrance	4,000	1,200,000	High
Theatre entrance	1,000	450,000	High
School entrance	1,250	225,000	High
School restroom door	1,250	225,000	High
Store or bank entrance	500	150,000	High
Office building restroom door	400	118,000	High
School corridor door	80	15,000	Average
Office building corridor door	75	22,000	Average
Store restroom door	60	18,000	Average
Residential			
Entrance	40	15,000	Low
Restroom door	25	9,000	Low
Corridor door	10	3,600	Low
Closet door	6	2,200	Low

NOTE: School classroom doors have approximately the same frequency as school restroom doors. We recommend that bearing hinges be used on all above categories other than "residential."

13.15 ASTM Specifications Applicable to Finish Hardware Requirements

Products Comply With:

ASTM B-117—Salt spray (fog) testing (paint test).

ASTM C-236—Test for thermal conductance and transmittance of built-up sections by means of the guarded hot box.

ASTM C-553—Specifications for mineral fiber blanket and felt insulation (industrial type).

ASTM D-610—Method of evaluating degree of rusting on painted steel surfaces.

ASTM D-714—Method of evaluating degree of blistering of paints.

ASTM D-1735—Method for water fog testing of organic coatings.

ASTM D-3359—Measuring adhesion by tape test (paint).

ASTM E-90—Recommended practice for laboratory measurement of airborne sound transmission loss of building partitions.

ASTM E-152—Fire tests of door assemblies.

ASTM E-283—Test for rate of air leakage through window.

ASTM E-413—Classification for determination of sound transmission class.

Foam Core Standards—Polystyrene/Polyurethane

ASTM C-165—Method for measuring compressive properties of thermal insulations.

ASTM C-177—Test method for steady-state heat flux measurements and thermal transmission properties by means of the guarded hot plate apparatus.

ASTM C-203—Test for breaking load and calculated flexural strength of preformed block-type thermal insulation.

ASTM C-272—Test for water absorption of core materials for structural sandwich constructions.

ASTM C-273—Shear test in flatwise plans of flat sandwich construction or sandwich cores.

ASTM C-303—Test method for density of preformed block-type thermal insulation.

ASTM C-518—Test method for steady-state heat flux measurements and thermal transmission properties by means of the heat flow meter apparatus.

ASTM C-355—Test for water vapor transmission of thick materials.

ASTM C-578—Specification for preformed, block-type cellular polystyrene thermal insulations.

ASTM D-732—Test for shear strength of plastics by punch tool.

ASTM D-1621—Test for compressive strength of rigid cellular plastics.

ASTM D-1622—Test for apparent density of rigid cellular plastics.

ASTM D-1623—Test for tensile and tensile adhesion properties of rigid cellular plastics.

ASTM D-2842—Test for water absorption of rigid cellular plastics.

ASTM D-2856—Test for open cell content of rigid cellular plastics by the air pycnometer.

ASTM D-2863—Measuring by minimum oxygen concentration to support candle-like combustion of plastics (oxygen index).

ASTM E-84—Test for surface burning characteristics of building materials.

ASTM E-96—Test methods for water vapor transmission of materials.

Steel & Galvanizing Standards

ASTM A-366—Specification for steel, carbon, cold-rolled sheet, commercial quality.

ASTM A-525—Specification for steel sheet, zinc-coated (galvanized) by the hot-dip process, general requirements.

ASTM A-526—Specification for steel sheet, zinc-coated (galvanized) by the hot-dip process, commercial quality.

ASTM A-568—Specification for steel, carbon, and high strength low-alloy hot-rolled strip, and cold-rolled sheet, general requirements.

ASTM A-569—Specification for steel, carbon (0.15 maximum percent), hot-rolled sheet and strip, commercial quality.

ASTM A-591—Specification for steel sheet, electrolytic zinc-coated.

ASTM A-620—Specification for steel sheet, carbon, cold-rolled, drawing quality, special killed.

ASTM A-642—Specification for steel sheet, zinc-coated (galvanized) by the hot-dip process, drawing quality, special killed.

ANSI/SDI 100—Recommended specifications for standard steel doors and frame.

ANSI A250.5-1994—Performance test procedure for steel door frames and frame anchors.

ANSI A123.1—Standard nomenclature for steel doors and steel door frames.

ANSI A224.1—Standard test procedure and acceptance criteria for prime-painted steel surfaces for steel doors and frames.

ANSI A250.4-1994—Test procedure and acceptance criteria for physical endurance for steel doors and hardware reinforcings.

A115 Series Of Door & Frame Preparation Standards

ANSI A115.1—Specifications for standard steel door and steel frame preparations for mortise locks 1-3/8" (35) and 1-3/4" (44) doors.

ANSI A115.2—Specifications for standard steel doors and frame preparation for bored or cylindrical locks for 1-3/8" (35) and 1-3/4" (44) doors.

ANSI A115.4—Specifications for standard steel doors and frame preparation for lever extension flush bolts.

ANSI A115.5—Specifications for steel frame preparation for 181 Series and 190 Series deadlock strikes.

ANSI A115.6—Specifications for standard steel door and steel frame preparation for preassembled door locks (unit lock).

ANSI A115.8—Specifications for door and frame preparation for floor closer center hung, single, or double acting.

ANSI A115.9—Specifications for hospital door roller latches.

ANSI A115.11—Specifications for standard steel door and frame preparation for mortise locks for 1-3/8" (35) doors.

ANSI A115.12—Specifications for standard steel door and steel frame preparation for offset intermediate pivot.

ANSI A115.13—Specifications for standard steel door and steel frame preparation for tubular deadlocks.

ANSI A115.14—Specifications for standard steel doors for open back strikes.

ANSI A2.2—Fire tests of door assemblies (UL 10B).

ANSI A155.1—Fire door frames UL 63 (outdated).

ANSI/NFPA 105—Installation of smoke and draft control door assemblies.

Section

14

Drywall, Metal Framing, and Plaster

Contents

14.0.0 Drywall systems
14.0.1 Non load-bearing partitions
14.0.2 Load-bearing partitions
14.0.3 High-performance sound control
14.0.4 Wall furring (partition details)
14.0.5 Non load-bearing ceilings
14.1.0 Wall furring (illustrations)
14.2.0 Partition construction details (illustrations)
14.3.0 Plumbing fixture attachment and electric outlet installation
14.4.0 Tub and shower details—single-layer details
14.5.0 Wall control joint details (illustrated)
14.6.0 Typical bath tub and swimming pool wall details
14.7.0 Soffit framing specifications
14.8.0 Shelf-wall specifications and illustrations
14.9.0 Chase-wall specifications and illustrations
14.10.0 Resilient channel partition specifications
14.11.0 Tall wall specifications and limiting heights
14.11.1 *L* over 120/240/360 explained
14.11.2 Structural stud specifications
14.11.3 Typical limiting heights of interior partitions
14.12.0 High-performance sound-control construction (illustrations)
14.13.0 Curtain wall construction (illustrations)
14.13.1 Typical curtain-wall limiting-height specifications
14.14.0 Super studs
14.14.1 Super stud section properties: 2½" (13.85 cm) by 4" (10.16 cm) studs
14.14.2 Super stud section properties: 4" (10.16 cm) by 8" (20.32 cm) studs
14.14.3 Super stud section properties: 8" (20.32 cm) by 16" (40.64 cm) studs
14.14.4 Super stud section properties: Terms and definitions
14.14.5 Super stud accessories
14.15.0 Plaster systems
14.15.1 Comparing conventional plaster, veneer plaster, and drywall systems
14.15.2 Lath and plaster installation procedures
14.15.3 Metal lath, hangers, channel, and stud specifications
14.15.4 Lath, framing, and furring accessories
14.16.0 Five levels of drywall-taping systems

14.0.0 Drywall Systems

Steel or wood studs, faced with gypsum panels (regular, fire rated, or vinyl faced) have dominated the construction industry, representing the most cost effective, light weight, and fire resistant means of creating interior walls. Specialty products, such as ½-inch (12 mm) thick cement board, sometimes referred to as *Wonder board* and *exterior-grade gypsum sheathing panels*, along with the development of heavier-gauge structural metal studs for curtain wall construction, has expanded the number of applications to which these products can be used.

14.0.1 Non Load-Bearing Partitions

Non-Load Bearing Partitions

☐ *Sound-deadening material** ■ *RC-1 ™ Resilient Channel***

Fire rating	Fire-rated construction: Detail & physical data	Fire-rated construction: Description & test no.	Acoustical performance: STC	Acoustical performance: Description & test no.	System reference
1 hr.	3 1/8"	Steel Stud—5/8" SHEETROCK brand gypsum panels, ULTRACODE core—1 5/8" studs 24" o.c.—panels vert appl & screw att with 1 1/2" Type S screws 8" o.c. perim, 12" o.c. field—joints stag & fin—**UL Des U496**			A
1 hr.	3 1/2"	Steel Stud—1/2" SHEETROCK brand gypsum panels, FIRECODE C core—2 1/2" studs 24" o.c.—single layer panels ea side appl vert & screw att—1 1/2" THERMAFIBER SAFB—joints fin—perimeter caulked—**UL Des U448** wt 5 width 3 1/2"	45 48	**TL-69-42** Based on 3 5/8" studs & 2" SAFB—**SA-800422**	B
1 hr. est	4 1/4"	Steel Stud—5/8" SHEETROCK brand gypsum panels, FIRECODE core—2 1/2" studs 24" o.c.—1 1/2" THERMAFIBER SAFB—2 layers—base layer 1/4" SHEETROCK brand gypsum panels screw att—5/8" facelayer screw att—joints fin—perimeter caulked—est. fire rating based on T-1174-OSU wt 7 width 4 1/4"	55 53	**CK-684-14** Based on 1/2" thick panels—**CK-684-13**	C
1 hr.	5 1/8"	Steel Stud—resil partition—1/2" SHEETROCK brand gypsum panels, FIRECODE C Core, or 5/8" SHEETROCK brand gypsum panels, FIRECODE core—3 5/8" studs 24" o.c.—3" THERMAFIBER SAFB 25" wide creased to fit cavity—RC-1 chan 24" o.c. screw att one side—panels vert appl & screw att—joints stag & fin—perimeter caulked—**UL Des U451** wt 6 width 5 1/8"	55 54	Based on 5/8" SHEETROCK brand gypsum panels, FIRECODE core & 25" wide creased SAFB—**SA-850415** Based on 5/8" SHEETROCK brand gypsum panels, FIRECODE core—**SA-850415**	D
1 hr. est	4"	Steel Stud—1/2" SHEETROCK brand gypsum panels, FIRECODE C core—2 1/2" studs 24" o.c.—single layer panels one side appl vert & screw att—1 1/2" THERMAFIBER SAFB—2 layers opp side—panels appl vert & screw att—joints stag & fin—perimeter caulked—est. fire rating based on T-3362-OSU wt 7 width 4"	50 41	**SA-800504** Based on same construction without SAFB—**TL-69-148**	E
1 hr.	3 5/8"	Steel Stud—2 layers 1/2" SHEETROCK brand gypsum panels ea side—1 5/8" studs 24" o.c.—panels appl vert & screw att—joints stag & fin—perimeter caulked—**U of C 9-21-64** wt 9 width 3 5/8"	55	Based on SHEETROCK brand gypsum panels FIRECODE C core, & 1 1/2" SAFB—**USG-840824**	F
1 hr.	4 7/8"	Steel Stud—5/8" SHEETROCK brand gypsum panels, FIRECODE core—3 5/8" studs 24" o.c.—single layer panels vert or horiz appl & screw att—joints stag & fin—perimeter caulked—**UL Des U465**—based on panels horiz appl—**GA-WP-1200** wt 6 width 4 7/8"	40 49 51	**USG-860808** Based on 3" SAFB in cavity—**SA-870717** Based on FIRECODE C core panels and 3" SAFB 25" wide, creased to fit cavity—**TL-90-166**	G
1 hr.	2 7/8"	Steel Stud—5/8" SHEETROCK brand gypsum panels, FIRECODE core—1 5/8" studs 24" o.c.—single layer panels vert appl & screw att 12" o.c.—joints fin—perimeter caulked —**U of C 7-31-62** wt 5 width 2 7/8"	38	**USG-860809**	H
1 hr.	3 3/4"	Steel Stud—5/8" SHEETROCK brand gypsum panels, FIRECODE core—2 1/2" studs 24" o.c.—1 1/2" THERMAFIBER SAFB—panels apply horiz & screw-att—joints opposite—vert joints unfin—horiz joints fin—**CEG 8-11-83**—rating also applies to assembly with 1/2" SHEETROCK brand gypsum panels, FIRECODE C core, joints fin—**CEG 5-9-84** wt 6 width 3 3/4"	47	**SA-831001**	I
1 hr.	10 3/4"	Steel Stud Chase Wall—5/8" SHEETROCK brand gypsum panels, FIRECODE core, ea side—1 5/8" studs 24" o.c. in 2 rows spaced 6 1/4" apart—5/8" gypsum panel gussets or steel run braces spanning chase screw att to studs—panels appl vert & screw att—joints stag & fin—**UL Des U420** wt 6 width 10 3/4"	52	Based on 3 1/2" insulation on one side—**TL-76-155**	J
1 hr. (truss 3 hr.)	14 1/4"	Steel Stud—5/8" SHEETROCK brand gypsum panels, FIRECODE C core, ea side—fireproofed steel truss —2 1/2" studs 24" o.c. in 2 rows spaced 8" apart—5/8" gypsum panel gussets spanning chase att to stud at qtr & ctr points—panels appl vert & screw att—joints stag & fin—**UL Des U805** wt 6 width 14 1/4"	N/A		K
2 hr.	5"	Steel Stud—3/4" SHEETROCK brand gypsum panels, ULTRACODE core, ea side—3 5/8" or 3 1/2" studs 24" o.c.—3" THERMAFIBER SAFB—panels vert appl & screw att 8" o.c. perim, 12" o.c. field—joints stag & fin—perimeter caulked—**UL Des U491**	50	**USG-910617**	L
2 hr.	3 5/8" 4 1/2" 5 5/8"	Steel Stud—2 layers 1/2" SHEETROCK brand gypsum panels, FIRECODE C core, ea side—1 5/8", 2 1/2" or 3 5/8" studs 24" o.c.—base layer appl vert, face layer appl vert or horiz, joints stag—base layer screw att—face layer strip lamin or screw att—joints fin—perimeter caulked—with or rating based on assembly without sound atten blankets—**UL Des U412** wt 10 width 4 1/2"	50 55 52 54	Based on 3 5/8" stud assembly without SAFB—**USG-840817** Based on 3 5/8" studs and 1 1/2" SAFB—**SA-800421** Based on lamin, face layer, 1 1/2" SAFB and 2 1/2" studs—**SA-860932** Based on 2 1/2" studs, screw att. face layer and 1 1/2" SAFB—**CK-654-40**	M
2 hr.	5"	Steel Stud—2 layers 5/8" SHEETROCK brand gypsum panels, FIRECODE core, plain or vinyl faced vert appl ea side—2 1/2" studs or 24" o.c.—base layers screw att—face layer lamin or screw att—joints stag & fin or unfin—perimeter caulked—**UL Des U411** wt 12 width 6 1/8"	48 56	Based on 3 5/8" studs and 5/8" SHEETROCK brand gypsum panels, FIRECODE C core—**BBN-770408** Based on 3 5/8" studs and 3" SAFB—**USG-840818**	N
2 hr	6 1/8"	Steel Stud—2 layers 5/8" SHEETROCK brand gypsum panels, FIRECODE core, ea side—2 1/2" studs 24" o.c.—panels appl horiz & joints stag—base and face layers screw att—joints fin—perimeter caulked—**GA-WP-1548** wt 12 width 5"	51 56	Based on 2 1/2" SAFB in cavity—**GA-WP-1548** Based on 2" SAFB in cavity—**USG-840819**	0

*Where thermal insulation is shown in assembly drawings, the specific product is required in the assembly to achieve the stated fire-rating. Fiberglass insulation cannot be substituted for THERMAFIBER Insulation.
**Use RC-Resilient Channel or equivalent.

14.0.1 Non Load-Bearing Partitions (Continued)

Fire rating	Fire-rated construction: Detail & physical data	Fire-rated construction: Description & test no.	Acoustical performance: STC	Acoustical performance: Description & test no.	System reference
2 hr.	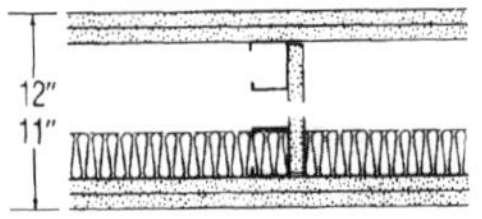 12"	Steel Stud Chase Wall—2 layers ⅝" SHEETROCK brand gypsum panels, FIRECODE core, ea side—1⅝" studs 24" o.c. in 2 rows spaced 6¼" apart—⅝" gypsum panel gussets or steel run braces spanning chase screw att to studs—panels appl vert & screw att—joints stag & fin—**UL Des U420** wt 13 width 12"	52 57	**TL-76-162** Based on 3½" insulation one side—**TL-76-156**	P
2 hr. est	11"	Steel Stud Chase Wall—2 layers ½" SHEETROCK brand gypsum panels, FIRECODE C core, ea side—1⅝" studs 24" o.c. in 2 rows spaced 5¾" apart—½" gypsum panel gussets spanning chase att to studs at qtr points—panels appl vert & screw att—1½" THERMAFIBER SAFB—joints stag & fin—perimeter caulked—est. fire rating based on **UL Des U412** wt 11 width 11"	55	**SA-860907**	Q
3 hr.	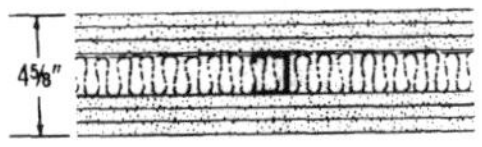 4⅝"	Steel Stud—3 layers ½" SHEETROCK brand gypsum panels, FIRECODE C core, ea side—1⅝" studs 24" o.c.—base layers appl vert—face layer appl horiz—panels screw att with joints stag and fin—perimeter caulked—rating based on assembly with or without SAFB—**UL Des U435** wt 13 width 4⅝"	59	Based on assembly with 1½" SAFB in cavity—**SA-830112**	R
3 hr.	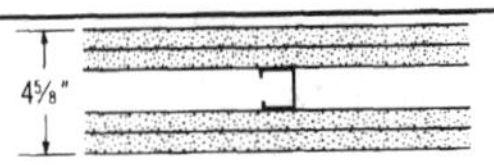 4⅝"	Steel stud—2 layers ¾" SHEETROCK brand gypsum panels, ULTRACODE core, ea side—1⅝" studs 24" o.c.—base layer app vert and att with 1¼" Type S screws 24" o.c., face layer att vert or horiz with 2¼" Type S screws 12" o.c.—att horiz joints with Type G screws midway betw framing (24" o.c.)—joints fin—perimeter caulked—**UL Des U435**			S
3 hr.	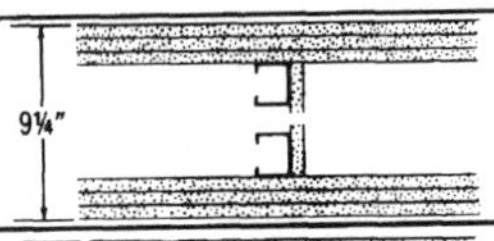 9¼"	Steel Stud—3 layers ½" SHEETROCK brand gypsum panels, FIRECODE C core, ea side—1⅝" studs 24" o.c. in 2 rows spaced 3" apart—steel truss member—gypsum panel gussets or steel run braces spanning chase screw att to studs—panels appl vert & screw att—joints stag & fin—2 hr. rating applies with 2 layers panels ea side—1 hr. rating applies with single layer ⅝" panels ea side—**UL Des U436** wt 13 width 9¼"	N/A		T
3 hr.	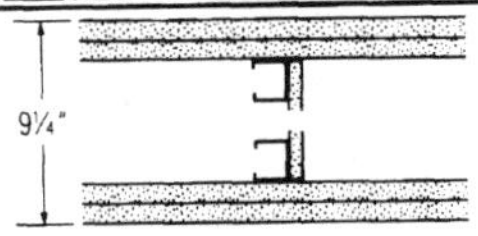 9¼"	Steel stud chase wall—2 layers ¾" SHEETROCK brand gypsum panels, ULTRACODE core, ea side—1⅝" studs 24" o.c. in two rows spaced 2" apart—steel truss member—gypsum panel gussets or stl run braces spanning chase screw-att to studs—base layer app vert and att with 1¼" Type S screws 24" o.c., face layer att vert or horiz with 2¼" Type S screws 12" o.c.—att horiz joints with Type G screws betw midway framing (24" o.c.)—joints stag & fin—**UL Des U436**			U
4 hr.	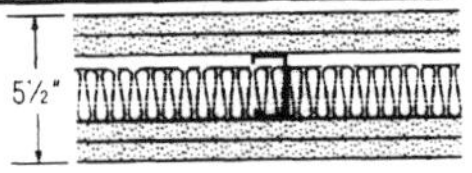 5½"	Steel Stud—2 layers ¾" SHEETROCK brand gypsum panels, ULTRACODE core, ea side—2½" studs 24" o.c.—2" THERMAFIBER SAFB—base layer app vert, joints stag & screw att 24" o.c.—face layer app vert or horiz & screw att 12" o.c.—att along horiz joints with Type G screws midway betw framing (24" o.c.)—joints fin—perimeter caulked—**UL Des U490**	56	**USG-910907**	V
4 hr.	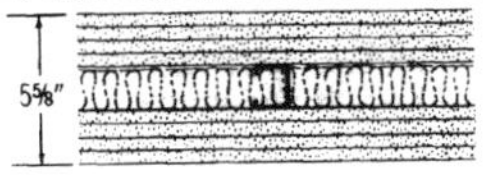 5⅝"	Steel Stud—4 layers ½" SHEETROCK brand gypsum panels, FIRECODE C core, ea side—1⅝" studs 24" o.c.—base layers appl vert—face layer appl horiz—panels screw att with joints stag & fin—perimeter caulked—rating based on assembly with or without sound atten fire blankets—**UL Des U435** wt 17 width 5⅝"	62	Based on assembly with 1½" SAFB in cavity—**SA-830113**	W

Steel stud 25 ga. will provide above fire and sound ratings.

By permission: United States Gypsum Corporation, Chicago, Illinois

14.0.2 Load-Bearing Partitions

Load-Bearing Partitions

Fire rating	Fire-rated construction: Detail & physical data	Fire-rated construction: Description & test no.	Acoustical performance: STC	Acoustical performance: Description & test no.	System reference
45 min.	4½"	½" SHEETROCK brand gypsum panels, FIRECODE C core—3½" 20 ga. structural studs 24" o.c.—panels appl vert & att with 1" Type S-12 screws 12" o.c.—joints fin—**load bearing up to 100% allowable stud axial—UL Des U425**	47	Based on engineering evaluation using 3" SAFB in cavity	A
1 hr.	4¾"	⅝" SHEETROCK brand gypsum panels, FIRECODE core—3½" 20 ga. structural studs 24" o.c.—panels appl vert & att with 1" Type S-12 Screws 12" o.c—joints fin—**load bearing up to 100% allowable stud axial load—UL Des U425**	40 41	**USG-810519** Based on 2" SAFB in cavity—**USG-810518**	B
1 hr.	6"	Dbl layer ½" SHEETROCK brand gypsum panels, FIRECODE C core—3½" 20 ga. structural studs 24" o.c.—1", 1½", 2", 3" THERMAFIBER SAFB—RC-1 chan one side spaced 24" o.c. screw-att to studs—panels appl vert with joints stag—base layer att with 1" Type S-12 screws 12" o.c.—face layer att with 1⅝" Type S-12 screws 12" o.c.—joints fin—rating also applies with IMPERIAL FIRECODE C base and veneer finish surface—**load bearing up to 100% allowable stud axial load—UL Des U440**	61 51	Based on 3½" 16 ga. structural studs, ⅝" thick panels, lateral bracing and 3" SAFB cavity—**SA-830628*** Based on 3½" 16 ga. structural studs and lateral bracing—**SA-840715**	C
1½ hr.	5½"	Dbl layer ½" SHEETROCK brand gypsum panels, FIRECODE C core—3½" 20 ga. structural studs 24" o.c.—panels appl vert—base layer att with 1" Type S-12 screws 12" o.c.—face layer att with 1⅝" Type S-12 screws 12" o.c.—joints fin—**load bearing up to 100% allowable stud axial load—UL Des U425**	49 49	Based on 2" SAFB—**USG-811009** Based on 2" SAFB and 6" 20 ga. structural studs—**USG-810940**	D
2 hr.	6"	Dbl layer ⅝" SHEETROCK brand gypsum panels, FIRECODE core—3½" 20 ga. structural studs 24" o.c.—panels appl vert—base layer att with 1" Type S-12 screws 12" o.c.—face layer att with 1⅝" Type S-12 screws 12" o.c.—joints fin—**load bearing up to 80% allowable stud axial load—UL Des U425**	48 49	Based on 2" SAFB in cavity—**USG-811006** Based on 2" SAFB and 6" 20 ga. structural studs—**USG-810937**	E

Fire rating	Fire-rated construction: Detail & physical data	Fire-rated construction: Description & test no.	Acoustical performance: STC	Acoustical performance: Description & test no.	System reference
3 hr.	7½"	Four layers ½" SHEETROCK brand gypsum panels, FIRECODE C core, ea side—3½" 20 ga. structural studs 24" o.c.—1", 1½", 2" or 3" THERMAFIBER SAFB optional—base layers appl vert with joints stag—base panels att with Type S-12 screws 48" o.c.—face layer appl vert or horiz with 2¼" Type S-12 screws 12" o.c. and 1½" Type G screws in panels—rating also applies with IMPERIAL FIRECODE C base and veneer finish surface—**load bearing up to 100% allowable stud axial load—UL Des U426**			F

*Assemblies with RC-1 Resilient Channel or equivalent require lateral bracing and offer estimated fire rating.

By permission: United States Gypsum Corporation, Chicago, Illinois

14.0.3 High-Performance Sound Control

High Performance Sound Control

Fire rating	Fire-Rated construction: Detail & physical data	Fire-Rated construction: Description & test no.	Acoustical performance: STC	Acoustical performance: Description & test no.	System reference
1 hr.	5¼" wt. 6	Resil Stud Drywall—½" SHEETROCK brand gypsum panels, FIRECODE C core—3⅝" 20 ga. structural studs 24" o.c.—3" THERMAFIBER SAFB—RC-1 chan one side spaced 24" o.c. screw-att to studs—single-layer gypsum panels screw-att to studs & RC-1 chan—panels appl vert with joints stag—joints fin—perimeter caulked—**UL Des U451**	50 54	**RAL-TL-87-156** (42 MTC) Based on ⅝" thick panels—**RAL-TL-83-216** (47 MTC)	A
1 hr.	7½" wt. 6	Resil Stud Drywall—½" SHEETROCK brand gypsum panels, FIRECODE C core—6" 20 ga. structural studs 24" o.c.—5" THERMAFIBER SAFB—RC-1 chan one side spaced 24" o.c.screw-att to studs—single-layer gypsum panels screw-att to studs & RC-1 chan—panels appl vert with joints stag—joints fin—perimeter caulked—**UL Des U451**	56 56	**RAL-TL-87-139** (48 MTC) Based on ⅝" thick panels—**RAL-TL-84-141** (50 MTC)	B
1½ hr.	8½" wt. 9	Resil Stud Drywall— ⅝" SHEETROCK brand gypsum panels. FIRECODE C core—6" 20 ga. structural studs 24" o.c.—5" THERMAFIBER SAFB—RC-1 chan one side spaced 24" o.c. screw-att to studs—2 layers gypsum panels crew-att to studs, 1 layer screw-att to chan—panels appl vert with joints stag—joints fin—perimeter caulked—**UL Des U452**	59	**RAL-TL-84-140** (54 MTC)	C
1½ hr.	6" wt. 8	Resil Stud Drywall—⅝" SHEETROCK brand gypsum panels, FIRECODE C core—3⅝" 20 ga. structural studs 24" o.c.—3" THERMAFIBER SAFB—RC-1 chan one side spaced 24" o.c. screw-att to studs—2 layers gypsum panels screw-att to studs, 1 layer screw-att to chan—panels appl vert with joints stag—joints fin—perimeter caulked—**UL Des U452**	58	**RAL-TL-83-215** (52 MTC)	D
2 hr.	8½" wt. 10	Resil Stud Drywall—½" SHEETROCK brand gypsum panels, FIRECODE C core—6" 20 ga. structural studs 24" o.c.—5" THERMAFIBER SAFB—RC-1 chan one side spaced 24" o.c. screw-att to studs—chan—double-layer gypsum panels screw-att to studs & RC-1 panels appl vert with joints stag—joints fin—perimeter caulked—**UL Des U454**	63 62	**RAL-TL-87-141** (59 MTC) Based on ⅝" thick panels—**RAL-TL-84-139** (58 MTC)	E
2 hr.	6¼" wt. 10	Resil Stud Drywall—½" SHEETROCK brand gypsum panels, FIRECODE C core—3⅝" 20 ga. structural studs 24" o.c.—3" THERMAFIBER SAFB—RC-1 chan one side spaced 24" o.c. screw-att to studs—2 layers gypsum panels screw-att to chan, 2 layers screw-att to chan—panels appl vert with joints stag—joints fin—perimeter caulked—**UL Des U454**	60 61	**RAL-TL-87-154** (54 MTC) Based on ⅝" thick panels—**RAL-TL-83-214** (57 MTC)	F
2 hr.	6" wt. 9	Resil Stud Drywall—½" SHEETROCK brand gypsum panels, FIRECODE C core—3⅝" 20 ga. structural studs 24" o.c.—3" THERMAFIBER SAFB—RC-1 chan one side spaced 24" o.c. screw-att to studs—single-layer gypsum panels screw-att to studs, 2-layers screw-att to chan—panels appl vert with joints stag—joints fin—perimeter caulked—**UL Des U453**	58 60 59	Estimated sound test (52 MTC) Based on ½" thick panels, 6" 20 ga. structural studs, 5" SAFB—**RAL-TL-87-140** (54 MTC) Based on ⅝" thick panels, 6" 20 ga. structural studs, 5" SAFB—**RAL-TL-84-136** (54 MTC)	G
3 hr.	6¾" wt. 12	Resil Stud Drywall—½" SHEETROCK brand gypsum panels, FIRECODE C core—3⅝" 20 ga. structural studs 24" o.c.—3" THERMAFIBER SAFB—RC-1 chan one side spaced 24" o.c. screw-att to studs—3 layers gypsum panels screw-att to studs, 2 layers screw-att to chan—panels appl vert with joints stag—joints fin—perimeter caulked—**UL Des U455**	61 62	**RAL-TL-87-153** (56 MTC) Based on ⅝" thick panels—**RAL-TL-83-213** (59 MTC)	H
3 hr.	9" wt. 12	Resil Stud Drywall—½" SHEETROCK brand gypsum panels, FIRECODE C core—6" 20 ga. structural studs 24" o.c.—5" THERMAFIBER SAFB—RC-1 chan one side spaced 24" o.c. screw-att to studs—3 layers gypsum panels screw-att to stud, 2 layers screw-att to chan—panels appl vert with joints stag—joints fin—perimeter caulked—**UL Des U455**	64 63 65	**RAL-TL-87-142** (59 MTC) Based on ⅝" thick panels—**RAL-TL-84-138** (59 MTC) Based on ⅝" thick panels, acoustical sealant bead between panels and studs, dabs 8" o.c. between panel layers on stud side—**RAL-TL-84-150** (60 MTC)	I

By permission: United States Gypsum Corporation, Chicago, Illinois

14.0.4 Wall Furring

Wall Furring

Detail & physical data	Description	Comments	System reference
1⅜"	Metal furring channels 24" o.c., ½" SHEETROCK brand gypsum panels, foil-back, screw-attached, joints finished	Provides good vapor resistance; no limiting height	A
1½"	SHEETROCK Z-furring channels applied vertically 24" o.c., THERMAFIBER fire safety FS-15 blankets between channels, ½" SHEETROCK brand gypsum panels, foil-back, screw-attached to channels, joints finished	Noncombustible system with mineral fiber insulation; suitable for up to 3" thick insulation; good vapor retarder; no limiting height	B
varies	Steel studs 24" o.c., set in runners, ½" SHEETROCK brand gypsum panels, foil-back, screw-attached to studs, joints finished	Free-standing; allows for pipe chase clearance; good vapor retarder	C
1½"	SHEETROCK Z-furring channels applied vertically 24" o.c., rigid plastic foam insulation between channels, ½" SHEETROCK brand gypsum panels, foil-back, applied vertically and screw-attached to channels, joints finished	Suitable for up to 3" thick insulation; no limiting height.	D

By permission: United States Gypsum Corporation, Chicago, Illinois

14.0.5 Nonload-Bearing Ceilings

Non-Load-Bearing Ceilings

Fire rating	Fire-rated construction: Detail & physical data	Fire-rated construction: Description & test no.	Acoustical performance: STC	Acoustical performance: IIC	Acoustical performance: Description & test no.	System reference
N/A		⅝″ SHEETROCK brand gypsum panels, FIRECODE core—1½″ chan 4′ o.c.—met fur chan 24″ o.c.—panels screw att 12″ o.c.—joints fin clg wt 3	N/A			A
1 hr. (beam 1 hr.)	9⅝″	½″ SHEETROCK brand gypsum panels, FIRECODE C core—7¼″ 18 ga. structural steel joists 24″ o.c.—dbl layer gypsum panel clg and ⅝″ T&G plywd flr att to joists with Type S-12 screws—dbl layer gypsum panels around beam—joints exp—**UL Des L524** clg wt 4	39		Based on 9½″ 16 ga. structural joists—**USG-760105**	B
			43		Based on 9½″ 16 ga. structural joists and 3″ SAFB*—**USG-760310**	
				56	Based on 9½″ 16 ga. structural joists and carpet & pad—**USG-760106**	
				60	Based on 9½″ 16 ga. structural joists and carpet & pad with 3″ SAFB* —**USG-760405**	
1½ hr.	27¼″	⅝″ SHEETROCK brand gypsum panels, FIRECODE C core—susp grid with main run 4′ o.c. and cross tees 2′ o.c.—gypsum panels screw-att below grid—joints stag and fin—min 1″ roof insul and ⅝″ gypsum bd on steel deck over bar joists—1-hr. rating based on assembly with ½″ thick panels—**UL Des P510** clg wt 4	N/A			C
2 hr. (beam 2 hr.)	13⅞″	½″ SHEETROCK brand gypsum panels, FIRECODE C core—furred or susp—met fur chan 24″ o.c.—panels att with 1″ Type S screws 12″ o.c.—joints exp or fin—2½″ conc on riblath or corrugated steel deck over bar joist—**UL Des G515** clg wt 3	N/A			D
2 hr. (beam 3 hr.)	21¼″	½″ SHEETROCK brand gypsum panels, FIRECODE C core—susp grid with main run 4′ o.c. and cross tees 2′ o.c.—gypsum panels screw-att below grid—joints fin—2½″ conc on riblath over bar joist—**UL Des G529** clg wt 3	N/A			E
2 hr.	9½″	⅝″ SHEETROCK brand gypsum panels, FIRECODE C core—met fur chan 24″ o.c.—panels att with 1″ Type S screws—joints fin—2″ prestressed reg or lightwt conc units with 6″ deep stems 48″ o.c.—**UL Des J502—UL Des J503** clg wt 3	N/A			F

*Insulation may affect fire rating. See SA-905.

Non-Load-Bearing Ceilings (cont.)

Fire rating	Fire-rated construction: Detail & physical data	Fire-rated construction: Description & test no.	Acoustical performance: STC	Acoustical performance: IIC	Acoustical performance: Description & test no.	System reference
3 hr. (beam 3 hr.)	21¼″	½″ SHEETROCK brand gypsum panels, FIRECODE C core—susp grid with main run 4′ o.c. and cross tees 2′ o.c.—gypsum panels screw-att below grid—joints fin—3¼″ conc on riblath over bar joist—rating also applies with ⅝″ panels and 2¾″ conc slab—**UL Des G529** clg wt 3	N/A			G
3 hr.	10¼″	⅝″ SHEETROCK brand gypsum panels, FIRECODE C core—met fur chan 24″ o.c.—panels att with 1″ Type S screws—joints fin—prestressed 2¾″ reg or 2½″ lightwt conc units with 6″ deep stems 48″ o.c.—**UL Des J502—UL Des J503—UL Des J504** clg wt 3		N/A		H
3 hr. (beam 3 hr.)	16″	⅝″ SHEETROCK brand gypsum panels, FIRECODE C core—met fur chan 24″ o.c.—panels att with 1″ Type S screws 12″ o.c.—joints exp or fin—3″ conc on corrugated steel deck or on riblath over bar joist—**UL Des G512** clg wt 3		N/A		I

By permission: United States Gypsum Corporation, Chicago, Illinois

14.1.0 Wall Furring (Illustrations)

Wall Furring

Interior and exterior walls are readily furred using ½" SHEETROCK brand Gypsum Panels, Foil-Back, screw-attached to steel framing erected vertically. In these systems, any of three different framing methods may be used to provide a vapor retarder, thermal insulation, and chase space for pipes, conduits and ducts.

With Metal Furring Channels

These furring channels, erected vertically 24" o.c., are fastened directly to interiors of exterior walls of monolithic concrete and virtually any type of masonry—brick, concrete block, tile. Channels may be furred using adjustable wall furring brackets and ¾" cold-rolled channels to provide additional space for pipes, conduits or ducts.

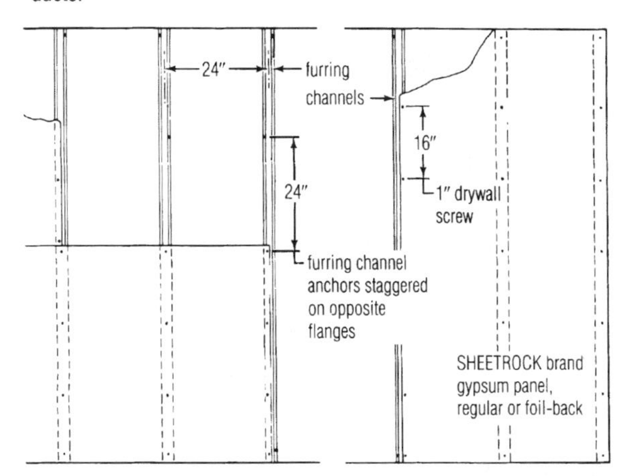

Perpendicular application *Parallel application*

With SHEETROCK Z-Furring Channels

In this assembly, SHEETROCK Z-Furring Channels are spaced 24" o.c. THERMAFIBER Fire Safety FS-15 Blanket or rigid foam insulation is friction-fit to interiors of exterior walls. Gypsum panels are screw-attached to channel flanges to provide a drywall surface isolated to a great degree from the masonry wall. In new construction and remodeling, this system provides a highly insulative self-furring solid backup for SHEETROCK brand Gypsum Panels. See construction details on page 26.

Thermal resistance (R) values for various assemblies are shown below.

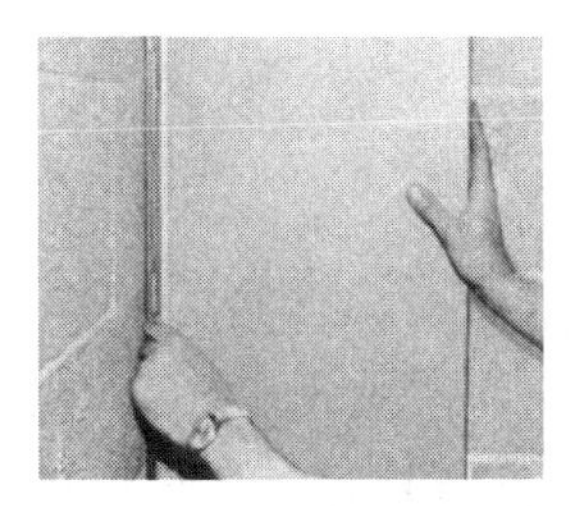

Installing insulation

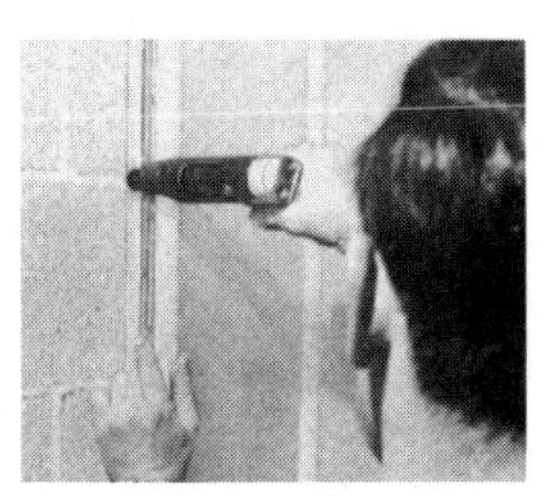

Attaching Z-furring channel

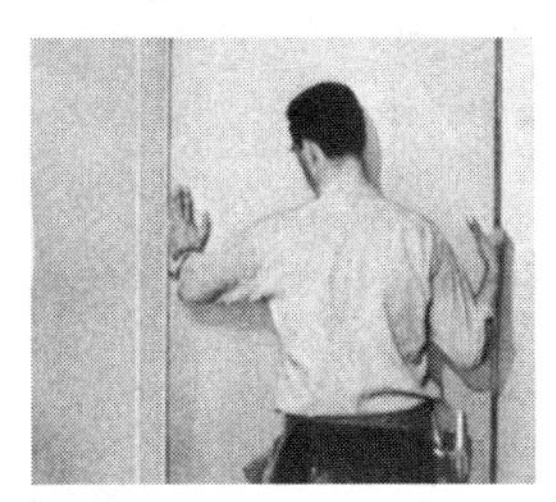

Erecting gypsum panel

Screw-attaching panel

Wall elevation

SHEETROCK Z-furring channels 24" o.c.
continuous wood nailer at opening
THERMAFIBER fire safety FS-15 blanket 2'-0" x 4'-0"
½" or ⅝" SHEETROCK brand gypsum panels applied vertically or horizontally

Design thermal resistance (R) values[1] with SHEETROCK Z-Furring Channel System

Wall construction	Nom. wall thickness	Uninsul. wall	wall[2] (no. insul.)	Wall insulated with—[2] THERMAFIBER Fire Safety FS-15 Blankets 1" (4.17)	1½" (6.00)	2" (8.00)	3" (12.00)	Rigid polystyrene 1" (5.00)	1½" (7.50)	2" (10.00)	3" (15.00)	Rigid urethane 1" (6.25)	1½" (9.38)	2" (12.50)	3" (18.75)
4" face brick & 8" cinder block	12"	3.01	4.38	7.63	9.46	11.46	15.46	8.46	10.96	13.46	18.46	9.71	12.84	15.96	22.21
4" face brick & 4" com. brick	8"	2.09	3.46	6.71	8.54	10.54	14.54	7.54	10.04	12.54	17.54	8.79	11.92	15.04	21.29
poured conc. (140 lb./cu. ft.)	8"	1.49	2.86	6.11	7.94	9.94	13.94	6.94	9.44	11.94	16.94	8.19	11.32	14.44	20.69
12" conc. block & 4" face brick	16"	2.57	3.94	7.19	9.02	11.02	15.02	8.02	10.52	13.02	18.02	9.27	12.40	15.52	21.77

(1) Resistances based on procedures and design values from 1981 ASHRAE Handbook of Fundamentals, winter conditions (15 mph wind) and neglect the effect of furring channels and fasteners. (2) Interior wall finish: ½" SHEETROCK brand Gypsum Panels, Foil-Back, (R-0.45). R-values for insulation shown in parentheses, based on 75 °F. mean temperature for insulation and components.

14.2.0 Partition Construction Details (Illustrations)

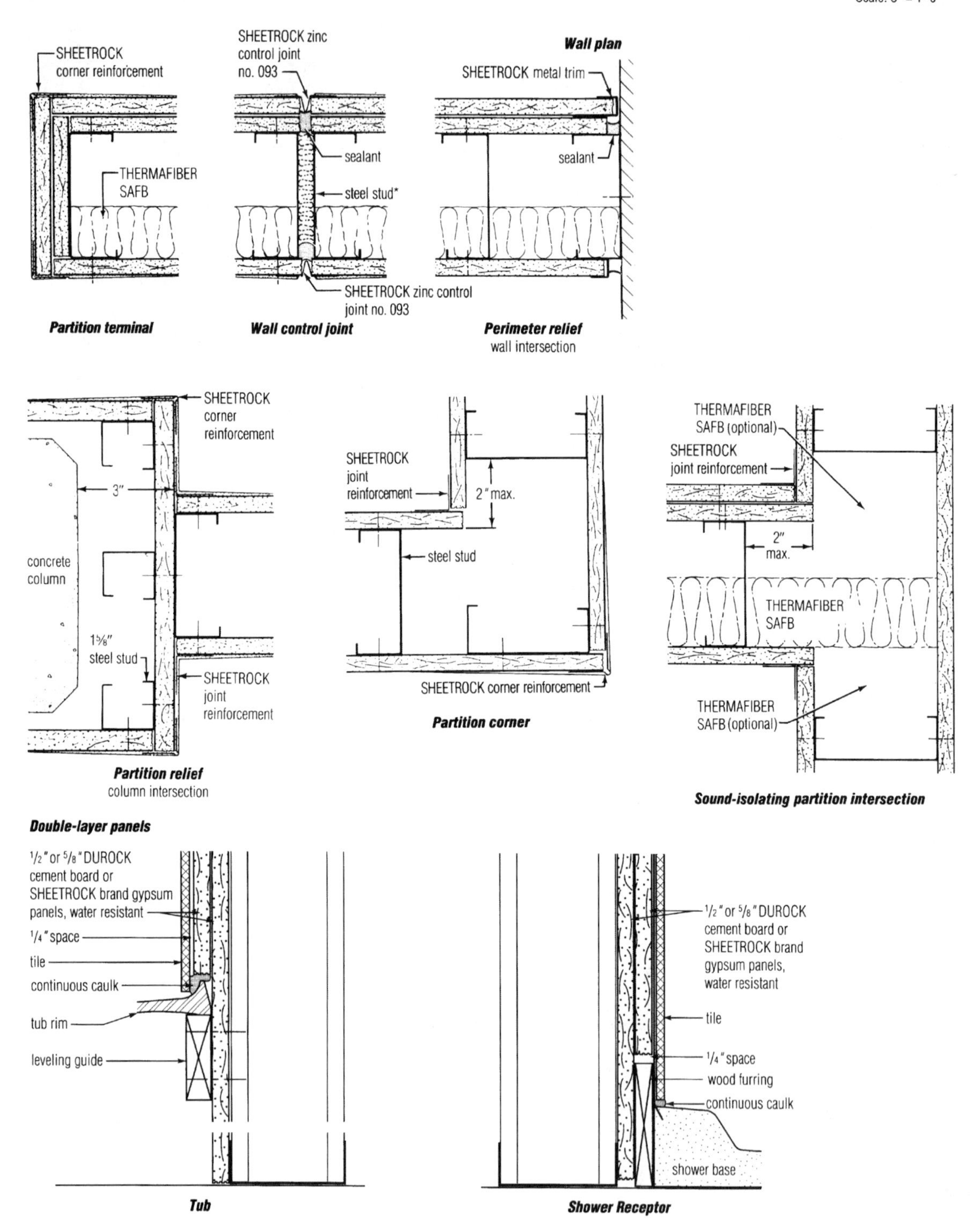

By permission: United States Gypsum Corporation, Chicago, Illinois

14.3.0 Plumbing Fixture Attachment and Electric Outlet Installation

Fixture Attachment

Scale: 3" = 1'-0"

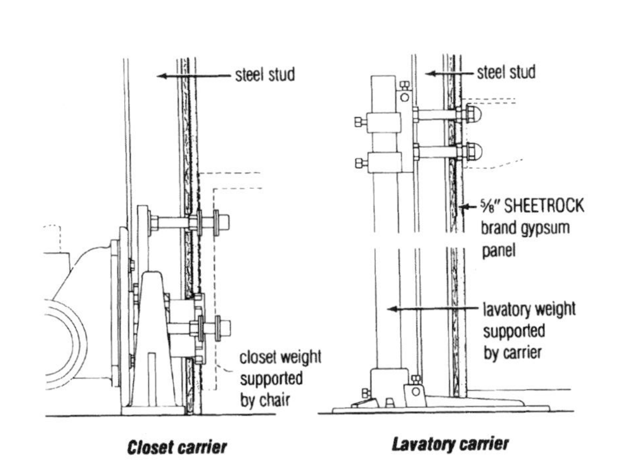

Closet carrier

Lavatory carrier

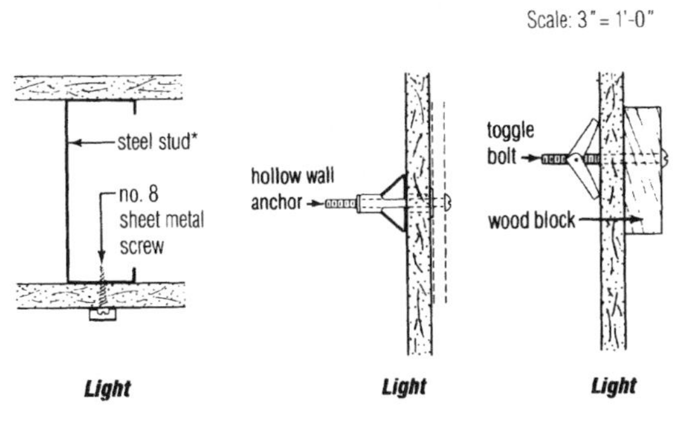

Light *Light* *Light*

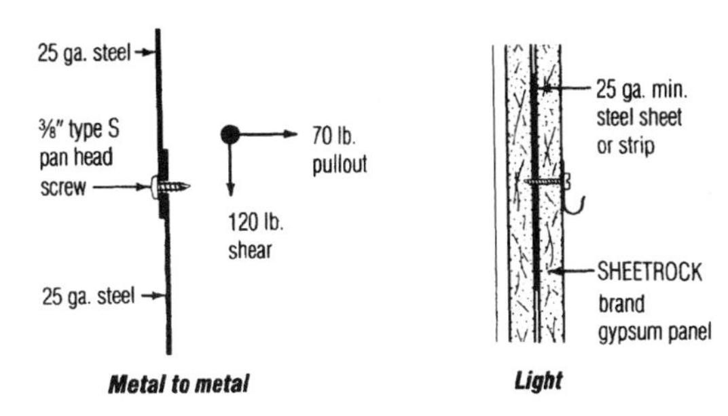

Metal to metal

Light

Load table

Fastener							
	Size		Base assembly	Allowable withdrawal resistance		Allowable shear resistance	
Type	in	mm		lb	N(1)	lb	N(1)
toggle bolt or hollow wall anchor	⅛	3.18	½" gypsum panel	20	89	40	178
	3/16	4.76		30	133	50	222
	¼	6.35		40	178	60	267
	⅛	3.18	½" gypsum panel & 25 ga. steel stud	70	311	100	445
	3/16	4.76		80	356	125	556
	¼	6.35		155	689	175	778
no. 8 sheet metal screw			½" gypsum panel & 25 ga.steel stud or 25 ga. steel insert	50	222	80	356
Type S bugle head screw				60	267	100	445
Type S-12 bugle head screw			½" gypsum panel & 20 ga. steel stud or 20 ga. steel insert	85	378	135	600
⅜" Type S pan head screw			25 ga. steel to 25 ga. steel	70	311	120	534
Type S-12			20 ga. steel to 20 ga. steel	53	235	133	591
two bolts welded to steel insert	3/16	4.76	see grab bar attachment below	175	778	200	890
	¼	6.35		200	890	250	1112
bolt welded to 1½" chan.	¼	6.35	see plumber's bracket below	200	890	250	1112

(1) Newtons

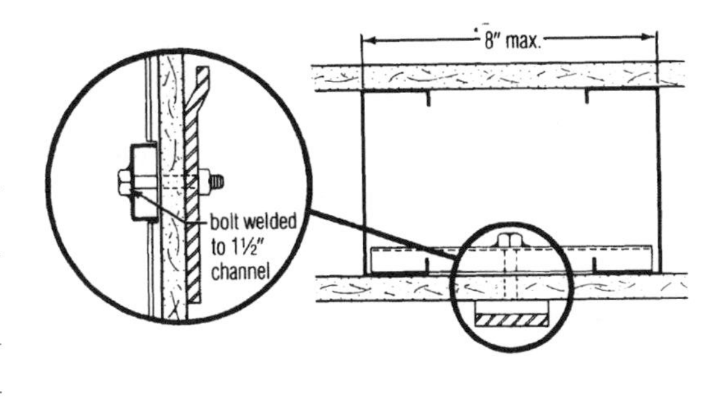

Outlet Boxes

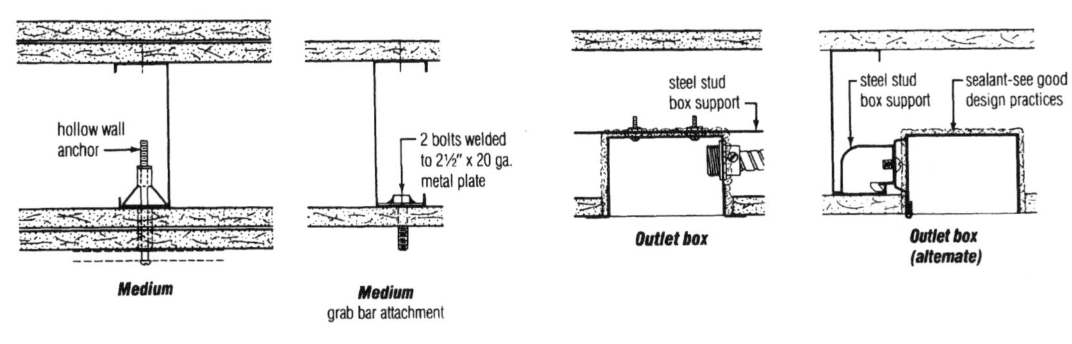

Medium

Medium
grab bar attachment

Outlet box

Outlet box (alternate)

By permission: United States Gypsum Corporation, Chicago, Illinois

14.4.0 Tub and Shower Details—Single-Layer Panels

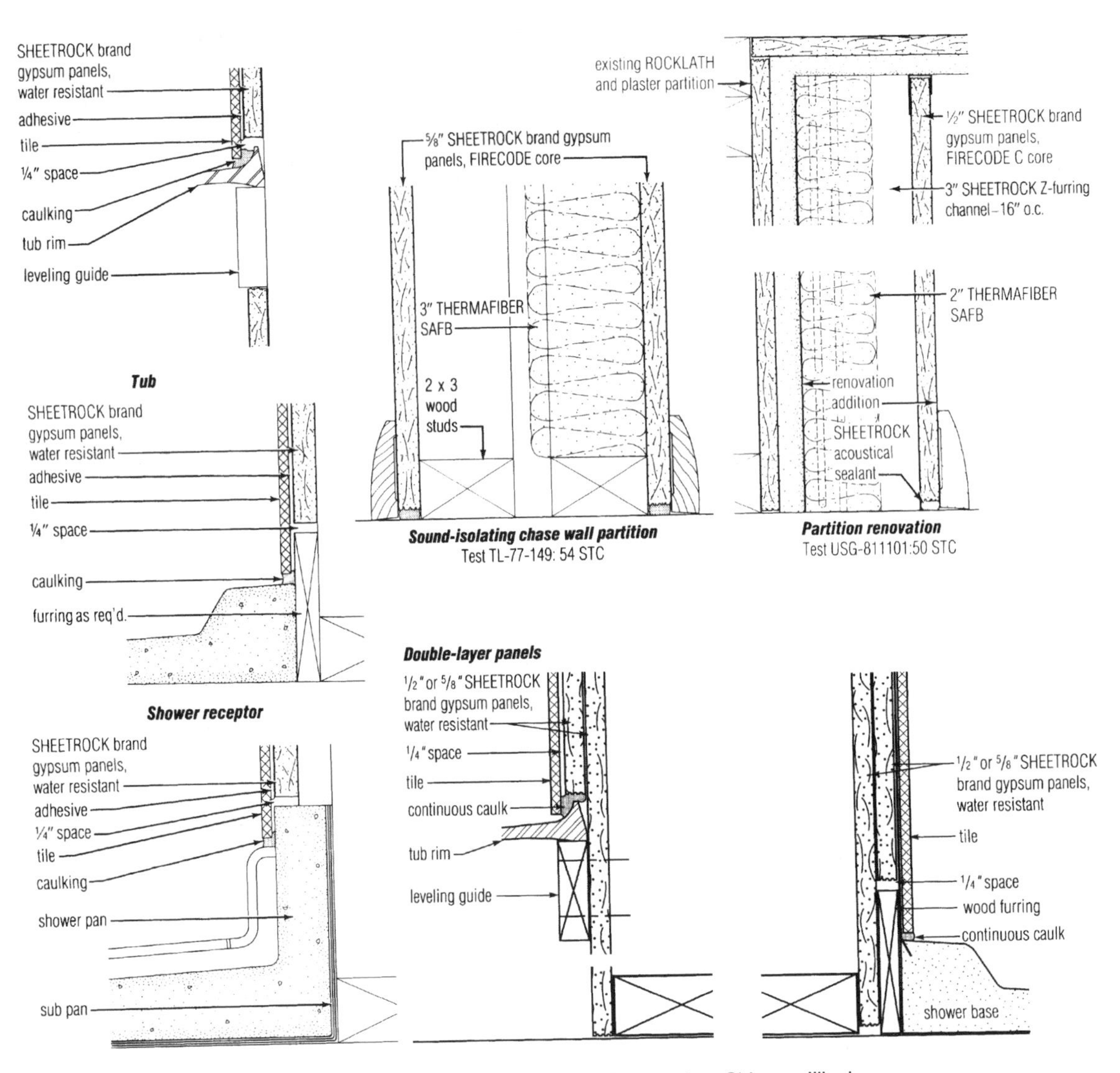

By permission: United States Gypsum Corporation, Chicago, Illinois

14.5.0 Wall Control Joint Details (Illustrated)

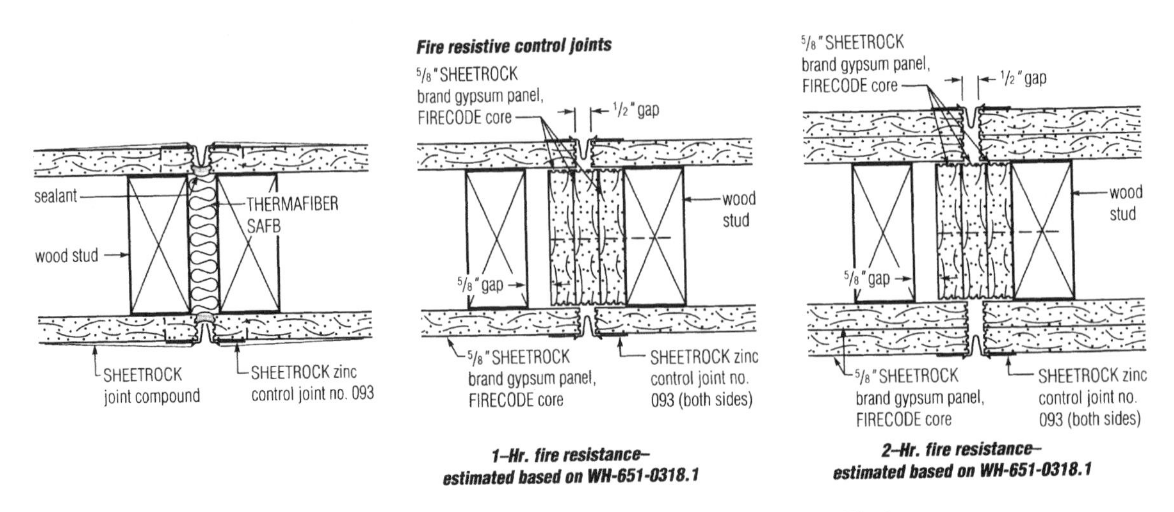

By permission: United States Gypsum Corporation, Chicago, Illinois

14.6.0 Typical Bath Tub and Swimming Pool Wall Details

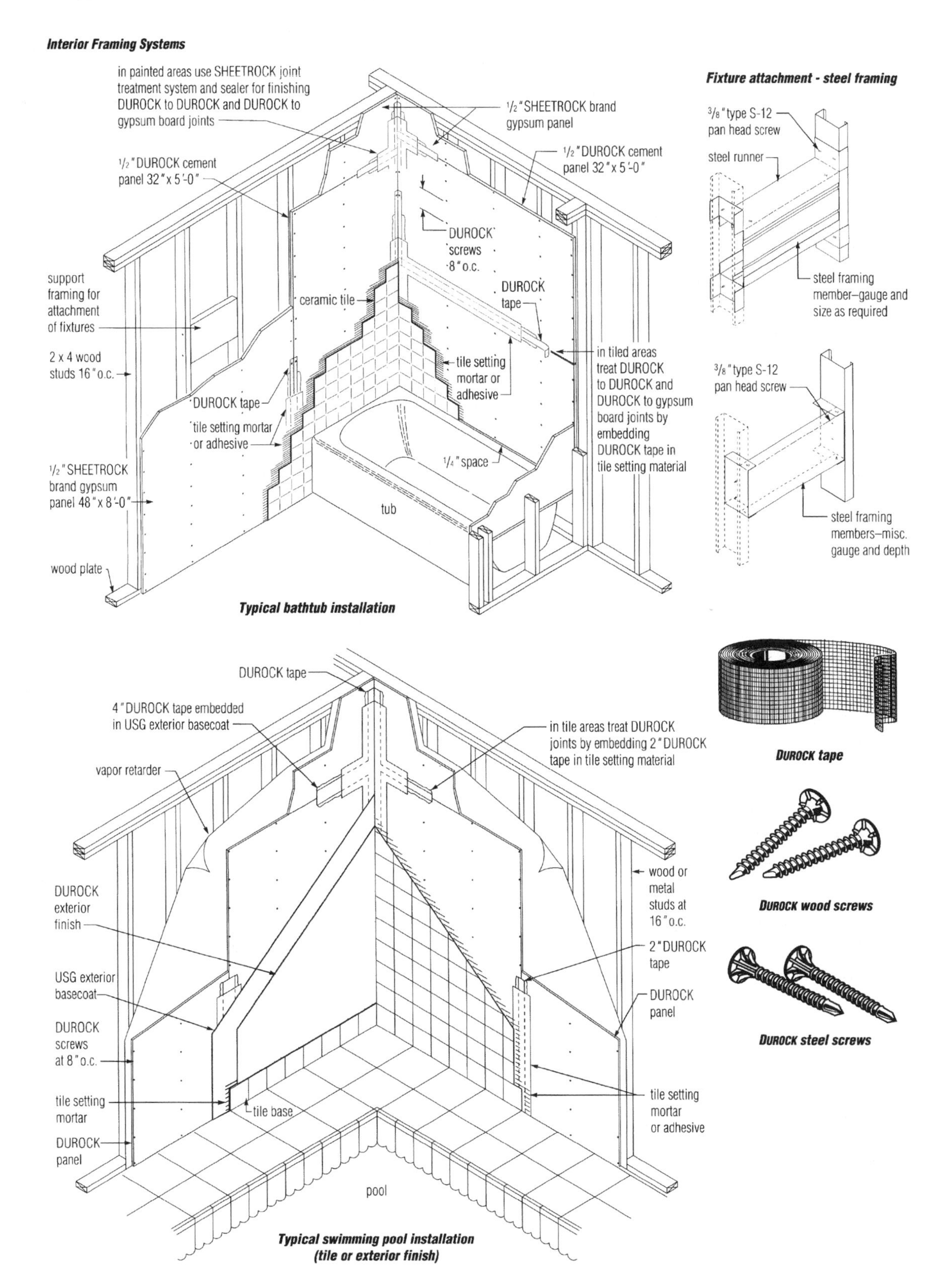

By permission: United States Gypsum Corporation, Chicago, Illinois

14.7.0 Soffit Framing Specifications

This assembly consists of galvanized steel channel runners and studs faced with Sheetrock brand Gypsum Panels, screw attached. It is a lightweight, fast and economical method of filling over cabinets or lockers and of housing overhead ducts, pipes or conduits. The braced system permits constructing soffits with depths of 48" (vertically) and widths to 72" (horizontally). The unbraced system is for soffits up to 24" × 24".

Maximum Width and Depth Dimensions(1)

Gypsum board thickness(2)		Steel stud size		Maximum width		Maximum depth for max. width shown	
In	mm	In	mm	In	mm	In	mm
½	12.7	1⅜	41.3	60	1500	48	1200
½	12.7	2½, 3⅛	63.5, 92.1	72	1800	36	900
⅝	15.9	1⅜	41.3	60	1500	30	800
⅝	15.9	2½, 3⅛	63.5, 92.1	72	1800	18	500

(1) The construction is not designed to support loads other than its own dead weight and should not be used where it may be subjected to excessive abuse.
(2) The double-layer system and ⅛" thick gypsum panels are not recommended for this construction.

14.8.0 Shelf-Wall Specifications and Illustrations

This system provides load-carrying walls for shelving in stores, offices, schools and other applications. Incorporating simple, quickly erected, economical steel stud components with Garcy shelf brackets, standards and accessories, the assembly offers advantages of steel stud-drywall construction plus structural strength to support shelving and merchandise.

In this assembly, 3⅝" steel studs spaced no more than 24" o.c. are securely fastened to floor and ceiling runners and surfaced with either single or double-layer Sheetrock brand Gypsum Panels. Slotted standards are screw-attached through gypsum board to studs or steel reinforcing inserted between layers.

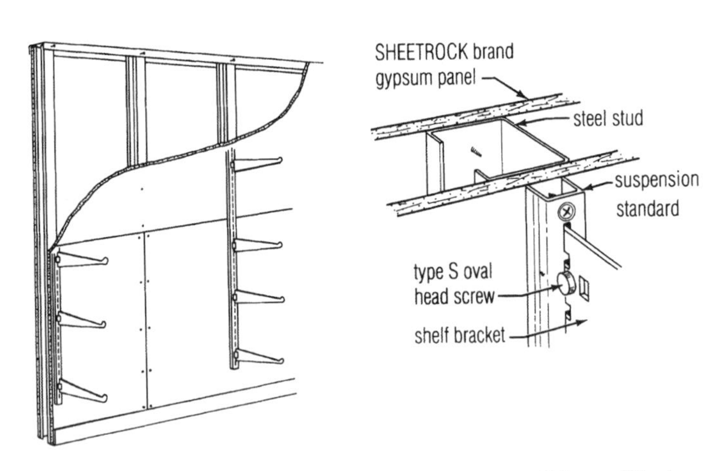

By permission: United States Gypsum Corporation, Chicago, Illinois

14.9.0 Chase-Wall Specifications and Illustrations

Typical limiting heights—Chase wall partitions

Stud width	Stud ga.	Stud spacing	Allow. defl.	One layer	Two layers
1⅝"	25	16"	L/120	15'3" f	15'3" f
			L/240	13'3" d	14'6" d
			L/360	11'6" d	12'9" d
		24"	L/120	12'6" f	12'6" f
			L/240	11'6" d	12'6" f
			L/360	10'0" d	11'0" d
2½"	25	16"	L/120	19'6" f	19'6" f
			L/240	17'6" d	19'0" d
			L/360	15'6" d	16'6" d
		24"	L/120	16'0" f	16'0" f
			L/240	15'6" d	16'0" f
			L/360	13'6" d	14'6" d
3⅝"	25	16"	L/120	23'6" f	23'6" f
			L/240	22'9" d	23'6" f
			L/360	19'9" d	21'3" d
		24"	L/120	19'3" f	19'3" f
			L/240	19'3" f	19'3" f
			L/360	17'3" d	18'6" d

Limiting height for ½" or ⅝" thick panels and 5 psf uniform load perpendicular to partition. Assemblies require vertical cross braces 4 ft. o.c. max. Use two-layer heights for multi-layer assemblies. Limiting criteria:d–deflection, f–bending stress. Consult local code authority for limiting criteria.

Chase walls provide vertical shafts where greater core widths are needed for pipe chase enclosures and other service installations. They consist of a double row of steel studs with gypsum panel cross braces between rows. Double-layer ½" SHEETROCK brand Gypsum

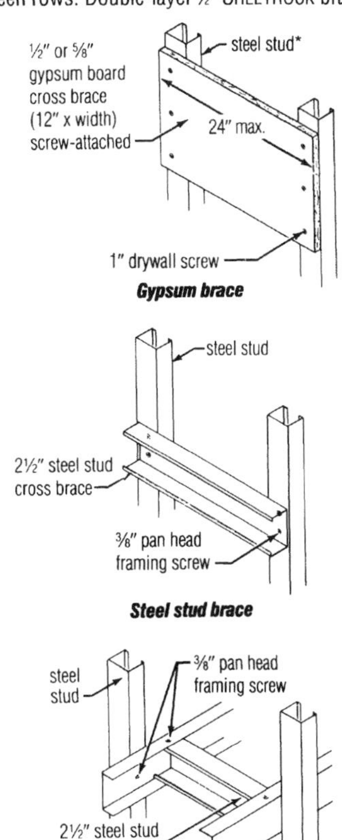

Gypsum brace

Steel stud brace

Steel stud & runner brace

By permission: United States Gypsum Corporation, Chicago, Illinois

14.10.0 Resilient Channel Partition Specifications

Resilient attachment of gypsum panels with RC-1 Resilient Channels or equivalent provides low-cost, highly efficient assemblies for increased privacy in corridor and party wall applications. The steel channels float the panels away from the studs and provide a spring action that decouples the board from the framing. When combined with THERMAFIBER SAFB in the framing cavity, highly effective sound attenuation is obtained.

In these thin, lightweight assemblies, horizontal RC-1 Resilient Channels (or equivalent), 24″ o.c., are screw-attached one side of 3⅝″ steel studs spaced 24″ o.c. and set in runners. Gypsum panels are screw-attached to these channels on one side and directly attached to the steel stud flanges on the opposite partition side. THERMAFIBER SAFB, 3″ thick and 25″ wide, are inserted and creased in the partition cavity. Because the blanket is wider than the cavity, it presses against the panels, thereby damping sound vibrations more effectively and offering 55 STC sound rating. (Use of a filler strip at the base may reduce STC rating.) Limiting heights for these assemblies are shown in the table below.

Limiting heights—resilient channel assemblies(1)

Stud width	Stud ga.	Stud spacing	Allow. defl.	One layer resilient partition
3⅝″	25	16″	L/120 L/240	16′7″f 13′4″d
		24″	L/120 L/240	13′6″f 11′8″d

(1) Limiting height for ⅝″ thick gypsum panels and 5-psf uniform load perpendicular to partition. Studs attached to top and bottom runners on resilient side. Limiting criteria: d—deflection, f—bending stress; consult local code authority for limiting criteria.

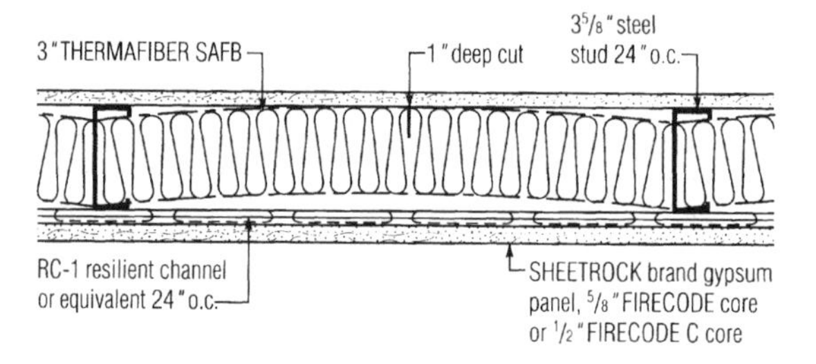

14.11.0 Tall Wall Specifications and Limiting Heights

Partitions exceeding 30' in height are considered tall. When these taller than normal partition heights are required, consideration must be given to length restrictions for manufacturing and shipping steel studs, scaffolding, stud placement, etc.

Use double structural studs back-to-back 24" o.c. The studs should be the maximum practical length so that the splice of one stud in each pair will occur at outer ⅙ of the span. The splice of the other stud will occur at the opposite end. Attach studs back to back with screws approximately 4' o.c. Attach each stud flange to top and bottom runner with ½" Type S-12 screws so that each pair of studs will have four screw attachments at each end. Attach 1½" 20 ga. V-bracing to stud flanges on each side assembly 12' o.c. for stud alignment and lateral bracing.

For 5 psf wind load, 20 ga. runner track is recommended. The fasteners should have a capacity of 300 lb. in single shear and bearing. For 10 psf wind load, 18 ga. runner track attached with fasteners with 400 lb. single shear and bearing is recommended.

Runner Attachment Spacing

Maximum wall height	Wind load	
	5 psf	10 psf
40'	24"	24"
48'	24"	20"
55'	24"	16"

Required Double Stud Sizes—Structural Studs

Maximum wall height	Wind load/deflection							
	5 psf L/240 size (in.) & ga.		L/360 Size (in.) & ga.		10 psf L/240 Size (in.) & ga.		L/360 Size (in.) & ga.	
35'	6	14 or	7¼	14 or	8	14 or	9¼	14 or
	7¼	18	8	16	9¼	16	11½	16
40'	7¼	14 or	8	14	9¼	14 or	11½	16 or
	8	16			11½	16		
45'	8	14	9¼	14	11½	16	13½	14
50'	9¼	14	11½	16	11½	14	13½	14
55'	11½	16	11½	14	13½	15	—	—

Conforms to 1986 AISI Specification for the Design of Cold-Formed Steel Structural Members.
Narrower flange is 1.552 in.; wider flange is 1.724 in, outside for all structural studs. See note on page 4.

Typical Limiting Heights—Structural Studs

Stud width(1)	Stud gauge	Wind load/deflection					
		5 psf		10 psf		15 psf	
		L/240	L/360	L/240	L/360	L/240	L/360
3½"	20	14'9"	13'0"	11'9"	10'3"	10'3"	9'0"
	18	16'3"	14'3"	13'0"	11'3"	11'3"	10'0"
	16	17'6"	15'3"	14'0"	12'3"	12'3"	10'9"
	14	18'9"	16'6"	15'0"	13'3"	13'3"	11'6"
4"	20	15'9"	14'0"	12'9"	11'3"	11'3"	9'9"
	18	17'3"	15'3"	14'0"	12'3"	12'3"	10'9"
	16	18'9"	16'6"	15'0"	13'3"	13'3"	11'6"
	14	20'0"	17'9"	16'3"	14'3"	14'3"	12'6"
6"	20	22'0"	19'3"	17'9"	15'6"	15'6"	13'6"
	18	24'0"	21'3"	19'3"	17'0"	17'0"	14'9"
	16	26'0"	23'0"	21'0"	18'6"	18'6"	16'0"
	14	28'0"	24'9"	22'6"	19'9"	19'9"	17'3"
8"	18	30'6"	26'9"	24'6"	21'6"	21'6"	18'9"
	16	33'0"	29'3"	26'6"	23'3"	23'3'	20'6"
	14	35'6"	31'3"	28'6"	25'0"	25'0"	22'0"

(1) Studs 24" o.c.

By permission: United States Gypsum Corporation, Chicago, Illinois

14.11.1 *L* over 120/240/360 Explained

Many of the tables included in this section make reference to *L*/120, *L*/240, and *L*/360. For those unfamiliar with these terms, the following explanation is of assistance in understanding the deflection specification included in these tables. The established rule is that a member should no deflect more than 1/360th of the length of its span, when the span is expressed in inches. To convert inches to centimeters, multiply by 2.54. *L* represents the length of the span, specifically, in the case of *L*/360, a 30-foot (9.144 meter) beam, and this beam should not deflect more than one inch (2.54 centimeters). If the criteria is *L*/240, then this 240-inch (609.6 cm), 20-foot (6.096 meter) beam shall not deflect more than one inch (2.54 cm).

14.11.2 Structural Stud Specifications

Typical Physical and Structural Properties[1]— Structural Studs (FY=40 ksi)

table 5

Size (in.) &	ga.	Weight[2] (lb/ft)	Weight[2] (kg/m)	Net area[3] (in^2)	AET (net effective area) (in^2)	Allow. Design steel thickness[3] (in)	bending moment about x x axis (K-in)	Lip width (in)	Major axis I_X (in^4)	Major axis S_X (in)	Major axis r_X (in)	Minor axis I_y (in^4)	Minor axis S_Y (in^3)	Minor axis r_y (in)	Full unreduced section modulus S_f (in^3)	Effective section modulus (M_c/S_f) S_C (in^3)	Q (column factor)	J (in^4)	C_W (in^5)	X_0 (in)
3-5/8	20	0.97	1.44	0.216	0.2136	0.0359	6.557	0.500	0.541	0.273	1.429	0.085	0.082c	0.621	0.302	0.236	0.752	0.0001	0.300700	1.357
3-5/8	18	1.24	1.85	0.285	0.2713	0.0478	9.247	0.500	0.708	0.385	1.423	0.111	0.106c	0.616	0.395	0.309	0.799	0.0003	0.387	1.345
3-5/8	16	1.59	2.37	0.368	0.3341	0.0598	11.678	0.625	0.893	0.486	1.411	0.147	0.146c	0.629	0.499	0.387	0.804	0.0005	0.5703	1.420
3-5/8	14	2.00	2.98	0.454	0.3917	0.0747	14.293	0.625	1.093	0.596	1.404	0.178	0.176c	0.622	0.611	0.466	0.802	0.0011	0.6833	1.406
4	20	1.02	1.52	0.228	0.1792	0.0359	7.464	0.500	0.673	0.311	1.556	0.091	0.084c	0.617	0.341	0.271	0.721	0.0001	0.3631	1.313
4	18	1.30	1.93	0.301	0.2576	0.0478	10.5	0.500	0.882	0.437	1.550	0.117	0.108c	0.611	0.447	0.355	0.803	0.0003	0.4679	1.301
4	16	1.67	2.48	0.388	0.3571	0.0598	13.302	0.625	1.115	0.554	1.539	0.157	0.150c	0.626	0.566	0.447	0.812	0.0006	0.6816	1.374
4	14	2.09	3.11	0.480	0.4833	0.0747	16.3	0.625	1.366	0.679	1.532	0.189	0.181c	0.619	0.693	0.539	0.811	0.0011	0.8176	1.359
6	20	1.27	1.89	0.300	0.2148	0.0359	12.93	0.500	1.787	0.539	2.253	0.112	0.088c	0.587	0.596	0.495	0.582	0.0002	0.8744	1.111
6	18	1.63	2.43	0.397	0.3107	0.0478	18.561	0.500	2.35	0.773	2.246	0.145	0.118c	0.581	0.785	0.659	0.653	0.0004	1.1309	1.099
6	16	2.08	3.10	0.508	0.4303	0.0598	23.759	0.625	2.99	0.99	2.243	0.195	0.163c	0.598	0.999	0.836	0.710	0.0007	1.5888	1.163
6	14	2.62	3.90	0.629	0.5858	0.0747	29.231	0.625	3.679	1.218	2.234	0.236	0.197c	0.591	1.229	1.017	0.767	0.0014	1.9148	1.148
7-1/4	18	1.84	2.74	0.457	0.2969	0.0478	24.361	0.500	3.732	1.015	2.663	0.157	0.118c	0.562	1.029	0.875	0.583	0.0004	1.7311	1.005
7-1/4	16	2.34	3.48	0.583	0.4304	0.0598	31.268	0.625	4.753	1.303	2.664	0.211	0.166c	0.579	1.311	1.121	0.637	0.0008	2.4067	1.064
7-1/4	14	2.95	4.39	0.720	0.6152	0.0747	35.529	0.625	5.857	1.605	2.654	0.256	0.203c	0.572	1.615	1.366	0.690	0.0015	2.9054	1.050
8	18	1.97	2.93	0.493	0.2937	0.0478	27.874	0.500	4.756	1.161	2.908	0.159	0.118c	0.550	1.187	1.018	0.547	0.0004	2.1644	0.956
8	16	2.50	3.72	0.628	0.456	0.0598	36.132	0.625	6.059	1.505	2.911	0.219	0.166c	0.568	1.513	1.306	0.600	0.0009	2.9966	1.013
8	14	3.15	4.69	0.779	0.6936	0.0747	44.557	0.625	7.473	1.856	2.901	0.265	0.205c	0.561	1.866	1.594	0.652	0.0017	3.6201	0.999
9-1/4	16	2.76	4.11	0.702	0.4146	0.0598	44.838	0.625	8.691	1.868	3.316	0.227	0.166c	0.550	1.875	1.647	0.546	0.0009	4.1512	0.938
9-1/4	14	3.48	5.18	0.872	0.6028	0.0747	55.351	0.625	10.73	2.306	3.306	0.278	0.206c	0.543	2.314	2.015	0.594	0.0018	5.0199	0.925
11-1/2	16	3.23	4.81	0.837	0.4355	0.0598	55.03	0.625	15.03	2.293	4.030	0.229	0.166c	0.521	2.606	2.326	0.470	0.0011	6.7915	0.830
11-1/2	14	4.07	6.06	1.040	0.5366	0.0747	77.138	0.625	18.58	3.214	4.018	0.292	0.207c	0.514	3.221	2.853	0.512	0.0021	8.2221	0.818
13-1/2	14	4.60	6.84	1.189	0.8562	0.0747	90.046	0.625	27.99	3.752	4.639	0.295	0.207c	0.491	4.134	3.704	0.456	0.0024	11.83235	0.743

Conforms to 1986 AISI Specification for the Design of Cold-Formed Steel Structural Members.(1) Narrower flange is 1.552 in.; wider flange is 1.724 in.outside width for all structural studs. See "Notice" on page 4. (2) Steel with corrosion-resistant coating. (3) Steel without coating.

By permission: United States Gypsum Corporation, Chicago, Illinois

14.11.3 Typical Limiting Heights of Interior Partitions

Stud width	Stud spacing	Allow. defl.	Partition, one layer	Partition, two layers	Furring, one layer
25 gauge stud (.0179 min.)					
1-5/8″	16″	L/120	10′9″ f	10′9″ d	10′3″ d
		L/240	9′6″ d	10′6″ d	8′3″ d
		L/360	8′3″ d	9′0″ d	7′3″ d
	24″	L/120	8′9″ f	8′9″ f	8′9″ f
		L/240	8′3″ d	8′9″ f	7′3″ d
		L/360	7′3″ d	8′0″ d	6′3″ d
2-1/2″	16″	L/120	13′9″ f	13′9″ f	13′9″ d*
		L/240	12′6″ d	13′6″ d	11′0″ d
		L/360	10′9″ d	11′9″ d	9′9″ d
	24″	L/120	11′3″ f	11′3″ f	11′3″ f
		L/240	10′9″ d	11′3″ f	9′9″ d
		L/360	9′6″ d	10′3″ d	8′6″ d
3-5/8″	16″	L/120	16′9″ f	16′9″ f	16′9″ f*
		L/240	16′0″ d	16′9″ f	14′6″ d*
		L/360	14′0″ d	14′9″ d	12′9″ d*
	24″	L/120	13′6″ f	13′6″ f	13′6″ f*
		L/240	13′6″ f	13′6″ f	12′9″ d*
		L/360	12′3″ d	13′0″ d	11′0″ d
4″	16″	L/120	17′3″ f	17′3″ f	17′3″ f*
		L/240	17′3″ d	17′3″ f	15′9″ d*
		L/360	15′0″ d	15′9″ d	13′9″ d*
	24″	L/120	14′3″ f	14′3″ f	14′3″ f*
		L/240	14′3″ f	14′3″ f	13′9″ d*
		L/360	13′0″ d	13′9″ d	12′0″ d
6″	16″	L/120	20′0″ f	20′0″ f	20′0″ f*
		L/240	20′0″ f	20′0″ f	20′0″ f*
		L/360	20′0″ f	20′0″ f	18′9″ f*
	24″	L/120	15′0″ v	15′0″ v	15′0″ v*
		L/240	15′0″ v	15′0″ v	15′0″ v*
		L/360	15′0″ v	15′0″ v	15′0″ v*
22 gauge stud (.0270 min.)					
2-1/2″	16″	L/120	16′6″ d	17′0″ f	15′3″ d*
		L/240	13′0″ d	14′0″ d	12′0″ d
		L/360	11′6″ d	12′3″ d	10′6″ d
	24″	L/120	14′0″ f	14′0″ f	13′3″ d*
		L/240	11′6″ d	12′3″ d	10′6″ d
		L/360	10′0″ d	10′6″ d	9′3″ d
3-5/8″	16″	L/120	21′9″ d	22′0″ f	20′3″ d*
		L/240	17′3″ d	18′0″ d	16′0″ d*
		L/360	15′0″ d	15′9″ d	14′0″ d*
	24″	L/120	18′0″ f	18′0″ f	17′9″ d*
		L/240	15′0″ d	15′9″ d	14′0″ d*
		L/360	13′0″ d	13′9″ d	12′3″ d*
4″	16″	L/120	23′3″ f	23′3″ f	21′9″ d*
		L/240	18′6″ d	19′3″ d	17′3″ d*
		L/360	16′3″ d	16′9″ d	15′0″ d*
	24″	L/120	19′0″ f	19′0″ f	19′0″ f*
		L/240	16′3″ d	16′9″ d	15′0″ d*
		L/360	14′0″ d	14′9″ d	13′3″ d*
6″	16″	L/120	29′0″ f	29′0″ f	29′0″ f*
		L/240	25′3″ d	26′0″ d	23′9″ d*
		L/360	22′0″ d	22′9″ d	20′9″ d*

Stud width	Stud spacing	Allow. defl.	Partition, one layer	Partition, two layers	Furring, one layer
22 gauge stud (.0270 min.)					
6″	24″	L/120	23′6″ f	23′6″ f	23′6″ f*
		L/240	22′0″ d	22′9″ d	20′9″ d*
		L/360	19′3″ d	19′9″ d	18′3″ d*
20 gauge stud (.0312 min.)					
2-1/2″	16″	L/120	17′4″f	17′11″f	16′6″d*
		L/240	13′10″d	16′1″d	13′0″d*
		L/360	12′0″d	14′0″d	11′6″d
	24″	L/120	14′7″f	14′7″f	14′6″d*
		L/240	12′0″d	13′5″f	11′6″d
		L/360	10′6″d	12′4″d	10′0″d
3-5/8″	16″	L/120	22′7″d	23′8″f	21′9″d*
		L/240	17′11″d	20′2″d	17′3″d*
		L/360	15′7″d	17′8″d	15′0″d*
	24″	L/120	19′4″f	19′4″f	19′0″d*
		L/240	15′7″d	17′8″f	15′0″d*
		L/360	13′8″d	15′6″d	13′3″d*
4″	16″	L/120	24′3″d	25′6″d	23′6″d*
		L/240	19′2″d	21′7″d	18′9″d*
		L/360	16′10″d	18′11″d	16′3″d*
	24″	L/120	20′9″f	20′9″f	20′6″d*
		L/240	16′10″d	18′11″d	16′3″d*
		L/360	14′8″d	16′6″d	14′3″d*
6″	16″	L/120	32′11″d	33′11″f	32′3″d*
		L/240	26′1″d	28′6″d	25′6″d*
		L/360	22′10″d	24′11″d	23′3″d*
	24″	L/120	25′3″f	25′3″f	28′0″d*
		L/240	22′10″d	24′11″d	22′3″d*
		L/360	19′11″d	21′10″d	19′6″d*
20 gauge joist (.0341 min.)					
3-5/8″	16″	L/120	24′0″ d	25′0″ d	23′0″ d*
		L/240	19′0″ d	19′9″ d	18′3″ d*
		L/360	16′9″ d	17′3″ d	16′0″ d*
	24″	L/120	21′0″ d	21′9″ d	20′3″ d*
		L/240	16′9″ d	17′3″ d	16′0″ d*
		L/360	14′6″ d	15′0″ d	14′0″ d*
4″	16″	L/120	25′9″ d	26′9″ d	24′9″ d*
		L/240	20′6″ d	21′3″ d	19′9″ d*
		L/360	18′0″ d	18′6″ d	17′3″ d
	24″	L/120	22′6″ d	23′3″ d	21′6″ d*
		L/240	18′0″ d	18′6″ d	17′3″ d*
		L/360	15′9″ d	16′3″ d	15′0″ d*

By permission: United States Gypsum Corporation, Chicago, Illinois

14.12.0 High-Performance Sound-Control Construction (Illustrations)

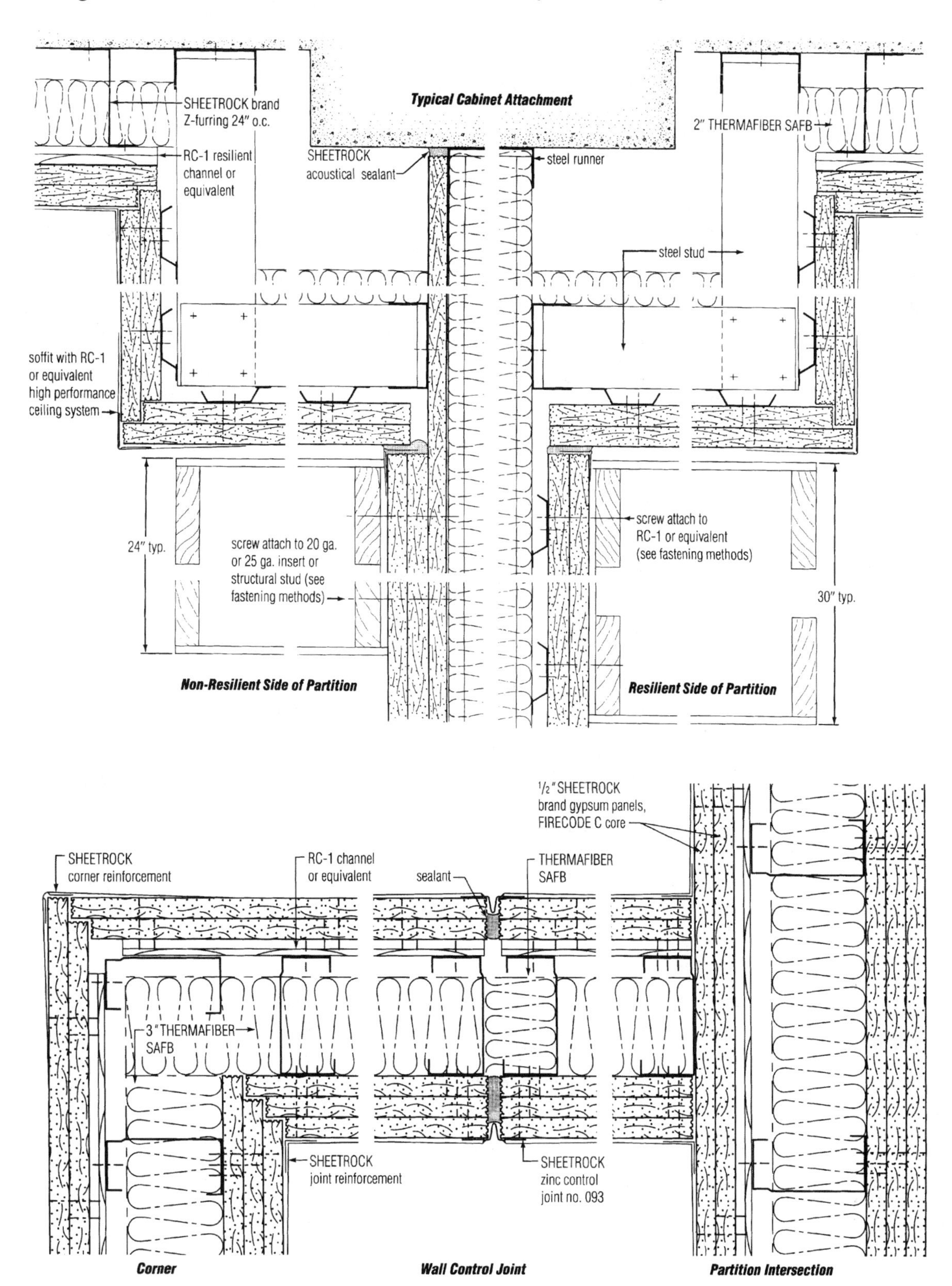

By permission: United States Gypsum Corporation, Chicago, Illinois

14.13.0 Curtain Wall Construction (Illustrations)

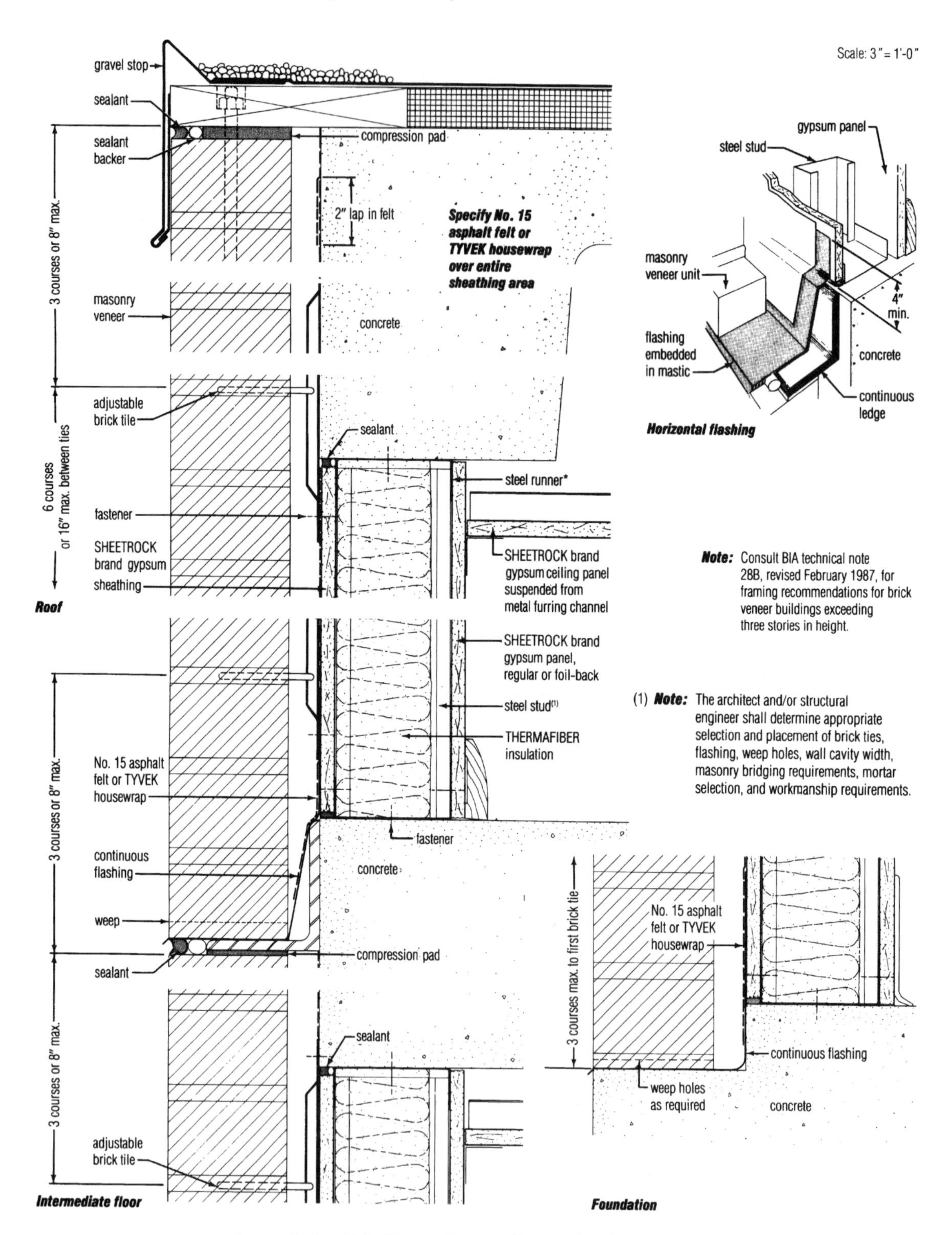

By permission: United States Gypsum Corporation, Chicago, Illinois

14.13.1 Typical Curtain-Wall Limiting-Height Specifications

Typical Curtain Wall Limiting Heights—Studs (20-ga.) (F_y = 33 ksi)

Maximum allowable simple span limiting heights calculated using stud properties[1] **stud properties only**

Design criteria		Deflection limitation (L/240)				Deflection limitation (L/360)				Deflection limitation (L/600)			
Wind load (psf)	Stud spacing (in o.c.)	2½″	3⅝″	4″	6″	2½″	3⅝″	4″	6″	2½″	3⅝″	4″	6″
		20 ga.	20 ga.	20 ga.	20 ga.	20 ga.	20 ga.	20 ga.	20 ga.	20 ga.	20 ga.	20 ga.	20 ga.
15 (80 mph)	12	9′1″	12′2″	13′2″	18′2″	8′0″	10′7″	11′6″	15′10″	6′8″	9′0″	9′8″	13′4″
	16	8′3″	11′1″	12′0″	16′7″	7′3″	9′8″	10′6″	14′6″	6′1″	8′2″	8′9″	12′2″
	24	7′3″	9′8″	10′6″	14′6″	6′3″	8′6″	9′1″	12′7″	5′3″	7′1″	7′8″	10′8″
20 (90 mph)	12	8′3″	11′1″	12′0″	16′7″	7′3″	9′8″	10′6″	14′6″	6′1″	8′2″	8′9″	12′2″
	16	7′6″	10′1″	10′10″	15′0″	6′7″	8′9″	9′6″	13′1″	5′7″	7′4″	8′0″	11′1″
	24	6′7″	8′9″	9′6″	13′1″	5′9″	7′8″	8′3″	11′6″	4′10″	6′6″	7′0″	9′8″
25 (100 mph)	12	7′8″	10′3″	11′1″	15′4″	6′8″	9′0″	9′8″	13′4″	5′8″	7′7″	8′2″	11′3″
	16	7′0″	9′3″	10′1″	14′0″	6′1″	8′2″	8′9″	12′2″	5′2″	6′10″	7′4″	10′3″
	24	6′1″	8′2″	8′9″	12′2″	5′3″	7′1″	7′8″	10′8″	4′6″	6′0″	6′6″	9′0″
30 (110 mph)	12	7′3″	9′8″	10′6″	14′6″	6′3″	8′6″	9′1″	12′7″	5′3″	7′2″	7′8″	10′8″
	16	6′7″	8′9″	9′6″	13′1″	5′9″	7′8″	8′3″	11′6″	4′10″	6′6″	7′0″	9′8″
	24	5′9″	7′8″	8′3″	11′6″	5′0″	6′8″	7′2″	10′0″	4′2″	5′8″	6′1″	8′6″
35 (120 mph)	12	6′10″	9′2″	9′10″	13′8″	6′0″	8′0″	8′8″	12′0″	5′1″	6′9″	7′3″	10′1″
	16	6′3″	8′3″	9′0″	12′6″	5′6″	7′3″	7′10″	10′10″	4′7″	6′2″	6′7″	9′2″
	24	5′6″	7′3″	7′10″	10′10″	4′9″	6′4″	6′10″	9′6″	4′0″	5′4″	5′9″	8′0″
40 (125 mph)	12	6′7″	8′9″	9′6″	13′1″	5′9″	7′8″	8′3″	11′6″	4′10″	6′6″	7′0″	9′8″
	16	6′0″	8′0″	8′7″	11′10″	5′2″	7′0″	7′6″	10′4″	4′4″	5′10″	6′4″	8′9″
	24	5′2″	7′0″	7′6″	10′4″	4′7″	6′1″	6′7″	9′1″	3′10″	5′1″	5′7″	7′8″

(1) Any independently supported exterior treatment over gypsum sheathing. Based on properties of studs alone with stress increased 33% for intermittent wind loading. Yield strength for studs and runners is 33 ksi.

Typical Curtain Wall Limiting Heights—Studs (F_y = 40 ksi)

Limiting heights calculated using stud properties[1] **stud properties only**

Design criteria		Stud spacing	Simple span limiting heights[1] for steel studs by size and gauge														
			3⅝″ stud				4″ stud				6″ stud				8″ stud		
Wind load	Deflection limitation	(in o.c.)	20 ga.	18 ga.	16 ga.	14 ga.	20 ga.	18 ga.	16 ga.	14 ga.	20 ga.	18 ga.	16 ga.	14 ga.	18 ga.	16 ga.	14 ga.
15 psf	L/240	12	13′4″	14′7″	15′9″	16′10″	14′4″	15′8″	16′11″	18′2″	19′10″	21′9″	23′7″	25′3″	27′6″	29′10″	32′0″
		16	12′1″	13′3″	14′4″	15′4″	13′0″	14′3″	15′5″	16′6″	18′0″	19′9″	21′5″	22′11″	25′0″	27′1″	29′1″
		24	10′7″	11′7″	12′6″	13′4″	11′4″	12′5″	13′5″	14′5″	15′9″	17′3″	18′8″	20′0″	21′10″	23′8″	25′4″
	L/360	12	11′8″	12′9″	13′9″	14′9″	12′6″	13′8″	14′10″	15′10″	17′4″	19′0″	20′7″	22′1″	24′0″	26′0″	27′11″
		16	10′7″	11′7″	12′6″	13′4″	11′4″	12′5″	13′5″	14′5″	15′9″	17′3″	18′8″	20′0″	21′10″	23′8″	25′4″
		24	9′3″	10′1″	10′11″	11′8″	9′11″	10′10″	11′9″	12′7″	13′9″	15′1″	16′4″	17′6″	19′1″	20′8″	22′2″
	L/600	12	9′1″	10′9″	11′7″	12′5″	10′7″	11′7″	12′6″	13′4″	14′7″	16′0″	17′4″	18′7″	20′3″	22′0″	23′7″
		16	8′11″	9′9″	10′6″	11′3″	9′7″	10′6″	11′4″	12′2″	13′3″	14′7″	15′9″	16′11″	18′5″	19′11″	21′5″
		24	7′9″	8′6″	9′2″	9′10″	8′5″	9′2″	9′11″	10′7″	11′7″	12′9″	13′9″	14′9″	16′1″	17′5″	18′8″
20 psf	L/240	12	12′1″	13′3″	14′4″	15′4″	13′0″	14′3″	15′5″	16′6″	18′0″	19′9″	21′5″	22′11″	25′0″	27′1″	29′0″
		16	11′0″	12′0″	13′0″	13′11″	11′10″	12′11″	14′0″	15′0″	16′5″	17′11″	19′5″	20′10″	22′8″	24′7″	26′5″
		24	9′7″	10′6″	11′4″	12′2″	10′4″	11′4″	12′3″	13′1″	14′4″	15′8″	17′0″	18′2″	19′10″	21′6″	23′1″
	L/360	12	10′7″	11′7″	12′7″	13′4″	11′4″	12′5″	13′5″	14′5″	15′9″	17′3″	18′8″	20′0″	21′10″	23′8″	25′4″
		16	9′7″	10′6″	11′4″	12′2″	10′4″	11′4″	12′3″	13′1″	14′4″	15′8″	17′0″	18′2″	19′10″	21′6″	23′1″
		24	8′5″	9′2″	9′11″	10′7″	9′0″	9′11″	10′8″	11′5″	12′6″	13′8″	14′10″	15′11″	17′4″	18′9″	20′2″
	L/600	12	8′11″	9′9″	10′6″	11′3″	9′7″	10′6″	11′4″	12′2″	13′3″	14′7″	15′9″	16′11″	18′5″	19′11″	21′5″
		16	8′1″	8′10″	9′7″	10′3″	8′9″	9′6″	10′4″	11′0″	12′1″	13′3″	14′4″	15′4″	16′9″	18′2″	19′5″
		24	7′1″	7′9″	8′4″	8′11″	7′7″	8′4″	9′0″	9′8″	10′7″	11′7″	12′6″	13′5″	14′7″	15′10″	17′0″
25 psf	L/240	12	11′3″	12′3″	13′3″	14′2″	12′1″	13′3″	14′4″	15′4″	16′9″	18′4″	19′10″	21′3″	23′2″	25′2″	26′11″
		16	10′2″	11′2″	12′1″	12′11″	11′0″	12′0″	13′0″	13′11″	15′2″	16′8″	18′0″	19′4″	21′1″	22′10″	24′6″
		24	8′11″	9′9″	10′6″	11′3″	9′7″	10′6″	11′4″	12′2″	13′3″	14′7″	15′9″	16′11″	18′5″	19′11″	21′5″
	L/360	12	9′10″	10′9″	11′7″	12′5″	10′7″	11′7″	12′6″	13′4″	14′7″	16′0″	17′4″	18′7″	20′3″	22′0″	23′7″
		16	8′11″	9′9″	10′6″	11′3″	9′7″	10′6″	11′4″	12′2″	13′3″	14′7″	15′9″	16′11″	18′5″	19′11″	21′5″
		24	7′9″	8′6″	9′2″	9′10″	8′5″	9′2″	9′11″	10′7″	11′7″	12′9″	13′9″	14′9″	16′1″	17′5″	18′8″
	L/600	12	8′3″	9′1″	9′9″	10′6″	8′11″	9′9″	10′6″	11′3″	12′4″	13′6″	14′8″	15′8″	17′1″	18′6″	19′10″
		16	7′6″	8′3″	8′11″	9′6″	8′1″	8′10″	9′7″	10′3″	11′2″	12′3″	13′4″	14′3″	15′6″	16′10″	18′1″
		24	6′7″	7′2″	7′9″	8′4″	7′1″	7′9″	8′4″	8′11″	9′9″	10′9″	11′7″	12′5″	13′7″	14′8″	15′9″
30 psf	L/240	12	10′7″	11′7″	12′8″	13′4″	11′4″	12′5″	13′5″	14′5″	15′9″	17′4″	18′8″	20′0″	21′10″	23′8″	25′4″
		16	9′7″	10′6″	11′4″	12′2″	10′4″	11′4″	12′3″	13′1″	14′5″	15′8″	17′0″	18′2″	19′10″	21′6″	23′1″
		24	8′5″	9′2″	9′11″	10′7″	9′0″	9′11″	10′8″	11′5″	12′6″	13′8″	14′10″	15′11″	17′4″	18′9″	20′2″
	L/360	12	9′3″	10′1″	10′11″	11′8″	9′11″	10′10″	11′9″	12′7″	13′9″	15′1″	16′4″	17′6″	19′1″	20′8″	22′2″
		16	8′5″	9′2″	9′11″	10′7″	9′0″	9′11″	10′8″	11′5″	12′6″	13′8″	14′10″	15′11″	17′4″	18′9″	20′2″
		24	7′4″	8′0″	8′8″	9′3″	7′11″	8′8″	9′4″	10′0″	10′11″	12′0″	13′0″	13′11″	15′2″	16′5″	17′7″
	L/600	12	7′9″	8′6″	9′2″	9′10″	8′5″	9′2″	9′11″	10′7″	11′7″	12′9″	13′9″	14′9″	16′1″	17′5″	18′8″
		16	7′1″	7′9″	8′4″	8′11″	7′7″	8′4″	9′0″	9′8″	10′7″	11′7″	12′6″	13′5″	14′7″	15′10″	17′0″
		24	6′2″	6′9″	7′4″	7′10″	6′9″	7′3″	7′10″	8′5″	9′3″	10′1″	10′11″	11′9″	12′9″	13′10″	14′10″
35 psf	L/240	12	10′1″	11′0″	11′10″	12′8″	10′10″	11′10″	12′9″	13′8″	15′0″	16′5″	17′9″	19′0″	20′9″	22′6″	24′1″
		16	9′1″	10′0″	10′9″	11′6″	9′10″	10′9″	11′7″	12′5″	13′7″	14′11″	16′2″	17′3″	18′10″	20′5″	21′11″
		24	8′0″	8′9″	9′5″	10′1″	8′7″	9′5″	10′2″	10′10″	11′10″	13′0″	14′1″	15′1″	16′5″	17′10″	19′2″
	L/360	12	8′9″	9′7″	10′4″	11′1″	9′5″	10′4″	11′2″	11′11″	13′1″	14′4″	15′6″	16′7″	18′1″	19′8″	21′1″
		16	8′0″	8′9″	9′5″	10′1″	8′7″	9′5″	10′2″	10′10″	11′10″	13′0″	14′1″	15′1″	16′5″	17′10″	19′2″
		24	7′0″	7′7″	8′3″	8′10″	7′6″	8′2″	8′10″	9′6″	10′4″	11′4″	12′4″	13′2″	14′4″	15′7″	16′9″
	L/600	12	7′5″	8′1″	8′9″	9′4″	8′0″	8′9″	9′5″	10′1″	11′0″	12′1″	13′1″	14′0″	15′3″	16′7″	17′9″
		16	6′9″	7′4″	7′11″	8′6″	7′3″	7′11″	8′7″	9′2″	10′0″	11′0″	11′11″	12′9″	13′11″	15′1″	16′2″
		24	5′10″	6′5″	6′11″	7′5″	6′4″	6′11″	7′6″	8′0″	8′9″	9′7″	10′5″	11′2″	12′1″	13′2″	14′1″
40 psf	L/240	12	9′7″	10′6″	11′4″	12′2″	10′4″	11′4″	12′3″	13′1″	14′4″	15′8″	17′0″	18′2″	19′10″	21′6″	23′1″
		16	8′9″	9′7″	10′4″	11′0″	9′5″	10′3″	11′1″	11′11″	13′0″	14′3″	15′5″	16′6″	18′0″	19′6″	20′11″
		24	7′7″	8′4″	9′0″	9′8″	8′2″	9′0″	9′8″	10′5″	11′4″	12′5″	13′6″	14′5″	15′9″	17′1″	18′4″
	L/360	12	8′5″	9′2″	9′11″	10′7″	9′0″	9′11″	10′8″	11′5″	12′6″	13′8″	14′10″	15′11″	17′4″	18′9″	20′2″
		16	7′7″	8′4″	9′0″	9′8″	8′2″	9′0″	9′8″	10′5″	11′4″	12′5″	13′6″	14′5″	5′9″	17′1″	18′4″
		24	6′8″	7′3″	7′10″	8′5″	7′2″	7′10″	8′6″	9′1″	9′11″	10′10″	11′9″	12′7″	13′9″	14′11″	16′0″
	L/600	12	7′1″	7′9″	8′4″	8′11″	7′7″	8′4″	9′0″	9′8″	10′7″	11′7″	12′6″	13′5″	14′7″	15′10″	17′0″
		16	6′5″	7′0″	7′7″	8′2″	6′11″	7′7″	8′2″	8′9″	9′7″	10′6″	11′4″	12′2″	13′3″	14′5″	15′5″
		24	5′7″	6′2″	6′8″	7′1″	6′1″	6′7″	7′2″	7′8″	8′4″	9′2″	9′11″	10′8″	11′7″	12′7″	13′6″

Conforms to 1986 AISI Specification for the Design of Cold-Formed Steel Structural Members.

(1) Any independently supported exterior treatment over gypsum sheathing. Based on properties of studs alone with stress increased 33% for wind loading. Yield strength for runners is 33 ksi.

IMPORTANT NOTE: U.S. Gypsum Company does not manufacture these steel framing members. The table above shows minimum limiting heights for "typical" curtain wall construction based on the typical physical and structural properties published in Tables 1 through 5 on pages 4 and 5. The physical and structural values that govern this table are suggested minimums and may vary by region and by manufacturer. Table is meant as a general guideline only. Request actual physical and structural property data from our local United States Gypsum Company representative or framing manufacturer.

By permission: United States Gypsum Corporation, Chicago, Illinois

14.14.0 Super Studs

General Information

◆ Mechanical Properties, Base Steel

Unless noted otherwise herein, structural framing components manufactured by **Super Stud** are formed from steel meeting the minimum requirements of the following specifications:

	ASTM A446 Grade D ($Fy_{(min)}$ = 50 KSI)	ASTM A446 Grade A ($Fy_{(min)}$ = 33 KSI)
Studs	12, 14 & 16 gage	18 & 20 gage
Track & Accessories	12 & 14 gage	16, 18 & 20 gage

Fy = Minimum Yield Point

Upon request, **Super Stud** will fulfill requests for any of our components manufactured from steel meeting the minimum requirements of ASTM A446, Grade B, $Fy_{(min)}$ = 37 KSI and Grade C, $Fy_{(min)}$ = 40 KSI.

ASTM A446, entitled *Standard Specification for Steel Sheet, Zinc Coated (Galvanized) by the Hot-Dip Process, Structural (Physical) Quality*, covers sheet steel of structural (physical) quality with zinc coating. Material of this quality is intended primarily for use in applications where mechanical or structural properties of the base metal are specified or required. These properties include those indicated by tension, hardness and other accepted mechanical requirements.

Steel manufactured in accordance with ASTM A446 meet the minimum ductility requirements of the AISI Specification. Steel ductility is critical to its formability during rolling or braking processes. Ductility also affects the performance of the cutting tips of screw fasteners used to connect steel framing.

◆ Galvanized Coating

Super Stud structural framing components are zinc coated (galvanized) in accordance with ASTM A525, G-60. ASTM A525 entitled *Specification for General Requirements for Steel Sheet, Zinc Coated (Galvanized)*, is listed as an applicable document in ASTM A446.

While a G-60 coating is our standard offering, **Super Stud** can also fulfill requests for G-90 coatings which provides 50 percent more zinc protection.

◆ Base Steel Thickness

The structural properties and load tables were prepared using the following base steel thicknesses:

20 gage: 0.0346 inch 18 gage: 0.0451 inch 16 gage: 0.0566 inch
14 gage: 0.0713 inch 12 gage: 0.1017 inch

In conformance with the AISI Specification, the actual delivered base steel thickness, individual measurement, must not be less than 95 percent of the values listed above.

◆ Identification

Super Stud's structural "C" studs can be furnished with a factory applied marking denoting company name and gage thickness of the section. It is complemented with a "Down" arrow to denote the indexed end to assure web knockout alignment. Studs manufactured from 33 KSI material are color coded with blue markings while 50 KSI steel is color coded red. The marking is provided upon written request of the purchaser.

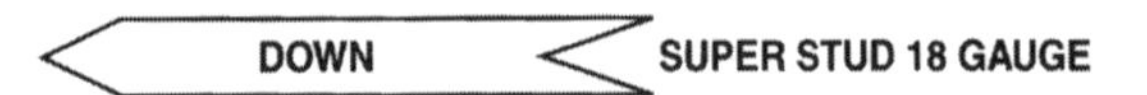

Studs (SSCW, SSC, SSJ, SJW & SSW)

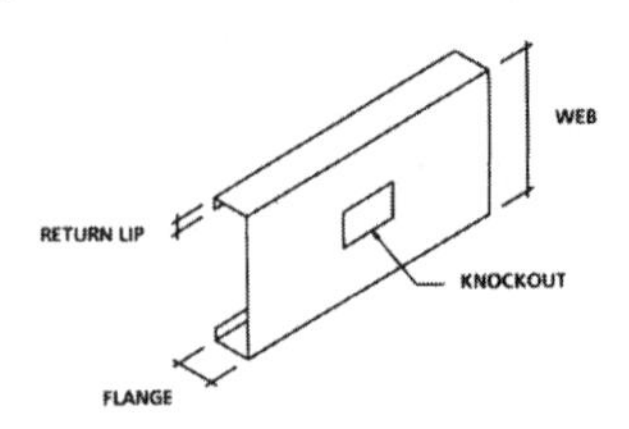

Studs serve as structural elements in the construction of exterior curtainwalls, soffits, load bearing walls and headers. They are also used in floor joist, roof rafter and truss frame assemblies.

"C" studs are defined by three basic components...the web, flange and return lip. The flange serves to stiffen the web while the return lip stabilizes the flange.

		Return Lip Length				
Section	Flange Width	20 Ga	18 Ga	16 Ga	14 Ga	12 Ga
SSCW	1-1/4"	0.375"	NA	NA	NA	NA
SSC	1-3/8"	0.437"	0.406"	0.437"	0.343"	NA
SSJ	1-5/8"	0.5"	0.531"	0.562"	0.5"	0.312"
SJW	2"	NA	0.625"	0.75"	0.687"	0.5"
SSW	2-1/2"	NA	0.687"	0.843"	0.937"	0.781"

◆ Web Knockouts

Studs are typically furnished with punched webs to facilitate the installation of conduit, piping and bridging. The knockout sizes and locations are defined below. Members may be furnished with unpunched webs upon request.

◆ Knockout Spacing

Section	First Knockout (Dim. A)	Typical Spacing (Dim. B)	Last Knockout (Dim. C)[Note 1]
All sections except those listed below	24"	24"	24" **min.**
8" SSJ - 12" SSJ 8" SJW - 14" SJW 8" SSW - 16" SSW	48"	48"	24" **min.**

Note 1: 24" Min. applies to millcut sections only. Does not apply to components saw cut from stock.

◆ Knockout Size

Web Depth	Knockout Length (Dim. D)	Knockout Height (Dim. E)
2-1/2"	1-1/2"	3/4"
3-5/8" to 6"	2"	1-1/2"
Over 6"	4"	2-1/2"

TAIL END
DIM. C
℄ LAST KNOCKOUT
DIM. B
DIM. D
℄
DIM. B
WEB
℄ FIRST KNOCKOUT
DIM. A
DIM. E
LEADING EDGE

By permission: Super Stud Building Products, Inc., Astoria, New York

14.14.1 Super Stud Section Properties: 2½" (13.85 cm) by 4" (10.16 cm) Studs

Section	Weight Plf	Ma K-in	Va Kip	Area in²	Ix in⁴	Sx. in³	Rx in	Iy in⁴	Sy. in³	Ry in	D in	JX1000 in⁴	Cw in⁶	Ro in	Xo in	Beta	h/t
2-1/2SSC20	0.689	3.263	1.037	0.202	0.208	0.165	1.013	0.056	0.060	0.526	0.501	0.081	0.086	1.666	-1.213	0.470	66.642
2-1/2SSJ20	0.762	3.557	1.037	0.224	0.237	0.188	1.028	0.088	0.082	0.625	0.628	0.089	0.145	1.920	-1.497	0.392	66.642
2-1/2SSC18	0.879	4.134	1.354	0.258	0.263	0.209	1.009	0.069	0.074	0.516	0.492	0.175	0.101	1.637	-1.181	0.480	50.432
2-1/2SSJ18	0.994	4.780	1.354	0.292	0.304	0.242	1.020	0.114	0.109	0.624	0.637	0.198	0.194	1.927	-1.511	0.385	50.432
2-1/2SSC16	1.099	7.694	2.510	0.323	0.323	0.257	1.001	0.085	0.093	0.514	0.500	0.345	0.130	1.643	-1.197	0.469	39.170
2-1/2SSJ16	1.243	8.897	2.510	0.365	0.373	0.297	1.011	0.141	0.137	0.622	0.645	0.390	0.250	1.935	-1.528	0.376	39.170
2-1/2SSC14	1.312	9.097	3.057	0.386	0.382	0.304	0.996	0.092	0.096	0.489	0.469	0.653	0.125	1.574	-1.117	0.497	30.063
2-1/2SSJ14	1.509	10.724	3.057	0.443	0.450	0.358	1.008	0.162	0.154	0.605	0.625	0.751	0.262	1.885	-1.474	0.389	30.063
3-1/2SSC20	0.806	4.992	1.046	0.237	0.452	0.253	1.381	0.063	0.060	0.515	0.431	0.095	0.168	1.827	-1.080	0.651	95.543
3-1/2SSJ20	0.880	5.424	1.046	0.259	0.511	0.286	1.406	0.099	0.082	0.618	0.547	0.103	0.273	2.043	-1.348	0.565	95.543
3-1/2SSC18	1.032	6.336	1.762	0.303	0.574	0.321	1.375	0.077	0.075	0.505	0.422	0.206	0.200	1.802	-1.050	0.661	72.605
3-1/2SSJ18	1.147	7.305	1.762	0.337	0.660	0.370	1.399	0.129	0.110	0.618	0.555	0.229	0.363	2.046	-1.360	0.558	72.605
3-1/2SSC16	1.291	11.856	3.417	0.379	0.709	0.396	1.367	0.096	0.094	0.503	0.429	0.405	0.254	1.803	-1.063	0.653	56.837
3-1/2SSJ16	1.436	13.670	3.417	0.422	0.815	0.457	1.390	0.160	0.139	0.616	0.563	0.451	0.463	2.050	-1.375	0.550	56.837
3-1/2SSC14	1.554	14.037	4.483	0.457	0.841	0.469	1.356	0.104	0.099	0.476	0.401	0.774	0.258	1.744	-0.998	0.679	44.088
3-1/2SSJ14	1.752	16.502	4.483	0.515	0.985	0.551	1.383	0.184	0.157	0.597	0.543	0.872	0.503	2.004	-1.321	0.565	44.088
3-5/8SSCW20	0.777	4.843	1.008	0.228	0.454	0.245	1.410	0.048	0.048	0.456	0.360	0.091	0.130	1.743	-.916	0.723	99.156
3-5/8SSC20	0.821	5.240	1.008	0.241	0.490	0.265	1.426	0.064	0.060	0.514	0.423	0.096	0.180	1.853	-1.066	0.669	99.156
3-5/8SSJ20	0.895	5.691	1.008	0.263	0.554	0.300	1.452	0.100	0.082	0.617	0.538	0.105	0.293	2.064	-1.331	0.584	99.156
3-5/8SSC18	1.052	6.652	1.762	0.309	0.623	0.337	1.420	0.078	0.075	0.503	0.415	0.210	0.215	1.828	-1.035	0.679	75.377
3-5/8SSJ18	1.167	7.662	1.762	0.343	0.716	0.388	1.445	0.130	0.111	0.616	0.546	0.232	0.390	2.067	-1.344	0.577	75.377
3-5/8SJW18	1.311	8.715	1.762	0.385	0.837	0.455	1.474	0.227	0.163	0.768	0.728	0.261	0.726	2.418	-1.757	0.472	75.377
3-5/8SSC16	1.315	12.454	3.417	0.387	0.770	0.416	1.411	0.097	0.094	0.502	0.422	0.413	0.274	1.828	-1.049	0.671	59.046
3-5/8SSJ16	1.460	14.346	3.417	0.429	0.884	0.479	1.436	0.162	0.139	0.615	0.554	0.458	0.496	2.070	-1.358	0.570	59.046
3-5/8SJW16	1.676	16.415	3.417	0.493	1.048	0.569	1.459	0.299	0.221	0.778	0.763	0.526	1.063	2.473	-1.840	0.447	59.046
3-5/8SSC18	1.052	6.652	1.762	0.309	0.623	0.337	1.420	0.078	0.075	0.503	0.415	0.210	0.215	1.828	-1.035	0.679	75.377
3-5/8SSJ18	1.167	7.662	1.762	0.343	0.716	0.388	1.445	0.130	0.111	0.616	0.546	0.232	0.390	2.067	-1.344	0.577	75.377
3-5/8SJW18	1.311	8.715	1.762	0.385	0.837	0.455	1.474	0.227	0.163	0.768	0.728	0.261	0.726	2.418	-1.757	0.472	75.377
3-5/8SJW12	2.723	27.148	6.339	0.800	1.672	0.907	1.446	0.407	0.284	0.713	0.683	2.759	1.160	2.291	-1.628	0.495	30.644
4SSCW20	0.821	5.558	0.909	0.241	0.572	0.281	1.540	0.049	0.048	0.451	0.342	0.096	0.161	1.830	-0.880	0.769	109.994
4SSC20	0.865	6.003	0.909	0.254	0.617	0.304	1.558	0.066	0.060	0.508	0.403	0.101	0.222	1.933	-1.025	0.719	109.994
4SSJ20	0.939	6.510	0.909	0.276	0.696	0.343	1.588	0.103	0.083	0.612	0.513	0.110	0.359	2.132	-1.285	0.637	109.994
4SSC18	1.109	7.627	1.762	0.326	0.785	0.386	1.552	0.081	0.075	0.497	0.394	0.221	0.266	1.910	-0.996	0.728	83.692
4SSJ18	1.224	8.759	1.762	0.360	0.899	0.443	1.581	0.135	0.111	0.612	0.522	0.244	0.475	2.134	-1.297	0.631	83.692
4SJW18	1.368	9.940	1.762	0.402	1.049	0.518	1.615	0.235	0.164	0.765	0.698	0.273	0.877	2.468	-1.702	0.524	83.692
4SSW18	1.541	10.852	1.762	0.453	1.235	0.611	1.652	0.413	0.237	0.955	0.929	0.307	1.598	2.924	-2.215	0.426	83.692
4SSC16	1.387	14.299	3.417	0.408	0.971	0.478	1.543	0.100	0.095	0.496	0.402	0.435	0.337	1.909	-1.008	0.721	65.671
4SSJ16	1.532	16.423	3.417	0.450	1.113	0.549	1.572	0.168	0.140	0.610	0.529	0.481	0.603	2.136	-1.310	0.624	65.671
4SJW16	1.749	18.764	3.417	0.514	1.317	0.650	1.601	0.310	0.223	0.776	0.732	0.549	1.269	2.519	-1.783	0.499	65.671
4SSW16	1.977	20.506	3.417	0.581	1.552	0.766	1.634	0.548	0.326	0.971	0.975	0.621	2.396	3.004	-2.326	0.400	65.671
4SSC14	1.676	16.960	5.196	0.492	1.153	0.566	1.530	0.108	0.101	0.468	0.375	0.835	0.347	1.853	-0.935	0.746	51.101
4SSJ14	1.873	19.851	5.196	0.550	1.346	0.663	1.564	0.192	0.161	0.591	0.511	0.933	0.662	2.092	-1.258	0.639	51.101
4SJW14	2.146	23.733	5.196	0.631	1.606	0.793	1.596	0.364	0.263	0.759	0.714	1.069	1.412	2.473	-1.731	0.510	51.101
4SSW14	2.510	27.674	5.196	0.738	1.931	0.955	1.618	0.700	0.433	0.974	1.001	1.250	3.303	3.043	-2.386	0.385	51.101

Flange Width: SSCW=1-1/4", SSC=1-3/8", SSJ=1-5/8", SJW=2" and SSW=2-1/2"

Reference Page 11 for Notes

By permission: Super Stud Building Products, Inc., Astoria, New York

14.14.2 Super Stud Section Properties: 4" (10.16 cm) by 8" (20.32 cm) Studs

Section	Weight Plf	Ma K-in	Va Kip	Area in^2	Ix in^4	Sx. in^3	Rx in	Iy in^4	Sy. in^3	Ry in	D in	JX1000 in^4	Cw in^6	Ro in	Xo in	Beta	h/t
4SJW12	2.853	31.051	7.102	0.838	2.103	1.037	1.584	0.421	0.292	0.709	0.655	2.890	1.427	2.342	-1.574	0.549	34.331
4SSW12	3.393	38.425	7.102	0.997	2.595	1.283	1.613	0.875	0.526	0.937	0.952	3.438	3.587	2.926	-2.255	0.406	34.331
5-1/2SSC20	1.042	9.365	0.652	0.306	1.313	0.474	2.071	0.072	0.061	0.485	0.337	0.122	0.443	2.307	-0.893	0.850	153.347
5-1/2SSJ20	1.115	10.104	0.652	0.328	1.465	0.529	2.114	0.114	0.083	0.590	0.435	0.131	0.705	2.469	-1.132	0.790	153.347
5-1/2SSC18	1.339	11.929	1.452	0.394	1.673	0.604	2.062	0.088	0.076	0.474	0.330	0.267	0.536	2.286	-0.866	0.857	116.951
5-1/2SSJ18	1.454	13.556	1.452	0.427	1.900	0.686	2.108	0.149	0.113	0.590	0.443	0.290	0.929	2.469	-1.142	0.786	116.951
5-1/2SSC16	1.676	22.448	2.902	0.493	2.078	0.750	2.054	0.110	0.096	0.473	0.337	0.526	0.678	2.282	-0.876	0.853	92.173
5-1/2SSJ16	1.821	25.510	2.902	0.535	2.360	0.852	2.100	0.185	0.142	0.589	0.450	0.571	1.174	2.467	-1.152	0.782	92.173
5-1/2SSC14	2.040	26.758	5.422	0.599	2.478	0.894	2.033	0.118	0.103	0.444	0.314	1.016	0.714	2.233	-0.809	0.869	72.139
5-1/2SSJ14	2.237	30.932	5.422	0.657	2.862	1.033	2.086	0.212	0.164	0.568	0.433	1.114	1.316	2.428	-1.103	0.794	72.139
6SSCW20	1.057	9.899	0.596	0.311	1.513	0.501	2.207	0.055	0.049	0.420	0.269	0.124	0.396	2.362	-0.729	0.905	167.798
6SSC20	1.101	10.598	0.596	0.323	1.619	0.536	2.237	0.074	0.061	0.478	0.320	0.129	0.537	2.443	-0.857	0.877	167.798
6SSJ20	1.174	11.415	0.596	0.345	1.801	0.597	2.284	0.117	0.084	0.582	0.414	0.138	0.853	2.597	-1.089	0.824	167.798
6SSC18	1.416	13.508	1.326	0.416	2.064	0.684	2.227	0.090	0.077	0.466	0.314	0.282	0.652	2.422	-0.830	0.882	128.038
6SSJ18	1.531	15.301	1.326	0.450	2.336	0.774	2.279	0.153	0.113	0.582	0.422	0.305	1.123	2.596	-1.099	0.821	128.038
6SJW18	1.675	17.174	1.326	0.492	2.686	0.891	2.336	0.269	0.167	0.739	0.575	0.334	2.013	2.855	-1.467	0.736	128.038
6SSW18	1.848	18.661	1.326	0.543	3.116	1.032	2.396	0.475	0.241	0.935	0.778	0.368	3.594	3.222	-1.941	0.637	128.038
6SSC16	1.773	25.440	2.648	0.521	2.566	0.850	2.219	0.113	0.096	0.465	0.320	0.556	0.823	2.418	-0.840	0.879	101.007
6SSJ16	1.917	28.817	2.648	0.563	2.904	0.963	2.270	0.190	0.142	0.581	0.428	0.602	1.418	2.593	-1.109	0.817	101.007
6SJW16	2.134	32.643	2.648	0.627	3.396	1.126	2.327	0.356	0.227	0.753	0.605	0.670	2.804	2.889	-1.538	0.717	101.007
6SSW16	2.362	35.484	2.648	0.694	3.947	1.305	2.384	0.632	0.332	0.954	0.821	0.741	5.132	3.280	-2.040	0.613	101.007
6SSC14	2.161	30.371	5.362	0.635	3.064	1.014	2.197	0.121	0.103	0.436	0.299	1.076	0.872	2.369	-0.774	0.893	79.151
6SSJ14	2.358	34.975	5.362	0.693	3.525	1.168	2.255	0.218	0.165	0.561	0.413	1.174	1.596	2.555	-1.061	0.828	79.151
6SJW14	2.631	41.211	5.362	0.773	4.150	1.376	2.317	0.417	0.269	0.734	0.589	1.310	3.202	2.850	-1.488	0.727	79.151
6SSW14	2.995	47.731	5.362	0.880	4.947	1.642	2.371	0.812	0.444	0.960	0.844	1.491	6.877	3.305	-2.093	0.599	79.151
6SJW12	3.545	54.215	11.030	1.042	5.462	1.811	2.290	0.481	0.309	0.679	0.537	3.592	3.466	2.740	-1.343	0.760	53.997
6SSW12	4.086	66.226	11.030	1.201	6.665	2.212	2.356	1.012	0.558	0.918	0.799	4.139	7.943	3.204	-1.967	0.623	53.997
7-1/4SSJ18	1.723	19.734	1.090	0.506	3.680	0.999	2.696	0.161	0.113	0.563	0.377	0.343	1.706	2.932	-1.006	0.882	155.754
7-1/4SJW18	1.867	22.027	1.090	0.549	4.197	1.141	2.766	0.284	0.167	0.720	0.518	0.372	3.031	3.162	-1.354	0.817	155.754
7-1/4SSJ16	2.158	37.221	2.173	0.634	4.581	1.243	2.688	0.200	0.143	0.562	0.384	0.677	2.151	2.927	-1.014	0.880	123.092
7-1/4SJW16	2.375	41.947	2.173	0.698	5.318	1.446	2.761	0.377	0.227	0.735	0.547	0.745	4.176	3.190	-1.420	0.802	123.092
7-1/4SSJ14	2.661	45.253	4.390	0.782	5.573	1.511	2.669	0.229	0.165	0.541	0.370	1.325	2.439	2.891	-0.969	0.888	96.683
7-1/4SJW14	2.934	52.973	4.390	0.862	6.508	1.769	2.747	0.441	0.269	0.715	0.532	1.461	4.808	3.153	-1.372	0.811	96.683
7-1/4SSJ12	3.588	58.558	11.030	1.054	7.224	1.956	2.617	0.240	0.172	0.477	0.326	3.635	2.556	2.789	-0.835	0.910	66.288
7-1/4SJW12	3.977	69.871	11.030	1.169	8.594	2.334	2.711	0.508	0.310	0.659	0.484	4.030	5.319	3.051	-1.234	0.836	66.288
8SSJ18	1.838	22.790	0.985	0.540	4.673	1.153	2.941	0.165	0.114	0.552	0.355	0.366	2.126	3.142	-0.958	0.907	172.384
8SJW18	1.982	25.181	0.985	0.582	5.307	1.312	3.019	0.292	0.167	0.708	0.489	0.395	3.764	3.360	-1.295	0.852	172.384
8SSW18	2.155	26.388	0.985	0.633	6.084	1.506	3.100	0.519	0.242	0.905	0.671	0.429	6.663	3.665	-1.734	0.776	172.384
8SSJ16	2.302	43.016	1.962	0.677	5.822	1.437	2.933	0.205	0.143	0.551	0.361	0.722	2.679	3.137	-0.966	0.905	136.343
8SJW16	2.519	48.314	1.962	0.740	6.729	1.663	3.015	0.387	0.227	0.723	0.517	0.790	5.162	3.385	-1.358	0.839	136.343
8SSW16	2.748	50.919	1.962	0.807	7.731	1.912	3.094	0.694	0.333	0.927	0.709	0.862	9.310	3.710	-1.824	0.758	136.343
8SSJ14	2.843	52.367	3.959	0.836	7.091	1.749	2.913	0.235	0.165	0.530	0.348	1.416	3.047	3.101	-0.921	0.912	107.202
8SJW14	3.116	60.981	3.959	0.916	8.242	2.037	3.000	0.454	0.270	0.704	0.503	1.552	5.964	3.349	-1.311	0.847	107.202
8SSW14	3.480	70.087	3.959	1.023	9.725	2.407	3.084	0.892	0.446	0.934	0.731	1.733	12.318	3.726	-1.872	0.748	107.202

Flange Width: SSCW=1-1/4", SSC=1-3/8", SSJ=1-5/8", SJW=2" and SSW=2-1/2"

Reference Page 11 for Notes

14.14.3 Super Stud Section Properties: 8" (20.32 cm) by 16" (40.64 cm) Studs

Section	Weight Plf	Ma K-in	Va Kip	Area in^2	Ix in^4	Sx. in^3	Rx in	Iy in^4	Sy. in^3	Ry in	D in	JX1000 in^4	Cw in^6	Ro in	Xo in	Beta	h/t
8SSJ12	3.848	68.023	11.030	1.131	9.223	2.272	2.856	0.246	0.173	0.466	0.307	3.898	3.218	3.000	-0.792	0.930	73.663
8SJW12	4.237	80.615	11.030	1.245	10.905	2.693	2.959	0.521	0.311	0.647	0.457	4.293	6.661	3.250	-1.178	0.869	73.663
8SSW12	4.778	97.324	11.030	1.404	13.137	3.251	3.059	1.110	0.562	0.889	0.691	4.841	14.658	3.636	-1.753	0.768	73.663
9-1/4SSJ18	2.030	25.485	0.849	0.597	6.672	1.429	3.344	0.170	0.114	0.534	0.324	0.404	2.946	3.501	-0.888	0.936	200.100
9-1/4SSW18	2.174	28.263	0.639	0.639	7.526	1.614	3.432	0.303	0.168	0.689	0.448	0.433	5.199	3.703	-1.208	0.894	200.100
9-1/4SSJ16	2.543	49.139	1.688	0.747	8.319	1.782	3.336	0.212	0.143	0.533	0.330	0.798	3.711	3.495	-0.895	0.934	158.428
9-1/4SJW16	2.760	54.733	1.688	0.811	9.551	2.048	3.432	0.403	0.228	0.705	0.474	0.866	7.090	3.726	-1.267	0.884	158.428
9-1/4SSJ14	3.147	65.089	3.403	0.925	10.151	2.174	3.313	0.243	0.166	0.512	0.318	1.567	4.237	3.459	-0.853	0.939	124.734
9-1/4SJW14	3.420	75.194	3.403	1.005	11.712	2.511	3.414	0.472	0.271	0.685	0.461	1.703	8.229	3.690	-1.223	0.890	124.734
9-1/4SSJ12	4.280	85.031	10.047	1.258	13.272	2.840	3.248	0.253	0.174	0.449	0.281	4.337	4.517	3.359	-0.730	0.953	85.954
9-1/4SJW12	4.670	99.757	10.047	1.372	15.547	3.332	3.366	0.540	0.313	0.627	0.420	4.731	9.295	3.595	-1.095	0.907	85.954
10SSJ16	2.687	52.165	1.558	0.790	10.094	2.003	3.575	0.216	0.143	0.523	0.314	0.843	4.424	3.714	-0.858	0.947	171.678
10SJW16	2.904	58.457	1.558	0.853	11.544	2.293	3.678	0.411	0.228	0.694	0.452	0.911	8.422	3.936	-1.219	0.904	171.678
10SSW16	3.133	62.108	1.558	0.921	13.131	2.611	3.777	0.740	0.334	0.896	0.626	0.983	15.084	4.219	-1.654	0.846	171.678
10SSJ14	3.329	73.245	3.138	0.978	12.329	2.446	3.550	0.247	0.166	0.502	0.303	1.658	5.060	3.677	-0.816	0.951	135.252
10SJW14	3.602	84.246	3.138	1.058	14.166	2.814	3.658	0.481	0.272	0.674	0.440	1.794	9.795	3.901	-1.176	0.909	135.252
10SSW14	3.965	96.031	3.138	1.165	16.549	3.290	3.768	0.952	0.448	0.904	0.646	1.975	19.827	4.231	-1.697	0.839	135.252
10SSJ12	4.540	95.981	9.253	1.334	16.167	3.206	3.481	0.257	0.174	0.439	0.268	4.600	5.417	3.577	-0.698	0.962	93.328
10SJW12	4.929	111.989	9.253	1.449	18.840	3.740	3.606	0.550	0.314	0.616	0.400	4.994	11.122	3.807	-1.051	0.924	93.328
10SSW12	5.470	133.419	9.253	1.607	22.419	4.456	3.735	1.182	0.566	0.858	0.610	5.542	23.993	4.146	-1.585	0.854	93.328
11-1/2SSJ16	2.976	58.517	1.350	0.875	14.323	2.477	4.047	0.222	0.144	0.504	0.286	0.934	6.066	4.154	-0.792	0.964	198.180
11-1/2SSW16	3.422	70.614	1.350	1.006	18.377	3.182	4.275	0.767	0.335	0.874	0.575	1.074	20.531	4.630	-1.548	0.888	198.180
11-1/2SSJ14	3.693	83.505	2.716	1.085	17.526	3.031	4.019	0.253	0.166	0.483	0.277	1.839	6.957	4.117	-0.753	0.967	156.290
11-1/2SSW14	4.329	109.335	2.716	1.272	23.181	4.014	4.268	0.989	0.449	0.882	0.595	2.156	26.910	4.639	-1.588	0.883	156.290
11-1/2SSJ12	5.059	119.570	7.990	1.487	23.105	3.994	3.942	0.264	0.175	0.421	0.246	5.126	7.493	4.016	-0.641	0.975	108.078
11-1/2SSW12	5.989	163.121	7.990	1.760	31.468	5.448	4.228	1.226	0.568	0.835	0.561	6.068	32.829	4.557	-1.480	0.895	108.078
12SSJ16	3.073	60.691	1.292	0.903	15.947	2.645	4.202	0.224	0.144	0.498	0.278	0.964	6.679	4.302	-0.773	0.968	207.014
12SJW16	3.289	68.622	1.292	0.967	18.067	2.998	4.323	0.429	0.229	0.666	0.403	1.032	12.644	4.512	-1.108	0.940	207.014
12SSW16	3.518	73.465	1.292	1.034	20.373	3.382	4.439	0.776	0.335	0.866	0.560	1.104	22.568	4.770	-1.516	0.899	207.014
12SSJ14	3.814	86.019	2.599	1.121	19.524	3.237	4.174	0.255	0.167	0.477	0.269	1.899	7.665	4.265	-0.734	0.970	163.303
12SJW14	4.087	103.953	2.599	1.201	22.208	3.685	4.300	0.501	0.272	0.646	0.392	2.035	14.762	4.477	-1.067	0.943	163.303
12SSW14	4.451	113.355	2.599	1.308	25.704	4.267	4.433	1.000	0.449	0.874	0.580	2.216	29.557	4.779	-1.555	0.894	163.303
12SSJ12	5.232	127.935	7.642	1.538	25.781	4.273	4.095	0.265	0.175	0.416	0.239	5.301	8.268	4.163	-0.624	0.978	112.994
12SJW12	5.621	147.361	7.642	1.652	29.674	4.922	4.238	0.572	0.315	0.588	0.357	5.695	16.927	4.383	-0.950	0.953	112.994
12SSW12	6.162	173.525	7.642	1.811	34.917	5.796	4.391	1.239	0.568	0.827	0.547	6.243	36.137	4.697	-1.449	0.905	112.994
14SJW16	3.675	79.140	1.104	1.080	26.524	3.331	4.956	0.443	0.229	0.641	0.363	1.153	17.868	5.100	-1.017	0.960	242.350
14SJW14	4.572	113.246	2.218	1.344	32.654	4.650	4.930	0.517	0.273	0.621	0.354	2.277	20.915	5.064	-0.978	0.963	191.353
14SSW14	4.936	129.723	2.218	1.451	37.477	5.339	5.083	1.038	0.450	0.846	0.526	2.458	41.626	5.349	-1.437	0.928	191.353
14SJW12	6.313	186.758	6.509	1.855	43.813	6.238	4.859	0.589	0.316	0.563	0.324	6.397	24.130	4.968	-0.868	0.970	132.660
14SSW12	6.854	217.663	6.509	2.014	51.038	7.270	5.034	1.284	0.570	0.798	0.497	6.944	51.238	5.269	-1.335	0.936	132.660
16SSW14	5.421	146.438	1.935	1.593	52.152	6.192	5.721	1.069	0.450	0.819	0.482	2.700	56.124	5.932	-1.336	0.949	219.404
16SSW12	7.546	251.869	5.669	2.218	71.191	8.879	5.666	1.321	0.571	0.772	0.456	7.646	69.405	5.851	-1.240	0.955	152.325

Flange Width: SSJ=1-5/8", SJW=2" and SSW=2-1/2"
Reference Page 11 for Notes

14.14.4 Super Stud Section Properties: Terms and Definitions

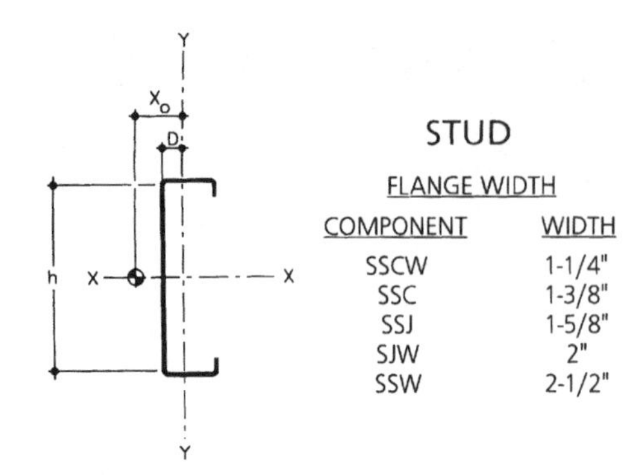

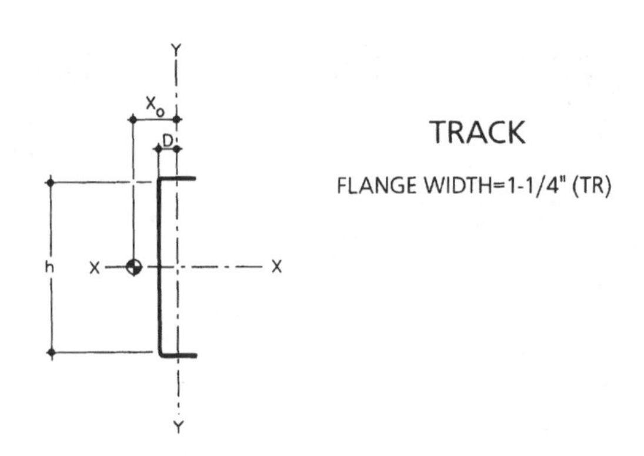

Terms and Definitions

Weight	Weight per lineal foot, Plf
Ma	Allowable bending moment of braced section, K-in
Va	Allowable shear force through an unpunched web, Kip
Area	Cross-sectional area of gross section, in^2
Ix, Iy	Moment of inertia of gross section about applicable axis, in^4
Sx_e, Sy_e	Section modulus of the effective section stressed at yield about the applicable axis, in^3
Rx, Ry	Radius of gyration of gross section about applicable axis, in
D	Distance from the Y axis to outside of web, in
Jx1000	St. Venant torsional constant, in^4, multiplied by 1,000
Cw	Torsional warping constant, in^6
Ro	Polar radius of gyration about the shear center, in
Xo	Distance from shear center to centroid along the principal axis, in
Beta	$1-(Xo/Ro)^2$
h/t	Flat web to thickness ratio

Notes:

1. Section properties were prepared in accordance with the American Iron and Steel Institute *Specification for the Design of Cold Formed Steel Structural Members*, 1986 edition with 1989 addendum.
2. Allowable bending moment, Ma, was calculated in accordance with AISI Section C3.1, Procedure 1, based on the initiation of yield in the effective section.
3. Bearing stiffeners are required for all components where the h/t ratio exceeds 200.
4. Sx_e and Sy_e are based on the effective section stressed at yield. Reference AISI Section B2.
5. Ma & Va are based on steel meeting the minimum requirements of the following specifications:

	ASTM A 446 Grade D ($Fy_{(min)}$= 50 KSI)	ASTM A446 Grade A ($Fy_{(min)}$= 33 KSI)
Studs	12, 14 & 16 gage	18 & 20 gage
Track & Accessories	12 & 14 gage	16, 18 & 20 gage

Fy = Minimum Yield Point

Upon request, **Super Stud** will fulfill requests for any of our components manufactured from steel meeting the minimum requirements of ASTM A446, Grade B, $Fy_{(min)}$=37 KSI and Grade C, $Fy_{(min)}$= 40 KSI.

6. The structural properties and load tables were prepared using the following base steel thicknesses:
20 gage: 0.0346 inch 18 gage: 0.0451 inch 16 gage: 0.0566 inch
14 gage: 0.0713 inch 12 gage: 0.1017 inch

In conformance with the AISI Specification, the actual delivered base steel thickness, individual measurement, must not be less than 95 percent of the values listed above.

STRUCTURAL PRODUCT MATRIX

SECTION/FLANGE WIDTH	2-1/2"	3-1/2"	3-5/8"	4"	5-1/2"	6"	7-1/4"	8"	9-1/4"	10"	11-1/2"	12"	14"	16"
SSCW / 1-1/4" FLG.			20 ga.	20 ga.		20 ga.								
SSC / 1-3/8" FLG.	20-14 ga.	20-14 ga.	20-14 ga.	20-14 ga.	20-14 ga.	20-14 ga.								
SSJ / 1-5/8" FLG.	20-14 ga.	20-14 ga.	20-14 ga.	20-14 ga.	20-14 ga.	20-12 ga.	18-12 ga.	18-12 ga.	18-12 ga.	16-12 ga.	16-12 ga.	16-12 ga.		
SJW / 2" FLG.			18-12 ga.	18-12 ga.	18-12 ga.	18-12 ga.	18-12 ga.	18-12 ga.	18-12 ga.	16-12 ga.	16-12 ga.	16-12 ga.	16-12 ga.	
SSW / 2-1/2" FLG.				18-12 ga.		18-12 ga.		18-12 ga.		16-12 ga.	16-12 ga.	16-12 ga.	14-12 ga.	14-12 ga.
TR / 1-1/4" FLG.	20-14 ga.	20-14 ga.	20-14 ga.	20-14 ga.	20-14 ga.	20-14 ga.	20-14 ga.	20-14 ga.	18-14 ga.	18-12 ga.	16-12 ga.	16-12 ga.	14-12 ga.	14-12 ga.
TW / 2" FLG.			20-14 ga.	20-14 ga.		20-12 ga.		18-12ga.		16-12ga				
DT / 2-1/2" FLG.			16-12 ga.	16-12 ga.		16-12 ga.		16-12ga.		16-12ga				

By permission: Super Stud Building Products, Inc., Astoria, New York

14.14.5 Super Stud Accessories

Track (TR & TW)

Track is used as an enclosure for studs in the construction of wall and joist assemblies. Track also serves as a structural component in the design of sill, head and jamb assemblies of framed openings.

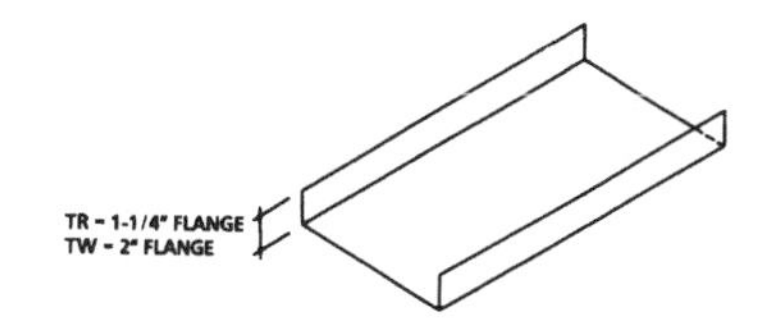

Deflection Clip (DC) (Patent Pending)

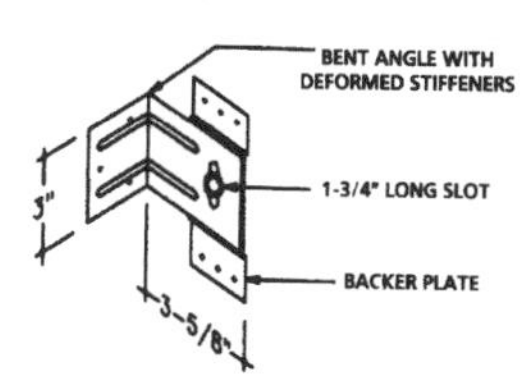

DC-500 & DC-850

For curtainwall construction, **Super Stud** has developed the industry's most advanced slide clip. Like competitive products, it provides lateral support for the framing member while it allows for vertical deflection of the primary frame. The difference lies in its ability to accommodate horizontal displacements between the face of support angles and an aligned stud wall.

Traditional slide clips require the installation of continuous support angles which must be accurately located to provide for a plumb and aligned wall. At best, this connection requires the installation of slotted adjustable angles requiring field attachment after they are placed in final position. At worst, the angle is attached out of position requiring the installation of scabbed stud pieces to make the transition from stud to angle. To simplify the installation, specify the **Super Stud** Deflection Clip. Eliminate the coordination headaches and expensive field modifications due to mis-aligned support angles.

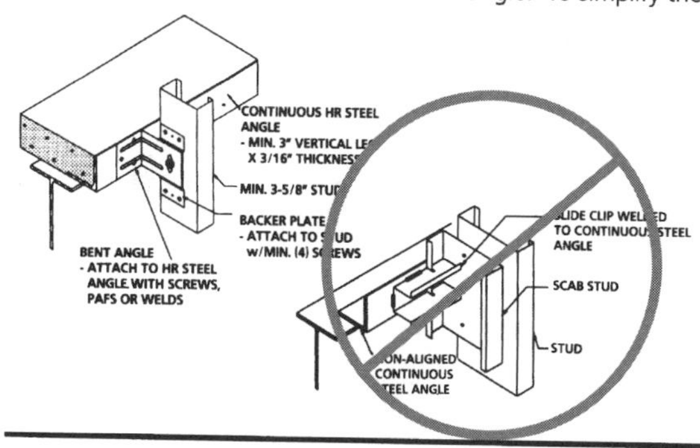

To reduce labor costs, **Super Stud** Deflection Clips may be installed in advance of the studs. Traditional clips which slide "inside" the flange must be installed simultaneously with the studs, an often difficult task for one person grappling to keep the stud positioned while attempting to field weld the clip to the support. The Deflection Clip, on the other hand, may be pre-attached to the angle at their required spacing before the installation of the studs.

Its unique offset, two piece construction allows for its direct attachment to the **web** of the stud. This eliminates flange "roll" resulting from the transfer of lateral load from the stud to conventional slide clips. Furthermore it is reversible and as such is not dependent on which direction the stud is turned.

The bent angle section of the clip is furnished with deformed stiffeners to increase its lateral load carrying capacities. It is provided pre-punched to accommodate either screw or powder actuated fastener attachment to the continuous angle. If desired, the clip may also be welded to the support.

The return leg provides ample surface for its **flush** attachment to the support angle. Some of our competitors clips do not provide the necessary offset required to accommodate the flange thickness of the stud. As a result, the clips are not installed flush to the surface causing the welds or mechanical fasteners to "bridge" the gap resulting in less predictable holding values.

Design with confidence. Our published allowable lateral load values were derived from the results of extensive, independent performance tests. The clip will provide for a maximum 3/4" of vertical primary frame movement and as much as 2" of horizontal displacement between the stud wall and support angle. For applications requiring displacements in excess of those listed above, **Super Stud** can modify the clip's construction to accommodate your needs.

Additional product information, including installation requirements, can be found in **Super Stud's** *Curtainwall Framing Systems Catalog*.

Deflection Strut (DS) (Patent Pending)

Manufactured employing the same techniques as the Deflection Clip, the Deflection Strut is used to make the transition of a curtainwall stud to a braced component of the primary frame. The tail end laps and connects to the structural component while the front end is screw attached to the stud.

Left and right side struts are available to correspond to the desired stud web position or where the installation requires two struts (i.e. where jambs attach to the structure).

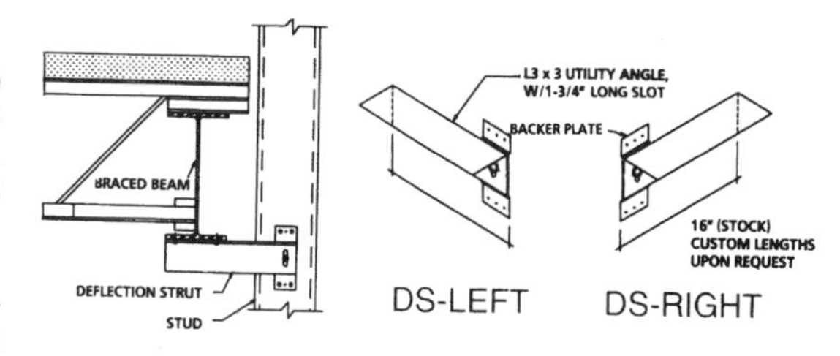

By permission: Super Stud Building Products, Inc., Astoria, New York

14.14.5 Super Stud Accessories (Continued)

Deflection Track (DT)

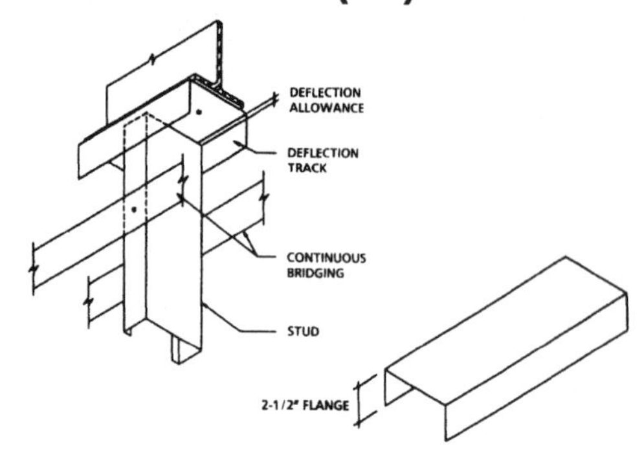

Thickness Offering: 12, 14 and 16 gage

Deflection Track is used to provide lateral support to an infill curtainwall assembly while it accommodates vertical deflections of the primary frame. As a result, the stud wall is not subjected to axial loads resulting from the deflection of the primary frame due to live loads, shrinkage, etc.

The track is typically furnished with 2-1/2" flanges. The extended leg provides for as much as 3/4" of primary frame deflection while it maintains a minimum bearing width of 1-1/2" at the end of the stud. For applications involving greater anticipated frame deflection, **Super Stud** can furnish flange lengths in excess of the standard 2-1/2" offering.

Utility Angle (UA)

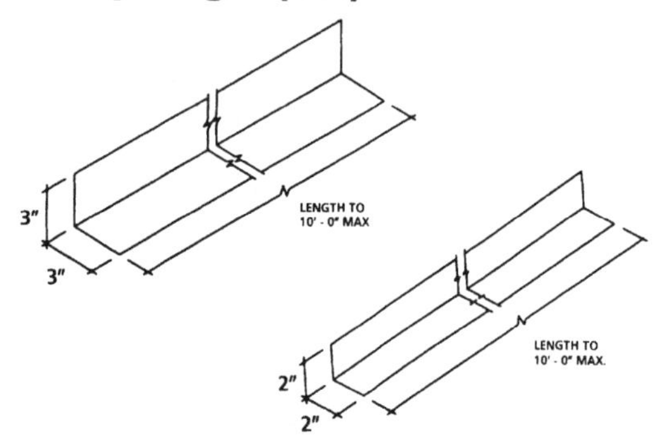

Thickness Offering: 20, 18, 16, 14 and 12 gage

Utility Angles, furnished in standard 10 foot lengths, are typically installed at corner framing conditions to provide continuous support for sheathing products.

Utility Angles are available in standard 2" or 3" leg dimensions. Non-standard angles with varying leg dimensions are also available.

Joist Hangers (JH)

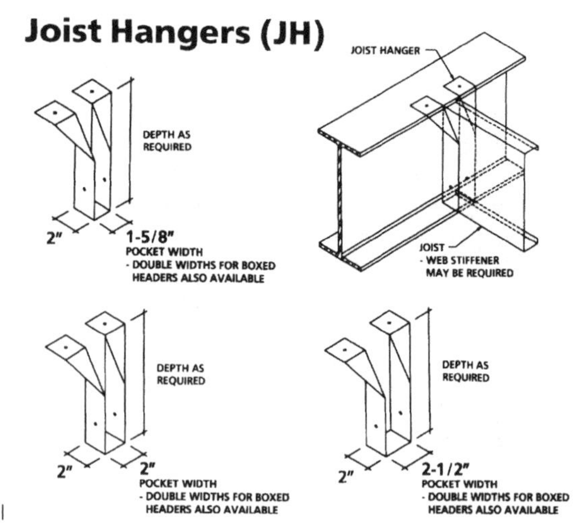

The joist hanger is used in applications where vertical alignment of the joist and the support structure is desired. The hanger, available in single or double widths, is used to attach joists or boxed headers to structural steel beams.

EZ Joist Support Clip (EZJSC)

(Patent Pending)

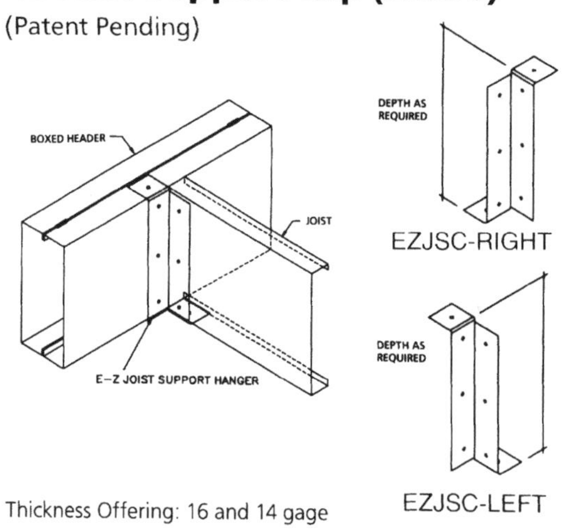

Thickness Offering: 16 and 14 gage

The EZ Joist Support Clip is distinguished by its return leg which positions the joist in place during installation. Simply rest the joist on the return leg and make the necessary screw attachments. It eliminates time consuming joist leveling and clamping steps required when a member is attached using traditional clip angles.

Left and right side clips are available to correspond to the desired joist web position and in depths to accommodate all "C" sections up to 12 inches in depth.

By permission: Super Stud Building Products, Inc., Astoria, New York

14.14.5 Super Stud Accessories (Continued)

Angle Clips (AC)

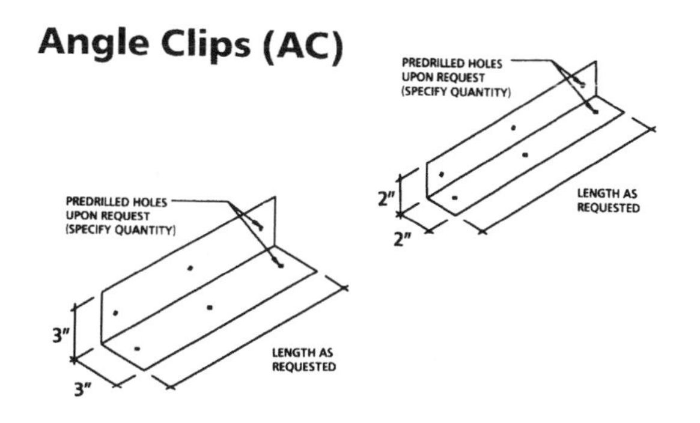

Thickness Offering: 20, 18, 16, 14 and 12 gage

Angle Clips serve a variety of functions in the construction of a steel framing system. They may be used to make attachments between framing members or, in the case of curtainwall construction, to transfer gravity and lateral loads from a stud to the primary frame.

Angle Clips are available in standard 2" or 3" leg dimensions. Non-standard angles with varying leg dimensions are also available.

Upon request, Angle Clips can be furnished pre-punched to facilitate screw attachments.

End Stiffener Clip (ESC)

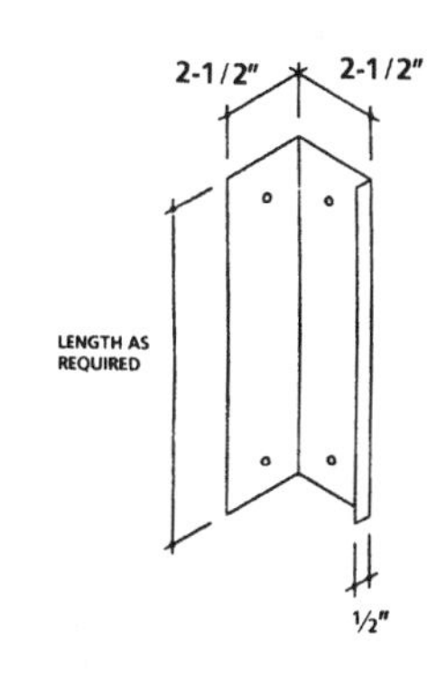

Thickness Offering: 16 and 14 gage

The End Stiffener provides web reinforcement while it functions as an attachment clip in the construction of a joist assembly. The clip is distinguished by its extended flange which provides ample surface area for attachment to a continuous steel track or wood ledger and its return lip which increases its stiffening capacity.

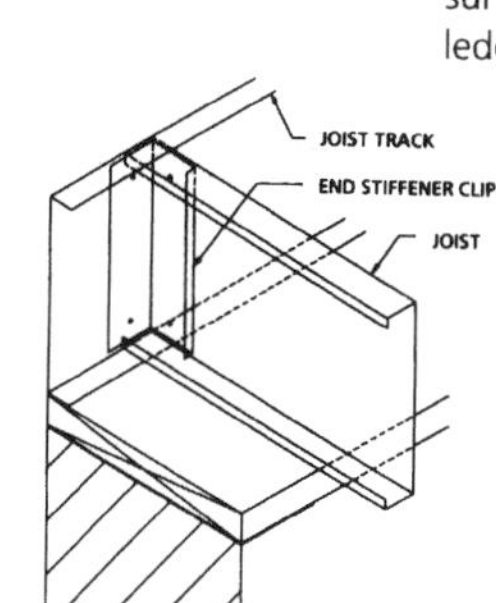

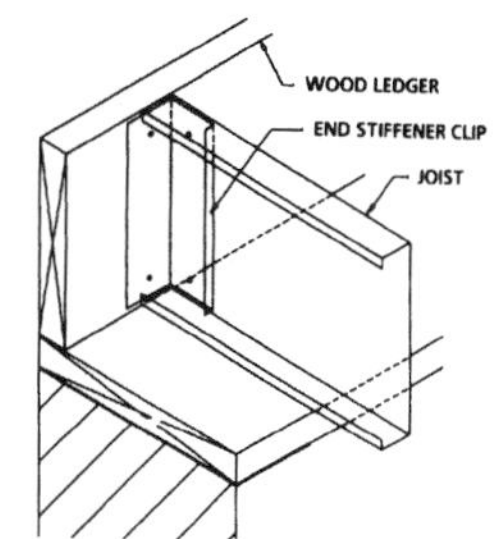

Web Stiffener (WS)

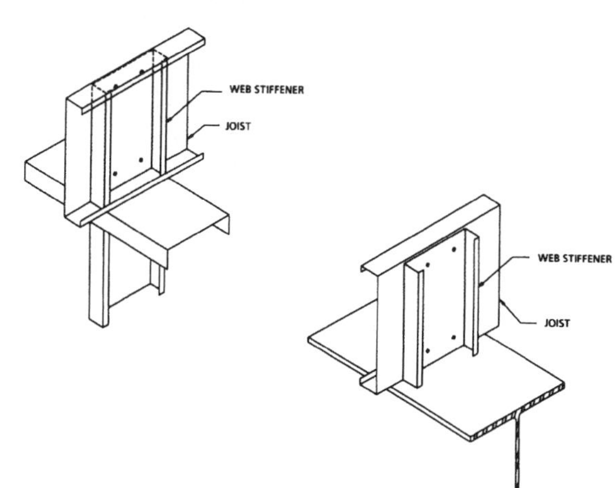

Thickness Offering: 18, 16 and 14 gage

Web Stiffeners provide web reinforcement for studs and joists at both end and intermediate support or concentrated load locations. The stiffener, placed either inside or outside the web, is manufactured on a custom order basis. Stiffeners located "inside" the web should be ordered to lengths equalling the depth of the member less 1/4" while those installed "outside" the web should match the depth of the member. The width of the stiffener should match the width of bearing above and/or below the stiffener.

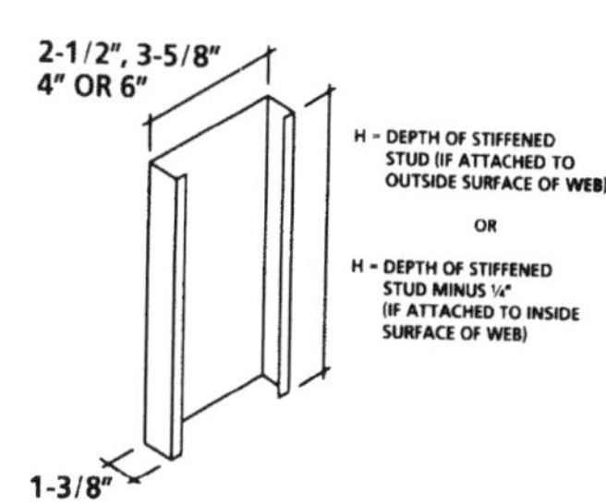

By permission: Super Stud Building Products, Inc., Astoria, New York

14.15.0 Plaster Systems

For years, the three-coat plaster system installed on expanded metal lath, attached to either wood or metal furring or studs, provided the ultimate in interior wall and ceiling construction. The smooth monlithic system created by plastering provided a relatively abuse-resistant surface; when decorative cornice or ceiling moldings were applied, an elegant room took shape.

However, the skills required to properly apply a three-coat plaster wall and their associated costs brought forth the development of veneer plaster systems in the 1960s. These systems took advantage of the large gypsum panels available to provide a smooth, stable foundation for a 1⁄16 inch (3 mm) to 1⁄8 inch (6 mm) application of plaster. The overall cost of this system is considerably lower than the conventional scratch-brown-finish coat. With a drying time of 48 hours versus 5 days for the regular three-coat system, production is greatly increased.

14.15.1 Comparing Conventional Plaster, Veneer Plaster, and Drywall Systems

Selecting a Plaster System

Because plaster systems provide more options in component selection than conventional drywall or masonry construction, plaster systems provide a much greater range of performance levels. The charts below compare conventional plaster, veneer plaster and drywall systems and list the distinctive characteristics of each.

Product Compatibility Selector

	Substrate						Finish Plaster								
Basecoat Plaster	**CMU**	**Mono. Conc.(1)**	**ML CH-FMG.**	**ML C-Studs**	**ROCKLATH Plaster Base**	**IMPERIAL Veneer Base**	**RED TOP Finish**	**STRUCTO-Gauge Lime**	**Keenes/ Lime**	**Gauging/ Lime**	**Keenes/ Lime/Sand**	**Gauging/ Lime/Sand**	**IMPERIAL Finish**	**DIAMOND Interior Finish**	**ORIENTAL Exterior Finish**
RED TOP & RED TOP Two Purpose (Sand)	✓	✓	✓	○	✓	○	✓	✓	✓	✓	✓	✓	✓	✓	○
RED TOP & RED TOP Two Purpose (Lightweight)	✓	✓	✓	○	✓	○	○	○	○	✓(3)	✓	✓	○	○	○
RED TOP Wood Fiber (Sand)	✓	✓	✓	○	✓	○	✓	✓	✓	✓	✓	✓	✓	✓	○
STRUCTO-BASE (Sand)	✓	✓	✓	✓	✓	○	✓	✓	✓	✓	✓	✓	✓	✓	○
STRUCTO-LITE	✓	✓	✓	○	✓	○	○	○	○	✓(3)	✓	✓	○	○	○
IMPERIAL Basecoat	✓	✓	○	○	○	✓	✓	✓	✓	✓	✓	✓	✓	✓	○
DIAMOND Veneer Basecoat	✓	✓	○	○	○	✓	✓	✓	✓	✓	✓	✓	✓	✓	○
IMPERIAL Finish	○	○	○	○	○	✓	—	—	—	—	—	—	—	—	—
DIAMOND Interior Finish	○	✓(2)	○	○	○	✓	—	—	—	—	—	—	—	—	—
DIAMOND Interior Finish (Electric Cable)	○	✓(2)	○	○	○	✓	—	—	—	—	—	—	—	—	—
Portland Cement/Lime/Sand (Stucco)	✓	✓	✓	○	○	○	○	○	○	○	○	○	○	○	✓

Notes: (1) A bonding agent must first be applied .(2) Job sanded. (3) Quality Gauging/not over metal lath.

Monolithic concrete to be treated with top quality bonding agent.
✓ = Acceptable
○ = Not Acceptable
— = Not Applicable

Comparing Conventional Plaster, Veneer Plaster and Drywall Systems*

System	**Characteristics**	**Comments**			
1. Conventional Plaster	Best system to attain a uniform, monolithic, blemish-free, smooth surface with good to excellent wear resistance based on the type of finish plaster. Ability to achieve intricate architectural details and ornamental shapes. High cost.				
2. Two Coat Veneer Plaster Systems IMPERIAL Basecoat Plaster (commercial application) or DIAMOND Veneer Basecoat Plaster (residential and light commercial applications) with finish plasters A-E below.	Provide distinct advantages over single coat veneer plaster and drywall systems. More monolithic surface with improved appearance under oblique lighting conditions. Ability to obtain truer wall surfaces, greater resistance to nail pops, joint ridging and joint shadowing/banding. Wider choice of finishing materials and texture options.	FINISH PLASTER RATING (No. 1 Best—No. 5 Acceptable) Productivity	Hardness	Workability	Ease to achieve smooth surface
A. IMPERIAL Finish	Ultimate in surface hardness and abrasion resistance. Easily textured. Low productivity and more difficult to achieve a smooth finish.	5	1	4	4
B. DIAMOND Interior Finish	Single bag, ready-to-use finish. Moderate strength. Acceptable workability. Extremely adaptable to textured finishes with or without the addition of aggregate. Satisfactory smooth finish.	2	3	3	3
C. STRUCTO-GAUGE Gauging Lime putty (1:1) or RED TOP Finish	Hardest dense putty finish. Moderate workability and ease of application. Excellent finish appearance.	2	4	2	2
D. Regular Gauging Lime Putty	Highest productivity. Best workability. Joinable, easiest to achieve a monolithic finish. Only moderate surface hardness.	1	5	1	1
E. RED TOP Keenes Cement Lime Putty and Sand	Ultimate choice for texturing. Unique, only retemperable material, allows extended time period for floating. Provides the ability for pigment addition to achieve colored textured surface also.	Due to unique nature Keenes is not rated with above finishes.			
3. One Coat IMPERIAL Finish Plaster	Monolithic, smooth or textured appearance. Ultimate in surface hardness. Direct to plaster base in a single coat veneer plaster system. Achieves high productivity due to compatibility with absorbent surface plaster base. Ready for further decoration in as little as 24 hours if completely dry.	Fast completion shortens construction time, brings in paying tenants faster, thus reducing interest paid on project construction loan.			
4. One Coat DIAMOND Interior Finish Plaster	Monolithic appearance. Moderate wear-resistant surface. Provides a wide range of texture types with or without the addition of aggregate. Ready for further decoration in as little as 24 hours if completely dry. Greatest coverage for single coat application over veneer plaster base. Lowest cost single coat veneer plaster system.	See comment IMPERIAL Finish.			
5. Gypsum Drywall	Relatively smooth surface with acceptable monolithic appearing surface under most conditions. Lowest cost. Resistant to light abrasion. Most susceptible to nail pops and joint photographing.				

*This table is meant to serve as a general guide to the selection of plaster systems. The information should not be construed as limiting materials or systems to specific types of construction.

By permission: United States Gypsum Corporation, Chicago, Illinois

14.15.2 Lath and Plaster Installation Procedures

Installation

Steel Studs—Space steel studs a maximum of 16″ o.c.

Metal Lath—Place Self-Furring Diamond Mesh Lath against studs and with end joints staggered in adjacent rows. Screw studs *through dimples only*, spacing screws 6″ o.c. Lap ends of lath at least 1″ between supports. Lap side (horizontal) joints at least ½″. Wire tie all side laps and end joints between supports together at intervals not exceeding 6″.

Basecoat—Mix STRUCTO-BASE Gypsum Plaster in a mechanical mixer to a uniform consistency. Scratch and brown coats shall be proportioned 2 cu. ft. of sand per 100 lbs. of plaster. Determine optimum batch material fluidity at the mixer by adjusting water usage to achieve the following slumps:

Machine application ....1½″ maximum, 1″ preferred.
Hand application......................¾″ maximum, ½″ preferred.

Slump Determination Procedure—Place a wetted 2″ IDx4″ high cylinder on base plate. Gradually fill cylinder with material, puddling occasionally. When full, strike-off flush with top of cylinder. Slowly raise cylinder and allow material to slump. Position empty cylinder beside material on the base plate (do not disturb) and place a rule on cylinder top to overhang material. Measurement from rule to material indicates slump.

Scratch Coat—For hand application, apply scratch coat with sufficient material and pressure to form good full keys and then cross rake.

For machine application, maintain sufficient angle in the spray pattern to develop full keys on the back of the lath and to prevent excessive material blow-through. Where leveling by trowel is necessary to remove high spots, cross rake for sufficient bond with subsequent brown coat.

Allow scratch coat to set and *partially* dry before application of brown coat.

Brown Coat—Apply brown coat after scratch coat has set firm and hard (maintaining proper "green state" or dampness). When applying the brown coat by hand application, use sufficient pressure to ensure proper bonding to the scratch coat application. Bring out to grounds (allow 1/16″ for finish coat) and straighten to a true surface with rod and darby with only limited use of additional water. Leave surface rough and open to receive finish coat. Minimum thickness of scratch and brown coats (basecoat plaster) shall be 11/16″ measured from the face of the lath.

For machine application of scratch and brown coats, consult the manufacturer of the particular plaster application machine for maximum length of hose and maximum vertical lift.

Finish Coat—Brown coat must be partially dry (green state) to receive finish coat. The following finishes are recommended and listed in descending order of hardness and abrasion resistance:

1 **IMPERIAL Finish Plaster**
2 **DIAMOND Interior Finish Plaster**
3 **RED TOP Finish Plaster**
4 **STRUCTO-GAUGE Gauging Plaster with (Type N or Type S) lime**
5 **CHAMPION, STAR, RED TOP, or Quality Gauging Plaster with (Type N or Type S) lime**
6 **Keenes Cement with (Type N or Type S) lime for a sand float finish**

A full specification for application of plaster finish coats can be found in the General Lathing and Plastering Specifications on page 34 (Part 3.14D).

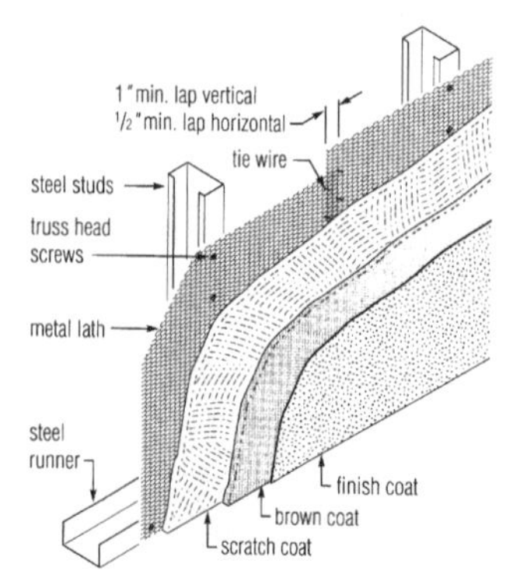

By permission: United States Gypsum Corporation, Chicago, Illinois

14.15.3 Metal Lath, Hangers, Channel, and Stud Specifications

Max. Frame Spacing–Metal Lath(1)

Type of lath(2)	Weight		Maximum allowable spacing									
			Vertical supports (wall) Wood		Metal Solid Partitions		Others(4)		Horizontal supports (ceiling) Wood or Concrete		Metal	
	lbs./yd²	kg/m²	in.	mm	in.	mm	in.	mm	in.	mm	in.	mm
Diamond Mesh	2.5	1.4	16	406	16	406	12	305	(5)		(5)	
Diamond Mesh(3)	3.4	1.8	16	406	16	406	16	406	16	406	13.5	343
1/8" Z-Rib	2.75	1.5	16	406	16	406	16	406	16	406	12	305
1/8" Z-Rib	3.4	1.8	19	483	24	610	19	483	19	483	19	483
3/8" Rib	3.4	1.8	24	610			24	610	24	610	24	610
3/4" Rib	4.0	2.2	24	610			24	610	24	610	24	610

(1) For spacing on fire-rated constructions, see test reports. (2) All types made from copper alloy steel containing from 0.20% to 0.25% pure copper, and painted with rust-inhibitive black asphaltum paint. Sheet size is 27" x 96". (3) Available in both copper alloy and galvanized steel. (4) Including vertical furring. (5) Not recommended except for fireproofing of steel shapes.

Metal Lath General Limitations

1 Metal lath products should not be used with magnesium oxychloride cement stuccos or stuccos containing calcium chloride additives.

2 In ceiling assemblies, certain precautions concerning construction, insulation and ventilation are necessary for good performance. A min. of 1/2 sq. in. net free vent area is recommended per sq. ft. of horizontal surface in plenum or other space.

Metal Lath Selector

Type of lath	Recommended Applications					
	Ornamental contour	Over int. substrate	Over ext. substrate(1)	Nail-on/tie-on flat ceiling	Solid partitions	Concrete centering
Diamond Mesh	X			X(3)	X(5)	
Self-Furring		X	X(2)	X(4)		
4-Mesh Z-Riblath				X		
3/8" Riblath					X	X

(1) For example: gypsum sheathing, replastering existing work, column fireproofing.
(2) 3.4 lb/yd² galvanized lath.
(3) For tie-on only: supports 16" o.c. max.
(4) For nail-on only: supports 16" o.c. max.
(5) Supports 16" o.c. max.

Technical Data

Frame and Fastener Spacing–ROCKLATH Plaster Base

Type framing	Base thickness		Fastener(1)	Max. frame spacing		Max. fastener spacing	
	in	mm		in	mm	in	mm
Wood	3/8	9.5	Nails–13 ga. 1 1/8" long, 19/64" flat head, blued.	16	406	5	127
			Staples—16 ga. galv. flattened wire, flat crown 7/16" wide, 1" divergent legs				
	1/2	12.7	Nails—13 ga. 1 1/4" long, 19/64" flat head blued	24	610	4	102
			Staples–16 ga. galv. flattened wire, flat crown 7/16" wide, 1" divergent legs				
Steel Stud	3/8	9.5	1" drywall screws	16	406	12	305
	1/2	12.7		24	610	6	152
Metal Furring-DWC-20 DWC-25	3/8	9.5	1" drywall screws	16	406	12	305
	1/2	12.7		24	610	6	152

(1) Metric; fastener dimensions: 19/64" = 7.5mm; 7/16" = 11.1mm; 1" = 25.4mm; 1 1/8" = 28.6mm; 1 1/4" = 31.8mm.

Technical Data

Thermal Coefficient of Expansion (unrestrained)
[Inches/inch/°F (40 –100 °F)]

Sanded Gypsum Plaster (100:2, 100:3)	7.0×10^{-6}
Wood Fiber Plaster (sanded 100:1)	8.0×10^{-6}
Gypsum Lath	9.0×10^{-6}

Max. Spacing–Main Runner–Carrying Channels

Main runner c. r. channel size		Max. c. to c. spacing of main runners		Max. spacing of hangers along runners	
in	mm	ft	mm	ft	mm
3/4	19.1	3	914	2	610
3/4	19.1	2 1/4	686	3(1)	914
1 1/2	38.1	4	1219	3	914
1 1/2	38.1	3 1/2	1067	3 1/2	1067
1 1/2	38.1	3	914	4	1219
2	50.8	4	1219	5	1524
2	50.8	2 1/2	762	6	1829
2	50.8	2	610	7	2134

(1) For concrete joist construction only—where 8 ga. wire may be inserted in joist before concrete is poured.

Max. Spacing–Cross-Furring Members

Cross-furring size	Max. c. to c. spacing of cross-furring		Main runner or support spacing	
	in	mm	ft	mm
3/4" (19.1mm) C.R. Channel	24	610	3	914
3/4" (19.1mm) C.R. Channel	19	483	3 1/2	1067
3/4" (19.1mm) C.R. Channel	16	406	4	1219
1" (25.4mm) H.R. Channel	24	610	4	1219
1" (25.4mm) H.R. Channel	19	483	4 1/2	1372
1" (25.4mm) H.R. Channel	12	305	5	1524
3/8" (9.5mm) Pencil Rod(1)	19	483	2	610
3/8" (9.5mm) Pencil Rod(1)	12	305	2 1/2	762

(1) Primary usage is on furred ceiling members.

Support Area–Hangers

Hanger size and type	Typical ceiling area per hanger		Maximum tensile load
	ft²	m²	lbs.(2)
9 ga. galvanized wire	12.5	1.2	340
8 ga. galvanized wire	16	1.5	408
3/16" (4.8mm) mild steel rod(1)	20	1.9	546
1/4" (6.4mm) mild steel rod(1)	22.5	2.1	972
3/16" x 1" (4.8mm x 25.4mm) mild steel flat(1)	25	2.3	3712

(1) Where severe moisture conditions may occur, rods galvanized or painted with rust-inhibitive paint, or galvanized straps are recommended. (2) Based on minimum yield 33,000 psi.

Section Properties–Studs and Channels

Item	Gauge	Width	Depth	Steel thick. in.*	I_x in⁴	S_x in³	F_c ksi
Steel Stud							
1 5/8"	25	1 5/8"	—	0.0188	0.038	0.040	19.8
2 1/2"	25	2 1/2"	—	0.0188	0.101	0.071	19.8
3 5/8"	25	3 5/8"	—	0.0188	0.239	0.113	19.8
4"	25	4"	—	0.0188	0.302	0.123	19.8
6"	25	6"	—	0.0188	0.773	0.184	19.8
Metal Furring Channel							
DWC-25 (hemmed)	25	—	7/8"	.0188	.0096	.0247	13.9
DWC-20 (unhemmed)	20	—	7/8"	.0344	.0165	.0355	18.8
Cold-Rolled Channel							
3/4"	16	—	3/4"	0.0566	0.007	0.019	19.8
1 1/2"	16	—	1 1/2"	0.0566	0.039	0.052	19.8
2"	16	—	2"	0.0566	0.083	0.083	19.8

*Base steel design thickness

By permission: United States Gypsum Corporation, Chicago, Illinois

14.15.4 Lath, Framing, and Furring Accessories

Lath Accessories

1-A Expanded Corner Bead
Easily flexed for irregular corners. Reinforces close to nose of bead, made with 2⅞" wide expanded flanges, in galvanized steel, or zinc alloy for exterior use.

Double-X Corner Bead
Ideal for structural tile and rough masonry, adjusts easily for plaster depth on columns. Perforated stiffening ribs along expanded flange.

4-A Flexible Corner Bead
Ideal for curved edges (archways, telephone niches, etc.). Versatile and economical as an "all purpose" corner bead. Snipping flanges lets you bend this bead to any curvature radius.

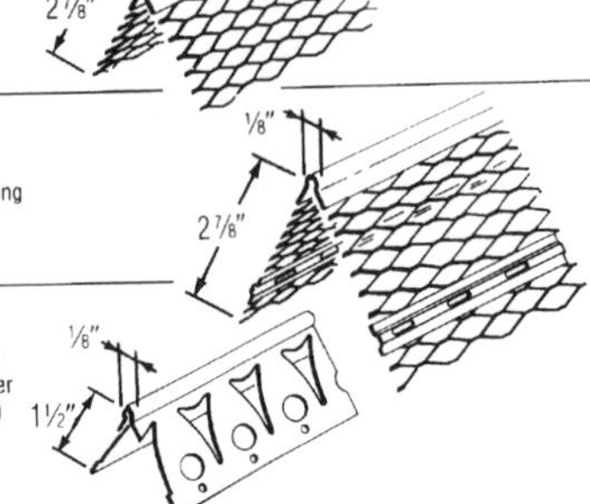

Cornerite
Strips of Diamond Mesh Lath for reinforcement. Available as painted or galvanized steel: Cornerite, bent in the center to a 100° angle, reinforces interior angles between unlapped metal lath, and between masonry constructions (to reduce plaster cracking), and nonferrous lath. Sizes: 2"x2"x96", 3"x3"x96".

Striplath, a flat strip, reinforces joints of nonmetallic and/or dissimilar plaster lathing/bases; also spans pipe chases. Sizes: 4"x96", 6"x96".

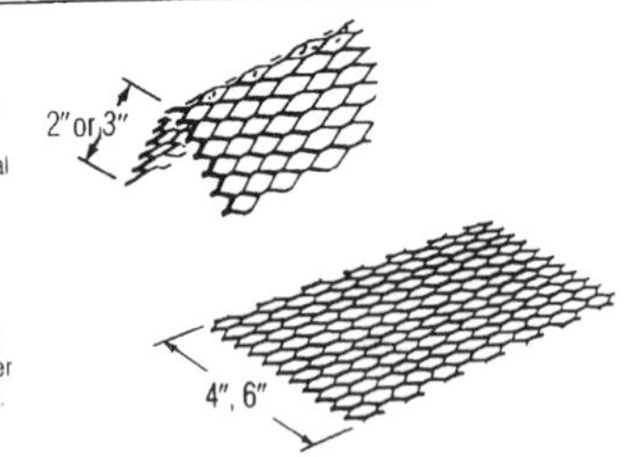

Casing Beads
Use ¾" casing beads with metal lath, ⅝" beads with all masonry units. When flange is applied under ROCKLATH Plaster Base, use ⅞" beads; over ROCKLATH Base, ½" beads. Made from corrosion-resistant galvanized steel or zinc alloy for exterior applications. #66 Square Edge Sizes with 1¼" short, solid flange: ¼", ⅜" ½", ¾", ⅞". #66 Square Edge Sizes with 3⅛" long, expanded flange: ¼", ⅜", ½", ⅝", ¾", ⅞", 1", 1¼". Length: 10'.

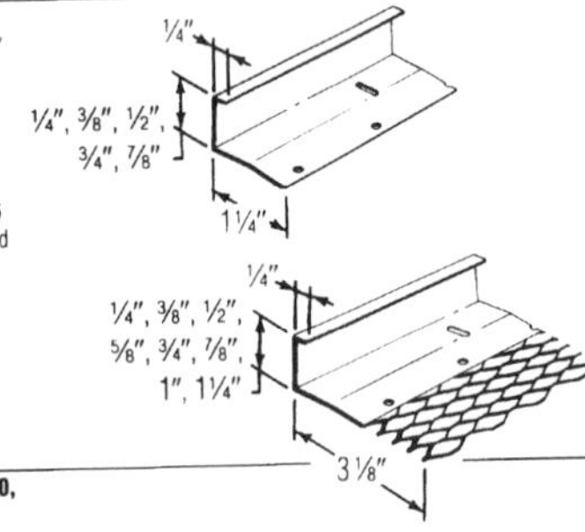

SHEETROCK Zinc Control Joints Nos. 50, 75, 100
Relieves plaster expansion/contraction stresses in large areas. Used from floor to ceiling in long partition runs, and from header to ceiling above door frames. Plastic tape, removed after plastering, protects a ¼"x ½" deep slot. Roll-formed from zinc, it is corrosion-resistant for both interior and exterior use with gypsum or portland cement plaster. Sizes, grounds: No. 50, ½"; No. 75, ¾"; No. 100, 1" (for exterior stucco curtain walls). Length: 10'. Limitations: adequate protection must be provided behind the control joint to maintain sound and/or fire ratings. Zinc control joint should not be used with magnesium oxychloride cement stuccos or stuccos containing calcium chloride additives.

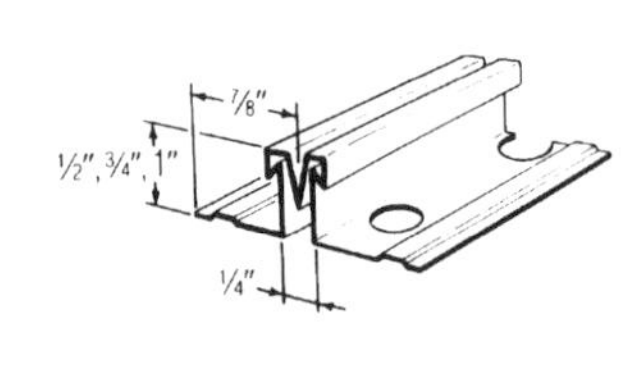

Double V Expansion Joint
Provides stress relief to control cracking in large plastered areas. Made with expanded flanges of corrosion-resistant galvanized steel, or zinc for exterior use. Grounds: ½", ¾". Lengths: 10'.

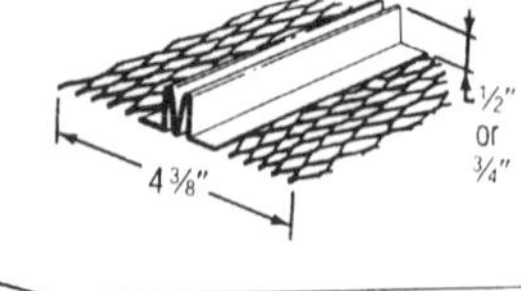

BRIDJOINT Field Clip B-1
Supports and aligns ⅜" thick ROCKLATH base ends where joints do not fall directly over 16" o.c. max. framing members: designed for use with ⅜" ROCKLATH Plaster Base.

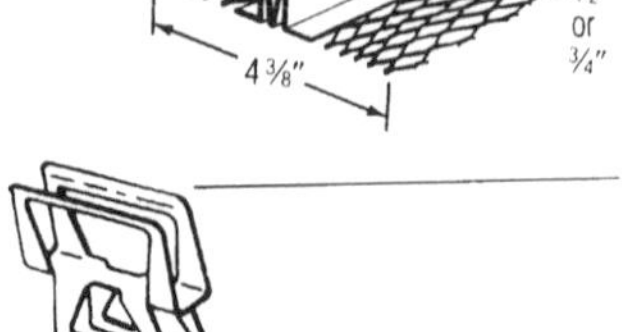

Framing and Furring Accessories

Steel ST-Studs and CR-Runners
Channel-shaped, roll-formed, with corrosion-resistant coating. Secure, rigid screw or clip attachment of the gypsum base utilizes the full structural contribution of the lath and plaster membrane. Stud widths: 1⅝" (for 25 ga. only), 2½", 3⅝", 4", 6"; stud styles: 25 ga., 22 ga., 20 ga. Stud lengths: 8' to 16'. Runners come in stud widths, 10' length only.

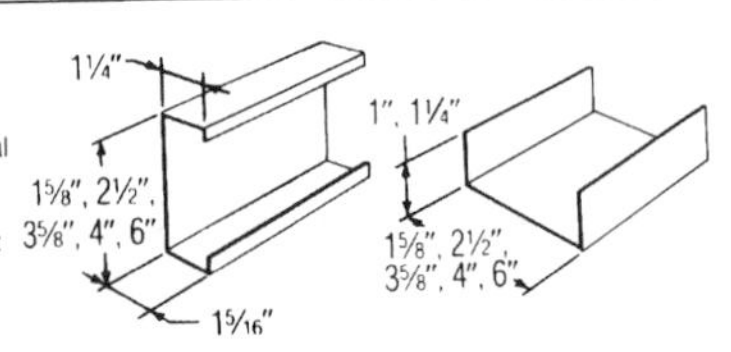

Cold-Rolled Channels
Made of 16 ga. steel. Used for furring, suspended ceilings, partitions, ornamental lathing. Available either galvanized or black asphaltum painted. Sizes: ¾", with ½" flange; 1½" and 2" with 17/32" flange. Lengths: 16' and 20'.

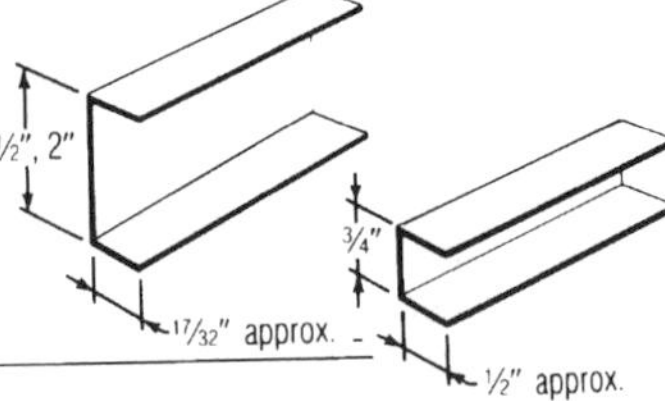

Metal Furring Channels
Hat-shaped channels for ceiling and wall furring. Roll-formed from two gauges of corrosion-resistant steel. DWC-25 for screw attachment of ROCKLATH Base. DWC-20 for greater spans, load-carrying capacity and exterior furring. Products comply with ASTM C645. Face width: 1¼", depth: ⅞"; length: 12'.

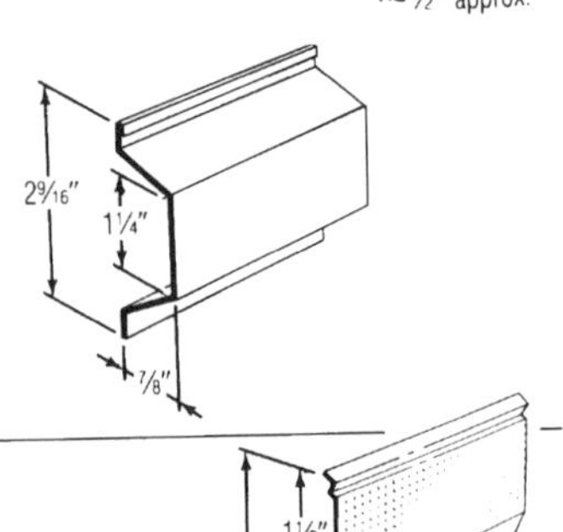

RC-1™ Resilient Channel
Part of the family of SHEETROCK Metal Products. Corrosion-resistant steel channel for resilient attachment of ROCKLATH Base to wood and steel framing. Prepunched holes in the flange facilitate screw fastening to framing members with 1¼" Type W or S screws. Not suitable for use with more than 2 layers of ⅝" thick gypsum panels.

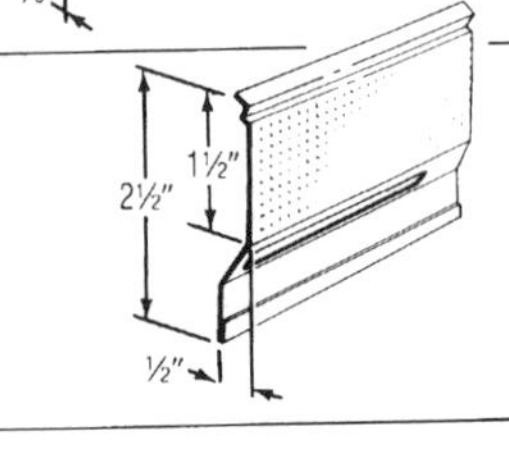

SHEETROCK Z-Furring Channels
Mechanically attach THERMAFIBER insulation and rigid foam insulations and ROCKLATH Base to interior of masonry walls. Made of corrosion-resistant, hot-dip galvanized steel. Furring depths: 1", 1½", 2", 3"; length: 8'6".

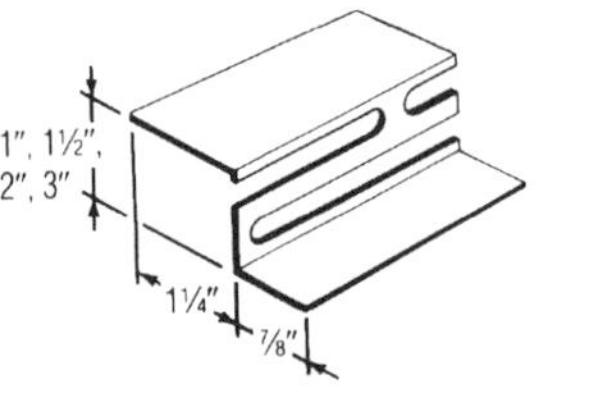

Adjustable Wall Furring Brackets
Used in braced furring systems for interior or exterior masonry walls. Made of 20 ga. galvanized steel, and attached to steel studs. Furring depth: up to 2¼" plus stud width.

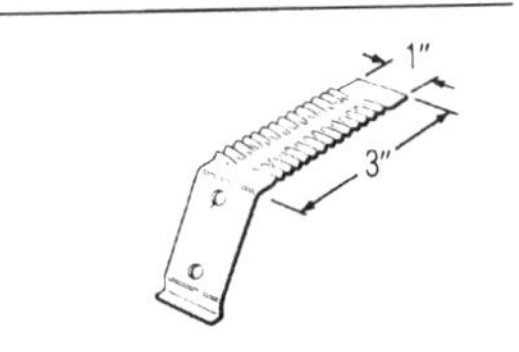

Metal Furring Channel Clip
Galvanized 1½" wire used to attach metal furring channels to 1½" cold-rolled channel ceiling grillwork.

Tie Wire and Hanger Wire
Tie wire is 18 ga. galvanized steel wire used in drywall/plaster to tie furring channel to runners and in plaster to tie metal lath to channel. Hanger wire is 8 ga. galvanized steel for hanging 1½" runner channels in drywall/plaster suspended ceilings.

By permission: United States Gypsum Corporation, Chicago, Illinois

14.16.0 Five Levels of Drywall-Taping Systems

- *Level 1* All joints and interior angles shall have tape embedded in joint compound. The surface shall be free of excess joint compound. Tool marks and ridges are acceptable. Suggested location of Level 1 taping: fire- and smoke-taped baffles above suspended ceilings and elsewhere where gypsum board is concealed from public view.
- *Level 2* All joints and interior angles shall have tape embedded in the joint compound and shall receive one separate coat of joint compound applied over all joints, angles, and fastener heads and accessories. The surface shall be free of excess joint compound. Tool marks and ridges are acceptable. Suggested location of Level 2 taping: Substrates that receive tile or paneling in excess of ¼ inch (8 mm) thickness.
- *Level 3* All joints and interior angles shall have tape embedded in the joint compound and shall receive two separate coats of joint compound applied over all joints, angles, fastener heads, and accessories. The surface shall be free of excess joint compound and all surfaces shall be smooth and free of tool marks and ridges. Suggested location of Level 3 taping: Areas scheduled to receive heavy-textured finishes (hand or spray applied), paneling less than ¼ inch (8 mm) thickness, or Class III vinyl wall coverings.
- *Level 4* All joints and interior angles shall have tape embedded in the joint compound and shall receive three separate coats of joint compound applied over all joints, angles, fastener heads, and accessories. Surfaces shall be free of excess joint compound and all surfaces shall be free of tool marks and ridges. Suggested location of Level 4 taping: Areas to receive paint coatings, paneling less than ¼ inch (8 mm) thickness, and where vinyl or wall fabric wall coverings will be applied.
- *Level 5* All joints and interior angles shall have tape embedded in the joint compound and shall receive three separate coats of joint compound applied over all joints, angles, fastener heads, and accessories. Provide a thin skim coat of joint compound (or other material manufactured expressly for this purpose) over the entire surface. The finished surface shall be free of excess joint compound and all surfaces shall be smooth of tool marks and ridges. Suggested location of Level 5 taping: Areas scheduled to be lightly by cove- and washing-type light fixtures.

Section

15

Flooring

Contents

15.0.0 Most frequently encountered flooring materials
15.1.0 Wood flooring (types)
15.2.0 Resilient flooring
15.3.0 Carpet construction and materials
15.3.1 Computing square yards and square meters of carpet (6 to 81 feet in length from rolls 9, 12, and 15 feet wide)
15.4.0 Seamless flooring
15.5.0 Stone veneer flooring
15.5.1 Thinset/mortar-bed stone veneer installation diagrammed
15.6.0 Terrazo floors
15.6.1 Terrazo floor components

15.0.0 Most Frequently Encountered Flooring Materials

Materials for floor coverings range from painted concrete to custom-made ceramic tiles or carpeting. This section deals primarily with those materials most frequently encountered on construction projects: wood flooring, resilient flooring and carpet, and secondarily, with less frequently used materials (stone veneer, seamless flooring, and terrazo).

15.1.0 Wood Flooring (Types)

The species of wood most commonly used for flooring is oak (red and white) and maple. Yellow birch and sweet birch are used on occasion, as are more exotic and costly species (such as pecan, walnut, cherry, ash, hickory, and teak).

- *Oak* Available in two grades of quartered sawed and five grades of plain sawed, generally milled as tongue-and-groove, oak flooring is sold in plank, strip, block, or parquet form.
- *Maple* Obtained from the sugar maple or rock maple trees, this wood is exceptionally hard and finds wide usage in gymnasium floors. Its resistance to abrasion and ability to take an excellent finish makes it desirable for all wood floor applications where heavy wear will be anticipated.
- *Acrylic-impregnated hardwood* Radiation polymerization of hardwood flooring replaces the air in the wood cells with a liquid polymer using a vacuum and pressure process. The liquid polymer can be colored or clear. The resultant finish will greatly improve the wood's resistance to wear.

15.2.0 Resilient Flooring

Vinyl Composition Tile (VCT)

The two types of vinyl composition tile are available in several thicknesses:

- *Type 1* Smooth surface
- *Type 2* Embossed surface

A thoroughly blended composition of thermoplastic binders, fillers, and pigment is used. The thermoplastic binder is polyvinyl chloride resin or a copolymer resin made by copolymerizing vinyl chloride with other monomeric materials. The size is usually 12" × 12" (304.8 mm). The difference between length and width shall be no greater than 0.020" (0.51 mm) for any size of square tile. Thickness will be either ⅛" (3.18 mm), 3/32" (2.38 mm), 0.080" (2.03 mm), 1/16" (1.59 mm).

Solid Vinyl Tile

Solid vinyl tiles are available in two types and three classes:

- *Type 1* Smooth surface
  - ~ *Class A* Monolithic
  - ~ *Class B* Multi-layered
- *Type 2* Embossed surface
  - ~ *Class A* Monolothic
  - ~ *Class B* Multi-layered
- *Class C* Class A or B with a permanently bonded coating.

Materials of Construction

- *Class A* Contains a constant composition throughout the tile thickness.
- *Class B* Contains layers of either Material I or Material II or combinations thereof.
- *Class C* Any construction of Class A or Class B that has a permanently bonded protective coating of Material III.

- *Material I* Vinyl plastic composed a binder stabilized against heat and polyvinyl chloride or a co-polymer of vinyl chloride (not less than 85% of which shall be polyvinyl chloride). The vinyl resin must be at least 60%, by weight, of the binder.
- *Material II (Translucent)* A transparent vinyl plastic containing resins, each one of which shall be polyvinyl chloride or a coploymer of vinyl chloride, not less than 85% of which is vinyl chloride. The vinyl resin must be at least 60%, by weight, of the binder.
- *Material III* A clear or transparent layer specifically formulated to function as a top coat to enhance the flooring material. This coating is composed of, but not restricted to, conventional vinyl resins of plasticizers. The size is generally 12" × 12" (304.8 mm) × (304.8 mm) with the same tolerances as VCT.

The nominal thicknesses an be 1/8" (3.18 mm), 0.100" (2.54 mm), 0.080" (2.03 mm), 0.0625" (1.59 mm), 0.050" (1.27 mm), 0.039" (1.00 mm).

Rubber floor tiles are made of 100% virgin synthetic rubber with a slip-retardant additive. This type of flooring has high strength as a result of its elasticity and resilience. Base thickness for heavy-duty wear is 0.130" (3.38 mm) and 0.100" (2.54 mm) for light-duty use.

15.3.0 Carpet Construction and Materials

Construction is the amount of pile packed into a given volume of carpet is translated into ounces of yarn for unit volume and depends upon the following:

- *Pitch* The number of warp lines of yarn in a 27" width. The higher the "pitch", the more dense the carpet.
- *Stitch* The number of lengthwise yarn tufts contained in a 1" area. More stitches per inch results in a more-dense carpet face.
- *Pile height* A measurement from the back of the pile to the front or top of the pile. High pile does not wear well; low pile does not wear well. Medium pile is the better service pile.

Weight per yard, expressed in ounces per yard, is the total weight of the pile yard, plus backings and coatings.

Materials of construction

- *Wool* Soft, good serviceabiliy, and resilient. The highest priced of the carpet materials.
- *Acrylics* Wool-like appearance; average durability, abrasion resistance, and stain resistance.
- *Polyester* Good abrasion resistance; feels like wool; susceptible to oil-based stains.
- *Olefin* Also referred to as *polypropylene*, often is used for indoor-outdoor carpet. Resistant to fading and staining; good abrasion resistance, resilience not good.
- *Nylon* Excellent abrasion resistance; easy to clean; and very good crush and stain resistance.

The backing material on all types of carpet can be either:

- *Primary backing* The material to which surface yarns are attached and constructed of jute cotton, or a synthetic.
- *Secondary backing* A material laminated to the primary backing to improve resiliency and add stability. It can be either jute or a woven or nonwoven synthetic material.
- *Separate padding* A cushioning material, separate from the carpet, that can be constructed of jute, foam rubber, plastic, or felted cattle hair.

15.3.1 Computing Square Yards and Square Meters of Carpet (6 to 81 Feet in Length From Rolls 9, 12, and 15 Feet Wide)

LENGTH	9 FEET			12 FEET			15 FEET		
FT IN	SQ FT	SQ YDS	SQ M	SQ FT	SQ YDS	SQ M	SQ FT	SQ YDS	SQ M
6-00	**54.0**	6.00	5.02	**72.0**	8.00	6.69	**90.0**	10.00	8.36
6-01	**54.8**	6.08	5.09	**73.0**	8.11	6.78	**91.3**	10.14	8.48
6-02	**55.5**	6.17	5.16	**74.0**	8.22	6.87	**92.5**	10.28	8.59
6-03	**56.3**	6.25	5.23	**75.0**	8.33	6.97	**93.8**	10.42	8.71
6-04	**57.0**	6.33	5.30	**76.0**	8.44	7.06	**95.0**	10.56	8.83
6-05	**57.8**	6.42	5.37	**77.0**	8.56	7.15	**96.3**	10.69	8.94
6-06	**58.5**	6.50	5.43	**78.0**	8.67	7.25	**97.5**	10.83	9.06
6-07	**59.3**	6.58	5.50	**79.0**	8.78	7.34	**98.8**	10.97	9.17
6-08	**60.0**	6.67	5.57	**80.0**	8.89	7.43	**100.0**	11.11	9.29
6-09	**60.8**	6.75	5.64	**81.0**	9.00	7.52	**101.3**	11.25	9.41
6-10	**61.5**	6.83	5.71	**82.0**	9.11	7.62	**102.5**	11.39	9.52
6-11	**62.3**	6.92	5.78	**83.0**	9.22	7.71	**103.8**	11.53	9.64
7-00	**63.0**	7.00	5.85	**84.0**	9.33	7.80	**105.0**	11.67	9.76
7-01	**63.8**	7.08	5.92	**85.0**	9.44	7.90	**106.3**	11.81	9.87
7-02	**64.5**	7.17	5.99	**86.0**	9.56	7.99	**107.5**	11.94	9.99
7-03	**65.3**	7.25	6.06	**87.0**	9.67	8.08	**108.8**	12.08	10.10
7-04	**66.0**	7.33	6.13	**88.0**	9.78	8.17	**110.0**	12.22	10.22
7-05	**66.8**	7.42	6.20	**89.0**	9.89	8.27	**111.3**	12.36	10.33
7-06	**67.5**	7.50	6.27	**90.0**	10.00	8.36	**112.5**	12.50	10.45
7-07	**68.3**	7.58	6.34	**91.0**	10.11	8.45	**113.8**	12.64	10.57
7-08	**69.0**	7.67	6.41	**92.0**	10.22	8.55	**115.0**	12.78	10.68
7-09	**69.8**	7.75	6.48	**93.0**	10.33	8.64	**116.3**	12.92	10.80
7-10	**70.5**	7.83	6.55	**94.0**	10.44	8.73	**117.5**	13.06	10.92
7-11	**71.3**	7.92	6.62	**95.0**	10.56	8.83	**118.8**	13.19	11.03
8-00	**72.0**	8.00	6.69	**96.0**	10.67	8.92	**120.0**	13.33	11.15
8-01	**72.8**	8.08	6.76	**97.0**	10.78	9.01	**121.3**	13.47	11.26
8-02	**73.5**	8.17	6.83	**98.0**	10.89	9.10	**122.5**	13.61	11.38
8-03	**74.3**	8.25	6.90	**99.0**	11.00	9.20	**123.8**	13.75	11.50
8-04	**75.0**	8.33	6.97	**100.0**	11.11	9.29	**125.0**	13.89	11.61
8-05	**75.8**	8.42	7.04	**101.0**	11.22	9.38	**126.3**	14.03	11.73
8-06	**76.5**	8.50	7.11	**102.0**	11.33	9.48	**127.5**	14.17	11.84
8-07	**77.3**	8.58	7.18	**103.0**	11.44	9.57	**128.8**	14.31	11.96
8-08	**78.0**	8.67	7.25	**104.0**	11.56	9.66	**130.0**	14.44	12.08
8-09	**78.8**	8.75	7.32	**105.0**	11.67	9.75	**131.3**	14.58	12.19
8-10	**79.5**	8.83	7.39	**106.0**	11.78	9.85	**132.5**	14.72	12.31
8-11	**80.3**	8.92	7.46	**107.0**	11.89	9.94	**133.8**	14.86	12.43
9-00	**81.0**	9.00	7.52	**108.0**	12.00	10.03	**135.0**	15.00	12.54
9-01	**81.8**	9.08	7.60	**109.0**	12.11	10.13	**136.3**	15.14	12.66
9-02	**82.5**	9.17	7.66	**110.0**	12.22	10.22	**137.5**	15.28	12.77
9-03	**83.3**	9.25	7.73	**111.0**	12.33	10.31	**138.8**	15.42	12.89
9-04	**84.0**	9.33	7.80	**112.0**	12.44	10.40	**140.0**	15.56	13.01
9-05	**84.8**	9.42	7.87	**113.0**	12.56	10.50	**141.3**	15.69	13.12
9-06	**85.5**	9.50	7.94	**114.0**	12.67	10.59	**142.5**	15.83	13.24
9-07	**86.3**	9.58	8.01	**115.0**	12.78	10.68	**143.8**	15.97	13.35
9-08	**87.0**	9.67	8.08	**116.0**	12.89	10.78	**145.0**	16.11	13.47
9-09	**87.8**	9.75	8.15	**117.0**	13.00	10.87	**146.3**	16.25	13.59
9-10	**88.5**	9.83	8.22	**118.0**	13.11	10.96	**147.5**	16.39	13.70
9-11	**89.3**	9.92	8.29	**119.0**	13.22	11.05	**148.8**	16.53	13.82

15.3.1 Computing Square Yards and Square Meters of Carpet (6 to 81 Feet in Length From Rolls 9, 12, and 15 Feet Wide) (Continued)

LENGTH	9 FEET			12 FEET			15 FEET		
FT IN	SQ FT	SQ YDS	SQ M	SQ FT	SQ YDS	SQ M	SQ FT	SQ YDS	SQ M
10-00	**90.0**	10.00	8.36	**120.0**	13.33	11.15	**150.0**	16.67	13.94
10-01	**90.8**	10.08	8.43	**121.0**	13.44	11.24	**151.3**	16.81	14.05
10-02	**91.5**	10.17	8.50	**122.0**	13.56	11.33	**152.5**	16.94	14.17
10-03	**92.3**	10.25	8.57	**123.0**	13.67	11.43	**153.8**	17.08	14.28
10-04	**93.0**	10.33	8.64	**124.0**	13.78	11.52	**155.0**	17.22	14.40
10-05	**93.8**	10.42	8.71	**125.0**	13.89	11.61	**156.3**	17.36	14.52
10-06	**94.5**	10.50	8.78	**126.0**	14.00	11.71	**157.5**	17.50	14.63
10-07	**95.3**	10.58	8.85	**127.0**	14.11	11.80	**158.8**	17.64	14.75
10-08	**96.0**	10.67	8.92	**128.0**	14.22	11.89	**160.0**	17.78	14.86
10-09	**96.8**	10.75	8.99	**129.0**	14.33	11.98	**161.3**	17.92	14.98
10-10	**97.5**	10.83	9.06	**130.0**	14.44	12.08	**162.5**	18.06	15.10
10-11	**98.3**	10.92	9.13	**131.0**	14.56	12.17	**163.8**	18.19	15.21
11-00	**99.0**	11.00	9.20	**132.0**	14.67	12.26	**165.0**	18.33	15.33
11-01	**99.8**	11.08	9.27	**133.0**	14.78	12.36	**166.3**	18.47	15.44
11-02	**100.5**	11.17	9.34	**134.0**	14.89	12.45	**167.5**	18.61	15.56
11-03	**101.3**	11.25	9.41	**135.0**	15.00	12.54	**168.8**	18.75	15.68
11-04	**102.0**	11.33	9.48	**136.0**	15.11	12.63	**170.0**	18.89	15.79
11-05	**102.8**	11.42	9.55	**137.0**	15.22	12.73	**171.3**	19.03	15.91
11-06	**103.5**	11.50	9.62	**138.0**	15.33	12.82	**172.5**	19.17	16.03
11-07	**104.3**	11.58	9.69	**139.0**	15.44	12.91	**173.8**	19.31	16.14
11-08	**105.0**	11.67	9.76	**140.0**	15.56	13.01	**175.0**	19.44	16.26
11-09	**105.8**	11.75	9.82	**141.0**	15.67	13.10	**176.3**	19.58	16.37
11-10	**106.5**	11.83	9.89	**142.0**	15.78	13.19	**177.5**	19.72	16.49
11-11	**107.3**	11.92	9.96	**143.0**	15.89	13.28	**178.8**	19.86	16.61
12-00	**108.0**	12.00	10.03	**144.0**	16.00	13.38	**180.0**	20.00	16.72
12-01	**108.8**	12.08	10.10	**145.0**	16.11	13.47	**181.3**	20.14	16.84
12-02	**109.5**	12.17	10.17	**146.0**	16.22	13.56	**182.5**	20.28	16.95
12-03	**110.3**	12.25	10.24	**147.0**	16.33	13.66	**183.8**	20.42	17.07
12-04	**111.0**	12.33	10.31	**148.0**	16.44	13.75	**185.0**	20.56	17.19
12-05	**111.8**	12.42	10.38	**149.0**	16.56	13.84	**186.3**	20.69	17.30
12-06	**112.5**	12.50	10.45	**150.0**	16.67	13.93	**187.5**	20.83	17.42
12-07	**113.3**	12.58	10.52	**151.0**	16.78	14.03	**188.8**	20.97	17.53
12-08	**114.0**	12.67	10.59	**152.0**	16.89	14.12	**190.0**	21.11	17.65
12-09	**114.8**	12.75	10.66	**153.0**	17.00	14.21	**191.3**	21.25	17.77
12-10	**115.5**	12.83	10.73	**154.0**	17.11	14.31	**192.5**	21.39	17.88
12-11	**116.3**	12.92	10.80	**155.0**	17.22	14.40	**193.8**	21.53	18.00
13-00	**117.0**	13.00	10.87	**156.0**	17.33	14.49	**195.0**	21.67	18.12
13-01	**117.8**	13.08	10.94	**157.0**	17.44	14.59	**196.3**	21.81	18.23
13-02	**118.5**	13.17	11.01	**158.0**	17.56	14.68	**197.5**	21.94	18.35
13-03	**119.3**	13.25	11.08	**159.0**	17.67	14.77	**198.8**	22.08	18.46
13-04	**120.0**	13.33	11.15	**160.0**	17.78	14.86	**200.0**	22.22	18.58
13-05	**120.8**	13.42	11.22	**161.0**	17.89	14.96	**201.3**	22.36	18.70
13-06	**121.5**	13.50	11.29	**162.0**	18.00	15.05	**202.5**	22.50	18.81
13-07	**122.3**	13.58	11.36	**163.0**	18.11	15.14	**203.8**	22.64	18.93
13-08	**123.0**	13.67	11.43	**164.0**	18.22	15.24	**205.0**	22.78	19.04
13-09	**123.8**	13.75	11.50	**165.0**	18.33	15.33	**206.3**	22.92	19.16
13-10	**124.5**	13.83	11.57	**166.0**	18.44	15.42	**207.5**	23.06	19.28
13-11	**125.3**	13.92	11.64	**167.0**	18.56	15.52	**208.8**	23.19	19.39

15.3.1 Computing Square Yards and Square Meters of Carpet (6 to 81 Feet in Length From Rolls 9, 12, and 15 Feet Wide) (Continued)

LENGTH	9 FEET			12 FEET			15 FEET		
FT. IN.	SQ FT	SQ YDS	SQ M	SQ FT	SQ YDS	SQ M	SQ FT	SQ YDS	SQ M
14-00	126.0	14.00	11.71	168.0	18.67	15.61	210.0	23.33	19.51
14-01	126.8	14.08	11.78	169.0	18.78	15.70	211.3	23.47	19.63
14-02	127.5	14.17	11.84	170.0	18.89	15.79	212.5	23.61	19.74
14-03	128.3	14.25	11.91	171.0	19.00	15.89	213.8	23.75	19.86
14-04	129.0	14.33	11.98	172.0	19.11	15.98	215.0	23.89	19.97
14-05	129.8	14.42	12.05	173.0	19.22	16.07	216.3	24.03	20.09
14-06	130.5	14.50	12.12	174.0	19.33	16.16	217.5	24.17	20.21
14-07	131.3	14.58	12.19	175.0	19.44	16.26	218.8	24.31	20.32
14-08	132.0	14.67	12.26	176.0	19.56	16.35	220.0	24.44	20.44
14-09	132.8	14.75	12.33	177.0	19.67	16.44	221.3	24.58	20.55
14-10	133.5	14.83	12.40	178.0	19.78	16.54	222.5	24.72	20.67
14-11	134.3	14.92	12.47	179.0	19.89	16.63	223.8	24.86	20.79
15-00	135.0	15.00	12.54	180.0	20.00	16.72	225.0	25.00	20.90
15-01	135.8	15.08	12.61	181.0	20.11	16.81	226.3	25.14	21.02
15-02	136.5	15.17	12.68	182.0	20.22	16.91	227.5	25.28	21.13
15-03	137.3	15.25	12.75	183.0	20.33	17.00	228.8	25.42	21.25
15-04	138.0	15.33	12.82	184.0	20.44	17.09	230.0	25.56	21.37
15-05	138.8	15.42	12.89	185.0	20.56	17.19	231.3	25.69	21.48
15-06	139.5	15.50	12.96	186.0	20.67	17.28	232.5	25.83	21.60
15-07	140.3	15.58	13.03	187.0	20.78	17.37	233.8	25.97	21.72
15-08	141.0	15.67	13.10	188.0	20.89	17.47	235.0	26.11	21.83
15-09	141.8	15.75	13.17	189.0	21.00	17.56	236.3	26.25	21.95
15-10	142.5	15.83	13.24	190.0	21.11	17.65	237.5	26.39	22.06
15-11	143.3	15.92	13.31	191.0	21.22	17.74	238.8	26.53	22.18
16-00	144.0	16.00	13.38	192.0	21.33	17.84	240.0	26.67	22.30
16-01	144.8	16.08	13.45	193.0	21.44	17.93	241.3	26.81	22.41
16-02	145.5	16.17	13.52	194.0	21.56	18.02	242.5	26.94	22.53
16-03	146.3	16.25	13.59	195.0	21.67	18.12	243.8	27.08	22.64
16-04	147.0	16.33	13.66	196.0	21.78	18.21	245.0	27.22	22.76
16-05	147.8	16.42	13.73	197.0	21.89	18.30	246.3	27.36	22.88
16-06	148.5	16.50	13.80	198.0	22.00	18.39	247.5	27.50	22.99
16-07	149.3	16.58	13.87	199.0	22.11	18.49	248.8	27.64	23.11
16-08	150.0	16.67	13.94	200.0	22.22	18.58	250.0	27.78	23.23
16-09	150.8	16.75	14.00	201.0	22.33	18.67	251.3	27.92	23.34
16-10	151.5	16.83	14.07	202.0	22.44	18.77	252.5	28.06	23.46
16-11	152.3	16.92	14.14	203.0	22.56	18.86	253.8	28.19	23.57
17-00	153.0	17.00	14.21	204.0	22.67	18.95	255.0	28.33	23.69
17-01	153.8	17.08	14.28	205.0	22.78	19.04	256.3	28.47	23.81
17-02	154.5	17.17	14.35	206.0	22.89	19.14	257.5	28.61	23.92
17-03	155.3	17.25	14.42	207.0	23.00	19.23	258.8	28.75	24.04
17-04	156.0	17.33	14.49	208.0	23.11	19.32	260.0	28.89	24.15
17-05	156.8	17.42	14.56	209.0	23.22	19.42	261.3	29.03	24.27
17-06	157.5	17.50	14.63	210.0	23.33	19.51	262.5	29.17	24.39
17-07	158.3	17.58	14.70	211.0	23.44	19.60	263.8	29.31	24.50
17-08	159.0	17.67	14.77	212.0	23.56	19.69	265.0	29.44	24.62
17-09	159.8	17.75	14.84	213.0	23.67	19.79	266.3	29.58	24.73
17-10	160.5	17.83	14.91	214.0	23.78	19.88	267.5	29.72	24.85
17-11	161.3	17.92	14.98	215.0	23.89	19.97	268.8	29.86	24.97

15.3.1 Computing Square Yards and Square Meters of Carpet (6 to 81 Feet in Length From Rolls 9, 12, and 15 Feet Wide) (Continued)

LENGTH	9 FEET			12 FEET			15 FEET		
FT. IN.	SQ FT	SQ YDS	SQ M	SQ FT	SQ YDS	SQ M	SQ FT	SQ YDS	SQ M
18-00	162.0	18.00	15.05	216.0	24.00	20.07	270.0	30.00	25.08
18-01	162.8	18.08	15.12	217.0	24.11	20.16	271.3	30.14	25.20
18-02	163.5	18.17	15.19	218.0	24.22	20.25	272.5	30.28	25.32
18-03	164.3	18.25	15.26	219.0	24.33	20.35	273.8	30.42	25.43
18-04	165.0	18.33	15.33	220.0	24.44	20.44	275.0	30.56	25.55
18-05	165.8	18.42	15.40	221.0	24.56	20.53	276.3	30.69	25.66
18-06	166.5	18.50	15.47	222.0	24.67	20.62	277.5	30.83	25.78
18-07	167.3	18.58	15.54	223.0	24.78	20.72	278.8	30.97	25.90
18-08	168.0	18.67	15.61	224.0	24.89	20.81	280.0	31.11	26.01
18-09	168.8	18.75	15.68	225.0	25.00	20.90	281.3	31.25	26.13
18-10	169.5	18.83	15.75	226.0	25.11	21.00	282.5	31.39	26.24
18-11	170.3	18.92	15.82	227.0	25.22	21.09	283.8	31.53	26.36
19-00	171.0	19.00	15.89	228.0	25.33	21.18	285.0	31.67	26.48
19-01	171.8	19.08	15.96	229.0	25.44	21.27	286.3	31.81	26.59
19-02	172.5	19.17	16.03	230.0	25.56	21.37	287.5	31.94	26.71
19-03	173.3	19.25	16.09	231.0	25.67	21.46	288.8	32.08	26.83
19-04	174.0	19.33	16.16	232.0	25.78	21.55	290.0	32.22	26.94
19-05	174.8	19.42	16.23	233.0	25.89	21.65	291.3	32.36	27.06
19-06	175.5	19.50	16.30	234.0	26.00	21.74	292.5	32.50	27.17
19-07	176.3	19.58	16.37	235.0	26.11	21.83	293.8	32.64	27.29
19-08	177.0	19.67	16.44	236.0	26.22	21.92	295.0	32.78	27.41
19-09	177.8	19.75	16.51	237.0	26.33	22.02	296.3	32.92	27.52
19-10	178.5	19.83	16.58	238.0	26.44	22.11	297.5	33.06	27.64
19-11	179.3	19.92	16.65	239.0	26.56	22.20	298.8	33.19	27.75
20-00	180.0	20.00	16.72	240.0	26.67	22.30	300.0	33.33	27.87
20-01	180.8	20.08	16.79	241.0	26.78	22.39	301.3	33.47	27.98
20-02	181.5	20.17	16.86	242.0	26.89	22.48	302.5	33.61	28.10
20-03	182.3	20.25	16.93	243.0	27.00	22.57	303.8	33.75	28.22
20-04	183.0	20.33	17.00	244.0	27.11	22.67	305.0	33.89	28.33
20-05	183.8	20.42	17.07	245.0	27.22	22.76	306.3	34.03	28.45
20-06	184.5	20.50	17.14	246.0	27.33	22.85	307.5	34.17	28.57
20-07	185.3	20.58	17.21	247.0	27.44	22.95	308.8	34.31	28.68
20-08	186.0	20.67	17.28	248.0	27.56	23.04	310.0	34.44	28.80
20-09	186.8	20.75	17.35	249.0	27.67	23.13	311.3	34.58	28.92
20-10	187.5	20.83	17.42	250.0	27.78	23.23	312.5	34.72	29.03
20-11	188.3	20.92	17.49	251.0	27.89	23.32	313.8	34.86	29.15
21-00	189.0	21.00	17.55	252.0	28.00	23.41	315.0	35.00	29.26
21-01	189.8	21.08	17.63	253.0	28.11	23.50	316.3	35.14	29.38
21-02	190.5	21.17	17.70	254.0	28.22	23.60	317.5	35.28	29.50
21-03	191.3	21.25	17.77	255.0	28.33	23.69	318.8	35.42	29.61
21-04	192.0	21.33	17.84	256.0	28.44	23.78	320.0	35.56	29.73
21-05	192.8	21.42	17.91	257.0	28.56	23.88	321.3	35.69	29.84
21-06	193.5	21.50	17.98	258.0	28.67	23.97	322.5	35.83	29.96
21-07	194.3	21.58	18.05	259.0	28.78	24.06	323.8	35.97	30.08
21-08	195.0	21.67	18.12	260.0	28.89	24.16	325.0	36.11	30.19
21-09	195.8	21.75	18.19	261.0	29.00	24.25	326.3	36.25	30.31
21-10	196.5	21.83	18.25	262.0	29.11	24.34	327.5	36.39	30.43
21-11	197.3	21.92	18.32	263.0	29.22	24.43	328.8	36.53	30.54

15.3.1 Computing Square Yards and Square Meters of Carpet (6 to 81 Feet in Length From Rolls 9, 12, and 15 Feet Wide) (Continued)

LENGTH	9 FEET			12 FEET			15 FEET		
FT. IN.	SQ FT	SQ YDS	SQ M	SQ FT	SQ YDS	SQ M	SQ FT	SQ YDS	SQ M
22-00	198.0	22.00	18.39	264.0	29.33	24.53	330.0	36.67	30.66
22-01	198.8	22.08	18.46	265.0	29.44	24.62	331.3	36.81	30.77
22-02	199.5	22.17	18.53	266.0	29.56	24.71	332.5	36.94	30.89
22-03	200.3	22.25	18.60	267.0	29.67	24.80	333.8	37.08	31.01
22-04	201.0	22.33	18.67	268.0	29.78	24.90	335.0	37.22	31.12
22-05	201.8	22.42	18.74	269.0	29.89	24.99	336.3	37.36	31.24
22-06	202.5	22.50	18.81	270.0	30.00	25.08	337.5	37.50	31.35
22-07	203.3	22.58	18.88	271.0	30.11	25.18	338.8	37.64	31.47
22-08	204.0	22.67	18.95	272.0	30.22	25.27	340.0	37.78	31.59
22-09	204.8	22.75	19.02	273.0	30.33	25.36	341.3	37.92	31.70
22-10	205.5	22.83	19.09	274.0	30.44	25.45	342.5	38.06	31.82
22-11	206.3	22.92	19.16	275.0	30.56	25.55	343.8	38.19	31.93
23-00	207.0	23.00	19.23	276.0	30.67	25.64	345.0	38.33	32.05
23-01	207.8	23.08	19.30	277.0	30.78	25.73	346.3	38.47	32.17
23-02	208.5	23.17	19.37	278.0	30.89	25.83	347.5	38.61	32.28
23-03	209.3	23.25	19.44	279.0	31.00	25.92	348.8	38.75	32.40
23-04	210.0	23.33	19.51	280.0	31.11	26.01	350.0	38.89	32.52
23-05	210.8	23.42	19.58	281.0	31.22	26.11	351.3	39.03	32.63
23-06	211.5	23.50	19.65	282.0	31.33	26.20	352.5	39.17	32.75
23-07	212.3	23.58	19.72	283.0	31.44	26.29	353.8	39.31	32.86
23-08	213.0	23.67	19.79	284.0	31.56	26.38	355.0	39.44	32.98
23-09	213.8	23.75	19.86	285.0	31.67	26.48	356.3	39.58	33.10
23-10	214.5	23.83	19.93	286.0	31.78	26.57	357.5	39.72	33.21
23-11	215.3	23.92	20.00	287.0	31.89	26.66	358.8	39.86	33.33
24-00	216.0	24.00	20.07	288.0	32.00	26.76	360.0	40.00	33.44
24-01	216.8	24.08	20.14	289.0	32.11	26.85	361.3	40.14	33.56
24-02	217.5	24.17	20.21	290.0	32.22	26.94	362.5	40.28	33.68
24-03	218.3	24.25	20.28	291.0	32.33	27.03	363.8	40.42	33.79
24-04	219.0	24.33	20.35	292.0	32.44	27.13	365.0	40.56	33.91
24-05	219.8	24.42	20.41	293.0	32.56	27.22	366.3	40.69	34.03
24-06	220.5	24.50	20.48	294.0	32.67	27.31	367.5	40.83	34.14
24-07	221.3	24.58	20.55	295.0	32.78	27.41	368.8	40.97	34.26
24-08	222.0	24.67	20.62	296.0	32.89	27.50	370.0	41.11	34.37
24-09	222.8	24.75	20.69	297.0	33.00	27.59	371.3	41.25	34.49
24-10	223.5	24.83	20.76	298.0	33.11	27.68	372.5	41.39	34.61
24-11	224.3	24.92	20.83	299.0	33.22	27.78	373.8	41.53	34.72
25-00	225.0	25.00	20.90	300.0	33.33	27.87	375.0	41.67	34.84
25-01	225.8	25.08	20.97	301.0	33.44	27.96	376.3	41.81	34.95
25-02	226.5	25.17	21.04	302.0	33.56	28.06	377.5	41.94	35.07
25-03	227.3	25.25	21.11	303.0	33.67	28.15	378.8	42.08	35.19
25-04	228.0	25.33	21.18	304.0	33.78	28.24	380.0	42.22	35.30
25-05	228.8	25.42	21.25	305.0	33.89	28.33	381.3	42.36	35.42
25-06	229.5	25.50	21.32	306.0	34.00	28.43	382.5	42.50	35.53
25-07	230.3	25.58	21.39	307.0	34.11	28.52	383.8	42.64	35.65
25-08	231.0	25.67	21.46	308.0	34.22	28.61	385.0	42.78	35.77
25-09	231.8	25.75	21.53	309.0	34.33	28.71	386.3	42.92	35.88
25-10	232.5	25.83	21.60	310.0	34.44	28.80	387.5	43.06	36.00
25-11	233.3	25.92	21.67	311.0	34.56	28.89	388.8	43.19	36.12

15.3.1 Computing Square Yards and Square Meters of Carpet (6 to 81 Feet in Length From Rolls 9, 12, and 15 Feet Wide) (Continued)

LENGTH	9 FEET			12 FEET			15 FEET		
FT. IN.	SQ FT	SQ YDS	SQ M	SQ FT	SQ YDS	SQ M	SQ FT	SQ YDS	SQ M
26-00	234.0	26.00	21.74	312.0	34.67	28.99	390.0	43.33	36.23
26-01	234.8	26.08	21.81	313.0	34.78	29.08	391.3	43.47	36.35
26-02	235.5	26.17	21.88	314.0	34.89	29.17	392.5	43.61	36.46
26-03	236.3	26.25	21.95	315.0	35.00	29.26	393.8	43.75	36.58
26-04	237.0	26.33	22.02	316.0	35.11	29.36	395.0	43.89	36.70
26-05	237.8	26.42	22.09	317.0	35.22	29.45	396.3	44.03	36.81
26-06	238.5	26.50	22.16	318.0	35.33	29.54	397.5	44.17	36.93
26-07	239.3	26.58	22.23	319.0	35.44	29.64	398.8	44.31	37.04
26-08	240.0	26.67	22.30	320.0	35.56	29.73	400.0	44.44	37.16
26-09	240.8	26.75	22.37	321.0	35.67	29.82	401.3	44.58	37.28
26-10	241.5	26.83	22.44	322.0	35.78	29.91	402.5	44.72	37.39
26-11	242.3	26.92	22.51	323.0	35.89	30.01	403.8	44.86	37.51
27-00	243.0	27.00	22.57	324.0	36.00	30.10	405.0	45.00	37.63
27-01	243.8	27.08	22.64	325.0	36.11	30.19	406.3	45.14	37.74
27-02	244.5	27.17	22.71	326.0	36.22	30.29	407.5	45.28	37.86
27-03	245.3	27.25	22.78	327.0	36.33	30.38	408.8	45.42	37.97
27-04	246.0	27.33	22.85	328.0	36.44	30.47	410.0	45.56	38.09
27-05	246.8	27.42	22.92	329.0	36.56	30.56	411.3	45.69	38.21
27-06	247.5	27.50	22.99	330.0	36.67	30.66	412.5	45.83	38.32
27-07	248.3	27.58	23.06	331.0	36.78	30.75	413.8	45.97	38.44
27-08	249.0	27.67	23.13	332.0	36.89	30.84	415.0	46.11	38.55
27-09	249.8	27.75	23.20	333.0	37.00	30.94	416.3	46.25	38.67
27-10	250.5	27.83	23.27	334.0	37.11	31.03	417.5	46.39	38.79
27-11	251.3	27.92	23.34	335.0	37.22	31.12	418.8	46.53	38.90
28-00	252.0	28.00	23.41	336.0	37.33	31.21	420.0	46.67	39.02
28-01	252.8	28.08	23.48	337.0	37.44	31.31	421.3	46.81	39.13
28-02	253.5	28.17	23.55	338.0	37.56	31.40	422.5	46.94	39.25
28-03	254.3	28.25	23.62	339.0	37.67	31.49	423.8	47.08	39.37
28-04	255.0	28.33	23.69	340.0	37.78	31.59	425.0	47.22	39.48
28-05	255.8	28.42	23.76	341.0	37.89	31.68	426.3	47.36	39.60
28-06	256.5	28.50	23.83	342.0	38.00	31.77	427.5	47.50	39.72
28-07	257.3	28.58	23.90	343.0	38.11	31.87	428.8	47.64	39.83
28-08	258.0	28.67	23.97	344.0	38.22	31.96	430.0	47.78	39.95
28-09	258.8	28.75	24.04	345.0	38.33	32.05	431.3	47.92	40.06
28-10	259.5	28.83	24.11	346.0	38.44	32.14	432.5	48.06	40.18
28-11	260.3	28.92	24.18	347.0	38.56	32.24	433.8	48.19	40.30
29-00	261.0	29.00	24.25	348.0	38.67	32.33	435.0	48.33	40.41
29-01	261.8	29.08	24.32	349.0	38.78	32.42	436.3	48.47	40.53
29-02	262.5	29.17	24.39	350.0	38.89	32.52	437.5	48.61	40.64
29-03	263.3	29.25	24.46	351.0	39.00	32.61	438.8	48.75	40.76
29-04	264.0	29.33	24.53	352.0	39.11	32.70	440.0	48.89	40.88
29-05	264.8	29.42	24.60	353.0	39.22	32.79	441.3	49.03	40.99
29-06	265.5	29.50	24.67	354.0	39.33	32.89	442.5	49.17	41.11
29-07	266.3	29.58	24.73	355.0	39.44	32.98	443.8	49.31	41.23
29-08	267.0	29.67	24.80	356.0	39.56	33.07	445.0	49.44	41.34
29-09	267.8	29.75	24.87	357.0	39.67	32.17	446.3	49.58	41.46
29-10	268.5	29.83	24.94	358.0	39.78	33.26	447.5	49.72	41.57
29-11	269.3	29.92	25.01	359.0	39.89	33.35	448.8	49.86	41.69

15.3.1 Computing Square Yards and Square Meters of Carpet (6 to 81 Feet in Length From Rolls 9, 12, and 15 Feet Wide) (Continued)

LENGTH	9 FEET			12 FEET			15 FEET		
FT. IN.	SQ FT	SQ YDS	SQ M	SQ FT	SQ YDS	SQ M	SQ FT	SQ YDS	SQ M
30-00	270.0	30.00	25.08	360.0	40.00	33.44	450.0	50.00	41.81
30-01	270.8	30.08	25.15	361.0	40.11	33.54	451.3	50.14	41.92
30-02	271.5	30.17	25.22	362.0	40.22	33.63	452.5	50.28	42.04
30-03	272.3	30.25	25.29	363.0	40.33	33.72	453.8	50.42	42.15
30-04	273.0	30.33	25.36	364.0	40.44	33.82	455.0	50.56	42.27
30-05	273.8	30.42	25.43	365.0	40.56	33.91	456.3	50.69	42.39
30-06	274.5	30.50	25.50	366.0	40.67	34.00	457.5	50.83	42.50
30-07	275.3	30.58	25.57	367.0	40.78	34.09	458.8	50.97	42.62
30-08	276.0	30.67	25.64	368.0	40.89	34.19	460.0	51.11	42.73
30-09	276.8	30.75	25.71	369.0	41.00	34.28	461.3	51.25	42.85
30-10	277.5	30.83	25.78	370.0	41.11	34.37	462.5	51.39	42.97
30-11	278.3	30.92	25.85	371.0	41.22	34.47	463.8	51.53	43.08
31-00	279.0	31.00	25.92	372.0	41.33	34.56	465.0	51.67	43.20
31-01	279.8	31.08	25.99	373.0	41.44	34.65	466.3	51.81	43.32
31-02	280.5	31.17	26.06	374.0	41.56	34.75	467.5	51.94	43.43
31-03	281.3	31.25	26.13	375.0	41.67	34.84	468.8	52.08	43.55
31-04	282.0	31.33	26.20	376.0	41.78	34.93	470.0	52.22	43.66
31-05	282.8	31.42	26.27	377.0	41.89	35.02	471.3	52.36	43.78
31-06	283.5	31.50	26.34	378.0	42.00	35.12	472.5	52.50	43.90
31-07	284.3	31.58	26.41	379.0	42.11	35.21	473.8	52.64	44.01
31-08	285.0	31.67	26.48	380.0	42.22	35.30	475.0	52.78	44.13
31-09	285.8	31.75	26.55	381.0	42.33	35.40	476.3	52.92	44.24
31-10	286.5	31.83	26.62	382.0	42.44	35.49	477.5	53.06	44.36
31-11	287.3	31.92	26.69	383.0	42.56	35.58	478.8	53.19	44.48
32-00	288.0	32.00	26.76	384.0	42.67	35.68	480.0	53.33	44.59
32-01	288.8	32.08	26.83	385.0	42.78	35.77	481.3	53.47	44.71
32-02	289.5	32.17	26.89	386.0	42.89	35.86	482.5	53.61	44.83
32-03	290.3	32.25	26.96	387.0	43.00	35.95	483.8	53.75	44.94
32-04	291.0	32.33	27.03	388.0	43.11	36.05	485.0	53.89	45.06
32-05	291.8	32.42	27.10	389.0	43.22	36.14	486.3	54.03	45.17
32-06	292.5	32.50	27.17	390.0	43.33	36.23	487.5	54.17	45.29
32-07	293.3	32.58	27.24	391.0	43.44	36.32	488.8	54.31	45.41
32-08	294.0	32.67	27.31	392.0	43.56	36.42	490.0	54.44	45.52
32-09	294.8	32.75	27.38	393.0	43.67	36.51	491.3	54.58	45.64
32-10	295.5	32.83	27.45	394.0	43.78	36.60	492.5	54.72	45.75
32-11	296.3	32.92	27.52	395.0	43.89	36.70	493.8	54.86	45.87
33-00	297.0	33.00	27.59	396.0	44.00	36.79	495.0	55.00	45.99
33-01	297.8	33.08	27.66	397.0	44.11	36.88	496.3	55.14	46.10
33-02	298.5	33.17	27.73	398.0	44.22	36.97	497.5	55.28	46.22
33-03	299.3	33.25	27.80	399.0	44.33	37.07	498.8	55.42	46.33
33-04	300.0	33.33	27.87	400.0	44.44	37.16	500.0	55.56	46.45
33-05	300.8	33.42	27.94	401.0	44.56	37.25	501.3	55.69	46.57
33-06	301.5	33.50	28.01	402.0	44.67	37.35	502.5	55.83	46.68
33-07	302.3	33.58	28.08	403.0	44.78	37.44	503.8	55.97	46.80
33-08	303.0	33.67	28.15	404.0	44.89	37.53	505.0	56.11	46.92
33-09	303.8	33.75	28.22	405.0	45.00	37.63	506.3	56.25	47.03
33-10	304.5	33.83	28.29	406.0	45.11	37.72	507.5	56.39	47.15
33-11	305.3	33.92	28.36	407.0	45.22	37.81	508.8	56.53	47.26

15.3.1 Computing Square Yards and Square Meters of Carpet (6 to 81 Feet in Length From Rolls 9, 12, and 15 Feet Wide) (Continued)

LENGTH	9 FEET			12 FEET			15 FEET		
FT. IN.	SQ FT	SQ YDS	SQ M	SQ FT	SQ YDS	SQ M	SQ FT	SQ YDS	SQ M
34-00	306.0	34.00	28.43	408.0	45.33	37.90	510.0	56.67	47.38
34-01	306.8	34.08	28.50	409.0	45.44	38.00	511.3	56.81	47.50
34-02	307.5	34.17	28.57	410.0	45.56	38.09	512.5	56.94	47.61
34-03	308.3	34.25	28.64	411.0	45.67	38.18	513.8	57.08	47.73
34-04	309.0	34.33	28.71	412.0	45.78	38.28	515.0	57.22	47.84
34-05	309.8	34.42	28.78	413.0	45.89	38.37	516.3	57.36	47.96
34-06	310.5	34.50	28.85	414.0	46.00	38.46	517.5	57.50	48.08
34-07	311.3	34.58	28.92	415.0	46.11	38.55	518.8	57.64	48.19
34-08	312.0	34.67	28.99	416.0	46.22	38.65	520.0	57.78	48.31
34-09	312.8	34.75	29.05	417.0	46.33	38.74	521.3	57.92	48.43
34-10	313.5	34.83	29.12	418.0	46.44	38.83	522.5	58.06	48.52
34-11	314.3	34.92	29.19	419.0	46.56	38.93	523.8	58.19	48.66
35-00	315.0	35.00	29.26	420.0	46.67	39.02	525.0	58.33	48.77
35-01	315.8	35.08	29.33	421.0	46.78	39.11	526.3	58.47	48.89
35-02	316.5	35.17	29.40	422.0	46.89	39.20	527.5	58.61	49.01
35-03	317.3	35.25	29.47	423.0	47.00	39.30	528.8	58.75	49.12
35-04	318.0	35.33	29.54	424.0	47.11	39.39	530.0	58.89	49.24
35-05	318.8	35.42	29.61	425.0	47.22	39.48	531.3	59.03	49.35
35-06	319.5	35.50	29.68	426.0	47.33	39.58	532.5	59.17	49.47
35-07	320.3	35.58	29.75	427.0	47.44	39.67	533.8	59.31	49.59
35-08	321.0	35.67	29.82	428.0	47.56	39.76	535.0	59.44	49.70
35-09	321.8	35.75	29.89	429.0	47.67	39.85	536.3	59.58	49.82
35-10	322.5	35.83	29.96	430.0	47.78	39.95	537.5	59.72	49.93
35-11	323.3	35.92	30.03	431.0	47.89	40.04	538.8	59.86	50.05
36-00	324.0	36.00	30.10	432.0	48.00	40.13	540.0	60.00	50.17
36-01	324.8	36.08	30.17	433.0	48.11	40.23	541.3	60.14	50.28
36-02	325.5	36.17	30.24	434.0	48.22	40.32	542.5	60.28	50.40
36-03	326.3	36.25	30.31	435.0	48.33	40.41	543.8	60.42	50.52
36-04	327.0	36.33	30.38	436.0	48.44	40.51	545.0	60.56	50.63
36-05	327.8	36.42	30.45	437.0	48.56	40.60	546.3	60.69	50.75
36-06	328.5	36.50	30.52	438.0	48.67	40.69	547.5	60.83	50.86
36-07	329.3	36.58	30.59	439.0	48.78	40.78	548.8	60.97	50.98
36-08	330.0	36.67	30.66	440.0	48.89	40.88	550.0	61.11	51.10
36-09	330.8	36.75	30.73	441.0	49.00	40.97	551.3	61.25	51.21
36-10	331.5	36.83	30.80	442.0	49.11	41.06	552.5	61.39	51.33
36-11	332.3	36.92	30.87	443.0	49.22	41.16	553.8	61.53	51.44
37-00	333.0	37.00	30.94	444.0	49.33	41.25	555.0	61.67	51.56
37-01	333.8	37.08	31.01	445.0	49.44	41.34	556.3	61.81	51.68
37-02	334.5	37.17	31.08	446.0	49.56	41.43	557.5	61.94	51.79
37-03	335.3	37.25	31.15	447.0	49.67	41.53	558.8	62.08	51.91
37-04	336.0	37.33	31.21	448.0	49.78	41.62	560.0	62.22	52.03
37-05	336.8	37.42	31.28	449.0	49.89	41.71	561.3	62.36	52.14
37-06	337.5	37.50	31.35	450.0	50.00	41.81	562.5	62.50	52.26
37-07	338.3	37.58	31.42	451.0	50.11	41.90	563.8	62.64	52.37
37-08	339.0	37.67	31.49	452.0	50.22	42.00	565.0	62.78	52.49
37-09	339.8	37.75	31.56	453.0	50.33	42.08	566.3	62.92	52.61
37-10	340.5	37.83	31.63	454.0	50.44	42.18	567.5	63.06	52.72
37-11	341.3	37.92	31.70	455.0	50.56	42.27	568.8	63.19	52.84

15.3.1 Computing Square Yards and Square Meters of Carpet (6 to 81 Feet in Length From Rolls 9, 12, and 15 Feet Wide) (Continued)

LENGTH	9 FEET			12 FEET			15 FEET		
FT. IN.	SQ FT	SQ YDS	SQ M	SQ FT	SQ YDS	SQ M	SQ FT	SQ YDS	SQ M
38-00	342.0	38.00	31.77	456.0	50.67	42.36	570.0	63.33	52.95
38-01	342.8	38.08	31.84	457.0	50.78	42.46	571.3	63.47	53.07
38-02	343.5	38.17	31.91	458.0	50.89	42.55	572.5	63.61	53.19
38-03	344.3	38.25	31.98	459.0	51.00	42.64	573.8	63.75	53.30
38-04	345.0	38.33	32.05	460.0	51.11	42.73	575.0	63.89	53.42
38-05	345.8	38.42	32.12	461.0	51.22	42.83	576.3	64.03	53.53
38-06	346.5	38.50	32.19	462.0	51.33	42.92	577.5	64.17	53.65
38-07	347.3	38.58	32.26	463.0	51.44	43.01	578.8	64.31	53.77
38-08	348.0	38.67	32.33	464.0	51.56	43.11	580.0	64.44	53.88
38-09	348.8	38.75	32.40	465.0	51.67	43.20	581.3	64.58	54.00
38-10	349.5	38.83	32.47	466.0	51.78	43.29	582.5	64.72	54.17
38-11	350.3	38.92	32.54	467.0	51.89	43.39	583.8	64.86	54.23
39-00	351.0	39.00	32.61	468.0	52.00	43.48	585.0	65.00	54.35
39-01	351.8	39.08	32.68	469.0	52.11	43.57	586.3	65.14	54.46
39-02	352.5	39.17	32.75	470.0	52.22	43.66	587.5	65.28	54.58
39-03	353.3	39.25	32.82	471.0	52.33	43.76	588.8	65.42	54.70
39-04	354.0	39.33	32.89	472.0	52.44	43.85	590.0	65.56	54.81
39-05	354.8	39.42	32.96	473.0	52.56	43.94	591.3	65.69	54.93
39-06	355.5	39.50	33.03	474.0	52.67	44.04	592.5	65.83	55.04
39-07	356.3	39.58	33.10	475.0	52.78	44.13	593.8	65.97	55.16
39-08	357.0	39.67	33.17	476.0	52.89	44.22	595.0	66.11	55.28
39-09	357.8	39.75	33.24	477.0	53.00	44.31	596.3	66.25	55.39
39-10	358.5	39.83	33.31	478.0	53.11	44.41	597.5	66.39	55.51
39-11	359.3	39.92	33.37	479.0	53.22	44.50	598.8	66.53	55.63
40-00	360.0	40.00	33.44	480.0	53.33	44.59	600.0	66.67	55.74
40-01	360.8	40.08	33.51	481.0	53.44	44.69	601.3	66.81	55.86
40-02	361.5	40.17	33.58	482.0	53.56	44.78	602.5	66.94	55.97
40-03	362.3	40.25	33.65	483.0	53.67	44.87	603.8	67.08	56.09
40-04	363.0	40.33	33.72	484.0	53.78	44.96	605.0	67.22	56.21
40-05	363.8	40.42	33.79	485.0	53.89	45.06	606.3	67.36	56.32
40-06	364.5	40.50	33.86	486.0	54.00	45.15	607.5	67.50	56.44
40-07	365.3	40.58	33.93	487.0	54.11	45.24	608.8	67.64	56.55
40-08	366.0	40.67	34.00	488.0	54.22	45.34	610.0	67.78	56.67
40-09	366.8	40.75	34.07	489.0	54.33	45.43	611.3	67.92	56.79
40-10	367.5	40.83	34.14	490.0	54.44	45.52	612.5	68.06	56.90
40-11	368.3	40.92	34.21	491.0	54.56	45.61	613.8	68.19	57.02
41-00	369.0	41.00	34.28	492.0	54.67	45.71	615.0	68.33	57.13
41-01	369.8	41.08	34.35	493.0	54.78	45.80	616.3	68.47	57.25
41-02	370.5	41.17	34.42	494.0	54.89	45.89	617.5	68.61	57.37
41-03	371.3	41.25	34.49	495.0	55.00	45.99	618.8	68.75	57.48
41-04	372.0	41.33	34.56	496.0	55.11	46.08	620.0	68.89	57.60
41-05	372.8	41.42	34.63	497.0	55.22	46.17	621.3	69.03	57.72
41-06	373.5	41.50	34.70	498.0	55.33	46.27	622.5	69.17	57.83
41-07	374.3	41.58	34.77	499.0	55.44	46.36	623.8	69.31	57.95
41-08	375.0	41.67	34.84	500.0	55.56	46.45	625.0	69.44	58.06
41-09	375.8	41.75	34.91	501.0	55.67	46.54	626.3	69.58	58.18
41-10	376.5	41.83	34.98	502.0	55.78	46.64	627.5	69.72	58.30
41-11	377.3	41.92	35.05	503.0	55.89	46.73	628.8	69.86	58.41

15.3.1 Computing Square Yards and Square Meters of Carpet (6 to 81 Feet in Length From Rolls 9, 12, and 15 Feet Wide) (Continued)

LENGTH	9 FEET			12 FEET			15 FEET		
FT. IN.	SQ FT	SQ YDS	SQ M	SQ FT	SQ YDS	SQ M	SQ FT	SQ YDS	SQ M
42-00	378.0	42.00	35.12	504.0	56.00	46.82	630.0	70.00	58.53
42-01	378.8	42.08	35.19	505.0	56.11	46.92	631.3	70.14	58.64
42-02	379.5	42.17	35.26	506.0	56.22	47.01	632.5	70.28	58.76
42-03	380.3	42.25	35.33	507.0	56.33	47.10	633.8	70.42	58.88
42-04	381.0	42.33	35.40	508.0	56.44	47.19	635.0	70.56	58.99
42-05	381.8	42.42	35.47	509.0	56.56	47.29	636.3	70.69	59.11
42-06	382.5	42.50	35.53	510.0	56.67	47.38	637.5	70.83	59.23
42-07	383.3	42.58	35.60	511.0	56.78	47.47	638.8	70.97	59.34
42-08	384.0	42.67	35.67	512.0	56.89	47.57	640.0	71.11	59.46
42-09	384.8	42.75	35.74	513.0	57.00	47.66	641.3	71.25	59.57
42-10	385.5	42.83	35.81	514.0	57.11	47.75	642.5	71.39	59.69
42-11	386.3	42.92	35.88	515.0	57.22	47.84	643.8	71.53	59.81
43-00	387.0	43.00	35.95	516.0	57.33	47.94	645.0	71.67	59.92
43-01	387.8	43.08	36.02	517.0	57.44	48.03	646.3	71.81	60.04
43-02	388.5	43.17	36.09	518.0	57.56	48.12	647.5	71.94	60.15
43-03	389.3	43.25	36.16	519.0	57.67	48.22	648.8	72.08	60.27
43-04	390.0	43.33	36.23	520.0	57.78	48.31	650.0	72.22	60.39
43-05	390.8	43.42	36.30	521.0	57.89	48.40	651.3	72.36	60.50
43-06	391.5	43.50	36.37	522.0	58.00	48.49	652.5	72.50	60.62
43-07	392.3	43.58	36.44	523.0	58.11	48.59	653.8	72.64	60.73
43-08	393.0	43.67	36.51	524.0	58.22	48.68	655.0	72.78	60.85
43-09	393.8	43.75	36.58	525.0	58.33	48.77	656.3	72.92	60.97
43-10	394.5	43.83	36.65	526.0	58.44	48.87	657.5	73.06	61.08
43-11	395.3	43.92	36.72	527.0	58.56	48.96	658.8	73.19	61.20
44-00	396.0	44.00	36.79	528.0	58.67	49.05	660.0	73.33	61.32
44-01	396.8	44.08	36.86	529.0	58.78	49.15	661.3	73.47	61.43
44-02	397.5	44.17	36.93	530.0	58.89	49.24	662.5	73.61	61.55
44-03	398.3	44.25	37.00	531.0	59.00	49.33	663.8	73.75	61.66
44-04	399.0	44.33	37.07	532.0	59.11	49.42	665.0	73.89	61.78
44-05	399.8	44.42	37.14	533.0	59.22	49.52	666.3	74.03	61.90
44-06	400.5	44.50	37.21	534.0	59.33	49.61	667.5	74.17	62.01
44-07	401.3	44.58	37.28	535.0	59.44	49.70	668.8	74.31	62.13
44-08	402.0	44.67	37.35	536.0	59.56	49.80	670.0	74.44	62.24
44-09	402.8	44.75	37.42	537.0	59.67	49.89	671.3	74.58	62.36
44-10	403.5	44.83	37.49	538.0	59.78	49.98	672.5	74.72	62.48
44-11	404.3	44.92	37.56	539.0	59.89	50.07	673.8	74.86	62.59
45-00	405.0	45.00	37.63	540.0	60.00	50.17	675.0	75.00	62.71
45-01	405.8	45.08	37.69	541.0	60.11	50.26	676.3	75.14	62.83
45-02	406.5	45.17	37.76	542.0	60.22	50.35	677.5	75.28	62.94
45-03	407.3	45.25	37.83	543.0	60.33	50.45	678.8	75.42	63.06
45-04	408.0	45.33	37.90	544.0	60.44	50.54	680.0	75.56	63.17
45-05	408.8	45.42	37.97	545.0	60.56	50.63	681.3	75.69	63.29
45-06	409.5	45.50	38.04	546.0	60.67	50.72	682.5	75.83	63.41
45-07	410.3	45.58	38.11	547.0	60.78	50.82	683.8	75.97	63.52
45-08	411.0	45.67	38.18	548.0	60.89	50.91	685.0	76.11	63.64
45-09	411.8	45.75	38.25	549.0	61.00	51.00	686.3	76.25	63.75
45-10	412.5	45.83	38.32	550.0	61.11	51.10	687.5	76.39	63.87
45-11	413.3	45.92	38.39	551.0	61.22	51.19	688.8	76.53	63.99

15.3.1 Computing Square Yards and Square Meters of Carpet (6 to 81 Feet in Length From Rolls 9, 12, and 15 Feet Wide) (Continued)

LENGTH	9 FEET			12 FEET			15 FEET		
FT. IN.	SQ FT	SQ YDS	SQ M	SQ FT	SQ YDS	SQ M	SQ FT	SQ YDS	SQ M
46-00	414.0	46.00	38.46	552.0	61.33	51.28	690.0	76.67	64.10
46-01	414.8	46.08	38.53	553.0	61.44	51.37	691.3	76.81	64.22
46-02	415.5	46.17	38.60	554.0	61.56	51.47	692.5	76.94	64.33
46-03	416.3	46.25	38.67	555.0	61.67	51.56	693.8	77.08	64.45
46-04	417.0	46.33	38.74	556.0	61.78	51.65	695.0	77.22	64.57
46-05	417.8	46.42	38.81	557.0	61.89	51.75	696.3	77.36	64.68
46-06	418.5	46.50	38.88	558.0	62.00	51.84	697.5	77.50	64.80
46-07	419.3	46.58	38.95	559.0	62.11	51.93	698.8	77.64	64.92
46-08	420.0	46.67	39.02	560.0	62.22	52.03	700.0	77.78	65.03
46-09	420.8	46.75	39.09	561.0	62.33	52.12	701.3	77.92	65.15
46-10	421.5	46.83	39.16	562.0	62.44	52.21	702.5	78.06	65.26
46-11	422.3	46.92	39.23	563.0	62.56	52.30	703.8	78.19	65.38
47-00	423.0	47.00	39.30	564.0	62.67	52.40	705.0	78.33	65.50
47-01	423.8	47.08	39.37	565.0	62.78	52.49	706.3	78.47	65.61
47-02	424.5	47.17	39.44	566.0	62.89	52.58	707.5	78.61	65.73
47-03	425.3	47.25	39.51	567.0	63.00	52.68	708.8	78.75	65.84
47-04	426.0	47.33	39.58	568.0	63.11	52.77	710.0	78.89	65.92
47-05	426.8	47.42	39.65	569.0	63.22	52.86	711.3	79.03	66.08
47-06	427.5	47.50	39.72	570.0	63.33	52.95	712.5	79.17	66.19
47-07	428.3	47.58	39.79	571.0	63.44	53.05	713.8	79.31	66.30
47-08	429.0	47.67	39.85	572.0	63.56	53.14	715.0	79.44	66.43
47-09	429.8	47.75	39.92	573.0	63.67	53.23	716.3	79.58	66.54
47-10	430.5	47.83	39.99	574.0	63.78	53.33	717.5	79.72	66.66
47-11	431.3	47.92	40.06	575.0	63.89	53.42	718.8	79.86	66.77
48-00	432.0	48.00	40.13	576.0	64.00	53.51	720.0	80.00	66.89
48-01	432.8	48.08	40.20	577.0	64.11	53.60	721.3	80.14	67.01
48-02	433.5	48.17	40.27	578.0	64.22	53.70	722.5	80.28	67.12
48-03	434.3	48.25	40.34	579.0	64.33	53.79	723.8	80.42	67.24
48-04	435.0	48.33	40.41	580.0	64.44	53.88	725.0	80.56	67.35
48-05	435.8	48.42	40.48	581.0	64.56	53.98	726.3	80.69	67.47
48-06	436.5	48.50	40.55	582.0	64.67	54.07	727.5	80.83	67.59
48-07	437.3	48.58	40.62	583.0	64.78	54.16	728.8	80.97	67.70
48-08	438.0	48.67	40.69	584.0	64.89	54.25	730.0	81.11	67.82
48-09	438.8	48.75	40.76	585.0	65.00	54.35	731.3	81.25	67.93
48-10	439.5	48.83	40.83	586.0	65.11	54.44	732.5	81.39	68.05
48-11	440.3	48.92	40.90	587.0	65.22	54.53	733.8	81.53	68.17
49-00	441.0	49.00	40.97	588.0	65.33	54.63	735.0	81.67	68.28
49-01	441.8	49.08	41.04	589.0	65.44	54.72	736.3	81.81	68.40
49-02	442.5	49.17	41.11	590.0	65.56	54.81	737.5	81.94	68.52
49-03	443.3	49.25	41.18	591.0	65.67	54.91	738.8	82.08	68.63
49-04	444.0	49.33	41.25	592.0	65.78	55.00	740.0	82.22	68.75
49-05	444.8	49.42	41.32	593.0	65.89	55.09	741.3	82.36	68.86
49-06	445.5	49.50	41.39	594.0	66.00	55.18	742.5	82.50	68.98
49-07	446.3	49.58	41.46	595.0	66.11	55.28	743.8	82.64	69.10
49-08	447.0	49.67	41.53	596.0	66.22	55.37	745.0	82.78	69.21
49-09	447.8	49.75	41.60	597.0	66.33	55.46	746.3	82.92	69.33
49-10	448.5	49.83	41.67	598.0	66.44	55.56	747.5	83.06	69.44
49-11	449.3	49.92	41.74	599.0	66.56	55.65	748.8	83.19	69.56

15.3.1 Computing Square Yards and Square Meters of Carpet (6 to 81 Feet in Length From Rolls 9, 12, and 15 Feet Wide) (Continued)

LENGTH	9 FEET			12 FEET			15 FEET		
FT. IN.	SQ FT	SQ YDS	SQ M	SQ FT	SQ YDS	SQ M	SQ FT	SQ YDS	SQ M
50-00	450.0	50.00	41.81	600.0	66.67	55.74	750.0	83.33	69.68
50-01	450.8	50.08	41.88	601.0	66.78	55.83	751.3	83.47	69.79
50-02	451.5	50.17	41.95	602.0	66.89	55.93	752.5	83.61	69.91
50-03	452.3	50.25	42.01	603.0	67.00	56.02	753.8	83.75	70.03
50-04	453.0	50.33	42.08	604.0	67.11	56.11	755.0	83.89	70.14
50-05	453.8	50.42	42.15	605.0	67.22	56.21	756.3	84.03	70.26
50-06	454.5	50.50	42.22	606.0	67.33	56.30	757.5	84.17	70.37
50-07	455.3	50.58	42.29	607.0	67.44	56.39	758.8	84.31	70.49
50-08	456.0	50.67	42.36	608.0	67.56	56.48	760.0	84.44	70.61
50-09	456.8	50.75	42.43	609.0	67.67	56.58	761.3	84.58	70.72
50-10	457.5	50.83	42.50	610.0	67.78	56.67	762.5	84.72	70.84
50-11	458.3	50.92	42.57	611.0	67.89	56.76	763.8	84.86	70.95
51-00	459.0	51.00	42.64	612.0	68.00	56.86	765.0	85.00	71.07
51-01	459.8	51.08	42.71	613.0	68.11	56.95	766.3	85.14	71.19
51-02	460.5	51.17	42.78	614.0	68.22	57.04	767.5	85.28	71.30
51-03	461.3	51.25	42.85	615.0	68.33	57.13	768.8	85.42	71.42
51-04	462.0	51.33	42.92	616.0	68.44	57.23	770.0	85.56	71.53
51-05	462.8	51.42	42.99	617.0	68.56	57.32	771.3	85.69	71.65
51-06	463.5	51.50	43.06	618.0	68.67	57.41	772.5	85.83	71.77
51-07	464.3	51.58	43.13	619.0	68.78	57.51	773.8	85.97	71.88
51-08	465.0	51.67	43.20	620.0	68.89	57.60	775.0	86.11	72.00
51-09	465.8	51.75	43.27	621.0	69.00	57.69	776.3	86.25	72.12
51-10	466.5	51.83	43.34	622.0	69.11	57.79	777.5	86.39	72.23
51-11	467.3	51.92	43.41	623.0	69.22	57.88	778.8	86.53	72.35
52-00	468.0	52.00	43.48	624.0	69.33	57.97	780.0	86.67	72.46
52-01	468.8	52.08	43.55	625.0	69.44	58.06	781.3	86.81	72.58
52-02	469.5	52.17	43.62	626.0	69.56	58.16	782.5	86.94	72.70
52-03	470.3	52.25	43.69	627.0	69.67	58.25	783.8	87.08	72.81
52-04	471.0	52.33	43.76	628.0	69.78	58.34	785.0	87.22	72.93
52-05	471.8	52.42	43.83	629.0	69.89	58.44	786.3	87.36	73.04
52-06	472.5	52.50	43.90	630.0	70.00	58.53	787.5	87.50	73.16
52-07	473.3	52.58	43.97	631.0	70.11	58.62	788.8	87.64	73.28
52-08	474.0	52.67	44.04	632.0	70.22	58.71	790.0	87.78	73.39
52-09	474.8	52.75	44.11	633.0	70.33	58.81	791.3	87.92	73.51
52-10	475.5	52.83	44.17	634.0	70.44	58.90	792.5	88.06	73.63
52-11	476.3	52.92	44.24	635.0	70.56	58.99	793.8	88.19	73.74
53-00	477.0	53.00	44.31	636.0	70.67	59.09	795.0	88.33	73.86
53-01	477.8	53.08	44.38	637.0	70.78	59.18	796.3	88.47	73.97
53-02	478.5	53.17	44.45	638.0	70.89	59.27	797.5	88.61	74.09
53-03	479.3	53.25	44.52	639.0	71.00	59.36	798.8	88.75	74.21
53-04	480.0	53.33	44.59	640.0	71.11	59.46	800.0	88.89	74.32
53-05	480.8	53.42	44.66	641.0	71.22	59.55	801.3	89.03	74.44
53-06	481.5	53.50	44.73	642.0	71.33	59.64	802.5	89.17	74.55
53-07	482.3	53.58	44.80	643.0	71.44	59.74	803.8	89.31	74.67
53-08	483.0	53.67	44.87	644.0	71.56	59.83	805.0	89.44	74.79
53-09	483.8	53.75	44.94	645.0	71.67	59.99	806.3	89.58	74.90
53-10	484.5	53.83	45.01	646.0	71.78	60.01	807.5	89.72	75.02
53-11	485.3	53.92	45.08	647.0	71.89	60.11	808.8	89.86	75.13

15.3.1 Computing Square Yards and Square Meters of Carpet (6 to 81 Feet in Length From Rolls 9, 12, and 15 Feet Wide) (Continued)

LENGTH	9 FEET			12 FEET			15 FEET		
FT. IN.	SQ FT	SQ YDS	SQ M	SQ FT	SQ YDS	SQ M	SQ FT	SQ YDS	SQ M
54-00	486.0	54.00	45.15	648.0	72.00	60.20	810.0	90.00	75.25
54-01	486.8	54.08	45.22	649.0	72.11	60.29	811.3	90.14	75.37
54-02	487.5	54.17	45.29	650.0	72.22	60.39	812.5	90.28	75.48
54-03	488.3	54.25	45.36	651.0	72.33	60.48	813.8	90.42	75.60
54-04	489.0	54.33	45.43	652.0	72.44	60.57	815.0	90.56	75.72
54-05	489.8	54.42	45.50	653.0	72.56	60.67	816.3	90.69	75.83
54-06	490.5	54.50	45.57	654.0	72.67	60.76	817.5	90.83	75.95
54-07	491.3	54.58	45.64	655.0	72.78	60.85	818.8	90.97	76.06
54-08	492.0	54.67	45.71	656.0	72.89	60.94	820.0	91.11	76.18
54-09	492.8	54.75	45.78	657.0	73.00	61.04	821.3	91.25	76.30
54-10	493.5	54.83	45.85	658.0	73.11	61.13	822.5	91.39	76.41
54-11	494.3	54.92	45.92	659.0	73.22	61.22	823.8	91.53	76.53
55-00	495.0	55.00	45.99	660.0	73.33	61.32	825.0	91.67	76.64
55-01	495.8	55.08	46.06	661.0	73.44	61.41	826.3	91.81	76.76
55-02	496.5	55.17	46.13	662.0	73.56	61.50	827.5	91.94	76.88
55-03	497.3	55.25	46.20	663.0	73.67	61.59	828.8	92.08	76.99
55-04	498.0	55.33	46.27	664.0	73.78	61.69	830.0	92.22	77.11
55-05	498.8	55.42	46.33	665.0	73.89	61.79	831.3	92.36	77.23
55-06	499.5	55.50	46.40	666.0	74.00	61.87	832.5	92.50	77.34
55-07	500.3	55.58	46.47	667.0	74.11	61.97	833.8	92.64	77.46
55-08	501.0	55.67	46.54	668.0	74.22	62.06	835.0	92.78	77.57
55-09	501.8	55.75	46.61	669.0	74.33	62.15	836.3	92.92	77.69
55-10	502.5	55.83	46.68	670.0	74.44	62.24	837.5	93.06	77.81
55-11	503.3	55.92	46.75	671.0	74.56	62.34	838.8	93.19	77.92
56-00	504.0	56.00	46.82	672.0	74.67	62.43	840.0	93.33	78.04
56-01	504.8	56.08	46.89	673.0	74.78	62.52	841.3	93.47	78.15
56-02	505.5	56.17	46.96	674.0	74.89	62.62	842.5	93.61	78.27
56-03	506.3	56.25	47.03	675.0	75.00	62.71	843.8	93.75	78.39
56-04	507.0	56.33	47.10	676.0	75.11	62.80	845.0	93.89	78.50
56-05	507.8	56.42	47.17	677.0	75.22	62.90	846.3	94.03	78.62
56-06	508.5	56.50	47.24	678.0	75.33	62.99	847.5	94.17	78.73
56-07	509.3	56.58	47.31	679.0	75.44	63.08	848.8	94.31	78.85
56-08	510.0	56.67	47.38	680.0	75.56	63.17	850.0	94.44	78.97
56-09	510.8	56.75	47.45	681.0	75.67	63.27	851.3	94.58	79.08
56-10	511.5	56.83	47.52	682.0	75.78	63.36	852.5	94.72	79.20
56-11	512.3	56.92	47.59	683.0	75.89	63.45	853.8	94.86	79.32
57-00	513.0	57.00	47.66	684.0	76.00	63.55	855.0	95.00	79.43
57-01	513.8	57.08	47.73	685.0	76.11	63.64	856.3	95.14	79.55
57-02	514.5	57.17	47.80	686.0	76.22	63.73	857.5	95.28	79.66
57-03	515.3	57.25	47.87	687.0	76.33	63.82	858.8	95.42	79.78
57-04	516.0	57.33	47.94	688.0	76.44	63.92	860.0	95.56	79.90
57-05	516.8	57.42	48.01	689.0	76.56	64.01	861.3	95.69	80.01
57-06	517.5	57.50	48.08	690.0	76.67	64.10	862.5	95.83	80.13
57-07	518.3	57.58	48.15	691.0	76.78	64.20	863.8	95.97	80.24
57-08	519.0	57.67	48.22	692.0	76.89	64.29	865.0	96.11	80.36
57-09	519.8	57.75	48.29	693.0	77.00	64.38	866.3	96.25	80.48
57-10	520.5	57.83	48.36	694.0	77.11	64.47	867.5	96.39	80.59
57-11	521.3	57.92	48.43	695.0	77.22	64.57	868.8	96.53	80.71

15.3.1 Computing Square Yards and Square Meters of Carpet (6 to 81 Feet in Length From Rolls 9, 12, and 15 Feet Wide) (Continued)

LENGTH	9 FEET			12 FEET			15 FEET		
FT. IN.	SQ FT	SQ YDS	SQ M	SQ FT	SQ YDS	SQ M	SQ FT	SQ YDS	SQ M
58-00	522.0	58.00	48.49	696.0	77.33	64.66	870.0	96.67	80.83
58-01	522.8	58.08	48.56	697.0	77.44	64.75	871.3	96.81	80.94
58-02	523.5	58.17	48.63	698.0	77.56	64.85	872.5	96.94	81.06
58-03	524.3	58.25	48.70	699.0	77.67	64.94	873.8	97.08	81.17
58-04	525.0	58.33	48.77	700.0	77.78	65.03	875.0	97.22	81.29
58-05	525.8	58.42	48.84	701.0	77.89	65.12	876.3	97.36	81.41
58-06	526.5	58.50	48.91	702.0	78.00	65.22	877.5	97.50	81.52
58-07	527.3	58.58	48.98	703.0	78.11	65.31	878.8	97.64	81.64
58-08	528.0	58.67	49.05	704.0	78.22	65.40	880.0	97.78	81.75
58-09	528.8	58.75	49.12	705.0	78.33	65.50	881.3	97.92	81.87
58-10	529.5	58.83	49.19	706.0	78.44	65.59	882.5	98.06	81.99
58-11	530.3	58.92	49.26	707.0	78.56	65.68	883.8	98.19	82.10
59-00	531.0	59.00	49.33	708.0	78.67	65.77	885.0	98.33	82.22
59-01	531.8	59.08	49.40	709.0	78.78	65.87	886.3	98.47	82.33
59-02	532.5	59.17	49.47	710.0	78.89	65.96	887.5	98.61	82.45
59-03	533.3	59.25	49.54	711.0	79.00	66.05	888.8	98.75	82.57
59-04	534.0	59.33	49.61	712.0	79.11	66 15	890.0	98.89	82.68
59-05	534.8	59.42	49.68	713.0	79.22	66.24	891.3	99.03	82.80
59-06	535.5	59.50	49.75	714.0	79.33	66.33	892.5	99.17	82.92
59-07	536.3	59.58	49.82	715.0	79.44	66.43	893.8	99.31	83.03
59-08	537.0	59.67	49.89	716.0	79.56	66.52	895.0	99.44	83.15
59-09	537.8	59.75	49.96	717.0	79.67	66.61	896.3	99.58	83.26
59-10	538.5	59.83	50.03	718.0	79.78	66.70	897.5	99.72	83.38
59-11	539.3	59.92	50.10	719.0	79.89	66.80	898.8	99.86	83.50
60-00	540.0	60.00	50.17	720.0	80.00	66.89	900.0	100.00	83.61
60-01	540.8	60.08	50.24	721.0	80.11	66.98	901.3	100.14	83.73
60-02	541.5	60.17	50.31	722.0	80.22	67.08	902.5	100.28	83.84
60-03	542.3	60.25	50.38	723.0	80.33	67.17	903.8	100.42	83.96
60-04	543.0	60.33	50.45	724.0	80.44	67.26	905.0	100.56	84.08
60-05	543.8	60.42	50.52	725.0	80.56	67.35	906.3	100.69	84.19
60-06	544.5	60.50	50.59	726.0	80.67	67.45	907.5	100.83	84.31
60-07	545.3	60.58	50.65	727.0	80.78	67.54	908.8	100.97	84.43
60-08	546.0	60.67	50.72	728.0	80.89	67.63	910.0	101.11	84.54
60-09	546.8	60.75	50.79	729.0	81.00	67.73	911.3	101.25	84.66
60-10	547.5	60.83	50.86	730.0	81.11	67.82	912.5	101.39	84.77
60-11	548.3	60.92	50.93	731.0	81.22	67.91	913.8	101.53	84.89
61-00	549.0	61.00	51.00	732.0	81.33	68.00	915.0	101.67	85.01
61-01	549.8	61.08	51.07	733.0	81.44	68.10	916.3	101.81	85.12
61-02	550.5	61.17	51.14	734.0	81.56	68.19	917.5	101.94	85.24
61-03	551.3	61.25	51.21	735.0	81.67	68.28	918.8	102.08	85.35
61-04	552.0	61.33	51.28	736.0	81.78	68.38	920.0	102.22	85.47
61-05	552.8	61.42	51.35	737.0	81.89	68.47	921.3	102.36	85.59
61-06	553.5	61.50	51.42	738.0	82.00	68.56	922.5	102.50	85.70
61-07	554.3	61.58	51.49	739.0	82.11	68.65	923.8	102.64	85.82
61-08	555.0	61.67	51.56	740.0	82.22	68.75	925.0	102.78	85.93
61-09	555.8	61.75	51.63	741.0	82.33	68.84	926.3	102.92	86.05
61-10	556.5	61.83	51.70	742.0	82.44	68.93	927.5	103.06	86.17
61-11	557.3	61.92	51.77	743.0	82.56	69.03	928.8	103.19	86.28

15.3.1 Computing Square Yards and Square Meters of Carpet (6 to 81 Feet in Length From Rolls 9, 12, and 15 Feet Wide) (Continued)

LENGTH	9 FEET			12 FEET			15 FEET		
FT. IN.	SQ FT	SQ YDS	SQ M	SQ FT	SQ YDS	SQ M	SQ FT	SQ YDS	SQ M
62-00	558.0	62.00	51.84	744.0	82.67	69.12	930.0	103.33	86.40
62-01	558.8	62.08	51.91	745.0	82.78	69.21	931.3	103.47	86.52
62-02	559.5	62.17	51.98	746.0	82.89	69.31	932.5	103.61	86.63
62-03	560.3	62.25	52.05	747.0	83.00	69.40	933.8	103.75	86.75
62-04	561.0	62.33	52.12	748.0	83.11	69.49	935.0	103.89	86.86
62-05	561.8	62.42	52.19	749.0	83.22	69.58	936.3	104.03	86.98
62-06	562.5	62.50	52.26	750.0	83.33	69.68	937.5	104.17	87.10
62-07	563.3	62.58	52.33	751.0	83.44	69.77	938.8	104.31	87.21
62-08	564.0	62.67	52.40	752.0	83.56	69.86	940.0	104.44	87.33
62-09	564.8	62.75	52.47	753.0	83.67	69.96	941.3	104.58	87.44
62-10	565.5	62.83	52.54	754.0	83.78	70.05	942.5	104.72	87.56
62-11	566.3	62.92	52.61	755.0	83.89	70.14	943.8	104.86	87.68
63-00	567.0	63.00	52.68	756.0	84.00	70.23	945.0	105.00	87.79
63-01	567.8	63.08	52.75	757.0	84.11	70.33	946.3	105.14	87.91
63-02	568.5	63.17	52.81	758.0	84.22	70.42	947.5	105.28	88.03
63-03	569.3	63.25	52.88	759.0	84.33	70.51	948.8	105.42	88.14
63-04	570.0	63.33	52.95	760.0	84.44	70.61	950.0	105.56	88.26
63-05	570.8	63.42	53.02	761.0	84.56	70.70	951.3	105.69	88.37
63-06	571.5	63.50	53.09	762.0	84.67	70.79	952.5	105.83	88.49
63-07	572.3	63.58	53.16	763.0	84.78	70.88	953.8	105.97	88.61
63-08	573.0	63.67	53.23	764.0	84.89	70.98	955.0	106.11	88.72
63-09	573.8	63.75	53.30	765.0	85.00	71.07	956.3	106.25	88.84
63-10	574.5	63.83	53.37	766.0	85.11	71.16	957.5	106.39	88.95
63-11	575.3	63.92	53.44	767.0	85.22	71.26	958.8	106.53	89.07
64-00	576.0	64.00	53.51	768.0	85.33	71.35	960.0	106.67	89.19
64-01	576.8	64.08	53.58	769.0	85.44	71.44	961.3	106.81	89.30
64-02	577.5	64.17	53.65	770.0	85.56	71.53	962.5	106.94	89.42
64-03	578.3	64.25	53.72	771.0	85.67	71.63	963.8	107.08	89.53
64-04	579.0	64.33	53.79	772.0	85.78	71.72	965.0	107.22	89.65
64-05	579.8	64.42	53.86	773.0	85.89	71.81	966.3	107.36	89.77
64-06	580.5	64.50	53.93	774.0	86.00	71.91	967.5	107.50	89.88
64-07	581.3	64.58	54.00	775.0	86.11	72.00	968.8	107.64	90.00
64-08	582.0	64.67	54.07	776.0	86.22	72.09	970.0	107.78	90.12
64-09	582.8	64.75	54.14	777.0	86.33	72.19	971.3	107.92	90.23
64-10	583.5	64.83	54.21	778.0	86.44	72.28	972.5	108.06	90.35
64-11	584.3	64.92	54.28	779.0	86.56	72.37	973.8	108.19	90.46
65-00	585.0	65.00	54.35	780.0	86.67	72.46	975.0	108.33	90.58
65-01	585.8	65.08	54.42	781.0	86.78	72.56	976.3	108.47	90.70
65-02	586.5	65.17	54.49	782.0	86.89	72.65	977.5	108.61	90.81
65-03	587.3	65.25	54.56	783.0	87.00	72.74	978.8	108.75	90.93
65-04	588.0	65.33	54.63	784.0	87.11	72.84	980.0	108.89	91.04
65-05	588.8	65.42	54.70	785.0	87.22	72.93	981.3	109.03	91.16
65-06	589.5	65.50	54.77	786.0	87.33	73.02	982.5	109.17	91.28
65-07	590.3	65.58	54.84	787.0	87.44	73.11	983.8	109.31	91.39
65-08	591.0	65.67	54.91	788.0	87.56	73.21	985.0	109.44	91.51
65-09	591.8	65.75	54.97	789.0	87.67	73.30	986.3	109.58	91.63
65-10	592.5	65.83	55.04	790.0	87.78	73.39	987.5	109.72	91.74
65-11	593.3	65.92	55.11	791.0	87.89	73.49	988.8	109.86	91.86

15.3.1 Computing Square Yards and Square Meters of Carpet (6 to 81 Feet in Length From Rolls 9, 12, and 15 Feet Wide) (Continued)

LENGTH	9 FEET			12 FEET			15 FEET		
FT. IN.	SQ FT	SQ YDS	SQ M	SQ FT	SQ YDS	SQ M	SQ FT	SQ YDS	SQ M
66-00	594.0	66.00	55.18	792.0	88.00	73.58	990.0	110.00	91.97
66-01	594.8	66.08	55.25	793.0	88.11	73.67	991.3	110.14	92.09
66-02	595.5	66.17	55.32	794.0	88.22	73.76	992.5	110.28	92.21
66-03	596.3	66.25	55.39	795.0	88.33	73.86	993.8	110.42	92.32
66-04	597.0	66.33	55.46	796.0	88.44	73.95	995.0	110.56	92.44
66-05	597.8	66.42	55.53	797.0	88.56	74.04	996.3	110.69	92.55
66-06	598.5	66.50	55.60	798.0	88.67	74.14	997.5	110.83	92.67
66-07	599.3	66.58	55.67	799.0	88.78	74.23	998.8	110.97	92.79
66-08	600.0	66.67	55.74	800.0	88.89	74.32	1000.0	111.11	92.90
66-09	600.8	66.75	55.81	801.0	89.00	74.41	1001.3	111.25	93.02
66-10	601.5	66.83	55.88	802.0	89.11	74.51	1002.5	111.39	93.13
66-11	602.3	66.92	55.95	803.0	89.22	74.60	1003.8	111.53	93.25
67-00	603.0	67.00	56.02	804.0	89.33	74.69	1005.0	111.67	93.37
67-01	603.8	67.08	56.09	805.0	89.44	74.79	1006.3	111.81	93.48
67-02	604.5	67.17	56.16	806.0	89.56	74.88	1007.5	111.94	93.60
67-03	605.3	67.25	56.23	807.0	89.67	74.97	1008.8	112.08	93.72
67-04	606.0	67.33	56.30	808.0	89.78	75.07	1010.0	112.22	93.83
67-05	606.8	67.42	56.37	809.0	89.89	75.16	1011.3	112.36	93.95
67-06	607.5	67.50	56.44	810.0	90.00	75.25	1012.5	112.50	94.06
67-07	608.3	67.58	56.51	811.0	90.11	75.34	1013.8	112.64	94.18
67-08	609.0	67.67	56.58	812.0	90.22	75.44	1015.0	112.78	94.30
67-09	609.8	67.75	56.65	813.0	90.33	75.53	1016.3	112.92	94.41
67-10	610.5	67.83	56.72	814.0	90.44	75.62	1017.5	113.06	94.53
67-11	611.3	67.92	56.79	815.0	90.56	75.72	1018.8	113.19	94.64
68-00	612.0	68.00	56.86	816.0	90.67	75.81	1020.0	113.33	94.76
68-01	612.8	68.08	56.93	817.0	90.78	75.90	1021.3	113.47	94.88
68-02	613.5	68.17	57.00	818.0	90.89	75.99	1022.5	113.61	94.99
68-03	614.3	68.25	57.07	819.0	91.00	76.09	1023.8	113.75	95.11
68-04	615.0	68.33	57.13	820.0	91.11	76.18	1025.0	113.89	95.23
68-05	615.8	68.42	57.20	821.0	91.22	76.27	1026.3	114.03	95.34
68-06	616.5	68.50	57.27	822.0	91.33	76.37	1027.5	114.17	95.46
68-07	617.3	68.58	57.34	823.0	91.44	76.46	1028.8	114.31	95.57
68-08	618.0	68.67	57.41	824.0	91.56	76.55	1030.0	114.44	95.69
68-09	618.8	68.75	57.48	825.0	91.67	76.64	1031.3	114.58	95.81
68-10	619.5	68.83	57.55	826.0	91.78	76.74	1032.5	114.72	95.92
68-11	620.3	68.92	57.62	827.0	91.89	76.83	1033.8	114.86	96.04
69-00	621.0	69.00	57.69	828.0	92.00	76.92	1035.0	115.00	96.15
60-01	621.8	69.08	57.76	829.0	92.11	77.02	1036.3	115.14	96.27
69-02	622.5	69.17	57.83	830.0	92.22	77.11	1037.5	115.28	96.39
69-03	623.3	69.25	57.90	831.0	92.33	77.20	1038.8	115.42	96.50
69-04	624.0	69.33	57.97	832.0	92.44	77.29	1040.0	115.56	96.62
69-05	624.8	69.42	58.04	833.0	92.56	77.39	1041.3	115.69	96.73
69-06	625.5	69.50	58.11	834.0	92.67	77.48	1042.5	115.83	96.85
69-07	626.3	69.58	58.18	835.0	92.78	77.57	1043.8	115.97	96.97
69-08	627.0	69.67	58.25	836.0	92.89	77.67	1045.0	116.11	97.08
69-09	627.8	69.75	58.32	837.0	93.00	77.76	1046.3	116.25	97.20
69-10	628.5	69.83	58.39	838.0	93.11	77.85	1047.5	116.39	97.32
69-11	629.3	69.92	58.46	839.0	93.22	77.95	1048.8	116.53	97.43

15.3.1 Computing Square Yards and Square Meters of Carpet (6 to 81 Feet in Length From Rolls 9, 12, and 15 Feet Wide) (Continued)

LENGTH	9 FEET			12 FEET			15 FEET		
FT IN	SQ FT	SQ YDS	SQ M	SQ FT	SQ YDS	SQ M	SQ FT	SQ YDS	SQ M
70-00	630.0	70.00	58.53	840.0	93.33	78.04	1050.0	116.67	97.55
70-01	630.8	70.08	58.60	841.0	93.44	78.13	1051.3	116.81	97.66
70-02	631.5	70.17	58.67	842.0	93.56	78.22	1052.5	116.94	97.78
70-03	632.3	70.25	58.74	843.0	93.67	78.32	1053.8	117.08	97.90
70-04	633.0	70.33	58.81	844.0	93.78	78.41	1055.0	117.22	98.01
70-05	633.8	70.42	58.88	845.0	93.89	78.50	1056.3	117.36	98.13
70-06	634.5	70.50	58.95	846.0	94.00	78.60	1057.5	117.50	98.24
70-07	635.3	70.58	59.02	847.0	94.11	78.69	1058.8	117.64	98.36
70-08	636.0	70.67	59.09	848.0	94.22	78.78	1060.0	117.78	98.48
70-09	636.8	70.75	59.16	849.0	94.33	78.87	1061.3	117.92	98.59
70-10	637.5	70.83	59.23	850.0	94.44	78.97	1062.5	118.06	98.71
70-11	638.3	70.92	59.29	851.0	94.56	79.06	1063.8	118.19	98.83
71-00	639.0	71.00	59.36	852.0	94.67	79.15	1065.0	118.33	98.94
71-01	639.8	71.08	59.43	853.0	94.78	79.25	1066.3	118.47	99.06
71-02	640.5	71.17	59.50	854.0	94.89	79.34	1067.5	118.61	99.17
71-03	641.3	71.25	59.57	855.0	95.00	79.43	1068.8	118.75	99.29
71-04	642.0	71.33	59.64	856.0	95.11	79.52	1070.0	118.89	99.41
71-05	642.8	71.42	59.71	857.0	95.22	79.62	1071.3	119.03	99.52
71-06	643.5	71.50	59.78	858.0	95.33	79.71	1072.5	119.17	99.64
71-07	644.3	71.58	59.85	859.0	95.44	79.80	1073.8	119.31	99.75
71-08	645.0	71.67	59.92	860.0	95.56	79.90	1075.0	119.44	99.87
71-09	645.8	71.75	59.99	861.0	95.67	79.99	1076.3	119.58	99.99
71-10	646.5	71.83	60.06	862.0	95.78	80.08	1077.5	119.72	100.10
71-11	647.3	71.92	60.13	863.0	95.89	80.17	1078.8	119.86	100.22
72-00	648.0	72.00	60.20	864.0	96.00	80.27	1080.0	120.00	100.33
72-01	648.8	72.08	60.27	865.0	96.11	80.36	1081.3	120.14	100.45
72-02	649.5	72.17	60.34	866.0	96.22	80.45	1082.5	120.28	100.57
72-03	650.3	72.25	60.41	867.0	96.33	80.55	1083.8	120.42	100.68
72-04	651.0	72.33	60.48	868.0	96.44	80.64	1085.0	120.56	100.80
71-05	651.8	72.42	60.55	869.0	96.56	80.73	1086.3	120.69	100.92
72-06	652.5	72.50	60.62	870.0	96.67	80.83	1087.5	120.83	101.03
72-07	653.3	72.58	60.69	871.0	96.78	80.92	1088.8	120.97	101.15
72-08	654.0	72.67	60.76	872.0	96.89	81.01	1090.0	121.11	101.26
72-09	654.8	72.75	60.83	873.0	97.00	81.10	1091.3	121.25	101.38
72-10	655.5	72.83	60.90	874.0	97.11	81.20	1092.5	121.39	101.50
72-11	656.3	72.92	60.97	875.0	97.22	81.29	1093.8	121.53	101.61
73-00	657.0	73.00	61.04	876.0	97.33	81.38	1095.0	121.67	101.73
73-01	657.8	73.08	61.11	877.0	97.44	81.48	1096.3	121.81	101.84
73-02	658.5	73.17	61.18	878.0	97.56	81.57	1097.5	121.94	101.96
73-03	659.3	73.25	61.25	879.0	97.67	81.66	1098.8	122.08	102.08
73-04	660.0	73.33	61.32	880.0	97.78	81.75	1100.0	122.22	102.19
73-05	660.8	73.42	61.39	881.0	97.89	81.85	1101.3	122.36	102.31
73-06	661.5	73.50	61.45	882.0	98.00	81.94	1102.5	122.50	102.43
73-07	662.3	73.58	61.52	883.0	98.11	82.03	1103.8	122.64	102.54
73-08	663.0	73.67	61.60	884.0	98.22	82.13	1105.0	122.78	102.66
73-09	663.8	73.75	61.66	885.0	98.33	82.22	1106.3	122.92	102.77
73-10	664.5	73.83	61.73	886.0	98.44	82.31	1107.5	123.06	102.89
73-11	665.3	73.92	61.80	887.0	98.56	82.40	1108.8	123.19	103.01

15.3.1 Computing Square Yards and Square Meters of Carpet (6 to 81 Feet in Length From Rolls 9, 12, and 15 Feet Wide) (Continued)

LENGTH	9 FEET			12 FEET			15 FEET		
FT. IN.	SQ FT	SQ YDS	SQ M	SQ FT	SQ YDS	SQ M	SQ FT	SQ YDS	SQ M
74-00	666.0	74.00	61.87	888.0	98.67	82.50	1110.0	123.33	103.12
74-01	666.8	74.08	61.94	889.0	98.78	82.59	1111.3	123.47	103.24
74-02	667.5	74.17	62.01	890.0	98.89	82.68	1112.5	123.61	103.35
74-03	668.3	74.25	62.08	891.0	99.00	82.78	1113.8	123.75	103.47
74-04	669.0	74.33	62.15	892.0	99.11	82.87	1115.0	123.89	103.59
74-05	669.8	74.42	62.22	893.0	99.22	82.96	1116.3	124.03	103.70
74-06	670.5	74.50	62.29	894.0	99.33	83.05	1117.5	124.17	103.82
74-07	671.3	74.58	62.36	895.0	99.44	83.15	1118.8	124.31	103.93
74-08	672.0	74.67	62.43	896.0	99.56	83.24	1120.0	124.44	104.05
74-09	672.8	74.75	62.50	897.0	99.67	83.33	1121.3	124.58	104.17
74-10	673.5	74.83	62.57	898.0	99.78	83.43	1122.5	124.72	104.28
74-11	674.3	74.92	62.64	899.0	99.89	83.52	1123.8	124.86	104.40
75-00	675.0	75.00	62.71	900.0	100.00	83.61	1125.0	125.00	104.52
75-01	675.8	75.08	62.78	901.0	100.11	83.71	1126.3	125.14	104.63
75-02	676.5	75.17	62.85	902.0	100.22	83.80	1127.5	125.28	104.75
75-03	677.3	75.25	62.92	903.0	100.33	83.89	1128.8	125.42	104.86
75-04	678.0	75.33	62.99	904.0	100.44	83.98	1130.0	125.56	104.98
75-05	678.8	75.42	63.06	905.0	100.56	84.08	1131.3	125.69	105.10
75-06	679.5	75.50	63.13	906.0	100.67	84.17	1132.5	125.83	105.21
75-07	680.3	75.58	63.20	907.0	100.78	84.26	1133.8	125.97	105.33
75-08	681.0	75.67	63.27	908.0	100.89	84.36	1135.0	126.11	105.44
75-09	681.8	75.75	63.34	909.0	101.00	84.45	1136.3	126.25	105.56
75-10	682.5	75.83	63.41	910.0	101.11	84.54	1137.5	126.39	105.68
75-11	683.3	75.92	63.48	911.0	101.22	84.63	1138.8	126.53	105.79
76-00	684.0	76.00	63.55	912.0	101.33	84.73	1140.0	126.67	105.91
76-01	684.8	76.08	63.61	913.0	101.44	84.82	1141.3	126.81	106.03
76-02	685.5	76.17	63.69	914.0	101.56	84.91	1142.5	126.94	106.14
76-03	686.3	76.25	63.75	915.0	101.67	85.01	1143.8	127.08	106.26
76-04	687.0	76.33	63.82	916.0	101.78	85.10	1145.0	127.22	106.37
76-05	687.8	76.42	63.89	917.0	101.89	85.19	1146.3	127.36	106.49
76-06	688.5	76.50	63.96	918.0	102.00	85.28	1147.5	127.50	106.61
76-07	689.3	76.58	64.03	919.0	102.11	85.38	1148.8	127.64	106.72
76-08	690.0	76.67	64.10	920.0	102.22	85.47	1150.0	127.78	106.84
76-09	690.8	76.75	64.17	921.0	102.33	85.56	1151.3	127.92	106.95
76-10	691.5	76.83	64.24	922.0	102.44	85.66	1152.5	128.06	107.07
76-11	692.3	76.92	64.31	923.0	102.56	85.75	1153.8	128.19	107.19
77-00	693.0	77.00	64.38	924.0	102.67	85.84	1155.0	128.33	107.30
77-01	693.8	77.08	64.45	925.0	102.78	85.93	1156.3	128.47	107.42
77-02	694.5	77.17	64.52	926.0	102.89	86.03	1157.5	128.61	107.53
77-03	695.3	77.25	64.59	927.0	103.00	86.12	1158.8	128.75	107.65
77-04	696.0	77.33	64.66	928.0	103.11	86.21	1160.0	128.89	107.77
77-05	696.8	77.42	64.73	929.0	103.22	86.31	1161.3	129.03	107.88
77-06	697.5	77.50	64.80	930.0	103.33	86.40	1162.5	129.17	108.00
77-07	698.3	77.58	64.87	931.0	103.44	86.49	1163.8	129.31	108.12
77-08	699.0	77.67	64.94	932.0	103.56	86.59	1165.0	129.44	108.23
77-09	699.8	77.75	65.01	933.0	103.67	86.68	1166.3	129.58	108.35
77-10	700.5	77.83	65.08	934.0	103.78	86.77	1167.5	129.72	108.46
77-11	701.3	77.92	65.15	935.0	103.89	86.86	1168.8	129.86	108.58

15.3.1 Computing Square Yards and Square Meters of Carpet (6 to 81 Feet in Length From Rolls 9, 12, and 15 Feet Wide) (Continued)

LENGTH	9 FEET			12 FEET			15 FEET		
FT. IN	SQ FT	SQ YDS	SQ M	SQ FT	SQ YDS	SQ M	SQ FT	SQ YDS	SQ M
78-00	702.0	78.00	65.22	936.0	104.00	86.96	1170.0	130.00	108.70
78-01	702.8	78.08	65.29	937.0	104.11	87.05	1171.3	130.14	108.81
78-02	703.5	78.17	65.36	938.0	104.22	87.14	1172.5	130.28	108.93
78-03	704.3	78.25	65.43	939.0	104.33	87.24	1173.8	130.42	109.04
78-04	705.0	78.33	65.50	940.0	104.44	87.33	1175.0	130.56	109.16
78-05	705.8	78.42	65.57	941.0	104.56	87.42	1176.3	130.69	109.28
78-06	706.5	78.50	65.64	942.0	104.67	87.51	1177.5	130.83	109.39
78-07	707.3	78.58	65.71	943.0	104.78	87.61	1178.8	130.97	109.51
78-08	708.0	78.67	65.77	944.0	104.89	87.70	1180.0	131.11	109.63
78-09	708.8	78.75	65.84	945.0	105.00	87.79	1181.3	131.25	109.74
78-10	709.5	78.83	65.91	946.0	105.11	87.89	1182.5	131.39	109.86
78-11	710.3	78.92	65.98	947.0	105.22	87.98	1183.8	131.53	109.97
79-00	711.0	79.00	66.05	948.0	105.33	88.07	1185.0	131.67	110.09
79-01	711.8	79.08	66.12	949.0	105.44	88.16	1186.3	131.81	110.21
79-02	712.5	79.17	66.19	950.0	105.56	88.26	1187.5	131.94	110.32
79-03	713.3	79.25	66.26	951.0	105.67	88.35	1188.8	132.08	110.44
79-04	714.0	79.33	66.33	952.0	105.78	88.44	1190.0	132.22	110.55
79-05	714.8	79.42	66.40	953.0	105.89	88.54	1191.3	132.36	110.67
79-06	715.5	79.50	66.47	954.0	106.00	88.63	1192.5	132.50	110.79
79-07	716.3	79.58	66.54	955.0	106.11	88.72	1193.8	132.64	110.90
79-08	717.0	79.67	66.61	956.0	106.22	88.81	1195.0	132.78	111.02
79-09	717.8	79.75	66.68	957.0	106.33	88.91	1196.3	132.92	111.13
79-10	718.5	79.83	66.75	958.0	106.44	89.00	1197.5	133.06	111.25
79-11	719.3	79.92	66.82	959.0	106.56	89.09	1198.8	133.19	111.37
80-00	720.0	80.00	66.89	960.0	106.67	89.19	1200.0	133.33	111.48
80-01	720.8	80.08	66.96	961.0	106.78	89.28	1201.3	133.47	111.60
80-02	721.5	80.17	67.03	962.0	106.89	89.37	1202.5	133.61	111.72
80-03	722.3	80.25	67.10	963.0	107.00	89.47	1203.8	133.75	111.83
80-04	723.0	80.33	67.17	964.0	107.11	89.56	1205.0	133.89	111.95
80-05	723.8	80.42	67.24	965.0	107.22	89.65	1206.3	134.03	112.06
80-06	724.5	80.50	67.31	966.0	107.33	89.74	1207.5	134.17	112.18
80-07	725.3	80.58	67.38	967.0	107.44	89.84	1208.8	134.31	112.30
80-08	726.0	80.67	67.45	968.0	107.56	89.93	1210.0	134.44	112.41
80-09	726.8	80.75	67.52	969.0	107.67	90.02	1211.3	134.58	112.53
80-10	727.5	80.83	67.59	970.0	107.78	90.12	1212.5	134.72	112.64
80-11	728.3	80.92	67.66	971.0	107.89	90.21	1213.8	134.86	112.76
81-00	729.0	81.00	67.73	972.0	108.00	90.30	1215.0	135.00	112.88
81-01	729.8	81.08	67.80	973.0	108.11	90.39	1216.3	135.14	112.99
81-02	730.5	81.17	67.87	974.0	108.22	90.49	1217.5	135.28	113.11
81-03	731.3	81.25	67.93	975.0	108.33	90.58	1218.8	135.42	113.23
81-04	732.0	81.33	68.00	976.0	108.44	90.67	1220.0	135.56	113.34
81-05	732.8	81.42	68.07	977.0	108.56	90.77	1221.3	135.69	113.46
81-06	733.5	81.50	68.14	978.0	108.67	90.86	1222.5	135.83	113.57
81-07	734.3	81.58	68.21	979.0	108.78	90.95	1223.8	135.97	113.69
81-08	735.0	81.67	68.28	980.0	108.89	91.04	1225.0	136.11	113.81
81-09	735.8	81.75	68.35	981.0	109.00	91.14	1226.3	136.25	113.92
81-10	736.5	81.83	68.42	982.0	109.11	91.23	1227.5	136.39	114.04
81-11	737.3	81.92	68.49	983.0	109.22	91.32	1228.8	136.53	114.15

15.4.0 Seamless Flooring

A monolithic surface containing a resin matrix, fillers, and a decorative topping. The thermosetting or thermoplastic matrix can be either an epoxy, one- or two-part polyester, one- or two-part polyurethane, or a one- or two-part neoprene (polychloroprene) material.

15.5.0 Stone Veneer

Various types of thin stone veneer flooring materials are available for installation over concrete or wood subfloors using a thin-set or mortar-bed installation process.

15.5.1 Thinset/Mortar-Bed Stone Veneer Installation Diagrammed

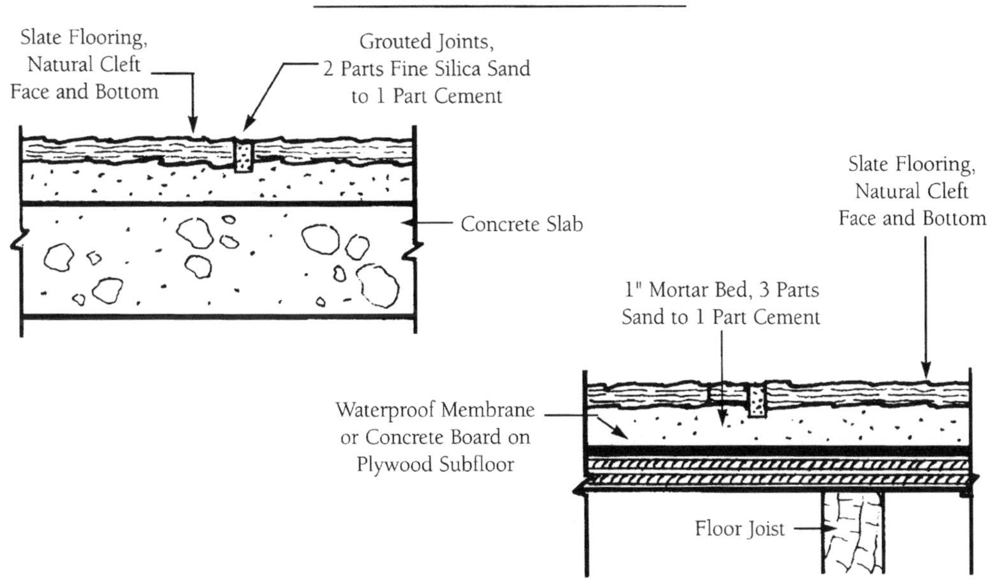

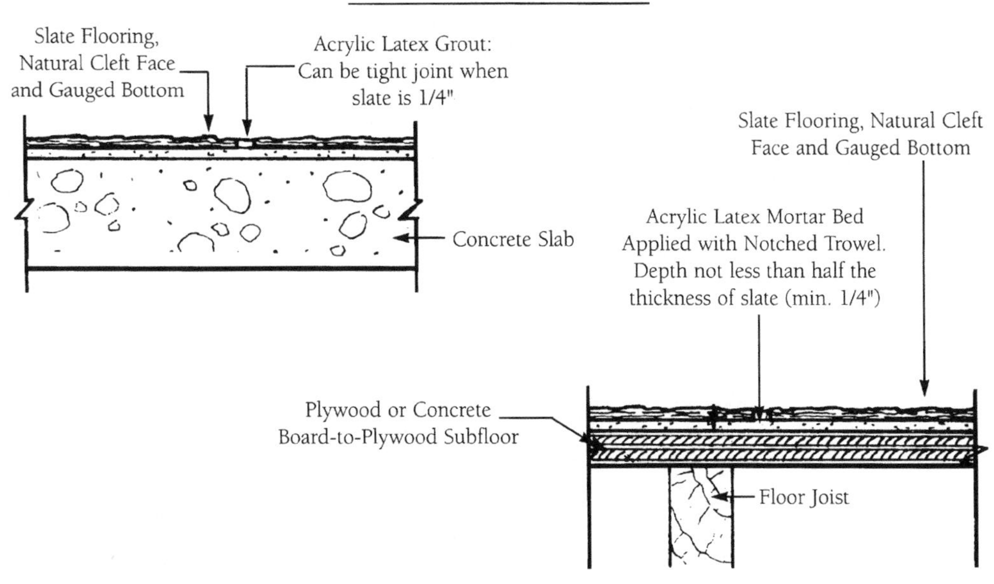

By permission: Buckingham-Virginia Slate Corp., Arvonia, Virginia

15.6.0 Terrazo Flooring

Derived from the Italian *terrace* or *terrazza*, this type of flooring is produced by embedding small pieces of marble in mortar. After curing, the surface is polished to a very smooth and shiny finish.

15.6.1 Terrazo Floor Components

TERRAZZO: Derived from the Italian "Terrace" or "Terrazza" and by definition over the centuries: "A form of mosaic flooring made by embedding small pieces of marble in mortar and polishing."

Today, the National Terrazzo and Mosaic Association (NTMA) defines this traditional material as follows: "Terrazzo consists of marble, granite, onyx or glass chips in portland cement, modified portland cement or resinous matrix. The terrazzo is poured, cured, ground and polished. Typically used as a finish for floors, stairs or walls, Terrazzo can be poured in place or precast."

"Rustic Terrazzo is a variation of where, in lieu of grinding and polishing, the surface is washed with water or otherwise treated to expose the chips. Quartz, quartzite and river bed aggregates can also be used."

"Mosaic is an artistic finish composed of small hand cut pieces of smalti, glass or marble called tessarae. The tessarae are mounted on paper by hand to form mosaic sheets. These sheets of mosaic are then set in mortar on the job site to create beautiful patterns, designs, and murals."

MARBLE CHIPS: Marble has been defined as a metamorphic rock formed by the recrystallization of limestone. However, in recent decades, marble has been redefined to include all calcareous rocks capable of taking a polish (such as onyx, travertine, and attractive serpentine rocks). Marble is quarried, selected to avoid off color or contaminated material, crushed and sized to yield marble chips for Terrazzo. Excellent domestic and imported marble chips are available for use in terrazzo in a wide range of colors and can be combined in infinite varieties to create color harmonies of every descriptions.

MARBLE CHIP SIZES: Marble chips are graded by number according to size in accordance with standards adopted by producers as follows:

Number	Passes screen (in inches)	Retained on screen (in inches)
0	⅛	1⁄16
1	¼	⅛
2	⅜	¼
3	½	⅜
4	⅝	½
5	¾	⅝
6	⅞	¾
7	1	⅞
8	1⅛	1

CUSTOMARY SIZES FOR TOPPINGS:

1. Standard: No. 1 and 2.
2. Intermediate: No. 1, 2, 3, and 4.
3. Venetian: No. 1, 2, 3, 4, and 5; and/or 6, 7, and 8.
4. Resinous: (¼ inch thickness) No. 1 and 0.
5. Resinous: (⅜ inch thickness) No. 1, 2, and 0.

NOTE: Marble chip quarries normally produce 0, 1, and 2 as separate sizes. Larger sizes are frequently grouped; for example #3–4 mixed and #7–8 mixed, and #4–7 mixed. #00 chips (1⁄16 to 1⁄32 inch size) are available for use in industrial floors.

SELECTING MARBLE CHIPS: It is highly desirable that color combinations be designated by NTMA plate numbers (NTMA Color Plates). In the absence of NTMA color plates, it is important that the size and color combinations be shown due to price differentials.

MATRICES: The matrix is the ingredient in a terrazzo floor which acts as a binder to hold the chips in position. There are three basic types of matrices: cementitious, modified-cementitious and resinous.

CEMENTITIOUS MATRICES: Portland Cement provides a good background for marble chips. It can be tinted to produce various colors. White cement is color controlled during manufacture. Gray Portland Cement may not be color controlled. For use in terrazzo, portland cement should exceed the minimum standards of ASTM C-150.

MINERAL COLOR PIGMENTS: Interior: Shall not exceed two pound per bag of portland cement. Exterior: Pigment shall not exceed ½ pound per bag of Portland Cement.

MODIFIED CEMENTITIOUS MATRICES: Polyacrylate Modified Cement: A composition resinous material which has proven to be an excellent binder for use in thin-set terrazzo. Minimum physical properties are stipulated in Polyacrylate Terrazzo specification.

RESINOUS MATRICES: EPOXY OR POLYESTER: A two component thermal setting resinous material which has proven to be an excellent binder for use in thin-set terrazzo. Minimum physical properties are stipulated in NTMA Terrazzo specifications.

DIVIDER STRIPS: White alloy of zinc, brass or plastic are used for function and aesthetics. Brass and plastic may have a reaction with some resinous materials and should be used only if deemed safe by the supplier of the resin.

The following are the most common types of strips available (in some systems, the strips act as control joints).

1¼ inch Standard Divider Strip with anchoring device. Available in white alloy of zinc or brass and 14, 16 or 18 B & S gauge. Extensively used in Sand Cushion, Bonded to Concrete, Structural and other types of cementitious terrazzo systems. Also used in monolithic terrazzo where slab has been recessed or sawn to create a weakened vertical plane. Available in 1½ inch and greater depths for Venetian Terrazzo control joints and special conditions.

1¼ inch Heavy Top Divider Strips with anchoring device. Available in white alloy of zinc or galvanized steel bottom section. Top section available in white alloy of zinc, brass or colored plastic. Width of the top section is ⅛, ¼, ⅜, or ½ inch. Basic use is the same for the 1¼ inch Standard Divider Strip. (Some plastic strips are 3/16 inch and 5/16 inch instead of ⅛, ¼ and ½ inch).

K or L Strips in standard gauges or with the heavy top feature for use in monolithic or resinous "thin-set" systems. Sizes vary according to the depth of the terrazzo topping. Can be attached to substrate with adhesive compatible with topping matrix.

CONTROL JOINTS: Double "L" strips (Angle strips) or straight strips positioned back to back are effective in allowing for anticipated shrinkage in the subfloor at construction joints. Double "L" (Angle strips) are used for Thin-set and Monolithic systems.

In Sand Cushion Terrazzo, the employment of the normal single divider strips, regardless of the gauge inserted in the Sand Cushion underbed up to five foot or less on centers, provides ample control of anticipated shrinkage that will take place when the terrazzo work is installed in accordance to these specifications as each divider picks up a minute amount of the contraction.

Construction joints in the structural slab have no bearing on the placement of divider strips in a Sand Cushion system due to the use of an isolation membrane.

NOTE: It is not this Association's intent to make expansion joint recommendations. Architects should specify expansion joints and indicate locations and details on the project drawings.

By permission: The National Terrazo & Mosaic Association Inc., Des Plaines, Illinois

Section

16

Painting

Contents

16.0.0 Generic paint formulations
16.1.0 Special-purpose coatings
16.2.0 Coating specifications for normal exposures (exterior)
16.3.0 Coating specifications for interior surfaces
16.4.0 Specifications for industrial exposure (light/moderate duty)
16.5.0 Coating specifications (industrial exposure and heavy-duty exposure)
16.6.0 Painting recommendations (immersion exposure)
16.7.0 Painting recommendations (low-temperature applications)
16.8.0 Painting recommendations (high-temperature exposure)
16.9.0 Recommended surface-preparation procedures for basic construction materials.
16.10.0 Preservative treatment for exterior woodwork
16.11.0 Myth of maintenance-free exterior coatings
16.12.0 Steel-structure painting procedures
16.12.1 SSPC specifications
16.12.2 SSPC grading of new and previously painted steel
16.12.3 Minimum surface preparation for various painting systems
16.12.4 Steel Structures Painting Council (SSPC) coating systems
16.13.0 Generic high-performance coatings for steel and concrete

Although surface preparation is the key to the proper application of any paint, a wide range of commercially produced products are available for every functional and aesthetic purpose.

16.0.0 Generic Paint Formulations

Water-Based Coatings

The first water-based coating contained styrene or styrene butadiene and was known as *latex paint.* These paints were for interior use only, but over the years, acrylic or acrylic ester resins were developed for exterior use. Other water-based paints are alkyds, vinyl or polyvinyl acetates and cement-based coatings.

Acrylic coatings are available as either opaque (colored) or clear. Methyl methacrylate is often used as a clear coating for concrete to provide weathering protection.

Water-based coatings have higher permeability to water vapor, making them suitable for application over moist, porous surfaces, such as wood, concrete, and masonry.

Solvent-Based Coatings

These coatings can be purchased as either clear or opaque materials. Clear solvent-based coatings use drying oils mixed with a resin and are generally referred to as varnishes. Various clear coatings will contain either:

- *Phenolics* Present good water and weathering characteristics. When mixed with tung oil, these varnishes are most durable for marine use. However, the relative dark color tends to darken with age and might preclude its use in some areas.
- *Shellacs* Shellac is a resin dissolved in spirit varnish, a volatile solvent. This coating is more often used as a sealer under a more-durable top coat.
- *Lacquers* Cellulose derivations in volatile spirits. They have some application in interior use, particularly for aesthetic considerations.
- *Silicon resins in a solvent solution of mineral spirits* This was once widely used as a masonry sealer. With a life span of 5 to 10 years, this coating has largely been replaced by acrylic coatings with a considerably longer life span.
- *Urethane* This is a one- or two-component, moisture-cured, solvent-based formulation with superior wear-resistance characteristics.

Opaque solvent-based coatings use alkyds as their principal binder and are available either water- or solvent-dispersed. When combined with an oil vehicle, these alkyd-oil coatings can be formulated to produce a flat, semi-gloss or gloss finish that is fast drying, flexible, durable, chalk resistant, and exhibits good color retention.

These coatings are not compatible with previous coatings that contain either lead or zinc. Alkyd-based paints could not be used to encapsulate lead-based paint because the new application will most likely cause blistering or peeling.

Chlorinated rubber coatings have good resistance to microorganisms, resistance to alkalis and acids and low permeability to water and water vapor.

Chlorosulfonated polyethylene coatings are resistant to chlorine, bromine, oxygen, ozone, and ultraviolet radiation.

Epoxy-ester coatings are made of epoxy resins and drying oils. These coatings exhibit resistance to chemical fumes and the marine environment. The polyamide-cured type is very abrasion resistant and will tolerate repeated scrubbing and washings. Bitumen epoxy coatings (both coal tar and asphalt types) are generally used for heavy-duty immersion service, such as below-grade structural steel, and underground tank and pipe coatings.

16.1.0 Special-Purpose Coatings

- Fire retardant or intumescent coatings.
- Reflective coatings to absorb the ultraviolet band of solar radiation and reflect it as visible light.
- Bituminous coatings of either water-based emulsions or solvent cut-back coal tar pitch or asphalt materials.

16.2.0 Coating Specifications for Normal Exposures (Exterior)

This table will help the specification writer select the best detailed specifications for normal exposures such as schools, hotels, apartments, stores, etc. as well as light, moderate, and heavy duty industrial specifications. It has been designed from the specification writer's point of view; starting with the information the specifier has—the material and the surface. The specifier can choose the coating's generic type, the finish desired, the surface preparation necessary, the appropriate primer, and the number of topcoats necessary to achieve a satisfactory coating system. Surface preparations shown are minimums and should be upgraded if necessary because of the service or environmental conditions.

Note: standard alkyd and epoxy coatings will chalk on exterior exposure.

Substrate/Area	Topcoat		Specifications for Normal Exposures		Minimum dft/ct		Product	
	Vehicle	Finish	Surface Preparation	Primers & Topcoats	Mils	Microns	Series	Page
Exterior Painting Recommendations—Normal Exposure								
drywall — exterior								
Drywall	acrylic latex	primer flat satin semi-gloss	S-W 8 or 12	1 ct: A-100 Exterior Latex Wood Primer 2 cts: A-100 Exterior Latex Flat, or 2 cts: LowTemp 35 Exterior Latex Flat, or 2 cts: A-100 Exterior Latex Satin, or 2 cts: LowTemp 35 Exterior Latex Satin, or 2 cts: A-100 Exterior Latex Gloss	1.4 1.3 1.5 1.3 1.3 1.3	35 32 37 32 32 32	B42 A6 B15 A82 B17 A8	27 26 91 26 91 26
masonry and cementitious surfaces								
Siding, Shingles	acrylic latex	primer flat satin semi-gloss	S-W 2, 4, 22, or 12	1 ct: A-100 Exterior Latex Wood Primer 2 cts: A-100 Exterior Latex Flat, or 2 cts: LowTemp 35 Exterior Latex Flat, or 2 cts: A-100 Exterior Latex Satin, or 2 cts: LowTemp 35 Exterior Latex Satin, or 2 cts: A-100 Exterior Latex Gloss	1.4 1.3 1.5 1.3 1.3 1.3	35 32 37 32 32 32	B42 A6 B15 A82 B17 A8	27 26 91 26 91 26
Concrete Masonry Units	latex acrylic latex	filler flat satin semi-gloss	S-W 3 or 12 or 12	1 ct: ProMar Interior/Exterior Latex Block Filler 2 cts: A-100 Exterior Latex Flat, or 2 cts: LowTemp 35 Exterior Latex Flat, or 2 cts: A-100 Exterior Latex Satin, or 2 cts: LowTemp 35 Exterior Latex Satin, or 2 cts: A-100 Exterior Latex Gloss	8.0 1.3 1.5 1.3 1.3 1.3	200 32 37 32 32 32	B25 A6 B15 A82 B17 A8	121 26 91 26 91 26
Concrete, Stucco, Brick	acrylic latex alkyd acrylic latex	primer primer flat flat satin semi-gloss	S-W 5, 22, 4, or 12	1 ct:: Loxon Exterior Acrylic Masonry Primer, or 1 ct: ProMar Masonry Conditioner 2 cts: Loxon Exterior Acrylic Masonry Coating, or 2 cts: A-100 Exterior Latex Flat, or 2 cts: LowTemp 35 Exterior Latex Flat, or 2 cts: A-100 Exterior Latex Satin, or 2 cts: LowTemp 35 Exterior Latex Satin, or 2 cts: A-100 Exterior Latex Gloss	3.1 2.2 3.6 1.3 1.5 1.3 1.3 1.3	77 55 90 32 37 32 32 32	A24 B46 A24 A6 B15 A82 B17 A8	92 122 92 26 91 26 91 26
	alkyd	primer gloss	S-W 5, 22, 4, or 12	1 ct: ProMar Masonry Conditioner 2 cts: SWP Exterior Gloss Oil Base Paint	2.2 2.1	55 70	B46 A2	122 30
Cementitious Hardboard	acrylic latex	primer flat flat satin semi-gloss	S-W 6 or 12	1 ct: Loxon Exterior Acrylic Masonry Primer 2 cts: Loxon Exterior Acrylic Masonry Coating, or 2 cts: A-100 Exterior Latex Flat, or 2 cts: LowTemp 35 Exterior Latex Flat, or 2 cts: A-100 Exterior Latex Satin, or 2 cts: LowTemp 35 Exterior Latex Satin, or 2 cts: A-100 Exterior Latex Gloss	3.1 3.6 1.3 1.5 1.3 1.3 1.3	77 90 32 37 32 32 32	A24 A24 A6 B15 A82 B17 A8	92 92 26 91 26 91 26
Concrete	acrylic stain or sealer	flat	S-W 5 or 12	1-2 cts: H&C Shield Plus Concrete Stain	none	none	-	74
Concrete	water repellent	none	S-W 5 or 12	1-2 cts: H&C HB-100 or HB-150 Water Repellent	none	none	-	73

16.2.0 Coating Specifications for Normal Exposures (Exterior) (Continued)

Substrate/Area	Topcoat		Specifications for Normal Exposures			Minimum dft/ct		Product	
	Vehicle	Finish	Surface Preparation	Primers & Topcoats		Mils	Microns	Series	Page
Exterior Painting Recommendations—Normal Exposure									
metal									
Aluminum Siding and trim	acrylic latex	flat	S-W 1 or 12	2 cts:	A-100 Exterior Latex Flat, or	1.3	32	A6	26
				2 cts:	LowTemp 35 Exterior Latex Flat, or	1.5	37	B15	91
		satin		2 cts:	A-100 Exterior Latex Satin, or	1.3	32	A82	26
				2 cts:	LowTemp 35 Exterior Latex Satin, or	1.3	32	B17	91
		semi-gloss		2 cts:	A-100 Exterior Latex Gloss	1.3	32	A8	26
Aluminum	acrylic	primer	SSPC-SP1	1 ct:	DTM Wash Primer	1.0	25	B66	59
	oleoresinous	aluminum		2 cts:	Silver-Brite Aluminum, B59S11	1.0	25	B59	130
Iron and Steel	alkyd	primer	SSPC-SP2	1 ct:	Kem Bond HS Universal Primer, or	5.0	125	B50	84
	acrylic latex			1 ct:	DTM Acrylic Primer/Finish	2.5	62	B66	57
		flat		2 cts:	A-100 Exterior Latex Flat, or	1.3	32	A6	26
				2 cts:	LowTemp 35 Exterior Latex Flat, or	1.5	37	B15	91
		satin		2 cts:	A-100 Exterior Latex Satin, or	1.3	32	A82	26
				2 cts:	LowTemp 35 Exterior Latex Satin, or	1.3	32	B17	91
		semi-gloss		2 cts:	A-100 Exterior Latex Gloss	1.3	32	A8	26
	alkyd	primer	SSPC-SP2	1 ct:	Kem Bond HS Universal Primer	5.0	125	B50	84
	oleoresinous	aluminum		2 cts:	Silver-Brite Aluminum, B59S11	1.0	25	B59	130
				2 cts:	Silver-Brite Rust Resistant Aluminum, B59S2	1.0	25	B59	131
Galvanized	acrylic latex	primer	SSPC-SP1	1 ct:	DTM Acrylic Primer/Finish (optional)	2.5	62	B66	57
		flat		2 cts:	A-100 Exterior Latex Flat, or	1.3	32	A6	26
				2 cts:	LowTemp 35 Exterior Latex Flat, or	1.5	37	B15	91
		satin		2 cts:	A-100 Exterior Latex Satin, or	1.3	32	A82	26
				2 cts:	LowTemp 35 Exterior Latex Satin, or	1.3	32	B17	91
		semi-gloss		2 cts:	A-100 Exterior Latex Gloss	1.3	32	A8	26
	acrylic	primer	SSPC-SP1	1 ct:	Galvite HS Primer	3.0	75	B50	72
	oleoresinous	aluminum		2 cts:	Silver-Brite Aluminum, B59S11	1.0	25	B59	130
wood									
Siding and Trim Paint	alkyd	primer	S-W 23 or 12	1 ct:	A-100 Exterior Oil Wood Primer	2.3	57	Y24	27
	acrylic latex			1 ct:	A-100 Exterior Latex Wood Primer	1.4	35	B42	27
		flat		2 cts:	A-100 Exterior Latex Flat, or	1.3	32	A6	26
				2 cts:	LowTemp 35 Exterior Latex Flat, or	1.5	37	B15	91
		satin		2 cts:	A-100 Exterior Latex Satin, or	1.3	32	A82	26
				2 cts:	LowTemp 35 Exterior Latex Satin, or	1.3	32	B17	91
		semi-gloss		2 cts:	A-100 Exterior Latex Gloss	1.3	32	A8	26
	alkyd	primer	S-W 23 or 12	1 ct:	A-100 Exterior Oil Wood Primer	2.3	57	Y24	27
		gloss		2 cts:	SWP Exterior Gloss Oil Base Paint	2.1	70	A2	139
Siding and Trim Stain	acrylic	stain—solid color	S-W 23 or 12	2 cts:	WoodScapes Solid Color Stain	2.0	50	A15	147
	polyurethane	stain—semi-transparent		2 cts:	WoodScapes Semi-Transparent	none	none	A15	146
Plywood Paint	acrylic latex	primer	S-W 23 or 12	1 ct:	A-100 Exterior Latex Wood Primer	1.4	35	B42	27
		flat		2 cts:	A-100 Exterior Latex Flat, or	1.3	32	A6	26
				2 cts:	LowTemp 35 Exterior Latex Flat, or	1.5	37	B15	91
		satin		2 cts:	A-100 Exterior Latex Satin, or	1.3	32	A82	26
				2 cts:	LowTemp 35 Exterior Latex Satin, or	1.3	32	B17	91
		semi-gloss		2 cts:	A-100 Exterior Latex Gloss	1.3	32	A8	26
Plywood Stain	acrylic	stain—solid color	S-W 23 or 12	2 cts:	WoodScapes Solid Color Stain	2.0	50	A15	147
	polyurethane	stain—semi-transparent		2 cts:	WoodScapes Semi-Transparent	none	none	A15	146

16.2.0 Coating Specifications for Normal Exposures (Exterior) (Continued)

Substrate/Area	Topcoat		Specifications for Normal Exposures		Minimum dft/ct		Product	
	Vehicle	Finish	Surface Preparation	Primers & Topcoats	Mils	Microns	Series	Page
Exterior Painting Recommendations—Normal Exposure								
wood								
	alkyd	varnish	S-W 23 or 12	2-3 cts: Exterior Varnish	1.8	45	A67	70
	alkyd	clear	S-W 23 or 12	1-2 cts: Cuprinol Clear Wood Preservative, or	none	none	-	52
				1-2 cts: Cuprinol Clear Deck & Wood Seal	none	none	-	50
	alkyd	clear	S-W 23 or 12	1-2 cts: Cuprinol Clear Deck & Siding Finish, or	none	none	-	51
				1-2 cts: Cuprinol Clear Deck & Wood Seal	none	none	-	50
	acrylic	flat	S-W 23 or 12	1-2 cts: Cuprinol Solid Color Deck Stain	2.0	50	-	53
vinyl siding								
Residential Siding	acrylic latex	flat	S-W 23 or 12	2 cts: A-100 Exterior Latex Flat, or	1.3	32	A6	26
				2 cts: LowTemp 35 Exterior Latex Flat, or	1.5	37	B15	91
		satin		2 cts: A-100 Exterior Latex Satin, or	1.3	32	A82	26
				2 cts: LowTemp 35 Exterior Latex Satin, or	1.3	32	B17	91
		semi-gloss		2 cts: A-100 Exterior Latex Gloss	1.3	32	A8	26
elastomeric coating systems — exterior								
Concrete, Stucco, Masonry	acrylic	primer	S-W 5, 22	1 ct: Loxon Exterior Acrylic Masonry Primer	3.1	77	A24	92
	elastomeric	flat		2 cts: Elastomeric Coating	4.8	120	A5	60

16.3.0 Coating Specifications for Interior Surfaces

acoustical tiles								
Perforated Fiberboard	latex	primer	S-W 8 or 12	1 ct: ProMar 200 or 400 Latex Wall Primer	1.1	27	B28	118
		flat		2 cts: ProMar 200 or 400 Int. Latex Flat Wall Paint, or	1.4	35	B30	114
	alkyd	flat		2 cts: ProMar 200 Int. Alkyd Flat Wall Paint	1.8	45	B32	109
Metal Pan Tiles	alkyd	primer	S-W 8 or 12	1 ct: Wall & Wood Primer	1.6	40	B49	143
	latex	flat		2 cts: ProMar 200 or 400 Int. Latex Flat Wall Paint, or	1.4	35	B30	114
	alkyd	flat		2 cts: ProMar 200 Int. Alkyd Flat Wall Paint	1.8	45	B32	109
drywall— interior								
Gypsum Board, Plaster Board	latex	primer	S-W 8 or 12	1 ct: ProMar Classic Latex Primer, or	1.6	40	B28	119
				1 ct: ProMar 200 or 400 Latex Wall Primer	1.1	27	B28	118
		flat		2 cts: ProMar 200 or 400 Int. Latex Flat Wall Paint, or	1.4	35	B30	114
		eg-shel		2 cts: ProMar 200 or 400 Int. Latex Eg-Shel, or	1.6	40	B20	115
		semi-gloss		2 cts: ProClassic Waterborne Semi-Gloss Enamel, or	1.3	32	B31	108
				2 cts: ProMar 200 or 400 Int. Latex Semi-Gloss, or	1.3	32	B31	116
		gloss		2 cts: ProClassic Waterborne Gloss Enamel, or	1.3	32	B21	108
				2 cts: ProMar 200 Int. Latex Gloss Enamel	1.5	37	B21	117
	latex texture	texture	S-W 8 or 12	Mixture of 1 gallon of ProMar Interior/Exterior Block Filler and 1 gallon of ProMar 200 or 400 Latex Flat	N/A	N/A	B25	121
					N/A	N/A	B30	114
	latex	primer	S-W 8 or 12	1 ct: ProMar Classic Latex Primer, or	1.6	40	B28	119
	alkyd			1 ct: ProMar 200 or 400 Latex Wall Primer	1.1	27	B28	2118
		flat		2 cts: ProMar 200 Int. Alkyd Flat, or	1.8	45	B32	109
		eg-shel		2 cts: ProMar 200 Int. Alkyd Eg-Shel, or	1.8	45	B33	110
		semi-gloss		2 cts: ProClassic Interior Alkyd Semi-Gloss, or			B34	107
				2 cts: ProClassic HS Interior Alkyd Semi-Gloss, or	1.7	42	B34	107
				2 cts: ProMar 200 or 400 Int. Alkyd Semi-Gloss, or	1.7	42	B34	111
		gloss		2 cts: ProMar 200 Int. Alkyd Gloss Enamel	1.6	40	B35	112

16.3.0 Coating Specifications for Interior Surfaces (Continued)

Substrate/Area	Topcoat Vehicle	Topcoat Finish	Specifications for Normal Exposures: Surface Preparation	Specifications for Normal Exposures: Primers & Topcoats	Minimum dft/ct: Mils	Minimum dft/ct: Microns	Product: Series	Product: Page
Interior Painting Recommendations—Normal Exposure								
drywall—interior, continued								
Stain resistant topcoat	latex	primer	S-W 8 or 12	1 ct: ProMar Classic Latex Primer, or	1.6	40	B28	119
				1 ct: ProMar 200 Interior Latex Wall Primer	1.1	27	B28	118
	acrylic	flat		2 cts: EverClean Interior Latex Flat, or	1.7	42	A96	69
		satin		2 cts: EverClean Interior Latex Satin, or	1.7	42	A97	69
		semi-gloss		2 cts: EverClean Interior Latex Semi-Gloss	1.3	32	A98	69
Low odor finishes	latex acrylic	primer	S-W 8 or 12	1 ct: HealthSpec Low Odor Int. Latex Primer	1.0	25	B11	76
		flat		2 cts: HealthSpec Low Odor Int. Latex Flat, or	1.5	37	B5	75
		eg-shel		2 cts: HealthSpec Low Odor Int. Latex Eg-Shel, or	1.5	37	B9	75
		semi-gloss		2 cts: HealthSpec Low Odor Int. Latex Semi-Gloss	1.5	37	B10	75
Ceilings	latex alkyd	primer	S-W 8 or 12	1 ct: ProMar 200 Interior Latex Wall Primer	1.1	27	B28	118
		flat		1 ct: Super Save Lite Hi-Tec Dryfall	2.0	50	B48	137
				1 ct: Dry Fall Flat White, or	4.0	100	B48	55
				1 ct: Super Save Lite Flat, or	3.0	75	B48	136
		semi-gloss		1 ct: Super Save Lite Semi-Gloss, or	3.0	75	B47	136
		gloss		1 ct: Super Save Lite Gloss	2.0	50	B47	136
masonry and cementitious surfaces								
Concrete, CMU Cement Board Block, Brick (unglazed)	latex	primer	S-W 5, 3, 4, or 12	1 ct: Loxon Interior Acrylic Masonry Primer, or	3.0	75	B28	93
		filler		1 ct: ProMar Interior/Exterior Block Filler	8.0	200	B25	121
		flat		2 cts: ProMar 200 or 400 Int. Latex Flat, or	1.4	35	B30	114
		eg-shel		2 cts: ProMar 200 or 400 Int. Latex Eg-Shel, or	1.6	40	B20	115
		semi-gloss		2 cts: ProClassic Waterborne Semi-Gloss Enamel, or	1.3	32	B31	108
				2 cts: ProMar 200 or 400 Int. Latex Semi-Gloss, or	1.3	32	B31	116
		gloss		2 cts: ProClassic Waterborne Gloss Enamel	1.3	32	B21	108
	latex	primer	S-W 5, 3, 4, or 12	1 ct: ProMar Classic Latex Primer, or	1.6	40	B28	119
				1 ct: ProMar 200 or 400 Latex Wall Primer, or	1.1	27	B28	118
		filler		1 ct: ProMar Interior/Exterior Block Filler	8.0	200	B25	121
	alkyd	flat		2 cts: ProMar 200 Int. Alkyd Flat, or	1.8	45	B32	109
		eg-shel		2 cts: ProMar 200 Int. Alkyd Eg-Shel, or	1.8	45	B33	110
		semi-gloss		2 cts: ProClassic Interior Alkyd Semi-Gloss, or	1.7	42	B34	107
				2 cts: ProClassic HS Interior Alkyd Semi-Gloss, or	2.3	57	B34	107
				2 cts: ProMar 200 or 400 Int. Alkyd Semi-Gloss, or	1.7	42	B34	111
		gloss		2 cts: ProMar 200 Int. Alkyd Gloss Enamel	1.6	40	B35	112
Low odor finishes	latex acrylic	primer	S-W 5, 3, 4, or 12	1 ct: HealthSpec Low Odor Int. Latex Primer	1.0	25	B11	76
		flat		2 cts: HealthSpec Low Odor Int. Latex Flat, or	1.5	37	B5	75
		eg-shel		2 cts: HealthSpec Low Odor Int. Latex Eg-Shel, or	1.5	37	B9	75
		semi-gloss		2 cts: HealthSpec Low Odor Int. Latex Semi-Gloss	1.5	37	B10	75
Concrete Floors	acrylic	gloss	S-W 5	1-2 cts: H&C Shield Plus Concrete Stain	none	none		74
metal								
Aluminum	acrylic latex	primer	S-W 1 or 12	1 ct: DTM Acrylic Primer/Finish	2.5	62	B66	57
		flat		2 cts: ProMar 200 or 400 Int. Latex Flat, or	1.4	35	B30	114
		eg-shel		2 cts: ProMar 200 or 400 Int. Latex Eg-Shel, or	1.6	40	B20	114
		semi-gloss		2 cts: ProClassic Waterborne Semi-Gloss Enamel, or	1.3	32	B31	108
				2 cts: ProMar 200 or 400 Int. Latex Semi-Gloss, or	1.3	32	B31	116
		gloss		2 cts: ProClassic Waterborne Gloss Enamel, or	1.3	32	B31	108
				2 cts: ProMar 200 Int. Latex Gloss Enamel	1.5	37	B21	117
	acrylic alkyd	primer	S-W 1 or 12	1 ct: DTM Acrylic Primer/Finish	2.5	62	B66	57
		flat		2 cts: ProMar 200 Int. Alkyd Flat, or	1.8	45	B32	109
		eg-shel		2 cts: ProMar 200 Int. Alkyd Eg-Shel, or	1.8	45	B33	110
		semi-gloss		2 cts: ProClassic HS Interior Alkyd Semi-Gloss, or	2.3	57	B34	107
				2 cts: ProClassic Interior Alkyd Semi-Gloss, or	1.7	42	B34	107
				2 cts: ProMar 200 or 400 Int. Alkyd Semi-Gloss, or	1.7	42	B34	111
		gloss		2 cts: ProMar 200 Int. Alkyd Gloss Enamel	1.6	40	B35	112

16.3.0 Coating Specifications for Interior Surfaces (Continued)

Substrate/Area	Topcoat		Specifications for Normal Exposures		Minimum dft/ct		Product	
	Vehicle	Finish	Surface Preparation	Primers & Topcoats	Mils	Microns	Series	Page
Interior Painting Recommendations—Normal Exposure								
metal, continued								
Galvanized Steel	acrylic latex	primer	S-W 1 or 12	1 ct: DTM Acrylic Primer/Finish	2.5	62	B66	57
		flat		2 cts: ProMar 200 or 400 Int. Latex Flat, or	1.4	35	B30	114
		eg-shel		2 cts: ProMar 200 or 400 Int. Latex Eg-Shel, or	1.6	40	B20	115
		semi-gloss		2 cts: ProClassic Waterborne Semi-Gloss Enamel, or	1.3	32	B31	108
				2 cts: ProMar 200 or 400 Int. Latex Semi-Gloss, or	1.3	32	B31	116
		gloss		2 cts: ProClassic Waterborne Gloss Enamel, or	1.3	32	B31	108
				2 cts: ProMar 200 Int. Latex Gloss Enamel	1.5	37	B21	117
Galvanized Steel	acrylic alkyd	primer	S-W 1 or 12	1 ct: Galvite HS Primer	3.0	75	B50	72
		flat		2 cts: ProMar 200 Int. Alkyd Flat, or	1.8	45	B32	109
		eg-shel		2 cts: ProMar 200 Int. Alkyd Eg-Shel, or	1.8	45	B33	110
		semi-gloss		2 cts: ProClassic HS Interior Alkyd Semi-Gloss, or	2.3	57	B34	107
				2 cts: ProClassic Interior Alkyd Semi-Gloss, or	1.7	42	B34	107
				2 cts: ProMar 200 or 400 Int. Alkyd Semi-Gloss, or	1.7	42	B34	111
		gloss		2 cts: ProMar 200 Int. Alkyd Gloss Enamel	1.6	40	B35	112
Steel and Iron	acrylic alkyd latex	primer	S-W 14 or 12	1 ct: DTM Acrylic Primer/Finish, or	2.5	62	B66	57
		primer		1 ct: Kem Kromik Universal Metal Primer	3.0	75	B50	89
		flat		2 cts: ProMar 200 or 400 Int. Latex Flat, or	1.4	35	B30	114
		eg-shel		2 cts: ProMar 200 or 400 Int. Latex Eg-Shel, or	1.6	40	B20	115
		semi-gloss		2 cts: ProClassic Waterborne Semi-Gloss Enamel, or	1.3	32	B31	108
				2 cts: ProMar 200 or 400 Int. Latex Semi-Gloss, or	1.3	32.5	B31	116
		gloss		2 cts: ProClassic Waterborne Gloss Enamel, or	1.3	32	B31	108
				2 cts: ProMar 200 Int. Latex Gloss Enamel	1.5	37.5	B21	117
	alkyd	primer	S-W 14 or 12	1 ct: Kem Bond HS Universal Primer	5.0	125	B50	84
		flat		2 cts: ProMar 200 Int. Alkyd Flat, or	1.8	45.0	B32	109
		eg-shel		2 cts: ProMar 200 Int. Alkyd Eg-Shel, or	1.8	45.0	B33	110
		semi-gloss		2 cts: ProClassic HS Interior Alkyd Semi-Gloss, or	2.3	57	B34	107
				2 cts: ProClassic Interior Alkyd Semi-Gloss, or	1.7	42	B34	107
				2 cts: ProMar 200 Int. Alkyd Semi-Gloss, or	1.7	42.5	B34	111
				2 cts: ProMar 400 Int. Alkyd Semi-Gloss, or	1.6	40.0	B34	111
		gloss		2 cts: ProMar 200 Int. Alkyd Gloss Enamel	1.6	40.0	B35	112
	alkyd	primer	S-W 14 or 12	1 ct: Kem Bond HS Universal Metal Primer	2.0	50	B50	84
		aluminum		2 cts: Silver-Brite Aluminum, B59S11, or	1.0	25	B59	130
				2 cts: Silver-Brite Rust Resistant Aluminum, B59S2	1.0	25	B59	130
wood								
Walls, Trim, Doors Windows, Ceilings	alkyd	primer	S-W 24 or 12	1 ct: ProMar Classic Latex Primer, or	1.6	40	B28	119
				1 ct: ProMar 200 Interior Enamel Undercoater, or	1.9	47	B49	113
				1 ct: Wall and Wood Primer	1.6	40	B49	143
		flat		2 cts: ProMar 200 Interior Alkyd Flat, or	1.8	45	B32	109
		eg-shel		2 cts: ProMar 200 Interior Alkyd Eg-Shel, or	1.8	45	B33	110
		semi-gloss		2 cts: ProClassic HS Interior Alkyd Semi-Gloss, or	2.3	57	B34	107
				2 cts: ProClassic Interior Alkyd Semi-Gloss, or	1.7	42	B34	107
				2 cts: ProMar 200 Interior Alkyd Semi-Gloss, or	1.7	42	B34	111
				2 cts: ProMar 400 Interior Alkyd Semi-Gloss, or	1.6	40	B34	111
		gloss		2 cts: ProMar 200 Interior Alkyd Gloss	1.6	40	B35	112
	alkyd	primer	S-W 24 or 12	1 ct: ProMar Classic Latex Primer, or	1.6	40	B28	119
				1 ct: ProMar 200 Interior Enamel Undercoater, or	1.9	47	B49	113
	latex			1 ct: Wall and Wood Primer	1.6	40	B49	143
		flat		2 cts: ProMar 200 or 400 Interior Latex Flat, or	1.4	35	B30	114
		eg-shel		2 cts: ProMar 200 or 400 Interior Latex Eg-Shel, or	1.6	40	B20	115
		semi-gloss		2 cts: ProClassic Waterborne Latex Semi-Gloss, or	1.3	32	B31	108
				2 cts: ProMar 200 or 400 Interior Latex Semi-Gloss, or	1.3	32	B31	116
		gloss		2 cts: ProMar 200 Interior Latex Gloss Enamel	1.5	37	B21	117

16.3.0 Coating Specifications for Interior Surfaces (Continued)

Substrate/Area	Topcoat: Vehicle	Topcoat: Finish	Specifications for Normal Exposures: Surface Preparation	Specifications for Normal Exposures: Primers & Topcoats		Minimum dft/ct: Mils	Minimum dft/ct: Microns	Product: Series	Product: Page
Interior Painting Recommendations—Normal Exposure									
wood, continued									
Stain resistant topcoat	latex	primer	S-W 24 or 12	1 ct:	ProMar Classic Latex Primer, or	1.6	40	B28	119
				1 ct:	ProMar 200 Interior Latex Wall Primer	1.1	27	B28	118
	acrylic	flat		2 cts:	EverClean Interior Latex Flat	1.7	42	A96	69
		satin		2 cts:	EverClean Interior Latex Satin	1.7	42	A97	69
		semi-gloss		2 cts:	EverClean Interior Latex Semi-Gloss	1.3	32	A98	69
Low odor finishes	latex	primer	S-W 24 or 12	1 ct:	HealthSpec Low Odor Int. Latex Primer	1.0	25	B11	76
	acrylic	flat		2 cts:	HealthSpec Low Odor Int. Latex Flat, or	1.5	37	B5	75
		eg-shel		2 cts:	HealthSpec Low Odor Int. Latex Eg-Shel, or	1.5	37	B9	75
		semi-gloss		2 cts:	HealthSpec Low Odor Int. Latex Semi-Gloss	1.5	37	B10	75
Ceilings	alkyd	primer	S-W 24 or 12	1 ct:	ProMar Classic Latex Primer, or	1.6	40	B28	119
				1 ct:	ProMar 200 Interior Enamel Undercoater, or	1.9	47	B49	113
				1 ct:	Wall and Wood Primer	1.6	40	B49	143
		flat		1 ct:	Super Save Lite Hi-Tec Dryfall	2.0	50	B48	137
				1 ct:	Dry Fall Flat White, or	4.0	100	B48	55
				1 ct:	Super Save Lite Flat, or	3.0	75	B48	136
		semi-gloss		1 ct:	Super Save Lite Semi-Gloss, or	3.0	75	B47	136
		gloss		1 ct:	Super Save Lite Gloss	2.0	50	B47	136
	alkyd	primer	S-W 24 or 12	1 ct:	ProMar Classic Latex Primer, or	1.6	40	B28	119
				1 ct:	ProMar 200 Interior Enamel Undercoater, or	1.9	47	B49	113
				1 ct:	Wall and Wood Primer	1.6	40	B49	143
	acrylic	flat		1-2 cts:	Waterborne Acrylic Dryfall Flat	4.0	100	B42	145
		eg-shel		1-2 cts:	Waterborne Acrylic Dryfall Eg-Shel	4.0	100	B42	145
Clear Finishes Varnishes	alkyd	stain	S-W 24 or 12	1 ct:	Oil Stain (omit if clear finish is desired)	none	none	A48	101
		sealer		1 ct:	ProMar Varnish Sanding Sealer (optional)	1.0	25	B26	123
		satin		2 cts:	Oil Base Varnish, Satin, or	1.3	32	A66	100
		gloss		2 cts:	Oil Base Varnish, Gloss	1.3	32	A66	100
	alkyd	stain	S-W 24 or 12	1 ct:	Oil Stain (omit if clear finish is desired)	none	none	A48	101
	polyurethane	satin		2 cts:	Polyurethane Varnish, Satin, or	1.7	42	A67	105
		gloss		2 cts:	Polyurethane Varnish, Gloss	1.7	42	A67	105
Floors	alkyd	gloss	S-W 24 or 12	1 ct:	Oil Stain (omit if clear finish is desired)	none	none	A48	101
	alkyd			2 cts:	Oil Base Varnish, or	1.3	32	A66	100
	polyurethane			2 cts:	Polyurethane Varnish	1.7	42	A67	105

16.4.0 Specifications for Industrial Exposure (Light/Moderate Duty)

Substrate/Area	Topcoat		Specifications for Normal Exposures		Minimum dft/ct		Product	
	Vehicle	Finish	Surface Preparation	Primers & Topcoats	Mils	Microns	Series	Page
steel and iron								
walls, joists, trim, doors, ducts, vents, structural items, miscellaneous	acrylic	primer	SSPC-SP2	1 ct: DTM Acrylic Primer/Finish	5.0	125	B66	57
		flat		2 cts: DTM Acrylic Primer/Finish	5.0	125	B66	57
	acrylic	primer	SSPC-SP2	1 ct: DTM Acrylic Primer/Finish	5.0	125	B66	57
		semi-gloss		2 cts: DTM Acrylic Semi-Gloss, or	4.0	100	B66	56
				2 cts: Metalatex Semi-Gloss, or	4.0	100	B42	99
		gloss		2 cts: DTM Acrylic Gloss	4.0	100	B66	56
	alkyd	semi-gloss	SSPC-SP2	2 cts: Direct-to-Metal Alkyd Semi-Gloss	5.0	125	B55	54
	alkyd	primer	SSPC-SP2	1 ct: Kem Kromik Universal Metal Primer, or	4.0	100	B50	89
				1 ct: Kem Bond HS, or	2.0	125	B50	84
				1 ct: High Solids Alkyd Metal Primer,or	5.0	125	B50	79
				1 ct: Kromik Metal Primer	4.0	100	E41	90
		gloss		2 cts: Industrial Enamel HS, or	4.0	100	B54Z	83
				2 cts: Industrial Enamel VOC	3.0	75	B54Z	83
	alkyd	primer	SSPC-SP2	1 ct: KemKromik Universal Metal Primer, or	4.0	100	B50	89
				1 ct: Kem Bond HS Primer, or	2.0	125	B50	84
				1 ct: High Solids Alkyd Metal Primer, or	5.0	125	B50	79
				1 ct: Kromik Metal Primer	4.0	100	E41	90
	silicone alkyd	gloss		2 cts: Silicone Alkyd Enamel Low VOC	4.0	100	B56	129
				2 cts: Steel-Master 9500 Silicone Alkyd	3.0	75	B56	134
	polyamide epoxy	semi-gloss	SSPC-SP2	1-2 cts: Surface Tolerant Epoxy Primer	8.0	200	B58	138
				1-2 cts: Surface Tolerant Epoxy Coating	6.0	150	B58	138
	moisture cured urethane	primer	SSPC-SP2	1 ct: Corothane I Zinc Primer	3.5	87	B65	48
				1 ct: Corothane I Mastic	3.5	87	B65	47
		gloss		1 ct: Corothane I Aromatic Finish, or	2.0	50	B65	45
				1 ct: Corothane I Aliphatic Finish	2.0	50	B65	43
	moisture cured urethane	primer	SSPC-SP2	1 ct: Corothane I Zinc Primer	3.5	87	B65	48
				1 ct: Corothane I Aliphatic Finish, or	2.0	50	B65	43
		gloss		1 ct: Corothane I Aluminum	3.0	75	B65	44
	water based epoxy	primer	SSPC-SP2	1 ct: Water Based Catalyzed Epoxy Primer	5.0	125	B70	144
		gloss		2 cts: Water Based Catalyzed Epoxy	3.0	75	B70	144
Ceilings	alkyd	primer	SSPC-SP2	1 ct: High Solids Alkyd Metal Primer	3.0	75	B50	79
		flat		1-2 cts: Super Save-Lite Hi-Tec Dryfall	1.5	37	B48	137
	alkyd	primer	SSPC-SP2	1 ct: Kem Bond HS Universal Primer	5.0	125	B50	84
		flat		1 ct: Dry Fall Flat White, or	4.0	100	B48	55
				1 ct: Super Save Lite Flat, or	3.0	75	B48	136
		semi-gloss		1 ct: Super Save Lite Semi-Gloss, or	3.0	75	B47	136
		gloss		1 ct: Super Save Lite Gloss, or	2.0	50	B47	136
	acrylic	flat		2 cts: DTM Acrylic Primer/Finish, or	5.0	125	B66	57
	epoxy ester	flat		1-2 cts: Galvite Epoxy Ester Dry Fall	4.0	100	B48	71
	alkyd	flat	SSPC-SP2	1-2 cts: Opti-Bond Multi-Surface Coating	3.5	87	B50	102
aluminum								
ducts, trim, miscellaneous	acrylic	semi-gloss	SSPC-SP1	2 cts: DTM Acrylic Semi-Gloss, or	4.0	100	B66	56
				2 cts: Metalatex Semi-Gloss, or	4.0	100	B42	99
		gloss		2 cts: DTM Acrylic Gloss	4.0	100	B66	56
	acrylic	primer	SSPC-SP1	1 ct: DTM Wash Primer	1.0	25	B71	59
	alkyd	gloss		2 cts: Industrial Enamel HS, or	4.0	100	B54	83
				2 cts: Industrial Enamel VOC	3.0	75	B54	83

16.4.0 Specifications for Industrial Exposure (Light/Moderate Duty) (Continued)

Substrate/Area	Topcoat Vehicle	Topcoat Finish	Surface Preparation	Primers & Topcoats		Minimum dft/ct Mils	Minimum dft/ct Microns	Product Series	Product Page
Painting Recommendations—Industrial Exposure, Light/Moderate duty exposure									
masonry									
walls,	acrylic	filler	S-W 5	1 ct: 2 cts:	Heavy Duty Block Filler (interior) Heavy Duty Block Filler (exterior)	10.0 10.0	250 250	B42 B42	77 77
	acrylic	filler flat semi-gloss gloss	S-W 5	1 ct: 2 cts: 2 cts: 2 cts: 2 cts:	Heavy Duty Block Filler DTM Acrylic Primer/Finish, or DTM Acrylic Semi-Gloss Coating, or Metalatex Semi-Gloss, or DTM Acrylic Gloss Coating	10.0 5.0 4.0 4.0 4.0	250 125 100 100 100	B42 B66 B66 B42 B66	77 57 56 99 56
	acrylic alkyd	primer gloss	S-W 5	1 ct: 2 cts: 2 cts:	ProMar Masonry Conditioner Industrial Enamel HS, or Industrial Enamel VOC	2.2 4.0 3.0	55 100 75	B46 B54Z B54Z	122 83 83
	acrylic silicone alkyd	filler gloss	S-W 5	1 ct: 2 cts:	Heavy Duty Block Filler Steel-Master 9500 Silicone Alkyd	10.0 3.0	250 75	B42 B56	77 134
	epoxy ester	filler	S-W 5	1 ct:	Epoxy Ester Masonry Filler/Sealer	10	250	B61	66
	acrylic	semi-gloss gloss	S-W 5	2 cts: 2 cts:	DTM Acrylic Semi-Gloss Coating, or DTM Acrylic Gloss Coating	4.0 4.0	100 100	B66 B66	56 56
	epoxy ester water based epoxy	filler gloss	S-W 5	1 ct: 1-2 cts:	Epoxy Ester Masonry Filler/Sealer Water Based Catalyzed Epoxy	10 3.0	250 75	B61 B70	66 144
	epoxy ester	filler	S-W 5	1-2 cts:	Sher-Crete Epoxy Ester Masonry Coating	10	250	B61	127
Ceilings	alkyd	primer flat	S-W 5	1 ct: 1-2 cts:	Epoxy Ester Masonry Filler Sealer Super Save-Lite Hi-Tec Dryfall	10 1.5	250 37	B61 B48	66 137
	epoxy ester alkyd	primer flat semi-gloss gloss	S-W 5	1 ct: 1 ct: 1 ct: 1 ct: 1 ct:	Epoxy Ester Masonry Filler Sealer Dry Fall Flat White, or Super Save Lite Flat, or Super Save Lite Semi-Gloss, or Super Save Lite Gloss, or	10 4.0 3.0 3.0 2.0	250 100 75 75 50	B61 B48 B48 B47 B47	66 55 136 136 136
	acrylic epoxy ester	filler flat	S-W 5	1 ct: 1-2 cts:	Heavy Duty Block Filler Galvite Epoxy Ester Dry Fall	10.0 4.0	250 100	B42 B48	77 71
	alkyd	flat	S-W 5	1-2 cts:	Opti-Bond Multi-Surface Coating	3.5	87	B50	102
Concrete Floors	water based epoxy system	primer gloss	S-W 5	1 ct: 1 ct:	ArmorSeal Water Based Epoxy Primer/Sealer ArmorSeal 700HS Water Based Epoxy	7.0 7.5	175 187	B70 B70	36 31
	waterbased epoxy system	primer gloss	S-W 5	1 ct: 2 cts:	ArmorSeal Floor-Plex 7100 (reduced) ArmorSeal Floor-Plex 7100	2.0 2.0	50 50	B70 B70	34 34
	solvent based epoxy	gloss	S-W 5	1-2 cts:	ArmorSeal 1000 HS Epoxy	4.5	112	B67	32

16.5.0 Coating Recommendations (Industrial Exposure and Heavy-Duty Exposure)

Substrate/Area	Vehicle	Finish	Surface Preparation	Primers & Topcoats		Mils	Microns	Series	Page
aluminum									
ducts, trim, miscellaneous	acrylic polyamide epoxy	primer gloss	SSPC-SP1	1 ct: 2 cts:	DTM Wash Primer Tile-Clad High Solids Epoxy	1.0 4.0	25 100	B71 B62	59 141
	acrylic epoxy	primer	SSPC-SP1	2 cts:	Water Based Catalyzed Epoxy	3.0	75	B70	144

16.5.0 Coating Recommendations (Industrial Exposure and Heavy-Duty Exposure) (Continued)

Substrate/Area	Topcoat Vehicle	Topcoat Finish	Surface Preparation	Primers & Topcoats	Minimum dft/ct Mils	Minimum dft/ct Microns	Product Series	Product Page
Painting Recommendations—Industrial Exposure, Heavy duty exposure								
aluminum, continued								
	acrylic polyamide epoxy	primer gloss	SSPC-SP1	1 ct: DTM Wash Primer 2 cts: Sher-Tile High Solids Epoxy	1.0 8.0	25 200	B71 B67	59 128
	polyamide epoxy	low sheen	SSPC-SP1	2 cts: Heavy Duty Epoxy	7.0	175	B67	78
	epoxy mastic	semi-gloss	SSPC-SP1	2 cts: Macropoxy High Solids Epoxy	6.0	150	B58	97
galvanized steel								
ducts, trim, miscellaneous	acrylic epoxy	primer	SSPC-SP1	2 cts: Water Based Catalyzed Epoxy	3.0	75	B70	144
	epoxy	low gloss gloss	SSPC-SP1	2 cts: Heavy Duty Epoxy, or 2 cts: Tile-Clad High Solids Epoxy, or 2 cts: Sher-Tile High Solids Epoxy, or	7.0 4.0 8.0	175 100 200	B67 B62 B67	78 141 128
	epoxy mastic	semi-gloss	SSPC-SP1	2 cts: Macropoxy High Solids Epoxy	6.0	150	B58	97
	epoxy polyurethane	primer gloss	SSPC-SP1	1 ct: Recoatable Epoxy Primer 1-2 cts: Poly-Lon 1900 Polyester Polyurethane	6.0 2.0	150 50	B67 B65	124 104
steel and iron								
walls, joists, trim, doors, ducts, vents structural items, miscellaneous	acrylic epoxy	primer gloss	SSPC-SP3	1 ct: Water Based Catalyzed Epoxy Primer 1-2 cts: Water Based Catalyzed Epoxy	5.0 3.0	125 75	B70 B70	144 144
	polyamide epoxy	primer low sheen gloss	SSPC-SP6	1 ct: Recoatable Epoxy Primer 2 cts: Heavy Duty Epoxy, or 2 cts: Tile-Clad High Solids Epoxy, or 2 cts: Sher-Tile High Solids Epoxy	6.0 7.0 4.0 8.0	150 175 100 200	B67 B67 B62 B67	124 78 141 128
	polyamide epoxy	primer semi-gloss	SSPC-SP2	1 ct: Surface Tolerant Epoxy Primer 1-2 cts: Surface Tolerant Epoxy Coating	8.0 6.0	200 150	B58 B58	138 138
	epoxy mastic	aluminum semi-gloss	SSPC-SP2 SSPC-SP2	1-2 cts: Macropoxy Aluminum, or 1-2 cts: Macropoxy High Solids Epoxy, or 1-2 cts: Epoxy Mastic Coating	6.0 6.0 10	150 150 250	B58 B58 B58	97 97 68
	epoxy	pre-primer primer semi-gloss gloss	SSPC-SP2	1 ct: Macropoxy 920 Pre-Prime 1 ct: Macropoxy Primer 1-2 cts: Macropoxy 646 Fast Cure Epoxy, or 1-2 cts: Macropoxy High Solids Epoxy	2.0 6.0 10.0 6.0	50 150 250 150	B58 B58 B58 B58	96 97 94 97
	epoxy polyurethane	primer gloss	SSPC-SP6	1 ct: Epolon II Rust Inhibitive Epoxy Primer 1-2 cts: Poly-Lon 1900 Polyester Polyurethane	2.0 2.0	50 50	B62 B65	62 104
	epoxy mastic polyurethane	aluminum primer gloss	SSPC-SP2	1 ct: Epoxy Mastic Aluminum II, or 1 ct: Macropoxy High Solids Primer 1-2 cts: Corothane II Polyurethane, or 1-2 cts: Hi-Solids Polyurethane	6.0 6.0 4.0 4.0	150 150 100 100	B62 B58 B65 B65	67 97 49 82
	zinc rich epoxy epoxy mastic	zinc primer low gloss semi-gloss gloss	SSPC-SP6	1 ct: Zinc Clad Primer 1-2 cts: Heavy Duty Epoxy, or 1-2 cts: Macropoxy High Solids Epoxy 1-2 cts: Sher-Tile Epoxy	5.0 7.0 6.0 8.0	125 175 150 200	B69 B67 B58 B67	148-151 78 97 128
	moisture cured urethane	primer matte gloss	SSPC-SP6	1 ct: Corothane I Zinc Primer 1 ct: Corothane I Mastic 1 ct: Corothane I Aliphatic Finish	3.5 2 2.0	87 50 50	B65 B65 B65	48 47 43

16.5.0 Coating Recommendations (Industrial, Exposure and Heavy-Duty Exposure) (Continued)

Substrate/Area	Topcoat		Specifications for Normal Exposures		Minimum dft/ct		Product	
	Vehicle	Finish	Surface Preparation	Primers & Topcoats	Mils	Microns	Series	Page
Painting Recommendations—Industrial Exposure, Heavy duty exposure								
steel and iron								
	zinc rich epoxy polyurethane	zinc primer semi-gloss gloss	SSPC-SP6	1 ct: Zinc Clad Primer 1-2 cts: Heavy Duty Epoxy 1 ct: Corothane II Polyurethane, or 1 ct: Hi-Solids Polyurethane	5.0 7.0 4.0 4.0	125 175 100 100	B69 B67 B65 B65	148-151 78 49 82
	epoxy polyester epoxy	primer gloss	SSPC-SP6	1 ct: Recoatable Epoxy Primer 1-2 cts: Armor-Tile HS Polyester Epoxy	6.0 4.0	150 100	B67 B67	124 37
	polyamide epoxy	primer semi-gloss	SSPC-SP6	1 ct: Epolon II Rust Inhibitive Epoxy Primer 1-2 cts: Epolon II Multi-Mil Epoxy	4.0 6	100 150	B67 B62	62 61
masonry								
walls	epoxy	filler semi-gloss gloss	brush blast	1 ct: Kem Cati-Coat Epoxy Filler/Sealer 1-2 cts: Heavy Duty Epoxy, or 1-2 cts: Sher-Tile HS Epoxy	30 7.0 8.0	750 175 200	B42 B67 B67	85 78 128
	epoxy polyurethane	filler satin/gloss gloss	brush blast	1 ct: Kem Cati-Coat Epoxy Filler/Sealer 1-2 cts: Corothane II Polyurethane, or 1-2 cts: Hi-Solids Polyurethane	30 4.0 4.0	750 100 100	B42 B65 B65	85 49 82
	moisture cured urethane	gloss	brush blast	1-2 cts: Corothane I Aliphatic Finish	2.0	50	B65	43
Concrete Floors	epoxy	primer gloss	brush blast	1 ct: ArmorSeal 33 Epoxy Primer/Sealer 1 ct: ArmorSeal 300 Heavy Duty Non-Skid	8.0 42.0	200 1050	B58 B67	28 29
	epoxy	primer gloss	brush blast	1 ct: ArmorSeal 33 Epoxy Primer/Sealer 1 ct: ArmorSeal 550SL Self Leveling Epoxy	8.0 30	200 750	B58 B58	28 30
	waterbased epoxy system	primer gloss	S-W 5	1 ct: ArmorSeal Floor-Plex 7100 (reduced) 2 cts: ArmorSeal Floor-Plex 7100	2.0 2.0	50 50	B70 B70	34 34

16.6.0 Painting Recommendations (Immersion Exposure)

Substrate/Area	Topcoat Vehicle	Topcoat Finish	Specifications for Normal Exposures: Surface Preparation	Specifications for Normal Exposures: Primers & Topcoats	Minimum dft/ct Mils	Minimum dft/ct Microns	Product Series	Product Page
steel								
non-potable water	coal tar epoxy	semi-gloss	SSPC-SP10	1-2 cts: Coal Tar Epoxy C-200, or 1 ct: Hi-Mil Sher-Tar Epoxy, or 1-2 cts: Corothane I Coal Tar	16.0 24.0 7.0	400 600 175	B69 B69 B65	41 80 46
non-potable water	epoxy epoxy	primer semi-gloss	SSPC-SP10	1 ct: Epoxide 52 Epoxy Primer, or 1 ct: Copoxy Shop Primer 2 cts: Hi-Solids Catalyzed Epoxy, or 2 cts: Tank Clad HS Epoxy	3.0 5.0 6.0 8.0	75 125 150 200	B67 B62 B62 B62	64 42 81 140
potable water	epoxy	semi-gloss	SSPC-SP10	2-3 cts: Hi-Solids Catalyzed Epoxy, or 2 cts: Tank Clad HS Epoxy see AWWA or NSF specifications for additional details	6.0 8.0	150 200	B62 B62	81 140
	epoxy amine	low sheen	SSPC-SP10	2 cts: Epoxide 33/34 Potable Water Epoxy	5.0	125	B62	63
concrete								
non-potable water	coal tar epoxy	semi-gloss	Brush Blast	1-2 cts: Coal Tar Epoxy C-200, or 1 ct: Hi-Mil Sher-Tar Epoxy	16.0 24.0	400 600	B69 B69	41 80
	epoxy system	low sheen	SSPC-SP10	1 ct: Kem Cati-Coat Epoxy Filler/Sealer 1 ct: EpoSeal 3040 Fairing and Sealing Compound† 2 cts: Epoxide 33/34 Potable Water Epoxy contact a representative for information on this product	10.0 5.0	250 125	B42 B62	85 63
Painting Recommendations—Immersion Exposure								
concrete, continued								
potable water	epoxy	semi-gloss	Brush Blast	2-3 cts: Hi-Solids Catalyzed Epoxy, or 2 cts: Tank Clad HS Epoxy see AWWA or NSF specifications for additional details	6.0 8.0	150 200	B62 B62	81 140

16.7.0 Painting Recommendations (Low-Temperature Applications)

Down to 40°F: Steel	polyamide epoxy	semi-gloss	SSPC-SP2	2 cts: Macropoxy 846 Winter Grade Epoxy	8.0	200	B58	95
Aluminum	polyamide epoxy	semi-gloss	SSPC-SP1	2 cts: Macropoxy 846 Winter Grade Epoxy	8.0	200	B58	95
Galvanized	polyamide epoxy	semi-gloss	SSPC-SP1	2 cts: Macropoxy 846 Winter Grade Epoxy	8.0	200	B58	95
Down to 35°F: Steel	epoxy amine	low sheen	SSPC-SP2	1-2 cts: Polar Epoxy Low Temperature Cure Epoxy	7.0	175	B62	103

16.8.0 Painting Recommendations (High-Temperature Exposure)

Steel									
up to 450°F	silicone acrylic	primer	SSPC-SP10	1 ct: 1 ct:	Kem Hi-Temp Heat Flex II 450 Zinc Dust Primer Kem Hi-Temp Heat-Flex II 450	1.5 1.5	37 37	B59 B59	87 87
	silicone acrylic	primer	SSPC-SP10	1 ct: 1 ct:	Kem Hi-Temp Heat-Flex II 450 Primer Kem Hi-Temp Heat-Flex II 450	1.5 1.5	37 37	B59 B59	87 87
up to 800°F	silicone	low luster	SSPC-SP10	2 cts:	Kem Hi-Temp Heat-Flex II 800	1.5	37	B59	88
up to 400°F interior/exterior	alkyd	aluminum	SSPC-SP6	2 cts:	Silver-Brite Aluminum, B59S11	1.5	37	B59	130
up to 700°F interior	alkyd	aluminum	SSPC-SP6	2 cts:	Silver-Brite Hi-Heat Resisting Aluminum, B59S3	0.5	12	B59	132
500°-1000°F interior/exterior	alkyd	aluminum	SSPC-SP6	2 cts:	Silver-Brite Hi-Heat Silicone Alkyd Aluminum B59S8	1.0	25	B59	133
Painting Recommendations—Traffic Marking Paints									
Concrete and Asphalt	latex	flat	SW	1 ct: 1 ct.	SetFast Acrylic Water Borne Traffic Paint Series TM226-White/TM225-Yellow, or SetFast Fast Dry Latex Traffic Marking Paint Series TM2136-White/TM2137-Yellow	7.0 8.5	175 212	TM TM	125 126

16.9.0 Recommended Surface-Preparation Procedures for Basic Construction Materials

Your responsibility, when writing a paint specification, is to understand the significant points in the task and include them in your specification. Details to be covered include: establish a central point from which the painting crew is to operate, provide parking space for painters' vehicles, proper identification, insurance, tools, etc. While these are important points that must be determined, they do not directly affect the paint job. Only those points pertaining to quality paint work will be covered here.

The scope of the paint job should be completely described, including everything that is to be cleaned and painted. DO NOT take anything for granted. Do not merely specify that "the surface should be sandblasted." Do specify the results you want to achieve and leave the choice of pressures, hose sizes, etc., up to the contractor. Allow the contractor to exercise initiative and ingenuity. You might get a better job at a lower price.

Write the specification in clear, precise, easy-to-understand language—so that all parties involved know what you mean. Be brief and to the point, do not confuse the reader. Remember, your primary objective is "a good paint job."

Surface Preparation

Coating performance is affected by proper product selection, surface preparation and application. Coating integrity and service life will be reduced because of improperly prepared surfaces. As high as 80% of all coatings failures can be directly attributed to inadequate surface preparation that affects coating adhesion. Selection and implementation of proper surface preparation ensures coating adhesion to the substrate and prolongs the service life of the coating system.

Selection of the proper method of surface preparation depends on the substrate, the environment, and the expected service life of the coating system. Economics, surface contamination, and the effect on the substrate will also influence the selection of surface preparation methods.

The surface must be dry and in sound condition. Remove oil, dust, dirt, loose rust, peeling paint or other contamination to ensure good adhesion.

Remove mildew before painting by washing with a solution of 1 quart liquid household bleach and 3 quarts of warm water. Apply the solution and scrub the mildewed area. Allow the solution to remain on the surface for 10 minutes. Rinse thoroughly with clean water and allow the surface to dry 48 hours before painting. Wear protective glasses or goggles, waterproof gloves, and protective clothing. Quickly wash off any of the mixture that comes in contact with your skin. Do not add detergents or ammonia to the bleach/water solution.

No exterior painting should be done immediately after a rain, during foggy weather, when rain is predicted, or when the temperature is below 50°F.

Aluminum S-W 1

Remove all oil, grease, dirt, oxide and other foreign material by cleaning per SSPC-SP1, Solvent Cleaning.

Asbestos Siding S-W 2

Remove all dust and dirt. If siding has been weathered and is porous, treat with Masonry Conditioner.

Block (Cinder and Concrete) S-W 3

Remove all loose mortar and foreign material. Surface must be free of laitance, concrete dust, dirt, form release agents, moisture curing membranes, loose cement, and hardeners. Concrete and mortar must be cured at least 30 days at 75°F. The pH of the surface should be between 6 and 9. On tilt-up and poured-in-place concrete, commercial detergents and abrasive blasting may be necessary to prepare the surface. Fill bug holes, air pockets, and other voids with a cement patching compound.

Brick S-W 4

Must be free of dirt, loose and excess mortar, and foreign material. All brick should be allowed to weather for at least one year followed by wire brushing to remove efflorescence. Treat the bare brick with one coat of Masonry Conditioner.

Concrete S-W 5

The following guides will help assure maximum performance of the coating system and satisfactory coating adhesion:

1. Cure—Concrete must be cured prior to coating application. Cured is defined as concrete poured and aged at a material temperature of at least 75°F for at least 30 days. The pH of the surface should be between 6 and 9.
2. Moisture—(Reference ASTM D4263) Concrete must be free of moisture as much as possible (moisture seldom drops below 15% in concrete). Test for moisture or dampness by taping the 4 edges of an 18 inch by 18 inch plastic sheet (4 mils thick) on the bare surface (an asphalt tile or other moisture impervious material will also do), sealing all of the edges. After a minimum of 16 hours, inspect for moisture, discoloration, or condensation on the concrete or the underside of the plastic. If moisture is present, the source must be located and the cause corrected prior to painting.
3. Temperature—Air, surface and material temperature must be at least 50°F (10°C) during the application and until the coating is cured.
4. Contamination—Remove all grease, dirt, loose paint, oil, tar, glaze, laitance, efflorescence, loose mortar, and cement by the recommendations A, B, C, or D, listed below.
5. Surface Condition—Hollow areas, bug holes, honeycombs, voids, fins, form marks, protrusions, or rough edges are to be ground or stoned to provide a smooth, continuous surface of suitable texture for proper adhesion of the coating. Imperfections may require filling with a material compatible with the Sherwin-Williams' coatings.
6. Concrete Treatment—Hardeners, sealers, form release agents, curing compounds, and other concrete treatments must be compatible with the coatings, or be removed.

16.9.0 Recommended Surface-Preparation Procedures for Basic Construction Materials (Continued)

Surface preparations for concrete

Method "A"—Blast Cleaning

(Reference ASTM D4259) Brush Blasting or Sweep Blasting—Includes dry blasting, water blasting, water blasting with abrasives, and vacuum blasting with abrasives.

1. Use 16 - 30 mesh sand and oil-free air.
2. Remove all surface contamination (ref. ASTM D4258). See Method "D" below.
3. Stand approximately 2 feet from the surface to be blasted.
4. Move nozzle at a uniform rate.
5. Laitance must be removed and bug holes opened.
6. Surface must be clean and dry (moisture check: ref. ASTM D4263) and exhibit a texture similar to that of medium grit sandpaper.
7. Vacuum or blow down and remove dust and loose particles from the surface (ref. ASTM D4258). See Method "D" below.

Method "B"—Acid Etching

1. Remove all surface contamination (ref. ASTM D4258)
2. Wet surface with clean water.
3. Apply a 10-15% Muriatic Acid or 50% Phosphoric Acid solution at the rate of one gallon per 75 square feet.
4. Scrub with a stiff brush.
5. Allow sufficient time for scrubbing until bubbling stops.
6. If no bubbling occurs, the surface is contaminated with grease, oil, or a concrete treatment which is interfering with proper etching. Remove the contamination with a suitable cleaner (ref. ASTM D4258, or Method "D" below) and then etch the surface.
7. Rinse the surface two or three times. Remove the acid/water mixture after each rinse.
8. Surface should have a texture similar to medium grit sandpaper.
9. It may be necessary to repeat this step several times if a suitable texture is not achieved with one etching. Bring the pH (ref. ASTM D4262) of the surface to neutral with a 3% solution of trisodium phosphate or similar alkali cleaner and flush with clean water to achieve a sound, clean surface.
10. Allow surface to dry and check for moisture (ref. ASTM D4263).

Method "C"—Power Tool Cleaning or Hand Tool Cleaning (ref. ASTM D4259)

1. Use needle guns or power grinders, equipped with a suitable grinding stone of appropriate size and hardness, which will remove concrete, loose mortar, fins, projections, and surface contaminants. Hand tools may also be used.
2. Vacuum or blow down and remove dust and loose particles from the surface (ref. ASTM D4258, or Method "D" below).
3. Test for moisture or dampness by taping the 4 edges of an 18 inch by 18 inch plastic sheet (4 mils thick) on the bare surface (an asphalt tile or other moisture impervious material will also do), sealing all of the edges. After a minimum of 16 hours, inspect for moisture, discoloration, or condensation on the concrete or the underside of the plastic. If moisture is present, the source must be located and the cause corrected prior to painting.

Method "D"—Surface Cleaning (ref. ASTM D4258)

The surface must be clean, free of contaminants, loose cement, mortar, oil, and grease. Broom cleaning, vacuum cleaning, air blast cleaning, water cleaning, and steam cleaning are suitable as outlined in ASTM D4258. Concrete curing compounds, form release agents, and concrete hardeners may not be compatible with recommended coatings. Check for compatibility by applying a test patch of the recommended coating system, covering at least 2 to 3 square feet. Allow to dry one week before testing adhesion per ASTM D3359. If the coating system is incompatible, surface preparation per methods outlined in ASTM D4259 are required.

Cement Composition Siding/Panels S-W 6

Remove all surface contamination by washing with an appropriate cleaner, rinse thoroughly and allow to dry. Existing peeled or checked paint should be scraped and sanded to a sound surface. Pressure clean, if needed, with a minimum of 2100 psi pressure to remove all dirt, dust, grease, oil, loose particles, laitance, foreign material, and peeling or defective coatings. Allow the surface to dry thoroughly. If the surface is new, test it for pH, many times the pH may be 10 or higher.

Copper S-W 7

Remove all oil, grease, dirt, oxide and other foreign material by cleaning per SSPC-SP 2, Hand Tool Cleaning.

Drywall—Interior and Exterior S-W 8

Must be clean and dry. All nail heads must be set and spackled. Joints must be taped and covered with a joint compound. Spackled nail heads and tape joints must be sanded smooth and all dust removed prior to painting. Exterior surfaces must be spackled with exterior grade compounds.

Exterior Composition Board (Hardboard) S-W 9

Some composition boards may exude a waxy material that must be removed with a solvent prior to coating. Whether factory primed or unprimed, exterior composition board siding (hardboard) must be cleaned thoroughly and primed with an alkyd primer.

Galvanized Metal S-W 10

Allow to weather a minimum of 6 months prior to coating. Clean per SSPC-SP1 using detergent and water or a degreasing cleaner, then prime as required. When weathering is not possible or the surface has been treated with chromates or silicates, first Solvent Clean per SSPC-SP1 and apply a test area, priming as required. Allow the coating to dry at least one week before testing. If adhesion is poor, Brush Blast per SSPC-SP7 is necessary to remove these treatments.

Plaster S-W 11

Must be allowed to dry thoroughly for at least 30 days before

16.9.0 Recommended Surface-Preparation Procedures for Basic Construction Materials (Continued)

painting. Room must be ventilated while drying; in cold, damp weather, rooms must be heated. Damaged areas must be repaired with an appropriate patching material. Bare plaster must be cured and hard. Textured, soft, porous, or powdery plaster should be treated with a solution of 1 pint household vinegar to 1 gallon of water. Repeat until the surface is hard, rinse with clear water and allow to dry.

Previously Coated Surfaces

Maintenance painting will frequently not permit or require complete removal of all old coatings prior to repainting. However, all surface contamination such as oil, grease, loose paint, mill scale dirt, foreign matter, rust, mold, mildew, mortar, efflorescence, and sealers must be removed to assure sound bonding to the tightly adhering old paint. Glossy surfaces of old paint films must be clean and dull before repainting. Thorough washing with an abrasive cleanser will clean and dull in one operation, or, wash thoroughly and dull by sanding. Spot prime any bare areas with an appropriate primer. Recognize that any surface preparation short of total removal of the old coating may compromise the service length of the system. Check for compatibility by applying a test patch of the recommended coating system, covering at least 2 to 3 square feet. Allow to dry one week before testing adhesion per ASTM D3359. If the coating system is incompatible, complete removal is required (per ASTM 4259, see Concrete, Method).

16.10.0 Preservative Treatment for Exterior Woodwork

Preservative Treatment for Exterior Woodwork

Modern technology has developed methods of treating certain species to extend their life when exposed to the elements. All lumber species used for exterior architectural woodwork, except species listed as "Resistant or very resistant" in the following tables (although it is desirable for those species) shall be treated with an industry tested and accepted formulation containing 3-iodo-2-propynyl butyl carbamate (IPBC) as its active ingredient according to manufacturer's directions.

Some domestic woods according to heartwood decay resistance:

Resistant or very resistant	Moderately resistant	Slightly or nonresistant
Cedars	Baldcypress (young growth) *	Ashes
Cherry, black	Douglas-fir	Basswood
Junipers	Pine, Eastern White *	Beech
White Oak	Pine, So. Longleaf *	Birches
Redwood, clear heart	Pine, Slash	Butternut
Walnut, black		Hemlocks
		Hickories
		Red Oak
		Pines (other than slash, longleaf, and E. white)
		Poplars
		Spruces
		True firs (western and eastern)

* - The southern and eastern pines and baldcypress are now largely second growth with a large proportion of sapwood. Substantial quantities of heartwood lumber of these species are not available.

Some imported woods according to heartwood decay resistance:

Resistant or very resistant	Moderately resistant	Slightly or nonresistant
Mahogany, American (Honduras)	Avodire	Obeche
Meranti **	European walnut	Mahogany, Philippine: Mayapis, White lauan
Teak	Mahogany, Philippine: Almon, Bagtikan, Red Lauan, Tangile	
	Sapele	

** - More than one species included, some of which may vary in resistance from that indicated.

DATA: U.S. Dept. of Agriculture, Forest Products Laboratory

16.11.0 Myth of Maintenance-Free Exterior Coatings

1. *What are the 20-year fluorocarbon paint coatings used on exterior aluminum members?*

   These coatings are high-molecular-weight polymers that have been formulated into a dispersion coating for application at the factory. Polyvinylidene fluoride (PVF2) is the base ingredient in these coatings. Other high-performance coatings are siliconized acrylics, siliconized polyesters, and other synthetic polymers.

2. *Are these coatings maintenance free?*

   No. Unless proper maintenance procedures are followed, these coated surfaces will degrade, over time, in the presence of atmospheric weathering and airborne pollutants.

3. *What specifically causes problems leading to degradation?*

   The collection of airborne dirt and chemical pollutants, in the presence of moisture, increases the potential for erosion, corrosion, loss of surface gloss, stainings, and discoloration.

4. *What is "chalking"?*

   Ultraviolet degradation of the resin vehicle and color in the coating results in loss of gloss and the formation of powder on the surface. This powder is referred to as *chalking*, a change in both the appearance and color of the coating. Regular maintenance can prevent chalking.

5. *When should the maintenance of exterior curtain walls begin?*

   As soon as possible after the installation and acceptance of the building by the owner so as to remove any dirt or pollutants caused during the construction process.

6. *What is AAMA 610.1?*

   The American Architectural Manufacturers Association (AAMA) developed AAMA 610.1, a procedure for the cleaning and maintenance of painted aluminum extrusions and curtain wall systems. These are general, not specific guidelines. AAMA suggests that owners hire experienced maintenance contractors for curtain wall cleaning, if they do not have such individuals on staff.

7. *What kind of cleaning cycles are considered adequate?*

   Exterior glazing is generally cleaned on a quarterly basis, depending upon the amount of atmospheric pollution in a specific geographic area. Curtain wall and exterior aluminum construction can be incorporated into the same schedule.

8. *Can the rundown from sealants contribute to the staining of aluminum with high performance coatings?*

   Yes. The oils and plasticizers in many caulking materials can bleed onto adjacent metal surfaces causing stains or discolorations.

9. *If a factory finish on a curtain wall is stained or discolored to the point where it needs to be re-coated, can a field applied coating be used to repair a factory applied coating?*

   In many cases—Yes. Coating manufacturers have developed a number of field applied airdried primers and finish coats for in-place coating repairs. The coating manufacturer or an approved applicator should be consulted for specifics.

16.12.0 Steel-Structure Painting Procedures

The authority on surface preparation and the subsequent painting of steel structures, the Steel Structures Painting Council, has developed a series of procedures that have become industry standards. The Steel Structures Painting Council developed specific surface-preparation procedures for the proper application of various types of coatings. Each surface-preparation procedure has been given an "SP" number, prefaced by their organization letters (SSPC). A particular procedure is referred to as *SSPC-SP* (and the number).

16.12.1 SSPC Specifications

SSPC specification	Description (summarized)
SP 1 Solvent Cleaning	Removal of oil, grease, dirt, soil, salts, and contaminants by cleaning with solvents, vapor, alkali, emulsion, or steam.
SP 2 Hand Tool Cleaning	Removal of loose rust, loose mill scale, and loose paint, by hand chipping, scraping, sanding, and wire brushing.
SP 3 Hand Tool Cleaning	Removal of loose rust, loose mill scale, and loose paint, by power-tool chipping, descaling, sanding, wire brushing, and grinding.
SP 5 White Metal Blasting	Removal of all visible rust, mill scale, paint, and foreign matter by blast cleaning by wheel or nozzle, dry or wet, using sand, grit, or shot.
SP 6 Commercial Blast Cleaning	Blast cleaning until at least $\frac{2}{3}$ of the surface area is free of all visible residues.
SP 7 Brush-off, Blast Cleaning	Blast cleaning of all, except tightly adhering residues of mill scale, rust, and coatings, exposing numerous evenly distributed flecks of underlying metal.
SP 8 Pickling	Complete removal of rust and mill scale by acid pickling, duplex pickling, or electrolytic pickling.
SP 10 Near-White Blast Cleaning	Blast cleaning to nearly white-metal cleanliness, until at least 95% of the surface area is free of all visible residues.
SP 11-87T Power-Tool Cleaning to Bare Metal	Complete removal of all rust, scales, and paint by power tools with resultant surface profile.

Note: SSPC does not have an SP 9 category.

16.12.2 SSPC Grading of New and Previously Painted Steel

Four surface conditions of new steel, with respect to its oxidation and rust formation, established by SSPC are:

- *Rust Grade A* A steel surface covered completely by adherent mill scale with little or no visible rust.
- *Rust Grade B* A steel surface covered with both mill scale and rust.
- *Rust Grade C* A steel surface completely covered with rust; little or no pitting is visible.
- *Rust Grade D* A steel surface completely covered with rust; pitting is visible.

Four conditions of previously painted steel construction are designated by SSPC for maintenance painting and are based upon the rust-grade classifications established by the Council, which range from:

- *Grade E* Non-deteriorated steel with 0 to 0.1% rust
- *Grade F* Slightly to moderately deteriorated steel with 0.1% to 1% rust

- *Grade G* Deteriorated steel with 1 to 10% rust
- *Grade H* Severely deteriorated steel with more than 10% rust and up to 100% rust

16.12.3 Minimum Surface Preparation for Various Painting Systems

According to the SSPC, certain minimum surface-preparation requirements are necessary for the application of various painting systems.

Painting System	Minimum Surface Preparation
Oil base	Hand tool cleaning (SSPC-SP2)
Alkyd	Commercial blast cleaning (SSPC-SP6 or pickling, SSPC-SP8)
Phenolic	Commercial blast cleaning (SSPC-SP6 or pickling, SSPC-SP8)
Vinyl	Commercial blast cleaning (SSPC-SP6 or pickling, SSPC-SP8)
Rust-Preventative Compounds	Solvent cleaning (SSPC-SP1 or nominal cleaning)
Asphalt Mastic	Commercial blast cleaning (SSPC-SP6 or pickling, SSPC-SP8)
Coal-Tar Coatings	Commercial blast cleaning (SSPC-SP6)
Coal-Tar Epoxy	Commercial blast cleaning (SSPC-SP6)
Zinc Rich	Commercial blast cleaning(SSPC-SP6)
Epoxy Polyamide	Commercial blast cleaning (SSPC-SP6 or pickling, SSPC-SP8)
Chlorinated Rubber	Commercial blast cleaning (SSPC-SP6 or pickling, SSPC-SP8)
Silicone Alkyd	Commercial blast cleaning (SSPC-SP6 or pickling, SSPC-SP8)
Urethane	Commercial blast cleaning (SSPC-SP6 or pickling, SSPC-SP8)
Latex	Commercial blast cleaning (SSPC-SP6 or pickling, SSPC-SP8)

16.12.4 Steel Structures Painting Council (SSPC) Coating Systems

SSPC-PS 1.04	Three-coat oil-alkyd (lead and chromate free) painting system for galvanized or nongalvanized steel (with zinc-dust/zinc-oxide linseed-oil primer
SSPC-PS 1.07	Three-coat oil-base red lead painting system
SSPC-PS 1.08	Four-coat oil-base red lead painting system
SSPC-PS 1.09	Three-coat oil-base zinc-oxide painting system (without lead or chromate pigment)
SSPC-PS 1.10	Four-coat oil-base zinc-oxide painting system (without lead or chromate pigment)
SSPC-PS 1.11	Three-coat oil-base red lead painting system
SSPC-PS 1.12	Three-coat oil-base zinc-chromate painting system
SSPC-PS 1.13	One-coat oil-base slow-drying maintenance painting system (without lead or chromate pigments)
SSPC-PS 2.03	Three-coat alkyd painting system with red lead-oxide primer)for weather exposure)
SSPC-PS 2.05	Three-coat alkyd painting system for unrusted galvanized steel (for weather protection)
SSPC-PS 4.01	Four-coat vinyl painting system with red lead primer (for salt-waste or chemical use)
SSPC-PS 4.02	Four-coat vinyl painting system (for fresh water, chemical, or corrosive atmospheres)
SSPC-PS 4.03	Three-coat vinyl painting system with wash primer (for salt-water and weather exposure)
SSPC-PS 4.04	Four-coat white or colored vinyl painting system (for fresh-water, chemical, or corrosive atmospheres)

SSPC-PS 4.05	Three-coat vinyl painting system with wash primer and vinyl alkyd finish coat (for atmospheric exposure)
SSPC-PS 8.01	One-coat rust-preventative painting system for thick-film compounds
SSPC-PS 9.01	Cold-applied asphalt mastic painting system with extra-thick film
SSPC-PS 10.01	Hot-applied coal-tar enamel painting system
SSPC-PS 10.02	Cold-applied coal-tar mastic painting system
SSPC-PS 11.02	Black (or dark red) coal-tar epoxy-polyamide painting system
SSPC-PS 12.01	One-coat zinc-rich painting system
SSPC-PS 13.10	Epoxy polyamide painting system
SSPC-PS 14.01	Steel-joist shop-painting system
SSPC-PS 15.01	Chlorinated-rubber painting system for salt-water immersion
SSPC-PS 15.02	Chlorinated-rubber painting system for fresh-water immersion
SSPC-PS 15.03	Chlorinated-rubber painting system for marine and industrial atmospheres
SSPC-PS 15.04	Chlorinated rubber painting system for field application over a shop-applied solvent-base inorganic zinc-rich primer
SSPC-PS 16.01	Silicone alkyd-base painting system for new steel
SSPC-PS 18.01	Three-coat latex painting system

16.13.0 Generic High-Performance Coatings for Steel and Concrete

The following formulations are a sampling of the types and ranges of high-performance coatings and their recommended service:

- *Polyurethane alkyd copolymer* Finish coat for pumps, motors, machinery, piping, and handrails, resulting in a high gloss that has excellent brush, roller, and spray characteristics. This finish exhibits excellent weathering capability and good abrasion resistance.
- *Epoxy polyamide* A 100% solid epoxy mastic that can be applied and cured underwater, providing protection against metal corrosion and erosion, and the deterioration of concrete and wood at (or below) the waterline. This type of coating is recommended for the repair of steel, concrete, or wood pilings; leaking tanks; boat hulls; and cracks in concrete; however, it is not recommended for immersion in (or exposure to) strong solvents or corrosive materials.
- *Aliphatic polyurethane* A two-part system that provides a satin finish coat on primed steel and exhibits very good resistance to splash and spillage of acids, alkalies, solvents, and salts. It has excellent abrasion-resistance qualities. This coating is used in chemical-processing, pulp and paper mills, and in the petro chemical industries.
- *Acrylic aliphatic polyurethane* Another two-part coating system that can be applied by brush, roller, or spray, and exhibits excellent weathering and abrasion-resistance characteristics. This coating is recommended as a finish coat over pigmented polyurethanes for exterior exposure where chemical resistance, gloss retention, and as excellent weathering characteristics are required. This coating can be used to provide a graffiti-free surface.
- *Elastomeric polyurethane* A two-component coating system that is utilized as a build coat overall compatible primer to provide a waterproof topping over concrete floors, decks, and walkways. A nonskid aggregate is often added to this coating to provide a slip-resistant surface.
- *Zinc-rich chlorinated rubber coating* Considered a "cold galvanizing" coating. When this coating is applied to a structural-steel member, the zinc metal in the coating bonds in much the same manner as hot-dip galvanizing. This single-component coating is an excellent material for the field touch-up of hot galvanized surfaces.
- *Thixotropic coal-tar coatings* A coal-tar-based material that can be applied in high-build layers by either brushing or rolling several coats to an 8-mil thickness. This coating is highly adaptable to application for underground or underwater usage.

- *Coal-tar epoxy polyamide* Providing excellent corrosion, chemical, and abrasion-resistance qualities, this coating is used in the sewage, water-treatment, chemical-processing industries, and on bridges and pilings, where steel and concrete structures are exposed to heavy-duty service conditions. Manufacturers, such as Carbonline and TNEMAC produce these, and other types of high-performance coatings.

Section

17

Elevators—Dumbwaiters

Contents

17.0.0 Basic elevator types, classifications, speeds, and capacities
17.1.0 Traction elevator installation (isometric)
17.1.1 Traction elevator (typical platform and sling assembly)
17.2.0 Traction gearless elevator installation (isometric)
17.3.0 Hydraulic elevator installation (isometric)
17.3.1 Hydraulic elevator (typical platform and sling assembly)
17.4.0 Hydraulic freight elevator installation (isometric)
17.5.0 Hoistway section (traction elevator)
17.6.0 Hoistway section (holeless, telescoping holeless, and conventional elevator)
17.7.0 Elevator machine-room configurations
17.8.0 Gearless elevator machine-room configuration
17.9.0 Hydraulic elevator preparation work (check list)
17.10.0 Dumbwaiter installation (isometrics)
17.11.0 Dumbwaiter (typical uses, standard sizes, and horsepower amperage)

17.0.0 Basic Elevator Types, Classifications, Speeds, and Capacities

THE ELEVATOR SYSTEM
HIGHLIGHTS

BASIC TYPES

Hydraulic (Oildraulic) • Traction

CAR CLASSIFICATION

Passenger • Freight

CAR SPEEDS

Oildraulic: 75 to 200 Feet Per Minute
Traction: 125 to 1200 Feet Per Minute

A very few extra tall buildings have cars which run at 1800 fpm.

TRAVEL

Oildraulic: Up to 65 Feet
Traction: Up to about 600 Feet

A few go higher

TYPICAL CAR CAPACITIES

Passenger: 1,000 to 3,500 Pounds
Freight: 1,000 to 75,000 Pounds

Only practicality limits capacity

MAJOR COMPONENTS

Hoistway • Car • Driving Machinery • Control Mechanism

17.1.0 Traction Elevator Installation (Isometric)

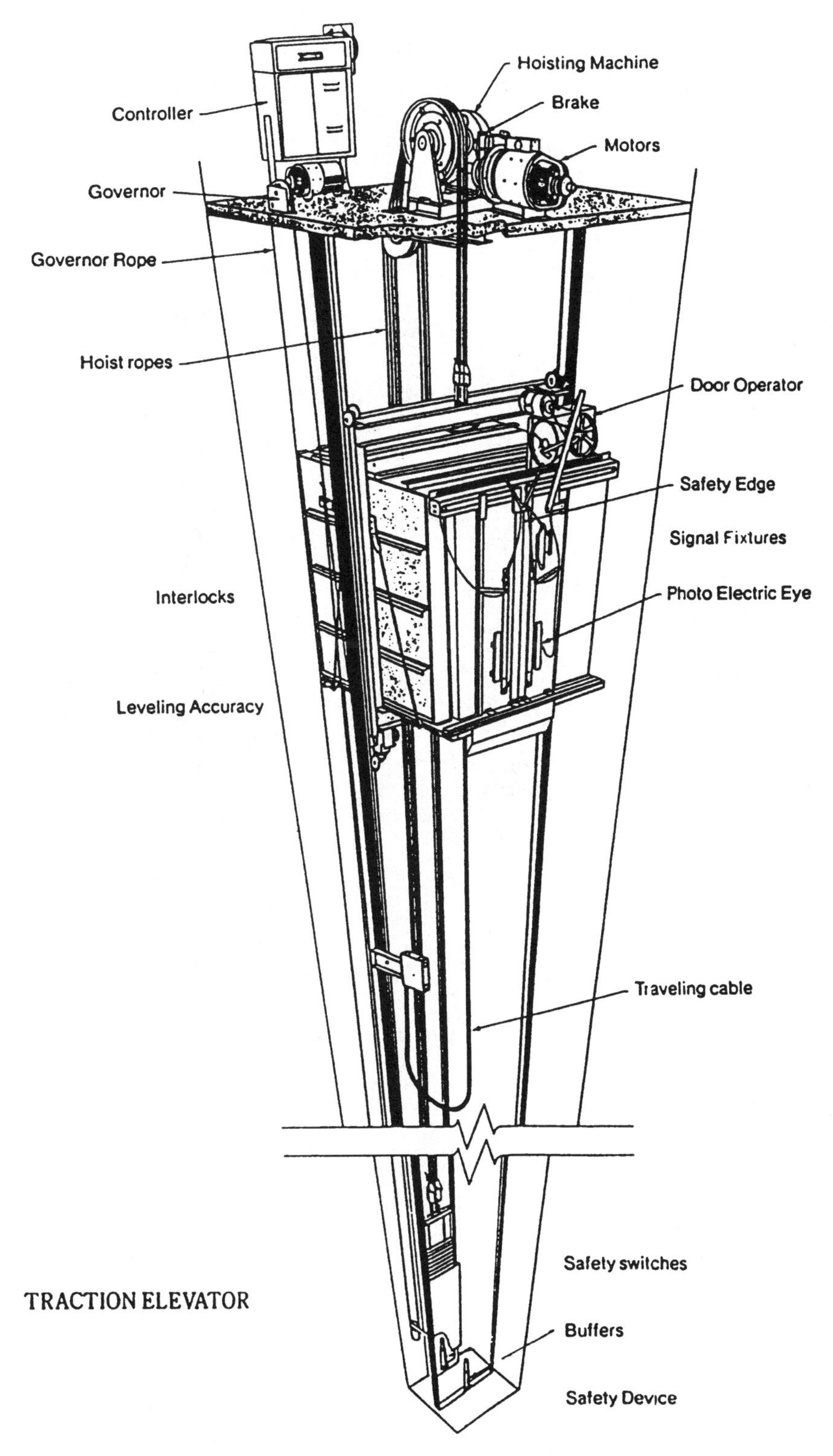

Reprinted with permission from Dover Elevator Company, Memphis, Tennessee

17.1.1 Traction Elevator (Typical Platform and Sling Assembly)

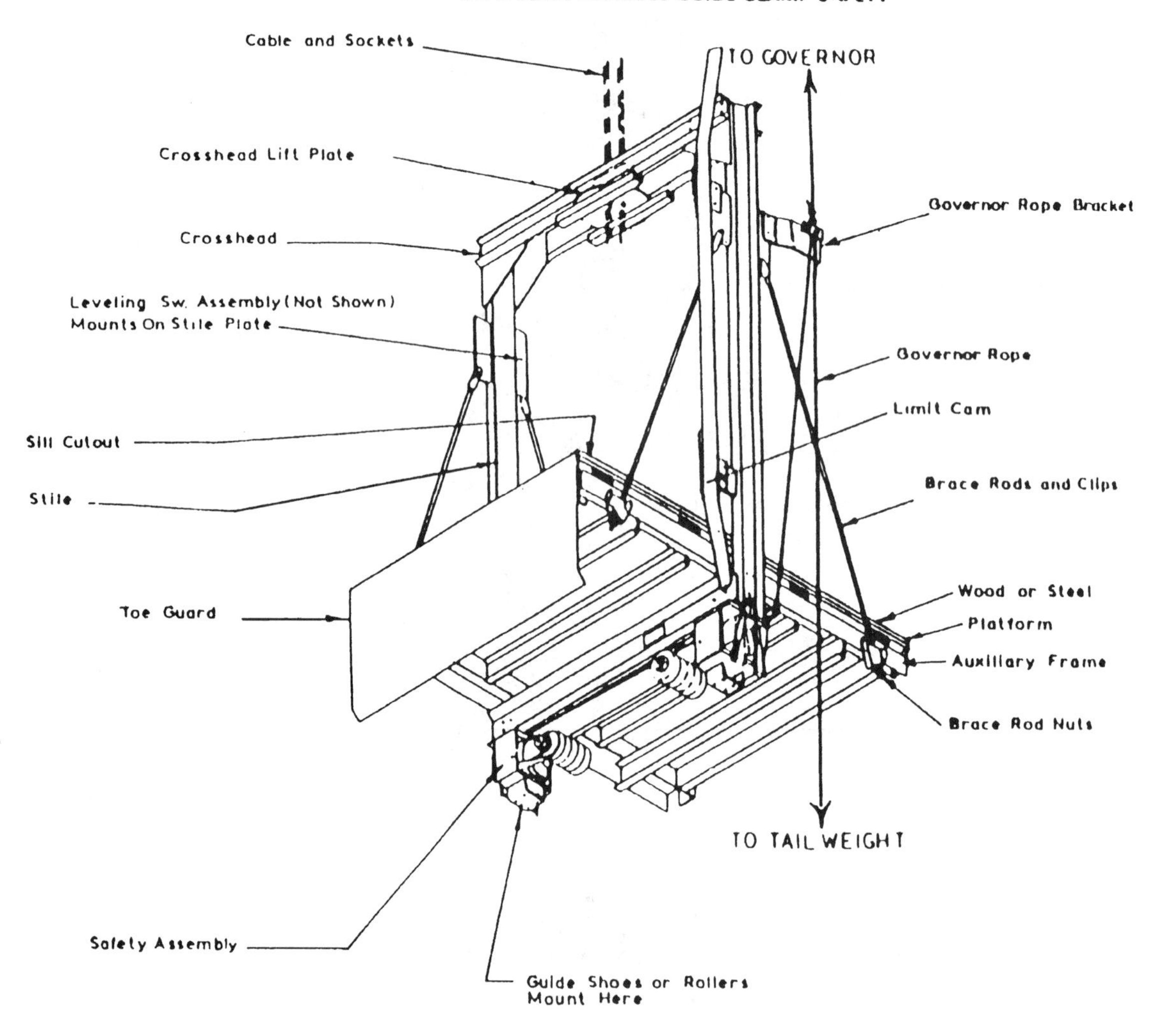

Reprinted with permission from Dover Elevator Company, Memphis, Tennessee

17.2.0 Traction Gearless Elevator Installation (Isometric)

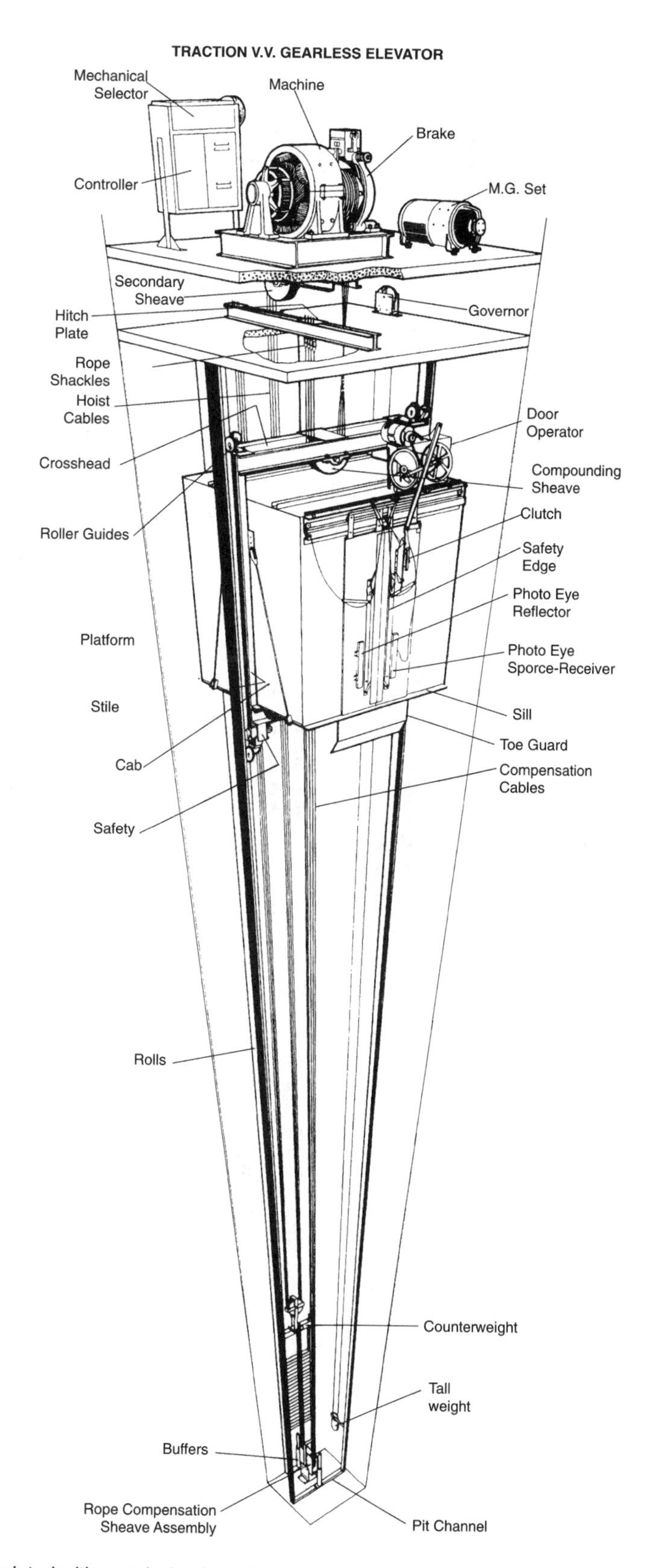

Reprinted with permission from Dover Elevator Company, Memphis, Tennessee

17.3.0 Hydraulic Elevator Installation (Isometric)

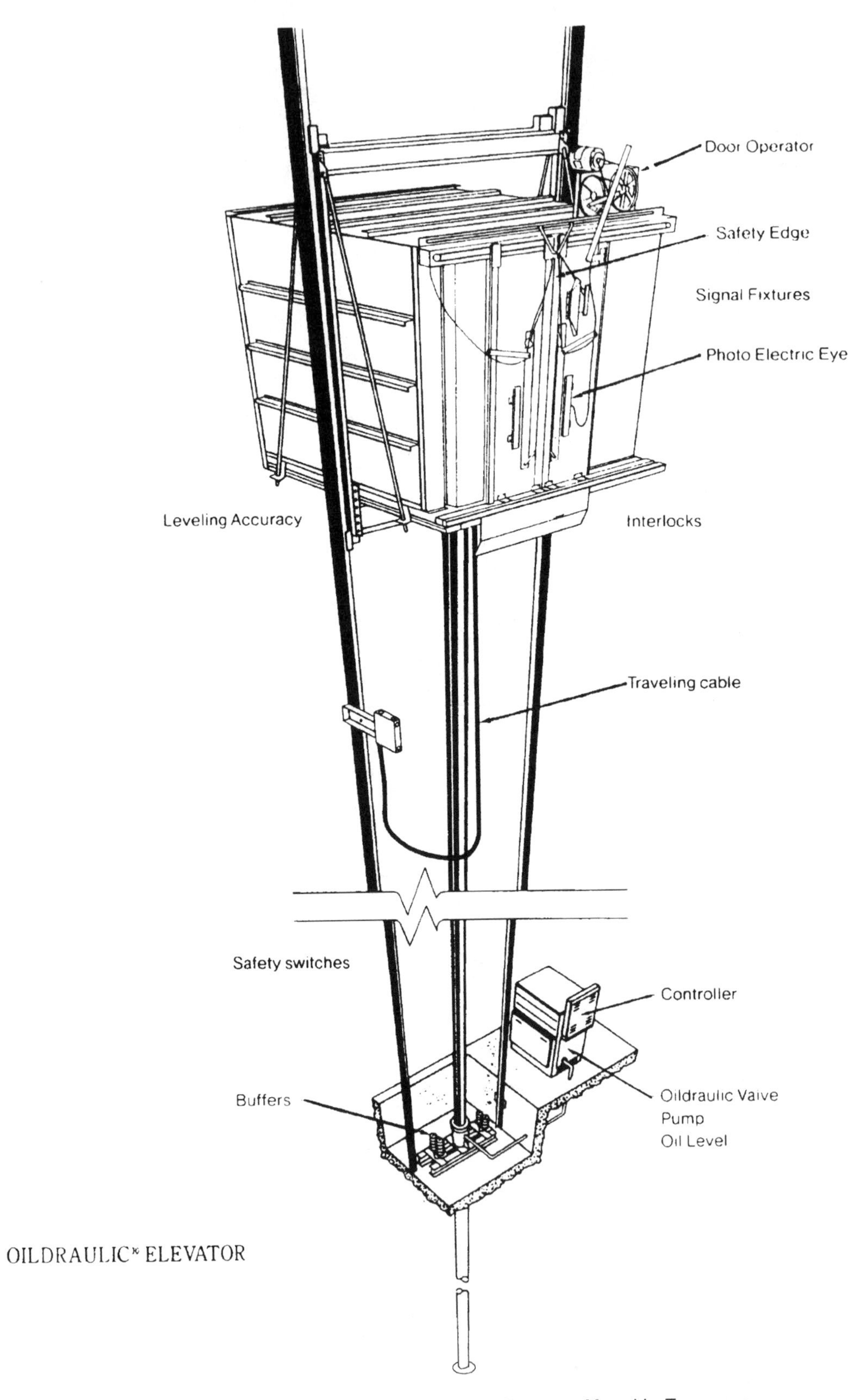

Reprinted with permission from Dover Elevator Company, Memphis, Tennessee

17.3.1 Hydraulic Elevator (Typical Platform and Sling Assembly)

TYPICAL PLATFORM & SLING
FOR
OILDRAULIC ELEVATOR

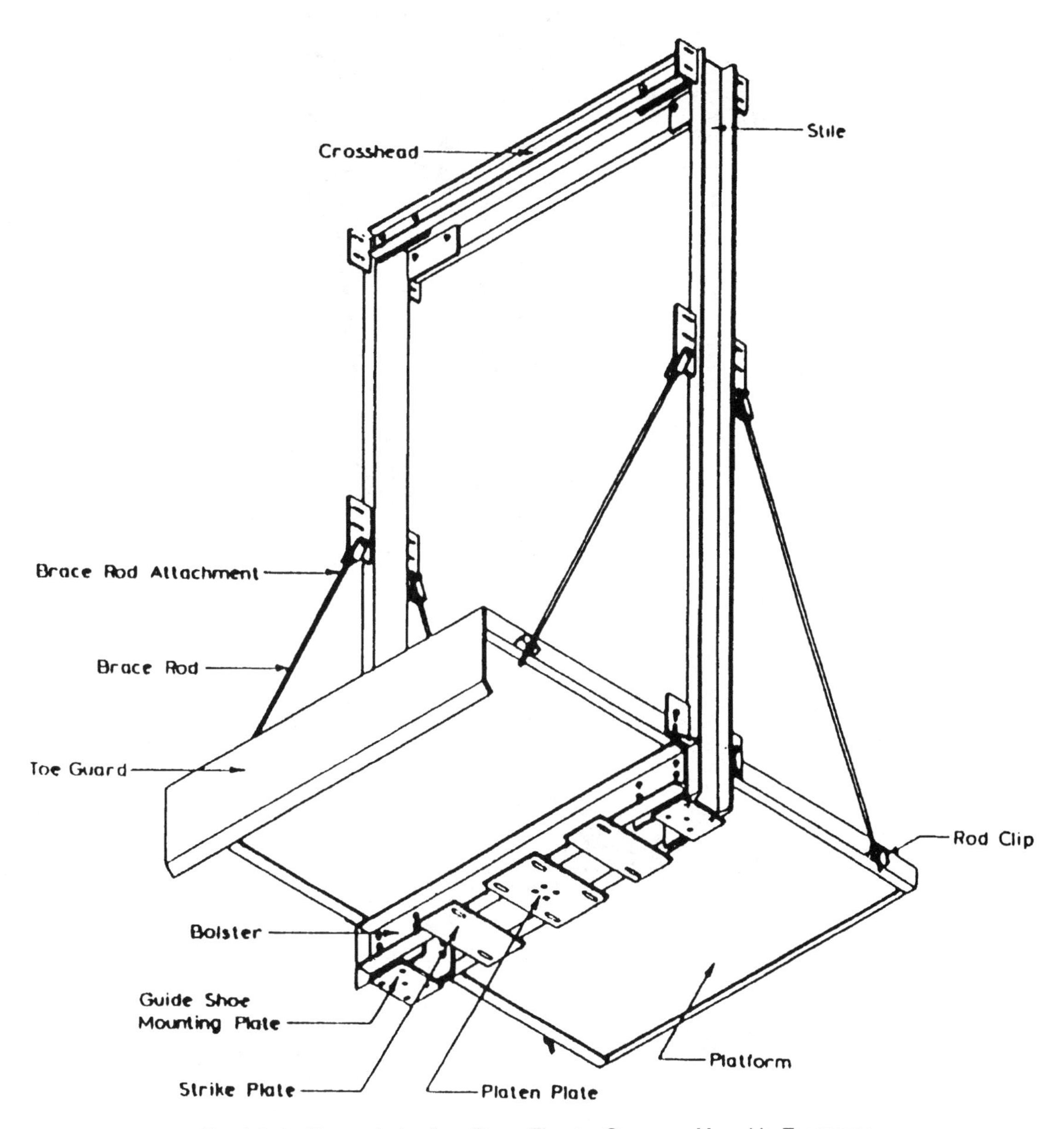

Reprinted with permission from Dover Elevator Company, Memphis, Tennessee

17.4.0 Hydraulic Freight Elevator Installation (Isometric)

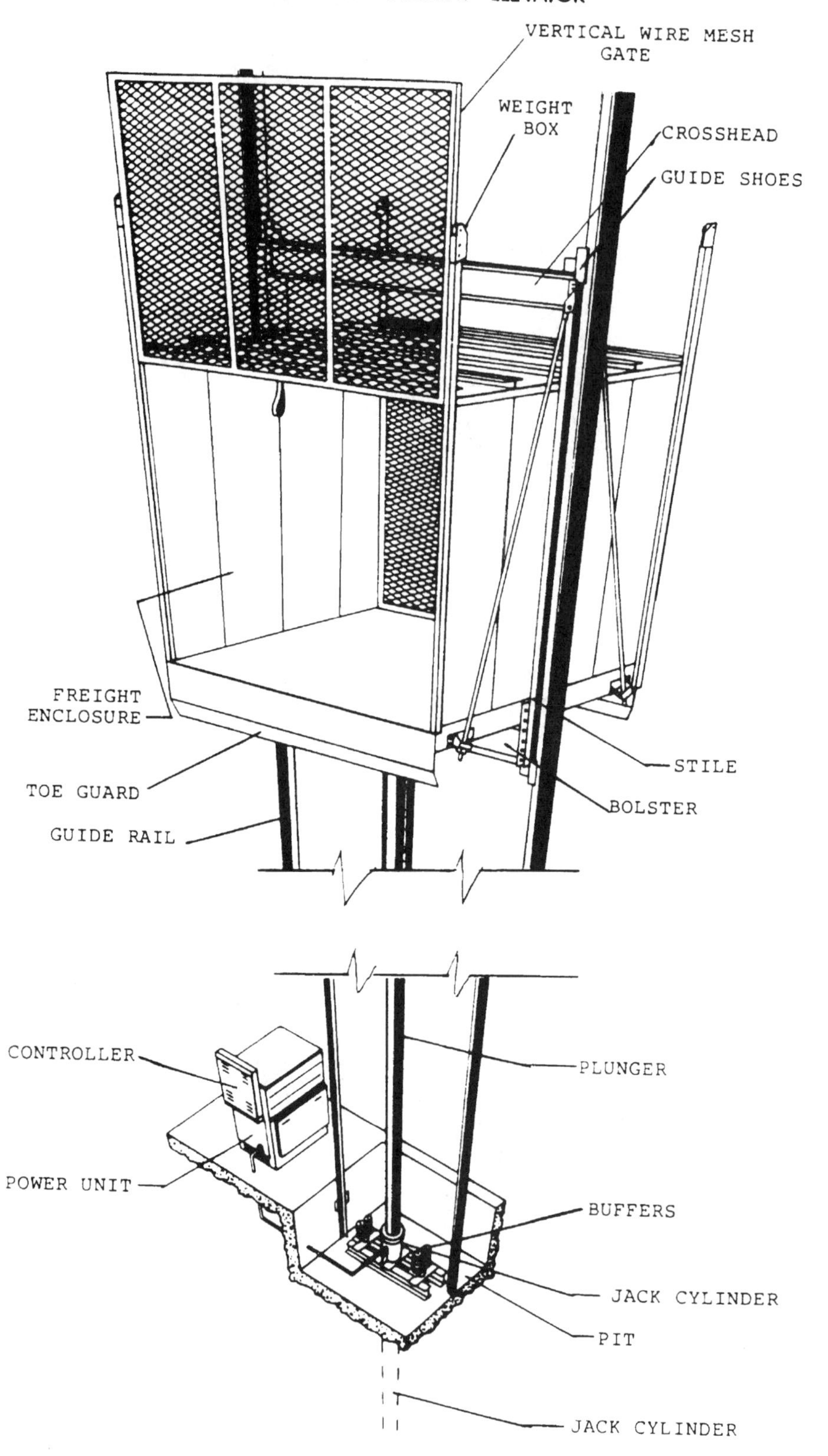

Reprinted with permission from Dover Elevator Company, Memphis, Tennessee

17.5.0 (Hoistway Section, Traction Elevator)

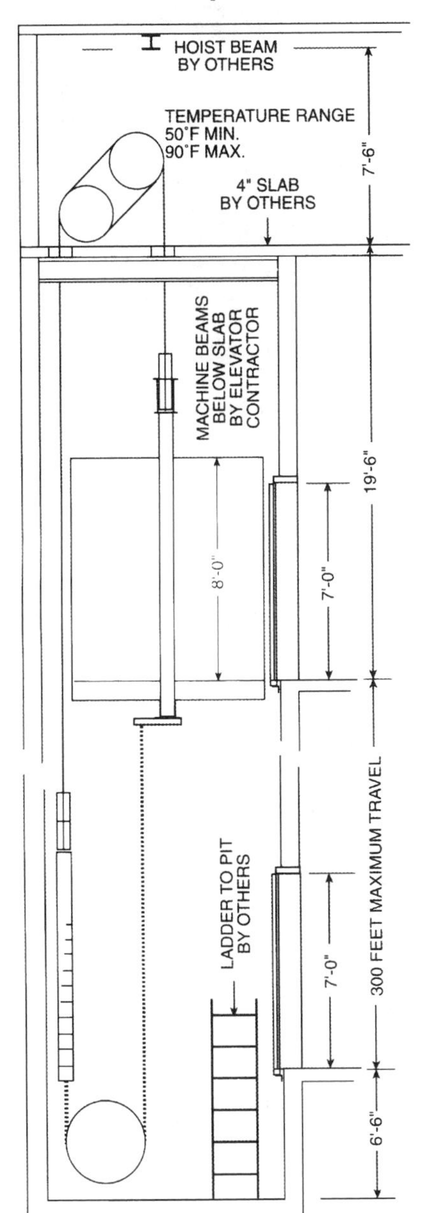

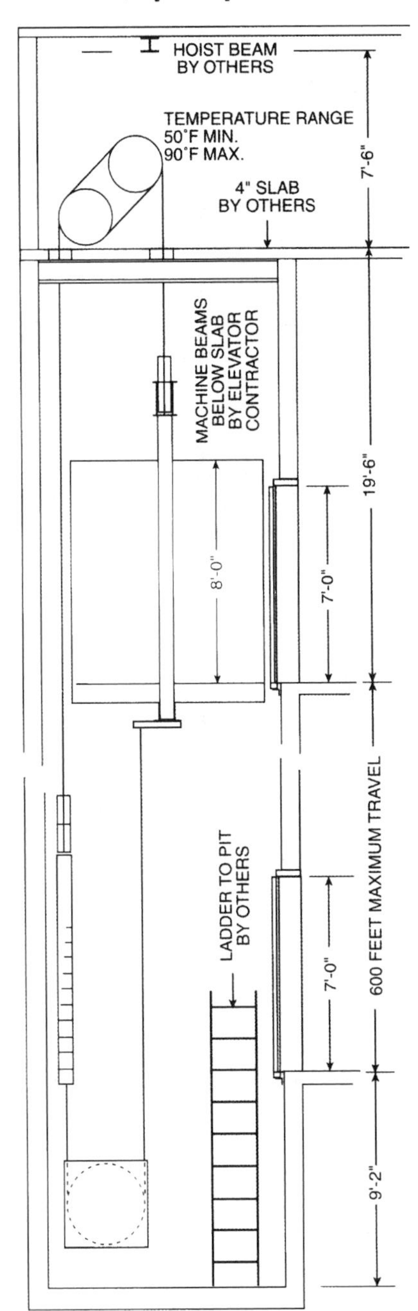

Reprinted with permission from Dover Elevator Company, Memphis, Tennessee

17.6.0 Hoistway Section (Holeless, Telescoping Holeless, and Conventional Elevator)

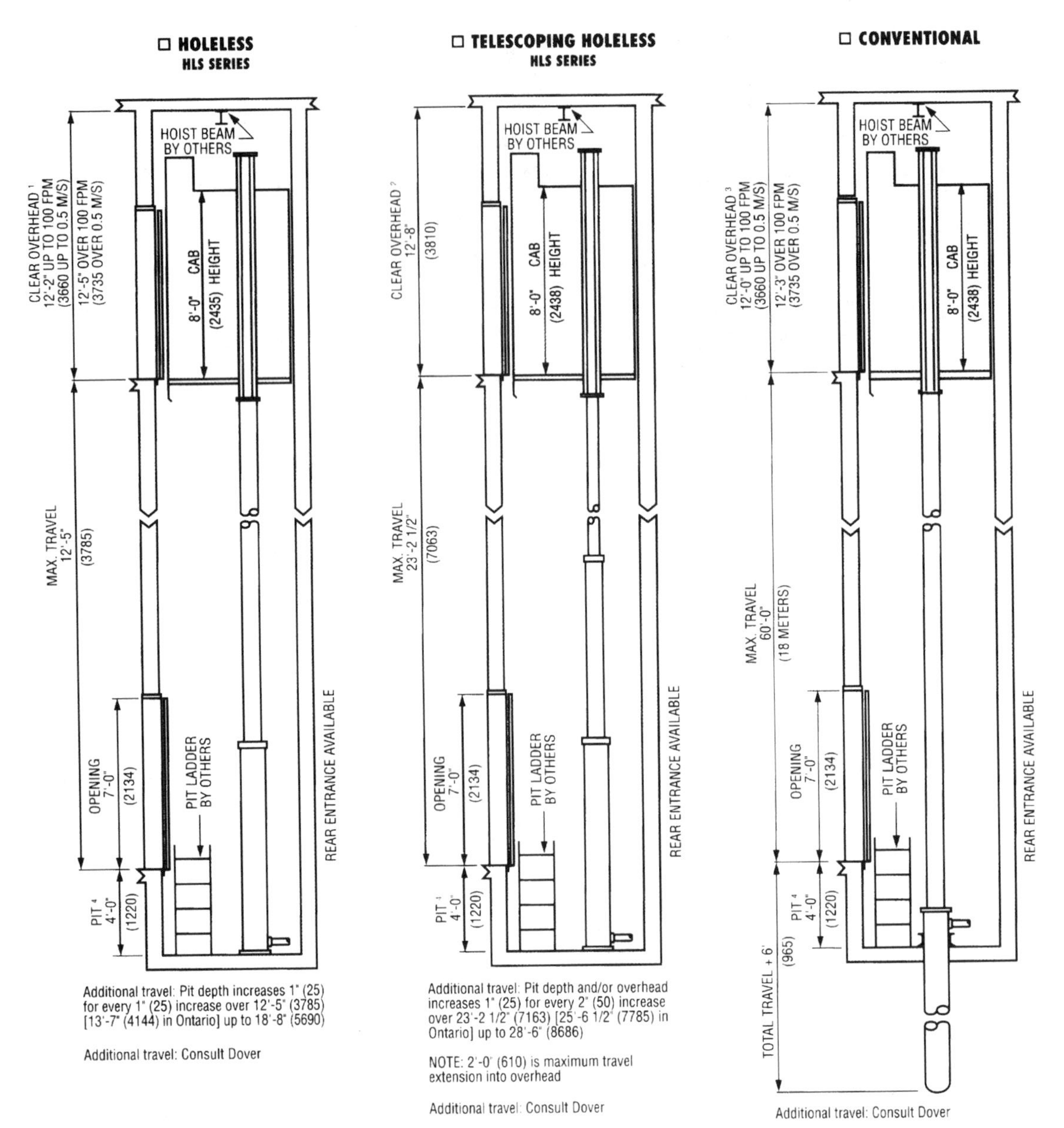

CLEAR OVERHEAD IN CANADA:

[1] 13'-1" up to 100 FPM (3988 up to 0.5 M/S)
13'-4" over 100 FPM (4064 over 0.5 M/S)

[2] 13'-7" (4140)

[3] 12'-1" up to 100 FPM (3685 up to 0.5 M/S)
12'-4" over 100 FPM (3760 over 0.5 M/S)

[4] PIT: 5'-2" (1580) in Ontario.

NOTE: All dimensions in parentheses are in millimeters unless otherwise indicated. Dimensional data shown here comply with the ASME A17.1 and CSA B44 Safety Code for Elevators. Local codes may vary from the national code. Consult your local Dover representative for details.

Reprinted with permission from Dover Elevator Company, Memphis, Tennessee

17.7.0 Elevator Machine-Room Configurations

All Dover pre-engineered Oildraulic elevators offer a wide variety of speeds, capacities, cab options and travel heights. The power unit size is determined by all of these options. To determine the correct machine room size for your application consult your local Dover representative.

The most desirable machine room location is on the lowest floor adjacent to the elevator hoistway. It may, however, be located remote from the hoistway if necessary.

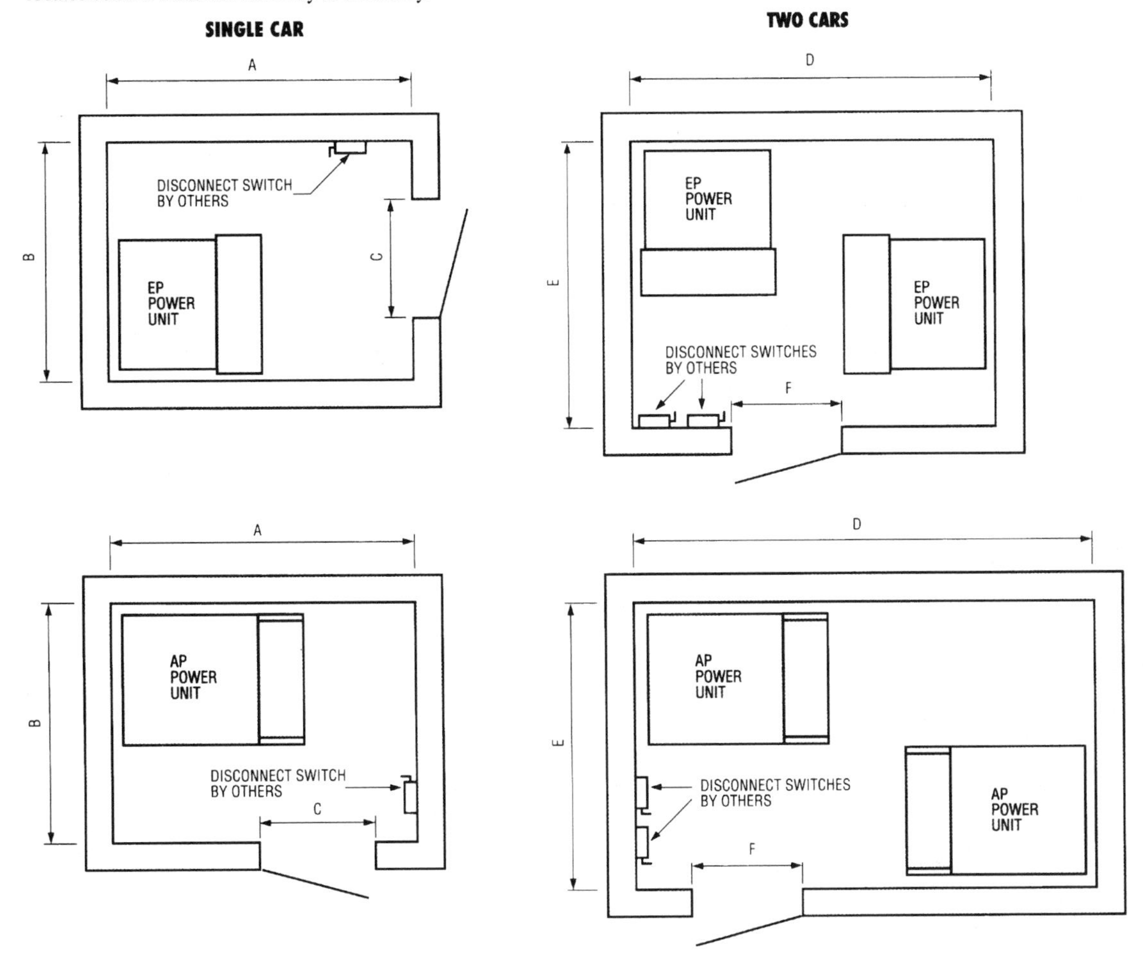

Dimensions will vary based on specific job requirements. Contact your local Dover Representative concerning your application.

Single Car Machine Room Dimensions				
Power Unit	EP1	EP2	AP1	AP2
A	6'-6" (1980)	6'-6" (1980)	8'-0" (2438)	10'-0" (3050)
B	5'-0" (1524)	6'-0" (1830)	5'-6" (1675)	5'-6" (1675)
C[1]	3'-0" (914)	3'-0" (914)	3'-0" (914)	3'-6" (1066)

NOTE: EP power units are submersible and AP units are dry. Each comes in two sizes.

[1]Clear Opening

Dimensions will vary based on specific job requirements. Contact your local Dover Representative concerning your application.

Two Car Machine Room Dimensions				
Power Unit	EP1	EP2	AP1	AP2
D	8'-0" (2438)	8'-6" (2590)	11'-0" (3355)	15'-0" (5475)
E	8'-0" (2438)	8'-6" (2590)	6'-6" (1980)	7'-6" (2286)
F[1]	3'-0" (914)	3'-0" (914)	3'-0" (914)	3'-6" (1066)

NOTE: EP power units are submersible and AP units are dry. Each comes in two sizes.

[1]Clear Opening

NOTE: All dimensions in parentheses are in millimeters unless otherwise indicated. Dimensional data shown here comply with the ASME A17.1 and CSA B44 Safety Code for Elevators. Local codes may vary from the national code. Consult your local Dover representative for details.

17.8.0 Gearless Elevator Machine-Room Configurations

Machine Room Dimensions				
	Capacity (In Pounds)			
Dimensions	2500	3000	3500	4000
A	8'-4"	8'-4"	8'-4"	9'-4"
B	6'-8"	7'-2"	7'-10"	7'-10"
C	17'-0"	17'-0"	17'-0"	19'-0"
D	22'-4"	23'-4"	24'-8"	24'-8"

Consult Dover on all 4,000 lb. capacity applications

Machine Room Plan View

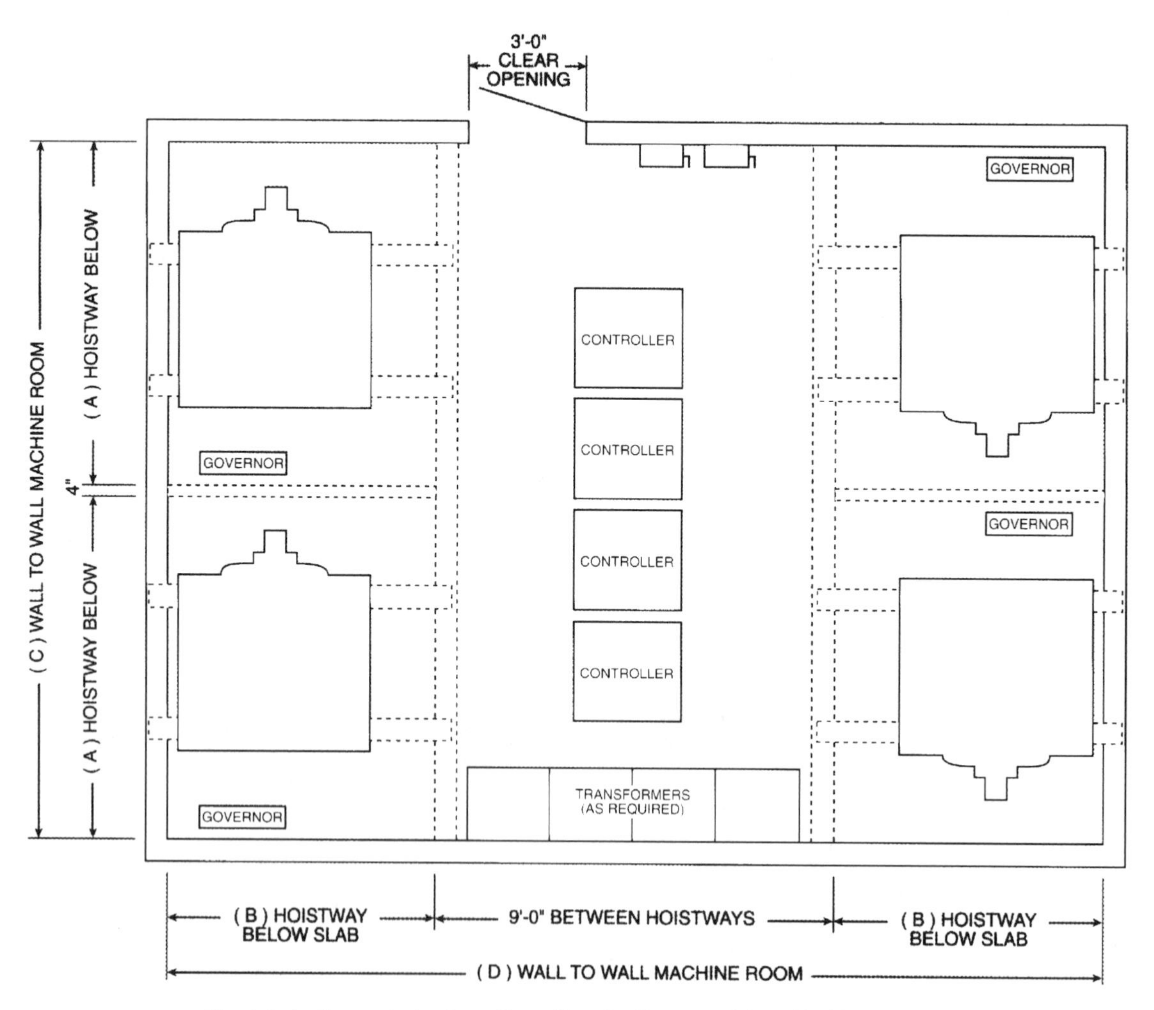

Reprinted with permission from Dover Elevator Company, Memphis, Tennessee

17.9.0 Hydraulic Elevator Preparation Work (Check List)

Hydraulic Passenger Elevators Preparatory Work

The following preparatory work is required in order to properly install the elevator equipment. The cost of this work is not included in the elevator proposal, since it is a part of the building construction.

1. A plumb and legal hoistway, properly framed and enclosed and including a pit of proper depth, and a pit ladder for each elevator. Drains, lights, access doors, waterproofing, and hoistway ventilation, as required.

2. Adequate supports and foundations to carry the loads of all equipment, including supports for guide rail brackets. Divider beam for rail bracket support as required.

3. An enclosed elevator equipment room with electrical work outlets, adequate lighting, and heating and ventilation sufficient to maintain the room at a temperature of 50°F minimum to 100°F maximum.

4. Proper trenching and backfilling for any underground piping or conduit.

5. Proper location of jack hole from building lines with adequate ingress and egress for mobile well drilling equipment, after final excavation and previous to the pouring of footings or foundation.

6. Cutting of walls, floor, etc. and removal of such obstructions as may be necessary for proper installation of the elevator. Setting of anchors and sleeves.

7. Grouting of door sills, hoistway frames, jack and signal fixtures after installation of elevator equipment.

8. Temporary enclosures, barricades, or other protection from open hoistways and elevator work area during the time the elevator is being installed.

9. Temporary elevator service prior to completion and acceptance of complete installation.

10. Complete connections from the electric power mains to each controller, including branch circuit protection devices.

11. Electric power of the same characteristics as the permanent supply, without charge, for construction, testing and adjusting.

12. Heat and smoke sensors as required in accordance with NFPA**#72E and ASME A17.1.***

13. All telephone wiring to machine room control panel, and installation of telephone instrument or other communication equipment in elevator cab with all connections to elevator trail cable and in machine room.

14. A standby power source, including necessary transfer switches and auxiliary contacts, where elevator operation from an alternate power supply is required.****

15. A means to automatically disconnect the main line power supply to the elevator prior to the application of water in the elevator machine room. This means shall not be self-resetting.

16. All painting, except as otherwise specified.

17. Adequate storage facilities for elevator equipment prior to and during installation.

* Refer to elevator layout drawings for details of each requirement.
** National Fire Protection Code.
*** Safety Code for Elevators and Escalators.
**** Contact your local Dover representative for design information.

17.10.0 Dumbwaiter Installation (Isometrics)

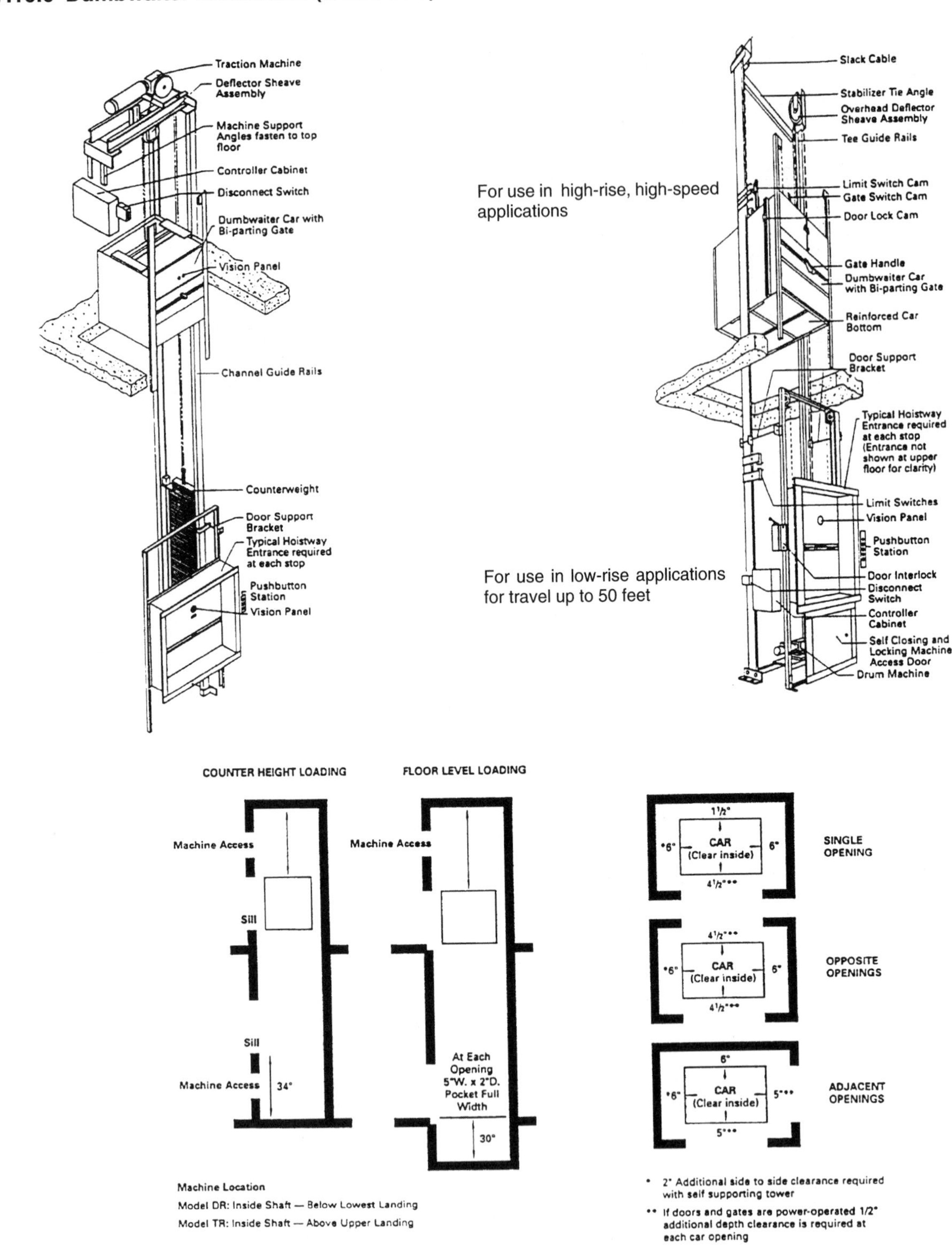

By permission MATOT, Bellwood, Illinois

17.11.0 Dumbwaiter (Typical Uses, Standard Sizes, and Horsepower Amperage)

TYPICAL USES OF DUMBWAITERS	Food Service-Carts	Food Service-Tray	Record, Money Handling Computer Service	Cleaning and Maintenance Supply	Light Supply and Material Handling	Bulk Supply and Material Handling	Pharmaceutical Specimen	Utility Cart Transportation	Mail and Correspondence
CAR SIZE (see guide below)	H	B-G	A-D	H-J	B-G	H-J	A-C	H-J	A-F
Hospital, Health Clinic	•		•	•	•	•	•	•	•
Financial			•	•	•	•			•
Manufacturing			•	•	•	•		•	
Hotel, Motel	•		•	•	•	•			•
Educational	•		•	•	•	•	•	•	•
Office Building			•	•	•	•			•
Government and Municipal			•	•	•	•		•	
Recreational, Club	•	•		•	•	•		•	
Stores and Shops		•	•	•	•	•		•	
Library		•	•	•	•	•		•	
Restaurant	•	•		•	•	•			
Nursing Home	•	•	•	•	•	•			
Ships and Marine Duty				•	•	•			•
Churches	•	•		•	•	•			
Laboratories			•	•	•	•	•		

In the "Typical Uses" chart, above, match the intended dumbwaiter use with type of building to find the letter code(s) for suggested car size(s). Then refer to the "Recommended Standard Sizes" chart, below, to find the dimensions of the cars that will meet your requirements. (Example: A dumbwaiter to carry cleaning and maintenance supplies in a hotel or motel has recommended car sizes of either H, I or J.)

HORSEPOWER AMPERAGE

DR (DRUM) DUMBWAITER CAPACITY	HP	FULL LOAD AMPS			
		3 PHASE		1 PHASE	
		230v	460v	110v	220v
up to 150 lbs	1	4.8	2.4	NOT RECOMMENDED	
150 – 300 lbs	2	5.9	3.0		
300 – 500 lbs	3	8.6	4.3		
TR (TRACTION) DUMBWAITER CAPACITY					
100 – 250 lbs	1	3.4	1.7	12.4	6.2
250 – 500 lbs	2	6.5	3.5	18.0	9.0

Use the Horsepower Amperage Chart only for determining the power supply feed circuit design, which is furnished by others.

- 3 phase starting amps = 5 times full load
- 1 phase starting amps = 6 times full load
- Information in this chart pertains to single speed equipment at 50 F.P.M.

RECOMMENDED STANDARD SIZES

STANDARD SIZE	INSIDE DIMENSIONS		
	WIDTH	DEPTH	HEIGHT
A	18	18	24
B	20	20	30
C	20	24	30
D	24	24	36
E	24	24	36
F	24	30	36
G	30	30	36
H	30	42	48
I	36	36	36
J	36	36	48

Dumbwaiters are restricted to a platform of 9 square feet, and a clear inside height of 4 feet by the *American National Standard Safety Code for Elevators, Dumbwaiters, Escalators and Moving Walks.*

By permission MATOT, Bellwood, Illinois

Section

18

Plumbing

Contents

18.0.0 Introduction to plumbing
18.1.0 Equivalent length (pipe, elbows, tees, and valves)
18.1.1 Equivalent length of pipe for 90-degree elbows (in feet)
18.2.0 Maximum capacity of gas pipe in cubic feet per hour
18.3.0 Iron and copper elbow-size equivalents
18.4.0 Water velocities (types of service)
18.5.0 Flow rates/demand for various plumbing fixtures
18.5.1 Hot-water demand for various fixtures
18.6.0 Head-of-water equivalents (in PSI)
18.7.0 Pipe sizes for horizontal rainwater piping
18.8.0 Velocity/flow in cast-iron sewer pipe of 2" (5.08 cm) and 3" (7.6 cm)
18.9.0 Expansion characteristics of metal and plastic pipe
18.9.1 Expansion characteristics of metal and plastic pipe in graph form
18.10.0 Size of roof drains for varying amounts of rainfall (in square feet)
18.11.0 Comparative costs of steam-condensate systems
18.12.0 Supports for pipe risers (illustrated)
18.12.1 Supports for horizontal pipe runs (illustrated)
18.13.0 Cast-iron pipe hub-barrel dimensions
18.14.0 Pipe diameters and trench widths (U.S. and metric sizes)
18.15.0 Pipe test plugs (illustrated)
18.16.0 Thrust pressures when hydrostatically testing soil pipe
18.17.0 Piping schematics (vent and stack installations)
18.17.1 Piping schematics (continuous- and looped-vent system)
18.17.2 Piping schematics (stacked fixture installation)
18.17.3 Piping schematics (roof drain and leader, hubless/hub pipe)
18.17.4 Piping schematics (battery of fixtures with a common vent)
18.17.5 Piping schematics (circuit venting/ wet venting)
18.17.6 Piping schematics (typical waste and vent installation)
18.18.0 Various city water-temperature and hardness figures
18.19.0 Abbreviations, definitions, and symbols that appear on plumbing drawings
18.19.1 Recommended symbols for plumbing on plumbing drawings
18.19.2 Symbols for pipe fittings and valves

18.0.0 Introduction to Plumbing

Leonardo DaVinci is credited with the design and installation of the first indoor plumbing system in Italy in the mid-sixteenth century. Other than the addition of sophisticated pumps on the supply side and designer fixtures on the other end, not much has changed, except for the materials of construction (gravity still plays as important role today as it did in 1550). A building plumbing system will generally consist of incoming domestic water service and distribution, above- and below-grade drainage systems for both sanitary and storm water, a venting system, pipe, fittings, valves, pumps, and fixtures.

18.1.0 Equivalent Length (Pipe, Elbows, Tees and Valves)

Find the nominal pipe size being used in the left-most column. For each fitting, read the value under the appropriate heading and add this to the length of piping. This allows total system pressure drop to be calculated. (This is valid for any fluid.)

PIPE SIZE	EQUIVALENT LENGTH OF STRAIGHT PIPE (FEET)				
	STANDARD ELBOW	STANDARD TEE	GATE VALV E FULL OPEN	GLOBE VALVE FULL OPEN	ANGLE VALVE FULL OPEN
1-1/2	4	9	0.9	41	21
2	5	11	1.2	54	27
2-/1/2	6	13	1.4	64	32
3	8	16	1.6	80	40
3-1/2	9	18	2.0	91	45
4	11	21	2.2	110	55
5	13	26	2.8	140	70
6	16	32	3.4	155	81
8	20	42	4.5	210	110
10	25	55	5.5	270	140
12	30	65	6.5	320	160
14	35	75	8.0	370	190

This table contains the number of feet of straight pipe usually allowed for standard fittings and valves.

18.1.1 Equivalent Length of Pipe for 90-Degree Elbows (In Feet)

Velocity, ft/s	Pipe Size														
	1/2	3/4	1	1-1/4	1-1/2	2	2-1/2	3	3-1/2	4	5	6	8	10	12
1	1.2	1.7	2.2	3.0	3.5	4.5	5.4	6.7	7.7	8.6	10.5	12.2	15.4	18.7	22.2
2	1.4	1.9	2.5	3.3	3.9	5.1	6.0	7.5	8.6	9.5	11.7	13.7	17.3	20.8	24.8
3	1.5	2.0	2.7	3.6	4.2	5.4	6.4	8.0	9.2	10.2	12.5	14.6	18.4	22.3	26.5
4	1.5	2.1	2.8	3.7	4.4	5.6	6.7	8.3	9.6	10.6	13.1	15.2	19.2	23.2	27.6
5	1.6	2.2	2.9	3.9	4.5	5.9	7.0	8.7	10.0	11.1	13.6	15.8	19.8	24.2	28.8
6	1.7	2.3	3.0	4.0	4.7	6.0	7.2	8.9	10.3	11.4	14.0	16.3	20.5	24.9	29.6
7	1.7	2.3	3.0	4.1	4.8	6.2	7.4	9.1	10.5	11.7	14.3	16.7	21.0	25.5	30.3
8	1.7	2.4	3.1	4.2	4.9	6.3	7.5	9.3	10.8	11.9	14.6	17.1	21.5	26.1	31.0
9	1.8	2.4	3.2	4.3	5.0	6.4	7.7	9.5	11.0	12.2	14.9	17.4	21.9	26.6	31.6
10	1.8	2.5	3.2	4.3	5.1	6.5	7.8	9.7	11.2	12.4	15.2	17.7	22.2	27.0	32.0

18.2.0 Maximum Capacity of Gas Pipe (in Cubic Feet Per Hour)

Nominal Iron Pipe Size, in.	Internal Diameter, in.	Length of Pipe, ft													
		10	20	30	40	50	60	70	80	90	100	125	150	175	200
1/4	0.364	32	22	18	15	14	12	11	11	10	9	8	8	7	6
3/8	0.493	72	49	40	34	30	27	25	23	22	21	18	17	15	14
1/2	0.622	132	92	73	63	56	50	46	43	40	38	34	31	28	26
3/4	0.824	278	190	152	130	115	105	96	90	84	79	72	64	59	55
1	1.049	520	350	285	245	215	195	180	170	160	150	130	120	110	100
1-1/4	1.380	1050	730	590	500	440	400	370	350	320	305	275	250	225	210
1-1/2	1.610	1600	1100	890	760	670	610	560	530	490	460	410	380	350	320
2	2.067	3050	2100	1650	1450	1270	1150	1050	990	930	870	780	710	650	610
2-1/2	2.469	4800	3300	2700	2300	2000	1850	1700	1600	1500	1400	1250	1130	1050	980
3	3.068	8500	5900	4700	4100	3600	3250	3000	2800	2600	2500	2200	2000	1850	1700
4	4.026	17,500	12,000	9700	8300	7400	6800	6200	5800	5400	5100	4500	4100	3800	3500

Notes: 1. Capacity is in cubic feet per hour at gas pressures of 0.5 psig or less and a pressure drop of 0.5 in. of water; Specific gravity = 0.60.

18.3.0 Iron and Copper Elbow-Size Equivalents

Fitting	Iron Pipe	Copper Tubing
Elbow, 90°	1.0	1.0
Elbow, 45°	0.7	0.7
Elbow, 90° long turn	0.5	0.5
Elbow, welded, 90°	0.5	0.5
Reduced coupling	0.4	0.4
Open return bend	1.0	1.0
Angle radiator valve	2.0	3.0
Radiator or convector	3.0	4.0
Boiler or heater	3.0	4.0
Open gate valve	0.5	0.7
Open globe valve	12.0	17.0

[a]See Table 4 for equivalent length of one elbow.
Source: Giesecke (1926) and Giesecke and Badgett (1931, 1932).

18.4.0 Water Velocities (Types of Service)

Type of Service	Velocity, ft/s	Reference
General service	4 to 10	a, b, c
City water	3 to 7	a, b
	2 to 5	c
Boiler feed	6 to 15	a, c
Pump suction and drain lines	4 to 7	a, b

[a]Crane Co. 1976. Flow of fluids through valves, fittings, and pipe. Technical Paper 410.
[b]*System Design Manual.* 1960. Carrier Air Conditioning Co., Syracuse, NY.
[c]*Piping Design and Engineering.* 1951. Grinnell Company, Inc., Cranston, RI.

Maximum Water Velocity to Minimize Erosion

Normal Operation, h/yr	Water Velocity, ft/s
1500	15
2000	14
3000	13
4000	12
6000	10

Source: *System Design Manual*, Carrier Air Conditioning Co., 1960.

18.5.0 Flow Rates/Demand for Various Plumbing Fixtures

Proper Flow and Pressure Required during Flow for Different Fixtures

Fixture	Flow Pressure[a]	Flow, gpm
Ordinary basin faucet	8	3.0
Self-closing basin faucet	12	2.5
Sink faucet—3/8 in.	10	4.5
Sink faucet—1/2 in.	5	4.5
Dishwasher	15-25	—[b]
Bathtub faucet	5	6.0
Laundry tube cock—1/4 in.	5	5.0
Shower	12	3-10
Ball cock for closet	15	3.0
Flush valve for closet	10-20	15-40[c]
Flush valve for urinal	15	15.0
Garden hose, 50 ft, and sill cock	30	5.0

[a]Flow pressure is the pressure (psig) in the pipe at the entrance to the particular fixture considered.
[b]Varies; see manufacturers' data.
[c]Wide range due to variation in design and type of flush valve closets.

Demand Weights of Fixtures in Fixture Units[a]

Fixture or Group[b]	Occupancy	Type of Supply Control	Weight in Fixture Units[c]
Water closet	Public	Flush valve	10
Water closet	Public	Flush tank	5
Pedestal urinal	Public	Flush valve	10
Stall or wall urinal	Public	Flush valve	5
Stall or wall urinal	Public	Flush tank	3
Lavatory	Public	Faucet	2
Bathtub	Public	Faucet	4
Shower head	Public	Mixing valve	4
Service sink	Office, etc	Faucet	3
Kitchen sink	Hotel or restaurant	Faucet	4
Water closet	Private	Flush valve	6
Water closet	Private	Flush tank	3
Lavatory	Private	Faucet	1
Bathtub	Private	Faucet	2
Shower head	Private	Mixing valve	2
Bathroom group	Private	Flush valve for closet	8
Bathroom group	Private	Flush tank for closet	6
Separate shower	Private	Mixing valve	2
Kitchen sink	Private	Faucet	2
Laundry trays (1 to 3)	Private	Faucet	3
Combination fixture	Private	Faucet	3

Note: See Hunter (1941).
[a]For supply outlets likely to impose continuous demands, estimate continuous supply separately, and add to total demand for fixtures.
[b]For fixtures not listed, weights may be assumed by comparing the fixture to a listed one using water in similar quantities and at similar rates.
[c]The given weights are for total demand. For fixtures with both hot and cold water supplies, the weights for maximum separate demands can be assumed to be 75% of the listed demand for the supply.

18.5.1 Hot-Water Demand for Various Fixtures

Hot water demand per fixture for various types of buildings in gph of water per fixture, calculated at a final temperature of 140 F. (Reprinted with permission of ASHRAE.)

Fixture	Apartment house	Club	Gymnasium	Hospital	Hotel	Industrial plant	Office building	Private residence	School	YMCA
Basins, private lavatory	2	2	2	2	2	2	2	2	2	2
Basins, public lavatory	4	6	8	6	8	12	6	—	15	8
Bath tubs[a]	20	20	30	20	20	—	—	20	—	30
Dishwashers[b]	15	50-150	—	50-150	50-200	20-100	—	15	20-100	20-100
Foot basins	3	3	12	3	3	12	—	3	3	12
Kitchen sinks	10	20	—	20	30	20	20	10	20	20
Laundry, stationary tubs	20	28	—	28	28	—	—	20	—	28
Pantry sinks	5	10	—	10	10	—	10	5	10	10
Showers	30	150	225	75	75	225	30	30	225	225
Service sinks	20	20	—	20	30	20	20	15	20	20
Hydrotherapeutic showers				400						
Hubbard baths				600						
Leg baths				100						
Arm baths				35						
Sitz baths				30						
Continuous-flow baths				165						
Circular wash sinks				20	20	30	20		30	
Semicircular wash sinks				10	10	15	10		15	
Demand factor	0.30	0.30	0.40	0.25	0.25	0.40	0.30	0.30	0.40	0.40
Storage capacity factor[c]	1.25	0.90	1.00	0.60	0.80	1.00	2.00	0.70	1.00	1.00

[a]Whirlpool baths require specific consideration based on capacity; they are not included in the bathtub category.
[b]Dishwasher requirements should be taken from this table or from manufacturers' data for the model to be used, if this is known.
[c]Ratio of storage tank capacity to probable maximum demand per hr. Storage capacity may be reduced where an unlimited supply of steam is available from a central street steam system or large boiler plant.

By permission of American Society of Heating, Refrigerating and Air-Conditioning Engineers, Inc. Atlanta, Georgia, from their *1993 ASHRAE Fundamentals Handbook*

18.6.0 Head-of-Water Equivalents (in PSI)

Head Ft.	0	1	2	3	4	5	6	7	8	9
0	....	0.433	0.866	1.299	1.732	2.165	2.598	3.031	3.464	3.987
10	4.330	4.763	5.196	5.629	6.062	6.495	6.928	7.361	7.794	8.277
20	8.660	9.093	9.526	9.959	10.392	10.825	11.258	11.691	12.124	12.557
30	12.990	13.423	13.856	14.289	14.722	15.155	15.588	16.021	16.454	16.887
40	17.320	17.753	18.186	18.619	19.052	19.485	19.918	20.351	20.784	21.217
50	21.650	22.083	22.516	22.949	23.382	23.815	24.248	24.681	25.114	25.547
60	25.980	26.413	26.846	27.279	27.712	28.145	28.578	29.011	29.444	29.877
70	30.310	30.743	31.176	31.609	32.042	32.475	32.908	33.341	33.774	34.207
80	34.640	35.073	35.506	35.939	36.372	36.805	37.238	37.671	38.104	38.537
90	38.970	39.403	39.836	40.269	40.702	41.135	41.568	42.001	42.436	42.867

18.7.0 Pipe Sizes for Horizontal Rainwater Piping

Size of Pipe in Inches 1/8" Slope	Maximum Rainfall in Inches per Hour				
	2	3	4	5	6
3	1644	1096	822	657	548
4	3760	2506	1880	1504	1253
5	6680	4453	3340	2672	2227
6	10700	7133	5350	4280	3566
8	23000	15330	11500	9200	7600
10	41400	27600	20700	16580	13800
11	66600	44400	33300	26650	22200
15	109000	72800	59500	47600	39650

Size of Pipe in Inches 1/4" Slope	Maximum Rainfall in Inches per Hour				
	2	3	4	5	6
3	2320	1546	1160	928	773
4	5300	3533	2650	2120	1766
5	9440	6293	4720	3776	3146
6	15100	10066	7550	6040	5033
8	32600	21733	16300	13040	10866
10	58400	38950	29200	23350	19450
11	94000	62600	47000	37600	31350
15	168000	112000	84000	67250	56000

Size of Pipe in Inches 1/2" Slope	Maximum Rainfall in Inches per Hour				
	2	3	4	5	6
3	3288	2295	1644	1310	1096
4	7520	5010	3760	3010	2500
5	13660	8900	6680	5320	4450
6	21400	13700	10700	8580	7140
8	46000	30650	23000	18400	15320
10	82800	55200	41400	33150	27600
11	133200	88800	66600	53200	44400
15	238000	158800	119000	95300	79250

By permission of Cast Iron Soil Pipe Institute

18.8.0 Velocity/Flow in Cast-Iron Sewer Pipe of 2" (5.08 cm) and 3" (7.6 cm)

Pipe Size (In.)	SLOPE		¼ FULL		½ FULL		¾ FULL		FULL	
	(In./Ft.)	(Ft./Ft.)	Velocity (Ft./Sec.)	Flow (Gal./Min.)	Velocity (Ft./Sec.)	Flow (Gal./Min.)	Velocity (Ft./Sec.)	Flow (Gal./Min.)	Velocity (Ft./Sec.)	Flow (Gal./Min.)
2.0	0.0120	0.0010	0.36	0.83	0.46	2.16	0.52	3.67	0.46	4.35
	0.0240	0.0020	0.51	1.18	0.66	3.06	0.74	5.18	0.66	6.15
	0.0360	0.0030	0.62	1.45	0.80	3.75	0.90	6.35	0.80	7.53
	0.0480	0.0040	0.72	1.67	0.93	4.33	1.04	7.33	0.93	8.69
	0.0600	0.0050	0.80	1.87	1.04	4.84	1.16	8.20	1.04	9.72
	0.0720	0.0060	0.88	2.04	1.13	5.30	1.27	8.98	1.13	10.65
	0.0840	0.0070	0.95	2.21	1.23	5.72	1.38	9.70	1.23	11.50
	0.0960	0.0080	1.01	2.36	1.31	6.12	1.47	10.37	1.31	12.29
	0.1080	0.0090	1.07	2.50	1.39	6.49	1.56	11.00	1.39	13.04
	0.1200	0.0100	1.13	2.64	1.47	6.84	1.64	11.59	1.47	13.75
	0.2400	0.0200	1.60	3.73	2.07	9.67	2.33	16.39	2.07	19.44
	0.3600	0.0300	1.96	4.57	2.54	11.85	2.85	20.07	2.54	23.81
	0.4800	0.0400	2.26	5.28	2.93	13.68	3.29	23.18	2.93	27.49
	0.6000	0.0500	2.53	5.90	3.28	15.29	3.68	25.92	3.28	30.74
	0.7200	0.0600	2.77	6.47	3.59	16.75	4.03	28.39	3.59	33.67
	0.8400	0.0700	2.99	6.98	3.88	18.10	4.35	30.66	3.88	36.37
	0.9600	0.0800	3.20	7.47	4.14	19.35	4.65	32.78	4.14	38.88
	1.0800	0.0900	3.39	7.92	4.40	20.52	4.93	34.77	4.40	41.24
	1.2000	0.1000	3.58	8.35	4.63	21.63	5.20	36.65	4.63	43.47
3.0	0.0120	0.0010	0.47	2.55	0.61	6.56	0.69	11.05	0.61	13.12
	0.0240	0.0020	0.67	3.61	0.86	9.28	0.97	15.63	0.86	18.55
	0.0360	0.0030	0.82	4.42	1.06	11.36	1.19	19.14	1.06	22.72
	0.0480	0.0040	0.95	5.11	1.22	13.12	1.37	22.10	1.22	26.24
	0.0600	0.0050	1.06	5.71	1.37	14.67	1.53	24.71	1.37	29.33
	0.0720	0.0060	1.16	6.25	1.50	16.07	1.68	27.07	1.50	32.13
	0.0840	0.0070	1.25	6.75	1.62	17.35	1.81	29.24	1.62	34.71
	0.0960	0.0080	1.34	7.22	1.73	18.55	1.94	31.26	1.73	37.11
	0.1080	0.0090	1.42	7.66	1.83	19.68	2.06	33.16	1.83	39.36
	0.1200	0.0100	1.50	8.07	1.93	20.74	2.17	34.95	1.93	41.49
	0.2400	0.0200	2.21	11.42	2.73	29.33	3.07	49.43	2.73	58.67
	0.3600	0.0300	2.60	13.98	3.35	35.93	3.76	60.53	3.35	71.86
	0.4800	0.0400	3.00	16.14	3.87	41.49	4.34	69.90	3.87	82.97
	0.6000	0.0500	3.35	18.05	4.32	46.38	4.85	78.15	4.32	92.77
	0.7200	0.0600	3.67	19.77	4.74	50.81	5.31	85.61	4.74	101.62
	0.8400	0.0700	3.96	21.36	5.12	54.88	5.74	92.47	5.12	109.76
	0.9600	0.0800	4.24	22.83	5.47	58.67	6.13	98.85	5.47	117.34
	1.0800	0.0900	4.50	24.22	5.80	62.23	6.51	104.85	5.80	124.46
	1.2000	0.1000	4.74	25.53	6.11	65.29	6.86	110.52	6.11	131.19

By permission of Cast Iron Soil Pipe Institute

18.9.0 Expansion Characteristics of Metal and Plastic Pipe

Expansion: Allowances for expansion and contraction of building materials are important design considerations. Material selection can create or prevent problems. Cast iron is in tune with building reactions to temperature. Its expansion is so close to that of steel and masonry that there is no need for costly expansion joints and special offsets. That is not always the case with other DWV materials.

Thermal expansion of various materials.

Material	Inches per inch 10^{-6} X per °F	Inches per 100′ of pipe per 100°F.	Ratio-assuming cast iron equals 1.00
Cast iron	6.2	0.745	1.00
Concrete	5.5	0.66	.89
Steel (mild)	6.5	0.780	1.05
Steel (stainless)	7.8	0.940	1.26
Copper	9.2	1.11	1.49
PVC (high impact)	55.6	6.68	8.95
ABS (type 1A)	56.2	6.75	9.05
Polyethylene (type 1)	94.5	11.4	15.30
Polyethylene (type 2)	83.3	10.0	13.40

Here is the *actual* increase in length for 50 feet of pipe and 70° temperature rise.

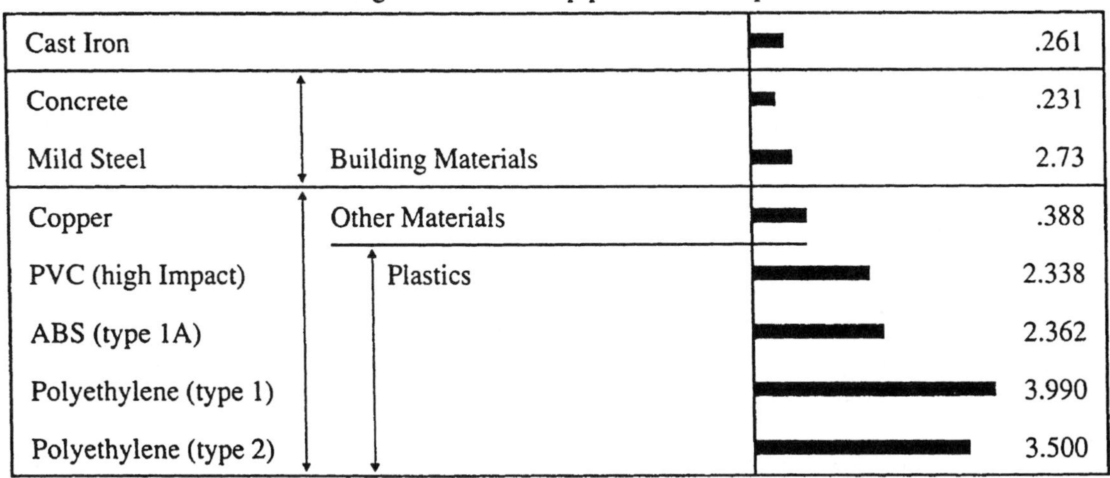

By permission of Cast Iron Soil Pipe Institute

18.9.1 Expansion Characteristics of Metal and Plastic Pipe in Graph Form

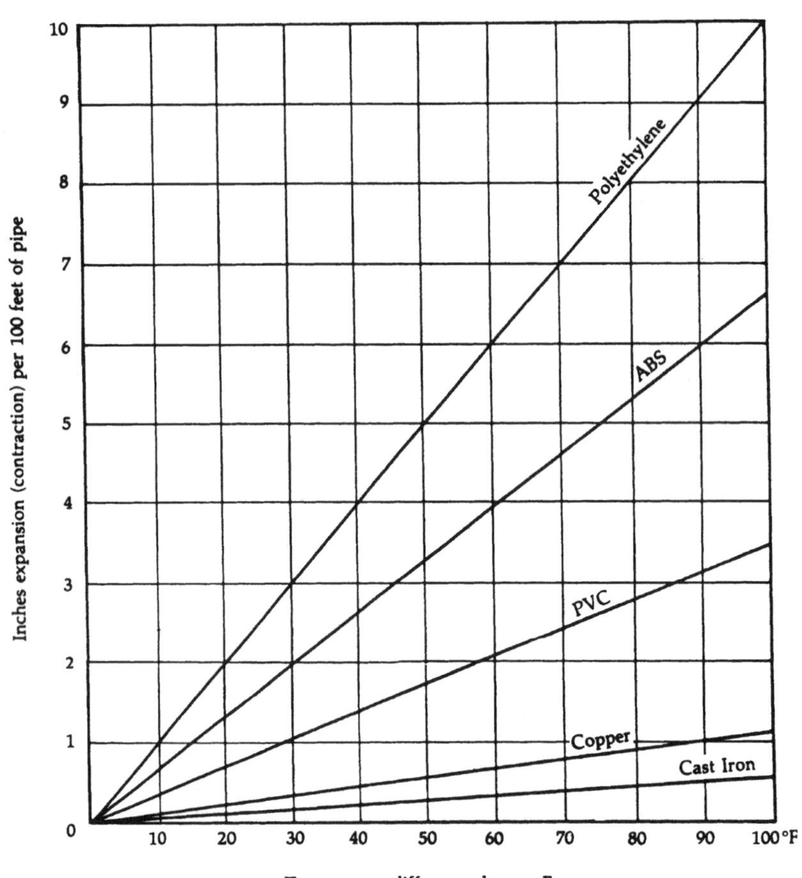

Example: Find the expansion allowance required for a 120 ft. run of ABS pipe in a concrete & masonry building and for a temperature difference of 90°F.

Answer: At a temperature difference of 90°F read from the chart, ABS expands 6″ and concrete expands ¾″.

$$(6 - \tfrac{3}{4}) \times \frac{120}{100} = 5\tfrac{1}{4} \times \frac{120}{100} = 6.3 \text{ inches}$$

By permission of Cast Iron Soil Pipe Institute, Chattanooga, Tn

18.10.0 Size of Roof Drains for Varying Amounts of Rainfall (in Square Feet)

Rain Fall in Inches	Size of Drain or Leader in Inches* 2	3	4	5	6	8
1	2880	8800	18400	34600	54000	116000
2	1440	4400	9200	17300	27000	58000
3	960	2930	6130	11530	17995	38660
4	720	2200	4600	8650	13500	29000
5	575	1760	3680	6920	10800	23200
6	480	1470	3070	5765	9000	19315
7	410	1260	2630	4945	7715	16570
8	360	1100	2300	4325	6750	14500
9	320	980	2045	3845	6000	12890
10	290	880	1840	3460	5400	11600
11	260	800	1675	3145	4910	10545
12	240	730	1530	2880	4500	9660

*Round, square or rectangular rainwater pipe may be used and are considered equivalent when closing a scribed circle quivalent to the leader diameter.

Source: Uniform Plumbing Code (IAPMO) 1985 Edition

By permission of Cast Iron Soil Pipe Institute

18.11.0 Comparative Costs of Steam-Condensate Lines

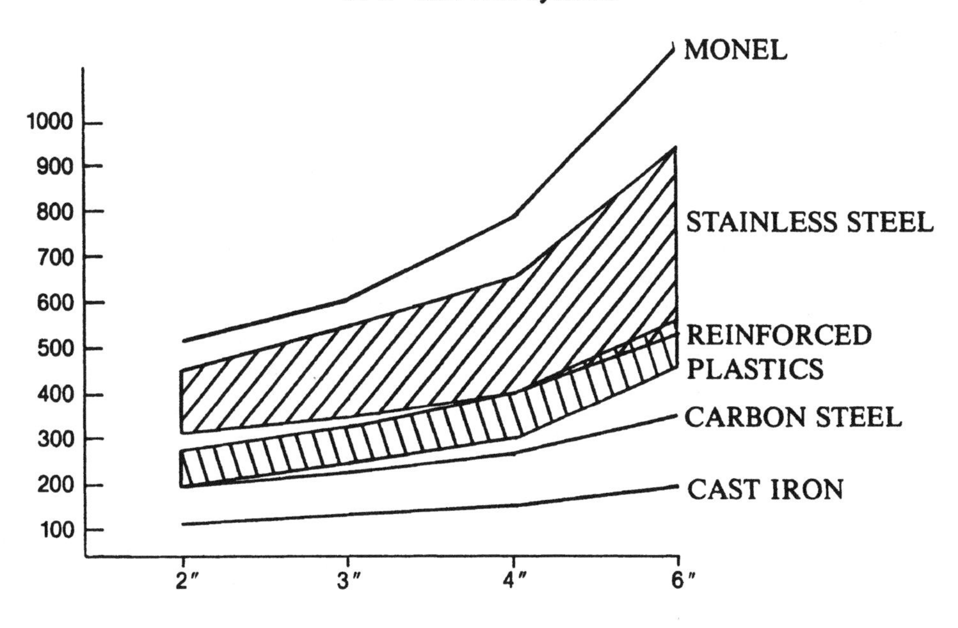

By permission of Cast Iron Soil Pipe Institute

18.12.0 Supports for Pipe Risers (Illustrated)

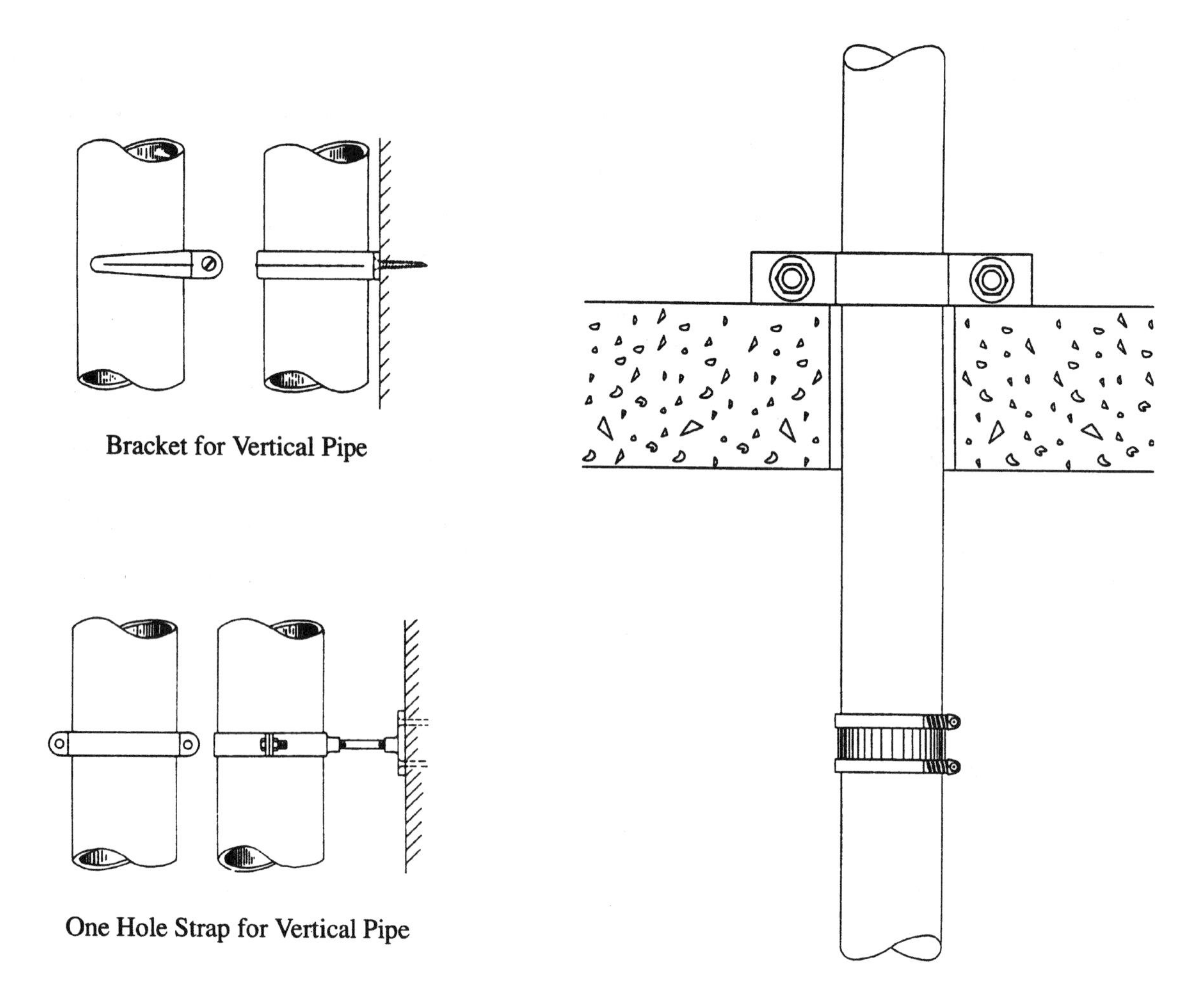

Bracket for Vertical Pipe

One Hole Strap for Vertical Pipe

Method of clamping
the Pipe at Each Floor,
Using a Friction Clamp
or Floor Clamp

By permission of Cast Iron Soil Pipe Institute

18.12.1 Supports for Horizontal Pipe Runs (Illustrated)

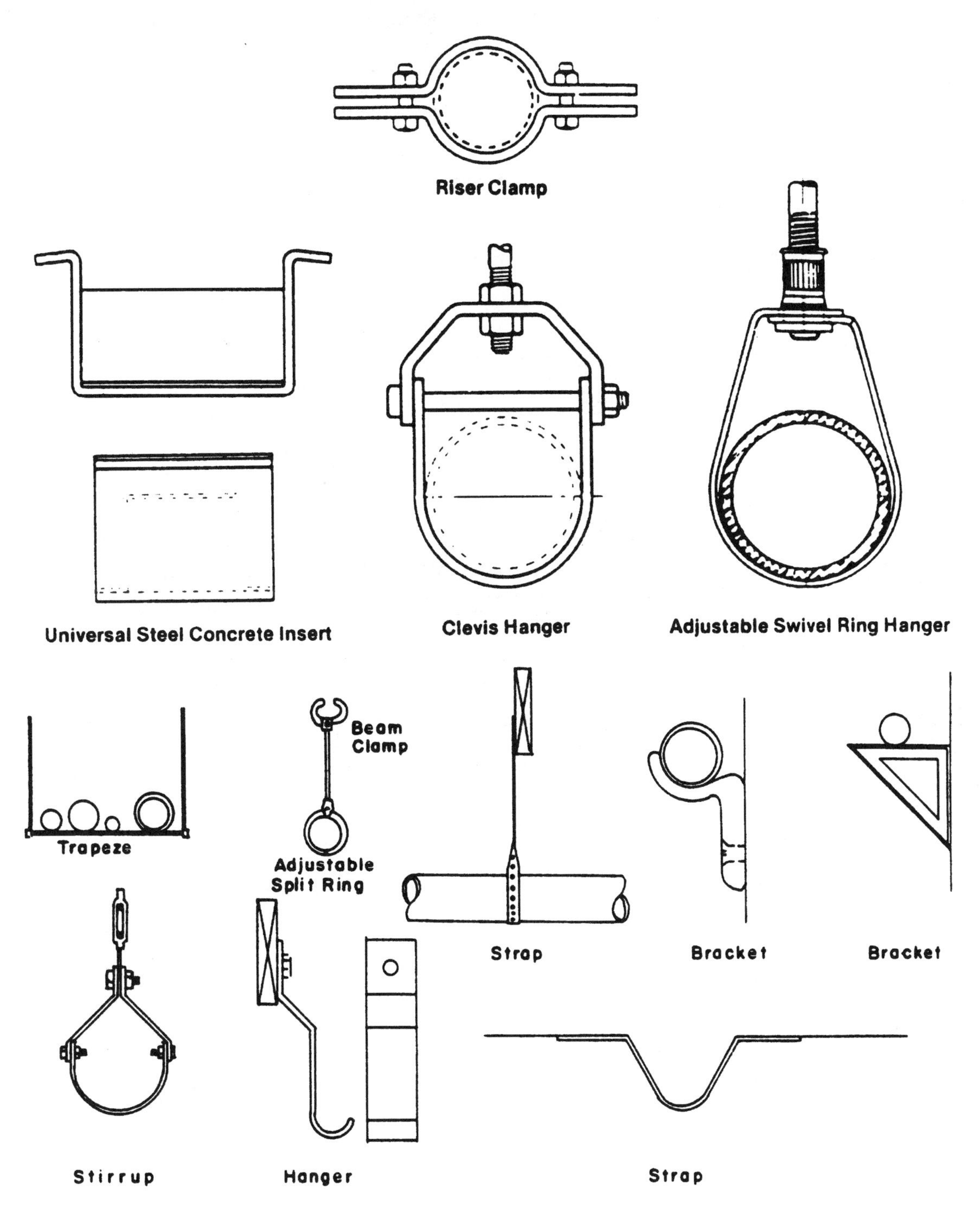

By permission of Cast Iron Soil Pipe Institute

18.13.0 Cast-Iron Pipe Hub-Barrel Dimensions

The following dimensions are given for use as convenient information on details of the hub barrel, and spigot, and are not requirements of this specification.

Outside Dimensions of Hub, Barrel, and Spigot for Detailing, in.

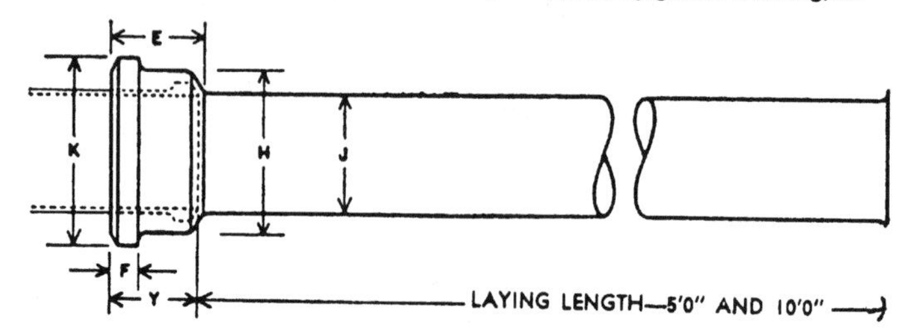

NOTE 1— in. = 25.4 mm.

Size (nominal ID)	Extra-Heavy Pipe 'XH'					
	K max	*H* max	*J*	*F*	*Y*	*E*
2	4 1/8	3 5/8	2 3/8	3/4	2 1/2	2 3/4
3	5 3/8	4 15/16	3 1/2	13/16	2 3/4	3 1/4
4	6 3/8	5 15/16	4 1/2	7/8	3	3 1/2
5	7 3/8	6 15/16	5 1/2	7/8	3	3 1/2
6	8 3/8	7 15/16	6 1/2	7/8	3	3 1/2
8	11 1/16	10 7/16	8 5/8	1 3/16	3 1/2	4 1/8
10	13 5/16	12 11/16	10 3/4	1 3/16	3 1/2	4 1/8
12	15 7/16	14 13/16	12 3/4	1 7/16	4 1/4	5
15	18 13/16	18 3/16	15 7/8	1 7/16	4 1/4	5

Size (nominal ID)	Service-Pipe 'SV'					
	K max	*H* max	*J*	*F*	*Y*	*E*
2	3 15/16	3 3/8	2 1/4	3/4	2 1/2	2 3/4
3	5	4 1/2	3 1/4	13/16	2 3/4	3 1/4
4	6	5 1/2	4 1/4	7/8	3	3 1/2
5	7	6 1/2	5 1/4	7/8	3	3 1/2
6	8	7 1/2	6 1/4	7/8	3	3 1/2
8	10 1/2	9 7/8	8 3/8	1 3/16	3 1/2	4 1/8
10	12 13/16	12 3/16	10 1/2	1 3/16	3 1/2	4 1/8
12	14 15/16	14 5/16	12 1/2	1 7/16	4 1/4	5
15	18 5/16	17 5/8	15 5/8	1 7/16	4 1/4	5

By permission of Cast Iron Soil Pipe Institute

18.14.0 Pipe Diameters and Trench Widths (U.S. and Metric Sizes)

Pipe Diameter (millimeters)	Trench Width (millimeters)	Pipe Diameter (millimeters)	Trench Width (millimeters)
100	470	1500	2500
150	540	1650	2800
200	600	1800	3000
250	680	1950	3200
300	800	2100	3400
375	910	2250	3600
450	1020	2400	3900
525	1100	2550	4100
600	1200	2700	4300
675	1300	2850	4500
825	1600	3000	4800
900	1700	3150	5000
1050	1900	3300	5200
1200	2100	3450	5400
1350	2300	3600	5600

NOTE: Trench widths based on 1.25 Bc + 300 where Bc is the outside diameter of the pipe in millimeters.

Pipe Diameter (inches)	Trench Width (feet)	Pipe Diameter (inches)	Trench Width (feet)
4	1.6	60	8.5
6	1.8	66	9.2
8	2.0	72	10.0
10	2.3	78	10.7
12	2.5	84	11.4
15	3.0	90	12.1
18	3.4	96	12.9
21	3.8	102	13.6
24	4.1	108	14.3
27	4.5	114	14.9
33	5.2	120	15.6
36	5.6	126	16.4
42	6.3	132	17.1
48	7.0	138	17.8
54	7.8	144	18.5

NOTE: Trench widths based on 1.25 Bc + 1 ft where Bc is the outside diameter of the pipe in inches.

18.15.0 Pipe Test Plugs (Illustrated)

Typical test plugs used for air/water tests.

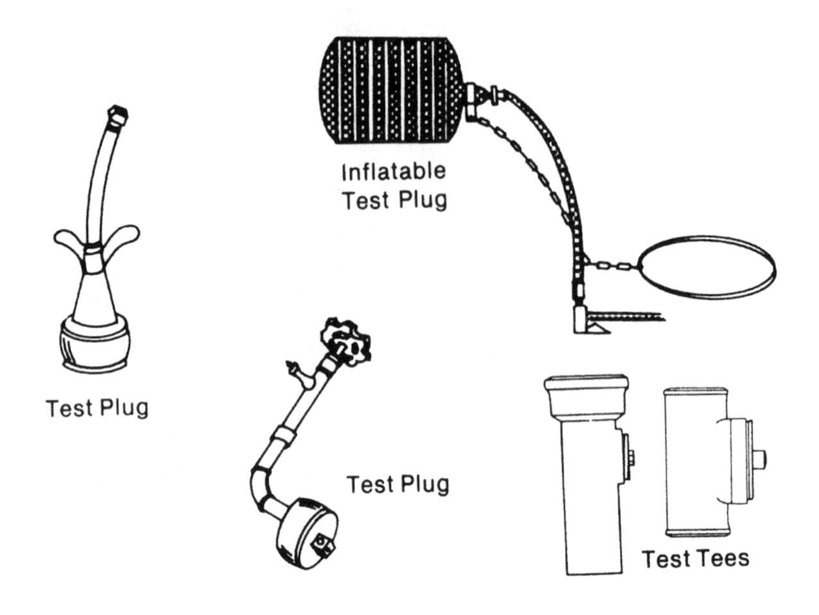

18.16.0 Thrust Pressures When Hydrostatically Testing Soil Pipe

PIPE SIZE		1½"	2"	3"	4"	5"	6"	8"	10"
HEAD, Feet of Water	PRESSURE PSI	THRUST lb.	THRUST lb.	THRUST lb.	THRUST lb.	THRUST lb.	THRUST lb.	THRUST lb.	THRUST lb.
10	4.3	12	19	38	65	95	134	237	377
20	8.7	25	38	77	131	192	271	480	762
30	13.0	37	56	115	196	287	405	717	1139
40	17.3	49	75	152	261	382	539	954	1515
50	21.7	62	94	191	327	479	676	1197	1900
60	26.0	74	113	229	392	574	810	1434	2277
70	30.3	86	132	267	457	668	944	1671	2654
80	34.7	99	151	306	523	765	1082	1914	3039
90	39.0	111	169	344	588	860	1216	2151	3416
100	43.4	123	188	382	654	957	1353	2394	3801
110	47.7	135	208	420	719	1052	1487	2631	4178
120	52.0	147	226	458	784	1147	1621	2868	4554
AREA, OD. in.^{2}		2.84	4.34	8.81	15.07	22.06	31.17	55.15	87.58

Thrust = Pressure x Area

By permission of Cast Iron Soil Pipe Institute

18.17.0 Piping Schematics (Vent and Stack Installations)

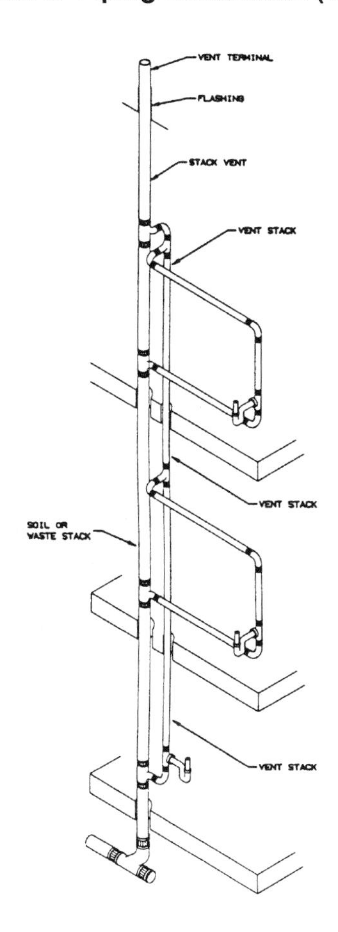

Vent Stack and Stack Vent

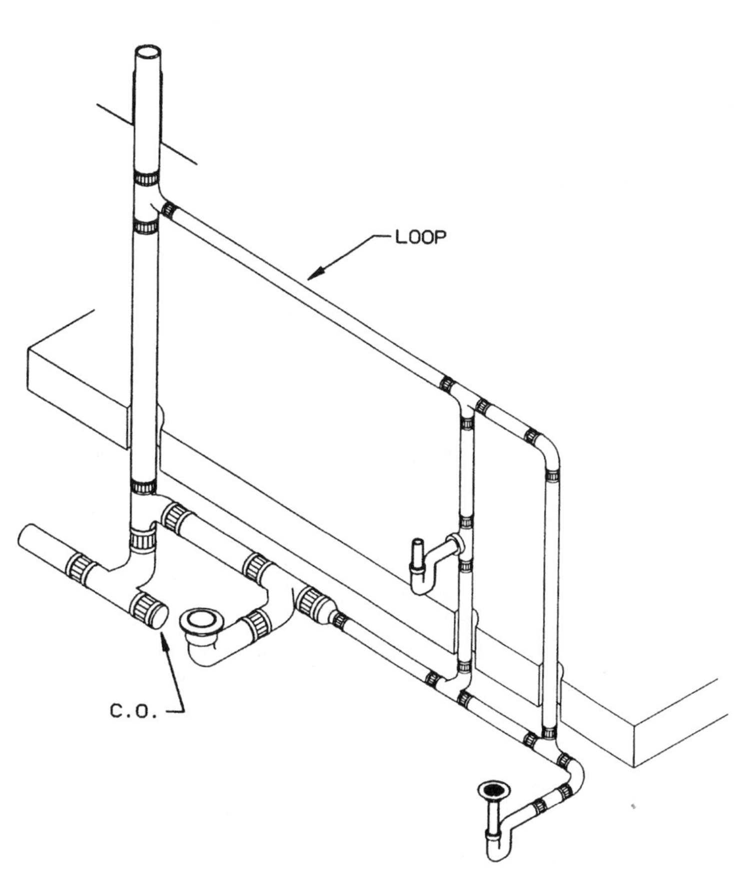

Loop Vent

By permission of Cast Iron Soil Pipe Institute

18.17.1 Piping Schematics (Continuous- and Looped-Vent System)

TYPICAL LAYOUTS

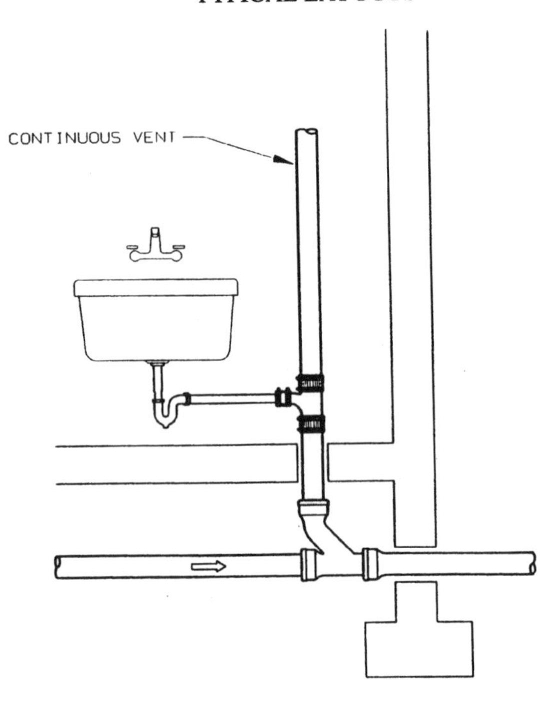

Continuous Vent

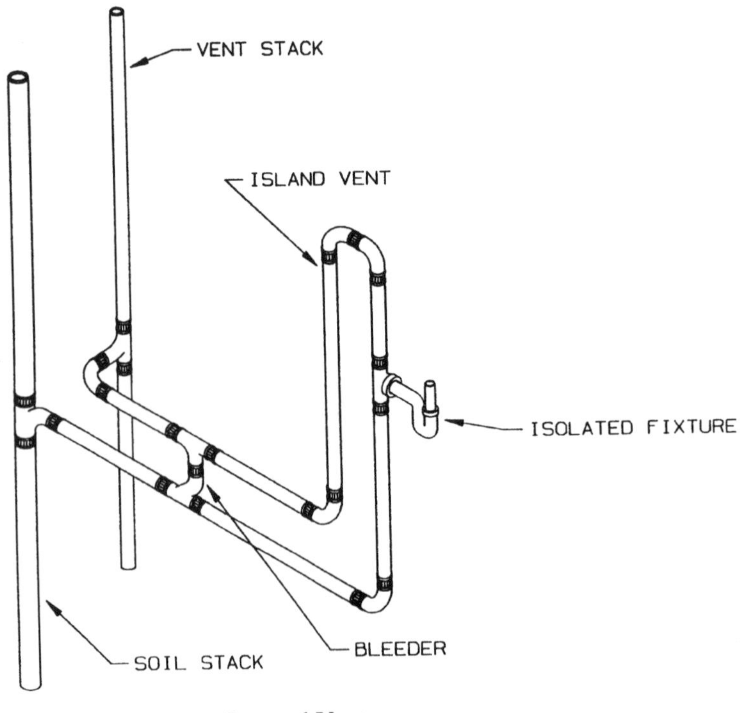

Looped Vent

By permission of Cast Iron Soil Pipe Institute

18.17.2 Piping Schematics (Stacked Fixture Installation)

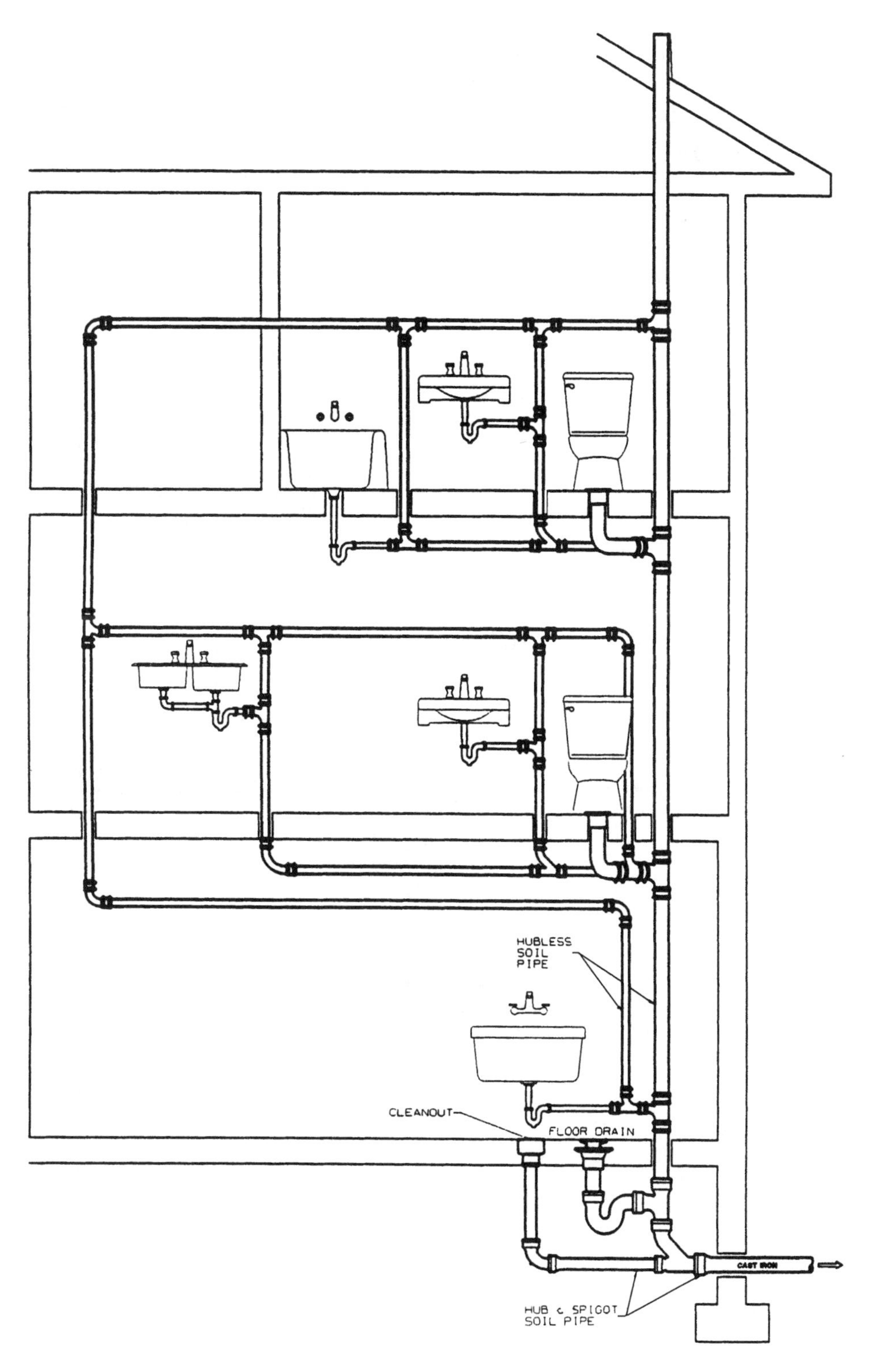

By permission of Cast Iron Soil Pipe Institute

18.17.3 Piping Schematics (Roof Drain and Leader, Hubless/Hub Pipe)

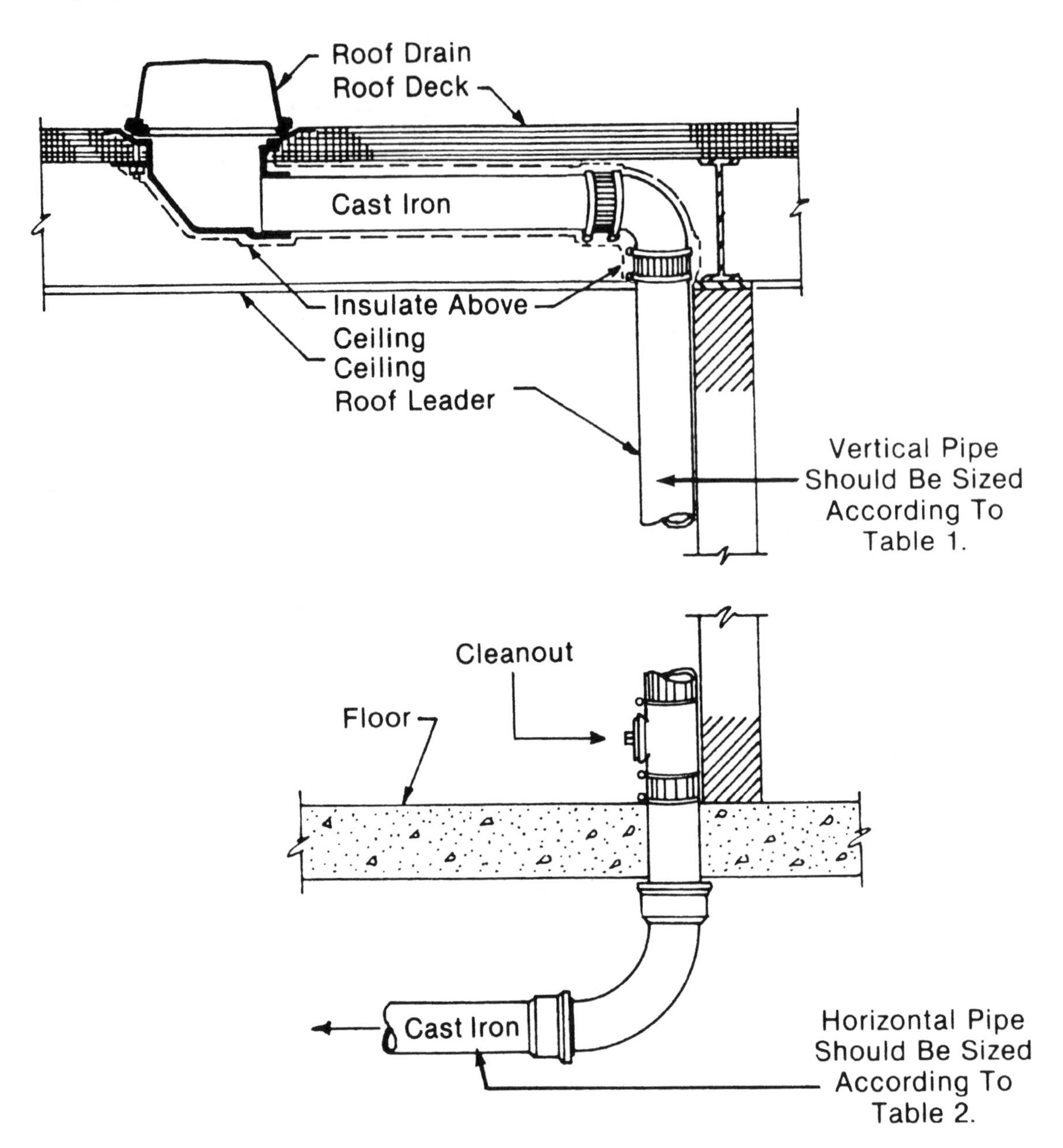

By permission of Cast Iron Soil Pipe Institute

18.17.4 Piping Schematics (Battery of Fixtures with a Common Vent)

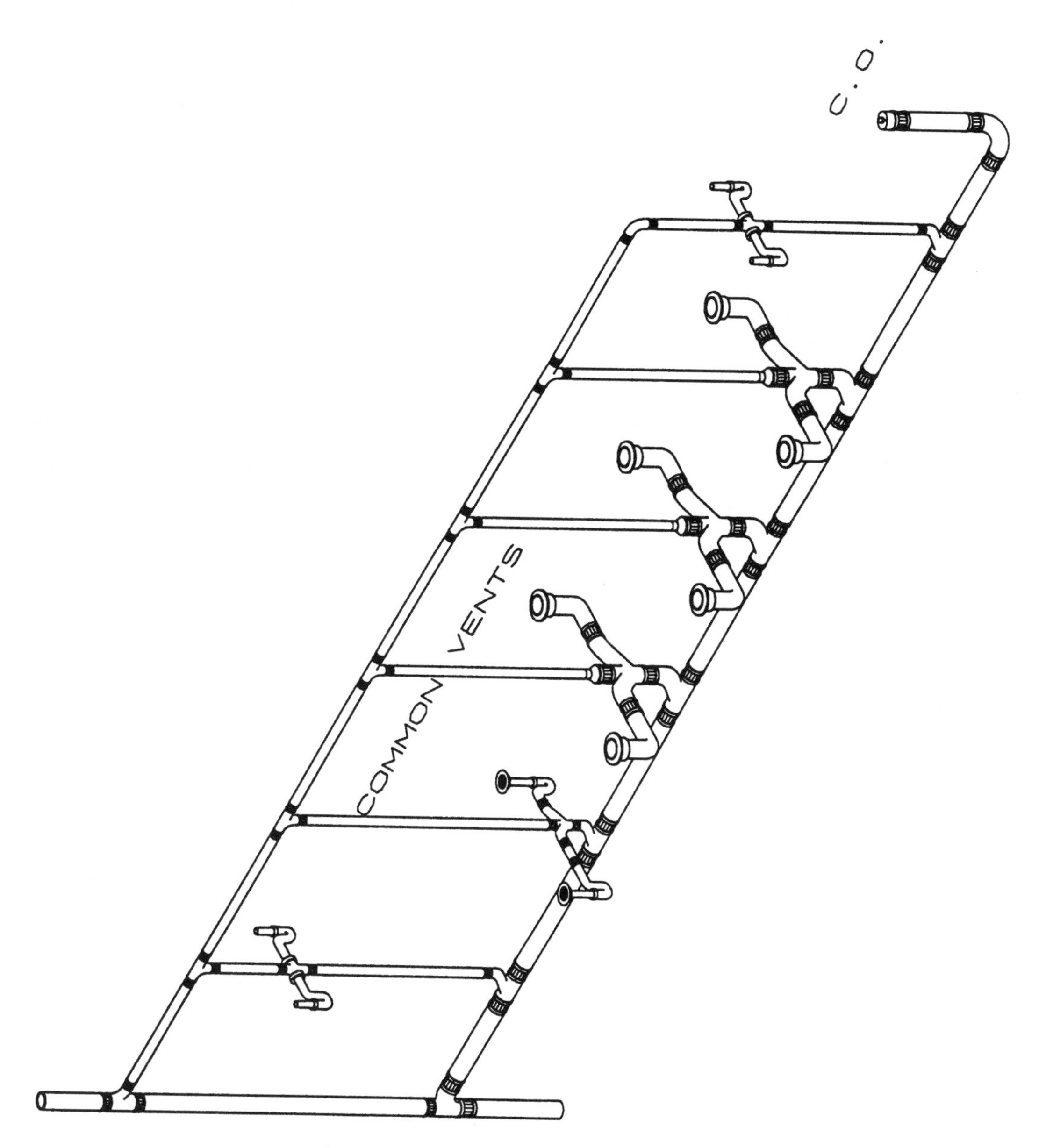

By permission of Cast Iron Soil Pipe Institute

18.17.5 Piping Schematics (Circuit Venting/Wet Venting)

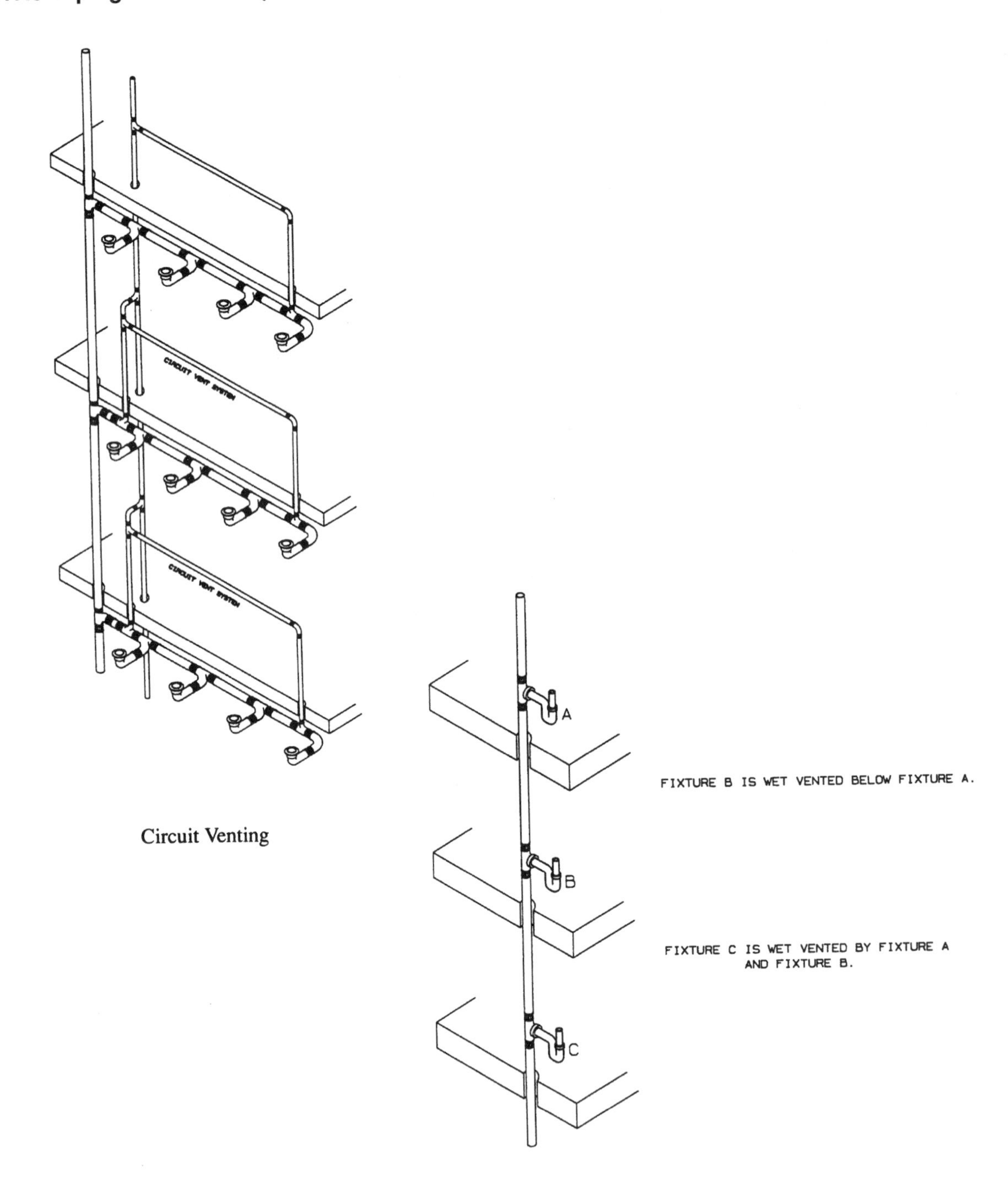

Circuit Venting

Wet Vent

By permission of Cast Iron Soil Pipe Institute

18.17.6 Piping Schematics (Typical Waste and Vent Installation)

Typical waste and vent pipe installation for plumbing fixtures

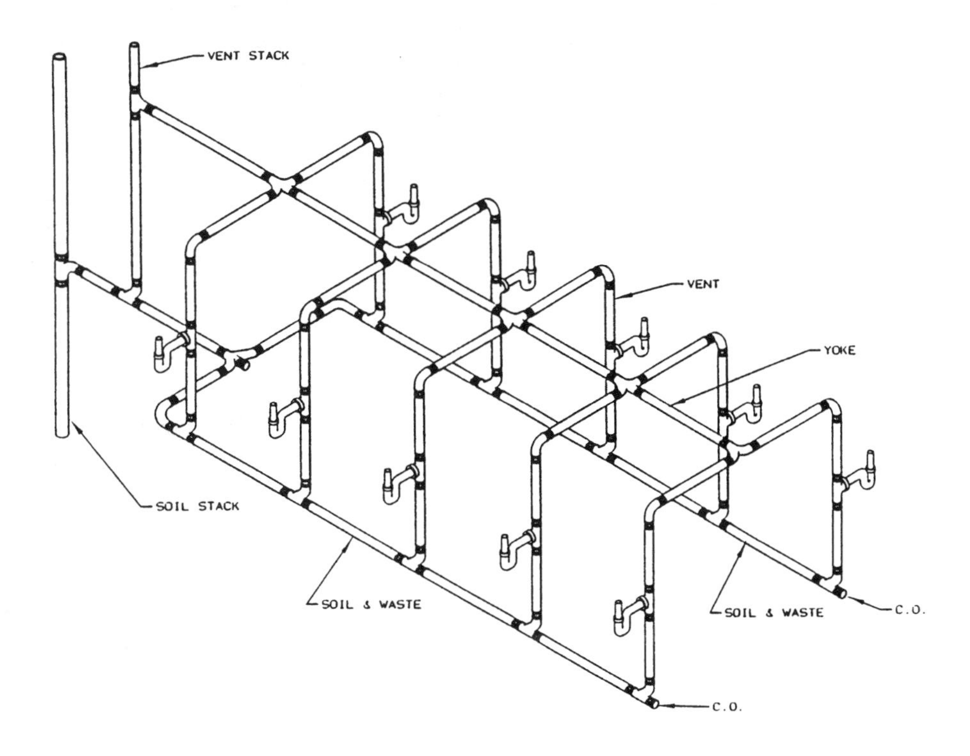

Drainage for a Battery of Fixtures with a Wide Pipe Space Available

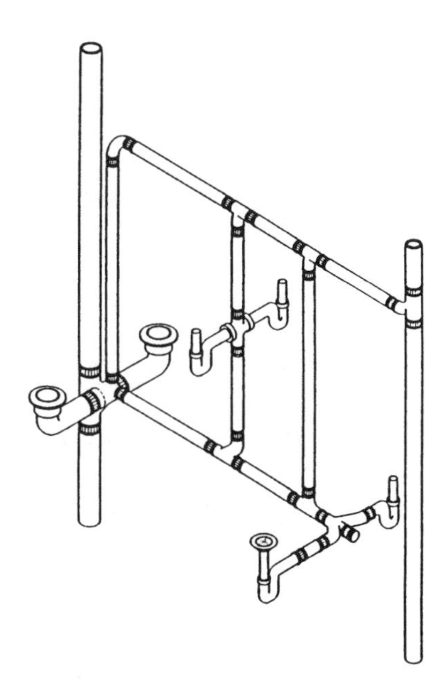

PIPING FOR TUB, LAVATORY & WATER CLOSET
EACH FIXTURE VENTED

Typical Piping Arrangement for a Water Closet, Lavatory and Tub. Piping may be either Hubless or Hub and Spigot.

By permission of Cast Iron Soil Pipe Institute

18.18.0 Various City Water-Temperature and Hardness Figures

City Water Data

State and City	Source of Supply	Maximum Water Temp. F	Hardness PPM
Alabama			
Anniston	W	70	104
Birmingham	S	85	43
Alaska			
Fairbanks	W	46	120
Ketchikan	S	44	4
Arizona			
Phoenix	W	81	210
Tucson	W	80	222
Arkansas			
Little Rock	WS	89	26
California			
Fresno	W	72	87
Los Angeles	WS	79	195
Sacramento	S	83	76
San Francisco	S	66	181
Colorado			
Denver	S	74	123
Pueblo	S	77	279
Connecticut			
Hartford	S	73	12
New Haven	S	76	46
Delaware			
Wilmington	S	83	48
District of Columbia			
Washington	S	84	162
Florida			
Jacksonville	WS	90	305
Miami	W	82	78
Georgia			
Atlanta	S	87	14
Savannah	W	85	120
Hawaii			
Honolulu	S	70	57
Idaho			
Boise	WS	65	71
Illinois			
Chicago	5	73	125
Peoria	W	67	386
Springfield	84	164	
Indiana			
Evansville	S	87	140
Fort Wayne	S	84	95
Indianapolis	WS	85	279
Iowa			
Des Moines	S	77	340
Dubuque	W	60	324
Sioux City	W	62	548
Kansas			
Kansas City	S	92	230
Kentucky			
Ashland	S	85	93
Louisville	S	85	104
Louisiana			
New Orleans	S	93	150
Shreveport	S	90	36
Maine			
Portland	S	70	12

City Water Data

State and City	Source of Supply	Maximum Water Temp. F	Hardness PPM
Maryland			
Baltimore	S	75	50
Massachusetts			
Cambridge	S	74	46
Holyoke	S	77	23
Michigan			
Detroit	S	78	100
Muskegon	S	71	153
Minnesota			
Duluth	S	58	54
Minneapolis	S	83	172
Mississippi			
Jackson	S	85	38
Meridian	WS	89	7
Missouri			
Springfield	WS	80	187
St. Louis	S	88	83
Montana			
Butte	WS	54	63
Helena	WS	57	96
Nebraska			
Lincoln	W	63	188
Omaha	S	85	135
New Hampshire			
Berlin	S	69	10
Nashua	W	70	25
Nevada			
Reno	S	63	114
New Jersey			
Atlantic City	WS	73	12
Newark	S	75	29
New Mexico			
Albuquerque	W	72	155
New York			
Albany	S	70	42
Buffalo	S	76	118
NewYork	WS	73	30
North Carolina			
Asheville	S	79	4
Wilmington	S	89	34
North Dakota			
Bismarck	S	80	172
Ohio			
Cincinnati	S	85	120
Cleveland	S	77	121
Oklahoma			
Oklahoma City	S	83	100
Tulsa	S	85	80
Oregon			
Portland	S	65	10
Pennsylvania			
Philadelphia	S	83	98
Pittsburgh	84	95	
Rhode Island			
Providence	S	71	26
South Carolina			
Charleston	S	85	18
Greenville	S	79	4

By permission of The Trane Company, LaCrosse, Wisconsin

18.19.0 Abbreviations, Definitions, and Symbols that Appear on Plumbing Drawings

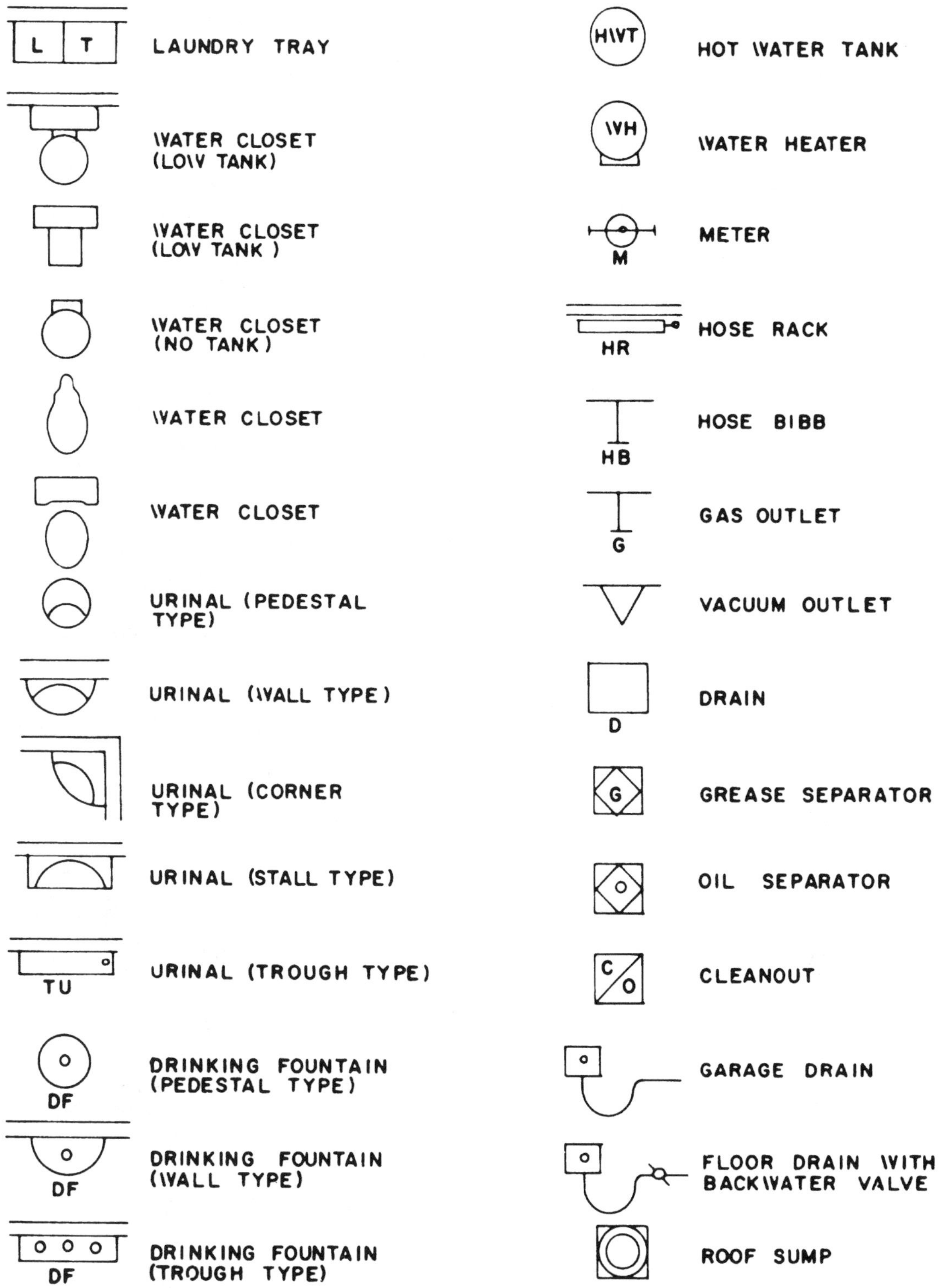

L T — LAUNDRY TRAY

WATER CLOSET (LOW TANK)

WATER CLOSET (LOW TANK)

WATER CLOSET (NO TANK)

WATER CLOSET

WATER CLOSET

URINAL (PEDESTAL TYPE)

URINAL (WALL TYPE)

URINAL (CORNER TYPE)

URINAL (STALL TYPE)

TU — URINAL (TROUGH TYPE)

DF — DRINKING FOUNTAIN (PEDESTAL TYPE)

DF — DRINKING FOUNTAIN (WALL TYPE)

DF — DRINKING FOUNTAIN (TROUGH TYPE)

HWT — HOT WATER TANK

WH — WATER HEATER

M — METER

HR — HOSE RACK

HB — HOSE BIBB

G — GAS OUTLET

VACUUM OUTLET

D — DRAIN

G — GREASE SEPARATOR

OIL SEPARATOR

C/O — CLEANOUT

GARAGE DRAIN

FLOOR DRAIN WITH BACKWATER VALVE

ROOF SUMP

By permission of Cast Iron Soil Pipe Institute

18.9.1 Recommended Symbols for Plumbing on Plumbing Drawings

Symbols for fixtures[1].

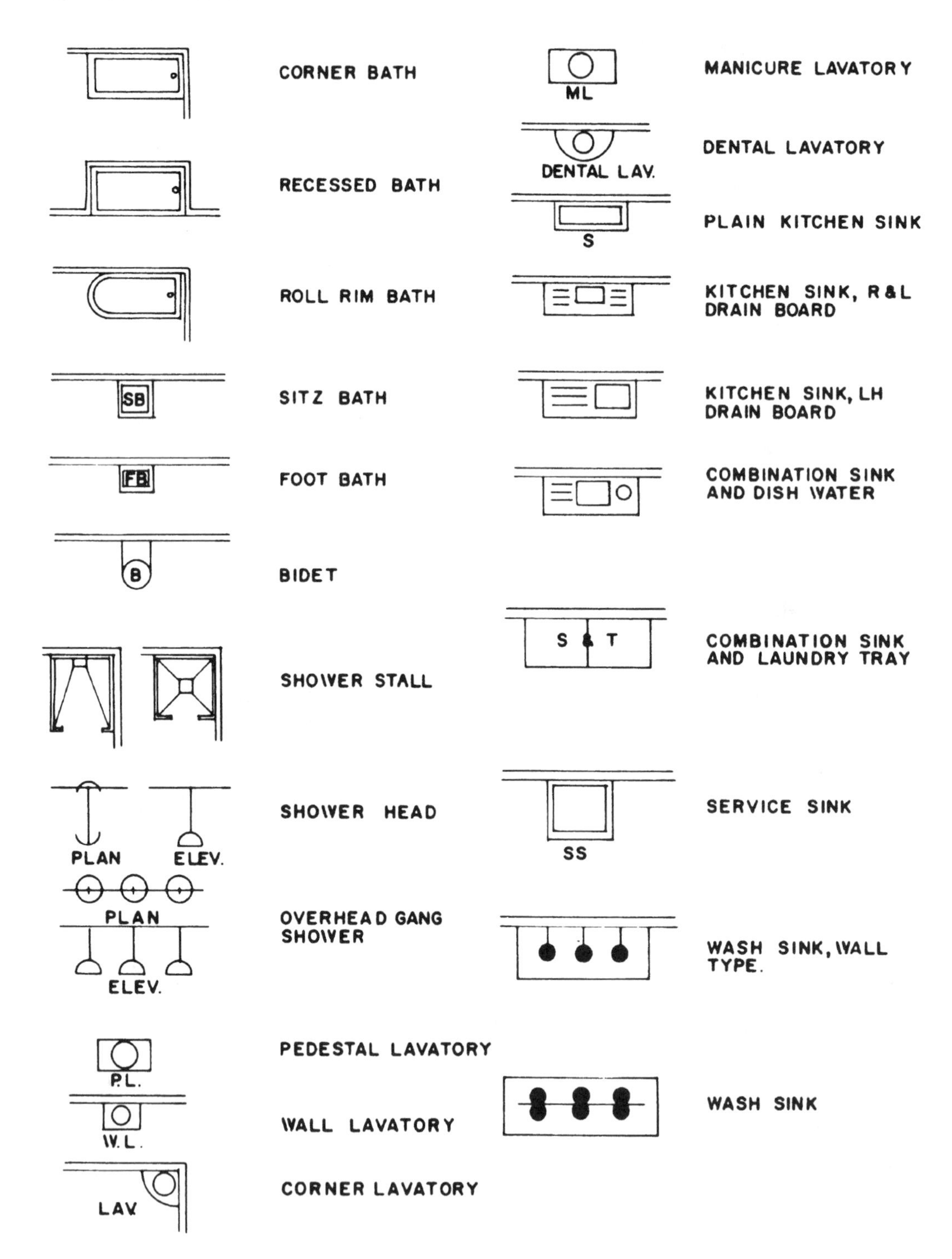

[1] Symbols adopted by the American National Standards Association (ANSI)

By permission of Cast Iron Soil Pipe Institute

18.19.1 Recommended Symbols for Plumbing on Plumbing Drawings (Continued)

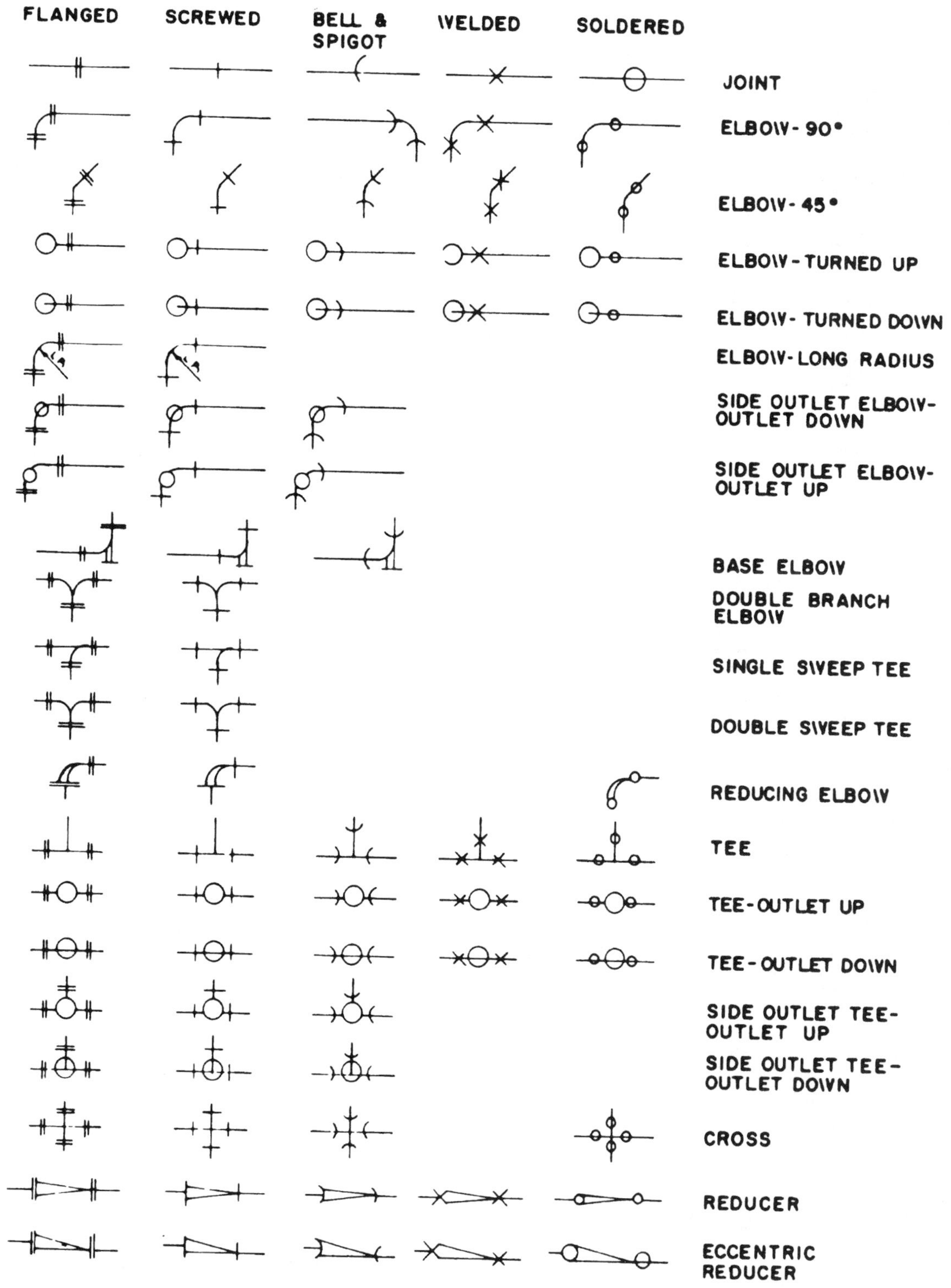

FLANGED SCREWED BELL & SPIGOT WELDED SOLDERED

JOINT

ELBOW - 90°

ELBOW - 45°

ELBOW - TURNED UP

ELBOW - TURNED DOWN

ELBOW - LONG RADIUS

SIDE OUTLET ELBOW - OUTLET DOWN

SIDE OUTLET ELBOW - OUTLET UP

BASE ELBOW

DOUBLE BRANCH ELBOW

SINGLE SWEEP TEE

DOUBLE SWEEP TEE

REDUCING ELBOW

TEE

TEE - OUTLET UP

TEE - OUTLET DOWN

SIDE OUTLET TEE - OUTLET UP

SIDE OUTLET TEE - OUTLET DOWN

CROSS

REDUCER

ECCENTRIC REDUCER

By permission of Cast Iron Soil Pipe Institute

18.19.2 Symbols for Pipe Fittings and Valves

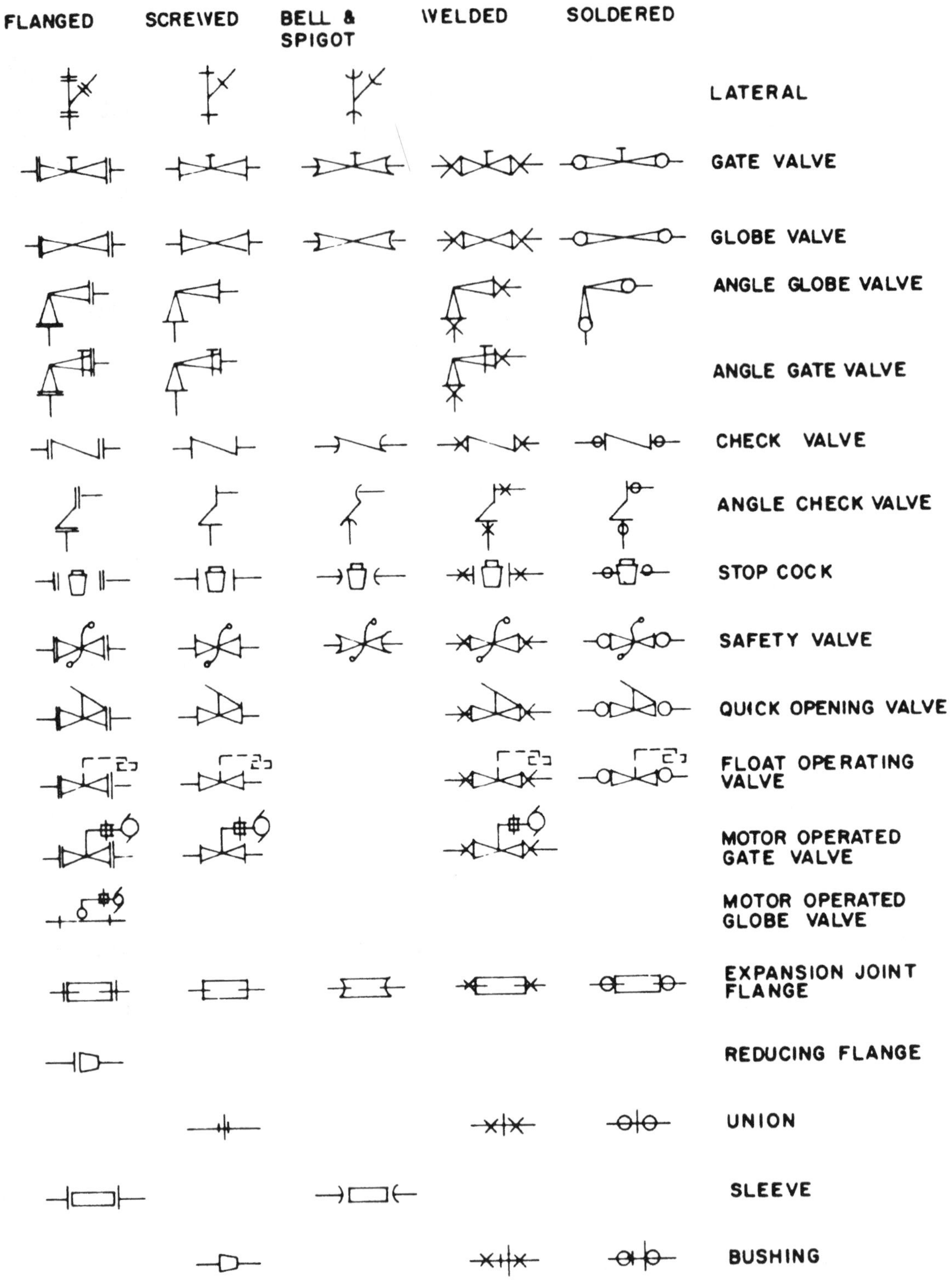

FLANGED SCREWED BELL & SPIGOT WELDED SOLDERED

LATERAL

GATE VALVE

GLOBE VALVE

ANGLE GLOBE VALVE

ANGLE GATE VALVE

CHECK VALVE

ANGLE CHECK VALVE

STOP COCK

SAFETY VALVE

QUICK OPENING VALVE

FLOAT OPERATING VALVE

MOTOR OPERATED GATE VALVE

MOTOR OPERATED GLOBE VALVE

EXPANSION JOINT FLANGE

REDUCING FLANGE

UNION

SLEEVE

BUSHING

By permission of Cast Iron Soil Pipe Institute

18.19.2 Symbols for Pipe Fittings and Valves (Continued)

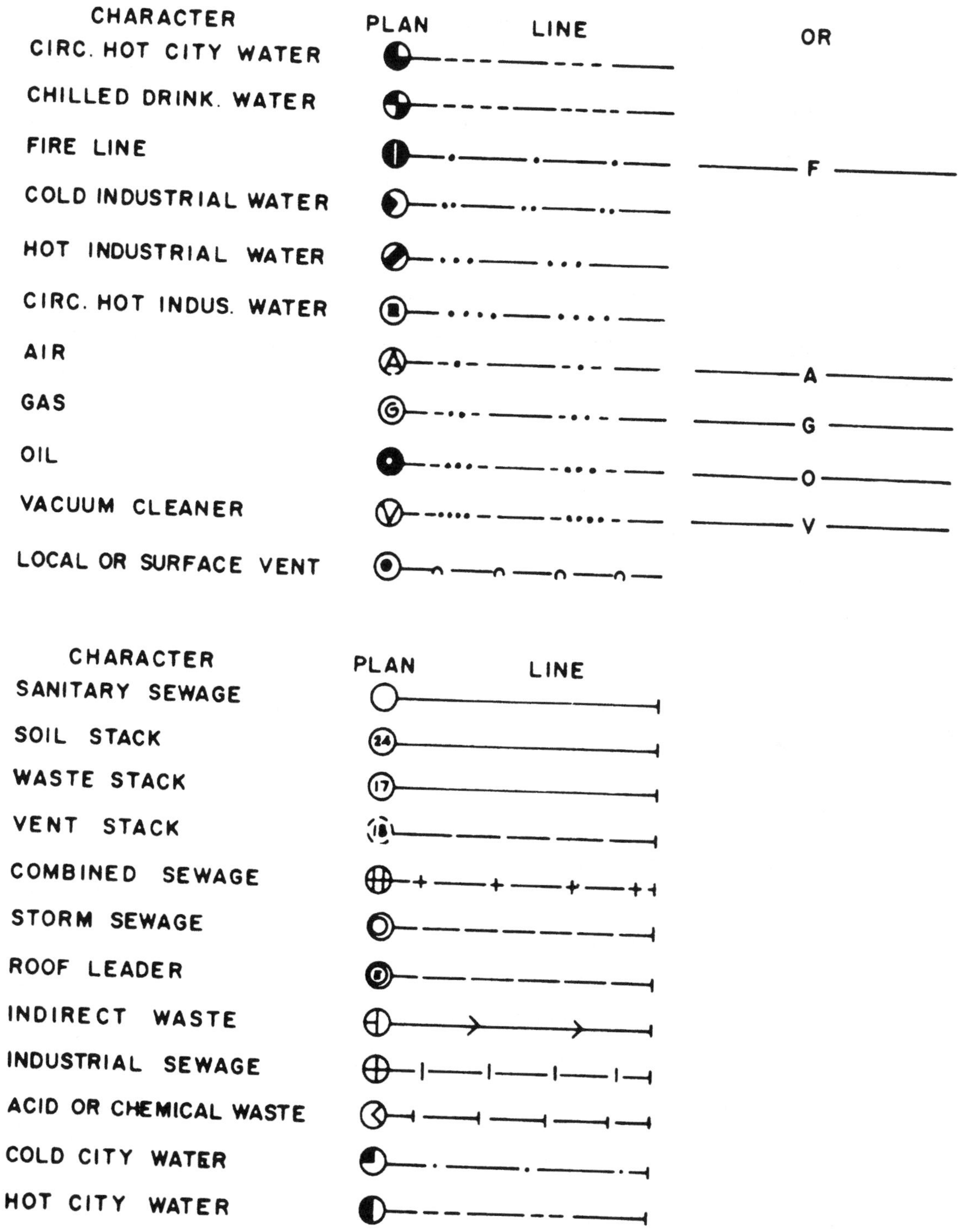

CHARACTER	PLAN	LINE	OR
CIRC. HOT CITY WATER			
CHILLED DRINK. WATER			
FIRE LINE			F
COLD INDUSTRIAL WATER			
HOT INDUSTRIAL WATER			
CIRC. HOT INDUS. WATER			
AIR	A		A
GAS	G		G
OIL			O
VACUUM CLEANER	V		V
LOCAL OR SURFACE VENT			

CHARACTER	PLAN	LINE
SANITARY SEWAGE		
SOIL STACK	24	
WASTE STACK	17	
VENT STACK	18	
COMBINED SEWAGE		
STORM SEWAGE		
ROOF LEADER		
INDIRECT WASTE		
INDUSTRIAL SEWAGE		
ACID OR CHEMICAL WASTE		
COLD CITY WATER		
HOT CITY WATER		

By permission of Cast Iron Soil Pipe Institute

Section

19

Fire Protection

Contents

19.9.0 Introduction to fire protection
19.1.0 Wet-pipe systems
19.2.0 Dry-pipe systems
19.3.0 Pre-action systems
19.4.0 Fire-cycle systems
19.5.0 Deluge systems
19.6.0 Standpipes
19.7.0 Sprinkler heads
19.8.0 Hose stations
19.9.0 Siamese connections
19.10.0 Light, ordinary, extra-hazard occupancy (defined
19.11.0 Illustration of grid versus looped system
19.12.0 Placement of sprinkler heads in relation to obstructions
19.13.0 Sprinkler-head placement requirements
19.14.0 Sprinkler-head requirements for various hazards
19.15.0 Temperature ratings of sprinklers, based on the distance from the heat source
19.16.0 Sprinkler maintenance schedules
19.17.0 Hangers for sprinkler pipes
19.18.0 Piping weights when filled with water
19.19.0 Seismic zones and piping modification requirements
19.20.0 Unacceptable pipe weld joints (illustrated)
19.21.0 Schematics of fire-department connections/water supply
19.22.0 Schematics of commercial cooking automatic nozzle installation
19.23.0 Contractor's material and test certification forms

19.0.0 Introduction to Fire Protection

Sprinklers and other fire-protection systems are available in many different variations, each designed for specific fire-suppression situations. Systems using water can be customized in many ways, but all maintain the same basic components. If local water systems cannot provide sufficient volume and/or pressure, water tanks are often installed to provide adequate flow, delivering the required amounts of water by gravity, air, or pump pressure.

Systems capable of delivering oxygen-starving foams or powders are frequently used if these particular agents are more effective in suppressing fire. Halogenated agents, halon, developed to replace water as the agent to extinguish fires without damaging sensitive equipment, is used in many computer rooms or other areas that contain delicate and valuable documents, fabrics, and relics. And portable fire extinguishers of various capacities are used in localized situations to extinguish combustible material, solvent oil, and electrical fires.

19.1.0 Wet-Pipe Systems

Wet-pipe systems are the most common systems used in commercial and industrial construction.

- *Advantages* Rapid response to fire control because the sprinkler pipes are always filled with water, relatively uncomplicated design, highly reliable.
- *Disadvantages* Cannot be used where systems are to be installed in a building that is not heated and where ambient temperatures are at (or below) freezing, unless an anti-freeze solution is added to the water in the system.

19.2.0 Dry-Pipe Systems

Dry-pipe systems are used where fire protection is required to be installed in unheated spaces, where ambient temperatures will dip below the freezing mark. This system is often used in low-hazard areas.

- *Advantages* Dry valves allow pressurized air to fill the piping until a sprinkler head requires that water enter the system; therefore, ambient freezing problems are eliminated.
- *Disadvantages* There is a delay in response time, which this requires more water to be delivered quickly. Therefore, piping sizes are generally much larger than those in a wet system. The dry system might also require the installation of more sprinkler heads than required in a wet system. The dry system will also need an air compressor (another piece of equipment that will require maintenance) to ensure that pressure is in the system at all times.

19.3.0 Pre-Action Systems

This form of sprinkler system is a two-stage system. The first stage is the alert and fire-notification phase. When detected, the presence of a fire activates the alarm (the first phase); the second phase is the sprinkler response.

- *Advantages* This system combines some of the advantages of a dry system, but adds a time delay to fill the lines with water prior to the opening of any sprinkler systems.
- *Disadvantages* There is a delay in delivering water through the sprinkler heads as the pre-action valves fills the pipes with water, following the sprinkler heads to open in response to the presence of fire.

19.4.0 Fire-Cycle Systems

This system functions in much the same manner as the pre-action system, except that it adds the installation of sensing devices to stop the flow of water when the fire has been extinguished.

- *Advantages* Minimizes water damage after the fire has been extinguished.
- *Disadvantages* Delay in delivering water to the required area(s).

19.5.0 Deluge Systems

The deluge system is similar to the pre-action system, except that all sprinkler heads are kept open when the system is activated.

- *Advantages* In high hazard areas, this system delivers water through all of the sprinkler heads in the system, instead of just opening selected heads in close proximity to the fire.
- *Disadvantages* The entire area is deluged with water—even if the fire is restricted to a much smaller area.

19.6.0 Standpipes

Standpipes are installed in high-rise buildings to create an internal water supply on upper floors so that firefighters can attach their hoses to connections on the standpipe and effectively fight the fire on that floor. Standpipes are generally located in or near stair enclosures so that firefighters can have a water supply available before entering the actual floor where the fire has occurred.

Standpipe systems are usually of a wet system design and operate through an up-feed pump to ensure proper volume and pressure on the upper floors of a multi-story building. However, water can be delivered through the standpipe system by the firefighter's pumper truck, connected to a city fire hydrant, delivering high-pressure, high-volume water through the standpipe via an external siamese connection.

19.7.0 Sprinkler Heads

Sprinkler heads are available in a variety of configurations and materials of construction, depending on their coverage and aesthetic requirements. Sprinkler heads are generally of two basic types:

- *Fusible* Heads with soldered metal links that keep the head closed until the temperature rises to the point where the metal reaches its melting point. The solder will then yield, allowing the sprinkler head to open. Fusible metal alloy pellets can also be used. The pellet will melt at a predetermined temperature and allow the sprinkler head to open.
- *Frangible* A breakable, transparent glass capsule containing a colored liquid that will expand to the point where the glass will break, allowing the sprinkler head to open. The liquid is color-coded so that visible inspection will confirm that the correct temperature-seeking head has been installed.

19.8.0 Hose Stations

There is often disagreement about the advantages of installing hose cabinets, complete with reeled hoses in strategic locations throughout a building. Often, these cabinets will contain only a valved, threaded connection, but no hose or reel. Unless the hoses are inspected and maintained properly by the building's owner, they could deteriorate, or even be removed from the cabinet. Many firefighters, not trusting the quality of cabinet fire hoses, bring their own hoses to the building to attach to hose connections in the cabinets.

Hose nozzles, when attached to these cabinet hoses, are available in adjustable fog, spray type, straight steam, smooth-bore, or combination solid stream and spray. Solid-stream nozzles, ranging in size from ¾ inch (1.90 cm) to 2 inches (5.08 cm) can deliver 120 to 560 gallons per minute, at pressures ranging from 40 to 10 pounds per square inch (psi).

19.9.0 Siamese Connections

A *siamese connection* is an external source, attached to the building, to which firefighters can attach their hoses from pumper trucks and pressurize the building's sprinkler system by drawing water from a city hydrant. The location of the siamese connection, number of connections, and size and type of thread pattern varies, depending upon local fire-marshall requirements.

19.10.0 Light, Ordinary, and Extra-Hazard Occupancy (Defined)

Light Hazard Occupancies include occupancies having conditions similar to:

Churches
Clubs
Eaves and overhangs, if combustible construction with no combustibles beneath
Educational
Hospitals
Institutional
Libraries, except large stack rooms
Museums
Nursing or convalescent homes
Office, including data processing
Residential
Restaurant seating area
Theaters and auditoriums excluding stages and prosceniums
Unused attics

Ordinary Hazard Occupancies (Group 1) include occupancies having conditions similar to:

Automobile parking and showrooms
Bakeries
Beverage manufacturing
Canneries
Dairy products manufacturing and processing
Electronic plants
Glass and glass products manufacturing
Laundries
Restaurant service areas.

Ordinary Hazard Occupancies (Group 2) include occupancies having conditions similar to:

Cereal mills
Chemical plants—ordinary
Confectionery products
Distilleries
Dry cleaners
Feed mills
Horse stables
Leather goods manufacturing
Libraries—large stack room areas
Machine shops
Metal working
Mercantile
Paper and pulp mills
Paper process plants
Piers and wharves
Post offices
Printing and publishing
Repair garages
Stages
Textile manufacturing
Tire manufacturing
Tobacco products manufacturing
Wood machining
Wood product assembly

Extra Hazard Occupancies (Group 1) include occupancies having conditions similar to:

Aircraft hangers (except as governed by NFPA 409)
Combustible hydraulic fluid use areas
Die casting
Metal extruding
Plywood and particle board manufacturing
Printing [using inks having flash points below 100°F (37.9°C)]
Rubber reclaiming, compounding, drying, milling, vulcanizing
Saw mills
Textile picking, opening, blending, garnetting, carding, combining of cotton, synthetics, wool shoddy, or burlap
Upholstering with plastic foams

Extra Hazard Occupancies (Group 2) include occupancies having conditions similar to:

Asphalt saturating
Flammable liquids spraying
Flow coating
Manufactured home or modular building assemblies (where finished enclosure is present and has combustible interiors)
Open oil quenching
Plastics processing
Solvent cleaning
Varnish and paint dipping

Other NFPA standards contain design criteria for fire control or fire suppression. While these may form the basis of design criteria, this standard describes the methods of design, installation, fabrication, calculation, and evaluation of water supplies that should be used for the specific design of the system.

Included among items requiring listing are sprinklers, some pipe and some fittings, hangers, alarm devices, valves controlling flow of water to sprinklers, valve tamper switches, and gauges.

Information regarding the highest temperature that may be encountered in any location in a particular installation may be obtained by use of a thermometer that will register the highest temperature encountered; it should be hung for several days in the location in question, with the plant in operation.

19.11.0 Illustration of Grid Versus Looped System

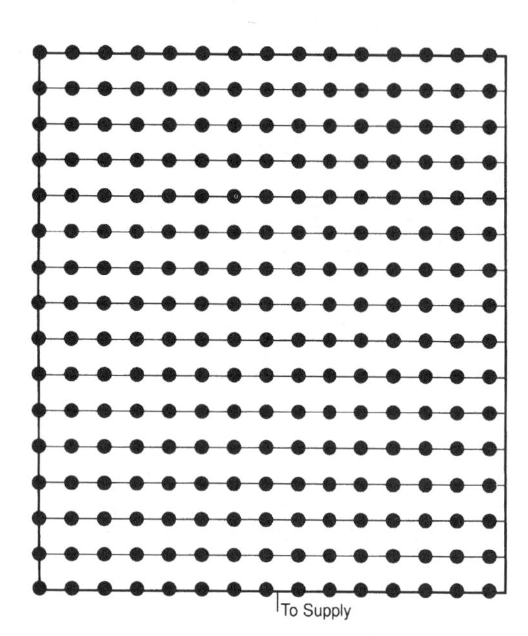

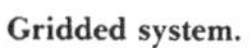

Gridded system.

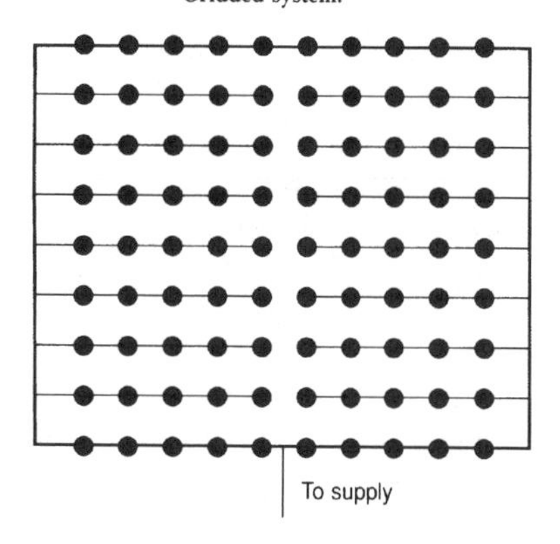

Looped system.

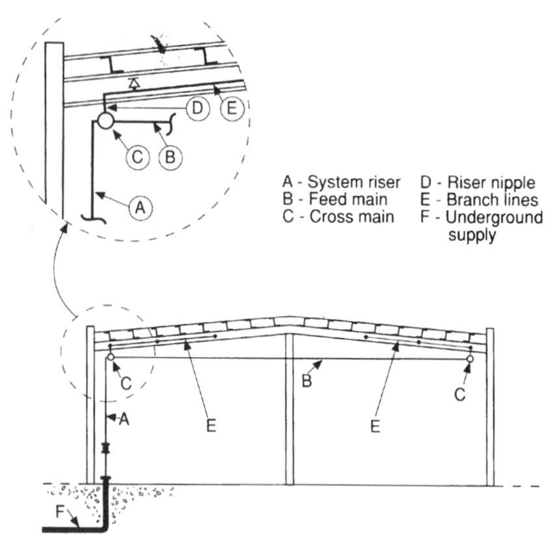

Building elevation showing parts of sprinkler piping system.

Dry Sprinkler. Under certain ambient conditions, wet pipe systems having dry-pendent (or upright) sprinklers may freeze due to heat loss by conduction. Therefore, due consideration should be given to the amount of heat maintained in the heated space, the length of the nipple in the heated space, and other relevant factors.

19.12.0 Placement of Sprinkler Heads in Relation to Obstructions

Position of Sprinklers in Relation to Obstruction Located Entirely Below the Sprinklers

Distance of Deflector above Bottom of Obstruction	Minimum Distance to Side of Obstruction, ft (m)
Less than 6 in. (152 mm)	1½ (0.5)
6 in. (152 mm) to less than 12 in. (305 mm)	3 (0.9)
12 in. (305 mm) to less than 18 in. (457 mm)	4 (1.2)
18 in. (457 mm) to less than 24 in. (610 mm)	5 (1.5)
24 in. (610 mm) to less than 30 in. (660 mm)	6 (1.8)

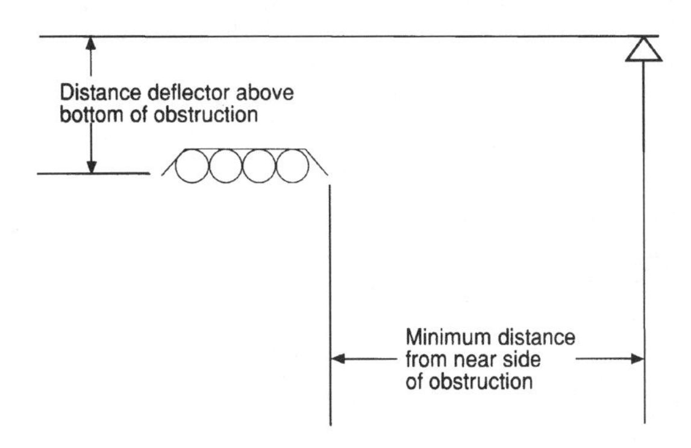

Position of sprinklers in relation to obstructions located entirely below the sprinklers. (To be used with Table 4-4.3.4.2.1.)

Exception: Where the obstruction is greater than 24 in. (610 mm) wide, one or more lines of sprinklers shall be installed below the obstruction.

(c) The obstruction shall not extend more than 12 in. (305 mm) to either side of the midpoint between sprinklers.

Exception: When the extensions of the obstruction exceed 12 in. (305 mm), one or more lines of sprinklers shall be installed below the obstruction.

(d) At least 18 in. (457 mm) clearance shall be maintained between the top of storage and the bottom of the obstruction.

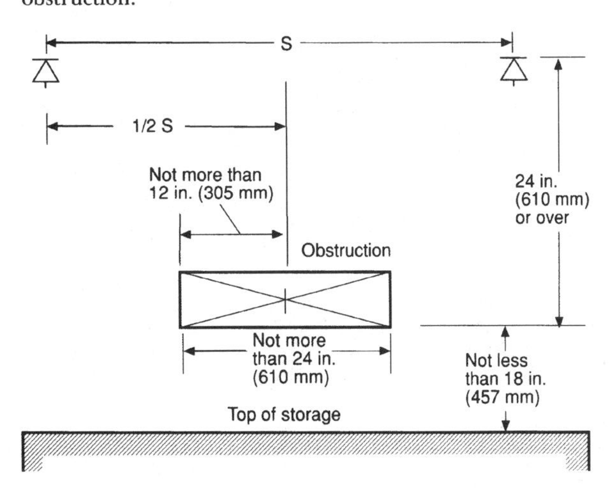

Position of sprinklers in relation to obstructions located 24 in. (610 mm) or more below deflectors.

19.13.0 Sprinkler-Head Placement Requirements

Position of Deflector when Located above Bottom of Obstruction

Distance from Sprinkler to Side of Obstruction	Maximum Allowable Distance of Deflector Above Bottom of Obstruction — Standard Sprinklers	Maximum Allowable Distance of Deflector Above Bottom of Obstruction — Extended Coverage Sprinklers
Less than 1 ft	0 in.	0 in.
1 ft to less than 1 ft 6 in.	1 in.	0 in.
1 ft 6 in. to less than 2 ft	1 in.	1 in.
2 ft to less than 2 ft 6 in.	2 in.	1 in.
2 ft 6 in. to less than 3 ft	3 in.	1 in.
3 ft to less than 3 ft 6 in.	4 in.	3 in.
3 ft 6 in. to less than 4 ft	6 in.	3 in.
4 ft to less than 4 ft 6 in.	7 in.	5 in.
4 ft 6 in. to less than 5 ft	9 in.	7 in.
5 ft to less than 5 ft 6 in.	11 in.	7 in.
5 ft 6 in. to less than 6 ft	14 in.	7 in.
6 ft to less than 6 ft 6 in.	N/A	9 in.
6 ft 6 in. to less than 7 ft	N/A	11 in.
7 ft and greater	N/A	14 in.

For SI Units: 1 in. = 25.4 mm; 1 ft = 0.3048 m

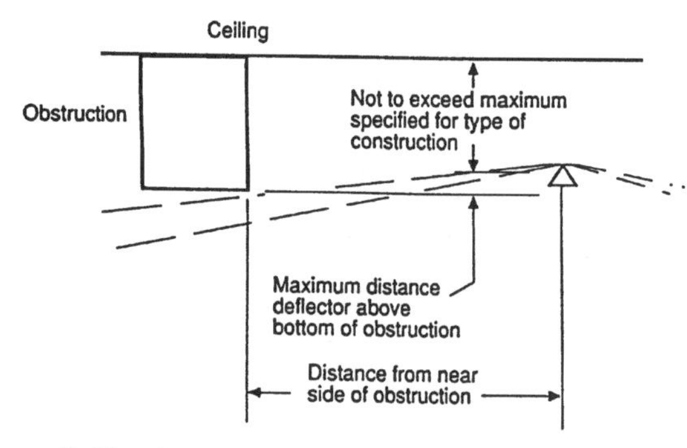

Position of deflector, upright, or pendent sprinkler when located above bottom of obstructions.

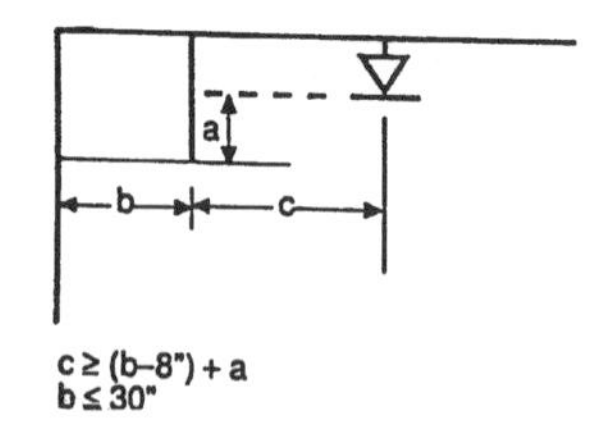

$c \geq (b-8") + a$
$b \leq 30"$

Horizontal obstructions against walls.

Under obstructed construction, the sprinkler deflector shall be located 1 to 6 in. (25.4 to 152 mm) below the structural members and a maximum distance of 22 in. (559 mm) below the ceiling/roof deck.

Exception No. 1: Sprinklers shall be permitted to be installed with the deflector at or above the bottom of the structural member to a maximum of 22 in. (559 mm) below the ceiling/roof deck where the sprinkler is installed in conformance.

Horizontal and Minimum Vertical Distances for Sprinklers

Horizontal Distance	Minimum Vertical Distance below Deflector
6 in. or less	3 in.
More than 6 in. to 9 in.	4 in.
More than 9 in. to 12 in.	6 in.
More than 12 in. to 15 in.	8 in.
More than 15 in. to 18 in.	9½ in.
More than 18 in. to 24 in.	12½ in.
More than 24 in. to 30 in.	15½ in.
More than 30 in.	18 in.

For SI Units: 1 in. = 25.4 mm.

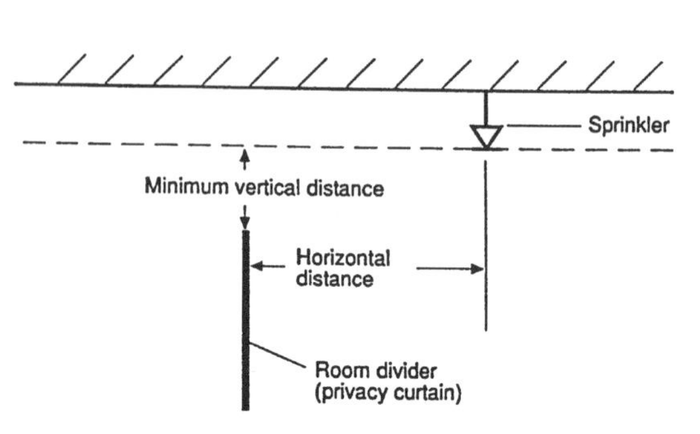

Sprinklers installed near privacy curtains, free-standing partitions, or room dividers.

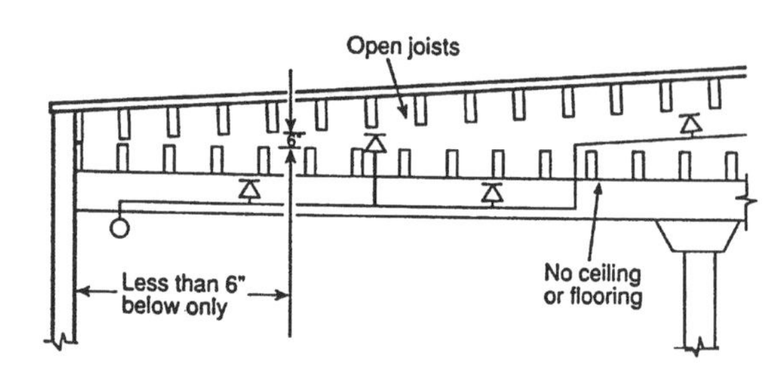

For SI Units: 1 in. = 25.4 mm.
Arrangement of sprinklers under two sets of open joists — no sheathing on lower joists.

Exception No. 2: Where sprinklers are installed in each bay of obstructed construction, deflectors shall be a minimum of 1 in. (25.4 mm) and a maximum of 12 in. (152 mm) below the ceiling.

Exception No. 3: Sprinklers shall only be permitted below composite wood joists where joist channels are firestopped to the full depth of the joists with material equivalent to the web construction so that individual channel areas do not exceed 300 sq ft (27.9 m²).

Exception No. 4: Deflectors of sprinklers under concrete tee construction with stems spaced less than 7½ ft (2.3 m) but more than 3 ft (0.9 m) on centers shall, regardless of the depth of the tee, be permitted to be located at or above the plane 1 in. (25.4 mm) below the bottom of the stems of the tees and shall comply with.*

19.14.0 Sprinkler-Head Requirements for Various Hazards

Hazard	Type of System	Minimum Operating Pressure,[1] psi (bar) 25 (1.7)	50 (3.4)	75 (5.2)	Hose Stream Demand gal/min (dm³/min)	Water Supply Duration, Hr
		Number Design Sprinklers				
Palletized[2] Storage						
Class I, II, and III commodities up to 25 ft (7.6 m) with maximum 10 ft (3.0 m) clearance to ceiling	Wet	15	Note 4	Note 4	500 (1900)	2
	Dry	25	Note 4	Note 4		
Class IV commodities up to 20 ft (6.1 m) with maximum 10 ft (3.0 m) clearance to ceiling	Wet	20	15	Note 4		
	Dry	Does not apply	Does not apply	Does not apply	500 (1900)	2
Unexpanded plastics up to 20 ft (6.1 m) with maximum 10 ft (3.0 m) clearance to ceiling	Wet	25	15	Note 4		
	Dry	Does not apply	Does not apply	Does not apply	500 (1900)	2
Expanded plastics commodities up to 18 ft (5.5 m) with maximum 8 ft (2.4 m) clearance to ceiling	Wet	Does not apply	15	Note 4	500 (1900)	2
	Dry	Does not apply	Does not apply	Does not apply		
Idle wood pallets up to 20 ft (6.1 m) with maximum 10 ft (3.0 m) clearance to ceiling	Wet	15	Note 4	Note 4		
	Dry	25	Note 4	Note 4	500 (1900)	1½
Solid Piled[2] Storage						
Class I, II, and III commodities up to 20 ft (6.1 m) with maximum 10 ft (3.0 m) clearance to ceiling	Wet	15	Note 4	Note 4		
	Dry	25	Note 4	Note 4	500 (1900)	1½
Class IV commodities and unexpanded plastics up to 20 ft (6.1 m) with maximum 10 ft (3.0 m) clearance to ceiling	Wet	Does not apply	15	Note 4		
	Dry	Does not apply	Does not apply	Does not apply	500 (1900)	1½
Double-Row Rack Storage[3] with Minimum 5.5 ft (1.7 m) Aisle Width and Multiple-Row Rack Storage with Minimum 8.0 ft (2.5 m) Aisle Width						
Class I and II commodities up to 25 ft (7.6 m) with maximum 5 ft (1.5 m) clearance to ceiling	Wet	20	Note 4	Note 4		
	Dry	30	Note 4	Note 4	500 (1900)	1½
Class I and II commodities up to 30 ft (9.2 m) with maximum 5 ft (1.5 m) clearance to ceiling	Wet	20 plus one level of in-rack sprinklers[5]	Note 4	Note 4		
	Dry	30 plus one level of in-rack sprinklers[5]	Note 4	Note 4	500 (1900)	1½
Class I, II, and III commodities up to 20 ft (6.1 m) with maximum 10 ft (3.0 m) clearance to ceiling	Wet	15	Note 4	Note 4		
	Dry	25	Note 4	Note 4	500 (1900)	1½
Class I, II, and III commodities up to 25 ft (7.6 m) with maximum 10 ft (3.0 m) clearance to ceiling	Wet	15 plus one level of in-rack sprinklers[5]	Note 4	Note 4	500 (1900)	1½
	Dry	25 plus one level of in-rack sprinklers[5]	Note 4	Note 4		
Class IV commodities up to 20 ft (6.1 m) with maximum 10 ft (3.0 m) clearance to ceiling	Wet	Does not apply	20	15		
	Dry	Does not apply	Does not apply	Does not apply	500 (1900)	2
Class IV commodities up to 25 ft (7.6 m) with maximum 10 ft clearance to ceiling	Wet	Does not apply	20 plus one level of in-rack sprinklers[5]	15 plus one level of in-rack sprinklers[5]		
	Dry	Does not apply	Does not apply	Does not apply	500 (1900)	2

19.15.0 Temperature Ratings of Sprinklers, Based on the Distance from the Heat Source

Temperature Ratings of Sprinklers Based on Distance from Heat Sources

Type of Heat Condition	Ordinary Degree Rating	Intermediate Degree Rating	High Degree Rating
1. Heating Ducts (a) Above	More than 2 ft 6 in.	2 ft 6 in. or less	—
(b) Side and Below	More than 1 ft 0 in.	1 ft 0 in. or less	—
(c) Diffuser Downward Discharge Horizontal Discharge	Any distance except as shown under Intermediate	*Downward:* Cylinder with 1 ft 0 in. radius from edge, extending 1 ft 0 in. below and 2 ft 6 in. above *Horizontal:* Semi-cylinder with 2 ft 6 in. radius in direction of flow, extending 1 ft 0 in. below and 2 ft 6 in. above	—
2. Unit Heater (a) Horizontal Discharge	—	*Discharge Side:* 7 ft 0 in. to 20 ft 0 in. radius pie-shaped cylinder [see Figure 4-3.1.3.2] extending 7 ft 0 in. above and 2 ft 0 in. below heater; also 7 ft 0 in. radius cylinder more than 7 ft 0 in. above unit heater	7 ft 0 in. radius cylinder extending 7 ft 0 in. above and 2 ft 0 in. below unit heater
(b) Vertical Downward Discharge [Note: For sprinklers below unit heater, see Figure 4-3.1.3.2.]	—	7 ft 0 in. radius cylinder extending upward from an elevation 7 ft 0 in. above unit heater	7 ft 0 in. radius cylinder extending from the top of the unit heater to an elevation 7 ft 0 in. above unit heater
3. Steam Mains (Uncovered) (a) Above	More than 2 ft 6 in.	2 ft 6 in. or less	—
(b) Side and Below	More than 1 ft 0 in.	1 ft 0 in. or less	—
(c) Blowoff Valve	More than 7 ft 0 in.	—	7 ft 0 in. or less

For SI Units: 1 in. = 25.4 mm; 1 ft = 0.3048 m.

Ratings of Sprinklers in Specified Locations

Location	Ordinary Degree Rating	Intermediate Degree Rating	High Degree Rating
Skylights	—	Glass or plastic	—
Attics	Ventilated	Unventilated	—
Peaked Roof: Metal or thin boards; concealed or not concealed; insulated or uninsulated	Ventilated	Unventilated	—
Flat Roof: Metal, not concealed; insulated or uninsulated	Ventilated or unventilated	Note: For uninsulated roof, climate and occupancy may necessitate intermediate sprinklers. Check on job.	—
Flat Roof: Metal; concealed; insulated or uninsulated	Ventilated	Unventilated	—
Show Windows	Ventilated	Unventilated	—

Note: A check of job condition by means of thermometers may be necessary.

Temperature Ratings, Classifications, and Color Codings

Max. Ceiling Temp. °F	Max. Ceiling Temp. °C	Temperature Rating °F	Temperature Rating °C	Temperature Classification	Color Code	Glass Bulb Colors
100	38	135 to 170	57 to 77	Ordinary	Uncolored or Black	Orange or Red
150	66	175 to 225	79 to 107	Intermediate	White	Yellow or Green
225	107	250 to 300	121 to 149	High	Blue	Blue
300	149	325 to 375	163 to 191	Extra High	Red	Purple
375	191	400 to 475	204 to 246	Very Extra High	Green	Black
475	246	500 to 575	260 to 302	Ultra High	Orange	Black
625	329	650	343	Ultra High	Orange	Black

19.16.0 Sprinkler Maintenance Schedules

Parts	Activity	Frequency
Flushing Piping	Test	5 years
Fire Department Connections	Inspection	Monthly
Control Valves	Inspection	Weekly—Sealed
	Inspection	Monthly—Locked
	Inspection	Monthly—Tamper Switch
	Maintenance	Yearly
Main Drain	Flow Test	Quarterly
Open Sprinklers	Test	Annual
Pressure Gauge	Calibration Test	
Sprinklers	Test	50 years
Sprinklers—High Temp	Test	5 years
Sprinklers—Residential	Test	20 years
Waterflow Alarms	Test	Quarterly
Preaction/Deluge Detection System	Test	Semiannually
Preaction/Deluge Systems	Test	Annually
Antifreeze Solution	Test	Annually
Cold Weather Valves	Open and Close Valves	Fall, Close; Spring, Open
Dry/Preaction/Deluge Systems		
Air Pressure and Water Pressure	Inspection	Weekly
Enclosure	Inspection	Daily—Cold Weather
Priming Water Level	Inspection	Quarterly
Low—Point Drains	Test	Fall
Dry Pipe Valves	Trip Test	Annual—Spring
Dry Pipe Valves	Full Flow Trip	3 years—Spring
Quick-Opening Devices	Test	Semi-annually

19.17.0 Hangers for Sprinkler Pipes

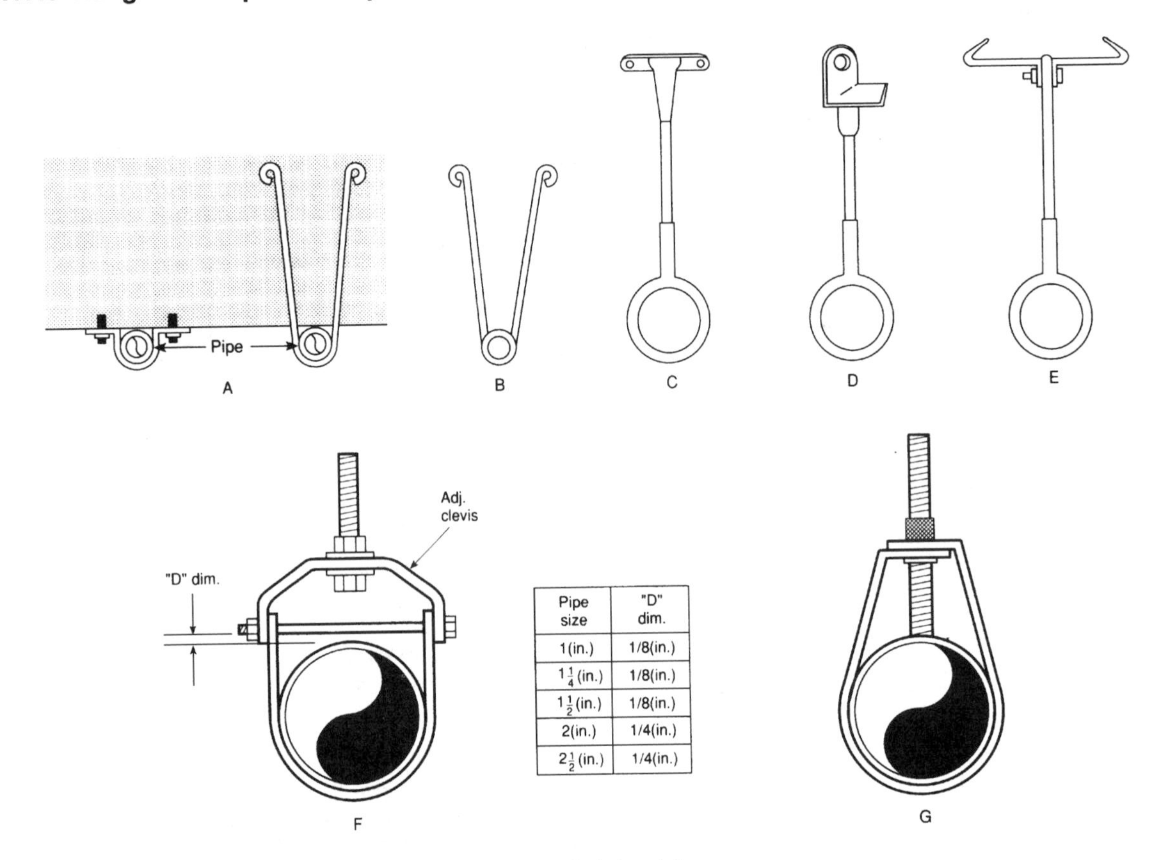

Pipe size	"D" dim.
1(in.)	1/8(in.)
$1\frac{1}{4}$(in.)	1/8(in.)
$1\frac{1}{2}$(in.)	1/8(in.)
2(in.)	1/4(in.)
$2\frac{1}{2}$(in.)	1/4(in.)

A U-type hangers for branch lines
B Wraparound U-hook
C Adjustable clip for branch lines
D Side beam adjustable hanger
E Adjustable coach screw clip for branch lines
F Clevis hanger
G Adjustable swivel loop hanger

Examples of acceptable hangers for end of line (or armover) pendent sprinklers.

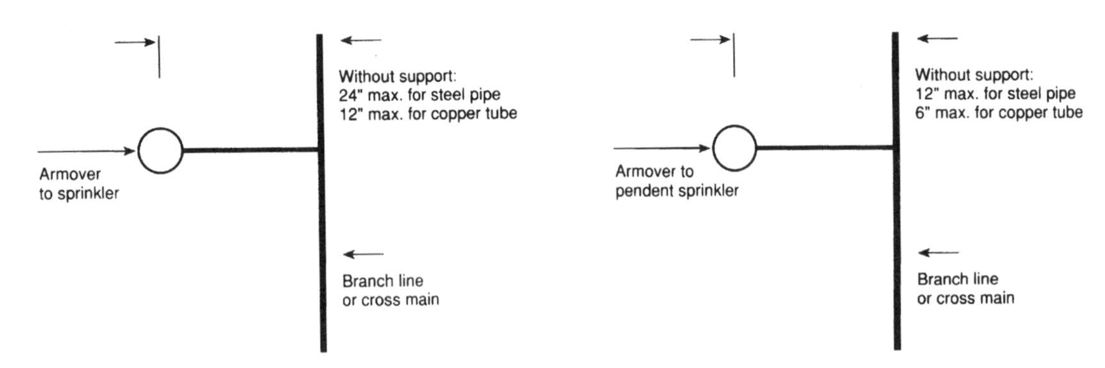

For SI Units: 1 in. = 25.4 mm; 1 ft = 0.3048 m.

Maximum length for unsupported armover.

For SI Units: 1 in. = 25.4 mm; 1 ft = 0.3048 m.
NOTE: The pendent sprinkler may be installed either directly in the fitting at the end of the armover or in a fitting at the bottom of a drop nipple.

Maximum length of unsupported armover where the maximum pressure exceeds 100 psi (6.9 bars) and a branch line above a ceiling supplies pendent sprinklers below the ceiling.

19.18.0 Piping Weights When Filled With Water

Piping Weights for Determining Horizontal Load

Schedule 40 Pipe (in.)	Weight of Water-Filled Pipe (lb per ft)	½ Weight of Water-Filled Pipe (lb per ft)
1	2.05	1.03
1¼	2.93	1.47
1½	3.61	1.81
2	5.13	2.57
2½	7.89	3.95
3	10.82	5.41
3½	13.48	6.74
4	16.40	8.20
5	23.47	11.74
6	31.69	15.85
8*	47.70	23.85
Schedule 10 Pipe (in.)		
1	1.81	0.91
1¼	2.52	1.26
1½	3.04	1.52
2	4.22	2.11
2½	5.89	2.95
3	7.94	3.97
3½	9.78	4.89
4	11.78	5.89
5	17.30	8.65
6	23.03	11.52
8	40.08	20.04

* Schedule 30
For SI Units: 1 in. = 25.4 mm; 1 lb = 0.45kg; 1 ft = 0.30 m.

19.19.0 Seismic Zones and Piping-Modification Requirements

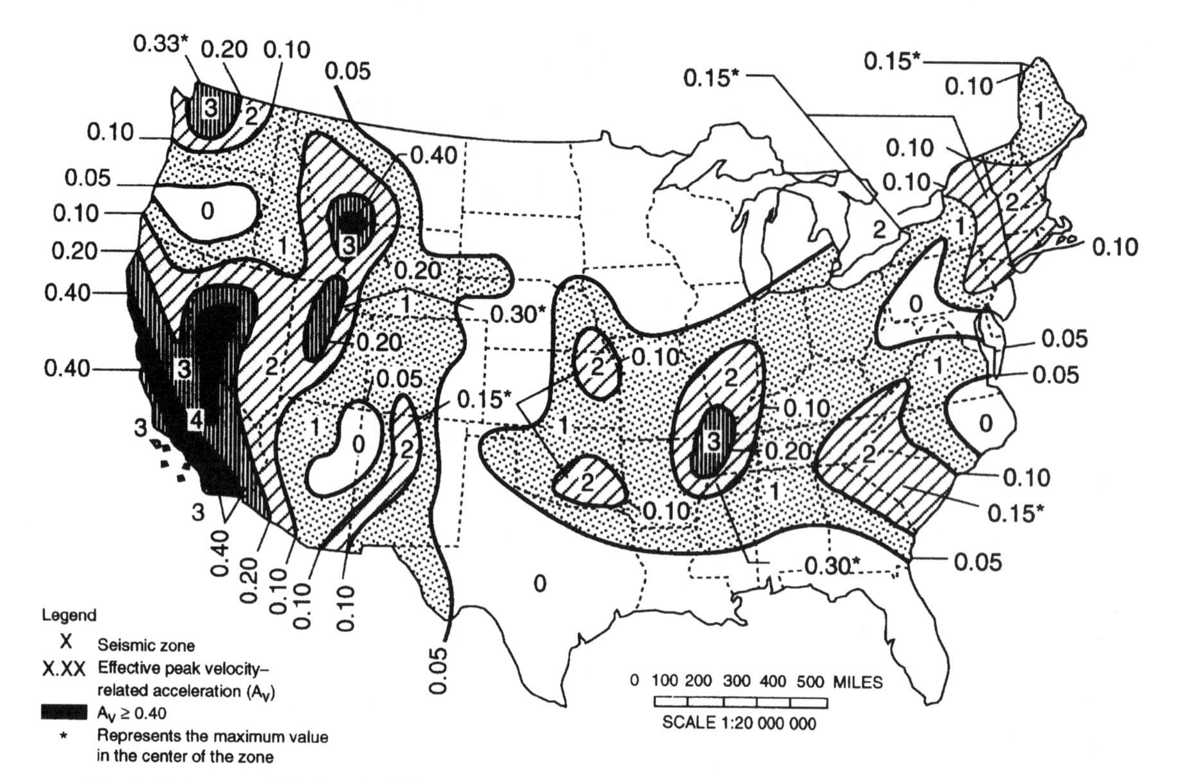

Map of seismic zones and effective peak velocity-related acceleration (A_v) for contiguous 48 states. Linear interpolation between contours is acceptable.

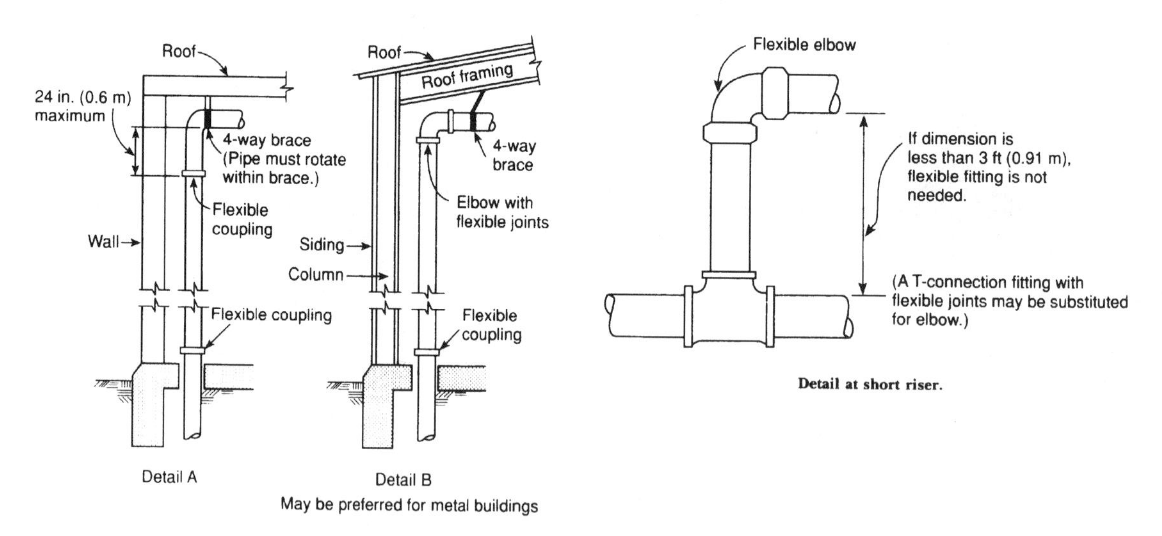

Detail at short riser.

Note to Detail A: The four-way brace should be attached above the upper flexible coupling required for the riser and preferably to the roof structure if suitable. The brace should not be attached directly to a plywood or metal deck.

Riser details.

19.20.0 Unacceptable Pipe Weld Joints (Illustrated)

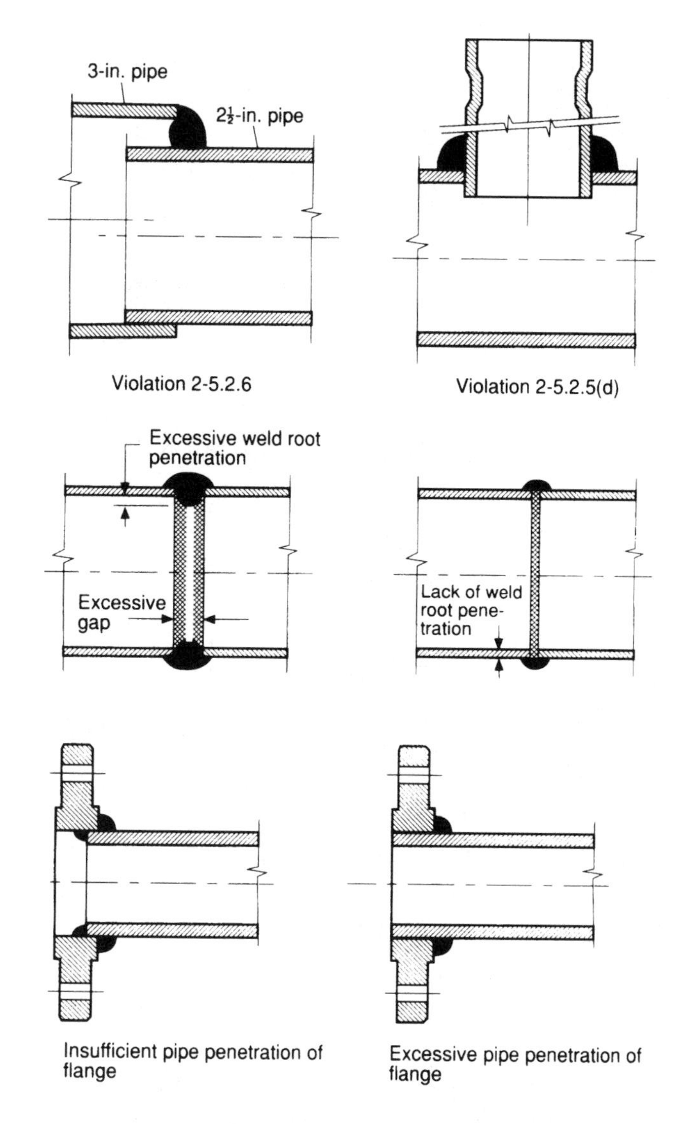

Unacceptable weld joints.

19.21.0 Schematics of Fire-Department Connections/Water Supply

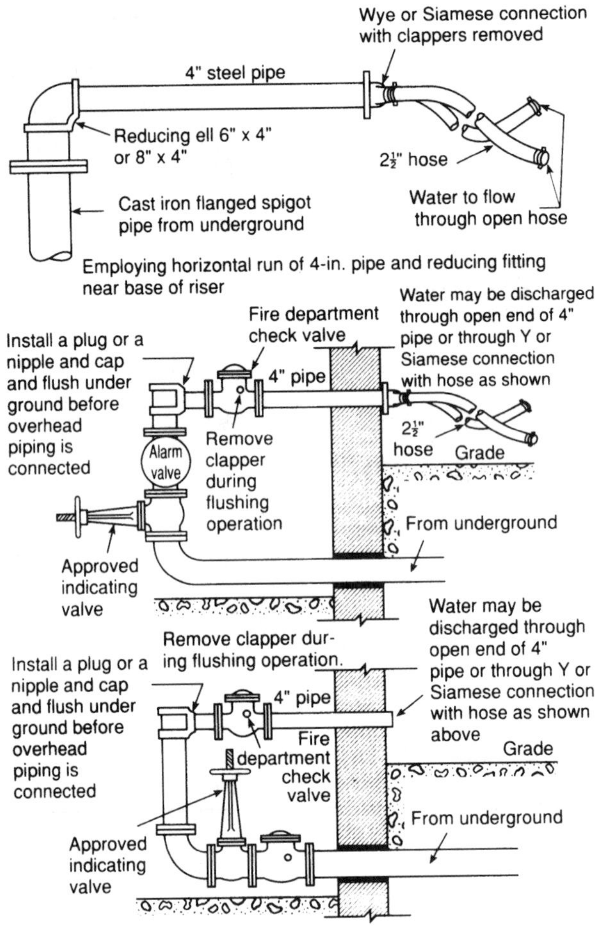

Methods of flushing water supply connections.

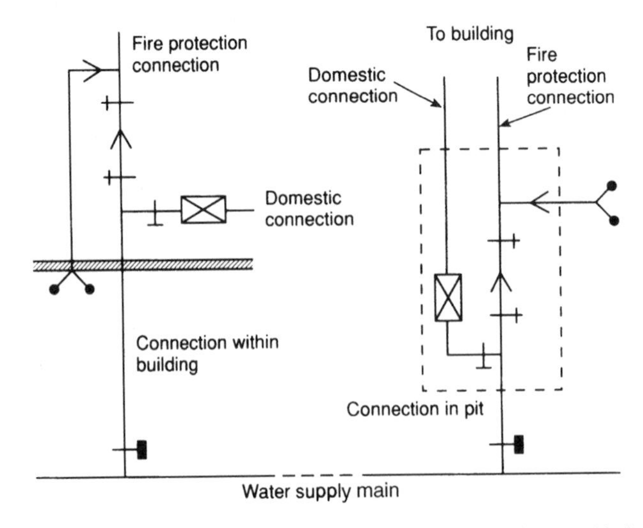

19.22.0 Schematics of Commercial Cooking Automatic Nozzle Installation

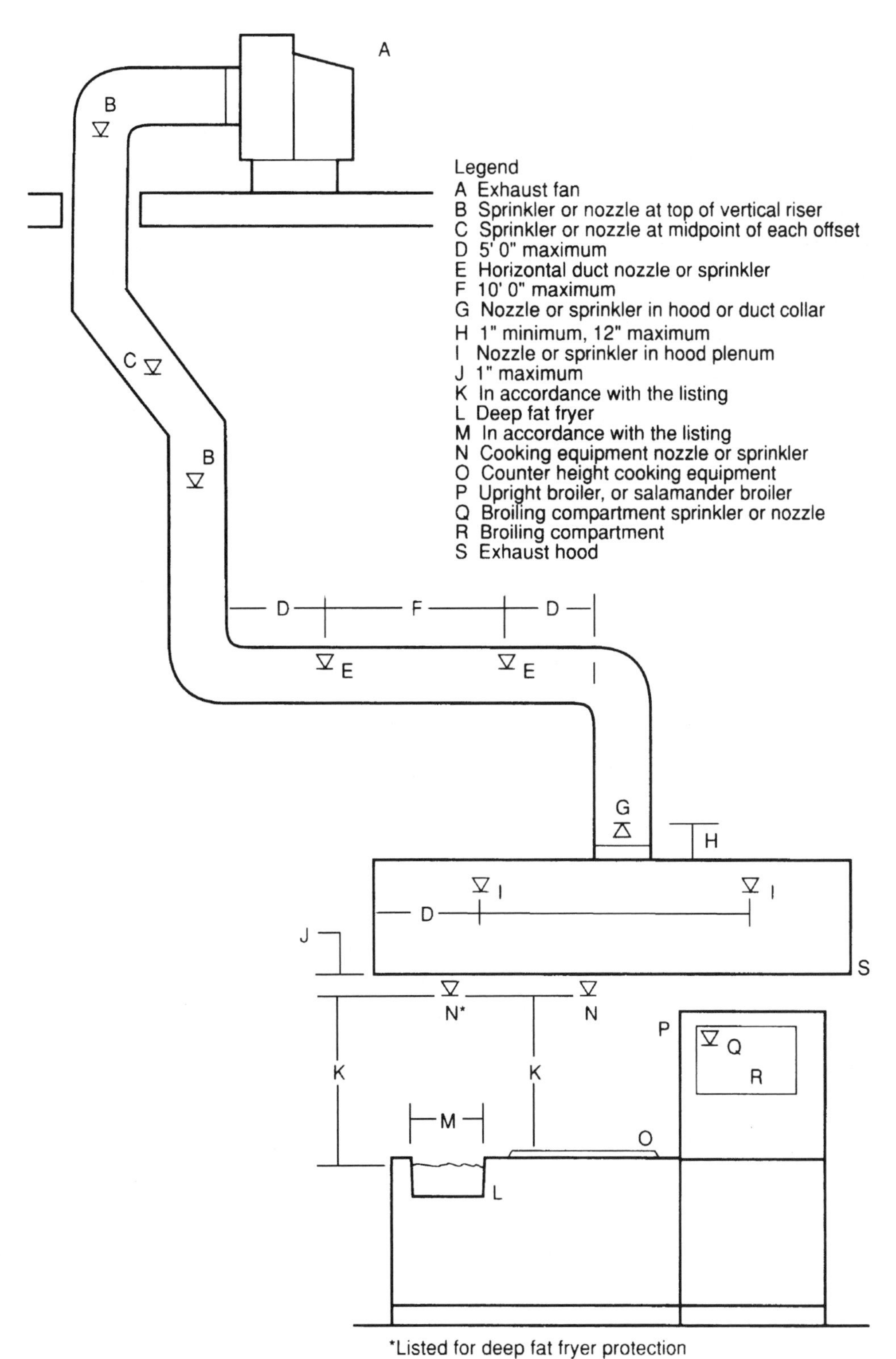

*Listed for deep fat fryer protection

Typical installation showing automatic sprinklers or automatic nozzles being used for the protection of commercial cooking equipment and ventilation systems.

19.23.0 Contractor's Material and Test Certification Forms

Contractor's Material and Test Certificate for Aboveground Piping

PROCEDURE
Upon completion of work, inspection and tests shall be made by the contractor's representative and witnessed by an owner's representative. All defects shall be corrected and system left in service before contractor's personnel finally leave the job.

A certificate shall be filled out and signed by both representatives. Copies shall be prepared for approving authorities, owners, and contractor. It is understood the owner's representative's signature in no way prejudices any claim against contractor for faulty material, poor workmanship, or failure to comply with approving authority's requirements or local ordinances.

PROPERTY NAME	DATE
PROPERTY ADDRESS	

PLANS	ACCEPTED BY APPROVING AUTHORITIES (NAMES)		
	ADDRESS		
	INSTALLATION CONFORMS TO ACCEPTED PLANS	☐ YES	☐ NO
	EQUIPMENT USED IS APPROVED IF NO, EXPLAIN DEVIATIONS	☐ YES	☐ NO
INSTRUCTIONS	HAS PERSON IN CHARGE OF FIRE EQUIPMENT BEEN INSTRUCTED AS TO LOCATION OF CONTROL VALVES AND CARE AND MAINTENANCE OF THIS NEW EQUIPMENT? IF NO, EXPLAIN	☐ YES	☐ NO
	HAVE COPIES OF THE FOLLOWING BEEN LEFT ON THE PREMISES:	☐ YES	☐ NO
	1. SYSTEM COMPONENTS INSTRUCTIONS	☐ YES	☐ NO
	2. CARE AND MAINTENANCE INSTRUCTIONS	☐ YES	☐ NO
	3. NFPA 25	☐ YES	☐ NO
LOCATION OF SYSTEM	SUPPLIES BUILDINGS		

SPRINKLERS	MAKE	MODEL	YEAR OF MANUFACTURE	ORIFICE SIZE	QUANTITY	TEMPERATURE RATING

PIPE AND FITTINGS
Type of Pipe __________
Type of Fittings __________

ALARM VALVE OR FLOW INDICATOR	ALARM DEVICE			MAXIMUM TIME TO OPERATE THROUGH TEST CONNECTION	
	TYPE	MAKE	MODEL	MIN.	SEC.

DRY PIPE OPERATING TEST	DRY VALVE			Q. O. D.		
	MAKE	MODEL	SERIAL NO.	MAKE	MODEL	SERIAL NO.

	TIME TO TRIP THROUGH TEST CONNECTION*		WATER PRESSURE	AIR PRESSURE	TRIP POINT AIR PRESSURE	TIME WATER REACHED TEST OUTLET*		ALARM OPERATED PROPERLY	
	MIN.	SEC.	PSI	PSI	PSI	MIN.	SEC.	YES	NO
Without Q.O.D.									
With Q.O.D.									

IF NO, EXPLAIN

*MEASURED FROM TIME INSPECTOR'S TEST CONNECTION IS OPENED.

19.23.0 Contractor's Material and Test Certification Forms (Continued)

DELUGE & PREACTION VALVES									
OPERATION	☐ PNEUMATIC	☐ ELECTRIC	☐ HYDRAULIC						
PIPING SUPERVISED	☐ YES	☐ NO	DETECTING MEDIA SUPERVISED	☐ YES	☐ NO				
DOES VALVE OPERATE FROM THE MANUAL TRIP AND/OR REMOTE CONTROL STATIONS	☐ YES	☐ NO							
IS THERE AN ACCESSIBLE FACILITY IN EACH CIRCUIT FOR TESTING	☐ YES	☐ NO	IF NO, EXPLAIN						
MAKE	MODEL	DOES EACH CIRCUIT OPERATE SUPERVISION LOSS ALARM		DOES EACH CIRCUIT OPERATE VALVE RELEASE		MAXIMUM TIME TO OPERATE RELEASE			
		YES	NO	YES	NO	MIN.	SEC.		

PRESSURE REDUCING VALVE TEST							
LOCATION & FLOOR	MAKE & MODEL	SETTING	STATIC PRESSURE		RESIDUAL PRESSURE (FLOWING)		FLOW RATE
			INLET (PSI)	OUTLET (PSI)	INLET (PSI)	OUTLET (PSI)	FLOW (GPM)

TEST DESCRIPTION

HYDROSTATIC: Hydrostatic tests shall be made at not less than 200 psi (13.6 bars) for two hours or 50 psi (3.4 bars) above static pressure in excess of 150 psi (10.2 bars) for two hours. Differential dry-pipe valve clappers shall be left open during test to prevent damage. All aboveground piping leakage shall be stopped.

PNEUMATIC: Establish 40 psi (2.7 bars) air pressure and measure drop, which shall not exceed 1-1/2 psi (0.1 bars) in 24 hours. Test pressure tanks at normal water level and air pressure and measure air pressure drop, which shall not exceed 1-1/2 psi (0.1 bars) in 24 hours.

TESTS			
ALL PIPING HYDROSTATICALLY TESTED AT	____ PSI FOR ____ HRS.		IF NO, STATE REASON
DRY PIPING PNEUMATICALLY TESTED	☐ YES	☐ NO	
EQUIPMENT OPERATES PROPERLY	☐ YES	☐ NO	
DO YOU CERTIFY AS THE SPRINKLER CONTRACTOR THAT ADDITIVES AND CORROSIVE CHEMICALS, SODIUM SILICATE OR DERIVATIVES OF SODIUM SILICATE, BRINE, OR OTHER CORROSIVE CHEMICALS WERE NOT USED FOR TESTING SYSTEMS OR STOPPING LEAKS?	☐ YES	☐ NO	
DRAIN TEST	READING OF GAGE LOCATED NEAR WATER SUPPLY TEST CONNECTION: ______ PSI		RESIDUAL PRESSURE WITH VALVE IN TEST CONNECTION OPEN WIDE ______ PSI
UNDERGROUND MAINS AND LEAD IN CONNECTIONS TO SYSTEM RISERS FLUSHED BEFORE CONNECTION MADE TO SPRINKLER PIPING.			
VERIFIED BY COPY OF THE U FORM NO. 85B	☐ YES	☐ NO	OTHER EXPLAIN
FLUSHED BY INSTALLER OF UNDERGROUND SPRINKLER PIPING	☐ YES	☐ NO	
IF POWDER DRIVEN FASTENERS ARE USED IN CONCRETE, HAS REPRESENTATIVE SAMPLE TESTING BEEN SATISFACTORILY COMPLETED?	☐ YES	☐ NO	IF NO, EXPLAIN

BLANK TESTING GASKETS		
NUMBER USED	LOCATIONS	NUMBER REMOVED

WELDING		
WELDED PIPING	☐ YES	☐ NO
IF YES. . .		
DO YOU CERTIFY AS THE SPRINKLER CONTRACTOR THAT WELDING PROCEDURES COMPLY WITH THE REQUIREMENTS OF AT LEAST AWS D10.9, LEVEL AR-3?	☐ YES	☐ NO
DO YOU CERTIFY THAT THE WELDING WAS PERFORMED BY WELDERS QUALIFIED IN COMPLIANCE WITH THE REQUIREMENTS OF AT LEAST AWS D10.9, LEVEL AR-3?	☐ YES	☐ NO
DO YOU CERTIFY THAT WELDING WAS CARRIED OUT IN COMPLIANCE WITH A DOCUMENTED QUALITY CONTROL PROCEDURE TO INSURE THAT ALL DISCS ARE RETRIEVED, THAT OPENINGS IN PIPING ARE SMOOTH, THAT SLAG AND OTHER WELDING RESIDUE ARE REMOVED, AND THAT THE INTERNAL DIAMETERS OF PIPING ARE NOT PENETRATED?	☐ YES	☐ NO

CUTOUTS (DISCS)		
DO YOU CERTIFY THAT YOU HAVE A CONTROL FEATURE TO ENSURE THAT ALL CUTOUTS (DISCS) ARE RETRIEVED?	☐ YES	☐ NO

19.23.0 Contractor's Material and Test Certification Forms (Continued)

<table>
<tr><td>HYDRAULIC DATA NAMEPLATE</td><td colspan="2">NAMEPLATE PROVIDED
☐ YES ☐ NO</td><td>IF NO, EXPLAIN</td></tr>
<tr><td rowspan="2">REMARKS</td><td colspan="3">DATE LEFT IN SERVICE WITH ALL CONTROL VALVES OPEN:</td></tr>
<tr><td colspan="3"></td></tr>
<tr><td rowspan="4">SIGNATURES</td><td colspan="3">NAME OF SPRINKLER CONTRACTOR</td></tr>
<tr><td colspan="3">TESTS WITNESSED BY</td></tr>
<tr><td>FOR PROPERTY OWNER (SIGNED)</td><td>TITLE</td><td>DATE</td></tr>
<tr><td>FOR SPRINKLER CONTRACTOR (SIGNED)</td><td>TITLE</td><td>DATE</td></tr>
<tr><td colspan="4">ADDITIONAL EXPLANATION AND NOTES</td></tr>
</table>

Section

20

Heating, Ventilating, and Air Conditioning

Contents

20.0.0 Introduction
20.1.0 Common boiler types
20.2.0 Hot-water boiler (schematic)
20.2.1 Exploded view of hot-water boiler
20.2.2 Hot-water boiler (parts list)
20.3.0 Typical steam boiler system
20.4.0 Summary of Federal EPA rules for boilers built/modified after June 9, 1989
20.5.0 Boiler feedback systems (illustrated)
20.6.0 Typical firetube boiler fuel consumption for No. 2 and No. 6 oil
20.7.0 Boiler economizer features and schematic
20.8.0 Boiler stack options
20.8.1 Typical stack construction
20.8.2 Stack expansion/contraction and installation concerns
20.9.0 Schematic of typical custom-built HVAC unit
20.10.0 Schematic of indirect evaporative precooling system
20.11.0 Heat-pump operation schematics
20.12.0 Air-cooled condenser and subcooling system (illustrated)
20.13.0 Variable air volume (VAV) systems diagrammed
20.13.1 Variable air volume (VAV) diagrams showing radiation heating, reheat and fan-powered systems
20.14.0 Single- and two-pipe cooling system diagrams
20.15.0 Two-pipe reverse main and three-pipe heating/cooling piping diagrams
20.16.0 Four-pipe systems with one- and two-coil piping diagrams
20.17.0 Shell and coil, and shell and tube condensers (illustrated and described)
20.18.0 Shell and tube evaporator (diagram and description)
20.19.0 Evaporative condenser (diagram and description)
20.20.0 Heating with a chiller (diagram and description)
20.21.0 Typical evaporative cooler (diagram and description)
20.22.0 Typical flow diagram of an ice-storage system
20.23.0 Types of humidifiers (illustrated and described)
20.24.0 Mechanical draft towers (illustrated and described)
20.25.0 Equivalent rectangular duct dimension tables
20.25.1 Equivalent spiral, flat, oval duct dimensions
20.26.0 Typical fan configurations
20.27.0 Rate of heat gain from selected office equipment
20.28.0 Thermal properties of common building materials

20.0.0 Introduction

The indoor work environment must be controlled and regulated to provide the occupants with a healthy and productive work area. The comfort zone for people in a work environment depends upon two factors: heat/air conditioning and humidity. The ideal indoor temperature should range from 65 degrees F (18 degrees C) to 75 degrees F (24.5 degrees C) and relative humidity levels should be between 30 and 50%. Basic HVAC systems all have common components: heat and/or cooling sources, a method by which the heating or cooling is distributed, terminal devices to disperse the heat or cooling, and a means to control the equipment.

20.1.0 Common Boiler Types

	CAST IRON	MEMBRANE WATERTUBE	ELECTRIC	FIREBOX	FIRETUBE	FLEXIBLE WATERTUBE	INDUSTRIAL WATERTUBE	VERTICAL FIRETUBE
Efficiency	Low	Medium	High	Medium	High	Medium	Medium	Low/Medium
Floor Space Required	Low	Very Low	Very Low	Medium	Medium/ High	Low	High	Very Low
Maintenance	Medium/High	Medium	Medium/High	Low	Low	Medium	High	Low
Initial Cost	Medium	Low/Medium	High	Low	Medium/ High	Low/Medium	High	Low
No. of Options Available	Low	Medium	Medium	Low/Medium	High	Medium	High	Low
Pressure Range	HW/LPS	HW/LPS HPS to 600 psig	HW/LPS HPS to 900 PSIG	HW/LPS	HW/LPS HPS to 350 psig	HW/LPS	High Temp HW HPS to 900 psig	HW/LPS HPS to 150 psig
Typical Sizes	To 200 hp	To 250 hp	To 300 hp	To 300 hp	To 1500 hp	To 250 hp		To 100 hp
Typical Applications	Heating/ Process	Heating/ Process	Heating/ Process	Heating	Heating/ Process	Heating	Process	Heating/ Process
Comments	Field Erectable					Field Erectable		

Scotch Marine - The Classic Firetube Boiler

The Scotch Marine style of boiler has become so popular in the last 40 years that it frequently is referred to simply as "a firetube boiler." Firetube boilers are available for low or high pressure steam, or for hot water applications. Firetube boilers are typically used for applications ranging from 15 to 1500 horsepower. A firetube boiler is a cylindrical vessel, with the flame in the furnace and the combustion gases inside the tubes. The furnace and tubes are within a larger vessel, which contains the water and steam.

The firetube construction provides some characteristics that differentiate it from other boiler types. Because of its vessel size, the firetube contains a large amount of water, allowing it to respond to load changes with minimum variation in steam pressure.

Stream pressure in a firetube boiler is generally limited to approximately 350 psig. To achieve higher pressure, it would be necessary to use a very thick shell and tube sheet material. For this reason, a watertube boiler is generally used if pressure above 350 psig desgn is needed.

Firetube boilers are usually built similar to a shell and tube heat exchanger. A large quantity of tubes results in more heating surface per boiler horsepower, which greatly improves heat transfer and efficiency.

Firetube boilers are rated in boiler horsepower (BHP), which should not be confused with other horsepower measurements.

The furnace and the banks of tubes are used to transfer heat to the water. Combustion occurs within the furnace and the flue gases are routed through the tubes to the stack outlet. Firetube boilers are available in two, three and four pass designs. A single "pass" is defined as the area where combustion gases travel the length of the boiler. Generally, boiler efficiencies increase with the number of passes.

Firetube boilers are available in either dryback or wetback design. In the dryback boiler, a refractory-lined chamber, outside of the vessel, is used to direct the combustion gases from the furnace to the tube banks. Easy access to all internal areas of the boiler including tubes, burner, furnace, and refractory, is available from either end of the boiler. This makes maintenance easier and reduces associated costs.

The wetback boiler design has a water cooled turn around chamber used to direct the flue gases from the furnace to the tube banks. The wetback design requires less refractory maintenance; however, internal pressure vessel maintenance, such as cleaning, is more difficult and costly. In addition, the wetback design is more prone to water side sludge buildup, because of the restricted flow areas near the turn around chamber.

By permission of Cleaver Brooks, Milwaukee, Wisconsin

20.2.0 Hot-Water Boiler (Schematic)

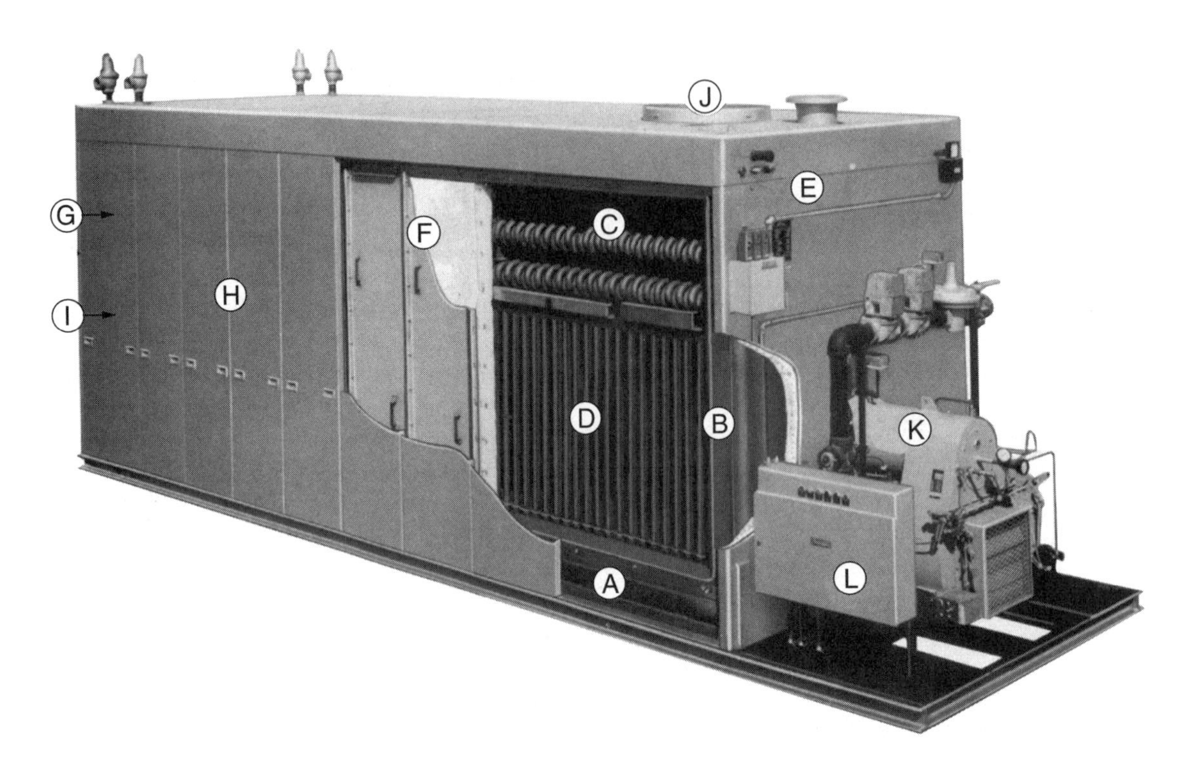

A. Heavy steel boiler frame, built and stamped in accordance with the appropriate ASME Boiler Code.

B. Large volume water leg downcomers promote rapid internal circulation and temperature equalization.

C. Bryan bent water tubes are flexible, individually replaceable without welding or rolling.

D. Internal water-cooled furnace with low heat release rate.

E. Water side interior accessible for cleanout and inspection, front and rear openings, upper and lower drums.

F. Boiler tube and furnace area access panels: heavy gauge steel-lined with high temperature ceramic fiber and insulation, bolted and tightly sealed to boiler frame.

G. Combustion chamber and burner head are completely accessible via manway in end of combustion chamber.

H. Heavy gauge steel boiler jacket with rust-resistant zinc coating and enamel finish. Insulated with fiberglass to insure exceptionally cool outer surface.

I. Rear flame observation port in access door at rear of boiler.

J. Minimum sized flue vent.

K. Forced draft, flame retention head-type burner. Efficient combustion of oil or gas, quiet operation.

L. Control panel: all controls installed with connections to terminal strip.

By permission of Bryan Steam Boiler, Peru, Indiana

20.2.1 Exploded View of Hot-Water Boiler

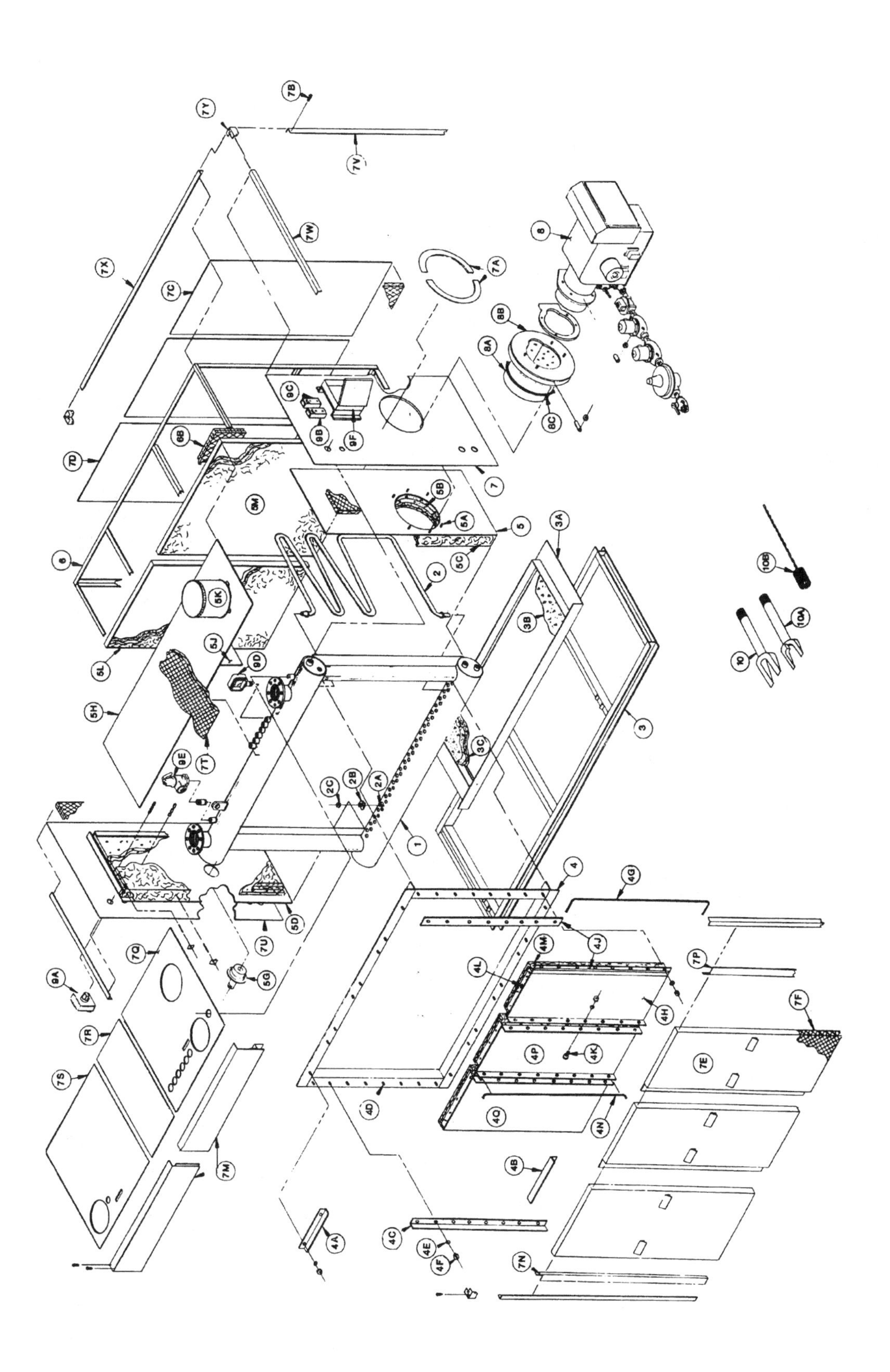

20.2.2 Hot-Water Boiler (Parts List)

DESCRIPTION
BOILER FRAME ASSEMBLY(Less Tubes)
BOILER TUBE ASSEMBLY
"A" Outside Tubes
"B" Inside Tubes
Tube Studs
Tube Clamps
Tube Nuts
BOILER BASE ASSEMBLY
Boiler Floor Pan Assembly
Floor Pan Insul.(Min. Fib.)Layers
Floor Pan Refr.(Castable) Bags
TUBE ACCESS PANEL ASSEMBLY
Tube Access Panel Frame
Panel Clamp, Top
Panel Clamp, Bottom
Panel Clamp, End
Panel Weld Stud - 3/8"-16 x 1-3/8"
Panel Washers - 3/8"
Panel Nuts - 3/8"-16
Rope Gasket (Ft.)
FRONT (HINGED) PANEL ASSEMBLY
Panel
Hinge and Shim
Bolts - 3/8"-16 x 1"
Washers - 3/8"
FRONT PANEL INSULATION ASSEMBLY
Insul.(Mineral Fib.)19" x 60-9/16"
Refractory(Cer.Fib) 29" x 60-9/16"
Rope Gasket (Ft.)
CENTER PANEL ASSEMBLY
Panel
Bolts - 3/8"-16 x 1"
Washers - 3/8"
Nuts - 3/8"-16
CENTER PANEL INSULATION ASSEMBLY
Insul.(Min. Fib.)17-5/8"x 60-9/16"
Refr.(Cer. Fib)30-1/8"x 60-9/16"
Rope Gasket (Ft.)
END PANEL ASSEMBLY "A"
Panel "A" - 20-15/16" Wide
Bolts - 3/8"-16 x 1"
Washers - 3/8"
Nuts - 3/8"-16
END PANEL INSULATION ASSEMBLY "A"
Insul.(Min. Fib)16-15/16"X60-9/16"
Refr.(Cer. Fib.)29-7/16"X 60-9/16"
Rope Gasket (Ft.)
END PANEL ASSEMBLY "B"
Panel "B" 24-1/8" Wide
Bolts 3/8"-16
Washers - 3/8"

ITEM NO.	DESCRIPTION
	Nuts - 3/8"
	END PANEL INSULATION ASSEMBLY "B"
	Insul.(Min. Fib)20-1/8"x 60-9/16"
	Refr.(Cer. Fib.)30-5/8"x 60-9/16"
	Rope Gasket (Ft.)
o	END PANEL ASSEMBLY "C"
	Panel "C"- 33-13/16" Wide
	Bolts - 3/8"-16 x 1"
	Washers - 3/8"
	Nuts - 3/8"
	END PANEL INSULATION ASSEMBLY "C"
	Insul.(Min. Fib.)29-13/16"X 60-9/16
	Refr.(Cer. Fib.)43-5/16"X 60-9/16"
	Rope Gasket (Ft.)
	FLUE COLLECTOR ASSEMBLY
5	FLUE COLLECTOR FRONT PLATE ASS'Y
	Angle Iron 14 Ft.
5A	Weld Studs-3/8"-16 x 2-1/2" Long
	FRONT PLATE INSULATION ASSEMBLY
5B	Insul.(Min. Fib.)28-1/2"x 70-5/8"
5C	Refr.(Cer. Fib.)40-1/2"x 70-5/8"
5D	FLUE COLLECTOR REAR PLATE ASS'Y
	Angle Iron 14 Ft.
	Weld Studs-3/8"-16 x 1-1/2" Long
	REAR PLATE INSULATION ASSEMBLY
5E	Insul.(Min. Fib.)28-1/2"x 70-5/8"
5F	Refr.(Cer. Fib.)40-1/2"x 70-5/8"
5G	Rear Access Plug/Site Port
5H	FLUE COLL. TOP PLATE ASSEMBLY
5J	VERTICAL TUBE BAFFLE ASSEMBLY
5K	FLUE EXTENSION/CONNECTION
5L	FLUE COLLECTOR SIDE ASSEMBLY "A"
	Side Panel "A" - 22-9/16" Wide
	SIDE PANEL "A" INSULATION ASS'Y
5M	Insul.(Cer. Fib.)22-9/16"x 70-3/4"
	Insul.(Cer. Fib.)23-1/16"x 70-3/4"
o	FLUE COLLECTOR SIDE ASSEMBLY "B"
	Side Panel "B" 44-1/4" Wide
	SIDE PANEL "B" INSUL. ASSEMBLY
	Insul. (Cer. Fib.)43-3/4"x 70-3/4"
o	FLUE COLLECTOR SIDE ASSEMBLY "C"
	Side Panel "C" 45-1/16" Wide
	SIDE PANEL "C" INSUL. ASSEMBLY
	Insul.(Cer. Fib.)45-1/16"x 70-3/4"
	Insul.(Cer. Fib.)45-9/16"x 70-3/4"
o	FLUE COLLECTOR SIDE ASSEMBLY "D"
	Side Panel "D" 53-7/8" Wide
	SIDE PANEL "D" INSUL. ASSEMBLY

ITEM NO.	DESCRIPTION
	Flue Coll. Side Assembly (Cont'd)
	Insul.(Cer. Fib.)53-7/8"x 70-3/4"
	Insul.(Cer. Fib.)54-3/8"x 70-3/4"
o	FLUE COLLECTOR SIDE ASSEMBLY "E"
	Side Panel "E" 57-1/8" Wide
	SIDE PANEL "E" INSULATION ASS'Y
	Insul.(Cer. Fib.)57-1/8"x 70-3/4"
	Insul.(Cer. Fib.)57-5/8"x 70-3/4"
6	JACKET FRAME ASSEMBLY
6A	JACKET INSULATION ASSEMBLY
6B	Insulation (Sq. Ft.)
	JACKET ASSEMBLY
	JACKET FRONT PANEL ASSEMBLY
7	Jacket Front
7A	Front Burner Plug Cover
7B	Jacket Screws
	JACKET SIDE ASSEMBLY
7C	38" Panel
7D	Modular Panel
	Jacket Screws
	JACKET ACCESS PANEL ASSEMBLY
7E	ACCESS PANEL ASSEMBLY "A"
	Panel "A" Front-30-9/16"x 74-1/2"
7F	PANEL "A" INSULATION ASSEMBLY
	Insul. (Fib'gls W/F)30-1/2" x 72"
o	ACCESS PANEL ASSEMBLY "B"
	Panel "B" 17-1/2 x 74-1/2"
	PANEL "B" INSULATION ASSEMBLY
	Insul. (Fib'gls W/F) 17-1/2" x 72"
o	ACCESS PANEL ASSEMBLY "C"
	Panel "C" 22-5/16" x 74-1/2"
	PANEL "C" INSULATION ASSEMBLY
	Insul. (Fib'gls W/F)17-1/2" x 72"
o	ACCESS PANEL "D"
	Panel "D" 27-1/8" x 72"
	PANEL "D" INSULATION ASSEMBLY
	Insulation (Fib'gls W/F)27" x 72"
o	ACCESS PANEL ASSEMBLY "E"
	Panel "E" 30-5/16" x 74-1/2"
	PANEL "E" INSULATION ASSEMBLY
	Insul. (Fib'gls W/ F)30-1/4" x 72"
	JACKET FILLER PANEL ASSEMBLY
7M	Filler Panel
	JACKET FILLER STRIP ASSEMBLY
7N	Filler Strip Left
7P	Filler Strip Right

ITEM NO.	DESCRIPTION
	JACKET TOP PANEL ASSEMBLY
7Q	Top Panel - Front
	Jacket Screws
7R	Top Panel - Center
	Jacket Screws
7S	Top Panel - Rear
	Jacket Screws
7T	JACKET TOP INSULATION (Sq. Ft.)
	JACKET REAR PANEL ASSEMBLY
7U	Rear Panel
	Rear Panel Access Plug Cover
	Jacket Screws
	JACKET REAR INSUL. ASSEMBLY
	Insulation (Fib'gls W/F)
	JACKET TRIM ASSEMBLY
7V	Vertical Moldings
7W	Horizontal Moldings (Ends)
7X	Horizontal Moldings (Sides)
7Y	Corner Moldings
	Jacket Screws
	FORCED DRAFT BURNER ASSEMBLY
8	Forced Draft Burner Assembly
8A	Forced Draft Burner Plug
8B	Forced Draft Burner Spacer
8C	Forced Draft Burner Gasket
9	WATER TRIM ASSEMBLY
9A	Low Water Cut-off
9B	High Limit
9C	Operator
9D	Pressure/Temperature Control
9E	Relief Valve
9F	Control Panel

By permission of Bryan Boilers, Peru, Indiana

20.3.0 Typical Steam Boiler System

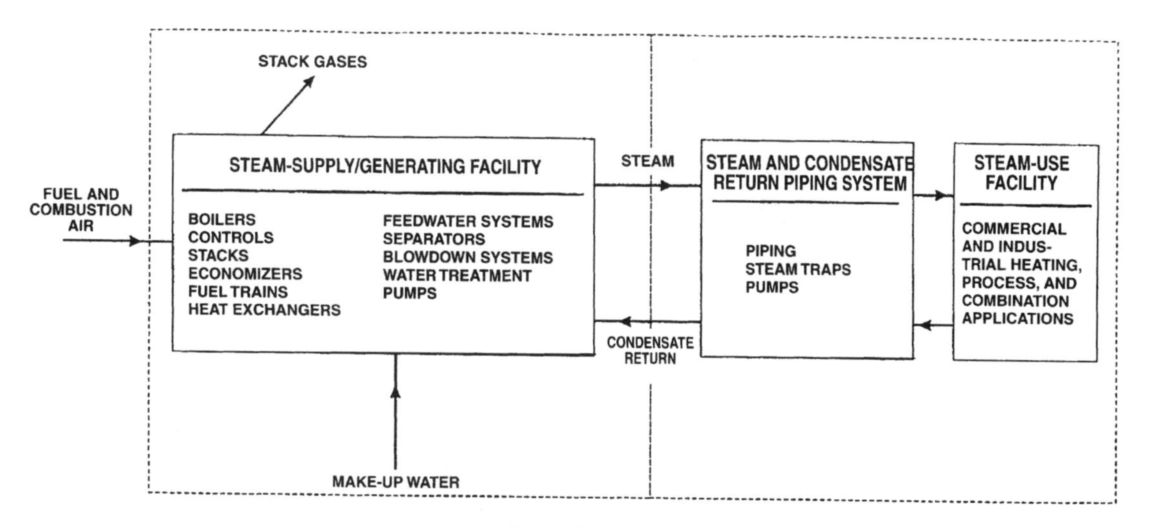

Typical Steam System

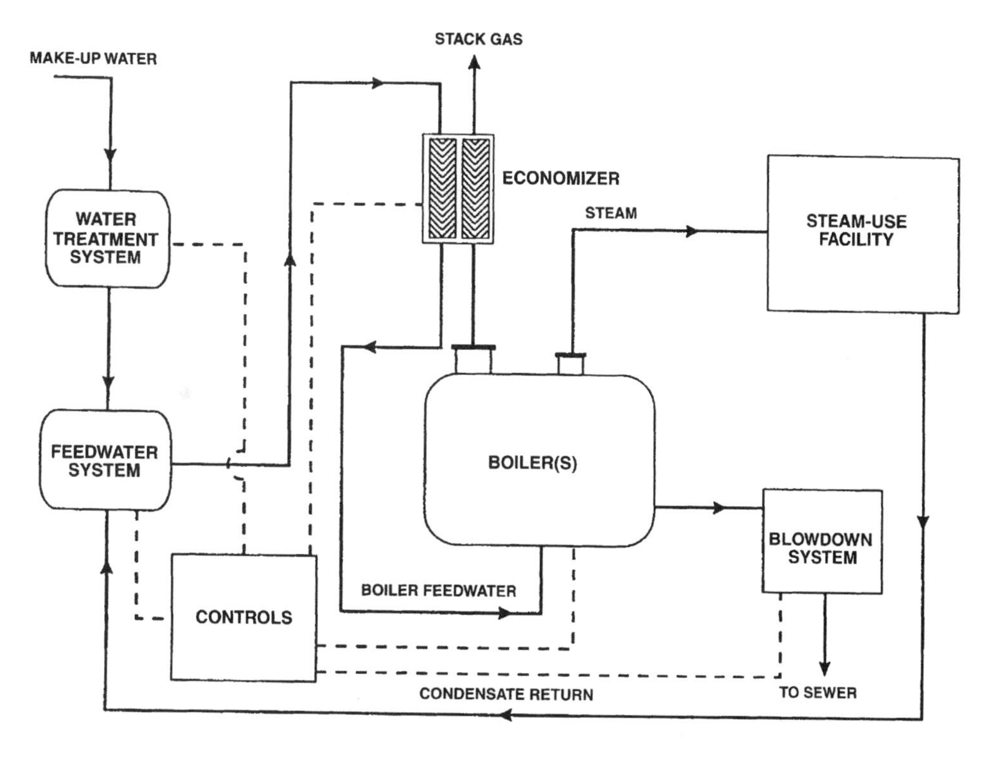

Schematic Diagram of a Generic Steam-Generating Facility

By permission of Cleaver Brooks, Milwaukee, Wisconsin

20.4.0 Summary of Federal EPA Rules for Boilers Built/Modified after June 9, 1989

RULES FOR SULFUR DIOXIDE (SO_2) EMISSIONS

1. Coal Firing

1.2 lb SO_2/MMBtu Limit all 10-100 MMBtu.
90% SO_2 reduction required if > 75 MMBtu and > 55% annual coal capacity.
Initial performance testing required within 180 days of start-up.
30 day rolling average used in calculations.
Continuous Emission Monitoring System (CEMS) required except:
Fuel analysis may be used (before cleanup equipment).
Units < 30 MMBtu may use supplier certificate for compliance.

2. Residual Oil Firing

Limit of 0.5 lb SO_2/MMBtu or 0.5% sulfur in fuel.
CEMS required to meet SO_2 limit except fuel analysis can be used as fired condition before cleanup equipment.
Fuel sulfur limit compliance can be:
Daily as fired fuel analysis.
As delivered (before used) fuel analysis.
Fuel supplier certificate for units < 30 MMBtu.
Initial performance testing and 30 day rolling average required except for supplier certificate.

3. Distillate Oil Firing (ASTM grades 1 and 2)

Limit 0.5% sulfur in fuel (required in ASTM standard).
Compliance by fuel supplier certificate.
No monitoring or initial testing required.

RULES FOR PARTICULATE MATTER (PM) EMISSIONS

1. General

Limits established only for units between 30-100 MMBtu.
All coal, wood and residual oil fired units > 30 MMBtu must meet opacity limit of 20%, except one 6 minute/hour opacity of 27%. CEMS required to monitor opacity.

2. Coal Firing

0.05 lb/MMBtu limit if > 30 MMBtu and > 90% annual coal capacity.
0.10 lb/MMBtu limit if > 30 MMBtu and < 90% annual coal capacity.
20% opacity (CEMS) and initial performance tests on both PM limit and opacity.

3. Wood Firing

0.10 lb/MMBtu limit if > 30 MMBtu and > 30% annual wood capacity.
0.30 lb/MMBtu limit if > 30 MMBtu and < 30% annual wood capacity.
Opacity limits and initial testing per above.

4. Oil Firing

All units > 30 MMBtu subject to opacity limit, only residual oil firing must use CEMS.
Initial performance testing required.

REPORTING REQUIREMENTS

Owners or operators of all affected units must submit information to the administrator, even if they are not subject to any emission limits or testing. Required reports include:
Information on unit size, fuels, start-up dates and other equipment information.
Initial performance test results, CEMS performance evaluation.
Quarterly reports on SO_2 and/or PM emission results, including variations from limits and corrective action taken.
For fuel supplies certificate, information on supplies and details of sampling and testing for coal and residual oil.
Records must be maintained for two years.

By permission of Cleaver Brooks, Milwaukee, Wisconsin

20.5.0 Boiler Feedback Systems (Illustrated)

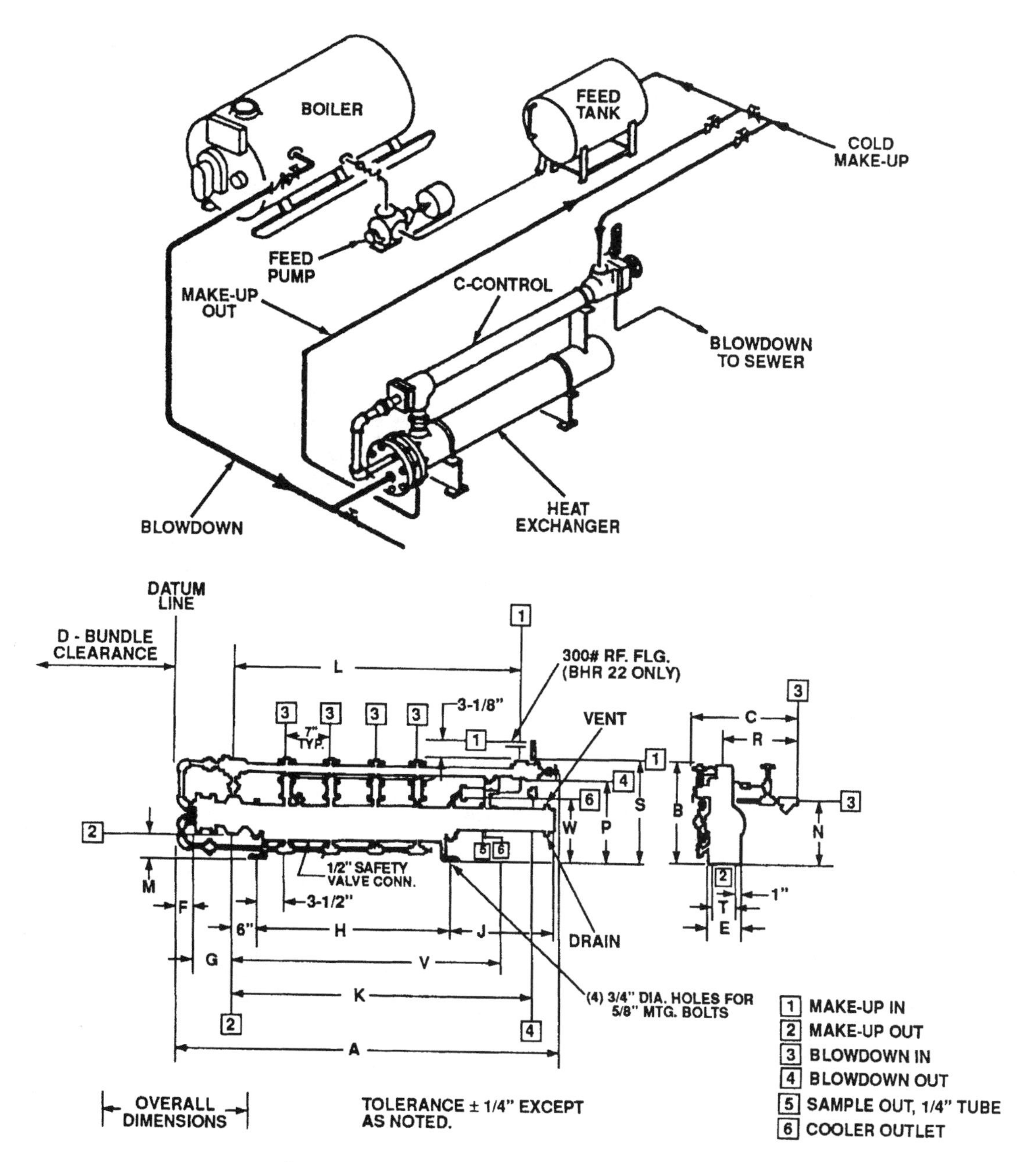

By permission of Cleaver Brooks, Milwaukee, Wisconsin

20.6.0 Typical Firetube Boiler Fuel Consumption for No. 2 and No. 6 Oil

Typical Firetube Boiler Fuel Consumption Rates - No. 6 Oil (gal/hr)[A]

AVERAGE OUTPUT	BOILER EFFICIENCY					
	86%	84%	82%	80%	78%	76%
BHP						
100	26	27	27	28	29	29
200	52	53	54	56	57	59
300	78	80	82	84	86	88
400	104	106	109	112	114	117
500	130	133	136	140	143	147
600	156	159	163	168	172	176
700	182	186	191	196	200	206
800	208	213	218	224	229	235
900	234	239	245	252	257	264
1000	260	266	272	280	286	294

A. Based on 150,000 Btu/gallon.

Typical Firetube Boiler Fuel Consumption Rates - No. 2 Oil (gal/hr)[A]

AVERAGE OUTPUT	BOILER EFFICIENCY					
	86%	84%	82%	80%	78%	76%
BHP						
100	28	28	29	30	31	31
200	56	57	58	60	61	63
300	83	85	87	90	92	94
400	111	114	117	120	123	126
500	139	142	146	149	153	157
600	167	171	175	179	184	189
700	195	199	204	209	215	220
800	222	228	233	239	245	252
900	250	256	262	269	276	283
1000	278	285	292	299	307	315

A. Based on 140,000 Btu/gallon.

By permission of Cleaver Brooks, Milwaukee, Wisconsin

20.7.0 Boiler Economizer Features and Schematic

Reduces Fuel Use and Cost:

- Recovers heat from flue gases that would otherwise be wasted.
- Heat is used to raise boiler feedwater temperature prior to entering the boiler.

Load Changes:

- Rapid changes in load demands can be met faster due to higher feedwater temperature.

Emissions:

- Reduced fuel-firing rates for any given steam output means reduced NOx emissions.

ASME Construction:

- Ensures high quality design and manufacturing standards.
- Provides safety and reliability.

High Efficiency Heat Exchanger:

- Provides continuous, high-frequency resistance welding.
- Provides uniform fin-to-tube contact for maximum heat transfer.
- Fin tubing offers up to 12 times the heat exchange surface of bare tubing of the same diameter.

Self-Drainng Design:

- Suitable for outdoor installation.

Low Pressure Drop:

- Provides gas side pressure drops of 0.8" WC or less.
- Permits use of smaller forced draft fans.
- Permits use of existing fans in almost all installations.

Gas Tight Combustion Stack:

- Provides inner casing of carbon steel.
- Provides outer casing of weather resistant, corrugated, galvanized carbon steel.
- Compact dimensions provide for easy installation.

Feedwater Preheating System:

- Controls cold end corrosion through all flow rates.
- Prevents the forming of corrosive acids in the economizer.
- Prevents stack corrosion.

By permission of Cleaver Brooks, Milwaukee, Wisconsin

20.7.0 Boiler Economizer Features and Schematic (Continued)

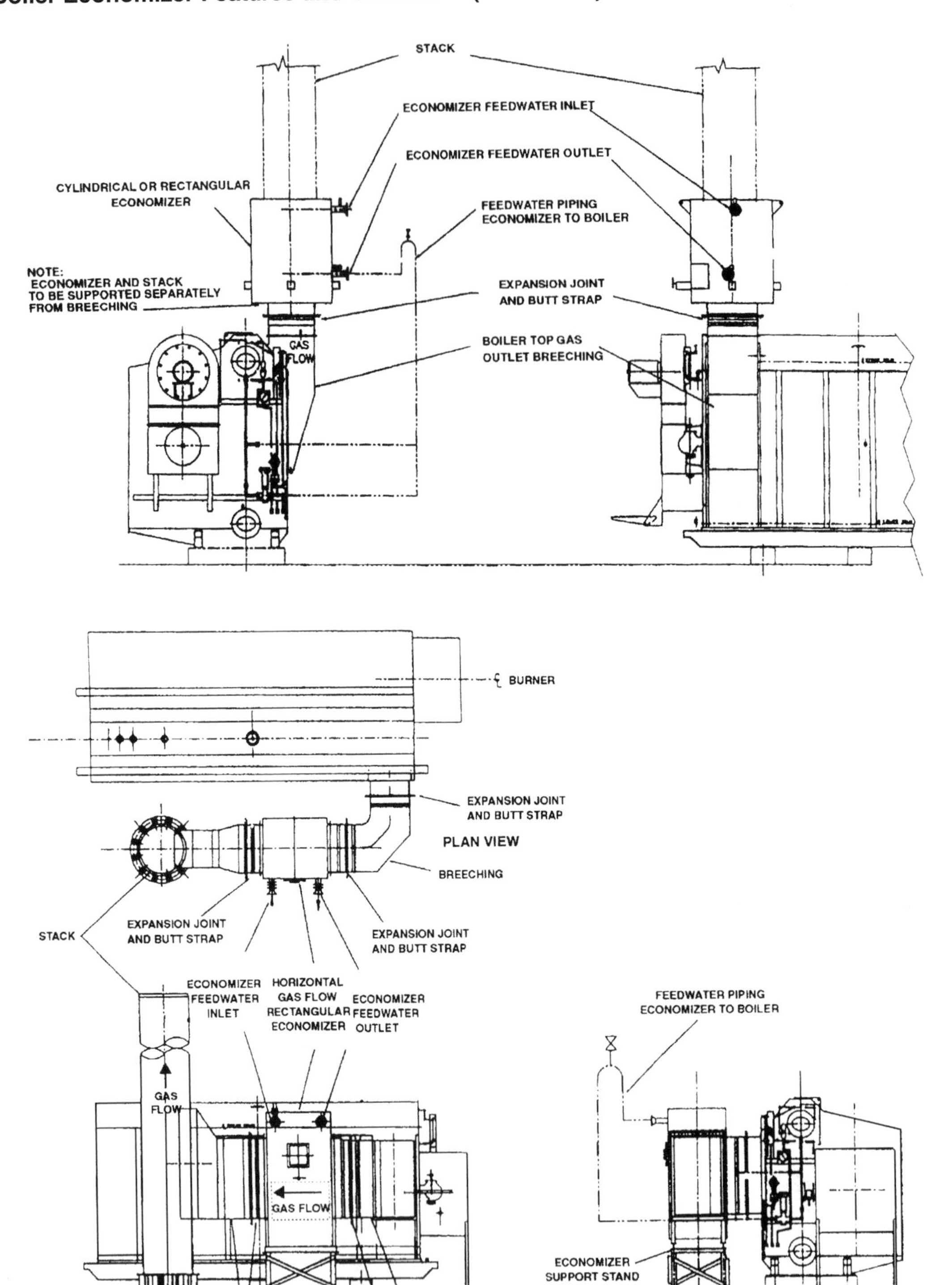

By permission of Cleaver Brooks, Milwaukee, Wisconsin

20.8.0 Boiler Stack Options

Stack Product Offering and Application Information

MODEL NO.	AMERI-VENT	CBS-I	CBS-II	CBS-III	ICBS
Description	Type "B" Gas Vent	Single Wall Stainless Steel	Double Wall Air Insulated	Triple Wall Air Insulated	Double wall either 1", 2" or 4" Material Insulated
Applications	AGA Listed Gas Appliances	Air/Product Containment Breeching Systems, Grease Duct	Boiler and Breeching Systems, Engine/Turbine Exhausts, Grease/Oven Exhausts, Air/Particle Containment		
Fuel Types	Natural or LP Gas	LP; Natural Gas; #2, #4^A, #5^A or #6^A Fuel Oil; Wood; Coal^A; Grease Vapors; Caustic Fumes; Particles			
Exhaust Pressures	Neutral or Negative	Positive, Neutral or Negative			
Exhaust Temp. Continuous/Intermittent	400°F Plus Ambient	Air Product Containment or 2000°F for Grease Duct	100 °F Continuous, 1400 °F Intermittent 1400 ° Continuous. 1800 °F Intermittent Grease Duct 500 °F Continuous		
Diameters	3" through 30"	6" through 48"			
Materials	Inner: .012" Aluminum 3" to 6" .014" Aluminum 7" to 18" .018" Aluminum 20" to 30"	Inner: Standard 304SS .035" All Diameters (Optional .035" 316SS Available)	Inner: .035" 20-ga 304SS Standard (Optional .035" 20-ga 316SS Available)	Inner: .035" 20-ga 304SS Standard (Optional .035" 20-ga 316SS Available)	Inner: .035" 20-ga 304SS Standard (Optional .035" 20-ga 316SS Available)
	Outer: 28-ga Galvanized Steel 3" to 30"	Outer: N/A	Outer: .025" 24-ga Aluminum Coated Steel 6" to 24" .034" 20-ga Aluminum Coated Steel 26" to 48" Optional 304 or 316SS Available)	Center: .025" 24-ga Aluminum Coated Steel 6" to 24" .034" 20-ga Aluminum Coated Steel 26" to 48" (Optional 304 or 316SS Available)	Insulation Material: Eleven Pound Fiber Insulation of 1", 2" or 4" thickness
				Outer: .025" 24-ga Aluminum Coated Steel 6" to 24" .034" 20-ga Aluminum Coated Steel 26" to 48" (Optional 304 or 316SS Available)	Outer: .025" 24-ga Aluminum CoatedSteel 6" to 24" .034" 20-ga Aluminum Coated Steel 26" to 48" (Optional 304 or 316SS Available)
Insulation	1/4" Air 3" through 6" 1/2" Air 7" through 30"	N/A	1" Air	Innerwall to Center = 1" Air Center to Outerwall = 1/2" Air	1" Material 2" Material or 4" Material
Application Ref. & Listings	Complies with one or more of the following: AGA; HUD; NBC; UBC; UMC; NMC; SMC; SBCCI; ICBO; BOCA; UL-103, 710, 411; ULC-S604; NFPA-85, A, B, D, 31, 34, 37, 54, 96, 211.				

A. Recommended 316 Stainless Steel.

By permission of Cleaver Brooks, Milwaukee, Wisconsin

20.8.1 Typical Stack Construction

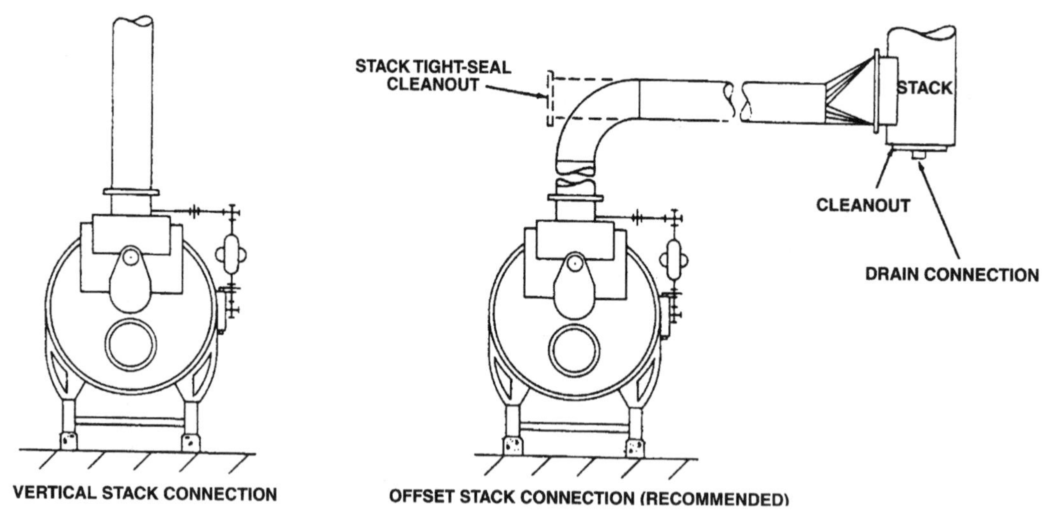

Typical Stack Locations

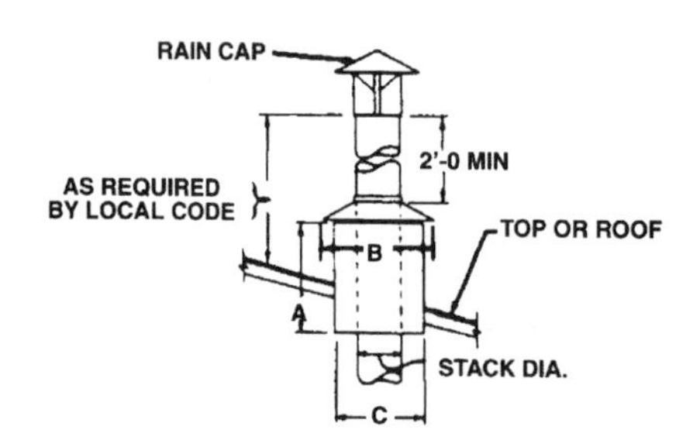

BOILER HP	STACK DIAMETER (IN.)	A (IN.)	B (IN.)	C (IN.)
15-20	6	15	15	12
25-40, 50A	8	20	20	16
50-60	10	25	25	20
70-100A, 125A	12	30	30	24
125-200	16	40	40	32
250-350	20	50	50	40
400-800	24	60	60	48

Typical Stack Construction

By permission of Cleaver Brooks, Milwaukee, Wisconsin

20.8.2 Stack Expansion/Contraction, and Installation Concerns

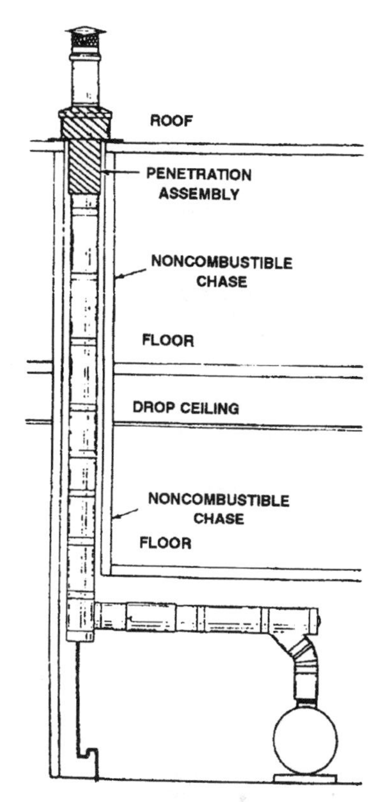

SYSTEMS EXTENDING THROUGH ANY STORY ABOVE THE BOILER ROOM REQUIRE A NONCOMBUSTIBLE CHASE ENCLOSURE FROM THE BOILER ROOM TO THE ROOF AND A PENETRATION ASSEMBLY OR ROOF SUPPORT ASSEMBLY AT THE ROOF LEVEL

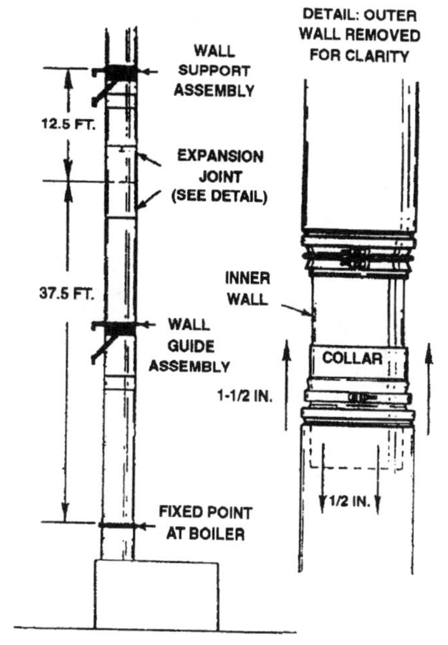

THERMAL EXPANSION OCCURS BETWEEN ANY TWO FIXED POINTS IN THE STACK SYSTEM. NOTE: DISTANCES SHOWN ON ILLUSTRATION ARE FOR EXAMPLE ONLY.

When stack systems are exposed to the heating and cooling of normal operation, the components will expand and contract. The systems are designed to adjust to this movement, provided the amount and direction of expansion is accurately calculated, and the system is correctly installed.

The amount of thermal expansion that will occur depends on the length of breeching, height of the stack, temperature of the flue gas, and arrangement of the system. Therefore, the following must be considered.

Thermal Expansion

The CBS/ICBS systems use two different parts to compensate for thermal expansion between two fixed points in the system. They are: expansion joints and bellows joints.

Each type of joint is:

- Designed to compensate for linear expansion only.
- Never used to correct for misalignment between components.
- Not load bearing. Therefore, these systems are usually between support or guide assemblies.

Expansion Direction

To determine expansion direction, it is necessary to understand how the expansion or bellows joint works. The expansion or bellows joint itself does not expand. In fact, the opposite happens. It compresses to absorb the movement of the parts expanding around it. This expansion movement occurs from fixed points in the system toward the expansion or bellows joint.

By permission of Cleaver Brooks, Milwaukee, Wisconsin

20.9.0 Schematic of a Typical Custom-Built HVAC Unit

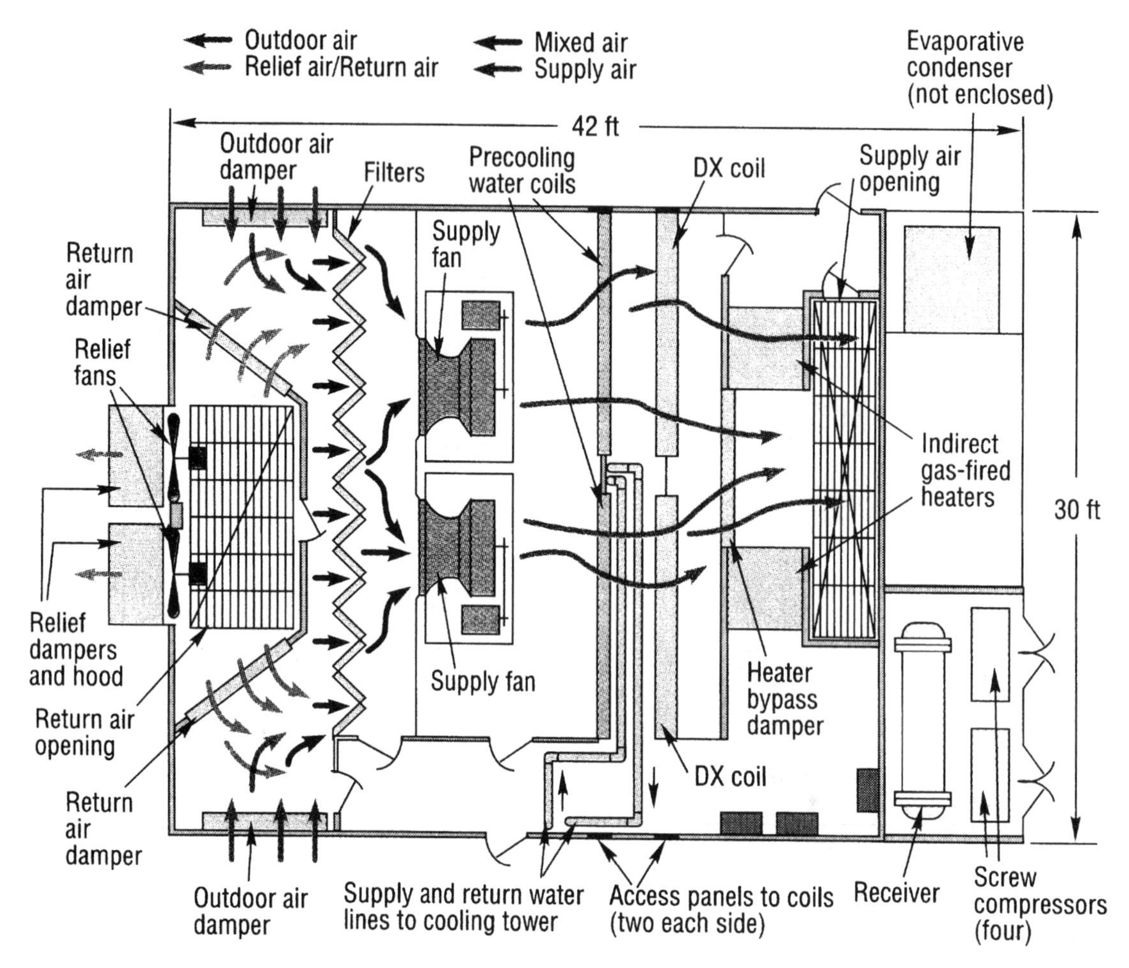

Reprinted by permission from *Heating/Piping/Air Conditioning* magazine, December 1996

20.10.0 Schematic of Indirect Evaporative Precooling System

Components include:

1. Stand along cooling tower
2. Water pump and piping
3. Water cooling coils
4. Centrifugal separator
5. Chemical treatment system

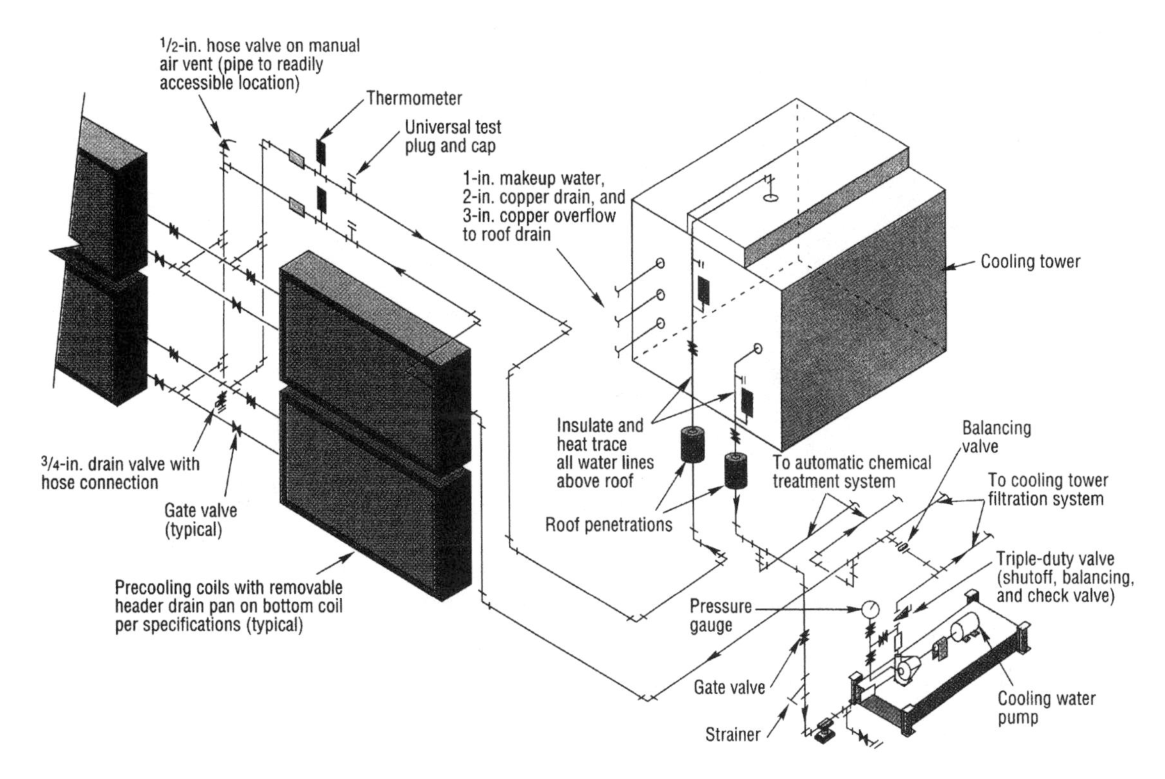

Reprinted by permission from Heating/Piping/Air Conditioning magazine, December 1996

20.11.0 Heat-Pump Operation Schematics

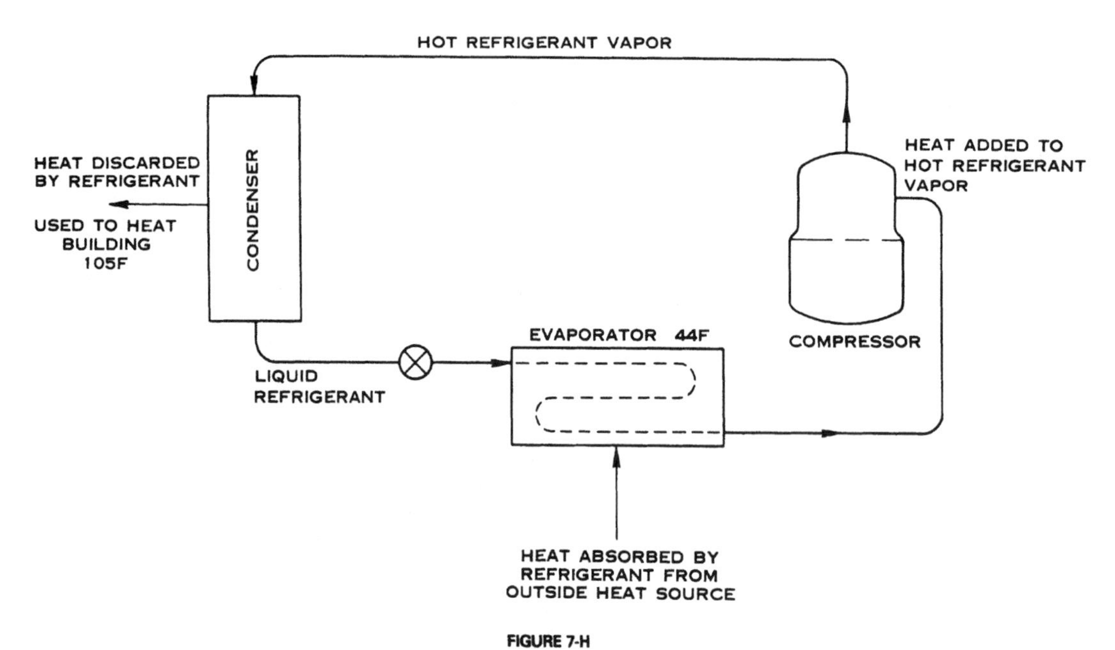

FIGURE 7-H
HEAT PUMP ON HEATING CYCLE

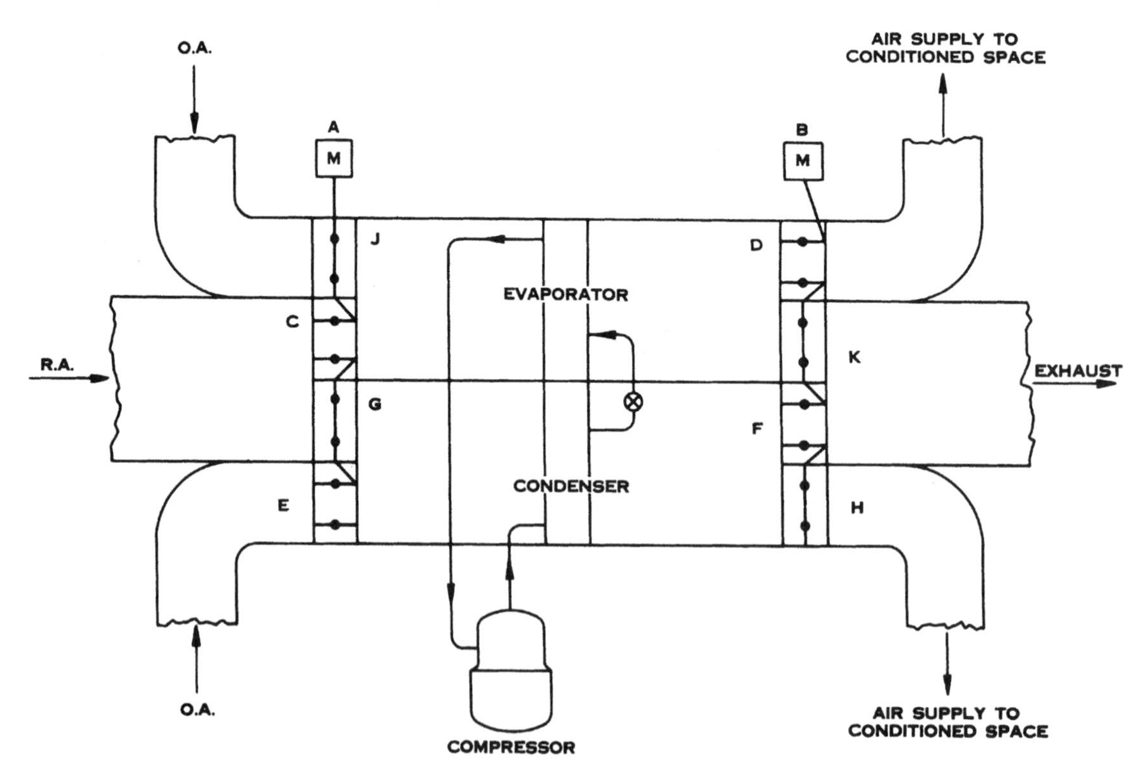

By permission of The Trane Company, LaCrosse, Wisconsin

20.12.0 Air-Cooled Condenser and Subcooling System (Illustrated)

System Piping Suggestions

If an air conditioning system with an air-cooled condenser will operate only when the outdoor temperature is above 40 F, a simple fan cycling or multilouvered damper control is usually adequate. The shutter control will follow the system load variations closely enough so there should be neither head pressure nor starting problems. The system piping can be simple, as illustrated in the piping diagram.

As will be noted, this system does not employ the conventional liquid receiver. The air condenser has sufficient volume to hold the charge on a system where the components are reasonably close together. Since the accumulator between the condensing circuit and the subcooler of the air condenser can handle a small variation in liquid volume, this would not be considered a critically charged system.

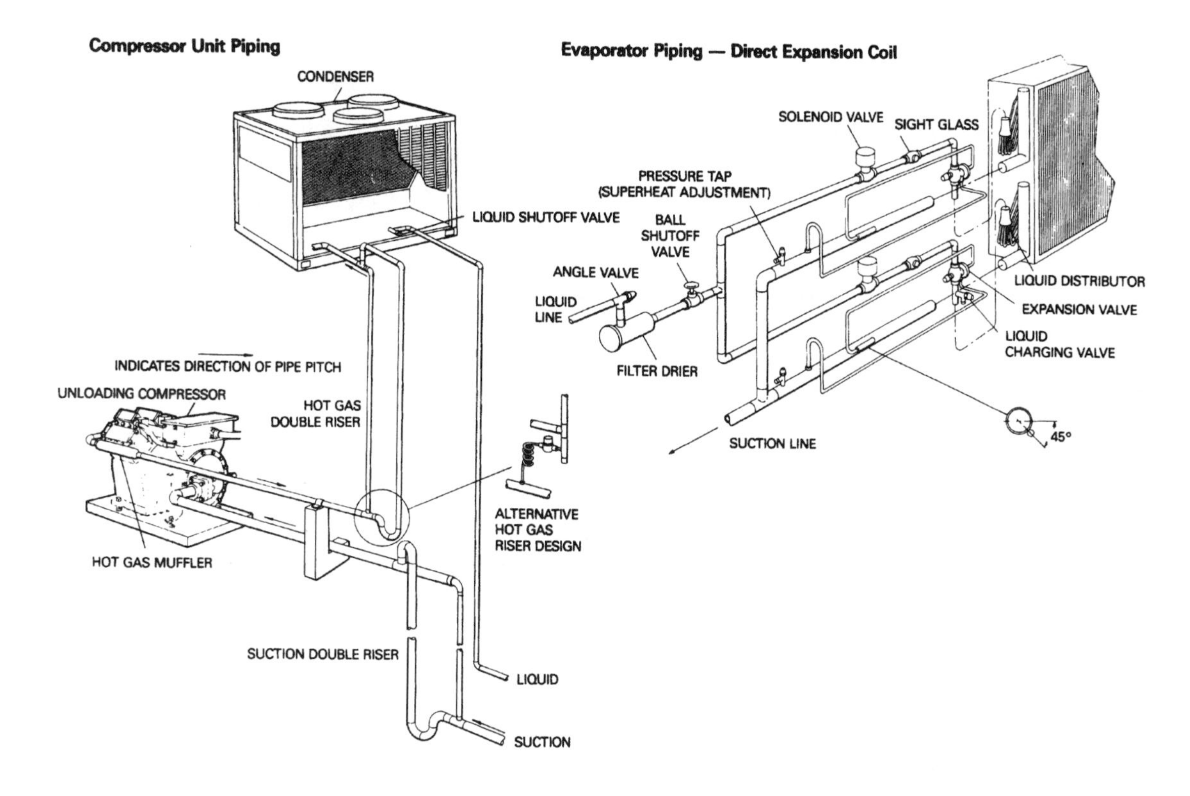

TYPICAL PIPING ARRANGEMENT OF SYSTEM WITH AIR-COOLED CONDENSER AND SUBCOOLING. NO HEAD PRESSURE CONTROL, OR HEAD PRESSURE CONTROL MAY BE WITH SHUTTERS.

By permission of The Trane Company, LaCrosse, Wisconsin

20.13.0 Variable Air Volume (VAV) Systems Diagrammed

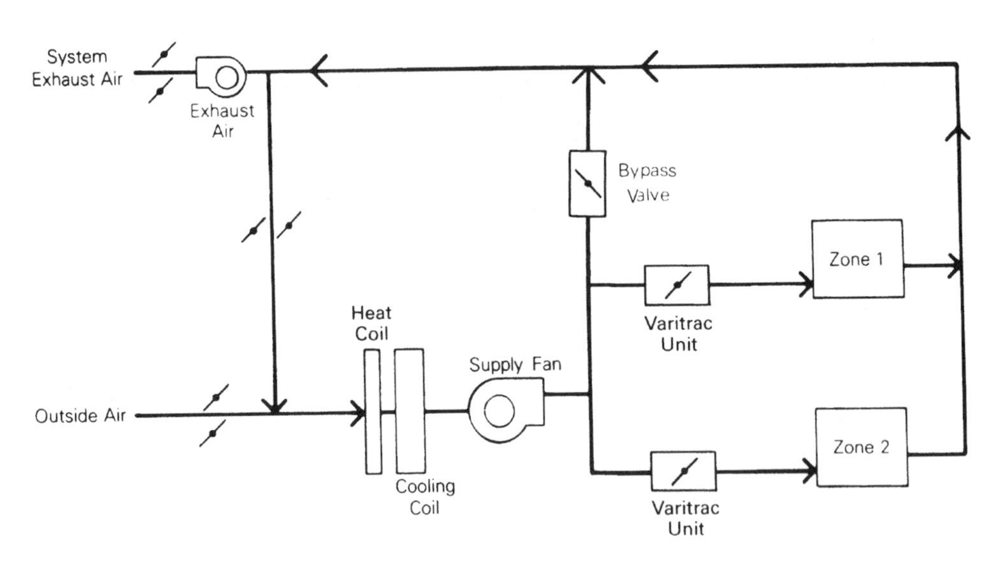

CHANGEOVER-BYPASS VARIABLE AIR VOLUME SYSTEM

The changeover-bypass variable air volume system offers perhaps the least expensive temperature control for a large number of zones when compared to other variable air volume systems. It is a flexible system in that it is relatively easy and inexpensive to subdivide a building into additional new zones should it become necessary after the initial building and system design have been completed. Operating and first cost savings are both possible through the ability to use building load diversity to not only reduce the installed system equipment size but also to reduce its energy use through more efficient operation of smaller pieces of equipment at part-load conditions.

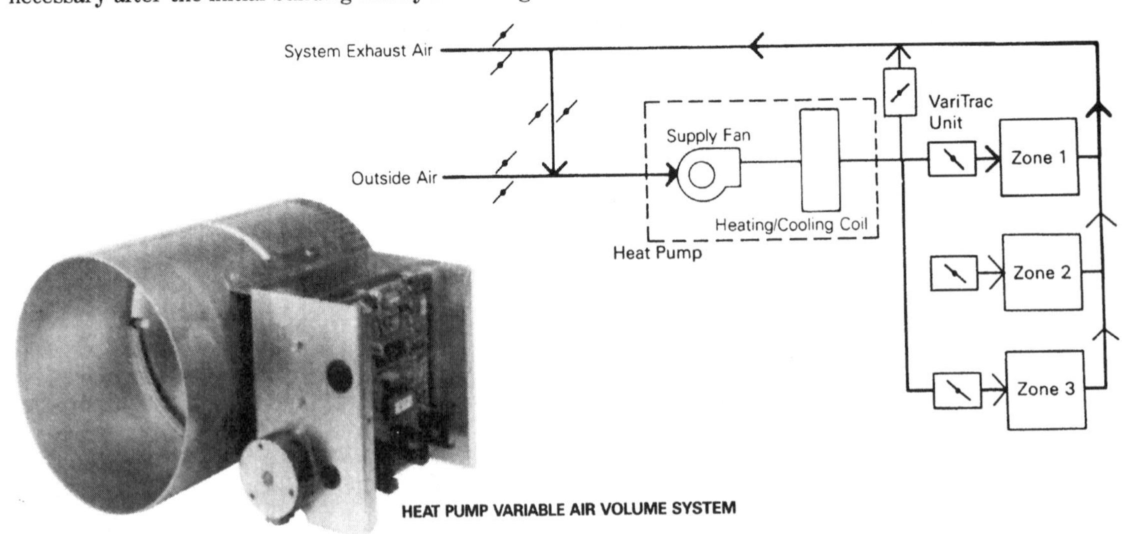

HEAT PUMP VARIABLE AIR VOLUME SYSTEM

By permission of The Trane Company, LaCrosse, Wisconsin

20.13.1 Variable Air Volume (VAV) Diagrams Showing Radiation Heating, Reheat and Fan-Powered Systems

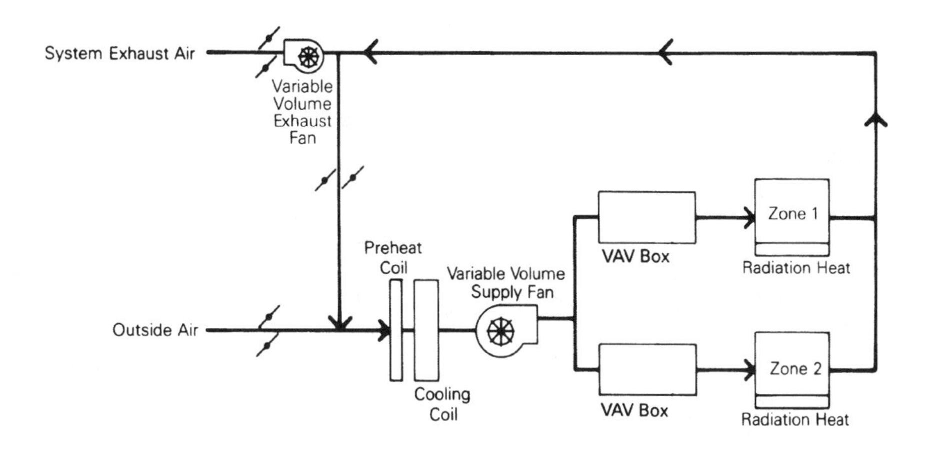

VARIABLE AIR VOLUME COOLING WITH PERIMETER RADIATION HEATING

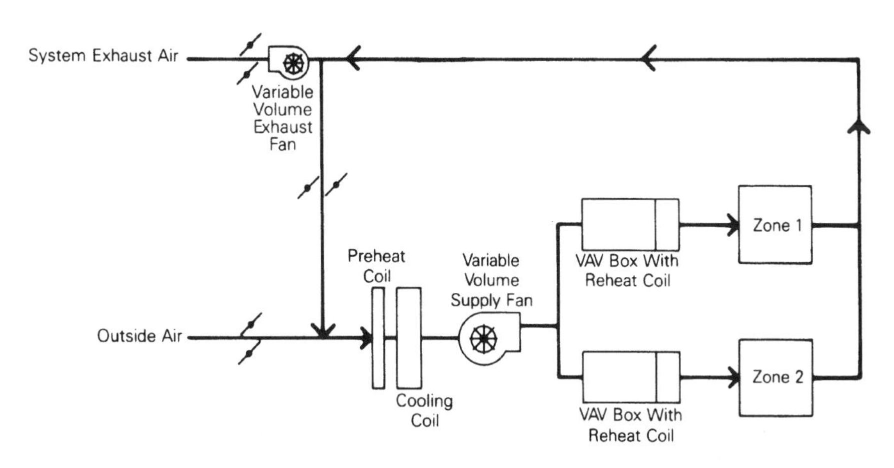

VARIABLE AIR VOLUME REHEAT SYSTEM

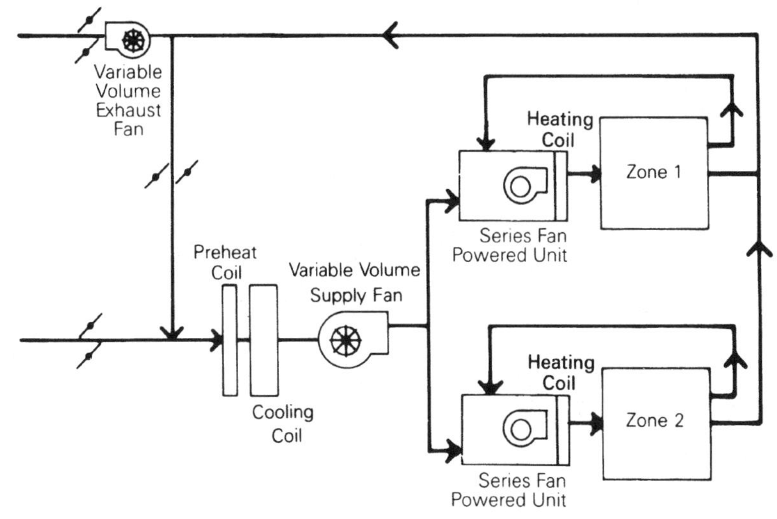

SERIES FAN POWERED VARIABLE AIR VOLUME SYSTEM

By permission of The Trane Company, LaCrosse, Wisconsin

20.14.0 Single- and Two-Pipe Cooling System Diagrams

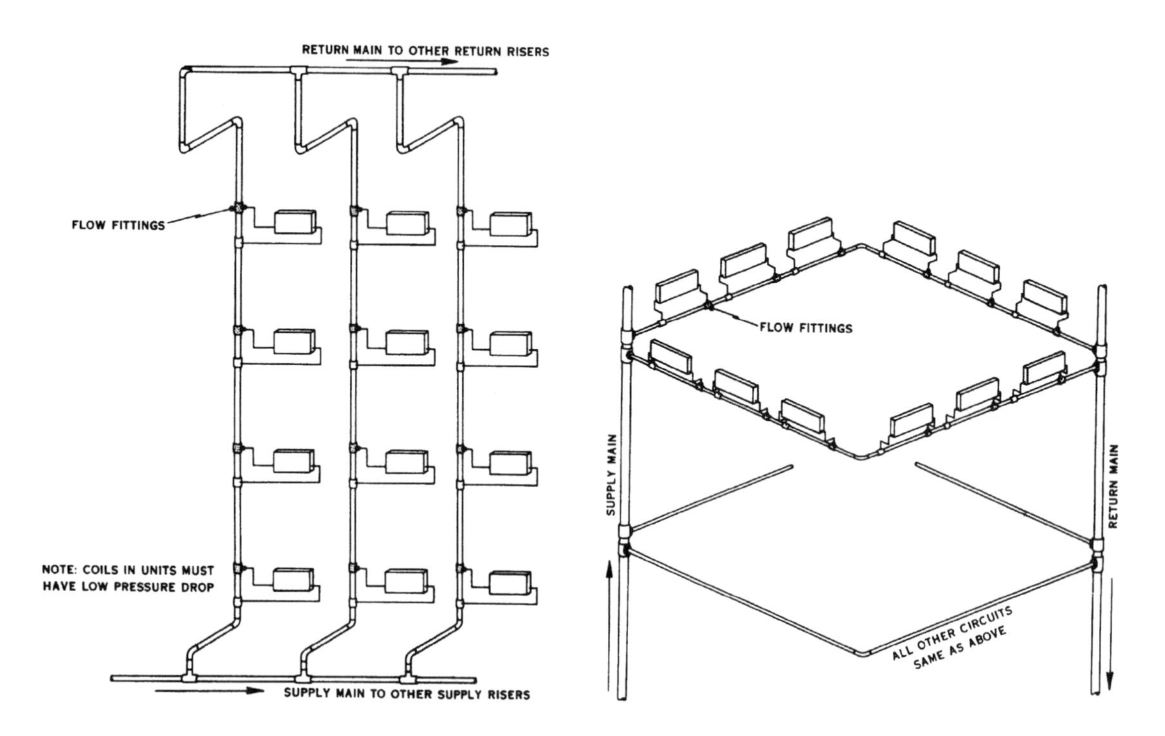

SINGLE-SUPPLY, SINGLE-RETURN RISER

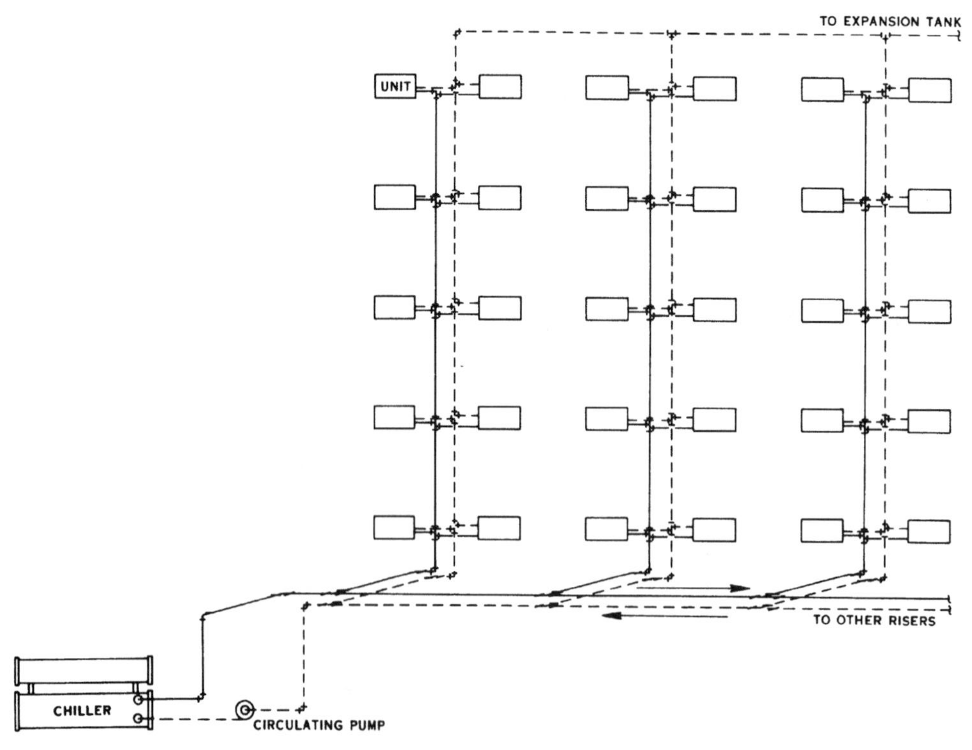

TWO-PIPE DIRECT RETURN MAINS AND RISERS

By permission of The Trane Company, LaCrosse, Wisconsin

20.15.0 Two-Pipe Reverse Main and Three-Pipe Heating/Cooling Piping Diagrams

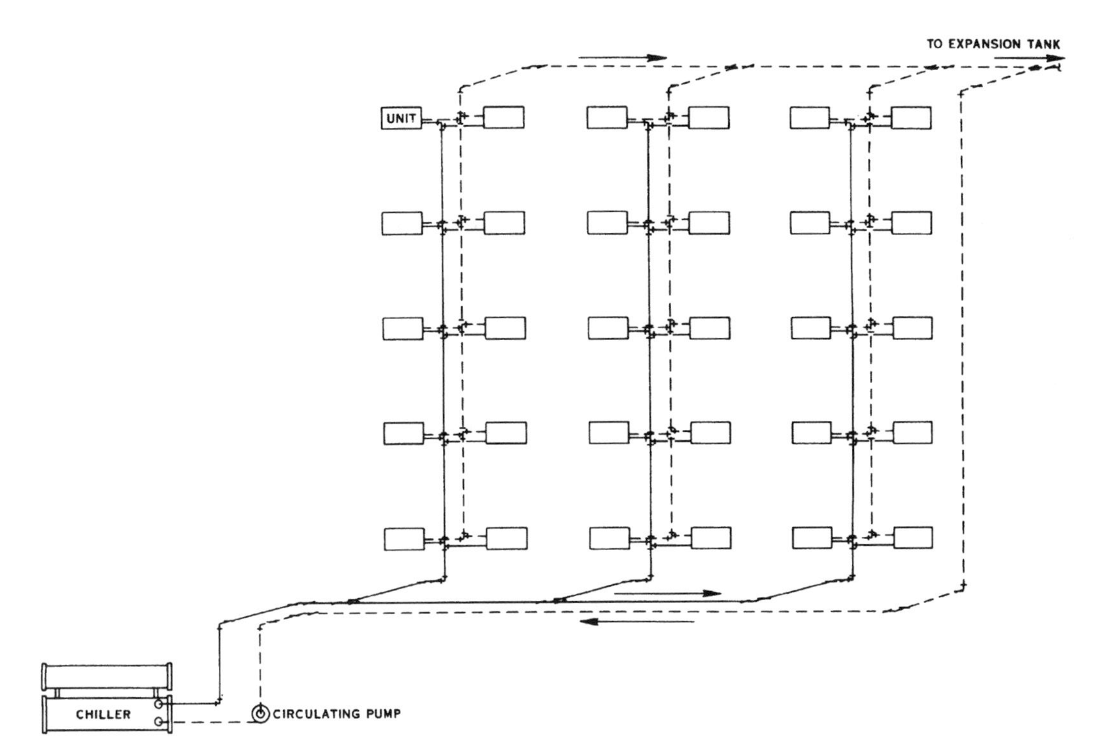

TWO-PIPE REVERSE RETURN MAINS AND RISERS

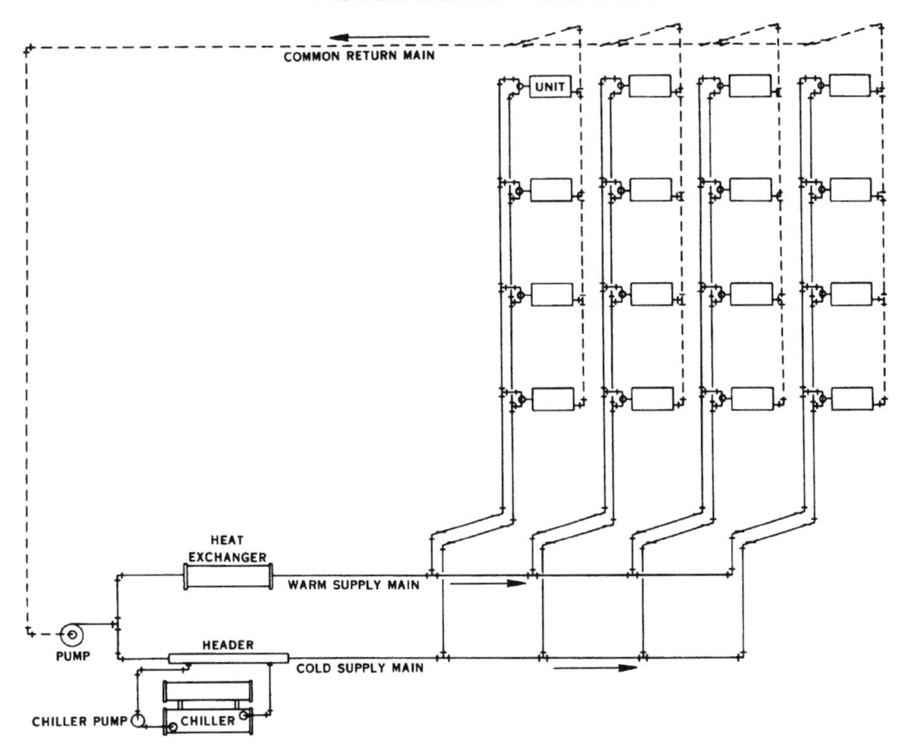

SIMPLE THREE-PIPE WATER DISTRIBUTION

By permission of The Trane Company, LaCrosse, Wisconsin

20.16.0 Four-Pipe Systems with One- and Two-Coil Piping Diagrams

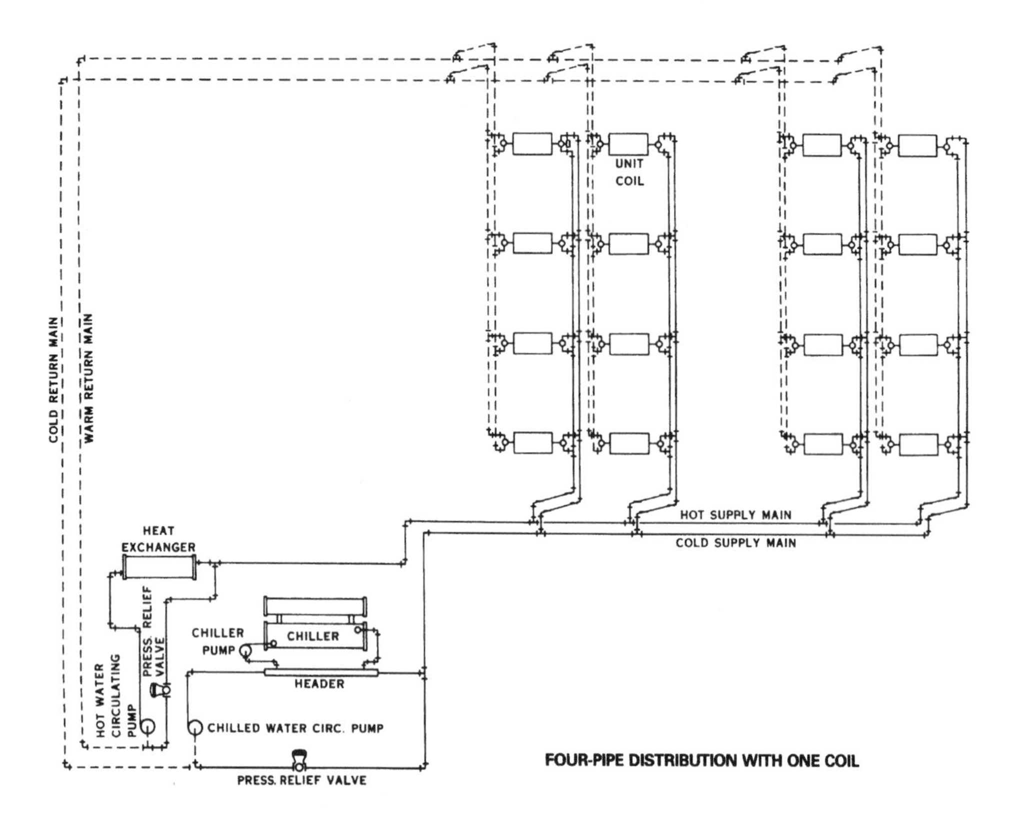

FOUR-PIPE DISTRIBUTION WITH ONE COIL

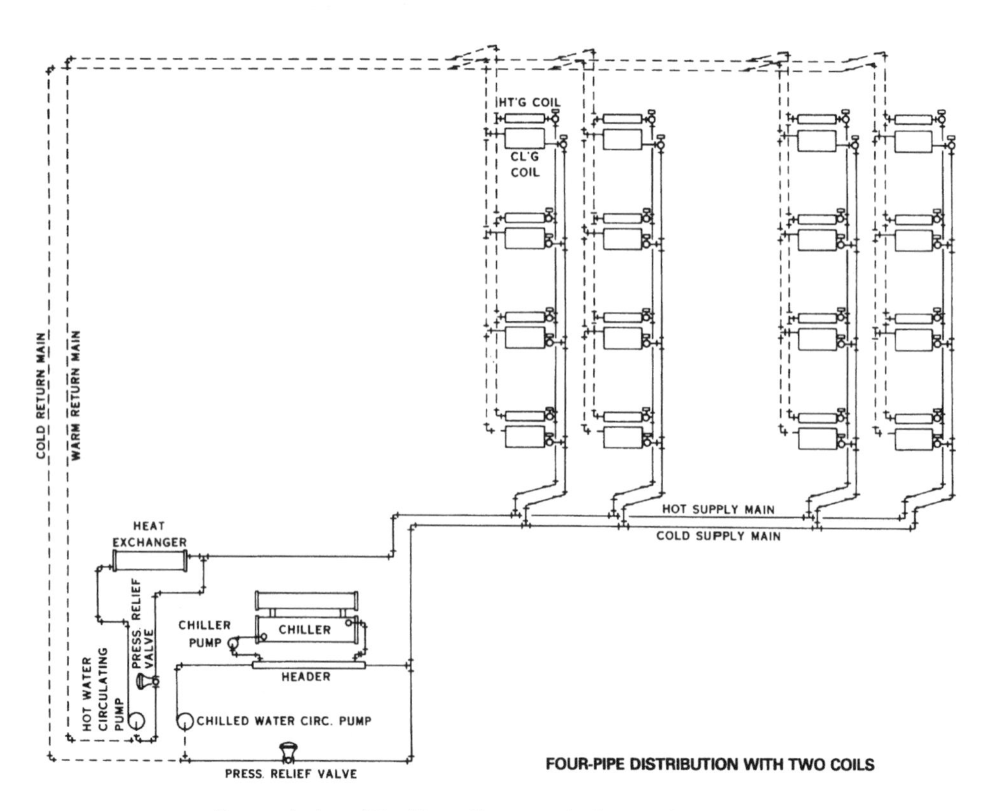

FOUR-PIPE DISTRIBUTION WITH TWO COILS

By permission of The Trane Company, LaCrosse, Wisconsin

20.17.0 Shell and Coil, and Shell and Tube Condensers (Illustrated and Described)

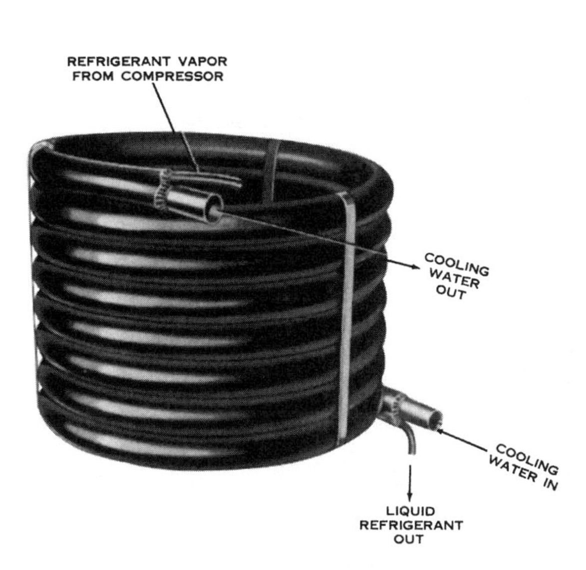

DOUBLE-TUBE CONDENSER

Shell-and-Coil Condenser

A shell-and-coil condenser is simply a continuous copper coil mounted inside a steel shell. Water flows through the coil and refrigerant vapor from the compressor is discharged inside the shell to condense on the outside of the cold tubes. In many designs, the shell also serves as a liquid receiver.

The shell-and-coil condenser has a low manufacturing cost but this is offset by the disadvantage that this type condenser is difficult to service in the field. If a leak develops in the coil, the head from the shell must be removed and the entire coil pulled from the shell in order to find and repair the leak. A continuous coil is a nuisance to clean whereas straight tubes are easy to clean with mechanical tube cleaners. Thus, with some types of fouled cooling water, it may be difficult to maintain a high rate of heat transfer with a shell-and-coil condenser.

Shell-and-Tube Condenser

The shell-and-tube condenser, permits a large amount of condensing surface to be installed in a comparatively small space. The condenser consists of a large number of ¾ or ⅝-inch tubes installed inside a steel shell. The water flows inside the tubes while the vapor flows outside, around the nest of tubes. The vapor condenses on the outside surface of the tubes and drips to the bottom of the condenser, which may be used as a receiver for the storage of liquid refrigerant. Shell-and-tube condensers are used for practically all water-cooled refrigeration systems.

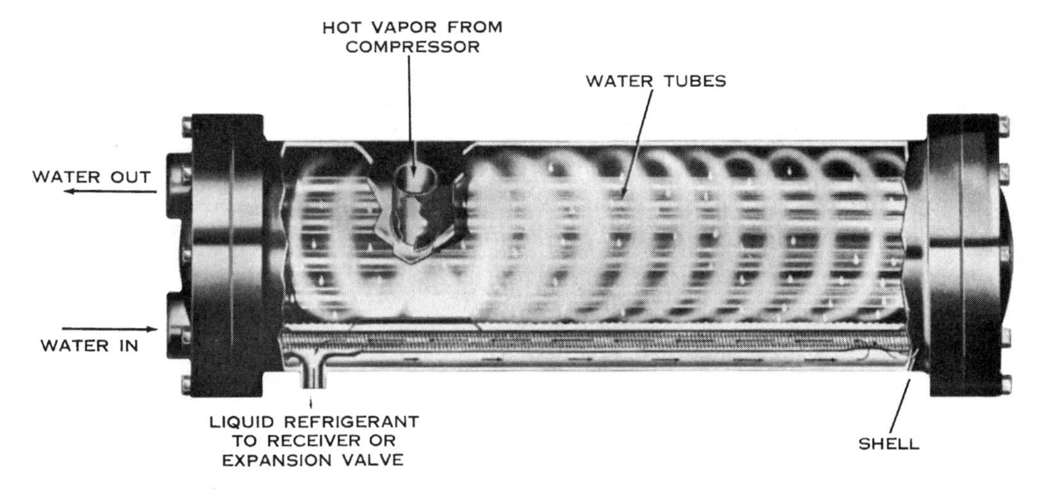

SHELL-AND-TUBE CONDENSER

By permission of The Trane Company, LaCrosse, Wisconsin

20.18.0 Shell and Tube Evaporator (Diagram and Description)

Shell-and-Tube Evaporators

There are two common types of shell-and-tube evaporators used to provide chilled water for air conditioning systems. There are the same two types as discussed previously with fin and tube coil evaporators; the **flooded** type and the **direct expansion** (dry) type. In the flooded type, the shell contains a tube bundle through which water to be chilled is pumped. Half to three-fourths of the tube bundle is immersed in liquid refrigerant, which boils because of the heat received from the water being cooled.

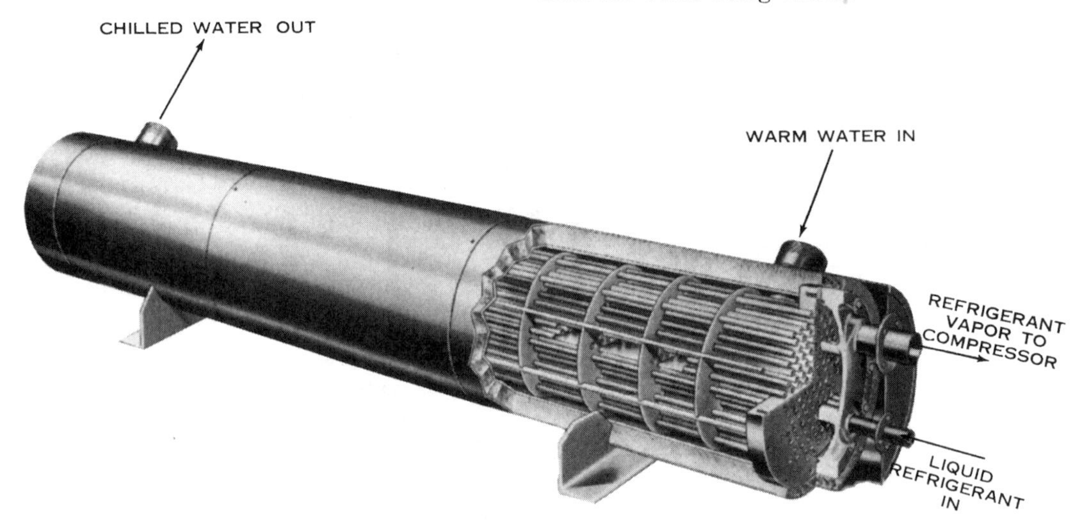

A DIRECT EXPANSION SHELL-AND-TUBE EVAPORATOR

By permission of The Trane Company, LaCrosse, Wisconsin

20.19.0 Evaporative Condenser (Diagram and Description)

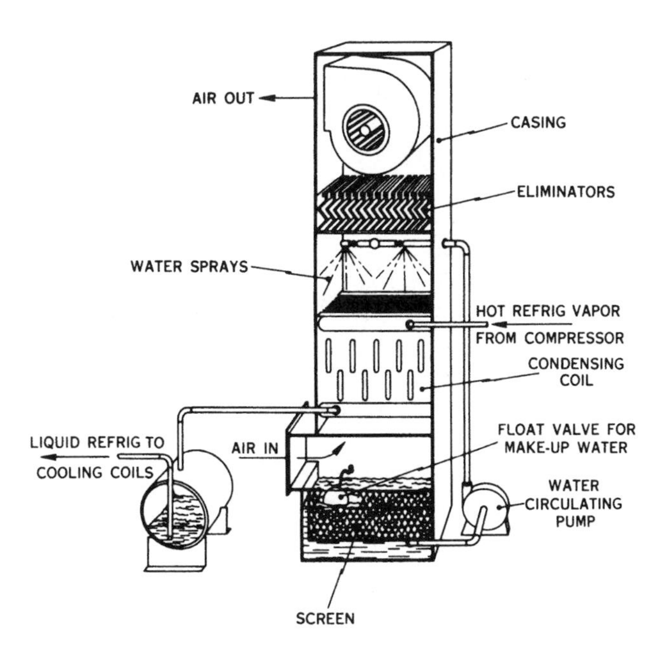

EVAPORATIVE CONDENSER

The evaporative condenser is a form of water-cooled condenser that offers a means of conserving water by combining the condenser and the cooling tower into one piece of equipment.

By permission of The Trane Company, LaCrosse, Wisconsin

20.20.0 Heating With a Chiller (Diagram and Description)

Freezing water to ice in the ice storage system removes 144 Btu per pound of ice generated by the chiller. This is why the ice-storage system designed with a heat-recovery loop can also make the chiller into a water-source heat pump for cold-weather heating.

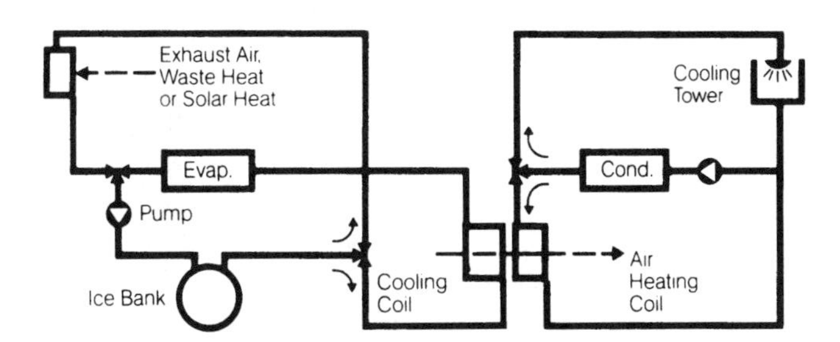

ICE STORAGE SYSTEM SCHEMATIC WITH HEAT RECOVERY LOOP

By permission of The Trane Company, LaCrosse, Wisconsin

20.21.0 Typical Evaporative Cooler (Diagram and Description)

A typical evaporative cooler is a metal housing with three sides containing porous material kept saturated with water. A pump lifts water from the sump in the bottom of the unit and delivers it to perforated troughs at the top of the unit. The fan draws outside air through the saturated material and discharges it directly into the conditioned space or into a duct system for distribution into several rooms. The porous material is generally spun glass fibers, aspen excelsior pads or tinsel made of copper or aluminum. The discharge line from the pump is usually plastic tubing although copper tubing or iron pipe are sometimes used. A float valve is normally provided to replenish the water evaporated into the air passing through the unit.

Generally, this valve is set to waste a fixed amount of water at all times. This ensures there will be a continual dilution of the natural minerals in the water that are left behind due to evaporation. This is commonly called "blowdown" and provides protection against a sticking float valve.

Variations in the above design are offered by several manufacturers for applications, primarily in dry climates with a low design wet bulb temperature.

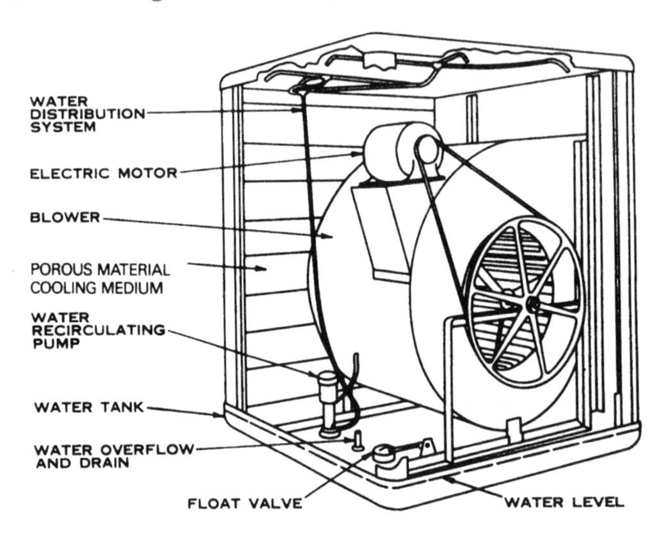

TYPICAL EVAPORATIVE COOLER

By permission of The Trane Company, LaCrosse, Wisconsin

20.22.0 Typical Flow Diagram for an Ice-Storage System

It is important to note that while making ice at night, the chiller must cool the water-glycol solution down to 26 F, rather than producing 44 F water required for conventional air conditioning systems.

This has the net effect of "derating" the nominal chiller capacity by a substantial amount (typically 25–30 percent). The compressor efficiency at this time is only slightly reduced because the lower nighttime outdoor ambient wet bulb temperatures result in cooler condenser water from the cooling tower which lowers the condensing temperature to keep the chiller operating efficiently. Similarly, chillers with air-cooled condensing also benefit from cooler outdoor ambient dry bulb temperatures to lower the system condensing temperature at night.

The temperature modulating valve in the bypass loop has the added advantage of providing excellent capacity control. During mild temperature days, typically in the spring and fall, the chiller will often be capable of providing all the necessary cooling capacity for the building without the use of cooling capacity from the ice storage system. When the building's actual cooling load is equal to or less than the chiller capacity at the time, all of the system coolant will flow through the bypass loop.

It is important that the coolant chosen be an ethylene glycol-based industrial coolant, such as Dowthern SR-1 or UCAR Thermofluid 17, which is specially formulated for low viscosity and good heat transfer properties. Either of these fluids contain a multi-component corrosion inhibitor which is effective with most materials of construction including aluminum, copper, silver solder and plastics. Further, they contain no anti-leak agents and produce no films to interfere with heat transfer efficiency. They also permit use of standard pumps, seals and air handling coils. It should be noted, however, that because of the slight difference in heat transfer properties between water and the mild glycol solution, the cooling coil capacities will need to be increased by approximately 5 percent. It is also important that the water and glycol solution be thoroughly mixed before the solution is placed into the system.

The use of ice storage system technology opens new doors to other economic opportunities in system design. These offer significant potential for not only first-cost savings but also operating cost savings that should be evaluated on a life cycle cost basis using a computerized economic analysis program such as the Trane Air Conditioning Economics (TRACE®) program.

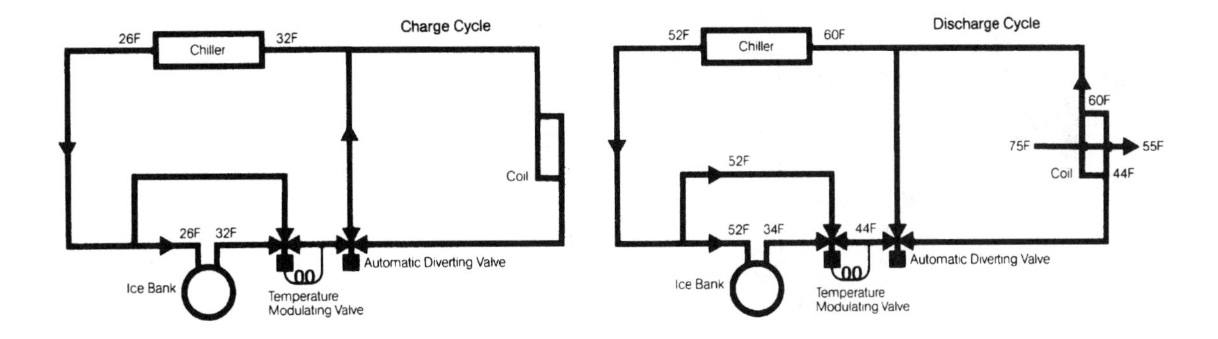

TYPICAL FLOW DIAGRAMS FOR A PARTIAL ICE STORAGE SYSTEM

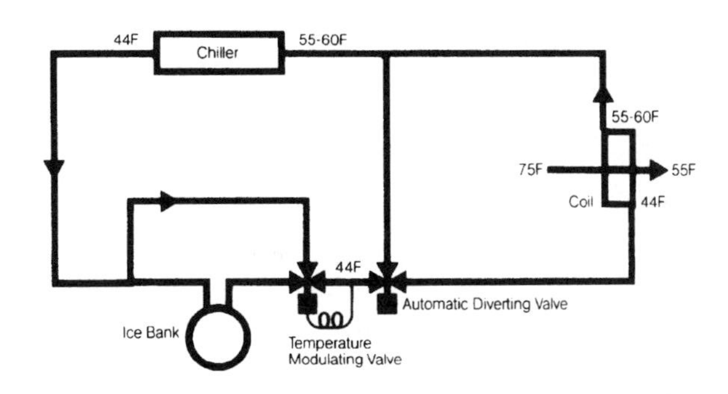

TYPICAL FLOW DIAGRAM FOR A PARTIAL ICE STORAGE SYSTEM — WITH ALL COOLANT THROUGH THE BYPASS LOOP

By permission of The Trane Company, LaCrosse, Wisconsin

20.23.0 Types of Humidifiers (Illustrated and Described)

Hot Element Humidifier

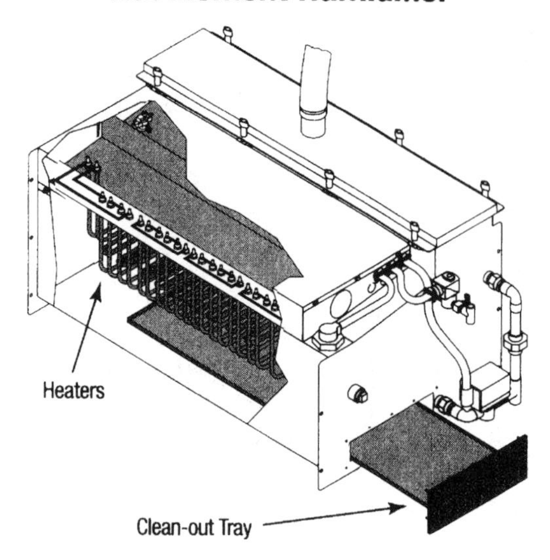

Resistance heating elements are submerged in an evaporating chamber full of water.

Atomizing Humidifier

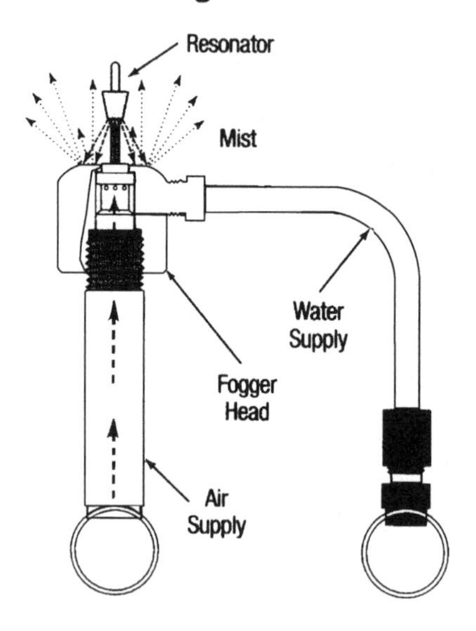

Compressed air is used to provide an ultrasonic shock wave to atomize the water into a mist, which is absorbed into the airstream.

Steam-to-Steam Humidifier

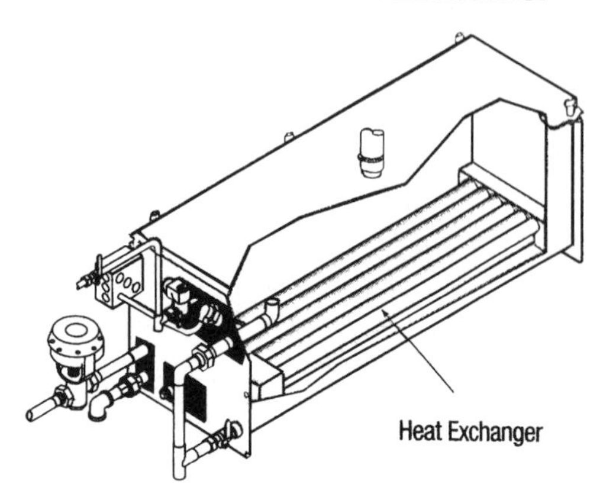

Humidification steam output is modulated to match load conditions by the use of a valve, controlling the flow to the heat exchanger.

Reprinted by permission from *Heating/Piping/Air Conditioning* magazine, December 1996

20.23.0 Types of Humidifiers (Illustrated and Described) (Continued)

Replaceable Cylinder Humidifier

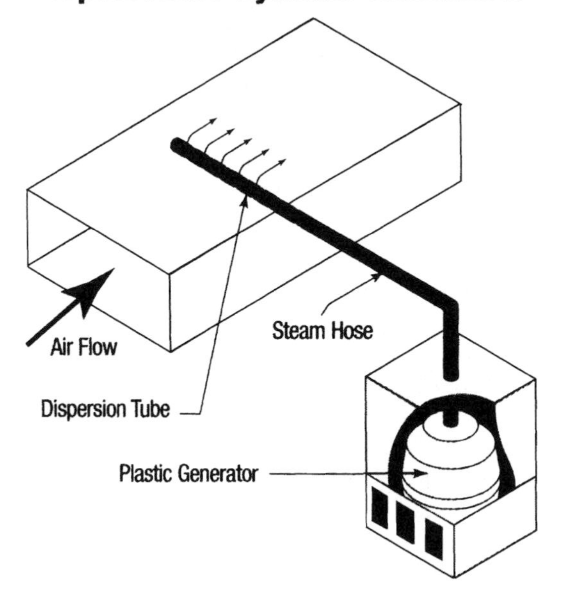

Consists of a replaceable plastic evaporator mounted in a cabinet along with a fill and drain, valve, and an electronic water-level detector. Electronic plates inside the plastic evaporator allow electric current to pass through the water, causing it to boil and create steam.

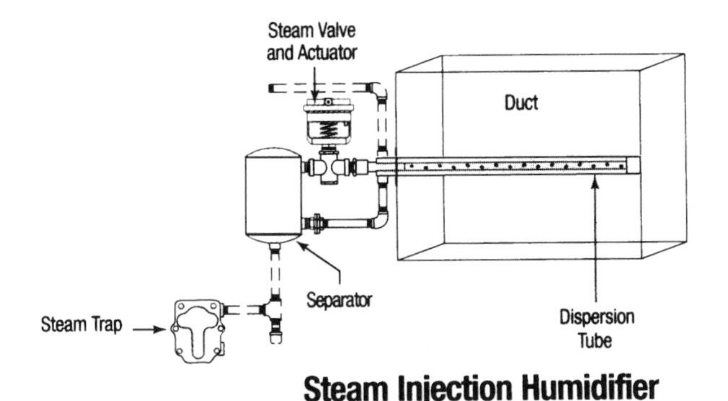

Steam Injection Humidifier

Direct steam injection is used in buildings with a boiler where the introduction of boiler chemicals into the air is not objectionable.

Reprinted by permission from *Heating/Piping/Air Conditioning* magazine, December 1996

20.24.0 Mechanical Draft Towers (Illustrated and Described)

Mechanical Draft

Mechanical draft towers use either single or multiple fans to prove flow of a known volume of air through the tower. Thus their thermal performance is considered to be more stable and is affected by fewer psychometric variables than that of the natural draft atmospheric towers. The presence of fans also provides a means of regulating air flow to compensate for changing atmospherc and load conditions through fan capacity modulation of speed and/or cycling.

Mechanical draft towers are categorized as either "induced draft" wherein a fan located in the exiting air stream draws air through the tower or "forced draft" in which the fan is located in the ambient air stream entering the tower, and the air is blown through the tower.

Induced Draft

An induced draft cooling tower is provided with a top-mounted fan that induces atmospheric air to flow up through the tower, as warm water falls downward. An induced draft tower may have only spray nozzles for water breakup or it may be filled with various slat and deck arrangements. There are several types of induced draft cooling towers.

In a counterflow induced draft tower, a top-mounted fan induces air to enter the sides of the tower and flow vertically upward as the water cascades through the tower. The counterflow tower is particularly well adapted to a restricted space as the discharge air is directed vertically upward, and the sides require only a minimum clearance for air intake area. The primary breakup of water may be by either pressure spray or by gravity from pressure filled fumes.

A doubleflow induced draft tower, has a top-mounted fan to induce air to flow across the fill material. The air is then turned vertically in the center of the tower. The distinguishing characteristics of a doubleflow induced draft tower are the two air intakes on opposite sides of the tower and the horizontal flow of air through the fill sections.

Comparing counterflow and doubleflow induced draft towers of equal capacity, the doubleflow tower would be somewhat wider but the height would be much less. Cooling towers must be braced against the wind. From a structural standpoint, therefore, it is much easier to design a doubleflow than a counterflow tower as the low silhouette of the doubleflow type offers much less resistance to the force of the winds.

Mechanical equipment for counterflow and doubleflow towers is mounted on top of the tower and is readily accessible for inspection and maintenance. The water distributing systems are completely open on top of the tower and can be inspected during operation. This makes it possible to adjust the float valves, and clean stopped-up nozzles while the towers are operating.

The crossflow induced draft tower is a modified version of the doubleflow induced draft tower. The fan in a crossflow cooling tower draws air through a horizontal opening and discharges the air vertically.

In some situations, an indoor location for the cooling tower may be desirable. An induced draft tower, of the counterflow or crossflow design, is generally selected for indoor installation. Two connections to the outside are usually required: one for drawing outdoor air into the tower, and the other for discharging it back to the outside. A centrifugal blower is often necessary for this application to overcome the static pressure of the ductwork. Many options are possible as to point of air entrance and air discharge. This flexibility is often important in designing an indoor installation. Primary water breakup is by pressure spray and fill of various types.

An indoor installation of an induced draft counterflow cooling tower is shown. In this particular case, air required for operation of the tower is being taken from the basement. As the cooling tower fan is therefore exhausting air from the building, the quantity of air exhausted must be included in sizing the outside air intake for the air conditioning system.

The induced draft cooling tower, for indoor installation, is usually a completely assembled packaged unit; but is so designed that it can be partially disassembled to permit passage through limited entrances. Indoor installations of cooling towers are becoming more popular. External space restrictions, architectural compatibility (including aesthetics), convenience for observation, diagnostics and maintenance all combine to favor an indoor location. The installation cost is somewhat higher than an outdoor location. Packaged towers are generally available in capacities to serve the cooling requirements of refrigeration plants in the 5 to 100 ton range.

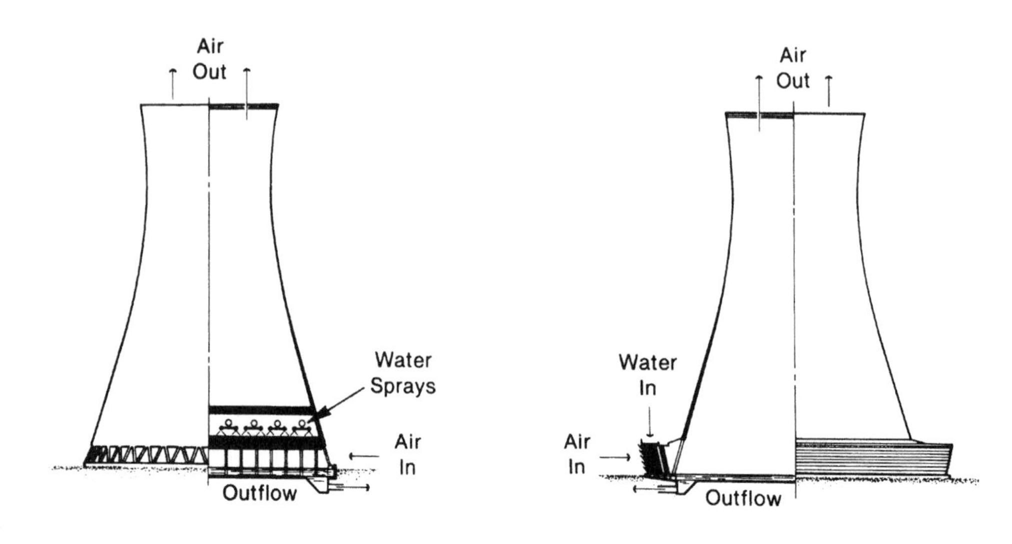

COUNTERFLOW NATURAL DRAFT TOWER

CROSSFLOW NATURAL DRAFT TOWER

By permission of The Trane Company, LaCrosse, Wisconsin

20.24.0 Mechanical Draft Towers (Illustrated and Described) (Continued)

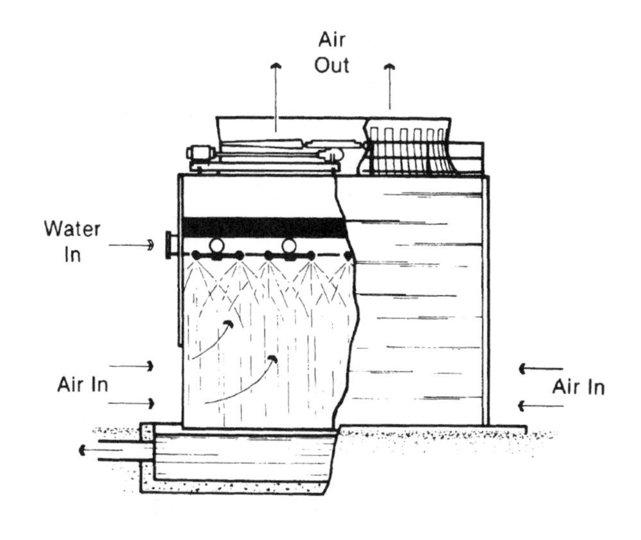

(A) SPRAY-FILLED, COUNTERFLOW COOLING TOWER

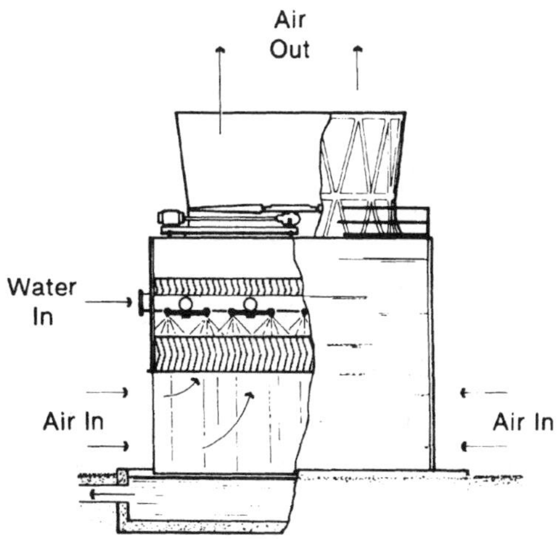

(B) COUNTERFLOW COOLING TOWER WITH FILL BANK

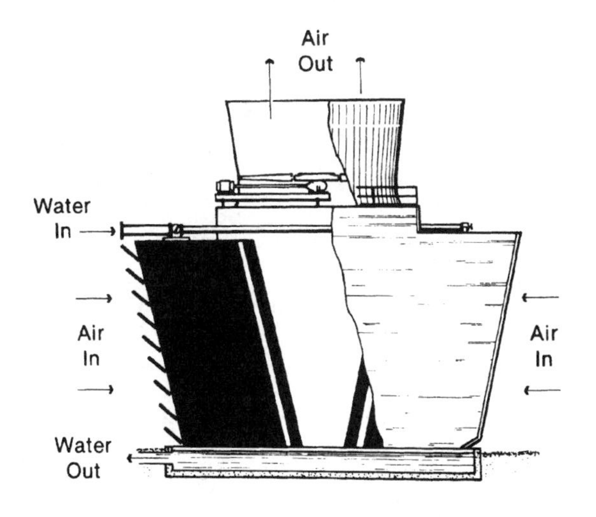

A DOUBLEFLOW INDUCED DRAFT COOLING TOWER

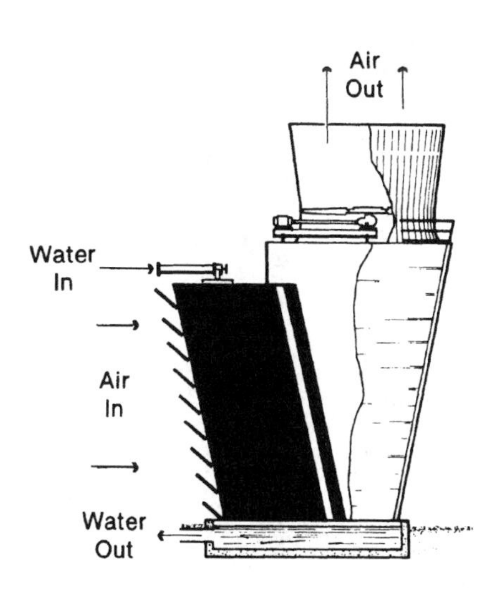

A SINGLE FLOW CROSS-FLOW INDUCED DRAFT COOLING TOWER

By permission of The Trane Company, LaCrosse, Wisconsin

20.25.0 Equivalent Rectangular Duct Dimension Tables

Duct Diameter, in.	Rectangular Size, in.	Aspect Ratio														
		1.00	1.25	1.50	1.75	2.00	2.25	2.50	2.75	3.00	3.50	4.00	5.00	6.00	7.00	8.00
6	Width	—	6													
	Height	—	5													
7	Width	6	8													
	Height	6	6													
8	Width	7	9	9	11											
	Height	7	7	6	6											
9	Width	8	9	11	11	12	14									
	Height	8	7	7	6	6	6									
10	Width	9	10	12	12	14	14	15	17							
	Height	9	8	8	7	7	6	6	6							
11	Width	10	11	12	14	14	16	18	17	18	21					
	Height	10	9	8	8	7	7	7	6	6	6					
12	Width	11	13	14	14	16	16	18	19	21	21	24				
	Height	11	10	9	8	8	7	7	7	7	6	6				
13	Width	12	14	15	16	18	18	20	19	21	25	24	30			
	Height	12	11	10	9	9	8	8	7	7	7	6	6			
14	Width	13	14	17	18	18	20	20	22	24	25	28	30	36		
	Height	13	11	11	10	9	9	8	8	8	7	7	6	6		
15	Width	14	15	17	18	20	20	23	25	24	28	28	35	36	42	
	Height	14	12	11	10	10	9	9	9	8	8	7	7	6	6	
16	Width	15	16	18	19	20	23	23	25	27	28	32	35	42	42	48
	Height	15	13	12	11	10	10	9	9	9	8	8	7	7	6	6
17	Width	16	18	20	21	22	25	25	28	27	32	32	35	42	49	48
	Height	16	14	13	12	11	11	10	10	9	9	8	7	7	7	6
18	Width	16	19	21	23	24	25	28	28	30	32	36	40	42	49	56
	Height	16	15	14	13	12	11	11	10	10	9	9	8	7	7	7
19	Width	17	20	21	23	24	27	28	30	30	35	36	40	48	49	56
	Height	17	16	14	13	12	12	11	11	10	10	9	8	8	7	7
20	Width	18	20	23	25	26	27	30	30	33	35	40	45	48	56	56
	Height	18	16	15	14	13	12	12	11	11	10	10	9	8	8	7
21	Width	19	21	24	26	28	29	30	33	33	39	40	45	54	56	64
	Height	19	17	16	15	14	13	12	12	11	11	10	9	9	8	8
22	Width	20	23	26	26	28	32	33	36	36	39	44	50	54	56	64
	Height	20	18	17	15	14	14	13	13	12	11	11	10	9	8	8
23	Width	21	24	26	28	30	32	35	36	39	42	44	50	54	63	64
	Height	21	19	17	16	15	14	14	13	13	12	11	10	9	9	8
24	Width	22	25	27	30	32	34	35	39	39	42	48	55	60	63	72
	Height	22	20	18	17	16	15	14	14	13	12	12	11	10	9	9
25	Width	23	25	29	30	32	36	38	39	42	46	48	55	60	70	72
	Height	23	20	19	17	16	16	15	14	14	13	12	11	10	10	9
26	Width	24	26	30	32	34	36	38	41	42	46	52	55	66	70	72
	Height	24	21	20	18	17	16	15	15	14	13	13	11	11	10	9
27	Width	25	28	30	33	36	38	40	41	45	49	52	60	66	70	80
	Height	25	22	20	19	18	17	16	15	15	14	13	12	11	10	10
28	Width	26	29	32	35	36	38	43	44	45	49	56	60	66	77	80
	Height	26	23	21	20	18	17	17	16	15	14	14	12	11	11	10
29	Width	27	30	33	35	38	41	43	44	48	53	56	65	72	77	88
	Height	27	24	22	20	19	18	17	16	16	15	14	13	12	11	11
30	Width	27	31	35	37	40	43	45	47	48	53	60	65	72	77	88
	Height	27	25	23	21	20	19	18	17	16	15	15	13	12	11	11
31	Width	28	31	35	39	40	43	45	50	51	56	60	70	78	84	88
	Height	28	25	23	22	20	19	18	18	17	16	15	14	13	12	11
32	Width	29	33	36	39	42	45	48	50	54	56	60	70	78	84	96
	Height	29	26	24	22	21	20	19	18	18	16	15	14	13	12	12
33	Width	30	34	38	40	44	47	50	52	54	60	64	75	78	91	96
	Height	30	27	25	23	22	21	20	19	18	17	16	15	13	13	12
34	Width	31	35	39	42	44	47	50	52	57	60	64	75	84	91	96
	Height	31	28	26	24	22	21	20	19	19	17	16	15	14	13	12
35	Width	32	36	39	42	46	50	53	55	57	63	68	75	84	91	104
	Height	32	29	26	24	23	22	21	20	19	18	17	15	14	13	13
36	Width	33	36	41	44	48	50	53	55	60	63	68	80	90	98	104
	Height	33	29	27	25	24	22	21	20	20	18	17	16	15	14	13
38	Width	35	39	44	47	50	54	58	61	63	67	72	85	96	105	112
	Height	35	31	29	27	25	24	23	22	21	19	18	17	16	15	14

*Shaded area not recommended.

By permission of American Society of Heating, Refrigerating and Air-Conditioning Engineers, Inc. Atlanta, Georgia, from their *1993 ASHRAE Fundamentals Handbook*

20.25.0 Equivalent Rectangular Duct Dimension Tables (Continued)

Duct Diameter, in.	Rectangular Size, in.	Aspect Ratio														
		1.00	1.25	1.50	1.75	2.00	2.25	2.50	2.75	3.00	3.50	4.00	5.00	6.00	7.00	8.00
40	Width	37	41	45	49	52	56	60	63	66	70	76	90	96	105	120
	Height	37	33	30	28	26	25	24	23	22	20	19	18	16	15	15
42	Width	38	43	48	51	56	59	63	66	69	74	80	90	102	112	120
	Height	38	34	32	29	28	26	25	24	23	21	20	18	17	16	15
44	Width	40	45	50	54	58	61	65	69	72	81	84	95	108	119	128
	Height	40	36	33	31	29	27	26	25	24	23	21	19	18	17	16
46	Width	42	48	53	56	60	65	68	72	75	84	88	100	114	126	136
	Height	42	38	35	32	30	29	27	26	25	24	22	20	19	18	17
48	Width	44	49	54	60	62	68	70	74	78	88	92	105	120	126	136
	Height	44	39	36	34	31	30	28	27	26	25	23	21	20	18	17
50	Width	46	51	57	61	66	70	75	77	81	91	96	110	120	133	144
	Height	46	41	38	35	33	31	30	28	27	26	24	22	20	19	18
52	Width	48	54	59	63	68	72	78	83	84	95	100	115	126	140	152
	Height	48	43	39	36	34	32	31	30	28	27	25	23	21	20	19
54	Width	49	55	62	67	70	77	80	85	90	98	104	120	132	147	160
	Height	49	44	41	38	35	34	32	31	30	28	26	24	22	21	20
56	Width	51	58	63	68	74	79	83	88	93	102	108	125	138	147	160
	Height	51	46	42	39	37	35	33	32	31	29	27	25	23	21	20
58	Width	53	60	66	70	76	81	85	91	96	105	112	130	144	154	168
	Height	53	48	44	40	38	36	34	33	32	30	28	26	24	22	21
60	Width	55	61	68	74	78	83	90	94	99	109	116	130	144	161	
	Height	55	49	45	42	39	37	36	34	33	31	29	26	24	23	
62	Width	57	64	71	75	82	88	93	96	102	112	120	135	150	168	
	Height	57	51	47	43	41	39	37	35	34	32	30	27	25	24	
64	Width	59	65	72	79	84	90	95	99	105	116	124	140	156		
	Height	59	52	48	45	42	40	38	36	35	33	31	28	26		
66	Width	60	68	75	81	86	92	98	105	108	119	128	145	162		
	Height	60	54	50	46	43	41	39	38	36	34	32	29	27		
68	Width	62	70	77	82	90	95	100	107	111	123	132	150	168		
	Height	62	56	51	47	45	42	40	39	37	35	33	30	28		
70	Width	64	71	80	86	92	99	105	110	114	126	136	155			
	Height	64	57	53	49	46	44	42	40	38	36	34	31			
72	Width	66	74	81	88	94	101	108	113	117	130	140	160			
	Height	66	59	54	50	47	45	43	41	39	37	35	32			
74	Width	68	76	84	91	98	104	110	116	123	133	144	165			
	Height	68	61	56	52	49	46	44	42	41	38	36	33			
76	Width	70	78	86	93	100	106	113	118	126	137	148	165			
	Height	70	62	57	53	50	47	45	43	42	39	37	33			
78	Width	71	80	89	95	102	110	115	121	129	140	152				
	Height	71	64	59	54	51	49	46	44	43	40	38				
80	Width	73	83	90	98	104	113	118	124	132	144	156				
	Height	73	66	60	56	52	50	47	45	44	41	39				
82	Width	75	84	93	100	108	115	123	129	135	147	160				
	Height	75	67	62	57	54	51	49	47	45	42	40				
84	Width	77	86	95	103	110	117	125	132	138	151	164				
	Height	77	69	63	59	55	52	50	48	46	43	41				
86	Width	79	88	98	105	112	119	128	135	141	154	168				
	Height	79	70	65	60	56	53	51	49	47	44	42				
88	Width	80	90	99	107	116	124	130	138	144	158					
	Height	80	72	66	61	58	55	52	50	48	45					
90	Width	82	93	102	110	118	126	133	140	147	161					
	Height	82	74	68	63	59	56	53	51	49	46					
92	Width	84	94	104	112	120	128	138	143	150	165					
	Height	84	75	69	64	60	57	55	52	50	47					
94	Width	86	96	107	116	124	131	140	146	153	168					
	Height	86	77	71	66	62	58	56	53	51	48					
96	Width	88	99	108	117	126	135	143	151	159						
	Height	88	79	72	67	63	60	57	55	53						
98	Width	90	100	111	119	128	137	145	154	162						
	Height	90	80	74	68	64	61	58	56	54						
100	Width	91	103	113	123	132	140	148	157	165						
	Height	91	82	75	70	66	62	59	57	55						
102	Width	93	105	116	124	134	142	153	160	168						
	Height	93	84	77	71	67	63	61	58	56						
104	Width	95	106	117	128	136	146	155	162							
	Height	95	85	78	73	68	65	62	59							

•Shaded area not recommended.

20.25.0 Equivalent Rectangular Duct Dimension Tables (Continued)

Duct Diameter, in.	Rectangular Size, in.	Aspect Ratio 1.00	1.25	1.50	1.75	2.00	2.25	2.50	2.75	3.00	3.50	4.00	5.00	6.00	7.00	8.00
106	Width	97	109	120	130	140	149	158	165							
	Height	97	87	80	74	70	66	63	60							
108	Width	99	110	122	131	142	151	160	168							
	Height	99	88	81	75	71	67	64	61							
110	Width	101	113	125	135	144	153	163								
	Height	101	90	83	77	72	68	65								
112	Width	102	115	126	137	146	158	165								
	Height	102	92	84	78	73	70	66								
114	Width	104	116	129	140	150	160									
	Height	104	93	86	80	75	71									
116	Width	106	119	131	142	152	162									
	Height	106	95	87	81	76	72									
118	Width	108	121	134	144	154	164									
	Height	108	97	89	82	77	73									
120	Width	110	123	135	147	158										
	Height	110	98	90	84	79										

*Shaded area not recommended.

By permission of American Society of Heating, Refrigerating and Air-Conditioning Engineers, Inc. Atlanta, Georgia, from their *1993 ASHRAE Fundamentals Handbook*

20.25.1 Equivalent Spiral, Flat, Oval Duct Dimensions

Duct Diameter, in.	Major Axis (a), in. / Minor Axis (b), in. 3	4	5	6	7	8	9	10	11	12	14	16	18	20	22	24
5	8															
5.5	9	7														
6	11	9														
6.5	12	10	8													
7	15	12	10	8												
7.5	19	13	—	9												
8	22	15	11	—												
8.5		18	13	11	10											
9		20	14	12	—	10										
9.5		21	18	14	12	—										
10			19	15	13	11										
10.5			21	17	15	13	12									
11				19	16	14	—	12								
11.5				20	18	16	14	—								
12				23	20	17	15	13								
12.5				25	21	—	—	15	14							
13				28	23	19	17	16	—	14						
13.5				30	—	21	18	—	16	—						
14				33	—	22	20	18	17	15						
14.5				36	—	24	22	19	—	17						
15				39	—	27	23	21	19	18						
16				45	—	30	—	24	22	20	17					
17				52	—	35	—	27	24	21	19					
18				59	—	39	—	30	—	25	22	19				
19						46	—	34	—	28	23	21				
20						50	—	38	—	31	27	24	21			
21						58	—	43	—	34	28	25	23			
22						65	—	48	—	37	31	29	26			
23						71	—	52	—	42	34	30	27			
24						77	—	57	—	45	38	33	29	26		
25								63	—	50	41	36	32	29		
26								70	—	56	45	38	34	31		
27								76	—	59	49	41	37	34		
28										65	52	46	40	36		
29										72	58	49	43	39	35	
30										78	61	54	46	40	38	
31										81	67	57	49	44	39	37
32											71	60	53	47	42	40
33											77	66	56	51	46	41
34												69	59	55	47	44
35												76	65	58	50	46
36												79	68	61	53	49
37													71	64	57	52
38													78	67	60	55
40														77	69	62
42															75	68
44															82	74

By permission of American Society of Heating, Refrigerating and Air-Conditioning Engineers, Inc. Atlanta, Georgia, from their *1993 ASHRAE Fundamentals Handbook*

20.26.0 Typical Fan Configurations

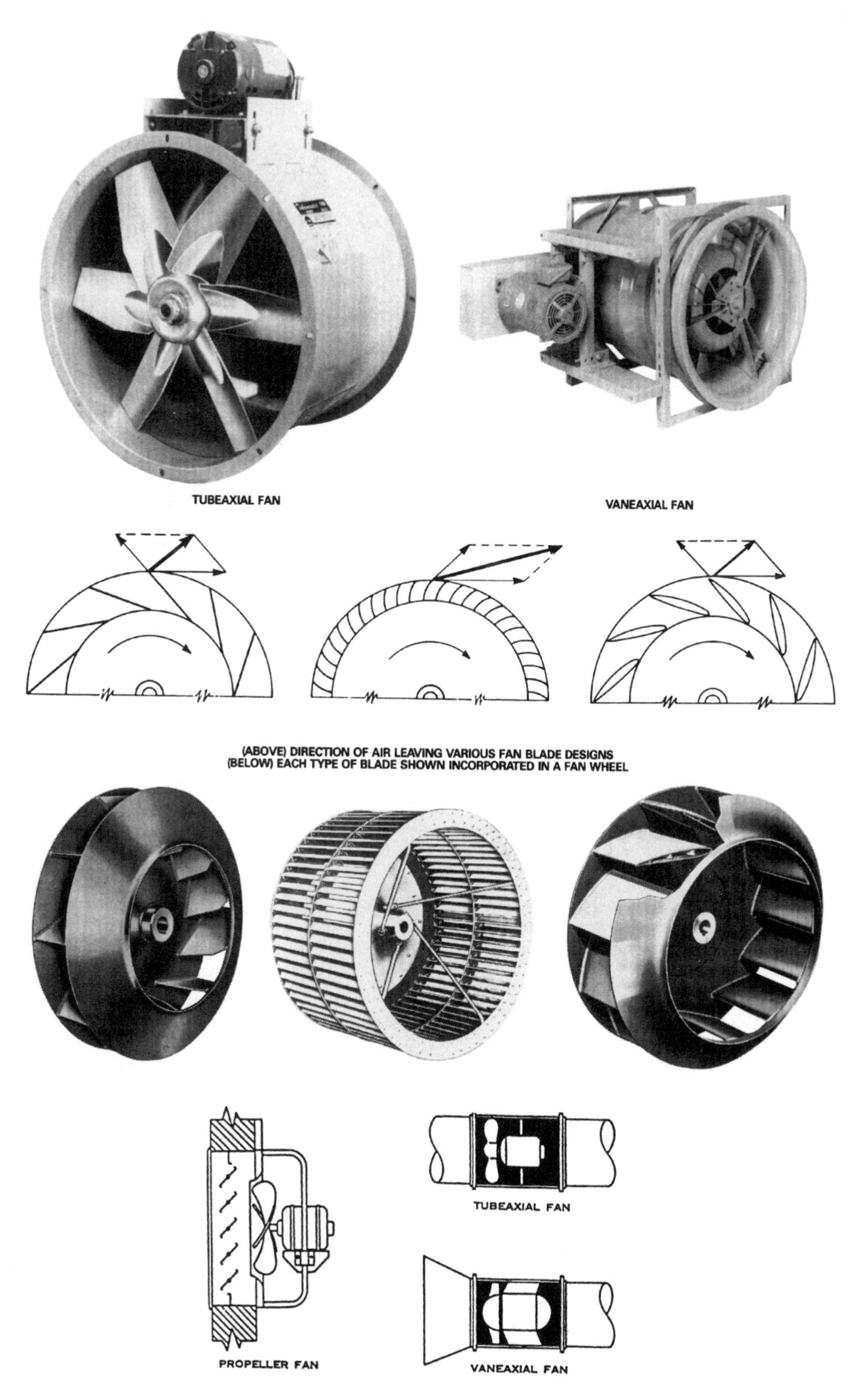

By permission of The Trane Company, La Crosse, Wisconsin

20.27.0 Rate of Heat Gain from Selected Office Equipment

Appliance	Size	Maximum Input Rating, Btu/h	Standby Input Rating, Btu/h	Recommended Rate of Heat Gain, Btu/h
Check processing workstation	12 pockets	16400	8410	8410
Computer devices				
Card puncher		2730 to 6140	2200 to 4800	2200 to 4800
Card reader		7510	5200	5200
Communication/transmission		6140 to 15700	5600 to 9600	5600 to 9600
Disk drives/mass storage		3410 to 34100	3412 to 22420	3412 to 22420
Magnetic ink reader		3280 to 16000	2600 to 14400	2600 to 14400
Microcomputer	16 to 640 KByte[a]	340 to 2050	300 to 1800	300 to 1800
Minicomputer		7500 to 15000	7500 to 15000	7500 to 15000
Optical reader		10240 to 20470	8000 to 17000	8000 to 17000
Plotters		256	128	214
Printers				
Letter quality	30 to 45 char/min	1200	600	1000
Line, high speed	5000 or more lines/min	4300 to 18100	2160 to 9040	2500 to 13000
Line, low speed	300 to 600 lines/min	1540	770	1280
Tape drives		4090 to 22200	3500 to 15000	3500 to 15000
Terminal		310 to 680	270 to 600	270 to 600
Copiers/Duplicators				
Blue print		3930 to 42700	1710 to 17100	3930 to 42700
Copiers (large)	30 to 67[a] copies/min	5800 to 22500	3070	5800 to 22500
Copiers (small)	6 to 30[a] copies/min	1570 to 5800	1020 to 3070	1570 to 5800
Feeder		100	—	100
Microfilm printer		1540	—	1540
Sorter/collator		200 to 2050	—	200 to 2050
Electronic equipment				
Cassette recorders/players		200	—	200
Receiver/tuner		340	—	340
Signal analyzer		90 to 2220	—	90 to 2220
Mailprocessing				
Folding machine		430	—	270
Inserting machine	3600 to 6800 pieces/h	2050 to 11300	—	1330 to 7340
Labeling machine	1500 to 30000 pieces/h	2050 to 22500	—	1330 to 14700
Postage meter		780	—	510
Wordprocessors/Typewriters				
Letter quality printer	30 to 45 char/min	1200	600	1000
Phototypesetter		5890	—	5180
Typewriter		270	—	230
Wordprocessor		340 to 2050	—	300 to 1800
Vending machines				
Cigarette		250	51 to 85	250
Cold food/beverage		3920 to 6550	—	1960 to 3280
Hot beverage		5890	—	2940
Snack		820 to 940	—	820 to 940
Miscellaneous				
Barcode printer		1500	—	1260
Cash registers		200	—	160
Coffee maker	10 cups	5120	—	3580 sensible 1540 latent
Microfiche reader		290	—	290
Microfilm reader		1770	—	1770
Microfilm reader/printer		3920	—	3920
Microwave oven	1 ft^3	2050	—	1360
Paper shredder		850 to 10240	—	680 to 8250
Water cooler	32 qt/h	2390	—	5970

[a] Input is not proportional to capacity.

By permission of American Society of Heating, Refrigerating and Air-Conditioning Engineers, Inc. Atlanta, Georgia, from their *1993 ASHRAE Fundamentals Handbook*

20.28.0 Thermal Properties of Common Building Materials

Description	Density, lb/ft³	Conductivity[b] (k), Btu·in / h·ft²·°F	Conductance (C), Btu / h·ft²·°F	Resistance[c] (R) Per Inch Thickness (1/k), °F·ft²·h / Btu·in	Resistance[c] (R) For Thickness Listed (1/C), °F·ft²·h / Btu	Specific Heat, Btu / lb·°F
Brick, fired clay *continued*	100	4.2–5.1	—	0.24–0.20	—	—
	90	3.6–4.3	—	0.28–0.24	—	—
	80	3.0–3.7	—	0.33–0.27	—	—
	70	2.5–3.1	—	0.40–0.33	—	—
Clay tile, hollow						
1 cell deep ... 3 in.	—	—	1.25	—	0.80	0.21
1 cell deep ... 4 in.	—	—	0.90	—	1.11	—
2 cells deep ... 6 in.	—	—	0.66	—	1.52	—
2 cells deep ... 8 in.	—	—	0.54	—	1.85	—
2 cells deep ... 10 in.	—	—	0.45	—	2.22	—
3 cells deep ... 12 in.	—	—	0.40	—	2.50	—
Concrete blocks[i]						
Limestone aggregate						
8 in., 36 lb, 138 lb/ft³ concrete, 2 cores	—	—	—	—	—	—
Same with perlite filled cores	—	—	0.48	—	2.1	—
12 in., 55 lb, 138 lb/ft³ concrete, 2 cores	—	—	—	—	—	—
Same with perlite filled cores	—	—	0.27	—	3.7	—
Normal weight aggregate (sand and gravel)						
8 in., 33-36 lb, 126-136 lb/ft³ concrete, 2 or 3 cores	—	—	0.90-1.03	—	1.11-0.97	0.22
Same with perlite filled cores	—	—	0.50	—	2.0	—
Same with verm. filled cores	—	—	0.52-0.73	—	1.92-1.37	—
12 in., 50 lb, 125 lb/ft³ concrete, 2 cores	—	—	0.81	—	1.23	0.22
Medium weight aggregate (combinations of normal weight and lightweight aggregate)						
8 in., 26-29 lb, 97-112 lb/ft³ concrete, 2 or 3 cores	—	—	0.58-0.78	—	1.71-1.28	—
Same with perlite filled cores	—	—	0.27-0.44	—	3.7-2.3	—
Same with verm. filled cores	—	—	0.30	—	3.3	—
Same with molded EPS (beads) filled cores	—	—	0.32	—	3.2	—
Same with molded EPS inserts in cores	—	—	0.37	—	2.7	—
Lightweight aggregate (expanded shale, clay, slate or slag, pumice)						
6 in., 16-17 lb 85-87 lb/ft³ concrete, 2 or 3 cores	—	—	0.52-0.61	—	1.93-1.65	—
Same with perlite filled cores	—	—	0.24	—	4.2	—
Same with verm. filled cores	—	—	0.33	—	3.0	—
8 in., 19-22 lb, 72-86 lb/ft³ concrete,	—	—	0.32-0.54	—	3.2-1.90	0.21
Same with perlite filled cores	—	—	0.15-0.23	—	6.8-4.4	—
Same with verm. filled cores	—	—	0.19-0.26	—	5.3-3.9	—
Same with molded EPS (beads) filled cores	—	—	0.21	—	4.8	—
Same with UF foam filled cores	—	—	0.22	—	4.5	—
Same with molded EPS inserts in cores	—	—	0.29	—	3.5	—
12 in., 32-36 lb, 80-90 lb/ft³ concrete, 2 or 3 cores	—	—	0.38-0.44	—	2.6-2.3	—
Same with perlite filled cores	—	—	0.11-0.16	—	9.2-6.3	—
Same with verm. filled cores	—	—	0.17	—	5.8	—
Stone, lime, or sand						
Quartzitic and sandstone	180	72	—	0.01	—	—
	160	43	—	0.02	—	—
	140	24	—	0.04	—	—
	120	13	—	0.08	—	0.19
Calcitic, dolomitic, limestone, marble, and granite	180	30	—	0.03	—	—
	160	22	—	0.05	—	—
	140	16	—	0.06	—	—
	120	11	—	0.09	—	0.19
	100	8	—	0.13	—	—
Gypsum partition tile						
3 by 12 by 30 in., solid	—	—	0.79	—	1.26	0.19
3 by 12 by 30 in., 4 cells	—	—	0.74	—	1.35	—
4 by 12 by 30 in., 3 cells	—	—	0.60	—	1.67	—
Concretes						
Sand and gravel or stone aggregate concretes (concretes with more than 50% quartz or quartzite sand have conductivities in the higher end of the range)	150	10.0-20.0	—	0.10-0.05	—	—
	140	9.0-18.0	—	0.11-0.06	—	0.19-0.24
	130	7.0-13.0	—	0.14-0.08	—	—
Limestone concretes	140	11.1	—	0.09	—	—
	120	7.9	—	0.13	—	—
	100	5.5	—	0.18	—	—
Gypsum-fiber concrete (87.5% gypsum, 12.5% wood chips)	51	1.66	—	0.60	—	0.21
Cement/lime, mortar, and stucco	120	9.7	—	0.10	—	—
	100	6.7	—	0.15	—	—
	80	4.5	—	0.22	—	—
Lightweight aggregate concretes						
Expanded shale, clay, or slate; expanded slags; cinders; pumice (with density up to 100 lb/ft³); and scoria (sanded concretes have conductivities in the higher end of the range)	120	6.4-9.1	—	0.16-0.11	—	—
	100	4.7-6.2	—	0.21-0.16	—	0.20
	80	3.3-4.1	—	0.30-0.24	—	0.20
	60	2.1-2.5	—	0.48-0.40	—	—
	40	1.3	—	0.78	—	—

By permission of American Society of Heating, Refrigerating and Air-Conditioning Engineers, Inc. Atlanta, Georgia, from their *1993 ASHRAE Fundamentals Handbook*

20.28.0 Thermal Properties of Common Building Materials (Continued)

Description	Density, lb/ft³	Conductivity[b] (k), Btu·in / h·ft²·°F	Conductance (C), Btu / h·ft²·°F	Resistance[c] (R): Per Inch Thickness (1/k), °F·ft²·h / Btu·in	Resistance[c] (R): For Thickness Listed (1/C), °F·ft²·h / Btu	Specific Heat, Btu / lb·°F
Expanded polystyrene, molded beads	1.0	0.26	—	3.85	—	—
	1.25	0.25	—	4.00	—	—
	1.5	0.24	—	4.17	—	—
	1.75	0.24	—	4.17	—	—
	2.0	0.23	—	4.35	—	—
Cellular polyurethane/polyisocyanurate[i] (CFC-11 exp.) (unfaced)	1.5	0.16–0.18	—	6.25–5.56	—	0.38
Cellular polyisocyanurate[i] (CFC-11 exp.)(gas-permeable facers)	1.5–2.5	0.16–0.18	—	6.25–5.56	—	0.22
Cellular polyisocyanurate[j] (CFC-11 exp.)(gas-impermeable facers)	2.0	0.14	—	7.04	—	0.22
Cellular phenolic (closed cell)(CFC-11, CFC-113 exp.)	3.0	0.12	—	8.20	—	—
Cellular phenolic (open cell)	1.8–2.2	0.23	—	4.40	—	—
Mineral fiber with resin binder	15.0	0.29	—	3.45	—	0.17
Mineral fiberboard, wet felted						
Core or roof insulation	16–17	0.34	—	2.94	—	—
Acoustical tile	18.0	0.35	—	2.86	—	0.19
Acoustical tile	21.0	0.37	—	2.70	—	—
Mineral fiberboard, wet molded						
Acoustical tile[k]	23.0	0.42	—	2.38	—	0.14
Wood or cane fiberboard						
Acoustical tile,[k] 0.5 in.	—	—	0.80	—	1.25	0.31
Acoustical tile[k] 0.75 in.	—	—	0.53	—	1.89	—
Interior finish (plank, tile)	15.0	0.35	—	2.86	—	0.32
Cement fiber slabs (shredded wood with Portland cement binder)	25–27.0	0.50–0.53	—	2.0–1.89	—	—
Cement fiber slabs (shredded wood with magnesia oxysulfide binder)	22.0	0.57	—	1.75	—	0.31
Loose Fill						
Cellulosic insulation (milled paper or wood pulp)	2.3–3.2	0.27–0.32	—	3.70–3.13	—	0.33
Perlite, expanded	2.0–4.1	0.27–0.31	—	3.7–3.3	—	0.26
	4.1–7.4	0.31–0.36	—	3.3–2.8	—	—
	7.4–11.0	0.36–0.42	—	2.8–2.4	—	—
Mineral fiber (rock, slag, or glass)[g]						
approx. 3.75–5 in.	0.6–2.0	—	—	—	11.0	0.17
approx. 6.5–8.75 in.	0.6–2.0	—	—	—	19.0	—
approx. 7.5–10 in.	0.6–2.0	—	—	—	22.0	—
approx. 10.25–13.75 in.	0.6–2.0	—	—	—	30.0	—
Mineral fiber (rock, slag, or glass)[g] approx. 3.5 in. (closed sidewall application)	2.0–3.5	—	—	—	12.0–14.0	—
Vermiculite, exfoliated	7.0–8.2	0.47	—	2.13	—	0.32
	4.0–6.0	0.44	—	2.27	—	—
Spray Applied						
Polyurethane foam	1.5–2.5	0.16–0.18	—	6.25–5.56	—	—
Ureaformaldehyde foam	0.7–1.6	0.22–0.28	—	4.55–3.57	—	—
Cellulosic fiber	3.5–6.0	0.29–0.34	—	3.45–2.94	—	—
Glass fiber	3.5–4.5	0.26–0.27	—	3.85–3.70	—	—
METALS (See Chapter 36, Table 3)						
ROOFING						
Asbestos-cement shingles	120	—	4.76	—	0.21	0.24
Asphalt roll roofing	70	—	6.50	—	0.15	0.36
Asphalt shingles	70	—	2.27	—	0.44	0.30
Built-up roofing 0.375 in.	70	—	3.00	—	0.33	0.35
Slate 0.5 in.	—	—	20.00	—	0.05	0.30
Wood shingles, plain and plastic film faced	—	—	1.06	—	0.94	0.31
PLASTERING MATERIALS						
Cement plaster, sand aggregate	116	5.0	—	0.20	—	0.20
Sand aggregate 0.375 in.	—	—	13.3	—	0.08	0.20
Sand aggregate 0.75 in.	—	—	6.66	—	0.15	0.20
Gypsum plaster:						
Lightweight aggregate 0.5 in.	45	—	3.12	—	0.32	—
Lightweight aggregate 0.625 in.	45	—	2.67	—	0.39	—
Lightweight aggregate on metal lath 0.75 in.	—	—	2.13	—	0.47	—
Perlite aggregate	45	1.5	—	0.67	—	0.32
Sand aggregate	105	5.6	—	0.18	—	0.20
Sand aggregate 0.5 in.	105	—	11.10	—	0.09	—
Sand aggregate 0.625 in.	105	—	9.10	—	0.11	—
Sand aggregate on metal lath 0.75 in.	—	—	7.70	—	0.13	—
Vermiculite aggregate	45	1.7	—	0.59	—	—
MASONRY MATERIALS						
Masonry Units						
Brick, fired clay	150	8.4–10.2	—	0.12–0.10	—	—
	140	7.4–9.0	—	0.14–0.11	—	—
	130	6.4–7.8	—	0.16–0.12	—	—
	120	5.6–6.8	—	0.18–0.15	—	0.19
	110	4.9–5.9	—	0.20–0.17	—	—

20.28.0 Thermal Properties of Common Building Materials (Continued)

Description	Density, lb/ft³	Conductivity[b] (k), Btu·in / h·ft²·°F	Conductance (C), Btu / h·ft²·°F	Resistance[c] (R) Per Inch Thickness (1/k), °F·ft²·h / Btu·in	For Thickness Listed (1/C), °F·ft²·h / Btu	Specific Heat, Btu / lb·°F
BUILDING BOARD						
Asbestos-cement board	120	4.0	—	0.25	—	0.24
Asbestos-cement board ....0.125 in.	120	—	33.00	—	0.03	
Asbestos-cement board ....0.25 in.	120	—	16.50	—	0.06	
Gypsum or plaster board ....0.375 in.	50	—	3.10	—	0.32	0.26
Gypsum or plaster board ....0.5 in.	50	—	2.22	—	0.45	
Gypsum or plaster board ....0.625 in.	50	—	1.78	—	0.56	
Plywood (Douglas Fir)[d]	34	0.80	—	1.25	—	0.29
Plywood (Douglas Fir) ....0.25 in.	34	—	3.20	—	0.31	
Plywood (Douglas Fir) ....0.375 in.	34	—	2.13	—	0.47	
Plywood (Douglas Fir) ....0.5 in.	34	—	1.60	—	0.62	
Plywood (Douglas Fir) ....0.625 in.	34	—	1.29	—	0.77	
Plywood or wood panels ....0.75 in.	34	—	1.07	—	0.93	0.29
Vegetable fiber board						
Sheathing, regular density[e] ....0.5 in.	18	—	0.76	—	1.32	0.31
....0.78125 in.	18	—	0.49	—	2.06	
Sheathing intermediate density[e] ....0.5 in.	22	—	0.92	—	1.09	0.31
Nail-base sheathing[e] ....0.5 in.	25	—	0.94	—	1.06	0.31
Shingle backer ....0.375 in.	18	—	1.06	—	0.94	0.31
Shingle backer ....0.3125 in.	18	—	1.28	—	0.78	
Sound deadening board ....0.5 in.	15	—	0.74	—	1.35	0.30
Tile and lay-in panels, plain or acoustic	18	0.40	—	2.50	—	0.14
....0.5 in.	18	—	0.80	—	1.25	
....0.75 in.	18	—	0.53	—	1.89	
Laminated paperboard	30	0.50	—	2.00	—	0.33
Homogeneous board from repulped paper	30	0.50	—	2.00	—	0.28
Hardboard[e]						
Medium density	50	0.73	—	1.37	—	0.31
High density, service-tempered grade and service grade	55	0.82	—	1.22	—	0.32
High density, standard-tempered grade	63	1.00	—	1.00	—	0.32
Particleboard[e]						
Low density	37	0.71	—	1.41	—	0.31
Medium density	50	0.94	—	1.06	—	0.31
High density	62.5	1.18	—	0.85	—	0.31
Underlayment ....0.625 in.	40	—	1.22	—	0.82	0.29
Waferboard	37	0.63	—	1.59	—	—
Wood subfloor ....0.75 in.	—	—	1.06	—	0.94	0.33
BUILDING MEMBRANE						
Vapor—permeable felt	—	—	16.70	—	0.06	
Vapor—seal, 2 layers of mopped 15-lb felt	—	—	8.35	—	0.12	
Vapor—seal, plastic film	—	—	—	—	Negl.	
FINISH FLOORING MATERIALS						
Carpet and fibrous pad	—	—	0.48	—	2.08	0.34
Carpet and rubber pad	—	—	0.81	—	1.23	0.33
Cork tile ....0.125 in.	—	—	3.60	—	0.28	0.48
Terrazzo ....1 in.	—	—	12.50	—	0.08	0.19
Tile—asphalt, linoleum, vinyl, rubber	—	—	20.00	—	0.05	0.30
vinyl asbestos						0.24
ceramic						0.19
Wood, hardwood finish ....0.75 in.	—	—	1.47	—	0.68	
INSULATING MATERIALS						
Blanket and Batt[f,g]						
Mineral fiber, fibrous form processed from rock, slag, or glass						
approx. 3–4 in.	0.4–2.0	—	0.091	—	11	
approx. 3.5 in.	0.4–2.0	—	0.077	—	13	
approx. 3.5 in.	1.2–1.6	—	0.067	—	15	
approx. 5.5–6.5 in.	0.4–2.0	—	0.053	—	19	
approx. 5.5 in.	0.6–1.0	—	0.048	—	21	
approx. 6–7.5 in.	0.4–2.0	—	0.045	—	22	
approx. 8.25–10 in.	0.4–2.0	—	0.033	—	30	
approx. 10–13 in.	0.4–2.0	—	0.026	—	38	
Board and Slabs						
Cellular glass	8.0	0.33	—	3.03	—	0.18
Glass fiber, organic bonded	4.0–9.0	0.25	—	4.00	—	0.23
Expanded perlite, organic bonded	1.0	0.36	—	2.78	—	0.30
Expanded rubber (rigid)	4.5	0.22	—	4.55	—	0.40
Expanded polystyrene, extruded (smooth skin surface) (CFC-12 exp.)	1.8–3.5	0.20	—	5.00	—	0.29
Expanded polystyrene, extruded (smooth skin surface) (HCFC-142b exp.)[h]	1.8–3.5	0.20	—	5.00	—	0.29

By permission of American Society of Heating, Refrigerating and Air-Conditioning Engineers, Inc. Atlanta, Georgia, from their *1993 ASHRAE Fundamentals Handbook*

Section

21

Electrical

Contents

21.0.0 Common electrical terminology
21.1.0 Conductor properties (AWG size 18 to 2000)
21.2.0 Maximum number of conductors in trade sizes of conduit or tubing THWN, THHN standard conductors
21.3.0 Percent of cross-section of conduit and tubing for conductors
21.4.0 Maximum number of concentric stranded conductors in trade sizes of conduit or tubing (RHW and RHH conductors with outer covering)
21.5.0 Dimensions of rubber and thermoplastic-covered conductors
21.6.0 Maximum number of conductors in trade sizes of conduit or tubing for TW, XHHW, RHW conductors
21.7.0 Maximum number of fixture wires in trade sizes of conduit or tubing
21.8.0 Conductor size increases from copper to aluminum
21.9.0 Minimum radii bends in conduit
21.10.0 Aluminum building wire nominal dimensions
21.11.0 Expansion characteristics of PVC rigid nonmetallic conduit
21.12.0 Maximum number of compact conductors in conduit or tubing
21.13.0 Maximum rating of motor-branch circuit, short-circuit, and ground-fault protection devices
21.14.0 Maximum number of conductors allowed in metal boxes
21.15.0 Electrical duct bank sizes for one to nine ducts
21.16.0 Minimum cover requirements for 0- to 600-volt conductors
21.17.0 Demand loads for various types of residential electrical appliances
21.18.0 General lighting loads by occupancy
21.19.0 Selection of overcurrent protection and switching devices
21.20.0 Size of equipment and raceway grounding conductors for 15- to 400-amp overcurrent devices
21.21.0 Enclosures for nonhazardous locations (NEMA designations)
21.22.0 Enclosures for hazardous locations (NEMA designations)
21.23.0 Motor-controller enclosure types (indoor and outdoor use)
21.24.0 Voltage-drop tables for 6- and 12-volt equipment
21.25.0 Seismic restraints and bracing
21.26.0 Full load current (in amperes) for single-phase, two-phase, three-phase, and direct-current motors

21.0.0 Common Electrical Terminology

Amp (A)
A measurement of the rate of flow of electrons along a wire. If electricity can be likened to plumbing, amps would be the same as gallons-per-second. Watts ÷Volts = Amps.

American Wire Gauge (AWG)
AWG refers to common wire sizes and ratings.

CO/ALR
15 or 20 A devices which can be used with copper or aluminum wire. Higher-rated devices appropriate for direct connection to aluminum or copper wire are marked "AL-CU".

Circuit
The path electricity follows as it moves along a conductor. Branch circuits distribute power to the parts of the home where it's needed.

Circuit Breaker
A resettable safety device that automatically stops electrical flow in a circuit when an overload or short circuit occurs. Either circuit breakers or fuses are located in the home's load center.

Conductor
A material capable of carrying electricity's energy. Opposite of Insulator.

Current
The rate of flow of electrons through a conductor, measured in Amps.

Electron
An invisible particle of negatively-charged matter that moves at the speed of light through an electrical circuit.

Fed Spec
Devices which comply with Federal Specifications such as W-C-596 for connecting devices and W-S-896 for switches. Fed Spec Standards for switches and connecting devices include NEMA Performance Standards.

Fuse
A non-resettable safety device that automatically stops electrical flow in a circuit when an overload or short circuit occurs. Either fuses or circuit breakers are located in the home's load center.

Ground
Refers literally to earth which has an electrical potential (voltage) of zero.

Ground Fault Circuit Interrupter (GFCI or GFI)
A safety device that senses shock hazard to a far greater degree than fuses or circuit breakers. Automatically stops electrical flow in a circuit.

Grounding Wire
The conductor used to connect the electrical equipment to ground (or earth) at the service entrance point, minimizing the potential for electrical shock. Either clad in green insulation or unclad.

Hospital Grade
UL-established criteria for devices used in hospitals. To obtain that listing and carry the Hospital Grade green dot identification, devices must pass many of the same tests as those included in NEMA Performance Standards and must go beyond in ability to withstand impact, crushing and continuous torture without loss of grounding path continuity. The highest grade attainable is Hospital Grade.

Hot Wire
The ungrounded conductor that carries electricity from the utility to a load center, or from a branch circuit to a receptacle or switch. It is normally clad in red or black insulation.

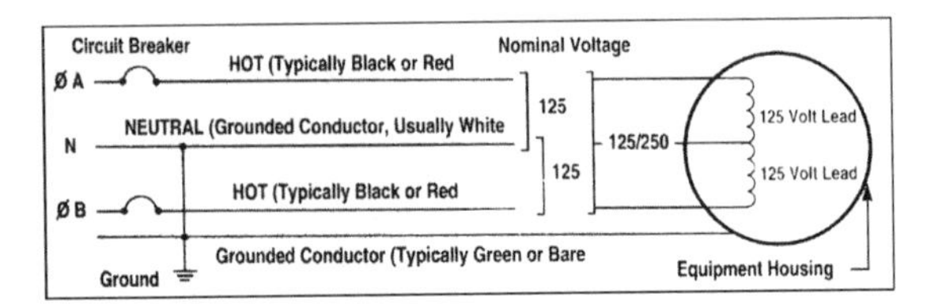

Insulation
A non-conductive covering that protects wires and other conductors of electricity.

Isolated Ground
In an isolated ground device, the grounding path is isolated from the device's mounting bracket. This "isolated ground" provides an electrical noise shield so that electromagnetic radiation waves will not turn into ground path noise which can disrupt sensitive electronics and can cause equipment malfunction.

Kilowatt (kw)
A thousand watts. (Watt is the measure of power that a electrical device consumes.) A kilowatt hour is the measurement most utilities use to measure electrical consumption. It indicates how many kilowatts are consumed for a full hour.

Knock-outs
Tabs that can be removed to make openings for wires and/or conduit in device and junction boxes or electrical panels.

Load Center
A home's fuse box or circuit breaker box. It divides the power into various branch circuits for distribution throughout the home.

21.1.0 Conductor Properties (AWG Size 18 to 2000)

Size AWG/ kcmil	Area Cir. Mils	Conductors				DC Resistance at 75°C (167°F)		
		Stranding		Overall		Copper		Aluminum
		Quantity	Diam. In.	Diam. In.	Area In.^{2}	Uncoated ohm/MFT	Coated ohm/MFT	ohm/MFT
18	1620	1	—	0.040	0.001	7.77	8.08	12.8
18	1620	7	0.015	0.046	0.002	7.95	8.45	13.1
16	2580	1	—	0.051	0.002	4.89	5.08	8.05
16	2580	7	0.019	0.058	0.003	4.99	5.29	8.21
14	4110	1	—	0.064	0.003	3.07	3.19	5.06
14	4110	7	0.024	0.073	0.004	3.14	3.26	5.17
12	6530	1	—	0.081	0.005	1.93	2.01	3.18
12	6530	7	0.030	0.092	0.006	1.98	2.05	3.25
10	10380	1	—	0.102	0.008	1.21	1.26	2.00
10	10380	7	0.038	0.116	0.011	1.24	1.29	2.04
8	16510	1	—	0.128	0.013	0.764	0.786	1.26
8	16510	7	0.049	0.146	0.017	0.778	0.809	1.28
6	26240	7	0.061	0.184	0.027	0.491	0.510	0.808
4	41740	7	0.077	0.232	0.042	0.308	0.321	0.508
3	52620	7	0.087	0.260	0.053	0.245	0.254	0.403
2	66360	7	0.097	0.292	0.067	0.194	0.201	0.319
1	83690	19	0.066	0.332	0.087	0.154	0.160	0.253
1/0	105600	19	0.074	0.373	0.109	0.122	0.127	0.201
2/0	133100	19	0.084	0.419	0.138	0.0967	0.101	0.159
3/0	167800	19	0.094	0.470	0.173	0.0766	0.0797	0.126
4/0	211600	19	0.106	0.528	0.219	0.0608	0.0626	0.100
250	—	37	0.082	0.575	0.260	0.0515	0.0535	0.0847
300	—	37	0.090	0.630	0.312	0.0429	0.0446	0.0707
350	—	37	0.097	0.681	0.364	0.0367	0.0382	0.0605
400	—	37	0.104	0.728	0.416	0.0321	0.0331	0.0529
500	—	37	0.116	0.813	0.519	0.0258	0.0265	0.0424
600	—	61	0.099	0.893	0.626	0.0214	0.0223	0.0353
700	—	61	0.107	0.964	0.730	0.0184	0.0189	0.0303
750	—	61	0.111	0.998	0.782	0.0171	0.0176	0.0282
800	—	61	0.114	1.03	0.834	0.0161	0.0166	0.0265
900	—	61	0.122	1.09	0.940	0.0143	0.0147	0.0235
1000	—	61	0.128	1.15	1.04	0.0129	0.0132	0.0212
1250	—	91	0.117	1.29	1.30	0.0103	0.0106	0.0169
1500	—	91	0.128	1.41	1.57	0.00858	0.00883	0.0141
1750	—	127	0.117	1.52	1.83	0.00735	0.00756	0.0121
2000	—	127	0.126	1.63	2.09	0.00643	0.00662	0.0106

These resistance values are valid ONLY for the parameters as given. Using conductors having coated strands, different stranding type, and especially, other temperatures, change the resistance.

Formula for temperature change: $R_2 = R_1 [1+\alpha(T_2-75)]$ where: $\alpha_{cu} = 0.00323$, $\alpha_{AL} = 0.00330$.

Conductors with compact and compressed stranding have about 9 percent and 3 percent, respectively, smaller bare conductor diameters than those shown. See Table 5A for actual compact cable dimensions.

The IACS conductivities used: bare copper = 100%, aluminum = 61%.

Class B stranding is listed as well as solid for some sizes. Its overall diameter and area is that of its circumscribing circle.

(FPN): The construction information is per NEMA WC8-1976 (Rev 5-1980). The resistance is calculated per National Bureau of Standards Handbook 100, dated 1966, and Handbook 109, dated 1972.

21.2.0 Maximum Number of Conductors in Trade Sizes of Conduit or Tubing THWN, THHN Standard Conductors

Conduit Trade Size (Inches)		½	¾	1	1¼	1½	2	2½	3	3½	4	5	6
Type Letters	Conductor Size AWG, kcmil												
THWN, THHN, FEP (14 through 2), FEPB (14 through 8), PFA (14 through 4/0) PFAH (14 through 4/0) Z (14 through 4/0) XHHW (4 through 500 kcmil)	14	13	24	39	69	94	154						
	12	10	18	29	51	70	114	164					
	10	6	11	18	32	44	73	104	160				
	8	3	5	9	16	22	36	51	79	106	136		
	6	1	4	6	11	15	26	37	57	76	98	154	
	4	1	2	4	7	9	16	22	35	47	60	94	137
	3	1	1	3	6	8	13	19	29	39	51	80	116
	2	1	1	3	5	7	11	16	25	33	43	67	97
	1		1	1	3	5	8	12	18	25	32	50	72
	1/0		1	1	3	4	7	10	15	21	27	42	61
	2/0		1	1	2	3	6	8	13	17	22	35	51
	3/0		1	1	1	3	5	7	11	14	18	29	42
	4/0		1	1	1	2	4	6	9	12	15	24	35
	250			1	1	1	3	4	7	10	12	20	28
	300			1	1	1	3	4	6	8	11	17	24
	350			1	1	1	2	3	5	7	9	15	21
	400				1	1	1	3	5	6	8	13	19
	500				1	1	1	2	4	5	7	11	16
	600				1	1	1	1	3	4	5	9	13
	700					1	1	1	3	4	5	8	11
	750					1	1	1	2	3	4	7	11
XHHW	6	1	3	5	9	13	21	30	47	63	81	128	185
	600				1	1	1	1	3	4	5	9	13
	700					1	1	1	3	4	5	7	11
	750					1	1	1	2	3	4	7	10

Note: This table is for concentric stranded conductors only. For cables with compact conductors, the dimensions in Table 5A shall be used.

21.3.0 Percent of Cross-Section of Conduit and Tubing for Conductors

Number of Conductors	1	2	3	4	Over 4
All conductor types except lead-covered	53	31	40	40	40
Lead-covered conductors	55	30	40	38	35

Note 1. See Tables 3A, 3B, and 3C for number of conductors all of the same size in trade sizes of conduit or tubing ½ inch through 6 inch.

Note 2. For conductors larger than 750 kcmil or for combinations of conductors of different sizes, use Tables 4 through 8, Chapter 9, for dimensions of conductors, conduit and tubing.

Note 3. Where the calculated number of conductors, all of the same size, includes a decimal fraction, the next higher whole number shall be used where this decimal is 0.8 or larger.

Note 4. When bare conductors are permitted by other sections of this Code, the dimensions for bare conductors in Table 8 of Chapter 9 shall be permitted.

Note 5. A multiconductor cable of two or more conductors shall be treated as a single conductor cable for calculating percentage conduit fill area. For cables that have elliptical cross section, the cross-sectional area calculation shall be based on using the major diameter of the ellipse as a circle diameter.

21.4.0 Maximum Number of Concentric Stranded Conductors in Trade Sizes of Conduit or Tubing (RHW and RHH Conductors with Outer Covering)

Conduit Trade Size (inches)		½	¾	1	1¼	1½	2	2½	3	3½	4	5	6
Type Letters	Conductor Size AWG, kcmil												
RHW,	14	3	6	10	18	25	41	58	90	121	155		
	12	3	5	9	15	21	35	50	77	103	132		
	10	2	4	7	13	18	29	41	64	86	110		
	8	1	2	4	7	9	16	22	35	47	60	94	137
RHH	6	1	1	2	5	6	11	15	24	32	41	64	93
	4	1	1	1	3	5	8	12	18	24	31	50	72
(with outer covering)	3	1	1	1	3	4	7	10	16	22	28	44	63
	2		1	1	3	4	6	9	14	19	24	38	56
	1		1	1	1	3	5	7	11	14	18	29	42
	1/0		1	1	1	2	4	6	9	12	16	25	37
	2/0			1	1	1	3	5	8	11	14	22	32
	3/0			1	1	1	3	4	7	9	12	19	28
	4/0			1	1	1	2	4	6	8	10	16	24
	250				1	1	1	3	5	6	8	13	19
	300				1	1	1	3	4	5	7	11	17
	350				1	1	1	2	4	5	6	10	15
	400				1	1	1	1	3	4	6	9	14
	500				1	1	1	1	3	4	5	8	11
	600					1	1	1	2	3	4	6	9
	700					1	1	1	1	3	3	6	8
	750						1	1	1	3	3	5	8

Note: This table is for concentric stranded conductors only.

21.5.0 Dimensions of Rubber- and Thermoplastic-Covered Conductors

	Types RFH-2, RH, RHH,*** RHW,*** SF-2		Types TF, THW,† TW		Types TFN, THHN, THWN		Types**** FEP, FEPB, FEPW, TFE, PF, PFA, PFAH, PGF, PTF, Z, ZF, ZFF		Type XHHW, ZW††		Types KF-1, KF-2, KFF-1, KFF-2	
Size AWG kcmil	Approx. Diam. Inches	Approx. Area Sq. In.	Approx. Diam. Inches	Approx. Area Sq. In.	Approx. Diam. Inches	Approx. Area Sq. In.	Approx. Diam. Inches	Approx. Area Sq. Inches	Approx. Diam. Inches	Approx. Area Sq. In.	Approx. Diam. Inches	Approx. Area Sq. In.
Col. 1	Col. 2	Col. 3	Col. 4	Col. 5	Col. 6	Col. 7	Col. 8	Col. 9	Col. 10	Col. 11	Col. 12	Col. 13
18	.146	.0167	.106	.0088	.089	.0062	.081	.0052	...	....	.065	.0033
16	.158	.0196	.118	.0109	.100	.0079	.092	.0066	...	....	.070	.0038
14	30 mils .171	.0230	.131	.0135	.105	.0087	.105 .105	.0087 .0087	...	....	.083	.0054
14	45 mils .204*	.0327*	...	....	...	....	... ...	....	...	....	...	....
14	....	....	.162†	.0206†	...	....	... ...	....	.129	.0131	...	....
12	30 mils .188	.0278	.148	.0172	.122	.0117	.121 .121	.0115 .0115	...	....	.102	.0082
12	45 mils .221*	.0384*	...	....	...	....	... ...	....	...	....	...	....
12	....	....	.179†	.0252†	...	....	... ...	....	.146	.0167	...	....
10	...... .242	.0460	.168	.0222	.153	.0184	.142 .142	.0158 .0158	...	....	.124	.0121
10	....	....	.199†	.0311†	...	....	... ...	....	.166	.0216	...	....
8	...... .328	.0845	.245	.0471	.218	.0373	.206 .186	.0333 .0272	...	....	...	....
8	....	....	.276†	.0598†	...	....	... ...	....	.241	.0456	...	....
6	.397	.1238	.323	.0819	.257	.0519	.244 .302	.0468 .0716	.282	.0625	...	....
4	.452	.1605	.372	.1087	.328	.0845	.292 .350	.0670 .0962	.328	.0845	...	....
3	.481	.1817	.401	.1263	.356	.0995	.320 .378	.0804 .1122	.356	.0995	...	....
2	.513	.2067	.433	.1473	.388	.1182	.352 .410	.0973 .1320	.388	.1182	...	....
1	.588	.2715	.508	.2027	.450*	.1590	.420 ...	.1385	.450	.1590	...	....
1/0	.629	.3107	.549	.2367	.491	.1893	.462 ...	.1676	.491	.1893	...	....
2/0	.675	.3578	.595	.2781	.537	.2265	.498 ...	.1948	.537	.2265	...	....
3/0	.727	.4151	.647	.3288	.588	.2715	.560 ...	.2463	.588	.2715	...	....
4/0	.785	.4840	.705	.3904	.646	.3278	.618 ...	.3000	.646	.3278	...	....

21.6.0 Maximum Number of Conductors in Trade Sizes of Conduit or Tubing for TW, XHHW, RHW Conductors

Type Letters	Conductor Size AWG, kcmil	½	¾	1	1¼	1½	2	2½	3	3½	4	5	6
TW, XHHW (14 through 8)	14	9	15	25	44	60	99	142					
	12	7	12	19	35	47	78	111	171				
	10	5	9	15	26	36	60	85	131	176			
	8	2	4	7	12	17	28	40	62	84	108		
RHW and RHH (without outer covering), THW	14	6	10	16	29	40	65	93	143	192			
	12	4	8	13	24	32	53	76	117	157			
	10	4	6	11	19	26	43	61	95	127	163		
	8	1	3	5	10	13	22	32	49	66	85	133	
TW,	6	1	2	4	7	10	16	23	36	48	62	97	141
	4	1	1	3	5	7	12	17	27	36	47	73	106
THW,	3	1	1	2	4	6	10	15	23	31	40	63	91
	2	1	1	2	4	5	9	13	20	27	34	54	78
	1		1	1	3	4	6	9	14	19	25	39	57
FEPB (6 through 2), RHW and RHH (without outer covering)	1/0		1	1	2	3	5	8	12	16	21	33	49
	2/0		1	1	1	3	5	7	10	14	18	29	41
	3/0		1	1	1	2	4	6	9	12	15	24	35
	4/0			1	1	1	3	5	7	10	13	20	29
	250			1	1	1	2	4	6	8	10	16	23
	300			1	1	1	2	3	5	7	9	14	20
	350				1	1	1	3	4	6	8	12	18
	400				1	1	1	2	4	5	7	11	16
	500				1	1	1	1	3	4	6	9	14
	600					1	1	1	3	4	5	7	11
	700					1	1	1	2	3	4	7	10
	750					1	1	1	2	3	4	6	9

Note: This table is for concentric stranded conductors only.

21.7.0 Maximum Number of Fixture Wires in Trade Sizes of Conduit or Tubing

(40 Percent Fill Based on Individual Diameters)

Conduit Trade Size (Inches)	½					¾					1					1¼					1½					2				
Wire Types	18	16	14	12	10	18	16	14	12	10	18	16	14	12	10	18	16	14	12	10	18	16	14	12	10	18	16	14	12	10
PTF, PTFF, PGFF, PGF, PFF, PF, PAF, PAFF, ZF, ZFF	23	18	14			40	31	24			65	50	39			115	90	70			157	122	95			257	200	156		
TFFN, TFN	19	15				34	26				55	43				97	76				132	104				216	169			
SF-1	16					29					47					83					114					186				
SFF-1	15					26					43					76					104					169				
TF	11	10				20	18				32	30				57	53				79	72				129	118			
RFH-1	11					20					32					57					79					129				
TFF	11	10				20	17				32	27				56	49				77	66				126	109			
AF	11	9	7	4	3	19	16	12	7	5	31	26	20	11	8	55	46	36	19	15	75	63	49	27	20	123	104	81	44	34
SFF-2	9	7	6			16	12	10			27	20	17			47	36	30			65	49	42			106	81	68		
SF-2	9	8	6			16	14	11			27	23	18			47	40	32			65	55	43			106	90	71		
FFH-2	9	7				15	12				25	19				44	34				60	46				99	75			
RFH-2	7	5				12	10				20	16				36	28				49	38				80	62			
KF-1, KFF-1, KF-2, KFF-2	36	32	22	14	9	64	55	39	25	17	103	89	63	41	28	182	158	111	73	49	248	216	152	100	67	406	353	248	163	110

21.8.0 Conductor Size Increases From Copper to Aluminum

When substituting aluminum, increase the sizes of conductors in accordance with the following table:

Copper Size Conductor	Minimum Substitute Aluminum Size Conductor
#2	2/0
#1	2/0
1/0	4/0
2/0	4/0
3/0	300 MCM
4/0	350 MCM
250 MCM	400 MCM
300 MCM	500 MCM
350 MCM	600 MCM
400 MCM	600 MCM
500 MCM	750 MCM

Utilization	Acceptable Types
Conductors #1 AWG and smaller	THWN, THHN, XHHW
Conductors 1/0 and larger in "dry" locations	THHW, THHN, XHHW
Conductors 1/0 and larger in "wet" locations	THHW-2, XHHW-2, THWN-2

21.9.0 Minimum Radii Bends in Conduit

Bends in conduit shall have minimum radii:

1. For primary feeder - - - - - - -15'-0", except where specifically indicated otherwise or where turning up at termination point.
2. For primary feeder turning up at termination point - - - - - - -4'-0"
3. For secondary feeder all bends. - - - -4'-0"
4. For communications and/or signal wiring all bends - - - - -4'-0"

21.10.0 Aluminum Building Wire Nominal Dimensions

Bare Conductor**			Type THW		Type THHN		Type XHHW		
Size AWG or kcmil	Number of Strands	Diam. Inches	Approx. Diam. Inches	Approx. Area Sq. In.	Approx. Diam. Inches	Approx. Area Sq. In.	Approx. Diam. Inches	Approx. Area Sq. In.	Size AWG or kcmil
8	7	.134	.255	.0510	—	—	.224	.0394	8
6	7	.169	.290	.0660	.240	.0452	.260	.0530	6
4	7	.213	.335	.0881	.305	.0730	.305	.0730	4
2	7	.268	.390	.1194	.360	.1017	.360	.1017	2
1	19	.299	.465	.1698	.415	.1352	.415	.1352	1
1/0	19	.336	.500	.1963	.450	.1590	.450	.1590	1/0
2/0	19	.376	.545	.2332	.495	.1924	.490	.1885	2/0
3/0	19	.423	.590	.2733	.540	.2290	.540	.2290	3/0
4/0	19	.475	.645	.3267	.595	.2780	.590	.2733	4/0
250	37	.520	.725	.4128	.670	.3525	.660	.3421	250
300	37	.570	.775	.4717	.720	.4071	.715	.4015	300
350	37	.616	.820	.5281	.770	.4656	.760	.4536	350
400	37	.659	.865	.5876	.815	.5216	.800	.5026	400
500	37	.736	.940	.6939	.885	.6151	.880	.6082	500
600	61	.813	1.050	.8659	.985	.7620	.980	.7542	600
700	61	.877	1.110	.9676	1.050	.8659	1.050	.8659	700
750	61	.908	1.150	1.0386	1.075	.9076	1.090	.9331	750
1000	61	1.060	1.285	1.2968	1.255	1.2370	1.230	1.1882	1000

* Dimensions are from industry sources
** Compact conductor per ASTM B 400

21.11.0 Expansion Characterisics of PVC Rigid Nonmetallic Conduit

Expansion Characteristics Of PVC Rigid Nonmetallic Conduit Coefficient Of Thermal Expansion = 3.38×10^{-5} In/In/°F

Temperature Change in Degrees F	Length Change in Inches per 100 ft. of PVC Conduit	Temperature Change in Degrees F	Length Change in Inches per 100 ft. of PVC Conduit	Temperature Change in Degrees F	Length Change in Inches per 100 ft. of PVC Conduit	Temperature Change in Degrees F	Length Change in Inches per 100 ft. of PVC Conduit
5	0.2	55	2.2	105	4.2	155	6.3
10	0.4	60	2.4	110	4.5	160	6.5
15	0.6	65	2.6	115	4.7	165	6.7
20	0.8	70	2.8	120	4.9	170	6.9
25	1.0	75	3.0	125	5.1	175	7.1
30	1.2	80	3.2	130	5.3	180	7.3
35	1.4	85	3.4	135	5.5	185	7.5
40	1.6	90	3.6	140	5.7	190	7.7
45	1.8	95	3.8	145	5.9	195	7.9
50	2.0	100	4.1	150	6.1	200	8.1

21.12.0 Maximum Number of Compact Conductors in Conduit or Tubing

Conductor Size AWG or kcmil	1 in. THW	1 in. THHN	1 in. XHHW	1¼ in. THW	1¼ in. THHN	1¼ in. XHHW	1½ in. THW	1½ in. THHN	1½ in. XHHW	2 in. THW	2 in. THHN	2 in. XHHW	2½ in. THW	2½ in. THHN	2½ in. XHHW	3 in. THW	3 in. THHN	3 in. XHHW	3½ in. THW	3½ in. THHN	3½ in. XHHW	4 in. THW	4 in. THHN	4 in. XHHW	Conductor Size AWG or kcmil
6	5	7	6	9	13	11	12	18	15																6
4	4	4	4	7	8	8	9	11	11	15	18	18													4
2	3	3	3	5	6	6	7	8	8	11	13	13													2
1				3	4	4	5	6	6	8	10	10	11	14	14										1
1/0				3	3	3	4	5	5	7	8	8	9	12	12										1/0
2/0					3	3	3	4	4	5	7	7	8	10	10	12	15	15							2/0
3/0							3	3	3	5	6	6	7	8	8	10	13	13							3/0
4/0								3	3	4	5	5	6	7	7	9	10	10	12	14	14				4/0
250										3	4	4	4	5	5	7	8	8	9	11	11				250
300										3	3	3	4	4	4	6	7	7	8	9	10	10	12	12	300
350											3	3	3	4	4	5	6	6	7	8	8	9	11	11	350
400													3	3	4	5	5	6	6	7	8	8	9	10	400
500														3	3	4	4	5	5	6	6	7	8	8	500
600																3	4	4	4	5	5	6	6	7	600
700																3	3	3	4	4	4	5	6	6	700
750																3	3	3	4	4	4	5	5	5	750
1000																			3	3	3	4	4	4	1000

21.13.0 Maximum Rating of Motor-Branch Circuit, Short-Circuit, and Ground-Fault Protection Devices

	Percent of Full-Load Current			
Type of Motor	Nontime Delay Fuse	Dual Element (Time-Delay) Fuse	Instantaneous Trip Breaker	* Inverse Time Breaker
Single-phase, all types				
No code letter	300	175	700	250
All ac single-phase and polyphase squirrel-cage and synchronous motors† with full-voltage, resistor or reactor starting:				
No code letter	300	175	700	250
Code letter F to V	300	175	700	250
Code letter B to E	250	175	700	200
Code letter A	150	150	700	150
All ac squirrel-cage and synchronous motors† with autotransformer starting:				
Not more than 30 amps				
No code letter	250	175	700	200
More than 30 amps				
No code letter	200	175	700	200
Code letter F to V	250	175	700	200
Code letter B to E	200	175	700	200
Code letter A	150	150	700	150
High-reactance squirrel-cage				
Not more than 30 amps				
No code letter	250	175	700	250
More than 30 amps				
No code letter	200	175	700	200
Wound-rotor —				
No code letter	150	150	700	150
Direct-current (constant voltage)				
No more than 50 hp				
No code letter	150	150	250	150
More than 50 hp				
No code letter	150	150	175	150

* The values given in the last column also cover the ratings of nonadjustable inverse time types of circuit breakers.

† Synchronous motors of the low-torque, low-speed type (usually 450 rpm or lower), such as are used to drive reciprocating compressors, pumps, etc. that start unloaded, do not require a fuse rating or circuit-breaker setting in excess of 200 percent of full-load current.

21.14.0 Maximum Number of Conductors Allowed in Metal Boxes

The maximum number of conductors permitted shall be computed using the volume per conductor listed in the table, with the deductions provided for, and these volume deductions shall be based on the largest conductor entering the box. Boxes described in the table have a larger cubic inch capacity than is designated in the table shall be permitted to have their cubic inch capacity marked as required by this section and the maximum number of conductors permitted shall be computed using the volume per conductor listed.

Metal Boxes

Box Dimension, Inches Trade Size or Type	Min. Cu. In. Cap.	Maximum Number of Conductors						
		No. 18	No. 16	No. 14	No. 12	No. 10	No. 8	No. 6
4 x 1¼ Round or Octagonal	12.5	8	7	6	5	5	4	2
4 x 1½ Round or Octagonal	15.5	10	8	7	6	6	5	3
4 x 2⅛ Round or Octagonal	21.5	14	12	10	9	8	7	4
4 x 1¼ Square	18.0	12	10	9	8	7	6	3
4 x 1½ Square	21.0	14	12	10	9	8	7	4
4 x 2⅛ Square	30.3	20	17	15	13	12	10	6
4¹¹⁄₁₆ x 1¼ Square	25.5	17	14	12	11	10	8	5
4¹¹⁄₁₆ x 1½ Square	29.5	19	16	14	13	11	9	5
4¹¹⁄₁₆ x 2⅛ Square	42.0	28	24	21	18	16	14	8
3 x 2 x 1½ Device	7.5	5	4	3	3	3	2	1
3 x 2 x 2 Device	10.0	6	5	5	4	4	3	2
3 x 2 x 2¼ Device	10.5	7	6	5	4	4	3	2
3 x 2 x 2½ Device	12.5	8	7	6	5	5	4	2
3 x 2 x 2¾ Device	14.0	9	8	7	6	5	4	2
3 x 2 x 3½ Device	18.0	12	10	9	8	7	6	3
4 x 2⅛ x 1½ Device	10.3	6	5	5	4	4	3	2
4 x 2⅛ x 1⅞ Device	13.0	8	7	6	5	5	4	2
4 x 2⅛ x 2⅛ Device	14.5	9	8	7	6	5	4	2
3¾ x 2 x 2½ Masonry Box/Gang	14.0	9	8	7	6	5	4	2
3¾ x 2 x 3½ Masonry Box/Gang	21.0	14	12	10	9	8	7	4
FS—Minimum Internal Depth 1¾ Single Cover/Gang	13.5	9	7	6	6	5	4	2
FD—Minimum Internal Depth 2⅜ Single Cover/Gang	18.0	12	10	9	8	7	6	3
FS—Minimum Internal Depth 1¾ Multiple Cover/Gang	18.0	12	10	9	8	7	6	3
FD—Minimum Internal Depth 2⅜ Multiple Cover/Gang	24.0	16	13	12	10	9	8	4

Volume Required per Conductor

Size of Conductor	Free Space Within Box for Each Conductor
No. 18	1.5 cubic inches
No. 16	1.75 cubic inches
No. 14	2. cubic inches
No. 12	2.25 cubic inches
No. 10	2.5 cubic inches
No. 8	3. cubic inches
No. 6	5. cubic inches

21.15.0 Electrical Duct Bank Sizes for One to Nine Ducts

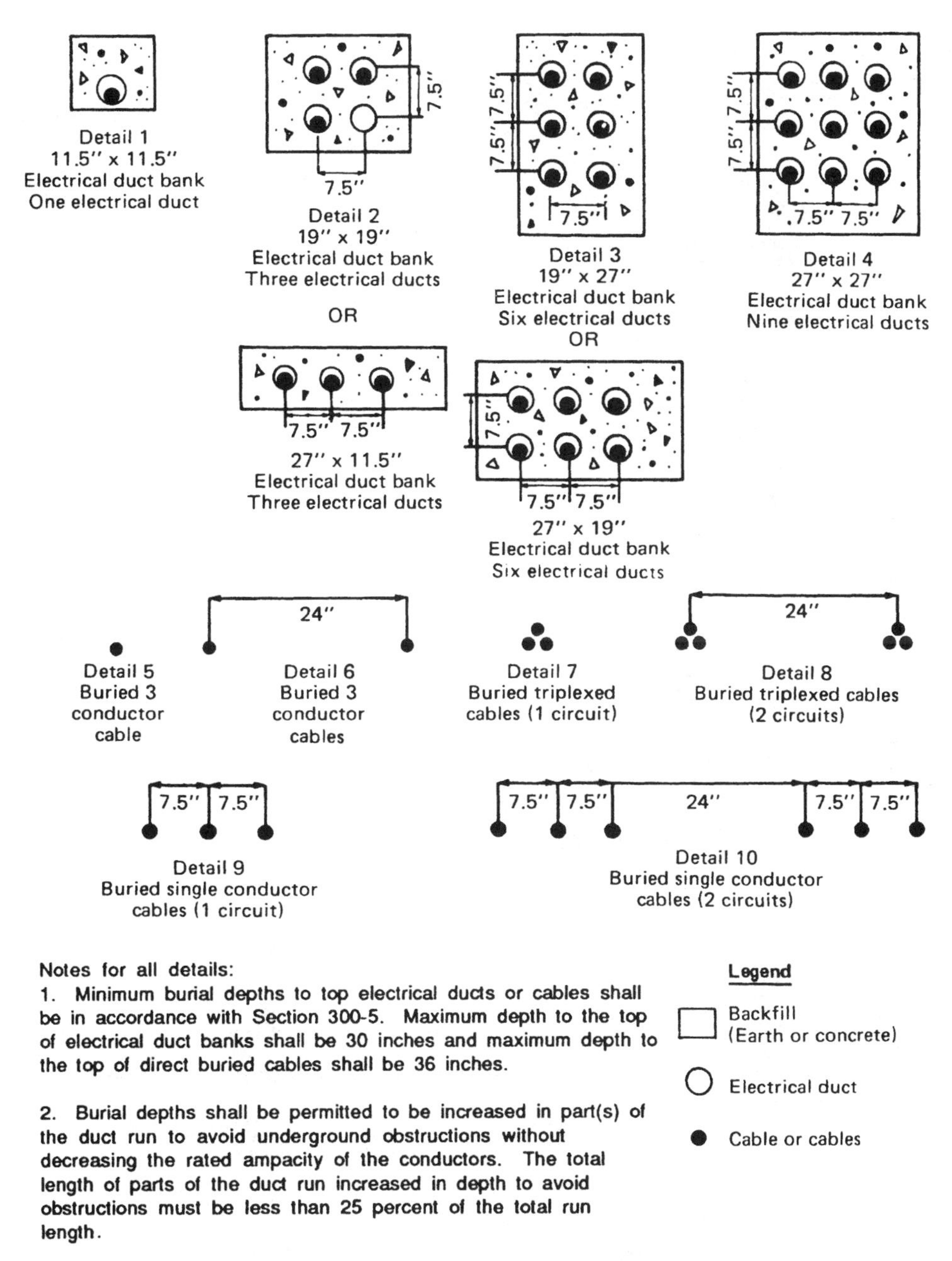

Notes for all details:

1. Minimum burial depths to top electrical ducts or cables shall be in accordance with Section 300-5. Maximum depth to the top of electrical duct banks shall be 30 inches and maximum depth to the top of direct buried cables shall be 36 inches.

2. Burial depths shall be permitted to be increased in part(s) of the duct run to avoid underground obstructions without decreasing the rated ampacity of the conductors. The total length of parts of the duct run increased in depth to avoid obstructions must be less than 25 percent of the total run length.

3. For SI units: one inch = 25.4 millimeters; one foot = 305 millimeters.

21.16.0 Minimum Cover Requirements for 0 to 600-Volt Conductors

(Cover is defined as the shortest distance measured between a point on the top surface of any direct buried conductor, cable, conduit or other raceway and the top surface of finished grade, concrete, or similar cover.)

Location of Wiring Method or Circuit	Type of Wiring Method or Circuit				
	Direct Burial Cables or Conductors	Rigid Metal Conduit or Intermediate Metal Conduit	Rigid Nonmetallic Conduit Approved for Direct Burial Without Concrete Encasement or Other Approved Raceways	Residential Branch Circuits Rated 120 Volts or less with GFCI Protection and Maximum Overcurrent Protection of 20 Amperes	Circuits for Control of Irrigation and Landscape Lighting Limited to Not More than 30 Volts and Installed with Type UF or in Other Identified Cable or Raceway
All Locations Not Specified Below	24	6	18	12	6
In Trench Below 2-Inch Thick Concrete or Equivalent	18	6	12	6	6
Under a Building	0 (In Raceway Only)	0	0	0 (In Raceway Only)	0 (In Raceway Only)
Under Minimum of 4-Inch Thick Concrete Exterior Slab with no vehicular traffic and the slab extending not less than 6 inches beyond the underground installation	18	4	4	6 (Direct Burial) 4 (In Raceway)	6 (Direct Burial) 4 (In Raceway)
Under Streets, Highways, Roads, Alleys, Driveways, and Parking Lots	24	24	24	24	24
One- and Two-Family Dwelling Driveways and Parking areas, and Used for No Other Purpose	18	18	18	12	18
In or Under Airport Runways Including Adjacent Areas Where Trespassing Prohibited	18	18	18	18	18
In Solid Rock Where Covered by Minimum of 2 Inches Concrete Extending Down to Rock	2 (In Raceway Only)	2	2	2 (In Raceway Only)	2 (In Raceway Only)

Note 1. For SI Units: one inch = 25.4 millimeters
Note 2. Raceways approved for burial only where concrete encased shall require concrete envelope not less than 2 inches thick.
Note 3. Lesser depths shall be permitted where cables and conductors rise for terminations or splices or where access is otherwise required.
Note 4. Where one of the conduit types listed in columns 1-3 is combined with one of the circuit types in columns 4 and 5, the shallower depth of burial shall be permitted.

21.17.0 Demand Loads for Various Types of Residential Electrical Appliances

Demand Factors for Household Electric Clothes Dryers

Number of Dryers	Demand Factor Percent
1	100
2	100
3	100
4	100
5	80
6	70
7	65
8	60
9	55
10	50
11-13	45
14-19	40
20-24	35
25-29	32.5
30-34	30
35-39	27.5
40 & over	25

Where two or more single-phase ranges are supplied by a three-phase, 4-wire feeder, the total load shall be computed on the basis of twice the maximum number connected between any two phases. kVA shall be considered equivalent to kW for loads computed under this section.

Demand Loads for Household Electric Ranges, Wall-Mounted Ovens, Counter-Mounted Cooking Units, and Other Household Cooking Appliances over 1¾ kW Rating. Column A to be used in all cases except as otherwise permitted

NUMBER OF APPLIANCES	Maximum Demand (See Notes)	Demand Factors Percent (See Note 3)	
	COLUMN A (Not over 12 kW Rating)	COLUMN B (Less than 3½ kW Rating)	COLUMN C (3½ kW to 8¾ kW Rating)
1	8 kW	80%	80%
2	11 kW	75%	65%
3	14 kW	70%	55%
4	17 kW	66%	50%
5	20 kW	62%	45%
6	21 kW	59%	43%
7	22 kW	56%	40%
8	23 kW	53%	36%
9	24 kW	51%	35%

21.18.0 General Lighting Loads by Occupancy*

Type of Occupancy	Unit Load per Sq. Ft. (Volt-Amperes)
Armories and Auditoriums	1
Banks	3½**
Barber Shops and Beauty Parlors	3
Churches	1
Clubs	2
Court Rooms	2
*Dwelling Units	3
Garages — Commercial (storage)	½
Hospitals	2
*Hotels and Motels, including apartment houses without provisions for cooking by tenants	2
Industrial Commercial (Loft) Buildings	2
Lodge Rooms	1½
Office Buildings	3½**
Restaurants	2
Schools	3
Stores	3
Warehouses (storage)	¼
In any of the above occupancies except one-family dwellings and individual dwelling units of two-family and multifamily dwellings:	
Assembly Halls and Auditoriums	1
Halls, Corridors, Closets, Stairways	½
Storage Spaces	¼

For SI units: one square foot = 0.093 square meter.

* All general use receptacle outlets of 20-ampere or less rating in one-family, two-family and multifamily dwellings and in guest rooms of hotels and motels [except those connected to the receptacle circuits specified shall be considered as outlets for general illumination, and no additional load calculations shall be required for such outlets.

** In addition a unit load of 1 volt-ampere square foot shall be included for general purpose receptacle outlets when the actual number of general-purpose receptacle outlets is unknown.

21.19.0 Selection of Overcurrent Protection and Switching Devices

Category of Application		Acceptable Device Types (See Legend Below)
Individually mounted service disconnect unit	(8–800 amps) (above 800 amps)	SW-QMQB/CF SW-BP/CF
Service disconnect unit in main switchboard	(0–800 amps) (above 800 amps)	SW-QMQB/CF SW-BP/CF
Feeder unit in main switchboard	(0–800 amps) (above 800 amps)	SW-QMQB/CF SW-BP/CF
Main or branch unit in 265/460 (277/480) volt distribution panel or power panel		SW-QMQB/CF except CLCB-MC if needed in order to meet the specified series connected rating of downstream lighting or appliance panel.
Main unit in 265/460 (277/480) volt lighting or appliance panel		CB-SMC, except CLCB-MC if needed in order to meet the specified series connected rating of the panel.
Branch unit in 265/460 (277/480) volt lighting or appliance panel		CB-SMC
Branch unit in 120/208 volt lighting or appliance panel.		CB-SMC
Main or branch unit in 120/208 volt distribution panel or power panel		SW-QMQB/CF, except CLCB if needed in order to meet the specified series connected rating of downstream lighting or appliance panel.
Main unit in 120/208 volt lighting or appliance panel		CB-SMC, except CLCB of needed in order to meet specified series connected rating of the panel.
Branch unit in panelette		CB-CMC
Main unit in metering assembly		QMQB/CF
Tenant main unit in metering assembly		CB-SMC
Individually mounted unit	(0–1200 amps)	SW-QMQB/CF except CLCB-MC if needed in order to meet the specified series connected rating of downstream lighting or appliance panel.
Individually mounted unit without overcurrent protection	(0–1200 amps) (above 1200 amps)	SW-QMQB SW-BP
Motor starting fusing		CF

Explanation of abbreviations used above is as follows:

ABBREVIATION	DESCRIPTION
SW-BP	Distribution switch; bolted pressure type.
SW-QMQB	Distribution switch; quick-make, quick-break type.
/	Fusible—fused with.
CF	Cartridge fuses.
CB-SMC	Circuit breaker, standard molded case type.
CB-CMC	Circuit breaker, compact molded case type.

21.20.0 Size of Equipment and Raceway Grounding Conductors for 15- to 400-Amp Overcurrent Devices

SIZING OF EQUIPMENT AND RACEWAY GROUNDING CONDUCTORS AND LOAD SIDE OF SERVICE BONDING JUMPERS

OVERCURRENT DEVICE FUSE OR TRIP SIZE (AMPS)	GROUNDING CONDUCTOR OR BONDING JUMPER SIZE--CU **	(AL)
15,20	#12	-
25-60	#10	-
70-100	#8	-
110-200	#4	(#4)
225-400	#2	(1/0)
500,600	* 2 x #1	(2 x 2/0)
700,800	* 2 x 1/0	(2 x 3/0)
1000	* 3 x 2/0	(3 x 4/0)
1200	* 4 x 3/0	(4 x 250) MCM
1600	* 5 x 4/0	(5 x 350) MCM
2000	* 6 x 250 MCM	(6 x 400) MCM
2500	* 7 x 350 MCM	(7 x 600) MCM
3000	* 8 x 400 MCM	(8 x 600) MCM
4000	* 11 x 500 MCM	(11 x750) MCM

* Adjust quantity (if needed) to match number of conduits in run.

** Where phase leg conductor ampacity exceeds overcurrent device, increase grounding conductor as if the overcurrent device size matched the phase leg ampacity.

CC. Grounding electrode conductors and conductors used for bonding on the supply side of the service device shall be sized in accordance with the following table:-

SIZING OF GROUNDING ELECTRODE CONDUCTORS AND MAIN (AND SUPPLY SIDE OF SERVICE) BONDING JUMPERS

SERVICE CONDUCTOR SIZE CABLE CU	(AL)	BUS	BROUNDING ELEC-TRODE CONDUCTOR SIZE--CU	(AL)	BONDING JUMPER SIZE----------- CU	(AL)
#2	(1/0)max.	100	#8	-	#8	-
3/0	(250)MCM max.	200	#4	#2	#4	(#2)
500	(700)MCM max.	400	1/0	3/0	1/0	(3/0)
2x350	(2x500)MCM max.	600	2/0	4/0	2/0	(4/0)
2x500	(2x700)MCM max.	800	2/0	4/0	2/0	(350)MCM
4x350	(4x500)MCM max.	1200	3/0	250MCM	250	(600)MCM
5x400	(5x600)MCM max.	1600	3/0	250MCM	400	(2x250)MCM
6x500	(6x750)MCM max.	2000	3/0	250MCM	500	(2x250)MCM
8x500	(8x750)MCM max.	3000	3/0	250MCM	2x500	(2x250)MCM
11x500	(11x750)MCM max.	4000	3/0	250MCM	2x500	(2x250)MCM

21.21.0 Enclosures for Nonhazardous Locations (NEMA Designations)

For a degree of protection against:	Designed to meet tests no. [1]	NEMA Type: For indoor use			Outdoor use		Indoor or outdoor		
		1	12	13	3R	3	4	4X	6P
Incidental contact with enclosed equipment	6.2	✓	✓	✓	✓	✓	✓	✓	✓
Falling dirt	6.2	✓	✓	✓	✓	✓	✓	✓	✓
Rust	6.8	✓	✓	✓	✓	✓	✓	✓	✓
Circulating dust, lint, fibers and flyings [2]	6.5.1.2 (2)		✓	✓		✓	✓	✓	✓
Windblown dust	6.5.1.1 (2)					✓	✓	✓	✓
Falling liquids and light splashing	6.3.2.2		✓	✓		✓	✓	✓	✓
Rain (test evaluated per 6.4.2.1)	6.4.2.1				✓	✓	✓	✓	✓
Rain (test evaluated per 6.4.2.2)	6.4.2.2					✓	✓	✓	✓
Snow and sleet	6.6.2.2				✓	✓	✓	✓	✓
Hosedown and splashing water	6.7						✓	✓	✓
Occasional prolonged submersion	6.11 (2)								✓
Oil and coolant seepage	6.3.2.2		✓	✓					
Oil or coolant spraying and splashing	6.12			✓					
Corrosive agents	6.9							✓	✓

[1] See below for abridged description of NEMA enclosure test requirements. Refer to NEMA Standards Publication No. 250 for complete test specifications.
[2] Non-hazardous materials, not Class III ignitable or combustible.

Rod Entry Test—A ⅛" diameter rod must not be able to enter enclosure except at locations where nearest live part is more than 4" from an opening—such opening shall not permit a ½" diameter rod to enter.

Drip Test—Water is dripped onto enclosure for 30 minutes from an overhead pan having uniformly spaced spouts, one every 20 square inches of pan area each spout having a drip rate of 20 drops per minute.
Evaluation 6.3.2.2: No water shall have entered enclosure.

Rain Test—Entire top and all exposed sides are sprayed with water at a pressure of 5 psi from nozzles for one hour at a rate to cause water to rise 18 inches in a straight-sided pan beneath the enclosure.
Evaluation 6.4.2.1: No water shall have reached live parts, insulation or mechanisms.
Evaluation 6.4.2.2: No water shall have entered enclosure.

Outdoor Dust Test (Alternate Method)—Enclosure and external mechanisms are subjected to a stream of water at 45 gallons per minute from a 1" diameter nozzle, directed at all joints from all angles from a distance of 10 to 12 feet. Test time is 48 seconds times the test length (height + width + depth of enclosure in feet), or a minimum of 5 minutes. No water shall enter enclosure.

Indoor Dust Test (Alternate Method)—Atomized water at a pressure of 30 psi is sprayed on all seams, joints and external operating mechanisms from a distanc of 12 to 15 inches at a rate of three gallons per hour. No less than five ounces of water per linear foot of test length (height + length + depth of enclosure) is applied. No water shall enter enclosure.

External Icing Test—Water is sprayed on enclosure for one hour in a cold room (2° C): then room temperature is lowered to approximately –5° C and water spray is controlled so as to cause ice to buil up at a rate of ¼" per hour until ¾" thick ice has formed on top surface of a 1" diameter metal test bar, then temperature is maintained at –5° C for 3 hours.
Evaluation 6.6.2.2: Equipment shall be undamaged after ice has melted (external mechanisms not required to be operable while ice-laden).

Hosedown Test—Enclosure and external mechanisms are subjected to a stream of water at 65 gallons per minute from a 1" diameter nozzle, directed at all joints from all angles from a distance of 10 to 12 feet. Test time is 48 seconds times the test length (height + width + depth of enclosure in feet), or a minimum of 5 minutes. No water shall enter enclosure.

Rust Resistance Test (Applicable only to enclosures incorporating external ferrous parts)—Enclosure is subjected to a salt spray (fog) for 24 hours, using water with five parts by weight of salt (NaCl), at 35°C, then rinsed and dried. There shall be no rust except where protection is impractical (e.g., machined mating surfaces, sliding surfaces of hinges, shafts, etc.)

Corrosion Protection—Sheet steel enclosures are evaluated per UL 50, Part 13 (test for equivalent protection as G-90 commercial zinc coated sheet steel). Other materials per UL 508, 5.9 or 5.10.

(2) Air Pressure Test (Alternate Method)—Enclosure is submerged in water at a pressure equal to water depth of six feet, for 24 hours. No water shall enter enclosure.

Oil Exclusion Test—Enclosure is subjected to a stream of test liquid for 30 minutes from a ⅜" diameter nozzle at two gallons a minute. Water with 0.1% wetting agent is directed from all angles from a distance of 12 to 18 inches, while any externally operated device is operated at 30 operations per minute. No test liquid shall enter the enclosure.

21.22.0 Enclosures for Hazardous Locations (NEMA Designations)

For a degree of protection against atmospheres typically containing: [3]	Designed to meet tests: [2]	Class (National Electrical Code)	NEMA Type 7, Class I Group:				NEMA Type 9, Class II Group:		
			A	B	C	D	E	F	G
Acetylene		I	✓						
Hydrogen, manufactured gas	Explosion test	I	✓	✓					
	Hydrostatic test								
Diethyl ether, ethylene, hydrogen sulfide		I			✓				
	Temperature test								
Acetone, butane, gasoline, propane, toluene		I			✓	✓			
Metal dusts and other combustible dusts with resistivity of less than 10^5 ohm-cm.		II					✓		
	Dust penetration test								
Carbon black, charcoal, coal or coke dusts with resistivity between 10^2–10^8 ohm-cm.	Temperature test with dust blanket	II						✓	
Combustible dusts with resistivity of 10^5 ohm-cm or greater		II							✓
Fibers, flyings	[4]	III							✓

[1] For indoor locations only unless cataloged with additional NEMA Type enclosure number(s) suitable for outdoor use as shown in table on Page 14. Some control devices (if so listed in the catalog) are suitable for Division 2 hazardous location use in enclosures for non-hazardous locations. For explanation of CLASSES, DIVISIONS and GROUPS, refer to the National Electrical Code.
Note: Classifications of hazardous locations are subject to the approval of the authority having jurisdiction. Refer to the National Electrical Code.

[2] See abridged description of test requirements below. For complete requirements, refer to UL Standard 698, compliance with which is required by NEMA enclosure standards.

[3] For listing of additional materials and information noting the properties of liquids, gases and solids, refer to NFPA 479M-1986, Classification of Gases, Vapors, and Dusts for Electrical Equipment in Hazardous (Classified) Locations.

[4] UL 698 does not include test requirements for Class III. Products that meet Class II, Group G requirements are acceptable for Class III.

21.23.0 Motor-Controller Enclosure Types (Indoor and Outdoor Use)

The table provides the basis for selecting enclosures for use in specific nonhazardous locations. The enclosures are not intended to protect against conditions such as condensation, icing, corrosion or contamination which may occur within the enclosure or enter via the conduit or unsealed openings. These internal conditions require special consideration by the installer and/or user.

Motor Controller Enclosure Selection Table

For Outdoor Use

Provides a Degree of Protection Against the Following Environmental Conditions	Enclosure Type Number †						
	3	3R	3S	4	4X	6	6P
Incidental contact with the enclosed equipment	X	X	X	X	X	X	X
Rain, snow and sleet	X	X	X	X	X	X	X
Sleet*	—	—	X	—	—	—	—
Windblown dust	X	—	X	X	X	X	X
Hosedown	—	—	—	X	X	X	X
Corrosive agents	—	—	—	—	X	—	X
Occasional temporary submersion	—	—	—	—	—	X	X
Occasional prolonged submersion	—	—	—	—	—	—	X

* Mechanism shall be operable when ice covered.

For Indoor Use

Provides a Degree of Protection Against the Following Environmental Conditions	Enclosure Type Number †										
	1	2	4	4X	5	6	6P	11	12	12K	13
Incidental contact with the enclosed equipment	X	X	X	X	X	X	X	X	X	X	X
Falling dirt	X	X	X	X	X	X	X	X	X	X	X
Falling liquids and light splashing	—	X	X	X	X	X	X	X	X	X	X
Circulating dust, lint, fibers and flyings	—	—	X	X	—	X	X	—	X	X	X
Settling airborne dust, lint, fibers and flyings	—	—	X	X	X	X	X	—	X	X	X
Hosedown and splashing water	—	—	X	X	—	X	X	—	—	—	—
Oil and coolant seepage	—	—	—	—	—	—	—	—	X	X	X
Oil or coolant spraying and splashing	—	—	—	—	—	—	—	—	—	—	X
Corrosive agents	—	—	—	X	—	—	X	X	—	—	—
Occasional temporary submersion	—	—	—	—	—	X	X	—	—	—	—
Occasional prolonged submersion	—	—	—	—	—	—	X	—	—	—	—

† Enclosure type number, except type number 1, shall be marked on the motor controller enclosure.

21.24.0 Voltage-Drop Tables for 6- and 12-Volt Equipment

The National Electrical Code limits voltage drop to a maximum of 5% of nominal. Thus, circuit runs must be of sufficient size to maintain operating voltage when remote fixtures and/or exit signs are connected to the emergency lighting equipment. The table below shows the length of wire run based on system voltage, wire gauge and total wattage on the run. To determine loads or lengths of wire runs not listed, divide the *known* value into the *constant* value at the bottom of the appropriate row.

Total Watts on Wire Run	6 Volt System				Total Watts on Wire Run	12 Volt System			
	Wire Gauge					Wire Gauge			
	12	10	8	6		12	10	8	6
	Length of Wire Run (Feet)					Length of Wire Run (Feet)			
6	94	150	238	379	6	378	600	955	1518
7	81	129	204	325	7	324	515	818	1301
8	70	112	179	284	8	283	450	716	1138
10	56	90	143	227	10	226	360	570	910
12	44	70	112	178	12	178	283	450	715
14	40	64	102	162	14	162	257	409	650
16	33	53	84	134	16	133	212	338	538
18	30	47	75	119	18	119	189	300	477
20	28	45	71	114	20	113	180	286	455
21	27	43	68	108	21	108	171	273	434
24	24	38	60	95	24	89	141	225	357
25	21	34	54	86	25	86	136	216	344
30	19	30	48	76	30	75	120	190	303
35	15	25	39	63	35	65	103	164	260
40	13	21	33	53	40	53	85	135	214
48	11	17	28	44	48	44	70	112	178
50	11	17	27	43	50	43	68	108	172
75	7	11	18	29	75	28	45	72	115
100	5	8	14	21	100	21	34	54	86
125	4	7	11	17	125	17	27	43	69
150	3	5	9	14	150	14	23	36	57
175	3	5	8	12	175	12	19	31	49
200	2	4	6	10	200	10	16	27	42
225	2	4	6	10	225	10	16	25	40
250	2	3	5	9	250	9	14	22	36
CONSTANT	567	901	1432	2277	CONSTANT	2267	3604	5730	9109

Example 1— A 12V system uses 8-gauge wire and will operate three 7W exit signs. Total watts on wire run is 21, length of run from table is 273'.

Longer Wire Runs

If loads are uniformly spaced along circuit path (equal watts, equal distances), the lengths in the table can be increased by certain values.

Number of fixtures	2	3	4	5
Multiplier	1.33	1.5	1.6	1.67

Example 2— Exit signs from example 1 will be uniformly spaced. Multiplier is 1.5 for three fixtures. Maximum permissible length of wire run is 273' X 1.5, or 409".

21.25.0 Seismic Restraints and Bracing

All seismic restraint and isolation devices, braces, and supports shall be capable of accepting without failure forces produced by seismic acceleration (expressed in multiples of the acceleration of gravity "G") based on the level above grade of the attachment of the equipment support system. For design purposes, the following acceleration levels shall be used:

DESIGN LEVEL OF ACCELERATION AT EQUIPMENT CENTER OF GRAVITY			
ELEVATION ABOVE GRADE	RIGIDLY FLOOR OR WALL MOUNTED EQUIPMENT	RESILIENTLY MOUNTED AND/OR SUPPORTED FROM CEILING OR STRUCTURE ABOVE	LIFE SAFETY EQUIPMENT (FIRE ALARM, HOSPITAL COMMUNICATIONS, EMERGENCY
BELOW GRADE UP TO 20 FEET ABOVE GRADE	0.125 "G"	0.500 "G"	1.000 "G"
21 FEET AND UP	0.500 "G"	0.750 "G"	

SEISMIC BRACING TABLE			
EQUIPMENT	ON CENTER SPACING		WITHIN EACH CHANGE OF DIRECTION
	TRANSVERSE	LONGITUDINAL	
CONDUIT	40 FEET	80 FEET	10 FEET OR 15 DIAMETERS

For all seismically supported trapeze supported conduit, the individual conduits shall be transversely and vertically restrained to the trapeze support at the designated restraint locations. Restrain at least every third trapeze hanger transversely and every fifth one longitudinally as well as the trapeze on both sides of every change of direction.

For overhead supported equipment, overstress of the building structure must not occur. Bracing may occur from:

1) Flanges of structural steel beams.

2) Upper truss chords in bar joists.

21.26.0 Full Load Current (In Amperes) for Single-Phase, Two-Phase, Three-Phase, and Direct-Current Motors

Full-Load Currents (in Amperes) for Single-Phase Alternating-Current Motors

The following values of full-load currents are for motors running at usual speeds and motors with normal torque characteristics. Motors built for especially low speeds or high torques may have higher full-load currents, and multispeed motors will have full-load current varying with speed, in which case the nameplate current ratings shall be used.

HP	115V	200V	208V	230V
1/6	4.4	2.5	2.4	2.2
1/4	5.8	3.3	3.2	2.9
1/3	7.2	4.1	4.0	3.6
1/2	9.8	5.6	5.4	4.9
3/4	13.8	7.9	7.6	6.9
1	16	9.2	8.8	8
1½	20	11.5	11	10
2	24	13.8	13.2	12
3	34	19.6	18.7	17
5	56	32.2	30.8	28
7½	80	46	44	40
10	100	57.5	55	50

21.26.0 Full Load Current (In Amperes) for Single-Phase, Two-Phase, Three-Phase, and Direct-Current Motors (Continued)

Full-Load Current* for Three-Phase Alternating-Current Motors

	Induction Type Squirrel-Cage and Wound-Rotor Amperes							Synchronous Type †Unity Power Factor Amperes			
HP	115V	200V	208V	230V	460V	575V	2300V	230V	460V	575V	2300V
½	4	2.3	2.2	2	1	.8					
¾	5.6	3.2	3.1	2.8	1.4	1.1					
1	7.2	4.1	4.0	3.6	1.8	1.4					
1½	10.4	6.0	5.7	5.2	2.6	2.1					
2	13.6	7.8	7.5	6.8	3.4	2.7					
3		11.0	10.6	9.6	4.8	3.9					
5		17.5	16.7	15.2	7.6	6.1					
7½		25.3	24.2	22	11	9					
10		32.2	30.8	28	14	11					
15		48.3	46.2	42	21	17					
20		62.1	59.4	54	27	22					
25		78.2	74.8	68	34	27		53	26	21	
30		92	88	80	40	32		63	32	26	
40		119.6	114.4	104	52	41		83	41	33	
50		149.5	143.0	130	65	52		104	52	42	
60		177.1	169.4	154	77	62	16	123	61	49	12
75		220.8	211.2	192	96	77	20	155	78	62	15
100		285.2	272.8	248	124	99	26	202	101	81	20
125		358.8	343.2	312	156	125	31	253	126	101	25
150		414	396.0	360	180	144	37	302	151	121	30
200		552	528.0	480	240	192	49	400	201	161	40

*These values of full-load current are for motors running at speeds usual for belted motors with normal torque characteristics. Motors built for especially low speeds or high torques may require more running current, and multispeed motors will have full-load current varying with speed, in which case the nameplate current rating shall be used.

†For 90 and 80 percent power factor the above figures shall be multiplied by 1.1 and 1.25 respectively.

The voltages listed are rated motor voltages. The currents listed shall be permitted for system voltage ranges of 110 to 120, 220 to 240, 440 to 480, and 550 to 600 volts.

21.26.0 Full Load Current (In Amperes) for Single-Phase, Two-Phase, Three-Phase, and Direct-Current Motors (Continued)

Full-Load Currents (in Amperes) for Direct-Current Motors

The following values of full-load currents* are for motors running at base speed.

	Armature Voltage Rating*					
HP	90V	120V	180V	240V	500V	550V
¼	4.0	3.1	2.0	1.6		
⅓	5.2	4.1	2.6	2.0		
½	6.8	5.4	3.4	2.7		
¾	9.6	7.6	4.8	3.8		
1	12.2	9.5	6.1	4.7		
1½		13.2	8.3	6.6		
2		17	10.8	8.5		
3		25	16	12.2		
5		40	27	20		
7½		58		29	13.6	12.2
10		76		38	18	16
15				55	27	24
20				72	34	31
25				89	43	38
30				106	51	46
40				140	67	61
50				173	83	75
60				206	99	90
75				255	123	111
100				341	164	148
125				425	205	185
150				506	246	222
200				675	330	294

* These are average direct-current quantities.

21.26.0 Full Load Current (In Amperes) for Single-Phase, Two-Phase, Three-Phase, and Direct-Current Motors (Continued)

Full-Load Current for Two-Phase Alternating-Current Motors (4-Wire)

The following values of full-load current are for motors running at speeds usual for belted motors and motors with normal torque characteristics. Motors built for especially low speeds or high torques may require more running current, and multispeed motors will have full-load current varying with speed, in which case the nameplate current rating shall be used. Current in the common conductor of a 2-phase, 3-wire system will be 1.41 times the value given.

The voltages listed are rated motor voltages. The currents listed shall be permitted for system voltage ranges of 110 to 120, 220 to 240, 440 to 480, and 550 to 600 volts.

HP	Induction Type Squirrel-Cage and Wound-Rotor Amperes				
	115V	230V	460V	575V	2300V
½	4	2	1	.8	
¾	4.8	2.4	1.2	1.0	
1	6.4	3.2	1.6	1.3	
1½	9	4.5	2.3	1.8	
2	11.8	5.9	3	2.4	
3		8.3	4.2	3.3	
5		13.2	6.6	5.3	
7½		19	9	8	
10		24	12	10	
15		36	18	14	
20		47	23	19	
25		59	29	24	
30		69	35	28	
40		90	45	36	
50		113	56	45	
60		133	67	53	14
75		166	83	66	18
100		218	109	87	23
125		270	135	108	28
150		312	156	125	32
200		416	208	167	43

Section

22

Metrification

Contents

22.0.0 Introduction to the 1975 Metric Conversion Act
22.1.0 What will change and what will remain the same
22.2.0 How metric units will apply in the construction industry
22.3.0 Metrification of pipe sizes
22.4.0 Metrification of standard lumber sizes
22.5.0 Metric rebar conversions
22.6.0 Metric conversion of ASTM diameter and wall thickness designations
22.7.0 Metric conversion scales (temperature and measurements)
22.8.0 Approximate metric conversions
22.9.0 Quick imperial (metric equivalents)
22.10.0 Metric conversion factors

22.0.0 Introduction to the 1975 Metric Conversion Act

As the federal government moves to convert the inch-pound units to the metric system, in accordance with the 1975 Metric Conversion Act, various parts of the construction industry will begin the conversion to this more universal method of measurement.

Metric units are often referred to as *SI units*, an abbreviation taken from the French Le Système International d'Unités. Another abbreviation that will be seen with more frequency is ISO - the International Standards Organization charged with supervising the establishment of a universal standards system. For everyday transactions it may be sufficient to gain only the basics of the metric system.

Name of metric unit	Symbol	Approximate size (length/pound)
meter	m	39½ inches
kilometer	km	0.6 mile
centimeter	cm	width of a paper clip
millimeter	mm	thickness of a dime
hectare	ha	2½ acres
square meter	m2	1.2 square yards
gram	g	weight of a paper clip
kilogram	kg	2.2 pounds
metric ton	t	long ton (2240 pounds)
liter	L	one quart and two ounces
milliliter	mL	⅕ teaspoon
kilopascal	kPa	atmospheric pressure is about 100 kPa

The Celsius temperature scale is used. Instead of referring to its measurement as *degree centigrade*, the term *degree Celsius* is the correct designation. Using this term, familiar points are:

- Water freezes at 0 degrees
- Water boils at 100 degrees
- Normal body temperature is 37 degrees (98.6 F)
- Comfortable room temperature 20 to 35 (68 to 77 F)

22.1.0 What Will Change and What Will Stay The Same?

Metric Module and Grid

What will change

- The basic building module, from 4 inches to 100 mm.
- The planning grid, from 2' × 2' to 600 × 600 mm.

What will stay the same

- A module and grid based on rounded, easy-to-use dimensions. The 100 mm module is the global standard.

Drawings

What will change

- Units, from feet and inches to millimeters for all building dimensions and to meters for site plans and civil engineering drawings. Unit designations are unnecessary: if there's no decimal point, it's millimeters; if there's a decimal point carried to one, two, or three places, it's meters. In accordance with ASTM E621, centimeters are not used in construction because (1) they are not consistent with the preferred use of multiples of 1000, (2) the order of magnitude between a millimeter and centimeter is only 10 and the use of both units would lead to confusion and require the use of unit designations, and (3) the millimeter is small enough to almost entirely eliminate decimal fractions from construction documents.

- Drawing scales, from inch-fractions-to-feet to true ratios. Preferred metric scales are:

  1:1 (full size)
  1:5 (close to 3" = 1'-0")
  1:10 (between 1"= 1'-0" and 1½" = 1'-0")
  1:20 (between ½" = 1'-0" and ¾" = 1'-0")
  1:50 (close to ¼" = 1'-0")
  1:100 (close to ⅛" = 1'-0")
  1:200 (close to 1⁄16" = 1'-0")
  1:500 (close to 1" = 40'-0")
  1:1000 (close to 1" = 80'-0")

As a means of comparison, inch-fraction scales may be converted to true ratios by multiplying a scale's divisor by 12; for example, for ¼" = 1'-0", multiply the 4 by 12 for a true ratio of 1:48.

- Drawing sizes, to ISO "A" series:

  A0 (1189 × 841 mm, 46.8 × 33.1 inches)
  A1 (841 × 594 mm, 33.1 × 23.4 inches)
  A2 (594 × 420 mm, 23.4 × 16.5 inches)
  A3 (420 × 297 mm, 16.5 × 11.7 inches)
  A4 (297× 210 mm, 11.7 × 8.3 inches)

  Of course, metric drawings can be made on any size paper.

  What will stay the same.

- Drawing contents

Never use dual units (both inch-pound and metric) on drawings. It increases dimensioning time, doubles the chance for errors, makes drawings more confusing, and only postpones the learning process. An exception is for construction documents meant to be viewed by the general public.

Specifications

What will change

- Units of measure, from feet and inches to millimeters for linear dimensions, from square feet to square meters for area, from cubic yards to cubic meters for volume (except use liters for fluid volumes), and from other inch-pound measures to metric measures as appropriate.

  What will stay the same

- Everything else in the specifications

Do not use dual units in specifications except when the use of an inch-pound measure serves to clarify an otherwise unfamiliar metric measure; then place the inch-pound unit in parentheses after the metric. For example, "7.5 kW (10 horsepower) motor." All unit conversions should be checked by a professional to ensure that rounding does not exceed allowable tolerances.

For more information, see the July–August 1994 issue of *Metric in Construction*.

Floor Loads

What will change

- Floor load designations, from "psf" to kilograms per square meter (kg/m^2) for everyday use and kilonewtons per square meter (kN/m^2) for structural calculations.

What will stay the same

- Floor load requirements

Kilograms per square meter often are used to designate floor loads because many live and dead loads (furniture, filing cabinets, construction materials, etc.) are measured in kilograms. However, kilonewtons per square meter or their equivalent, kilopascals, are the proper measure and should be used in structural calculations.

Construction Products

What will change

- Modular products: brick, block, drywall, plywood, suspended ceiling systems, and raised floor systems. They will undergo "hard" conversion; that is, their dimensions will change to fit the 100 mm module.
- Products that are custom-fabricated or formed for each job (for example, cabinets, stairs, handrails, ductwork, commercial doors and windows, structural steel systems, and concrete work). Such products usually can be made in any size, inch-pound or metric, with equal ease; therefore, for metric jobs, they simply will be fabricated or formed in metric.

What will stay the same

- All other products, since they are cut-to-fit at the jobsite (for example, framing lumber, woodwork, siding, wiring, piping, and roofing) or are not dimensionally sensitive (for example, fasteners, hardware, electrical components, plumbing fixtures, and HVAC equipment). Such products will just be "soft" converted—that is, relabeled in metric units. A 2¾" × 4½" wall switch face plate will be relabeled 70 × 115 mm and a 30 gallon tank, 114 L. Manufacturers eventually may convert the physical dimensions of many of these products to new rational "hard" metric sizes but only when it becomes convenient for them to do so.

"2-By-4" Studs and Other "2-By" Framing (Both Wood and Metal)

What will change

- Spacing, from 16" to 400 mm, and 24" to 600 mm.

What will stay the same

- Everything else.

"2-bys" are produced in "soft" fractional inch dimensions so there is no need to convert them to new rounded "hard" metric dimensions. 2-by-4s may keep their traditional name or perhaps they'll eventually be renamed 50 by 100 (mm), or, more exactly, 38 × 89.

Drywall, Plywood, and Other Sheet Goods

What will change

- Widths, from 4'-0" to 1200 mm.
- Heights, from 8'-0" to 2400 mm, 10'-0" to 3000 mm.

What will stay the same

- Thicknesses, so fire, acoustic, and thermal ratings won't have to be recalculated.

Metric drywall and plywood are readily available but may require longer lead times for ordering and may cost more in small amounts until their use becomes more common.

Batt Insulation

What will change

- Nominal width labels, from 16" to 16"/400 mm and 24" to 24"/600 mm.

What will stay the same

- Everything else.

Batts will not change in width; they'll just have a tighter "friction fit" when installed between metric-spaced framing members.

Doors

What will change

- Height, from 6'-8" to 2050 mm or 2100 mm and from 7'-0" to 2100 mm.
- Width, from 2'-6" to 750 mm, from 2'-8" to 800 mm, from 2'-10" to 850 mm, from 3'-0" to 900 mm or 950 mm, and from 3'-4" to 1000 mm.

What will stay the same

- Door thicknesses.
- Door materials and hardware.

For commercial work, doors and door frames can be ordered in any size since they normally are custom-fabricated.

Ceiling Systems

What will change

- Grids and lay-in ceiling tile, air diffusers and recessed lighting fixtures, from 2' × 2' to 600 × 600 mm and from 2' × 4' to 600 × 1200 mm.

What will stay the same

- Grid profiles, tile thicknesses, air diffuser capacities, florescent tubes, and means of suspension.

On federal building projects, metric recessed lighting fixtures may be specified if their total installed costs are estimated to be no more than for inch-pound fixtures.

Raised Floor Systems

What will change

- Grids and lay-in floor tile, from 2' × 2' to 600 × 600 mm.

What will stay the same

- Grid profiles, tile thicknesses, and means of support.

HVAC Controls

What will change

- Temperature units, from Fahrenheit to Celsius.

What will stay the same

- All other parts of the controls.

Controls are now digital so temperature conversions can be made with no difficulty.

Brick

What will change

- Standard brick, to 90 × 57 × 190 mm.
- Mortar joints, from ⅜" and ½" to 10 mm.
- Brick module, from 2' × 2' to 600 × 600 mm.

What will stay the same

- Brick and mortar composition.

Of the 100 or so brick sizes currently made, 5 to 10 are within a millimeter of a metric brick so the brick industry will have no trouble supplying metric brick.

For more information, see the March-April 1995 issue of *Metric in Construction*.

Concrete Block

What will change

- Block sizes, to 190 × 190 × 390 mm.
- Mortar joints, from ½" to 10 mm.
- Block module, from 2' × 2' to 600 × 600 mm.

What will stay the same

- Block and mortar composition.

On federal building projects, metric block may be specified if its total installed cost is estimated to be no more than for inch-pound block. The Construction Metrication Council recommends that, wherever possible, block walls be designed and specified in a manner that permits the use of either inch-pound or metric block, allowing the final decision to be made by the contractor.

Sheet Metal

What will change

- Designation, from "gage" to millimeters.

What will stay the same

- Thickness, which will be soft-converted to tenths of a millimeter.

In specifications, use millimeters only or millimeters with the gage in parentheses.

Concrete

What will change

- Strength designations, from "psi" to megapascals, rounded to the nearest 5 megapascals per ACI 318M as follows:

  2500 psi to 20 MPa
  3000 psi to 25 MPa
  3500 psi to 25 MPa

4000 psi to 30 MPa
4500 psi to 35 MPa
5000 psi to 35 MPa

Depending on exact usage, however, the above metric conversions may be more exact than those indicated.

What will stay the same

- Everything else.

For more information, see the November-December 1994 issue of *Metric in Construction.*

Rebar

What will change

- Rebar will not change in size but will be renamed per ASTM A615M-96a and ASTM A706M-96a as follows:

No. 3 to No. 10
No. 4 to No. 13
No. 5 to No. 16
No. 6 to No. 19
No. 7 to No. 22
No. 8 to No. 25
No. 9 to No. 29
No. 10 to No. 32
No. 11 to No. 36
No. 14 to No. 43
No. 18 to No. 57

What will stay the same

- Everything else.

For more information, see the July-August 1996 issue of *Metric in Construction.*

Glass

What will change

- Cut sheet dimensions, from feet and inches to millimeters.

What will stay the same

- Sheet thickness; sheet glass can be rolled to any dimension and often is rolled in millimeters now.

See ASTM C1036.

Pipe and Fittings

What will change

- Nominal pipe and fitting designations, from inches to millimeters

What will stay the same

- Pipe and fitting cross sections and threads.

Pipes and fittings are produced in "soft" decimal inch dimensions but are identified in nominal inch sizes a matter of convenience. A 2-inch pipe has neither an inside nor an outside diameter of 2 inches, a 1-inch fitting has no exact 1-inch dimension, and a ½-inch sprinkler head contains no ½-inch dimension anywhere; consequently, there is no need to "hard" convert pipes and fittings to rounded metric dimensions. Instead, they will not change size but simply be relabeled in metric as follows:

⅛" = 6 mm
³⁄₁₆" = 7 mm
1½" = 40 mm
2" = 50 mm

¼" = 8 mm	2½" = 65 mm
⅜" = 10 mm	3" = 75 mm
½" = 15 mm	3½" = 90 mm
⅝" = 18 mm	4" = 100 mm
¾" = 20 mm	4½" = 115 mm
1" = 25 mm	1" = 25 mm for all larger sizes
1¼" = 32 mm	

For more information, see the September-October 1993 issue of *Metric in Construction*.

Electrical Conduit

What will change

- Nominal conduit designations, from inches to millimeters.

What will stay the same

- Conduit cross sections.

Electrical conduit is similar to piping: it is produced in "soft" decimal inch dimensions but is identified in nominal inch sizes. Neither metallic nor nometallic conduit will change size; they will be relabeled in metric units as follows:

½" = 16 (mm)	2½" = 63 (mm)
¾" = 21 (mm)	3" = 78 (mm)
1" = 27 (mm)	3½" = 91 (mm)
1¼" = 35 (mm)	4" = 103 (mm)
1½" = 41 (mm)	5" = 129 (mm)
2" = 53 (mm)	6" = 155 (mm)

These new metric names were assigned by the National Electrical Manufacturers Association.

Electrical Wire

What will change

- Nothing at this time.

What will stay the same

- Existing American Wire Gage (AWG) sizes.

Structural Steel

What will change

- Section designations, from inches to millimeters and from pounds per foot to kilograms per meter, in accordance with ASTM A6M.
- Bolts—to metric diameters and threads per ASTM A325M and A490M.

What will stay the same

- Cross sections.

Like pipe and conduit, steel sections are produced in "soft" decimal inch dimensions (with actual depths varying by weight) but are named in rounded inch dimensions so there is no need to "hard" convert them to metric units. Rather, their names will be changed to metric designations, and rounded to the nearest 10 mm. Thus, a 10-inch section is relabeled as a 250-mm section and a 24-inch section is relabeled as a 610-mm section.

22.2.0 How Metric Units Will Apply in the Construction Industry

	Quantity	Unit	Symbol
Masonry	length	meter, millimeter	m, mm
	area	square meter	m^2
	mortar volume	cubic meter	m^3
Steel	length	meter, millimeter	m, mm
	mass	megagram (metric ton) kilogram	Mg (t) kg
	mass per unit length	kilogram per meter	kg/m
Carpentry	length	meter, millimeter	m, mm
Plastering	length	meter, millimeter	m, mm
	area	square meter	m^2
	water capacity	liter (cubic decimeter)	L (dm^3)
Glazing	length	meter, millimeter	m, mm
	area	square meter	m^2
Painting	length	meter, millimeter	m, mm
	area	square meter	m^2
	capacity	liter (cubic decimeter) milliliter (cubic centimeter)	L (dm^3) mL (cm^3)
Roofing	length	meter, millimeter	m, mm
	area	square meter	m^2
	slope	percent ratio of lengths	% mm/mm, m/m
Plumbing	length	meter, millimeter	m, mm
	mass	kilogram, gram	kg, g
	capacity	liter (cubic decimeter)	L (dm^3)
	pressure	kilopascal	kPa
Drainage	length	meter, millimeter	m, mm
	area	hectare (10 000 m2) square meter	ha m^2
	volume	cubic meter	m^3
	slope	percent ratio of lengths	% mm/mm, m/m
HVAC	length	meter, millimeter	m, mm
	volume (capacity)	cubic meter liter (cubic decimeter)	m^3 L (dm^3)
	air velocity	meter/second	m/s
	volume flow	cubic meter/second liter/second (cubic decimeter per second)	m^3/s L/s (dm^3/s)
	temperature	degree Celsius	°C
	force	newton, kilonewton	N, kN
	pressure	pascal, kilopascal	Pa, kPa
	energy	kilojoule, megajoule	kJ, MJ
	rate of heat flow	watt, kilowatt	W, kW
Electrical	length	millimeter, meter, kilometer	mm, m, km
	frequency	hertz	Hz
	power	watt, kilowatt	W, kW
	energy	magajoule kilowatt hour	MJ kWh
	electric current	ampere	A
	electric potential	volt, kilovolt	V, kV
	resistance	milliohm, ohm	mΩ, Ω

22.3.0 Metricification of Pipe Sizes

Pipe diameter sizes can be confusing because their designated size does not correspond to their actual size. For instance, a 2-inch steel pipe has an inside diameter of approximately 2⅛ inches and an outside diameter of about 2⅜ inches.

The *2 inch* designation is very similar to the 2" × 4" designation for wood studs, neither dimensions are "actual", but they are a convenient way to describe these items.

Pipe sizes are identified as *NPS* (*nominal pipe size*) and their conversion to metric would conform to ISO (International Standards Organization) criteria and are referred to as *DN* (*diameter nominal*). These designations would apply to all plumbing, mechanical, drainage, and miscellaneous pipe commonly used in civil works projects.

NPS size	DN size
⅛"	6 mm
3/16"	7 mm
¼"	8 mm
⅜"	10 mm
½"	15 mm
⅝"	18 mm
¾"	20 mm
1"	25 mm
1¼"	32 mm
1½"	40 mm
2"	50 mm
2½"	65 mm
3"	80 mm
3½"	90 mm
4"	100 mm
4½"	115 mm
5"	125 mm
6"	150 mm
8"	200 mm
10"	250 mm
12"	300 mm
14"	350 mm
16"	400 mm
18"	450 mm
20"	500 mm
24"	600 mm
28"	700 mm
30"	750 mm
32"	800 mm
36"	900 mm
40"	1000 mm
44"	1100 mm
48"	1200 mm
52"	1300 mm
56"	1400 mm
60"	1500 mm

For all pipe over 60 inches nominal, use 1 inch equals 25 mm.

22.4.0 Metrification of Standard Lumber Sizes

Metric units: ASTM Standard E 380 was used as the authoritative standard in developing the metric dimensions in this standard. Metric dimensions are calculated at 25.4 millimeters (mm) times the actual dimension in inches. The nearest mm is significant for dimensions greater than ⅛ inch, and the nearest 0.1 mm is significant for dimensions equal to or less than ⅛ inch.

The rounding rule for dimensions greater than 1/8 inch: If the digit in the tenths of mm position (the digit after the decimal point) is less than 5, drop all fractional mm digits; if it is greater than 5 or if it is 5 followed by at least one nonzero digit, round one mm higher; if 5 followed by only zeroes, retain the digit in the unit position (the digit before the decimal point) if it is even or increase it one mm if it is odd.

The rounding rule for dimensions equal to or less than 1/8 inch: If the digit in the hundredths of mm position (the second digit after the decimal point) is less than 5, drop all digits to the right of the tenth position; if greater than or it is 5 followed by at least one nonzero digit, round one-tenth mm higher; if 5 followed by only zeros, retain the digit in the tenths position if it is even or increase it one-tenth mm if it is odd.

In case of a dispute on size measurements, the conventional (inch) method of measurement shall take precedence.

22.5.0 Metric Rebar Conversions

A615 M-96a & A706M-96a Metric Bar Sizes	Nominal Diameter	A615-96a & A706-96a Inch-Pound Bar Sizes
#10	9.5 mm/0.375"	#3
#13	12.7 mm/0.500"	#4
#16	15.9 mm/0.625"	#5
#19	19.1 mm/0.750"	#6
#22	22.2 mm/0.875"	#7
#25	25.4 mm/1.000"	#8
#29	28.7 mm/1.128"	#9
#32	32.3 mm/1.270"	#10
#36	35.8 mm/1.410"	#11
#43	43.0 mm/1.693"	#14
#57	57.3 mm/2.257"	#18

22.6.0 Metric Conversion of ASTM Diameter and Wall Thickness Designations

METRIC CONVERSION OF ASTM DIAMETER DESIGNATIONS

in	mm	in	mm	in	mm	in	mm
6	150	30	750	57	1425	96	2400
8	200	33	825	60	1500	102	2550
10	250	36	900	63	1575	108	2700
12	300	39	975	66	1650	114	2850
15	375	42	1050	69	1725	120	3000
18	450	45	1125	72	1800	132	3300
21	525	48	1200	78	1950	144	3600
24	600	51	1275	84	2100	156	3900
27	675	54	1350	90	2250	168	4200

METRIC CONVERSION OF ASTM WALL THICKNESS DESIGNATIONS

in	mm	in	mm	in	mm	in	mm
1	25	3-1/8	79	5	125	8	200
1-1/2	38	3-1/4	82	5-1/4	131	8-1/2	213
2	50	3-1/2	88	5-1/2	138	9	225
2-1/4	56	3-3/4	94	5-3/4	144	9-1/2	238
2-3/8	59	3-7/8	98	6	150	10	250
2-1/2	63	4	100	6-1/4	156	10-1/2	263
2-5/8	66	4-1/8	103	6-1/2	163	11	275
2-3/4	69	4-1/4	106	6-3/4	169	11-1/2	288
2-7/8	72	4-1/2	113	7	175	12	300
3	75	4-3/4	119	7-1/2	188	12-1/2	313

22.7.0 Metric Conversion Scales (Temperature and Measurements)

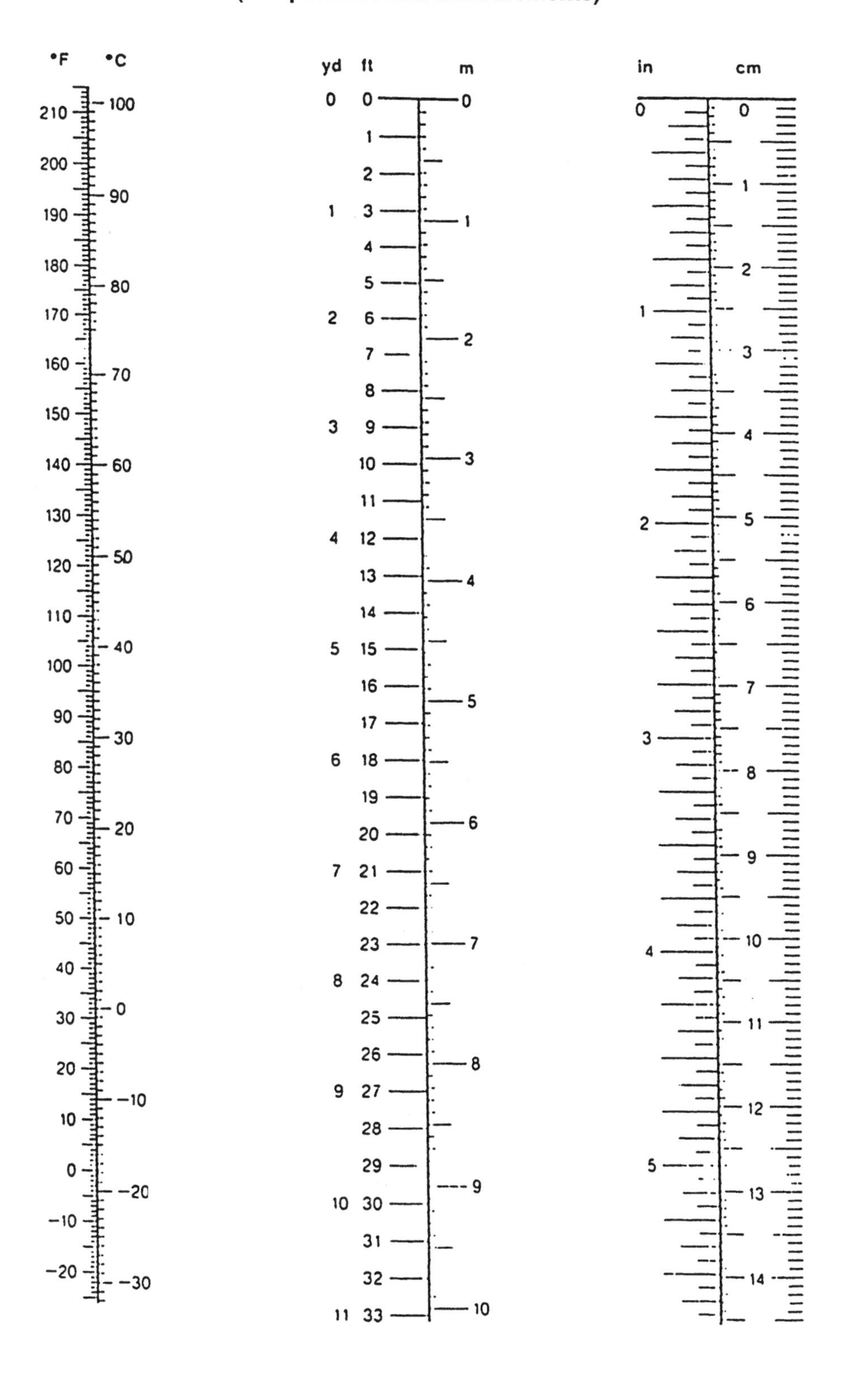

22.8.0 Approximate Metric Conversions

Symbol	*When You Know*	*Multiply by*	*To Find*	*Symbol*
		LENGTH		
mm	millimeters	0.04	inches	in
cm	centimeters	0.4	inches	in
m	meters	3.3	feet	ft
m	meters	1.1	yards	yd
km	kilometers	0.6	miles	mi
		AREA		
cm^2	square centimeters	0.16	square inches	in^2
m^2	square meters	1.2	square yards	yd^2
km^2	square kilometers	0.4	square miles	mi^2
ha	hectares (10,000 m^2)	2.5	acres	
		MASS (weight)		
g	grams	0.035	ounces	oz
kg	kilograms	2.2	pounds	lb
t	metric ton (1,000 kg)	1.1	short tons	
		VOLUME		
mL	milliliters	0.03	fluid ounces	fl oz
mL	milliliters	0.06	cubic inches	in^3
L	liters	2.1	pints	pt
L	liters	1.06	quarts	qt
L	liters	0.26	gallons	gal
m^3	cubic meters	35	cubic feet	ft^3
m^3	cubic meters	1.3	cubic yards	yd^3
		TEMPERATURE (exact)		
°C	degrees Celsius	multiply by 9/5, add 32	degrees Fahrenheit	°F

°C: -40, -20, 0, 20, 37, 60, 80, 100

°F: -40, 0, 32, 80, 98.6, 160, 212

water freezes — body temperature — water boils

U.S. Department of Commerce Technology Administration, Office of Metric Programs, Washington, DC 20230

22.8.0 Approximate Metric Conversion (Continued)

Symbol	*When You Know*	*Multiply by*	*To Find*	*Symbol*
		LENGTH		
in	inches	2.5	centimeters	cm
ft	feet	30	centimeters	cm
yd	yards	0.9	meters	m
mi	miles	1.6	kilometers	km
		AREA		
in^2	square inches	6.5	square centimeters	cm^2
ft^2	square feet	0.09	square meters	m^2
yd^2	square yards	0.8	square meters	m^2
mi^2	square miles	2.6	square kilometers	km^2
	acres	0.4	hectares	ha
		MASS (weight)		
oz	ounces	28	grams	g
lb	pounds	0.45	kilograms	kg
	short tons (2000 lb)	0.9	metric ton	t
		VOLUME		
tsp	teaspoons	5	milliliters	mL
Tbsp	tablespoons	15	milliliters	mL
in^3	cubic inches	16	milliliters	mL
fl oz	fluid ounces	30	milliliters	mL
c	cups	0.24	liters	L
pt	pints	0.47	liters	L
qt	quarts	0.95	liters	L
gal	gallons	3.8	liters	L
ft^3	cubic feet	0.03	cubic meters	m^3
yd^3	cubic yards	0.76	cubic meters	m^3
		TEMPERATURE (exact)		
°F	degrees Fahrenheit	subtract 32, multiply by 5/9	degrees Celsius	°C

United States Department of Commerce, Technology Administration,
National Institute of Standards and Technology, Metric Program, Gaithersburg, MD 20899

22.9.0 Quick Imperial (Metric Equivalents)

Distance

Imperial		Metric
1 inch	=	2.540 centimetres
1 foot	=	0.3048 metre
1 yard	=	0.9144 metre
1 rod	=	5.029 metres
1 mile	=	1.609 kilometres

Metric		Imperial
1 centimetre	=	0.3937 inch
1 decimetre	=	0.3281 foot
1 metre	=	3.281 feet
	=	1.094 yard
1 decametre	=	10.94 yards
1 kilometre	=	0.6214 mile

Weight

1 ounce (troy)	=	31.103 grams
1 ounce (avoir)	=	28.350 grams
1 pound (troy)	=	373.242 grams
1 pound (avoir)	=	453.592 grams
1 ton (short)	=	0.907 tonne*

1 gram	=	0.032 ounce (troy)
1 gram	=	0.035 ounce (avoir)
1 kilogram	=	2.679 pounds (troy)
1 kilogram	=	2.205 pounds (avoir)
1 tonne	=	1.102 ton (short)

*1 tonne = 1000 kilograms

Capacity

Imperial

1 pint	=	0.568 litre
1 gallon	=	4.546 litres
1 bushel	=	36.369 litres
1 litre	=	0.880 pint
1 litre	=	0.220 gallon
1 hectolitre	=	2.838 bushels

U.S.

1 pint (U.S.)	=	0.473 litre
1 quart (U.S.)	=	0.946 litre
1 gallon (U.S.)	=	3.785 litres
1 barrel (U.S.)	=	158.98 litres

Area

1 square inch	=	6.452 square centimetres
1 square foot	=	0.093 square metre
1 square yard	=	0.836 square metre
1 acre	=	0.405 hectare*
1 square mile	=	259.0 hectares
1 square mile	=	2.590 square kilometres
1 square centimetre	=	0.155 square inch
1 square metre	=	10.76 square feet
1 square metre	=	1.196 square yard
1 hectare	=	2.471 acres
1 square kilometre	=	0.386 square mile

***1 hectare = 1 square hectometre**

Volume

1 cubic inch	=	16.387 cubic centimetres
1 cubic foot	=	0.0283 cubic decimetres
1 cubic yard	=	0.765 cubic metre
1 cubic centimetre	=	0.061 cubic inch
1 cubic decimetre	=	35.314 cubic foot
1 cubic metre	=	1.308 cubic yard

22.10.0 Metric Conversion Factors

The following list provides the conversion relationship between U.S. customary units and SI (International System) units. The proper conversion procedure is to multiply the specified value on the left (primarily U.S. customary values) by the conversion factor exactly as given below and then round to the appropriate number of significant digits desired. For example, to convert 11.4 ft to meters: 11.4 × 0.3048 = 3.47472, which rounds to 3.47 meters. Do not round either value before performing the multiplication, as accuracy would be reduced. A complete guide to the SI system and its use can be found in ASTM E 380, Metric Practice.

To convert from	to	multiply by
Length		
inch (in.)	micron (μ)	25,400 E*
inch (in.)	centimeter (cm)	2.54 E
inch (in.)	meter (m)	0.0254 E
foot (ft)	meter (m)	0.3048 E
yard (yd)	meter (m)	0.9144
Area		
square foot (sq ft)	square meter (sq m)	0.09290304 E
square inch (sq in.)	square centimeter (sq cm)	6.452 E
square inch (sq in.)	square meter (sq m)	0.00064516 E
square yard (sq yd)	square meter (sq m)	0.8361274
Volume		
cubic inch (cu in.)	cubic centimeter (cu cm)	16.387064
cubic inch (cu in.)	cubic meter (cu m)	0.00001639
cubic foot (cu ft)	cubic meter (cu m)	0.02831685
cubic yard (cu yd)	cubic meter (cu m)	0.7645549
gallon (gal) Can. liquid	liter	4.546
gallon (gal) Can. liquid	cubic meter (cu m)	0.004546
gallon (gal) U.S. liquid**	liter	3.7854118
gallon (gal) U.S. liquid	cubic meter (cu m)	0.00378541
fluid ounce (fl oz)	milliliters (ml)	29.57353
fluid ounce (fl oz)	cubic meter (cu m)	0.00002957
Force		
kip (1000 lb)	kilogram (kg)	453.6
kip (1000 lb)	newton (N)	4,448.222
pound (lb) avoirdupois	kilogram (kg)	0.4535924
pound (lb)	newton (N)	4.448222
Pressure or stress		
kip per square inch (ksi)	megapascal (MPa)	6.894757
kip per square inch (ksi)	kilogram per square centimeter (kg/sq cm)	70.31
pound per square foot (psf)	kilogram per square meter (kg/sq m)	4.8824
pound per square foot (psf)	pascal (Pa)†	47.88
pound per square inch (psi)	kilogram per square centimeter (kg/sq cm)	0.07031
pound per square inch (psi)	pascal (Pa)†	6,894.757
pound per square inch (psi)	megapascal (MPa)	0.00689476
Mass (weight)		
pound (lb) avoirdupois	kilogram (kg)	0.4535924
ton, 2000 lb	kilogram (kg)	907.1848
grain	kilogram (kg)	0.0000648

To convert from	to	multiply by
Mass (weight) per length		
kip per linear foot (klf)	kilogram per meter (kg/m)	0.001488
pound per linear foot (plf)	kilogram per meter (kg/m)	1.488
Mass per volume (density)		
pound per cubic foot (pcf)	kilogram per cubic meter (kg/cu m)	16.01846
pound per cubic yard (lb/cu yd)	kilogram per cubic meter (kg/cu m)	0.5933
Temperature		
degree Fahrenheit (°F)	degree Celsius (°C)	$t_C = (t_F - 32)/1.8$
degree Fahrenheit (°F)	degree Kelvin (°K)	$t_K = (t_F + 459.7)/1.8$
degree Kelvin (°K)	degree Celsius (C°)	$t_C = t_K - 273.15$
Energy and heat		
British thermal unit (Btu)	joule (J)	1055.056
calorie (cal)	joule (J)	4.1868 E
Btu/°F · hr · ft²	W/m² · °K	5.678263
kilowatt-hour (kwh)	joule (J)	3,600,000. E
British thermal unit per pound (Btu/lb)	calories per gram (cal/g)	0.55556
British thermal unit per hour (Btu/hr)	watt (W)	0.2930711
Power		
horsepower (hp) (550 ft-lb/sec)	watt (W)	745.6999 E
Velocity		
mile per hour (mph)	kilometer per hour (km/hr)	1.60934
mile per hour (mph)	meter per second (m/s)	0.44704
Permeability		
darcy	centimeter per second (cm/sec)	0.000968
feet per day (ft/day)	centimeter per second (cm/sec)	0.000352

*E indicates that the factor given is exact.
**One U.S. gallon equals 0.8327 Canadian gallon.
†A pascal equals 1.000 newton per square meter.

Note:
One U.S. gallon of water weighs 8.34 pounds (U.S.) at 60°F.
One cubic foot of water weighs 62.4 pounds (U.S.).
One milliliter of water has a mass of 1 gram and has a volume of one cubic centimeter.
One U.S. bag of cement weighs 94 lb.

The prefixes and symbols listed below are commonly used to form names and symbols of the decimal multiples and submultiples of the SI units.

Multiplication Factor	Prefix	Symbol
$1,000,000,000 = 10^9$	giga	G
$1,000,000 = 10^6$	mega	M
$1,000 = 10^3$	kilo	k
$1 = 1$	—	—
$0.01 = 10^{-2}$	centi	c
$0.001 = 10^{-3}$	milli	m
$0.000001 = 10^{-6}$	micro	μ
$0.000000001 = 10^{-9}$	nano	n

Section

23

Useful Tables, Charts, and Formulas

Contents

23.0.0 Nails: penny designations ("d") and lengths (U.S. and metric)
23.1.0 Stainless steel sheets (thicknesses and weights)
23.2.0 Comparable thickness and weights of stainless steel, aluminum, and copper
23.3.0 Wire and sheet-metal gauges and weights
23.4.0 Weights and specific gravities of common materials
23.5.0 Useful formulas
23.6.0 Decimal equivalents of inches in feet and yards
23.7.0 Conversion of fractions to decimals
23.7.1 Decimals of a foot for each 1/32"
23.7.2 Decimals of an inch for each 1/64", with millimeter equivalents
23.8.0 Solutions of the right triangle
23.9.0 Area and other formulas
23.10.0 Volume of vertical cylindrical tanks (in gallons per foot of depth)
23.11.0 Volume of rectangular tank capacities (in U.S. gallons per foot of depth)
23.12.0 Capacity of horizontal cylindrical tanks
23.13.0 Round, tapered tank capacities
23.14.0 Circumferences and areas of circles

23.0.0 Nails: Penny Designation ("d") and Lengths (U.S. and Metric)

Nail - Penny Size	Length in Inches	Length in Millimeters
2d	1	25.40
3d	1 1/4	31.75
4d	1 1/2	38.10
5d	1 3/4	44.45
6d	2	50.80
7d	2 1/4	57.15
8d	2 1/2	63.50
9d	2 3/4	69.85
10d	3	76.20
12d	3 1/4	82.55
16d	3 1/2	88.90
20d	3 3/4	95.25
30d	4 1/2	114.30
40d	5	127.00
50d	5 1/2	139.70
60d	6	152.40

23.1.0 Stainless Steel Sheets (Thicknesses and Weights)

Gauge	Thickness		Weight	
	Inches	Mm.	lbs/ft2	kg/m2
8	0.17188	4.3658	7.2187	44.242
10	0.14063	3.5720	5.9062	28.834
11	0.1250	3.1750	5.1500	25.6312
12	0.10938	2.7783	4.5937	22.427
14	0.07813	1.9845	3.2812	16.019
16	0.06250	1.5875	2.6250	12.815
18	0.05000	1.2700	2.1000	10.252
20	0.03750	0.9525	1.5750	7.689
22	0.03125	0.7938	1.3125	6.409
24	0.02500	0.6350	1.0500	5.126
26	0.01875	0.4763	0.7875	3.845
28	0.01563	0.3970	0.6562	3.1816
Plates				
3/16"	0.1875	4.76	7.752	37.85
1/4"	0.25	6.35	10.336	50.46
5/16"	0.3125	7.94	12.920	63.08
3/8"	0.375	9.53	15.503	75.79
1/2"	0.50	12.70	20.671	100.92
5/8"	0.625	15.88	25.839	126.15
3/4"	0.75	19.05	31.007	151.38
1"	1.00	25.4	41.342	201.83

23.2.0 Comparable Thicknesses and Weights of Stainless Steel, Aluminum, and Copper

STAINLESS STEEL			ALUMINUM			COPPER		
Thickness (Inch)	Gauge (U.S. Standard)	Lb. sq. ft.	Thickness (Inch)	Gauge (B&S)	Lb. sq. ft.	Thickness (Inch)	Oz. sq. ft.	Lb. sq. ft.
.010	32	.420	.010	30	.141	.0108	8	.500
.0125	30	.525	.0126	28	.177	.0121	9	.563
						.0135	10	.625
.0156	28	.656	.0156		.220	.0148	11	.688
			.0179	25	.253	.0175	13	.813
.0187	26	.788						
.0219	25	.919	.020	24	.282	.021	16	1.000
.025	24	1.050	.0253	22	.352			
						.027	20	1.250
.031	22	1.313	.0313	—	.441	.032	24	1.500
.0375	20	1.575	.032	20	.451	.0337	28	1.750
			.0403	18	.563	.0431	32	2.000
			.0453	17	.100			
.050	18	2.100	.0506	16	.126			

Note that U.S. Standard Gauge (stainless sheet) is not directly comparable with the B&S Gauge (aluminum). A 20-gauge stainless averages .0375" thick; while a 20-gauge aluminum averages .032" thick; and 20-ounce copper is .027" thick. The higher strength of stainless steel permits use of thinner gauges than required for aluminum or copper, which makes stainless more competitive with aluminum on a weight-to-coverage basis and provides stainless with a substantial weight saving compared to copper. For example, 100 sq. ft. of .032" aluminum will weigh about 45 pounds, .021" (16-ounce) copper will weigh about 100 pounds, and .015" stainless will weigh about 66 pounds.

23.3.0 Wire and Sheet-Metal Gauges and Weights

Name of Gage	*United States Standard Gage		The United States Steel Wire Gage	American or Brown & Sharpe Wire Gage	New Birmingham Standard Sheet & Hoop Gage	British Imperial or English Legal Standard Wire Gage	Birmingham or Stubs Iron Wire Gage	Name of Gage
Principal Use	Uncoated Steel Sheets and Light Plates		Steel Wire except Music Wire	Non-Ferrous Sheets and Wire	Iron and Steel Sheets and Hoops	Wire	Strips, Bands, Hoops and Wire	Principal Use
Gage No.	Weight Oz. per Sq. Ft.	Approx. Thickness Inches	Thickness, Inches					Gage No.
7/0's			.4900		.6666	.500		7/0's
6/0's			.4615	.5800	.625	.464		6/0's
5/0's			.4305	.5165	.5883	.432	.550	5/0's
4/0's			.3938	.4600	.5416	.400	.454	4/0's
3/0's			.3625	.3648	.500	.372	.425	3/0's
2/0's			.3310	.3249	.4452	.348	.380	2/0's
1/0			.3065	.2893	.3964	.324	.340	1/0
1			.2830	.2576	.3532	.300	.300	1
2			.2625	.2294	.3147	.276	.284	2
3	160	.2391	.2437	.2043	.2804	.252	.259	3
4	150	.2242	.2253	.1819	.250	.232	.238	4
5	140	.2092	.2070	.1620	.2225	.212	.220	5
6	130	.1943	.1920	.1443	.1981	.192	.203	6
7	120	.1793	.1770	.1285	.1764	.176	.180	7
8	110	.1644	.1620	.1144	.1570	.160	.165	8
9	100	.1495	.1483	.1019	.1398	.144	.148	9
10	90	.1345	.1350	.0907	.1250	.128	.134	10
11	80	.1196	.1205	.0808	.1113	.116	.120	11
12	70	.1046	.1055	.0720	.0991	.104	.109	12
13	60	.0897	.0915	.0641	.0882	.092	.095	13
14	50	.0747	.0800	.0571	.0785	.080	.083	14
15	45	.0673	.0720	.0508	.0699	.072	.072	15
16	40	.0598	.0625	.0453	.0625	.064	.065	16
17	36	.0538	.0540	.0403	.0556	.056	.058	17
18	32	.0478	.0475	.0359	.0495	.048	.049	18
19	28	.0418	.0410	.0320	.0440	.040	.042	19
20	24	.0359	.0348	.0285	.0392	.036	.035	20
21	22	.0329	.0317	.0253	.0349	.032	.032	21
22	20	.0299	.0286	.0226	.0313	.028	.028	22
23	18	.0269	.0258	.0201	.0278	.024	.025	23
24	16	.0239	.0230	.0179	.0248	.022	.022	24
25	14	.0209	.0204	.0159	.0220	.020	.020	25
26	12	.0179	.0181	.0142	.0196	.018	.018	26
27	11	.0164	.0173	.0126	.0175	.0164	.016	27
28	10	.0149	.0162	.0113	.0156	.0148	.014	28
29	9	.0135	.0150	.0100	.0139	.0136	.013	29
30	8	.0120	.0140	.0089	.0123	.0124	.012	30
31	7	.0105	.0132	.0080	.0110	.0116	.010	31
32	6.5	.0097	.0128	.0071	.0098	.0108	.009	32
33	6	.0090	.0118	.0063	.0087	.0100	.008	33
34	5.5	.0082	.0104	.0056	.0077	.0092	.007	34
35	5	.0075	.0095	.0050	.0069	.0084	.005	35
36	4.5	.0067	.0090	.0045	.0061	.0076	.004	36
37	4.25	.0064	.0085	.0040	.0054	.0068		37
38	4	.0060	.0080	.0035	.0048	.0060		38
39			.0075	.0031	.0043	.0052		39
40			.0070		.0039	.0048		40

* U.S. Standard Gage is officially a weight gage, in oz. per sq. ft. as tabulated. The Approx. Thickness shown is the "Manufacturers' Standard" of the American Iron and Steel Institute, based on steel as weighing 501.81 lb. per cu. ft. (489.6 true weight plus 2.5 per cent for average over-run in area and thickness).

23.4.0 Weights and Specific Gravities of Common Materials

Substance	Weight Lb. per Cu. Ft.	Specific Gravity
METALS, ALLOYS, ORES		
Aluminum, cast, hammered	165	2.55-2.75
Brass, cast, rolled	534	8.4-8.7
Bronze, 7.9 to 14% Sn	509	7.4-8.9
Bronze, aluminum	481	7.7
Copper, cast, rolled	556	8.8-9.0
Copper ore, pyrites	262	4.1-4.3
Gold, cast, hammered	1205	19.25-19.3
Iron, cast, pig	450	7.2
Iron, wrought	485	7.6-7.9
Iron, spiegel-eisen	468	7.5
Iron, ferro-silicon	437	6.7-7.3
Iron ore, hematite	325	5.2
Iron ore, hematite in bank	160-180	....
Iron ore, hematite loose	130-160	....
Iron ore, limonite	237	3.6-4.0
Iron ore, magnetite	315	4.9-5.2
Iron slag	172	2.5-3.0
Lead	710	11.37
Lead ore, galena	465	7.3-7.6
Magnesium, alloys	112	1.74-1.83
Manganese	475	7.2-8.0
Manganese ore, pyrolusite	259	3.7-4.6
Mercury	849	13.6
Monel Metal	556	8.8-9.0
Nickel	565	8.9-9.2
Platinum, cast, hammered	1330	21.1-21.5
Silver, cast, hammered	656	10.4-10.6
Steel, rolled	490	7.85
Tin, cast, hammered	459	7.2-7.5
Tin ore, cassiterite	418	6.4-7.0
Zinc, cast, rolled	440	6.9-7.2
Zinc ore, blende	253	3.9-4.2
VARIOUS SOLIDS		
Cereals, oats ... bulk	32	....
Cereals, barley ... bulk	39	....
Cereals, corn, rye ... bulk	48	....
Cereals, wheat ... bulk	48	....
Hay and Straw ... bales	20	....
Cotton, Flax, Hemp	93	1.47-1.50
Fats	58	0.90-0.97
Flour, loose	28	0.40-0.50
Flour, pressed	47	0.70-0.80
Glass, common	156	2.40-2.60
Glass, plate or crown	161	2.45-2.72
Glass, crystal	184	2.90-3.00
Leather	59	0.86-1.02
Paper	58	0.70-1.15
Potatoes, piled	42	....
Rubber, caoutchouc	59	0.92-0.96
Rubber goods	94	1.0-2.0
Salt, granulated, piled	48	....
Saltpeter	67	....
Starch	96	1.53
Sulphur	125	1.93-2.07
Wool	82	1.32

Substance	Weight Lb. per Cu. Ft.	Specific Gravity
TIMBER, U. S. SEASONED		
Moisture Content by Weight:		
Seasoned timber 15 to 20%		
Green timber up to 50%		
Ash, white, red	40	0.62-0.65
Cedar, white, red	22	0.32-0.38
Chestnut	41	0.66
Cypress	30	0.48
Fir, Douglas spruce	32	0.51
Fir, eastern	25	0.40
Elm, white	45	0.72
Hemlock	29	0.42-0.52
Hickory	49	0.74-0.84
Locust	46	0.73
Maple, hard	43	0.68
Maple, white	33	0.53
Oak, chestnut	54	0.86
Oak, live	59	0.95
Oak, red, black	41	0.65
Oak, white	46	0.74
Pine, Oregon	32	0.51
Pine, red	30	0.48
Pine, white	26	0.41
Pine, yellow, long-leaf	44	0.70
Pine, yellow, short-leaf	38	0.61
Poplar	30	0.48
Redwood, California	26	0.42
Spruce, white, black	27	0.40-0.46
Walnut, black	38	0.61
Walnut, white	26	0.41
VARIOUS LIQUIDS		
Alcohol, 100%	49	0.79
Acids, muriatic 40%	75	1.20
Acids, nitric 91%	94	1.50
Acids, sulphuric 87%	112	1.80
Lye, soda 66%	106	1.70
Oils, vegetable	58	0.91-0.94
Oils, mineral, lubricants	57	0.90-0.93
Water, 4°C. max. density	62.428	1.0
Water, 100°C.	59.830	0.9584
Water, ice	56	0.88-0.92
Water, snow, fresh fallen	8	.125
Water, sea water	64	1.02-1.03
GASES		
Air, 0°C. 760 mm.	.08071	1.0
Ammonia	.0478	0.5920
Carbon dioxide	.1234	1.5291
Carbon monoxide	.0781	0.9673
Gas, illuminating	.028-.036	0.35-0.45
Gas, natural	.038-.039	0.47-0.48
Hydrogen	.00559	0.0693
Nitrogen	.0784	0.9714
Oxygen	.0892	1.1056

The specific gravities of solids and liquids refer to water at 4°C., those of gases to air at 0°C. and 760 mm. pressure. The weights per cubic foot are derived from average specific gravities, except where stated that weights are for bulk, heaped or loose material, etc.

23.4.0 Weights and Specific Gravities of Common Materials (Continued)

Substance	Weight Lb. per Cu. Ft.	Specific Gravity
ASHLAR MASONRY		
Granite, syenite, gneiss	165	2.3-3.0
Limestone, marble	160	2.3-2.8
Sandstone, bluestone	140	2.1-2.4
MORTAR RUBBLE MASONRY		
Granite, syenite, gneiss	155	2.2-2.8
Limestone, marble	150	2.2-2.6
Sandstone, bluestone	130	2.0-2.2
DRY RUBBLE MASONRY		
Granite, syenite, gneiss	130	1.9-2.3
Limestone, marble	125	1.9-2.1
Sandstone, bluestone	110	1.8-1.9
BRICK MASONRY		
Pressed brick	140	2.2-2.3
Common brick	120	1.8-2.0
Soft brick	100	1.5-1.7
CONCRETE MASONRY		
Cement, stone, sand	144	2.2-2.4
Cement, slag, etc.	130	1.9-2.3
Cement, cinder, etc.	100	1.5-1.7
VARIOUS BUILDING MATERIALS		
Ashes, cinders	40-45	....
Cement, portland, loose	90	....
Cement, portland, set	183	2.7-3.2
Lime, gypsum, loose	53-64	....
Mortar, set	103	1.4-1.9
Slags, bank slag	67-72	....
Slags, bank screenings	98-117	....
Slags, machine slag	96	....
Slags, slag sand	49-55	....
EARTH, ETC., EXCAVATED		
Clay, dry	63	....
Clay, damp, plastic	110	....
Clay and gravel, dry	100	....
Earth, dry, loose	76	....
Earth, dry, packed	95	....
Earth, moist, loose	78	....
Earth, moist, packed	96	....
Earth, mud, flowing	108	....
Earth, mud, packed	115	....
Riprap, limestone	80-85	....
Riprap, sandstone	90	....
Riprap, shale	105	....
Sand, gravel, dry, loose	90-105	....
Sand, gravel, dry, packed	100-120	....
Sand, gravel, wet	118-120	....
EXCAVATIONS IN WATER		
Sand or gravel	60	....
Sand or gravel and clay	65	....
Clay	80	....
River mud	90	....
Soil	70	....
Stone riprap	65	....

Substance	Weight Lb. per Cu. Ft.	Specific Gravity
MINERALS		
Asbestos	153	2.1-2.8
Barytes	281	4.50
Basalt	184	2.7-3.2
Bauxite	159	2.55
Borax	109	1.7-1.8
Chalk	137	1.8-2.6
Clay, marl	137	1.8-2.6
Dolomite	181	2.9
Feldspar, orthoclase	159	2.5-2.6
Gneiss, serpentine	159	2.4-2.7
Granite, syenite	175	2.5-3.1
Greenstone, trap	187	2.8-3.2
Gypsum, alabaster	159	2.3-2.8
Hornblende	187	3.0
Limestone, marble	165	2.5-2.8
Magnesite	187	3.0
Phosphate rock, apatite	200	3.2
Porphyry	172	2.6-2.9
Pumice, natural	40	0.37-0.90
Quartz, flint	165	2.5-2.8
Sandstone, bluestone	147	2.2-2.5
Shale, slate	175	2.7-2.9
Soapstone, talc	169	2.6-2.8
STONE, QUARRIED, PILED		
Basalt, granite, gneiss	96	....
Limestone, marble, quartz	95	....
Sandstone	82	....
Shale	92	....
Greenstone, hornblende	107	....
BITUMINOUS SUBSTANCES		
Asphaltum	81	1.1-1.5
Coal, anthracite	97	1.4-1.7
Coal, bituminous	84	1.2-1.5
Coal, lignite	78	1.1-1.4
Coal, peat, turf, dry	47	0.65-0.85
Coal, charcoal, pine	23	0.28-0.44
Coal, charcoal, oak	33	0.47-0.57
Coal, coke	75	1.0-1.4
Graphite	131	1.9-2.3
Paraffine	56	0.87-0.91
Petroleum	54	0.87
Petroleum, refined	50	0.79-0.82
Petroleum, benzine	46	0.73-0.75
Petroleum, gasoline	42	0.66-0.69
Pitch	69	1.07-1.15
Tar, bituminous	75	1.20
COAL AND COKE, PILED		
Coal, anthracite	47-58	....
Coal, bituminous, lignite	40-54	....
Coal, peat, turf	20-26	....
Coal, charcoal	10-14	....
Coal, coke	23-32	....

The specific gravities of solids and liquids refer to water at 4°C., those of gases to air at 0°C. and 760 mm. pressure. The weights per cubic foot are derived from average specific gravities, except where stated that weights are for bulk, heaped or loose material, etc.

23.5.0 Useful Formulas

Circumference of a circle = π × *diameter* or 3.1416 × *diameter*
Diameter of a circle = *circumference* × 0.31831
Area of a square = *length* × *width*
Area of a rectangle = *length* × *width*
Area of a parallelogram = *base* × *perpendicular height*
Area of a triangle = ½ *base* × *perpendicular weight*
Area of a circle = π *radius squared* or *diameter squared* × 0.7854
Area of an ellipse = *length* × *width* × 0.7854
Volume of a cube or rectangular prism = *length* × *width* × *height*
Volume of a triangular prism = *area of triangle* × *length*
Volume of a sphere = *diameter cubed* × 0.5236 (*diameter* × *diameter* × *diameter* × 0.5236)
Volume of a cone = π × *radius squared* × ⅓ *height*
Volume of a cylinder = π × *radius squared* × *height*
Length of one side of a square × 1.128 = *diameter of an equal circle*
Doubling the diameter of a pipe or cylinder increases its capacity 4 times
Pressure (in lb/sq in.) *of a column of water* = *height of the column* (in feet) × 0.434
Capacity of a pipe or tank (in U.S. gallons) = *diameter squared* (in inches) × *length* (in inches) × 0.0034
1 gal water = 8⅓ lb = 231 cu in.
1 cu ft water = 62½ lb = 7½ gal.

23.6.0 Decimal Equivalents of Inches in Feet and Yards

Inches	Feet	Yards
1	.0833	.0278
2	.1667	.0556
3	.2500	.0833
4	.333	.1111
5	.4166	.1389
6	.5000	.1667
7	.5833	.1944
8	.6667	.2222
9	.7500	.2500
10	.8333	.2778
11	.9166	.3056
12	1.000	.3333

23.7.0 Conversion of Fractions to Decimals

Fractions	Decimal	Fractions	Decimal
1/64	.015625	33/64	.515625
1/32	.03125	17/32	.53125
3/64	.046875	35/64	.546875
1/16	.0625	9/16	.5625
5/64	.078125	37/64	.578125
3/32	.09375	19/32	.59375
7/64	.109375	38/64	.609375
1/8	.125	5/8	.625
9/64	.140625	41/64	.640625
5/32	.15625	21/32	.65625
11/64	.1719	43/64	.67187
3/16	.1875	11/16	.6875
13/64	.2031	45/64	.70312
7/32	.2188	23/32	.71875
15/64	.234375	47/64	.734375
1/4	.25	3/4	.75
17/64	.265625	49/64	.765625
9/32	.28125	25/32	.78125
19/64	.296875	51/64	.796875
5/16	.3125	13/10	.8125
21/64	.328125	53/64	.828125
11/32	.34375	27/32	.84375
23/64	.359375	55/64	.859375
3/8	.375	7/8	.875
25/64	.398625	57/64	.890625
13/32	.40625	29/32	.90625
27/64	.421875	60/64	.921875
7/16	.4375	15/16	.9375
20/64	.453125	61/64	.953125
15/32	.46875	31/32	.96875
31/64	.484375	63/64	.984375
1/2	.50	1″	1.000000

By permission of Cast Iron Soil Pipe Institute

23.7.1 Decimals of a Foot for Each 1/32"

Inch	0	1	2	3	4	5
0	0	.0833	.1667	.2500	.3333	.4167
1/32	.0026	.0859	.1693	.2526	.3359	.4193
1/16	.0052	.0885	.1719	.2552	.3385	.4219
3/32	.0078	.0911	.1745	.2578	.3411	.4245
1/8	.0104	.0938	.1771	.2604	.3438	.4271
5/32	.0130	.0964	.1797	.2630	.3464	.4297
3/16	.0156	.0990	.1823	.2656	.3490	.4323
7/32	.0182	.1016	.1849	.2682	.3516	.4349
1/4	.0208	.1042	.1875	.2708	.3542	.4375
9/32	.0234	.1068	.1901	.2734	.3568	.4401
5/16	.0260	.1094	.1927	.2760	.3594	.4427
11/32	.0286	.1120	.1953	.2786	.3620	.4453
3/8	.0313	.1146	.1979	.2812	.3646	.4479
13/32	.0339	.1172	.2005	.2839	.3672	.4505
7/16	.0365	.1198	.2031	.2865	.3698	.4531
15/32	.0391	.1224	.2057	.2891	.3724	.4557
1/2	.0417	.1250	.2083	.2917	.3750	.4583
17/32	.0443	.1276	.2109	.2943	.3776	.4609
9/16	.0469	.1302	.2135	.2969	.3802	.4635
19/32	.0495	.1328	.2161	.2995	.3828	.4661
5/8	.0521	.1354	.2188	.3021	.3854	.4688
21/32	.0547	.1380	.2214	.3047	.3880	.4714
11/16	.0573	.1406	.2240	.3073	.3906	.4740
23/32	.0599	.1432	.2266	.3099	.3932	.4766
3/4	.0625	.1458	.2292	.3125	.3958	.4792
25/32	.0651	.1484	.2318	.3151	.3984	.4818
13/16	.0677	.1510	.2344	.3177	.4010	.4844
27/32	.0703	.1536	.2370	.3203	.4036	.4870
7/8	.0729	.1563	.2396	.3229	.4063	.4896
29/32	.0755	.1589	.2422	.3255	.4089	.4922
15/16	.0781	.1615	.2448	.3281	.4115	.4948
31/32	.0807	.1641	.2474	.3307	.4141	.4974

23.7.2 Decimals of an Inch for each 1/64", with Millimeter Equivalents

Fraction	1/64ths	Decimal	Millimeters (Approx.)	Fraction	1/64ths	Decimal	Millimeters (Approx.)
...	1	.015625	0.397	...	33	.515625	13.097
1/32	2	.03125	0.794	17/32	34	.53125	13.494
...	3	.046875	1.191	...	35	.546875	13.891
1/16	4	.0625	1.588	9/16	36	.5625	14.288
...	5	.078125	1.984	...	37	.578125	14.684
3/32	6	.09375	2.381	19/32	38	.59375	15.081
...	7	.109375	2.778	...	39	.609375	15.478
1/8	8	.125	3.175	5/8	40	.625	15.875
...	9	.140625	3.572	...	41	.640625	16.272
5/32	10	.15625	3.969	21/32	42	.65625	16.669
...	11	.171875	4.366	...	43	.671875	17.066
3/16	12	.1875	4.763	11/16	44	.6875	17.463
...	13	.203125	5.159	...	45	.703125	17.859
7/32	14	.21875	5.556	23/32	46	.71875	18.256
...	15	.234375	5.953	...	47	.734375	18.653
1/4	16	.250	6.350	3/4	48	.750	19.050
...	17	.265625	6.747	...	49	.765625	19.447
9/32	18	.28125	7.144	25/32	50	.78125	19.844
...	19	.296875	7.541	...	51	.796875	20.241
5/16	20	.3125	7.938	13/16	52	.8125	20.638
...	21	.328125	8.334	...	53	.828125	21.034
11/32	22	.34375	8.731	27/32	54	.84375	21.431
...	23	.359375	9.128	...	55	.859375	21.828
3/8	24	.375	9.525	7/8	56	.875	22.225
...	25	.390625	9.922	...	57	.890625	22.622
13/32	26	.40625	10.319	29/32	58	.90625	23.019
...	27	.421875	10.716	...	59	.921875	23.416
7/16	28	.4375	11.113	15/16	60	.9375	23.813
...	29	.453125	11.509	...	61	.953125	24.209
15/32	30	.46875	11.906	31/32	62	.96875	24.606
...	31	.484375	12.303	...	63	.984375	25.003
1/2	32	.500	12.700	1	64	1.000	25.400

23.8.0 Solutions of the Right Triangle

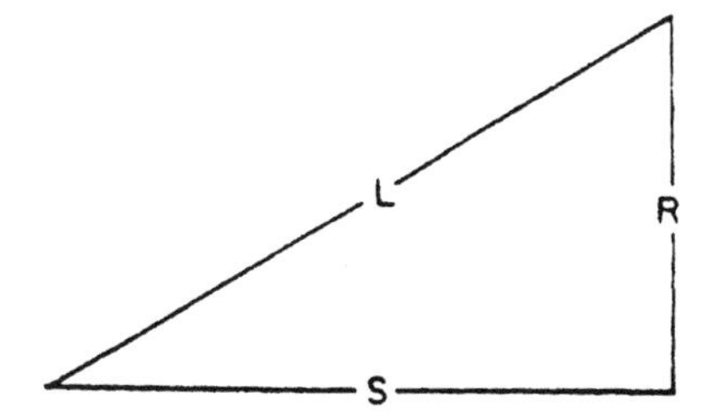

To find side	When you know side	Multiply side	For 45 Ells-By	For 22 1/2 Ells-By	For 67 1/2 Ells-By	For 72 Ells-By	For 60 Ells-By	For 80 Ells-By
L	S	S	1.4142	2.6131	1.08	1.05	1.1547	2.00
S	L	L	.707	.3826	.92	.95	.866	.50
R	S	S	1.000	2.4142	.414	.324	.5773	.1732
S	R	R	1.000	.4142	2.41	3.07	1.732	.5773
L	R	R	1.4142	1.0824	2.61	3.24	2.00	1.1547
R	L	L	.7071	.9239	.38	.31	.50	.866

By permission of Cast Iron Soil Pipe Institute

23.9.0 Area and Other Formulas

Parallelogram	*Area = base × distance between the two parallel sides*
Pyramid	*Area = ½ perimeter of base × slant height + area of base*
	Volume = area of base × ⅓ of the altitude
Rectangle	*Area = length × width*
Rectangular prisms	*Volume = width × height × length*
Sphere	*Area of surface = diameter × diameter × 3.1416*
	Side of inscribed cube = radius × 1.547
	Volume = diameter × diameter × diameter × 0.5236
Square	*Area = length × width*
Triangle	*Area = one half of height times base*
Trapezoid	*Area = one half of the sum of the parallel sides × height*
Cone	*Area of surface = one half of circumference of base × slant height + area of base*
	Volume = diameter × diameter × 0.7854 × one third of the altitude
Cube	*Volume = width × height × length*
Ellipse	*Area = short diameter × long diameter × 0.7854*
Cylinder	*Area of surface = diameter × 3.1416 × length + area of the two bases*
	Area of base = diameter × diameter × 0.7854
	Area of base = volume + length
	Length = volume + area of base
	Volume = length × area of base
	Capacity in gallons = volume in inches + 231
	Capacity of gallons = diameter × diameter × length × 0.0034
	Capacity in gallons = volume in feet × 7.48
Circle	*Circumference = diameter × 3.1416*
	Circumference = radius × 6.2832
	Diameter = radius × 2
	Diameter = square root of = (area + 0.7854)
	Diameter = square root of area × 1.1283

23.10.0 Volume of Vertical Cylindrical Tanks (in Gallons Per Foot of Depth)

Diameter in		U. S.	Diameter in		U. S.	Diameter in		U. S.
Feet	Inches	Gallons	Feet	Inches	Gallons	Feet	Inches	Gallons
1	0	5.875	3	6	71.97	6	0	211.5
1	1	6.895	3	7	75.44	6	3	220.5
1	2	7.997	3	8	78.99	6	6	248.2
1	3	9.180	3	9	82.62	6	9	267.7
1	4	10.44	3	10	86.33	7	0	287.9
1	5	11.79	3	11	90.13	7	3	308.8
1	6	13.22	4	0	94.00	7	6	330.5
1	7	14.73	4	1	97.96	7	9	352.9
1	8	16.32	4	2	102.0	8	0	376.0
1	9	17.99	4	3	106.1	8	3	399.9
1	10	19.75	4	4	110.3	8	6	424.5
1	11	21.58	4	5	114.6	8	9	449.8
2	0	23.50	4	6	119.0	9	0	475.9
2	1	25.50	4	7	123.4	9	3	502.7
2	2	27.58	4	8	127.9	9	6	530.2
2	3	29.74	4	9	132.6	9	9	558.5
2	4	31.99	4	10	137.3	10	0	587.5
2	5	34.31	4	11	142.0	10	3	617.3
2	6	36.72	5	0	146.9	10	6	647.7
2	7	39.21	5	1	151.8	10	9	679.0
2	8	41.78	5	2	156.8	11	0	710.9
2	9	44.43	5	3	161.9	11	3	743.6
2	10	47.16	5	4	167.1	11	6	777.0
2	11	49.98	5	5	172.4	11	9	811.1
3	0	52.88	5	6	177.7	12	0	846.0
3	1	55.86	5	7	183.2	12	3	881.6
3	2	58.92	5	8	188.7	12	6	918.0
3	3	62.06	5	9	194.2	12	9	955.1
3	4	65.28	5	10	199.9			
3	5	68.58	5	11	205.7			

By permission of Cast Iron Soil Pipe Institute

23.11.0 Volume of Rectangular Tank Capacities (in U.S. Gallons Per Foot of Depth)

Width	LENGTH OF TANK — IN FEET						
Feet	2	2 1/2	3	3 1/2	4	4 1/2	5
2	29.92	37.40	44.88	52.36	59.84	67.32	74.81
2 1/2	—	46.75	56.10	65.45	74.81	84.16	93.51
3	—	—	67.32	78.55	89.77	101.0	112.2
3 1/2	—	—	—	91.64	104.7	117.8	130.9
4	—	—	—	—	119.7	134.6	149.6
4 1/2	—	—	—	—	—	151.5	168.3
5	—	—	—	—	—	—	187.0
	5 1/2	6	6 1/2	7	7 1/2	8	8 1/2
2	82.29	89.77	97.25	104.7	112.2	119.7	127.2
2 1/2	102.9	112.2	121.6	130.9	140.3	149.6	159.0
3	123.4	134.6	145.9	157.1	168.3	179.5	190.8
3 1/2	144.0	157.1	170.2	183.3	196.4	209.5	222.5
4	164.6	179.5	194.5	209.5	224.4	239.4	254.3
4 1/2	185.1	202.0	218.8	235.6	252.5	269.3	286.1
5	205.7	224.4	243.1	261.8	280.5	299.2	317.9
5 1/2	226.3	246.9	267.4	288.0	308.6	329.1	349.7
6	—	269.3	291.7	314.2	336.6	359.1	381.5
6 1/2	—	—	316.1	340.4	364.7	389.0	413.3
7	—	—	—	366.5	392.7	418.9	445.1
7 1/2	—	—	—	—	420.8	448.8	476.9
8	—	—	—	—	—	478.8	508.7
8 1/2	—	—	—	—	—	—	540.5
	9	9 1/2	10	10 1/2	11	11 1/2	12
2	134.6	142.1	149.6	157.1	164.6	172.1	179.5
2 1/2	168.3	177.7	187.0	196.4	205.7	215.1	224.4
3	202.0	213.2	224.4	235.6	246.9	258.1	269.3
3 1/2	235.6	248.7	261.8	274.9	288.0	301.1	314.2
4	269.3	284.3	299.2	314.2	329.1	344.1	359.1
4 1/2	303.0	319.8	336.6	353.5	370.3	387.1	403.9
5	336.6	355.3	374.0	392.7	411.4	430.1	448.8
5 1/2	370.3	390.9	411.4	432.0	452.6	473.1	493.7
6	403.9	426.4	448.8	471.3	493.7	516.2	538.6
6 1/2	437.6	461.9	486.2	510.5	534.9	559.2	583.5
7	471.3	497.5	523.6	549.8	576.0	602.2	628.4
7 1/2	504.9	533.0	561.0	589.1	617.1	645.2	673.2
8	538.6	568.5	598.4	628.4	658.3	688.2	718.1
8 1/2	572.3	604.1	635.8	667.6	699.4	731.2	763.0
9	605.9	639.6	673.2	706.9	740.6	774.2	807.9
9 1/2	—	675.1	710.6	746.2	781.7	817.2	852.8
10	—	—	748.1	785.5	822.9	860.3	897.7
10 1/2	—	—	—	824.7	864.0	903.3	942.5
11	—	—	—	—	905.1	946.3	987.4
11 1/2	—	—	—	—	—	989.3	1032.0
12	—	—	—	—	—	—	1077.0

By permission of Cast Iron Soil Pipe Institute

23.12.0 Capacity of Horizontal Cylindrical Tanks

% Depth Filled	% of Capacity	% Depth Filled	% of Capacity	% Depth Filled	% of Capacity	% Depth Filled	% of Capacity
1	.20	26	20.73	51	51.27	76	81.50
2	.50	27	21.86	52	52.55	77	82.60
3	.90	28	23.00	53	53.81	78	83.68
4	1.34	29	24.07	54	55.08	79	84.74
5	1.87	30	25.31	55	56.34	80	85.77
6	2.45	31	26.48	56	57.60	81	86.77
7	3.07	32	27.66	57	58.86	82	87.76
8	3.74	33	28.84	58	60.11	83	88.73
9	4.45	34	30.03	59	61.36	84	89.68
10	5.20	35	31.19	60	62.61	85	90.60
11	5.98	36	32.44	61	63.86	86	91.50
12	6.80	37	33.66	62	65.10	87	92.36
13	7.64	38	34.90	63	66.34	88	93.20
14	8.50	39	36.14	64	67.56	89	94.02
15	9.40	40	37.36	65	68.81	90	94.80
16	10.32	41	38.64	66	69.97	91	95.50
17	11.27	42	39.89	67	71.16	92	96.26
18	12.24	43	41.14	68	72.34	93	96.93
19	13.23	44	42.40	69	73.52	94	97.55
20	14.23	45	43.66	70	74.69	95	98.13
21	15.26	46	44.92	71	75.93	96	98.66
22	16.32	47	46.19	72	77.00	97	99.10
23	17.40	48	47.45	73	78.14	98	99.50
24	18.50	49	48.73	74	79.27	99	99.80
25	19.61	50	50.00	75	80.39	100	100.00

By permission of Cast Iron Soil Pipe Institute

23.13.0 Round-Tapered Tank Capacities

$$Volume = \frac{h^3}{3} \frac{[(Area_{Top} + Area_{Base}) + \sqrt{(Area_{Top} + Area_{Base}}]}{231}$$

If inches are used.

$$Volume = \frac{h}{3} [(Area_{Base} + Area_{Top}) + \sqrt{(Area_{Base} + Area_{Top}}] \times 7.48$$

If feet are used.

Sample Problem

Let d be 12" (2 ft.)
D be 36" (3 ft.)
h be 48" (4 ft.)

Find volume in gallons.

$$Volume = \frac{48}{3} \frac{[(\pi \times 12^2) + (\pi \times 18^2) + \sqrt{\pi\ 12^2 \times 18^2}]}{231}$$

Where dimensions are in inches

$$Volume = \frac{4}{3} [(\pi \times 12^2) + (\pi \times 1½^2) + \sqrt{(\pi \times 1^2) \times (\pi \times ½^2)}] \times 7.48$$

Where dimensions are in feet

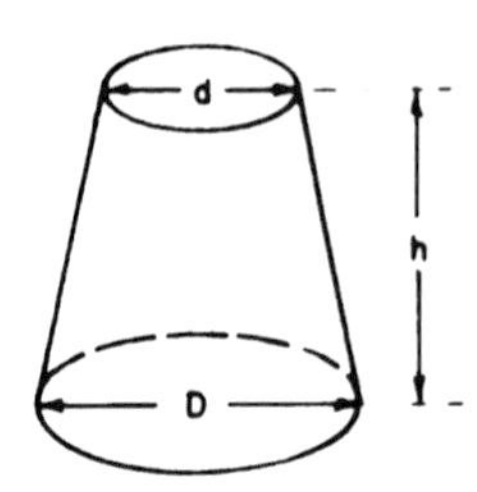

By permission of Cast Iron Soil Pipe Institute

23.14.0 Circumferences and Areas of Circles

Of One Inch				Of Inches or Feet					
Fract.	Decimal	Circ.	Area	Dia.	Circ.	Area	Dia.	Circ.	Area
1/64	.015625	.04909	.00019	1	3.1416	.7854	64	201.06	3216.99
1/32	.03125	.09818	.00077	2	6.2832	3.1416	65	204.20	3318.31
3/64	.046875	.14726	.00173	3	9.4248	7.0686	66	207.34	3421.19
1/16	.0625	.19635	.00307	4	12.5664	12.5664	67	210.49	3525.65
5/64	.078125	.24545	.00479	5	15.7080	19.635	68	213.63	3631.68
3/32	.09375	.29452	.00690	6	18.850	28.274	69	216.77	3739.28
7/64	.109375	.34363	.00939	7	21.991	38.485	70	219.91	3848.45
1/8	.125	.39270	.01227	8	25.133	50.266	71	223.05	3959.19
9/64	.140625	.44181	.01553	9	28.274	63.617	72	226.19	4071.50
5/32	.15625	.49087	.01917	10	31.416	78.540	73	229.34	4185.50
11/64	.171875	.53999	.02320	11	34.558	95.033	74	232.48	4300.84
3/16	.1875	.58.905	.02761	12	37.699	113.1	75	235.62	4417.86
13/64	.203125	.63817	.03241	13	40.841	132.73	76	238.76	4536.46
7/32	.21875	.68722	.03757	4	43.982	153.94	77	241.90	4656.63
15/64	.234375	.73635	.04314	15	47.124	176.71	78	245.04	4778.36
1/4	.25	.78540	.04909	16	50.265	201.06	79	248.19	4901.67
17/64	.265625	.83453	.05542	17	53.407	226.98	80	251.33	5026.55
9/32	.28125	.88357	.06213	18	56.549	254.47	81	254.47	5153.0
10/64	.296875	.93271	.06922	19	59.690	283.53	82	257.61	5281.02
5/16	.3125	.98175	.07670	20	63.832	314.16	83	260.75	5410.61
21/64	.328125	1.0309	.08456	21	65.973	346.36	84	263.89	5541.77
11/32	.34375	1.0799	.09281	22	69.115	380.13	85	267.04	5674.50
23/64	.35975	1.1291	.10144	23	72.257	415.48	86	270.18	5808.80
3/8	.375	1.1781	.11045	24	75.398	452.39	87	273.32	5944.68
25/64	.390625	1.2273	.11984	25	78.540	490.87	88	276.46	6082.12
13/32	.40625	1.2763	.12962	26	81.681	530.93	89	279.60	6221.14
27/64	.421875	1.3254	.13979	27	84.823	572.56	90	282.74	6361.71
7/16	.4375	1.3744	.15033	28	87.965	615.75	91	258.88	6503.88
29/64	.453125	1.4236	.16126	29	91.106	660.52	92	289.03	6647.61
15/32	.46875	1.4726	.17257	30	94.248	706.86	93	292.17	6792.91
31/64	.484375	1.5218	.18427	31	97.389	754.77	94	295.31	6939.78
1/2	.5	1.5708	.19635	32	100.53	804.25	95	298.45	7088.22

By permission of Cast Iron Soil Pipe Institute

23.14.0 Circumferences and Areas of Circles (Continued)

Of One Inch					Of Inches or Feet				
Fract.	Decimal	Circ.	Area	Dia.	Circ.	Area	Dia.	Circ.	Area
33/64	.515625	1.6199	.20880	33	103.67	855.30	96	301.59	7238.23
17/32	.53125	1.6690	.22166	34	106.81	907.92	97	304.73	7339.81
35/64	.546875	1.7181	.23489	35	109.96	962.11	98	307.88	7542.96
9/16	.5625	1.7671	.24850	36	113.10	1017.88	99	311.02	7697.69
37/64	.578125	1.8163	.26248	37	116.24	1075.21	100	314.16	7853.98
19/32	.59375	1.8653	.27688	38	119.38	1134.11	101	317.30	8011.85
30/64	.609375	1.9145	.29164	39	122.52	1194.59	102	320.44	8171.28
5/8	.625	1.9635	.30680	40	125.66	1256.64	103	323.58	8332.29
41/64	.640625	2.0127	.32232	41	128.81	1320.25	104	326.73	8494.87
21/32	.65625	2.0617	33824	42	131.95	1385.44	105	327.87	8659.01
43/64	.671875	2.1108	.35453	43	135.09	1452.20	106	333.01	8824.73
11/16	.6875	2.1598	.37122	44	138.23	1520.53	107	336.15	1992.02
45/64	.703125	2.2090	.38828	45	141.37	1590.43	108	339.29	9160.88
23/32	.71875	2.2580	.40574	46	144.51	1661.90	109	342.43	9331.32
47/64	.734375	2.3072	.42356	47	147.65	1734.94	110	345.58	9503.32
3/4	.75	2.3562	.44179	48	150.80	1809.56	111	348.72	9676.89
49/64	.765625	2.4050	.45253	49	153.94	1885.74	112	351.86	9853.03
23/32	.78125	2.4544	.47937	50	157.08	1963.50	113	355.0	10028.75
51/64	.796875	2.5036	.49872	51	160.22	2042.82	114	358.14	10207.03
13/16	.8125	2.5525	.51849	52	163.36	2123.72	115	361.28	10386.89
53/64	.828125	2.6017	.53862	53	166.50	2206.18	116	364.42	10568.32
27/32	.84375	2.6507	.55914	54	169.65	2290.22	117	367.57	10751.32
55/64	.859375	2.6999	.58003	55	172.79	2375.83	118	370.71	10935.88
7/8	.875	2.7489	.60123	56	175.93	2463.01	119	373.85	11122.02
57/64	.890625	2.7981	.62298	57	179.07	2551.76	120	376.99	11309 ’3
29/32	.90625	2.8471	.64504	58	182.21	2642.08	121	380.13	11499 01
59/64	.921875	2.8963	.66746	59	185.35	2733.97	122	383.27	11689.07
15/16	.9375	2.9452	.69029	60	188.50	2827.43	123	386.42	11882.29
61/64	.953125	2.9945	.71349	61	191.64	2922.47	124	389.56	12076.28
31/32	.96875	3.0434	.73708	62	194.78	3019.07	125	392.70	12271.85
63/64	.984375	3.0928	.76097	63	197.92	3117.25	126	395.84	12468.98

By permission of Cast Iron Soil Pipe Institute

Index

A

Acoustics, 320-335
  Decibel levels, common, 321
    Electrical transformers, 335
  Doors, 331
  Duct systems, 333
  Drywall partitions, 322, 323, 328, 329, 334
    Do's and dont's, 322
    High performance construction, 424, 429, 439
    Illustrations, 323
    STC ratings, 328
  Electrical installations, 334
    Transformers, 335
  Plumbing installations, 333
  Sound attenuation blankets, 327
  STC ratings, 321, 323, 328, 330
    Ceilings, 330
    Concrete, 323, 330
    Controlling, 327
    Doors, 331
    Floors, 323
    Suggested ratings, 329
    Wood, 323
  Stereo, effects on, 325-326
  Television, 325-326
  What is sound?, 320
Aluminum doors, 372, 373
  Revolving doors, 373
Aluminum windows, 374, 375
American Lumber Standards, 183-185
  Pressure treated lumber markings, 184
  Registered trademarks, 185
American Plywood Association, 224
  Grading guidelines, 224
  Roof sheathing specs, 239
  Siding, surfaces available, 228, 229
  Span tables, 231-238
  Sturdi-Floor, 239
  Trademarks, 225
American Softwood Standards, 196-200
ASTM, 41, 42, 150, 654
  Metric conversion-pipe, 654
  Reinforced concrete pipe, 41, 42
  Threaded fasteners, 150

B

Batter boards, 35
Bituminous paving, 28-32
    Calculating amounts, 29-31
      Metric, 30, 31
    Paving blocks, 32
    Pitfalls to avoid, 28
Boilers (see HVAC), 575-586
Brick, 102-107, 109-113, 116-120
  Arches, 107
  Corbeling, 119
  Estimating, 108
  Flashing details, 115-117
  Horizontal coursing, 109
  Modular/non-modular, 104
  Mortar, 98-101
    Additives, 98, 99
    Compressive strengths, 99, 100
    Testing, 99
    Types, 98
  Orientation, 112
  Pilasters, 114, 118

Brick *continued*
Positions in the wall, 104
Sizes, 102, 103
Traditional bonds, 105, 106
Vertical coursing, 110
Wall elevation details, 114, 119, 120,
Built-up roofs, 253-261, 278, 286
Three (3) ply, 253-255, 258, 259
Four (4) ply, 256-259

C

Caissons, 11
Capacity calculations, 673-675
Cylindrical tanks (horizontal, 675
Cylindrical tanks (vertical), 673
Rectangular tanks, 674
Round tanks, 676
Tapered tanks, 676
Carpet, 459-478
Construction/materials, 459
Measuring for, 460-478
Caulking (see Sealants), 302-312
Cedar shingles, 285-286
Cement, 72, 73, 75, 100
Chase walls, 433
Circles, 677, 678
Areas of, 677, 678
Circumferences, 677, 678
Column fireproofing, 294-299
COM-PLY, 235
Composite wood products, 233-235
Cellulosic fiberboard, 234
Hardboard, 234
Medium density fiberboard (MDF), 234
Oriented strand board (OSB), 234
Waferboard, 234
Concrete, 72-79
Acoustical properties, 323, 330
Admixtures, 74
Accelerating, 74
Air entrained, 74
Fly ash, 74
High range water reducer, 74
Multi-filament fibers, 74
Retarders, 74
Silica fume, 74
Water reducers, 74
Cellular, 72
Control joints, 73
Dowel spacing, 74
Spacing, 73, 74
Curing, 78
At 50 degrees F, 78
At 70 degrees F, 78
Procedures, 78
Ferrocement, 72
Filled soldier piles, 21
Forms, 75-77
Max. allowable tolerance, 75, 76
Commercial form ties, 77
Release agents, 76
General properties, 72
Gunite, 72
History, 72
Lightweight, 72
Mechanical/chemical reqmnts, 81
Mixing small batches, 75
By volume, 75
By weight, 75
Painting, 485, 488, 492, 494
Generic high performance coatings, 504, 505
Immersion exposure, 495
Industrial exposure, 494
Surface preparation, 497, 498
Pipe, 41-43
Portland cement, 72
High early, 73
Shrinkage factor, 73
Water content, 73
Types I-V, 72
Reinforcing steel, 79-95
Bar size/weights, 79
Specifications, 80
Corrugated steel supports, 95
Fabrication practices, 85-88
Metrication, 96, 653
Sequence for flat slabs, 93, 94
Two way flat plate, 93
Two way flat, 94
Storm/sanitary manholes, 67, 68
Supports, 89-95
Geometry/wire size, 90
Plastic, 92
Precast usage, 91
Typical types, 89
Ties, 77
Welded wire fabric WWF, 82-84
Common styles, 83
Identification, 84
Slumps, 75
Water, 73, 75
Chloride content, 75
Effect on shrinkage, 73
Conduits (see Electrical), 618, 620, 621, 623, 624,
Conductors (see Electrical), 617-621, 624, 626, 628, 632
Curtain walls, 440

D

Decimals, 667-670
Inch equivalents, 667, 670
Feet equivalents, 667, 669
Fractions, 668
Millimeters, 670
Yard equivalents, 667
Dewatering, 24-26
Draw downs, 24, 25

Protecting foundations, 26
Doors, 339-373
Aluminum, 372, 373
Revolving, 373
Types, sections, 372
Hardware, 388-418
Hinges, recommended number, 417
Types, 388, 389
Keying terminology, 415
Knob designs, 395
Lever handle, 396, 397
Locksets, latchsets, 390-392
Mortise locks, 393, 399, 401, 407, 408
Panic devices, 405-414
Strikes, 394
Hollow metal, 339-354
Classifications, 339
Hardware locations, 341
Installation problems, 342-351
Door binding, 346
Sagging, 346
Hinge binding, 350
Improper clearances, 345
Loose frame in drywall, 343.344
Paint problems, 353, 354
Reswagging hinges, 349
Thermal bow, 351
Twisted door, 347, 348
Metal thicknesses, 342
Reinforcing, 341
Standard opening sizes, 340
Storage/handling, 369-371
UL labels missing, 352, 353
Wood, 355-360, 363, 365-371
Construction details, 360-363
Electrostatic shield, 364
Finishing tips, 369-371
Fire-rated, 363-364
Construction, 363
Rating, 364
Glazing options, 367
Handling tips, 369-371
Installation, exterior, 368
Lead lined, 364
Laminate faced, 360-361
Mineral core, 361
Cut-away, 361
Particle core, 360
Cut-away, 360
Mineral core, 357
Cut-away, 357
Louver options, 367
Maintenance tips, 369-371
Matched veneers, 359
Ordering pre-machined, 365
Particle board, 356
Cut-away, 356
Special hardware, reinforcing requirements, 366
Sound retardant, 364
Stave core specs, 355
Cut-away, 355
Storage tips, 369-371
Telegraphing tolerance, 369
Veneers, 355-359
Appearance, 358
Description, 358
Matching, 359
Warp tolerance, 369
Drywall, 322, 323, 328, 329, 332-334, 419-455
Acoustical considerations, 322, 323, 329, 332-335
Chase walls, 433
Construction details, partitions, 428
Control joints, 430
Door frame installation, 343-344
Do's and dont's, 322, 323
Fireproofing, 294-298
Furring, 425, 427
Lath and plaster, 450-454
Installation procedures, 452
Metal lath, hangers, channels, 453
Systems, 450
Comparison of, 451
L over 120/240/360 explained, 436
Load bearing partitions, 423
Non-load bearing partitions, 421, 422
Ceilings, 426
Plumbing fixture attachment, 429
Tub and shower details, 430
Typical tub, pool details, 431
Resilient channels, partitions, 434
Shelf wall specs, illustrated, 432
Soffits, framing specs, 432
Sound control, 322, 323, 328, 329, 334, 424
Do's and dont's, 322
STC ratings, 328
Studs, 437, 442-449, 453
Curtain wall, 441
Construction details, 440
Plaster systems, 453
Specifications, 437, 443-3445
Structural, 437, 442-449
Super studs, 442-449
Accessories, 447-449
Section properties, 443-446
Tall walls, 435
Limiting heights, 438
Swimming pool applications, 431
Systems, 420
Tall wall construction, 435
Taping, five levels, 455
Dumbwaiters, 520, 521

E

Electrical, 615-641
Acoustical considerations, 334-335
Common terminology, 616

Electrical *continued*
- Conductors, 617-620, 621, 626, 632,
  - Aluminum, 621, 622
    - Size vs copper, 621
    - Nominal dimensions, 622
  - Conduit for, 618, 619
  - Cover requirements 0-600 volts, 628
  - Dimensions of rubber/plastic covered, 619
  - Grounding for 15-400 amp devices, 632
  - Maximum numbers in conduits, 618
    - RHH and RHW stranded, 619
    - TW, XHHW, RHW conductors, 620
  - Number allowed in metal boxes, 626
  - Properties, AWG size 18-2000, 617
- Conduits, 618-621, 623, 624
  - Minimum radii bends, 621
  - Non-metallic, expansion of, 623
  - Maximum number of conductors, 618-620
- Copper versus aluminum—sizes, 621
- Demand loads—residential appliances, 629
- Duct banks, 627
- Enclosures, 633-635
  - Hazardous, 634
  - Motor controllers, 635
  - Non-hazardous, 633
- Full load current for 1, 2, 3 phase motors, 638-641
- General lighting loads, 630
- Ground fault interrupters (GFI), 625
- Grounding 15-400 amps, 632
- Overcurrent protection, 631
- Seismic restraints/bracing, 637
- Voltage drops-6/12 volts, 636

Elevators, 507-519
- Basic types, 508
- Capacities, 508
- Holeless, 516
- Hydraulic, 512-514, 519
  - Freight elevator, 514
  - Hoistway section, 516
  - Installation, isometric, 514
  - Preparatory work, 519
  - Typical platform/sling, 513
- Machine rooms, 517, 518
- Telescoping, 516
- Traction, 508-511, 515, 518
  - Hoistway section, 515
  - Installation, isometric, 509
    - Gearless, 511
  - Typical platform/sling, 510

Exfiltration (pipe tests), 66

F

Finish Hardware, 387-418
- ASTM Specs, 418
- Construction key systems, 403
  - Removal core cylinders, 404
- Deadbolts, 402
- Door hinges (types), 388, 389
- Door knob designs, 395
- Finish symbol descriptions, 416
- Introduction to, 388
- Keying, terms, codes, designations, 415
- Lever handle designs, 396
  - Forged, wrought designs, 397
- Lockset, latchset configurations, 390–392
- Mortise, cases, hubs, parts, 393
  - Cylinders, 399
  - Miscellaneous cams, 401
  - Panic devices, 407, 408
- Panic devices, 405–413
  - Concealed, surface applied, 405, 406
  - Mortise lock devices, 407, 408
  - Outside trim, 413, 414
  - Rim devices, 409-412
- Removable core cylinders, 404
- Recommended number of hinges, 417
- Rosette, blocking rings, 400
- Strikes, 394
- Turn levers, 398

Fire Protection, 553-572
- Commercial kitchen schematic, 569
- Contractor's Material/Test Certifications, 570-572
- Fire department connection schematic, 568
- Grid system vs looped, 558
- Hazard classification, 556-557
  - Extra hazard, 557
  - Light, 556
  - Ordinary, 556
- Hose stations, 555
- Introduction to systems, 554, 555
  - Deluge system, 544
  - Dry pipe systems, 554
  - Fire-cycle systems, 554
  - Pre-action systems, 554
  - Wet pipe systems, 554
- Maintenance schedules, 563
- Pipe, 564, 565, 567,
  - Fire department connections, 568
  - Hangers, 564, 566
  - Unacceptable welds, 567
  - Weights filled with water, 565
- Siamese connections, 555
- Seismic zones/modifications, 566
- Standpipes, 555

Sprinkler heads, 555, 559-561
Fusible/frangible, 555
Placement requirements, 560
Depending upon hazard, 561
Relation to obstructions, 559
Temperature rating-distance to heat source, 562
Fireproofing, 288-300
Concrete column enclosure, 299
Dry system, 288
Applications guide, 290.291
Drywall, 294-296
2 & 3 hr columns, 294-296
3 hr at precast panel, 297
3 hr at 12" block wall, 298
Fireproof vs fire resistance, 288
Masonry—3 hour column, 300
Spray on, 288
Physical performance properties, 293
Typical specification, 289
Steel—4 accepted methods, 288
Terminolgy, 288
Trowel on, 288
Wet system, 288
Fire resistant ratings, 135, 178, 179, 274, 292, 294-299, 421, 424, 426
Concrete, 297, 299
Drywall, 294-298, 421, 424, 426
Floor-ceiling joists, 178, 179
Masonry, 135
Roof decks, 178, 274
Single ply membrane, 274
Spray fireproofing, 292
Wood doors, 363, 364
Flashings, 115-117, 251, 252, 267-270, 272, 282
Masonry, 115-117
Cavity walls, 117
Details, 115
Relieving angles, 116
Roof, 251, 252, 267-270, 272, 282
Copper/lead, 282
Locations, 252
Single ply membrane, 267-270, 272
Types, 251
Flooring, 452-481
Carpet, 459
Construction/materials, 459
Computing square yards, 460-478
Common types, 458
Resilient, 458, 459
Seamless, 479
Stone veneer, 479
Mortar set, 479
Thin set, 479
Terrazo, 480, 481
Components, 480, 481
Wood, 458
Fluid applied membrane roof, 250
Formulas, useful, 667, 671, 672
Areas, 672
Miscellaneous, 667
Right triangles, 671

G

Glass block, 124-126
Glazing, 363, 377, 379, 380, 385
Effects on heat gain, 377
Fire-rated, in doors, 363
Low-E, 379
Sloped, 385
Thermal movement, 380
Grout, 121

H

Handsignals for boom work, 27
Hardboard, 234
Hatches, roof, 279-281
High pressure laminates, 240-247
Crack avoidance, 243, 244
Low pressure laminates, 247
Post forming counter tops, 245-247
Common problems, 247
Manual forming techniqes, 246
Questions and answers, 240, 241
Stress crack avoidance, 243, 244
Tips to avoid warpage, 242
Holeless elevators, 516
HVAC, 573-613
Boilers, 575-586
Common types, 575
Economizers, 583, 584
Federal EPA rules, 580
Feedback system (illustrated), 581
Fire tube boiler fuel consumption, 582
Hot water(schematic), 576
Exploded view, 577
Parts list, 578
Stack options, 585
Expansion/contraction, 587
Typical installation, 586
Steam boiler, 579
Chillers—heating with coolers, 599
Condensors, 591, 598, 599,
Air-cooled/subcooling system, 591
Evaporative (diagram/description), 599
Shell and coil (diagram/ description), 597
Custom built HVAC unit, 588
Draft towers, mechanical, 603-605
Ducts, 606-608
Equivalent-round vs rectangular, 606-608
Equivalent-spiral, flat, oval, 608
Evaporators-shell and tube, 598
Shell and tube (diagram/ description), 598

HVAC—types *continued*
Evaporative cooler(diagram/ description), 600
Fans-typical configurations, 609
Heat pump operation schematic, 590
Heat loss/gain from office equipment, 610
Humidifiers—types, 602, 603
Ice storage systems, 600, 601
Indirect evaporative pre-cooling system, 589
Piping systems, 594-596
Four pipe with one-two coils, 596
Single/two pipe cooling, 594
Three pipe heat/cooling, 595
Two pipe reverse main, 595
Thermal properties—common materials , 611-613
Variable air volume (VAV), 592, 593
Systems diagrammed, 592
With heat, re-heat, fan powered units, 592, 593
Hydraulic elevators, 512-514, 516-519

I

Infiltration tests (pipe), 65
Insulation, 252, 260, 261, 269, 271, 272, 274, 320, 323, 324, 326-328, 421-424, 427, 611-613
Acoustical, 320, 323, 324, 326-328, 333, 334
Common building products, 611-613
Drywall partitions, 421-424, 427, 440
Roof, 252, 260, 261, 269, 271, 272, 274

J

Joists, 158-168
Characteristics, 158
DLW series, 162, 164
Girders, 165-168
Connection details, 166
Moment connections, 167
Specifying, 168
What are they?, 165
K series, 158-161
Specifications, 160
Top chord extensions, 161
LH series, 162, 163

L

Laminated veneer lumber (LVL), 234
Low pressure laminates, 247
Lumber, 182-221
American Lumber Standards, 183-185
Stamp marking, 183
Trademarks, 184
Wood preservatives, 183
Laminated veneer, 234
Metrication, 653
Oriented strand, 235
Parallel strand, 235
Softwoods, 196-200
American Softwood Standards, 196-200
Boards and timbers, 196
Shiplap, centermatch, 197
Siding, 199
Worked lumber, 198
Southern pine, 201-216
Birdsmouths, 205
Decking, 213
Framing, 210-212
Grading, 201, 213-216
Decking, 213
Finish boards, 214
Stamps, 201
Two inch dimensional , 215
Wood preservatives, 216
Header connections, 209
Industry abbreviations, 220-221
Inspection bureau, 182
Joist span tables, 202
Wet service joist, 203
Load tables, 209
Properties, 207
Rafter spans, 205
Oriented strand, 235
Western wood products, 182-195
Design values, 190-192
Grade stamps, 187
Scaffold planks, 196
Species of wood, 188
Standard sizes, 193, 194
Adjustment factors, 192
Western Wood Products Assoc., 182

M

Masonry, 98-135
Acoustical properties, 322
Bearing areas running bond, 120
Block, 110, 114-115, 118, 135
Fire ratings, 135
Vertical coursing, 110
Pilasters, 118
Brick, 102-120
Arches—illustrated, 107
Beam details, 113
Bond patterns, 105, 106
Illustrated, 105
Explained, 106
Corbeling limitations, 119
Corner details, 113
Coursing, 109, 110
Horizontal, 109
Vertical, 110
Orientation, 112

Caulking details—relieving angles, 116
Cavity walls, 309, 310, 314
Compressive strengths, 99-101
Mortar types, 99
Cement types, 100
Allowable for masonry, 101
Estimating, 108
Fireproofing column, 300
3 hr rating at block wall, 298
Flashing details, 115-117
At relieving angles, 116
Base flashing, 115
Cavity walls, 116
Double unit base, 115
Integral pilaster, 115
Single unit, 115
Fire resistance, 135
1-4 hour ratings, 135
Foundations—depth of backfill, 102
Nomenclature, 102
Positions in the wall, 104
Sizes, 103, 104
Modular, 104
Non-modular, 104
Position in the wall, 104
Glass block, 124-126
Typical head/sill details, 124
Typical panel anchorage, 125
Typical sill details, 126
Grout, 121
Proportions by weight, 121
Proportions by volume, 121
Jamb details, 113
History, 98
Mortar, 98-100
Additives, 98, 99
Compressive strengths, 99
using cement, 100
Joint details, 313
Testing, 99
Time to repoint?, 313
Inspection of joints, 313
Tips, 313-314
Types, 98
Parapet details, 114
Pilaster details, 114, 118
Block, 118
Brick, 118
Preventing failures, 134
Ladur, 133
Parapet walls, 114
Sealants, 312
Physical properties, 128
Repointing, 313, 314
Reinforcement, 127-134
Seismic ladur, 134
Ties, types, 127, 131
Truss, 130
Veneer anchors, 132
Seismic, 133
Metrication, 5, 30, 31, 96, 230, 643-659
Asphalt, 30, 31
Construction industry application, 651
Conversion tables, 656-659
Lumber sizes, 653
Metric Conversion Act-1975, 644
Millimeter equivalents, 670
Nails, sizes, 662
Pipe sizes, 652
ASTM diameter pipe, 654
Plywood panels, 230
Reinforcing steel, 96, 653
Sieve sizes, 5
Temperature conversions, 655
What will/will not change, 644-650
Moisture content (lumber), 186

N

Nails, size–U.S./metric, 662
NEMA designations, 633, 634

O

Oriented strand board (OSB), 234
Oriented strand lumber (OSL), 235
OSHA, 7-11
Slope requirements, 9-11
Soil classifications, 7, 8

P

Painting, 484-505
Exterior coating specs., 485-487
Generic formulations, 484
High performance, 504, 505
High temperature application, 496
Immersion exposure, 495
Industrial exposure specs., 491, 492
Heavy duty exposure, 492-494
Interior coating specs, 487-490
Low temperature applications, 495
Maintenance free coatings?, 501
Preservative treatment, wood, 500
Special purpose coatings, 484
Structural steel coatings, 502-505
Procedures, 502
Steel Structures Painting Council (SSPC), 502, 504
Coating systems, 503, 504
For new, previously painted surfaces, 502
Minimum surface preparation, 503
SSPC specifications, 502
Surface preparation, 497-499
Parallel strand lumber (PSL), 235
Parking garage checklists, 317-318
Piles, 11-22
Concrete, 12
Concrete filled, 21

Piles *continued*
Deadmen, 20
H piles, 18
Rock bearing, 19
Drilled into, 22
Sheet, 12
Soldier, 12, 20-22
Steel, 11
Timber, 11
Wood lagging, 22
Driving rigs, 12-16
Battered configuration, 16
Components of, 12
Hammer types, 16-17
Lead types, 12-15
Piping, 24-64
Bedding materials, 34
Cast iron pipe, 44, 49, 536, 537, 538
Crushing loads, 47
Diameter vs trench width, 537
Equivalent sizes, 47
Hub—barrel dimensions, 536
Slope required, 48
Thrust pressures, 538
Trench requirements, 44, 45, 537
Concrete pipe, 36
Rigid/pvc pipe, 37
Depth of cover, 46
Shields, 39, 40
Width requirements, 38
Trench shields, 39, 40
Assembly of, 39
Typical joining methods, 49
Working pressures, 46
Concrete pipe, 41-43
Reinforced concrete pipe, 36, 42
Specifications, 41, 42
Trench requirements, 36, 38
Corrugated metal, 60-63
Alum/galvanized specs., 60-63
Arch height cover, 61
Specifications 12"-144", 62
Ductile iron, 50-57
Assembly tips, 54, 55
Installation, 54, 55
Specifications, 54-57
Weights per foot, 50-53
Exfiltration tests, 66
Infiltration tests, 65
Installation—general, 24
PVC, 58-60
Deflection, 59
Expansion/contraction, 59, 64
Trench requirements, 37
Metrication—sizes, 652
Wall thickness, 654
Plaster systems, 450-454
Installation procedures, 452
Metal lath, accessories, 453, 454
Systems, 450
Types, 451
Plumbing, 523-551
Abbreviations, symbols, 547-551
Equivalent length vs, nominal, 524, 525
Pipe, elbows, tees, valves, 524
90 degree elbows, 525
Expansion of pipe, 531, 532
Metal/plastic, 531
In graph form, 532
Gas piping, 525
Fixtures, 527, 528, 541, 543
Flow rates, typical, 527
Hot water demand, 528
Schematic, stacked installation , 541
Schematic, stacked installation, common vent, 543
Head-of-water equivalents (in PSI), 528
Pipe sizes, 524-526, 529, 533, 536, 537
Actual vs nominal, 524, 525
Cast iron hub—barrel, 536
Velocity/flow rates, 530
Diameter/ trench width required, 537
Metric, 537
For horizontal rainwater, 529
Roof drains & rainfall, 533
Pipe supports, 534, 535,
Horizontal runs, 535
Risers, 534
Schematic piping drawings, 539-545
Continuous/looped vents, 540
Roof drain/leaders, 542
Stacked fixtures, 541
Fixtures with common vent, 543
Vent and stacks, 539
Circuit and wet venting, 544
Waste and vent, 545
Water velocities (types of service), 526
Steam—condensate systems costs, 533
Symbols–plumbing drawings, 548, 549
Pipe fitting and valves, 550-551
Velocity/flow cast iron sewer pipe, 530
Test plugs, 538
Thrust pressures, 538
Water temperatures-various cities, 546
Plywood, 224-239, 249
American Plywood Association, 224
Registered trademarks, 225
Roof sheathing specs., 232

Sidings, APA ratings, 228, 229
Sturdi-floor ratings, 239
Composite wood products, 233-234
Cellulosic fiberboard, 234
COMPLY, 235
Hardboard, 234
Medium density fibreboard, 234
Waferboard, 234
Oriented strand board (OSB) p 234
Dimensional changes, 238
Dimensions, U.S./Metric, 230
Exposure ratings, 227
Fastening schedules, 232, 236-239
Framing, for, 233-236, 239
Grading guidelines, 224
Ideal fabrication conditions, 237
Installation tips, 224
Moisture content vs warpage, 235
U.S. moisture content zones, 238
Span tables, 231, 232, 236,
Roof sheathing, 232
Sheathing, subfloors, 231
Specialty panels, 224, 225
Species, classification, 227
Group numbers, 227
Sturdi-floor, 239
Types and applications, 224
Underlayment, 236-238
Veneer grades, 226
Plywood panels, 230

R

Reinforcement, 127-134
Concrete, 79-95
Masonry, 127-134
Materials/properties, 128
Seismic, 133, 134
Ladur/comb, 134
Veneer anchors, 133
Ties (types), 127, 131
Truss and ladur reinforcement, 130
Unstable conditions, 134
Veneer anchors, 132
Wall anchorage details, 129
Ties, 131
Tile walls, 122, 123
standard cladding shapes, 123
Wall sections, 114-117
Waterpoofing cavity walls, 314
Wall systems, 111
Dual framing system, 111
Load bearing—shear wall , 111
Prefab curtain wall, 111
Structural skin, 111
Resilient flooring, 458, 459
Roofing, 250–286
Built-up, 250, 253-257
Fire vents, 278
Flashing details, 252, 260, 261
Four ply, gravel surface-3" per foot, 256
Four ply, smooth surface-3" per foot , 257
Four ply, modified bitumen, 259
Three ply on insulation, 253
Three ply-modified bitumen, 258
Three ply on nailable deck, 254
Three ply on lightweight fill, 255
Checklist to avoid leaks, 286
Copper/lead coated material specs, 282
Flashing details, 251, 252, 260, 261, 267-271
For built-up roofs, 260, 261
Single ply, 267-271
Cap, 268
Counterflashing, 270
Curb, 267, 269
Expansion joint, 271
Gutter/roof drain, 272
Reglet, 268
Types and locations, 252
Vertical pipe, 269
Fluid applied roofs, 250
Metal sheet panels, 250
Roof hatchs, 279-281
With ladder, 279
With ships ladder, 280
With stair access, 281
Single ply roofing, 262-277
Acceptable roof deck chart, 273
Ballasted roof stone spec., 266
Cap/reglet, 267
Counterflashing, 270
Curb flashing, 267, 269
Expansion joint flashing, 271
Fire vents, 278
Gutter/roof drain flashing, 272
Leak investigation, 276, 277
Preventative maintenance, 275
Securement data, 262, 263
Splicing cement guide, 265
UL specifications, 274
Wind speed map, 264
Slate roofs, 283, 284
Exposure to weather, 283
Installation procedures, 284
Standard size shingles, 283
Frequently used types, 250
Shingles/shakes, 250, 285, 286
Grade facsimile labels, 285
Installation diagrams, 285
Installation tips, 285
Maintenance tips, 286

S

Sealants, 302-318
Acceptable air seal applications, 310

Sealants *continued*
Adhesion test procedures, 311
Illustrations, 311
Best barriers to water entry, 312
Butt joints, 305
Cavity wall waterproofing, 314
Cementitious, properties, 303
Composite waterproofing, 315
Dow Corning designs, 307
Estimating, 307
Exterior wall applications, 309
Joint filling compounds, 302
Mortar joints, 313, 314
Details, 313
Inspections, 313
Time to repoint?, 313
Proper steps to take, 314
Non-cementitious, properties, 303
Parapet wall details, 312
Illustrations, 312
Parking garage checklists, 317, 318
Proper applications, 302, 303, 308-310
Properties of various types, 303, 305
Silicone, 305, 307
Temperature vs performance, 306
Time to repoint?, 313
Types, advantages, disadvantages, 303-305
Neutral, 304
Organics, 304
Polysulfides, 304
Polyurethanes, 304
Silicones, 304
Unacceptable air seal applications, 310
Seamless flooring, 479
Seismic, 133, 134, 637
Electrical bracing, 637
Fire protection modifications, 566
Masonry anchors, reinforcing, 133, 134
Zones, 566
Sheet piles, 17
Silicone sealants, 307-311
Single ply membrane roofs, 262-277
Sitework, 2-32
Asphalt paving, 28-32
Amount per cubic meter, 31
Amount per linear foot, 30
Amount per linear meter, 30
Paving blocks, 32
Pitfalls to avoid, 28
Tons req'd per mile, 31
Batter boards, 35
Bedding materials, 34, 35
Backfill, 34
Terminology, 34
Types I-V, 34, 35
Boring logs, 6, 7
Interpretation, 6
Classification terminolgy, 7
Caissons, 11
Dewatering, 23-25
Drawdown, 25
Groundwater, 26
Waterflow, 24
Glossary of terms, 2, 3
Hand signals for booms, 27
Investigations, 1
OSHA, 7-10
Soil classifications, 7, 8
Slope requirements, 9, 10
Piles, 11-22
H piles, 18
Sheet piles, 17
Typical sections, 18
Soldier piles, 19-22
Concrete filled, 21
Drilled into rock, 22
Rock bearing, 19
With deadmen, 20
With wood lagging, 22
Types, 11
Concrete, 12
Sheet, 12
Soldier, 12
Steel, 11
Timber, 11.
Pile driving rigs, 12-16
Battered configuration, 16
Hammer types, 16
Drop, 16
Double acting, 16
Single acting, 16
Vibratory, 17
Lead types, 12-15
Fixed, 13
Swing lead, 14
Typical parts, 12
Soils, 2-7
Classifications, 3
OSHA, 7-10
Glossary of terms, 2
Grain size, 4
Sieve size—U.S./Metric, 5
Test borings, 6, 7
Soffits, 432
Sound transmission coefficient (STC), 321, 323, 328-331
Southern Pine Inspection Bureau, 288-293
Span tables, 169, 177, 202-205, 213, 232, 236-238
Metal deck, 169, 177
Plywood, 231, 232, 236-238
Southern pine, 204
Wood joists, 202, 203
Wood rafters, 205
Spray fireproofing, 289-293
Sprinklers (see Fire Protection), 554-572
Stainless steel, 662, 663

Thickness, sheets, 662, 663
Weights, sheets, 662, 663
Steel, Structural, 138-179
ASTM specifications, 138
A36, A529, A441, A572, 138
A242, A588, A852, A514, 138
Composition of steel, 138
Details, 144-146
Base plate details, 145
Beam framing, moment splices, 144
Beam over column, 144
Moment splice at ridge, 144
Shelf angles with adjustment, 146
Tie rods and anchors, 146
Fasteners, 148-157
ASTM A325/A490 bolts, 150
Bolts, 148, 149, 152, 153, 155-157
Diameters/hole dimensions, 152
Dimensions of finished hex, 155
Head shapes, 148
Markings, 153
Tension control (TC), 156-157
Installation, 156
Specifications, 157
Weights, 149
Capscrews/boltsnut markings, 153
Nuts, 151, 153, 154
Dimensions, 154
Markings (heavy), 153
Properties, 151
History, 138
Joists, 158-168
Camber for standard types, 163
Characteristics, 158
CS Series, 158
DLH Series, 158, 162-164
DLW, 164
Joist Girders, 158, 165-168
Configurations, 165
Connection details, 166
Moment connections, 167
Specifying joist girders, 168
What are they?, 165
K Series, 158-160
Bridging, 158, 159
Open web, 160
Top chord/extended end, 161
LH Series, 158, 162-163
Standard/non-standard types, 163
Bearing details, 164
Standard/non standard types , 163
Bearing details, 164
SLH Series, 158
Floor—ceiling fire resistance, 174
Metal decking, 169-178
Cellular floor deck, 173
Composite deck details, 171
Fire resistance ratings—roof, 178
Lapping steel deck, 170
Non-composite deck details, 171
Openings, 175, 176
Reinforcing details, 175
Six inch penetration, 176
Pour stop selection table, 172
Spans recommended, 169, 177
Maximum spans, 177
Painting, 489, 493, 498, 502-505
High performance coatings, 504, 505
Steel Structures Painting Council, 502-504
Surface preparation, 498
Shapes, 139-141
W4 to W12, 139
W12 to W18, 140
W18 to W36, 141
Standard mill practices, 142, 143
Camber, 142
Shape tolerances, 143
Weld symbols, 147
Storm sewer, 67-69
Castings, manholes, 69
Components, 68
Sections, 67
Structural studs, 437, 442-449

T

Terrazo, 480, 481
Test borings, 6, 7
Thermal properties—common building materials, 611-613
Tile, 122, 123, 430
Cladding, 123
Framing for, 430
Shapes, 123
Systems, 122
Traction elevators, 509-511, 515
Trenching, 9, 10, 34-40, 44, 45, 537
Bedding materials, 34, 35
Cast iron pipe, 44, 45
Concrete pipe, 36
OSHA slope requirements, 9, 10
PVC pipe, 37
Trench shields, 39, 40

U

Underwriters Lab (UL), 292, 296-298, 352, 353
Labels off doors/frames, 352, 353
Fireproofing material ratings, 292

Underwriters Lab (UL) *continued*
Drywall ratings, 295-297
2 hr UL X519, 295
3 hour col. ULX518, 515, 296
3 hour col. ULU904, 297
3 hour block ULX515, 298
Single ply roof membrane, 274

V

Variable Air Volume (VAV), 592, 593

W

Waferboard, 234
Weights, 662-666
Aluminum, 663
Common materials, 665, 666
Copper, 663
Sheet metal, 664
Stainless steel, 662, 663
Welded wire fabric, 82-84
Windows, 374-378
ANSI standards, 376
Aluminum, 374
Glazing selections—heat gain, 377
NWWDA standards, 376
Air-infiltration standards, 377
Ordering wood clad windows, 378
Steel, 374
Vinyl, 374
Wood, 375
Wire, 664
Wood/Lumber, 182-221
Acoustical performance, 323-324
American Lumber Standards Committee, 183-185
Pressure treated, 184
Trademarks, 185
Wood preservatives, 183, 500
American Softwood Standards, 196-200
Boards and timbers, 196
Finish/floor/ceilings, 200
Siding (19% moisture content), 199
Shiplap/centermatch, 197
Worked lumber, 198
Flooring, 239, 236-238, 323, 324, 458
Acoustical performance, 323-324
Framing for, 239
Underlayment for, 236
Recommended spans , 236
Knots, how to measure, 217
Lumber industry abbreviations, 220-221
Moisture content, 186, 199
Painting, 484-499
Generic formulations, 484
High temperatures, 496
Immersion exposures, 495
Industrial exposures, 491-494
Interior applications, 487-490
Low temperatures, 495
Normal exposures, 485-487
Preservative treatment, 500
Surface preparation, 497-499
Softwoods/hardwoods, 182, 188, 218, 219
Commercial names—softwoods, 218, 219
Southern Pine Inspection Bureau, 182, 201-216
Decking, 213
Grade stamp markings, 210
Decking, 213
Finish and boards, 214
Two inch dimension, 215
Span tables, 202-205
Birdsmouth data, 205
For various species, 204
Joists, 202, 211
Framing details, 211, 212
Rafters, 205
Conversion diagram, 206
Framing details, 210
Maximum spans, 205
Standard sizes, 208
Structural framing, 207, 209-212
Connection details, 209
Header load tables, 209
Introduction to, 182
Properties, 207
Wet service joists, 203
Veneers, 355-359
Western Wood Products, 182, 187-195
Design values, 190
Adjustments to, 191, 192
Grade stamps, 187, 189
Guide to, 187
Introduction to, 182
Species included, 188
Species identification, 189
Scaffolding-size, design values, 195
Standard sizes-finish and select, 193
Standard sizes common boards, studs, 194
Windows, 375
Aluminum, 374-375
NWWDA standards, 376
Air infiltration standards, 377
Wood windows, 376
Performance grades, 376
Plastic, 374-375
Steel, 374-375
Wood clad–ordering, 378
Wood preservatives, 183, 216
ALSC standards, 183
Southern pine, 216
WWPA, 187–189

About the Author

Sidney M. Levy is a construction consultant in Chestertown, Maryland, with more than 40 years' experience in the industry. He is the author of 11 books, including several devoted to international construction. Mr. Levy is the author of *Project Management in Construction*, published in both English and Spanish editions.